ROWAN UNIVERSITY
LIBRARY
201 MULLICA HILL RD.
GLASSBORO, NJ 08028-1701

Handbook of
PHOTONICS

Editor-in-Chief

Mool C. Gupta

CRC Press
Boca Raton   New York

Library of Congress Cataloging-in-Publication Data

Handbook of photonics / editor-in-chief, Mool C. Gupta.
    p.   cm.
    Includes bibliographical references and index.
    ISBN 0-8493-8909-7
    1. Photonics.   I. Gupta, M. C. (Mool Chand)
   TA1520.H37 1996
   621.36—dc20   96-43592
                                                                                                                                 CIP

    This book contains information obtained from authentic and highly regarded sources. Reprinted material is quoted with permission, and sources are indicated. A wide variety of references are listed. Reasonable efforts have been made to publish reliable data and information, but the author and the publisher cannot assume responsibility for the validity of all materials or for the consequences of their use.

    Neither this book nor any part may be reproduced or transmitted in any form or by any means, electronic or mechanical, including photocopying, microfilming, and recording, or by any information storage or retrieval system, without prior permission in writing from the publisher.

    All rights reserved. Authorization to photocopy items for internal or personal use, or the personal or internal use of specific clients, may be granted by CRC Press LLC, provided that $.50 per page photocopied is paid directly to Copyright Clearance Center, 27 Congress Street, Salem, MA 01970 USA. The fee code for users of the Transactional Reporting Service is ISBN 0-8493-8909-7/97/$0.00 + $.50. The fee is subject to change without notice. For organizations that have been granted a photocopy license by the CCC, a separate system of payment has been arranged.

    The consent of CRC Press does not extend to copying for general distribution, for promotion, for creating new works, or for resale. Specific permission must be obtained in writing from CRC Press for such copying.

    Direct all inquiries to CRC Press LLC, 2000 Corporate Blvd., N.W., Boca Raton, Florida 33431.

© 1997 by CRC Press LLC

No claim to original U. S. Government works
International Standard Book Number 0-8493-8909-7
Library of Congress Card Number 96-43592
Printed in the United States of America 1 2 3 4 5 6 7 8 9 0
Printed on acid-free paper

Preface

The **Handbook of Photonics** is intended to serve as a reference source for introductory material and collection of published data for research and teaching in the field of photonics. The subject has grown rapidly in the past decade and plays a key role in many segments of industry and academic research. The book provides a comprehensive source of information for scientists, engineers, and students working in this field with an extensive list of references for further details.

The book is divided into three sections, respectively, photonic materials, photonic devices and optics, and photonic systems. The photonic material section contains chapters on semiconductor materials, ferroelectric and nonlinear materials, and nonlinear organic materials. The devices section includes information on semiconductor diode lasers and detectors, solid state lasers, modulators, fibers, binary optics, gradient optics, and guided wave devices. The photonic systems section contains information on optical signal processing, optical communication, optical storage, and electronic displays. With the rapid growth of the field it has not been possible to cover every aspect of photonics in a single volume. Yet we have tried to cover many of the important areas with the latest information and a large collection of data in a single source.

I would like to thank my wife and children for their support and for taking on additional tasks. My thanks also go to Erwin Cohen at CRC Press for getting this project started. I would like to thank Eastman Kodak Company for providing an environment where such work can be done in demanding times. Additional thanks go to all the contributors and the editorial board for their input and share of work.

Editor-in-Chief

Mool C. Gupta is a group leader at the Research Laboratories of Eastman Kodak Company and an adjunct associate professor at Cornell University. In 1982 he joined Kodak Research Laboratories as a senior scientist. Prior to joining Kodak he was a senior scientist at Jet Propulsion Laboratory. He received his Ph.D. degree in physics from Washington State University and was a postdoctoral fellow at Cornell University and a senior research fellow at the California Institute of Technology.

Dr. Gupta has spent over 25 years in research and development with over 70 technical publications and 20 patents. He has worked in the areas of optical information storage, guided wave devices, lasers using nonlinear frequency conversion, electro-optic and GaN semiconductor materials, processes, devices, and applications. He has taught courses on optoelectronic materials, devices, and applications at Cornell University, University of Rochester and as a short course at various Materials Research Society meetings. He was a conference chair for the 1996 SPIE meeting on Nonlinear Frequency Conversion and editor of the SPIE proceeding. He is a member of OSA, SPIE, MRS, and IEEE societies.

Editorial Board

Robert Byer
Stanford University
Palo Alto, California

Hyatt Gibbs
Optical Science Center
University of Arizona
Tucson, Arizona

L. P. Kaminow
Bell Laboratories
Murray Hill, New Jersey

P. K. Tien
Bell Laboratories
Murray Hill, New Jersey

Amnon Yariv
California Institute of Technology
Pasadena, California

Contributors

Ilesanmi Adesida
University of Illinois
Urbana, Illinois

Robert W. Boyd
University of Rochester
Rochester, New York

Charles F. Brucker
Eastman Kodak Company
Rochester, New York

David Casasent
Carnegie Mellon University
Pittsburgh, Pennsylvania

Lap Tak Cheng
E.I. Du Pont Nemours & Company, Inc.
Experimental Station
Wilmington, Delaware

James J. Coleman
University of Illinois
Urbana, Illinois

Michael W. Farn
Massachusetts Institute of Technology/
  Lincoln Laboratory
Lexington, Massachusetts

Mool C. Gupta
Eastman Kodak Company
Rochester, New York

James Harrison
Schwartz Electro-Optics, Inc.
Concord, Massachusetts

Seppo Honkanen
Optonex Ltd.
Espoo, Finland

Carlo Infante
CBI Technology Consultants
Scottsdale, Arizona

Jürgen Jahns
Fernuniversität–G.H. Hagen
Hagen, Germany

John N. Lee
U.S. Naval Research Laboratory
Washington, D.C.

Chia-Yen Li
University of California
Los Angeles, California

H. Luo
State University of New York
Buffalo, New York

Terry W. McDaniel
IBM Corporation
San Jose, California

H. Morkoç
University of Illinois at
  Urbana-Champaign
Urbana, Illinois

S. Iraj Najafi
École Polytechnique
Montreal, Quebec, Canada

D. A. Nolan
Corning, Inc.
Corning, New York

A. Petrou
State University of New York
Buffalo, New York

Andrew J. Stentz
Bell Laboratories
Murray Hill, New Jersey

G. W. Wicks
University of Rochester
Rochester, New York

Alan E. Wilner
University of Southern California
Los Angeles, California

Yuhuan Xu
University of California
Los Angeles, California

G. L. Yip
McGill University
Montreal, Quebec, Canada

John J. Zayhowski
Massachusetts Institute of Technology/
   Lincoln Laboratory
Lexington, Massachusetts

Contents

INTRODUCTION *Mool C. Gupta* xiii

PART A PHOTONIC MATERIALS

SECTION I Semiconductors

1. III-V Semiconductor Materials *G. W. Wicks* 5
2. Optical Properties and Optoelectronic Applications of II-VI Semiconductor Heterostructures *H. Luo and A. Petrou* 24
3. GaN and Silicon Carbide as Optoelectronic Materials *H. Morkoç* 49

SECTION II Oxide Materials

4. Ferroelectric Materials *Chia-Yen Li and Yuhuan Xu* 87
5. Nonlinear Optics *Andrew J. Stentz and Robert W. Boyd* 125

SECTION III Organic Materials

6. Second-Order Nonlinear Optical Properties of Organic Materials
   Lap Tak Cheng 155

PART B PHOTONIC DEVICES AND OPTICS

SECTION IV Devices

7. Optoelectronic Devices *Ilesanmi Adesida and James J. Coleman* 291
8. Miniature Solid-State Lasers *J. J. Zayhowski and J. Harrison* 326
9. Optical Modulators *John N. Lee* 393
10. Optical Fibers *D. A. Nolan* 435

SECTION V Optics

11	Binary Optics *Michael W. Farn*	459
12	Gradient-Index Glass Waveguide Devices *S. Iraj Najafi and Seppo Honkanen*	502
13	Design Methodology for Guided-Wave Photonic Devices *G. L. Yip*	530

PART C PHOTONIC SYSTEMS

14	Analog Coherent Optical Processing *David Casasent*	591
15	Optical Digital Computing *Jürgen Jahns*	605
16	Optical Communications *Alan E. Willner*	624
17	Optical Data Storage *Charles F. Brucker, Terry W. McDaniel, and Mool C. Gupta*	719
18	Electronic Displays *Carlo Infante*	768

Index 799

Introduction

General Introduction

The area of photonics reflects the synergy between optics and electronics and also shows the tie between optical materials, devices and systems. The subject of photonics plays a key role in many segments of industry such as optical communication, information storage, electronic display, signal processing, electronic imaging, etc.

The subject of classical optics is full of major discoveries by well-known contributors such as Sir Isaac Newton, Maxwells, Einstein, Fourier, Bragg, Planck, Gauss, and many more. The invention of lasers, semiconductor materials and optical devices which are based on it, guided wave optics in optical fibers and in other materials, and nonlinear optical phenomena have opened many new areas of devices and systems that are of practical interest and continue to impact our lives.

Photonic Materials

Semiconductor materials play a major role in the photonics area as they are used for light generation, for detection, modulation, and fabrication of monolithic optoelectronic devices. In the last decade significant improvements in epitaxial growth of materials using molecular beam epitaxy and metal-organic chemical vapor deposition have occurred. With these techniques, layered structures (super lattices or quantum-well structures) which show novel properties have been grown. III–V semiconductor materials have provided lasers of wavelength around 650 nm for optical storage and also around 1.55 μm for optical communication. Due to bandgap limitation of III–V materials it would be difficult to get blue/green or ultraviolet laser sources. But recent developments in II–VI and III–N semiconductor materials hold promises for achieving visible and ultraviolet semiconductor lasers useful for optical storage, printing, display, and many more new applications. The first three chapters in this handbook provide introductory information on general properties of these materials followed by a tabulation of data for various design applications.

Ferroelectric materials show properties such as birefringence, electro-optic effect, nonlinear optical effects, photorefractive effect, photoelastic effect, etc. These properties have been used for applications such as light modulators, beam scanning, nonlinear frequency conversion, holographic storage, optical display, etc. Chapter 4 discusses the basic properties of ferroelectric materials. Chapter 5 discusses the nonlinear optics and provides data on properties of various nonlinear materials.

Nonlinear optical properties of organic materials are of importance because of optical and electro-optical device applications. These include frequency conversion and electro-optic modulation. The low dielectric constant of organic solids coupled with their ultra-fast electro-optic response and large nonlinearity have driven much of the materials research. The real significance of such materials will be measured by the transparency, processibility, and stability of these materials. Chapter 6 describes second-order nonlinear optical effects in organic single crystals and polymers with an extensive list of structural and other optical properties of various organic materials.

Photonic Devices and Optics

Optoelectronic devices such as semiconductor LEDs, lasers, detectors, and optoelectronic integrated circuits along with optical fibers provide the pathway for optical communication. Significant progress has been made in achieving high laser powers (watts) and low threshold currents. Because of intramodal dispersion in optical fibers, semiconductor laser sources with very narrow linewidths have been fabricated. Also, direct modulation of semiconductor lasers to frequencies greater than 10 GHz and without significant chirping has been achieved for optical communication applications. Recently efficient blue LEDs have been commercialized and current injected blue/green semiconductor lasers have been demonstrated. Semiconductor lasers are being used in optical communication, optical data storage, high speed printing, electronic displays, optical signal processing, pump source for solid state lasers and fiber amplifiers and many other applications. Availability of laser arrays, particularly surface emitting devices, would open new applications of semiconductor lasers. Chapter 7 discusses the basic properties of semiconductor lasers and detectors.

Significant progress has been made in the area of lasers, particularly in achieving extremely high peak powers, femtosecond pulse widths, and wide spectral range with tunability. Recent progress in miniature solid state lasers has been impressive due to the availability of semiconductor lasers as pump source. Diode-pumped miniature solid state lasers offer an efficient, compact and robust means of generating diffraction limited, single frequency radiation. Chapter 8 presents the fundamental concepts and formulae for the design and utilization of miniature solid state lasers.

Optical modulators are required in optical communication, optical storage, signal processing, and optical sensing. The modulation is achieved by principles such as electro-optic, acousto-optic, magneto-optic, and micromechanical means. Electro-optic modulators can be used for modulation frequencies in excess of 20 GHz. Optical modulators are of interest for display and holographic storage as spatial light modulators. Arrays of either incoherent light sources or laser diodes with individual element addressing has been employed for spatial light modulation. Spatial light modulators of 1024×1024 pixels are possible. Chapter 9 provides discussions on various optical modulators and their performance comparison.

Optical fibers play a key role in optical communication and in sensor applications. The optical power attenuation has been reduced close to theoretical values of 0.2 dB/km at 1.55 µm and dispersion shifted fibers have achieved low optical loss and zero dispersion at 1.55 µm wavelength. Areas of research have included new index profiles for transmission in the 1.55 µm wavelength window, rare earth-doped fibers for optical amplifiers and fiber laser, demonstration of soliton transmission in fiber and observation of a number of nonlinear effects in fibers. Chapter 10 reviews the fundamental properties of optical fibers and their fabrication.

Binary optics-based devices rely on diffraction for the manipulation of light while conventional optical elements rely on refraction or reflection. Modern-day binary optical devices are usually multilevel or continuous surface reliefs rather than strictly binary. Binary optical components are usually manufactured using semiconductor fabrication technology, in particular photolithography and micromachining. The optical properties of binary optics are solely the result of the surface profile of the element. Binary optical elements are used to correct chromatic or residual monochromatic aberrations, construct arrays of micro-optics, and reshape and homogenize laser beams. Binary optics find applications in many areas such as in medical areas for bifocal intraocular lenses, in optical storage for optical heads, microoptics for semiconductor lasers, and grating beam splitters, etc. Chapter 11 describes the principles and design of binary optical components along with various applications.

Optical materials containing a distribution of refractive indices are called gradient index materials and are used in miniature optical systems. These include planar microlenses, optical power dividers, wavelength selective couplers, etc. The devices are fabricated by ion exchange process in glass and hence are relatively inexpensive passive components. Many passive and active components can be made by ion exchange processes to fabricate guided wave devices. Diffusion or an ion exchange process is used to alter the refractive index. Chapter 12 provides information on the fabrication and application of gradient-index optics. Chapter 13 describes the methodology of analyzing waveguide properties and verification with experimental characterization. Illustrative examples are provided where both the design and measured device performance parameters are compared.

Photonic Systems

Optical systems are of interest for information processing using digital or analog methods. Optics with large bandwidth and interconnection capabilities can alleviate limitations of all-electronic computers. Optical processing has several advantages over electronic processing, which must usually be done serially and is limited in speed by the broadening of pulses in interconnecting wires and is limited in density by "cross talk" between those wires. Optical systems capable of handling very large quantities of data await the development of digital optical logic elements with low switching threshold. It is not clear yet how the merging of electronic and optical technology will be achieved. But analog optical signal processing represents a rapidly maturing area with a number of hardware modules soon to be completed and tested. Laser diode arrays, an input spatial light modulator, and a set of computer generated hologram filters allow fabrication of a general purpose optical image processor. Optical processors can produce many operations such as Fourier transform and optical correlation, etc. Chapters 14 and 15 provide an introduction to analog and digital computing with various application areas.

Because of the ultra-wide bandwidth (>10 THz) available with optical frequencies, currently data transmission at rates of >10 Gb/s have been practically achieved with optical communication systems. This progress has been achieved due to the availability of essential components such as low loss optical fibers, 1.55 μm wavelength laser sources and detectors, filters, directional couplers, switches, isolators, modulators, etc. Optical amplifiers are playing a major role in optical communication as they can be used to compensate for signal attenuation resulting from distribution, transmission, or component insertion loss. The deployment of optical fibers is at an astounding rate with the increase in demand for video and audio data transmission requirements. Further progress in higher data rate transfer over longer distance can be expected. Optical fibers will be deployed widely in local areas for the distribution of a wide variety of broadband services; major population centers are being connected by long haul fiber systems. Multiplexing will help relieve congestion as information super highway traffic grows. Chapter 16 gives a thorough overview of the optical communication components, systems, applications, and future trends.

Optical information storage is playing an increasingly major role in audio and video digital storage of information. The noncontacting nature of optical recording along with the exceptional reliability of data are the key attributes of optical recording. Storage capacity over Gbytes on a 5.25-inch disk surface with data rates of >10 Mb/s are routinely achievable. Terabytes of information can be stored using library systems. Dramatic improvements in optical storage technology can be expected in storage capacity and access time and much of the improvements will come through the use of shorter wavelength light, miniaturized optical heads, and 3-D storage. Holographic storage provides a 3-dimensional optical storage where terabytes of information can be stored with data rates of Gb/s. The other novel features of holographic storage are fast access time (microseconds), parallel processing of pages, and no moving parts. Advances in materials research is needed for holographic storage to be competitive with other methods of information storage and have to achieve practical storage capacity and data rates which are theoretically possible. Chapter 17 provides an overview of optical storage systems and recording media along with future trends.

A number of different display technologies have achieved a great technological and market success over the years. Performance of major display technologies such as cathode ray tubes, light emitting diodes, liquid crystal displays, plasma display panels, electroluminescent displays, etc. have been continuously improved. Older technologies such as the cathode ray tube continue to be perfected, but new technologies such as active matrix liquid crystals achieve new performance levels. Chapter 18 provides a brief introduction to electronic displays and future trends.

The need for continous improvements in (a) optical communication data rates, (b) information storage capacity with higher data rates and fast access time, (c) high resolution printing, scanning, and display of images, (d) signal processing at faster rates, (e) biomedical applications, (f) sensors, and (g) integration of optical and electronic components, and many more applications will continue to provide new challenges to scientists and engineers working in the field of photonics.

References

Bass, M., Van Stryland, E.W., Williams, D. R., and Wolfe, W. L. (eds.) 1995. *Handbook of Optics,* Vol. 1 and 2., McGraw Hill, NY.

Howard, J. N. (ed.) 1986. *Optics Today,* American Institute of Physics, NY.

Miller, S. E. and Kaminow, I. P. (eds.) 1988. *Optical Fiber Telecommunications II,* Academic Press, NY.

Saleh, B. E. A. and Teich, M. C. 1991. *Fundamentals of Photonics,* John Wiley & Sons, NY.

Yariv, A. 1985. *Optoelectronics,* 3rd ed. Holt, Rinehart and Winston, NY.

Mool C. Gupta
Editor-in-Chief

Handbook of
PHOTONICS

PART A
PHOTONIC MATERIALS

SECTION I
Semiconductors

1
III-V Semiconductor Materials

G. W. Wicks
University of Rochester

1.1 Introduction 5
1.2 Fundamentals 5
    Optical Processes and Relationships among Constants • Quantum Wells and Strained Materials
1.3 Data on Specific Systems 7
    AlGaAs Material System • GaInAsP/InP Material System • AlGaInAs/InP Material System • AlGaInP Material System • GaSb Material System • GaAsP Material System

1.1 Introduction

The purpose of this chapter is to provide a source of data on the properties of III–V semiconductors that are relevant to optoelectronic applications. The family of III–V materials discussed in this chapter consists of compounds of the column III elements, aluminum, gallium, and indium, and the column V elements, phosphorus, arsenic, and antimony. The III–V nitrides and bismuthides are excluded from consideration here. The three column III and three column V elements can be grouped into 9 binary, 18 ternary and 15 quaternary compounds. Not all of these possible combinations have been successfully grown, and only about a quarter of them have important optoelectronic applications. This subset of the III–V family that has important optoelectronic applications is the material covered in this chapter.

This chapter organizes the III–V family into material systems, i.e., sets of materials that are compatible with each other and can be used to construct useful epitaxial heterostructures. The important III–V material systems are listed in Table 1.1.

1.2 Fundamentals

Optical Processes and Relationships among Constants

The main optical processes of importance in optoelectronic applications are reflection, waveguiding, diffraction, absorption, emission, and electrooptic and nonlinear optical effects. The main aspects of reflection, waveguiding, diffraction, absorption, and emission are expressed with the refractive index (n), absorption coefficient (α), and emission energy or direct bandgap energy (E_Γ), respectively, and are tabulated in this chapter. In material systems that have an indirect bandgap (E_X or E_L) that can become smaller than E_Γ, the indirect bandgap is also tabulated. Only

TABLE 1.1 III–V Material Systems with Important Optoelectronic Applications

Material System	Substrate	Lattice-matched Members	Important Strained Members	Main Optoelectronic Applications
AlGaAs	GaAs	GaAs $Al_xGa_{1-x}As$ $0 \leq x \leq 1$ AlAs	$Ga_{1-x}In_xAs$ $0 \leq x \leq 0.25$	Emitters and modulators: $0.75\mu m \leq \lambda \leq 1.1\mu m$, Detectors: $0.4\ \mu m \leq \lambda \leq 1.1\ \mu m$ Saturable absorbers: $\lambda \sim 0.8\text{--}0.9\ \mu m$
GaInAsP/InP	InP	$Ga_{0.47}In_{0.53}As$ $Ga_xIn_{1-x}As_yP_{1-y}$ $x = 0.47\ y;\ 0 \leq y \leq 1$ InP	$Ga_{1-x}In_xAs$ $0.4 \leq x \leq 0.6$ $InAs_xP_{1-x}$ $0 \leq x \leq 0.2$	Optoelectronic devices at $\lambda = 1.3\ \mu m$ and $1.55\ \mu m$
AlGaInAs/InP	InP	$Ga_{0.47}In_{0.53}As$ $(Al_xGa_{1-x})_{0.47}In_{0.53}As$ $0 \leq x \leq 1$ $Al_{0.48}In_{0.52}As$	$Ga_{1-x}In_xAs$ $0.4 \leq x \leq 0.6$	Optoelectronic devices at $\lambda = 1.3\ \mu m$ and $1.55\ \mu m$
AlGaInP	GaAs	GaAs $Ga_{0.5}In_{0.5}P$ $(Al_xGa_{1-x})_{0.5}In_{0.5}P$ $0 \leq x \leq 1$ $Al_{0.5}In_{0.5}P$	$Ga_{1-x}In_xAs$ $0 \leq x \leq 0.25$ $Ga_{1-x}In_xP$ $0.4 \leq x \leq 0.6$	Red emitters
AlGaAsSb/ GaInAsSb/ GaSb	GaSb	GaSb $Al_xGa_{1-x}As_ySb_{1-y}$ $x = 12\ y;\ 0 \leq x \leq 1$ $Ga_{1-x}In_xAs_{1-y}Sb_y$ $x = 1.1\ y;\ 0 \leq x \leq 1$		Emitters and detectors: $\lambda \sim 2\text{--}3\ \mu m$
GaAsP	GaAs or GaP	GaAs (on GaAs substrates); GaP (on GaP substrates	GaAsP	Visible LED's

those direct gap materials, i.e., materials with E_Γ smaller than E_X or E_L, can be efficient light emitters. Electrooptic effects and optical nonlinearities are discussed in other chapters of this handbook.

Other optical properties more relevant to solid state physics than to optoelectronic applications will not be discussed here. Examples of optical data that are not included in this chapter are critical point energies above the fundamental bandgap, Raman phonon energies, mid-IR properties, and low temperature (4K) data. These data can be found in other standard references (Madelung, 1982).

In many cases, low temperature data on these materials is easier to find and is more complete. Room temperature data, however, are more relevant to optoelectronic applications. Thus, this chapter provides room temperature data on the spectral and material composition dependence of the optical constants and emission energies. Not discussed here are the effects of doping and optical intensity on the optical constants and emission energy.

Data is provided here on the optical constants, n and α, and bandgaps. Rather than listing the optical constants n and α, some references list n and k (the extinction coefficient), or ϵ_1 and ϵ_2 (the real and imaginary parts of the dielectric function). These pairs of optical constants are all interrelated. Any of the three pairs of optical constants can be determined from either of the other two pairs. The relationships among these optical constants are given in many common references (Pankove, 1975). For most common optoelectronic applications, the most convenient form of the optical constants is n and α, thus, it is this pair of optical constants that is tabulated here.

Quantum Wells and Strained Materials

The optical properties of a semiconductor are altered by quantum size effects when at least one of the dimensions of the material is less than a few hundred Å. At present, the main structure in which quantum size effects are important are quantum wells—structures consisting of a thin well material sandwiched between two layers of a barrier material. Bulk material, i.e., material with large enough dimensions so that quantum size effects are negligible, are simpler cases to tabulate than quantum wells. The optical characteristics of bulk material depend mainly on photon energy and material composition, and are comprehensively documented here. The optical properties of quantum wells depend on photon energy—two materials compositions (well and barrier materials), and the thicknesses of the two materials. The additional parameters affecting the quantum wells make it impractical to completely catalogue optical properties for all possible quantum well structures. Only general trends for quantum wells are discussed here.

The present documentation of materials is mainly limited to bulk materials that lattice match an available substrate. Lattice mismatch produces strain. If the strain exceeds a critical value, it will be relieved via the production of dislocations or cracks, drastically degrading material quality. The critical amount of strain is determined by the magnitude of the lattice mismatch and the thickness of the layer. Thick layers need to be accurately lattice matched but thin layers, such as those in quantum wells, can be quite severely mismatched without suffering a degradation in quality. By relaxing the lattice matching requirement, strained quantum wells present the device designer with an additional adjustable parameter, an expanded range of usable material compositions. Additionally, in many cases the alteration of the bandstructure resulting from strain can be used to advantage. For these reasons strained quantum wells, incorporated into epitaxial structures with thicker nominally lattice matched layers, are becoming increasingly used in such optoelectronic devices as diode lasers. Although not exhaustively catalogued here, expanded ranges of bandgaps attainable with strained quantum wells are mentioned here, where relevant.

1.3 Data on Specific Systems

AlGaAs Material System

General Characteristics and Optoelectronic Applications

The materials technology and optical characteristics of the $Al_xGa_{1-x}As$ material system is the best understood of all the III–V materials. A primary reason for the large interest in this material system is the fact that it is a *ternary* material the lattice matches an available substrate, GaAs, over its whole composition range. Other material systems must to resort to the more difficult *quaternary* materials to have continuous composition ranges that lattice match substrates. The only other ternaries that similarly lattice match substrates are AlGaSb and AlGaP, but these materials are less useful than AlGaAs. AlGaSb is either indirect or nearly indirect over its whole composition range, and AlGaP is an indirect gap material over its whole composition range. AlGaAs is *direct* over nearly half of its composition range.

The AlGaAs material system is the prototype III–V semiconductor where new physics and device concepts are usually initially explored before being moved into other, more difficult material systems. AlGaAs has widespread applications in electronics and optoelectronics. Electronics applications include advanced devices, such as high electronic mobility transistors (HEMTs), heterojunction bipolar transistors (HBTs) and resonant tunneling diodes. Optoelectronic applications include diode lasers, solar cells and high speed photodetectors, optical modulators, and saturable absorbers. AlGaAs, like many of the III–V systems, has a certain relationship between electronic and optical properties that is extremely useful for optoelectronic applications; namely, the compositions of AlGaAs with larger bandgap energies also have smaller refractive indices and lower optical absorption. The relationship between bandgap and refractive index enables double heterostructure waveguides to confine injected electrons and holes together with the optical mode, an extremely useful characteristic for laser diodes. The relationship between bandgap and transparency allows the construction of high bandgap, transparent windows to let light in or out of a heterostructure, while limiting access of electrons to non-radiative recombination centers at surfaces. In a typical optoelectronic application, the active region of the device would consist of GaAs (or strained GaInAs); cladding/electron confinement layers of waveguides and transparent window layers of photodetectors employ AlGaAs.

Bandgaps, Refractive Index and Absorption Coefficient

The ternary, $Al_xGa_{1-x}As$, is a direct gap material (like GaAs) over the gallium rich portion of its composition range and indirect (like AlAs) over the aluminum rich portion of its composition range. Several bandgap-composition curves have been published. One of the most accurate results gives the direct-indirect crossover composition at room temperature to be at a composition of $x_c = 0.43$ and a bandgap energy near 2 eV (Chang et al., 1991). In its direct gap range, dependence of the composition, x, on the T = 300K bandgap is given by (Chang et al., 1991)

$$x = 0.702(E(eV) - 1.424), \quad \text{for } x \leq 0.43.$$

Rearranging this expression, the dependence of the direct gap on composition is obtained:

$$E_\Gamma(eV) = 1.424 + 1.4245x, \quad \text{for } x < 0.43.$$

A second reference (Guzzi et al., 1992) cites a value of $x_c = 0.40$ and gives expressions for the direct and indirect room temperature bandgaps over the whole composition range of the ternary. This information is given in the following expressions and in Fig. 1.1:

$$E_\Gamma(eV) = 1.423 + 1.36x + 0.22x^2$$

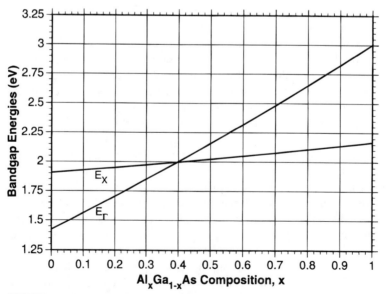

FIGURE 1.1 Room temperature values of the lowest direct (E_Γ) and indirect (E_x) bandgaps vs. $Al_xGa_{1-x}As$ composition. *Source:* Guzzi, M., Grilli, E., Oggioni, S., et al. 1992. *Phys. Rev. B* 45:10951. With permission.

and

$$E_X(\text{eV}) = 1.906 + 0.207x + 0.055x^2 \quad \text{for } 0 \leq x \leq 1.$$

The above two expressions for $E_\Gamma(x)$ agree within 7 meV for the direct gap compositions of $\text{Al}_x\text{Ga}_{1-x}\text{As}$, i.e., x values less than 0.40.

The compositional dependence of the refractive index spectra and absorption spectra (or equivalently, n and k spectra or ϵ_1 and ϵ_2 spectra) have been measured (Marple, 1964; Sell, 1974; Aspnes et al., 1986) and modeled (Adachi, 1988; Jenkins, 1990) by several authors. A fairly accurate model is that of Jenkins (Jenkins, 1990) which is an extension of the model of Adachi (1988). Figures 1.2 and 1.3, displaying the compositional dependence of the refractive index and absorption spectra, respectively, are based mainly on the Jenkins model.

Additional Data

Strained Quantum Wells. The long wavelength limit of emitters and detectors that employ lattice-matched AlGaAs is set by the GaAs bandgap wavelength at $\lambda \sim 870$ nm. The use of strained, pseudomorphic $\text{Ga}_{1-x}\text{In}_x\text{As}$ quantum wells extends the wavelength to beyond 1 µm (Feketa et al., 1986; Choi and Wang, 1990), thereby encompassing the important wavelengths of $\lambda = 980$ nm (for pumping erbium-doped fiber amplifiers) and $\lambda = 1.06$ µm (for YAG replacements). Typical InAs mole fractions for these longer wavelength applications are as large as $x \sim 0.2 - 0.25$.

Thermal Effects. The variation of the bandgap energy of GaAs with temperature can be fit with the Varshni expression (Thurmond, 1975).

$$E_g(T) = 1.519 - \frac{5.405 \times 10^{-4} T^2}{T + 204} \text{ eV}, \quad 0 < T(\text{K}) < 1000.$$

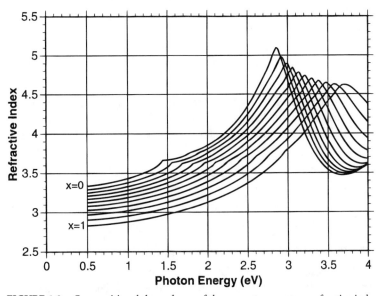

FIGURE 1.2 Compositional dependence of the room temperature refractive index spectrum of $\text{Al}_x\text{Ga}_{1-x}\text{As}$. The eleven curves represent compositions from $x = 0$ to $x = 1$ in equal increments of $\Delta x = 0.1$.

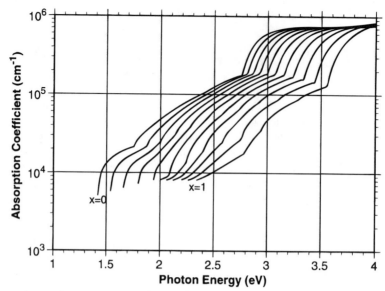

FIGURE 1.3 Compositional dependence of the room temperature above-bandgap absorption spectrum of $Al_xGa_{1-x}As$. The eleven curves represent compositions from $x = 0$ to $x = 1$ in equal increments of $\Delta x = 0.1$.

Near room temperature, this expression indicates that the GaAs bandgap shrinks as the material is heated at a rate of 0.45 meV per degree:

$$\frac{dE_g}{dT} = -4.5 \times 10^{-4} \frac{eV}{K}, \qquad T \sim 300K.$$

Equivalently, near room temperature the heating-induced lengthening of the GaAs bandgap wavelength is given by

$$\frac{d\lambda_g}{dT} = 2.8 \frac{\text{Å}}{K}, \qquad T \sim 300K.$$

A second report (Kirillov and Metz, 1983) on the rate of bandgap change with temperature gives values 30% larger than those cited above.

The thermal variation of the refractive index of GaAs is given by (Cardona, 1961)

$$\frac{1}{n}\frac{dn}{dT} = 1.1 \times 10^{-4} \, K^{-1}.$$

GaInAsP/InP Material System

General Characteristics and Optoelectronic Applications

The general $Ga_xIn_{1-x}As_yP_{1-y}$ quaternary has compositions specified by the parameters x and y that can independently take on any values between 0 and 1. The most important of these compositions are those that lattice match InP, denoted here as GaInAsP/InP. The GaInAsP/InP quaternary can be viewed as a mixture of two simpler materials, $Ga_{0.47}In_{0.53}As$ and InP, each of which lattice matches InP:

III–V Semiconductor Materials

$$Ga_xIn_{1-x}As_yP_{1-y} = (Ga_{0.47}In_{0.53}As)_y(InP)_{1-y}.$$

By comparing the Ga subscript on both sides of the above relationship, it is easily seen that the $Ga_xIn_{1-x}As_yP_{1-y}$ compositions that lattice match InP are specified by

$$x = 0.47y \quad (0, \leq y \leq 1).$$

The lowest bandgap, highest refractive index mixture of these two materials is the endpoint composition, $Ga_{0.47}In_{0.53}As$; the highest bandgap, lowest refractive index mixture of these two materials is the other endpoint composition, InP. All other compositions of GaInAsP/InP have bandgaps and refractive indices (and most other properties) intermediate to those of $Ga_{0.47}In_{0.53}As$ and InP.

Similar to many of the III–V systems, GaInAsP/InP materials with higher bandgaps have lower refractive indices. As discussed earlier (see AlGaAs+ Material System: General Characteristics and Optoelectronic Applications) this useful characteristic enables optical and electrical confinement in the same layer of a double heterostructure. Important applications of the GaInAsP/InP material system involve optoelectronic devices for optical communications systems based on 1.3 or 1.55 μm wavelengths. Diode lasers operating at either of these two wavelengths typically consist of InP cladding/electron confinement layers with GaInAsP cores and may or may not contain $Ga_{0.47}In_{0.53}As$ quantum wells. Infrared detectors with $Ga_{0.47}In_{0.53}As$ active layers cover wavelengths shorter than the bandgap wavelength of $Ga_{0.47}In_{0.53}As$, namely 1.65 μm, thus including the important communications wavelengths of 1.3 and 1.55 μm.

There are two other material systems closely related to GaInAsP/InP. AlGaInAs/InP, another quaternary that lattice matches InP, spans a similar range of bandgaps and refractive indices. This system is covered in the section on AlGaInAs/InP. The $Ga_xIn_{1-x}As_yP_{1-y}$ quaternary can also lattice match GaAs (with $x = 0.48 y + 0.52; 0 \leq y \leq 1$). However, it is not especially important because its range of bandgaps and optical constants are also covered by the well-developed AlGaAs/GaAs material system.

Direct Bandgap, Refractive Index and Absorption Coefficient

As discussed above, the compositions of the quaternary $Ga_xIn_{1-x}As_yP_{1-y}$/InP can be pictured as mixtures of InP and $Ga_{0.47}In_{0.53}As$. Because both of these constituents of $Ga_xIn_{1-x}As_yP_{1-y}$/InP are direct gap, the quaternary itself is direct gap over the whole range of compositions that lattice matches InP. The room temperature direct bandgap energy vs. composition is given by Nahory et al., 1978).

$$E_g(y) = 1.35 - 0.72y + 0.12y^2 \text{ (eV)} \quad \text{for } x = 0.47y; 0 \leq y \leq 1.$$

This relationship is depicted in Fig. 1.4.

Fairly complete refractive index data (Chandra et al., 1981) and models (Jensen and Torabi, 1983) exist for photon energies below the bandgap. The modeled spectral dependence of the room temperature refractive index is shown in Fig. 1.5.

Data on above-bandgap absorption of GaInAsP is scarce. An absorption spectrum of one composition of GaInAsP, with a bandgap wavelength of $\lambda_g = 1.31$ μm, has been reported (Yokouchi et al., 1992) and is graphed in Fig. 1.6. along with that of the two end point compositions, InP, and of $Ga_{0.47}In_{0.53}As$ (Backer et al., 1988) (Newman, 1958; Cardona, 1961, 1965; Petti and Turner, 1965).

A simple model of above-gap absorption, valid within a few tenths of an eV of the fundamental gap, uses the expression

$$\alpha(E) = \text{constant} \times (E - E_\Gamma)^{1/2}.$$

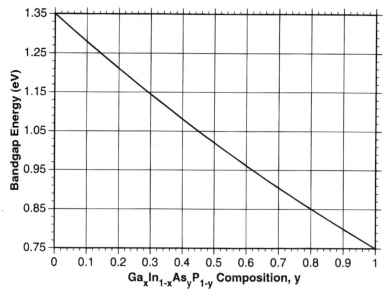

FIGURE 1.4 Room temperature bandgap of $Ga_xIn_{1-x}As_yP_{1-y}$ for compositions lattice matched to InP, i.e., $x = 0.47\,y$ and $0 \leq y \leq 1$.

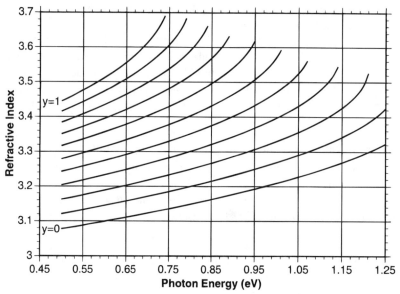

FIGURE 1.5 Room temperature refractive index of $Ga_xIn_{1-x}As_yP_{1-y}$ for compositions lattice matched to InP, i.e., $x = 0.47\,y$ and $0 \leq y \leq 1$.

The resulting fits of the three absorption spectra in Fig. 6 are

$$\alpha_{\text{GaInAs}} = 22900(\text{cm}^{-1}\,\text{eV}^{-1/2})(E - 0.705\,\text{eV})^{1/2},$$

$$\alpha_{\text{GaInAsP}} = 36600(\text{cm}^{-1}\,\text{eV}^{-1/2})(E - 0.938\,\text{eV})^{1/2},$$

$$\alpha_{\text{InP}} = 54500(\text{cm}^{-1}\,\text{eV}^{-1/2})(E - 1.354\,\text{eV})^{1/2}$$

In the absence of other absorption data, absorption spectra of compositions near that with $\lambda_g = 1.31$ μm can be estimated by inserting the appropriate direct bandgap into the fit expression from Figure 1.6 and converting bandgap wavelength to bandgap energy:

III–V Semiconductor Materials

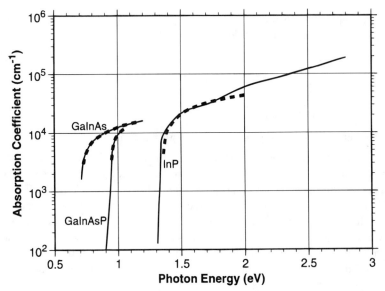

FIGURE 1.6 Room temperature absorption spectra of $Ga_{0.47}In_{0.53}As$, GaInAsP/InP ($\lambda_g = 1.3$ μm), and InP. The solid line represents the experimental data. The dashed line is the fit to the above gap absorption with the expression, $\alpha = \text{const} \times (E - E_g)^{1/2}$, as discussed in the text.

$$E_g = \frac{1.2398 \; \mu\text{m} \cdot \text{eV}}{\lambda_g}$$

Note that the α fit for the sample labeled $\lambda_g = 1.31$ μm in Fig. 1.6 indicates that its bandgap energy is approximately 0.938 eV, corresponding to a bandgap wavelength closer to 1.32 μm than the cited value of 1.31 μm.

Additional Data

Thermal Effects. The heating-induced shrinkage of the bandgap has been reported for the two endpoint compositions and one intermediate composition of GaInAsP/InP. The Varshni expression for the full temperature dependence of the $Ga_{0.47}In_{0.53}As$ bandgap (Yu and Kuphal, 1984) is

$$E_g(T) = 0.814 - \frac{4.906 \times 10^{-4} T^2}{T + 301} \; \text{eV} \quad (T \text{ in degrees Kelvin}).$$

Evaluating the derivative of this expression at room temperature gives

$$\frac{dE_g}{dT} = -3.7 \times 10^{-4} \; \frac{\text{eV}}{\text{K}}. \quad (Ga_{0.47}In_{0.53}As)$$

The rate of bandgap shift with temperature has been measured (Turner et al., 1964) for InP:

$$\frac{dE_g}{dT} = -2.9 \times 10^{-4} \; \frac{\text{eV}}{\text{K}}. \quad (\text{InP})$$

A second source (Kirillov and Merz, 1983) gives this temperature coefficient of InP as -5.3×10^{-4} eV/K.

Finally, the temperature coefficient of the bandgap of a quaternary composition, $Ga_{0.28}In_{0.72}As_{0.59}P_{0.41}$, has been reported (Madelon and Dore, 1981):

$$\frac{dE_g}{dT} = -3.9 \times 10^{-4} \frac{eV}{K} \quad (Ga_{0.28}In_{0.72}As_{0.59}P_{0.41})$$

AlGaInAs/InP Material System

General Characteristics and Optoelectronic Applications

The range of compositions of the general quaternary, $Al_xGa_yIn_{1-x-y}As$, can produce lattice constants as large as that of InAs ($x = y = 0$) or as small as that of GaAs ($x + y \approx 1$). The most useful members of this quaternary are the compositions that produce a lattice constant equal to that of InP. The InP-lattice-matched compositions of this quaternary are denoted here as AlGaInAs/InP. The AlGaInAs/InP quaternary can be pictured as consisting of mixtures of two ternaries, $Al_{0.48}In_{0.52}As$ and $Ga_{0.47}In_{0.53}As$, each of which lattices matches InP. Because the InAs mole fraction varies only from 0.52 to 0.53 (or, more precisely, from 0.523 to 0.532) over this whole composition range, it is a good approximation to take it to be constant at 0.53. This approximation leads to writing the compositions of AlGaInAs that lattice match InP as

$$(Al_xGa_{1-x})_{0.47}In_{0.53}As, \quad 0 \leq x \leq 1.$$

AlGaInAs/InP and GaInAsP/InP have very similar properties. Both lattice match InP. Both quaternaries have the same low bandgap end point composition, namely, the ternary $Ga_{0.47}In_{0.53}As$, with a room temperature bandgap of 0.75 eV (Fig. 1.7). The upper limits of the two quaternaries' bandgaps are similar. The high bandgap member of AlGaInAs/InP is $Al_{0.48}In_{0.52}As$ with a bandgap of 1.45 eV; the high bandgap member of GaInAsP/InP is InP with a bandgap of 1.35 eV. The choice between the two material systems is usually determined by the epitaxial growth technique. AlGaInAs is the easier system for molecular beam epitaxy (MBE), whereas GaInAsP is easier for the other epitaxial techniques.

Applications of AlGaInAs/InP are similar to those of GaInAsP/InP (see GaInAsP/InP Material System: General Characteristics and Optoelectronic Applications).

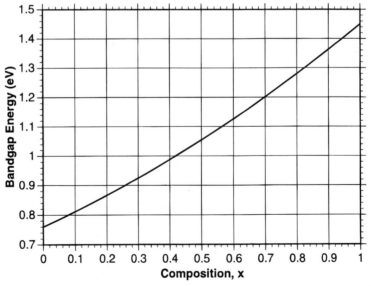

FIGURE 1.7 Compositional dependence of the room temperature bandgap of $(Al_xGa_{1-x})_{0.47}In_{0.53}As$.

Direct Bandgap, Refractive Index and Absorption Coefficient

The quaternary is a direct bandgap material over the whole range of compositions that lattice match InP. Its room temperature bandgap (Olego et al., 1982) is given by:

$$E_g = 0.76 + 0.49x + 0.20x^2.$$

The room temperature refractive index at photon energies below the bandgap has been measured for compositions of $x \geq 0.3$ (Mondry et al., 1992) and $x = 0$ (Chandra et al., 1981), and is graphed in Fig. 1.8.

Absorption data is scarce for the AlGaInAs quaternary, except for the binary end point, $Ga_{0.47}In_{0.53}As$. Because this ternary is common to the GaInAsP/InP material system, its absorption spectrum is displayed along with that of GaInAsP and InP in Fig. 1.6. The absorption spectra of other compositions of the AlGaInAs quaternary can be estimated by comparison with materials of similar direct bandgaps in the GaInAsP/InP system.

AlGaInP Material System

General Characteristics and Optoelectronic Applications

The range of compositions of the general quaternary AlGaInP can produce lattice constants as large as that of InP or as small as that of GaP. Compositions which produce an intermediate lattice constant equal to that of GaAs are the most important members of this quaternary. The GaAs-lattice-matched compositions of this quaternary are denoted here as AlGaInP/GaAs. The AlGaInP/GaAs quaternary can be thought of as consisting of mixtures of two ternaries, $Al_{0.52}In_{0.48}P$ and $Ga_{0.52}In_{0.48}P$, each of which lattice match GaAs. Because the InP mole fractions of these two ternaries are equal (to at least two significant digits) at 0.52, their mixture (the compositions of AlGaInP that lattice match GaAs) can be written as

$$(Al_xGa_{1-x})_{0.52}In_{0.48}P, \quad 0 \leq x \leq 1.$$

The main application of AlGaInP/GaAs is in the construction of red diode lasers. As discussed in the section on AlGaAs Material System (General Characteristics and Optoelectronic Applica-

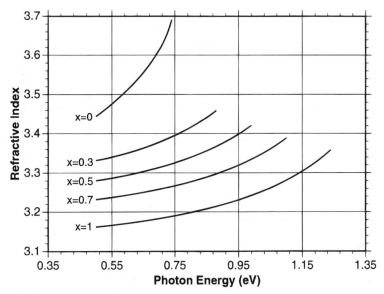

FIGURE 1.8 Compositional dependence of the room temperature refractive index spectrum of $(Al_xGa_{1-x})_{0.47}In_{0.53}As$.

tions) the bandgap and refractive index trends are such that electrical and optical confinement are possible in a single layer of a double heterostructure. Cladding/electron confinement layers are constructed of the aluminum rich compositions, often $Al_{0.52}In_{0.48}P$. Core layers consist of low aluminum compositions, e.g., $(Al_{0.2}Ga_{0.8})_{0.52}In_{0.48}P$. The active quantum well often consists of (lattice matched) $Ga_{0.52}In_{0.48}P$ or strained $Ga_{1-x}In_xP$ with InP mole fractions, x, of 0.6–0.65.

Bandgaps, Refractive Index, and Absorption Coefficient

The most important member of the quaternary $(Al_xGa_{1-x})_{0.52}In_{0.48}P$ is the ternary end point, $Ga_{0.52}In_{0.48}P$. Certain MOCVD epitaxial conditions cause this ternary to grow in an ordered CuPt-type structure (Kondow and Minagawa, 1988). Other MOCVD conditions as well as MBE growth (Wicks et al., 1991) produce a random ternary. The bandgap of the ordered structure is 90 meV lower than that of the random ternary. The possibility of ordering and its effects on the bandgap energy across the composition range of the quaternary $(Al_xGa_{1-x})_{0.52}In_{0.48}P$ are not known. The data reported here apply to the random alloys.

The lower bandgap compositions of the quaternary are direct, like $Ga_{0.52}In_{0.48}P$; the higher bandgap compositions are indirect, like $Al_{0.52}In_{0.48}P$. The dependence of the bandgap energies on composition can be approximated as linear interpolations between the two ternary end points. The room temperature direct gap, E_Γ, of the (random) quaternary has been reported (Kato et al., 1994):

$$E_\Gamma(eV) = 1.89 + 0.64x.$$

The room temperature indirect gap involving the X minimum of conduction band and Γ maximum of the valence band, E_X, can be approximated by a linear interpolation between the values reported for $Ga_{0.52}In_{0.48}P$ (Nelson and Holonyak, 1976) and $Al_{0.52}In_{0.48}P$ (Onton and Chicotka, 1970):

$$E_X(eV) = 2.25 + 0.09x.$$

The direct/indirect crossover is at the composition $x_c \sim 0.65$, and a room temperature bandgap of $E_g^c \sim 2.3$ eV. The compositional variations of these two bandgaps are graphed in Fig. 1.9.

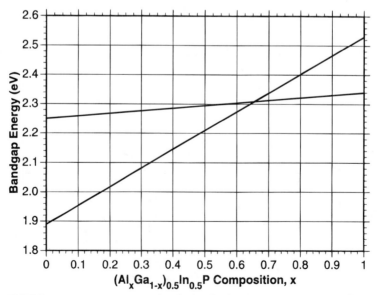

FIGURE 1.9 Room temperature bandgaps of $(Al_xGa_{1-x})_{0.52}In_{0.48}P$. Data applies to random alloy materials.

III–V Semiconductor Materials

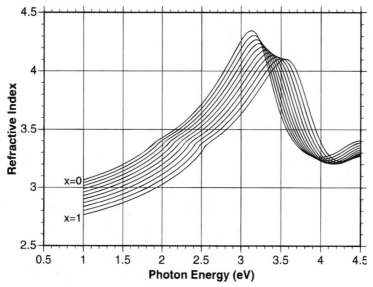

FIGURE 1.10 Modeled room temperature refractive index spectra of $(Al_xGa_{1-x})_{0.52}In_{0.48}P$. *Source:* Kato, H., Adachi, S., Nakanishi, H. et al. 1994. *Jpn. J. Appl. Phys.* 33(186). With permission.

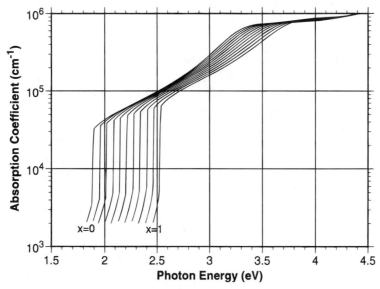

FIGURE 1.11 Modeled room temperature absorption spectra of $(Al_xGa_{1-x})_{0.52}In_{0.48}P$.

The refractive index of $(Al_xGa_{1-x})_{0.52}In_{0.48}P$ has been experimentally measured and modeled (Kato et al., 1994).[29] The room temperature indexes are graphed in Fig.1.10.

The model used to generate the refractive index data of $(Al_xGa_{1-x})_{0.52}In_{0.48}P$ in Fig. 1.10 can also be applied to generate absorption coefficient data. (Kato et al., 1994). The modeled above-bandgap absorption coefficients of the quaternary are exhibited in Fig. 1.11.

GaSb Material System

General Characteristics and Optoelectronic Applications

The GaSb material system is more complicated than the others listed in this chapter in that it consists of two quaternaries, AlGaAsSb and GaInAsSb, each of which lattice matches GaSb substrates. Each quaternary can be envisioned as a mixture of GaSb and a GaSb-lattice-matched ternary:

$$Al_xGa_{1-x}As_ySb_{1-y} = (GaSb)_{1-x}(AlAs_{0.083}Sb_{0.917})_x$$

and

$$Ga_{1-x}In_xAs_ySb_{1-y} = (GaSb)_{1-x}(InAs_{0.911}Sb_{0.089})_x.$$

The alloying of GaSb with $AlAs_{0.083}Sb_{0.917}$ increases the AlGaAsSb's bandgap above that of GaSb, whereas the alloying of GaSb with $InAs_{0.911}Sb_{0.089}$ decreases GaInAsSb's bandgap (at least the direct bandgap, E_Γ) below that of GaSb (see Fig. 1.12). Equating the As subscript on both sides of the above expressions gives the GaSb lattice matching conditions, $y = 0.083x$ and $y = 0.911x$, for $Al_xGa_{1-x}As_ySb_{1-y}$ and $Ga_{1-x}In_xAs_ySb_{1-y}$, respectively.

Common optoelectronic applications of these GaSb-based materials involve detectors and emitters operating in the 2–3 μm wavelength region. In such applications, GaInAsSb is often the active layer of the laser or detector, and AlGaAsSb is the cladding/electron confinement layer of the laser or the window/cap layer of the detector.

Because the bandgaps of GaSb and InAs differ by only 0.6%, AlGaAsSb and GaInAsSb can also be lattice matched to InAs substrates. The range of bandgaps of GaInAsSb/InAs is smaller than that of GaInAsSb/GaSb, and the former material is generally used more rarely than the latter. Additionally AlGaAsSb and GaInAsSb can lattice match InP, but are rarely employed because of competition from the GaInAsP/InP and AlGaInAs/InP systems.

Bandgaps, Refractive Indices, and Absorption Coefficients

GaSb is just barely a direct bandgap material, with the conduction band minimum at the L-point only a few tens of meV above that at the Γ point. Because this energy difference is on the order of k_BT at room temperature, a sizable fraction of the conduction band electrons populate indirect minima, making the material unsuitable as the active layer for an efficient room temperature laser. This difficulty is rapidly alleviated by a rising E_L and a falling E_Γ as the composition moves away from GaSb in the GaInAsSb/GaSb quaternary system. The GaInAsSb/GaSb quaternary remains direct over its whole composition range, spanning direct gaps from that of GaSb, $E_\Gamma = 0.72$ eV, down to that of $InAs_{0.911}Sb_{0.089}$, $E_\Gamma \sim 0.3$ eV.

The AlGaAsSb/GaSb quaternary is indirect or nearly indirect over its whole bandgap range. Its bandgap ranges from that of the barely direct gap of GaSb, $E_\Gamma = 0.72$ eV, to that of $AlAs_{0.08}Sb_{0.92}$, $E_X \sim 1.6$ eV. Moving away from the GaSb end point, the bandgap of the AlGaAsSb/GaSb quaternary rises from that of GaSb, becoming indirect around 1 eV when E_Γ crosses E_L at an aluminum composition just above $x = 0.2$. At yet higher aluminum concentrations, $x \sim 0.45$, the indirect nature changes as E_X becomes the lowest lying bandgap

The compositional dependence of the bandgaps of AlGaAsSb/GaSb and GaInAsSb/GaSb have been modeled (Adachi, 1987) and are graphed in Fig. 1.12.

Although the modeling is somewhat involved, the results of the modeling can be fit with simple quadratic expressions with errors of typically only a few meV. The fits are listed in Table 1.2.

The room temperature refractive index of AlGaAsSb/GaSb has been measured (Alibert et al., 1991) and modeled (Adachi, 1987; Alibert et al., 1991) at room temperature. The two models agree reasonably well with each other and with available experimental data. We use the first (Adachi, 1987) of these two models here, as it also applies to GaInAsSb/GaSb. The room temperature below-gap refractive indices of these two quaternaries that lattice match GaSb are graphed in Figs. 1.13 and 1.14.

As do most of the III–V materials, the larger bandgap AlGaAsSb/GaSb quaternaries have smaller refractive indices.

III–V Semiconductor Materials

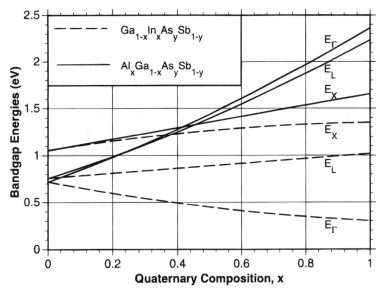

FIGURE 1.12 Modeled room temperature bandgaps vs. compositions of $Al_xGa_{1-x}As_ySb_{1-y}$ and $Ga_{1-x}In_xAs_ySb_{1-y}$ lattice matched to GaSb. Note that the $x = 0$ binary endpoint, GaSb, is common to both quaternaries. *Source:* Adachi, S. 1987. *J. Appl. Phys.* 61, 4869. With permission.

TABLE 1.2 Quadratic Fits to Modeled Room Temperature Bandgaps of AlGaAsSb/GaSb and GaInAsSb/GaSb

	$Al_xGa_{1-x}As_ySb_{1-y}$ $0 \leq x \leq 1; y = 0.083x$	$Ga_{1-x}In_yAs_ySb_{1-x}$ $0 \leq x \leq 1; y = 0.911x$
E_Γ (eV)	$0.717 + 1.233\,x + 0.414\,x^2$	$0.721 - 0.670\,x + 0.0252\,x^2$
E_X (eV)	$1.051 + 0.612\,x - 0.0937\,x^2$	$1.057 + 0.537\,x - 0.0242\,x^2$
E_L (eV)	$0.756 + 1.075\,x + 0.0404\,x^2$	$0.760 + 0.261\,x - 0.00190\,x^2$

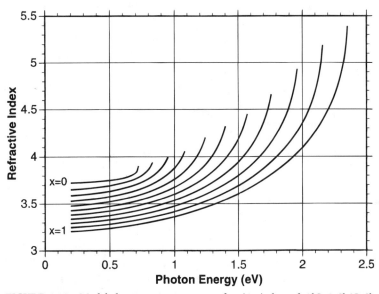

FIGURE 1.13 Modeled room temperature refractive index of AlGaAsSb/GaSb. Below-bandgap photon energies are displayed. *Source:* Adachi, S. 1987. *J. Appl. Phys.* 61, 4869. With permission.

The behavior of the refractive index of GaInAsSb/GaSb is anomalous in that the larger bandgap materials have larger refractive indices, at least at below gap photon energies. The cause of this anomaly is, perhaps, related to the aspect of the band structure that causes both the X and L bandgaps to have a compositional trend is opposite to that of the fundamental gap at Γ (see Fig. 1.12). A practical consequence of this index anomaly affects the design of a typical diode laser structure (double heterostructure or quantum well heterostructure). In most III–V material systems the cladding layer of the waveguide (lower refractive index) also functions as an electron confinement layer (larger bandgap). This scheme will not work in the GaInAsSb system, because the trend toward high bandgaps is not accompanied by the trend toward low refractive indices. Fortunately, the problem can be solved by the constructing laser cladding/electron confinement layers of AlGaAsSb, which have a higher bandgap and lower refractive index than those of the GaInAsSb core and quantum well.

Very little has been published on the room temperature absorption of the GaSb-based materials. A few experimental data points exist (Becker et al., 1961) only for GaSb itself, as shown in Fig. 1.15.

Fitting the simple expression, $\alpha \propto (E - E_g)^{1/2}$ to the three above gap data points of Fig. 1.15 gives the approximation:

$$\alpha_{GaSb} = 22600(\text{cm}^{-1}\ \text{eV}^{-1/2})(E - 0.71\ \text{eV})^{1/2}.$$

In the absence of more complete data, an estimate of the absorption coefficient of other materials in the AlGaAsSb/GaSb or GaInAsSb/GaSb systems could be made by replacing the 0.71 eV in the above expression with the relevant E_Γ bandgap.

GaAsP Material System

General Characteristics and Optoelectronic Applications

GaAs$_x$P$_{1-x}$ has a band structure similar to that of Al$_x$Ga$_{1-x}$As in that it is a direct gap material in its GaAs-rich composition range and crosses over to an indirect bandgap in its GaAs-deficient compositions. It has long been the dominant material for visible light emitting diodes, although this is in the process of changing as newer, higher performance materials become available. The

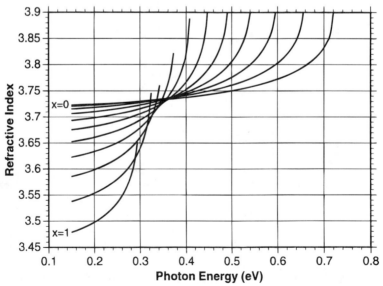

FIGURE 1.14 Modeled room temperature refractive index in GaInAsSb/GaSb. Below bandgap photon energies are displayed.

III–V Semiconductor Materials

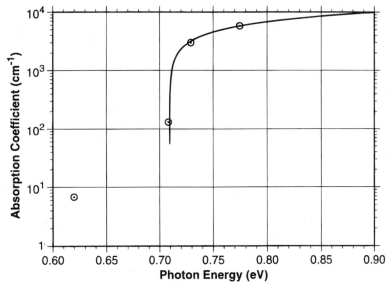

FIGURE 1.15 Room temperature absorption spectrum of GaSb. The circles are experimental data points, the line is a fit to the three highest energy data points, as discussed in the text.

GaAsP approach to visible LEDs involves simple epitaxial growth techniques, thus, minimizing cost, but relatively low performance results because of lattice matching and band structure factors. Only the two binary end points, GaAs and GaP, lattice match available substrates; the general quaternary is grown mismatched on either GaAs or GaP substrates, resulting in dislocations. The direct, lower bandgap compositions are used for red LEDs; the higher compositions (GaP) are used for yellow-green LEDs despite their indirect bandgap. Nitrogen doping of the indirect gap GaAsP compositions increases their radiative efficiencies to modest levels ($10^{-4} - 10^{-3}$) (Campbell

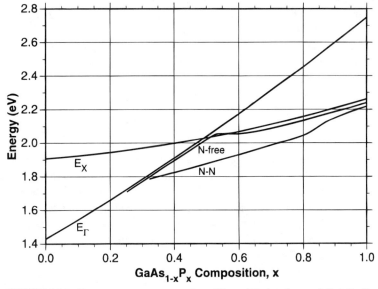

FIGURE 1.16 Room temperature energies of E_Γ and E_X bandgaps of GaAsP. Also displayed are the electroluminescence peak energies of nitrogen-free GaAsP and the electroluminescnece peak involving nitrogen-nitrogen pairs (N-N) of nitrogen doped GaAsP.

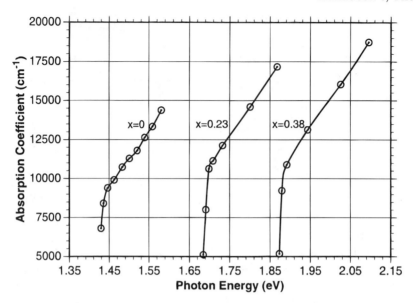

FIGURE 1.17 Compositional dependence of the room temperature absorption spectra near the fundamental absorption edge of $GaAs_{1-x}P_x$.

et al., 1974). Strong electron confinement by the nitrogen isoelectronic impurity relaxes the k-selection rules, enabling light emission from the indirect material. Emission efficiencies, however, are still at least an order of magnitude lower than that of direct material.

Bandgaps, Electroluminescence Energies and Absorption Spectra

The direct bandgaps of a series of 11 samples of $GaAs_{1-x}P_x$ were measured by Thompson et al., (1966) in an early work. In this paper, the data is fit with an expression produce is a value for the GaAs bandgap that is 15 meV or so larger than the presently accepted value. Although the GaAs bandgap produced by the original fit appears to be slightly too large, the original data itself does produce the correct value of the GaAs bandgap. The error in the fit appears to result from errors in the measurements of the compositions of the ternary samples. A more reliable fitting expression can be obtained from the original data by weighting the binary end points more heavily than the ternary points, because there can be no error in binary composition measurements. This fitting procedure produces the following result:

$$E_g(eV) = 1.428 + 1.125x + 1.952x^2.$$

The crossover of the direct and indirect bandgaps of $GaAs_xP_{1-x}$ is near a composition of $x = 0.5$ and a bandgap of 2 eV. At compositions greater than $x = 0.5$, the X indirect bandgap is the lowest. The Γ (direct) and the X bandgaps are displayed as functions of composition in Fig. 1.16. Also displayed in the figure are the peak electroluminescence energies (Craford and Holonyak, 1976) of N-doped and N-free GaAsP, useful information in the analysis of LEDs.

Optical absorption spectra are available for only a few compositions (Hasehawa et al., 1984) and are displayed in Fig. 1.17.

References

Adachi, S. 1987. *Appl. Phys.* 61, 4869.
Adachi, S. 1988. *Phys. Rev.* B 38, 12345.

Alibert, C., Skouri, M., Joullie, A., Benouna, M., and Sadiq, S. 1991. *J. Appl. Phys.* 69, 3208.
Aspnes, D. E., Kelso, S. M., Logan, R. A., and Bhat. R. J. 1986 *J. Appl. Phys.* 60, 754.
Bacher, F. R., Blakemore, J. S., Ebner, J. T., and Arthur, J. S. 1988. *Phys. Rev.* B 37, 2551.
Becker, W. M., Ramdas, A. K., and Fan, H. Y. 1961. *J. Appl. Phys. Suppl.* 32, 2094.
Campbell, J. C., Holonyak, N., Jr., Craford, M. G., and Keune, D. L. 1974. *J. Appl. Phys.* 45, 4543.
Cardona, M., 1961. In *Proc. Int. Conf. on Semiconductor Physics, Prague, 1960,* p. 388.
Cardona, M., 1961. *J. Appl. Phys.* 32, 958.
Cardona, M., 1965. *J. Appl. Phys.* 36, 2181.
Chandra, P., Coldren, L. A., and Strege, K. E. 1981. *Electron, Lett.* 17, 6.
Chang, K. H., Lee, C. P., Wu, J. S., Liu, D. G., Liou, D. C., Wang, M. H., Chen, L. J., and Marais, M. A. 1991. *J. Appl. Phys.* 70, 4877.
Choi, H. K. and Wang, C. A. 1990. *Appl. Phys. Lett.* 57, 321.
Craford, M. G. and Holonyak, N., Jr. In *Optical Properties of Solids New Developments,* ed. Seraphin, B. O. Chapt. 5, North-Holland, Amsterdam, The Netherlands, 1976.
Feketa, D., Chan, K. T., Ballantyne, J. M., and Eastman, L. F. 1986. *Appl. Phys. Lett.* 49, 1659.
Guzzi, M., Grilli, E., Oggioni, S., Stachli, J. L., Bosio, C., and Pavesi, L. 1992. *Phys. Rev.* B 45, 10951.
Hasehawa, S., Tanaka, A., and Sukegawa, T. 1984. *J. Appl. Phys.* 55, 3188.
Jenkins, D. W., 1990. *J. Appl. Phys.* 68, 1848.
Jensen, B. and Torabi, A. 1983. *J. Appl. Phys.* 54, 3623.
Kato, H., Adachi, S., Nakanishi, H., and Ohtsuka, K. 1994. *Jpn. J. Appl. Phys.* 33, 186.
Kirillov, D. and Merz, J. L. 1983. *J. Appl. Phys.* 54, 4104.
Kondow, M. and Minagawa, S. 1988. *J. Appl. Phys.* 64, 793.
Madelon, R. and Dore, M. 1981. *Solid State Commun.* 39, 639.
Madelung, O. ed. 1982. *Landholt-Bornstein, New Series, Group III*. Volume 17, Springer-Verlag, Berlin, Germany.
Marple, D. T. F., 1964. *J. Appl. Phys.* 35, 1241.
Mondry, M. J., Babic, D. I., Bowers, J. E., and Coldren, L. A. 1992. *IEEE Phontonics Technol. Lett.* 4, 627.
Nahory, R. R., Pollack, M. A., Johnston, W. D., Jr., and Barnes, R. L. 1978. *Appl. Phys. Lett.* 33, 659.
Nelson, R. J. and Holonyak, N., Jr., 1976. *J. Phys. Chem. Solids* 37, 629.
Newman, R. 1958. *Phys. Rev.* 111, 1518.
Olego, D., Chang, T. Y., Silberg, E., Caridi, E. A., and Pinczuk, P. 1982. *Appl. Phys. Lett.* 41, 476.
Onton, A. and Chicotka, R. J., 1970. *J. Appl. Phys.* 41, 4205.
Pankove, J. I. 1975. *Optical Processes in Semiconductors,* Dover Publications, New York.
Pettit, G. D. and Turner, W. J. 1965. *J. Appl. Phys.* 36, 2081.
Sell, D. D., Casey, H. C., and Weeht, K. W. 1974. *J. Appl. Phys.* 45, 2650.
Thompson, A. G., Cardona, M., Shaklee, K. L., and Woolley, J. C. 1966. *Phys. Rev.* 146, 601.
Thurmond, C. D. 1975. *J. Electrochem. Soc.* 122, 1133.
Turner, W. J., Reese, W. E., and Pettit, G. D. 1964. *Phys. Rev.* 136, A1467.
Wicks, G. W., Koch, M. W., Varriano, J. A., Johnson, F. G., Wie, C. R., Kim, H. M., and Colombo, P. 1991. *Appl. Phys. Lett.* 59, 342.
Yokouchi, N., Uchida, T., Miyamoto, T., Inaba, Y., Koyama, F., and Iga, K. 1992. *Jpn. J. Appl. Phys.* 1255.
Yu, P. W. and Kuphal, E. 1984. *Solid State Commun.* 49, 907.

2
Optical Properties and Optoelectronic Applications of II-VI Semiconductor Heterostructures

H. Luo
State University of New York at Buffalo

A. Petrou
State University of New York at Buffalo

2.1 Introduction 24
2.2 Overview 25
2.3 Quantum Confinement in Wide-Gap II-VI Semiconductors Heterostructures 27
2.4 ZnSe-Based Blue-Green Light Emitting Devices 31
  Quantum Well Structures for Blue-Green Lasers • Doping of ZnSe • ZnSe-Based Laser Diodes • The Lasing Mechanism in ZnSe-Based Laser Diodes • Future Research and Development
2.5 Type-II Heterostructures 38
2.6 HgTe/CdTe Type-III Heterostructures 39
2.7 Diluted Magnetic Semiconductor Heterostructures 42
  Magnetic Field Induced Type-I to Type-II Transition • Wave Function Mapping • Spin Superlattices

2.1 Introduction

Semiconductor heterostructures of II-VI compounds offer a wide range of properties which can be exploited for fundamental studies and for device applications. We will review recent studies addressing these issues, emphasizing ZnSe-based structures for blue-green light emitting devices. Other systems, such as type-II and diluted magnetic semiconductor heterostructures and narrow-gap type-III heterostructures, will be discussed.

Semiconductor applications have long been dominated by electronic devices, i.e., transistor-based integrated circuits. However, optoelectronic devices and photonic applications are becoming increasingly important—for example, optical data storage and communication, laser printing, flat panel displays, and so on. II-VI semiconductors are of interest, primarily, because of their wide range of optoelectronic properties. This is exemplified by the wide spectrum of energy gaps E_g of these materials, ranging from 0 to over 3.8 eV, as listed in Table 2.1. Studies of most II-VI

TABLE 2.1 Energy Gaps for II-VI Compounds for Liquid Helium (LHe), Liquid Nitrogen (LN$_2$), and Room Temperature (RT). (Mn-Based Chalcogenides Are Included)

Compound	T	E$_g$(eV)	Compound	T	E$_g$(eV)
ZnS	LHe	3.84	HgS	LHe	<0
	LN$_2$	3.78		LN$_2$	~0
	RT	3.68		RT	>0
ZnSe	LHe	2.82	HgSe	LHe	−0.27
	LN$_2$	2.80		LN$_2$	−0.205
	RT	2.71		RT	+0.10
ZnTe	LHe	2.40	HgTe	LHe	−0.300
	LN$_2$	2.38		LN$_2$	−0.252
	RT	2.28		RT	−0.146
CdS	LHe	2.56	MnS	LHe	~3.6
	LN$_2$	2.55		LN$_2$	~3.5
	RT	2.45		RT	~3.4
CdSe	LHe	1.83	MnSe	LHe	3.3
	LN$_2$	1.82		LN$_2$	3.14
	RT	1.75		RT	2.9
CdTe	LHe	1.606	MnTe	LHe	3.2
	LN$_2$	1.596		LN$_2$	3.05
	RT	1.528		RT	2.9

semiconductors can be traced back to the 1960s, concentrating on the fundamental properties of bulk materials. Most of the excitement in these materials, however, came from heterostructures after the introduction of molecular beam epitaxy (MBE), as demonstrated in the area of blue-green laser diodes (Haase et al., 1991). Considering the current interests in these materials and their application prospects, we will focus on heterostructures of II-VI semiconductors.

II-VI semiconductor heterostructures are unique in many ways. First, the spectrum of energy gaps allows the exploration of optical and optoelectronic phenomena and device applications covering wavelengths ranging from the far infrared to the ultraviolet. Furthermore, II-VI semiconductor heterostructures involve a wide range of band offsets, which adds variety and flexibility to band gap engineering. The II-VIs also have stronger polarity compared to other well studied semiconductors, e.g., III-Vs, Si and Ge, providing testing ground for various effects related to lattice distortion and vibration. Magnetic ions, such as Mn^{++} and Fe^{++}, can also be easily incorporated into II-VIs, which have been traditionally referred to diluted magnetic semiconductors (although a concentration of magnetic ions as large as 100% can be achieved with MBE).

In this paper, we will review various optical and optoelectronic studies involving II-VI semiconductor heterostructures and the development of II-VI optoelectronic devices. In light of the recent rapid progress in the fabrication of ZnSe-based blue-green laser diodes and their importance to the future of II-VI materials, we will focus our attention on the progress and remaining problems in this area.

2.2 Overview

Modern growth techniques, such as MBE and metal-organic chemical vapor phase deposition (MOCVD), have had a major impact on the evolution of the general field of II-VI semiconductors in several important ways. First, it was demonstrated that the II-VIs can be grown in monolithic multilayer structures, making it possible to form a variety of quantum well systems involving type-I, type-II and type-III (a zero-gap and open-gap combination) band alignments. Because of the various lattice constants of the II-VIs, both lattice matched systems and strained layer structures have been achieved. Second, it was demonstrated that epitaxy can be used to form new phases which do not exist naturally in the bulk. All II-VI compounds except HgS are

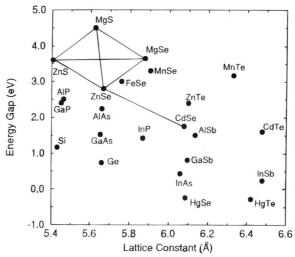

FIGURE 2.1 Energy gaps and lattice parameters for cubic group IV, III-V and II-VI semiconductors.

tetrahedrally coordinated in their stable bulk phases, crystallizing either in the cubic (zincblende) or in hexagonal (wurtzite) structures. It was demonstrated that epitaxy allows one to grow zincblende phases of CdSe (Samarth et al., 1989), MnTe (Durbin et al., 1989), MnSe (Kolodziejski et al., 1986) and FeSe (Jonker et al., 1988)—all of which have electronic and structural properties closely related to the common II-VIs. These new phases can, therefore, be alloyed and combined with the zinc blende II-VIs, thus, dramatically extending the scope of the physical phenomena that can be investigated, as will be discussed below.

Lattice constants and band gaps are essential as guides in designing various heterostructures. In Fig. 2.1, we show the relationship between energy gaps and lattice parameters of II-VI compounds. As evident in the figure, the II-VIs cover a wide range of energy gaps. It should be pointed out that II-VIs resemble III-V compounds in many ways. A useful guide can be obtained from a portion of the Periodic Table containing these elements, with their tetrahedral radii r_i in tetrahedrally-bonded compounds, shown in Table 2.2. We note that, for any zinc blende or diamond-structure crystal, the lattice constant is given with reasonable accuracy by $(4/\sqrt{13})(r_i + r_j)$, where r_i and r_j are the tetrahedral radii of nearest-neighbor atoms (e.g., of Zn and Se in ZnSe). Thus, Table 2.2 also provides a very useful and compact guide for selecting lattice matched systems not only between II-VI compounds themselves, but between the II-VIs and either III-V or Group-IV materials. It is also clear that, except for the elements in the first row (i.e., Al, Si, P, and S), elements in the same row have the same tetrahedral radii. This means that II-VI compounds will have approximately the same lattice constant as the III-V counterpart, which involves elements closest in the Periodic Table (Table 2.2) to the II-VI elements. For example,

TABLE 2.2 Tetrahedral Radii r_i for Group II, III, IV, V, and VI Elements (in Å)

II	III	IV	V	VI
	Al	Si	P	S
	1.230 Å	1.173 Å	1.128 Å	1.127 Å
Zn	Ga	Ge	As	Se
1.225 Å	1.225 Å	1.225 Å	1.225 Å	1.225 Å
Cd	In	α-Sn	Sb	Te
1.405 å	1.405 Å	1.405 Å	1.405 Å	1.405 Å
Hg				
1.402 Å				

ZnSe has a lattice constant very close to that of GaAs, noting that Ga and As are the Group-III and Group-V elements closest to Zn and Se, respectively.

Although the offset of a given semiconductor heterostructure is often a complicated matter, the type of band alignment is often the same between the counterparts of II-VI and III-VI compounds (defined by the closest elements in the Periodic Table, as was used above). For instance, ZnCdSe/ZnSe and GaInAs/GaAs are type-I systems, and ZnTe/CdSe and GaSb/InAs have type-II band alignments.

Furthermore, there are also analogies in doping II-VI and III-V compounds. Doping II-VIs has been a more difficult task in most cases. For a given II-VI compound, the relative difficulty of n-type and p-type doping is similar to that for its III-V counterpart. For example, in ZnSe n-type doping is much easier than p-type doping, which is the same as for GaAs. All of these similarities can be traced back to the atomic states of these elements that affect the absolute positions of the conduction and the valence bands, which involve complicated arguments. From a practical point of view, however, such similarities can be used as simple guides for exploring new heterostructures and for transferring techniques used in dealing with existing compounds to their lesser known counterparts.

2.3 Quantum Confinement in Wide-Gap II-VI Semiconductor Heterostructures

Research and development of II-VI semiconductor heterostructures has long been driven by the possibility of optoelectronic applications in the blue region. Two systems, namely, ZnSe- and ZnTe-based quantum well structures, have attracted most of the attention and have experienced some degree of success. We will focus on the quantum confinement effects in wide-gap II-VI semiconductor quantum wells and will present a qualitative description of the reason for quantum confinement being strong in some systems and weak in others.

Modification of electronic states of bulk materials using semiconductor heterostructures provides the possibility of tailoring the electronic and optical properties of the structures. Excitonic effects in wide gap II-VI semiconductors are particularly strong, because of the ionic nature of the bonds in these materials (which results in smaller dielectric constants and consequently stronger Coulomb interaction) and of the large effective masses of the carriers that lead to more tightly bound excitons (i.e., smaller Bohr radii). One of the most important properties that can be achieved in quantum well structures is the quantum confinement of carriers. It is responsible for a list of characteristics in type-I quantum wells, which have been utilized for various purposes, such as the increase of optical transition probability, enhancement of the exciton binding energy, and quantum confined stark effect (QCSE), to name a few.

The degree of confinement is determined by the depths of the quantum wells in the conduction and the valence bands, which in turn are derived from the energy gap difference between the two materials comprising the heterostructure and their alignment (or offset). The band alignment of wide gap II-VI semiconductors relative to the vacuum level is shown in Fig. 2.2. Examining the positions of the edges in the conduction and the valence bands for the various compounds, one finds a small conduction band offset when two compounds share the same cation, and a small valence band offset when there is a common anion. In other words, although the "common anion rule" has been shown to be inaccurate, it does provide a very good rule of thumb in II-VI systems.

Most type-I quantum wells consisting of II-VI compounds and alloys that were fabricated before 1990 did not have sufficient well depths to confine either electrons, or holes at room temperature. One of the extensively studied-systems is ZnMnSe/ZnSe (a common anion system) (Kolodziejski et al., 1985), which showed reasonable confinement effects of electrons and holes at low temperatures. The lack of deep wells in the valence band was evidenced by the temperature dependence of exciton

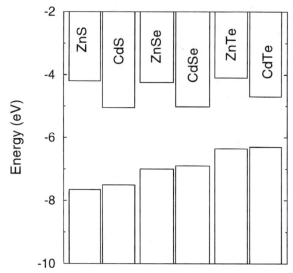

FIGURE 2.2 Band alignment for wide-gap II-VI semiconductors on an absolute scale. Note the relatively good agreement with the common-anion and common-cation rule.

absorption in this system. The strong exciton absorption at low temperatures deteriorates rapidly as the temperature is increased and totally disappears at room temperature. Attempts to observe stimulated emission in these structures revealed the same temperature behavior (Bylsma et al., 1985).

The reason for the temperature dependence of excitons originates from the strong electron-LO phonon coupling (i.e., Fröhlich interaction). The LO phonon energies are typically larger than the exciton binding energies. In the case of ZnSe, the LO phonon energy is 32 meV, whereas the exciton binding energy is 18 meV. Thus at high temperatures, all excitons dissociate into electron-hole pairs by LO phonons. The confinement of electrons and holes increases the exciton binding energy E_{ex}. When E_{ex} becomes larger than the LO phonon energy, the exciton dissociation rate will be greatly reduced. The enhancement of the Coulomb interaction, however, strongly depends on the height of the confining potentials in both the conduction and the valence bands. With the band alignments shown in Fig. 2.2, it appears difficult to find a ternary system in which wells in both the conduction and the valence bands are sufficiently deep. The first room temperature exciton absorption was observed in ZnCdTe/ZnTe quantum wells. Although it is one of the common anion systems, a strong carrier confinement effect was observed when well width was reduced to values less than that of the exciton Bohr radius. The same effect was soon observed in another common anion system, namely, ZnCdSe/ZnSe quantum wells (Ding et al., 1990).

The effect of carrier confinement in ZnCdSe/ZnSe quantum wells was demonstrated by the study of the exciton absorption as a function of well width and temperature. In Fig. 2.3, the absorption coefficient is plotted for two multiquantum well structures, one with a well width of 200 Å and the other 35 Å, both having a Cd concentration of 24%. The confinement effect can be seen from the blue shift of the exciton peak when the well width is decreased. The temperature dependence of the exciton absorption coefficients α in the two systems is also drastically different. The coefficient α in the 200 Å wells shows a bulk like behavior, disappearing when the temperature approaches 200 K. In the 35 Å wells, however, the exciton absorption survives at much higher temperatures and is observable at temperatures as high as 400 K. This indicates that the excitons in the 35 Å wells have a binding energy greater than the LO phonon energy.

The linewidth broadening of the quasi-2D case is plotted in Fig. 2.4 as a function of temperature, together with the 3D case in the one-phonon scattering model (see Chemla et al., 1988)

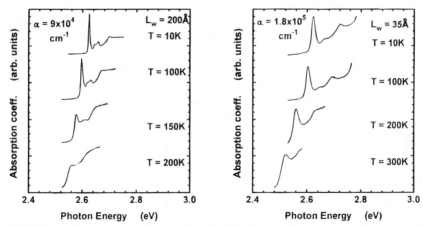

FIGURE 2.3 Absorption coefficient of two ZnCdSe/ZnSe multi-quantum well samples with well thicknesses L_w = 200 Å (left panel) and L_w = 35 Å (right panel), as a function of temperature.

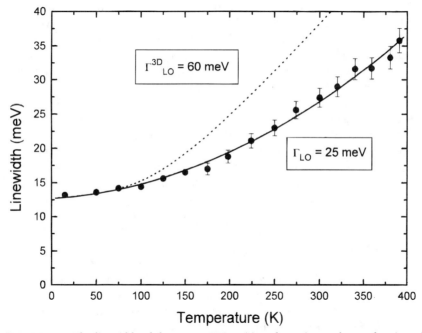

FIGURE 2.4 The linewidth of the n = 1 HH exciton absorption peak as a function of temperature of a multiquantum well sample with L_w = 30 Å. The Fröhlich interaction parameter obtained from a theoretical fit (the solid line) is Γ_{LO} = 28 meV. For comparison, the behavior of bulk ZnSe is shown by the dashed line.

$$\Gamma(T) = \Gamma_0 + \frac{\Gamma_{LO}}{e^{\hbar\omega_{LO}/kT} - 1}, \qquad (2.1)$$

where Γ_{LO} is the extrapolated linewidth at $T = 0$, and Γ_{LO} a phenomenological parameter, which describes the strength of the Frölich interaction. The value of Γ_{LO} is estimated from the data to be 25 meV, which is significantly less than the bulk value of 60 meV. The quasi-2D nature is further studied by magnetoabsorption experiments in fields up to 24 Tesla. Additional information is provided by the simultaneous observation of both 1s (n = 1) heavy-hole (HH) exciton and 2s (n = 1) HH exciton absorption peaks. The low field behavior of the two peaks is very similar.

At fields greater than 10 Tesla, the 2s exciton peak shows a much larger field dependence. At $B = 23.5$ Tesla, the field-induced shift of the 2s exciton peak is six times larger than that of the 1s state, which is consistent with the results of 2-D excitons (Akimoto and Hasegawa, 1966). The advantage of observing the 2s exciton peak is that it allows an estimate of the exciton binding energy. The difference between the 1s and the 2s exciton transition energies is 0.75 Rydberg^{3D} in the 3D case, and 0.88 Rydberg^{2D} for 2D excitons. Therefore, the exciton binding energy E_{ex} of the observed quasi-2D exciton can be estimated from $0.75 < (E_{2s} - E_{1s})/E_{ex} < 0.88$. The observed value of $E_{2s} - E_{1s}$ is 31 meV. Thus, the exciton binding energy in the quasi-2D case should be 40 meV $> E_{ex} >$ 35 meV, which is larger than the LO phonon energy of 32 meV.

Detailed studies are needed to illustrate why some structures demonstrate large confining potentials for carriers (such as ZnCdSe/ZnSe and ZnCdTe/ZnTe), whereas other structures show a weaker confinement effect (e.g., ZnMnSe/ZnSe and ZnSe/ZnSSe). Most of the observations can be qualitatively understood by considering the effects of strain in these systems. It can be seen in Fig. 2.1 that all common-anion or common-cation ternary combinations of wide-gap II-VIs will be lattice-mismatched. Because the offsets can be very small in these two systems, the offset introduced by strain can be significant. When a layer is under compressive strain (in the plane of the layer, as is the case in quantum well structures), the conduction band will move to a higher energy. Meanwhile, the heavy hole and the light hole bands split, with the heavy hole moving up in energy and the light hole to a lower energy. On the other hand, when the layer is under tensile strain, all three bands mentioned above move in opposite directions compared to the case of compressive strain. In all wide-gap II-VI compounds, the heavy-hole to conduction-band transition is stronger than that of the light hole to the conduction band. Thus, the effects of strain on the heavy hole and the conduction band are particularly important.

When the deformation potentials of the two materials involved in a quantum well structure are similar, the offset for any given band (e.g., the heavy-hole band) is determined mainly by the lattice mismatch and is not sensitive to whether the wells or the barriers are strained. We will use an example in which the wells are strained. For common-anion quantum wells (i.e., shallow wells in the valence band), it will be desirable to have the layer of the well under compressive strain, thus, moving the heavy-hole band edge up in the well, increasing the well depth for the heavy-hole quantum well. On the other hand, in dealing with shallow quantum wells in the conduction band in the case of common-cation systems, a tensile strain will lower the conduction band edge in the well.

For common anion structures, such as ZnMnSe/ZnSe and ZnMnTe/ZnTe quantum wells, the lattice constants of the well layers (ZnSe and ZnTe) are smaller than that of the barriers. Thus, the quantum well layers are under tensile strain. As a result, the effect of strain reduces the already small band offset for the heavy holes. This is consistent with the lack of success in achieving sufficient quantum confinement in these quantum wells. For ZnCdSe/ZnSe and ZnCdTe/ZnTe quantum wells, which have common anions, the quantum well layers (ZnCdSe and ZnCdTe) have a larger lattice constant than that in the barriers. Therefore, the heavy hole confining potentials will be larger due to the compressive strain, which is responsible for the observations of room temperature exciton absorption in such quantum wells, as discussed earlier. The same effect is also true for common-cation quantum wells. For example, in the ZnSe/ZnSSe system, the well layers (i.e., ZnSe), have the larger lattice constant of the two constituents. The compressive strain in the ZnSe layers, therefore, lifts its conduction-band edge, reducing the well depth in the conduction band, which is consistent with experimental observations.

In the examples discussed above, we have assumed that the barriers are lattice-matched to the buffer layers and only the wells are strained. The effect, however, remains qualitatively the same when the condition is changed to the opposite limit where the barriers are strained and the wells are unstrained. Furthermore, the same conclusion can be extended to situations where both the wells and the barriers are strained, which can be caused, for example, by a buffer layer with a lattice constant different from those in either the wells or the barriers.

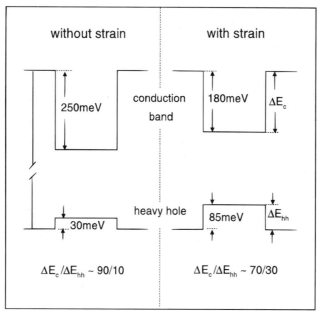

FIGURE 2.5 Band alignment in the ZnSe/Zn$_{0.75}$Cd$_{0.25}$Se quantum well system. Note the effect of strain, leading to an increased band offset in the valence band.

To illustrate the significance of this effect, we examine the case of a Zn$_{0.75}$Cd$_{0.25}$Se/ZnSe quantum well. The lattice constant of CdSe is about 7% larger than that of ZnSe. Detailed studies indicate that the effect of strain, in fact, provides a major portion of the band discontinuity for the heavy holes (more than the initial offset in strain-free structures). This is illustrated in Fig. 2.5, which shows band offsets in the conduction and the valence bands of a Zn$_{0.75}$Cd$_{0.25}$Se/ZnSe quantum well in two cases, namely, with strain absent (Fig. 2.5a) and included (Fig. 2.5b). Before considering the effect of strain, the valence band offset is estimated to be 30 meV. With the band gap difference, this would lead to a conduction-band to valence-band offset ratio of 9:1. Such a ratio is not favorable for room temperature lasing because the hole-confining potential is too shallow. With the effect of strain, this ratio changes to 7:3, which is much closer to that in GaAs/GaAlAs quantum wells. Furthermore, the ratio should remain qualitatively the same for different Cd concentrations because the energy gap, the offset, and the effect of strain all have close to linear dependence on the Cd concentration.

2.4 ZnSe-Based Blue-Green Light Emitting Devices

All wide-gap II-VI semiconductors have direct energy gaps and large effective masses (thus, large density of states), which result in intrinsically efficient light absorption and emission. Therefore, they are ideally suited for optical studies of quasi-2D electronic confinement in quantum well systems and possible photonic device applications. The quest for fabricating short wavelength semiconductor lasers and light emitting diodes started long before the growth of wide-gap semiconductor heterostructures became a reality. Among II-VI semiconductors, ZnSe appears to be well-suited for blue light emitting devices not only because of its band gap (2.7 eV at room temperature), but also because its lattice constant is very close to that of GaAs. Although there had been low-temperature observations of lasing action in bulk ZnSe with various pumping schemes, it was very clear that, for a practical device, one would have to employ a suitable quantum well structure. One of the consequences of the large effective mass in ZnSe is that, for a specific quantum well configuration (i.e., for a given well depth and width), the ground state is much better localized in wide gap quantum wells. The disadvantage for the ZnSe system, however, is that there is no common Group-II element which forms a compound with Se in the zincblende structure (as an example, CdSe has the wurtzite structure in the bulk). As a result, it is difficult to form a common-anion heterostructure in the form of ZnXSe/ZnSe (similar to AlGaAs/GaAs), where X is a Group-II element.

Due to the lack of a suitable candidate of Group-II elements, one of the compounds investigated was ZnMnSe, where Mn (being a transition metal) is introduced to substitute for Zn atoms. Because the MnSe has a rock-salt structure in the bulk, a zincblende phase of ZnMnSe can be formed in the bulk only at low Mn concentrations. However, cubic ZnMnSe can be grown by MBE for the whole composition range. The ZnMnSe/ZnSe heterostructures studied first involved compositions higher than 10% Mn, in which case the ZnMnSe and ZnSe layers form barriers and wells, respectively (Kolodziejski et al., 1985). Because of the attractive wavelength of the light emission in these quantum wells, lasing experiments with optical pumping were carried out, and were reasonably successful at low temperatures (Bylsma et al., 1985). The insufficient carrier confinement discussed earlier results in optical characteristics that deteriorate rapidly as the temperature is raised. When ZnMnSe is used as barriers, the internal transitions in Mn^{++} in the yellow region of the spectrum were also detrimental to achieving strong luminescence from the quantum wells in the blue, due to relaxation of the exciton to the localized states in Mn^{++}. In common-cation system, namely, ZnSSe/ZnSe quantum wells, in which ZnSe is the well, the conduction-band offset is small. Despite some early success in achieving lasing action (Cammack et al., 1987; Suemune et al., 1989; Zmudzinski et al., 1990; Sun et al., 1991), it was recognized that such a system was not suitable for room temperature applications. As a result, early experiments in this area were plagued with a number of problems, some of which are intrinsic to wide-gap materials in general, such as doping, whereas others are material-specific, e.g., finding a heterostructure suitable for fabricating the active region of a short-wavelength laser. Therefore, there was no report of lasing action in the blue which showed promise for realistic device applications (CW lasing action at room temperature) until 1990. The interest in blue lasers was revived by a series of breakthroughs since 1990. The demonstration of lasing action in ZnSe/ZnCdSe quantum wells identified a viable active-region configuration for the fabrication of blue lasers (Jeon et al., 1990), and p-type doping of ZnSe by atomic nitrogen was successfully carried out (Park et al., 1990; Ohkawa et al., 1991). These results were immediately followed by the successful fabrication of the first laser diode (Haase et al., 1991). In this section, we will focus on ZnSe-based blue-green laser structures, specific achievements that led to the current status of these devices, and describe obstacles that still exist.

Quantum Well Structures for Blue-Green Lasers

As already mentioned, early choices of ZnSe-based quantum wells, such as ZnSe/ZnMnSe, ZnSe/ZnSeS, and so on, were shown to have confining potentials that are very shallow either for electrons (e.g., ZnSe/ZnSeS) or for holes (e.g., ZnSe/ZnMnSe). Even when these wells are sufficiently deep to provide some confinement at low temperatures, thermal excitation degrades the degree of localization at room temperature below what would be required for efficient radiative recombination. It is important to notice that the condition for exciton absorption was also shown later to apply to lasing action. In other words, quantum well depth and width suitable for exciton absorption at room temperature are also optimum for lasing action, although the number of wells needed for lasing can be more than that needed for observing exciton absorption due to wave guiding (which can be necessary when the structure is only ZnCdSe/ZnSe, where the quantum wells are used for both carrier and light confinement). After exhausting all alloys based on II-VI zincblende binary compounds, it became clear that other alloys involving compounds that do not occur in the zincblende structure in nature need to be explored. One such possibility involves ZnCdSe alloys. The CdSe component has a wurtzite structure in the bulk. The first problem in such an approach is that of structural integrity, namely, whether the compound can be grown in the zinc blende phase with reasonable crystalline quality. It was demonstrated that zinc blende CdSe can indeed be grown by MBE on (100) GaAs substrates, although the density of stacking faults turns out to be rather high (Samarth et al., 1989), because CdSe is heavily lattice mismatched to GaAs ($\sim$ 7% mismatch). Because the natural structure of CdSe is not zinc blende, the crystal quality does not improve significantly at thicknesses beyond the critical thickness, in contrast to other "natural" zinc blende compounds, such as ZnTe or CdTe on GaAs. Significantly, however, it was demonstrated that when cubic CdSe is grown on ZnTe buffer layers (which is nearly lattice-matched to CdSe), its crystal quality improves dramatically (Luo et al., 1991). These initial efforts in perfecting the structure of CdSe led to the subsequent MBE growth of ZnCdSe alloys. It was shown that such alloys—especially at lower Cd concentrations—can be grown with an impressively high degree of perfection (Samarth et al., 1990a). Furthermore, the optical properties of this new family of alloys were favorable for light-emitting devices, including "clean" PL spectra, with relatively narrow band-edge luminescence and with no significant luminescence below the band edge. With the successful growth of ZnCdSe alloys and ZnCdSe/ZnSe quantum well and superlattice structures, the first room temperature lasing action in a ZnSe/ZnCdSe superlattice was reported in 1990 using optical pumping (Jeon et al., 1990).

Doping of ZnSe

The active region of any type of laser is unquestionably the most important component for its operation. At the same time, the formation of a suitable p-n junction configuration is a crucial link in the series of steps leading to the ultimate fabrication of laser diodes. Problems with doping ZnSe, especially with p-type doping, constituted a major and long-standing bottleneck. Part of the difficulty in doping any wide-gap semiconductor arises intrinsically from the size of the energy gap itself. The Fermi level for an undoped semiconductor lies in the middle of the energy gap. Thus, in trying to dope a wide-gap semiconductor, one is essentially shifting the Fermi level by half of the energy gap to the edge of either the conduction band (for n-type doping) or the valence band (for p-type doping). For ZnSe, which has an energy gap of 2.7 eV at room temperature, this corresponds to an energy change of 1.35 eV, which is enough to promote compensation through defect formation.

Although doping of II-VIs is in general difficult, n-type doping of ZnSe has been considerably easier, compared to p-type doping. This can be partly attributed to the relatively low position of the conduction band with respect to the vacuum level. Two promising dopants, Ga and Cl, have been the most exhaustively studied elements in this connection. The doping level of Ga in ZnSe can easily reach 10^{17} cm^{-3}. However, attempts to increase the carrier concentration by introducing higher concentrations of Ga during growth usually result in either saturation or even a decrease of the actual carrier concentration (Venkatesan et al., 1989), which is believed as due to the so-called "self-compensation" effect. When Cl is introduced (usually from a ZnCl$_2$ source in MBE growth) as the n-type dopant, doping levels can be successfully raised to above 10^{19} cm^{-3} (Ohkawa et al., 1987). Deep levels associated with Ga and Cl donors were compared by deep level transient spectroscopy (Karczewski et al., 1994). These results indicate that Cl is far superior to Ga as an n-type dopant, because no deep levels directly associated with Cl atoms can be detected, in contrast with Ga-doping, which results in a significant concentration of defects involving Ga atoms.

As mentioned earlier, p-type doping of ZnSe has been and, to some extent, still remains a major obstacle. The problem encountered in p-type doping of ZnSe stems from the strong lattice relaxation associated with most—and especially the heavier—group-V dopant candidates, which leads to the formation of deep centers and consequently to heavy compensation (Chadi et al., 1989). The discovery of incorporating nitrogen (by using a plasma source in MBE growth) made it possible to dope ZnSe p-type with hole concentrations close to what is needed for a laser diode. Although the use of the plasma source, as well as other means of exciting nitrogen gas, such as by microwaves, still awaits a thorough investigation, it is clear that the breakup of N_2 molecules into single N atoms is responsible for the success of this doping process. The highest net hole concentrations reached by this method are generally around 2×10^{18} cm^{-3}. As additional nitrogen is incorporated in ZnSe beyond the above concentration, compensation becomes a problem that, as yet, has not been overcome. It is widely believed that at the 10^{18} cm^{-3} level, nitrogen begins to form complexes responsible for this self-limiting behavior. Although the doping level is sufficiently high for the operation of laser diodes, in practice, higher doping levels are still very desirable. This is because all metal contacts to p-type ZnSe create high Schottky barriers, since their work functions are without exception much higher than the valence band of ZnSe. If the hole concentration in ZnSe could be further increased, the Schottky barriers would eventually become so narrow that tunneling would be sufficient for injecting the holes into the p-ZnSe. Thus, further improvement of p-type doping of ZnSe remains critical to developing practical blue laser diodes.

Doping of ZnSe-based alloys involved in current laser diodes, namely, ZnSSe, ZnMgSe, and ZnMgSSe, remains a problem, especially in the case of nitrogen doping. The mechanisms responsible are yet to be identified. Because the structures of such diodes consist mostly of these alloys, better doping schemes are needed.

ZnSe-Based Laser Diodes

Although significant progress has been made in recent years, there are still problems in fabricating practical ZnSe-based laser diodes, which are all, in one way or another, related to the very high lasing threshold. A reduced threshold will, undoubtedly, prolong the lifetime of the device. Three key factors are particularly important in reducing the threshold current: 1) quantum efficiency of the active region determined by the confinement of charge carriers in the quantum wells; 2) light confinement for maximum overlap between light in the laser cavity and carriers in the quantum wells; and 3) contact resistance. Below, we will discuss the efforts made in these three areas.

Electron and Light Confinement

The degree of quantum confinement is affected by two factors: the height of the confining potential and the well width. Because of the large difference in the ZnSe and CdSe lattice constants, there is a practical limit for the Cd concentration in the ZnCdSe quantum well region if one wishes to avoid excessive dislocation densities. Furthermore, it was demonstrated that, as Cd concentration increases, the overall crystalline quality of ZnCdSe deteriorates. Finally, a high Cd content will also reduce the energy gap of the active region. With all these factors considered, the Cd composition yielding the best results is currently around 25 to 30%. As mentioned earlier, the ZnSe/ZnCdSe system exhibits strong carrier confinement when the well width is reduced below the exciton Bohr radius. This includes not only an increase in the exciton binding energy, but also an enhancement of the oscillator strength due to high spatial overlap of the electron and hole states. It was demonstrated that the lasing threshold is closely related to such efficient confinement of carriers. The quantum well width in current laser diodes is around 50Å. With the introduction of S and Mg as alloy constituents, the well depth can be further increased in both the conduction band (by adding Mg) and in the valence band (with S). Thus, in principle, the system ZnMgSeS/ZnCdSe can achieve large confining potentials for both electrons and holes. Other considerations, involving trade-offs between electron confinement and light confinement, and lattice matching (see below), suggest that the optimal configuration of quantum wells currently in use is most commonly in the form of ZnSeS/ZnCdSe (Gaines et al., 1993).

Light confinement in the ZnSe-based lasing structures is also a crucial issue in achieving a low operating threshold, as in the case of III-V semiconductor lasers. Although ZnSe/ZnCdSe

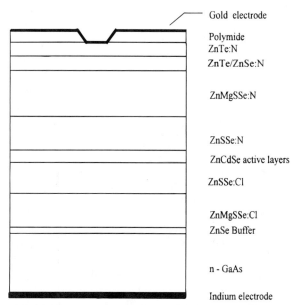

FIGURE 2.6 Schematic diagram of a separate-confinement laser diode structure involving ZnMgSeS.

quantum wells themselves, have a significant waveguiding effect (ZnSe has a lower index of refraction than ZnCdSe), materials with an even lower refractive index are needed to sandwich the quantum well active region to further improve light confinement. The fact that the index of refraction varies inversely with the energy gap has been used in selecting the II-VI materials to optimize light confinement in the present structures. The first II-VI blue laser diode fabricated at the 3M Company consisted of ZnSe/ZnCdSe quantum wells for the active region, sandwiched between ZnSeS layers for the purpose of optical confinement. A more recent structure demonstrated by North American Philips Laboratories, shown in Fig. 2.6, involved $Zn_{1-x}Mg_xSe_{1-y}S_y$ (Gaines et al., 1993), a quaternary alloy first grown by Sony to achieve separate confinement. This alloy not only has the largest energy gap among the ZnSe-based alloys (thus, ideal for the cladding layer), but also can be tailored to match the lattice constant with ZnSeS. Thus, allows the use of ZnSeS, lattice matched as barriers. We note here that the use of S in the barriers not only improves the structural integrity of the system by reducing stacking faults and dislocations through a better lattice match, but also increases the valence band offset between ZnSeS and ZnCdSe and, thus, the confinement of holes. This is achieved via two independent but additive mechanisms: the presence of S lowers the valence band of the barrier, and the reduction of the barrier lattice parameter due to the presence of S increases compression of the ZnCdSe layer, moving the valence band of the quantum well upward.

Ohmic Contacts

As mentioned earlier, ohmic contacts to wide-gap semiconductors constitute a major problem because ideally one would wish to use metals with work functions above the bottom of the conduction band on the n-type side of the p-n junction and below the top of the valence band on the p-type side, to avoid formation of Schottky barriers. As the gap increases, metals with requisite work functions are, thus, harder and harder to find. When such metals do not exist, there are two methods that can be used to overcome the unavoidable Schottky barriers. One of them is to heavily dope the semiconductor layer to which the contact is to be made, such that the depletion layer associated with the Schottky barrier is sufficiently thin to promote carrier injection by tunneling. For n-ZnSe, this approach does indeed provide contacts that appear to be ohmic, owing to the high doping levels achievable with Cl-doping and the relatively small

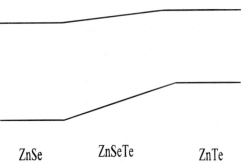

FIGURE 2.7 Band profile of linearly graded ZnSeTe region. The graded region is used to reduce Schottky barriers at p-ZnSe/Au contacts.

difference between the bottom of the conduction band and the metal work function. The dominant problem is making an ohmic contact to p-ZnSe, because there is no metal that has a work function near the top of the valence band. The problem is further compounded by the inability to attain high doping levels in p-type ZnSe. A solution to this problem is to use a graded alloy, so as to move the top of the valence band close to the metal work function, as was done in the case of n-type GaAs/Au contacts by using GaInAs (Woodall et al., 1981). For p-type ZnSe, the alloy ZnSeTe was used (because of the high valence band position of ZnTe, as discussed earlier; see also Fig. 2.2) to raise the valence band edge of the overall structure on its p-doped side close to the work function of Au, as shown in Fig. 2.7. It is also significant in this context that ZnTe can be doped p-type to a much higher level than ZnSe. Together, these effects result in a dramatically reduced contact resistance (Fan et al., 1992). It was also shown that, by growing HgSe on p-ZnSe, the contact resistance can be reduced in a similar way (Lansari et al., 1992).

A recent theoretical study indicates that the profile of the graded region is also very crucial to the contact resistance (Yang et al., 1995). So far, all of the graded-alloy regions have used a linear profile, as shown in Fig. 2.7. By changing to a parabolic grading profile, it was shown that the length of the residual barrier can be further reduced. With the same doping level, the grading length (i.e., the thickness of the graded region) can, therefore, be much smaller (by almost a factor of 5) for the parabolic grading than for the linear case. This is additionally advantageous because ZnTe has a large lattice mismatch with ZnSe, and parabolic grading reduces the thickness of the part in the graded region where a high density of dislocations is expected.

The Lasing Mechanism in ZnSe-Based Laser Diodes

Semiconductor lasers that have been well studied, such as III-V semiconductor lasers, involve strongly interacting states. In such cases, the large Bohr radii of the ground-state (1S) excitons lead to strong overlap, as enough excitons are generated to reach the lasing threshold. Thus, the excitonic nature of the underlying recombination transition is replaced by a plasma of electrons and holes, and stimulated emission arises from recombining such electrons and holes in the plasma. This situation is naturally linked to the exciton population and to the size of the excitons. In principle, if the Bohr radius of the exciton is reduced, the mechanism would revert to the recombination of individual excitons. The large effective masses in ZnSe result in Bohr radii that are much smaller than those in GaAs and are further reduced by quantum confinement. It was shown that in ZnSe/ZnCdSe quantum well lasers exhibiting strong confinement, exciton recombination remains even under lasing condition, especially at low temperatures (Ding et al., 1992). There have been recent studies showing that in certain ZnSe/ZnCdSe quantum well lasers the mechanism involves an electron-hole plasma (see, for example, Cingolani et al., 1994). However, in all the studies reported so far which showed the electron-hole plasma as the active medium, there is substantially less confinement (either wide quantum wells or shallow wells corresponding to small Cd concentrations) compared to the quantum wells used by Ding et al. (1992). Thus to some extent the question still remains unresolved. The results available seem to suggest that when Bohr radii close to those in bulk ZnSe favor the electron-hole plasma situation. On the other hand, a reduced Bohr radius in narrow and deep ZnSe/ZnCdSe quantum wells, with their strong confining effects, can tip the scales from the band-to-band lasing involving an electron-hole plasma to excitonic lasing involving discrete exciton levels. Considerable systematic experimental and theoretical work is needed to identify the mechanisms and the exact conditions to which these mechanisms correspond.

Future Research and Development

Although significant improvements have been achieved in a remarkably short time, current ZnSe-based blue-green lasers have one outstanding problem—their short operating lifetime at room temperature. There appear to be several areas, not necessarily independent of one another, where improvements are still needed. The short lifetime itself has been shown to be related to defects which propagate (irreversibly) as the temperature is raised during the operation of the laser (Guha et al., 1993a; Hua et al., 1994). One can therefore categorize the problem as consisting of two issues: defect formation during growth and defect expansion due to the heat that is generated because of the high threshold current. Since defect densities in even closely lattice-matched systems are still high ($> 10^3$ cm^{-2}), which causes failures of laser diodes, it appears that improvement in growth and other procedures related to material quality of ZnSe-based systems is needed. It has been demonstrated that MBE-grown GaAs buffer layers on GaAs substrates greatly reduce the density of dislocations (Xie et al., 1992; Guha et al., 1993b), which improve the lifetimes of laser diodes. Homoepitaxy is another area that has been studied in this connection (Ohishi et al., 1988; Ohkawa et al., 1992; Harsch et al., 1994). The key to such an approach is the availability of high quality ZnSe substrates. Recent reports by the collaboration between the North Carolina State University and Eagle-Picher have demonstrated the successful growth (by seeded physical vapor transport technique) of undoped ZnSe substrates. X-ray diffraction rocking curves with full width at half maximum as low as 11 arc sec are observed on ZnSe substrates grown in this way, which is comparable to that of GaAs substrates (Harsch et al., 1994). Attempts in doping these substrates are also underway and have shown moderate success with carrier concentrations reaching high 10^{17} cm^{-3} (Harsch et al., 1994). However, such substrates are currently not commercially available, and therefore, the advantages of using ZnSe substrates, especially for laser diodes, have not been fully explored. The second issue, that of high threshold, is a complicated one, involving several possible factors, such as still less than perfect contact resistance, the overall design of the laser structure with regard to both carrier and optical confinement, and device fabrication. Thus, all three major achievements that are emphasized in this article—quantum well structure, doping, and ohmic contact—all have room for improvement.

There is yet another question which is much harder to address at this stage, namely, whether the heat generated by the lowest possible threshold current for room temperature laser operation can itself generate defects in an initially defect-free structure. The answer to this, however, cannot be properly arrived at before the problems discussed above are fully resolved. Finally, as more attention is directed toward GaN-based, wide gap structures, a comparison between GaN-based and ZnSe-based devices seems appropriate. Although there are advantages to GaN-based light emitting diodes (e.g., extremely long lifetimes) (Nakamura et al., 1994), it is much more difficult to compare laser diodes fabricated from the two materials, because the requirements for LEDs and for laser diodes are significantly different and because a practical GaN laser diode remains to be fabricated. The issue of finding a suitable substrate, the thermal mismatch with such a substrate, and the growth of GaN-based heterostructures are all major problems. Although GaN belongs to the well-studied III-V family, the nitrides differ in many ways from the phosphides, arsenides, and antimonides, and it may very well be the case that integration with the standard III-V semiconductors will, in the end, prove much easier for ZnSe-based systems than for the nitrides. It is very clear that both ZnSe-based and GaN-based systems still face significant difficulties. Therefore, it appears much more appropriate to explore both systems to assess their potential.

2.5 Type-II Heterostructures

Because of the large offset between Se and Te compounds, as shown in Fig. 2.2, it is possible to fabricate type-II structures with extremely large offsets both in the conduction and the valence bands. One of the type-II heterostructures in the II-VI semiconductor family that is fairly well understood is the ZnSe/ZnTe short period superlattice (Kobayashi et al., 1986). Due to the large lattice mismatch (around 7%), individual layers in this structure have to be very thin (a few monolayers) to keep the density of dislocations low. The valence band offset in this case is close to 1 eV. Optical experiments were carried out on such structures, and demonstrated the type-II band alignment. There has not been significant effort in utilizing this structure itself for optoelectronic applications. The large value of the valence band offset, however, proved to be useful for forming ohmic contacts, as discussed earlier.

Another type-II structure in the II-VI family is CdSe/ZnTe (Luo et al., 1991; Yu et al., 1991), which is the exact analog of GaSb/InAs. As expected from Table 2.1, it is also a nearly lattice-matched system. Here, we will describe two fundamentally important phenomena. The first is structural, involving a switch of atoms at the interface (Kemner et al., 1992). Either the cations or the anions on the two sides of the interface exchange positions, which results in a few monolayers of CdTe and ZnSe at the interface.

Because of the large offsets in both the conduction and the valence bands, the wide energy gaps, and the absence of the X-point in the energy range of interest, CdSe/ZnTe heterostructures provide an ideal system for fundamental studies of optical properties of type-II alignment. The band alignment of this type-II system is schematically shown in Fig. 2.8. It had been a common belief that type-II superlattices are optically less active compared to type-I superlattices and quantum wells, because the type-II transitions are spatially indirect (i.e., the initial and the final states are localized in different layers of the superlattice). However, the studies of quasilocalized states above the barriers indicated that there are direct (type-I) transitions in type-II superlattices that involve both localized and quasilocalized states, as indicated by arrows in Fig. 2.8 (Luo and

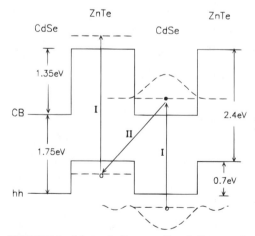

FIGURE 2.8 Schematic diagram of band alignment for a type-II CdSe/ZnTe superlattice. Wave functions of both localized and quasilocalized states are shown by the dashed lines. Arrows indicate optical transitions.

Furdyna, 1992). Optical studies of this system revealed indeed that type-I excitons can form in type-II superlattices, which results in large oscillator strengths (Luo et al., 1993a). It was observed that the absorption coefficients corresponding to such type-I transitions can be as large as 5×10^4 cm^{-1}, comparable to those in type-I quantum wells with similar energy gaps.

Furthermore, there are two types of type-I excitons in a type-II superlattice, respectively localized in the two layers (i.e., CdSe and ZnTe layers). One type of such excitons involves a confined electron and a localized above-barrier hole, both in the CdSe layers. The other kind consists of a confined hole and a localized above-barrier electron, both in the ZnTe layers. The behavior of both excitons shown resemble type-I excitons in type-I quantum wells, such as the confinement energy that increases with a reduction of layer thickness (Zhang et al., 1993). The dynamic properties of such excitons, however, dramatically differ from their counterparts in type-I systems. Because the localized above-barrier states are not the ground state, an electron (or a hole) in such a state will quickly relax to lower energy states in the same band. Such a relaxation process (typically in the sub-picosecond range) is several orders of magnitude faster than optical recombination processes which are on the order of a few nanoseconds. As a result, the type-I transitions can be observed easily only in measurements involving the excitation process, such as absorption or reflectivity experiments, but not in measurements of optical recombination, e.g., photoluminescence.

It is important to point out that the observed type-I transitions in type-II CdSe/ZnTe superlattices, which originate from the presence of quasilocalized states in the barriers, should be present in all type-II systems, such as InAs/GaSb superlattices. These transitions should be taken into account in analyzing the optical properties of GaAs/AlGaAs superlattices, where the band alignments at the Γ-point and the X-point can be of different types (Lew Yan Voon et al., 1993).

2.6 HgTe/CdTe Type-III Heterostructures

As mentioned in the Introduction, there have been review articles (see, for example, Meyer et al., 1993), in which HgTe/CdTe systems are discussed in detail. Here, we will briefly describe the unique properties of type-III superlattices and their potential for device applications.

The band structure of HgTe is characterized by a symmetry-induced zero energy gap, i.e., both the conduction and the valence bands belong to the Γ_8 bands that are degenerate at the Γ-point, as shown in Fig. 2.9b. The light-hole band is now the conduction band. The Γ_6 band, which is typically the conduction band in an open-gap semiconductor, is now in the valence band. The band diagram consisting of a zero-gap compound, e.g., HgTe, and an open-gap compound is referred to as the type-III alignment.

It is worth noting that not only the band structures in HgTe/CdTe are qualitatively different, but the band gaps and the effective masses are also drastically different. In addition to the new properties which can be expected from such a novel electronic configuration, type-III superlattices are of interest from the applied point of view, because the effective energy gap (the energy difference between the valence and conduction subbands allowed by the superlattice geometry) and the effective masses can be tuned over a wide range. We will discuss these two aspects—the physical properties and the capabilities for infrared detector applications.

The band structure of a HgTe/CdTe superlattice is illustrated in Fig. 2.10 for two combinations of well and barrier dimensions, showing two qualitatively distinct regions of the band structure (Figs. 2.10a and 2.10b). The difference in band structure is determined primarily by the well width, which, to some extent, is similar to Hg concentration in CdHgTe. For wide wells, the HH1 subband (originating from the bulk Γ_8 band) lying above the E_1 subbands (originating from Γ_6) at the center of the Brillouin zone. We will refer to superlattices corresponding to the wide-well limit as semimetallic. Note that—unlike bulk HgTe and its alloys HgCdTe with less than 10% Hg, which are zero-gap semiconductors—the superlattices in this range behave as true semimetals,

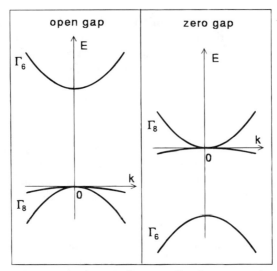

FIGURE 2.9 Schematic diagrams of band structures of an open-gap (left panel) and a zero-gap (right panel) semiconductor. A heterostructure consisting of the two materials is referred to as having type-III band alignment.

i.e., they have a significant overlap of the conduction and the valence bands. As expected, the widths of the bands displayed in the figures vary inversely as a function of barrier width.

As the well width decreases, the E1 band of the superlattice is gradually pushed upward, eventually ending up above the HH1 band. In this region, the superlattice resembles in many ways, the alloy $Hg_{1-x}Cd_xTe$ for $x > 0.15$, and we shall refer to it as semiconducting. The gap between the Γ_6-like E1 conduction band and the Γ_8-like HH1 valence band continues to increase as the well width decreases. One should note that, for any constant barrier width, a decrease of the well width represents a physical increase in the Cd content of the superlattice system as a whole and, in this sense, the behavior of the superlattice is similar to that of the $Hg_{1-x}Cd_xTe$ alloy with increasing x.

This similarity notwithstanding, there are major differences between $Hg_{1-x}Cd_xTe$ alloys and HgTe/CdTe superlattices. Most obvious is the anisotropy (represented by the two directions, the growth direction, z, and the in-plane direction, x) in the band structure diagram. Note that, in the valence band, the E vs. k dispersion in the growth direction is very flat, indicating that the hole mass characterizing motion along z is extremely heavy, whereas the mass in-plane is very light. Second, we see that, for x-motion the effective mass of the holes is nonparabolic. Such nonparabolicity (where a mass can increase, e.g., from 0.01 m_0 to a value two orders of magnitude larger in a span of only a few tens of meV) is not found in any other known material. A closely related feature is the mass broadening effect, where one must take into account a spectrum of masses (for different k-values) to properly describe the behavior of carriers in a given subband. These novel properties and the effects that follow are unique to type-III superlattices and do not occur in any other material.

The ability to tune the energy gap of HgTe/CdTe superlattices by varying the superlattice parameters (especially the well width) makes such structures of interest for infrared and far infrared detector applications, as an alternative to HgCdTe alloys. We note here two advantages of the superlattice in this context. First, it can be argued that the superlattice, grown by deposition of binary compounds, lends itself to better composition control and avoids band-gap broadening effects due to alloy fluctuations. Second, and probably more important, the large electron and hole mass in the z-direction (already discussed in connection with Fig. 2.10) guarantees a reduction in the background noise arising from dark current, which is expected to flow in any photovoltaic

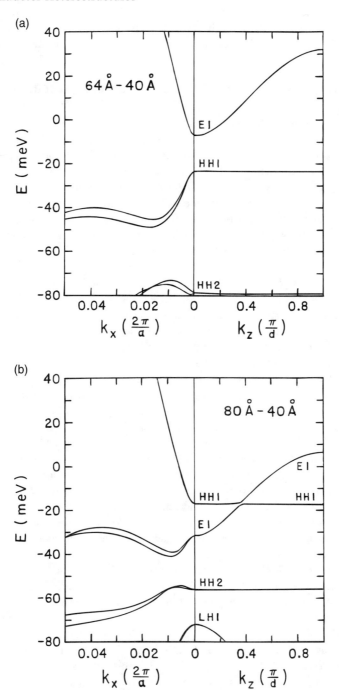

FIGURE 2.10 Energy band structures for two representative cases of HgTe/CdTe superlattices: (a) shows an open-gap semiconductor system, and (b) corresponds to a semimetallic system. Note that there is a true overlap between the E1 and the HH1 bands at the center of the Brillouin zone.

device. Whether these advantages will, in the end, compensate for the comparatively greater cost of producing HgTe/CdTe superlattices over HgCdTe alloys still remains unresolved.

2.7 Diluted Magnetic Semiconductor Heterostructures

Transition metals, such as Mn, Fe, Co, and so on, can be alloyed into II-VI semiconductors, forming what is referred to as diluted magnetic semiconductors (DMSs) (Furdyna, 1988). The exchange interaction between free carriers and localized electrons in these magnetic ions (sp-d exchange interaction) leads to a wide range of unique properties, combining semiconductor physics and magnetism. For example, the sp-d exchange interaction leads to enormous band Zeeman splittings (two orders of magnitude larger than those in nonmagnetic II-V materials). The efforts in utilizing this property for practical applications have been very limited, because in most cases the large splittings are present only when the temperature is low. Nevertheless, such efforts played an important role in the studies not only of DMSs, but semiconductors, in general.

In bulk DMSs, the transition metal concentrations are typically low because of the different crystal structures of the compounds involved, which is the reason for the use of the word *diluted*. Modern crystal growth techniques, e.g., MBE, have greatly extended the composition range of the magnetic component to 100% for the cases of Mn and Fe (Kolodziejski et al., 1986; Durbin et al., 1989; Jonker et al., 1988), for example MnTe. Despite the extended composition range, we will still use the name DMS for historic reasons. The large composition of magnetic ions has revealed new magnetic properties, such as novel forms of long range antiferromagnetic order, which had not been possible in bulk samples because of the limited magnetic ion compositions. Although most recent studies in this area have focused on II-VI DMS heterostructures, efforts have been made to introduce magnetic ions, i.e., Mn^{++}, into III-V semiconductors using the technique of MBE (Munekata et al., 1989). Both structural and magnetic properties have been studied (Krol et al., 1993; von Molnar et al., 1991), which greatly enriched the family of DMSs. The introduction of magnetic ions in heterostructures has exciting implications because of its multicomponent nature —semiconductor physics, magnetism and quantum confinement of charge carriers— and the sp-d exchange interaction. In this section, we will focus our attention on II-VI DMS heterostructures, namely, their optical properties (originating from the sp-d exchange interaction) and examples of applying such properties to the understanding of semiconductor heterostructures in general.

The most interesting DMS heterostructures involve a combination of DMS and non-DMS layers, such as ZnMnSe/ZnSe and ZnFeSe/ZnSe. Because of the drastically different electron and hole Zeeman splittings in DMS and non-DMS compounds, band offsets can be significantly changed by an externally applied magnetic field, to the extent that even the type of the band alignment can be changed (e.g., from type-I to type-II). The advantage of such a choice is twofold. First, one can have a continuous range of offset values in one sample, rather than growing a number of samples, as is often done for non-DMS heterostructures. Therefore, the properties of a given structure can be studied over a wide parameter range without ambiguity concerning alloy compositions and structural dimensions, which is often encountered in dealing with different samples. The second is the accurate mapping of the wave function for the carrier states. For example, in optical studies, one can quantitatively determine the shape of wave functions by measuring the Zeeman splitting of a given transition that reflects the location of the confined states and their penetration into the adjacent layers. These unique effects have led to a series of novel observations, some of which strictly belong to DMS heterostructures (such as long range antiferromagnetic order observed by Giebultowicz et al., 1992, and magnetic polarons by Yakovlev et al., 1990), whereas others can be applied to non-DMS heterostructures (e.g., III-V heterostructures). We will discuss selected examples of general interest in the following.

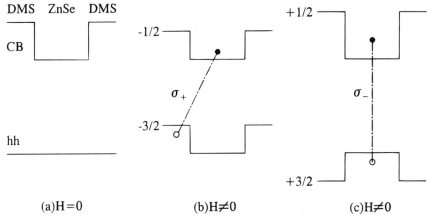

FIGURE 2.11 Schematic diagram of the conduction and the valence band alignment in ZnSe/Zn$_{0.9}$Fe$_{0.1}$Se quantum well structures. (a) B = 0. (b) B ≠ 0, m$_j$ = −1/2 electrons, m$_j$ = −3/2 holes. (c) B ≠ 0, m$_j$ = +1/2 electrons, m$_j$ = +3/2 holes. The dot-dashed lines indicate the allowed heavy hole exciton transitions in the Faraday geometry.

Magnetic Field Induced Type-I to Type-II Transition

As discussed above, band edges in DMS layers can be tuned over a wide energy range by an applied magnetic field, which in some cases can exceed the zero-field offsets. As a result, it is possible to convert the types of the band alignment of a heterostructure that consists of DMS and non-DMS layers. As an example, consider the conduction band and the heavy-hole valence band of a single non-DMS quantum well with DMS barriers, as shown in Fig. 2.11. The structure corresponds to a type-I band alignment at $B = 0$, with a deep conduction band well and a shallow well in the valence band (e.g., a common-anion system). Note that the heavy-hole spin splitting is typically much larger than that in the conduction band (Furdyna, 1988). Upon applying a magnetic field, one spin orientation (m$_j$ = +1/2 electrons, m$_j$ = +3/2 holes) will experience a deepening of the wells in both the conduction and the valence bands (enhancing the type-I nature). For the other spin states (m$_j$ = −1/2 electrons, m$_j$ = −3/2 holes), however, the enormous Zeeman splitting will, at some field, exceed the valence band offset, and the original valence band well will be transformed into a barrier, as shown in Fig. 2.11 for $B \neq 0$.

The conversion from type-I to type-II is manifested by a dramatic change in optical properties, because, in a type-II structure, the carriers involved in the strongest optical transitions are now physically separated in different layers. The phenomenon of such a conversion was first observed in ZnFeSe/ZnSe (Liu et al., 1989), and later in CdMnTe/CdTe (Deleporte et al., 1990).

It is important to note that this phenomenon is particularly important to the determination of band offsets. The Zeeman splitting of a given DMS layer can be measured accurately. Thus, the splitting at the field that corresponds to the type-I to type-II transition can provide direct information concerning the zero-field offset. Such a direct measurement of the band offsets, using simple optical techniques, has been a challenge for non-DMS heterostructures (e.g., III-V heterostructures).

Wave Function Mapping

The degree of confinement of charge carriers is of great importance for fundamental properties and device applications, as discussed earlier. The shape of the wave function, characterized by its localization in the quantum well and by its penetration into the adjacent layers, determines the confinement characteristics of a given state. Most of the experimental evidence concerning wave-function shapes (thus the degree of confinement) has been indirect, which includes measurements of the transition energies or deviation of the exciton states from the 3-D hydrogenic model. For low-lying states (such as the ground state), such estimates can provide reasonably accurate distribution of wave functions, but the uncertainty increases dramatically for higher energy states.

One can improve the situation by exploiting the fact that, in heterostructures consisting of DMS and non-DMS layers, the Zeeman splitting of a given state will reflect its probability distribution over these two media. In other words, the Zeeman splitting will be determined effectively by how many localized magnetic moments the electron "sees". It can, thus, be used as a tool for mapping the wave function (or probability), providing spatial identification of electronic states. This technique has been essential in observing several novel effects.

One such observation involved type-I excitons in type-II superlattices (Luo et al., 1993a). Type-II heterostructures have been studied for a long time and have been considered optically less active because of the spatial separation of the initial and final states. The strong transitions observed in the type-II system CdSe/ZnTe discussed earlier have been ellusive for several years after the first observation. The final identification of the origins of the optical transitions was made in CdSe/ZnTe structures, in which Mn^{++} was introduced either in the CdSe or ZnTe layers.

Another nonintuitive phenomenon is the quasilocalized states in a single quantum barrier and related structures. A single barrier is characterized by a continuous energy spectrum, which appears to be uninteresting in optical studies, presumably due to lack of discrete states. It was shown, however, quasilocalized states in a single barrier behave similarly to a single quantum well (Luo et al., 1993b). Such a similarity is characterized by quasilocalization of wave functions in the barrier region (measured by $|A|^2 + |B|^2$, A and B being the coefficients of the plane waves in the barrier region, normalized to the probability of the incoming wave, $|I|^2$) and peaks in the density of states, which resemble the discrete states in a quantum well, as shown in Fig. 2.12. The ability to spatially identify the optical transitions greatly simplified the analysis of such quasilocalized states. The structures used consist of ZaMnSe layers (barriers) sandwiched between ZnSe. The absorption peaks, in a system with a *continuous energy spectrum* between the quasilocalized states in the valence and the conduction bands, exhibited large Zeeman splitting (comparable to that in a ZnMnSe epilayer with the same Mn concentration) (Luo et al., 1993b). Thus, one can unambiguously identify that the transitions occur in the barrier region.

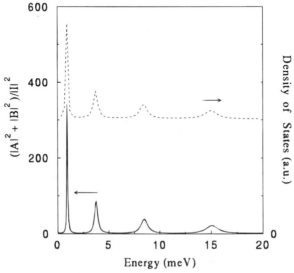

FIGURE 2.12 Quasilocalization and density of states associated with a single quantum barrier.

Spin Superlattices

A "spin superlattice" is a superlattice in which carriers with opposite spin states are confined in different layers of the structure. To achieve such spin-modulation, one must look for a structure having a small energy gap difference between the constituent layers and small band offsets in the absence of a magnetic field. In other words, an ideal system would be one in which both the conduction and the valence bands are flat without an applied field. When a magnetic field is applied, the large Zeeman splitting of the band edges in the DMS layers results in band offsets so as to produce a spatial separation of the spin-up and the spin-down states, as shown in Fig. 2.13. The band offsets are created by the magnetic field. Layers of the superlattice that behave as quantum wells for one spin state will be the barriers for the other spin state in both the conduction and the valence bands. The spatial separation of the spin states can be monitored by optical experiments. Two systems have been used for this purpose, namely, ZnFeSe/ZnSe and ZnMnSe/ZnSe, both of which exhibited spin superlattice behavior (Chou et al., 1991; Dai et al., 1991).

Concluding Remarks

The future of II-VI semiconductor heterostructures relies strongly on the development of device applications, which are not necessarily limited to currently emphasized ZnSe-based laser diodes. Other optoelectronic applications, e.g., modulators, that have fewer restrictions on operating power (and thus present fewer problems to the lifetime issue), will add to the variety. The efforts in fabricating ZnTe- and ZnSe-based modulators have already cleared the way for more sophisticated optoelectronic devices (Wang et al., 1993; Partovi et al., 1991). Despite the significant advances in the fabrication of blue-green laser diodes, device processing of II-VIs is still in its infancy. Many techniques commonly used for Si and III-V materials, such as ion implantation, have so far not been adequately applied to II-VI semiconductors.

The fundamental studies of II-VI heterostructures are still far behind what has been accomplished in their III-V and Si-based counterparts. This is particularly true for structures of lower dimensions, namely, quantum wires and dots. Effects related to growth processes, such as spontaneous chemical ordering and the formation of quantum dots on lattice mismatched substrates or buffer layers, also present opportunities for future studies. There are numerous problems which have not been properly addressed because of the limited progress in device applications. Such problems include the surface structures of II-VI compounds, surface electronic states, and growth

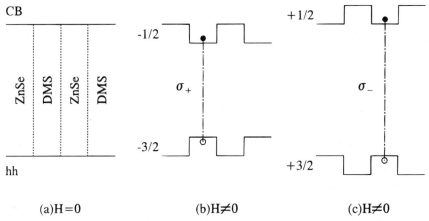

FIGURE 2.13 Schematic diagram of the conduction and the valence band alignment in spin superlattices. (a) B = 0. (b) B ≠ 0, m_j = −1/2 electrons, m_j = −3/2 holes. (c) B ≠ 0, m_j = +1/2 electrons, = +3/2 holes. The vertical dot-dashed lines represent the allowed heavy-hole exciton transitions in the Faraday geometry.

and characterization of oxide layers. The development in these areas, however, will, to a great extent, depend on the progress in device applications.

Theoretical studies have traditionally concentrated on III-V and Si-related problems. Recent activities in the area of blue-green laser diodes have stimulated a great deal of interest in II-VI materials, which has added strength to the overall effort in this area and will continue to be important in the future. For example, a thorough understanding of stabilities of the II-VIs and defect formation related to doping will provide important input to device fabrication. II-VI semiconductor heterostructures have experienced a great deal of development in the last few years, which, in turn, stimulated worldwide interest. The field as a whole, however, faces tremendous technological challenges.

Acknowledgments

The authors would like to acknowledge the support of the Center for Electronic and Electrooptical Materials (CEEM) at SUNY, NSF (DMR 9223054) and ONR/DoD under the MFEL program. We would also like to thank E. H. Lee, H. C. Chang, and M. S. Salib for their help in preparing this manuscript.

References

Akimoto, O. and Hasegawa, H. 1966. *J. Phys. Soc.* Japan 22, 181.
Bylsma, R., Becker, W., Bonsett, T., Kolodziejski, L.A., Gunshor, R.L., Yamanishi, M., and Datta, S. 1985. *Appl. Phys. Lett.* 47, 1039.
Cammack, D. A., Dalby, R., Corneliassen, H., and Khurgin, J. 1987. *J. Appl. Phys.* 62, 3071.
Chadi, J. and Chang, K. J. 1989. *Appl. Phys. Lett.* 55, 575.
Chemla, D., Schmitt-Rink, S., and Miller, D. A. B. 1988. *Optical Nonlinearities and Instabilities in Semiconductors,* Haug, H., ed., p. 83, *Academic Press, New York.* p.83
Chou, W. C., Petrou, A., Warnock, J., and Jonker, B. T. 1991. *Phys. Rev. Lett.* 67, 3820.
Cingolani, R., Rinaldi, R., Calcagnile, L., Prete, P., Sciacovelle, P., Tapfer, L., Vanzetti, L., Mula, G., Bassani, F., Sorba, L., and Franciosi, A. 1994. *Phys. Rev. B* 49, 16769.
Dai, N., Luo, H., Zhang, F. C., Samarth, N., Dobrowolska, M., and Furdyna, J. K. 1991, *Phys. Rev. Lett.* 67, 3824.
Deleporte, E., Berroir, J. M., Bastard, G., Delalande, C., Hong, J. M., and Chang, L. L. 1990. *Phys. Rev. B* 42 (1990), 5891.
Ding, J., Pelekanos, N., Nurmikko, A. V., Luo, H., Samarth, N., and Furdyna, J. K. 1990. *Appl. Phys. Lett.* 57, 2885.
Ding, J., Jeon, H., Ishihara, T., Hagerott, M., Nurmikko, A. V., Luo, H, Samarth, N., and Furdyna, J. K. 1992. *Phys. Rev. Lett.* 69, 1707.
Durbin, S. M. 1989. *Appl. Phys. Lett.* 55, 2087.
Fan, Y., Han, J., He, L., Saraie, J., Gunshor, R. L., Hagerott, M., Jeon, H., Nurmikko, A. V., Hua, G. C., and Otsuka, N. 1992. *Appl. Phys. Lett.* 61, 3160.
Furdyna, J. K. 1988. *J. Appl. Phys.* 64, R29.
Gaines, J., Drenten, R., Haberern, K., Marshall, T., Mensz, P., and Petruzzelo, J. 1993. *Appl. Phys. Lett.* 62, 2462.
Giebultowicz, T. M., Samarth, N., Luo, H., Furdyna, J. K., Klosowski, P., and Rhyne, J.J. 1992. *Phys. Rev. B* 46, 12076.
Guha, S., DePuydt, J. M., Haase, M. A, Qiu, J., and Cheng, H. 1993a. *Appl. Phys. Lett.* 63, 3107.
Guha, S., Munekata, H., and Chang, L. L. 1993b. *J. Appl. Phys.* 73, 2294.
Haase, M. A., Qiu, J., DePuydt, J. M., and Cheng, H. 1991. *Appl. Phys. Lett.* 59, 1272.
Harsch, W. C., Cantwell, G., and Schetzina, J. F. 1994. In *Proceedings of the International Workshop on ZnSe-Based Blue-Green Laser Structures.*

Hua, G. C., Otsuka, N., Grillo, D. C., Fan, Y., Han, J., Ringle, M. D., Gunshor, R. L., Hovinen, M., and Nurmikko, A. V. 1994. *Appl. Phys. Lett.* 65, 1331.
Jeon, H., Ding, J., Nurmikko, A. V., Luo, H., Samarth, N., Furdyna, J. K., Bonner, W. A., and Nahory, R. E. 1990. *Appl. Phys. Lett.* 57, 2413.
Jonker, B. T., Krebs, J. J., Qadri, S. B., Prinz, G. A., Volkening, F. A., and Koon, N.C. 1988. *J. Appl. Phys.* 63, 3303.
Karczewski, G., Hu, B., Yin, A., Luo, H., and Furdyna, J. K. 1994. *J. Appl. Phys.* 75, 7382.
Kemner, K. M., Bunker, B. A., Luo, H., Samarth, N., Furdyna, J. K., Weidmann, M. R., and Newman, K. E. 1992. *Phys. Rev.* B 46, 7272.
Kobayashi, M., Mino, N., Katagiri, H., Kimura, R., Konagai, M. 1986. *J. Appl. Phys.* 60, 773.
Kolodziejski, L. A., Gunshor, R. L., Bonsett, T. C., Venkatasubramaniam, R., Datta, S., Bylsma, R. B., Becker, W. M., and Otsuka, N. 1985. *Appl. Phys. Lett.* 47, 169.
Kolodziejski, L. A., Gunshot, R. L., Otsuka, N., Gu, B. P., Hefetz, Y., and Nurmikko, A. V. 1986. *Appl. Phys. Lett.* 48, 1482.
Krol, A., Soo, Y. L., Huang, S., Ming, Z. H., and Kao, Y. H. 1993. *Phys. Rev.* B 47, 7187.
Lansari, Y., Ren, J., Sneed, B., Bowers, K. A., Cook, Jr., J. W., and Schetzina, J. F. 1992. *Appl. Phys. Lett.* 61, 2554.
Lee, D., Zucker, J., Johnson, A. M., Feldman, R. D., and Austin, R. F. 1990. *Appl. Phys. Lett.* 57, 1132.
Lew Yan Voon, L. C., Ram-Mohan, L. R., Luo, H., and Furdyna, J. K. 1993. *Phys. Rev.* B 47, 6585.
Liu, X., Petrou, A., Warnock, J., Jonker, B. T., Prinz, G. A., and Krebs, J. J. 1989. *Phys. Rev. Lett.* 63, 2280.
Luo, H., Samarth, N., Zhang, F. C., Pareek, A. Dobrowolska, M., Furdyna, J. K., and Ostsuka, N. 1991. *Appl. Phys. Lett.* 58, 1783.
Luo, H. and Furdyna, J. K. 1992. *Bull. Am. Phys. Soc.* 37, 659.
Luo, H., Chou, W. C., Samarth, N., Petrou, A., and Furdyna, J. K., 1993a. *Solid State Commun.* 85, 691.
Luo, H., Dai, N., Zhang, F. C., Samarth, N., Dobrowolska, M., Furdyna, J. K, Parks, C., and Ramdas, A. K. 1993b. *Phys. Rev. Lett.* 70, 1307.
Meyer, J. R., Hoffman, C. A., Myers, T. H., and Giles, N. C. 1993. *Handbook on Semiconductors,* Mahajan, S., ed., Vol. III, 2nd edition, Elsevier, Amsterdam.
Munekata, H. et al. 1989. *Phys. Rev. Lett.* 63, 1849.
Myers, T. H., Meyer, J. R., and Hoffman, C. A. 1993. *Quantum Wells and Superlattices for Long Wavelength Infrared Detectors,* Manasreh, M. O., ed., Artech House, Boston.
Nakamura, S., Mukai, T., and Senoh, M. 1994. *Appl. Phys. Lett.* 64, 1697.
Park, R. M., Troffer, M. T., Rouleau, C. M., DePuydt, J. M., and Haase, M. A. 1990. *Appl. Phys. Lett.* 57, 2127.
Ohishi, M., Ohmori, K., Fujii, Y., and Saito, H. 1988. *J. Crystal Growth* 86, 375.
Ohkawa, O., Mitsuyu, T., and Yamazaki, O. 1987. *J. Appl. Phys.* 62, 3216.
Ohkawa, K., Karasawa, T., and Mitsuyu, T. 1991. *Jpn. J. Appl. Phys.* 30, L152.
Ohkawa, K., Ueno, A., and Mitsuyu, T. 1992. *J. Crystal Growth* 117, 375.
Partovi, A., Glass, A. M., Olson, D. H., Feldman, R. D., Austin, R. F., Lee, D., Johnson, A. M., and Miller, D. A. B. 1991. *Appl. Phys. Lett.* 58, 334.
Pelekanos, N. T., Ding, J., Hagerott, M., Nurmikko, A. V., Luo, H., Samarth, N., and Furdyna, J. K. 1992. *Phys. Rev.* B 45, 6037.
Samarth, N., Luo, H., Furdyna, J. K., Qadri, S. B., Lee, Y. R., Ramdas, A. K., and Otsuka, N. 1989. *Appl. Phys. Lett.* 54, 2680.
Samarth, N., Luo, H., Furdyna, J. K., Qadri, S. B., Lee, Y. R., Ramdas, A. K., and Otsuka, N. 1990a. *J. Electronic Materials* 19, 543.
Samarth, N., Luo, H., Furdyna, J. K., Alonso, R. G., Lee, Y. R., Ramdas, A. K., Qadri, S. B., and Otsuka, N. 1990b. *Appl. Phys. Lett.* 56, 1163.
Suemune, I., Yamada, K., Masato, H., Kan, Y., and Yamanishi M. 1989. *Appl. Phys. Lett.* 54, 981.

Sun, G., Shahzad, K., Khurgin, J., and Gaines, J. 1991. Conf. Lasers and Electrooptics, Baltimore.

Venkatesan, S., Pierret, R. F., Qiu, J., Kobayashi, M., Gunshor, R. L., and Kolodziejski, L. A. 1989. *J. Appl. Phys.* 66, 3656.

von Molnar, S. et al. 1991. *J. Magn. Magn. Mater.* 93, 356.

Wang, S. Y., Kanakami, Y., Simpson, J., Stewart, H., Prior, K. A., and Cavenett, B. C. 1993. *Appl. Phys. Lett.* 62, 1715.

Woodall, J. M., Freeout, J. L., Pettit, G. D., Jackson, T., and Kirchner, P. 1981. *J. Vac. Sci. Technol.* 19, 626.

Xie, W., Grillo, D. C., Gunshor, R. L., Kobayashi, M., Jeon, H., Ding, J., Nurmikko, A.V., Hua, G. C., and Otsuka, N. 1992. *Appl. Phys. Lett.* 60, 1999.

Yakovlev, D. R., Ossau, W., Landwehr, G., Bicknell-Tassius, R. N., Waag, A., and Uraltsev, I. N. 1990. *Solid State Commun.* 76, 325.

Yang, G. L., Luo, H., Lowandowski, L., and Furdyna, J. K. 1995. *Phys. Stat. Sol.* (h) **187**, 435.

Yu, E. T., Phillips, M. C., McCaldin, J. O., and McGill, T. C. 1991. *J. Vac. Sci. Technol.* B 9, 2233.

Zmudzinski, C. A., Guan, Y., and Zory, P. S. 1990. *IEEE Photonics and Techn. Letters* 2, 94.

Zhang, F. C., Luo, H., Dai, N., Samarth, N., Dobrowolska, M., and Furdyna, J. K. 1993. *Phys. Rev. B.* 47, 3806.

3
GaN and Silicon Carbide as Optoelectronic Materials

3.1	Introduction	49
3.2	SiC	51
	SiC Substrate Crystal Growth • SiC Thin Film Epitaxy • Dopant Considerations • Ohmic Contacts to SiC	
3.3	SiC Light Emitting Diodes	56
3.4	SiC Photodiodes	57
3.5	Remaining Issues in SiC and Summary	58
3.6	III–V Nitride Semiconductors	59
	Nitride Crystal Growth • Substrates for Nitride Epitaxy • Buffer Layers for Nitride Heteroepitaxy on Sapphire • Polytypism in the III–V Nitrides • Fundamental Properties of GaN, AlN, and InN • Electrical Porperties of Undoped Nitride Thin Films • Properties of Doped GaN	
3.7	Ohmic Contacts to GaN	69
3.8	Properties of Nitride Alloys	69
3.9	GaN-Based LEDs	72
3.10	Toward a GaN Laser	75
3.11	Detectors	80
3.12	Discussion	80

H. Morkoç
University of Illinois at Urbana-Champaign

3.1 Introduction

Recent advances in the science and art of the wide bandgap semiconductors silicon carbide and gallium nitride are reviewed with an emphasis on optoelectronics. The III–V nitrides have the potential for operating as high efficiency emitters and detectors in the blue-green, blue, and ultraviolet region of the optical spectrum, whereas SiC is conducive for the blue and ultraviolet region of the optical spectrum. In fact, SiC and extremely bright GaN light emitting diodes are currently in the marketplace. These technologies, combined with the relatively mature red and yellow emitters, comprise the three primary colors of the visible spectrum which will enable full color displays to be fabricated from semiconductor technology. For optical recording applications, storage capacity increases geometrically as the probe wavelength is reduced, a major impetus for short wavelength semiconductor laser development. In the past several years, each of these material systems has been the subject of significant research breakthroughs which have produced growing optimism that the future for these materials is now.

The recent surge of activity in wide bandgap semiconductors has arisen from the need for electronic devices capable of operating at high power levels, high temperatures, and caustic environments, and, separately, the need for optical materials, especially emitters, which are active in the blue and ultraviolet wavelengths. Electronics based on the existing semiconductor device technologies of Si and GaAs cannot tolerate elevated temperatures or chemically hostile environments. The wide bandgap semiconductors SiC and GaN, and perhaps sometime in the future, diamond, with their excellent thermal conductivities, large breakdown fields, and resistance to chemical attack, will be the materials of choice for these applications. In the optical device arena, the ever increasing need for higher density optical storage and full color display technologies are driving researchers to develop wide bandgap semiconductor emitter technologies capable of shorter wavelength operation.

SiC has emerged as the strong candidate for high temperature and high power device applications due to the availability of high quality SiC substrates, advances in chemical vapor deposition (CVD) growth of epitaxial structures, and the ability to easily dope the material in both n- and p-types. The large Si-C bonding energy makes SiC resistant to chemical attack and radiation, ensuring its stability at high temperatures. In addition, SiC has a large avalanche breakdown field, excellent thermal conductivity, and a predicted high electron saturation velocity which make it ideal for high power operation. Metal semiconductor and metal-oxide-semiconductor transistors with outstanding high temperature performance have already been demonstrated. With the recent introduction of a controllable 4H polytype exhibiting large electron mobilities, SiC is certain to attract more attention for high power electronic applications.

Industries, such as aerospace, electric power, automobile, petroleum and others, have continuously provided the impetus pushing the development of fringe technologies which are tolerant of increasingly high temperatures and hostile environments. Beyond SiC, other materials such as the III–V nitrides are potentially capable of increased power and higher temperature operation, due to their even larger predicted electron velocities, bandgaps, and lower ohmic contact resistances. A suitable semiconductor technology would allow bulky and hazardous aircraft hydraulics to be replaced with actuators controlled with heat tolerant electronics. In satellites, heat radiators could be reduced in size resulting in considerable weight reduction. Other bulky power electronic components in power transmitters could also be replaced with more reliable and compact solid state amplifiers. Both SiC and GaN are characterized by having not just large bandgaps but also small lattice constants. Figure 3.1 shows the lattice constant and bandgap of various polytypes of SiC and both hexagonal and wurtzite phases of the nitrides.

In the field of optical devices, several trends are pushing research into new materials. The ever increasing need for denser optical storage media is driving the development of shorter wavelength semiconductor laser technology due to the fact that the diffraction-limited, optical storage density increases geometrically as the probe laser wavelength is reduced. Towards this end, yellow lasers based on InGaAlP heterostructures have been successfully demonstrated. However, this material system is limited to 650 nm. Toward still shorter wavelengths, the recently demonstrated ZnSe-based laser technology is capable of operating in the green and blue wavelengths, and AlGaN can, potentially, lase in the ultraviolet.

Wide bandgap emitters are also bringing semiconductor technology to full color displays. For the first time, all three primary colors can be generated using semiconductor technology, which promises full introduction of the reliability, compactness, and other desirable attributes of semiconductors to an important technological market.

In this review, we concentrate on device-oriented research and applications of SiC and GaN. The first section covers SiC and describes recent progress in SiC substrates, epitaxy, processing, and the devices to which these advances have led. The next section examines GaN along with its

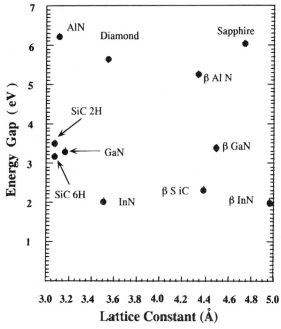

FIGURE 3.1 Zinc blende and wurtzite SiC and GaN lattice constants vs. the energy gap.

alloys with AlN and InN, discussing advances in epitaxial techniques, doping, and device technology.

3.2 SiC

SiC was one of the first semiconductors discovered, and its large cohesive energy caused some to mistake it for an element. Its high electron saturation velocity, wide bandgap, and high thermal conductivity, among other properties, make it highly attractive for emitters and detectors in the blue region of the spectrum and beyond. An appreciation of the potential of SiC for electronic applications can be gained by examining Table 3.1 which compares relevant material properties of SiC and GaN with Si and GaAs, the two most popular semiconductor device technologies, and GaP and diamond, two other contenders for high temperature applications, the former also for emitters up to green photon energies.

SiC is the most prominent of a family of close-packed materials which exhibit a one-dimensional polymorphism called polytypism. The SiC polytypes are differentiated by the stacking sequence of the tetrahedrally bonded Si-C bilayers, such that the individual bond lengths and local atomic environments are nearly identical, whereas the overall symmetry of the crystal is determined by the stacking periodicity. A shorthand has been developed to catalogue the literally infinite number of possible crystal structures. Each SiC bilayer, while maintaining the tetrahedral bonding scheme of the crystal, can be situated in one of three possible positions with respect to the lattice. These are each arbitrarily assigned the notation A, B, or C.

Depending on the stacking order, the bonding between Si and C atoms in adjacent bilayer planes is either of a zinc blende (cubic) or wurtzite (hexagonal) nature. Zinc blende bonds are rotated 60° with respect to nearest neighbors whereas hexagonal bonds are mirror images. Each type of bond provides a slightly altered atomic environment, making some lattice sites inequivalent in polytypes, with mixed bonding schemes, and reducing the overall crystal symmetry. These

TABLE 3.1 Important Parameters of Semiconductors of Interest for Conventional Electronics and Emerging High Temperature Electronics

Property	Si	GaAs	GaP	3C SiC (6H SiC)	Diamond	GaN
Bandgap (eV) at 300K	1.1	1.4	2.3	2.2 (2.9)	5.5	3.39
Maximum operating temperature (K)	600?	760?	1250?	1200 (1580) sublimes	1400(?) phase change	
Melting point (K)	1690	1510	1740	>2100		
Physical stability	Good	Fair	Fair	excellent	very good	good
Electron mobility R.T., cm^2/V-s	1400	8500	350	1000 (600)	2200	900
Hole mobility R.T., cm^2/V-s	600	400	100	40	1600	50?
Breakdown voltage E_b, 10^6/V/cm	0.3	0.4	–	4	10	5?
Thermal conductivity K, W/cm-C	1.5	0.5	0.8	5	20	1.3
Sat. elec. drift vel. v(sat), 10^7 cm/s	1	2	–	2	2.7	2.7
Dielectric const., K	11.8	12.8	11.1	9.7	5.5	9

effects are important when considering the substitutional impurity incorporation and electronic transport properties of SiC.

If the stacking is ABCABC ..., the purely cubic zinc blende structure is realized, which is commonly abbreviated as 3C SiC (or beta SiC). The number 3 refers to the three bilayer periodicity of the stacking, and the letter C denotes the overall cubic symmetry of the crystal. 3C SiC is the only cubic polytype. The purely wurtzite ABAB.... stacking sequence has also been observed and is abbreviated as 2H SiC, reflecting its two bilayer stacking periodicity and hexagonal symmetry. All of the other polytypes are mixtures of the fundamental zinc blende and wurtzite bonds. Some common hexagonal polytypes with more complex stacking sequences are 4H and 6H SiC. 4H SiC is composed equally of cubic and hexagonal bonds whereas 6H SiC is two thirds cubic, despite the overall hexagonal crystal symmetry of each. The family of hexagonal polytypes is collectively referred to as alpha SiC. Rhombohedral structures, such as 15R and 21R, have also been documented (Tagodzinski, Arnold, 1960).

The different polytypes have widely ranging physical properties. 3C SiC has the highest electron mobility and saturation velocity, resulting from the increased symmetry of the cubic crystal which reduces phonon scattering. The bandgaps differ widely among the polytypes, ranging from 2.3 eV for 3C SiC to 2.9 eV in 6H SiC to 3.3 eV for 2H SiC. Among the SiC polytypes, 6H is most easily prepared and has been the most extensively studied, whereas the 3C and 4H polytypes, particularly the latter, are attracting more attention. The polytypism of SiC makes it nontrivial to grow single phase material, but it also offers some potential advantages if crystal growth methods can be developed sufficiently to capitalize on the possibility of polytype (homo/hetero)-junctions. Figure 3.2 shows one such possibility, a 6H/3C SiC interface. Such a junction incorporates the advantages of heterojunction band offsets while maintaining a completely charge-free, lattice-matched, coherent interface.

Powell et al. (1990) have shown that 3C SiC can be grown on well-oriented basal plane (0001) 6H substrates. In this case, the heteroepitaxy is terrace-controlled, and the adatoms are free to choose the most energetically favorable stacking sequence in the direction perpendicular to the substrate surface. Proper control of the growth conditions, then, gives rise to the heteroepitaxy of 3C SiC. When the substrate is miscut several degrees away from the basal plane towards (11$\overline{2}$0), steps are prevalent on the growth surface. If the step spacing is less than the surface diffusion

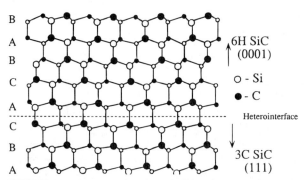

FIGURE 3.2 An example of a 3C SiC polytype and 6H SiC polytype heterojunction. When the plane of the interface is normal to the stacking direction, a coherent, lattice-matched heterojunction is possible.

length of the adatoms, the resultant growth is step controlled, allowing the underlying 6H material to influence the epitaxy, and homoepitaxial 6H SiC is grown. Researchers envision a SiC heterojunction bipolar transistor incorporating all of the high-power, high temperature capabilities of SiC, with the advantages of a wide bandgap emitter.

SiC Substrate Crystal Growth

One major advantage that SiC enjoys over other wide bandgap competitors is an established, commercialized process for the growth of high quality substrate material. 6H SiC substrates have been prepared in the laboratory by three separate processes (Levin et al., 1978; Tairov and Tvorskov, 1981; Ikeda et al., 1979), namely, thermal decomposition, growth from a carbon enriched Si melt, and sublimation. The sublimation growth technique is presently used to grow commercial substrates. One-inch diameter 6H SiC wafers have been available for some years. However, efforts to increase the substrate area have not yet reached the marketplace. Substrates are available with both n- and p-type conductivity over a wide range, but a semi-insulating SiC substrate growth technology is actively being pursued.

In sublimation growth, SiC is transported in the vapor phase to a SiC seed crystal which is held at a lower temperature. Typical growth parameters are 1800 °C for the seed crystal compared with a source temperature of 2000 °C. A thermal gradient of 20 K/cm across the growing crystal results in a growth rate of 0.7 mm/hr. For larger growth rates, higher source and seed crystal temperatures (2200 °C and 2300 °C) in an ambient pressure of roughly 5 torr have been successful. In the latter approach, the sublimated SiC clusters must be diffused though porous graphite under carefully controlled thermal and pressure gradients to form high quality single crystal 6H SiC. Reduced defect densities were noted when the source was situated below the seed crystal. Sublimation is also a suitable technique for epitaxial growth of SiC.

One serious problem faced by crystal growers is the formation of micropipes. Micropipes are voids which propagate through the entire length of the boule. They are observed to nucleate at the seed crystal and always have hexagonal cross-sections. Researchers have had some success eliminating micropipes by preparing seed crystals in which epitaxial layers are grown on the A face prior to sublimation growth of the bulk crystal.

In one process, the diced substrates undergo mechanical polishing involving a sequence of SiC and diamond pastes (Matsunami, private communication). The wafers are typically characterized by both optical microscopy to determine etch pit densities (EPD) and reflection mode Fourier transform infrared spectroscopy to study the width of the LO phonon signal. Typical EPD values are 10^4–10^5 cm^{-2}. Although efforts to improve substrate quality continue, it is encouraging to note that the present quality is nearly as good as early GaAs substrates. No chemical polishing technique has yet been published in the open literature.

SiC Thin Film Epitaxy

Early SiC epitaxy was achieved using liquid phase epitaxy (LPE). LPE can be performed at lower temperatures (1500–1700 °C) compared to sublimation. The first efforts suffered from contamination of C-saturated molten Si coming from the conventional graphite crucibles (Ikeda et al., 1979; Muench and Kurzinger, 1978). Dmitriev et al. (1985) circumvented the problem by introducing a graphite-free technique in which the Si melt is suspended in an electromagnetic field.

Chemical vapor deposition (CVD) has replaced LPE and sublimation as the growth method of choice. Both low and atmospheric pressure CVD have been successfully applied to SiC epitaxy. Low pressure is preferred when the deposition is over large area substrates, such as Si. However, because typical SiC substrates are only one inch in diameter, atmospheric pressure CVD provides adequate uniformity. Several groups have also used gas source, molecular beam epitaxy (GSMBE) for SiC growth, especially those interested in exploring Si and SiC technologies together. In this case, GSMBE allows lower growth temperatures which restrict dopant diffusion in Si.

Many workers have attempted to grow SiC heteroepitaxially on Si, taking advantage of the lower cost and larger area of these substrates and the potential for Si-SiC device integration. These advantages are offset by the large thermal and lattice mismatch between the two materials, which has made high quality material extremely difficult to achieve. Nevertheless, work continues on improving the quality of SiC/Si. Nishino (Nishino et al., 1983) developed a technique in which the clean Si surface is exposed to a carbon-containing gas at or above the growth temperature. This results in forming a thin monocrystalline 3C SiC layer which serves as a template for epitaxial growth. Nearly all of the SiC grown on Si substrates is cubic (Nishino et al., 1980; Golecki et al., 1992; Furumura et al., 1988; Suzuki et al., 1992; Sugil et al., 1990; Fuyuki et al., 1989; Zhou et al., 1993; Paisley et al., Wahab et al., 1992).

In addition to threading dislocations, two types of defects dominate SiC/Si epilayers. Planar defects such as microtwins and stacking faults are commonly observed (Nutt et al., 1987). This is not surprising given the polytypism of SiC. A stacking fault is simply a temporary disruption of the bulk stacking sequence. When growing on the zinc blende (100) surface, a coherent interface such as that shown in Fig. 3.2 is no longer possible because a bilayer plane must be terminated. Therefore, in SiC grown on (100) Si, the stacking faults are accompanied by dislocations, and, together, they relieve the lattice and thermal mismatch strain.

The other prevalent defects are double positioning (or, equivalently, inversion or antiphase domain) boundaries (DPB) (Shibahara, 1986; Pirouz, 1987) which arise from the reduced symmetry of SiC compared to the Si substrate. In the earliest stages of growth, the Si and C sublattices are interchangeable with regard to the Si substrate. When the Si and C sublattices are exchanged, the resulting 3C SiC crystal is rotated 90 °. When islands having opposite sublattice orientations nucleate together at later stages of the growth, the boundaries between the two domains necessarily have Si-Si and C-C bonding which increases the total energy of the crystal and alters the atomic charge distribution, thereby, scattering carriers. Inversion domain boundaries are seen as bands of mottled contrast in transmission electron micrographs.

As SiC substrates have become more widely available, more researchers are choosing to grow SiC homoepitaxially. Greatly reduced defect densities have been reported in 3C SiC epilayers grown on 6H SiC substrates (Powell et al., 1990), however DPBs continue to be observed by some workers (Kong et al., 1988). A group (Matsunami, private communication) using sublimation growth on in-house 6H SiC boules has succeeded in growing 3C SiC devoid of DPBs. In this approach the 6H (01$\bar{1}$4) surface is used for GSMBE epitaxial growth.

Dopant Considerations

One of the major limitations in developing wide bandgap semiconductors has been finding suitable shallow dopants. Although it is not ideal, achieving ambipolar doping in SiC has proved much easier than in the GaN and ZnSe systems. Nitrogen is the most popular n-type impurity whereas Al is favored for p-type doping. Dopants may be introduced during epitaxy or, later, using ion implantation. For CVD, ammonia and triethylaluminum (TMA) have proven to be suitable dopant source gases for n- and p-type doping, respectively (Kong et al., 1988).

When N doping is introduced during the growth, carrier concentrations as high as 10^{18} cm^{-3} can be realized. Efforts to further increase N incorporation resulted in polycrystalline material, probably driven by SiN$_x$ formation. Ion implantation with subsequent Ar annealing has yielded electron concentrations as large as 3×10^{19} cm^{-3} at a N dose of 5×10^{20} cm^{-3} (Cooper and Melloch, private communication).

Acceptor p-type doping is a recognized problem in SiC, although considerable progress has been made. All of the acceptor impurities thus far investigated, namely, Al, B, Ga, and Sc, form deep levels and are difficult to activate, generally requiring high temperature annealing. The depth of the acceptor levels also leads to strong variation of hole concentration with temperature, which considerably complicates device design and operation. Al is somewhat difficult to incorporate into the SiC lattice, and high carrier concentrations are difficult to achieve. Researchers at Kyoto University (Matsunami, private communication) have obtained p-type carrier concentrations in the 10^{19}–10^{20} cm^{-3} range using TMA in a CVD process on the Si face of 6H SiC. In contrast, growth on the C face permitted only 2×10^{18} cm^{-3} p-type doping. The carrier concentration could be easily controlled down to the low 10^{16} cm^{-3} range. On the upper end, the observed hole concentration became nonlinear as a function of TMA flow above 10^{19} cm^{-3}.

In general, background N causes unintentionally doped crystals to be n-type. In the best 6H SiC samples (the most developed of the polytypes), background carrier concentrations in the mid 10^{15} cm^{-3} range have been achieved (Palmour et al., 1991b). Further improvements should be possible as sources of nitrogen contamination are eliminated.

An excellent review of the optical and electrical properties of doped SiC has recently been published by Pensl and Choyke (1993). One important point made by the authors is that dopants can occupy either hexagonal or cubic sites in the more complex SiC polytypes. These different environments give rise to different binding energies and care must be taken when deconvolving the separate contributions from Hall data. Analyzing several samples, Pensl and Choyke showed that the relative abundance of the various N dopant levels corresponded to the ratio of available binding sites, that is, in 4H SiC an equal number of donors occupy cubic sites and hexagonal sites. In 6H SiC, the ratio is 2:1, reflecting the fact that two-thirds of the Γ sites have cubic bonding. In 6H SiC, the measured ionization energy of the hexagonal site was 85.5 meV whereas the energy of the cubic site was 125 meV. The experimental resolution was insufficient to resolve the two separate cubic donor energy levels. For 4H SiC, the hexagonal (h) and cubic (k) binding energies were measured to be 45 meV and 100 meV, respectively. In 3C SiC, a value of 48 meV was determined. Typical compensation values were one to two orders of magnitude below the observed electron concentration. Similar measurements for Al-doped SiC yielded an acceptor ionization energy of roughly 200 meV for each of the three most common SiC polytypes. These values are all smaller than those measured optically due to a reduction in the average electron energy when donor spacing is small.

Ion implantation plays a major role in commercial SiC technology. Due to the excellent stability of SiC, the material lends itself well to high temperature annealing for implantation-related damage removal. In 1970, Marsh and Dunlap (1970) characterized the first ion-implanted SiC junctions which were formed by implanting n-type dopant into a p-type substrate at room temperature. More modern approaches utilize a process in which the target material is heated during implantation (Ghezzo et al., 1992).

Ohmic Contacts to SiC

Many vital parameters of semiconductor devices, including speed and high power performance, depend strongly on the ohmic contact resistance. The issue is especially important in wide bandgap semiconductor systems because of the large Schottky barrier heights involved. Although the problem can often be alleviated in other semiconductors by the choice of a metal with an optimal barrier height, little flexibility is available in SiC due to Fermi level pinning at the surface. The barrier heights of Au, Pt, Ti, Hf, and Co are all within several hundred meV of one another. The best values for SiC remain as large as the upper 10^{-4} ohm-cm^2 and low 10^{-3} ohm-cm^2 ranges for reliable n- and p-type contacts, respectively. The contact problem is exacerbated by the high-temperature, high-power operation desired for SiC. Many contact schemes prove unreliable due to the formation of carbides and silicides which degrade the contact over time. The maximum operating temperatures of present day devices are limited by the ohmic contacts, not by the material properties of SiC. It is the overwhelming consensus of the SiC device community that ohmic contacts, along with micropipes, are a critical element in need of improvement.

Most of the contact schemes have been demonstrated to be stable up to 450 °C, but SiC, in the applications envisioned, will require reliable contacts at 600 °C and above. To achieve this performance, it is generally assumed that refractory metals will be required. Mixtures of Au with refractory metals have been tried (Naumor et al., 1987; McMullin et al., 1990; Anikin et al., 1990; and Gardner et al., 1991). However, when annealed above 450 °C, the metal diffused into the SiC or out through the capping material. Silicides with both refractory (TaSi$_2$) (Zeller, 1991) and nonrefractory metals have also been explored (Palmour et al., 1988a). WSi$_2$ contacts gave contact resistances of 4×10^{-4} ohm-cm^2 but delaminated at 600 °C.

An alternative approach is to evaporate a carbide-forming metal onto SiC, followed by thermal annealing with the aim of causing a chemical reaction. The contact is, then, passivated by a top noble metal layer. Carbide-based contacts, generally, adhere better than silicides. Au/Ta/SiC contacts having resistances as low as 1×10^{-5} ohm-cm^2 were achieved, but degraded, when annealed in air at 900 °C (Kelner et al., 1989). Informal results indicate that ohmic contacts with specific resistivies of high 10^{-6} ohm-cm^2 in heavily n-type SiC have been achieved.

3.3 SiC Light Emitting Diodes

Despite its indirect bandgap, SiC can be made to electroluminesce across the entire visible spectrum by adding various impurities. Although conventional GaP-based LEDs can provide red and green light at better efficiencies and lower cost than SiC, a niche remains for blue LEDs to be used in full color, display applications. Whereas the direct bandgap GaN and ZnSe semiconductors are potentially more efficient LEDs, they are not yet commercial. Brander et al. (1969) were the first to fabricate a blue SiC LED and since then other groups have followed suit (Hoffman et al., 1983; Edmond et al., 1993; and Hong et al., 1992).

Cree Research (Edmond et al., 1993) presently markets the best performing SiC blue LEDs, which have a pure blue emission centered at 470 nm (Fig. 3.3). Groups at Siemens AG, Sanyo, and Sharp have also developed prototype devices, and some have reached the marketplace. Cree's devices are reported to radiate 18.3 μW at a 25 mA (3 V) forward bias with a spectral half-width of 69 nm. Due to the indirect bandgap, the efficiency of these devices is only 0.02–0.03%, but that is partly compensated for by the ability to drive the SiC LEDs at higher currents. At 50 mA, 36 μW of output has been achieved. LEDs run at 50 mA show a typical degradation of 10–15%

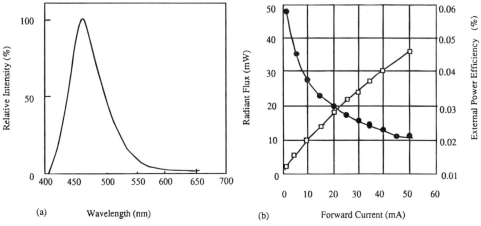

FIGURE 3.3 Cree Research, Inc. SiC LEDs. (a) Pure blue output spectrum centered at 470 nm. (b) LED efficiency as a function of bias. After Edmond et al., 1993.

over 10,000 hours, significantly less than GaP LEDs, and, once again, attesting to the durability of SiC.

The main light-producing mechanisms have been identified as (1) donor to acceptor (DA) pair recombination ($\sim$ 480 nm), (2) bound exciton recombination at localized Al centers ($\sim$ 455 nm), and (3) free exciton recombination ($\sim$ 425 nm). The key to reaching shorter wavelengths in SiC LEDs is reducing the background N contamination, which would shift the emission away from DA recombination towards exciton-related luminescence. On the other hand, the efficiency decrease at higher power levels is believed to result, in part, from a saturation of the DA pair levels (Edmond et al., 1993). The authors speculate that an increased DA pair density could result in an overall increase in the high-power efficiency of SiC LEDs.

3.4 SiC Photodiodes

Wide bandgap semiconductors are desirable for ultraviolet (uv) optical detection applications because of their insensitivity to longer wavelengths and their very small dark current even at elevated temperatures. The low dark current values have enabled SiC photodiodes to exceed Si UV detector sensitivities by four orders of magnitude (Palmour et al., 1993). Many of the applications for SiC *uv* detection involve hostile environments, such as in situ combustion monitoring and satellite based missile plume detection, where the ruggedness of SiC is important. Other applications capitalize on the sensitivity of SiC detectors, such as air quality monitoring, gas sensing, and personal UV exposure dosimetry. Among the wide bandgap semiconductors, SiC once again leads the way, being the first to reach the marketplace (Palmour et al., 1993). Fig. 3.4 shows a typical SiC UV photodiode device structure (Brown et al., 1993) fabricated from commercial 6H substrates on which a p-n junction was grown epitaxially. Photodiodes reported by Edmond et al. (1993) have demonstrated extremely small dark currents, as little as 10^{-11} A at -1 V and 200 °C (Fig. 3.5), representing a considerable improvement over previous efforts (Campbell and Chang, 1967; Strite and Morkoç, 1992a). The same devices exhibited near unity peak responsivities between 268 nm and 299 nm at temperatures as high as 450 °C (Fig. 3.6). However, as is apparent in Fig. 6, the responsivity falls off quickly with decreasing wavelength.

Brown et al. (1993) have attempted to push the operation of *uv* SiC photodiodes to still shorter wavelengths. As the detection wavelength is decreased, a larger proportion of the absorption occurs near the semiconductor surface, and surface recombination becomes more important. By thinning the top n$^+$ layer outside of the mesa contact, a responsivity of 50 mA/W was achieved

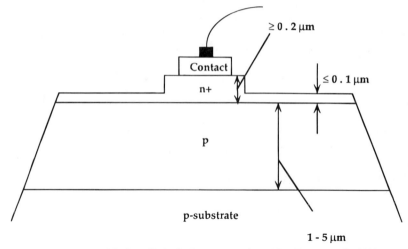

FIGURE 3.4 SiC photodiode device cross section. After Brown et al., 1993.

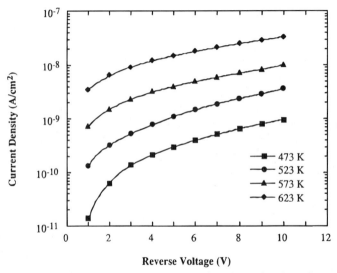

FIGURE 3.5 SiC photodiode dark current density as a function of temperature and reverse bias. After Edmond et al., 1993.

at 200 nm. Further improvement might be realized by adding a SiO_2 cap layer, which is transparent down to nearly 125 nm, to serve as a *uv* window, while, simultaneously, passivating the SiC surface.

3.5 Remaining Issues in SiC and Summary

SiC device technology has made great strides and has reached the marketplace. Nevertheless, many aspects of the substrate and material quality, epitaxial growth, device processing, and oxide quality need to be improved. Availability of SiC substrates has made possible the recent advances in SiC and GaN material and device quality. Currently, researchers are very active in trying to improve substrate quality by eliminating micropipes, lowering EPD densities, and developing improved surface polishing processes. Efforts are already underway to obviate the micropipe problem. Among the approaches is the use of the A face which so far has led to inferior electronic properties. The micropipe nemesis may simply be eliminated by getting rid of surface contaminants during bulk growth. Research 2-inch substrates are available in limited quantities; 1 3/16-inch

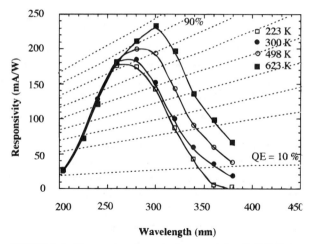

FIGURE 3.6 Temperature dependent responsivity of a SiC photodiode. After Edmond et al., 1993.

substrates and also the 4H substrates have reached the marketplace. Impressive room and cryogenic temperature electron mobilities, possible with the 4H polytype along with shallower donor activation energies, must be exploited. As larger SiC substrates are developed, economies of scale will be possible, making SiC devices practical for a wider range of applications.

The quality of epitaxial SiC has benefited from SiC substrates and improved CVD growth techniques and understanding. Both 6H and 3C SiC can now be deposited on 6H substrates, although double positioning boundaries and stacking faults remain a problem to be addressed. It would be useful for heterojunction device designs if 6H SiC could be grown on the 3C polytype. There is also a need to find methods of increasing dopant incorporation and activation in epitaxial SiC processes while reducing Al dopant diffusion.

In the SiC device arena, improvements in ohmic contact, packaging, and oxide technologies will allow better, more practical devices for a wider range of applications. FET devices require increased doping in the source and drain regions and better oxide interfaces. To allow long term operation at high temperatures, it is the consensus that some sort of encapsulation must be developed. Integration will require better packaging, contacts, insulators, and interconnects, all of which must be capable of long term, reliable performance at the ambient temperatures and power levels which will be demanded of SiC devices. On the positive side, all of these problems were faced and solved by pioneering Si researchers in the early sixties. It is now the quest of the nineties to bring about similar improvements in SiC.

3.6 III-V Nitride Semiconductors

The III-V nitrides have long been viewed as a promising system for optoelectronic applications in the blue and UV wavelengths and, more recently, as a high-power, high-temperature semiconductor with electronic properties superior to SiC. However, progress in the nitrides has been much slower than in SiC and ZnSe, and only recently have practical devices been realized.

Because nitride development presently lags behind that of ZnSe, many research groups have overlooked the long term advantages of the nitrides for laser and photodetector applications. The wurtzite nitride polytypes form a continuous alloy system whose direct room temperature bandgaps range from 6.2 eV in AlN, to 3.4 eV in GaN and to 1.9 eV in InN. Although ZnSe-based laser devices are limited to the visible wavelengths by their relatively smaller band gaps, lasers based on AlGaN quantum wells (QW) could conceivably operate at energies up to 4 eV. The extremely high thermal conductivities and superior stability of the nitrides and their substrates should, eventually, allow higher power operation with less degradation. GaN and AlN have a smaller lattice mismatch than any of the ZnSe alloys which would allow greater range and flexibility in heterostructure design. The growth of the nitrides is simpler because only ternary alloys are necessary. Nitride composition can be modulated by controlling group III fluxes as opposed to the higher vapor pressure group II and VI elements in effect in II-VI nitrides. Finally, ohmic contacts have been demonstrated for both polarities of GaN, meaning that GaN-based devices will not be limited by contact resistance, presently with ZnSe-based lasers.

Early research pinpointed a number of problems with the nitrides which hindered efforts to grow high quality material and fashion it into devices. These were (1) lack of a good substrate material, (2) large n-type background carrier concentrations, (3) an inability to dope p-type GaN, and (4) lack of a suitable etchant. In the past five years, great progress has been made in each of these areas. Optimized buffer layers on sapphire substrates and, more recently, the availability of better lattice-matched and thermally matched SiC substrates, has allowed greatly improved layers to be grown. GaN background carrier concentrations have been reduced as low as 4×10^{16} cm^{-3} in high quality, thin films. An understanding of the Mg acceptor impurity has allowed GaN to be doped to hole concentrations as high as 3×10^{18} cm^{-3}. Finally, reactive ion etching has proven to be a reliable and convenient method of etching nitride materials. These advances in material quality and processing have allowed researchers to demonstrate the first GaN p-n junction. LED, giving rise to optimism in many circles that a GaN-based laser will soon follow.

We have already reviewed the III-V nitrides from a historical perspective (Strite and Morkoç, 1992a). In this section, we only briefly cover the fundamental physical properties of the nitrides and their alloys, preferring instead to focus on current epitaxial growth techniques and devices which are at the forefront of today's research. Modern MBE and MOVPE growth techniques, including new developments in plasma based sources and substrate materials, are covered. In addition, new devices, processing techniques, and predictions in GaN-based devices are reviewed.

Nitride Crystal Growth

Nearly every crystal growth technique, substrate type, and orientation have been tried in an effort to grow high quality III-V nitride thin films. In recent times, researchers have successfully taken advantage of the newer molecular beam epitaxy (MBE) and metalloorganic vapor phase epitaxy (MOVPE) techniques which have yielded greatly improved film quality.

Maruska and Tietjen (1969) grew the first single crystal epitaxial GaN thin films by vapor transport. In their method, HCl vapor flowed over a Ga melt, causing the formation of GaCl which was transported downstream. At the substrate, the GaCl mixed with NH$_3$, resulting in the following chemical reaction (Ban, 1972).

$$GaCl + NH_3 \rightarrow GaN + HCl + H_2 \qquad (3.1)$$

The growth rates were quite large (0.5 μm/min) which allowed deposit of extremely thick films. However, GaN grown by vapor transport had very high background n-type carrier concentrations, typically 10^{19} cm^{-3}.

Maruska and Tietjens' approach was an early version of the modern day MOVPE GaN growth technique. In MOVPE, trimethylgallium (TMG), triethylaluminum (TMA), and trimethylindium (TMI) react with NH$_x$ at a substrate heated to roughly 1000 °C (Hashimoto et al., 1984; Sasaki and Matsuoka, 1988; Khan et al., 1991c; and Nakamura et al., 1991a). Nakamura et al. (1991) designed an atmospheric pressure MOVPE reactor specifically for nitride growth which has been

GaN and Silicon Carbide as Optoelectronic Materials 61

highly successful (Fig. 3.7 (a)). Reactant gases, diluted with H_2, enter the growth area through a quartz nozzle which directs the flow across the rotating substrate. The key aspect of the Nakamura et al. (1991) design is a downward subflow of He and N_2 which they claim improves the interaction of the reactant gases with the substrate (Fig. 3.7 (b)).

A disadvantage of the MOVPE approach is the high substrate temperature necessary to thermally dissociate the NH_3. Due to thermal mismatch with all available substrates, postgrowth cooling introduces significant amounts of strain and defects in the nitride film. In addition, the high growth temperatures may encourage other undesirable effects, such as dopant and group III metal desorption and segregation.

To reduce substrate temperatures, many groups have begun exploring an MBE approach in which the reactive nitrogen is supplied by microwave plasma excitation (Paisley et al., 1989c; Strite et al., 1991; and Lei et al., 1991). This has been made possible by the commercial development of compact electron cyclotron resonance (ECR) microwave plasma sources, like the Wavemat MPDR 610 and ASTEX CECR. These sources use a coaxial cavity geometry to efficiently couple

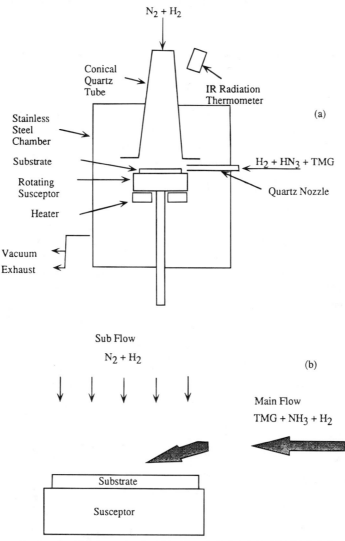

FIGURE 3.7 (a) Schematic of Nichia Chemical GaN MOVPE. (b) Sub flow of $N_2 + H_2$ from above which is claimed to improve interaction of reactant gases. After Nakamura et al., 1991a.

microwave energy (2.45 GHz) into the nitrogen discharge region. The plasma stream is diffusive and neutral, providing both atomic nitrogen and, to some extent, detrimental ionic nitrogen radicals to the growth surface. An ECR source is a considerable improvement over conventional RF sources in that it is compact enough to be installed into the MBE source flange, gaining a direct line of sight to the substrate (Fig. 3.8).

Researchers have learned that ECR sources, as they are presently designed for MBE use, have an undesirable trade-off (Lei et al., 1991). In normal use, the GaN growth rate is limited by the available flux of low kinetic energy reactive nitrogen species. This growth rate has proven to be quite small, on the order of 500 Å/hr (Lin, 1993d), although material quality is beginning to rival that of MOVPE (Lin et al., 1993a). If higher growth rates are attempted, Ga droplets form on the surface, leading to highly n-type GaN. The growth rate can be increased when the ECR is run at higher microwave power, but, in this regime, more energetic ions are created which degrade material quality, probably by introducing point defects. The GaN, in this case, is semi-insulating due to compensating deep levels. For ECR-based MBE to continue to develop, present day ECR designs must be upgraded or else the MBE source flange must be modified to permit installing larger diameter ECRs.

Substrates for Nitride Epitaxy

Probably what has hindered nitride researchers most in their quest for improved material quality is the lack of a suitable substrate material that is both lattice-matched and thermally matched to GaN. Efforts are ongoing to grow bulk GaN crystals for substrates, but, at present, researchers have no choice but to grow nitrides heteroepitaxially. Many different substrates have been tried, and the community has come to favor basal plane sapphire as a substrate. However, substrates like SiC, MgO, and ZnO, which have thermal and/or lattice matches superior to the sapphire, are increasingly available and should become popular in the near future. Table 3.2 compares the relevant properties of GaN and AlN with those of the most popular substrate materials.

The preference for sapphire substrates in nitride heteroepitaxy can be attributed to their wide availability, hexagonal symmetry, ease of handling, simple pregrowth cleaning requirements, and stability at the elevated temperatures required for nitride MOVPE. However, as can be seen in Table VII, the lattice and thermal mismatch between sapphire and GaN is quite large, resulting in a considerable strain field upon post growth cooling (Amano et al., 1988a; Naniwae et al., 1990). Nevertheless, with a suitable buffer layer, sapphire substrates have proven extremely serviceable. Presently, the best GaN material and devices have been realized on basal plane sapphire substrates.

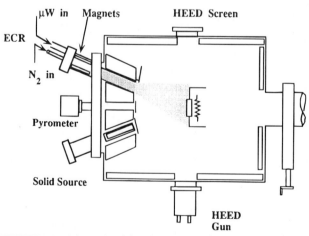

FIGURE 3.8 Schematic of the GaN MBE with ECR nitrogen source available to the author.

TABLE 3.2 Properties of GaN and AlN Vis a Vis Those of the Substrate Materials on Which These Wide Bandgap Materials Have Been Deposited by Various Growth Techniques

Substrate Material	Lattice Parameters	Thermal Conductivity	Coefficients of Thermal Expansion
GaN	a = 3.189 Å	1.3 W/cmK	5.59×10^{-6}/K
	c = 5.185 Å		3.17×10^{-6}/K
AlN	a = 3.112 Å	2.0 W/cmK	4.2×10^{-6}/K
	c = 4.982 Å		5.3×10^{-6}/K
6H SiC	a = 3.08 Å	4.9 W/cmK	4.2×10^{-6}/K
	c = 15.12 Å		4.68×10^{-6}/K
Sapphire	a = 4.758 Å	3.5 W/cmK	7.5×10^{-6}/K
	c = 12.99 Å		8.5×10^{-6}/K
ZnO	a = 3.252 Å		2.9×10^{-6}/K
	c = 5.213 Å		4.75×10^{-6}/K
Si	a = 5.4301 Å	1.5 W/cm	3.59×10^{-6}/K
GaAs	a = 5.6533 Å	0.5 W/cmK	6×10^{-6}/K
3C SiC	a = 4.36 Å	4.9 W/cmK	
MgO	a = 4.216 Å		10.5×10^{-6}/K

From Landolt-Börnstein, Vol. 17, Springer-Verlag, New York, 1982.

Buffer Layers for Nitride Heteroepitaxy on Sapphire

Because researchers, at present, have no choice for nitrides but heteroepitaxy, the optimization of the initial growth conditions is critical. Much progress has been made by introducing an initial buffer layer, especially on sapphire substrates, but also on SiC (Lin et al., 1993), GaAs (Amano et al., 1988; Naniwae et al., 1990; and Strite et al., 1992b) and Si (Amano et al., 1988a; Naniwae et al., 1990; and Lei et al., 1992).

Early efforts to grow GaN on sapphire substrates had only moderate success. The epilayers were often cracked after postgrowth cooling as a result of thermal strain (Grimmeis and Monemar, 1970). Yoshida et al. (1983) reported a substantial improvement in GaN/sapphire when the epitaxy was initiated with an AlN buffer layer. Akasaki and co-workers (Amano et al., 1986; Koide et al., 1988; Amano et al., 1988b; and Akasaki et al., 1989) have employed low temperature AlN buffer layers and studied, in detail, their role in the quality of the GaN layers grown. Akasaki et al. (1989) observed that the AlN grown at low temperature is initially amorphous, and, during subsequent growth, the buffer layer converts to single crystal. In their GaN grown with AlN buffer layers, a decrease of two orders of magnitude in the background carrier concentration was accompanied by a factor of ten in increased mobility. Bandgap photoluminescence was more intense by two orders of magnitude, midgap emission was suppressed, and the X-ray diffraction peak was four times narrower. As a result, the AlN buffer layer was quickly adopted by most workers in the field.

Recently, however, Nakamura (1991b) has grown GaN of equal or better quality by using a low temperature GaN buffer layer instead. Kuznia et al. (1993) have compared both low temperature GaN and AlN buffer layers to determine which is superior. Low-energy electron diffraction patterns confirmed that noncrystalline buffer layers convert to single crystals when actual growth conditions were simulated by a 1000 °C annealing. GaN film quality was observed to be a strong function of buffer layer thickness, with optimal thicknesses of 250 Å and 500 Å for GaN and AlN buffer layers, respectively. Both low-temperature GaN and AlN buffer layers improved the bulk GaN quality with AlN buffer layers providing slightly superior results in this study.

Polytypism in the III-V Nitrides

Like SiC, the III-V nitrides have been observed to nucleate in multiple tetrahedrally bonded polytypes. GaN, AlN, and InN are all most commonly observed as the wurtzite (2H) polytype

(Fig. 3.9 (a)), but each can also crystallize in a metastable zinc blende structure (Mizutu et al., 1986; Petror et al., 1992; Strite, 1993) (Figure 3.9 (b)). In general, wurtzite material grows on hexagonal substrates whereas zinc blende can be grown on cubic substrates. The exception is the sapphire (0001) (Humphreys, 1990; Nakamura, 1991c) zinc blende (111) faces, which are normal to the stacking direction and are, therefore, polytype neutral.

Fundamental Properties of GaN, AlN, and InN

Workers have labored for 25 years and have been able to determine many of the fundamental physical properties of the III-V nitride semiconductors. In Tables 3.3a, b, and c we tabulate the most important properties of each semiconductor. The more interested reader is referred to a longer discussion of the fundamental physical properties of these materials (1992a).

Electrical Properties of Undoped Nitride Thin Films

Control of III-V nitride electrical properties has traditionally been, and still is, one of the greatest challenges facing nitride researchers. All unintentionally doped GaN and InN, reported to date, suffer from n-type background carrier concentrations, except in films of poor quality which are compensated by deep levels. With improved crystal growth techniques, researchers in several leading laboratories have succeeded in reducing the background electron concentration to 10^{16} cm^{-3}. Nakamura et al. (1991c) have reported the highest GaN bulk mobility μ_n = 600 and 1500 cm^2/Vs at 300 K and 77K, respectively, in an undoped sample having n = 4 × 10^{16} cm^{-3}. InN films have proven much more difficult to grow and, generally, have background electron

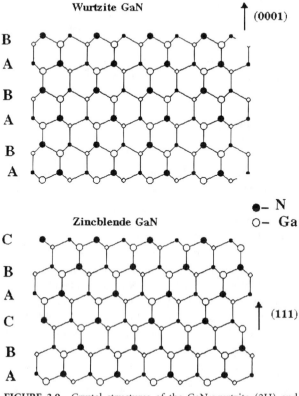

FIGURE 3.9 Crystal structures of the GaN wurtzite (2H) and zinc blende (3C) polytypes.

TABLE 3.3 Properties of GaN(a), AlN(b), and InN(c)

Wurtzite Polytype		
Bandgap energy	$E_g (300K) = 3.39$ eV	$E_g (1.6K) = 3.50$ eV
Temperature coefficient	$\frac{dE_g}{dT} = -6.0 \times 10^{-4}$ eV/K	
Pressure coefficient	$\frac{dE_g}{dP} = 4.2 \times 10^{-3}$ eV/kbar	
Lattice constants	$a = 3.189$ Å	
Thermal expansion	$\frac{\Delta a}{a} = 5.59 \times 10^{-6}$/K	$\frac{\Delta c}{c} = 3.17 \times 10^{-6}$/K
Thermal conductivity	$\kappa = 1.3$ W/cmK	
Index of refraction	$n(1\ eV) = 2.33$	$n(3.38\ eV) = 2.67$
Dielectric constants	$\epsilon_r \approx 9$	$\epsilon_\infty = 5.35$
Zincblende polytype		
Bandgap energy	$E_g (300K) = 3.2\text{–}3.3$ eV	
Lattice constant	$a = 4.52$ Å	
Index of refraction	$n(3\ eV) = 2.5$	
Bandgap energy	$E_g (300K) = 6.2$ eV	$E_g (5K) = 6.28$ eV
Lattice constants	$a = 3.112$ Å, $c = 4.982$ Å	
Thermal expansion	$\frac{\Delta a}{a} = 4.2 \times 10^{-6}$/K	$\frac{\Delta c}{c} = 5.3 \times 10^{-6}$/K
Thermal conductivity	$\kappa = 2$ W/cmK	
Index of refraction	$n(3eV) = 2.15 \pm 0.05$	
Dielectric constants	$\epsilon_r \approx 8.5 \pm 0.2$	$\epsilon_\infty = 4.68\text{–}4.84$
Zincblende Polytype		
Bandgap energy	$E_g (300K) = 5.11$ eV, theory	
Lattice constant	$a = 4.38$ Å	
Bandgap energy	$E_g (300K) = 1.89$ eV	
Temperature coefficient	$\frac{dE_g}{dT} = -1.8 \times 10^{-4}$ eV/K	
Lattice constants	$a = 3.548$ Å	$c = 5.760$ Å
Index of refraction	$n = 2.80\text{–}3.05$	
Dielectric constants	$\epsilon_r \approx$	
Zincblende Polytype		
Bandgap energy	$E_g (300K) = 2.2$ eV, theory	
Lattice constant	$a = 4.98$ Å	

concentrations in excess of 10^{18} cm^{-3}. AlN is always observed to be insulating, even when doped, most likely because its donor, acceptor and defect levels lie deep within the bandgap.

Even in the purest samples, the conductivity behavior described above persists, leading researchers to conclude that native defects, most likely nitrogen vacancies, are responsible. Evidence supporting nitrogen vacancies as the culprit has been accumulating. Jenkins and Dow (1989) calculated the native defect levels in InN, GaN, and AlN and found that the nitrogen vacancy defect is be a donor state with binding energies of 10 meV, 40 meV, and 1 eV in InN, GaN, and AlN, respectively. Powell et al. (1993) observed a strong correlation between the nitrogen beam pressure and the resultant carrier concentration. In that experiment, the N/Ga flux ratio was increased by a factor of eight resulting in an increase by 6 orders of magnitude in the GaN resistivity. Tansley and Egan (1993) have recently published a review of native defects in the nitrides to which the interested reader is referred for a more complete treatment.

Properties of Doped GaN

The achievement of p-type doping has been the major catalyst in the resurgent interest in nitrides. On the other hand, improved epitaxial techniques have reduced the background electron concentration in unintentionally doped GaN, necessitating new research into n-type dopants. Until recently, efforts to dope p-type GaN have led to compensated high resistivity material. Akasaki et al. (Amano et al., 1989; Akasaki et al., 1991a) made the initial breakthrough when they observed that compensated Mg-doped GaN could be converted into conductive p-type material by low-energy electron beam irradiation (LEEBI). Nakamura et al. (1992a; 1992b; 1992d) have since improved upon those results, using LEEBI to achieve GaN with $p = 3 \times 10^{18}$ cm^{-3} and a resistivity of 0.2 ohm-cm. They soon discovered that thermal annealing at 700 °C under an N_2 ambient converted the material to p-type equally well (Fig. 3.10). The process was observed to be reversible, with the GaN reverting to insulating compensated material when annealed under NH_3. Hydrogen was, thus, identified as the critical compensating agent. Workers using hydrogen-free growth techniques, notably ECR based MBE, have been able to achieve p-type conductivity in as-grown wurtzite (Molnar) and zinc blende (Lin, 1993b) GaN. Due to the deep binding energy (150–200 meV) of Mg, acceptor activation ratios of only 10^{-2}–10^{-3} are typically achieved requiring very large Mg incorporation to obtain high doping levels. Van Vechten et al. (1992) have proposed a plausible model describing H acceptor compensation in GaN in which Mg-H defect complexes are converted to conventional acceptor impurities by annealing or LEEBI.

Recent theoretical insights have provided a viable explanation for the success of the Mg acceptor in GaN compared to the other group II metals, which continue to compensate GaN, even after LEEBI or annealing (Amano et al., 1988c). Due to the strong binding of the nitrogen anion, III-V nitrides are considerably more ionic than typical III-V semiconductors. Their calculated band structures resemble II-VI semiconductors in many ways, including a large splitting between the

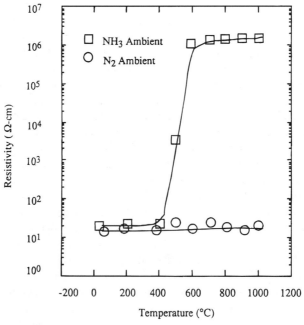

FIGURE 3.10 Mg-doped GaN resistivity as a function of annealing temperature under N_2 and NH_3 ambients. Hydrogen from the NH_3 recompensates the Mg acceptors, making the GaN insulating. The process is reversible; compensated GaN can be made conductive by reannealing under N_2 ambient. After Nakamura et al., 1992a.

GaN and Silicon Carbide as Optoelectronic Materials 67

upper and lower valence bands (LVB) as shown in Fig. 3.11 (Fiorentini, 1993). The Ga 3d core level energies have been predicted and observed (Fiorentini, 1993) to overlap in energy with the N_{2s}-like LVB states because the LVBs are deeper in GaN compared to GaAs, a more typical III-V semiconductor. The resulting energy resonance makes the Ga 3d electrons strongly hybridize with both the upper and lower valence band s and p levels. Such a hybridization is predicted to

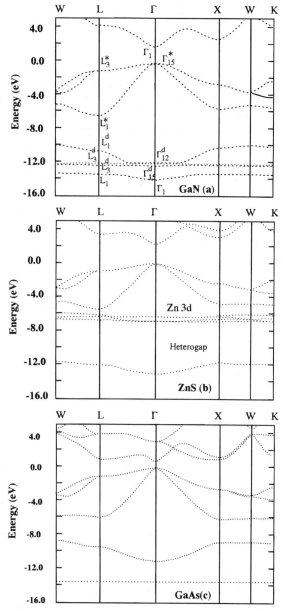

FIGURE 3.11 Calculated GaN (a) band structure compared with ZnS (b) and GaAs (c). The splitting between the upper and lower VBs is much larger in GaN than GaAs as a result of its higher ionicity. The GaN LVB also has considerably more structure because the Ga 3d levels hybridize with the other valence-band orbitals. After Fiorentini et al., 1993.

have a profound influence on the properties of GaN, including such quantities as the bandgap, lattice constant, acceptor levels and valence-band heterojunction offsets (Martin et al., 1994). In the cases of ZnS and ZnSe, it is known that potential acceptors, such as Cu, have the p-d acceptor level raised due to antibonding repulsion between Cu_{3d} and $Se_{4p}(S_{3p})$ resulting in a deep level, whereas impurities without d electron resonances form shallow acceptors (Wei and Zunger, 1988). Mg has no d electrons and turns out to be sufficiently shallow for room temperature p-type doping of GaN. On the other hand, Zn, Cd, and Hg, all of which have d electrons, form deep levels in GaN. These observations are consistent with the theory.

Nakamura et al. (1992c) have reported the properties of Si- and Ge-doped GaN grown by MOVPE. The observed carrier concentrations of Si-doped GaN were in the range 10^{17}–2×10^{19} cm^{-3} whereas Ge doping was observed to produce material having electron concentrations of 7×10^{16}–10^{19} cm^{-3}. A linear variation in the electron concentration as a function of both the SiH_4 and GeH_4 flow rates was observed across the entire experimental range (Fig. 3.12a and b). Ge incorporation was roughly an order of magnitude less efficient than Si as judged by the factor of 10 larger GeH_4 flow rates required to obtain similar electron concentrations. Goldenberg (1993) observed higher n-type GaN conductivities as NH_3 flow was increased during growth. He postulated that passivation of acceptors in his material leads to improved electrical characteristics.

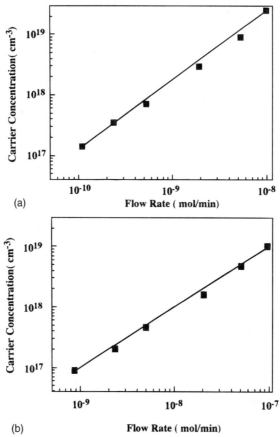

FIGURE 3.12 MOVPE (a) Si and (b) Ge incorporation rates as a function of hydride flow. Both dopants are well behaved; active donor incorporation rates are linear functions of gas flow. However, GeH_4 requires higher flow by a factor of 10 to obtain the same doping level. After Nakamura et al., 1992c.

3.7 Ohmic Contacts to GaN

Ohmic contacts remain a limiting factor in nearly all wide bandgap devices. Early results for GaN indicate that ohmic contacts can easily be formed to n- material. For many years, workers have made Hall measurements on samples contacted with soldered In contacts. Foresi and Moustakas (1993) have published the first investigation of GaN contact resistance. Contact resistances to n-type GaN of 10^{-4} and 10^{-3} ohm-cm^2 were realized using Al and Au metallizations, respectively. Importantly, Schottky barrier heights were found to be dependent on the metal work function, indicating that the surface Fermi energy of GaN was unpinned in contrast to both SiC and ZnSe. Khan et al. (1993) used Ti/Au to contact n-type GaN and measured a contact resistance of 7.8×10^{-5} ohm-cm^2. Nakamura et al. (1991e) have used Au (and later Au/Ni (1993a)) and Al as p- and n-type contacts, respectively, in their LED structures. Although contact resistances were not reported, an operating voltage of 4V at 20 mA forward bias is clear evidence that reasonable contact resistances were obtained.

Recently Lin et al. (1994b) have obtained extremely good ohmic contact in *n*-type GaN layers grown on sapphire substrates. Using the Ti/Al metallization scheme, they were able to obtain specific contact resistivities as low as 8×10^{-6} ohm-cm^2 with annealing at 900 °C for 30 seconds. The GaN sample used for this investigation was grown on sapphire and was, nominally, 1 μm thick. The electron concentration was 1×10^{17} cm^{-3} with a room temperature electron mobility of 100 cm^2/Vs.

Now, we will speculate on the nature of the reactions responsible for these low-resistance contacts. Typically, low-barrier Schottky contacts coupled with intermediate or graded bandgap interface material and/or tunneling are responsible for low-resistance contacts. The first type of interface requires a semiconductor compound with a bandgap of intermediate value, thus, eliminating both of the possible simple binary compounds of the reaction, AlN and TiN. AlN has a bandgap (6.2 eV) which is larger than GaN's whereas TiN's bandgap is too low and behaves metallically. Other, more complicated ternary and quaternary Ga, Ti, and Al nitride compounds are possible low-barrier Schottky contact materials, but further microstructure and phase identification of the interface will be necessary to link them with the observed low-contact resistance found in the work of Lin et al. (1994b).

For the second type mechanism, tunneling, to be applicable, the GaN at the metal/GaN interface must become heavily doped during annealing. One plausible process for this to occur involves the solid phase reaction between Ti and GaN, forming TiN. Suppose N is extracted from GaN without decomposing the GaN structure (i.e., N out-diffusion from the GaN lattice), then, an accumulation of N vacancies would be created in GaN near the junction. Because N vacancies in GaN act as donors, this region would be heavily doped n-GaN, which provides the configuration needed for tunneling contacts. We note that only two monolayers of TiN are needed to be formed to generate a 100Å layer of GaN with an electron density of 10^{20} cm^{-3}. Further investigation is needed to clarify these observations and refine the ohmic contact technology on GaN.

3.8 Properties of Nitride Alloys

Most modern device designs are optimized through heterostructures. Thus, the properties of nitride alloys and heterojunctions will be of increasing importance as more researchers focus on fabricating devices. Although the bandgap and lattice constants, as a function of alloy composition, have been measured for both AlGaN and InGaN, more work is clearly needed to determine the dopability of AlN alloys and the heterojunction band lineups.

AlN and GaN are reasonably well latticed-matched (2.4%) and, for many devices, only small amounts of AlN are needed in the GaN lattice to provide sufficient carrier and optical field confinement. This is a major advantage of the nitrides over II-VI materials which require quaternaries whose compositions are extremely difficult to control. Yoshida et al. (1982) performed a comprehensive study of AlGaN across the entire compositional range. The bandgap and lattice constant dependencies were found to be reasonably linear. An important point for future device considerations arises from their observations that the resistivity of unintentionally doped AlGaN increased rapidly with increasing AlN mole fraction, becoming insulating above 20% (Fig. 3.13). This behavior almost certainly reflects an increase of the native defect ionization energies, as AlN is added. It is not yet known how the most popular dopant impurities, Si and Mg, behave as a function of AlN mole fraction, but they can also be expected to move deeper into the forbidden gap. Devices, such as lasers, which depend critically on the overall device series resistance, will probably be restricted by the ability to dope high mole fraction AlGaN, especially when low-resistivity p-type material is required. We view this as a major outstanding issue which must be addressed before the potential of GaN, with respect to the other wide bandgap semiconductors, can be fully evaluated. Fortunately, the emergence of InGaN, coupled with the fact that good optical field confinement can be obtained with low AlN mole fraction AlGaN, mitigates this problem enormously, and the potential appears great for laser development in this material system.

As touched upon above, InGaN is expected to find use in GaN-based LEDs and laser structures as a strained QW material which emits in the violet or blue wavelengths. The most important properties of InGaN have been studied. Osamura et al. (1972) measured the bandgap across the entire compositional range and found a smooth variation with some bowing (Fig. 3.14). Nagatomo et al. (1989) have reported that the $In_x Ga_{1-x}N$ lattice constant varies linearly with InN mole fraction up to at least $x = 0.42$. The InN lattice constant is 11% larger than GaN's which places strict limits on the InN content and thickness of InGaN layers.

Recent reports have begun to fulfill the promise of InGaN. The major challenge of growing high quality InN-containing alloys is to find a compromise in the growth temperature, because InN is unstable at typical GaN deposition temperatures. Yoshimoto et al. (1991) first obtained InGaN layers of sufficient quality to observe a photoluminescent signal. Those layers were grown by MOVPE at 800 °C at higher than usual NH_3 and TMI flow rates. Significant In desorption

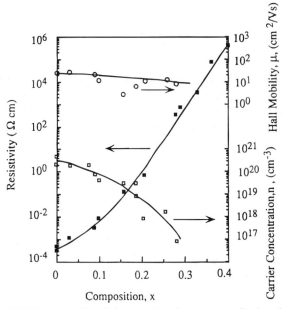

FIGURE 3.13 Electrical properties of unintentionally doped AlGaN as a function of AlN mole fraction. Resistivity increases quickly with AlN content, becoming insulating above $x = 0.2$. After Yoshida et al., 1982, p. 128.

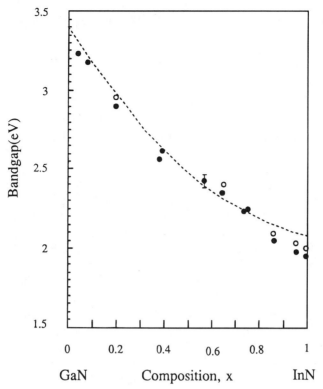

FIGURE 3.14 InGaN bandgap vs. InN mole fraction. After Osamura et al., 1972.

occurred despite the lower than usual growth temperature. InN mole fractions as large as 23% were achieved by these workers. Nakamura et al. (1992d) have since improved on this work, incorporating as much as 30% InN, and expanded the study of InGaN to Si (1993b) and Cd (1993c) doping. Recently, Nakamura et al. have reported InGaN double-heterostructure LEDs which emit both blue and violet light (1993d).

Less attention has been paid to the InAlN alloy system (Kobuta et al., 1989) and As and P alloys of GaN (Igarashi and Okada, 1988). InAlN could, conceivably, be used as a lattice-matched barrier material for GaN. The group V materials have been considered as a way to increase the lattice constant of AlN to lattice-match it to GaN. There has not yet been a report of GaN_{1-x} $(As\ or\ P)_x$ alloys having more than several percent As or P incorporation, so little can be said about their physical properties.

Recently, there has been a growing interest in solid solutions of AlN and SiC because of their excellent thermal and lattice matches (Tables 2 and 3). AlN-SiC alloys can, potentially, span the entire bandgap range of the two materials, i. e., 3.2 eV for 2H SiC to 6.2 eV for wurtzite AlN. Researchers face the challenge of a quaternary alloy in which phase separation occurs easily due to the large degree of polytypism in the component materials. The first report of continuous AlN_xSiC_{1-x} solid solutions appeared (Cutler et al.) in 1978. Hot pressing, liquid phase epitaxy, and, most recently, MOVPE (Jenkins et al.) have been used to form AlN-SiC alloys. Its properties, such as the lattice parameters, linear expansion coefficients, and micro hardness, have been studied across the entire composition range (Rafaniello, 1981). The transition from a direct to an indirect bandgap occurs at AlN = 0.7 (Nurnagomedov et al., 1989). Both p- and n-type conductivity have been observed in unintentionally doped AlN-SiC alloys at low AlN mole fractions (Nurmagomedov et al., 1986; Dimitriev et al., 1991). More work directed towards the electrical properties of AlN-SiC at higher AlN mole fractions is required to determine whether conductive material can be made.

3.9 GaN-Based LEDs

The first GaN LED was reported over twenty years ago (Pankove et al., 1971). Due to the inability, then, to dope GaN p-type, these devices were not conventional p-n junction LEDs, but rather MIS structures. Only recently, when Amano et al. (1990) first obtained p-type GaN by LEEBI, was the first p-n junction GaN LED realized. Soon after, these same workers introduced AlGaN as a barrier material (Akasaki and Amano, 1991).

Nakamura and co-workers (1991e) at Nichia Chemical Industries, Ltd. have seized the vanguard, having reported the most impressive GaN-based LEDs to date. A schematic diagram of a double-heterojunction LED with an InGaN light emitting region straddled by GaN n and p layers is shown in Fig. 3.15, where the light can be collected from the top and/or bottom because the sapphire substrate is transparent to the emission wavelength. At 10 mA forward bias, their LEDs emitted in the blue at 430 nm with a FWHM of 55 nm. Only a 4V bias was necessary to reach a forward current of 20 mA. The durability of GaN was also evident, with this device being operated at currents up to 100 mA. Nakamura et al. (1993a; 1993d) have also developed blue (440 nm) and violet (411–420 nm) double-heterostructure LEDs with InGaN QWs. In the blue device, output power up to 125 µW with a room temperature efficiency of 0.22% was observed at 20 mA forward bias. The electroluminescence had a sharp FWHM of about 30 nm, the best value yet reported for a GaN based LED.

A sample Nichia Chemical InGaN LED, provided by Mr. Nakamura, was analyzed in the author's laboratory by Dr. Boris Sverdlov. In a series of figures below, the performance of this particular LED will be documented and, when appropriate, will be compared to the other commercially available blue diode, SiC LED. Shown in Fig. 3.16 are the spectra of the InGaN double-heterostructure diode for forward current levels in the range of 1–20 mA with strong emission at 450 nm. This particular transition represents emission from the conduction band of InGaN to the Zn acceptor level. At forward currents above 5 mA, a weak, but noticeable, higher energy peak is apparent. Though more experiments are necessary, this particular emission path is tentatively attributed to the band edge emission in the InGaN quantum well.

The output power and external quantum efficiency versus the injection level CW are shown in Fig. 3.17. The LED measured exhibited astounding power levels up to about 2.5 mW at 30 mA with an associated efficiency of 3%. A maximum efficiency of 4.4% was observed at a current level of 2 mA. These power levels are comparable to the venerable InAlGaP and AlGaAs based LEDs (Nakamura, private communication). The Nichia LED also shows excellent current voltage characteristics,

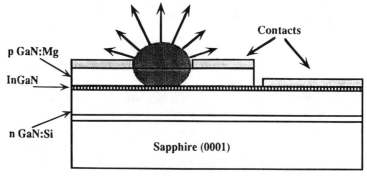

FIGURE 3.15 A schematic cross-sectional view of an InGaN/GaN double-heterojunction LED on sapphire. The insulating nature of the substrate necessitates surface-oriented electrodes, and the light can be collected from the top or the bottom surface as the substrate is transparent to the wavelength of emission.

GaN and Silicon Carbide as Optoelectronic Materials 73

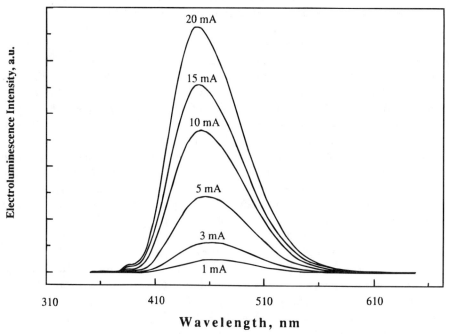

FIGURE 3.16 Light output spectrum for 10- and 20-mA forward current for the Nichia Chemical GaN p-n junction LED. After Nakamura et al., 1991.

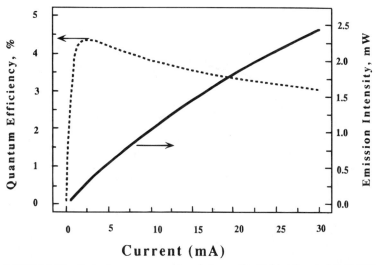

FIGURE 3.17 Output power comparison between the Nichia Chemical GaN LED and a commercial Sanyo' SiC LED. After Nakamura et al., 1991e. The present output powers are about 1mW and 20–30 µW for InGaN/GaN and SiC LEDs, respectively.

indicative of low ohmic contact resistance and excellent junction properties as characterized by the light and forward voltage vs. injection current diagram of Fig. 3.18. The sharp forward turn on is about what would be expected from the bandgap of the material system used. Shown in Fig. 3.19 are the output power levels of a Nichia Chemical InGaN LED (solid squares), a Nichia Chemical GaN LED (open squares), a Cree Research SiC LED (solid circles), and a commercial Sanyo SiC LED (open triangles). The recent LED by Nichia produces an astonishing power level of almost 3 mW at a forward current of 30 mA. This compares with 20 µW and 8 µW reported for SiC LED produced by Cree Research and Sanyo, respectively. In addition, the recent InGaN LED exhibits quantum efficiencies of about 3% at the maximum power level of almost 3 mW as compared to a figure that is about two orders of magnitude smaller for SiC LEDs. The Nichia LED is able to produce a brightness level of over 2 cd which, again, is simply outstanding.

GaN LEDs of the classic MIS type, emitting at 440–480 nm, and of the p-n junction type have been tested for brightness and longevity for their application in displays. The GaN MIS LEDs produced brightness of about 200 candelas for 10 mA forward current, which is comparable to that of the GaP-based LEDs with 20 mA emitting in the green (Akasaki, private communication). Considering the weak response of the human eye to blue as opposed to green, the GaN LEDs are excellent. The associated forward voltage is typically about 7V. The mean time to failure figures obtained so far in MIS type LEDs are excellent, with better performance expected from the p-n junction devices. The light intensity degrades to about 80% of its initial value in 10^4 hours. An additional attribute of these LEDs is that devices surviving the initial premature degradation exhibit the aforementioned long lifetime, making it very convenient to weed out the troublesome devices.

The p-n homojunction LEDs, emitting at wavelengths of 440–480 nm exhibit a 5 V forward voltage. They exhibit external efficiencies of about η_{ext} = 1.5%. With a forward current of 30 mA (5V), p-n junctions produce about 1.5 mW optical power. In AlGaN/GaN/AlGaN p-n DH structures, the output power goes up to 4 mW at a forward current of 30 mA with an external efficiency η_{ext} = 0.5%. Even shorter wavelengths are possible, i.e., 368 nm, which result in even higher density storage provided that coherent radiation sources can be obtained.

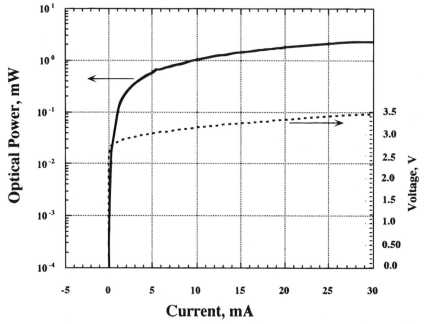

FIGURE 3.18 Light output intensity as a function of forward bias current for a double-heterostructure p-n junction LED with an InGaN QW active region.

GaN and Silicon Carbide as Optoelectronic Materials 75

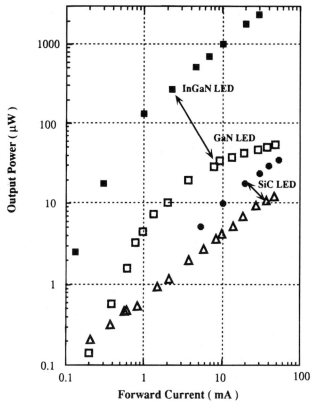

FIGURE 3.19 Output power comparison between the Nichia chemical InGaN LED (solid squares), Nichia GaN LED (open squares), a commercial Sanyo SiC (open triangles) and a Cree SiC LED (solid circles) in part after Nakamura et al., 1991e.

3.10 Toward a GaN Laser

The Holy Grail of GaN research remains the realization of a GaN laser, which will be the shortest wavelength semiconductor laser ever demonstrated. Optically pumped stimulated emission from GaN was first observed over twenty years ago (Dingle et al., 1971). Since then, researchers have been working toward both vertical cavity surface emitting lasers (SEL) and conventional separate confinement heterostructure, edge-emitting lasers in GaN.

Analogous to the conventional III-V compound semiconductors, such as AlGaAs/GaAs, the nitride system lends itself well to the formation of heterostructures favorable to the confinement of carriers and light for laser operation. Device designers have available to them a plethora of possible structures employing AlGaN, GaN, and InGaN. Because both the refractive indices and bandgaps are steep functions of composition, ternaries with small fractions are sufficient to provide one with separate confinement and graded-index, separate confinement, laser heterostructures with minimum lattice mismatches.

Khan and coworkers at APA Optics have reported several experiments directed toward an eventual SEL structure. They were the first to report the optical properties of nitride QWs (Khan et al., 1990). In that study, $Al_{0.14}Ga_{0.86}N$ was used as the barrier material for GaN QWs of thicknesses between 100 Å and 300 Å (Fig. 3.20). The photoluminescent energy shifts as a function of GaN QW thickness were in agreement with the predictions of a simple model fit to roughly equal valence- and conduction-band offsets (Kolbas, private communication). This was followed by a demonstration of vertical cavity lasing in which the stimulated emission was perpendicular

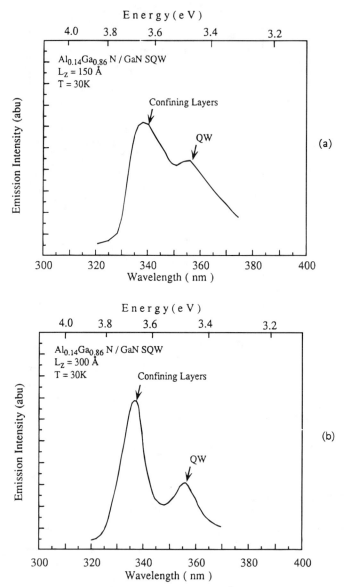

FIGURE 3.20 Photoluminescence from a 300 Å GaN QW. After Khan et al., 1990.

to the growth plane (Khan et al., 1991a). A third paper (Khan et al., 1991b) describes an 18-period AlGaN/GaN superlattice designed to operate as a quarter-wave mirror. Peak reflectivities of 80% at 442 nm and 95% at 375 nm were achieved. These studies have laid the groundwork for a GaN SEL, but critical issues, such as overall series resistance of the AlGaN sections and p-type material, still need to be addressed. If designed and executed properly, an SEL may actually be favorable for GaN because no cleaving is necessary.

Carrier confinement of both electrons and holes in the same spatial region is crucial for laser operation, and this requires Type I heterojunctions between GaN and its alloys with AlN and its alloys with InN. To date, no direct measurements of band alignment have been reported in the literature. The author and colleagues have recently performed photoemission spectroscopic measurements demonstrating Type I alignment between wurtzite GaN and AlN, and measured the valence-band offset $\Delta E_v = 0.9 \pm 0.5$ eV. This is roughly a 70:30 conduction:valence split of

GaN and Silicon Carbide as Optoelectronic Materials

the bandgap difference, reminiscent of GaAs/AlAs and agreeing with the local density function calculation of Lambrecht and Segall at Case Western University. GaN/AlN are found appropriate for confining both electrons and holes, with the large band offsets providing excellent barriers. Judging from the excellent LED results, the same appears to hold for the GaN/InGaN system.

One critical parameter for laser design is good knowledge of the index of refraction as a function of alloy composition at the lasing wavelength. Relying on observations of the AlGaAs alloy, to a first approximation, one can simply apply a rigid shift to the refractive index vs. energy obtained for GaN to move the band edge to the band edge of the alloy (Lin et al., 1992c). The results so obtained are in excellent agreement with those reported for $Al_{0.1}Ga_{0.9}N$ by Amano et al. (1993) and those available for AlN. The same method can be employed for the InGaN ternary, albeit with less accuracy, allowing the preliminary determination of the optimum waveguide structure for potential lasers which are actively pursued at various laboratories. Combining the data on InGaN with those for AlGaN and GaN allows one to calculate the confinement factor for a range of possible laser SCH waveguides utilizing InGaN and GaN for the quantum well, GaN, InGaN, and AlGaN for the waveguide layers, and AlGaN and GaN for the cladding layers.

The energy-dependent refractive indices of GaN and its In- and Al-containing ternaries are shown in Fig. 3.21. The GaN data have been measured in the author's laboratory. The AlGaN data have been arrived at by a method discussed above. The refractive indices so obtained for $Al_{0.1}Ga_{0.9}N$ and the AlN end point and the measured values agree, lending credibility to the method employed. The same method, although somewhat questionable compared to that for AlGaN, has also been employed with In-containing ternaries, as shown in Fig. 3.21. Using these refractive indices, confinement factors for DH and SCH laser structures, based on GaN and $Al_{0.29}Ga_{0.71}N$ structures, are shown in Fig. 3.22. To show the effectiveness of waveguides employing

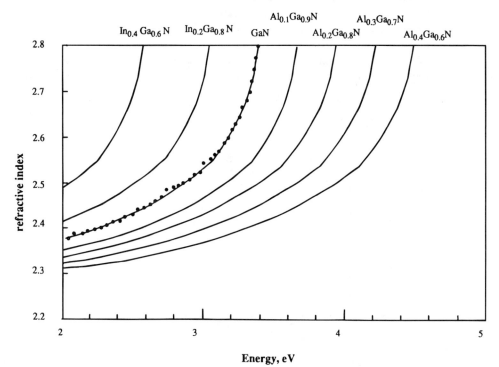

FIGURE 3.21 Refractive indices of important III-V nitrides as a function of energy, measured for GaN and extrapolated for AlGaN and InGaN.

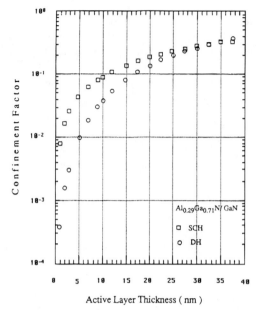

FIGURE 3.22 Confinement factor as a function of the active layer thickness for GaN lasers in the double-heterojunction and separate confinement, heterojunction laser configurations.

$Al_{0.1}Ga_{0.9}N$ cladding layers, a GaN light pipe, and an InGaN active lasing medium, the confinement factor for a separate structure can be evaluated. The result of such an exercise, in the form of the modal loss as a function of the waveguide thickness, designed for 480 nm at room temperature, is shown in Fig. 3.23. The use of the InGaN as the carrier-confining layer in an SCH or a GRIN-SCH does not affect the waveguide properties, compared to GaN active layers, because the active layer is very thin. The ternary InGaN can also be used as the waveguide. The results for confinement factors for operation at 370 nm (GaN active layer) and 480 nm ($In_{0.4}Ga_{0.6}N$ active layer) are tabulated in Table 3.4. We also include data for a SCH blue-green laser based on the combination of wide-bandgap II-VI compounds which have recently gained considerable attention. As Table 3.4 indicates, the confinement factor values for (In)(Al)GaN heterostructures are extremely encour-

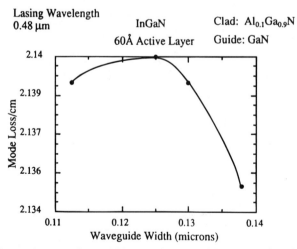

FIGURE 3.23 The model less as a function of the waveguide width in a separate confinement laser structure designed to emit at 480 nm.

TABLE 3.4 Confinement Factor Γ for Various Separate Confinement Waveguide Structures Based on GaN and ZnSe (the last item) Material Systems

Active Layer	Cladding	Waveguide	d_w (μ)	Γ (%)
GaN	$Al_{0.1}Ga_{0.9}N$	$Al_{0.03}Ga_{0.97}N$	0.07	3.763
60 Å	2 μm	$n = 2.7$	0.06	3.763
$n = 2.8$	$n = 2.58$		0.05	3.761
$In_{0.4}Ga_{0.6}N$	GaN	$In_{0.2}Ga_{0.8}N$	0.10	2.685
60Å	2 μm	$n = 2.5$	0.09	2.691
$n = 2.85$	$n = 2.42$		0.08	2.680
$In_{0.4}Ga_{0.6}N$	GaN	$In_{0.3}Ga_{0.7}N$	0.07	3.662
60Å	2 μm	$n = 2.6$	0.06	3.712
$n = 2.85$	$n = 2.42$		0.05	3.661
$In_{0.4}Ga_{0.6}N$	$Al_{0.1}Ga_{0.9}N$	$In_{0.2}Ga_{0.8}N$	0.13	2.139
60Å	2 μm	$n = 2.5$	0.12	2.139
$n = 2.85$	$n = 2.38$		0.11	2.139
ZnCdSe	ZnMgSSe	ZnSSe	0.12	1.671
60Å	2 μm	$n = 2.7$	0.11	1.679
$n = 2.78$	$n = 2.625$		0.10	1.676

Note: The wavelengths for which the calculations have been carried out are 370 nm for the GaN active layer, 480 nm for the $In_{0.4}Ga_{0.6}N$ active layer, and 520 nm for the CdZnSe active layer.

aging, and, pending improvements in the materials quality, one can optimistically expect lasers to be fabricated in this material system. The confinement factors obtainable with the nitrides are comparable to the venerable AlGaAs/GaAs case.

The injection level required to reach the lasing condition or to render the active layer nonabsorbing at the wavelength of emission in GaAs is about 1.6×10^{18} cm^{-3}. At this injection level, the separation of the quasi-Fermi levels equals the bandgap of the active layer (Morkoç et al., 1993). The lower the injection level, the lower the injection current required to reach the lasing condition or transparency. Then, it follows that, in semiconductors with large density of states, the transparency condition is reached at higher injection levels. Naturally, the band with the larger density of states or effective mass (the valence band in all the semiconductors) dominates. Using the hexagonal band parameters, the injection level for transparency in GaN-based lasers with GaN active layers is about 6×10^{18} cm^{-3}. Simply put, all else being equal, the transparency current for GaN lasers is about 3–4 times larger than in GaAs-based lasers. Considering the inferior materials quality in GaN compared to GaAs, even much larger currents would be required. Borrowing from the GaAs-based lasers, where the concept of coherent strain has been successfully used, one can use a InGaN active layer in the cubic phase to reduce the injection level required. This active layer not only has lower density of states due to the smaller carrier effective mass, by virtue of the coherent compressive strain, it also reduces the density of states of the heavy-hole band. Consequently, injection levels about 30–50% of those required for GaN active layers should be possible with strained InGaN active layers. Whether this concept can directly be applied to the wurtzite phase also requires the knowledge of the change of the band structures under coherent stress.

On the experimental side, Utilizing $Al_{0.1}Ga_{0.9}N$/GaN separate confinement laser structures, Amano et al. (1993) were able to reduce the optical pumping required for lasing at room temperature to about one-twelfth that required for GaN homojunction structures. It was noted that the refractive index difference between $Al_{0.1}Ga_{0.9}N$-clad and GaN waveguide layers is about 0.19 at 0.37 μm, which is sufficiently large for efficient wave guiding with good confinement. The authors measured the refractive indices of the aforementioned semiconductors by ellipsometry, as a function of wavelength from 600 nm down to that corresponding to the respective band edges of the aforementioned materials.

3.11 Detectors

Interest in GaN based detectors stems, in part, from the fact that Earth's atmosphere is opaque at a wavelength of about 200 nm. Consequently, space to space communication can take place without the blinding radiation from Earth. An additional application of UV detectors, in general, and Al(Ga)N detectors, in particular, is in furnaces and nuclear power plants. There are also other applications, such as combustion sensing and control for aircraft engines. A particular nitride-based detector, now attracting considerable activity, is the AlN photoconductive detector.

Khan et al. (1992) have reported on their GaN detectors with responsivity up to a wavelength of 365 nm, with an estimated gain of about 6×10^3. The absorbing GaN layer was about 0.8-μm thick and was grown on a sapphire substrate following a 0.1-μm thick AlN buffer layer. Measurements were conducted in the wavelength range of 200–365 nm with a responsivity of 2000 A/W at 365 nm. The sharp absorption edge was characterized with a drop of three orders of magnitude in responsivity over just 10 nm, from 365 nm to 375 nm. The responsivity was observed to be a linear function of the incident optical power over five orders of magnitude attesting to the quality of the GaN absorber.

Reitz and Khan (1992, unpublished) described the fabrication and investigation of AlGaN photoresistors sensitive to radiation with $\lambda < 360$ nm. Layers had a sharp band edge cut-off and a large responsivity in the UV which is nearly independent of wavelength. Devices operating at several Hertz performed quite well. The typical values of the dark resistance were about 0.5 MΩ for the devices with 1 mm^2 area. The dark current increases from 10^{-11}A to 10^{-6}A as the temperature increased from 20 °C to 220 °C. The photoresponse was linear in a photocurrent range of 10^{-7}A to 10^{-3}A. Photoconductors have also been investigated by Zook and Goldenberg (1992, unpublished). In this particular investigation, it was found that the photoconductive response of GaN decreased with increasing frequency. Also, increased intensity lowered the photoconductive gain.

3.12 Discussion

Wide-bandgap semiconductors discussed in this review have applications in blue displays and recording/reading. In particular, SiC boasts a very large thermal conductivity of about 5 W/cmK, and both SiC and GaN have large cohesive energies. To some extent, the high thermal conductivity of SiC can also be utilized for GaN by growing the epitaxial layers of GaN on SiC substrates. One must keep in mind that even the relatively thin GaN buffer layers can reduce the thermal conductivity of the overall structure.

Important advantages of GaN over SiC are its direct bandgap and heterostructures with allied AlN and InN binaries and their ternaries. Preliminary results also indicate that ohmic contacts on n-type GaN are much better than on SiC. Fortuitously large bandgap disparities between AlN, GaN, and InN render this particular semiconductor system excellent for obtaining heterointerfaces with good carrier confinement. Though heterojunctions can also be obtained in SiC, it requires exploitation of the many polytypes of SiC which are not necessarily easy to obtain and control.

On the optical emitter side, the issues are distinctly different. SiC's nemesis is its reliance on its indirect bandgap or localized impurities for light emission. Although excellent progress has been made in SiC based LEDs, with commercially available brightness levels of 10–20 millicandelas at about 440, GaN LEDs are quite competitive. GaN has a direct bandgap, and, as a result, the LEDs are very bright. Brightness levels about 1 cd and 200 mcd in InGaN/GaN DH p/n junction and MIS-type LEDs, respectively, have been obtained. In p/n junction GaN LEDs, efficiencies and power levels approaching 2% and 5 mW have been achieved. In addition, the lifetime of GaN LEDs is about several times 10^4 hours and certainly competitive with SiC. Failures occur early and devices surviving the first few hours tend to exhibit excellent longevity. The prevailing explanation for the excellent lifetime figures in GaN LEDs rests on the resilience of GaN against defect generation during growth and operation. The high threshold for damage generation in GaN, and also in SiC, is responsible for excellent lifetimes.

Although LEDs are fine for display purposes, high density compact disks require blue and even shorter wavelength coherent sources provided by lasers. The availability of recorder media would facilitate not just reading but also recording on compact disks by consumers. The critical technology for this capability is inexpensive high-power, short-wavelength lasers. Although blue lasers can result in a four-fold increase in disk density, UV lasers will permit additional increases. Such UV lasers are possible with GaN.

Judging from the extreme brightness of GaN and GaN/InGaN-based LEDs and favorably high threshold for defect generation, GaN-based heterostructures with their direct bandgap and favorable wave guiding properties may prove to be good contenders for the blue laser market. Confinement factors comparable to, if not better than, those available in the ZnMgSSe-based semiconductor system can be obtained. In terms of coherent optical emitters, the next few years will produce many exciting developments in the GaN arena.

Acknowledgments

The author benefited from Drs. S. Strite, M. E. Lin, B. Sverdlov, and Messrs. G. B. Gao and M. Burns in connection with previous manuscript projects. Funding from the Air Force Office of Scientific Research and Office of Naval Research under grant numbers F49620-92-J-0221 and N00014-89-J-1780, respectively, is greatly appreciated. I benefited immensely from participating in the High Temperature Electronics panel sanctioned by AFOSR under the leadership of Dr. Jerry Witt. I would like to express my appreciation to Ms. Susan Burns for assisting in the preparation of this manuscript and to Greg Martin for careful reading of the manuscript. Last, but not least, I would also like to acknowledge the enthusiasm and support of Max Yoder of ONR and G. L. Witt of AFOSR, in every imaginable way.

References

Akasaki, I., Amano, H., Koide, Y., Hiramatsu, K., and Sawaki, N., *J. Cryst. Growth*, 98, 209, 1989.
Akasaki, I., Amano, H., Kito, M., and Hiramatsu, K., *J. Lumin.*, 48/49, 666, 1991a.
Akasaki, I., and Amano, H., Mater. Res. Soc. Fall Meeting, Boston, MA, November 1991b.
Akasaki, I., private Communication.
Amano, H., Sawaki, N., Akasaki, I., and Toyoda, Y., *Appl. Phys. Lett.*, 48, 353, 1986.
Amano, H., Hiramatsu, K., and Akasaki, I., *Jpn. J. Appl. Phys.*, 27, L1384, 1988a.
Amano, H., Akasaki, I., Hiramatsu, K., Koide, N., and Sawaki, N., *Thin Solid Films*, 163, 415, 1988b.
Amano, H., Akasaki, I., Kozawa, T., Hiramatsu, K., Sawaki, N., and Ikeda, K., *J. Lumin.*, 40/41, 121, 1988c.
Amano, H., Kito, M., Hiramatsu, K., and Akasaki, I., *Jpn. J. Appl. Phys.*, 28, L2112, 1989; ibid., *Inst. Phys. Conf. Ser.*, 106, 725, 1990.
Amano, H., Watanabe, N., Koide, N., and Akasaki, I., *Jpn. J. Appl. Phys.*, 32, L1000, 1993.
Anikin, M. M., Rastegaeva, M. G., Syrkin, A. L., and Chuiko, I. in., *Proc. ICACSC* 90, 294, 1990.
Ban, V. S., *J. Electrochem. Soc.*, 119, 761, 1972.
Brander, R. W., and Sutton, R. P., *J. Phys. D2*, 309, 1969.
Brown, D. M., Downey, E. T., Ghezzo, M., Kretchner, J. W., Saia, R. J., Liu, Y. S., Edmond, J. A., Gati, G., Pimbley, J. M., and Schneider, W. E., *IEEE Trans. Electron. Dev.*, ED-40, 325, 1993.
Campbell, R. B., and Chang, H. C., *Solid State Electron.*, 63, 949, 1967.
Cutler, J. B., Viller, P. D., Rafaniello, W., Park, H.K., Thompson, D. P., and Jack, K. H., *Nature*, 275, 434, 1978.

Cooper, J. A., and Melloch, M., Purdue University, private communication.
Dingle, D., Shaklee, K. L., Leheny, R. F., and Zetterstrom, R. B., *Appl. Phys. Lett.,* 19, 5, 1971.
Dmitriev, V. A., Elfimov, L. B., Lin'kov, I. Yu., Morozenko, Ya. Y., Nikitina, I. P., Cheinokov, V. E., Cherenkov, A. E., and Chernov, M. A., *Sov. Tech. Phys. Lett.,* 17, 214, 1991.
Dmitriev, V. A., Ivanov, P. A., Korkin, I. in., Morozenko, Y. in., Popov, I. in., Sidorova, T. A., Strslchuk, A. M., Chelnokov, in. E., *Sov. Phys. Tech. Lett.,* 11, 98, 1985.
Edmond, J. A., Kong, H. S., and Carter, C. H., Jr., *Physica,* B185, 453, 1993.
Fiorentini, Y., Methfessel, M., and Scheffler, M., *Phys. Rev.,* B47, 13353, 1993.
Foresi, J. S., and Moustakas, T. D., *Appl. Phys. Lett.,* 62, 2859, 1993.
Furumura, Y., Doki, M., Meino, F., Eshita, T., Suzuki, T., and Maeda, M., *J. Electrochem. Soc.,* 135, 1255, 1988.
Fuyuki, T., Nakayama, M., Yoshinobu, T., and Matsunami, H., *J. Crystal Growth,* 95, 461, 1989.
Gardner, C. T., Cooper, J. A., Jr., Melloch, M. R., Palmour, J. W., and Carter, C. H., Jr., in 4th Int. Conf. on Amorphous and Crystalline Silicon Carbide and other IV-IV Materials, Santa Clara, CA, Oct. 10–11, 1991.
Ghezzo, M., Brown, D. M., Downey, E., Kretchmer, J., Hennessy, W., Polla, D. L., and Bakhru, H., *IEEE Electron. Dev. Lett.,* EDL-13, 639, 1992.
Goldenberg, B., APS March Meeting, Seattle, 1993.
Golecki, I., Reidinger, F., and Marti, J., *Appl. Phys. Lett.,* 60, 1703, 1992.
Grimmeiss, H. G., and Monemar, B., *J. Appl. Phys.,* 41, 4054, 1970.
Hashimoto, M., Amano, H., Sawaki, N., and Akasaki, I., *J. Cryst. Crowth,* 68, 163, 1984.
Hoffman, L., Ziegler, G., Theis, D., and Weyrich, C., *J. Appl. Phys.,* 53, 6962, 1982; Hoffman, L., Ziegler, G., Theis, D., and Weyrich, C., *IEEE Trans. Electron. Dev.,* ED-30, 277, 1983.
Hong, J. W., Shin, N. F., Jen, T. S., Ning, S. L., and Chang, C. Y., *IEEE Electron. Dev. Lett.,* 13, 375, 1992.
Humphreys, T. P., Sukow, C. A., Nemanich, R. J., Posthill, J. B., Rudder, R. A., Hattangaddy, S. V., and Markunas, R. J., *Mater. Res. Soc. Symp. Proc.,* 162, 531, 1990.
Igarashi, O., and Okada, Y., *Jpn. J. Appl. Phys.,* 27, 790, 1988.
Ikeda, M., Hayakawa, T., Yamagiva, Z., Matsunami, H., and Tanaka, T., *J. Appl. Phys.,* 50, 8215, 1979.
Jagodzinski, H., and Arnold, H., in *Silicon Carbide, A High Temperature Semiconductor,* edited O'Connor, J. R., and Smiltens, J., Eds., Pergamon, New York, 1960, pp. 136–145.
Jenkins, D. W., and Dow, J. D., *Phys. Rev.,* B39, 3317, 1989.
Jenkins, I., Irvine, K. C., Spencer, M. C., Dmitriev, V., and Chen, N., Semiannual Progress Report for ONR contract #N00014-92-J- 1136.
Kelner, G., Binari, Z., Sleger, K., Kong, H., *IEEE Electron. Dev. Lett.,* EDL-8, 428, 1987; Kelner, G., Shur, M. S., Binari, S., Sleger, K. J., and Kong, H., *IEEE Trans. Electron. Dev.,* ED-36, 1045, 1989.
Khan, M. A., Skogman, R. A., Yan Hove, J. M., Krishnankutty, S., and Kolbas, R. M., *Appl. Phys. Lett.,* 56, 1257 (1990)
Khan, M.A., Olson, D. T., Van Hove, J. M., and Kuznia, J. N., *Appl. Phys. Lett.,* 58, 1515, 1991a.
Khan, M.A., Kuznia, J. N., Van Hove, J. M., and Olson, D. T., *Appl. Phys. Lett.,* 59, 1449, 1991b.
Khan, M. A., Kuznia, J. N., Van Hove, J. M., Olsen, D. T., Krishnankutty, S., and Kolbas, R. M., *Appl. Phys. Lett.,* 58, 526, 1991c.
Khan, M. A., Kuznia, J. N., Olson, D. T., Van Howe, J. M., Blasingame, M., and Reitz, L., *Appl. Phys. Lett.* 60, 2917, 1992.
Khan, M. A., Kuznia, J. N., Bhattarai, A. R., and Olson, D. T., *Appl. Phys. Lett.,* 62, 1786, 1993.
Koide, Y., Itoh, N., Itoh, X., Sawaki, N., and Akasaki, I., *Jpn. J. Appl. Phys.,* 27, 1156, 1988.
Kolbas, R., private communication.
Kong, H. S., Glass, J. T., and Davis, R. F., *J. Mater. Res.,* 4, 204, 1989 Kong, H. S., Jiang, B. L., Glass, J. T., Rozgonyi, G. A., and Moore, K. L., *J. Appl. Phys.,* 63, 2645, 1988.
Kubota, K., Kobayashi, Y., and Fujimoto, K., *J. Appl. Phys.,* 66, 2984, 1989.

Kuznia, N., Khan, M. A., and Olson, D. T., *J. Appl. Phys.*, 73, 4700, 1993.
Levin, V. I., Tairov, Y. M., Travazhdyan, M. G., Tsvetkov, in. F., and Chernov, M. A., *Sov. Phys. Izvestia*, 14, 830; 1978.
Lei, T., Fanciulli, M., Molnar, R. J., Moustakas, T. D., Graham, R. J., and Scanlon, J., *Appl. Phys. Lett.*, 59, 944, 1991.
Lei, T., Moustakas, T. D., Craham, R. J., He, Y., and Berkowitz, S. J., *J. Appl. Phys.*, 71, 4933, 1992.
Lin, M. E., Sverdlov, B., Zhou, G. L., and Morkoç, H., *Appl. Phys. Lett.*, 62, 3479, 1993a.
Lin, M. E., Xue, C., Zhou, G. L., Greene, J. E., and Morkoç, H., *Appl. Phys. Lett.*, 63, 932, 1993b.
Lin, M. E., Sverdlov, B. N., Strite, S., Morkoç, H., and Drakin, A. E., *Electron. Lett.* 29 (11), 1019–1021, 1993c.
Lin, M. E., Sverdlov, B. N., and Morkoc, H., *Appl. Phys. Lett.*, 63d, 3625, 1993.
Lin, M. E., Ma, Fan, Z., Allen, L., and Morkoç, H., *Appl. Phys. Lett.*, in 64, 887, 1994.
Marsh, O. J., and Dunlap, H. L., *Radiation Effect.*, 6, 301, 1970.
Martin, G., Strite, S., Betchkaren, A., Agarwal, A., Rockett, A., W. R. L. Lambrecht, B. Segell and H. Morkoç. *Appl. Phys. Lett.*, 65, 610, 1994.
Maruska, H. P., and Tietjen, J. J., *Appl. Phys. Lett.*, 15, 327, 1969.
Matsunami, H., Kyoto University, private communication.
McMullin, P. G., Spitznagel, J. A., Szedon, J. R., and Costello, J. A., *Proc. ICACSC* 90, 191, 1990.
Mizuta, M., Fujieda, S., Matsumoto, Y., and Kawamura, T., *Jpn. J. Appl. Phys.*, 25, L945, 1986.
Molnar, R. J., Lei, T., and Moustakas, T. D., *Appl. Phys. Lett.*, 62, 72, 1993.
Morkoç, H., Sverdlov, B., and Gao, G. B., *Proc. IEEE*, 81, 492, 1993.
Muench, W., and Kurzinger, W., *Solid State Electron.*, 21, 1129, 1978.
Nagatomo, T., Kuboyama, T., Minamino, H., and Omoto, O., *Jpn. J. Appl. Phys.*, 28, L1334, 1989.
Nakamura, S., Harada, Y., and Seno, M., *Appl. Phys. Lett.*, 58, 2021, 1991a.
Nakamura, S., *Jpn. J. Appl. Phys.*, 30, L1705, 1991b.
Nakamura, S., Harada, Y., and Seno, M., *Appl. Phys. Lett.*, 58, 2021, 1991c.
Nakamura, S., Seno, M., and Mukai, T., *Jpn. J. Appl. Phys.*, 30, L1708, 1991d.
Nakamura, S., Mukai, T., and Seno, M., *Jpn. J. Appl. Phys.*, 30, L1998, 1991e.
Nakamura, S., Iwasa, N., Seno, M., and Mukai, T., *Jpn. J. Appl. Phys.*, 31, 1258, 1992a.
Nakamura, S., Mukai, T., Seno, M., and Iwasa, N., *Jpn. J. Appl. Phys.*, 31, L139, 1992b.
Nakamura, S., Mukai, T., and Seno, M., *Jpn. J. Appl. Phys.*, 31, 195, 1992c.
Nakamura, S., and Mukai, T., *Jpn. J. Appl. Phys.*, 31, L1457, 1992d.
Nakamura, S., Seno, M., and Mukai, T., *Appl. Phys. Lett.*, 62, 2390, 1993a.
Nakamura, S., Mukai, T., and Seno, M., *Jpn. J. Appl. Phys.*, 32, L16, 1993b.
Nakamura, S., Iwasa, N., and Nagahama, S., *Jpn. J. Appl. Phys.*, 32, L338, 1993c.
Nakamura, S., Seno, M., and Mukai, T., *Jpn. J. Appl. Phys.*, 32, L8, 1993d.
Nakamura, S., private Communication.
Naniwae, K., Itoh, S., Amano, H., Itoh, K., Hiramatsu, K., and Akasaki, I., *J. Cryst. Growth*, 99, 381, 1990.
Naumov, A. in., Nikitin, Z. in., Ostroumov, A. T., and Inodakov, Y. A., *Sov. Phys. Semicond.*, 21, 377, 1987.
Nishino, S., Hazuki, Y., Matsumani, H., and Tanaka, T., *J. Electrochem. Soc.*, 127, 2674, 1980.
Nishino, S., Powell, J. A., and Will, H. A., *Appl. Phys. Lett.*, 42, 460, 1983.
Nurmagomedov, S. A., Pikhtin, A. N., Rasbegaev, V. N., Safaraliev, C. K., Tairov, Yu, M., and Tzvetkov, V. F., *Sov. Tech. Phys. Lett.*, 12, 431, 1986.
Nurmagomedov, S. A., Pikhtin, A. N., Rasbegaev, V. N., Safaraliev, C. K., Tairov, Yu, M., and Tzvetkov, V. F., *Sov. Phys. Semicond.*, 23, 100, 1989.
Nutt, S. R., Smith, D. J., Kim, H. J., and Davis, R. F., *Appl. Phys. Lett.*, 50, 203, 1987.
Osamura, K., Nakajima, K., Murakami, Y., Shingu, P. H., and Otsuki, A., *Solid State Comm.*, 11, 617, 1972.
Paisley, M. J., Sitar, Z., Carter, Jr., C. H., and Davis, R. F., *Proc. SPIE*, 8, 877, 1988.

Paisley, M. J., Sitar, Z., Posthill, J. B., and Davis, R. F., *J. Vac. Sci. Technol.,* A7, 701, 1989c.
Palmour, J. W., Kong, H. S., and Davis, R. F., *Appl. Phys. Lett.,* 51, 2028, 1987; Palmour, J.W., Kong, H. S., and Davis, R. F., *J. Appl. Phys.,* 64, 2168, 1988a.
Palmour, J. W., Edmond, J. A., Kong, H. S., and Carter, C. H., Jr. in 4th Int. Conf. Amorphous and Crystalline Silicon Carbide and other IV-IV Materials, Santa Clara, CA, Oct. 10–11, 1991b.
Palmour, J. W., Edmond, J. A., Kong, H. S., and Carter, C. H., Jr., *Physica,* B185, 461, 1993.
Pankove, J. I., Miller, E. A., and Berkeyheiser, J. E., *RCA Rev.,* 32, 383, 1971.
Pensl, G., and Choyke, W. J., *Physica,* B185, 264, 1993.
Petrov, I., Mojab, E., Powell, R. C., Creene, J. E., Hultman, L., and Sundgren, J. –E., *Appl. Phys. Lett.,* 60, 2491, 1992.
Pirouz, P., Chorey, C. M., and Powell, J. A., *Appl. Phys. Lett.,* 50, 221, 1987; Chorey, C. M., Pirouz, P., Powell, J. A., and Mitchell, T. E., in *Semiconductor-Based Heterostructures: Interfacial Structure and Stability,* Green M. L., et al., Eds., Philadelphia, PA, The Metallurgical Soc. 1987, p. 115.
Powell, J. A., Larkin, D. J., Matus, L. G., Choyke, W. J., Bradshaw, J. L., Henderson, L., Yoganathan, M., Yang, J., and Pirouz, P., *Appl. Phys. Lett.,* 56, 1353, 1990; Powell, J.A., Larkin, D. J., Matus, L. G., Choyke, W. J., Bradshaw, J. L., Henderson, L., Yoganathan, M., Yang, J., and Pirouz, P., *Appl. Phys. Lett.,* 56, 1442, 1990.
Powell, R. C., Lee, N. E., Kim, Y.-W., and Greene, J. E., *J. Appl. Phys.,* 73, 189, 1993.
Rafaniello, W., Cho, K., and Yirkar, A.Y., *J. Mater. Sci.,* 16, 3479, 1981.
Reitz, L., and Khan, M. A., presented at the 1st Wide Bandgap Nitride Workshop, April 13–14, 1992, St. Louis, MO, unpublished.
Sasaki, T., and Matsuoka, T., *J. Appl. Phys.,* 64, 4531, 1988.
Shibahara, K., Nishino, S., and Matsunami, H., *J. Cryst. Growth,* 78, 538, 1986.
Strite, S., Ruan, J., Li, Z., Manning, N., Salvador, A., Chen, H., Smith, D. J., Choyke, W. J., and Morkoç, H., *J. Vac. Sci. Technol.,* B9, 1924, 1991.
Strite, S. T., and Morkoç, H., *J. Vac. Sci. Technol.,* B 10, 1237, 1992a.
Strite, S., Mui, D. S. L., Martin, G., Li, Z., Smith, D. J., and Morkoç, H., *Inst. Phys. Conf. Ser.,* 120, 89, 1992b.
Strite, S. T., Chandrasekhar, D., Smith, D. J., Sariel, J., Chen, H., Teraguchi, N., and Morkoç, H., *J. Cryst. Growth,* 127, 204, 1993.
Sugii, T., Aoyama, T., and Ito, T., *J. Electrochem. Soc.,* 137, 989, 1990.
Suzuki, A., Furukawa, K., Fujii, Y., Shigeta, M., and Nakajima, S., *Springer Proc. Phys.,* 56, 101, 1992.
Tairov, Y. M., and Tsverkov, in. R., *J. Cryst. Growth,* 52, 146, 1981.
Tansley, T. L., and Egan, R. J., *Phys. Rev.,* B45, 10942, 1993.
Van Vechten, J. A., Zook, J. D., Horning, R. D., and Goldenberg, B., *Jpn. J. Appl. Phys.,* 31, 3662, 1992.
Wahab, Q., Hultman, L., Sundgren, J.-E., and Willander, M., *Mat. Sci. Eng.,* B11, 61, 1992.
Wei, S.-H., and Zunger, A., *Phys. Rev.,* B37, 8958, 1988.
Yoshida, S., Misawa, S., and Gonda, S., *J. Appl. Phys.,* 53, 6844, 1982.
Yoshida, S., Misawa, S., and Gonda, S., *J. Vac. Sci. Technol.,* B1, 250, 1983, ibid, *Appl. Phys. Lett.,* 42, 427, 1983.
Yoshimoto, N., Matsuoka, T., Sasaki, T., and Katsui, A., *Appl. Phys. Lett.,* 59, 2251, 1991.
Zeller, M. Z., in Trans. 1st Int. High Temperature Electron. Conf., June 16–20, Albuquergue, NM, 1991.
Zhou, G.L., Ma, A., Lin, M.E., Shen, T.C., Allen, L.H., and Morkoç, H., *J. Crystal Growth,* 134, 167, 1993.
Zook, D., and Goldenberg, B., presented at the 1 Wide Bandgap Nitride Workshop, April 13–14, 1992, St. Louis, MO, unpublished.

SECTION II
Oxide Materials

4
Ferroelectric Materials

Chia-Yen Li
*University of California,
Los Angeles*

Yuhuan Xu
*University of California,
Los Angeles*

4.1 Introduction 87
4.2 Structure, Synthesis, and General Properties of
     Ferroelectric Materials 92
     Material Structure and Preparation Method of Ferroelectric Crystals
     • Ferroelectric Thin Films
4.3 Optical Properties of Ferroelectric Materials 101
     Electrooptic Effect • Nonlinear Optical Properties • Photorefractive
     Properties • Pyroelectric Effect • Ferroelectric Structure-Related Optical Properties • PLZT
4.4 Optical Applications of Ferroelectric Materials 115
     Electrooptic Applications • Nonlinear Optical Applications • Photorefractive Applications • Pyroelectric Applications • Ferroelectric Structure-Related Applications

4.1 Introduction

Ferroelectric materials have long been used for various kinds of photonic applications such as spatial light modulation, optical frequency multiplication, light-beam amplification, integrated optics, optical storage, and others. These applications are based on the phenomena of optical birefringence, electrooptic effects, nonlinear optical effects, photorefractive effects, and electroelastic effects in ferroelectric materials. Progress in the study of ferroelectric materials (including crystal, ceramics, and thin films) has played a significant role in the development of laser techniques and optical communications. This chapter, first, reviews basic concepts and characteristics of ferroelectric materials that are important for photonic applications. Important ferroelectric materials are grouped into several families according to their crystallographic structures. Their structures are described in the second section of this chapter. General properties and growth techniques for these ferroelectric materials are also mentioned in this section. Although these are not optical properties, they are important for selecting materials for applications. Optical properties, such as electrooptic, nonlinear optic, photorefractive, pyroelectric and ferroelectric structure-related, of ferroelectric materials are discussed in the third section. Photonic applications of ferroelectric materials are covered in the fourth section of this chapter.

Ferroelectricity was discovered in Rochelle salt in 1921 by Valasek. A ferroelectric crystal exhibits a spontaneous polarization P_s in a certain temperature range and the direction of P_s can be reversed by an external electric field. From a physical point of view, ferroelectric crystals are those crystals which possess one or more ferroelectric phases. The 'ferroelectric phase' is a particular state exhibiting spontaneous polarization which can be reversed by an external field. A reversal of polarization is considered as a special case of the polarization reorientation. From a crystallographic point of view, ferroelectricity can be found in polar crystals. A polar crystal is a piezoelectric (without centrosymmetry) crystal whose point-group symmetry has a unique rotational axis, but does not have any mirror perpendicular to this axis. Along a unique rotational axis, the atomic arrangement at one end is different from that at the other (opposite end). Therefore, they display spontaneous polarization. Polar crystals, which can be found in ten point groups, are 1, 2, *m*, *mm*2, 4, 4*mm*, 3, 3*m*, 6, and 6*mm*. (see Table 4.1)

Spontaneous polarization P_s is defined by the value of the dipole moment per unit volume or by the value of the charge per unit area on the surface perpendicular to the axis of spontaneous polarization. Because electric and optical properties are strongly correlated with crystal structure, the axis of spontaneous polarization is usually a crystal axis. A crystal exhibiting spontaneous polarization can be visualized as composed of negative and positive ions. At a certain temperature, these ions are at their equilibrium positions, at which the free energy of the crystal is at a minimum and the center of positive charge does not coincide with the center of negative charge. We may visualize each pair of positive and negative ions as an electric dipole and the spontaneous polarization as the ensemble of these dipoles. Usually, a spontaneous polarization can be written in the form

$$P_s = \left[\iiint \mu dv \right] / \text{volume} \tag{4.1}$$

where μ is the dipole moment per unit volume. This formula implies that a polycrystalline material (ceramic or a composite) may have a remanent polarization.

In general, uniform alignment of electric dipoles occurs only in certain regions of a crystal. Such regions with uniform polarization are called ferroelectric domains. The interface between two domains is called the domain wall. A ferroelectric crystal always has a polydomain structure prior to poling by electric field. A strong field may reverse the spontaneous polarization of a domain. This dynamic process of domain reversal is called 'domain switching'. The process of switching depends on temperature and on the applied electric field. When the direction of the applied external field E is opposite to the orientation of polarization of a domain, a new domain, with its polarization parallel to the field, may occur in the old domain by nucleation and growth.

TABLE 4.1 Symbols of the 32 Point Groups in Crystallography

Crystal System	International Notation	Schönflies' Notation	Remarks†	Crystal System	International Notation	Schönflies' Notation	Remarks†
Triclinic	1	C_1	* +	Trigonal (Rhombohedral)	3	C_3	* +
	$\bar{1}$	$C_i(S_2)$	—		$\bar{3}$	$C_{3i}(S_6)$	—
					3*m*	C_{3v}	* +
Monoclinic	2	C_2	* +		32	D_3	*
	$m(\bar{2})$	$C_s(C_{1h})$	* +		$\bar{3}m$	D_{3d}	—
	2/*m*	C_{2h}	—				
Orthorhombic	2*mm*	C_{2v}	* +	Hexagonal	6	C_6	* +
	222	$D_2(V)$	*		$\bar{6}$	C_{3h}	*
	mmm	$D_{2h}(V_h)$	—		6*mm*	C_{6v}	*+
					6/*m*	C_{6h}	—
Tetragonal	4	C_4	* +		622	D_6	*
	$\bar{4}$	S_4	*		$\bar{6}m2$	D_{6h}	*
	$\bar{4}2m$	$D_{2d}(V_d)$	*		6/*mmm*	D_{6h}	—
	422	D_4	*				
	4*mm*	C_{4v}	* +	Cubic	23	T	*
	4/*m*	C_{4h}	—		$\bar{4}3m$	T_d	*
	4/*mmm*	D_{4h}	—		*m*3	T_h	—
					43	O	—
					*m*3*m*	O_h	—

†'*' Implies that piezoelectric effects may be exhibited, and '+' implies that pyroelectric and ferroelectric effects may be exhibited.

Ferroelectric Materials

An important characteristic of ferroelectrics is the ferroelectric hysteresis loop. A typical ferroelectric hysteresis loop is shown in Fig. 4.1. As indicated in Fig. 4.1, the polarization P has two stable states at a certain value of the applied field E. If we first apply a small field across, we will have a linear relationship between P and E, because the field is not large enough to switch any domain and the crystal will behave as a normal dielectric material (paraelectric). This case corresponds to the segment OA of the curves in Fig. 4.1. As the electric field strength increases, a number of the domains, which are polarized opposite to the direction of the field, will be switched over into the field direction, and the polarization will increase rapidly (segment AB) until all the domains are aligned in the field direction (segment BC). This is a state of saturation in which the crystal is composed of one single domain.

As the field strength decreases from the saturation state, the polarization will decrease. When the field is reduced to zero, some of the domains will still remain in the previous field direction and the crystal will exhibit a remanent polarization P_r. The extrapolation of the linear segment BC of the curve back to the polarization axis (at the point E on the vertical axis in Fig. 4.1) represents the value of the spontaneous polarization P_s.

The remanent polarization P_r in the crystal cannot be removed until the applied field in the opposite direction reaches a certain value (at the point F in the figure). The strength of the field required to reduce the polarization P back to zero is called the 'coercive field strength' E_c. Further increasing the field strength in the opposite direction will cause a complete alignment of the dipoles in this direction and the cycle can be completed by reversing the field direction once again. Thus, the relationship between P and E is represented by a hysteresis loop (CDFGHC), as shown in Fig. 4.1.

Another important characteristic of ferroelectrics is the phase-transition temperature called the Curie point Θ_C. When the temperature decreases through the Curie point, a ferroelectric crystal undergoes a structural change from a paraelectric phase to a ferroelectric phase. When the temperature is above Θ_C, the crystal does not exhibit ferroelectricity. On the other hand, when the temperature is below Θ_C, the crystal exhibits ferroelectricity.

When the temperature is in the vicinity of the Curie point, thermodynamic properties (such as dielectric, elastic, optical and thermal) of a ferroelectric crystal show anomalies and the structure of the crystal changes. For example, the dielectric constant in most ferroelectric crystals has an

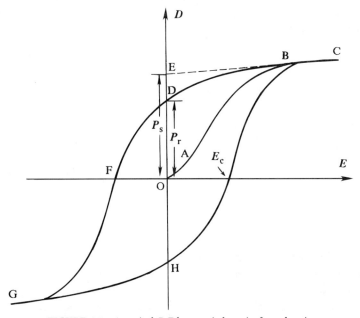

FIGURE 4.1 A typical *P-E* hysteresis loop in ferroelectrics.

abnormally large value (up to 10^4–10^5) near their Curie points. In most ferroelectrics, the temperature dependence of the dielectric constant above the Curie point (in the paraelectric phase) can be described fairly accurately by a simple law called the Curie-Weiss law:

$$\epsilon = \epsilon_o + C/(T - T_C) \qquad (T > T_C), \tag{4.2}$$

where C is the Curie-Weiss constant and T_c the Curie-Weiss temperature. T_c differs from the Curie point Θ_C. In the case of a first-order phase transition, $T_c < \Theta_C$, whereas for the second-order phase transition, $T_c = \Theta_C$. Usually, the temperature-independent term ϵ_o can be neglected, because it is much smaller than the term $C \mid (T - T_c)$, when T is near T_c.

The pyroelectric effect is the effect of temperature T on the value of spontaneous polarization P_s. The relation between P_s and T can be described by the pyroelectric coefficient. If the change in temperature ΔT in a crystal is gradual and small in magnitude, the change in the spontaneous polarization vector is given by

$$\Delta P_s = p \Delta T, \tag{4.3}$$

where p is the pyroelectric coefficient vector with three components:

$$p_m = \partial P_{sm}/\partial T, \qquad (m = 1, 2, 3). \tag{4.4}$$

The unit of the pyroelectric coefficient is $Cm^{-2}K^{-1}$ (or $\mu Ccm^{-2}K^{-1}$). A pyroelectric material with high Curie temperature can be used as an infrared detector.

Ferroelectric materials are important in the application of electrooptics and nonlinear optics because of their unique crystal structures. According to their microscopic symmetries, solid ferroelectrics are classified as optically isotropic and optically anisotropic. An example of the former type is ferroelectric ceramics whose isotropic behavior is due to the random orientation of the grains they possess. Based on their symmetrical point groups, the latter type of ferroelectrics can be further classified as optical uniaxial crystals and optically biaxial crystals. In the visible range, the optical relative dielectric constant ϵ/ϵ_o is equal to n^2, where n is the optical refractive index. If a coordinate system (X, Y, Z) is chosen to coincide with the three principle axes of a crystal, considering the crystal as an anisotropic medium,

$$\epsilon_x/\epsilon_o = n_1^2,$$
$$\epsilon_y/\epsilon_o = n_2^2,$$

and

$$\epsilon_z/\epsilon_o = n_3^2, (\epsilon_0 \text{ is the dielectric constant of free space}).$$

The optical anisotropy of a crystal is characterized by an index ellipsoid (or indicatrix) defined as

$$\frac{x^2}{n_1^2} + \frac{y^2}{n_2^2} + \frac{z^2}{n_3^2} = 1, \tag{4.5}$$

where n_1, n_2, and n_3 are the principal refractive indices. Given a direction in a crystal, two possible linear polarized rays (eigenmodes) may exist when an optical beam propagates in the crystal. Each ray has a unique direction of polarization which is parallel to the direction of electric displacement D and a corresponding refractive index (i.e., a phase velocity). The mutually orthogonal polarization directions and the indices of the two rays (one called "ordinary", the other called "extraordinary") can be described by the index ellipsoid. Along the principal axes, D and electric field E are parallel. The existence of two rays with different refractive indices is called *birefringence*.

Optical properties of a substance may be affected by an external mechanical stress, due to the effect of photoelasticity, which is also called the elastooptic or piezooptic effect and may occur in either ferroelectrics and nonferroelectrics. For crystals belonging to a noncentral, symmetric point group, the refractive indices are affected by external electric fields. This is the electrooptic effect. Both of these effects are produced by a distortion of the indicatrix caused by either mechanical stresses or electrical fields.

A ferroelectric crystal may exhibit both linear electrooptic (Pöckels effect) and quadratic (Kerr effect) effects. The former can be found only in optically anisotropic materials. The latter exists also in optically isotropic media, such as glasses and liquids.

We define the impermeability tensor β_{ij} as

$$\beta_{ij} = [\epsilon^{-1}]_{ij} = (-1)^{i+j}\Delta^{\beta}_{ij}/\Delta^{\beta}, \tag{4.6}$$

where ϵ^{-1} is the inverse of the dielectric tensor ϵ, Δ^{β} is an algebraic determinant of β_{ij}, and Δ^{β}_{ij} represents a remanent subdeterminant in which the term β_{ij} is removed. The propagation of light radiation in a crystal can be described completely in terms of β_{ij}. If we substitute x, y, and z in Eq. (4.5) by x_1, x_2, and x_3, respectively, Eq. (4.6) can be written in the form

$$\sum \epsilon_o \beta_{ij} x_i x_j = 1 \qquad (i, j = 1, 2, 3), \tag{4.7}$$

where $1/n_1^2$, $1/n_2^2$, and $1/n_3^2$ are the principal values of the impermeability tensor β_{ij}.

According to the quantum theory of solids, the optical dielectric permeability tensor depends on the distribution of electric charges in the crystal. The action of an external field results in a redistribution of the bound charges and a slight deformation of the ionic lattice. This result is reflected in a change in the optical permeability tensor. This is the electrooptic effect. Conventionally, the electrooptic coefficients are defined as

$$\begin{aligned}\beta_{ij}(E) - \beta_{ij}(0) &= \Delta\beta_{ij} \\ &= r_{ijk}E_k + s_{ijkl}E_k E_l \\ &= f_{ijk}P_k + g_{ijkl}P_k P_l \qquad (k, l = 1, 2, 3),\end{aligned} \tag{4.8}$$

where E is the applied electric field and P is the polarization field vector. Constants r_{ijk} and s_{ijkl} are the linear and the quadratic electrooptic coefficients, respectively, and constants f_{ijk} and g_{ijkl} are the linear and quadratic polarization-optic coefficients, respectively. It should be noted that higher order terms in Eq. (4.8) have been neglected because higher order effects are relatively much smaller than the linear and quadratic terms in most applications. The coefficients in Eq. (4.8) are related to each other as follows:

$$f_{ijk} = r_{ijk}/(\epsilon_k - \epsilon_o) \tag{4.9}$$

$$g_{ijkl} = s_{ijkl}/(\epsilon_k - \epsilon_o)(\epsilon_l - \epsilon_o),$$

where ϵ_k and ϵ_l are the principal optical dielectric constants. The values of electrooptic and polarization-optic coefficients of important ferroelectric materials are tabulated in section 4.3.

All of the ferroelectric crystals exhibit nonlinear polarization effects. Nonlinear optic effects have been used in laser frequency multiplication and optical parametric amplification. The optical nonlinearity results from an anharmonic response of the bound electrons driven by optical frequency fields of the laser radiation. The nonlinear behavior of the induced polarization of a medium under intense laser beam radiation can be expressed by the following power series expansion:

$$P = P_o + \chi^{(1)} \cdot E + \chi^{(2)} \cdot E \cdot E + \chi^{(3)} \cdot E \cdot E \cdot E + \cdots \qquad (4.10)$$

where P_0 is the static polarization and $\chi^{(i)}$ are the ith-order nonlinear optical susceptibilities (they are tensors of order $i + 1$). The second term on the right of Eq. (4.10) corresponds to a linear effect. The third term is the response for the second-harmonic generation (frequency doubling), for sum- and difference-frequency generation, and for parametric amplification and oscillation. The third term is responsible for such phenomena as third-harmonic generation, Raman scattering, Brillouin scattering, self-focusing, and optical phase conjugation. The magnitude of $\chi^{(i)}$ decreases by orders as i increases which means that higher order nonlinear effects are generally much weaker compared to those of lower orders. Electrooptic and photorefractive effects can be considered special cases of nonlinear optical phenomena. They are discussed separately in this chapter. Thus, we limit our discussion of nonlinear optical effects to second-order nonlinear effects at optical frequencies. Data related to the second-order nonlinear optical properties of ferroelectric materials will be given in later section. More detail about the theory of nonlinear optical processes can be found in other chapters, especially Chapter 12, of this handbook.

Photorefractivity was first discovered as an optical damage effect in $LiNbO_3$. The photorefractive effect is a photo-induced change in the refractive index of a medium. The effect is similar to optical phase conjugation which is a third-order nonlinear optical effect. Therefore, the photorefractive effect is also referred to as self-pumped phase conjugation. In general, the mechanism of the photorefractive effect is considered to be the spatial modulation of photocurrents by a nonuniform illumination. The photocurrents are generated by the transportation (i.e., drift, diffusion, and trapping) of excited charge carriers. In ferroelectric materials, impurity centers provide ground states and recombination sites for charge carriers (charge donors and acceptors). Due to the nonuniform illumination, a space charge field is produced that modulates the refractive index via the electrooptic effect.

Photorefractive effects have been successfully applied to real-time interferometry, beam deflection, image and signal processing and many other areas. Photorefractive effects in ferroelectric materials are discussed in later section.

4.2 Structure, Synthesis, and General Properties of Ferroelectric Materials

This section describes the structure, synthesis, and general properties of ferroelectric materials, which are classified by families. The families include perovskites, lithium tantalate and lithium niobate, tungsten-bronze-type niobates, water-soluble crystals, potassium titanyl phosphate, miscellaneous inorganic materials, polymers, composites, and ferroelectric liquid crystals. The first three families are often called oxygen octahedra ferroelectrics. The latter three families are excluded from the discussion because of space limitations. Liquid crystals are important in field of display (for more details, see Chapter 18) because of their unique electrooptic properties. Detailed description of liquid crytals is beyond the scope of this chapter. The attention of this section is directed mostly to oxide ferroelectric materials which are practical and promising for optical applications. Ferroelectric crystals (including single crystals and ceramics) and thin films are discussed separately in this section. Table 4.2 lists the names, chemical formulas, abbreviations, and crystal growth methods of the important ferroelectric materials. General properties of these materials, such as point group, phase-transition temperature, spontaneous polarization, density, and melting point are summarized in Table 4.3.

TABLE 4.2 List of Ferroelectric Materials and Their Crystal Growth Methods

Family	Ferroelectric Material	Chemical Formula	Abbrev.	Growth Method
Perovskite type	Barium titanate	$BaTiO_3$	—	Remeika method
				Top seed pulling method
	Potassium niobate	$KNbO_3$	—	Spontaneous nucleation and slow cooling
				Top seed solution growth
				Kyropoulos pulling
	Potassium tantalate	$KTaO_3$	—	The same as $KNbO_3$
	Potassium tantalate niobate	$KTa_{1-x}Nb_xO_3$	KTN	Kyropoulos technique
				Top seed solution growth
	Lead lanthanum zirconate titanate (in the form of ceramics)	$Pb_{1-x}(Zr_yTi_{1-y})_{1-0.25x}V^B_{0.25x}O_3$	PLZT	Chemical coprecipitation of powder and subsequent hot-pressing in oxygen environment
Lithium niobate family	Lithium niobate	$LiNbO_3$	—	Czochralski's technique
	Lithium tantalate	$LiTaO_3$	—	Czochralski's technique
Tungsten-bronze type	Barium strontium niobate	$Ba_{5x}Sr_{5(1-x)}Nb_{10}O_{30}$	SBN	Czochralski's method
	Barium sodium niobate	$Ba_{5x}Na_{5(1-x)}Nb_{10}O_{30}$	BNN	Czochralski's method
	Potassium lithium niobate	$K_3Li_2Nb_5O_{15}$	KLN	Kyropoulos method
	Potassium sodium strontium niobate	$(K_xNa_{1-x})_{0.4}(Sr_yBa_{1-y})_{0.8}Nb_2O_6$	KNSBN	Czochralski's technique
KDP family	Potassium dihydrogen phosphate	KH_2PO_4	KDP	Water solution temperature reduction method
	Potassium dihydrogen arsenate	KH_2AsO_4	KDA	The same as KDP
	Rubidium dihydrogen phosphate	RbH_2PO_4	RDP	The same as KDP
TGS type	Triglycine sulphate	$(NH_2CH_2COOH)_3 \cdot H_2SO_4$	TGS	Temperature reduction method
	Triglycine selenate	$(NH_2CH_2COOH)_3 \cdot H_2SeO_4$	TGSe	The same as TGS
KTP family	Potassium titanyl phosphate	$KTiOPO_4$	KTP	Top seed flux growth
Bismuth titanate	Bismuth titanate	$Bi_4Ti_3O_{12}$	—	Flux-growth method
Rare earth molybdate	Gadolinium molybdate	$\beta\text{-}Gd_2(MoO)_3$	GMO	Pulling from melt
Lead germanium oxide	Lead germanium oxide	$5PbO \cdot 3GeO_2$ or $Pb_5Ge_3O_{11}$	—	Czochralski's technique
				Bridgman's technique
Antimony sulphoiodide	Antimony sulphoiodide	SbSI		Vapor phase growth

Source: Xu, Y. *Ferroelectric Materials And Their Applications*, Elsevier Science Publishers B.V., Amsterdam, The Netherlands, 1991. With permission.

TABLE 4.3 General Physical Properties of Ferroelectric Materials

Chemical Formula	Point Group*	Phase Transition Temperature (°C)	Spontaneous Polarization (μC/cm²)	Density (g/cm³)	Melting Point (°C)
BaTiO$_3$	$m3m \to \mathbf{4mm} \to mm2 \to 3m$	120, 5, −90	26	6.02	1618
KNbO$_3$	$m3m \to 4mm \to \mathbf{mm2} \to 3m$	435, 225, −10	30		1050
KTaO$_3$	$m3m \to 4mm \to mm2 \to 3m$				
KTa$_{1-x}$Nb$_x$O$_3$	$m3m \to 4mm \to mm2 \to 3m$				
Pb$_{1-x}$La$_x$(Zr$_y$Ti$_{1-y}$)$_{1-0.25x}$V$^B_{0.25x}$O$_3$	See fig. 4.X for phase diagram				
LiNbO$_3$	$\bar{3}m \to \mathbf{3m}$	1210	71	4.64	1240
LiTaO$_3$	$\bar{3}m \to \mathbf{3m}$	665	50	7.45	1650
Ba$_{0.4}$Sr$_{0.6}$Nb$_2$O$_6$	$(4/m)mm \to \mathbf{4mm} \to m$	75, −213	32	~5.4	~1480
Ba$_2$NaNb$_5$O$_{15}$	$(4/m)mm \to 4mm \to \mathbf{mm2}$	560, 300	40	5.40	~1450
K$_3$Li$_2$Nb$_5$O$_{15}$	$(4/m)mm \to \mathbf{4mm}$	430	~40		1250
(K$_x$Na$_{1-x}$)$_{0.4}$(Sr$_y$Ba$_{1-y}$)$_{0.8}$Nb$_2$O$_6$	$(4/m)mm \to \mathbf{4mm}$		~30		
KH$_2$PO$_4$ (KDP)	$\bar{4}2m \to \mathbf{mm2}$	−150	−4.8	2.34	Decomposes at 180 °C
KH$_2$AsO$_4$	$\bar{4}2m \to \mathbf{mm2}$	−176			
RbH$_2$PO$_4$	$\bar{4}2m \to \mathbf{mm2}$	−126			
(NH$_2$CH$_2$COOH)$_3$·H$_2$SO$_4$ (TGS)	$2/m \to \mathbf{2}$	49	2.8	1.69	
(NH$_2$CH$_2$COOH)$_3$·H$_2$SeO$_4$	$2/m \to \mathbf{2}$	26			
KTiOPO$_4$	$mmm \to \mathbf{mm2}$	943	~17		
Bi$_4$Ti$_3$O$_{12}$	$(4/m)mm \to \mathbf{m}$	675	50, a-axis 4, c-axis	6.1	
β-Gd$_2$(MoO$_4$)$_3$	$\bar{4}2m \to \mathbf{mm2}$	159	0.17	7.33	1175
5PbO·3GeO$_2$, or Pb$_5$Ge$_3$O$_{11}$	$\bar{6} \to \mathbf{3}$	177	4.8	5.25	738
SbSI	$mmm \to \mathbf{mm2}$	22	25 (0 °C)		

*Point groups in bold are point groups at room temperature

Source: Xu, Y. *Ferroelectric Materials And Their Applications*, Elsevier Science Publishers B.V., Amsterdam, The Netherlands, 1991. With permission.

Material Structure and Preparation Method of Ferroelectric Crystals

Perosvkite-type Ferroelectrics

Perovskite is the name of the mineral calcium titanate ($CaTiO_3$). Perovskites have the general chemical formula ABO_3, where O is oxygen, A represents a cation with larger ionic radius, and B is a cation with a smaller ionic radius (in $BaTiO_3$, A is Ba and B is Ti). A perovskite is essentially a three-dimensional network of BO_6-octahedra. Most of the ferroelectrics with perovskite-type structure are compounds with either $A^{2+}B^{4+}O_3^{2-}$ or $A^{1+}B^{5+}O_3^{2-}$ -type formulas. In optical applications, $BaTiO_3$, $KNbO_3$, and $KTa_{1-x}Nb_xO_3$ (KTN) are important crystals in this class. In the case of $BaTiO_3$, the prototype crystal structure is cubic above the Curie temperature of 120°C, with Ba^{2+} ions at the cube corners, O^{2-} ions at the face center and a Ti^{4+} ion at the body center. Below the Curie temperature, the structure is slightly deformed, with Ba^{2+} and Ti^{4+} ions displaced relative to the O^{2-} ions, thereby creating a dipole. Thus, we may visualize each pair of positive and negative ions as an electric dipole, and the spontaneous polarization is due to the collective effect of these dipoles. $KNbO_3$ is biaxial with a perovskite structure, and its orthorhombic phase of symmetry $mm2$ exists between -5 °C and 225 °C. KTN is a completely miscible ferroelectric solid solution of $KNbO_3$ and $KTaO_3$. The phase-transition sequence is similar to that of $KNbO_3$.

One of the most important achievements in the field of ferroelectric ceramics is the development of transparent ferroelectrics of lead lanthanum zirconate titanate (PLZT). The composition of transparent ferroelectric ceramics is mainly that of stoichiometric lead zirconate titanate (PZT) with additional chemical modifiers, such as La, Bi, Sn, and Ba. Among them, La was found to be significantly better than others in optical homogeneity. The PLZT materials have a wide range of homogeneous compositions based on the completely miscible $PbZrO_3$-$PbTiO_3$ solid-solution system. The solubility of La is up to 35% in the $PbTiO_3$-rich phase. Therefore, it is possible to tailor a specific property of PLZT by designing the chemical composition. The chemical formula is listed in Table 4.2 where lanthanum ions replace lead ions on the A sites of the Perovskite structure and the charge balance is maintained by creating lattice site vacancies (represented by V^B) on B sites. The concentration of La, x (at .%), may vary from 2 to 30. The ratio y of Zr/(Zr + Ti) may take any value contineously from 0 to 1. The composition of PLZT is usually represented by the notation '$x/(1-y)$', e.g., 8/65/35 represents the composition $Pb_{0.92}La_{0.08}(Zr_{0.65}Ti_{0.35})_{0.98}O_3$. The preparation of PLZT ceramics involves chemical coprecipitation (solution reaction process) for preparing powder materials and subsequent hot-pressing in an oxygen atomosphere. The wet chemical processing of powder materials is used to achieve excellent compositional homogeneity. The advantage of hot-pressing is that high-density, low-porosity ceramic pieces can be made by this process. The phase diagram of the PLZT system at room temperature is shown in Fig. 4.2. From the phase diagram, we can see that excess La results in an undesirable reduction in optical transparency because it produces a heterogeneous mixed phase of PLZT, $La_2Zr_2O_7$, and $La_2Ti_2O_7$.

Lithium Niobate and Lithium Tantalate

Lithium niobate and lithium tantalate have similar structures. Their structures are in form of ABO_3 composed of oxygen octahedra. The neighboring oxygen octahedra are connected to each other through an oxygen ion that serves as a common tie. The slightly deformed oxygen octahedra are aligned along the c-axis (threefold axis). The cation arrangement follows the sequence Nb (or Ta), vacancy, Li, Nb (or Ta), vacancy, Li, and so on. The symmetry of both crystals belongs to the point group $3m$ in the trigonal ferroelectric phase at room temperatures. The symmetry changes to the point group $\bar{3}m$ in the paraelectric phase above the Curie temperature.

Ferroelectric-paraelectric phase transition in both $LiNbO_3$ and $LiTaO_3$ results from the displacement of Li^+ and Nb^{5+} (or Ta^{5+}) ions. The direction of spontaneous polarization is the same as that of ion displacement, which coincides with the c-axis.

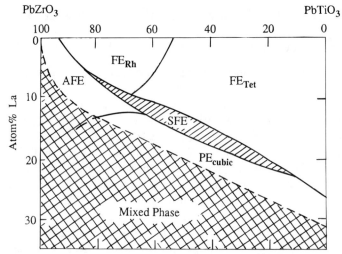

FIGURE 4.2 Room temperature phase diagram of the PLZT system. FE_{Tet}: ferroelectric tetragonal phase; FE_{Rh}: ferroelectric rhombohedral phase; AFE: antiferroelectric phase; PE_{cubic}: paraelectric cubic phase; SFE: ferroelectric phase having a slim loop (after Haertling, 1987).

Tungsten-Bronze-Type Niobate Crystals

Tungsten-bronze (TB) ferroelectric niobate crystals contain oxygen octahedra and maintain a crystallographic structure similar to that of the tetragonal tungsten-bronze-type K_xWO_3 and Na_xWO_3 ($x < 1$). There are three types of sites: A_1, A_2, and C. The interstitial site of the oxygen octahedron has two types, B_1 and B_2, each having different symmetry positions. A tetragonal unit cell includes two A_1 sites, four A_2 sites, four C sites, two B_1 sites, eight B_2 sites and thirty oxygen ions (at the corners of the octahedra). The site occupancy formula may, thus, be written as $(A_1)_2(A_2)_4(C)_4(B_1)_2(B_2)_8O_{30}$. The sites A_1, A_2, C, B_1, and B_2 can be either partially or fully occupied by different cations. In the case of niobates or tantalates, the B_1 and B_2 sites are occupied by Na^{5+} or Ta^{5+}, whereas the A_1, A_2, and C sites are occupied by either alkaline earth ions or alkaline ions, or both of them. The A and C sites may be fully or partially occupied. The number of occupied sites depends on the valence and the number of these cations, in accordance with the requirements of electroneutrality, and on their size in accordance with crystallographic constraints.

Barium sodium niobate (BNN, also known as 'banana'), strontium barium niobate (SBN), potassium lithium niobate (KLN), and potassium sodium strontium barium niobate (KNSBN) are important ferroelectrics in this class for optical applications. The chemical formula of BNN can be written as $Ba_2NaNb_5O_{15}$ in its ideal state, where the A_1 sites are fully occupied by Na^+ ions and the A_2 sites are fully occupied by Ba^{2+} ions. The properties of BNN can be modified by incoporating alkaline, alkaline earth ions other than Ba^{2+}, and Na^+ or Pb ions. The general formula of SBN is $Sr_{1-x}Ba_xNb_2O_6$, where $0.2 < x < 0.8$. The KLN ($K_3Li_2Nb_5O_{15}$) crystal is a fully occupied TB-type structure in which all of the A_1 and A_2 sites are occupied by the larger K^+ ions and all of the C sites are occupied by the smaller Li^+ ions. Some of the Nb^{5+} ions can be replaced by Ta^{5+} ions. The solid solution crystal is $K_3Li_2(Ta_xNb_{1-x})O_{15}$, i.e., potassium lithium tantalate niobate (KLTN). KNSBN, $(K_xNa_{1-x})_{0.4}(Sr_yBa_{1-y})_{0.8}Nb_2O_6$ ($0.50 < x < 0.75$, $0.30 < y < 0.90$), has a TB-type structure in which the A sites are fully occupied.

Water-soluble Ferroelectric Crystals

Well known water-soluble ferroelectric crystals in optical and infrared (pyroelectric) applications include the KDP and TGS families, respectively. The structural framework of potassium dihydrogen phosphate (KH_2PO_4) (KDP) is composed of two interpenetrating sets of body-centered sublattices of PO$_4$ tetrahedrons and is also composed of two interpenetrating sets of body-centered sublattices of K$^+$ ions. K$^+$ and P^{5+} ions are alternately arranged in different layers perpendicular to the c-axis and spaced from each other by a distance of $c/4$. Each phosphous ion is surrounded by four oxygen ions at the corner of a tetrahedron. Each PO$_4$ tetrahedron is linked to four other PO$_4$ groups, spaced $c/4$ apart along the c-axis. Thus, the linkage is such that there is an 'O-H-O' hydrogen bond between the 'upper' oxygen of one PO$_4$ group and one 'lower' oxygen of the neighboring PO$_4$ group, each hydrogen lying nearly perpendicular to the c-axis. Only two hydrogen atoms are located near each PO$_4$ group, therefore, forming (HPO$_4$)$^-$ ions as a group. Spontaneous polarization in KDP is along either $+c$-axis or $-c$-axis. Other crystals with a KDP type structure have the common chemical formula XH$_2$YO$_4$, where X represents K, Rb, or Cs and Y represents P or As, respectively. The hydrogen (H) in these compounds can be replaced entirely or partially by deuterium D. The chemical formula of deuterated KDP (DKDP or KD*P) is K(H$_{1-x}$D$_x$)$_2$PO$_4$. The physical properties of DKDP depend heavily on the amount of deuterium (value of x).

KDP is a water-soluble crystal. Large KDP single crystals with high optical quality can be easily grown. These crystals have fair mechanical properties. KDP does not contain any water molecules nor does it effloresce, but it decomposes at 180 °C. The major shortcoming of KDP is its deliquescence. However, this can be overcome by sealing or by coating a protective film on the surface.

At room temperature, the symmetry axis of triglycine sulfate (TGS) is in the direction of the b-axis of a monoclinic cell (twofold symmetry axis). Isomorphous crystals of the TGS family include triglycine selenate (NH$_2$CH$_2$COOH)$_3$·H$_2$SeO$_4$ (TGSe) and triglycine fluoberyllate, (NH$_2$CH$_2$COOH)$_3$H$_2$BeF$_4$ (TGFB). Better properties can be found in TGS-TGSe and TGS-TGFB mixed crystals. As in the KDP family, all of the hydrogen atoms in TGS can be replaced by deuterium atoms, and the deuterated crystal has the formula (ND$_2$CD$_2$COOD)$_3$·D$_2$SO$_4$ (DTGS). The main shortcoming of TGS for applications is the depolarization owing to its low Curie point (49°C). This disadvantage can be improved by incorporating α-alanine into TGS. The resulting doped crystal has the chemical formula (NH$_2$CH$_2$COOH)$_{3(1-x)}$(CH$_3$CHNH$_2$COOH)$_3$·H$_2$SO$_4$ (LATGS) where x is, generally, between 2% to 35%. The pyroelectric properties of TGS can be futher improved by doping phosphoric acid or arsenic acid into LATGS [Fang et al., 1983; 1989].

Potassium Titanyl Phosphate Family

Orhtorhombic crystals of potassium titanyl phosphate KTiOPO$_4$ (KTP) and its isomorphs with Rb (RTP) and Tl (TTP) represent another class of ferroelectric materials with unique combination of physical properties. Solid solutions exist in the MTiOPO$_4$ (KTP) structure where M can be K, Rb, Ti, Cs (partial), NH$_4$, or any combination of these ions, and solid solutions exist between MTiOPO$_4$ and MTiOAsO$_4$. It is ferroelectric because the symmetry between *mmm* (symmetry above the phase-transition temperature) and *mm*2 permits the existence of the ferroelectric state, and shows a net polarization. In the case of the transition from *mmm* to *mm*2 in K$_x$Rb$_{1-x}$TiOPO$_4$, the Rb (and K) ions are displaced along the c-axis, and Ti atoms are displaced so as to create chains of alternating long and short Ti-O bonds. At room temperature the T$_i$ ions are displaced in the T$_i$O$_b$ octahedra ($\Delta Z = 0.30$Å) and K ions undergo a large displacement ($\Delta Z = 1.66$Å). The spontaneous polarization which is along <011> and <0T1> directions is contributed by the displacements of K$^+$ and T$_i^{4+}$ ions. Ferroelectric domains in K$_x$Rb$_{1-x}$TiOPO$_4$ have not been observed. It is believed that this material grows to a single domain state (Bierlein et al., 1976 and 1989).

Miscellaneous Inorganic Ferroelectric Crystals

Several important ferroelectric crystals for optical applications which do not belong to the families mentioned previously are introduced in this part. These ferroelectrics have rather unique optical properties due to their crystal structures. These crystals are bismuth titanate (Bi$_4$Ti$_3$O$_{12}$), gadolinium molybdate ($Gd_2(MoO_4)_3$), and lead germanium oxide ($Pb_5Ge_3O_{11}$). Their structures are described following and their optical properties are discussed in a later section.

Bismuth titanate is one of the bismuth compounds with alternating layers of oxygen octahedra (a layer of the pseudo-perovskite structure) and Bi_2O_2. These bismuth compounds have the general chemical formula

$$(Bi_2O_2)^{2+}(A_{m-1}B_mO_{3m+1})^{2-}, \quad (m = 1, 2, \ldots, 5)$$

where A and B represent the ions with suitable chemical valences and ionic radii. Usually A represents Bi, Pb, Ba, Sr, Ca, Na, K, and rare earth elements; B represents Ti, Nb, Ta, W, Mo, Fe, Co, Cr, etc. Most of the compounds are ferroelectrics at room temperature with Curie points in the range of 300–700°C. Bismuth titanate is a typical representative of these compounds. Below T_c (675°C), the $Bi_4Ti_3O_{12}$ crystal belongs to a ferroelectric monoclinic phase with the symmetry of point group m. Because the monoclinic unit cell of bismuth titanate closely approximates an orthorhombic unit cell, a 'pseudo-orthorhombic' crystal system is always used to describe this crystal. The lattice parameters are $a = 5.41$ Å, $b = 5.445$ Å, $c = 32.8$ Å, and $\beta = 90°$. The switching characteristic of the polarization is rather unusual in this crystal. In this material, the polarization direction is inclined at a small angle (~4.5°) to the monoclinic a-axis in the a–c plane and can be resolved into two independently reversible components, a large one of about 50 $\mu C/cm^2$ along the a-axis and a small one of about 4 $\mu C/cm^2$ along the c-axis. These two components can be switched independently by an external field. Along the c-axis, two opposite polarization directions correspond to two kinds of domains, which can be distinguished optically.

β-$Gd_2(MoO_4)_3$ (β-GMO) is composed of three different sets of $(MoO_4)^{2-}$ tetrahedra arranged along the c-axis and ordered layer by layer. Gd^{3+} ions occupy approximately the center of the GdO_7 polyhedron. The lattice parameters of ferroelectric β-GMO are $a = 10.39$ Å, $b = 10.42$ Å, and $c = 10.70$ Å at room temperature. In the ferroelectric phase (orthorhombic), the polar axis of the crystal is parallel to the c-axis, and the spontaneous polarization is so small that its value is only 0.17–0.20 $\mu C/cm^2$ at room temperature. The spontaneous polarization originates in the displacements of Gd^{3+} ions and $(MoO_4)^{2-}$ sublattices, the polarizations caused by different Gd^{3+} ions cancel out mostly, and, thus, the resulting spontaneous polarization is very small. In the orthorhombic phase (temperature below T_c), if a compressive stress is applied to the crystal along the b-axis, the a-axis and b-axis (the difference in magnitude between the lattice parameters a and b is only 0.3%) will be interchanged, whereas the electric polarity will be reversed because there are small displacements of $(MoO_4)^{2+}$ tetrahedra caused by the applied stress. In this case, the state of spontaneous polarization is coupled with mechanical shear strain, whereas the positive or negative sign can be reversed by an external stress. This effect is called ferroelastic. β-GMO exhibits both ferroelectricity and ferroelasticity at room temperature.

$Pb_5Ge_3O_{11}$ is structurally composed of two different groups, (Ge_2O_7) and $Pb_5(GeO_4)$, where (GeO_4) is a tetrahedron with Ge occupying the center position. When viewed along the c-axis, the group tetrahedron (GeO_4) and the group (Ge_2O_7) are arranged in alternating layers. The (Ge_2O_7) group is composed of two (GeO_4) tetrahedrons sharing a corner (oxygen atom). In the ferroelectric phase, the crystal possesses two kind of antimorphic structures corresponding to displacements of Pb atoms along the $+c$- and $-c$-directions, where spontaneous polarization occurs. As the spontaneous polarization reverses, the group GeO_4 turns around in the a–b plane. These two antimorphic structures correspond to two kinds of ferroelectric domains with antiparallel polarization.

Ferroelectric Thin Films

Ferroelectric thin films have been extensively investigated for a wide variety of electrical and optical applications (Francombe, 1972). There are several reasons for the increasing importance of ferroelectric thin films: (1) The trend toward miniaturization of electronic and optical components has led to the development of thin film ferroelectric devices performing the same functions with only a fraction of the volume of devices based on bulk ceramic or single crystal elements. (2) Thin films have added designing advantages, such as small volume, high geometrical flexibility, and convenient integration with semiconductor ICs or optical ICs compared to bulk materials. (3) Lower fabrication cost compared to single crystals. (4) New areas of application are being indentified that utilize new device concepts, exploiting properties that are unique to both thin films and ferroelectric materials.

Many fabrication techniques of thin film, including physical techniques (such as evaporation, sputtering, liquid phase epitaxy, etc.) and chemical techniques (such as chemical vapor deposition (CVD), metalloorganic deposition (MOD), metalloorganic chemical vapor deposition (MOCVD) (Chour et al., 1993), and sol-gel processing) have been used to make different ferroelectric thin films on various substrates including sapphire, silicon, GaAs, fused silica, Pt-coated Al_2O_3, single crystal MgO and $MgAl_2O_4$, and single crystals of several ferroelectric crystals, etc. Table 4.4 summarizes various deposition techniques that have been used to fabricate different ferroelectric thin films. For a review of various techniques for preparing ferroelectric thin films, valuable papers have been published by Roy et al., (1990), Francombe et al., (1990), and Mansingh (1990).

Because ferroelectric films have almost the same properties as ferroelectric ceramics or single crystal materials, i.e., large dielectric permittivity, ferroelectricity, piezoelectricity, pyroelectricity, and nonlinear optical effects, they have been utilized to make various devices. Potential optical applications of ferroelectric thin films, such as pyroelectric infrared detectors, ferroelectric/photoconductive displays, electrooptic waveguide devices or optical modulators, optical memories, and frequency doublers for diode lasers. Examples of electronic and optical devices where ferroelectric films are key elements are shown in Fig. 4.3.

For the application of ferroelectric thin films in electronic and optical devices, bulk ferroelectric properties must be achieved in thin films. Optical clarity is an additional requirement for electrooptic device application. Therefore, a high quality ferroelectric thin film should possess the following properties: a stoichiometric composition, a dense and crystalline microstructure, a single crystal film (or, at least, a preferentially grain-oriented polycrystalline film) and a uniform large area. R. W. Roy and his co-workers compared principal features of different techniques for ferroelectric thin film preparation (Roy, 1990). Physical vapor deposition processes (such as sputtering which is very good for epitaxy) have been the most widely investigated deposition techniques for

TABLE 4.4 Comparison of Salient Features of Common Methods of Ferroelectric Thin Film Preparation. Numerical Scale for Epitaxy and Stoichiometry: 1-worst, 10-best (After R. W. Roy et al., 1990)

Method	Rate Å/min.	Epitaxy 1–10	Stoichiometry 1–10	Temperature(°C) Substrate	Anneal	Devices†	Cost	Miscellaneous Problems
RF Sputter	5–50	8	3	RT–700	500–700	1–6	H	negative ions
Magnetron sputter	50–300	5	5	RT	500	1–6	H	target surface
RF magnetron	50–100	9	5	RT	500	1–6	H	
Ion beam sputter	20–100	9	8	RT	500	1–6	H	uniformity
Evaporation	100–1000	7	4	RT	500	1–6	H	rate control
Laser deposition	50–1000	9	6	RT	500	1,3,5	H	debris, uniformity
Sol-Gel	1000Å/C	2–8	9	RT	450–750	1–5	L	multiple coating
MOD	3000Å/C	2	9	RT	500–800	1,3,5	L	high $T_{(anneal)}$
MOCVD	50–1000	5	7	400–800	600	1–5	H	high $T_{(substrate)}$

† Device number: 1-Capacitor, 2-Memory cell, 3-Actuator, 4-Electrooptic, 5-Pyroelector, 6-SAW.

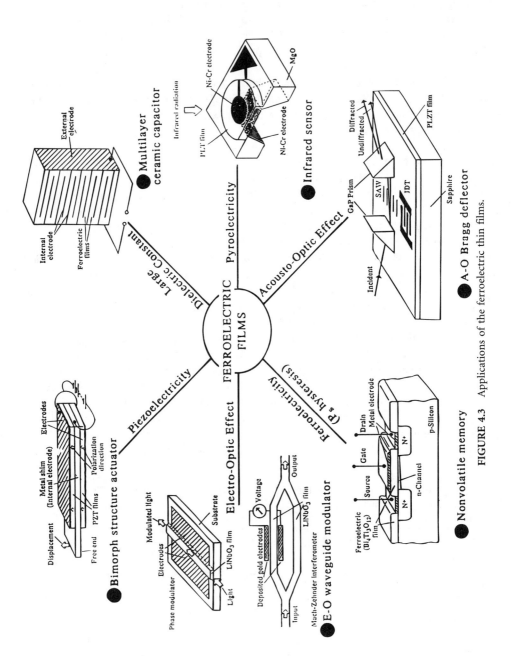

FIGURE 4.3 Applications of the ferroelectric thin films.

Ferroelectric Materials

preparing ferroelectric thin films. However, the application of ferroelectric thin films prepared by sputtering have been hindered by the difficulties associated with control of the chemical and physical properties of the film (i.e., stoichiometry, crystal structure, microstructure, and uniformity over large areas). Chemical vapor deposition (CVD) techniques have excellent potential as a production method for device-quality ferroelectric thin films, especially when organometallic precursors are used. The CVD process can be extremely reproducible once conditions are established to produce a film with a particular composition and crystal structure. An additional advantage is that scale-up of a CVD process from laboratory to production is easy. A limitation to the use of the CVD methods for ferroelectric thin film applications is the difficulty associated with controlling the stoichiometry of the complex compositions of typical ferroelectric materials. Lower deposition temperatures provide the opportunity to integrate the ferroelectric thin film with nonrefractory electrode materials compatible with semiconductor processes. The temperature of the CVD deposition process has been significantly decreased of a plasma-assisted CVD system. Sol-gel processing is a relatively new method for fabricating ferroelectric thin films. The advantages of sol-gel processing are good homogeneity, ease of composition control, low processing temperature, large area thin films and cost lower than other techniques. A general review of the sol-gel process for fabricating ferroelectric thin films is presented in (Xu, 1992). It is too early to conclude which one among the techniques listed in the table is the best.

4.3 Optical Properties of Ferroelectric Materials

Most of the oxide ferroelectric materials are transparent in the visible and infrared range where most of the photonic application wavelengths are covered. Several optical properties, such as pyroelectric, electrooptic, nonlinear optic, and photorefractive properties of ferroelectric materials, which are important for photonic applications, are discussed in this section. Some ferroelectric materials, such as $LiNbO_3$ and $LiTaO_3$, have been used for acoustooptic applications. However the acoustooptic effects are not discussed in this chapter. For more details of acoustooptic effects, see Chapters 9 and 11 of this handbook.

Electrooptic Effect

Linear electrooptic effects of ferroelectric materials have been successfully applied in many optical devices, such as amplitude modulators and phase modulators. The electrooptic effect can be described by the third-rank tensor of linear electrooptic coefficient r_{ijk}. Using simplified subscript indices, r_{ijk} is contracted to $r_{\alpha k}$, where the first and the second indices are replaced by a single index running from 1 to 6, while the third index is fixed. These matrix elements (6×3) do not have the usual tensor transformative or multiplicative properties. The simplified linear electrooptic coefficient tensors of some of the point groups are shown following as illustrative examples.

Point group $\bar{4}2m$ (e.g., KDP)

$$\begin{pmatrix} 0 & 0 & 0 \\ 0 & 0 & 0 \\ 0 & 0 & 0 \\ r_{41} & 0 & 0 \\ 0 & r_{41} & 0 \\ 0 & 0 & r_{63} \end{pmatrix}$$

Point group $4mm$ (e.g., $BaTiO_3$, SBN)

$$\begin{pmatrix} 0 & 0 & r_{13} \\ 0 & 0 & r_{13} \\ 0 & 0 & r_{33} \\ 0 & r_{51} & 0 \\ r_{51} & 0 & 0 \\ 0 & 0 & 0 \end{pmatrix}$$

Point group 3m (e.g., LiNbO$_3$, LiTaO$_3$) Point group 6mm (e.g., poled ceramic PLZT)

$$\begin{pmatrix} 0 & -r_{22} & r_{13} \\ 0 & r_{22} & r_{13} \\ 0 & 0 & r_{33} \\ 0 & r_{51} & 0 \\ r_{51} & 0 & 0 \\ -r_{22} & 0 & 0 \end{pmatrix} \qquad \begin{pmatrix} 0 & 0 & r_{13} \\ 0 & 0 & r_{13} \\ 0 & 0 & r_{33} \\ 0 & r_{51} & 0 \\ r_{51} & 0 & 0 \\ 0 & 0 & 0 \end{pmatrix}$$

When a linear polarized beam enters an electrooptic medium, the light intensity of the transmitted beam in the absence of light attenuation is given by

$$I = I_o \sin^2\left(\frac{\Delta\phi}{2}\right) \tag{4.11}$$

where the phase shift $\Delta\phi$ between ordinary and extraordinary waves is expressed as

$$\Delta\phi = \frac{2\pi l}{\lambda}\left(\Delta n - \frac{n^3}{2} r_{eff} E_3\right) \tag{4.12}$$

where Δn is the static birefringence and the second term in the parenthesis on the right-hand side of the equation is the electric field-induced birefringence (r_{eff} is the effective electrooptic coefficient). An important parameter for EO modulator applications is the *half-wave voltage* $V_{\lambda/2}$ (or V_π) which is the voltage required for a phase retardation of a half wavelength or $\Delta\phi = \pi$. It can be derived from Eq. (4.12) and is expressed as

$$V_{\lambda/2} = V_\pi = \frac{\lambda}{n^3 r_{eff}} \cdot \frac{d}{L} \tag{4.13}$$

where d is the spacing between two electrodes and L is the interaction length (sample dimension along the z-axis). The term $n^3 r_{eff}$ is considered a figure of merit for the half-wave voltage.

When an electric field is applied in the direction of the optical axis (z-axis), in a uniaxial crystal, the propagation of the light beam is perpendicular to this axis (transverse electrooptic effect). In this case, $d = L$, and the half-wave voltage V_π is given by

$$V_\pi = \lambda/n_3^3 r_c \tag{4.14}$$

where $r_c = r_{33} - (n^3_1/n_3^3)r_{13}$ is called the transverse effective electrooptic coefficient. Thus, by applying a known electric field in certain directions in a crystal, one can measure the electrooptic birefringence induced by the field and determine the electrooptic coefficient r_{ijk}. Table 4.5 lists the linear electrooptic coefficients and the half-wave voltage of several ferroelectric materials.

In the case of the quadratic electrooptic effect, the field-induced birefringence of electrooptic crystals and transparent ferroelectric ceramics is given by

$$\Delta n = -1/2(n^3 s_{13} E_3^2) \tag{4.15}$$

where s_{13} is the Kerr electrooptic coefficient, sometimes called the quadratic electrooptic coefficient.

TABLE 4.5 Electrooptic Properties of Ferroelectric Materials

Material	Point Group	r_{ij} (10^{-12} m/V)	V_π^T (kV)	V_π^L (kV)	λ (μm)	n	TR (μm)
BaTiO$_3$	4mm	$108\left(r_{33} - (n_a/n_c)^3\, r_{13}^T\right)$	–	–	0.546	2.44 (n_o)	0.45–0.7
		$23\left(r_{33} - (n_a/n_c)^3\, r_{13}^S\right)$	–	–		2.37 (n_e)	
		1640 (r_{51}^T)	–	–			
		820 (r_{51}^S)	–	–			
			0.48	–	0.67	2.42 (n_o)	
						2.36 (n_e)	
KNbO$_3$	mm2	64 (r_{33}^T)	–	–		2.279 (n_1)	
		28 (r_{13}^T)	–	–		2.329 (n_2)	
		380 (r_{42}^T)	–	–		2.167 (n_3)	
		25 (r_{33}^S)	–	–			
		10 (r_{13}^S)	–	–			
KTN(65/35)	4mm	500 ($|r_{33} - r_{13}|$)	0.11	–	0.546	2.318 (n_o)	0.4–6
		16000 (r_{51}, r_{42})	–	–		2.27 (n_e)	
PLZT(La/Zr/Ti):							
7/62/38	6mm	433 (r_c)¶¶	~0.07	–	0.546	2.5–3.0	
8/65/35 (10μm)††		523¶(r_c')§§	–	–	–	–	–
8/65/35 (3μm)††		612¶ (r_c')§§	–	–	–	–	–
LiNbO$_3$	3m	8.6 (r_{13}^S)	–	–	0.633	2.286 (n_o)	0.4–5
		30.8 (r_{33}^S)	–	–		2.200 (n_e)	
		28 (r_{51}^S)	2.8‡	–			
		3.4 (r_{22}^S)	–	–			
		7 (r_{22}^T)	–	–			
		21 ($r_c'^S$)§§	–	–			
		19 ($r_c'^T$)§§	–	–			
LiTaO$_3$	3m	7 (r_{13}^S)	2.7‡	–	0.633	2.176 (n_o)	0.4–2.9
		30.3 (r_{33}^S)	–	–		2.180 (n_e)	and
		20 (r_{51}^S)	–	–			3.2–4.0
		24 ($r_c'^S$)§§	–	–			
Sr$_{0.25}$Ba$_{0.75}$Nb$_2$O$_6$	4mm	41 ($r_c'^S$)§§	1.34‡	–	0.633	2.3144 (n_o)	–
						2.2596 (n_e)	
Sr$_{0.50}$Ba$_{0.50}$Nb$_2$O$_6$	4mm	205($r_c'^T$)§§	0.25‡	–	0.633	2.3123 (n_o)	–
		90 ($r_c'^S$)§§	–	–		2.2734 (n_e)	
Sr$_{0.75}$Ba$_{0.25}$Nb$_2$O$_6$	4mm	1300 (r_{33}^T)	0.037‡	–	0.633	2.3117 (n_o)	–
		66 (r_{13}^T)	–	–		2.2987 (n_e)	
		40 (r_{51}^T)	–	–			
		1400 ($r_c'^T$)§§	–	–			
		1066 ($r_c'^S$)§§	–	–			
Ba$_2$NaNb$_5$O$_{15}$	mm2	370 ($|n_3^3 r_{33} - n_1^3 r_{14}|$)	1.72¶	–	0.633	2.326 (n_1)	0.4–5
		400 ($|n_3^3 r_{33} - n_2^3 r_{23}|$)	1.57§	–		2.324 (n_2)	
						2.221 (n_3)	
K$_3$Li$_2$Nb$_5$O$_{15}$	4mm	790 ($|n_e^3 r_{33}|$)	0.93‡	–	0.633	2.277 (n_o)	0.4–5
		105 ($|n_o^3 r_{13}|$)	–	–		2.163 (n_e)	
KSr$_2$Nb$_5$O$_6$	4mm	130 (r_c')§§	0.5‡	–	0.633	2.28 (n_e)	–
		220 (r_{33})	–	–			
		130 (r_{11})	–	–			
KNSBN	4mm	59 (r_c)	0.864	–	0.633	2.3066 (n_o)	
($x = 0.5$, $y = 0.6$)§§						2.2490 (n_e)	
KH$_2$PO$_4$ (KDP)	$\bar{4}2m$	−10.5 (r_{63}^T)	17.6	8.8	0.656	1.5064 (n_o)	0.2–1.55
		9.7 (r_{63}^S)	19.1	9.6		1.4664 (n_e)	
		8.6 (r_{41}^T)	22.4	11.2			
		−10.3 ± 0.1 (r_{63}^T)	–	7.65	0.546	1.5120 (n_o)	
		8.77 ± 0.14 (r_{41}^T)	–	–		1.4683 (n_e)	
KD$_2$PO$_4$ (KD*P)	$\bar{4}2m$	−26.4 (r_{63}^T)	7.2	3.6	0.546	1.5079 (n_o)	0.2–2.15
		17.2 (r_{63}^S)	11.1	5.6		1.4683 (n_e)	
		8.8 (r_{41}^T)	23.9	12.0			

TABLE 4.5 Electrooptic Properties of Ferroelectric Materials—*(continued)*

Material	Point Group	r_{ij} (10^{-12}m/V)	V_π^T(kV)	V_π^L(kV)	λ (μm)	n	TR (μm)
NH$_4$H$_2$PO$_4$ (ADP)	$\bar{4}2m$	-8.5 (r_{63}^T)	21.2	10.6	0.656	1.52098 (n_o)	0.19–1.4
		5.5 (r_{63}^S)	32.8	10.4		1.5721 (n_e)	
		24.5 (r_{41}^T)	7.74	—			
KH$_2$AsO$_4$ (KDA)	$\bar{4}2m$	-10.9 (r_{63}^T)	17.0	18.5	0.656	1.5632 (n_o)	0.246–?
		12.5 (r_{41}^T)	15.5	7.8		1.5721 (n_e)	
		-10.9 ± 0.1 (r_{63}^T)	—	6.43	0.546	1.5707 (n_o)	
		12.5 ± 0.4 (r_{41}^T)	—	—		1.5206 (n_e)	
RbH$_2$PO$_4$ (RDA)	$\bar{4}2m$	13.0 (r_{63}^T)	14.3	7.2	0.656	1.56 (n_o)	
Pb$_5$Ge$_3$O$_{11}$	3	5.3 (r_{11}^T)	—	—	0.633?	2.16 (n_o)	
		2 ± 0.5 (r_{22}^T)	—	—		2.13 (n_e)	
		5.3 ± 0.4 (r_c^T)	—	—			
KTiOPO$_4$ (KTP)	$mm2$	9.5 ± 0.5 (r_{13})	—	—	0.633	1.7634 (n_1)	0.35–4.5
		15.7 ± 0.8 (r_{23})	—	—		1.7717 (n_2)	
		36.3 ± 1.8 (r_{33})	—	—		1.8639 (n_3)	
		7.3 ± 0.7 (r_{51})	—	—			
		9.3 ± 0.9 (r_{42})	—	—			

†r_{ij}: linear electrooptice coefficients at constant stress (r_{ij}^T) and constant strain (r_{ij}^S); V_π^T, V_π^L: transverse and logitudinal half-wave voltage; λ:wavelength; n:refractive index; TR:transmission range.

‡$E \parallel x_3$.

¶$E \parallel x_3$ and beam $\parallel x_2$.

§$E \parallel x_3$ and beam $\parallel x_3$.

§§$r_c' = r_{33}(n_e \mid n_o)^3 - r_{13}$

¶¶$r_c' = r_{33}(n_3 \mid n_1)^3 - r_{13}$

††Grain size.

Source: Xu, Y. *Ferroelectric Materials And Their Applications*, Elsevier Science Publishers B.V., Amsterdam, The Netherlands, 1991. With permission.

The quadratic electrooptic coefficients of some perovskite-type materials can be found in (Xu, 1991).

Nonlinear Optical Properties

For a second-order nonlinear optical effect, the instantaneous polarization P_i can be expressed in terms of the Fourier components of the electric field $E_i(\omega)$ by the basic equation

$$P_i(\omega_1 \pm \omega_2) = \sum_{jk} \epsilon_o d_{ijk} E_j(\omega_1) E_k(\omega_2) \tag{4.16}$$

where the third-rank tensor d_{ijk} has second-order nonlinear dielectric susceptibility, also called a nonlinear optical coefficient. The second-order nonlinear optical effects resulting from Eq. (4.16) are the optical sum-frequency generation or difference-frequency generation effect for $\omega_1 \neq \omega_2$ and the second-harmonic generation or optical rectification effect, respectively, for $\omega_1 = \omega_2$. A variety of differences in definition and notations of second-order nonlinear optical susceptibility can be found in the literature. For example, d_{ijk} can be equal to $\chi_{ijk}^{(2)}$ or, more often, equal to $1/2\chi_{ijk}^{(2)}$ depending on the definitions. Detailed discussions can be found in Chapter 12.

Similar to electrooptic coefficients, using simplified subscript, index d_{ijk} is contracted to $d_{i\alpha}$. The latter has only 18 components. According to the crystal symmetry, the number of independent components can be further reduced. The contracted d_{ij} tensor for some point groups is show following.

Ferroelectric Materials

Point group $\bar{4}2m$

$$\begin{pmatrix} 0 & 0 & 0 & d_{14} & 0 & 0 \\ 0 & 0 & 0 & 0 & d_{14} & 0 \\ 0 & 0 & 0 & 0 & 0 & d_{36} \end{pmatrix}$$

Point group $4mm$

$$\begin{pmatrix} 0 & 0 & 0 & 0 & d_{15} & 0 \\ 0 & 0 & 0 & d_{24} & 0 & 0 \\ d_{31} & d_{31} & d_{33} & 0 & 0 & 0 \end{pmatrix}$$

Point group $3m$

$$\begin{pmatrix} 0 & 0 & 0 & 0 & d_{15} & -d_{22} \\ -d_{22} & d_{22} & 0 & d_{15} & 0 & 0 \\ d_{31} & d_{31} & d_{33} & 0 & 0 & 0 \end{pmatrix}$$

Point group $mm2$

$$\begin{pmatrix} 0 & 0 & 0 & 0 & d_{15} & 0 \\ 0 & 0 & 0 & d_{24} & 0 & 0 \\ d_{31} & d_{32} & d_{33} & 0 & 0 & 0 \end{pmatrix}$$

Nonlinear optic coefficients of some ferroelectric crystals are listed in Table 4.6.

To obtain a significant second-harmonic wave, the phase velocity of the fundamental-frequency radiation must be equal to that of the second-harmonic radiation. This means that the refractive index at the fundamental wavelength must be equal to that at half the fundamental wavelength.

In the normal dispersion region, we usually have $n(2\omega) > n(\omega)$. Fortunately, however, by using the different phase velocities of ordinary and extraordinary rays propagating in special directions in a birefringent crystal, the condition of $n(2\omega) = n(\omega)$ can be satisfied. When the medium is a negative uniaxial crystal ($n_o > n_e$), the ordinary ray is used as the incident beam (the fundamental-frequency beam) and the extraordinary ray is used for frequency doubling. The condition of phase matching is satisfied when the angle between the normal to the wavefront and the optical axis of the crystal θ_k satisfies the following equation:

$$n_o^{2\omega} n_e^{2\omega} / [(n_o^{2\omega})^2 \sin^2\theta_k + (n_e^{2\omega})^2 \cos^2\theta_k]^{1/2} = n_o^{\omega}, \tag{4.17}$$

The solution of this equation is given by

$$\sin^2\theta_m = \frac{(n_e^{2\omega})^2[(n_o^{2\omega})^2 - (n_o^{\omega})^2]}{(n_o^{\omega})^2[(n_o^{2\omega})^2 - (n_e^{2\omega})^2]}, \tag{4.18}$$

where θ_m is usually called the phase-matching angle, i.e., the angle at which phase matching is satisfied. For the ideal case of monochromic Gaussian laser beams, neglecting attenuation, the second-harmonic intensity is given by

$$I(2\omega) = \frac{2\omega^2}{\pi\epsilon_o n_\omega^2 n_{2\omega} w_o} \cdot d_{eff}^2 \cdot I^2(\omega) \cdot l^2 \cdot \frac{\sin^2\left(\frac{\pi l}{2l_c}\right)}{\left(\frac{\pi l}{2l_c}\right)^2} \tag{4.19}$$

where l is the interaction length, w_o the beam waist, c the speed of light, and l_c the SHG coherence length ($l_c = \lambda/[4(n_{2\omega} - n_\omega)]$). It can be seen from Eq. (4.19) that the second-harmonic intensity is proportional to the square of the effective d_{ij} coefficient of the material.

Photorefractive Properties

Photorefraction of ferroelectric materials depends upon the exposure time of the light beam, the light intensity, and the material temperature. The mechanism of the photorefractive effect in ferroelectric crystals can be described as follows: when the crystal is illuminated, donor ions (structural defects and impurity ions) are ionized and photoelectrons are ejected. These photocarriers (mainly electrons), in turn, change the internal polarization field in the ferroelectric crystal and, subsequently, drift toward the high-field direction in this altered internal field. Carriers

TABLE 4.6 Second-Order Nonlinear Optical Properties (second harmonic coefficients) of Ferroelectric Materials†

Material	Point Group	d_{ij} (10^{-12} m/V)	λ (μm)	n_ω	$n_{2\omega}$
BaTiO$_3$	4mm	$d_{15} = 17.0 \pm 1.8$	1.0642	2.3175	2.4760
		$d_{31} = 15.7 \pm 1.8$	1.0642	2.3379	2.4128
		$d_{33} = 6.8 \pm 1.0$	1.0642	2.2970	2.4128
KNbO$_3$	mm2	$d_{31} = -12.88 \pm 1.03$	1.0642	2.2574	2.2029
		$d_{32} = +11.34 \pm 1.03$	1.0642	2.2200	2.2029
		$d_{33} = -19.58 \pm 1.03$	1.0642	2.1196	2.2029
		$d_{15} = -12.36 \pm 2.0$	1.0642	2.1885	2.3807
		$d_{24} = 11.85 \pm 2.0$	1.0642	2.1698	2.3224
LiNbO$_3$ (Li/Nb=0.946) (congruent-melt)	3m	$d_{31} = -4.76$	1.064	–	–
		$d_{22} = +2.3$	1.064	–	–
		$d_{33} = -29.7$	1.064	–	–
LiTaO$_3$	3m	$d_{31} = -1.07 \pm 0.2$	1.0582	2.1366	2.2089
		$d_{22} = +1.76 \pm 0.2$	1.0582	2.1366	2.2043
		$d_{33} = -16.4 \pm 2$	1.0582	2.1406	2.2089
SrBaNb$_5$O$_{15}$	4mm	$d_{31} = 4.31 \pm 1.32$	1.0642	2.2506	2.3092
		$d_{33} = 11.3 \pm 3.3$	1.0642	2.2138	2.3092
		$d_{33} = 5.98 \pm 2$	1.0642	2.2322	2.3583
Ba$_2$NaNb$_5$O$_{15}$	mm2	$d_{31} = -12.8 \pm 1.28$	1.0642	2.2570	2.2502
		$d_{32} = -12.8 \pm 0.64$	1.0642	2.2584	2.2502
		$d_{33} = -17.6 \pm 1.28$	1.0642	2.1700	2.2502
		$d_{15} = -12.8 \pm 0.64$	1.0642	2.2135	2.3656
		$d_{24} = -12.8 \pm 0.64$	1.0642	2.2142	2.3673
K$_3$Li$_2$Nb$_5$O$_{15}$	4mm	$d_{31} = 6.18 \pm 1.28$	1.0642	2.2057	2.1980
		$d_{33} = 11.2 \pm 1.6$	1.0642	2.1113	2.1980
		$d_{31} = 5.45 \pm 0.54$	1.0642	2.1585	2.3297
PbNb$_4$O$_{11}$	mm2	$d_{31} = +6.5 \pm 0.97$	1.0642	2.2979	2.4396
		$d_{32} = -5.87 \pm 0.88$	1.0642	2.3010	2.4396
		$d_{33} = -8.88 \pm 1.32$	1.0642	2.3254	2.4396
		$d_{15} = +5.89 \pm 0.64$	1.0642	2.3115	2.4113
KH$_2$PO$_4$ (KDP)	$\bar{4}$2m	$d_{14} = 0.43 \pm 0.03$	0.6943	1.4856	1.5335
		$d_{14} = 0.44 \pm 0.003$	1.064	1.5124	1.4768
		$d_{36} = 0.47 \pm 15\%$	0.6943	1.5058	1.4874
		$d_{36} = 0.44$	1.064	1.4942	1.4708
KD$_2$PO$_4$ (KD*P)	$\bar{4}$2m	$d_{14} = 0.342 \pm 0.02$	0.6943	1.4830	1.5285
		$d_{14} = 0.37 \pm 0.012$	1.0582	1.4789	1.5085
		$d_{36} = 0.34 \pm 0.01$	0.6943	1.5022	1.4855
		$d_{36} = 0.38 \pm 0.016$	1.0582	1.4978	1.4689
NH$_4$H$_2$PO$_4$ (ADP)	$\bar{4}$2m	$d_{14} = 0.40 \pm 0.02$	0.6943	1.4973	1.5498
		$d_{14} = 0.40 \pm 0.02$	1.0582	1.4874	1.5277
		$d_{36} = 0.42 \pm 0.03$	0.6943	1.5193	1.5004
		$d_{36} = 0.41 \pm 0.02$	1.0582	1.5067	1.4816
ND$_4$D$_2$PO$_4$ (AD*P)	$\bar{4}$2m	$d_{36} = 0.495 \pm 0.07$	0.6943	1.5138	1.4926
KH$_2$AsO$_4$ (KDA)	$\bar{4}$2m	$d_{14} = 0.39 \pm 0.045$	0.6943	1.538	1.606
		$d_{14} = 0.46 \pm 0.02$	1.06	1.554	1.521
		$d_{36} = 0.45 \pm 0.045$	0.6943	1.562	1.549
		$d_{36} = 0.43 \pm 0.025$	1.06	1.531	1.572
RbH$_2$PO$_4$ (RDP)	$\bar{4}$2m	$d_{14} = 0.49 \pm 0.07$	1.0642	1.4813	1.5106
		$d_{36} = 0.414 \pm 0.045$	0.6943	1.4969	1.5020
		$d_{36} = 0.38 \pm 0.04$	1.0642	1.4926	1.4811
RbH$_2$AsO$_4$ (RDA)	$\bar{4}$2m	$d_{36} = 0.47 \pm 0.05$	0.6943	1.5543	1.5531
(NH$_2$CH$_2$COOH)$_3$·H$_2$SO$_4$ (TGS)	2	$d_{23} = 0.32$	0.6943	1.567	1.618
Gd$_2$(MoO$_4$)$_3$	mm2	$d_{31} = -2.49 \pm 0.37$	1.064	1.8146	1.9102
		$d_{32} = +2.42 \pm 0.36$	1.064	1.8142	1.9102
		$d_{33} = -0.044 \pm 0.008$	1.064	1.8637	1.9102

TABLE 4.6 Second-Order Nonlinear Optical Properties (second harmonic coefficients) of Ferroelectric Materials†—(continued)

Material	Point Group	d_{ij} (10^{-12} m/V)	λ (μm)	n_ω	$n_{2\omega}$
		$d_{15} = -2.62 \pm 0.4$	1.064	1.8386	1.8549
		$d_{24} = +2.58 \pm 0.39$	1.064	1.8384	1.8545
Tb$_2$(MoO$_4$)$_3$	mm2	$d_{31} = -2.99 \pm 0.35$	1.064	1.8226	1.9185
		$d_{32} = +2.22 \pm 0.33$	1.064	1,8222	1.9185
		$d_{33} = -0.11 \pm 0.03$	1.064	1.8704	1.9185
		$d_{15} = -2.52 \pm 0.38$	1.064	1.8459	1.8649
		$d_{24} = +2.55 \pm 0.35$	1.064	1.8458	1.8645
Pb$_5$Ge$_3$O$_{11}$	3	$d_{11} = 0.96 \pm 0.16$	1.064	–	–
		$d_{22} = -2.1 \pm 0.3$	1.064	–	–
		$d_{31} = +0.51 \pm 0.07$	1.064	–	–
		$d_{33} = -0.79 \pm 0.12$	1.064	–	–
K$_x$Rb$_{1-x}$TiOPO$_4$	mm2	$d_{31} = 6.5$	1.064	1.743	1.891
(KTP will small amount of		$d_{32} = 5.0$	1.064	1.753	1.891
Rb)		$d_{33} = 13.7$	1.064	1.837	1.891
		$d_{15} = 6.1$	1.064	1.790	1.781
		$d_{24} = 7.6$	1.064	1.795	1.790

†d_{ij}: second harmonic coefficients; λ: wavelength; n_ω: refractive index at fundamental wavelength; $n_{2\omega}$: refractive index at second-harmonic wavelength.

Source: Singh, S., Nonlinear Optical Properties/Radiation Damage in CRC *Handbook of Laser Science and Technology,* Vol. 3., Optical Materials, Part I, M. J. Weber, Ed. CRC Press, Boca Raton, FL, 1986, p. 64–90.

propagate in the direction of the polarization axis, and, when they reach the unilluminated regions, they are trapped again. Thereby, an inhomogeneous electric field is built up at specific locations in the crystal. Due to the electrooptical effect, the refractive indices at these locations in the crystal are changed by this space-charge field.

The mechanism of photoexcitation in different materials is not the same, and different photofrefractive species are involved. Intrinsic defects, impurities, and dopants may affect the photofractve properties dramatically. Doping of multivalent ions has a dramatic effect on the photorefractivity of ferroelectric crystals. For example, ions, such as $Fe^{2+}-Fe^{3+}$, Cu^+-Cu^{2+}, $Mn^{2+}-Mn^{3+}$, $Ce^{3+}-Ce^{4+}$, etc., were found to cause photorefractive effects in ferroelectric crystals. For Fe^{2+}-doped LiNbO$_3$ crystals, one electron of Fe^{2+} ion can be excited to the conduction band by light illumination. The reaction is

$$Fe^{2+} + h\nu \rightarrow Fe^{3+} + e^- \tag{4.20}$$

where Fe^{3+} ions plays the role of a trapping center for one electron. Experiments show that the doped BaTiO$_3$ crystals are more sensitive to photorefractive effects than the pure crystals. For example, the holographic efficiency reached the maximum for 0.07 at.% Fe-doped BaTiO$_3$ (Godefroy et al., 1986). The dopants act as donor-acceptor traps via intervalent exchange, such as $Fe^{2+} \leftrightarrow Fe^{3+}$ (Staebler and Philips, 1974).

Defects in a crystal depend on a number of processing factors, such as starting materials, crystal growth conditions, and the chemical and poling treatments after growth. The study of the defect chemistry of photorefractive materials is important in improving material performance. Defect properties and the photorefractive effect in BaTiO$_3$ are discussed in (Godefroy, 1989; Jullien et al., 1989; and Wechsler et al., 1991). A review of defect chemistry of nonlinear oxide crystals can be found in ref (Morris, 1991).

Photorefraction of ferroelectric crystals depends upon the exposure time t, the light intensity I, and the sample temperature T. This phenomenon has been investigated by the polarized-optical method (Chen, 1969) with automatic compensation for the phase difference between ordinary and extraordinary waves. Time dependence of the birefringence change $\delta \Delta n = \delta (n_e - n_o)$ [or $\Delta B = \Delta(n_e - n_o)$] has been recorded and the steady-state values of photorefraction $\delta \Delta n_s$ have been determined.

Photorefraction in LiNbO$_3$ and LiTaO$_3$ crystal originates in the bulk photovoltaic effect, when neglecting other comparatively weak effects. In the illuminated region, there arises a photovoltaic current $G\alpha I$ determined by the Glass constant G, the absorbed light energy αI (α is the absorption coefficient), and the reverse current σE_i, where σ is the conductivity of the illuminated region. The current density is given by

$$j = G\alpha I - \sigma E_i. \tag{4.21}$$

The E_i field causes the birefringence change $\delta \Delta n$ via the linear electrooptic effect (Pöckels effect); hence, E_i can be found by measuring $\delta \Delta n$:

$$E_i = \delta \Delta n / m \tag{4.22}$$

where $m = 1/2(r_{33}n_e^3 - r_{13}n_o^3)$ (r_{33} and r_{13} are the electrooptic coefficients, where n_e and n_o are the extraordinary and ordinary refractive indices). From Eq. 4.21, Poisson's equation, and the the continuity equation, the kinetics of photorefractivity can be described by the exponent with a Maxwell's relaxation time $\tau_M = \kappa \epsilon_o / \sigma$ (κ is the relative dielectric constant of the crystal and ϵ_0 is the permittivity of vacuum), i.e.,

$$\delta \Delta n = \delta \Delta n_s [1 - \exp(-t/\tau_M)] \tag{4.23}$$

where

$$\delta \Delta n_s = mG\alpha I/\sigma \tag{4.24}$$

The rates of increase of both $\delta \Delta n$ and $\delta \Delta n_s$ depend on I. The conductivity of the illuminated region can be expressed as $\sigma = \sigma_d + \sigma_{ph}$, where σ_d is the dark conductivity and $\sigma_{ph} = \beta I$ is the photoconductivity (5.92). Then, from Eq. 4.24,

$$\delta \Delta n_s = mG\alpha I/(\sigma_d + \beta I) \tag{4.25}$$

As seen from Eq. 4.25, the linear increase of $\delta \Delta n_s$ versus I takes place when $\sigma_d \gg \sigma_{ph}$. At large I, when $\sigma_d \ll \sigma_{ph}$, it follows that $\delta \Delta n_s =$ constant.

The relaxation time $\tau_{ph} = \epsilon \epsilon_o (\sigma_d + \sigma_{ph})^{-1}$, after which the steady state of photorefraction is reached, depends on both σ_d and σ_{ph}, whereas the photorefractive dark relaxation time $\tau_d = \epsilon \epsilon_o \sigma_d^{-1}$ depends only on σ_d. To compare τ_{ph} with τ_d, the ratio of σ_d and σ_{ph} can be estimated.

A useful method of characterizing photorefractive behavior is two-beam coupling. When the refractive index grating pattern forms in a sample by the interference of the two coherent laser beams, Bragg self-diffraction occurs, leading to energy exchange between the two incident beams called attenuated beam I_a and pump beam I_p. The result is that the attenuated beam is amplified at the expense of the pumping beam. The transmission of the signal beam is given by

$$I_a(L)/I_a(0) = \frac{[I_a(0) + I_p(0)]\exp[(\Gamma - \alpha)L]}{[I_p(0) + I_a(0)\exp \Gamma L]} \tag{4.26}$$

where α is the absorption coefficient, L is the interaction length, and Γ is the effective exponential gain coefficient given by

$$\Gamma = [2\pi n^3/\lambda \cos \theta] r_{\text{eff}} \, \text{Im}(E_{sc}) \cos 2\theta \tag{4.27}$$

where λ is the wavelength of the incidental beams, n is the refractive index, r_{eff} is the effective

Ferroelectric Materials

electrooptic coefficient, Im(E_{sc}) is the imaginary part of the amplitude of the space charge electric field, and 2θ is the angle between the incidental beams.

Pyroelectric Effect

The pyroelectric effect can be found only in polar crystals. This effect can be used for detecting infrared light. The mechanism of pyroelectric infrared detection is explained as follows: Usually, the surface charge (due to spontaneous polarization) of a pyroelectric crystal is neutralized by free carriers from the interior and exterior of the crystal. The average time needed for interior free carriers to neutralize surface charge is $\tau = \epsilon/\sigma$, where ϵ is the dielectric constant and σ is the conductivity of the crystal. The value of τ ranges from 1 second to 1000 seconds for most of the pyroelectric materials. When an infrared beam with modulated frequency f is irradiated on a pyroelectric material, temperature, spontaneous polarization, and, thus, the surface charge of the material will be modulated with frequency f. If $f > 1/\tau$, free carriers from the interior of the material do not have enough time to compensate for the alternating surface charge. An alternating voltage perpendicular to the direction of spontaneous polarization will arise across the crystal. When the crystal is connected to a load R by electrodes on the surfaces, the signal voltage can be expressed as

$$\Delta V = AR\left(\frac{dP_s}{dt}\right) = AR\left(\frac{dP_s}{dT}\right)\left(\frac{dT}{dt}\right), \qquad (4.28)$$

where A is the area of electrodes, dP_s/dt is the spontaneous polarization change rate dT/dt is temperature change rate, and dP_s/dT is, by definition, the pyroelectric coefficient p. When the temperature change ΔT is small, the pyroelectric coefficient can be considered to be a constant. Then, the magnitude of signal ΔV is proportional to dT/dt.

Ideal materials for pyroelectric applications should have large temperature change rates, which means materials should have a large IR aborption coefficient, a small specific heat, low density, and small physical dimensions. High Curie temperatures, high pyroelectric coefficients p as well as low dielectric constants and low dielectric loss tan δ, especially for high frequency applications, are desirable material properties for pyroelectric infrared detectors. Table 4.7 lists pyroelectric properties of some ferroelectric materials.

TABLE 4.7 Pyroelectric Properties of Some Ferroelectric Materials

Materials	Pyroelectric Coefficient p (nC/cm^{2}K)	Dielectric Constant ϵ/ϵ_o	Specific Heat (J/cm^{3}K)	Thermal Conductivity (W/cmK)
BaTiO$_3$	20	160 (ϵ_{33}/ϵ_o) 4100 (ϵ_{11}/ϵ_o)	3.0	9×10^{-3}
PbTiO$_3$ (poled ceramics)	60	~200	3.1	–
PLZT (8/65/35) (poled ceramics)	170	~3800	2.6	–
LiNbO$_3$	4	30 (ϵ_{33}/ϵ_o) 75 (ϵ_{11}/ϵ_o)	2.8	–
LiTaO$_3$	19	44 (ϵ_{33}/ϵ_o)	3.19	–
Sr$_{0.5}$Ba$_{0.5}$Nb$_2$O$_6$	60	~500	2.34	–
Sr$_{0.75}$Ba$_{0.25}$Nb$_2$O$_6$	310	~1700	–	–
TGS (or modified TGS)	20–35	25–50	1.6–2.48	6.8×10^{-3}
TGS/15% TGSe (mixture crystals)	50	28	–	–
Pb$_5$Ge$_3$O$_{11}$	9.5	40 (ϵ_{33}/ϵ_o) 22 (ϵ_{11}/ϵ_o)	–	–

Source: Xu, Y. *Ferroelectrc Materials And Their Applications*, p. 30. Elsevier Science Publishers B. V., Amsterdam, The Netherlands, 1991. With permission.

Ferroelectric Structure-Related Optical Properties

Unique optical properties are found in some ferroelectric materials because of their material structures. Transparent PLZT ceramics are notable examples to illustrate the close relationship between materiel structure and optical properties. In PLZT ceramics, optical properties are strongly dependent upon the compositions (crystal structure) as well as microstructures (grain morphology). The miscellaneous inorganic ferroelectric crystals mentioned before have rather unique optical properties. Their variable optical anisotropic properties make them extremely attractive in applications, such as memory and display devices, electrooptic modulators, etc. For example, in $Bi_4Ti_3O_{12}$, the a- and c-axis components of polarization may be reversed independently through 180° so that the optical indicatrix may be rotated in the a–c plane in a variety of ways. The 180° reversal of ferroelectric domain in $Gd_2(MoO_4)_3$ allows a rotation of the optical indicatrix through 90° in the plane perpendicular to the polarization vector, thereby producing a large change of birefringence for light propagating along the polar axis. In $Pb_5Ge_3O_{11}$, the sense of spontaneous polarization is directly related to the sense of optical rotary power. Optical properties of the above four ferroelectric materials are discussed following.

PLZT

The most outstanding feature of PLZT materials is their high optical transparency. The transparency is a function of La concentration as well as the Zr/Ti ratio. The maximum transparency is achieve with compositions along the FE-PE phase boundary shown in Fig. 4.2. After poling, a PLZT ceramic with a composition in the ferroelectric phase region exhibit piezoelectricity and optical birefringence. The eletrooptic properties of transparent PLZT ceramics are closely related to the ferroelectric domain states which, in turn, depend on the compositions of PLZT ceramics (Land and Holland, 1970, Land, and Thacher, 1969; Thacher and Land, 1971). Typical P-E hysteresis loops obtained from different PLZT ceramic compositions are shown in Fig. 4.4. Four different shapes of the loops have the following characters: (a) ferroelectric linear memory. Normal linear electrooptic effects are generally observed in these kinds of PLZT material. The polarization and electrical domains cannot be reversed due to high E_c in this class; (b) ferroelectric switchable

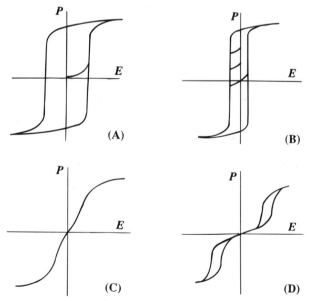

FIGURE 4.4 Typical hysteresis loops obtained from different PLZT compositions. (A) FE linear memory, (B) FE switchable memory, (C) SFE nonmemory, and (D) AFE nonmemory (after Haertling, 1987).

memory. The polarization and electrical domains can be reversed by an external electric field due to the low E_c of the materials; (c) slim-loop nonmemory. The slim loop ferroelectrics do not possess permenent polarization but do exhibit substantial induced polarization when subjected to an electric field. This electrically induced polarization effect is actually an electrically induced phase change from the paraelectric, optically isotropic state to the ferroelectric, optically anisotropic state. The electrically induced birefringence of such a material is a quadratic function of the applied field. The quadratic electrooptic effect is, thus, found in this class of materials; (d) antiferroelectric nonmemory. The antiferroelectrics are essentially nonpolar and nonferroelectric in their natural state, but revert to a ferroelectric state when subjected to a high external electric field. Similar to the slim-loop ferroelectrics electrooptic effects in these materials are only noticeable while the electric field is on.

Domains are important in ferroelectric and electrooptic ceramics because they affect electrical switching behavior and light scattering characteristics (refraction at domain walls) according to their size, number, and wall mobility under stress. No domains are seen in a material exhibiting quadratic electrooptic effect, because it is cubic and nonferroelectric in its normal state.

The electrooptic properties of PLZT materials are closely related to their ferroelectric properties. It was found that modification of the ferroelectric polarization by an electric field causes a change in the optical properties of the ceramics. Moreover, the magnitude of the observed electrooptic effect is dependent on both the strength and the direction of the electric field. In this sense, we say that the optical properties of a ferroelectric ceramic are electrically controllable. There are three unique, electrically controlled optical phenomena found in transparent ferroelectric PLZT ceramics related to the domain structure: (1) electrically controllable birefringence; (2) electrically controllable light scattering; and (3) electrically controllable surface deformation. These are manifestations of the changes in the optical properties caused by a change of P_r, i.e., a change in the state of alignment of the domains, which is controlled by an external field. In the case of ceramics having fine-grained structures (average single-domain grain size 1–2 μm), the main effect is a variable birefringence. In the case of ceramics having a coarse-grained structure (average grain size large than 3 μm), the main effect is a variable intensity of scattered light. The variable surface deformation effect exists in both fine-grained and coarsed-grained ceramics. Another interesting effect occurring in PLZT is the photoassisted domain-switching effect. We will explain these effects following.

(1) Electrically Controlled Birefringence. The principle involved in the electrical control of birefringence can be explained by Fig. 4.5. In the upper part of this figure is the normalized effective birefringence of a PLZT ceramic as a function of the partially remanent polarization P_r (when $E = 0$). Different values of P_r controlled by a negative pulse are shown in the lower part of the figure. In the figure, we can see that, as P_r changes from $+P_R$ to zero and then from zero to $-P_R$, Δn changes from the maximum to the minimum and, then, from minimum to maximum. It should be noted that the intensity of transmitted beam described by Eq. 4.11 is still valid when a light beam passes through transparent PLZT ceramic. However, the relationship between birefringence and the electric field is more complex than that in single crystals.

The behavior of electrically controlled birefringence is dependent on the composition of PLZT. The composition dependency is depicted in Fig. 4.6, where three regions, A, B, and C, divided by dot lines, exhibit different relationships between birefringence and electric field. The behavior of electrically controlled birefringence at different compositions includes memory effects (region A), linear effects (region B), and quadratic effects (region C). The three types of behavior of the electrically controlled birefringence as well as the ferroelectric hysteresis loop corresponding to regions A, B, and C are shown in Fig. 4.7.

The ceramics with composition in region A possess a low coercive field, a square hysteresis loop, and large piezoelectric and electrooptic coefficients. Important properties in this class include the effective birefringence at saturation polarization $\Delta n(P_R)$, the variation range of Δn, i.e.,

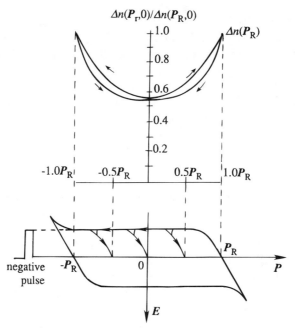

FIGURE 4.5 Relationship between the normalized effective birefringence and the remanent polarization P_r (when $E = 0$). Composition of PLZT: 2/65/35; grain size: 1 μm; light wavelength: 0.656 μm; $\Delta n\ (P_r = 0) = 0.012$.

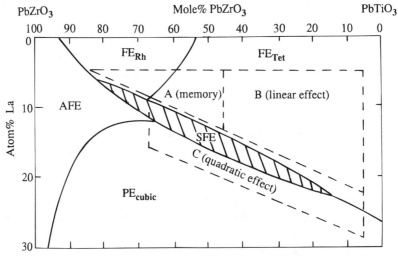

FIGURE 4.6 Various regions possessing different electrooptic characteristics denoted in the phase diagram of PLZT (after Haertling, 1987).

$\Delta n(P_R) - \Delta n(P_b)$, where $\Delta n(P_b)$ is the value of Δn at the lowest point in the Δn versus P curve and the variation range of remanent polarization, i.e., $P_R - P_b$. The electrooptic memory characteristics of several PLZT materials with different compositions, processing conditions, and grain sizes can be found in [Land and Thacher,1969; Thacher and Land 1971.].

The ceramics with composition in region B possess a high coercive field and exhibit a linear electrooptic effect at saturation polarization. A transverse effective linear electrooptic coefficient r_c can be used to describe the linear electrooptic effect. When the direction of the external electric field is perpendicular to the propagation direction of the light beam, r_c is determined by

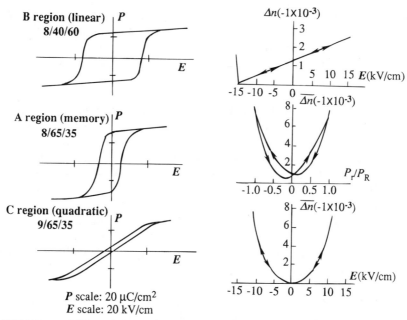

FIGURE 4.7 Different hysteresis loops and electrooptic characteristics of PLZT with compositions corresponding to the A, B, and C regions in Fig. 4.6 (after Haertling, 1987).

$$\Delta n = -\frac{1}{2} n_1^3 r_c E_3 \tag{4.29}$$

where n_1 is the refractive index in a direction perpendicular to the external field E_3.

The ceramics with composition in the region C have a very small (almost zero) coercive field. When an external electric field is applied to such a material, a ferroelectric phase exhibiting birefringence is induced. When the external electric field is removed, the ceramic returns to its isotropic state (no birefringence). In addition, the ceramics in region C exhibit a quadratic electrooptic effect, i.e., the birefringence Δn is proportional to E^2. Using the transverse effective quadratic electrooptic coefficient R or g, this effect can be described by the following equation:

$$\Delta n = -\frac{1}{2} n_1^3 R E_3^2, \tag{4.30}$$

or

$$\Delta n = -\frac{1}{2} n_1^3 g P_3^2$$

Linear and quadratic electrooptic coeffients of PLZT ceramics with different compositions can be found in (Haertling, 1987)

(2) Electrically Controlled Scattering. Transparent ferroelectric ceramics with coarse grains (e.g., $Pb_{1-x}Bi_x(Zr,Ti)O_3$ ceramics with grain size larger than 2 μm or PLZT ceramics with grain size larger than 3–4 μm) have electrically controlled light scattering (Land and Holland 1970). Contrary to birefringence, light scattering is preponderant in coarse-grained ceramics because a large number of scattering centers are formed by ferroelectric domains in a coarse-grained ceramic and these scattering centers disorder the light polarization state to dissolve birefringence. When

the direction of polarization is perpendicular to the surface of a ceramic specimen (parallel to the propagation direction of incident light), the maximum amount of light is permitted to pass through the ceramic; when the direction of the polarization is parallel to the speciment surface (perpendicular to the propagation direction of light), a very small amount of light is permitted to pass through the ceramic [Smith and Land 1972].

By the virtue of the ferro-antiferroelectric phase transition induced by an external field, another electrically controlled light scattering effect different from the previous one can be demonstrated. Using an electric field, the state of a PLZT ceramic can be changed from its original state without scattering (antiferroelectric phase) to a state with light scattering (ferroelectric state). The original state is restored by applying a reverse electric field. Only a small field (several kV/cm^2) is required to induce this electrically controlled light scattering and the accompanying phase transition (Kumada et al., 1974).

When a field is applied longitudinally (i.e., along the direction of light beam) to a PLZT material exhibiting a quadratic birefringence effect, an effect termed 'scattering depolarization' is observed (Haertling and McCampbell, 1972). This effect produced a depolarization of the incoming light, leading to the 'ON' state, when the field is applied, and to the 'OFF' state, when the field is removed.

(3) Electrically Controlled Surface Deformation. In PLZT ceramics with a rhombohedral crystal phase, a partial strain can be caused by partial domain switching, and relative deformation occurs on those locations of the surface where the domains have been switched. Consequently, uneven patterns on the surface are formed by these deformations. A light beam will be refracted or scattered by these uneven patterns on the surface (Land and Thacher, 1969; Thacher and Land, 1971).

Surface deformation can be caused by the strain due to domain reorientation or/and phase transition when the polarization is switched. As P_r increases (i.e., the number of switched domains increases), a relative strain S_3 changes from the minimum (where P_r is near zero) to the maximum value. The variation of S_3 is a function of the normalized remanent polarization P_r/P_R. Because the variation of P_r can be controlled by an electric field, the variation of surface deformation can also be controlled by the field.

(4) Photoassisted Domain-Switching Effect. Ferroelectric domains in a PLZT ceramic can be switched either by an external electric field or by an optical beam under an applied field. The later is called the photoassisted domain-switching effect. The mechanism of this effect is as follows: When a UV light of energy equal to or greater than the PLZT band gap of 3.35 eV is applied, carriers are photoexcited across the bandgap or from trapping centers in the bandgap to the conduction band. The photoexcited carriers are, then, diffused or drift under the influence of an applied field. Some of the carriers are moved beyond the penetration depth of light and are retrapped to establish a space-charge field resulting in a transient photocurrent. The space-charge field, in turn, modulates the applied field and assists in the domain-switching process by effectively lowering the E_c of the materials. Carriers remaining in the conduction band contribute to a steady-state photovoltaic current which is driven by the bulk photovoltaic effect. These two competing processes lead to various degrees of domain switching which is activated at different electric field levels when exposed to UV light of varying intensity. Thus, photographic images may be stored in a PLZT ceramic using this effect when an ordinary photographic negative is used to provide a variable intensity light source while an appropriate d.c. voltage is applied.

Bismuth Titanate

The unusual switching characteristic of polarization in bismuth titanate is applied in ferroelectric memory and display systems. The crystal is an optical biaxial crystal and exhibits a large anisotropy in the values of refractive indices with respect to the principal axes of the optical indicatrix $X = 2.5984$, $Y = 2.6094$, and $Z = 2.7000$, where both X and Y are perpendicular to the b-axis in the monoclinic lattic, the angle between the Y-direction and the c-axis is 29°, and the Z-direction is parallel to the b-axis. There is an angle of 50° between the two Y-directions of the two optical indicatrices, which are related to the two domains in which the c-components of polarization P_s are opposite. As the reversal of the c-component of P_s leads to a rotation of the optical indicatrix major axes (in the a–c plane) from +25° to −25° relative to the c-axis, domains in the crystal can be switched by an external field and the domain pattern can be displayed optically.

Gadolinium Molybdate

The use of the reversal characteristic of ferroelectricity-ferroelasticity as well the birefringence of β-GMO crystal has greatly contributed to developing electrooptic devices. β-GMO is optically transparent in the range 0.3–5 μm. At room temperature, the crystal is optically a biaxial crystal, and the optical indicatrix along the principle axes are $n_b = 1.8362 - 0.000204$, $n_b = 1.8362 + 0.000204$, and $n_c = 1.8888$ at the wavelength of 0.633 μm. Further, $n_c = 1.9065$, $n_b = 1.8518$, and the birefringence $\Delta n_{bc} = 0.0547$ at the wavelength of 0.546 μm. Reversal of the c-axis polarization interchanges the a-axis and b-axis with a resulting change of birefringence of $2\Delta n \approx 8 \times 10^{-4}$ along the c-axis. The a-axis and b-axis are interchanged when c is reversed across a domain wall. The domain wall can be moved by changing the applied voltage.

Lead Germanium Oxide

The reversal of the ferroelectric domain and the reversal of optical rotation in the lead germanium oxide crystal are of great interest. The two kinds of ferroelectric domains mentioned before with opposite optical rotations can be observed when polarized light is incident on the optical uniaxial $Pb_5Ge_3O_{11}$ crystal along the c-axis. When an external field is applied to the crystal to reverse the spontaneous polarization, a reversal of the optical rotation results. An optical rotating power of 5°35′ mm^{-1} about the c-axis of a $Pb_5Ge_3O_{11}$ crystal has been measured at room temperature at a laser wavelength of 0.6328 μm. In addition, the $Pb_5Ge_3O_{11}$ crystal exhibits a first-order electrogyration effect with optical rotatory power proportional to the spontaneous polarization (Uesu et al., 1990). Rotatory strength R is a result of the occurrence of P_s in the ferroelectric phase, and the spontaneous rotatory strength R is found to be proportional to P_s in the case of $Pb_5Ge_3O_{11}$. In the ferroelectric phase, the optical rotation of $Pb_5Ge_3O_{11}$ depends on temperature. In the paraelectric phase, optical rotation can be caused by an external electric field applied on the crystal; this effect is called 'electro-optical rotation'. The derivative of the optical rotational power ρ_3 with respect to the field E_3 along the c-axis ($d\rho_3/dE_3$) follows the Curie–Weiss law, with regard to its temperature dependence, and is proportional to the value of the permittivity ϵ_{33} (Konac et al., 1978). It is also known that displays of nonlinear optical effects and the signs of its nonlinear optical coefficients d_{31} and d_{33} can be changed when the spontaneous polarization is reversed (d_{31} and d_{33} vanish at temperatures above T_C). However, signs of the nonlinear optical coefficients d_{11} and d_{22} are independent of the P_s reversal, and they are constants above T_C (Miller et al., 1974).

4.4. Optical Applications of Ferroelectric Materials

Ferroelectric materials have many unique optical properties which make them very promising in various optical applications. In this section, optical applications of the ferroelectric materials discussed in section 4.2 are reviewed. These applications are grouped by their functional effects, i.e., pyroelectric, electrooptic, nonlinear optic, photorefractive, and ferroelectric structure-related effects. Material considerations are emphasized in the discussion. For readers needing more information and referencing of optical applications of ferroelectric materials, refer to the last section of this chapter for suggested sources.

Electrooptic Applications

Important optical devices, such as electroopic modulators, crystal light valves, and electrooptic light switches (including Q-switches), have been fabricated using linear (or nonlinear) electrooptic effects of electrooptic crystals. Ferroelectric crystals have long been recognized as one the most important material families in electrooptic applications because of their large electrooptic coefficients. The oxygen octahedra ferroelectrics, KDP and KTP families, are the important ferroelectric families for electrooptic applications. The electrooptic applications of PLZT ceramics are discussed in the above section because the applications are related to the structures and the compositions of PLZT ceramics.

General considerations of using a material for electrooptic (EO) modulators are the magnitude of half-wave voltage and power comsumption, optical transit-time limitations which affect the operating frequency, availability of large optical quality crystal, laser damage threshold, and optical transmission range. Material comparisons have been reviewed by Günter (Günter, 1987) in terms of the figure of merit for the half-wave voltage and by Kaminov and Turner (Kaminov and Turner, 1966) using the figure of merit with respect to minimizing the reactive driving power of an EO modulator. Other considerations are discussed in (Wemple and DiDomenico, 1972; Nye, 1957.).

Among ferroelectric perovskite crystals, $BaTiO_3$, $KNbO_3$, and KTN have very large electrooptic coefficients which give them high figures of merit as EO modulators. In addition, KTN exhibits a high saturation polarization, low dielectric loss, good infrared transmission up to 5 μm, and a low driving voltage for modulation. Therefore, KTN crystals have been considered very attractive materials for many optical applications, such as electrooptic and acoustooptic light modulators and deflectors, second-harmonic generators, parametric oscillators, variable delay lines, and holographic data storage.

In view of their large linear electrooptical coefficients and because large-sized crystals of high optical qualtity can be easily grown, $LiNbO_3$ and $LiTaO_3$ have been widely used in electrooptical modulation devices, such as linear EO modulators, travelling-wave modulators, waveguide modulators and EO Q-switches.

$LiNbO_3$ is generally considered to be the state-of-art, single-crystal, electrooptical material. Numerous modulators and multielement modulator circuits have been fabricated with $LiNbO_3$. A review paper by Thylen discusses the developments in devices for telecommunications (Thylen, 1988). Another paper by Abouelleil and Leonberger describes various waveguides made of $LiNbO_3$ (Abouelleil and Leonberger, 1989). Even though the coefficient of r_{22} is small, the electrooptic effect of r_{22} is frequently used because its temperature coefficient is the smallest among all other candidates. Contrary to the application of photorefractive effects, a material with a high laser-damage threshold is desired in electrooptical modulation and nonlinear optical applications. Therefore, it is necessary to avoid defects and impurities in the material. Pernicious ions (e.g., iron or rare earth ions) were found to be responsible for the laser damage of $LiNbO_3$.

With the rapid development of optical integrated circuits, ferroelectric thin films of $LiNbO_3$ and $LiTaO_3$ are under intensive study to achieve compatibility with the ICs. Active optical waveguides based on $LiNbO_3$ and $LiTaO_3$ thin films can be fabricated by various ion exchange techniques. These waveguides can be used as polarizing waveguides, Bragg gratings, frequency translators, chirp grating lenses, ring resonators, planar lenses and Mach–Zehnder interference modulators. A review paper by Abouelleil and Leonberger describes various waveguides made of $LiNbO_3$ (Abouelleil and Leonberger, 1989).

Several TB-type niobate crystals are important in electrooptic applications. They are BNN and its derivatives, SBN, KLN, KLTN and KNSBN crystals.

Electrooptic properties of BNN can be further improved by substituting all or a part of Ba^{2+} and Na^+ ions by other alkaline, alkaline earth ions, and/or Pb ions. For example, substitution of Na by K or of both Na and Ba by Li, K, Sr, and Pb. Some substitued crystals with large electrooptic coefficients and low half-wave voltages required for electrooptic modulators and deflectors can be found in (Burns et al., 1972; Watanabe et al., 1970). The othorhombic TB-type crystal $Pb_2KNb_5O_{15}$ (abbreviated as PKN) exhibits large linear electrooptic effects and also large quadratic electrooptic effects (g_{13} = 0.011 m^4/C^2 and g_{33} = 0.10 m^4/C^2). A half-wave voltage as low as 90 V at approximately 80 °C was found in a nonstoichiometric composition of $Sr_2KNb_5O_{15}$ (abbreviated as KSN) crystal (Clarke and Ainger, 1974). Some BNN derivatives, such as $(Ba_{1-x}Sr_x)_2$-$NaNb_5O_{15}$, $Ba_2LiNb_5O_{15}$, $(Ba_{1-x}Sr_x)_2LiNb_5O_{15}$ and $(Ba_{1-x}Mg_x)_2NaNb_5O_{15}$, are very useful in electrooptic modulators, optical parametric oscillators, and second-harmonic generators. Some BNN derivatives, such as high quality crystals of $K_{0.8}Na_{0.2}Ba_2Nb_5O_{15}$ and $Ba_2LiNb_5O_{15}$, can be poled even more easily than BNN itself (Kramer, et al., 1975; Matthes, 1972.)

SBN crystals have very large electrooptic coefficients. An electrooptic modulator made by using crystal with r_c = 269 × 10^{-12} m/V and T_c = 85 °C has been reported (Nomura, et al., 1974). For a light beam traversing along a direction perpendicular to the c-axis under a driving electric voltage 30 V, this modulator was found to operate effectively at several hundred kHz.

Both KLN and KLTN crystals have very high laser-damage thresholds. Their electrooptic properties can be found in (Xu, 1991). A KLTN crystal with the composition $K_3Li_2(Ta_{0.33}Nb_{0.67})_5O_{15}$ has a half-wave voltage $V_{\lambda/2}$ of 450 V and its figure of merit as an electrooptic modulator material is three times larger than that of BNN crystal.

KNSBN (also abbreviated as BSKNN) crystals are used as electrooptic materials. KNSBN crystals have a low half-wave voltage and a large effective electrooptic constant $r_c = r_{33} - (n_a \mid n_c)^3 r_{13}$. The low half-wave voltage of KNSBN results in values as high as 730 × 10^{-12} m/V, which is the figure of merit for electrooptic modulator materials. The threshold power for laser damage of KNSBN crystals is found to be very high. The high value of $n_o^3 r_c$ and a high threshold energy for laser damage show that KNSBN is better than $LiNbO_3$ and is, thus, an excellent material for optical switches, waveguide modulators, etc. in integrated optics.

KDP family crystals are transparent in the wavelength range from 0.2 to 1.5 μm covering the ultraviolet, visible, and near infrared regions. Because the infrared absorption is caused by vibration of hydrogen atoms and the vibration frequency of deuterium is lower than that of hydrogen (vibration frequency is approximately inversely proportional to the square root of the ion mass), the infrared absorption wavelength limit in deuterated crystals can be extended by a factor of $\sqrt{2}$.

The excellent linear electrooptic effect of KDP family crystals is extensively used in many applications (Salvo, 1991). Using the effect of a KDP crystal, an ultrafast optical shutter with a very short switching time (10^{-10}s) has been fabricated. In laser technology, KDP crystals can be used in combination with a Q-switched laser to generate a high-power pulse laser. In addition, a binary system, digital, light-beam deflector can be made by combining a KDP electrooptic switch and a birefringent crystal (e.g., calcite or sodium nitrate), and this device can be used in optical information storage. Another important application of KDP is electrooptic modulation. The configuration of a longitudinal modulator is rather simple; however, high voltages are needed for modulation. A transverse modulator made of two identical pieces of a KDP crystal plate for compensation is used to cancel the natura phase retardation and the temperature effect caused by natural birefringence. The configuration of transverse modulator is more complex than that of a longitudinal modulator, but it has certain advantages, such as low half-wave voltage and small drive power. Using the linear electrooptic effect, a KDP solid-state, light-value display tube has been designed. The image is 'written' into a KDP plate in the tube by a modulated sweep electron beam and, later, 'read' by a polarized light beam illuminating the plate. This display can be used in large screen TV, in light-valve display devices having photoconductive addressing capability, and hologram input devices. Moreover, using the electrooptic effect of KDP family crystals, electrooptic prism deflectors, interdigital electrode light deflectors, and adjustable light filters have been successfully demonstrated.

Among deuterated KDP family crystals, KD*P (DKDP) crystals have excellent electrooptic properties (Wipps and Bye, 1974). Another attractive crystal is CsD_2AsO_4 (DCsDA). The DCsDA crystals are well known for having the strongest linear electrooptic effect and the lowest half-wave voltage of longitudinal modulation in the KDP family (Adhav, 1969).

The KTP family of crystals has large linear electrooptic coefficients and low dielectric constants. Their distinctive properties makes this family of crystals very attractive in various electrooptic applications such as modulators and Q-switches. KTP has an electrooptic waveguide modulator figure of merit that is nearly double that of any other inorganic material and, hence, is very useful in active integrated optical devices. By using ion exchange processes (using Rb ion to replace K ion), both planar and channel optical waveguides with a wide range of Δn were fabricated in KTP, and a simple waveguide modulator was demonstrated (Bierlein et al., 1987).

Nonlinear Optical Applications

The nonlinear optical response of a material can give rise to exchange of energy between a number of optical beams at different frequencies. For second-order effects, important applications of this phenomenon are second-harmonic generation (SHG), sum- and difference-frequency generation, and parametric oscillation. In parametric oscillation, two waves with frequencies of ω_1 and ω_2 are simultaneously generated by a nonlinear medium pumped by a strong input beam at ω_3, where $\omega_3 = \omega_1 + \omega_2$.

As stated previously, electrooptic effects can be regarded as special cases of nonlinear optical effects in the limit where one of the field components is of nearly zero frequency compared to optical frequencies. Therefore, the materials with large electrooptic coefficients, generally, also exhibit strong nonlinear optical effects.

From the point of view of nonlinear optical application, the growth of large crystals with good optical quality is the key issue. The requirements for optical homogeneity of nonlinear optical crystals are very stringent. All factors, such as concentration variance and stress distribution, that may cause fluctuation in the refractive index of a crystal should be reduced to the minimum.

Second-harmonic generation is the most widely used nonlinear optical effect. Crystals used for frequency doubling should have the following characteristics: (1) have large nonlinear optical coefficients, $d_{ij}^{2\omega}$; (2) can meet the requirement of phase matching; (3) have wide transmission range low absorption coefficients at both fundamental and second-harmonic wavelengths; and (4) have high laser-damage threshold.

In the past, ferroelectric crystals from the KDP family, such as KDP and ADP, have been widely used for second-harmonic generation. However, their conversion efficiencies are quite low. Since the first growth of large $LiNbO_3$ crystals in 1965, it has been found that the nonlinear optical coefficients of $LiNbO_3$ are one order of magnitude higher than that of the KDP family crystals. One drawback of this crystal is its low laser-damage threshold. Later on, the successful growth of some TB-type ferroelectric crystals, such as BNN and KTP crystals, has led to the discovery of the SHG performance of these crystals better than that of $LiNbO_3$. $KNbO_3$ were also found to be a excellent crystal for SHG.

$KNbO_3$ crystals have large nonlinear coefficients $d_{ij}(2\omega)$ (see Table 4.6). In this crystal, second-harmonic generation with the best phase matching and a SHG conversion efficiency near 100% (as in the case with barium sodium niobate) can be obtained. It has been shown, in a pulsed mode operation, that milliwatts of a coherent, dark-blue, 430-nm, second-harmonic wave can be generated by using $KNbO_3$ crystals and with suitable single-mode lasers (Günter, 1987). This crystal is also very suitable for frequency summation of short wavelength radiations under noncritical phase-matching conditions.

$LiNbO_3$ crystal is important in the field of nonlinear optics. Because $LiNbO_3$ exhibits a large birefringence ($\Delta n = n_o - n_e \approx 0.08$) from the visible to infrared region, this crystal is widely used as an optical frequency doubler with phase matching achieved by using the nonlinear coefficient d_{31}. The phase-matching angle is relatively large, the temperature coefficient of birefringence negative, and the temperature coefficient of dispersion positive. Thus, by varying temperature to alter the refractive indices, optimum phase matching with an angle of 90° can be obtained. $LiTaO_3$ cannot be used as a frequency-doubling crystal because its birefringence is too small to obtain phase matching for SHG.

BNN is regarded as one of the best crystals for optical frequency multiplication. In addition, BNN crystal parametric oscillators are able to function in the wavelength range from 0.98 μm to 1.06 μm. Polycrystalline ferroelectric thin films of SBN (grain size approximately 2–3 μm) with approximately the same nonlinear properties as SBN single crystal have been reported (Zhdanov et al. 1980).

The KTP crystal family has very large nonlinear optical coefficients. The transparent range of a KTP crystal (or $K_xRb_{1-x} TiOPO_4$) extends from 0.35 μm to 4.5 μm. KTP crystals also have thermally stable phase-matching properties useful for SHG.

Photorefractive Applications

In the 1980's, facinating feasibilies of coherent optics were demonstrated with the help of photorefractive materials. For information storage applications, a number of volume phase holograms are superimposed at the same site with different angles according to the Bragg condition; extremely large storage densities up to 10^{10} bit/cm^3 can be obtained. The degenerate four-wave mixing yields phase-conjugated waves. In special cases, the pumping beams can even be dispensed with and self-pumped phase-conjugated mirrors can be realized. Photorefractive effects have been successfully applied to real-time interferometry, beam deflection, image and signal processing (e.g., convolution, correlation edge enhancement, inversion, substraction, amplification, or incoherent-to-coherent conversion), and many other areas. A number of ferroelectric crystals have been used in photorefractive applications . Among them are the perovskite-type ferroelectrics, tungsten-bronze type ferroelectrics, $LiNbO_3$, and $LiTaO_3$.

The photorefractive properties of $KNbO_3$ crystals, such as photoconductivity, beam coupling gain, and response time, can be enhanced by the electrochemical reduction method. Because the conductivities of $KNbO_3$ are much larger in crystals compared to $LiNbO_3$, the relaxation times are much shorter in $KNbO_3$, typically, a few milliseconds in the reduced crystals and up to a few seconds in the oxidized ones. A resolution of up to 10 line pairs per mm was achieved for a laser beam modulated by a white light image. However, the diffraction efficiencies are limited to a few percent for a 1-mm crystal thickness as a trade-off for their high conductivities. Iron doping is also known to increase the gain coefficient Γ of $KNbO_3$ (Medrano et al., 1989).

$LiNbO_3$ and $LiTaO_3$ crystals were the early ferroelectric materials in which the photorefractive, light-induced, refractive index change was discovered. On the one hand, photorefraction is the basic mechanism of optical phase recording. On the other hand, it is a serious hindrance in applications of ferroelectric crystals in the fields of electrooptics and nonlinear optics. Due to this fact, the mechanism of photorefraction, as well as its application to holography, have been intensively investigated.

Fe-doped $LiNbO_3$ crystals display a strong photorefractive effect and have been used in holographic storages for more than twenty years. It has been found that the sensitivity of holographic recording in Fe-doped crystals depends only on the concentration of Fe^{2+} ions and that it is independent of the concentration of Fe^{3+} ions and the ratio of Li/Nb. Doped $LiNbO_3$ (e.g., Fe as well as Ce-doped $LiNbO_3$) provides high diffraction eficiency with faster erasure characteristics and has been utilized in real-time holographic optical storages (Xu, et al., 1989).

A pure SBN crystal exhibits a small photorefractive effect. However, Ce-doped SBN crystal shows a diffraction efficiency above 90%—a characteristic describing high-sensitivity required for laser holography. In this crystal, holograms can be recorded, read, and erased with an He-Ne laser. Ce-doped SBN and Cu-doped SBN crystals possess large laser-beam coupling capabilities, important for energy transfer and large phase-conjugate reflectivity (Rakuljic et al., 1986; Liu et al., 1987). Self-pumped phase conjugation in SBN, Cu-doped SBN, and Ce-doped SBN crystals has also been realized.

When ions with different chemical valences are used as dopants in KNSBN crystals, semiconducting ferroelectric (or photoferroelectric) KNSBN can be obtained. Such crystals have excellent controllable photoconductive and photorefractive effects. KNSBN crystal doped with transition-metal or rare-earth ions have been grown by Xu and Chen (Xu et al., 1989; Chen and Xu, 1989). These crystals show sensitive photorefractive properties and possess great potential for applications in holographic memories, image storage devices, and optical phase-conjugation devices (Neurgaonkar et al., 1987; Rodriguez et al., 1987). The photorefractive sensitivity S_{n2} (i.e., the refractive index change per unit incident-energy density) of doped KNSBN is larger than that of undoped KNSBN. The photoresponse time in doped KNSBN crystals is shorter than that in undoped KNSBN. Optical phase conjugation in KNSBN crystals has also been investigated. A comparison of the intensity dependency of the photoinduced birefringency change at 578 nm in Nd-doped KNSBN, Ce-doped SBN, $BaTiO_3$ and Fe-doped $KNbO_3$ can be found in ref. (Xu, 1989). Self-pumped phase conjugation in Fe-doped KNSBN crystals has been observed. Phase-conjugated reflectivities up to 38% and phase-conjugated images of very high quality were observed. Holographic associative memory has also been demonstrated with a plate of Co-doped KNSBN crystal (Xu et al., 1990a; Yuan et al., 1990).

Pyroelectric Applications

Infrared detection is the main application of pyroelectric materials. Until now, some ferroelectric materials have been used for infrared detectors. They are TGS, $LiNbO_3$ and $LiTaO_3$, some TB-type crystals, and PLZT ceramics. $Pb_5Ge_3O_{11}$ crystals have great potential in infrared applications.

TGS crystals are the mostly widely used pyroelectric material in infrared detectors. TGS has a large pyroelectric coefficient of 2 to 3.5×10^{-2} $\mu Ccm^{-2}K^{-1}$ at room temperature and the largest response sensitivity D^* among the known pyroelectric materials. In doped TGS, D^* reaches a value of 2.5×10^9 $cmHz^{1/2}W^{-1}$. However, there are certain shortcomings of TGS for pyroelectric applications. The main problem is the depolarization owing to its low Curie point (in fact, even at room temperature, partial depolarization occurs). To solve the depolarization problem, TGS crystals were grown from an aqueous TGS solution with the addition of α-alanine (the amount of the doping of L-alanine is between 2% to 35% of glycine) (Fang et al., 1989). The crystal grown from the doped solution is an α-alanine-doped TGS (abbreviated as LATGS). In LATGS crystal, an 'internal bias field' exists and its ferroelectric hysteresis loop shifts to one side. Thus, at temperatures below T_c, LATGS maintains its saturation polarization state without any external field. LATGS crystals remain polarizaed even after repeated cycles of heat treatment between Tc and room temperature.

Although the pyroelectric coefficients in $LiTaO_3$ and $LiNbO_3$ are smaller than those in many other pyroelectric materials, the Curie temperatures in both crystals are, in contrast, higher than those of most of the other pyroelectric crystals. Moreover, they have stable physical and chemical properties, which some pyroelectric materials lack, especially the water-soluble crystals. Therefore, both crystals, especially $LiTaO_3$, which has a higher pyroelectric coefficient and a lower dielectric constant, are still very attractive as materials for infrared detectors. An infrared detector employing $LiTaO_3$ crystal is able to withstand high-energy infrared radiation and has a fast response time of 0.5 ns (Stokowski, 1976; Putley, 1980). Picosecond response time has been demonstrated by using $LiTaO_3$ crystals doped with Cu^{2+} impurities in detecting 10 ps pulses from a mode-lock Nd:glass laser (Auston and Glass, 1972). By combining a $LiTaO_3$ plate and a Si-CCD (charge coupling device), a infrared image sensor CCD has been developed (Okuyama, 1989).

SBN shows strong pyroelectric effects. To be considered for use as infrared pyroelectric detectors, a large pyroelectric coefficient, a small specific heat, and fairly low relative permittivity are required. The dielectric and pyroelectric properties at room temperature are reported in (Glass, 1969; Glass

1970; Lang, 1974; Liu, 1978). Using the pyroelectric effect in SBN, infrared detectors have been fabricated. The characteristics of these detectors (at room temperature) can be improved by changing the ratio of Sr/Ba. These detectors have high stability in air without a protective window which is necessary in the case of TGS. This material is, therefore, widely used for fast detectors. The pyroelectric effect in thin SBN film has been reviewed by Zook and Liu (Zook and Liu, 1978). The main shortcoming of SBN is that its dielectric constant is not low enough for application at high frequency. Its high-frequency figure of merit $p/\rho c_p \kappa$ (where p is the pyroelectric coefficient, ρ the density, c_p the specific heat, and κ the dielectric constant) is lower than that of TGS by one order of magnitude. Thus, SBN thin-film detectors are mainly used for small area, low-frequency applications. For pyroelectric applications, SBN crystals are frequently modified by doping with Pb and La. For example, doping with 0.01% Pb improves the pyroelectric properties of $Ba_{0.50}Sr_{0.50}Nb_2O_6$ (Maciolek and Liu, 1973) and $Ba_{0.52}Sr_{0.48}Nb_2O_6$ over undoped SBN; laser damages are also reduced. La-doped SBN has larger pyroelectric coefficients and dielectric constants and lower Curie temperatures.

Strontium sodium lithium niobate $Sr_4NaLiNb_{10}O_{30}$ (SNLN) is a tetragonal TB-type crystal transparent in the wavelength range from 0.39–5 μm. The pyroelectric coefficient of SNLN crystal is about 7×10^{-2} μC/cm^{2}K, which is slightly higher than that of SBN with Sr/Ba=1. Another TB-type crystal, KNSBN (also abbreviated as BSKNN) is also used as a pyroelectric material.

The PLZT ceramic with a composition of 8/65/35 has a pyroelectric coefficient of 17×10^{-2} μC/cm^{2}K. The pyroelectric coefficient of PLZT increases with increasing amount of La and is higher than most coefficients reported for pyroelectric materials except SBN. The advantages of using PLZT ceramics as pyroelectric materials are their high Curie temperatures and a low fabrication cost. However, in the PLZT system, the composition which has a larger pyroelectric coefficient also has a larger dielectric constant and dielectric loss (Liu and Kyonka, 1974). This is unfavorable to its pyroelectric voltage sensitivity.

$Pb_5Ge_3O_{11}$ crystals have excellent pyroelectric properties with a pyroelectric coefficient of 9.5×10^{-3} μC/cm^{2}K. Due to the crystal's low dielectric permittivity, a rather high pyroelectric voltage sensitivity can be achieved and this presents great potential for infrared detector applications [Luff et al., 1974].

Ferroelectric Structure-Related Applications

Electrooptic Applications of PLZT

Since the development of the first transparent ferroelectric PLZT ceramic in 1971, the application of PLZT ceramics in electrooptic devices has been evolving at a steady pace. Over 20 years of research and development efforts on these materials have resulted in the characterization of their properties and a general understanding of their many interactive phenomena. A summary of the various phenomena related to the electrooptic effect and device criteria for application of this group of materials was given in (Haertling, 1987).

Successful application of the PLZT ceramics, which have been developed, are mainly in the category of optical shutters. Applications involving optical modulators, image storage, and displays are presently being developed as an outgrowth of optical shutter technology. Continuous research efforts on image storage-display devices show a promising future for these materials in new areas of applications. The successful fabrication of PLZT thin films with useful electrooptic properties has added a new dimension to the possibilities for unique device applications in the emerging fields of integrated optics and optical communications. The application of PLZT thin films to inegrated optics was first explored by Chen and Marzwell (Chen and Marzwell, 1975) in 1975.

One category of shutter devices may be grouped under the general heading of spatial light modulators, or, in short, SLMs. For these device applications, PLZT technology continues to be the front runner because of the high speed, high contrast and resolution, good thermal stability, single surface electrode technology, availability of high-quality materials, and coherent modulation.

The SLMs can be one-two, or three-dimensional. A one-dimensional SLM is commonly referred to as a linear gate array (LGA). The LGAs have been used for high-speed optical data recording, document identification, and color printing. The two-dimensional SLMs can be used for holographic page composers. Three-dimensional SLMs have been used for 3-D stereo viewing systems. Their applications include flight simulators, computer graphics, home entertainment, and industrial scanning electron microscopy.

From the P-E characteristic point of view, PLZT ceramics suitable for switchable memory devices possess permanent polarization (P_R); the electrical and/or optical information is stored in the ceramic by the incremental switching of polarization on traversing the hysteresis loop. Desirable characteristics of materials for such switchable memories are large P_R and low E_c (the coercive field). Therefore, PLZT ceramics with compositions which exhibit hysteresis loops similar to the one shown in Fig. 4.4 (B) are desirable for switchable memory devices. For nonswitchable memory (linear memory), the materials are permanently polarized to saturation remanence and left to remain in that state. Optical information is extracted from the ceramic by the action of an electric field which causes slight changes (linear) in the optical birefiringence of the ceramic, but the polarization and electric domain reversal in the material do not occur. Desirable characteristics of materials for nonswitchable menmory devices are high E_c and large change in birefringence for a given applied electric field.

From the compositional point of view, the PLZT ceramics with compositions in region A of Fig. 4.6 can be used to make light valves, optical storage-display devices, ferroelectric picture devices, and optical spectral filters. For ceramics with compositions in region B, birefringence is a linear function of the electric field. These materials are used mainly for linear electrooptic modulators and transient optical switches. The quadratic electrooptic effect found in materials in region C can be utilized in quadratic electrooptic modulators.

Other Ferroelectric Structure-Related Applications

Bismuth titanate crystals are presently used in making ferroelectric optical memory devices. The simplest optical memory is conceptually an electrically addressed light valve. The light valve consists of a polarizer, a $Bi_4Ti_3O_{12}$ crystal plate, and an analyzer set for extinction (Cummins and Luke, 1971). The incident beam is usually arranged to propagate along the b-axis of $Bi_4Ti_3O_{12}$ initially polarized parallel to the major or minor axis of the indicatrix. Reversal of the c-axis polarization by the application of a transverse electric field results in the rotation of the optical axis through 50 ° in the a–c plane. This is close to the optimum condition of 45 ° for a maximum transmission of light throught the analyzer. However, at present, $Bi_4Ti_8O_{12}$ crystal with a large a–c face area is not easy grow from the flux. Progress in overcoming this difficulty has been made by epitaxially depositing large a–c crystal face films onto single crystals of MgO and $MgAl_2O_4$ (Wu et al., 1972; Wu et al., 1973).

Gadolinium molybdate single crystals have been used in light valve devices with a longitudinally applied field and transparent electrodes. The crystal is placed between two crossed polarizers and the natural phase retardation introduced by this crystal is compensated for with a quarter-wave plate. The electrodes are used to provide a field across the element, just exceeding the coercive field, to reverse the c-axis polarization. Several shortcomings exist when applications of the light-valve arrays of β-GMO are considered. First, because the coercive field of the crystal is rather high, the switching speed is still of the order of microseconds even with a high switching field. Second, solving the problem of mechanical cross-talk between elements in GMO due to the large strain associated with polarization reversal requires using separate crystal plates for rows and columns. Therefore, β-GMO crystals are more suitable for applications in memory or page-composer devices than in display devices.

A ferroelectric-photoconductive image storage device with optical write-in and read-out has been made by using the reversal effect of the ferroelectric domain and optical rotation in $Pb_5Ge_3O_{11}$ (Cummins and Luke, 1973). This device can achieve a storage resolving power of 57 lines/mm.

Due to the small optical rotation of $Pb_5Ge_3O_{11}$, the efficiency of read-out is very low in applications involving ferroelectric displays or light valves. In addition, the switching speed of the device is also low, between several microseconds to tens of microseconds in fields of the order of 10 kV/cm. However, because of its unique optical properties, $Pb_5Ge_3O_{11}$ crystal is still of great interest. In the near future it is hoped that scientists may find a new promising crystal in this family for applications.

References

Abouelleil, M. M., and Leonberger, F. J. *J. Am. Ceram. Soc.*, 72, 1311, 1989.
Adhav, R. S. *J. Opt. Soc. Am.*, 59, 414, 1969.
Auston, D. H., and Glass, A. M., *Apply. Phys. Lett.*, 20, 398, 1972.
Bierlein, J. D., and Gier, T. E., U.S. Patent 3,949,323 1976; Zumsteg, F. C., Bierlein, J. D., and Gier, T. E., *Apply. Phys. Lett.*, 47, 4980, 1976; Bierlein, J. D., and Vanherzeele, H., *J. Opt. Soc. Am.*, B/Vol. 6, 622, 1989.
Bierlein, J. D., Ferretti, A., Brixner, L. H., and Hsu, W. Y., *Apply. Phys. Lett.*, 50, 1216, 1987.
Burns, G., Giess, E. A., and O'Kane, D. F., U.S. Patent 3,640,865, 1972.
Chen, F. S., *J. Appl. Phys.*, 40, 3389, 1969.
Chen, H. C., and Xu, Y. H., *J. Cryst. Growth*, 96, 357, 1989.
Chen, D., and Marzwell, N., ONR Report No. 0395–1, NITS No. AD-A009102, 1975.
Chour, K. W., Wang, G., and Xu, R., *Proc. MRS Fall Meeting, Boston: MOCVD of Electronic Ceramics*, Seshu Desu, Ed., 1993.
Clarke, R., and Ainger, F. W., *Ferroelectrics*, 8, 101, 1974.
Cummins, S. E., and Luke, T. E., *IEEE Trans. Electron. Dev.*, ED-18, 761, 1971.
Cummins, S. E., and Luke, T. E., *Proc. IEEE* 61, 1039, 1973.
Fang, C. S., Yao, X., Chen, Z. X., Bhalla, A. S., and Cross, L. E., *Mater. Sci. Lett.*, 22, 134, 1983.
Fang, C. S., Wang, M., and Zhuo, H. S., *Ferroelectrics*, 91, 349, 1989a.
Fang, C. S., Wang, M., and Zhuo, H. S., *Ferroelectrics*, 91:373, 1989b.
Francome, M. H., *Thin Solid Films*, 13, 413, 1972.
Francombe, M. H., and Krishnaswamy, S. V., 1990. *Ferroelectric thin films*, Mat. Res. Soc. Symp. Proc. E. R. Myers, and A. I. Kingon, Eds., 1990, Vol. 200, p. 179.
Fridkin, V. M., *Photoelectrics*. Springer, Berlin, 1979.
Glass, A. M., *J. Appl. Phys.*, 40, 4699, 1969.
Glass, A. M., *J. Appl. Phys.*, erratum 41, 2268, 1970.
Godefroy, G., Ormancey, G., Jullien, P., Ousi, W., and Semanou, Y., *IEEE ISAF*, 1986, pp. 12–15.
Godefroy, G., *Ferroelectrics*, 92, 205, 1989.
Günter, P. Electro optical Effects in Dielectric Crystals. *Ferroelectrics* 75, 5–23, 1987.
Haertling, G. H. *Ferroelectrics*, 75, 25, 1987.
Haertling, G. H., and McCampbell, C., *Proc. IEEE*, 60, 450, 1972.
Jullien, P., Maillard, A., Ormancey, G., Lahlafi, A., and Matull, R., *Ferroelectrics*, 94, 81, 1989.
Kaminov, I. P., and Turner, E. H., *Proc. IEEE*, 54, 1374, 1966.
Konac, C., Fousek, J., and Kursten, H. D., *Ferroelectrics*, 21, 347, 1978.
Kramer, V., Matthes, H., and Marshall, A., *J. Mat. Sci.* 10, 547, 1975.
Kumada, A., Toda, G., and Otomo, Y., *Ferroelectrics*, 7, 367, 1974.
Land, C. E., and Holland, R., *IEEE Spectrum* 7(2), 71, 1970.
Land, C. E., and Thacher, P., *Proc. IEEE* 57, 751, 1969.
Lang, S. B., *Ferroelectrics and Related Phenomena, Vol. 2. Sourcebook of Pyroelectricity*, Lefkowitz, I., and Taylor, G. W., Eds., Gordon and Breach, New York, 1974.
Liu, W. H., Qiu, Y. S., Zhang, H. J., Dai, J. H., Wang, P. Y., and Xu, L. Y., *Opt. Commun.* 64, 81, 1987.
Liu, S. T., *Ferroelectrics*, 22, 709, 1978.
Liu, S. T., and Kyonka, J., *Ferroelectrics*, 7, 167, 1974.

Luff, D., Lane, R., Brown, K. R., and Marshallsay, H. J., *Trans. & J. Br. Ceram. Soc.*, 73, 215, 1974.
Maciolek, R. B., and Liu, S. T., *J. Electron. Mater.* 2, 191, 1973.
Mansingh, A., *Ferroelectrics*, 102, 69, 1990.
Matthes, H., *J. Cryst. Growth*, 15, 157, 1972.
Medrano, C., Voit, E., Amrhein, P., and Günter, P., *Ferroelectrics*, 92, 289, 1989.
Miller, R. C., Nordland, W. A., and Ballman, A. A., *Ferroelectrics*, 7, 109, 1974.
Morris, P. A., 1991. Defect Chemistry of Nonlinear Optical Oxide Crystals, in *Materials for Nonlinear Optics: Chemical Perspectives*, S. R. Marder, J. E. Sohn, and G. D. Stucky, Eds., American Chemical Society, Washington, DC, 1991, 380–393.
Neurgaonkar, R. R., Cory, W. K., Oliver, J. R., Clark, W. W. III, Wood, G. L., Miller, M. J., and Sharp, E. J. *J. Cryst. Growth*, 84, 629, 1987.
Nomura, S., Kojima, H., Hattori, Y., and Kotsuka H., *Jpn. J. Apply. Phys.*, 13, 1185, 1974.
Nye, J. F., *Physical Properties of Crystals*, Clarendon Press, Oxford, 1957.
Okuyama, M., Togami, Y., Ohnishi, J. and Hamakawa, Y., *Ferroelectrics*, 91, 127, 1989.
Putley, E. H., *Infrared Physics* 20, 149, 1980.
Rakuljic, G. A., Yariv, A., and Neurgaonkar, R. R., *Opt. Eng.*, 25, 1212, 1986.
Rodriguez, J., Siahmakoun, A., Salamo, G. J., Miller, M. J., Clark, W. W. III, Wood, G. L., Sharp, E. J., Neurgaonkar, R. R., *Appl. Opt.* 26, 1732, 1987.
Roy, R. W., Etzold, K. F., and Cuomo, J. J., *Ferroelectric thin films*, Mat. Res. Soc. Symp. Proc. Vol. 200, E. R. Myers and A. I. Kingon. Eds., 1990, Vol. 200, p. 141.
Salvo, G. J. *IEEE Trans. Electron. Dev.* ED-18, 748, 1971.
Smith, W. D., and Land, C. E.,*Apply. Phys. Lett.*, 20, 169, 1972. Land, C. E., and Smith, W. D., *Apply. Phys. Lett.*, 23, 57, 1973.
Staebler, D. L., and Philips, W., *Appl. Opt.*, 13, 788, 1974.
Stokowski, S. E., *Apply. Phys. Lett.*, 29, 393, 1976.
Thacher, P., and Land, C. E., *Wescon Technical Report* 15, 31/2-2-31/1-11, 1971.
Thylen, L., *J. Lightwave Technol.*, 6, 847, 1988.
Uesu, Y., Okada, N., Inoue, M., and Hara, S., *Ferroelectrics*, 107, 33, 1990.
Watanabe, A., Sato, Y., Yano, T., and Kitahiro, I., *J. Phys. Soc. Jpn.*, 28, Suppl. 93, 1970.
Wechsler, B. A., Rytz, D., Klein, M. B., and Schwartz, R. N., 1991. Defect Properties and the Photorefractive Effect in Barium Titanate, in Materials for Nonlinear Optics: Chemical Perspectives, Marder, S. R., Sohn, J. E., and Stucky G. D., Eds., American Chemical Society, Washington, DC, 1991, pp. 394–409.
Wemple, S. H. and DiDomenico, pp. 264–380. in *Apply. Solid State Science* 3, R. Wolfe, Ed., Academic Press, 1972.
Wipps, P. W., and Bye, K. L., *Ferroelectrics*, 7, 183, 1974.
Wu, S. Y., Takei, W. J., Francombe, M. H., and Cummins, S. E., *Ferroelectrics*, 3, 217, 1972.
Wu, S. Y., Takei, W. J., Francombe, M. H., *Apply. Phys. Lett.*, 22, 26, 1973.
Xu, Y., Yuan, Y., Yu, Y. L., Xu, K. B., Xu, Y. H. and Zhu, D. R., *Proc. SPIE*, 1220, 94, 1990c.
Xu, K. B., Xu, H. Y., Yuan, Y., Hong, J., and Xu, Y. H., *Proc. SPIE*, 1078, 331, 1989.
Xu, Y., and Mackenzie, J. D., Ferroelectric Thin Films Prepared by Sol-Gel Processing, *Integrated Ferroelectrics*, 1, 17–42, 1992.
Xu, Y., *Ferroelectric Materials And Their Applications*, Elesevier Science Publishers B.V., Amsterdam, The Netherlands, 1991.
Xu, Y., Huang, Z. Q., Li, W., Wang, H., Zhu, D. R., and Chen, H. C., *Ferroelectrics*, 92, 211, 1989.
Yuan, Y., Xu, H. Y., Yu, Y. L., Xu, K. B., Xu, Y. H., Zhu, D. R., Chen, H. C., and Zhang, Q. L., *Proc. SPIE*, 1220, 116, 1990c.
Zhdanov, V. G., Kostsov, E. G., Malinovsky, V. K., Pokrovsky, and Sterelyukhina, L. N., *Ferroelectrics*, 29, 219, 1980.
Zook, J. D., and Liu, S. T., *J. Appl. Phys.*, 49, 4604, 1978.

5

Nonlinear Optics

5.1 Introduction.. 125
5.2 Nonlinear Optical Susceptibility................................. 126
    Definition of Nonlinear Optical Susceptibility • Symmetry Properties of Nonlinear Optical Susceptibility
5.3 Second-Order Nonlinearities...................................... 129
    Nonlinear Wave Equation • Sum-Frequency Generation • Difference-Frequency Generation • Phase-Matching Considerations • Materials with Second-Order Nonlinear Susceptibilities
5.4 Third-Order Nonlinearities... 137
    Third-Harmonic Generation • Four-Wave Mixing • Degenerate Four-Wave Mixing and Phase Conjugation • Intensity-Dependent Refractive Index • Self-Focusing • Tensor Nature of the Third-Order Susceptibility • Two-Photon Absorption • Enhancement of Nonlinear Interactions in Optical Waveguides • Soliton Propagation • Prototypical Nonlinear Optical Switches • Materials with Third-Order Nonlinear Susceptibilities
5.5 Stimulated Light Scattering.. 146
    General Features of Stimulated Light Scattering • Stimulated Raman Scattering • Stimulated Brillouin Scattering

Andrew J. Stentz
Bell Laboratories

Robert W. Boyd
University of Rochester

5.1 Introduction

Nonlinear optical phenomena occur when the response of a material system to an applied optical field depends nonlinearly on the strength of the applied field. The subsequent interaction of the optical field with this nonlinear response generates a myriad of interesting physical processes. In some applications, these nonlinear processes are merely limitations to the performance of linear systems, whereas in photonics, nonlinear processes promise to be much more. Applications of nonlinear optics include the frequency doubling of semiconductor lasers, the generation of ultrashort laser pulses, dispersion compensation in communication systems, and all-optical switching. In this chapter, we provide a brief summary of this field, emphasizing important results and qualitative features. A more complete, pedagogical introduction to nonlinear optics is presented in Boyd (1992). The reader is also directed to the textbooks by Shen (1984), Butcher and Cotter (1990), and Zernike and Midwinter (1973).

Two systems of units are commonly used in nonlinear optics. The SI (or mks) system is typically used in applied nonlinear optics and photonics and is the primary system of units used in this handbook. The Gaussian systems of units is often used in fundamental studies of nonlinear optics, and most of the original papers in nonlinear optics are written with these units. To facilitate transformation between these systems of units, every equation quoted in this chapter can be interpreted as an SI equation as written or as a Gaussian-system equation by omitting the prefactors (often ϵ_0 or $4\pi\epsilon_0$) that appear in square brackets. For example, Eq. 2.1 is correct as written as an SI equation and becomes $\tilde{P}(t) = \chi^{(1)} \tilde{E}(t)$ in the Gaussian system.

5.2 Nonlinear Optical Susceptibility

In conventional linear optics, it is assumed that the applied optical field is sufficiently weak that the induced polarization $\tilde{P}(t)$ depends linearly on the applied electric field strength $\tilde{E}(t)$ such that

$$\tilde{P}(t) = [\epsilon_0]\chi^{(1)}\tilde{E}(t) \tag{5.1}$$

where the constant of proportionality $\chi^{(1)}$ is the linear susceptibility. When the applied field is sufficiently strong, one can often accurately describe the induced polarization as a power series in the applied electric field strength such that

$$\tilde{P}(t) = [\epsilon_0][\chi^{(1)}\tilde{E}(t) + \chi^{(2)}\tilde{E}^2(t) + \chi^{(3)}\tilde{E}^3(t) + \cdots] \tag{5.2}$$

where $\chi^{(2)}$ and $\chi^{(3)}$ are known, respectively, as the second- and third-order nonlinear optical susceptibilities. The two lowest order nonlinear terms, which are displayed explicitly in Eq. 5.2, are responsible for most nonlinear optical interactions. For simplicity, we have neglected the vector nature of the polarization and the electric field and the tensor nature of the optical susceptibility in writing Eqs. 5.1 and 5.2. As the strength and nature of nonlinear optical interactions are dependent on the magnitude and form of the nonlinear susceptibility tensor, in the following section, we present a proper definition of susceptibility and describe the symmetry properties that are reflected in its form.

Definition of Nonlinear Optical Susceptibility

If we assume that the electric field $\tilde{\mathbf{E}}(\mathbf{r},t)$ can be decomposed into a sum of discrete frequency components $\mathbf{E}(\omega_n)$ as

$$\tilde{\mathbf{E}}(\mathbf{r}, t) = \sum_n \mathbf{E}(\omega_n)e^{-i\omega_n t} + c.c., \tag{5.3}$$

then, we can define the components of the second-order nonlinear susceptibility tensor as

$$P_i(\omega_n + \omega_m) = [\epsilon_0] \sum_{jk} \sum_{(nm)} \chi^{(2)}_{ijk}(\omega_n + \omega_m, \omega_n, \omega_m)E_j(\omega_n)E_k(\omega_m) \tag{5.4}$$

where i,j,k refer to the Cartesian coordinates, and the notation (nm) indicates that the sum over n and m is to be made with the sum $\omega_n + \omega_m$ held constant. In general, the summation over n and m may be performed to obtain the result

$$P_i(\omega_n + \omega_m) = [\epsilon_0]D \sum_{jkl} \chi^{(2)}_{ijk}(\omega_n + \omega_m, \omega_n, \omega_m)E_j(\omega_n)E_k(\omega_m) \tag{5.5}$$

where D is known as the degeneracy factor and is equal to the number of distinct permutations of the frequencies ω_n and ω_m. Similarly, we define the components of the third-order nonlinear susceptibility tensor as

$$P_i(\omega_0 + \omega_n + \omega_m) = [\epsilon_0]D \sum_{jkl} \chi^{(3)}_{ijkl}(\omega_0 + \omega_n + \omega_m, \omega_0, \omega_n, \omega_m)E_j(\omega_0)E_k(\omega_n)E_l(\omega_m). \tag{5.6}$$

To completely describe all second-order nonlinear interactions with the nonlinear susceptibility described in Eq. 5.5, one would need to know 324 complex numbers. This figure accounts for all permutations of the Cartesian coordinates and of the frequency components for both positive and negative frequencies. Fortunately, the symmetries summarized in the following section greatly reduce the number of independent components.

Symmetry Properties of Nonlinear Optical Susceptibility

General symmetry properties of nonlinear susceptibility are reviewed in Table 5.1. Detailed discussions of their derivations are given by Boyd (1992). In addition to the symmetry properties found in Table 5.1, the spatial symmetries of a given material may place additional constraints on the allowed form of the nonlinear susceptibility tensor. See Boyd (1992) for the symmetry properties of various crystal point groups.

The symmetry with the most striking impact on nonlinear susceptibility is inversion symmetry. When a material possesses a center of inversion (i.e., is centrosymmetric), the second-order susceptibility is identically zero. This is the case for all gases, liquids, and amorphous solids as well as for some crystal classes. Inversion symmetry also has a dramatic impact on the number of independent elements in the third-order nonlinear susceptibility, reducing the number of possible independent elements from 81 to 3.

The conditions for the validity of Kleinman's symmetry are so frequently met that a contracted notation for the second-order nonlinear susceptibility under these conditions has been developed. We introduce this notation by defining the tensor d_{ijk} as

$$d_{ijk} = \frac{1}{2} \chi^{(2)}_{ijk}. \tag{5.7}$$

Under the conditions of Kleinman symmetry and, in general, for second-harmonic generation, the tensor d_{ijk} is symmetric in its last two indices. Therefore, we can introduce the contracted matrix d_{il} according to the prescription

jk:	11	22	33	23,32	31,13	12,21
l:	1	2	3	4	5	6

The nonlinear susceptibility then takes the following form:

$$d_{il} = \begin{bmatrix} d_{11} & d_{12} & d_{13} & d_{14} & d_{15} & d_{16} \\ d_{21} & d_{22} & d_{23} & d_{24} & d_{25} & d_{26} \\ d_{31} & d_{32} & d_{33} & d_{34} & d_{35} & d_{36} \end{bmatrix}. \tag{5.8}$$

When the Kleinman symmetry condition is valid, only 10 of these elements are independent and the matrix takes the following general form:

$$d_{il} = \begin{bmatrix} d_{11} & d_{12} & d_{13} & d_{14} & d_{15} & d_{16} \\ d_{16} & d_{22} & d_{23} & d_{24} & d_{14} & d_{12} \\ d_{15} & d_{24} & d_{33} & d_{23} & d_{13} & d_{14} \end{bmatrix}. \tag{5.9}$$

TABLE 5.1 Symmetry Properties of Nonlinear Optical Susceptibility

Property	Conditions of Validity	Transformation That Preserves Numerical Value of χ	Example with Second-Order Susceptibility
Reality of fields	Always true	Change sign of all frequency components and take complex conjugate	$\chi^{(2)}_{ijk}(-\omega_n-\omega_m,-\omega_n,-\omega_m) = \chi^{(2)}_{ijk}(\omega_n+\omega_m,\omega_n,\omega_m)^\star$
Intrinsic permutation symmetry	Always true	Simultaneously interchange last two frequency components and last two cartesian coordinates	$\chi^{(2)}_{ijk}(\omega_n+\omega_m,\omega_n,\omega_m) = \chi^{(2)}_{ikj}(\omega_n+\omega_m,\omega_m,\omega_n)$
Reality of nonlinear susceptibility tensor	Lossless medium	Take complex conjugate	$\chi^{(2)}_{ijk}(\omega_n+\omega_m,\omega_n,\omega_m) = \chi^{(2)}_{ijk}(\omega_n+\omega_m,\omega_n,\omega_m)^\star$
Full permutation symmetry	Lossless medium	Simultaneously interchange any frequency components and the corresponding cartesian coordinates	$\chi^{(2)}_{ijk}(\omega_n+\omega_m,\omega_n,\omega_m) = \chi^{(2)}_{jki}(-\omega_n,\omega_m,-\omega_n-\omega_m)$ and, therefore, with the above results, $\chi^{(2)}_{jki}(\omega_n+\omega_m,\omega_n,\omega_m) = \chi^{(2)}_{jki}(\omega_n,-\omega_m,\omega_n+\omega_m) = \chi^{(2)}_{jki}(\omega_n+\omega_m,\omega_n,\omega_m)$
Kleinman's symmetry	χ is independent of frequency and medium is lossless (e.g. $\omega_{opt.} \ll \omega_{reson.}$)	Interchange cartesian coordinates without interchanging frequency components	
Inversion symmetry	Medium is centrosymmetric	All values of second-order susceptibility are zero	$\chi^{(2)}_{ijk}(\omega_n+\omega_m,\omega_n,\omega_m) = 0$

5.3 Second-Order Nonlinearities

Second-order nonlinearities are commonly employed to generate new frequency components from one or more incident laser beams. For example, frequency-doubling the output of near-infrared semiconductor lasers is a viable option for the generation of compact blue-green laser sources. Optical parametric oscillators based on difference-frequency generation are routinely used to generate frequency-tunable radiation throughout the optical spectrum. In this section, we review the basic equations governing sum- and difference-frequency generation and second-harmonic generation.

Nonlinear Wave Equation

Propagation of light in a nonlinear, nonmagnetic medium with no currents and no free charges is governed by the nonlinear wave equation

$$\nabla \times \nabla \times \tilde{\mathbf{E}} + \frac{1}{[\epsilon_0]c^2} \frac{\partial^2}{\partial t^2} \tilde{\mathbf{D}}^{(1)} = -\frac{4\pi}{[4\pi\epsilon_0]} \frac{1}{c^2} \frac{\partial^2}{\partial t^2} \tilde{\mathbf{P}}^{NL}, \qquad (5.10)$$

where we have split the polarization **P** into its linear and nonlinear parts such that

$$\tilde{\mathbf{P}} = \tilde{\mathbf{P}}^{(1)} + \tilde{\mathbf{P}}^{NL}, \qquad (5.11)$$

and where we have introduced the linear displacement field $\mathbf{D}^{(1)}$ as

$$\tilde{\mathbf{D}}^{(1)} = [\epsilon_0]\tilde{\mathbf{E}} + \frac{4\pi}{[4\pi]} \tilde{\mathbf{P}}^{(1)}. \qquad (5.12)$$

In a dispersive medium, each frequency component needs to be considered separately, and, therefore, we decompose the fields into their discrete frequency components as

$$\tilde{\mathbf{E}}(\mathbf{r}, t) = \sum_n \mathbf{E}_n(\mathbf{r})e^{-i\omega_n t} + c.c., \qquad (5.13a)$$

$$\tilde{\mathbf{D}}^{(1)}(\mathbf{r}, t) = \sum_n \mathbf{D}_n^{(1)}(\mathbf{r})e^{-i\omega_n t} + c.c., \qquad (5.13b)$$

and

$$\tilde{\mathbf{P}}^{NL}(\mathbf{r}, t) = \sum_n \mathbf{P}_n^{NL}(\mathbf{r})e^{-i\omega_n t} + c.c.. \qquad (5.13c)$$

If the medium is also lossless, $\mathbf{D}^{(1)}$ and $\mathbf{E}_n$ are related by a real, frequency-dependent dielectric tensor as

$$\tilde{\mathbf{D}}_n^{(1)}(\mathbf{r}, t) = \boldsymbol{\epsilon}^{(1)}(\omega_n) \cdot \tilde{\mathbf{E}}_n(\mathbf{r}, t). \qquad (5.14)$$

Therefore, under most conditions, the nonlinear wave equation reduces to the form

$$\nabla^2 \mathbf{E}_n(\mathbf{r}) + \frac{1}{[\epsilon_0]c^2} \omega_n^2 \epsilon^{(1)}(\omega_n) \cdot \mathbf{E}_n(\mathbf{r}) = -\frac{4\pi}{[4\pi\epsilon_0]} \frac{1}{c^2} \omega_n^2 \mathbf{P}_n^{NL}. \tag{5.15}$$

Sum-Frequency Generation

Sum-frequency generation, depicted schematically in Fig. 5.1, is a common nonlinear optical process used to produce light at the sum frequency of the incident optical waves. If the incident beams are of the same frequency, the process is known as second-harmonic generation. The coupled-wave equations that govern this interaction and their solutions in a few illustrative regimes are presented below.

Coupled-Wave Equations

If we consider sum-frequency generation in a lossless, nonlinear medium with monochromatic plane waves, we may assume that the electric fields and nonlinear polarizations take the forms

$$\tilde{E}_i(z, t) = A_i e^{i(k_i z - \omega_i t)} + c.c., \tag{5.16}$$

where A_i is known as the slowly varying envelope of the electric field and

$$k_i = \frac{n_i \omega}{c}, \quad n_i = \left(\frac{\epsilon^{(1)}(\omega_i)}{[\epsilon_0]} \right)^{1/2}. \tag{5.17}$$

Similarly, we may assume that the nonlinear polarizations takes the form

$$\tilde{P}_i(z, t) = P_i e^{-i\omega_i t} + c.c.. \tag{5.18}$$

By considering which terms of the nonlinear polarization oscillate at the frequency $\omega_3 = \omega_1 + \omega_2$, we find from Eqs. 5.6–5.9 that the contribution to the nonlinear polarization that leads to sum-frequency generation is of the form

$$P_3 = [\epsilon_0] 4 d A_1 A_2 e^{i(k_1 + k_2)z}. \tag{5.19}$$

Using Eqs. 5.15 through 5.19, we obtain the nonlinear wave equation

$$\frac{d^2 A_3}{dz^2} + 2ik_3 \frac{dA_3}{dz} = -\frac{16\pi}{[4\pi]} \frac{d\omega_3^2}{c^2} A_1 A_2 e^{i(k_1 + k_2 - k_3)z}. \tag{5.20}$$

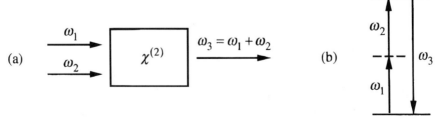

FIGURE 5.1 Sum-frequency generation. (a) Geometry of the interaction. (b) Energy-level description.

Nonlinear Optics

In most cases, we can neglect the first term on the left-hand side of the this equation because it is much smaller than the second term. This simplification is known as the slowly varying amplitude approximation. We, thereby, obtain the following equation for the spatial evolution of the sum-frequency field A_3:

$$\frac{dA_3}{dz} = \frac{8\pi}{[4\pi]} \frac{id\omega_3^3}{k_3 c^2} A_1 A_2 e^{i\Delta kz} \tag{5.21}$$

where the quantity

$$\Delta k = k_1 + k_2 - k_3 \tag{5.22}$$

is known as the wave vector mismatch. Similarly the coupled-wave equations for the other two optical fields are given by

$$\frac{dA_1}{dz} = \frac{8\pi}{[4\pi]} \frac{id\omega_1^3}{k_1 c^2} A_3 A_2^* e^{-i\Delta kz} \tag{5.23a}$$

and

$$\frac{dA_2}{dz} = \frac{8\pi}{[4\pi]} \frac{id\omega_2^3}{k_2 c^2} A_3 A_1^* e^{-i\Delta kz}. \tag{5.23b}$$

The Manley-Rowe Relations

If we now consider the spatial evolution of the intensities of the three interacting waves, it can easily be shown from Eqs. 5.21 through 5.23 that the following relationships hold for sum-frequency generation:

$$\frac{d}{dz}\left(\frac{I_1}{\omega_1}\right) = \frac{d}{dz}\left(\frac{I_2}{\omega_2}\right) = -\frac{d}{dz}\left(\frac{I_3}{\omega_3}\right) \tag{5.24}$$

where the intensities are given by

$$I_i = \frac{[4\pi\epsilon_0]}{2\pi} n_i c |A_i|^2 \tag{5.25}$$

Equations 5.24 are known as the Manley-Rowe relations. They tell us that the rate at which a photon at ω_1 is destroyed equals the rate at which a photon at ω_2 is destroyed which equals the rate at which a photon at frequency ω_3 is created. The relations can also be useful in deriving expressions for the spatial evolution of the intensities during nonlinear interactions.

Undepleted Pump Regime

The general solution of the three coupled equations for sum-frequency generation was derived by Armstrong et al. (1962). Here, we present the solution of the equations in a few specific cases to illustrate the general properties of the nonlinear interaction. When the efficiency of sum-frequency generation is sufficiently small that depletion of the incident waves can be neglected, the spatial evolution of the intensity of the sum-frequency wave is given by

$$I_3 = \left[\frac{1}{64\pi^3\epsilon_0}\right] \frac{512\pi^5 d^2 I_1 I_2}{n_1 n_2 n_3 \lambda_3^2 c} z^2 \mathrm{sinc}^2\left(\frac{\Delta k z}{2}\right). \tag{5.26}$$

Note the dependence on the wave vector mismatch Δk. For distances z greater than $2/\Delta k$, the sum-frequency field becomes out of phase with its driving polarization, and the power of the sum-frequency field flows back into the incident fields. For this reason, the coherence length of the interaction is defined as

$$L_C = \frac{2}{\Delta k}. \tag{5.27}$$

Upconversion

We now consider the case where one of the incident beams is weak and other is sufficiently strong that we can treat its amplitude as constant. This case is known as upconversion. Taking A_2 as the strong field amplitude, we find that the coupled-amplitude equations governing the evolution of A_1 and A_3 are

$$\frac{dA_1}{dz} = \kappa_1 A_3 e^{-i\Delta k z} \tag{5.28a}$$

and

$$\frac{dA_3}{dz} = \kappa_3 A_1 e^{i\Delta k z} \tag{5.28b}$$

where

$$\kappa_1 = \frac{8\pi}{[4\pi]} \frac{i\omega_1^2 d}{k_1 c^2} A_2^* \tag{5.29}$$

and

$$\kappa_3 = \frac{8\pi}{[4\pi]} \frac{i\omega_3^2 d}{k_3 c^2} A_2.$$

The general solution to these equations is given by

$$A_1(z) = \left[A_1(0)\cos gz + \left(\frac{\kappa_1}{g} A_3(0) + \frac{i\Delta k}{2g} A_1(0)\right)\sin gz\right] e^{-i\Delta k z/2} \tag{5.30a}$$

and

$$A_3(z) = \left[A_3(0)\cos gz + \left(\frac{\kappa_3}{g} A_1(0) - \frac{i\Delta k}{2g} A_3(0)\right)\sin gz\right] e^{i\Delta k z/2} \tag{5.30b}$$

where g is the positive square root of the expression

$$g = \sqrt{-\kappa_1\kappa_3 + \frac{1}{4}\Delta k^2}. \qquad (5.31)$$

Consider the special case in which there is no sum-frequency field incident upon the nonlinear medium. In this case, the evolution of the intensity of sum-frequency field is given by

$$I_3(z) = \frac{n_3}{n_1} I_1(0) \frac{|\kappa_3|^2}{g^2} \sin^2 gz. \qquad (5.32)$$

Note that the maximum sum-frequency intensity is generated when $\Delta k = 0$ and that the maximum decreases by the factor $\kappa_1\kappa_3/\left(\kappa_1\kappa_3 - \frac{1}{4}\Delta\kappa^2\right)$ for nonzero wave vector mismatch. The evolution of the intensity of the sum-frequency field is illustrated in Fig. 5.2. Note the sinusoidal flow of energy into and then out of the sum-frequency field. The characteristic length of evolution of the interaction is g^{-1} so that the frequency of the oscillations increases with the wave vector mismatch.

Second-Harmonic Generation

Second-harmonic generation displays the same basic features as illustrated above. See the work of Armstrong et al. (1962) for the general plane-wave solutions. In practice, however, to achieve the highest conversion efficiency, it is necessary to consider detailed aspects of the problem such as the need for a small positive wave vector mismatch and the spatial walk-off of the focused fundamental and second-harmonic beams. A detailed review of this problem is beyond the scope of this chapter, and readers are referred to the text by Zernike and Midwinter (1973) and to the exhaustive treatment by Boyd and Kleinman (1968).

Difference-Frequency Generation

The process of difference-frequency generation is schematically illustrated in Figure 5.3. Some aspects of the evolution of the difference-frequency wave are qualitatively different from the processes treated above. The following treatment illustrates those differences.

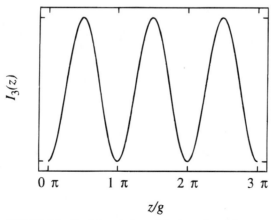

FIGURE 5.2 Spatial evolution of the intensity of the sum-frequency wave in the undepleted pump regime.

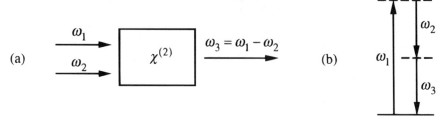

FIGURE 5.3 Difference-frequency generation. (a) Geometry of the interaction. (b) Energy-level description.

Coupled-Wave Equations

An treatment analogous to that outlined in section 3.2.1 for sum-frequency generation leads to the following coupled-wave equations for the spatial evolution of the difference-frequency field:

$$\frac{dA_1}{dz} = \frac{8\pi}{[4\pi]} \frac{id\omega_1^2}{k_1 c^2} A_3 A_2 e^{-i\Delta k z}, \tag{5.33a}$$

$$\frac{dA_2}{dz} = \frac{8\pi}{[4\pi]} \frac{id\omega_2^2}{k_2 c^2} A_1 A_3^* e^{i\Delta k z}, \tag{5.33b}$$

and

$$\frac{dA_3}{dz} = \frac{8\pi}{[4\pi]} \frac{id\omega_3^2}{k_3 c^2} A_1 A_2^* e^{i\Delta k z} \tag{5.33c}$$

where

$$\Delta k = k_1 - k_2 - k_3. \tag{5.34}$$

Undepleted Pump Regime

If we assume that the applied pump field A_1 is sufficiently strong that it can be treated as constant, the general solutions to Equations 5.33 are

$$A_2(z) = \left[A_2(0) \left(\cosh gz - \frac{i\Delta k}{2g} \sinh gz \right) + \frac{\kappa_2}{g} A_3^*(0) \sinh gz \right] e^{i\Delta k z/2} \tag{5.35a}$$

and

$$A_3(z) = \left[A_3(0) \left(\cosh gz - \frac{i\Delta k}{2g} \sinh gz \right) + \frac{\kappa_3}{g} A_2^*(0) \sinh gz \right] e^{i\Delta k z/2} \tag{5.35b}$$

where

$$g = \left[\kappa_2 \kappa_3^* - \left(\frac{\Delta k}{2} \right)^2 \right]^{1/2} \tag{5.36}$$

Nonlinear Optics

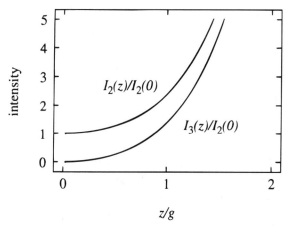

FIGURE 5.4 Spatial evolution of the intensities I_2 and I_3 for difference-frequency generation in the undepleted pump regime.

and

$$\kappa_j = \frac{8\pi}{[4\pi]} \frac{id\omega_j^2 A_1}{k_j c^2}. \tag{5.37}$$

The spatial evolution of the intensity of the A_2 and A_3 fields for the boundary condition $A_3(0) = 0$ and zero wave vector mismatch is shown in Fig. 5.4. Notice that both fields experience monotonic growth. This is qualitatively different from the oscillatory behavior of sum-frequency generation.

Processes such as difference-frequency generation, in which the initial and final quantum-mechanical states are identical, are known as parametric processes. For this reason, the amplification of the field A_2 as illustrated in Fig. 5.4 is often referred to as parametric amplification. If the purpose of the nonlinear process is to amplify the A_2 field, this field is referred to as the signal wave, and the field A_3 is called the idler wave. If an optical resonator that is highly reflective at frequency ω_2 is placed around the nonlinear medium, oscillation can occur due to the gain provided by parametric amplification. This type of a device, shown schematically in Fig. 5.5, is known as an optical parametric oscillator (OPO). The efficiency of an OPO can be increased at the expense of simplicity and stability if the device is made resonant at both the signal and idler frequencies. For a review of the current status of OPOs, see the review article by Tang et al. (1992).

Phase-Matching Considerations

As illustrated in the above discussion, the conversion efficiency of nonlinear mixing processes is critically sensitive to the amount of wave vector mismatch. To maximize the conversion efficiency,

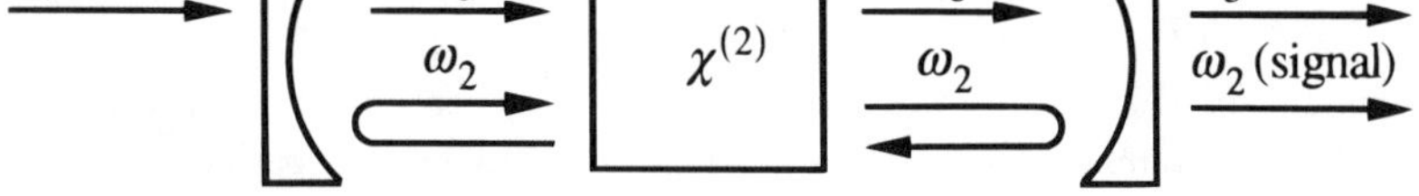

FIGURE 5.5 Schematic illustration of an optical parametric oscillator.

one usually needs to satisfy the phase-matching condition $\Delta k = 0$. For example, for the case of sum-frequency generation, the phase-matching condition $\Delta k = 0$ implies that

$$n_1\omega_1 + n_2\omega_2 = n_3\omega_3 \qquad (5.38)$$

where

$$\omega_1 + \omega_2 = \omega_3. \qquad (5.39)$$

Because most optically transparent materials exhibit normal dispersion (i.e., their refractive indices monotonically increase with frequency), this phase-matching requirement is impossible to satisfy without using special techniques of the sort described below.

Angle Tuning

Angle-tuning is by far the most common method employed to achieve phase matching. With this method, optical fields of different polarizations propagate in a birefringent crystal at the orientation necessary to achieve phase matching. For example, in the case of sum-frequency generation, if the optical fields with the lower two frequencies ω_1 and ω_2 are polarized so that they experience the ordinary refractive index of a uniaxial crystal and the sum-frequency field ω_3 is polarized so that it experiences the extraordinary index, the phase-matching conditions is given by

$$n_1^o\omega_1 + n_2^o\omega_2 = n_3^e\omega_3. \qquad (5.40)$$

This condition can be satisfied if the crystal is a negative, uniaxial crystal such that $n_e < n_o$. The case (assumed in Eq. 5.40) in which the lower frequency waves have the same polarization is known as type I phase matching, and the case where the lower frequency waves are orthogonally polarized is known as type II phase matching.

Of course, only rarely will the principal axis of a crystal provide the refractive indices necessary to achieve phase matching. In practice, the angular orientation of the crystal must be adjusted. In the case of a uniaxial crystal, the extraordinary polarization experiences the refractive index $n_e(\theta)$ defined by the relationship

$$\frac{1}{n_e(\theta)^2} = \frac{\sin^2\theta}{\overline{n}_e^2} + \frac{\cos^2\theta}{n_o^2} \qquad (5.41)$$

where θ is the angle between the propagation vector and the optic axis. By properly adjusting the angle θ, the value of $n_e(\theta)$ can be altered to satisfy the phase-matching condition $\Delta k = 0$.

Temperature Tuning

If the phase matching angle θ of Eq. 5.41 has a value other than 90°, the Poynting and the propagation vectors of extraordinary waves propagating within the crystal are not parallel. This effect leads to a spatial walk-off of the interacting beams and can sufficiently decrease the efficiency of the nonlinear mixing process as to make angle tuning impractical. In some cases, this problem can be overcome by using temperature to tune the birefringence of the crystal.

Quasi Phase Matching

Another technique for achieving efficient sum-frequency generation is known as quasi phase matching. With this technique, parameters of the nonlinear medium are periodically altered along the length of the medium to prevent the flow of energy from the sum-frequency field. Consider the sinusoidal evolution of the sum-frequency field with nonzero wave vector mismatch, as illustrated in Fig. 5.2. If the sign of the wave vector mismatch or the sign of the nonlinear susceptibility is periodically changed at multiples of $\pi/2g$, energy will monotonically flow from the pump fields into the sum-frequency field. This technique was first proposed by Armstrong et al. (1962). The recent work of Suhara and Nishihara (1990) has shown that the optimum solution is to sinusoidally modify the nonlinear susceptibility. This may be done with lithium niobate, for example. Great strides have been made in recent years in implementing this scheme in nonlinear guided-wave devices. See the work of Lim et al. (1989) as an example.

Cherenkov Radiation in Waveguides

Another technique specific to nonlinear waveguides has been very effective in overcoming phase-matching limitations and efficiently generating second-harmonic radiation. With this technique, the fundamental wave is propagated in a nonlinear waveguide with a phase velocity that is less than the phase velocity of the second-harmonic wave freely propagating in the substrate. This situation leads to the efficient generation of second-harmonic radiation in the form of Cherenkov radiation into the substrate. For more information on this technique, see the early work of Tien et al. (1970) and the review by Stegeman (1992).

Materials with Second-Order Nonlinear Susceptibilities

Properties of common second-order nonlinear materials can be found in Table 5.2. The nonlinear susceptibilities are listed in the contracted notation described above. Most of the values for the inorganic materials are taken from Singh (1986). BBO, LBO, and KTP are the most recently developed inorganic materials listed in the table and have become quite popular for use in optical parametric oscillators. See Eimerl et al. (1987), Chen et al. (1989), and Bierlein et al. (1989) for more information on these materials. Although they are not yet commercially available, several organic materials are included in the table. Single crystals of these materials were grown to characterize their susceptibility tensors. The values for the organic materials are taken from Bierlein et al. (1990), Kerkoc et al. (1989), Ledoux et al. (1990), and Singh (1986). The dispersion coefficients for most of the materials found in Table 5.2 are listed in Table 5.3. These coefficients allow for the calculation of phase-matching angles as described above.

5.4. Third-Order Nonlinearities

Third-Harmonic Generation

Third-harmonic generation is the process that leads to the creation of a wave at frequency 3ω in response to an applied wave at frequency ω. The source of the third-harmonic wave is the nonlinear polarization

$$P(3\omega) = [\epsilon_0]\chi^{(3)}(3\omega, \omega, \omega, \omega)E(\omega)^3. \tag{5.42}$$

Third harmonic generation is often used as a means of measuring the value of $\chi^{(3)}$ of optical materials.

Four-Wave Mixing

Four-wave mixing is the name given to a process in which input waves at frequencies ω_1, ω_2, and ω_3 interact in a nonlinear material by means of the nonlinear susceptibility $\chi^{(3)}(\omega_4, \omega_3, \omega_2, \omega_1)$ to create an output wave at frequency $\omega_4 = \omega_1 + \omega_2 + \omega_3$.

TABLE 5.2 Properties of Common Second-Order Nonlinear Optical Materials

Material	Point Group	d_{il} (pm/V)	$\lambda_{meas.}$ (μm)	Transmission Range (μm)
Inorganics				
Ammonium dihydrogen phosphate (ADP) $NH_4H_2PO_4$	$\overline{4}2m$	$d_{36} = 0.53$	1.064	0.2–1.1
Barium borate (BBO) β-BaB_2O_4	$3m$	$d_{11} = 1.8$	1.064	0.2–2.2
Barium sodium niobate $Ba_2NaNb_5O_{15}$	$mm2$	$d_{15} = -13$ $d_{24} = -13$ $d_{31} = -13$ $d_{32} = -13$ $d_{33} = -18$	1.064	0.25–5
Cadmium germanium arsenide $CdGeAs_2$	$\overline{4}2m$	$d_{36} = 460$	10.6	2.5–18
Cadmium selenide CdSe	$6mm$	$d_{15} = 31$ $d_{31} = -29$ $d_{33} = 55$	10.6	0.75–8
Cadmium sulfide CdS	$6mm$	$d_{15} = 14$ $d_{31} = -13$ $d_{33} = 26$	1.058	0.55–15
Gallium arsenide GaAs	$\overline{4}3m$	$d_{14} = 230$ $d_{36} = 210$	1.058	0.9–17
Lithium iodate $LiIO_3$	6	$d_{31} = -5.0$ $d_{33} = -5.2$	1.064	0.3–>5.6
Lithium niobate $LiNbO_3$	$3m$	$d_{22} = 2.6$ $d_{31} = -4.9$ $d_{33} = -44$	1.058	0.35–4.5
Lithium triborate (LBO) LiB_3O_5	$mm2$	$d_{31} = -1.2$ $d_{32} = 1.2$	1.079	0.16–2.6
Potassium dideuterium phosphate (KD*P) KD_2PO_4	$\overline{4}2m$	$d_{14} = 0.37$ $d_{36} = 0.38$	1.058	0.2–1.6
Potassium dihydrogen phosphate (KDP) KH_2PO_4	$\overline{4}2m$	$d_{14} = 0.44$ $d_{36} = 0.44$	1.064	0.2–1.4
Potassium titanyl phosphate (KTP) $KTiOPO_4$	mm	$d_{15} = 6.1$ $d_{24} = 7.6$ $d_{31} = 6.5$ $d_{32} = 5.0$ $d_{33} = 14$	1.064	0.35–4.5
Silver gallium selenide $AgGaSe_2$	$\overline{4}2m$	$d_{36} = 50$	10.6	0.75–20
Organics				
DAN 4-(N,N-dimethylamino)-3-acetamidonitrobenzene 4-(CH_3NCH_3)-3-$(CH_2CONH_2)NO_2C_6H_3$	2	$d_{21} = 1.5$ $d_{22} = 5.2$ $d_{23} = 50$ $d_{25} = 1.5$	1.064	0.49–2.2

TABLE 5.2 Properties of Common Second-Order Nonlinear Optical Materials—(*continued*)

Material	Point Group	d_{il} (pm/V)	$\lambda_{meas.}$ (μm)	Transmission Range (μm)
MAP methyl- (2,4-dinitrophenyl)- amino-propanoate $C_{10}H_{12}N_3O_6$	2	$d_{21} = 2.4$ $d_{22} = 12$ $d_{23} = 11$ $d_{25} = -0.35$	1.064	0.55–2.3
MNA 2-methyl-4-nitroaniline $CH_3\text{-}NH_2\text{-}NO_2\text{-}C_6H_4$	m	$d_{11} = 160$ $d_{24} = 24$	1.064	0.48–2
MMONs 3-methyl-4-methoxy- 4'-nitrostilbene $3\text{-}CH_3\text{-}4\text{-}CH_3OC_6H_5C =$ $CH(C_6H_4NO_2\text{-}4)$	mm2	$d_{24} = 71$ $d_{32} = 41$ $d_{33} = 184$	1.064	0.51–1.6
NPP N-4-nitrophenyl-(L)-prolinol $4\text{-}NO_2C_6H_4C_5H_8NO_3$	2	$d_{21} = 57$ $d_{22} = 19$	1.34	0.52–1.6
POM 3-methyl-4-nitropyridine-1- oxide $NO_2CH_3NOC_5H_4$	222	$d_{36} = 6.4$	1.064	0.49–2.1

(See text for references.)

Degenerate Four-Wave Mixing and Phase Conjugation

Degenerate four-wave mixing refers to the interaction of four waves all of the same frequency by means of the nonlinear susceptibility $\chi^{(3)}(\omega, \omega, \omega, -\omega)$. When the four waves interact in the geometry of Fig. 5.6, this process leads to optical phase conjugation, that is, the generation of an output wave, (the "conjugate" wave) whose complex amplitude A_4 is proportional to the complex conjugate of that of the signal wave A_3. In particular, a simple model of this interaction shows that (Yariv and Pepper, 1977)

$$A_4 = \frac{i\kappa}{|\kappa|} A_3^* \tan|\kappa|L \tag{5.43a}$$

where

$$\kappa = \frac{12\pi}{[4\pi]} \frac{\omega}{nc} \chi^{(3)} A_1 A_2. \tag{5.43b}$$

The importance of being able to generate a wave proportional to the complex conjugate of another wave is that this is exactly the transformation required to remove the effects of aberrations from certain optical systems. Optical phase conjugation thus, is, useful in aberration correction and other optical signal processing applications.

Intensity-Dependent Refractive Index

The process of degenerate four-wave mixing can be understood conceptually as a process in which the refractive index of a material becomes modified in proportion to the local intensity $I = [4\pi\epsilon_0](nc/2\pi)|E|^2$ of the light field according to

TABLE 5.3 Dispersion Parameters of Common Second-Order Nonlinear Optical Materials

Material	Dispersion Eq. #*	Index	A	B	C	D	E
Inorganics							
ADP	2	n_o	1.0	1.28196	0.01069		
		n_e	1.0	1.15607	0.00890		
AgGaSe$_2$	3	n_o	3.9362	2.9113	0.38821	1.7954	40
		n_e	3.3132	3.3616	0.38201	1.7677	40
Ba$_2$NaNb$_5$O$_{15}$	2	n_x	1.0	3.94655	0.040179		
		n_y	1.0	3.95233	0.040252		
		n_z	1.0	3.60287	0.032149		
BBO	1	n_o	2.7405	0.0184	0.0179	0.0155	
		n_e	2.3730	0.0128	0.0156	0.0044	
CdGeAs$_2$	3	n_o	10.1064	2.2988	1.0872	1.6247	1370
		n_e	11.8018	1.2152	2.6971	1.6922	1370
CdS	1	n_o	5.235	0.1819	0.1651		
		n_e	5.239	0.2076	0.1651		
CdSe	3	n_o	4.2243	1.7680	0.2270	3.1200	3380
		n_e	4.2009	1.8875	0.2171	3.6461	3629
GaAs	4	n	3.5	4.5×10^9	1.4×10^5	2.45×10^4	269
KDP	2	n_o	1.0	1.24361	0.00959		
		n_e	1.0	1.12854	0.00841		
KD*P	3	n_o	1.661824	0.585337	0.016017	0.691221	30
		n_e	1.687522	0.447488	0.017039	0.596216	30
KTP	2	n_x	2.1146	0.89188	0.20861	0.01320	
		n_y	2.1518	0.87862	0.21801	0.01327	
		n_z	2.3136	1.00012	0.23831	0.01679	
LBO	1	n_x	2.4517	0.01177	0.00921	0.00960	
		n_y	2.5279	0.01652	−0.005459	0.01137	
		n_z	2.5818	0.01414	0.01186	0.01457	
LiIO$_3$	2	n_o	1.0	2.40109	0.021865		
		n_e	1.0	1.91359	0.01940		
LiNbO$_3$	1	n_o	4.9048	0.11768	0.04750	0.027169	
		n_e	4.5820	0.099169	0.044432	0.021950	
Organics							
DAN	2	n_x	2.1390	0.1474	0.1355		
		n_y	2.3290	0.3072	0.1547		
		n_z	2.5379	0.7196	0.1759		
MAP	2	n_x	2.7523	0.6079	0.1606	0.05361	
		n_y	2.3100	0.2258	0.17988	0.01886	
		n_z	2.1713	0.10305	0.16951	0.01667	
MMONS	2	n_x	1.987	0.314	0.363		
		n_y	2.184	0.405	0.403		
		n_z	2.507	1.130	0.421		
NPP	2	n_x	2.3532	1.1299	0.1678	−0.0392	
		n_y	2.8137	0.3655	0.2030	0.0816	
		n_z	2.1268	0.0527	0.155	0.0608	
POM	2	n_x	2.5521	0.7962	0.1289	0.0941	
		n_y	2.4315	0.3556	0.1276	0.0579	
		n_z	2.4529	0.1641	0.1280	0.0	

*Dispersion Eq. #   Dispersion Equation

1. $n^2 = A + \dfrac{B}{(\lambda^2 - C)} - D\lambda^2$ (λ in μm)
2. $n^2 = A + \dfrac{B\lambda^2}{(\lambda^2 - C)} - D\lambda^2$ (λ in μm)
3. $n^2 = A + \dfrac{B}{(\lambda^2 - C)} + \dfrac{D\lambda^2}{(\lambda^2 - E)}$ (λ in μm)
4. $n^2 = A + \dfrac{B}{D^2 - \gamma^2} + \dfrac{E^2}{E^2 - \gamma^2}$ (γ in cm^{-1})

(See text for references.)

Nonlinear Optics

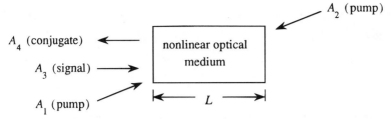

FIGURE 5.6 Geometry of degenerate four-wave mixing that leads to optical phase conjugation.

$$n = n_o + n_2 I \tag{5.44}$$

where

$$n_2 = \left[\frac{1}{16\pi^2 \epsilon_0}\right] \frac{12\pi^2}{n_o^2 c} \chi^{(3)}(\omega, \omega, \omega, -\omega). \tag{5.45}$$

This conceptual picture can be used to obtain an understanding of the nature of the degenerate four-wave mixing process illustrated in Fig. 5.6. One can imagine the signal wave A_3 interfering with one of the pump waves (A_1) to create a spatially varying intensity distribution which modulates the refractive index in accordance with Eq. (5.45) to form a transient grating in the material. The other pump wave, then, scatters from the grating to create the output, conjugate wave. Typical values of n_2 arising from various physical mechanisms are listed in Table 5.4.

Self-Focusing

Under many circumstances, the nonlinear refractive index coefficient n_2 is positive, and, in this case, the propagation of light through such a material can be unstable to a process known as self-focusing. In this process, rays of light are brought to a focus inside the material as a consequence of the lensing action induced by the tendency of the material to have a larger refractive index near the center of the beam where the intensity is largest. Self-focusing can occur only if the tendency of a beam to form a focus overcomes the tendency of the beam to expand due to diffraction effects, and a simple model of the self-focusing process predicts that this can occur only if the power in a beam exceeds the critical power whose value is given by

$$P_{cr} = \frac{\pi(0.61)^2}{8n_o n_2} \lambda^2. \tag{5.46}$$

TABLE 5.4 Typical Values of the Nonlinear Refractive Index

Physical Mechanism	n_2 (cm^2/W)	Response Time (sec)
Thermal effects	10^{-6}	10^{-3}
Electrostriction	10^{-14}	10^{-9}
Molecular orientation	10^{-14}	10^{-12}
Electronic polarization	10^{-16}	10^{-15}

For $P \gg P_{cr}$, the distance from the entrance of the nonlinear optical medium to the point of the self-focus is given by

$$z_f = \frac{2n_0}{0.61} \frac{w_0^2}{\lambda} \frac{1}{(P/P_{cr})^{1/2}} \tag{5.47}$$

where w_0 is the beam waist radius at the entrance face of the nonlinear medium.

Tensor Nature of the Third-Order Susceptibility

The third-order susceptibility is a fourth-rank tensor and, thus, possesses 81 components, each of which is independent in the most general case. However, for degenerate four-wave mixing in an isotropic material, $\chi^{(3)}(\omega,\omega,\omega,-\omega)$ possesses only two independent components and the nonlinear polarization can be expressed as (Maker and Terhune, 1965)

$$\mathbf{P} = [\epsilon_o]\left[A(\mathbf{E}\cdot\mathbf{E}^*)\mathbf{E} + \frac{1}{2} B(\mathbf{E}\cdot\mathbf{E})\mathbf{E}^* \right] \tag{5.48}$$

where $A = 6\chi^{(3)}_{1122}$ and $B = 6\chi^{(3)}_{1221}$.

Two-Photon Absorption

At sufficiently large laser intensity, the usual law of linear absorption becomes modified to

$$\frac{dI}{dz} = -\alpha I - \beta I^2 \tag{5.49}$$

where α is the usual linear absorption coefficient and β is a new constant that describes processes in which two photons are simultaneously removed from the light beam. Two-photon absorption can be understood as arising from the imaginary part of $\chi^{(3)}(\omega,\omega,\omega,-\omega)$.

Enhancement of Nonlinear Interactions in Optical Waveguides

The nonlinear susceptibility of many waveguide materials is orders of magnitude smaller than the susceptibility of common bulk nonlinear materials. Nonetheless, nonlinear effects in many waveguides can often be observed at modest power levels due to the small mode size and low loss of the waveguides. The strength of third-order nonlinear interactions depends on the product of the intensity and the effective interaction length. For a focused Gaussian beam, this product is given by

$$(IL_{\text{eff}})_{\text{gaussian}} = \left(\frac{P}{\pi w_0^2}\right)\left(\frac{\pi w_0^2}{\lambda}\right) = \frac{P}{\lambda} \tag{5.50}$$

where w_0 is the beam waist radius. Notice that the product is independent of w_0. For a waveguide, this product is given by

$$(IL_{\text{eff}})_{\text{waveguide}} = \left(\frac{P}{\pi w_0^2}\right)\left(\frac{1}{\alpha}\right) = \frac{P}{\pi w_0^2 \alpha} \tag{5.51}$$

Nonlinear Optics

where w_o is the mode radius and L_{eff} is taken as one Beer's length. The ratio of these two product is

$$\frac{(IL_{eff})_{waveguide}}{(IL_{eff})_{Gaussian}} = \frac{\lambda}{\pi w_0^2 \alpha}. \tag{5.52}$$

This is the enhancement factor for the strength of nonlinear interactions occurring in a waveguide rather than in a bulk geometry. In some cases, this enhancement factor is exceedingly large. For example, near the minimum-loss wavelength of 1.55 μm in optical fibers, the enhancement factor is $\sim 10^9$. This huge enhancement causes nonlinear optical processes, like stimulated Raman scattering, to limit the power levels of fiber communication systems. In the next section, we discuss a situation in which nonlinear optics can be used to enhance the performance of communication systems.

Soliton Propagation

The propagation of pulses of picosecond duration in optical fibers is often governed by the nonlinear Schrodinger equation shown below:

$$\frac{\partial A}{\partial z} = -\frac{i}{2}\beta_2 \frac{\partial^2 A}{\partial \tau^2} + i\gamma |A|^2 A. \tag{5.53}$$

Here, A is the slowly varying envelope of the electric field, z is the longitudinal coordinate, τ is the local time defined by $\tau = t - z v_g$ where v_g is the group velocity, γ is a nonlinear response parameter given by $\gamma = \{[4\pi\epsilon_0]/2\pi\} n_2 n_o \omega_o$, and β_2 is the group velocity dispersion parameter defined by

$$\beta_2 = \left(\frac{\partial^2 \beta}{\partial \omega^2}\right)\bigg|_{\omega=\omega_o} \tag{5.54}$$

where β is the wavenumber of the fundamental mode of the fiber. The first term on the right-hand side of Eq. 5.53 treats the effects of group-velocity dispersion. At the minimum-loss wavelength of standard communication fibers, pulses experience anomalous group-velocity dispersion. In this regime, the higher frequency components of the pulse propagate at a faster group velocity than do lower frequency components, and, thus, the pulse broadens and becomes down-chirped. This effect can be balanced against the effects of self-phase modulation which are treated by the second term on the right-hand side of Eq. 5.53. Shown in Fig. 5.7 is the nonlinear phase shift experienced by a pulse propagating under the influence of self-phase modulation in a medium with a positive nonlinear refractive index. This nonlinear phase modulation induces an instantaneous frequency shift given by

$$\delta\omega = -\frac{\partial \phi_{ne}}{\partial \tau}. \tag{5.55}$$

The leading edge of the pulse is shifted to lower frequencies whereas the trailing edge is upshifted. This effect can counteract the pulse-broadening effects of group-velocity dispersion for the proper pulse shape and intensity. The result is a pulse known as the fundamental bright soliton which has the form

$$A = \sqrt{\frac{1}{L_D \gamma}} \operatorname{sech}\left(\frac{\tau}{\tau_p}\right) e^{iz/2L_D} \tag{5.56}$$

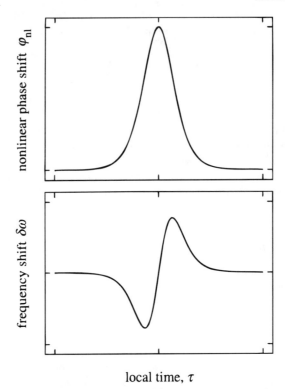

FIGURE 5.7 Nonlinear phase shift and instantaneous frequency shift induced by self-phase modulation.

where

$$L_D = \frac{\tau_p^2}{|\beta_2|}. \tag{5.57}$$

Notice that this pulse propagates without changing shape and without becoming chirped. In addition, the pulse accumulates a uniform nonlinear phase shift across its profile. This feature makes solitons natural bits of information for the all-optical switching schemes discussed in the next section. For a derivation of the above equations and for an extensive discussion of nonlinear fiber optics, see Agrawal (1989). See the work of Mollenauer (1993) as an example of recent soliton-transmission experiments.

Prototypical Nonlinear Optical Switches

Four examples of nonlinear optical switches are depicted in Fig. 5.8. For simplicity, the first example is a nonlinear Mach–Zehnder interferometer. One arm of the interferometer is constructed with a highly nonlinear material to produce an intensity-dependent nonlinear phase shift and, thereby, nonlinear switching. Nonlinear common-path interferometers are also being investigated (Blow et al., 1989). In particularly, nonlinear loop mirrors (i.e., nonlinear Sagnac interferometers constructed with all-fiber optic components) are being investigated as all-optical switches in communication systems (Mollenauer et al., 1993) and have been used to mode lock fiber lasers (Duling, 1991). The nonlinear directional coupler depicted in Fig. 5.8b is another promising waveguide switch. In this device, evanescent coupling occurs between two nonlinear waveguides. Nonlinear directional coupling has been demonstrated in glass fibers (Gusovskii et al., 1985), in

Nonlinear Optics

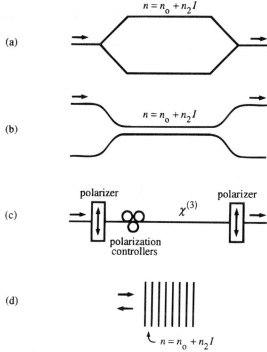

FIGURE 5.8 Four examples of nonlinear optical switches: (a) nonlinear Mach–Zehnder interferometer, (b) nonlinear directional coupler, (c) nonlinear polarization evolution switch, and (d) nonlinear Bragg reflector.

polymers (Townsend et al., 1989), and in semiconductors (Jin et al., 1990). The switch depicted in Fig. 5.8c is based on nonlinear polarization evolution. The input light is passed through a linear polarizer and, then, polarization controllers to produce elliptically polarized light. Due to the tensor nature of the third-order susceptibility, the light, then, undergoes nonlinear polarization evolution, creating an intensity-dependent transmission through the output polarizer. This device has recently been used to mode lock fiber lasers (Matsas et al., 1992). Depicted in Fig. 5.8d is a nonlinear Bragg reflector, constructed with nonlinear material to make the reflectivity intensity-dependent. For more details, see Ehrlich (1990).

Two-photon absorption and nonlinear index saturation limit the performance of the nonlinear switches described above. To identify materials suited for nonlinear switching applications, the following figures of merit have been developed:

$$W = \frac{\Delta n_{\max}}{\alpha \lambda}, \qquad T = \frac{2\beta\lambda}{n_2} \tag{5.58}$$

where α and β are the linear and two-photon absorption coefficients, respectively, as defined in Eq. 5.49, and $\Delta n_{\max}$ is the limiting value of the nonlinear index change. Useful optical switching requires $W > 2$ and $T > 1$. See Stegeman (1987), Mizrahi (1989), and Stegeman (1992) for more details.

Materials with Third-Order Nonlinear Susceptibilities

Parameters of some common third-order nonlinear materials are listed in Table 5.5. In each case, the nonlinear refractive index experienced by linearly polarized light is listed. The values of the

TABLE 5.5 Properties of Common Third-Order Nonlinear Materials

Material	n_o	n_2 (10^{-16} cm²/W)	$\lambda_{meas.}$ (μm)	Pulse Duration (picoseconds)
Acetone $(CH_3)_2$ CO	1.35	24	1.064	10
Borosilicate glass (BK-7)	1.52	3.4	1.064	125
Carbon disulfide CS_2	1.63	514	1.064	10
		290	1.064	1000
Carbon tetrachloride CCl_4	1.45	15	1.064	10
Cadmium sulfide CdS	2.34	500	1.064	30
Diamond C	2.42	12.6	0.545	4000
Gallium arsenide GaAs	3.47	−3300	1.064	30
Germanium Ge	4.00	2800	10.6	30
Lithium flouride LiF	1.39	1.05	1.064	125
Nitrobenzene $C_6H_5NO_2$	1.54	685	1.064	10
Potassium titanyl phosphate (KTP) $KTiOPO_4$	1.78	31	1.064	30
Sapphire Al_2O_3	1.76	3.1	1.064	30
Silica, fused SiO_2	1.458	2.7	1.064	125
Yttrium aluminum garnet (YAG) $Y_3Al_5O_{12}$	1.83	7.2	1.064	150

(See text for references.)

parameters for the liquids and the large-bandgap crystals are from Smith (1986), the semiconductor parameters are from Sheik-Bahae et al. (1991), and the parameters for the glasses are from Milam and Weber (1976).

5.5 Stimulated Light Scattering

General Features of Stimulated Light Scattering

Light can be removed from an incident laser beam either by spontaneous or by stimulated light scattering. Spontaneous light scattering is the dominant process at low intensities, whereas stimulated scattering dominates at higher intensities (greater than approximately 100 MW/cm²). The distinction between these processes is that spontaneous light scattering involves the interaction of light with material excitations created either by thermal excitation or by quantum mechanical zero-point fluctuations, whereas stimulated light scattering involves material excitations created or enhanced by the presence of the light field within the material. The connection between stimulated and spontaneous light scattering was first elucidated by Hellwarth (1963), who showed, theoretically, that the rate at which photons are scattered into some particular mode S (which, for historical reasons, is known as the Stokes mode) can be represented as

$$\frac{dm_S}{dt} = Dm_L(m_S + 1) \quad (5.59)$$

where m_L and m_S are the number of photons per mode for the laser and Stokes fields, respectively, and D is a constant of proportionality. In the factor $m_S + 1$, the m_S and the 1 represent the stimulated and spontaneous contributions to light scattering, respectively. It is generally believed that Equation (5.59) is valid for all light scattering processes and, thus, that a stimulated version of every spontaneous light scattering process must exist. Table 5.6 shows the properties of some typical light scattering processes. Note that stimulated Brillouin scattering (SBS) has highest gain coefficient of these processes, but has a rather slow response time of $\sim$ 1 ns. Thus, SBS tends to be the dominant scattering process for laser pulses sufficiently long to excite SBS. For shorter

Nonlinear Optics

TABLE 5.6 Typical Light Scattering Processes

Process	Material Excitation	Frequency Shift	Response Time	Gain (cm/MW)
Raman	Molecular vibrations (optical phonons)	30 THz	1 ps	5×10^{-3}
Brillouin	Sound waves (acoustic phonons)	3 GHz	1 ns	10^{-2}
Rayleigh	Density variations	0	10 ns	10^{-4}
Rayleigh wing	Molecular orientation	0	1 ps	10^{-3}

pulses, stimulated Raman scattering (SRS) tends to be the dominant process. For these reasons, the most important stimulated light scattering processes are SBS and SRS, and the ensuing discussion will be restricted to these processes.

Boyd (1992) has presented a pedagogical description of stimulated light scattering.

Stimulated Raman Scattering

Raman scattering is light scattering by the interaction of light with the vibrational degree of freedom of a molecule. The scattered light is shifted in frequency by a vibrational frequency of the molecule that scatters the light. Raman scattering can be described by the energy level diagram of Figure 5.9, which shows a laser photon of frequency ω_L being scattered into a photon of frequency $\omega_S = \omega_L - \omega_V$ with the molecule being excited from its ground state $v = 0$ to its first excited vibrational state $v = 1$. These states differ in energy by $\hbar\omega_V$.

Stimulated Raman scattering can be studied either in the amplifier configuration of Figure 5.10 or the generator configuration of Figure 5.10b. In the amplifier configuration, both the laser wave and a (typically weak) Stokes wave are applied to the optical medium and the Stokes wave experiences exponential growth according to

$$|A^S(z)|^2 = |A^S(0)|^2 e^{Gz} \tag{5.60}$$

where the gain coefficient G can be represented as

$$G = \frac{24\pi}{[4\pi]} i \frac{\omega_s}{n_s c} \chi_R(\omega_S) |A_L|^2, \tag{5.61}$$

where $\chi_R(\omega_S) \equiv \chi^{(3)}(\omega_S = \omega_S + \omega_L - \omega_L)$ is the Raman susceptibility. The gain coefficient can often be represented in terms of molecular parameters as

FIGURE 5.9 Energy-level description of stimulated Raman scattering.

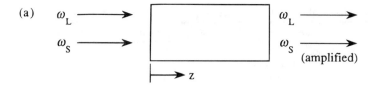

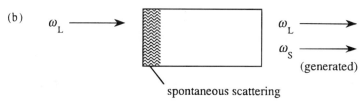

FIGURE 5.10 (a) Amplifier and (b) generator configuration for stimulated Raman scattering.

$$G = \frac{2\pi N}{n_S cm\gamma} \frac{\omega_S}{\omega_v} \left(\frac{\partial \alpha}{\partial q}\right)_0^2 |A_L|^2, \qquad (5.62)$$

where N is the number density of molecules, m is the reduced mass associated with the vibrational degree of freedom, γ is the width of the Raman resonance, and $(\partial\alpha/\partial q)_0$ is a measure of how rapidly the electronic polarizability α changes with the vibrational coordinate q. Analogous considerations show that a weak wave at the anti-Stokes frequency $\omega_a = \omega_L + \omega_V$ experiences attenuation according to

$$|A^a(z)|^2 = |A^a(0)|^2 e^{-Gz}. \qquad (5.63)$$

Stokes amplification (anti-Stokes attenuation) is an example of an induced gain (absorption) process. The presence of gain, that is, does not depend on the fulfillment of a phase-matching condition, and the value of the gain coefficient G does not depend critically on the angle between the laser and Stokes beams (which for simplicity is shown as zero in Fig. 5.10).

The properties of some common Raman materials are shown in Table 5.7.

While the Raman amplifier configuration of Fig. 5.10a lends itself to a simple theoretical description, most applications of SRS use the generator configuration of Fig. 5.10b. In this case, no Stokes beam is applied externally; instead, SRS is initiated by light produced at the Stokes frequency by spontaneous Raman scattering, which occurs as a consequence of quantum fluctuations of the material medium. The theoretical analysis of this situation is quite complicated, although detailed models of the initiation of SRS have been presented by Raymer and Mostowski

TABLE 5.7 Properties of Some Raman Media

Substance	Frequency Shift (cm^{-1})	Gain Factor G/I_L (cm/GW)
Benzene	992	2.8
Nitrobenzene	1345	2.1
SiO$_2$	467	0.8
H$_2$ gas (P > 10 atm)	4155	1.5
CS$_2$	656	24

Source: Kaiser and Maier, 1972; Penzkopfer et al., 1979.

(1981). As a good approximation, SRS in the generator configuration is a threshold process, and the condition for the occurrence of SRS is given by

$$GL > (GL)_{th}, \qquad (5.64)$$

where L is the length of the interaction region and the value $(GL)_{th}$ of the product GL at the threshold for SRS is approximately 30.

One application of SRS is as a frequency shifter for light fields. As an example, recall that radiation at a wavelength of 1.06 μm can be generated with high efficiency by the Nd:YAG laser. This radiation can be shifted to different wavelengths through SRS. Note that molecular hydrogen has the largest frequency shift of any known material (see Table 5.7) and, thus, is particularly useful for Raman frequency shifting.

Stimulated Brillouin Scattering

Brillouin scattering is the scattering of light from sound waves. Like SRS, stimulated Brillouin scattering (SBS) is an induced gain process, i.e., the Stokes wave intensity I_S experiences expotential growth of the form

$$I_S(z) = I_S(0)e^{gI_Lz}. \qquad (5.65)$$

Nonetheless, unlike SRS, SBS cannot occur in the exact forward direction. The reason is that the acoustic wave must obey the dispersion relation $\Omega = v|\mathbf{q}|$, where Ω is the angular frequency of the acoustic wave and $\mathbf{q}$ is its wavevector, and consequently, in the forward direction, only a zero-frequency photon can couple to the laser and Stokes waves so that $\omega_S = \omega_L - \Omega$ and $\mathbf{k}_S = \mathbf{k}_i - \mathbf{q}$. Consequently, SBS is usually observed only in the backward and near backward directions, in the configuration illustrated in Fig. 5.11. In the backward direction, the frequency shift associated with the SBS process is given by

$$\Omega = 2\omega \frac{v}{c/n}. \qquad (5.66)$$

where v denotes the velocity of sound in the material.

Detailed consideration (Boyd, 1992) of the coupling between the acoustic and optical waves through use of Maxwell's equation and the equation of hydrodynamics shows that the gain factor g of Equation (5.64) is given by

$$g = \frac{\gamma^2 \omega^2}{[\epsilon_0^2] nvc^3 \rho_0 \Gamma_B}, \qquad (5.67)$$

where $\gamma \equiv \rho(\partial\epsilon/\partial\rho)$ is the electrostrictive coupling coefficient, ρ_0 is the material density, and Γ_B is the width (full width at half-maximum in angular frequency units) of the Brillouin resonance.

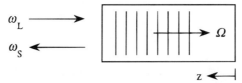

FIGURE 5.11 Stimulated Brillouin scattering involves scattering of the optical wave ω_L from a retreating sound wave to produce a backward Stokes wave.

TABLE 5.8 Properties of Several Materials Used for SBS*

Material	Frequency Shift $\Omega/2\pi$ (MHz)	Linewidth $\Gamma/2\pi$ (MHz)	Gain Factor g (cm/GW)
CS_2	5,850	52.3	150
CCl_4	4,390	520	6
Methanol	4,250	250	13
Ethanol	4,550	353	12
H_2O	5,690	317	4.8
Optical glasses	11,000–17,000	10–106	4–25
Fused silica	17,000	78	4.5

*Values quoted for a wavelength of 0.694 μm.

Source: Kaiser and Maier, 1972.

Some of the properties of materials commonly used for SBS are listed in Table 5.8

Like SRS, SBS can be excited either in an amplifier configuration (in which both laser and Stokes waves are externally applied to the interaction region) or a generator configuration (in which only a laser wave is applied externally). Most applications involve a generator configuration. In a generator configuration, SBS displays a threshold behavior; the threshold for the occurrence of SBS is given by (Boyd, Rzazewski, and Narum, 1990)

$$gI_L L \geq (gI_L L)_{th} \tag{5.68}$$

where $gI_L L)_{th} \approx 25 - 30$.

One application of SBS is in the field of optical phase conjugation. Detailed considerations of the transverse structure of the laser and Stokes fields show (Zel'dovich et al., 1972; Boyd and Grynberg, 1992) that, under many circumstances, the generated Stokes wave is a phase-conjugate replica of the incident laser wave, that is $E_S^{out} \propto (E_L^{in})^*$. The generation of phase-conjugate wavefronts is important because, under certain situations, phase conjugation can be used to remove the influence of aberrations from optical systems.

Another application of SBS is laser pulse compression. Under appropriate conditions, an appreciable fraction of the laser pulse energy can be transferred to an output Stokes pulse that is considerable shorter than the incident laser pulse. Pulse compression by a factor of 10 is possible by means of this effect. The origin of this effect is purely kinematic. The leading edge of the Stokes pulse is more efficiently amplified than the trailing edge because the laser and Stokes pulses are conterpropagating and the trailing edge interacts with a portion of the pump pulse that that has been depleted by the leading edge of the Stokes pulse. One necessary condition for the occurrence of pulse shortening is that the longitudinal extent of the interaction region must exceed the physical length of the laser pulse.

Acknowledgments

The authors would like to acknowledge the assistance of George Fischer in preparing the tables on organic materials.

References

Agrawal, G. P., *Nonlinear Fiber Optics*, 1st ed., Academic Press, San Diego, 1989.

Armstrong, J. A., Bloembergen, N., Ducuing, J., and Pershan, P. S., Interactions between light waves in a nonlinear dielectric, *Phys. Rev.*, 127, 1918, 1962.

Bierlein, J. D., Cheng L. K., Wang, Y., and Tam, W., Linear and nonlinear optical properties of 3-methyl-4-methoxy-4'-nitrostilbene single crystals, *Appl. Phys. Lett.*, 56, 423, 1990.

Bierlein, J. D., and Vanherzeele, H., Potassium titanyl phosphate: properties and new applications, *J. Opt. Soc. Am. B,* 6, 622, 1989.

Blow, K. J., Doran, N. J., and Nayar, B. K., Experimental demonstration of optical soliton switching in an all-fiber nonlinear Sagnac interferometer, *Opt. Lett.,* 14, 754, 1989.

Boyd, G. D., and Kleinman, D. A., Parametric interaction of focused Gaussian light beams, *J. Appl. Phys.,* 39, 3597, 1968.

Boyd, R. W., *Nonlinear Optics,* 1st ed., Academic Press, Boston, 1992.

Boyd, R. W., and Grynberg, G., *Contemporary Nonlinear Optics,* 1st ed., Academic Press, San Diego, Chapter 3, 1992.

Boyd, R. W., Rzazewski, K., and Narum, P., Noise initiation of stimulated Brillouin scattering, *Phys. Rev. A,* 42, 5514, 1990.

Butcher, P. N., and Cotter, D., *The Elements of Nonlinear Optics,* Cambridge University Press, Cambridge, 1990.

Chen, C., Wu, Y., Jiang, A., Wu, B., You, G., Li, R., and Lin, S., New nonlinear-optical crystal: LiB_3O_5, *J. Opt. Soc. Am. B,* 6, 616, 1989.

Duling, I. N., All-fiber ring soliton laser mode locked with a nonlinear mirror, *Opt. Lett.,* 16, 539, 1991.

Ehrlich, J. E., Assanto, G., and Stegeman, G. L, All-optical tuning of waveguide nonlinear distributed feedback gratings, *Appl. Phys. Lett.,* 56, 602, 1990.

Eimerl, D., Davis, L., Velsko, S., Graham, E. K., and Zalkin, A., Optical, mechanical, and thermal properties of barium borate, *J. Appl. Phys.,* 62, 1968, 1987.

Gusovskii, D. D., Dianov, E. M., Maier, A. A., Neustruev, V. B., Shklovskii, E. I., and Shcherbakov, I. A., Nonlinear light transfer in tunnel-coupled optical waveguides, *Sov. J. Quantum Electron.,* 15, 1523, 1985.

Hellwarth, R. W., Theory of stimulated Raman scattering, *Phys. Rev.,* 130, 1850, 1963.

Jin, R., Sokoloff, J. P., Harten, P. A., Chuang, C. L., Lee, S. G., Warren, M., Gibbs, H. M., Peyghambarian, Ultrafast modulation with subpicosecond recovery time in a GaAs/AlGaAs nonlinear directional coupler, *Appl. Phys. Lett.,* 56, 993, 1990.

Kaiser, W. and Maier, M. *Laser Handbook,* North-Holland, Amsterdam, Vol. 2, part E2, 1972.

Kerkoc, P., Zgonik, M., Sutter, K., Bosshard, Ch., and Gunter, P., Optical and nonlinear optical properties of 4-(N,N-dimethylamino)-3-acetamidonitrobenzene single crystals, *Appl. Phys. Lett.,* 54, 2062, 1989.

Ledoux, I., Lepers, C., Perigaud, A., Bandan, J., and Zyss, J., Linear and nonlinear optical properties of N-4-nitrophenyl-l-prolinol single crystals, *Opt. Commun.,* 80, 149, 1990.

Lim, E. J., Fejer, M. M., and Byer, R. L., Blue light generation by frequency doubling in periodically poled lithium niobate channel waveguide, *Electron. Lett.,* 25, 731, 1989.

Maker, P. D., and Terhune, R. W., Study of optical effects due to an induced polarization third order in the electric field strength, *Phys. Rev.,* 137, A801, 1965.

Matsas, V. J., Newson, T. P., Richardson, D. J., and Payne, D. N., Self-starting passively mode-locked fibre ring soliton laser exploiting nonlinear polarisation rotation, *Electron Lett.,* 28, 1391, 1992.

Milam, D., and Weber, M. J., Measurement of nonlinear refractive-index coefficients using time-resolved interferometry: Application to optical materials for high-power neodymium lasers, *J. Appl. Phys.,* 47, 2497, 1976.

Mizrahi, V., DeLong, K. W., Stegeman, G. I., Saifi, M. A., and Andrejco, M. J., Two-photon absorption as a limitation to all-optical switching, *Opt. Lett.,* 14, 1140, 1989.

Mollenauer, L. F., Lichtman, E., Neubelt, M. J., and Harvey, G. T., Demonstration, using sliding-frequency guiding filters, of error-free soliton transmission over more than 20 Mm at 10 Gbits/s, single channel, and over more than 13 Mm at 20 Gbit/s in a two-channel wdm, *Electron. Lett.,* 29, 910, 1993.

Penzkopfer, A., Laubereau, A., Kaiser, W., High intensity Raman interactions, *Prog. Quantum Electron.*, 6, 55, 1979.

Raymer, M. G., and Mostowski, J., Stimulated raman scattering: Unified treatment of spontaneous initiation and spatial propagation, *Phys. Rev. A*, 24, 1980, 1981.

Sheik–Bahae, M., Hutchings, D. C., Hagan, D. J., and Van Stryland, E. W., Dispersion of bound electronic nonlinear refraction in solids, *IEEE J. Quantum Electron.*, 27, 1296, 1991.

Shen, Y. R., *The Principles of Nonlinear Optics*, John Wiley & Sons, New York, 1984.

Singh, S., *Handbook of Laser Science and Technology*, CRC Press, Boca Raton, FL, Volume III, Section 1.1, 1986.

Smith, W. L., *Handbook of Laser Science and Technology*, CRC Press, Boca Raton, FL. Volume III, Section 1.3, 1986.

Stegeman, G. I., *Contemporary Nonlinear Optics*, 1st ed., Academic Press, San Diego, 1992, Chapter 1.

Stegeman, G. I., Parameter trade-offs in nonlinear directional couplers: Two level saturable nonlinear media, *Opt. Comm.*, 63, 281, 1987.

Suhara, T., and Nishihara, H., (1990). Theoretical analysis of waveguide second-harmonic generation phase matched with uniform and chirped gratings, *IEEE J. Quant. Electron.*, 26, 1265, 1990.

Tang, C. L., Bosenberg, W. R., Ukachi, T., Lane, R. J., and Cheng, L. K., Optical parametric oscillators, *Proc. IEEE*, 80, 365, 1992.

Tien, P. K., Ulrich, R., and Martin, R. J., Optical second-harmonic generation in form of coherent Cerenkov radiation from a thin-film waveguide, *Appl. Phys. Lett.*, 17, 447, 1970.

Townsend, P. D., Jackel, J. L., Baker, G. L., Shelburne, J. A., and Etemad, S., Observation of nonlinear optical transmission and switching phenomena in polydiacetylene-based directional couplers, *Appl. Phys. Lett.*, 55, 1829, 1989.

Yariv, A., and Pepper, D. M., Amplified reflection, phase conjugation, and oscillation in degenerate four-wave mixing, *Opt. Lett.*, 1, 16, 1977.

Zel'dovich, B. Y., Popovichev, V. I., Ragulsky, V. V., and Faizullov, F. S., Connection between the wave fronts of the reflected and excited light in stimulated Mandel'shtam-Brillouin scattering, *JETP Lett.*, 15, 109, 1972.

Zernike, F., and Midwinter, J. E., *Applied Nonlinear Optics*, 2nd ed., John Wiley & Sons, New York, 1973.

SECTION III
Organic Materials

6

Second-Order Nonlinear Optical Properties of Organic Materials

6.1	Introduction	155
6.2	Second-order Nonlinear Optical Effects	156
	Macroscopic Susceptibility • Microscopic Susceptibility	
6.3	Experimental Techniques	164
	Second-Harmonic Generation • Electric-Field Poling • Electro-Optic Modulation • DC Electric-Field-Induced Second-Harmonic Generation • Solvatochromic Method • Hyper-Rayleigh Scattering • Data Reduction at Infinite Dilution	
6.4	Molecular Studies	173
	Electron Donors and Acceptors • Substitution Pattern • Nonlinearity and Transparency Trade-Off • The π System	
6.5	Single Crystal Studies	229
	Single Crystals for Frequency Conversion • Single Crystals for Electro-Optics	
6.6	Poled-Polymer Studies	239
	Property Considerations • Polymers	
6.7	Concluding Remarks	271

Lap Tak Cheng
E.I. Du Pont de Nemours &
Company, Inc., Experimental
Station

6.1 Introduction

The nonlinear optical properties of organic materials have been extensively investigated during the past two decades. Early results of fundamental studies quickly led to intense activities motivated by potential applications, which include optical and electro-optical devices offering both performance and economy. Owing to the molecular nature of organic solids, investigations on properties, such as their optical nonlinearity, are amenable to a natural partitioning. The basic nonlinear optical polarization can be studied at the molecular level. Great effort has been devoted to understanding the structure and property relationships, the enhancement of molecular nonlinearities, and the optimization of trade-offs among relevant properties. Effective organization of the microscopic nonlinearity into macroscopic responses represents a significant challenge. Organizational approaches based on a single crystal, a quasi-crystalline thin film, and a glassy matrix are all pursued with encouraging results. As the stringent and variable material requirements for different applications become apparent, work on material optimization has emerged as the focus of much recent research.

In both molecular and bulk studies, the extreme flexibility of the organic approach has generated a large volume of published data. This article aims to condense much of the published information on second-order, optically nonlinear, organic materials into a single convenient reference. Inorganic, organometallic, and biological materials are not included. A brief introduction to second-order nonlinear optical effects is included to define relevant quantities. Common experimental techniques used in characterizing molecular and bulk materials are reviewed. Molecular results are reviewed to illustrate structure and property trends. Property trade-offs, as they impact key applications, are discussed. Published information on well-characterized noncentrosymmetric organic single crystals is surveyed. Their potential application for frequency conversion and electro-optic modulation are examined. Poled polymeric materials hold the best promise for realizing applications. The diverse material approaches are summarized, and key stability issues are discussed.

6.2 Second-Order Nonlinear Optical Effects

Macroscopic Susceptibility

Many authoritative treatments exist on the subject of nonlinear optics (Bloembergen, 1965; Shen, 1984) and its adaptation to organic materials (Chemla, 1987; Prasad and Williams, 1991). To facilitate discussion, we begin with an abbreviated introduction to nonlinear optics. Only charge neutral and nonmagnetic media will be considered. Under the dipole approximation, Maxwell equations and the associated constitutive equations are reduced to the following relationships between the electric field vector E, the electric displacement vector D, and the electric polarization vector P:

$$\nabla \times \nabla \times E = -\frac{1}{c^2}\frac{\partial^2}{\partial t^2} D \tag{6.1}$$

and

$$D = E + 4\pi P = \epsilon \cdot E, \tag{6.2}$$

where c is the speed of light and ϵ is the dielectric constant. Gaussian cgs units are used. Eqs. (6.1) and (6.2) can be combined to yield the familiar wave equation (3) with plane wave solutions, which are expressed as one half of the sum of two complex conjugates plus a static field in Eq. (6.4). E_0^ω is time independent but carries an arbitrary phase, $E_0^\omega = \xi(r)e^{i\varphi}$.

$$\nabla^2 E - \frac{\epsilon}{c^2}\frac{\partial^2 E}{\partial t^2} = 0, \tag{6.3}$$

and

$$E(r, t) = \frac{1}{2}\left[E_0^\omega e^{i(\omega t - kr)} + E_0^\omega e^{-i(\omega t - kr)}\right] + E^0. \tag{6.4}$$

Substituting Eq. (6.4) in Eq. (6.3) yields the dispersion relationship defining the wave vector $k = \omega\sqrt{\epsilon}/c = \omega n/c$, where n is the index of refraction.

The nonlinear optical response of a medium perturbed by an optical field is formally incorporated into Maxwell's equations by introducing a field-dependent electric polarization P^{NL}, which can be expanded as a power series of the electric field in Fourier space:

$$P(\omega) = P^{(1)}(\omega) + P^{NL}(\omega) = P^{(1)}(\omega) + P^{(2)}(\omega) + P^{(3)}(\omega) + \cdots \quad (6.5)$$

$$P^{(n)}(\omega) = \chi^{(n)}(-\omega; \omega_1, \omega_2, \ldots \omega_n) : E(\omega_1) E(\omega_2) \cdots E(\omega_n), \quad (6.6)$$

where $\chi^{(n)}$ defines the nonlinear susceptibility tensors of rank $n + 1$. From Eq. (6.2), the linear susceptibility $\chi^{(1)}$ is related to the dielectric tensor by $\epsilon = n^2 = (1 + 4\pi\chi^{(1)})$. The frequency arguments in Eq. (6.6) must sum to zero for energy conservation. A tensor of rank n can have 3^n number of independent components. Take the second-order nonlinear susceptibility $\chi^{(2)}$, for instance. Being a third-rank tensor, it has 27 components, $\chi^{(2)}_{IJK}$, due to the permutation of its three indices I, J, and K, each of which takes the values X, Y, and Z (1, 2, and 3 are used in Tables 2 and 3). The permutation of the frequency arguments, ω, ω_1, and ω_2, further increases the number to 81. Fortunately, various symmetry considerations greatly limit the number of independent and nonzero components. For nonabsorbing media, i.e., the susceptibilities are real quantities, relationships such as $\chi^{(2)}_{IJK}(-\omega; \omega_1, \omega_2) = \chi^{(2)*}_{JKI}(-\omega_1; -\omega_2, \omega)$ can be deduced from microscopic considerations, reducing the number to 27. The symmetrization procedure, assumed in the definition of the susceptibilities, also insures that the components are invariant upon permutation of the index and frequency pairs (j, ω_1) and (k, ω_2) and further reduces the number to 18. If some of the frequency arguments are the same, as in the case of second-harmonic generation (SHG), their associated indices J and K can be freely permuted. If the frequency dispersion of the material can be ignored, the permutation of the indices becomes independent of the frequency arguments. Because inversion symmetry dictates that $P^{(2)} = \chi^{(2)} E_1 E_2$ changes to $-P^{(2)} = \chi^{(2)}(-E_1)(-E_2)$, $\chi^{(2)}$ must be zero for all centrosymmetric materials. This is also true for other even order nonlinear susceptibilities. Moreover, the susceptibility must have the same symmetry elements as the material, further limiting the nonzero components. For SHG, a different notation is often used to describe the susceptibility:

$$d_{IJK}(-2\omega; \omega, \omega) = \frac{1}{2} \chi^{(2)}_{IJK}(-2\omega; \omega, \omega). \quad (6.7)$$

Because the last two indices interchange freely, it is a symmetric tensor. The notation for the d coefficients is further simplified by a contraction of the last two indices: $d_{IJJ} = d_{I1}$, $d_{I23} = d_{I4}$, $d_{I13} = d_{I5}$, and $d_{I12} = d_{I6}$.

The equation for propagation in a nonlinear medium can be obtained by substituting Eq. (6.5) in Eqs. (6.1) and (6.2):

$$\nabla \times \nabla \times E(k, \omega) - \frac{\omega^2}{c^2} \epsilon \cdot E(k, \omega) = \frac{4\pi\omega'^2}{c^2} P^{NL}(k', \omega'). \quad (6.8)$$

Because P^{NL} contains two or more electric field factors, a set of $(n + 1)$ coupled equations is needed for the nth order nonlinear optical process. Take the second-order process of sum frequency generation (SFG) for instance. Three waves, with different wave vectors and frequencies given by k, k_1, k_2 and $\omega = \omega_1 + \omega_2$, can transfer energy among each other as they propagate in the nonlinear medium according to the following coupled wave equations:

$$[\nabla \times \nabla \times - \frac{\omega^2}{c^2} \epsilon \cdot]E(k, \omega) = \frac{4\pi\omega^2}{c^2} \chi^{(2)}(-\omega; \omega_1, \omega_2)E_1(k_1, \omega_1)E_2(k_2, \omega_2), \quad (6.9a)$$

$$[\nabla \times \nabla \times - \frac{\omega_1^2}{c^2} \epsilon_1 \cdot]E_1(k_1, \omega_1) = \frac{4\pi\omega_1^2}{c^2} \chi^{(2)}(-\omega_1; -\omega_2, \omega)E_2^*(k_2, \omega_2)E(k, \omega), \quad (6.9b)$$

and

$$[\nabla \times \nabla \times - \frac{\omega_2^2}{c^2} \epsilon_2 \cdot]E_2(k_2, \omega_2) = \frac{4\pi\omega_2^2}{c^2} \chi^{(2)}(-\omega_2; \omega, -\omega_1)E(k, \omega)E_1^*(k_1, \omega_1). \quad (6.9c)$$

As before, conservation of energy dictates that the nonlinear susceptibilities, $\chi^{(2)}$'s, which are the coupling coefficients, must be equal in a nonabsorbing medium. Energy gain in the high-frequency wave must equal the energy loss from the two low-frequency waves.

The solution for Eqs. (6.9a–6.9c) (Armstrong et al., 1962) will not be given here. Instead, we will examine the commonly encountered condition where $E(k, 2\omega)$ is much less than the input fields and consider the case of SHG where $\omega_1 = \omega_2 = \omega$. In the remainder of this section, we will limit our discussion to Z propagating plane waves, with the input face of the medium fixed at $Z = 0$, and will specialize the case where all waves are polarized in the X direction. Substituting Eq. (6.4) in Eq. (6.9a) and collecting 2ω terms gives

$$\nabla^2 E_X^{2\omega} + \frac{(2\omega)^2 \epsilon}{c^2} E_X^{2\omega} = \frac{-4\pi(2\omega)^2}{c^2} P^{NL}(2\omega)$$

$$= \frac{-4\pi(2\omega)^2}{c^2} \frac{1}{4} \chi_{XXX}^{(2)}(-2\omega; \omega, \omega)E_X^\omega(Z)E_X^\omega(Z)[e^{i(2\omega-2k_\omega Z)} + \text{c.c.}], \quad (6.10)$$

where

$$\nabla^2 E_X^{2\omega} = \frac{1}{2} \nabla^2 [E_X^{2\omega}(Z)e^{i(2\omega - k_{2\omega}Z)} + \text{c.c.} + E_X^0]$$

$$= \frac{1}{2} \left[\frac{\partial^2}{\partial Z^2} - i2k_{2\omega} \frac{\partial}{\partial Z} - k_{2\omega}^2 \right] E_X^{2\omega}(Z)e_i^{(2\omega - k_{2\omega} Z)} + \text{c.c.} \quad (6.11)$$

"c.c." stands for "complex conjugate". Because $k_{2\omega}^2 = (2\omega)^2 \epsilon / c^2$, assuming $E_X^{2\omega}(Z)$ varies slowly with Z, i.e., $\partial^2/\partial Z^2 E_X^{2\omega}(Z) \ll k^{2\omega} \partial/\partial Z E_X^{2\omega}(Z)$, Eq. (6.10) is approximated by

$$ik_{2\omega} \frac{\partial E_X^{2\omega}(Z)}{\partial Z} e^{i(2\omega - k_{2\omega}Z)} + \text{c.c.} = \frac{\pi(2\omega)^2}{c^2} \chi_{XXX}^{(2)}(-2\omega; \omega, \omega)E_X^\omega(Z)E_X^\omega(Z)e^{i(2\omega - 2k_\omega Z)} \quad (6.12)$$

$$+ \text{c.c.}$$

or

$$\frac{\partial E_X^{2\omega}(Z)}{\partial Z} = -\frac{i\pi(2\omega)^2}{k_{2\omega}c^2} \chi_{XXX}^{(2)}(-2\omega; \omega, \omega)E_X^\omega E_X^\omega e^{i(\Delta k Z)} \quad (6.13)$$

where $\Delta k = k_{2\omega} - 2k_\omega$, which is known as the phase mismatch of the nonlinear process. Integration yields the solution for Eq. (6.13):

$$E_X^{2\omega}(Z) = E_X^{2\omega}(0) + \frac{\pi(2\omega)^2}{k_{2\omega}c^2\Delta k} \chi_{XXX}^{(2)}(-2\omega; \omega, \omega)E_X^\omega E_X^\omega(1 - e^{i(\Delta kZ)}). \qquad (6.14)$$

If the harmonic field is zero at the input face, i.e., $E_X^{2\omega}(0) = 0$, the harmonic intensity is proportional to

$$E_X^{2\omega}(Z)E_X^{*2\omega}(Z) = \left[\frac{\pi(2\omega)^2}{k_{2\omega}c^2}\chi_{XXX}^{(2)}(-2\omega; \omega, \omega)E_X^\omega E_X^{*\omega}\right]^2\left[\frac{\sin(\Delta kZ/2)}{\Delta kZ/2}\right]^2 Z^2. \qquad (6.15)$$

The second square bracket in Eq. (6.15) is an oscillatory function of Z, peaking strongly at $\Delta k = 0$, which is known as the phase-matching condition. For $\Delta k \neq 0$, the fundamental and the harmonic waves propagate with different phase velocities, leading to an oscillatory harmonic intensity due to constructive and destructive interferences. At a phase-mismatched condition, half of the propagation distance between adjacent harmonic intensity maxima is known as the coherence length, $l_c = \pi/\Delta k = \lambda/4(n_{2\omega} - n_\omega)$, over which the two waves propagate out of phase by π. Due to normal material dispersion, phase matching cannot be achieved along arbitary propagative directions. In fact, for $\chi_{XXX}^{(2)}$ and other diagonal tensor elements, phase matching is never satisfied (unless anomalous dispersion occurs at wavelengths near electronic or vibrational transitions). However, if the material possesses optical birefringence, phase matching may be achieved for certain combinations of wavelength and polarization along specific directions. When phases match, the harmonic intensity is proportional to the square of the interaction length.

The static component of the input field gives rise to a phenomenon known as the Pockels or electro-optic (EO) effect. If $\omega_1 = \omega$ and $\omega_2 = 0$, substituting Eq. (6.4) to Eq. (6.9a) and collecting ω terms yields

$$\frac{\partial E_X^\omega(Z)}{\partial Z} = -\frac{i4\pi(\omega)^2}{k_\omega c^2} P^{NL}(\omega)$$

$$= -\frac{i4\pi(\omega)^2}{k_\omega c^2} \chi_{XXX}^{(2)}(-\omega; \omega, 0)E_X^\omega(Z)E_X^0. \qquad (6.16)$$

Note that $\Delta k = 0$ is always satisfied. Integration yields

$$E_X^\omega(Z) = E_X^\omega(0)e^{-\Delta_\phi}$$

where

$$\Delta\phi = 4\pi\omega^2\chi_{XXX}^{(2)}(-\omega; \omega, 0)E_X^0 Z/k_\omega c^2. \qquad (6.17)$$

Therefore, the result of the nonlinear interaction is a field and propagation length-dependent phase shift, which is equivalent to a change in the refractive index. Historically, the electro-optic effect is defined as

$$\Delta\left(\frac{1}{n^2}\right)_{XX} = r_{XXX}E_X^0, \qquad (6.18)$$

which gives

$$\Delta n_X = \frac{-n_X^3 r_{XXX}E_X^0}{2}. \qquad (6.19)$$

The change in refractive index results in a phase shift $\Delta\delta = -kn_X^3 r_{XXX}E_X^0 Z/2$, where $\Delta k_\omega = k\Delta n$

is used. Therefore, equating the two phase shifts, i.e., $\Delta\delta = \Delta\phi$, leads to the relationship between the two definitions:

$$r_{XXX} = \frac{-8\pi\chi^{(2)}_{XXX}(-\omega;\omega,0)}{n_X^2 n_X^2}. \tag{6.20}$$

Because the first two frequencies are the same, the first two indices of the electro-optic coefficients are often given in contracted notation analogous to the SHG d coefficients. A useful quantity is the half-wave voltage V_π obtained by setting $\Delta\delta$ or $\Delta\phi$ equal to π:

$$V_\pi = \frac{\lambda}{n_X^3 r_{XXX}} \frac{h}{Z} \tag{6.21}$$

where Z is the interaction length, h is the distance between the electrodes, and $E_\pi^0 = V_\pi/h$. Because $h = Z$ for the longitudinal geometry being discussed, V_π is independent of the device dimensions and can be decreased only by increasing the material figure of merit, $n_X^3 r_{XXX}$. For transverse geometry, V_π can be reduced by lowering h/Z as well as increasing $n_X^2 n_y r_{XXY}$. Other polarization combinations and propagation directions can be similarly considered for both SHG and EO effects.

Microscopic Susceptibility

Analogous to Eq. (6.5), a molecular nonlinear electric polarization p^{NL} can be defined for the microscopic response in molecular materials:

$$p(\omega) = \mu + \alpha(\omega) \cdot E(\omega) + p^{NL}(\omega), \tag{6.22}$$

and

$$p^{NL}(\omega) = \beta(-\omega;\omega_1,\omega_2):E(\omega_1)E(\omega_2) + \gamma(-\omega;\omega_1,\omega_2,\omega_3):E(\omega_1)E(\omega_2)E(\omega_3) + \cdots \tag{6.23}$$

where μ is the permanent molecular dipole moment and α, β, and γ are the linear polarizability, quadratic, and cubic hyperpolarizability tensors, respectively. When properly modified to account for intermolecular interactions and statistical ensemble averages, the three microscopic polarizabilities, determine the bulk optical properties of the molecular solid. Respectively, they determine bulk refractive indexes, EO and other three-wave mixing susceptibilities, and susceptibilities for four-wave mixing processes, such as third-harmonic generation (THG) and dc electric-field induced, second-harmonic generation (EFISH). Symmetry considerations associated with the bulk susceptibilities apply similarly to the hyperpolarizabilities. For instance, the lowest order asymmetric response β vanishes for molecules with inversion symmetry.

For single crystals, components of the macroscopic second-order susceptibility are given by an oriented gas model (Zyss and Oudar, 1982) which relates components of the hyperpolarizability β_{ijk} of a molecule located at a nonequivalent crystallographic site S with s equivalent positions, by the expression

$$\chi^{(2)}_{IJK} = \frac{1}{V} f_I^{2\omega} f_J^\omega f_K^\omega \sum_S \sum_s \sum_{ijk} \cos\theta_{Ii}^{(S,s)} \cos\theta_{Jj}^{(S,s)} \cos\theta_{Kk}^{(S,s)} \beta_{ijk}, \tag{6.24}$$

where V is the unit-cell volume, fs are local field factors, and $\cos\theta$s are the directional cosines determined by the relative orientation of the crystallographic axes (I, J, K) and the molecular axes (i, j, k). For dipolar liquids or glassy matrices whose centrosymmetry is broken by an external poling electric field along Z, the transient or quasi-static $\chi^{(2)}_{IJK}$ is given by

$$\chi^{(2)}_{IJK} = N f_I^{2\omega} f_J^{\omega} f_K^{\omega} \sum_{ijk} \langle \cos\theta_{Ii} \cos\theta_{Jj} \cos\theta_{Kk} \rangle \beta_{ijk}, \tag{6.25}$$

where N is the number density and $\langle \ \rangle$ denotes the thermodynamic orientational average. For molecules with a dominant hyperpolarizability tensor component β_{zzz} along the dipole axis z, Eq. (6.25) becomes

$$\chi^{(2)}_{ZZZ} = N f_Z^{2\omega} f_Z^{\omega} f_Z^{\omega} \langle \cos^3\theta \rangle \beta_{zzz} \tag{6.26}$$

and

$$\chi^{(2)}_{ZXX} = N f_Z^{2\omega} f_X^{\omega} f_X^{\omega} \langle \cos\theta \sin^2\theta \cos^2\phi \rangle \beta_{xxx}$$
$$= \frac{1}{2} N f_Z^{2\omega} f_X^{\omega} f_X^{\omega} (\langle \cos\theta \rangle - \langle \cos^3\theta \rangle) \beta_{xxx}, \tag{6.27}$$

where θ and ϕ are polar angles. For isotropic media, the thermodynamic averages are given by the first- and third-order Langevin functions of $a = \mu E^0/kT$:

$$\langle \cos\theta \rangle = \frac{a}{3} - \frac{a^3}{45} + \frac{2a^5}{945} - \frac{2a^7}{9450} + \cdots$$

and

$$\langle \cos^3\theta \rangle = \frac{a}{5} - \frac{a^3}{105} + \frac{8a^5}{9450} + \cdots, \tag{6.28}$$

where E^0 is the poling field. Therefore for small α,

$$\chi^{(2)}_{ZZZ} = 3\chi^{(2)}_{ZXX} = N f_Z^{2\omega} f_Z^{\omega} f_Z^{\omega} \frac{\mu E^0 \beta_{xxx}}{5kT}. \tag{6.29}$$

If axial order is additionally present in the medium (such as in liquid crystalline materials), the thermodynamic averages are modified, enhancing the diagonal susceptibility by as much as a factor of five and reducing the off-diagonal susceptibility to null.

The frequency-dependent local field corrections are given by the Lorenz–Lorentz or Onsager models. The Onsager model takes into account the dipolar reaction field and must be used for polar liquids. In mixtures, the Onsager local fields are also species dependent. Expressions for the local field factors are [Bottcher, 1973]

$$\text{Lorenz–Lorentz model: } f^0 = f^\omega = \frac{\epsilon + 2}{3}; \tag{6.30}$$

$$\text{Onsager model: } f_i^0 = \frac{\epsilon(n_i^2 + 2)}{2\epsilon + n_i^2}, \tag{6.31}$$

$$f_i^\omega = \frac{n^2(\omega)[n_i^2(\omega) + 2]}{2n^2(\omega) + n_i^2(\omega)} \qquad (6.32)$$

where $n(\omega)$ is the refractive index of the mixture and $n_i(\omega)$ is the refractive index of species i.

The quantum mechanical expressions for the hyperpolarizabilities can be obtained from time-dependent perturbation theory. Under the perturbation of an external field E, the time evolution of the density matrix at thermal equilibrium, $\rho = \Sigma_m e^{-E_m/kT}|m\rangle\langle m|$, is governed by the Liouville equation:

$$\frac{\partial \rho}{\partial t} = \frac{1}{i\hbar}[H, \rho] \qquad (6.33)$$

where $|m\rangle$ and E_m are the eigen states and energies of the unperturbed system. The Hamiltonian $H = H_0 + \mu \cdot E$ includes a dipolar interaction term $\mu \cdot E = qr \cdot E$ which couples the electronic position r with charge q to the perturbation. Because a solution to Eq. (6.33) can be obtained by successive approximations of the density matrix

$$\rho = \rho^{(0)} + \rho^{(1)} + \rho^{(2)} + \rho^{(3)} + \cdots, \qquad (6.34)$$

the expectation values of the molecular polarizability and hyperpolarizabilities are given by

$$\langle p \rangle = Tr[\rho p] = \sum_{n=0} \langle p^{(n)} \rangle = \sum_{n=0} Tr[\rho^{(n)} p] = \langle \mu \rangle + \langle \alpha \rangle + \langle \beta \rangle + \langle \gamma \rangle + \cdots \qquad (6.35)$$

Substituting Eq. (6.34) in Eq. (6.33) and collecting terms of the appropriate orders yields

$$\frac{\partial \rho_{mm'}^{(n)}}{\partial t} = \frac{1}{i\hbar}[(H_0, \rho^{(n)})_{mm'} + (\mu \cdot E, \rho^{(n-1)})_{mm'}] \qquad (6.36)$$

or

$$\rho_{mm'}^{(n)}(\omega) = \frac{(\mu \cdot E, \rho^{(n-1)})_{mm'}}{\hbar(\omega - \omega_{mm'})}. \qquad (6.37)$$

At normal temperature, all molecules are assumed to be in their ground state, i.e., $\rho_{gg}^{(0)} = 1$ and using the field definition given by Eq. (6.4) and combining Eqs. (6.37) and (6.35) yields the quantum mechanical expressions for α_{ij} and β_{ijk}:

$$\alpha_{ij}(\omega) = \frac{1}{\hbar} \sum_m \left[\frac{\langle g|\mu_j|m\rangle\langle m|\mu_i|g\rangle}{(\omega_{gm} - \omega)} + \frac{\langle g|\mu_i|m\rangle\langle m|\mu_j|g\rangle}{(\omega_{gm} + \omega)} \right], \qquad (6.38)$$

and

$$\beta_{ijk}(-\omega_3; \omega_1, \omega_2) = \frac{-1}{2\hbar^2} \sum_l \sum_m \left[\frac{\langle g|\mu_i|l\rangle\langle l|\mu_j|m\rangle\langle m|\mu_k|g\rangle}{(\omega_{gl} + \omega_3)(\omega_{gm} + \omega_2)} + \frac{\langle g|\mu_i|l\rangle\langle l|\mu_j|m\rangle\langle m|\mu_k|g\rangle}{(\omega_{gl} - \omega_3)(\omega_{gm} - \omega_2)} \right.$$
$$\left. + \frac{\langle g|\mu_k|l\rangle\langle l|\mu_j|m\rangle\langle m|\mu_i|g\rangle}{(\omega_{gl} + \omega_2)(\omega_{gm} + \omega_3)} + \frac{\langle g|\mu_k|l\rangle\langle l|\mu_j|m\rangle\langle m|\mu_i|g\rangle}{(\omega_{gl} - \omega_2)(\omega_{gm} - \omega_3)} \right.$$

$$+ \frac{\langle g|\mu_i|l\rangle\langle l|\mu_k|m\rangle\langle m|\mu_j|g\rangle}{(\omega_{gl}+\omega_3)(\omega_{gm}+\omega_1)} + \frac{\langle g|\mu_i|l\rangle\langle l|\mu_k|m\rangle\langle m|\mu_j|g\rangle}{(\omega_{gl}-\omega_3)(\omega_{gm}-\omega_1)}$$

$$+ \frac{\langle g|\mu_j|l\rangle\langle l|\mu_k|m\rangle\langle m|\mu_i|g\rangle}{(\omega_{gl}+\omega_1)(\omega_{gm}+\omega_3)} + \frac{\langle g|\mu_j|l\rangle\langle l|\mu_k|m\rangle\langle m|\mu_i|g\rangle}{(\omega_{gl}-\omega_1)(\omega_{gm}-\omega_3)}$$

$$+ \frac{\langle g|\mu_k|l\rangle\langle l|\mu_i|m\rangle\langle m|\mu_j|g\rangle}{(\omega_{gl}-\omega_2)(\omega_{gm}+\omega_1)} + \frac{\langle g|\mu_k|l\rangle\langle l|\mu_i|m\rangle\langle m|\mu_j|g\rangle}{(\omega_{gl}+\omega_2)(\omega_{gm}-\omega_1)}$$

$$+ \left. \frac{\langle g|\mu_j|l\rangle\langle l|\mu_i|m\rangle\langle m|\mu_k|g\rangle}{(\omega_{gl}-\omega_1)(\omega_{gm}+\omega_2)} + \frac{\langle g|\mu_j|l\rangle\langle l|\mu_i|m\rangle\langle m|\mu_k|g\rangle}{(\omega_{gl}+\omega_1)(\omega_{gm}-\omega_2)} \right]. \quad (39)$$

It is evident that the full electronic structure of a molecule is needed to determine its nonlinear optical response. However, as a result of the energy denominators, transitions which poccess large transitional moments and low energy gaps are expected to contribute strongly. Depending on the input optical frequency, the resonances that exist in the frequency dependence of the hyperpolarizability can be exploited to single out the contribution of individual states. Further insight can be obtained by specializing Eq. (6.39) to a two-level system (Oudar and Chemla, 1977b), i.e., $l = m = e$, and limiting the sums to just the ground (g) and excited state (e). The diagonal tensor element is given by

$$\beta_{zzz}(-\omega_3; \omega_1, \omega_2) = \frac{\omega_{ge}^2(3\omega_{ge}^2 + \omega_1\omega_2 - \omega_3^2)}{\hbar^2(\omega_{ge}^2 - \omega_1^2)(\omega_{ge}^2 - \omega_2^2)(\omega_{ge}^2 - \omega_3^2)}$$

$$\times [\langle g|\mu_z|e\rangle(\langle e|\mu_z|e\rangle - \langle g|\mu_z|g\rangle)\langle e|\mu_z|g\rangle]$$

$$= \frac{\omega_{ge}^2(3\omega_{ge}^2 + \omega_1\omega_2 - \omega_3^2)\mu_{ge}^2\Delta\mu_{ge}}{\hbar^2(\omega_{ge}^2 - \omega_1^2)(\omega_{ge}^2 - \omega_2^2)(\omega_{ge}^2 - \omega_3^2)} \quad (6.40)$$

where μ_{ge} and $\Delta\mu_{ge}$ denote the transitional moment and the change in dipole moment between the two states. Therefore, it is evident that a strongly contributing state must possess both a strong coupling and a large redistribution of charge. These states are referred to as charge-transfer (CT) states. In the special case of SHG, i.e, $\omega_3 = 2\omega_1 = 2\omega_2 = 2\omega$, the two-level approximation for $\beta_{zzz}(-2\omega; \omega, \omega)$ is

$$\beta_{zzz}(-2\omega; \omega, \omega) = \frac{3\omega_{ge}^2\mu_{ge}^2\Delta\mu_{ge}}{\hbar^2(\omega_{ge}^2 - \omega^2)(\omega_{ge}^2 - 4\omega^2)}, \quad (6.41)$$

and, for the electro-optic effect,

$$\beta_{zzz}(-\omega; \omega, 0) = \frac{(3\omega_{ge}^2 - \omega^2)\mu_{ge}^2\Delta\mu_{ge}}{\hbar^2(\omega_{ge}^2 - \omega^2)^2}. \quad (6.42)$$

Therefore, from Eqs. (6.20) and (6.24), the susceptibilities for electrooptic and second-harmonic generation for the two-level system are connected by dispersion and local field factors:

$$r_{IJK}(-\omega^1, \omega_1, 0) = \frac{-8\pi}{n_I^2 n_J^2} \frac{f_I^{\omega_1} f_J^{\omega_1} f_K^0}{f_I^{2\omega_2} f_J^{\omega_2} f_K^{\omega_2}} \frac{(3\omega_{eg}^2 - \omega_2^2)(\omega_{eg}^2 - \omega_2^2)(\omega_{eg}^2 - 4\omega_1^2)}{3\omega_{eg}^2(\omega_{eg}^2 - \omega_1^2)^2}$$

$$\times \chi_{IJK}^{(2)}(-2\omega_2, \omega_2, \omega_2) \quad (6.43)$$

Given the theoretical formulation outlined above, full scale calculations of molecular hyperpolarizabilities and crystalline susceptibilities can be performed at various levels of approximations. The advantage of the theoretical approach is that it permits inquiry into structure-property relationships without undertaking elaborate synthesis and experimental characterization. It also affords information which is not accessible to measurements. However, the success of various computational approaches must be evaluated with experimental data. In the next section, we will review some common experimental techniques for studying nonlinear optical materials

6.3 Experimental Techniques

Second-Harmonic Generation

Second-harmonic generation is, perhaps, the most commonly used technique for studying second-order nonlinear optical materials. It involves detecting second-harmonic intensity when a monochromatic optical field traverses a noncentrosymmetric material. The polarization states of the fundamental and harmonic waves and the sample orientation determine which one or combination of the tensor components $\chi_{IJK}^{(2)}(-2\omega;\omega,\omega)$ are being measured. As is clear from Eq. (6.15), the harmonic intensity is determined by both the susceptibility and the phase mismatch of the material. Because Eq. (6.15) contains an oscillatory term of ΔkZ,

$$I(2\omega) \propto (\chi_{IJK}^{(2)} I_\omega)^2 \frac{\sin^2(\Delta kZ/2)}{(\Delta kZ/2)^2}, \tag{6.44}$$

either a continuous variation of the phase mismatch Δk or the optical path length Z can lead to an oscillatory harmonic signal, which can then be fitted to extract the two unknowns. The phase mismatch within the nonlinear medium can be varied by rotating a plane-parallel sample about an axis perpendicular to the beam path, and the path length can be varied by translating a wedge-shaped sample along the wedge at normal incidence. For the former, the resulting oscillatory signal is known as Maker fringes and the technique is called the Maker fringe method. The latter is known as the wedge method.

The intensity of the Maker fringes is nonperiodic and oscillatory in incident angle θ and is caused by the angular dependence of the phase mismatch,

$$\Delta k(\theta) = \frac{\pi}{l_c(\theta)} = \frac{4\pi(n_\omega \cos\theta_\omega - n_{2\omega} \cos\theta_{2\omega})}{\lambda}, \tag{6.45}$$

where θ_ω and $\theta_{2\omega}$ are the fundamental and harmonic refractive angles inside the sample related to θ by Snell's law. For small angular deviation from normal incidence, Eq. (6.45) can be expanded about $\theta = 0$ to give

$$\Delta k = \frac{1}{\cos\theta_\omega l_c(\theta = 0)}. \tag{6.46}$$

For the wedge method, no angular dependence needs to be considered, and the signal is a much simpler periodic function of the translation X and the wedge angle ϕ:

$$I(2\omega) \propto (\chi_{IJK}^{(2)} l_c I_\omega)^2 \sin^2 \frac{\pi X \tan\phi}{2 l_c}. \tag{6.47}$$

However, wedged samples with appropriate orientation must be fabricated to determine each $\chi_{IJK}^{(2)}$.

Electric-Field Poling

SHG can be carried out in a medium in which a polar order is established by the poling of dipoles with a dc electric field. For a liquid in which dipolar orientation is rapid, accentricity can be created and destroyed instantaneously with the switching of the field. This situation is encountered in characterizing molecular hyperpolarizability and will be discussed in section 3.4. For a glassy or viscoelastic medium, the polar order responds to the dc electric field in time scales limited by dipolar relaxation times. A commonly encountered situation is the poling of nonlinear optical chromophores in a polymeric matrix (Havinga and Van Pelt, 1979; Meredith et al., 1982; Singer et al., 1986). This type of material, referred to as a poled polymer, will be reviewed in section 6.6. Here the poling and SHG characterization of these polymers are discussed.

To maximize the poling field, chromophore-containing polymers are typically poled in a thin film. The thin films are deposited by spin coating a polymer solution on conducting substrates, such as indium tin oxide coated glass or a silicone wafer. The dried films are typically a few μm thick and two layers are, sometimes, used to prevent pinholes. The poling electric field is produced either by a corona discharge or by charged electrodes in direct contact with the polymer surfaces. For poling with electrodes, semitransparent thin metal electrodes (10–20 nm of Au or Al) are used to permit optical experiments. To control the dipolar relaxation time, the polymer sample is mounted on a heating stage equipped with optical access. SHG provides a means of monitoring the poling and relaxation of the chromophores.

The second-order nonlinearity of a poled medium is given by Eqs. (6.26) and (6.27). There are only two nonzero SHG tensor components ($\chi^{(2)}_{ZZZ}$ and $\chi^{(2)}_{ZXX}$). Because the poling field is along the thin film normal Z coupling to $\chi^{(2)}_{ZZZ}$ is achieved by tilting the sample at a 45° angle with respect to the laser beam, which is polarized in the plane of incidence. Assuming uniformly induced nonlinearity, the magnitude of $\chi^{(2)}_{ZZZ}$ is determined by referencing the harmonic intensity to a standard material with known susceptibility and coherence length. Because the typical film thickness is much less than its coherence length, i.e., $Z \ll l_c$, the oscillatory term in Eq. (6.15) can be expanded to give an intensity independent of the coherence length:

$$I(2\omega) \propto (\chi^{(2)}_{ZZZ} \sin \varphi I_\omega Z/n_\omega)^2 \frac{\sin^2(\Delta k Z/2)}{(\Delta k Z/2)^2} \approx (\chi^{(2)}_{ZZZ} \sin \varphi I_\omega Z/n_\omega)^2, \qquad (6.48)$$

where φ is the external angle between the film normal and the propagative direction. In addition, harmonic absorption can be additionally accounted for by the expression

$$I(2\omega) \propto \frac{e^{-2\alpha_{2\omega}Z} - 2e^{-\alpha_{2\omega}Z} + 1}{n_\omega^2 \alpha_{2\omega}^2} (\chi^{(2)}_{ZZZ} \sin \varphi I_\omega)^2 \qquad (6.49)$$

where $\alpha_{2\omega}$ is the absorption coefficient. Note that the harmonic intensity becomes independent of Z and is limited, instead, by the harmonic absorption length of the film.

There are several advantages of using a corona for poling (Mortazavi et al., 1989). Foremost is its convenience: electrode deposition is not required. The nonconducting polymer surface also prevents catastrophic electrical breakdown, allowing the study of imperfect samples. Under normal circumstances, a much higher electric field (several MV/cm) can be obtained with a corona because the surface voltage is determined only by the intrinsic dielectric strength of the polymer. When electrodes are used, catastrophic breakdown is usually initiated at a surface imperfection, limiting the achievable voltage in routine experiments. Unlike contact poling with electrodes and a power supply which can supply the required currrent, a corona source is current-limited. The ion deposition rate of a corona is determined by its wire voltage, which cannot be increased indefinitely without corona instability. Therefore, a high poling field may not be sustainable as the ionic conductivity of the polymer increases rapidly with temperatures. For some materials, effective poling with corona charging at high temperatures is not possible. Other disadvantages include the reactivity of the corona (Hampsch et al., 1990; Dao et al., 1993), which can be controlled with an inert atmosphere, and the spatial nonuniformity of the poling field.

Electro-Optic Modulation

Many techniques have been developed to measure the electro-optic properties of crystalline and polymeric materials. They are mostly interferometric to maximize sensitivity. The sample is placed either in the optical path of a Mach–Zehnder or Michelson interferometer (Sigelle and Hierle, 1981), or the interference between reflected waves from the sample surfaces is used (Teng and Man, 1990). The sensitivity of these thin film reflection techniques can be much enhanced by fabricating highly finished etalons (Yankelevich et al., 1991). The electro-optic phase shift is, then, detected from the intensity change at a particular interference fringe with a lock-in amplifier. For thin film samples with good optical quality, a prism coupling technique (Horsthuis and Krijnen, 1989; Gallo et al., 1993) can be used to determine the phase shift by monitoring the modulation of sharply defined synchronous angle for the guided wave. Here we outline a Mach–Zehnder technique which is suitable for both thin film and bulk samples.

A Mach–Zehnder interferometer is arranged by splitting a well-collimated laser beam with a 50% partial reflector into two optical branches (see Fig. 1). The light polarizations and the electric field direction depend on the r_{IJK} being measured. The electroded sample with proper orientation is placed at one branch of the interferometer. A reference phase modulator with accurately known half-wave voltage is placed in the other optical branch. The two branches are recombined with high and partial reflectors and are arranged to copropagate. Neutral density filters are used to achieve the same intensities for the two branches. The interference fringes are magnified with a telescope. A slit, small compared to the fringe spacing, spatially selects part of a fringe for detection. The amplified outputs of a dual-channel signal generator with phase and amplitude controls provide the modulating voltage. The intensity modulation, detected by a lock-in amplifier, is the result of the total phase shifts generated at both branches:

$$I_m \propto \cos(\Delta\phi_S - \Delta\phi_R). \tag{6.50}$$

The phases and amplitudes of the two channels are adjusted so that the phase shift from the

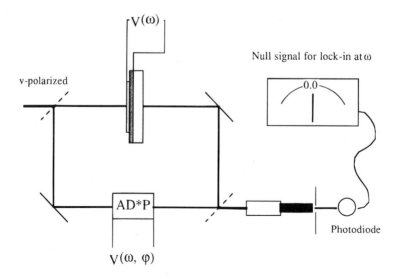

FIGURE 6.1 Schematic of a Mach–Zehnder interferometer.

sample is precisely cancelled by that from the reference modulator, thus, nulling the modulated signal. The half wave voltage of the sample is determined by

$$V_\pi^S = V^S V_\pi^R / V^R \qquad (6.51)$$

where V^S and V^R are the modulating voltages for the sample and reference at null, respectively. From the known index of refraction, wavelength, and sample dimensions, r_{IJK} can be calculated by Eq. (6.21). Potential errors with this technique include possible piezoelectric contributions and over estimation due to etalon effects. For thin samples, certain r_{IJK} cannot be determined independently with this technique. In the case of a poled polymer, only r_{ZZX} can be measured. At the weak poling field limit, $r_{ZZZ} = 3r_{ZZX}$ is often assumed.

DC Electric-Field-Induced, Second-Harmonic Generation

The study of molecular hyperpolarizability is most conveniently carried out in dilute solutions (Levine and Bethea, 1974). Nevertheless, because the molecular ground state can be influenced by the solvent environment, results are often solvent dependent. Additionally, only quantities which have nonzero thermodynamic averages can be obtained. For odd order optical polarizabilities, such as α and γ, only the scalar parts of the tensors, denoted as α_Z and γ_Z, are measured:

$$\alpha_Z = \frac{1}{3}(\alpha_{ii} + \alpha_{jj} + \alpha_{kk}), \qquad (6.52)$$

and

$$\gamma_Z = \frac{1}{5}(\gamma_{iiii} + \gamma_{jjjj} + \gamma_{kkkk} + \gamma_{iiii} + 2\gamma_{iijj} - 2\gamma_{iikk} + 2\gamma_{jjkk}). \qquad (6.53)$$

The random orientation of molecules in solution precludes any even order optical response. Bulk second-harmonic generation is not allowed by symmetry, for instance. If the molecule of interest possesses a permanent dipole moment and if the external electric field contains a slow or dc component which induces dipolar alignment, a third-order process known as dc electric-field-induced, second-harmonic generation (EFISH) gives access to the hyperpolarizability β. As before, letting the external field be

$$E_i = E_i^0 + \frac{1}{2} E_i^\omega [e^{iex} + c.c.], \qquad (6.54)$$

the induced molecular polarization oscillating at 2ω, obtained by inserting E_i in Eq. (6.23) and collecting terms at 2ω, is expressed as

$$p_i^{NL}(2\omega) = \frac{1}{2} \beta_{ijk}(-2\omega, \omega, \omega) E_j^\omega E_k^\omega + \frac{3}{2} \gamma_{ijkl}(-2\omega, \omega, \omega, 0) E_j^\omega E_k^\omega E_l^0. \qquad (6.55)$$

For a Z polarized external field, the macroscopic response is obtained by suming the molecular and local field contributions and carrying out the orientational average:

$$P_Z^{NL}(2\omega) = Nf^{2\omega}f^\omega f^\omega <\frac{1}{2}\beta_{ZZZ}(-2\omega, \omega, \omega) + \frac{3f^0}{2}\gamma_{ZZZ}(-2\omega, \omega, \omega, 0)E_Z^0> E_Z^\omega E_Z^\omega. \qquad (6.56)$$

From Eq. (6.28) and keeping only terms that are third order in field, Eq. (6.56) becomes

$$P_Z^{NL}(2\omega) = Nf^{2\omega}f^{\omega}f^{\omega}f^0 \left[\frac{1}{2}\frac{\mu_Z\beta_Z(-2\omega, \omega, \omega)}{5kT} + \frac{3}{2}\gamma_Z(-2\omega, \omega, \omega, 0)\right]E_Z^{\omega}E_Z^{\omega}E_Z^0, \qquad (6.57)$$

where γ_z is given by Eq. (6.53) and

$$\beta_Z(-2\omega, \omega, \omega) = \beta_{ZZZ} + \frac{1}{3}(\beta_{ZXX} + \beta_{ZYY} + 2\beta_{XXZ} + 2\beta_{YYZ}) \qquad (6.58)$$

is the vectorial projection of the hyperpolarizability tensor along the dipole moment. Eq. (6.57) defines a third-order susceptibility $\chi_{\text{EFISH}}^{(3)}$ which is given by

$$\chi_{\text{EFISH}}^{(3)}(-2\omega, \omega, \omega, 0) = Nf^{2\omega}f^{\omega}f^{\omega}f^0\gamma_{\text{EFISH}}E_Z^{\omega}E_Z^{\omega}E_Z^0 \qquad (6.59)$$

where

$$\gamma_{\text{EFISH}} = \frac{\mu\beta_{\mu}(-2\omega, \omega, \omega)}{5kT} + \gamma^e(-2\omega, \omega, \omega, 0). \qquad (6.60)$$

Note that, in Eqs. (6.59) and (6.60), the degeneracy factors of 1/2 and 3/2 have been absorbed into the definition of β_{μ} and γ^e. This rather arbitrary definition has been widely used by many experimentalists and results from omitting the degeneracy factors in Eq. (6.55). Additional confusion arises in the literature when different polarization expansions, as in Eq. (6.23), and different field definitions, as in Eq. (6.54), are used. Therefore, care must be exercised when experimental values are compared to theory (Li et al., 1992), to each other (Willetts et al., 1992), and among different techniques (Levine and Bethea, 1975; Burland et al., 1991).

The experimental setup for the EFISH and third-harmonic generation (THG) experiments typically consists of a high-peak-power laser. Flash lamp-pumped Nd:YAG lasers with a good gaussian mode are often used. The Q-switched nanosecond pulses at 1.064 μm wavelength are usually amplified to pulse energies of 0.5–1 joule. A gas Raman cell filled with methane or hydrogen is often used to generate Stokes radiation at 1.53 or 1.91 μm. Any one of these radiations can serve as the fundamental frequency for the EFISH or THG experiments. For 1.91 μm laser, the harmonics occur at 954 nm and 636 nm, respectively. Because of the presence of resonances in the susceptibilities, see Eq. (6.39), it is desirable to use the longest convenient radiation. For most lightly colored compounds with absorption edges below 500 nm, results obtained at 1.91 μm are minimally influenced by the resonant excitation frequencies of the molecule and can be considered to be approximately the intrinsic (electronic) nonlinearity associated with a specific molecular structure.

As in SHG, both the susceptibility and coherence length have to be determined. Several sample cell designs have been used. The most commonly used is a wedged cell equipped with electrodes (see Fig. 2). For EFISH, a high voltage, typically a few kV/mm, is applied and the second-harmonic intensity is collected as a function of the cell translation X perpendicular to the optical path. As in a solid sample in a SHG experiment, this translation leads to a change in the propagation length inside the liquid. The wedge angle ϕ and the dispersion of the liquid determine the periodicity of an oscillatory signal given by

$$I(2\omega) \propto [(\chi_S^{\text{EFISH}}l_S - \chi_G^{\text{EFISH}}l_G)I(\omega)]^2 \sin^2\frac{X\pi\tan\phi}{2l_S}, \qquad (6.61)$$

where the subscripts denote window glass (G) and sample solution (S). Note that the amplitude of the signal is given by the nonlinearity difference between the sample and the window material.

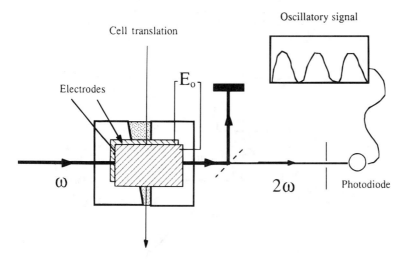

FIGURE 6.2 Schematic of an EFISH measurement with a wedged cell.

Therefore, from the signal periodicity and intensity and the known property of the window, both the susceptibility and coherence length can be obtained. To avoid interference of the EFISH signal arising from the window material and the air, a large optical slab extending well beyond the liquid and electrode regions is usually needed. The focusing of the incoming radiation also needs to be well controlled to achieve good fringe contrast. Ideally, the laser beam should have a small beam waist with minimum divergence. For THG, the wedged cell can give rise to complicated interference fringes due to signals arising from the window material and the air (Meredith et al., 1983). The tight and collimated focusing mentioned for EFISH tends to exacerbate this interference.

An alternative technique has been used for the simultaneous THG and EFISH experiments (Cheng et al., 1991b). The harmonic amplitudes and coherence lengths are measured in separate cells (see Fig. 3). The gaussian propagation of a tightly focused laser beam across a single interface can be approximated by using a thick glass window and a long path-length liquid cell. (see insert A of Fig. 3). Harmonic generation from the outer two interfaces of the cell can be ignored due to the rapid beam divergence, thus, minimizing possible air contributions which are known to be important in THG and, to a lesser extent, in EFISH measurements. To carry out THG and EFISH concurrently, electrodes are attached at the glass/liquid interface. The liquid cell is partitioned into two chambers, one chamber containing the organic solution and the other containing a standard liquid with known nonlinearity. Comparative measurement is performed by alternately translating the two liquid chambers into the focal region of the laser beam. Because the signal arises from the discontinuities associated with a single interface, no variable or oscillatory interference effects need to be considered. The ratio $\mathcal{R}$ of the harmonic intensities I_S and I_R from the two chambers is determined. Its relationship to the material properties is given by

$$\sqrt{\mathcal{R}} = \sqrt{I_S/I_R} \propto \frac{1 - R_S}{1 - R_R},$$

$$R_S \propto \frac{\chi_S^{(3)} l_S}{\chi_G^{(3)} l_G}, \tag{6.62}$$

and

$$R_R \propto \frac{\chi_R^3 l_R}{\chi_G^{(3)} l_G},$$

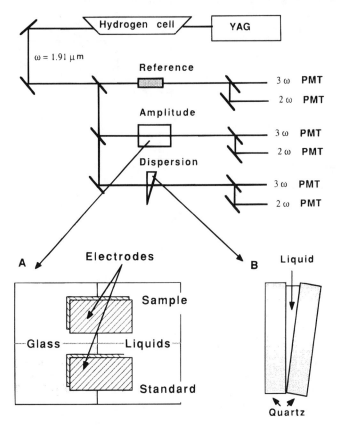

FIGURE 6.3 Schematic of an EFISH/THG measurement with a single interface cell.

where the $\chi^{(3)}$'s and l's are the THG or EFISH susceptibilities and coherence lengths for the window glass (G), sample (S), or reference (R) liquids, respectively.

The coherence lengths for both THG and SHG are measured separately with a wedged cell constructed with matching crystalline quartz windows (see insert B of Fig. 3). The usual oscillatory SHG and THG signals yield the coherence lengths. The disadvantages of this technique include a relatively large sample size and potential error when harmonic absorption becomes significant. For liquids with low absorption at the fundamental wavelength, harmonic absorption is not an issue because the radiation can impinge at the interface from the liquid side.

In both the wedged cell and the single-interface methods, the EFISH signal results from the nonlinearity difference between the two media across the interfaces. This is due to the presence of a dc electric field parallel to the interface in both materials. An alternative method (Uemiya et al., 1993), which is similar to the SHG study of poled polymers to be discussed in section 3.2, avoids the window EFISH signal by applying the electric field perpendicularly to the interfaces. Figure 4 shows a cell made of two glass plates separated by a spacer of length L. The electric field is applied through two semitransparent electrodes coated on the inside surface of the glass plates. Because the electric field exists only in the liquid medium, the signal is determined only by the nonlinearity of the liquid. An oscillatory harmonic signal, the Maker fringes, arises when the cell is rotated about the axis perpendicular to the dc and optical fields. The second-harmonic intensity is given by

$$I(2\omega) \propto T(\theta)[\chi_S^{EFISH} \sin\theta E^0 I(\omega) \sin\psi(\theta)]^2$$

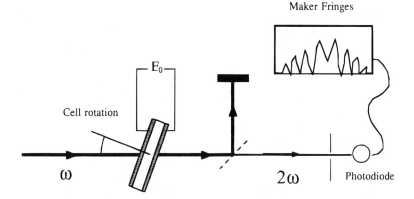

FIGURE 6.4 Schematic of an EFISH measurement with a thin-film geometry.

and

$$\psi(\theta) = 2\pi L/\lambda (n_S^\omega \cos \theta_S^\omega - n_S^{2\omega} \cos \theta_S^{2\omega}) \quad (6.63)$$

where ns are the fundamental and harmonic indices of the solution, $T(\theta)$ is the angle-dependent transmission factor, and θ is the incident angle of the fundamental wavelength.

Solvatochromic Method

By the two-level approximation, Eq. (6.41), the diagonal hyperpolarizability tensor element is connected to spectroscopic quantities. Therefore, when compared with results obtained from EFISH, spectroscopic measurements can be used to evaluate the contribution of a particular transition to the hyperpolarizability (Buckley et al., 1986; Paley and Harris, 1989). For molecules possessing a single dominant low-energy transition which is CT in character, the spectroscopic method provides a quick assessment of the molecular nonlinearity. Because spectroscopic equipment is widely available, this method is particularly valuable in synthetic research.

The transition moment and energy can be readily determined from the UV vis absorption spectrum in a suitable solvent, which, for comparison, should be the same as used in an EFISH measurement. The ground state dipole moment μ_g can be obtained by the standard capacitance bridge technique. The excited state dipole moment μ_g or the change in dipole moment $\Delta\mu$ can be determined by measuring the absorption solvatochromism of the molecule in solutions of different polarity. $\Delta\mu$ is most conveniently extracted from the solvatochromic shifts $\Delta\omega$, observed in the absorption spectra using solvents with closely matching refractive indices but quite different dielectric constants ϵ, by (Mataga, 1970) the expression

$$\Delta\omega = \frac{2\mu_g(\mu_g - \mu_e)}{a^3 c\hbar} \Delta\left(\frac{\epsilon - 1}{\epsilon + 2}\right) \quad (6.64)$$

where c is the speed of light and a is the Onsager radius of the solvent cavity which the molecule occupies. Typically, a is estimated to be near the molecular length but remains a source of uncertainty in this measurement. The sign of β_{zzz} can also be readily determined from the sign of $\Delta\omega$. A large $\Delta\omega$ alone does indicate high hyperpolarizability because it may result from a large ground state dipole moment. Therefore, μ_g must be independently measured or estimated. There are other well established techniques for measuring $\Delta\mu$, but one must keep in mind that, for nonlinear optics, the dipole moment of the Frank–Condon state is of interest. Results obtained from vibrationally relaxed fluorescing states or strongly charge-separated states, such as the twisted-CT states are of no consequence. For molecules whose ground state geometry is highly solvent dependent, the solvatochromic method cannot be used because the hyperpolarizability itself changes substantially in different solvents.

Hyper-Rayleigh Scattering

Although the orientational average of the second-order polarizability, $\langle \beta_{ijk} \rangle$ vanishes in liquids, its fluctuation $\langle \beta_{ijk}^* \beta_{lmn} \rangle$ does not. This fluctuation in nonlinear polarization gives rise to a phenomenon known as hyper-Rayleigh scattering (Terhune et al., 1965), in which photons from an incident laser are scattered into photons of the second-harmonic. Within the dipole approximation, the theory of hyper-Rayleigh scattering gives the following expression for the scattered second-harmonic intensity of a binary solution (Kielich et al., 1975):

$$I_i(2\omega) \propto \left(\frac{2\omega}{c}\right)^4 \sum_{n=1}^{2} \left\langle \sum_{p=1}^{N^n} \sum_{q=1}^{N^n} \beta_{ijk}^{p,n*} \beta_{lmn}^{q,n} \right\rangle (f^\omega)^4 (f^{2\omega})^2 I_{jk,lm}^2(\omega), \tag{6.65}$$

where n denotes the solvent and solute species, $\beta_{ijk}^{p,n}$'s denote the *ijk* components of the hyperpolarizability tensor of molecule p of species n, and $\langle \rangle$ indicates the orientational average. The harmonic scattering intensity is proportional to the fourth power of the fundamental frequency (as in Rayleigh scattering) and to the square of the fundamental intensity (as in SHG). By neglecting coherent scattering, i.e., any spatial and orientational correlations, for polarized (VV) and depolarized (VH) scattering, Eq. (6.65) simplifies to (Clays and Persoon, 1991)

$$I_Z^{VV}(2\omega) \propto \left(\frac{2\omega}{c}\right)^4 (f^\omega)^4 (f^{2\omega})^2 I_Z^2(\omega) [N_1 \langle \beta_{ZZZ}^1 \beta_{ZZZ}^1 \rangle + N_2 \langle \beta_{ZZZ}^2 \beta_{ZZZ}^2 \rangle] \tag{6.66}$$

and

$$I_X^{VH}(2\omega) \propto \left(\frac{2\omega}{c}\right)^4 (f^\omega)^4 (f^{2\omega})^2 I_Z^2(\omega) [N_1 \langle \beta_{XZZ}^1 \beta_{XZZ}^1 \rangle + N_2 \langle \beta_{XZZ}^2 \beta_{XZZ}^2 \rangle], \tag{6.67}$$

where subscripts 1 (2) denote solvent (solute) and $\langle \beta\beta \rangle$s depend on the point-group symmetry of the molecules. For point group C_{2v}, for example (Zyss et al., 1993),

$$\langle \beta_{ZZZ} \beta_{ZZZ} \rangle = \frac{1}{7} \beta_{iii} \beta_{iii} + \frac{9}{35} \beta_{ijj} \beta_{ijj} + \frac{6}{35} \beta_{iii} \beta_{ijj}, \tag{6.68}$$

and

$$\langle \beta_{XZZ} \beta_{XZZ} \rangle = \frac{1}{35} \beta_{iii} \beta_{iii} + \frac{11}{105} \beta_{ijj} \beta_{ijj} - \frac{2}{105} \beta_{iii} \beta_{ijj}. \tag{6.69}$$

In addition, the depolarization ratio allows separating the two tensor components by the expression

$$\frac{\langle \beta_{ZZZ} \beta_{ZZZ} \rangle}{\langle \beta_{XZZ} \beta_{XZZ} \rangle} = \frac{15 + 21\kappa^2 + 30\kappa}{3 + 11\kappa^2 - 2\kappa}, \tag{6.70}$$

where $\kappa = \beta_{XZZ}/\beta_{ZZZ}$.

The key advantage of hyper-Rayleigh scattering is its access to the hyperpolarizability of molecules which lack a permanent dipole moment. This includes charged species as well as molecules with T_d, D_n, or D_{nh} point-group symmetry, where n is an odd number ≥ 3. It also yields tensor components which are not available from EFISH. Currently, it is difficult to the measurement wavelength to the nonresonant regime by using long wavelength (≈ 2 μm) fundamental radiation due to the λ^{-4} dependence on scattering efficiency and the lack of a good photodetector at the harmonic wavelength near 1 μm.

Data Reduction at Infinite Dilution

In addition to the nonlinear optical measurements such as EFISH, THG, or hyper-Rayleigh scattering, other physical and optical quantities, such as the density, refractive indices at several wavelengths, and dielectric constant are usually needed to obtain molecular parameters. These quantities, together with the susceptibilities and coherence lengths, are usually determined for a series of solutions in graded concentrations. From the solution properties, the relevant molecular properties, such as the permanent dipole moment and appropriate tensor projections of all three polarizabilities in Eq. (6.23), can be calculated according to well-accepted approximations of liquid dielectric behavior. For EFISH and THG in a binary solution, for instance, the following equations relate the molecular and solution quantities:

$$\frac{n^2(\omega) - 1}{4\pi} = \sum_{i=1}^{2} N_i \alpha_i^\omega f_i^\omega, \tag{6.71}$$

$$\frac{\epsilon - 1}{4\pi} = \sum_{i=1}^{2} N_i \left(\alpha_i^0 f_i^0 + \frac{\mu_i^2 f_i^0}{3kT} \right), \tag{6.72}$$

$$\chi_{THG}^{(3)} = \sum_{i=1}^{2} N_i (f_i^\omega)^3 f_i^{3\omega} \gamma_i^{THG}, \tag{6.73}$$

and

$$\chi_{EFISH}^{(3)} = \sum_{i=1}^{2} N_i (f_i^\omega)^2 f_i^{2\omega} f_i^0 \left(\gamma_i^e + \frac{\mu_i \beta_{\mu,i}}{5kT} \right), \tag{6.74}$$

where i denotes the solvent and solute species, α^0 is the static polarizability, and the local fields are given by Eqs. (6.31) and (6.32) with $f_i^* = \epsilon(2\epsilon + 1)(n_i^2 + 2)^2/3(2\epsilon + n_i^2)^2$. The concentration-dependent data can be analyzed in the infinite dilution limit (Singer and Garito, 1981). This involves rewriting Eqs. (6.71–6.74) as functions of mole fraction, differentiating with respect to mole fraction, and dropping all terms proportional to mole fraction. At this limit, the initial slopes and the solvent values of the measured quantities determine the results. When one or more molecular parameters are complex, as in the case of optical absorption, a nonlinear least-squares fit of the concentration-dependent data can be used to obtain both the real and imaginary contributions (Kajzar et al., 1987).

6.4 Molecular Studies

Since the early recognition that charge-transfer interaction at the molecular level holds the key to large optical nonlinearity, organic molecules possessing extraordinarily large hyperpolarizabilities have been successfully developed. Molecular engineering, focusing on the structural motif of "donor-π-acceptor" conjugated molecules, has yielded an increase in hyperpolarizability of over four orders of magnitude since the first interest in urea for nonlinear optics. Table 1 lists most of the measurement results in the literature. It is organized to facilitate discussion of structure-property trends and is in approximate ascending order according to the size of the π system, acceptor strength, and donor strength. Where substituents are not specified in the molecular structure, hydrogen is implied. Multiple measurement results reported for the same molecule are listed in chronological order.

The majority of the results was obtained from EFISH measurements. β_μ as defined in Eq. (6.60) is given together with the available dipole moment and λ_{max}, which denotes the absorption maximum of the lowest energy, prominent transition. The solvent and fundamental wavelength used for the measurement are also given because they may influence measurement results. Where μ is not available, $\mu\beta_\mu$ is given. The quantity $\beta_0 = \beta_\mu(1 - \lambda_{max}^2/\lambda^2)(1 - 4\lambda_{max}^2/\lambda^2)$, which is derived from the 2-level approximation, is sometimes given. Results obtained from solvatochromic analysis and hyper-Rayleigh scattering are indicated with (S) and (HR), respectively. Typical experimental uncertainties in μ and β_μ are ± 10–15%. When comparing results from different laboratories, additional discrepancies can be expected due to differing methodologies. Unfortunately, confusion caused by different theoretical conventions used by various authors is difficult to eliminate. In cases where the difference is beyond the bounds of reasonable uncertainties due to the wavelengths or solvents used, readers are requested to consult the original references for clarification.

Electron Donors and Acceptors

Mono- and *para* disubstituted benzene and stilbene results can be used to evaluate the effectiveness of various donor and acceptor groups in inducing hyperpolarizability. A wide range of electron donating and withdrawing substituents have been examined. They can be ranked in increasing effectiveness; for simple acceptors,

$$SO_2CH_3 < CN < COH < SO_2CF_3 < COCF_3 < NO < NO_2 < CHC(CN)_2 < CCNC(CN)_2,$$

and, for simple donors,

$$CH_3 < Br < OH < OC_6H_5 < OCH_3 < SH < SCH_3 < NH_2 < NHCH_3 < N(CH_3)_2.$$

From monosubstituted derivatives, it is evident that acceptors are considerably more effective than donors. The ineffectiveness of donors may be attributed to two factors. Other than the -$N(CH_3)_2$ group and the bridged structure in julolidine, the connecting atom in all of the donors, including -NH_2, is known to be sp^3 hybridized. The tetrahedral geometry does not allow efficient overlap between the lone pair-containing donor orbital and the π system. In addition, these donor substitutions result only in minimal extensions of the overall conjugation. The proper hybridization and longer extension of the conjugation may be reasons for the effectiveness of the acceptors. This view is supported by the large β_μ of benzylidene malonitrile ($C_6H_5CHC(CN)_2$) and its extended structure. However, the substantially higher dipole moments of the acceptor substituted benzenes do not mean a greater polarization of the benzene π system because the sulfonyl, cyano, carbonyl, and nitro group dipoles are substantial [e.g., μ(nitrohexane) = 3.6 D, and μ(acetonitrile) = 3.4 D]. The cooperative effects of donor and acceptor groups are evident in *para* disubstituted benzene (Levine, 1976) and 4,4' disubstituted stilbene derivatives (Oudar, 1977a). A significant increase in μ and β_μ values over the sum of the monosubstituted fragments is observed. This enhancement results from the CT interactions between the donor and acceptor groups, leading to polarization of both the ground and excited states. From simple vector additivity

TABLE 6.1 Experimental Results of Molecular Studies: λ_{max}, Absorption Peak Wavelength of Lowest Energy Strong Transition; μ, Dipole Moment; $\beta_\mu(\lambda)$, EFISH Determined First Hyperpolarizability at a Fundamental Wavelength Given in Microns; β_0, as Defined in Section 6.4; S Indicates Solvatochromic Result; HR Indicates Hyper-Rayleigh Scattering Results

Structure	Solvent	λ_{max} (nm)	μ (10^{-18} esu)	$\beta_\mu(\lambda)^{ref.}$ (10^{-30} esu)
Unconjugated derivatives				
(N-methyl structure)	chloroform	<240	3.5	0.5 (1.9)[156]
(cyanamide structure)		<200	4.6	2.3 (1.06)[37]
K⁺ tricyanomethanide	H₂O			7 (1.06HR)[265]
Benzene derivatives				
X = CH₃	neat		0.34	0.18 (1.06)[141]
X = OH	neat	260	0.38	<0.2 (1.9)[46]
	neat			0.17 (1.06)[193]
	neat			0.17 (1.06)[191]
X = SH	p-dioxane	270	1.5	<0.2 (1.9)[46]
X = OCH₃	neat		1.5	<0.2 (1.9)[46]
X = SCH₃	neat	275	1.4	<0.2 (1.9)[46]
X = NH₂	neat		1.4	<0.2 (1.9)[46]
	neat		1.3	0.89 (1.06)[141]
	neat		1.5	1.2 (1.06)[193]
	neat			1.1 (1.06)[190]
	neat		1.5	1.2 (1.06)[191]
	neat	284	1.5	0.55 (1.9)[46]
X = N(CH₃)₂	neat			1.8 (1.06)[193]
	neat		1.8	1.3 (1.06)[191]
	neat	293	1.6	1.1 (1.9)[46]

TABLE 6.1 Experimental Results of Molecular Studies—*(continued)*

Structure	Solvent	λ_{max} (nm)	μ (10^{-18} esu)	$\beta_\mu(\lambda)^{ref.}$ (10^{-30} esu)
X = F	neat		1.5	0.44 (1.06)[141]
	neat			0.53 (1.06)[193]
	neat		1.6	0.70 (1.06)[191]
	neat	266	0.63	<0.2 (1.9)[46]
X = Cl	neat		1.5	0.28 (1.06)[141]
	neat			0.22 (1.06)[193]
	neat		1.7	0.33 (1.06)[191]
	neat	272	0.44	<0.2 (1.9)[46]
X = Br	neat		1.4	0.04 (1.06)[141]
	neat			0.02 (1.06)[193]
	neat		1.7	0.2 (1.06)[191]
	neat	274	0.81	<0.2 (1.9)[46]
X = I	neat		1.1	0.28 (1.06)[141]
	neat			0.7 (1.06)[193]
	neat		1.7	0.46 (1.06)[191]
	neat	304	0.96	<0.2 (1.9)[46]
X = SO$_2$CH$_3$	p-dioxane		1.1	<0.2 (1.9)[46]
X = SO$_2$F	neat		1.7	0.3 (1.9)[46]
X = CN	neat			0.51 (1.06)[193]
	neat			0.48 (1.06)[191]
	neat	230	4.1	0.36 (1.9)[46]
X = COH	neat	320	3.9	0.8 (1.9)[46]
X = COCF$_3$	p-dioxane	300	2.8	1.3 (1.9)[46]
X = NO	p-dioxane	340/745	3.3	1.7 (1.9)[46]
X = NO$_2$	neat		3.1	2.0 (1.06)[145]
	neat			2.3 (1.06)[193]
	neat		4.0	2.2 (1.06)[190]
	neat		4.0	2.3 (1.06)[191]
	hexane		4.2	1.1 (1.06)[232]
	neat	268	4.0	1.9 (1.9)[46]
	p-dioxane	303	4.8	3.1 (1.9)[46]
X = C$_2$H(CN)$_2$	p-dioxane	310	1.6	1.3 (1.9)[46]

Second-Order Nonlinear Optical Properties of Organic Materials

X	Y	Solvent	λ	μβ	β
X = SO$_2$CH$_3$	Y = OH	p-dioxane	290	3.4	1.3 (1.9)^{46}
X = SO$_2$CH$_3$	Y = N(CH$_3$)$_2$	chloroform		μβ = 26 × 10^{-48}	(1.9)^{264}
X = CN	Y = CH$_3$	DMSO	232	4.4	2.9 (1.9)^{56}
		neat		4.4	0.7 (1.9)^{46}
X = CN	Y = Cl	p-dioxane		2.3	0.8 (1.9)^{46}
X = CN	Y = Br	p-dioxane		2.4	1.1 (1.9)^{46}
		p-dioxane	240	2.5	2.2 (1.59)^{87}
X = CN	Y = OC$_6$H$_5$	p-dioxane	247	4.1	1.2 (1.9)^{46}
X = CN	Y = OCH$_3$	DMSO	248	4.8	4.8 (1.9)^{56}
		p-dioxane		4.8	1.9 (1.9)^{46}
X = CN	Y = SCH$_3$	p-dioxane	269	4.4	2.8 (1.9)^{46}
X = CN	Y = NH$_2$	DMSO	270	5.9	13 (1.9)^{56}
		p-dioxane	271	5.0	3.1 (1.9)^{46}
		p-dioxane	297	6.4	2.4 (1.59)^{87}
X = CN	Y = N(CH$_3$)$_2$	DMSO	290	6.6	14 (1.9)^{56}
		p-dioxane	290	5.6	5.0 (1.9)^{46}
		CCl$_4$		μβ = 44 × 10^{-48}	(1.06)^{211}
		chloroform	292	μβ = 50 × 10^{-48}	(1.06)^{211}
		H$_2$CCl$_2$	292	μβ = 53 × 10^{-48}	(1.06)^{211}
		H$_3$CCN	291	μβ = 48 × 10^{-48}	(1.06)^{211}
		H$_3$COH	295	μβ = 47 × 10^{-48}	(1.06)^{211}
X = COH	Y = CH$_3$	neat		3.0	1.7 (1.9)^{46}
X = COH	Y = OC$_6$H$_5$	neat	269	2.8	1.9 (1.9)^{46}
X = COH	Y = OCH$_3$	neat	269	3.5	2.2 (1.9)^{46}
X = COH	Y = SCH$_3$	neat	310	3.1	2.6 (1.9)^{46}
X = COH	Y = N(CH$_3$)$_2$	p-dioxane	326	5.1	6.3 (1.9)^{46}
X = SO$_2$C$_3$H$_7$	Y = OCH$_3$	chloroform	290	5.4	3.3 (1.9)^{46}
X = SC$_6$H$_5$	Y = NH$_2$	acetone	320	3.0	14 (1.06)^{16}
X = COCH$_3$	Y = NH$_2$	DMSO	310	4.5	2.4 (1.9)^{56}
X = COCF$_3$	Y = OC$_6$H$_5$	p-dioxane	292	3.5	3.6 (1.9)^{46}
X = COCF$_3$	Y = OCH$_3$	p-dioxane	292	4.0	3.6 (1.9)^{46}
X = COCF$_3$	Y = N(CH$_3$)$_2$	p-dioxane	356	5.9	10 (1.9)^{46}
X = NO	Y = N(CH$_3$)$_2$	p-dioxane	407	6.2	12 (1.9)^{46}
X = NO$_2$	Y = CH$_3$	DMSO	280	4.0	8 (1.9)^{56}
		p-dioxane	272	4.2	2.1 (1.9)^{46}
X = NO$_2$	Y = F	neat		2.7	2.1 (1.06)^{141}
X = NO$_2$	Y = Br	p-dioxane	274	3.0	3.3 (1.9)^{46}
X = NO$_2$	Y = OH	p-dioxane	304	5.0	3.0 (1.9)^{46}
X = NO$_2$	Y = OC$_6$H$_5$	p-dioxane	294	4.2	4.0 (1.9)^{46}

TABLE 6.1 Experimental Results of Molecular Studies—*(continued)*

Structure		Solvent	λ_{max} (nm)	μ (10^{-18} esu)	β_μ (λ)[ref.] (10^{-30} esu)
X = NO_2	Y = OCH_3	DMSO	314	4.5	16 (1.9)[56]
		p-dioxane		4.9	6 (1.06)[98]
				5.3	4 (1.9S)[195]
		p-dioxane	302	4.6	5.1 (1.9)[46]
X = NO_2	Y = SCH_3				6.7 (1.9S)[195]
		p-dioxane	322	4.4	6.1 (1.9)[46]
		acetone	335	4.4	20 (1.06)[16]
X = NO_2	Y = SC_6H_5	acetone	334	4.3	19 (1.06)[16]
X = NO_2	Y = N_2H_3	p-dioxane	366	6.3	7.6 (1.9)[46]
X = NO_2	Y = NH_2	methanol			36 (1.06)[193]
		melt		7.2	21 (1.06)[140]
		methanol		6.2	35 (1.06)[190]
		DMSO	378	6.1	47 (1.9)[56]
		p-dioxane			9.6 (1.91)[250]
		p-dioxane			12 (1.37)[250]
		p-dioxane			17 (1.06)[250]
		p-dioxane			25 (0.91)[250]
		p-dioxane			40 (0.83)[250]
		DMSO	384	$\mu\beta = 75 \times 10^{-48}$	(1.58)[231]
				7.1	9.6 (1.9S)[195]
					23 (1.06HR)[48]
		chloroform			9.2 (1.9)[46]
		acetone	365	6.2	16 (1.06)[237]
		p-dioxane	354	7.0	17 (1.06)[237]
		chloroform	348	6.4	29 (1.06)[237]
		H_3CCN	366	6.2	32 (1.06)[237]
		methanol	370	6.1	38 (1.06)[237]
		NMP	386	6.8	(1.9)[107]
		p-dioxane		$\mu\beta = 69 \times 10^{-48}$	20 (1.06)[117]
		chloroform	347	6.2	
X = NO_2	Y = $NH(NH_2)$	DMSO	378	$\mu\beta = 10 \times 10^{-47}$	(1.36)[231]
X = NO_2	Y = $N(CH_3)_2$	DMSO	418	6.8	52 (1.9)[56]
		chloroform		$\mu\beta = 14 \times 10^{-47}$	(1.9)[264]
		acetone	388	6.9	26 (1.06)[16]
		DMSO	404	$\mu\beta = 14 \times 10^{-47}$	(1.36)[231]
		acetone	376	6.4	12 (1.9)[46]
X = NO_2	Y = CN	p-dioxane		0.9	0.6 (1.9)[46]
X = NO_2	Y = COH	p-dioxane	376	2.5	0.2 (1.9)[46]

		λ (nm)	μβ or β	β (ω)
X = C₂H(CN)₂, Y = OCH₃	p-dioxane	345	5.5	9.8 (1.9)[46]
X = C₂H(CN)₂, Y = N(CH₃)₂	DMSO	440	μβ = 27 × 10⁻⁴⁷	(1.36)[231]
	DMSO	440	μβ = 62 × 10⁻⁴⁷	(1.06)[101]
			7.1	12 (1.9S)[195]
X = C₂H(CN)₂	chloroform	420	7.8	32 (1.9)[46]
X = C₂H(CN)₂, Y = N(C₂H₅)₂	p-dioxane	419	μβ = 30 × 10⁻⁴⁷	(1.9)[106]
X = C₂H(CN)₂, Y = NC₅H₁₀	chloroform	440	8.5	17 (β₀)[177]
X = C₂H(CN)₂, Y = N(4-CH₃C₆H₄)₂	chloroform	452	7.6	25 (β₀)[177]
X = C₂(CN)₃, Y = NH₂	CH₂Cl₂	498	7.8	39 (1.9)[46]
X = C₂(CN)₃, Y = N(CH₃)₂	CH₂Cl₂	516	8.2	50 (1.9)[46]
	DMSO	528	μβ = 85 × 10⁻⁴⁷	(1.36)[231]
	chloroform	458	5.7	14 (1.9)[43]
	CH₂Cl₂		8.0	44 (1.9)[46]
	chloroform		8.3	51 (1.9)[43]
	CH₂Cl₂	556	1.6	60 (1.9)[46]
	DMSO	561	μβ = 73 × 10⁻⁴⁷	(1.36)[231]
	DMSO	458	μβ = 36 × 10⁻⁴⁷	(1.36)[231]
	DMSO	602	μβ = 12 × 10⁻⁴⁶	(1.58)[231]
	p-dioxane	505	μβ = 71 × 10⁻⁴⁷	(1.36)[203]
		397	β = 171 × 10⁻³⁰	(1.06S)[200]

TABLE 6.1 Experimental Results of Molecular Studies—*(continued)*

Structure	Solvent	λ_{max} (nm)	μ (10^{-18} esu)	$\beta_\mu(\lambda)^{ref.}$ (10^{-30} esu)
ortho-X-nitrobenzene				
X = CH$_3$	neat		3.9	1.0 (1.9)[46]
X = F	neat		5.0	1.8 (1.9)[141]
X = Br	p-dioxane		4.0	0.4 (1.9)[46]
X = OH	p-dioxane	348	3.4	1.2 (1.9)[46]
X = OCH$_3$	neat	318	3.8	1.4 (1.9)[46]
X = NH$_2$	melt		5.0	6.4 (1.32)[140]
	acetone		4.3	10 (1.06)[190]
X = CN	p-dioxane		4.1	2.5 (1.9)[46]
X = COH	p-dioxane		5.5	1.2 (1.9)[46]
	p-dioxane		4.0	0.8 (1.9)[46]
meta-X-nitrobenzene				
X = CH$_3$	neat		3.9	1.5 (1.9)[46]
X = F	neat		3.6	−1.6 (1.06)[141]
X = Cl	benzene		3.4	1.6 (1.06)[191]
X = Br	benzene		3.4	1.2 (1.06)[191]
X = OH	p-dioxane		3.4	1.0 (1.9)[46]
X = OCH$_3$	p-dioxane		3.6	0.8 (1.9)[46]
X = NH$_2$	p-dioxane	326	3.9	1.6 (1.9)[46]
	melt		5.5	4.2 (1.32)[140]
	acetone	396	4.9	6 (1.06)[190]
X = CN	p-dioxane		4.7	1.9 (1.9)[46]
X = COH	p-dioxane		3.7	0.8 (1.9)[46]
X = NO$_2$	p-dioxane		2.8	1.7 (1.9)[46]
	benzene		3.9	1.2 (1.06)[191]
2-methyl-4-nitroaniline	p-dioxane			9.5 (1.91)[250]
	p-dioxane			13 (1.37)[250]
	p-dioxane			17 (1.06)[250]
	p-dioxane			27 (0.91)[250]
	p-dioxane			45 (0.83)[250]

Solvent	λ (nm)	μ (D)	β × 10³⁰ (esu)	Structure
p-dioxane	361	6.2	8.7 (1.9)[46]	
p-dioxane		7.4	18 (1.06)[25]	
p-dioxane			9.1 (1.32)[25]	
p-dioxane			7.6 (1.91)[25]	
p-dioxane		5.7	13 (1.06)[262]	4-amino-3-chloronitrobenzene
p-dioxane	350	5.9	6.8 (1.9)[46]	4-amino-3-methoxynitrobenzene
p-dioxane		6.0	8.7 (1.9)[46]	3-acetamido-4-(dimethylamino)nitrobenzene
p-dioxane		7.4	18 (1.06)[262]	
p-dioxane		8.1	9.5 (1.06S)[195]	
p-dioxane	240	3.5	1.4 (1.59)[87]	4-bromo-2,3,5,6-tetrafluorobenzonitrile
p-dioxane	274	6.5	2.6 (1.59)[87]	4-amino-2,3,5,6-tetrafluorobenzonitrile
p-dioxane	304	4.4	2.5 (1.9)[46]	4-methoxy-2-fluoronitrobenzene
p-dioxane	304	4.9	2.6 (1.9)[46]	4-methoxy-2,5-difluoronitrobenzene
p-dioxane	270	4.2	1.7 (1.9)[46]	4-methoxy-2,3,5,6-tetrafluoronitrobenzene

TABLE 6.1 Experimental Results of Molecular Studies—*(continued)*

Structure	Solvent	λ_{max} (nm)	μ (10^{-18} esu)	β_μ (λ)$^{ref.}$ (10^{-30} esu)
F, F, F, F, NO$_2$, H$_2$N (tetrafluoro nitroaniline)	*p*-dioxane	319	6.0	5.5 (1.59)[87]
NO$_2$, NO$_2$, H$_2$N	acetone		5.5	21 (1.06)[190]
	p-dioxane		$\mu\beta = 49 \times 10^{-48}$	(1.9)[107]
NO$_2$, NO$_2$, N, H$_3$CO$_2$C	acetone		5.6	22 (1.06)[190]

Styrene derivatives

X—⌬=⌬—Y

	Solvent	λ_{max} (nm)	μ (10^{-18} esu)	β_μ (λ)$^{ref.}$ (10^{-30} esu)
X = CN, Y = OCH$_3$	chloroform	304	4.2	7.0 (1.9)[45]
X = CN, Y = N(CH$_3$)$_2$	chloroform	364	6.0	23 (1.9)[45]
X = COH, Y = Br	chloroform	298	2.0	6.5 (1.9)[45]
X = COH, Y = OCH$_3$	DMSO	320	$\mu\beta = 80 \times 10^{-48}$	(1.9)[55]
	chloroform	318	4.2	11 (1.9)[45]
X = COH, Y = N(CH$_3$)$_2$	DMSO	387	$\mu\beta = 37 \times 10^{-47}$	(1.9)[55]
	chloroform	384	$\mu\beta = 32 \times 10^{-47}$	(1.34)[14]
	chloroform	384	5.6	30 (1.9)[45]
X = COCH$_3$, Y = OCH$_3$	chloroform	316	4.0	8.9 (1.9)[45]
X = COCF$_3$, Y = N(CH$_3$)$_2$	chloroform		6.6	64 (1.9)[43]
X = NO$_2$, Y = H	chloroform	312	3.8	8.0 (1.9)[45]
X = NO$_2$, Y = OH	chloroform	352	5.1	18 (1.9)[45]
X = NO$_2$, Y = OCH$_3$	*p*-dioxane		5.5	12 (1.06)[98]
	chloroform	352	4.6	17 (1.9)[45]
X = NO$_2$, Y = N(CH$_3$)$_2$	chloroform	438	7.9	220 (1.06)[189]
	chloroform		6.5	50 (1.9)[45]
X = CHC(CN)$_2$, Y = N(CH$_3$)$_2$	DMSO	500	$\mu\beta = 36 \times 10^{-46}$	(1.06)[102]
	DMSO	500	Re($\mu\beta$) = 34×10^{-46}	(1.06)[101]
			Im($\mu\beta$) = 36×10^{-46}	(1.06)[101]

Compound	Solvent	λ (nm)	μ (D)	μβ	β(ω)[ref]
(julolidine-CH=CH-C6H4-Y); X = N(CH$_3$)$_2$, Y = NO$_2$	chloroform	438	5.9		35 (1.9)[45]
X = COH	chloroform		6.3		48 (1.9)[43]
X = NO$_2$	chloroform		7.1		90 (1.9)[43]
(nitro-benzylidene-amine, isomer A)	p-dioxane			μβ = 15 × 10^{-47}	(1.9)[107]
(nitro-benzylidene-amine, isomer B)	p-dioxane			μβ = 16 × 10^{-47}	(1.9)[107]

Biphenyl derivatives

X	Y	Solvent	λ (nm)	μ (D)	β(ω)[ref]
CN	H	p-dioxane	272	4.0	1.9 (1.9)[45]
COCH$_3$	H	p-dioxane	280	3.1	2.0 (1.9)[45]
NO$_2$	H	p-dioxane	304	3.8	4.1 (1.9)[45]
SO$_2$CH$_3$	N(CH$_3$)$_2$	chloroform	340	6.0	13 (1.9)[45]
CN	OH	p-dioxane	292	4.8	6.3 (1.9)[45]
COCH$_3$	OCH$_3$	p-dioxane	304	3.4	4.9 (1.9)[45]
NO$_2$	Br	p-dioxane	306	2.7	4.4 (1.9)[45]
NO$_2$	OH	p-dioxane	334	4.9	7.7 (1.9)[45]
NO$_2$	OCH$_3$	p-dioxane	332	4.5	9.2 (1.9)[45]
NO$_2$	NH$_2$	chloroform	372	5.0	24 (1.9)[45]
NO$_2$	NH$_2$	NMP		7.8	24 (1.9)[45]
NO$_2$	N(CH$_3$)$_2$	chloroform	390	5.5	50 (1.9)[45]
NO$_2$	N(CH$_3$)$_2$	toluene	401	7.5	89 (1.06)[271]
NO$_2$	N(C$_6$H$_{13}$)$_2$	C$_6$H$_{11}$CH$_3$	400	7.5	73 (1.06)[271]

TABLE 6.1 Experimental Results of Molecular Studies—(*continued*)

Structure	Solvent	λ_{max} (nm)	μ (10^{-18} esu)	$\beta_\mu(\lambda)^{ref.}$ (10^{-30} esu)
Fluorene derivatives				
X = CN, Y = H	*p*-dioxane	284	3.9	3.0 (1.9)[45]
X = NO$_2$, Y = H	*p*-dioxane	328	4.1	5.1 (1.9)[45]
X = NO$_2$, Y = Br	*p*-dioxane	330	2.8	6.0 (1.9)[45]
X = NO$_2$, Y = OCH$_3$	*p*-dioxane	356	4.7	11 (1.9)[45]
X = NO$_2$, Y = N(CH$_3$)$_2$	*p*-dioxane	410	5.6	40 (1.9)[45]
	chloroform	417	6.0	55 (1.9)[45]
Diphenyl ethers and derivatives				
CH$_3$O–C$_6$H$_4$–O–C$_6$H$_4$–SO–	chloroform	252	5.2	6.3 (1.9)[43]
H$_2$N–C$_6$H$_4$–O–C$_6$H$_4$–NO$_2$	*p*-dioxane	296	6.2	15 (1.06)[210]
	p-dioxane		4.9	4.5 (1.9)[43]
(CH$_3$)$_2$N–C$_6$H$_4$–O–C$_6$H$_4$–NO$_2$	*p*-dioxane	294	5.2	5.3 (1.9)[43]
CH$_3$O–C$_6$H$_4$–S–C$_6$H$_4$–SO–	chloroform	295	5.3	8.9 (1.9)[43]
H$_2$N–C$_6$H$_4$–S–C$_6$H$_4$–NO$_2$	benzene	341	6.0	26 (1.06)[210]
	acetone		5.8	28 (1.06)[16]
	p-dioxane	334	5	8.7 (1.9)[43]

Compound	Solvent	λ (nm)	μ	β (ω)
H₂N–C₆H₄–SO₂–C₆H₄–NO₂	acetone	332	8	19 (1.06)[16]
H₂N–C₆H₄–SO₂–C₆H₄–NH₂	acetone	295	7.8	8 (1.06)[16]
H₂N–C₆H₄–Se–C₆H₄–NO₂	p-dioxane		5.9	27 (1.06)[210]
H₂N–C₆H₄–Te–C₆H₄–NO₂	p-dioxane		5.6	27 (1.06)[210]
n = 1	chloroform	320	6	16 (1.06)[164]
n = 2	chloroform	334	7	22 (1.06)[164]
n = 3	chloroform	385	6.8	38 (1.06)[164]
X = SO₂CH₃, Y = H	p-dioxane	310	5.3	3.8 (1.06)[32]
X = SO₂CH₃, Y = OCH₃	p-dioxane	310	5.9	11 (1.06)[32]
X = SO₂CH₃, Y = CH₃S	p-dioxane	320	5.2	16 (1.06)[32]
X = SO₂CH₃, Y = NH₂	chloroform	338	6.5	13 (1.9)[45]
X = SO₂CH₃, Y = N(CH₃)₂	p-dioxane	358	7.5	5.6 (1.06)[32]
X = SO₂CF₃, Y = OCH₃	p-dioxane	327	6.2	21 (1.06)[32]
X = SO₂CF₃, Y = N(CH₃)₂	p-dioxane	388	8.4	40 (1.06)[32]
X = CO₂CH₃, Y = SCH₃	chloroform	328	2.9	8 (1.9)[45]
X = CO₂CH₃, Y = NH₂	chloroform	332	3.8	15 (1.9)[45]
X = COCH₃, Y = SCH₃	chloroform	336	3.7	9.8 (1.9)[45]
X = COCH₃, Y = NH₂	chloroform	334	3.3	12 (1.9)[45]

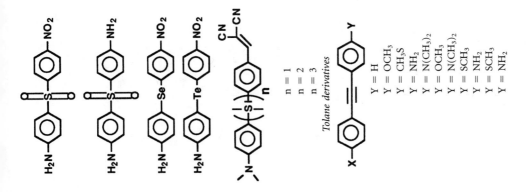

Tolane derivatives

TABLE 6.1 Experimental Results of Molecular Studies—*(continued)*

Structure	Solvent	λ_{max} (nm)	μ (10^{-18} esu)	$\beta_\mu (\lambda)^{ref.}$ (10^{-30} esu)
X = COC$_6$H$_5$, Y = NH$_2$	chloroform	352	3.7	19 (1.9)^{45}
X = CN, Y = SCH$_3$	chloroform	333	4.0	15 (1.9)^{45}
X = CN, Y = NH$_2$	chloroform	342	5.2	20 (1.9)^{45}
X = CN, Y = NHCH$_3$	chloroform	358	5.7	27 (1.9)^{45}
X = CN, Y = N(CH$_3$)$_2$	chloroform	372	6.1	29 (1.9)^{45}
X = NO$_2$, Y = H	p-dioxane	326	4.6	16 (1.06)^{32}
X = NO$_2$, Y = Br	chloroform	335	3.0	10 (1.9)^{45}
X = NO$_2$, Y = OCH$_3$	p-dioxane	356	4.4	14 (1.9)^{45}
X = NO$_2$, Y = SCH$_3$	chloroform	362	4.0	20 (1.9)^{45}
X = NO$_2$, Y = NH$_2$	chloroform	380	5.5	24 (1.9)^{45}
	NMP	410	5.5	40 (1.9)^{45}
X = NO$_2$, Y = NHCH$_3$	chloroform	400	5.7	46 (1.9)^{45}
X = NO$_2$, Y = N(CH$_3$)$_2$	chloroform	415	6.1	46 (1.9)^{45}
	p-dioxane	402	7.1	102 (1.06)^{32}
	chloroform	416	6.6	33 (β_0)^{177}
X = NO$_2$, Y = N(C$_6$H$_5$)$_2$	chloroform	418	4.8	28 (β_0)^{177}

Stilbene derivatives

Structure	Solvent	λ_{max} (nm)	μ (10^{-18} esu)	$\beta_\mu (\lambda)^{ref.}$ (10^{-30} esu)
X = H, Y = OCH$_3$	benzene	320	2.8	6.1 (1.06)^{175}
X = H, Y = NH$_2$	p-dioxane	332	2.1	12 (1.06)^{189}
	benzene		2.2	7.4 (1.9)^{46}
X = H, Y = N(CH$_3$)$_2$	p-dioxane	340	2.4	29 (1.06)^{189}
	benzene		2.1	10 (1.9)^{46}
X = Cl, Y = H	chloroform		1.5	3.6 (1.06)^{189}
X = Cl, Y = N(CH$_3$)$_2$	p-dioxane	323	4.0	42 (1.06)^{189}
X = CF$_3$, Y = OCH$_3$		326	4.3	12 (1.06)^{244}
		4.2	16 (1.06)^{175}	
X = CF$_3$, Y = OH	p-dioxane	327	4.7	12 (1.06)^{244}
X = SO$_2$CH$_3$, Y = H	p-dioxane	364	4.4	58 (1.06)^{32}
X = SO$_2$CH$_3$, Y = OCH$_3$	chloroform	336	6.5	10 (1.9)^{46}
	p-dioxane	335	6.1	9.1 (1.06)^{32}
X = SO$_2$CH$_3$, Y = SC$_6$H$_5$	p-dioxane	344	4.4	19 (1.06)^{32}
X = SO$_2$CH$_3$, Y = N(CH$_3$)$_2$	p-dioxane	376	6.9	66 (1.06)^{32}

X	Y	solvent	λ	μβ	β (ω)ref
$X = SO_2CH_3$	$Y = N(CH_2C_2H_3)_2$	chloroform	391	$\mu\beta = 57 \times 10^{-47}$	$(1.9)^{264}$
$X = SO_2CF_3$	$Y = OCH_3$	p-dioxane	347	7.8	$14 (1.9)^{46}$
		p-dioxane	350	6.6	$34 (1.06)^{32}$
$X = SO_2C_6F_{13}$	$Y = N(CH_3)_2$	chloroform		8.0	$59 (1.9)^{43}$
$X = COCF_3$	$Y = OCH_3$	p-dioxane	368	4.2	$16 (1.9)^{46}$
$X = COH$	$Y = N(CH_3)_2$	chloroform	360	3.5	$24 (1.9)^{254}$
$X = CN$	$Y = OH$	p-dioxane	344	4.5	$13 (1.9)^{46}$
$X = CN$	$Y = OCH_3$	DMSO	342	$\mu\beta = 82 \times 10^{-47}$	$(1.9)^{55}$
		chloroform	340	3.8	$19 (1.9)^{46}$
$X = CN$	$Y = N(CH_3)_2$	DMSO	390	$\mu\beta = 98 \times 10^{-48}$	$(1.9)^{55}$
		chloroform	382	5.7	$36 (1.9)^{46}$
$X = NO_2$	$Y = H$	benzene		4.6	$29 (1.06)^{189}$
		p-dioxane	345	4.2	$11 (1.9)^{46}$
$X = NO_2$	$Y = CH_3$	DMSO	368	$\mu\beta = 20 \times 10^{-48}$	$(1.9)^{55}$
		p-dioxane	351	4.7	$15 (1.9)^{46}$
$X = NO_2$	$Y = Cl$	chloroform		3.1	$39 (1.06)^{189}$
$X = NO_2$	$Y = Br$	p-dioxane	344	3.2	$14 (1.9)^{46}$
		chloroform	356	3.4	$18 (1.9)^{46}$
$X = NO_2$	$Y = OH$	chloroform			$93 (1.06HR)^{48}$
		p-dioxane	370	5.5	$17 (1.9)^{46}$
$X = NO_2$	$Y = OC_6H_5$	p-dioxane	350	4.6	$18 (1.9)^{46}$
$X = NO_2$	$Y = OCH_3$	p-dioxane		5.7	$81 (1.06)^{98}$
		chloroform			$105 (1.06HR)^{48}$
		p-dioxane	364	4.5	$28 (1.9)^{46}$
		chloroform	370	4.5	$34 (1.9)^{46}$
$X = NO_2$	$Y = SCH_3$	p-dioxane	364	4.5	$60 (1.06)^{32}$
		p-dioxane	374	4.3	$26 (1.9)^{46}$
		chloroform	380	4.3	$34 (1.9)^{46}$
$X = NO_2$	$Y = NH_2$	p-dioxane	378	5.1	$68 (1.06)^{32}$
		acetone		7.5	$260 (1.06)^{189}$
		chloroform	402	5.1	$40 (1.9)^{46}$
$X = NO_2$	$Y = N(CH_3)_2$	acetone		7.4	$450 (1.06)^{193}$
		DMSO	447	$\mu\beta = 42 \times 10^{-46}$	$(1.9)^{55}$
		DMSO	453	$\mu\beta = 76 \times 10^{-47}$	$(1.36)^{231}$
				7.1	$323 (1.06S)^{195}$
		chloroform	427	6.6	$73 (1.9)^{46}$
		NMP		7.2	$70 (1.9)^{46}$
		p-dioxane		$\mu\beta = 58 \times 10^{-47}$	$(1.9)^{107}$
$X = NO_2$	$Y = N(CH_2C_2H_3)_2$	chloroform	438	6.7	$42 (\beta_0)^{177}$
$X = NO_2$	$Y = N(C_6H_5)_2$	chloroform	452	$\mu\beta = 57 \times 10^{-47}$	$(1.9)^{264}$
		chloroform	436	4.8	$37 (\beta_0)^{177}$

TABLE 6.1 Experimental Results of Molecular Studies—(*continued*)

Structure	Solvent	λ_{max} (nm)	μ (10^{-18} esu)	$\beta_\mu(\lambda)$ ref. (10^{-30} esu)
X = NO$_2$, Y = COOCH$_3$	CH$_2$Cl$_2$	350	4.0	4 (1.9)[46]
X = NO$_2$, Y = COH	p-dioxane	352	4.1	6 (1.9)[46]
X = CHC(CN)$_2$, Y = N(CH$_3$)$_2$	chloroform	485	7.8	210 (1.9)[43]
X = CHC(CN)$_2$, Y = N(C$_2$H$_5$)$_2$	CH$_2$Cl$_2$	468	8.2	180 (1.58)[54]
	p-dioxane		$\mu\beta = 11 \times 10^{-46}$	(1.9)[106]
X = Br, Y = OCH$_3$	p-dioxane	325	4.0	25 (1.9)[46]
[julolidine-CH=CH-C$_6$H$_4$-NO$_2$]	chloroform	438	7	96 (1.9)[46]
[2-NO$_2$-C$_6$H$_4$-CH=CH-C$_6$H$_4$-OCH$_3$ (o,o)]	chloroform	360	3.8	4.4 (1.9)[46]
[2-NO$_2$-C$_6$H$_4$-CH=CH-C$_6$H$_4$-OCH$_3$ (o,m)]	chloroform	370	3.7	1.6 (1.9)[46]
[2-NO$_2$-C$_6$H$_4$-CH=CH-C$_6$H$_4$-OCH$_3$ (o,p)]	chloroform	390	3.5	3.8 (1.9)[46]
[3-NO$_2$-C$_6$H$_4$-CH=CH-C$_6$H$_4$-OCH$_3$ (m,o)]	chloroform	320	4.4	5.5 (1.9)[46]
[3-NO$_2$-C$_6$H$_4$-CH=CH-C$_6$H$_4$-OCH$_3$ (m,m)]	chloroform	292	3.9	4.5 (1.9)[46]
[3-NO$_2$-C$_6$H$_4$-CH=CH-C$_6$H$_4$-OCH$_3$ (m,p)]	p-dioxane	318	3.9	5.3 (1.9)[46]

chloroform	362	5.0	22 (1.9)[46]
chloroform	352	4.0	21 (1.9)[46]
chloroform	346	4.6	12 (1.9)[46]
chloroform	346	3.4	14 (1.9)[46]
	322	3.1	9.2 (1.06)[175]
	320	4.6	6.8 (1.06)[175]
	322	3.4	10 (1.06)[175]
	324	3.5	12 (1.06)[175]
	322	3.3	16 (1.06)[175]

TABLE 6.1 Experimental Results of Molecular Studies—(continued)

Structure	Solvent	λ_{max} (nm)	μ (10^{-18} esu)	β_μ (λ)$^{ref.}$ (10^{-30} esu)
[H₃CO–C₆H₄–CH=CH–C₆F₄–CF₃]		334	4.0	23 (1.06)[175]
[H₃CO–C₆H₄–CH=CH–C₆H₃(CF₃)₂]		326	4.2	11 (1.06)[175]
[X–C₆H₄–C(CN)=CH–C₆H₄–CF₃]				
X = H	p-dioxane	314	4.7	4.5 (1.06)[244]
X = Cl	p-dioxane	318	4.2	5.1 (1.06)[244]
X = Br	p-dioxane	320	4.6	8.1 (1.06)[244]
X = CH₃	p-dioxane	323	4.8	7.4 (1.06)[244]
X = OCH₃	p-dioxane	340	5.3	15 (1.06)[244]
X = OH	p-dioxane	348	5.5	16 (1.06)[244]
X = SCH₃	p-dioxane	362	5.0	16 (1.06)[244]
X = N(CH₃)₂	p-dioxane	410	6.4	29 (1.06)[244]
[H₃CO–C₆H₃(CH₃)–CH=CH–C₆H₄–NO₂]	p-dioxane	366	5.2	26 (1.9)[46]
[H₃CO–C₆H₃(OCH₃)–CH=CH–C₆H₄–NO₂]	p-dioxane	380	4.7	23 (1.9)[46]
[H₃CO–C₆H₃(F)–CH=CH–C₆H₄–NO₂]	p-dioxane	363	4.1	18 (1.9)[46]
[H₃CO–C₆H₃(OCH₃)–CH=CH–C₆H₄–NO₂]	p-dioxane	395	4.8	32 (1.9)[46]

Solvent	λ (nm)	μ	β (×10⁻³⁰)
p-dioxane	361	5.3	21 (1.9)[46]
p-dioxane	340	4.6	8 (1.9)[46]
p-dioxane	382	4.1	2.1 (1.9)[46]
p-dioxane	355	4	10 (1.9)[46]
p-dioxane	354	2	5 (1.9)[46]
p-dioxane	378	5.0	12 (1.9)[255]
p-dioxane	384	4.7	22 (1.9)[46]
p-dioxane	466	7.0	57 (1.9)[255]
p-dioxane		μβ = 66 × 10⁻⁴⁷	(1.9)[107]
chloroform		4.1	15 (1.9)[43]

TABLE 6.1 Experimental Results of Molecular Studies—*(continued)*

Structure	Solvent	λ_{max} (nm)	μ (10^{-18} esu)	$\beta_\mu(\lambda)^{ref.}$ (10^{-30} esu)
	chloroform		6.2	45 (1.9)[43]
	p-dioxane	404	5.6	25 (1.9)[46]
	p-dioxane	390	3.1	11 (1.9)[46]
Aza- and azobenzene derivatives				
	p-dioxane	346	4.4	4.9 (1.9)[46]
	p-dioxane	351	4.7	15 (1.9)[46]
	p-dioxane	376	4.4	14 (1.9)[46]
	DMSO	458	$\mu\beta = 50 \times 10^{-47}$	(1.36)[231]
	p-dioxane	349	5.4	6.6 (1.9)[46]

Structure: X–C₆H₄–N=N–C₆H₄–Y (azobenzene derivative)

Substituents	Solvent	λ	μβ or β	β (ω) ref
X = SO₂CH₃, Y = N(C₄H₉)₂	chloroform	461	$\mu\beta = 51 \times 10^{-47}$	(1.9)[264]
X = NO₂, Y = NH₂	p-dioxane	420	5.8	29 (1.9)[46]
X = NO₂, Y = N(CH₃)₂	DMSO	470	$\mu\beta = 77 \times 10^{-47}$	(1.36)[231]
	choloroform	498	$\mu\beta = 13 \times 10^{-46}$	(1.9)[264]
X = NO₂, Y = N(C₂H₅)₂	chloroform	480	7.7	40 (β₀)[177]
X = NO₂, Y = N(C₂H₅)(C₂H₄)OH	chloroform	494	8.0	50 (β₀)[177]
	CH₂Cl₂	480	8.9	90 (1.57)[54]
	p-dioxane	455	7.0	49 (1.9)[46]
	DMSO	508	$\mu\beta = 11 \times 10^{-46}$	(1.36)[231]
X = NO₂, Y = N(C₆H₅)₂	chloroform	486	5.9	54 (β₀)[177]
X = CHC(CN)₂, Y = N(CH₃)₂	DMSO	492	$\mu\beta = 27 \times 10^{-46}$	(1.36)[231]
X = C₂(CN)₃, Y = N(C₂H₅)₂	DMSO		$\mu\beta = 41 \times 10^{-46}$	(1.58)[264]
X = C₂(CN)₃, Y = N(C₂H₅)(C₂H₄)OH	CH₂Cl₂	513	10	190 (1.57)[54]

Pyridine derivatives

Structure	Solvent	λ	μβ or β	β (ω) ref
H₂N–pyridine–NO₂	acetone	376	6.5	3.7 (1.9)[46]
(prolinol)N–pyridine–NO₂	p-dioxane		7.2	18 (1.06)[25]
	p-dioxane			11 (1.32)[25]
	p-dioxane			11 (1.91)[25]
H₄₅C₂₂–N–pyridine–NO₂	p-dioxane	357	5.5	17 (1.06)[262]
cyclooctyl-N–pyridine–NO₂	p-dioxane		6.7	13 (1.06)[25]
	p-dioxane			9.3 (1.32)[25]
	p-dioxane			6.0 (1.91)[25]
(1-phenylethyl)N–pyridine–NO₂	p-dioxane	361	6.8	22 (1.06)[25]
	p-dioxane			12 (1.32)[25]
	p-dioxane			11 (1.91)[25]
H₃CO–pyridine–NO₂	p-dioxane		6.1	15 (1.06)[262]
	p-dioxane		3.5	2.2 (1.9)[46]

TABLE 6.1 Experimental Results of Molecular Studies—*(continued)*

Structure	Solvent	λ_{max} (nm)	μ (10^{-18} esu)	$\beta_\mu(\lambda)^{ref.}$ (10^{-30} esu)
H₃CO–⟨⟩–N⁺–NO₂	*p*-dioxane		3.5	2.2 (1.9)[46]
Br–⟨⟩–CH=CH–⟨⟩N	chloroform		0.9	10 (1.9)[46]
CH₃O–⟨⟩–CH=CH–⟨⟩N	chloroform	335	3.8	16 (1.9)[43]
O₂N–⟨⟩–CH=CH–⟨⟩N	chloroform		1.3	8 (1.9)[46]
F₃C–⟨⟩–C(CN)=CH–⟨⟩N	*p*-dioxane	312	6.2	4.3 (1.06)[244]
F₃C–⟨⟩–CH=C(CN)–⟨⟩N	*p*-dioxane	309	5.5	4.4 (1.06)[244]
F₃C–⟨⟩–CH=C(CN)–⟨⟩N	*p*-dioxane	307	5.1	4.2 (1.06)[244]
Coumarin derivatives				
coumarin structure	chloroform		5	15 (1.9)[43]

	chloroform	5	8 (1.9)[43]
	chloroform	7.3	30 (1.9)[43]
	chloroform	8.8	50 (1.9)[43]
	chloroform	5.8	9.5 (1.9)[43]
	chloroform	7.3	15 (1.9)[43]
	p-dioxane	3.6	<1 (1.9)[43]
	p-dioxane	4	<1 (1.9)[43]
	p-dioxane	2.6	<1 (1.9)[43]
	p-dioxane	3	2.7 (1.9)[43]

Other polycyclic aromatic derivatives

X = H
X = SH
X = COOH

TABLE 6.1 Experimental Results of Molecular Studies—*(continued)*

Structure	Solvent	λ_{max} (nm)	μ (10^{-18} esu)	$\beta_\mu(\lambda)^{ref.}$ (10^{-30} esu)
X = H, Y = O	*p*-dioxane	340		4.8 (1.34)[28]
X = OCH$_3$, Y = O	*p*-dioxane	312		13 (1.34)[28]
X = H, Y = S	*p*-dioxane	438		11 (1.34)[28]
X = OCH$_3$, Y = S	*p*-dioxane	438		21 (1.34)[28]
(di-tert-butyl phenol pyridine, OH)		333	4.0	$\beta_0 = 12(S)^7$
(di-tert-butyl phenol pyridine, OH)		356	6.2	$\beta_0 = 11(S)^7$
(aminonaphthalene nitro)	DMSO		$\mu\beta = 23 \times 10^{-47}$	(1.36)[231]
(bis-dimethylamino styryl dinitrobenzene)	chloroform		7.6	36 (1.9)[43]
(bis-dimethylamino diphenyl dicyanoethylene)	chloroform		7.3	20 (1.9)[43]

CCl$_4$	590		580 (1.06HR)[291]
chloroform	400	4	10 (1.9)[43]
chloroform	400	6.7	31 (1.9)[43]
chloroform	423	7	45 (1.9)[43]
p-dioxane		6	5 (1.9)[43]
p-dioxane		8	11 (1.9)[43]
p-dioxane	575	$\mu\beta = 30 \times 10^{-47}$	(≈ 2)[226]

TABLE 6.1 Experimental Results of Molecular Studies—*(continued)*

Structure	Solvent	λ_{max} (nm)	μ (10^{-18} esu)	β_μ $(\lambda)^{ref.}$ (10^{-30} esu)
	chloroform		6.5	35 $(1.9)^{43}$
	NMP	364	7	18 $(1.9)^{43}$
	NMP	406	8	30 $(1.9)^{43}$
	chloroform	394	8	52 $(1.9)^{43}$
	p-dioxane		$\mu\beta = 13 \times 10^{-46}$	$(1.9)^{107}$
	chloroform	352	7.6	1 $(1.9)^{156}$
	chloroform	372	$\mu\beta = 30 \times 10^{-48}$	$(1.34)^{14}$
Polyene derivatives				
n = 1 X = COH Y = N(CH$_3$)$_2$	chloroform	284	6.3	3.3 $(1.9)^{156}$
n = 2 X = COH Y = N(C$_2$H$_5$)$_2$	chloroform	363	6.5	20 $(1.9)^{156}$
n = 3 X = COH Y = N(CH$_3$)$_2$	chloroform	422	6.9	53 $(1.9)^{156}$

n = 1	X = CHC(CN)$_2$	Y = N(CH$_3$)$_2$	chloroform	374	8.9	6.1 (1.9)[156]
n = 2	X = CHC(CN)$_2$	Y = N(C$_2$H$_5$)$_2$	chloroform	476	10.7	45 (1.9)[156]
n = 3	X = CHC(CN)$_2$	Y = N(CH$_3$)$_2$	chloroform	550	9.9	211 (1.9)[156]
n = 1	X = NO$_2$	Y = N(CH$_3$)$_2$	chloroform		6.3	4.8 (1.9)[43]
n = 2	X = NO$_2$	Y = N(CH$_3$)$_2$	chloroform		6.7	21 (1.9)[43]
n = 3	X = NO$_2$	Y = N(CH$_3$)$_2$	chloroform		8.4	73 (1.9)[43]
n = 3	X = SO$_2$CF$_3$	Y = N(CH$_3$)$_2$	chloroform		9.8	40 (1.9)[43]
			chloroform	456	$\mu\beta = 12 \times 10^{-46}$	(1.34)[14]
			chloroform	466	$\mu\beta = 22 \times 10^{-46}$	(1.34)[14]
			chloroform	500	$\mu\beta = 73 \times 10^{-46}$	(1.34)[14]
			DMSO	380	$\mu\beta = 23 \times 10^{-47}$	(1.06)[101]
			DMSO	470	Re($\mu\beta$) = 13 × 10^{-46} Im($\mu\beta$) = 15 × 10^{-46}	(1.06)[101] (1.06)[101]
			DMSO	480	Re($\mu\beta$) = 14 × 10^{-46} Im($\mu\beta$) = 35 × 10^{-46}	(1.06)[101] (1.06)[101]
			DMSO	440	Re($\mu\beta$) = 15 × 10^{-46} Im($\mu\beta$) = 17 × 10^{-46}	(1.06)[101] (1.06)[101]

TABLE 6.1 Experimental Results of Molecular Studies—*(continued)*

Structure	Solvent	λ_{max} (nm)	$\mu\beta$ (10^{-18} esu)	$\beta_\mu(\lambda)^{ref.}$ (10^{-30} esu)
(polyene-COH)	$Cl_2CHCHCl_2$	470	$Re(\mu\beta) = 10 \times 10^{-47}$ $Im(\mu\beta) = 44 \times 10^{-46}$	$(1.06)^{101}$ $(1.06)^{101}$
	chloroform	476	$\mu\beta = 96 \times 10^{-47}$	$(1.9)^{53}$
(polyene-NO$_2$)	$Cl_2CHCHCl_2$	500	$Re(\mu\beta) = -20 \times 10^{-46}$ $Im(\mu\beta) = 25 \times 10^{-46}$	$(1.06)^{101}$ $(1.06)^{101}$
(polyene-CN/CN)	$Cl_2CHCHCl_2$	570	$Re(\mu\beta) = 35 \times 10^{-46}$ $Im(\mu\beta) = 47 \times 10^{-46}$	$(1.06)^{101}$ $(1.06)^{101}$
	chloroform	566	$\mu\beta = 44 \times 10^{-46}$	$(1.9)^{53}$
(polyene-CO$_2$C$_2$H$_5$/CN)	$Cl_2CHCHCl_2$	510	$Re(\mu\beta) = -12 \times 10^{-45}$ $Im(\mu\beta) = 73 \times 10^{-46}$	$(1.06)^{101}$ $(1.06)^{101}$
(polyene)	chloroform	502	$\mu\beta = 15 \times 10^{-46}$	$(1.9)^{53}$

α-Phenylpolyene Derivatives

(X—C$_6$H$_4$—[CH=CH]$_n$—Y)

	Structure	Solvent	λ_{max} (nm)	$\mu\beta$	$\beta_\mu(\lambda)^{ref.}$
$n = 2$	X = COH, Y = OCH$_3$	chloroform	350	4.3	$28\ (1.9)^{45}$
$n = 3$	X = COH, Y = OCH$_3$	chloroform	376	4.6	$42\ (1.9)^{45}$
$n = 2$	X = COH, Y = N(CH$_3$)$_2$	chloroform	412	6.0	$52\ (1.9)^{45}$
$n = 3$	X = COH, Y = N(CH$_3$)$_2$	chloroform	434	6.3	$88\ (1.9)^{45}$
		chloroform		6.6	$105\ (1.9)^{43}$
$n = 4$	X = COH, Y = N(CH$_3$)$_2$	chloroform		8.0	$138\ (1.9)^{43}$
$n = 2$	X = COCF$_3$, Y = N(CH$_3$)$_2$	chloroform		6.6	$126\ (1.9)^{43}$

n = 2	X = NO₂	Y = OCH₃		DMSO	370	$\mu\beta = 81 \times 10^{-48}$	$(1.06)^{102}$
				chloroform		4.6	$42\ (1.9)^{45}$
n = 3	X = NO₂	Y = OCH₃		DMSO	400	$\mu\beta = 30 \times 10^{-47}$	$(1.06)^{102}$
n = 2	X = NO₂	Y = N(CH₃)₂		acetone		8.8	$630\ (1.06)^{189}$
				DMSO	460	$\mu\beta = 17 \times 10^{-46}$	$(1.06)^{102}$
				chloroform	466	6.5	$140\ (1.9)^{254}$
n = 3	X = NO₂	Y = N(CH₃)₂		DMSO	490	$\mu\beta = 55 \times 10^{-46}$	$(1.06)^{102}$
				chloroform	487	6.6	$240\ (1.9)^{254}$
n = 4	X = NO₂	Y = N(CH₃)₂		chloroform	502	7.6	$280\ (1.9)^{254}$
n = 2	X = CHC(CN)₂	Y = N(CH₃)₂		DMSO	500	$\mu\beta = 36 \times 10^{-46}$	$(1.06)^{102}$
				chloroform		9.0	$163\ (1.9)^{45}$
n = 3	X = CHC(CN)₂	Y = N(CH₃)₂		chloroform	520	8.8	$432\ (1.9)^{43}$

n = 3	X = COH	chloroform		7.1	$162\ (1.9)^{43}$
n = 3	X = NO₂	chloroform		7.8	$287\ (1.9)^{43}$
n = 4	X = NO₂	chloroform		$\mu\beta = 26 \times 10^{-46}$	$(1.9)^{43}$
n = 3	X = CHC(CN)₂	chloroform		8.7	$485\ (1.9)^{43}$

	chloroform	450	$\mu\beta = 20 \times 10^{-46}$	$(1.34)^{14}$
	chloroform	461	$\mu\beta = 42 \times 10^{-46}$	$(1.34)^{14}$
	chloroform	498	$\mu\beta = 89 \times 10^{-46}$	$(1.34)^{14}$

TABLE 6.1 Experimental Results of Molecular Studies—*(continued)*

	Structure		Solvent	λ_{max} (nm)	μ (10^{-18} esu)	β_μ $(\lambda)^{ref.}$ (10^{-30} esu)
	Diphenylpolyene derivatives					
$n = 2$	X = CN	Y = OCH$_3$	chloroform	360	4.3	27 (1.9)^{45}
$n = 3$	X = CN	Y = OCH$_3$	chloroform	380	4.6	40 (1.9)^{45}
$n = 2$	X = NO$_2$	Y = Br	chloroform	378	3.5	21 (1.9)^{45}
$n = 3$	X = NO$_2$	Y = Br	chloroform	400	3.8	35 (1.9)^{45}
$n = 2$	X = NO$_2$	Y = OCH$_3$	p-dioxane		6.0	135 (1.06)^{98}
			chloroform	397	4.8	47 (1.9)^{45}
$n = 3$	X = NO$_2$	Y = OCH$_3$	p-dioxane		6.6	274 (1.06)^{98}
			chloroform	414	5.1	76 (1.9)^{45}
$n = 4$	X = NO$_2$	Y = OCH$_3$	p-dioxane		6.7	367 (1.06)^{98}
			chloroform	430	5.8	55 (1.9)^{45}
$n = 5$	X = NO$_2$	Y = OCH$_3$	p-dioxane		7.0	623 (1.06)^{98}
$n = 2$	X = NO$_2$	Y = SCH$_3$	chloroform	398	4.5	101 (1.9)^{45}
$n = 2$	X = NO$_2$	Y = N(CH$_3$)$_2$	chloroform	442	7.6	107 (1.9)^{45}
			p-dioxane		$\mu\beta = 75 \times 10^{-47}$	(1.9)^{107}
$n = 3$	X = NO$_2$	Y = N(CH$_3$)$_2$	chloroform	458	8.2	131 (1.9)^{45}
$n = 4$	X = NO$_2$	Y = N(CH$_3$)$_2$	chloroform	464	9	190 (1.9)^{45}
$n = 2$	X = C$_2$HCN$_2$	Y = N(CH$_3$)$_2$	DMSO	481	$\mu\beta = 13 \times 10^{-46}$	(1.36)^{231}
$n = 2$	X = CN	Y = OCH$_3$	chloroform	354	3.8	4.5 (1.9)^{45}
$n = 3$	X = CN	Y = OCH$_3$	chloroform	376	3.8	7.1 (1.9)^{45}
$n = 2$	X = NO$_2$	Y = OCH$_3$	chloroform	376	3.7	6.4 (1.9)^{45}
$n = 3$	X = NO$_2$	Y = OCH$_3$	chloroform	392	4.1	11 (1.9)^{45}
$n = 2$	X = OCH$_3$	Y = CN	chloroform	356	3.9	2.6 (1.9)^{45}

n	X	Y	Solvent	λ	value	β (ω)[ref]
n = 3	X = OCH$_3$	Y = CN	chloroform	378	3.9	4.3 (1.9)[45]
n = 2	X = CN	Y = OCH$_3$	chloroform	358	4.9	4.3 (1.9)[45]
n = 2	X = OCH$_3$	Y = NO$_2$	chloroform	376	3.8	4.9 (1.9)[45]
n = 3	X = OCH$_3$	Y = NO$_2$	chloroform	392	3.8	11 (1.9)[45]
n = 2	X = NO$_2$	Y = OCH$_3$	chloroform	380	4.3	17 (1.9)[45]
n = 3	X = NO$_2$	Y = OCH$_3$	chloroform	412	4.8	56 (1.9)[45]

(NMe$_2$–stilbene–NO$_2$ derivatives)

Solvent	λ	value	β (ω)[ref]
p-dioxane	480	7.1	98 (1.9)[255]
p-dioxane		$\mu\beta = 13 \times 10^{-46}$	(1.9)[107]

(dinitro derivative)

Solvent	λ	value	β (ω)[ref]
chloroform		6.2	71 (1.9)[255]

α,ω-Diphenylpolyene derivatives

n	X	Y	Solvent	λ	value	β (ω)[ref]
n = 2	X = CN	Y = SCH$_3$	chloroform	330	3.7	17 (1.9)[45]
n = 2	X = CN	Y = NH$_2$	NMP	388	$\mu\beta = 11 \times 10^{-47}$	(1.9)[45]
n = 2	X = NO$_2$	Y = SCH$_3$	chloroform	338	3.9	17 (1.9)[45]
n = 2	X = NO$_2$	Y = NH$_2$	chloroform	334	6.3	28 (1.9)[45]
n = 2	X = NO$_2$	Y = NH$_2$	NMP	416	$\mu\beta = 24 \times 10^{-47}$	(1.9)[45]
n = 3	X = NO$_2$	Y = NH$_2$	NMP	440	$\mu\beta = 41 \times 10^{-47}$	(1.9)[45]

Cumulene derivatives

X	Y	Solvent	λ	μβ	(ω)[ref]
X = NO$_2$	Y = H	p-dioxane	442	$\mu\beta = 15 \times 10^{-47}$	(1.9)[61]
X = NO$_2$	Y = H	p-dioxane	442	$\mu\beta = 66 \times 10^{-47}$	(1.06)[61]
X = NO$_2$	Y = CH$_3$	p-dioxane	448	$\mu\beta = 60 \times 10^{-47}$	(1.06)[61]
X = NO$_2$	Y = OCH$_3$	p-dioxane	459	$\mu\beta = 85 \times 10^{-47}$	(1.06)[61]
X = NO$_2$	Y = O(CH$_2$)$_{11}$CH$_3$	p-dioxane	461	$\mu\beta = 28 \times 10^{-47}$	(1.9)[61]

TABLE 6.1 Experimental Results of Molecular Studies—(continued)

Structure		Solvent	λ_{max} (nm)	μ (10^{-18} esu)	$\beta_\mu(\lambda)^{ref.}$ (10^{-30} esu)
α,ω-Polyphenyl derivatives					
$n = 3$	X = NO$_2$, Y = OCH$_3$	p-dioxane	340	5.0	11 (1.9)[45]
$n = 3$	X = NO$_2$, Y = NH$_2$	NMP	360	7.8	24 (1.9)[45]
$n = 4$	X = NO$_2$, Y = NH$_2$	NMP	344	7.6	16 (1.9)[45]
Pyrrol derivatives					
		p-dioxane	376	5.8	11 (1.06)[244]
		p-dioxane	381	7.2	12 (1.06)[244]
		p-dioxane	410	5.5	26 (1.9)[43]
	X = OCH$_3$	DMSO	485	1.4	32 (1.9S)[64]
	X = N(CH$_3$)$_2$	DMSO	553	1.2	439 (1.9S)[64]
	X = NH, Y = CH$_3$	DMSO	539	3.7	20 (1.9S)[64]

Structure	Solvent	λ_{max}	μ	β (ω)
X = S, Y = H	DMSO	493	2.5	10 (1.9S)[64]
(benzodithiole derivative)	DMSO	553	2.5	578 (1.9S)[64]
Furan derivatives				
furan–CH=CH–C₆H₄–CF₃/CN	p-dioxane	345	5.7	7.7 (1.06)[244]
furan–CH=CH–C₆H₄–CF₃/CN	p-dioxane	314	5.4	7.1 (1.06)[244]
O_2N–furan–CH=CH–C₆H₄–NMe₂	chloroform	478	6.9	83 (1.9)[45]
O_2N–furan–CH=CH–julolidine	chloroform		5.9	173 (1.9)[43]
O_2N–furan–(CH=CH)₂–C₆H₄–NMe₂	chloroform	488	7.2	113 (1.9)[45]
O_2N–C₆H₄–furan–C₆H₄–OCH₃	chloroform	400	5.2	40 (1.9)[45]
Thiophene derivatives				
thiophene–CH=CH–NO₂	chloroform	351	5.2	20 (1.06)[117]
thiophene–(CH=CH)₂–NO₂	chloroform	382	5.4	67 (1.06)[117]
thiophene–CH=CH–C₆H₄–CF₃/CN	p-dioxane	346	6.8	6.6 (1.06)[244]

TABLE 6.1 Experimental Results of Molecular Studies—*(continued)*

Structure	Solvent	λ_{max} (nm)	μ (10^{-18} esu)	β_μ (λ)[ref.] (10^{-30} esu)
	p-dioxane	321	6.7	5.4 (1.06)[244]
	chloroform		5.8	7.5 (1.9)[43]
	chloroform		9.0	21 (1.9)[43]
	chloroform		8.8	21 (1.9)[43]
	chloroform		5.9	23 (1.9)[43]
	in PMMA	510	$\mu\beta_0 = 60 \times 10^{-47}$ (from EO at 633 nm)[165]	
	in PMMA	499	$\mu\beta_0 = 26 \times 10^{-47}$ (from EO at 633 nm)[165]	
	chloroform		3.7	30 (1.9)[43]
	chloroform	492	7.4	98 (1.9)[45]
	p-dioxane	478	$\mu\beta = 60 \times 10^{-47}$	(1.9)[203]

Second-Order Nonlinear Optical Properties of Organic Materials

Solvent	λ	μβ	ref
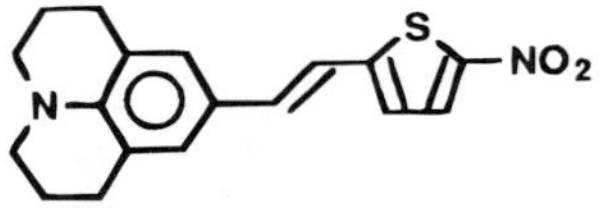 chloroform		7.0	197 (1.9)[43]
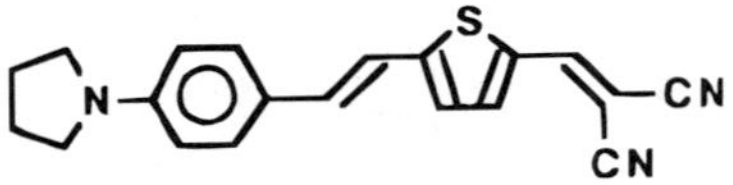 p-dioxane	584	μβ = 26 × 10⁻⁴⁶	(1.9)[106]
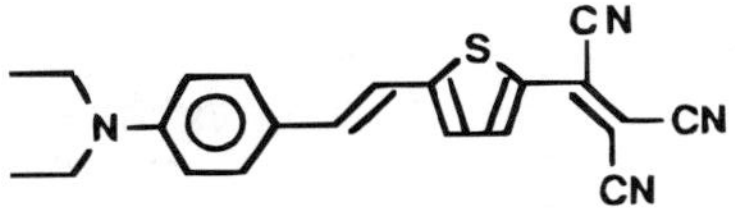 p-dioxane	640	μβ = 62 × 10⁻⁴⁶	(1.9)[203]
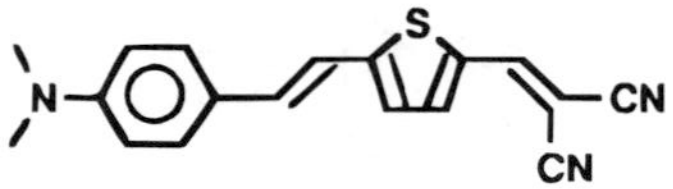 chloroform		7.5	161 (1.9)[43]
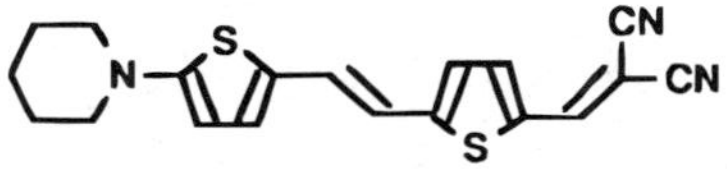 chloroform		7.6	250 (1.9)[72]
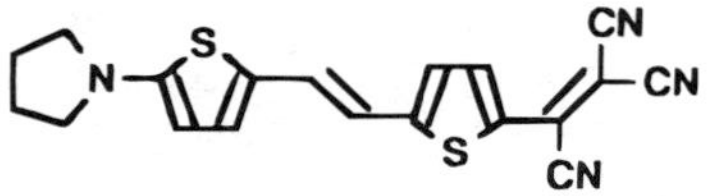 p-dioxane	718	μβ = 69 × 10⁻⁴⁶	(1.9)[203]
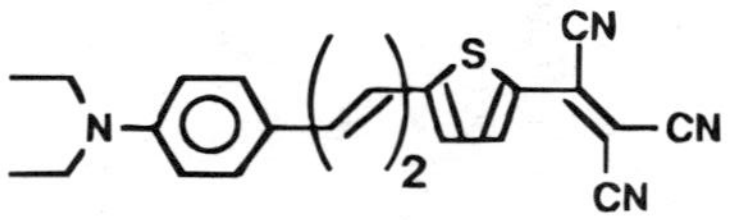 p-dioxane	662	μβ = 91 × 10⁻⁴⁶	(1.9)[203]
p-dioxane	513	μβ = 13 × 10⁻⁴⁶	(1.9)[106]
p-dioxane	547	μβ = 23 × 10⁻⁴⁶	(1.9)[106]
p-dioxane	556	μβ = 38 × 10⁻⁴⁶	(1.9)[106]
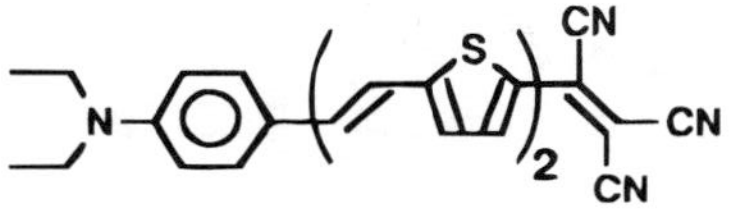 p-dioxane	653	μβ = 74 × 10⁻⁴⁶	(1.9)[203]

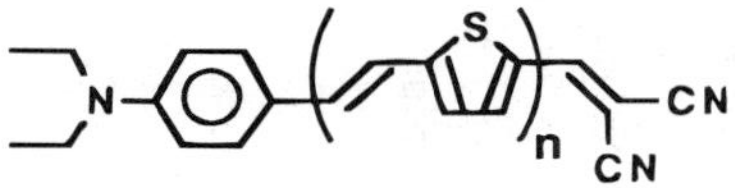

Note: $\mu\beta$ values given in esu; wavelengths in nm.

(structures shown; n = 1, n = 2, n = 3 for one series)

TABLE 6.1 Experimental Results of Molecular Studies—(continued)

Structure		Solvent	λ_{max} (nm)	μ (10^{-18} esu)	$\beta_\mu(\lambda)$ ref. (10^{-30} esu)
X–(S)ₙ–Y					
n = 1, X = NO₂	Y = OCH₃	CCl₄	349	5.5	7.6 (1.06)[271]
n = 2, X = NO₂	Y = H	C₆H₁₁CH₃	378	5.0	19 (1.06)[271]
n = 2, X = NO₂	Y = OCH₃	C₆H₁₁CH₃	411	6.0	41 (1.06)[271]
n = 2, X = NO₂	Y = OCH₃	CCl₄	418	6.0	46 (1.06)[271]
n = 2, X = NO₂	Y = SCH₃	C₆H₁₁CH₃	394	5.0	49 (1.06)[271]
n = 2, X = NO₂	Y = N(CH₃)₂	C₆H₁₁CH₃	472	8.0	319 (1.06)[271]
n = 3, X = NO₂	Y = OCH₃	CCl₄	452	6.0	162 (1.06)[271]

Azole derivatives

(pyrazole with X-phenyl-N and Y substituent, Z on phenyl)

X = NO₂	Y = CH₃OC₆H₄	Z = H	p-dioxane	354	6.6	31 (1.06)[166]
X = NO₂	Y = C₆H₅CH–CH-	Z = H	p-dioxane	346	6.1	32 (1.06)[166]
X = NO₂	Y = p-CH₃OC₆H₄CH–CH-	Z = H	p-dioxane	368	5.7	40 (1.06)[166]
X = SO₂CH₃	Y = p-CH₃OC₆H₄CH–CH-	Z = H	p-dioxane	330	5.6	16 (1.06)[166]
X = NO₂	Y = p-CH₃SC₆H₄CH–CH-	Z = NO₂	p-dioxane	354	4.1	42 (1.06)[166]
X = CH₃O	Y = p-O₂NC₆H₄CH–CH-	Z = H	p-dioxane	370	7.0	48 (1.06)[166]

| X = NO₂ | Y = p-CH₃OC₆H₄C≡C- | Z = H | p-dioxane | 340 | 5.6 | 20 (1.06)[166] |
| X = NO₂ | Y = p-CH₃OC₆H₄CH–CH- | Z = H | p-dioxane | 358 | 5.1 | 63 (1.06)[166] |

X = NO₂	Y = CH₃OC₆H₄	Z = H	p-dioxane	306	6.3	13 (1.06)[166]
X = NO₂	Y = C₆H₅CH–CH-	Z = H	p-dioxane	298	4.6	7 (1.06)[166]
X = NO₂	Y = p-CH₃OC₆H₄CH–CH-	Z = H	p-dioxane	302	5.5	14 (1.06)[166]

X = NO$_2$	Y = p-CH$_3$SC$_6$H$_4$CH=CH-	Z = NO$_2$	p-dioxane	312	3.3	12 (1.06)[166]
X = NO$_2$	Y = p-(CH$_2$)$_6$N	Z = N	chloroform	476	8.3	79 (1.9)[176]
X = NO$_2$	Y = p-CH$_3$OC$_6$H$_4$C≡C-	Z = N	p-dioxane	344	8.0	53 (1.06)[176]
X = NO$_2$	Y = p-(CH$_2$)$_5$N	Z = N	chloroform	438	7.2	46 (1.9)[176]
X = O$_2$NC$_6$H$_4$C≡C-	Y = p-CH$_3$OC$_6$H$_4$	Z = N	p-dioxane	400	8.1	69 (1.06)[176]
X = NO$_2$	Y = p-CH$_3$OC$_6$H$_4$S	Z = N	p-dioxane	398	6.6	56 (1.06)[176]
X = NO$_2$	Y = p-CH$_3$OC$_6$H$_4$	Z = N	p-dioxane	410	7.0	52 (1.06)[176]
X = NO$_2$	Y = p-C$_4$H$_9$CH(C$_2$H$_5$)CH$_2$OC$_6$H$_4$	Z = N	chloroform	416	6.3	25 (1.9)[176]
X = NO$_2$	Y = p-CH$_3$OC$_6$H$_4$	Z = N	chloroform	412	6.4	20 (1.9)[176]
X = SO$_2$C$_4$F$_9$	Y = p-C$_4$H$_9$CH(C$_2$H$_5$)CH$_2$OC$_6$H$_4$	Z = N	chloroform	384	6.5	14 (1.9)[176]
X = SO$_2$C$_6$H$_5$	Y = p-CH$_3$OC$_6$H$_4$	Z = N	p-dioxane	362	8.0	10 (1.06)[176]
X = O$_2$NC$_6$H$_4$C≡-	Y = p-CH$_3$OC$_6$H$_4$	Z = N	p-dioxane	378	7.1	64 (1.06)[176]
X = SO$_2$C$_6$H$_4$CF$_3$	Y = p-CH$_3$OC$_6$H$_4$	Z = N	chloroform	370	8.0	32 (1.06)[176]
X = NO$_2$	Y = p-CH$_3$OC$_6$H$_4$	Z = O	p-dioxane	390	6.3	47 (1.06)[176]
X = SO$_2$C$_4$F$_9$	Y = p-CH$_3$OC$_6$H$_4$	Z = O	p-dioxane	378	7.9	31 (1.06)[176]
X = SO$_2$C$_6$H$_5$	Y = p-CH$_3$OC$_6$H$_4$	Z = O	p-dioxane	358	7.0	17 (1.06)[176]
X = SO$_2$CH$_3$	Y = p-CH$_3$OC$_6$H$_4$	Z = O	p-dioxane	352	7.0	15 (1.06)[176]
X = C$_6$F$_{16}$	Y = p-CH$_3$OC$_6$H$_4$	Z = O	p-dioxane	344	5.7	13 (1.06)[176]
X = CF$_3$	Y = p-CH$_3$OC$_6$H$_4$	Z = O	p-dioxane	338	5.3	9 (1.06)[176]
X = NO$_2$	Y = p-CH$_3$OC$_6$H$_4$	Z = S	p-dioxane	400	8.1	21 (β$_0$)[176]
			chloroform	564	6.2	71 (1.9)[43]
			chloroform	578	6.1	83 (1.9)[43]
			chloroform	596	6.5	75 (1.9)[43]
			CH$_2$Cl$_2$	579	8.5	260 (1.58)[54]

TABLE 6.1 Experimental Results of Molecular Studies—*(continued)*

Structure	Solvent	λ_{max} (nm)	μ (10^{-18} esu)	$\beta_\mu(\lambda)^{ref.}$ (10^{-30} esu)
	chloroform	582	9.0	52 (β_0)[177]
	chloroform	582	6.9	68 (β_0)[177]
	CH_2Cl_2	645	10	530 (1.58)[54]
	chloroform	634	6.9	100 (1.9)[43]
	chloroform	670	6.9	130 (1.9)[43]
Azulene derivatives				
	chloroform		1.2	<1 (1.9)[43]
	chloroform		2.5	<1 (1.9)[43]
	chloroform		4	<1 (1.9)[43]

Pentafulvene derivatives

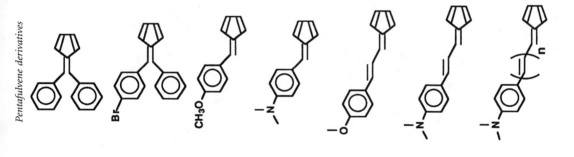

	chloroform		1	7 (1.9)[43]
	chloroform		1	5 (1.9)[43]
	chloroform		2	10 (1.9)[43]
	chloroform DMSO	420	3 $\mu\beta = 44 \times 10^{-47}$	30 (1.9)[43] (1.06)[102]
	DMSO	370	$\mu\beta = 70 \times 10^{-48}$	(1.06)[102]
	chloroform DMSO	450	3.6 $\mu\beta = 20 \times 10^{-46}$	74 (1.9)[43] (1.06)[102]
	chloroform		$\mu\beta = 73 \times 10^{-47}$	(1.9)[43]

TABLE 6.1 Experimental Results of Molecular Studies—(continued)

Structure	Solvent	λ_{max} (nm)	μ (10^{-18} esu)	$\beta_\mu(\lambda)^{ref.}$ (10^{-30} esu)
	chloroform		4.0	0.9 (1.9)[43]
	chloroform		3.5	4.6 (1.9)[43]
	methanol	444	26	−100 (1.3)[144]
	pyridine	606	17	−210 (1.3)[144]
	DMSO	570	8	1000 (1.9)[57]
	chloroform	590	4.0	190 (1.9)[153]
	chloroform	580	4.0	79 (1.9)[153]
	chloroform	610	4.3	91 (1.9)[153]
X = H	chloroform	428	1.5	5.9 (1.9)[154]
X = Br	chloroform	432	1.3	13 (1.9)[154]

X = OCH$_3$	chloroform	469	2.4	17 (1.9)[154]
X = OCH$_3$; *ortho*: OCH$_3$	chloroform	498	3.4	19 (1.9)[154]
X = SCH$_3$	chloroform	476	2.3	17 (1.9)[154]
X = NH$_2$	chloroform	—	3.3	38 (1.9)[43]
X = N(CH$_3$)$_2$	chloroform	558	3.9	78 (1.9)[154]
X = CH-CH-C$_6$H$_4$-OCH$_3$	chloroform	497	2.6	48 (1.9)[154]
X = CH-CH-C$_6$H$_4$-N(CH$_3$)$_2$	chloroform	512	3.7	116 (1.9)[154]
	DMSO	470	$\mu\beta = 74 \times 10^{-47}$	(1.06)[104]
	DMSO	550	$\mathrm{Re}(\mu\beta) = 48 \times 10^{-45}$ $\mathrm{Im}(\mu\beta) = 61 \times 10^{-45}$	(1.06)[104] (1.06)[104]
	DMSO	550	$\mathrm{Re}(\mu\beta) = 48 \times 10^{-45}$ $\mathrm{Im}(\mu\beta) = 61 \times 10^{-45}$	(1.06)[104] (1.06)[104]
Y = O	DMSO	390	$\mu\beta = 12 \times 10^{-47}$	(1.06)[103]
Y = S	DMSO	480	$\mathrm{Re}(\mu\beta) = 11 \times 10^{-46}$ $\mathrm{Im}(\mu\beta) = 31 \times 10^{-47}$	(1.06)[103] (1.06)[103]
	DMSO	530	$\mathrm{Re}(\mu\beta) = 21 \times 10^{-45}$ $\mathrm{Im}(\mu\beta) = 30 \times 10^{-45}$	(1.06)[104] (1.06)[104]

X = N
X = S

TABLE 6.1 Experimental Results of Molecular Studies—(*continued*)

Structure		Solvent	λ_{max} (nm)	μ (10^{-18} esu)	$\beta_\mu(\lambda)^{ref.}$ (10^{-30} esu)
	n = 2, X = C(CH$_3$)$_2$	DMSO		Re($\mu\beta$) = 17 × 10^{-45}	(1.06)[104]
				Im($\mu\beta$) = 24 × 10^{-45}	(1.06)[104]
	n = 3, X = S	DMSO		Re($\mu\beta$) = 13 × 10^{-45}	(1.06)[104]
				Im($\mu\beta$) = 23 × 10^{-45}	(1.06)[104]
	n = 3, X = C–CH$_2$	DMSO		Re($\mu\beta$) = 20 × 10^{-45}	(1.06)[104]
				Im($\mu\beta$) = 27 × 10^{-45}	(1.06)[104]
	n = 3, X = O	chloroform		6.9	360 (1.9)[43]
	n = 1, X = S	chloroform		7.4	−46 (1.9)[43]
	n = 2, X = S	chloroform		8.9	52 (1.9)[43]
	n = 3, X = S	chloroform		9.1	300 (1.9)[43]
	n = 1	DMSO		Re($\mu\beta$) = 13 × 10^{-45}	(1.06)[104]
				Im($\mu\beta$) = 15 × 10^{-45}	(1.06)[104]
	n = 3	DMSO		Re($\mu\beta$) = 27 × 10^{-45}	(1.06)[104]
				Im($\mu\beta$) = 49 × 10^{-45}	(1.06)[104]

	n	X	Y	Z	Solvent	λ (nm)	μβ or β	β₀ / ref
	0	O	CH₃	—	chloroform	510	$\mu\beta = -37 \times 10^{-47}$	$(1.9)^{60}$
	0	S	C₂H₅	—	chloroform	526	$\mu\beta = -60 \times 10^{-47}$	$(1.9)^{60}$
	0	O	4-CH₃C₆H₄	H	DMSO	470	$\mu\beta = 63 \times 10^{-47}$	$(1.06)^{103}$
	0	O	CH₃	CH₃	chloroform	470	6.2	20 $(\beta_0)^{177}$
	0	O	CH₃	CH₃	chloroform	486	4.5	40 $(\beta_0)^{177}$
	1	O	CH₃	C₂H₅	chloroform	464	5.1	47 $(1.9)^{43}$
	1	O	C₂H₅	CH₃	DMSO	540	$Re(\mu\beta) = 13 \times 10^{-46}$	$(1.06)^{103}$
	1	O	4-CH₃C₆H₄	C₆H₅	DMSO	540	$Im(\mu\beta) = 52 \times 10^{-46}$	$(1.06)^{103}$
	1	O	CH₃	C₆H₅	chloroform	558	7.1	91 $(\beta_0)^{177}$
	2	O	CH₃	C₂H₅	chloroform	560	4.0	97 $(\beta_0)^{177}$
	3	O	CH₃	C₂H₅	chloroform	532	5.1	173 $(1.9)^{43}$
	0	S	CH₃	C₂H₅	chloroform	554	5.8	338 $(1.9)^{43}$
	0	S	CH₃	H	chloroform	572	5.3	660 $(1.9)^{43}$
	0	S	C₂H₅	C₂H₅	DMSO	500	$Re(\mu\beta) = 12 \times 10^{-46}$	$(1.06)^{103}$
	0	S	C₂H₅	C₂H₅	DMSO	500	$Im(\mu\beta) = 60 \times 10^{-47}$	
	0	S	CH₃	C₂H₅	chloroform	494	6.4	20 $(\beta_0)^{177}$
	0	S	4-CH₃C₆H₄	C₆H₅	chloroform	484	5.4	68 $(1.9)^{155}$
	0	S	C₂H₅	C₆H₅	chloroform	500	7.1	20 $(\beta_0)^{177}$
	0	S	4-CH₃C₆H₄	C₆H₅	chloroform	518	5.2	45 $(\beta_0)^{177}$
	0	S	CH₃	C₆H₅	chloroform	502	7.6	25 $(\beta_0)^{177}$
	0	S	CH₃	4-NO₂C₆H₄	chloroform	520	5.7	57 $(\beta_0)^{177}$
	1	S	C₂H₅	C₆H₅	chloroform	588	5.5	58 $(1.9)^{43}$
	1	S	CH₃	C₆H₅	chloroform	594	9.0	80 $(1.9)^{43}$
	1	S	4-CH₃C₆H₄	C₂H₅	chloroform	572	8.3	94 $(\beta_0)^{177}$
	1	S	CH₃	CH₃	chloroform	604	6.5	119 $(\beta_0)^{177}$
	1	S	CH₃	CH₃	chloroform	—	5.7	256 $(1.9)^{155}$
	2	S	CH₃	C₂H₅	chloroform	—	6.2	636 $(1.9)^{155}$
	2	S	CH₃	CH₃	chloroform	—	6.6	1490 $(1.9)^{155}$
	3	S	CH₃	CH₃	chloroform	624	4.2	299 $(1.9)^{43}$

TABLE 6.1 Experimental Results of Molecular Studies—(*continued*)

Structure	Solvent	λ_{max} (nm)	μ (10^{-18} esu)	$\beta_\mu (\lambda)^{ref.}$ (10^{-30} esu)
n = 2	chloroform		6.8	630 (1.9)[43]
n = 3	chloroform		7.2	771 (1.9)[43]
n = 4	chloroform		7.5	893 (1.9)[43]
	chloroform		$\mu\beta = 71 \times 10^{-46}$	(1.9)[43]
n = 0, X = O	chloroform	480	5.7	64 (1.9)[43]
n = 1, X = O	chloroform	574	5.9	213 (1.9)[43]
n = 2, X = O	chloroform	616	6.5	504 (1.9)[43]
n = 3, X = O	chloroform		7.1	935 (1.9)[43]

n = 0, X = S	chloroform	522	7.0	87 (1.9)[155]
n = 1, X = S	chloroform	614	6.6	355 (1.9)[155]
n = 2, X = S	chloroform	680	6.3	1141 (1.9)[155]
n = 3, X = S	chloroform	686	8.8	2169 (1.9)[155]
	chloroform		$\mu\beta = 19 \times 10^{-45}$	(1.9)[43]
	chloroform		$\mu\beta = 12 \times 10^{-45}$	(1.9)[43]
	chloroform	680	$\mu\beta = 35 \times 10^{-46}$	(1.9)[155]
	chloroform		7.9	11 (1.9)[43]
	chloroform		7.5	13 (1.9)[43]

TABLE 6.1 Experimental Results of Molecular Studies—(continued)

Structure	Solvent	λ_{max} (nm)	μ (10^{-18} esu)	$\beta_\mu(\lambda)^{ref.}$ (10^{-30} esu)
[structure 1]	chloroform		5.5	14 (1.9)[43]
[structure 2, X = OCH$_3$; X = N(CH$_3$)$_2$]	chloroform chloroform		2.7 5.6	171 (1.9)[43] 339 (1.9)[43]
[structure 3, n = 0, X = C$_2$H$_5$; n = 1, X = C$_2$H$_5$; n = 1, X = C$_6$H$_5$]	chloroform chloroform chloroform		$\mu\beta = 37 \times 10^{-46}$ $\mu\beta = 71 \times 10^{-46}$ $\mu\beta = 86 \times 10^{-46}$	(1.9)[43] (1.9)[43] (1.9)[74]
[structure 4]	chloroform		11	280 (1.9)[43]

chloroform	478	8.3		37 (1.9)[155]
chloroform	530	8.6		140 (1.9)[155]
chloroform	562	8.7		362 (1.9)[155]
chloroform	582	8.9		918 (1.9)[155]
chloroform			$\mu\beta = 67 \times 10^{-45}$	(1.9)[43]
chloroform	504	9.5		51 (1.9)[155]
chloroform	586	9.1		180 (1.9)[155]
chloroform	620	9.0		656 (1.9)[155]
chloroform	640	9.8		995 (1.9)[155]
chloroform	647		$\mu\beta = 29 \times 10^{-45}$	(1.9)[155]
chloroform		6.5		46 (1.9)[43]

n = 0
n = 1
n = 2
n = 3

n = 0
n = 1
n = 2
n = 3

TABLE 6.1 Experimental Results of Molecular Studies—(continued)

Structure	Solvent	λ_{max} (nm)	μ (10^{-18} esu)	$\beta_\mu(\lambda)^{ref.}$ (10^{-30} esu)
n = 0	chloroform		$\mu\beta = 25 \times 10^{-46}$	$(1.9)^{72}$
n = 1	chloroform		$\mu\beta = 41 \times 10^{-46}$	$(1.9)^{72}$
	chloroform		4.4	$500 \, (1.9)^{43}$
n = 0, X = C$_2$H$_5$	chloroform	490	3.7	$38 \, (\beta_0)^{177}$
n = 0, X = 4-CH$_3$C$_5$H$_4$	chloroform	498	2.0	$64 \, (\beta_0)^{177}$
n = 0, X = CH$_3$	chloroform		3.7	$58 \, (1.9)^{43}$
n = 1, X = C$_2$H$_5$	chloroform	552	4.4	$92 \, (\beta_0)^{177}$
n = 1, X = 4-CH$_3$C$_5$H$_4$	chloroform	544	2.0	$93 \, (\beta_0)^{177}$
n = 1, X = CH$_3$	chloroform		3.6	$191 \, (1.9)^{43}$
n = 2, X = CH$_3$	chloroform		4.9	$266 \, (1.9)^{43}$
n = 3, X = CH$_3$	chloroform		5.3	$347 \, (1.9)^{43}$
n = 0	chloroform		4.0	$104 \, (1.9)^{43}$
n = 1	chloroform		4.5	$215 \, (1.9)^{43}$

chloroform	670	6.1	84 (1.9)[43]
chloroform	680	6.6	118 (1.9)[43]
DMSO	620	6.8	163 (1.58)[34] −118 (0.95)[34]
chloroform		5.0	98 (1.9)[43]
chloroform		6.0	264 (1.9)[43]
chloroform		6.9	466 (1.9)[43]
chloroform		6.5	750 (1.9)[43]
chloroform		6.3	124 (1.9)[43]
chloroform		6.1	268 (1.9)[43]
chloroform		6.8	534 (1.9)[43]

TABLE 6.1 Experimental Results of Molecular Studies—*(continued)*

Structure	Solvent	λ_{max} (nm)	μ (10^{-18} esu)	$\beta_\mu (\lambda)^{ref.}$ (10^{-30} esu)
	chloroform		5.0	98 (1.9)[43]
	chloroform		5.6	64 (1.9)[43]
	chloroform		6.7	97 (1.9)[43]
	chloroform		8.0	89 (1.9)[43]
X = OCH$_3$	chloroform		6.0	165 (1.9)[43]
X = N(CH$_3$)$_2$	chloroform		6.0	1024 (1.9)[43]

chloroform		7.8		169 (1.9)[43]
chloroform		7.7		83 (1.9)[43]
chloroform	696	8.1		432 (1.9)[43]
chloroform	548	9.5		60 (β_0)[177]
chloroform	550	7.2		72 (β_0)[177]
chloroform		5.5		164 (1.9)[43]
chloroform		8.7		129 (1.9)[43]
chloroform		7.0		87 (1.9)[43]

TABLE 6.1 Experimental Results of Molecular Studies—(*continued*)

Structure	Solvent	λ_{max} (nm)	μ (10^{-18} esu)	$\beta_\mu(\lambda)^{ref.}$ (10^{-30} esu)
	chloroform		7.5	93 (1.9)[43]
	chloroform		8.6	82 (1.9)[43]
	chloroform		7.9	102 (1.9)[43]
	chloroform		8.5	95 (1.9)[43]

of the monosubstituted fragments, the charge-transfer contribution to μ is typically less than 1 D for benzene and up to 2 D for stilbene derivatives. Typically, β_μ is enhanced 6 to 8 times.

Substitution Pattern

Charge-transfer interactions between substituents are often depicted by resonance structures which connect charge-neutral and charge-separated states with alternating single and double bonds. By this picture, *para* and *ortho* substitution patterns allow interaction between donor and acceptor groups whereas *meta* does not for benzene derivatives. For stilbenes, resonance interaction is allowed for 2(donor)-2'(acceptor), 4-4', 2-4', and 4-2', but is not allowed for 2-3', 3-2', 3-3', 3-4', and 4-3' substitutions. To what extent is such a picture reliable in predicting hyperpolarizability? For benzenes, *meta* and *ortho* compounds are much less nonlinear than the *para* compounds and, in most cases, their β_μ values are even less than the vector sums of their monosubstituted derivatives. In addition, *ortho* derivatives are not always superior to the *meta* derivatives, contrary to expectations based on resonance structures. The importance of factors other than resonance structures in determining β_μ values is further illustrated by results obtained for the nitrostilbene derivatives. Most unexpected are the relatively large β_μ values found for the 3-4' and the very low value for the 2-2' derivatives. A second unexpected result is found in the nonreciprocity of β_μ values upon interchanging the donor and acceptor positions. Intuitively, CT interactions, if effective for 2-4' and 3-4', should also be effective for 4-2' and 4-3' substitution patterns.

To understand these observations, two factors which are undoubtedly important must be considered. First, the dominant portion of the molecular dipole moment of nitrostilbene derivatives is due to the polar nature of the nitro group itself, apart from the relatively small mesomeric contribution due to donor acceptor interaction between the substituents. Therefore the molecular dipole direction is principally determined by the orientation of the nitro group. Second, because the EFISH-determined β_μ is a vectorial projection of the hyperpolarizability tensor along the molecular dipole direction, it samples the dominant tensor components only if the dipole direction coincides with the molecular CT axis. For *meta* and *ortho* nitrobenzene derivatives and $-3'$ and $-2'$ nitrostilbene derivatives, whose CT and dipole directions are angularly displaced, β_μ values are expected to decrease. With these simple geometric considerations, one can easily rationalize the low β_μ values for these compounds which exhibit angular displacements greater than 60 degrees. The comparable values for all stilbenes with 4'-nitro substitution are also partly explained by the good alignment of molecular dipoles with CT axes. Because the group moment of the methoxy donor is much smaller than that of the nitro acceptor, their substitution patterns have only a minor effect on the molecular dipole directions. This picture is confirmed by the much increased value for the 3',5'-dinitro derivative in which the CT and dipolar axes are well aligned. Large components of the β tensor, which are not fully accessed by EFISH because of geometric considerations, should also be present in all of the 2'- and 3'-nitrostilbene derivatives. Hyper-Rayleigh scattering can confirm this expectation.

Substitution patterns with multiple donors and acceptors have also been examined. Generally, no significant enhancement can be achieved with additional substituents. This may be the result of a saturation phenomenon well known for perturbation interactions, in which the mixing between a donor and an acceptor orbital stabilizes the lower energy orbital which becomes less effective when mixing with addition donors due to the increased energy splitting. More importantly, from understandings developed earlier, steric factors and redirection of the ground-state dipole moment often dominate the outcome, when additional acceptor substituents are involved. Although the 3',5'-dinitrostilbene derivative leads to significant improvement over the 3'-nitro derivative, the 2',4'-dinitro derivative is found to offer lower β_μ than the 4'-nitro analog. Because donors have minimal effects on the dipole direction, 2,4- donor-4'- acceptor substituted stilbenes are the only examples where enhancement over that of the 4-4' derivative is realized through multiple substituents. Heteroatom and side-group substitution generally have a detrimental effect on molecular nonlinearity probably due to the electronic localization effects on the substituted sites.

Nonlinearity and Transparency Trade-Off

According to Eq. (6.41), a trade-off exists between nonlinearity and transparency, and it can be optimized by enhancing $\mu_{ge}^2 \Delta\mu_{ge}$ for a given ω_{ge} or λ_{max}. However, because the oscillator strength $f_{ge} \propto \omega_{ge}\mu_{ge}^2$ is not known to be much greater than unity, it is likely that large hyperpolarizabilities are to be associated with low energy transitions due to the strong cubic dependence on λ_{max}. Spectroscopic and β_μ results from Table 1 can be used to explore this issue. For a given π system, β_μ and λ_{max} are tuned with combinations of donor and acceptor groups of increasing strengths, and results obtained for several molecular classes are given in logarithmic plots in Fig. 5. Within the uncertainty of some resonance enhancement for the high λ_{max} species, strong power law dependences are found, with exponents ranging from 4.7 to 8.7. These values are considerably larger than the two-level prediction and appear to increase with the extension of the π system. Therefore, for a given λ_{max}, a better trade-off can be realized by extending the π system. Hyperpolarizability arising from substituents with strong inductive but weak resonance contributions, such as cyano, fluorinated sulfonyl, and fluorinated sulfonylsulfimide groups have been shown to offer an improved trade-off (Cheng et al., 1990).

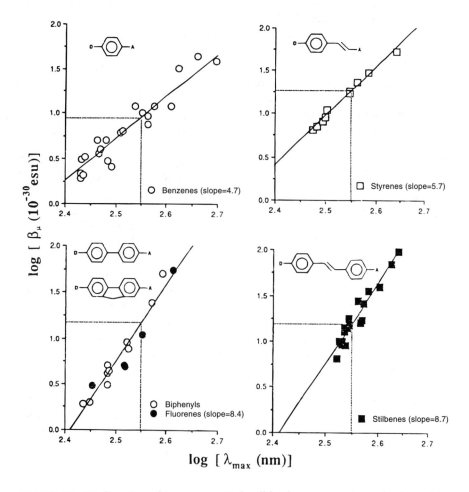

FIGURE 6.5 Nonlinearity and transparency trade-off for donor-π-acceptor molecular systems.

The π System

Although hyperpolarizability is enhanced with increasing donor-acceptor strength and an appropriate substitution pattern, the most significant increase resulted from conjugation extension. Although the expectation of enhanced hyperpolarizabilities for extended π systems is quite reasonable, considering the increased numbers of electrons and the reduced energy gaps, the structural features of a π system, such as length, planarity, bond alternation, and aromaticity, are also expected to play important roles. Because styrene has a conjugation length intermediate between those of benzene and stilbene, the β_μ values of its derivatives fall between those of benzene and stilbene derivatives. When comparing styrene, biphenyl, and stilbene derivatives (Cheng et al., 1991a), the efficacy of the vinyl unit compared to the phenyl unit becomes evident. A two carbon atom extension of the former results in about equal enhancement as that of a four atom extension in the latter. Clearly there is an advantage in molecular volume with the vinyl unit. It is interesting to note that the superior accepting strength of the dicyanovinyl and tricyanovinyl groups arises largely from the vinyl extension of conjugation length. When the donor and acceptor positions are interchanged in the α,ω-substituted styrene derivatives, significant differences are found in their properties. Although the CT band position and bandwidth are very similar, the oscillator strength, for example, of 4-N, N-dimethylaminophenyl β-nitrostyrene is found to be 1.6 times higher than that of 4-nitrophenyl β-N, N-dimethylaminostyrene. This is consistent, within the two-level approximation, with the lower β_μ value found for the latter derivative.

Planarity and Bond Alternation

The importance of planarity can be addressed by comparing donor-acceptor 4,4′-substituted biphenyl, whose phenyl linkage is free to rotate, and 4,4′-substituted fluorene which has bridged planar structures. The fluorenes are found to be more nonlinear but the difference is small and decreases with the strength of donor and acceptor groups whose interaction may impose significant double-bond character on the phenyl linkage resulting in a more planar structure in solution. Given an empirical $\cos^2\theta$ dependence on the extinction coefficient for the π-π* transition, β_μ is expected to have a similar dependence within the two-level approximation and is therefore quite tolerant to small torsional angles. Using such a model, the torsional angle of the biphenyl derivatives can be estimated from β_μ and λ_{max} measurements. Such an analysis, indeed, shows an end group dependent torsional angle for the donor-acceptor substituted biphenyl derivatives (Cheng et al., 1991a).

Stilbene derivatives have β_μ values significantly higher than the tolane derivatives, despite comparable physical length. This indicates an important difference between the olefinic and acetylenic bridges. Although the π orbitals of both linkages are coplanar with the phenyl rings, the p orbitals of the sp hybridized acetylenic carbons, which are electron-rich due to the low energy of the s orbitals, may lead to less effective π delocalization because of the orbital energy mismatch with the p orbitals of the sp^2 hybridized phenyl carbons. In addition, due to the short triple bond length, p orbital overlap is maximized in the acetylenic resonance structure. These factors are likely to result in localization of electron density at the phenyl rings. Such a difference of course does not exist in the *trans*-stilbene structure where all carbons are sp^2 hybridized, allowing effective delocalization of all π electrons. Structural analysis (Stiegman et al., 1991) on tolane and α,ω-diphenylpolyne derivatives has shown that the quinoidal character is present only in the phenyl fragments and that the acetylenic linkages remain largely unchanged in the ground state. Therefore, the π system is poorly delocalized in the ground state, and the extent of charge transfer in the ground and excited states is relatively independent of the acetylenic linkage. The acetylene linkage acts as a rather passive conduit to transfer charge into what are, ultimately, highly localized orbitals at the nitrophenyl and aniline fragments.

Length Dependence

Further extensions of aromatic conjugations lead to substantial reduction of nonlinearity in the α,ω-polyphenyl derivatives (Cheng et al., 1991a) and only moderate enhancement in the phenylvinyl-extended stilbene derivatives. The properties of the α,ω-diphenylpolyne derivatives are equally disappointing, yielding only a modest increase for the triyne structure. This is in sharp contrast to the rapid rise of β_μ values for vinyl extensions in the α,ω-polyene (Marder et al., 1993), α-phenylpolyene (Cheng et al., 1991a), and α,ω-diphenylpolyene (Huijts and Hesselink, 1989a; Barzoukas et al., 1989a) series, which show power law dependences. Clearly, physical extension of the π system alone does not insure higher hyperpolarizability. The substantial aromatic stabilization of extended aromatic structures and the strong bond alternation in the acetylenic fragments, as discussed above, are seen as inhibiting factors for charge delocalization, resulting in the low β_μ values of these structures. Torsional effects in the α,ω-polyphenyl derivatives lead to a rapid decrease in β_μ for that series. For the polyene derivatives, because the neutral and other charge-separated resonance structures formally consist of similar π systems of alternating single and double bonds, they are more effectively mixed in the ground and excited states based on energy considerations. Unhampered by aromaticity, the donor-acceptor interaction can significantly reduce the bond length alternation and approach the cyanine limit with reduced band gap and increased oscillator strength. Data for aldehyde- and dicyanovinyl-terminated polyene with tertiary amino donors show that significantly stronger conjugation length power law exponents than those of the phenyl-terminated polyenes were found (Marder et al., 1993).

Aromaticity

However, the chemical stabilities of polyene derivatives are not sufficient for practical devices. A balance between aromaticity, which tends to increase chemical stability, and nonlinearity needs to be considered. One approach is to consider structures, in which certain aromatic fragments are replaced with either quinoid- or polyene-like moieties. When one of the phenyl units in conventional diphenyl polyene derivatives is replaced with a five member heterocyclic, such as pyrole, furan, thiophene, and azole, which have reduced aromatic characteristics and are, thus, more polyene-like, higher β_μ and λ_{max} values are found. The aromatic hindrance to charge-transfer can be further minimized with quinoidal moieties where both the charge-neutral and charge-separated resonance structures have an unchanged degree of aromaticity. An odd merocyanine, N, N-dimethylaminoindolaniline, provides an interesting case study (Marder et al., 1991). The two rings interchange their electronic characteristics between aromatic and quinoidal forms upon charge transfer, thus, minimizing the energetic barrier for charge redistribution in the excited state. This structural feature results in a highly polarizable ground state and a low energy CT band, both of which contribute to the extraordinary high β_μ value for the relatively small structure.

This approach of using a quinoid-aromatic driving force to optimize hyperpolarizability is extended to other merocyanine-type derivatives containing highly effective cyclic acceptors, such as barbituric acid, thiobarbituric acid (Marder et al., 1994), 1,3-indandione, phenyl-isoxazolone (Marder et al., 1994), 1,3-indandimalononitrile, and other heterocyclic acceptors. In conjugation with powerful tertiary amine donors, these acceptors yield chromophores possessing some of the highest hyperpolarizability values ever reported. The effectiveness of the barbituric acid and thiobarbituric acid acceptors is believe to be the result of the strong aromatic driving force of the heterocyclic ring through the amide linkage which has a substantial carbon-nitrogen double bond character. Providing the correct balance between aromatic and quinoid elements, phenylpolyene derivatives are found to be more effective than polyene derivatives. When thiophene units are introduced into the conjugation, materials with reasonable chemical stability have been reported (Jen et al., 1993).

At present, there is no shortage of molecular materials with sufficient microscopic nonlinearity for numerous interesting applications. The challenge is to optimize various property trade-offs, including optical transparency, thermal, chemical, photochemical and electrochemical stabilities, or to develop device schemes with tolerant device requirements. There remains the challenge of effective organization and fabrication. In the remainder of this article, two well studied approaches to organizing the microscopic response into bulk nonlinearity are reviewed. The single crystal approach which offers high number density but only marginal control will be followed by the poled-polymer approach which suffers from the quasi-static nature of its macroscopic responses.

6.5 Single Crystal Studies

The organization of an optically nonlinear chromophore in a single crystalline form offers maximum number density and a potentially high degree of alignment. However, the crystallographic forces involved for an optimal crystal structure for nonlinear optical applications are difficult to control. Therefore, common practice in the search for single crystal materials has relied on near random but minor modifications of an optical nonlinear molecular unit and exploiting the polymorphism of a molecular crystal by crystallizing derivatives in different solvents. A large data base in the noncentrosymmetry of crystalline powders has been accumulated based on the Kurtz powder SHG test (Kurtz and Perry, 1968). Numerous crystals have been further developed, and their single crystal optical and physical properties have been studied. Table 2 summarizes much of the published single crystal results. Only materials which fit the current discussion of donor-acceptor substituted conjugates are included. New materials whose single crystal properties are less well characterized are excluded. Table 2 organizes single crystal materials according to increasing size of the π system, the acceptor strength, and the donor strength. Because the material requirements are quite different for frequency conversion and electrode-optic modulation applications, crystals suitable for each of the applications will be discussed separately. For frequency conversion, our discussion includes only crystals which have sufficient transparency for the frequency doubling of the Nd:YAG laser.

Single Crystals for Frequency Conversion

Transparency Constraint

The efficient frequency doubling of diode lasers will have significant technological implications in imaging and optical recording. For example, the ability to use 400 nm light, instead of 800 nm for optical recording, will increase the storage capacity by four times. Due to the typically low power of these lasers, materials with high optical nonlinearity are desired for efficient frequency doubling. The high nonlinearity of organics and the potential flexibility in material and device engineering have generated much interest in the frequency conversion with single organic crystals. However, because the origination of optical nonlinearity in organic materials is dominated by low energy CT optical excitations, the trade-off between nonlinearity and optical transparency (see Fig. 5) prevents the use of the highly nonlinear organic molecules. For second-harmonic generation (SHG), a large optical window between the fundamental and the harmonic waves is required, making it quite a challenge to find efficient organic materials for SHG in the blue region of the optical spectrum. The near infrared overtone absorption of C-H and O-H stretches also limits the use of organics for parametric downconversion processes.

Within a certain transparency requirement, such as that for harmonic generation in the blue-green region (as given by the dashed boxes in Fig. 5), β is limited to between 1 (for blue) to 3 (for green) times that of **p**-nitroaniline (*p*-NA) (see Table 1). These materials are either conjugated derivatives with weak donors and acceptors or small π systems with somewhat stronger substituents. In particular, cyano, ketone, fluorinated ketone, sulfone, and fluorinated sulfone substituted benzenes and styrenes are found to yield mostly colorless materials suitable for SHG of short wavelength diode lasers. Nitrobenzene and stilbene derivatives are mostly yellow materials with

TABLE 6.2 Single Crystal Results on Organic Nonlinear Optical Materials: n_i, refractive index at 633 nm; $\lambda_{NC}(\theta)$, θ-noncritical phase-matching wavelength; $\lambda_{NC}(\lambda)$, λ-noncritical phase-matching wavelength; and at 1.064 μm: DT(τ), optical damage threshold (pulse duration); d_{eff}, effective nonlinearity (phase-matching type); $\Delta\theta l$, angular acceptance; ΔTl, temperature acceptance; $d\theta_{pm}/dT$, temperature tuning of phase-matching angle; and ρ, walk-off angle (phase-matching type)

Structure and Nomenclature (acronym)	Point Group	SHG $d_{ij}(\lambda)$ & EO $r_{ij}(\lambda)$ (pm/v) (μm)	Cut-Off λ (nm)	Ref.	Properties
Urea	m	$d_{36}(1.06) \approx 1.3$ $r_{41}(0.63) \approx 1.9$ $r_{63}(0.63) \approx 0.83$	200 & 1800	212, 80, 18, 21, 172	DT(10 ns) = 5 GW/cm². $n_o = 1.485$, $n_e = 1.567$.
p,p'-dihydroxydiphenyl sulfone (DHDPS)	mm2	$d_{33}(1.06) \approx 7$ $d_{32}(1.06) \approx 0.4$ $d_{11}(1.06) \approx 3$	300 & 1500	275	Relatively high hardness, nonhygroscopic, and photochemically stable. Crystals cleave at input power >400 MW/cm², $n_x = 2.009$, $n_y = 2.000$, $n_z = 1.921$. $\lambda_{NC}(\theta) = 865$ nm.
3-methoxy-4-hydroxy-benzaldehyde (MHBA)	2	$d_{13}(1.06) \approx 13$ $d_{11}(1.06) \approx 9.8$ $d_{12}(1.06) \approx 3.9$ $d_{14}(1.06) \approx 3.2$	370	248, 286	DT(10 ns) = 2 GW/cm². $n_x = 1.545$, $n_y = 1.685$, $n_z = 1.780$. $\Delta\theta l = 0.9$ mrad-cm. $\rho = 6°$.
8-(4'-acetylphenyl)-1,4-dioxa-8-azaspiro[4,5]decane (APDA)	mm2	$d_{33}(1.06) \approx 50$ $d_{32}(1.06) \approx 7$	384	217	$n_x = 1.56$, $n_y = 1.66$, $n_z = 1.68$. $d_{eff}(1.06) \approx 14.9$ pm/V.
5-nitrouracil (5NU)	222	$d_{14}(1.06) \approx 8.7$	410 & 1550	201	Relatively transparent at near IR, DT(10 ns) = 3 GW/cm². $n_x = 1.569$, $n_y = 1.901$, $n_z = 1.707$. $\Delta\theta l = 8$ mrad-cm, $\rho = 7°$, $\lambda_{NC}(\lambda) = 1440$ nm.
m-aminophenol (mAP)	mm2	$d_{33}(1.06) \approx 3.3$ $d_{32}(1.06) \approx 2.4$ $d_{31}(1.06) \approx 0.7$	320 & 1700	36	poor mechanical strength, $n_x = 1.659$, $n_y = 1.765$, $n_z = 1.578$.
m-chloronitrobenzene (mCNB)	mm2	$d_{33}(1.06) \approx 7.8$ $d_{32}(1.06) \approx 4$ $d_{31}(1.06) \approx 4.5$	400 & 2000	36	Cleaves easily, not phase-matchable for SHG at 1.064 μm, $n_x = 1.676$, $n_y = 1.684$, $n_z = 1.649$.
m-bromonitrobenzene (mBNB)	mm2	$d_{33}(1.06) \approx 8$ $d_{32}(1.06) \approx 4.5$ $d_{31}(1.06) \approx 4$	420 & 2100	36	$\lambda_{NC}(\theta)$ at 1.064 μm for solid solution of mC$_{0.95}$Br$_{0.05}$NB, $n_x = 1.649$, $n_y = 1.729$, $n_z = 1.678$.

TABLE 6.2 Single Crystal Results on Organic Nonlinear Optical Materials—(continued)

Structure and Nomenclature (acronym)	Point Group	SHG $d_{ij}(\lambda)$ & EO $r_{ij}(\lambda)$ (pm/v) (μm)	Cut-Off λ (nm)	Ref.	Properties
4-nitrophenol sodium dihydrate (NPNa)	$mm2$	type I $d_{eff}(1.06) \approx 8$	515	167	Vickers hardness: 34, good thermal conductivity.
m-nitroaniline (mNA)	$mm2$	$d_{33}(1.06) \approx 20$ $d_{32}(1.06) \approx 1.5$ $d_{31}(1.06) \approx 20$ $r_{33}(0.63) \approx 17$ $r_{23}(0.63) \approx 0.1$ $r_{13}(0.63) \approx 7.5$	500 & 1900	256, 36, 240	Cleaves easily, melt grown into channel waveguide (unfavorable orientation for SHG), $n_x = 1.752$, $n_y = 1.715$, $n_z = 1.665$.
2-methyl-4-nitroaniline (MNA)	m	$d_{11}(1.06) \approx 167$ $d_{12}(1.06) \approx 25$ $r_{11}(0.63) \approx 67$	500 & 1900	143, 170, 181, 150	Melt grown into channel waveguide (orientation controlled by electric or temperature gradients). DT(20 ns) = 0.2 GW/cm². $n_x = 2.001$, $n_y = 1.658$, $n_z = 1.435$, near optimum molecular alignment for EO.
4-(N,N-dimethylamino)-3-acetamidonitrobenzene (DAN)	2	$d_{21}(1.06) \approx 1.5$ $d_{22}(1.06) \approx 5.2$ $d_{23}(1.06) \approx 50$ $d_{25}(1.06) \approx 1.5$ $r_{11}(0.63) \approx 13$	485 & 2270	121, 122, 123	Fiber waveguide allows full use of nonzero d_{ij}, DT(15 ns) = 80 MW/cm² $d_{eff} \approx 35.5$ (I) & 9 (II) pm/v. $n_x = 1.539$, $n_y = 1.682$, $n_z = 1.949$. $\Delta\theta l \approx 1.5$ mrad-cm. $\rho = 7°$, and $\Delta Tl \approx 1.9$ °C-cm.
N-(4-nitrophenyl)-(s)-prolinol (NPP)	2	$d_{22}(1.06) \approx 28$ $d_{21}(1.06) \approx 85$	500 & 2000	289, 138	Near optimum molecular alignment for largest d_{21}. $n_x = 2.066$, $n_y = 1.876$, $n_z = 1.478$. $\lambda_{NC}(\theta) = 1.15$ μm using d_{21} $\lambda_{NC}(\lambda) = 1.5$ μm.
2-methyl-4-nitro-N-methylaniline (MNMA)	$mm2$	$d_{33}(1.06) \approx 2.6$ $d_{31}(1.06) \approx 13$ $d_{15}(1.06) \approx 12$ $r_{13}(0.63) \approx 8$ $r_{33}(0.63) \approx 7.5$	510	245	$n_x = 2.148$, $n_x = 1.520$.
N-(4-nitrophenyl)-N-aminoacetonitrile (NPAN)	$mm2$	$d_{33}(1.06) \approx 27$ $d_{32}(1.06) \approx 57$ $d_{33}(1.34) \approx 24$ $d_{32}(1.34) \approx 48$	500	171, 15	Near optimum molecular alignment for largest d_{32}. $\Delta\theta l \approx 2$ mrad-cm. $\Delta Tl \approx 5$ °C-cm.
L-N-(5-nitro-2-pyridyl)leucinol (NPLO)	2	type I: $d_{eff}(1.06) \approx 37$ type II: $d_{eff}(1.06) \approx 3$	480	263	Vickers hardness: 34, nonhygroscopic, DT(8 ns) = 6 GW/cm². $n_x = 1.457$, $n_y = 1.631$, $n_z = 1.933$. $\rho = 0.22$ (I) & 0.24 (I) mrad.

TABLE 6.2 Single Crystal Results on Organic Nonlinear Optical Materials—(*continued*)

Structure and Nomenclature (acronym)	Point Group	SHG $d_{ij}(\lambda)$ & EO $r_{ij}(\lambda)$ (pm/v) (μm)	Cut-Off λ (nm)	Ref.	Properties
3,5-dimethyl-1-(4-nitrophenyl) pyrazole (DMNP)	mm2	$d_{33}(0.84) \approx 29$ $d_{32}(0.84) \approx 90$	450	83, 100	(100) oriented core fibers allow full use of d_{32}, $\lambda_{NC}(\theta) = 944$ nm using d_{32}.
m-dinitrobenzene (*m*DB)	mm2	$d_{33}(1.06) \approx 0.7$ $d_{32}(1.06) \approx 2.7$ $d_{31}(1.06) \approx 1.8$	400 & 2200	20	$n_x = 1.738$, $n_y = 1.680$, $n_z = 1.483$.
mehtyl-(2,4-dinitrophenyl)-aminopropanoate (MAP)	2	$d_{22}(1.06) \approx 18$ $d_{21}(1.06) \approx 17$ $d_{23}(1.06) \approx 3.7$	500 & 2000	192	DT(10 ns) = 3 GW/cm^2. $d_{eff} \approx 3.8$ (I) & 8.8 (II) pm/v. $n_x = 1.531$, $n_y = 1.653$, $n_z = 1.935$. $\Delta\theta l \approx 1.5$ mrad-cm. $\rho = 11.5°$(I) & $2.4°$(I).
3-methyl-4-nitropyridine-*N*-oxide (POM)	222	$d_{14}(1.06) \approx 10$ $r_{52}(0.63) \approx 5.2$	450 & 2100	288, 230	DT(20 ps) = 2 GW/cm^2. $d_{eff} \approx 7.9$ (I) & 4.0 (II) pm/v. $n_x = 1.663$, $n_y = 1.829$, $n_z = 1.625$. $\rho = 6.3°$(I) & $1.4°$(I).
2-amino-5-nitropyridinium-dihydrogen phosphate (2A5NPDP)	mm2	$d_{33}(1.06) \approx 12$ $d_{15}(1.06) \approx 7$ $d_{24}(1.06) \approx 1$	420 & 2000	135	$n_x = 1.752$, $n_y = 1.715$, $n_z = 1.665$. $\lambda_{NC}(\theta) = 1084$ & 1129 nm. $\lambda_{NC}(\lambda) = 1340$ nm. not phase-matchable for type-II SHG at 1.064 μm. Type-I $d_{eff} \approx 2$ pm/v. $d\theta_{pm}/dT = 2.4'/°$C at 1.3 μm.
(−)2-(α-methylbenzylamino)-5-nitropyridine (MBANP)	2	$d_{22}(1.06) \approx 60^{58}$ $d_{22}(1.06) \approx 35^{57}$	430 & 1500	12, 11, 134	DT(425 ns) = 1 GW/cm^2. $n_y = 1.813$, $n_c = 1.676$. $d_{eff} \approx 1.2 \times$ LiIO$_3$ ≈ 6 pm/v.
2-*N*-cyclooctylamino-5-nitropyridine (COANP)	mm2	$d_{33}(1.06) \approx 14$ $d_{32}(1.06) \approx 32$ $d_{31}(1.06) \approx 15$ $r_{33}(0.63) \approx 15$ $r_{13}(0.63) \approx 3.4$ $r_{23}(0.63) \approx 13$	470	77, 26, 27	$\lambda_{NC}(\theta) = 1023$ nm using d_{32}. $\lambda_{NC}(\theta) = 1413$ nm using d_{31}. $n_x = 1.68$, $n_y = 1.78$, $n_z = 1.64$. $d_{eff} \approx 24$ pm/v, $r_{33}(0.52) \approx 28$, $r_{33}(1.06) \approx 7.7$. $r_{13}(0.52) \approx 6.8$, $r_{13}(1.06) \approx 0.9$. $r_{23}(0.52) \approx 26$, $r_{23}(1.06) \approx 6.3$.
2-adamantylamino-5-nitropyridine (AANP)	mm2	$d_{33}(1.06) \approx 60$ $d_{31}(1.06) \approx 80$	460	257	At 533 nm: $n_x = 1.77$, $n_y = 1.61$, $n_z = 1.86$. At 1.06 μm: $n_x = 1.67$, $n_y = 1.59$, $n_z = 1.71$.

TABLE 6.2 Single Crystal Results on Organic Nonlinear Optical Materials—(continued)

Structure and Nomenclature (acronym)	Point Group	SHG $d_{ij}(\lambda)$ & EO $r_{ij}(\lambda)$ (pm/v) (μm)	Cut-Off λ (nm)	Ref.	Properties
N-(4-nitro-2-pyridinyl)-(s)-phenylalaninol (NPPA)	2	$d_{22}(1.06) \approx 2.6$ $d_{21}(1.06) \approx 0.4$ $d_{23}(1.06) \approx 31$ $d_{16}(1.06) \approx 0.5$ $d_{34}(1.06) \approx 25$	480	261, 247	$\lambda_{NC}(\theta)$ for SHG and SFG within absorption edge, calculated $d_{eff} \approx 31$ pm/v. $n_x = 1.524$, $n_y = 1.694$, $n_z = 1.907$.
2-(N-prolinol)-5-nitropyridine (PNP)	2	$d_{22}(1.06) \approx 17$ $d_{21}(1.06) \approx 48$ $r_{22}(0.63) \approx 13$ $r_{12}(0.63) \approx 13$	490 & 2080	246, 27	$\lambda_{NC}(\theta) = 1020$ nm using d_{21}. $d_{eff} \approx 47$ pm/v. $n_x = 1.990$, $n_y = 1.788$, $n_z = 1.467$. $\rho = 7°$, $r_{22}(0.52) \approx 28$, $r_{12}(0.52) \approx 20$. $r_{22}(1.06) \approx 8$, $r_{12}(1.06) \approx 9$.
2-dicyanovinylanisole (DIVA)	2	$d_{22}(1.06) \approx 10$		78	$d_{eff}(1.06) \approx 20$ pm/V, n = 1.65.
3-(1,1-dicyanoethenyl)-1-phenyl-4,5-dihydro-1H-pyrazole (DCNP)	m	$r_{33}(0.63) \approx 87$	700 & 1600	3	High melting point: 190 °C, near optimum molecular alignment for EO, $n_x = 1.9$, $n_z = 2.7$.
ω-(p-methoxyphenyl) benzofulvene (MPBF)	mm2	$d_{eff}(1.06) \approx 7$	>450	133	$d_{eff} \approx 7$ pm/v. $\Delta\theta l = 0.32$ mrad-cm. $\Delta Tl \approx 4.6$ °C-cm.
2-cyano-3-(2-methoxyphenyl)-2-propenoic acid methyl ester (CMP-methyl)	2	$d_{22}(1.06) \approx 29$	410	183	$n_y = 1.85$
p-methylbenzal-1,3-dimethylbarbituric acid	2	$d_{eff}(1.06) \approx 8$	460	132	Vickers hardness: 25.5.

TABLE 6.2 Single Crystal Results on Organic Nonlinear Optical Materials—(*continued*)

Structure and Nomenclature (acronym)	Point Group	SHG $d_{ij}(\lambda)$ & EO $r_{ij}(\lambda)$ (pm/v) (μm)	Cut-Off λ (nm)	Ref.	Properties
4-Br-4′-methoxychalcone	m	$d_{33}(1.06) \approx 6$ $d_{13}(1.06) \approx 27$	420	285	$n_x = 1.55$, $n_y = 1.47$, $n_z = 1.90$.
4-ethoxy-4′-methoxychalcone	mm2	$d_{eff}(1.06) \approx 5.7$	430	129	Low hardness, $DT(1\ ns) > 30\ GW/cm^2$. At 532 nm: $n_x = 1.493$, $n_y = 1.710$, $n_z = 1.983$. At 1.06 μm: $n_x = 1.477$, $n_y = 1.663$, $n_z = 1.850$. $d_{eff} \approx 3.5\ (I)\ \&\ 5.7\ (II)$ pm/v.
4-methyl-2-thienylchalcone	2	$d_{eff}(1.06) \approx 7$	430	128	Low hardness, $n_x = 1.648$, $n_y = 1.696$, $n_z = 1.775$. $d_{eff} \approx 7$ pm/v. $\Delta\theta l = 0.9$ mrad-cm. $\rho = 3.6°$. $\Delta Tl = 2.2$ °C-cm.
3-methyl-4-methoxy-4′-nitrostilbene (MMONS)	mm2	$d_{33}(1.06) \approx 184$ $d_{32}(1.06) \approx 41$ $d_{24}(1.06) \approx 71$ $r_{33}(0.63) \approx 40$	515 & 2000	22, 220	$\lambda_{NC}(\theta) = 1028$ nm using d_{24}. $d_{eff} \approx 43$ pm/v[71]. $n_x = 1.569$, $n_y = 1.693$, $n_z = 2.129$. $\rho = 9.6°$. $\Delta Tl = 0.17$ °C-cm.
4-nitro-4′-methylbenzylidiene aniline (NMBA)	m	$d_{11}(1.06) \approx 139$ $d_{33}(1.06) \approx 0.6$ $d_{31}(1.06) \approx 41$ $r_{11}(0.63) \approx 25$	480 & 1600	8, 9	Near optimum molecular alignment for EO, $\lambda_{NC}(\theta)$ for SHG within absorption edge, $\lambda_{NC}(\lambda) = 1500$ nm. $n_x = 1.951$, $n_y = 1.657$, $n_z = 1.510$. $d\Delta n/dT = 15.8 \times 10^{-5}$ K^{-1}. $d_{eff} \approx 2$ pm/v.
4′-nitrobenzylidene-3-acetamino-4-methoxyaniline (MNBA)	m	$d_{11}(1.06) \approx 175$ $d_{31}(1.06) \approx 2$ $d_{33}(1.06) \approx 2$ $r_{11}(0.63) \approx 29$ $r_{13}(0.63) \approx 0.5$ $r_{33}(0.63) \approx 2.4$	505	131	Near optimum molecular alignment for EO, $n_x = 2.024$, $n_y = 1.648$, $n_z = 1.583$.
Cyanostilbazolium p-toluenesulfonate complex	mm2	$d_{33}(1.06) \approx 21$	415	258	Melting point: 279 °C, $n_z = 1.775$.
4′-dimethyamino-N-methyl-4-stilbazolium methyl sulfate (SPCD)	mm2	$r_{33}(0.63) \approx 430$	600	282	Only thin-film cystal reported, $n_y = 1.31$, $n_z = 1.55$.

TABLE 6.2 Single Crystal Results on Organic Nonlinear Optical Materials—(continued)

Structure and Nomenclature (acronym)	Point Group	SHG $d_{ij}(\lambda)$ & EO $r_{ij}(\lambda)$ (pm/v) (μm)	Cut-Off λ (nm)	Ref.	Properties
4'-dimethyamino-N-methyl-4-stilbazolium tosulate (DAST)	m	$d_{11}(1.91) \approx 600$ $d_{22}(1.91) \approx 100$ $d_{12}(1.91) \approx 30$ $r_{33}(0.82) \approx 400$	700 & 2000	157, 198	Near optimum molecular alignment for EO. At 820 nm: $n_x = 2.216$, $n_y = 1.66$, $n_z = 1.65$. Dielectric constant: $\epsilon_{ab} = 5.1$, $\epsilon_c = 3.1$.

a cut-off at around 500 nm suitable for SHG of the Nd:YAG laser (see Table 1). Other conjugation systems, such as vinyl (Nogami et al., 1989), tolane (Burland et al., 1992), and cumulene derivatives (Ermer et al, 1991), offer similar nonlinearity and transparency trade-offs.

An estimate of the potential nonlinearity for optimized single organic crystals suitable for second-harmonic generation is useful. The macroscopic susceptibility d_{IJK} can be estimated by using the oriented gas model according to Eq. (6.24). Due to the normal dispersion of materials, phase-matched harmonic conversion can be simply achieved only between waves of differing polarizations. Therefore, the diagonal susceptibility tensor elements are not useful. To optimize the potentially phase-matchable nondiagonal tensor elements, the molecular units must be aligned according to certain geometric guidelines depending on the tensorial nature of their hyperpolarizabilities (Zyss and Ouder, 1982). For prototypical CT chromophores, if one assumes β_{zzz} dominates, the optimal structures for optimization of the phase-matchable d_{IJ}'s belong to the point group 2, m, and mm2. The directional cosines are given by $\cos\theta \sin^2\theta$, $\sin\theta \cos^2\theta$, and $\sin^2\psi \cos\theta \sin^2\theta$ ($\sin^2\psi \cos\theta \sin^2\theta$), respectively, where ($\psi$, θ) are the Euler angles which fix the orientation of the molecular CT axis against the crystalline polar axis. They are optimized at (ψ, 54.74°), (ψ, 35.26°), and (90°, 54.74°) (0°, 54.74°) respectively. Take point group mm2, for example. An optimized crystal structure yields birefringence phase-matchable coefficients expressed by

$$d_{ZYY} = \frac{1}{2} Nf_Z^{2\omega} f_Y^{\omega} f_Y^{\omega} \frac{2}{3\sqrt{3}} \beta_{zzz} \qquad (6.75)$$

and

$$d_{ZXX} = \frac{1}{2} Nf_Z^{2\omega} f_X^{\omega} f_X^{\omega} \frac{2}{3\sqrt{3}} \beta_{zzz}.$$

Given a typical number density, N, of 5×10^{21} cm^{-3}, local field factor of 5, and a hyperpolarizability of 10×10^{-30} esu (about that of p-NA), $d_{zyy} = d_{zxx}$ is estimated to be about 48×10^{-9} esu (20 pm/V). At the wavelength of interest, resonance enhancement of the hyperpolarizability is likely to significantly increase the SHG d coefficients. Therefore, even for moderately nonlinear organic molecules (due to the transparency requirement for frequency conversion), d coefficients several times that of the best inorganic materials can be expected.

Phase-Matching Properties

Several general remarks can be made concerning phase matching in organic nonlinear optical crystals. Due to the strongly anisotropic polarizability of molecules possessing π systems with one or two dimensional CT transitions, their acentric crystals, which often resulted from polar stacking of the π system, will most likely be highly birefringent and therefore phase-matchable throughout their transparency range. This differs from inorganics where many members lack sufficient birefringence for phase matching in the blue-green region. In fact, the large birefringence of well-aligned organic crystals has long been recognized as a drawback in frequency conversion application due to the small angular acceptance and excessive fundamental and harmonic beam walk-offs which limit the use of tight focusing and long interaction length for improving conversion efficiency.

To avoid walk-off and increase angular aperture, angular noncritical phase matching (θ-NCPM) by propagation along principal dielectric axes is desired. Although NCPM SHG is limited to a few discrete wavelengths determined by crystal birefringence and dispersion, sum frequency generation (SFG) allows for some tuning with a tunable laser. For organics, this process is, however, often limited by material absorption. With the compensating effect of a CT transfer band at the blue and UV and vibronic absorption in the near IR, organic crystals usually allow for wavelength quasi-noncritical phase matching (λ-NCPM) in the near IR (1.5 μm) (Zyss et al., 1981). This is an advantage for SHG of ultrashort and tunable IR sources (Zyss et al., 1981). The relatively large and anisotropic thermal expansion of molecular crystals, their low melting points, and potential thermochromic properties can lead to poor temperature bandwidths (ΔTl) for frequency conversion processes. In fact, small overtone absorption of near IR fundamental in combination with poor ΔTl have been suggested as the source of saturation in conversion efficiency at high pump power. On the other hand, the relatively large temperature dependence of the phase-matching angle ($d\theta_{pm}/dT$) can be viewed as an advantage (Kotler et al., 1992; Morita and Vidakovic, 1992) because it allows for temperature tuning of a NCPM wavelength.

Physical Properties

The physical properties of organic molecular crystals are generally less desirable. Because they are held together by dispersion force, their mechanical strengths are considerably less than that of the inorganics which have Vickers hardnesses, typically, in the low tens and melting points often below 150°C. This often leads to difficulty in fabrication and handling. One approach is to use an ionic nonlinear moiety to form salts (Kotler et al., 1992) which are expected to have improved mechanical strength due to ionic force and improved thermal conductivity. One such ionic crystal, deuterate NPNa, has been used for the intracavity frequency doubling of diode-pumped CW Nd:YVO$_4$ laser (Minemoto et al., 1993). This is an application which requires well-polished surfaces and high thermal loading. An impressive 4.4 mW of green light has been reported. The ionic approach may, however, give rise to undesirable hygroscopicity.

Chemical and photochemical stability must also be considered for each material but, generally, due to the limited size and relatively high oxidation potential of materials which satisfy the transparency requirement, they have reasonably good stability. Reported optical damage thresholds (see Table 2) are typically in the order of 0.1–1 GW/cm^2 for 10 ns pulses at 1.064 μm and much lower for 533 nm. These values are about an order of magnitude below that of the inorganics (except for those prone to photorefractive damage) and long-term, multishot damage data are generally lacking. For short pulses, the damage thresholds are usually considerably higher. Degradation due to chemical decomposition is potentially an issue due to the low melting point of many organic crystals.

Crystals

Over 60% of the entries in Table 2 are nitrobenzene or nitropyridine derivatives. The majority of them have amino donors. Although these molecular structures insure a sizable molecular nonlinearity, they also have limited transparency in the blue region. NPP, PNP, NPAN, and, probably, DMNP crystallize in near optimum crystal structures for frequency conversion. Their figures of merit for frequency conversion (d^2/n^3) are nearly two orders of magnitude higher than that of the best inorganic materials. It is interesting to note the large difference in macroscopic nonlinearity between NPP and PNP because the two materials have very similar molecular and crystalline structures (Twieg and Dirk, 1986). The blue shift of the optical spectrum in PNP (20 nm in solution) leads to a predicted (by the two-level model) decrease in macroscopic nonlinearity on the order of < 25%. Molecular studies of *p*-nitroanisole and its 5-nitropyridine derivative have found a 50% decrease in hyperpolarizability accompanying the spectral blue shift as predicted by Dewar's rule (Dewar, 1950). Evidently, the 50% decrease in macroscopic nonlinearity for the amino derivative agrees with molecular results. The comparison between NPP and PNP represents a rare opportunity in which the nonlinearity and transparency trade-off is evident in the solid state. The highly nonlinear and well-developed NPP has, however, not received wide acceptance as a replacement for inorganics despite its commercially availability. Secondary properties, as mentioned above, must be considered.

Unlike SHG, optical parametric oscillation (OPO) using an organic single crystal has been reported only for urea (Rosker et al., 1985) and NPP (Josse et al., 1992). Urea is of interest primarily due to its UV transparency. Due to the low nonlinearity of urea, however, high pump power is needed for stable operation at several times threshold. Multishot laser damage has prevented practical devices. For NPP, the large nonlinearity has allowed demonstration of an extremely low-threshold nanosecond OPO. This OPO is pumped at an awkward wavelength of 593 nm. For CT organics, the presence of a CT band leads to a rapidly deteriorating optical damage threshold with decreasing wavelength. This and the vibronic absorption cut-off prevent high-power pumping with convenient sources, such as the fundamental and harmonics of Nd:YAG laser. Significant overtone absorption of the near IR idler is also expected.

Of particular interest in frequency conversion of low-power lasers, single crystal fibers of *m*NA, MNA, DAN, and DMNP can be melt grown in glass capillaries. For *m*NA, the crystal orientation is unfavorable for frequency conversion (Tomaru et al., 1984). Control of the crystal orientation in the case of MNA has been demonstrated (Murayama et al., 1992). For DAN and DMNP, the optical x axis lies along the fiber axis, allowing the use of the large d_{II} (Kerkoc et al., 1989; Harada et al., 1990). Voids between crystal core and cladding have led to poor phase matching in DAN fibers (Kerkoc et al., 1989). The DMNP fiber is reported to have good optical quality. Simultaneous generation of red, green, and blue light in a crystal core fiber of DMNP by Cerenkov phase-matched SHG and SFM of 890 nm and 1300 nm diodes has been demonstrated (Hyuga et al., 1993). The X-propagating NCPM wavelength of DMNP is at 944 nm which is attractive from the viewpoint of direct blue light generation from diode or 946 nm Nd:YAG outputs. However, the 472-nm NCPM second-harmonic is quite close to the absorption cut-off. Noncritical intracavity sum frequency mixing of an 844-nm diode and diode-pumped 1.064 μm YAG was recently reported (Hyuga et al., 1993).

DHDPS, MHBA, APDA, and the chalcone derivatives are relatively recent materials developed for their blue transparency. They are CT derivatives with reduced strength acceptors in the form of ketone and sulfone groups. The chalcones clearly have two dimensional hyperpolarizabilities and include a more extended styrene conjugation. Their absorption cut-offs are red shifted from that of MHBA and APDA. Although their crystal structures are not optimum for frequency conversion, sizable nonlinear coefficients are found. Increasing attention is given to improving transparency, and many recent activities have been focusing on reducing the CT transfer interaction by using weak acceptors such as CN (Nogami et al., 1989), CO_2R (Nogami et al., 1989), COR (Cockerham et al., 1991; Kawamata and Inoke, 1991; 1992), and COR_2, (Itoh, 1989; Herning, 1989), weaker donors such as I, Br (Zhang et al., 1990; Canon, 1989; Chemical, 1990), CH_3 (Bailey et al., 1989), and OCH_3 (Zhang et al., 1990), or no donor at all (Sakai et al., 1990). Smaller π systems, such as methylene derivatives, are being investigated (Nogami et al., 1989). Although most of these investigations have not gone beyond scouting studies with the Kurtz powder SHG test, judging from the fruitful activities on nitroaniline derivatives, well-characterized crystals with perhaps more moderate nonlinearity, but significantly improved transparency, can be expected.

Single Crystals for Electro-Optics

The low dielectric constant of organic solids coupled with the ultrafast electronic nature of their electro-optic responses have driven much of the material research. However, investigation of single crystalline materials has not been as intense as in the case of polymeric materials which will be reviewed in the next section. In fact, detailed investigations of the electro-optic properties of organic single crystals are far fewer than those for harmonic generation. Even fewer crystals have been developed primarily for their electro-optic properties.

Material Requirements

A recent review (Bosshard et al., 1993) has identified only a dozen crystals whose r coefficients have been measured. The low number of investigations is unexpected because the adaptation of hyperpolarizabilities to electro-optic applications is most suited for the organic approach. The material property requirements for electro-optics are less stringent than for frequency conversion. The transparency requirement, which limits the available nonlinearity to a few times that of the inorganics for frequency conversion, is much relaxed for electro-optics because, typically, only radiation in the near IR (830 nm, 1.3 μm, or 1.5 μm) is involved, and, often, only a single wavelength is used, requiring transparency at a much narrower spectral region. This tolerant device requirement, thus, allows the usage of larger π systems in the molecular engineering. Long-wavelength absorbing materials can be developed to exploit both the nonlinearity-transparency trade-off and the potential resonance enhancement. In addition, because this process is always phase-matched, the optimized crystalline structure calls for parallel alignment of the CT axis for all molecules of the unit cell, which is a much more efficient way of utilizing hyperpolarizability (note that, in Eq. (6.75), only a factor of $2/3\sqrt{3}$ of β_{zzz} is utilized to satisfy phase matching). Unlike frequency conversion, where high optical density is necessary for efficiency, Eq. (6.17) shows that the electro-optic phase shift is independent of the optical intensity, thus, greatly moderating the material requirement for optical damage and photochemical stability.

Material Status

Despite the arguments above, only four of the characterized materials listed in Table 2 have conjugations extending beyond a single phenyl ring. It may be that the more colored materials are difficult to screen with the powder SHG test which is most conveniently carried out with 1.06 μm lasers. A recently published technique (Kiguchi et al., 1993), which uses a prism-coupled evanescent field to minimize harmonic absorption, may be helpful. Moreover, good quality single crystals with an extended π system are difficult to obtain due to the high degree of flexibility and isomorphism in their molecular structures. Other disincentives include the difficulty of device fabrication. Research on molecular crystal microfabrication, such as creating optical waveguides, is only beginning. Despite the limited effort, current materials have shown considerable promise. Materials with figures of merit n^3r, comparable to the best of the inorganics, have been identified. When the low dielectric constant and ultrafast response are factored in, organics offers unmatched device figures of merit.

Again, the oriented gas model for one-dimensional nonlinearity can be used to estimate the electro-optic coefficients of optimized materials. The optimal crystal structures for electro-optics (Zyss and Oudar, 1982) belong to point groups 1, 2, m, and $mm2$. Their directional cosines are given by 1, $\cos^3\theta$, $\sin^3\theta$, and $\cos^3\theta$, respectively, where (ψ, θ) are optimized at (ψ, θ), (ψ, $0°$), (ψ, $90°$), and (ψ, $0°$), respectively. The birefringence nonphase-matchable coefficient for all cases is simply given by

$$d_{ZZZ} = \frac{1}{2} N f_Z^{2\omega} f_Z^{\omega} f_Z^{\omega} \beta_{zzz}. \qquad (6.76)$$

Combining Eqs. (6.43) and (6.76) and, assuming that the local field factors cancel, yields

$$r_{ZZ} = \frac{-8N\pi\beta_{zzz}}{n_g^4} \frac{[3 - (\lambda_{max}/\lambda_{EO})^2][1 - 4(\lambda_{max}/\lambda_{SHG})^2][1 - (\lambda_{max}/\lambda_{SHG})^2]}{3[1 - (\lambda_{max}/\lambda_{EO})^2]^2}. \quad (6.77)$$

Assuming that $N = 5 \times 10^{21}$ cm^{-3}, $n_g = 1.6$, $\lambda_{max} = 630$ nm, and $\beta_{zzz} = 1 \times 10^{-27}$ esu (1.9 µm EFISH results from Table 1), r_{33} at 1.0 µm is estimated to be 2.5×10^{-5} esu (10,400 pm/V), which is over an order of magnitude higher then the best known values. Therefore, in contrast to the nearly optimized material status for crystals suitable for frequency conversion, much higher nonlinearity can be expected from future crystals for electro-optics.

Crystals

Among the crystals whose electro-optic properties have been studied, only MNA (Lipscomb et al., 1981), DCNP (Allen et al., 1988), NMBA (Bailey et al., 1992), MNBA (Knopfle et al., 1994), and DAST (Marder et al., 1989; Perry et al., 1991) have near ideal crystal structures, yielding greater than 80% of the alignment optimized electro-optic coefficients according to the oriented gas model. In addition, the molecular hyperpolarizability of MNA is within 75% of the highest measured value for CT benzene derivatives; that of DCNP, due to the potent tertiary amine donor and bis-cyano acceptor, is among the highest of the push-pull butadiene derivatives; and that of DAST is also among the highest values reported for a "stilbene" type phenyl-=-phenyl conjugation. Therefore, their electro-optic coefficients ($r_{11} = 67$ pm/V for MNA, $r_{33} = 87$ pm/V for DCNP, and $r_{33} = 400$ pm/V for DAST) can be considered as substantially optimized for their molecular classes.

NMBA gives a rather low r_{11} of 25 pm/V because NMBA lacks an effective donor and the aza substitution within the conjugation also contributes to a low hyperpolarizability. MNBA also has a somewhat reduced hyperpolarizability due to the aza and acetamino substitution (see Table 1 for the effects of the latter substituent). SPCD (Yoshimura, 1987) contains the same cationic chromophore as DAST but the crystalline packing of SPCD is much less efficient (about 55% of optimum). The high r_{33} value of SPCD in comparison to DAST is likely due to resonance enhancement because the measurement was carried out near the optical cut-off. MMONS (Bierlein et al., 1990) gives an attractive r_{33} value of 40 pm/V despite an inefficient (55%) crystal structure.

Numerous other factors affects the device potential of an electro-optic material. For single crystal materials, their crystal growth properties are of key importance. As with other inorganic materials, single crystal devices are expensive to produce because of fabrication difficulties. The cost of fabrication is increased by the small supply of available samples. The extreme difficulty of growing MNA has, undoubtedly, hampered its development since its early discovery. Nevertheless significant progress has been reported in growing single crystal fibers and thin films with some degree of orientation control. SPCD also appears to be difficult to grow. Only micron thin film crystals of mm dimensions have been reported. DAST crystallizes quite readily. Crystals with cm dimensions have been grown, although not free of internal defects. Large crystals of DCNP, MMONS, NMBA, and MNBA can be readily grown by slow evaporation of solutions. The sizable coefficients and availability of large crystals should permit more detailed studies of other device-related issues.

6.6 Poled-Polymer Studies

Property Considerations

To exhibit second-order nonlinear optical effects, the isotropic symmetry in amorphous polymers must be broken. For a poled polymer, an acentricity is imparted with a poling electric field. If a polar alignment of the dipolar species in the polymer occurs during poling, this acentricity can be maintained, upon the removal of the poling field, for a time period determined by the orientation relaxation time of the dipoles. For maximum alignment, poling can be carried out near the polymer glass-transition temperature (T_g) where the dipolar orientation time is short. The polar alignment is then dynamically "frozen" into the polymer by cooling under the poling field. At the reduced temperature, the relaxation of the polar order slows considerably. If this relaxation can be reduced to an acceptable minimum, an amorphous material with quasi-permanent, second-order nonlinearity is obtained.

Poled polymers represents an attractive alternative to single crystal materials for nonlinear optics. The poled-polymer approach offers more rational controls in the organization of molecular nonlinearity. Macroscopic activity is no longer dependent on intractable crystallization forces. When chromophores are incorporated into a polymeric matrix, optically nonlinear thin films and fibers can be easily obtained, allowing for large area coverage as well as long interaction length. Existing microfabrication technology on polymers, such as lithographic and reactive ion etching techniques, may also be extended to these materials. By covalently linking the chromophores into a polymeric network, high number density and, thus, high nonlinearity is possible. The optical quality of the polymer can also be optimized through compositional designs. Together with the ultrafast nature of the nonlinear optical response, these advantages have led many to believe that poled polymers are the most promising organic nonlinar optical materials to realize wide-spread application.

The following considerations are necessary during the design of a poled polymer. Beginning with the choice of a suitable chromophore, application requirements for nonlinear optical, linear optical, chemical, photochemical, and other properties are considered. At a minimum, the chromophore and the polymer must be optically transparent at the application wavelengths. To optimized poled nonlinearity, the vectorial product of dipole moment and hyperpolarizability $\mu\beta_\mu$ and the chromophore number density must be maximized. The polymer matrix is generally chosen for synthetic accessibility and certain desired polymer properties, such as good optical quality, dielectric strength, and mechanical properties. A high T_g is also sought because of the strong correlation between temporal stability of the poled nonlinearity and T_g. A high T_g is also required for compatibility with well-established manufacturing steps in the electronic industry. Of practical importance, the polymer must also be processible into thin film and multilayer structures. Together, these properties represent a set of rather stringent material requirements and, despite much progress, they have yet to be fully met by any existing material. To help assess current materials, some of these requirements will be discussed further.

Number Density

The simplest way of incorporating chromophores into a polymer is by doping (Singer et al., 1986). The chromophore and polymer, thus, coexist in a guest-host environment. The maximum dopant concentration depends on solubility which, in turn, depends on the polar nature of the guest and host. Typically, less than 20 wt% (number density between 2–3×10^{20} cm^{-3}) can be achieved without phase separation. If the chromophores are covalently linked to the polymer backbone, much higher chromophore content is possible (Esselin et al., 1988). Depending on the relative molecular weight of the chromophoric and polymeric fragments, a chromophore content in excess of 80 wt% is possible in some homopolymers. However, copolymerization is often necessary to maintain solubility and increase molecular weight. To insure a high degree of cross-linking, multifunctionalized chromophores and cross-linking groups are used for cross-linked networks and this reduces the number density (high number density can be maintained if the cross-linking groups are functionalized directly on the chromophore). Typically, a number density of 5–15×10^{20} cm^{-3} is achieved by covalent linking.

Poling

Although the dielectric strengths of high molecular weight polymers are as high as several MV/cm, typical poling fields reported on the electrode poling of highly functionalized nonlinear optical polymers are around 0.5–1 MV/cm. Although corona poling may achieve a higher poling field, its current limiting characteristic often prevents its use for high T_g materials due to their ionic conductivity at poling temperatures. For cross-linked materials, high ionic conductivity is also observed during much of the thermosetting reaction. Although ionic conductivity may not be intrinsic and may be eliminated by purification, poling with high fields at temperatures in excess of 250°C remains a difficult task. In addition, the issue of electrochemical degradation of the chromophore under high field and high current poling conditions has not been fully addressed. It may limit the poling time and temperature and the choice of chromophore. As alternatives to thermally activated corona or contact poling, other techniques have been reported including the use of a plasticizing gas (Barry and Soane, 1991) or photo-induced cis-trans isomerization to assist orientational mobility (Sekkat and Dumont, 1992; Blanchard and Mitchell, 1993). These low-temperature techniques may offer potential advantages including higher poling efficiency through the Boltzmann factor in Eq. (6.29) and high poling fields in the of poling of material whose ionic conductivity is excessive at temperatures near its T_g. Poling by charge carrier trapping (Yitzchaik et al., 1991) and charge carrier injection with electron beam (Yang et al., 1994) have also been reported. Deeply trapped carriers may improve the temporal stability of the polar order. However, the generality of these techniques has not been established.

Potential Nonlinearity

As before, the oriented gas model provides an estimate of the potential nonlinearity of poled polymers. For polymers containing molecules with one-dimensional and two-level nonlinearity, their second-harmonic generation coefficients are given by Eqs. (6.26) and (6.29). For SHG applications, given typical values for number density = 1×10^{21} cm^{-3}, local field factors = 5, order parameter $\langle \cos^3\theta \rangle$ = 0.2, and β_μ(1.06 µm) = 50 × 10^{-30} esu for optimized optically transparent molecules, a $d_{ZZZ} \approx 2.5 \times 10^{-8}$ esu (10 pm/v) can be expected. This value is quite modest when compared to the best nonphase-matchable coefficients of inorganic single crystals (for example, d_{33} = 34 pm/v for LiNbO$_3$). As seen in the single crystal studies, the transparency requirement greatly limits the available nonlinearity for the organic approach. Given the lower number density and lower orientational order, the poled-polymer approach becomes particularly challenging. All material parameters must be optimized to develop competitive poled polymers for SHG. Higher d coefficients can be obtained by exploiting resonance enhancement. Phase matching by anomalous dispersion (Cahill et al., 1989), that is, having the CT transition situated between the fundamental and harmonic wavelengths, has been attempted to use more colored chromophores with high nonlinearity. Other approaches, such as the use of noncollinear phase-matching with Cherenkov radiation (Sugihara et al., 1991) and surface emitted SHG with counter-propagating fundamental waves (Otomo et al., 1993) have also been reported.

TABLE 6.3 Experimental Results of Poled Polymer Studies: M_w, molecular weight; ρ#, chromophore number density; α(wavelength), optical propagation loss; τ_1 and τ_2, relaxation time in $d_{eff}(t)/d_{eff}(0)$ = $Ae^{-t/\tau_1} + Be^{-t/\tau_2}$.

Structure and Nomenclature (Ref)	Properties
Guest-host polymer composites (4-[N-ethyl-N-(2-hydroxyethyl)]amino-4'-nitroazobenzene) (DR1) doped in poly-(methylmethacrylate) (PMMA) (Singer et al., 1986)	$T_g \approx 100$ °C. $\lambda_{max} \approx 470$ nm. $\rho^\# = 2.4 \times 10^{20}$ cm^{-3}. low dielectric constant: ϵ = 3.6. contact poled with 62 V/µm at 100 °C. d_{33}(1.58 µm) = 2.5 pm/V. Novelty: first study of amorphous guest-host sytem.

TABLE 6.3 Experimental Results of Poled Polymer Studies—(*continued*)

Structure and Nomenclature (Ref)	Properties
4-(4'-cyanophenylazo)-N,N-bis-(methoxycarbonylmethyl)-aniline doped in copolymer of vinylidene flouride and trifluoroethylene (Foraflon®, 70:30 mol %) (Pantelis et al., 1988; Hill et al., 1987)	$T_g \approx 100$ °C. $\lambda_{max} = 400$ nm. Up to 10 wt% doping. $\alpha(1 \ \mu m) \approx -1.5$ dB/cm. Corona poled at 25 °C. d_{33} (1.06 μm) up to 2.6 pm/V. Stable nonlinearity after 300 days at ambient condition. Novelty: ferroelectric host polymer provides a stable internal field of 150V/μm.
4-N,N-dimethylamino-4'-nitrostilbene doped in thermosetting epoxy (EPO-TEK® 301-2) (Hubbard et al., 1989)	Precure at 80 °C prior to poling. $\lambda_{max} \approx 430$ nm. $\rho^{\#} = 0.2 \times 10^{20}$ cm^{-3}. Contact poled at 60 V/μm. Temporal stability: τ_1 (25 °C) = 7 days and τ_2 (25 °C) = 72 days. Novelty: use of crosslinked polymer as host.
p-nitrophenol doped in type A gelatin (Ho et al., 1992)	$T_g \approx 60$–70 °C. $\lambda_{max} < 350$ nm. Up to 35 wt% doping. Spin coating from aqueous solution. Contact poled at 40 V/μm. r_{33} (633 nm) = 10–40 pm/V. 40% activity remains after 5 days. Novelty: use of cross-linked biopolymer as host.
(dicyanomethylene)-2-methyl-6-(p-dimethylaminostyryl-4H-pyran (DCM) doped in polyimide (Amoco Ultradel®) (Ermer et al., 1992)	$T_g = 220$ °C $\lambda_{max} = 474$ nm. Doping level: 20 wt%. n (830 nm) = 1.651. α (830 nm) ≈ -1.5 dB/cm. Multilayers contact poled with 312 V/ $\bar{\mu}$ 190 °C. r_{33} (830 nm) = 3.4 pm/V. Low poling field at NLO layer due to high conductivity. Novelty: use of high T_g host polymer.
2,4,5-triarylimidazole derivative (lophine) doped in polyimide (Ultem®) (Stahelin et al., 1992a)	$T_g = 210$-150 °C for 0–35 wt%. $\lambda_{max} = 410$ nm. Chromophore: $\mu_g = 7 \times 10^{-18}$, $\beta_0 = 18 \times 10^{-30}$ (esu). High loading, up to 35 wt%. For 0–35 wt% doping, corona poling gives d_{33} (1.047 μm) = 6–17 pm/V. For 20 wt% doping, corona poling gives $T_g = 180$ °C, $d_{33} = 10.5$ pm/v, τ (80 °C) = 1.5 year. Novelty: use of thermally stable chromophore.

TABLE 6.3 Experimental Results of Poled Polymer Studies—(*continued*)

Structure and Nomenclature (Ref)	Properties
DR1 doped in phenylsiloxane polymer (Allied Signal Accuglass 204®) (Jeng et al., 1992a)	λ_{max} = 493 nm. n (533 nm) = 1.628. Doping level: 35 mg dye in 2g of A204. Corona poled at 200 °C for 10 min. d_{33} (1.06 μm) = 11.43 pm/V. d_{33} (40 hrs, 100 °C) = 3.8 pm/V. Novelty: Use of sol-gel as host polymer.
Side-Chain acrylate polymers Methylmethacrylate (MMA)/DR1 functionalized methacrylate copolymer (Esselin et al., 1988)	T_g = 128–134 °C for 1–19 mol %. λ_{max} = 470 nm. $\rho^{\#}$: up to 7.5 × 10^{20} cm^{-3}. Contact poled with 90 V/μm at 130 °C. d_{33} (1.064 μm) = 3–58 pm/V for 1–19 mol%. Novelty: high $\rho^{\#}$.
MMA/4-dicyanovinyl-4'-(N,N-dialkylamino)azobenzene functionalized methacrylate copolymer (Singer et al., 1988)	T_g = 127 °C. $\rho^{\#}$ = 8 × 10^{20} cm^{-3}. n (800 nm) = 1.58. Corona poled above T_g. d_{33} (1.58 μm) = 21 pm/V. r_{33} (799 nm) = 15 pm/V. 90% of activity remains stable after 35 days at ambient conditions.
Poly-4-(4'-nitrophenylazo)-N-methyl-N-(2-acroyloxyethyl)aniline (Hill et al., 1989; 1988)	T_g = 105 °C. λ_{max} ≈ 470 nm. M_W = 4.9 × 10^3. Contact poling at 190 V/μm gives d_{33} (1.06 μm) = 55 pm/V. Contact poling at 20 V/μm gives r_{33} (633 nm) = 30 pm/V. Activity remains stable for over 2 years at ambient conditions.
MMA/4-(N-ethyl-N-2-methacroyloxyethoxy)-2-methyl-4'-nitroazobenzene copolymer (Ore et al., 1989)	λ_{max} = 486 nm. M_w = 10 × 10^3. $\rho^{\#}$ = 9.2 × 10^{20} cm^{-3}. Corona poled d_{33} (1.06 μm) = 41 pm/V. Stable nonlinearity after 75 days at ambient conditions.
Poly-N-(2-methacroyloxyethyl)-N-methyl-4'-nitroaniline (Hayashi et al., 1992)	T_g = 100 °C. λ_{max} = 390 nm. M_w = 11–17 × 10^3. n (633 nm) = 1.66. Corona poled d_{33} (1.58 μm) = 30 pm/V. Activity decays to 65% in 40 days at ambient conditions.

TABLE 6.3 Experimental Results of Poled Polymer Studies—(*continued*)

Structure and Nomenclature (Ref)	Properties
MMA/4-*N*-(2-methacroyloxyethyl)-*N*-ethyl-4′-aminophenylazo-4″-nitroazobenzene copolymer (Amano et al., 1990)	$T_g \approx 100$ °C. $\lambda_{max} = 500$ nm. $\rho^{\#} = 4.3 \times 10^{20}$ cm^{-3}. Corona poled at 100 °C. d_{33} (1.06 μm) = 142 pm/V. d_{33} (1.7 μm) = 70 pm/V.
MMA/2,5-dimethyl-4-*N*-(2-methacroyloxyethyl)-*N*-ethyl-4′-aminophenylazo-4″-dicyanoazobenzene copolymer (Shuto et al., 1991)	$T_g \approx 140$ °C. $\lambda_{max} = 510$ nm. $\rho^{\#} = 4 \times 10^{20}$ cm^{-3}. α (633 nm) = -50 dB/cm due to absorption. Corona poling with 200 V/μm at 140 °C gives d_{33} (1.06 μm) = 417 pm/V. Contact poling at 150 V/μm at 140 °C gives r_{33} (633 nm) = 40 pm/V. Excellent temporal stability at 80 °C.

MMA/4-(*N*-methyl-*N*-2-methacroyloxyethoxy)-4′-cyanoabenzene homo- and copolymers (S'heeren et al., 1993a)

Corona poled near T_g.
d_{33} values are measured 10 days after poling at 1.064 μm.

dye (mol%)	T_g (°C)	Mw (×10^3)	d_{33} (pm/V)
12	129	59	5.2
16	128	44	21
32	134	59	68
44	128	39	45
49	125	98	32
72	124	36	31
100	181	15	26

MMA/4-(*N*-methyl-*N*-2-methacroyloxyethoxy)-4′-nitrostilbene co-polymers (S'heeren et al., 1993c)

X = H 1
   CN 2

Corona poled near T_g.
d_{33} values are measured several days after poling at 1.064 μm.

	dye (mol%)	T_g (°C)	Mw (×10^3)	d_{33} (pm/V)
1	11	122	69	18
	19	120	70	31
	31	124	69	44
	39	126	66	15
	52	120	48	66
2	9	124	56	22
	18	131	62	41
	33	126	60	20

TABLE 6.3 Experimental Results of Poled Polymer Studies—*(continued)*

Structure and Nomenclature (Ref)	Properties
MMA/*N*-(3-methacryloxyalkyl)-7-diethylaminocoumarin-3-carboxamide (76:23 mol%) copolymer (Mortazavi et al., 1991)	T_g = 135 °C. λ_{max} = 410 nm. $\rho^{\#}$ = 9.8 × 10²⁰ cm⁻³. M_w = 89 × 10³. Corona poled for SHG and contact poled with 100 V/μm for EO at 160 °C. d_{33} (1.06 μm) = 13 pm/V. r_{33} (477–1100 nm) = 2–12 pm/V. d_{33} decays by 25% at 100 °C in 50 hours.
Isobornyl methacrylate/*N*-(3-methacryloxyalkyl)-7-diethylaminocoumarin-3-carboxamide (76:23 mol%) copolymer (Mortazavi et al., 1991)	T_g = 170 °C. λ_{max} = 410 nm. M_w = 50 × 10³. Corona poled at 200 °C. d_{33} (1.06 μm) = 11 pm/V. d_{33} decays by 10% at 100 °C and by 40% at 140 °C in 50 hours.
Poly(*p*-*N*-(2-methacroyloxyethyl)-*N*-ethylaminobenzall-1-3-diethyl (or diphenyl thiobarbituric acid) (Cheng and Tan, 1993)	Corona poled near T_g. d_{33} and r_{33} values are measured at 1.064 μm. At 100 °C d_{33} of A(B) decays by 80% (10%) in 150 mins (15 days)

	T_g (°C)	M_w (×10³)	d_{33} (pm/V)	r_{33} (pm/V)
A	125	74	33	14
B	194	60	40	18

Structure and Nomenclature (Ref)	Properties
Poly-4-(6-acroyloxyhexyloxy)-4′-nitrostilbene (Huijts et al., 1989b)	T_g = 65 °C. λ_{max} = 380 nm. $\rho^{\#}$: 18 × 10²⁰ cm⁻³. n (633 nm) = 1.62. α (1.3 μm) ≈ −1.5 dB/cm. Contact poled with 23 V/μm at 65 °C. r_{33} (633 nm) = 0.9 pm/V.
Poly-4′-*N*-(6-methacroyloxyhexyl)-*N*-methyl-amino-4-methylsulfonylazobenzene (Robello et al., 1991a; 1992)	T_g = 99 °C. λ_{max} = 446 nm. $\rho^{\#}$ = 17 × 10²⁰ cm⁻³. M_w = 89 × 10³. n (633 nm) = 1.76. α (830 nm) = −1 dB/cm. Contact poled with 90V/μm at 100 °C. r_{33} (633 nm) = 39 pm/V; r_{33} (860 nm) = 13 pm/V.
MMA/4′-(6-methacroyloxyhexyloxy)-4-methylsulfonylstilbene (55:45 wt%) copolymer (Rikken et al., 1991; Seppen et al., 1991)	T_g = 117 °C. λ_{max} = 355 nm. Low absorption at 420 nm. Corona poled with 120 V/μm at 100 °C. d_{33} (820 nm) = 9 pm/V. Activity decreases to a quasi-stable 70% value after 120 days at ambient. Waveguide formation by UV photobleaching.

TABLE 6.3 Experimental Results of Poled Polymer Studies—*(continued)*

Structure and Nomenclature (Ref)	Properties
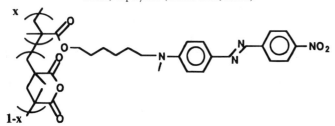 MMA/4′-(6-methacroyloxyhexylsulfonyl)-4-N,N dimethylamino-biphenyl (50:50 wt%) copolymer (Rikken et al., 1992)	$T_g \approx 100$ °C. $\lambda_{max} = 340$ nm. Corona poled with 120 V/μm at 100 °C. d_{33} (820 nm) = 45 pm/V. Poor temporal stability at 60 °C.
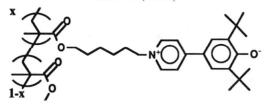 MMA/4-(4′-nitrophenylazo)-N-methyl-N-(6-methacroyloxyhexyl)aniline (81:19 mol%) copolymer (Muller et al., 1992a)	$T_g = 104$ °C. $\lambda_{max} \approx 470$ nm. $M_w = 100 \times 10^3$. Thermally decomposes at 267 °C. Poled with 110 V/μm. $r_{33} - r_{13}$ (1.3 μm) = 9 pm/V.
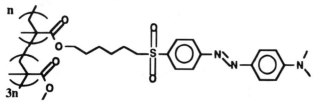 Methacrylic anhydride/4-(4′-nitrophenylazo)-N-methyl-N-(6-methacroyloxyhexyl)aniline (33:67 mol%) copolymer (Strohriegl, 1993; Muller et al., 1992a)	$T_g = 90$ °C. $\lambda_{max} \approx 470$ nm. Thermally decomposes at 227 °C. Poled with 110 V/μm. $r_{33} - r_{13}$ (1.5 μm) = 19 pm/V.
MMA/[2,6-di-*tert*-butyl-4-(1-ω-methacroyloxyalyl)-4-pyridino]phenolates (85:15 mol%) copolymer (Combellas et al., 1992)	$T_g = 130$ °C. $\lambda_{max} \approx 525$ nm. Contains zwitterionic chromophore with $\beta_0 = -30 \times 10^{-30}$ esu. Soluble in THF.
MMA/4′-(6-methacroyloxyhexylsulfonyl)-4-N,N-dimethylamino-azobenzene (25 mol%) copolymer (Xu et al., 1993)	$T_g = 124$ °C. $\lambda_{max} = 437$ nm. $M_w = 17 \times 10^3$. Corona poled at 125 °C for 45 mins. d_{33} (1.06 μm) = 100 pm/V. Activity stabilizes at 95% value after 10 days at ambient conditions.

Second-Order Nonlinear Optical Properties of Organic Materials

TABLE 6.3 Experimental Results of Poled Polymer Studies—(*continued*)

Structure and Nomenclature (Ref)	Properties

Functionalized homo- and copolymers of isocyanatoethyl methacrylate, MMA, and dimethylaminoethyl methacrylate. (Cheng et al., 1992)

Corona poled near T_g.
d_{33} values are measured at 1.064 μm.
d_{33} of A & C are stable at room temperature. Polymer B is stable at 80 °C.
$\alpha(1.06\ \mu m) \approx -(1.5 - 3)$ dB/cm.

(x, y, z)	T_g (°C)	$\rho^{\#}$ (10^{20}/cm^3)	d_{33} (pm/V)
A(1, 1, 0)	120	6.8	19
B(1, 0, 0)	130	13	44
C(1, 0, 0)	140	13	88
C(1, 0, 1)	130	9.6	121
C(1, 0, .2)	135	12	90

Side-Chain Non-acrylate Polymers

Poly-4-[*N*-(4′-nitrophenyl)amino-methyl]ethylene (Eich et al., 1989b)

$T_g = 125$ °C.
$\lambda_{max} \approx 390$ nm.
Thermally decomposes at 260 °C.
Corona poled at 140 °C.
d_{33} (1.06μm) = 31 pm/V.
d_{33} relaxes to 19 pm/V in 5 days.

Poly-4-[*N*-methyl-N-(4′-nitrophenyl)amino-methyl]styrene (Hayashi, 1991)

$T_g = 103$ °C.
$\lambda_{max} = 393$ nm.
$M_w = 12 \times 10^3$.
$n(633\ nm) = 1.73$.
$\alpha(633\ nm) = -10$ dB/cm.
Corona poled at 110 °C.
d_{33} (1.06 μm) = 28 pm/V which stabilizes to 18 pm/V in 5 months at ambient conditions.

Poly-4-[N-(4′-nitrophenyl)amino-methyl]styrene (Hayashi et al., 1992)

$T_g = 123$ °C.
$\lambda_{max} = 383$ nm.
$M_w = 15 \times 10^3$.
$n(633\ nm) = 1.70$.
Corona poled at T_g.
$d_{33}(1.06\ \mu m) = 10$ pm/V
Low d_{33} due to poling-induced decomposition.

Poly-4-[*N*-(4′-cyanophenyl)amino-methyl]styrene (Hayashi et al., 1992)

$T_g = 132$ °C.
$\lambda_{max} = 288$ nm.
$M_w = 15 \times 10^3$.
$n(633\ nm) = 1.65$.
Corona poled at T_g.
$d_{33}(1.06\ \mu m) = 1$ pm/V.

Styrene/*p*-[4-nitro-4′-(N-ethyl-N-2-oxyethyl)azobenzene] methylstyrene copolymer (88:12 mol%) (Ye et al., 1987)

$T_g = 110$ °C.
$\lambda_{max} \approx 470$ nm.
Contact poled with 30 V/μm at $\approx$ 110 °C.
d_{33} (1.06 μm) = 1.1 pm/V.
Stable activity at room temperature.

TABLE 6.3 Experimental Results of Poled Polymer Studies—(*continued*)

Structure and Nomenclature (Ref)	Properties
Styrene/*N*-(4-nitrophenyl)-*S*-prolinoxy methylstyrene copolymer (64:36 mol%) (Ye et al., 1989)	T_g = 110 °C. λ_{max} ≈ 380 nm. Contact poled with 70 V/μm at ≈ 110 °C. d_{33} (1.06 μm) = 1.6 pm/V. Low d_{33} due to material impurity.
p-hydroxystyrene/*N*-(4-nitrophenyl)-*S*-prolinoxy styrene copolymer (10:90 mol%) (Ye et al., 1989)	T_g = 96 °C. λ_{max} ≈ 380 nm. $\rho^\#$ = 2.3 × 10^{21} cm^{-3}. Corona poled at T_g. d_{33} (1.06 μm) = 33 pm/V. T_g increases to 146 °C at 16 mol% functionalization.
p-hydroxystyrene/*p*-[4-(2,2-dicyanovinyl)-4'-(N-ethyl-N-2-oxyethyl)azobenzene]styrene copolymer (Ye et al., 1992)	Corona poled at 117 °C for 15 hours. Thermal and electrochemical decomposition observed at 147 °C.
Poly-(2,6-dimethylbromo-1,4-phenylene oxide) partially functionalized with *N*-(4-nitrophenyl)-*S*-prolinol (Dai et al., 1990)	T_g = 170 °C. $\rho^\#$ = 2.6 × 10^{21} cm^{-3}. n (633 nm) = 1.584. Corona poled at 190 °C for 30 min. α = −1 dB/cm. Temporal stability: τ_1(25 °C) = 0.3 days and τ_2(25 °C) = 39 days.
p-nitroaniline functionalized polyimide (Lin et al., 1992)	T_g = 236 °C. λ_{max} = 390 nm. $\rho^\#$ = 7 × 10^{20} cm^{-3}. Films are cured and corona poled at 240 °C d_{33}(1.06 μm) = 5.4 pm/V. d_{33}(24 hrs, 85 °C) ≈ 5 pm/V and stable.
Poly-(4-nitro-4'-(vinylooxyethyloxy)azobenzene, 1, and other poly-(vinyl ethers) (S'heeren et al., 1992)	

TABLE 6.3 Experimental Results of Poled Polymer Studies—(*continued*)

Structure and Nomenclature (Ref)	Properties
Poly-(4-cyano-4'-(vinylooxyethyloxy) azobenzene 2	
Poly-(4-dicyanovinyl-4'-(vinylooxyethyloxy) benzene 3	
Poly-(4-cyano-4-carbomethoxyvinyl-4'-(vinylooxyethyloxy) benzene 4	
Polyvinylalcohol/N-ethyl-N-methylamino nitroaniline copolymer (Sasaki, 1993)	$T_g = 120\ °C$. $\lambda_{max} \approx 380$ nm. Corona poled at T_g. d_{33} (1.06 μm) = 10 pm/V. d_{33} relaxes to 7 pm/V in 40 days.
DR1/Poly(maleic anhydride-co-propylene copolymer (16:84 mol%) (Bauer et al., 1993)	$T_g = 180\ °C$. $\lambda_{max} \approx 470$ nm. Corona poled at 185 °C. r_{33}(780 nm) = 6pm/V. With 3 °C/min heating rate, r_{33} is stable up to 100 °C.

	T_g (°C)	Mw (×10³)	λ_{max} (nm)	dye (mol%)
A	170	54	468	92
B	180			96

Azo dye/Poly(maleic anhydride-co-styrene or co-norbornadiene copolymer
(Ahlheim and Lehr, 1994)
Main-Chain Polymers

TABLE 6.3 Experimental Results of Poled Polymer Studies—*(continued)*

Structure and Nomenclature (Ref)	Properties
Vinylidene cyanide/vinyl acetate copolymer (50:50 mol%) (Azumai et al., 1990; Eich et al., 1988; Sato et al., 1987)	T_g = 180 °C. M_w = 470 × 10³. n(2.94 μm) = 1.434. α(2.94 μm) = 1.4 mm⁻¹. Corona poled at 180 °C for 2 hours. d_{33}(1.064 μm) = 0.3 pm/V[Eich]. d_{eff}(1.94 μm) = 117 pm/V[Azumai]. Note: large discrepancy in SHG d coefficients.
Aromatic polyurea (Nagamori et al., 1992; Kajikawa et al., 1991)	Thin film prepared by vapor deposition polymerization. T_g = > 150 °C. $\lambda_{cut-off}$ = 400 nm. Corona poled at 180 °C for 3 minutes. d_{33}(1.06 μm) = 1.7 pm/V. Negligible relaxation of activity over 2 months at ambient conditions. Activity remains stable after short-term heating to 200 °C.
N-phenylated polyurea (Nalwa et al., 1993a; Nalwa et al., 1993b; Azumai et al., 1990; Sato et al., 1987)	T_g = 123 °C. λ_{max} = 253 nm. n(633 nm) = 1.577. α(633 nm) = −1.2 dB/cm. Corona poled at 130 °C for 1 hour. d_{33}(1.06 μm) = 5.5 pm/V. 90% of activity remains stable after 40 days at ambient.
Epoxy polymers containing 4-amino-4'-nitroazobenzene (Teraoka et al., 1991]	Polymers prepared by melt condensation at 150 °C under N_2. d_{33} values are measured at 1.064 μm during corona poling.

	1	2
T_g(°C)	86	77
λ_{max}(nm)	392	412
Mw(×10³)	7	5
Poling temp. (°C)	100	85
d_{33}(pm/V)	31	12
τ(hour)	24000	770
β	0.32	0.34

Epoxy polymers containing p-aminonitrobenzene (Gadret et al., 1991]	T_g = 77 °C. λ_{max} = 480 nm. Corona poled after precuring. d_{33}(1.06 μm) = 25 pm/V. Stable activity at ambient conditions for >20 days. Activity decreases rapidly to zero when heated above 80 °C.

TABLE 6.3 Experimental Results of Poled Polymer Studies—(*continued*)

Structure and Nomenclature (Ref)	Properties
Epoxy polymer containing 4-amino-4′nitrotolane (Jungbauer et al., 1991)	T_g = 125 °C. λ_{max} = 418 nm. n(633 nm) = 1.71. Corona poled at 135 °C for 1 hour. d_{33}(1.06 μm) = 89 pm/V. r_{13}(633 nm) = 8 pm/V. Stable birefringence at ambient conditions. Birfringence decays with τ = 448 hour and β = 0.45 at 100 °C.
Epoxy polymers containing 4-amino-4′methylsulfonyltolane [Twieg et al., 1992]	Polymers prepared by melt condensation at 150 °C. d_{33} values are measured at 1.064 μm during corona poling. |	1	2
---	---	---
T_g (°C)	128	180
λ_{max} (nm)	366	366
Mw (×10³)	28	
d_{33} (pm/V)	15	30	
Epoxy polymers containing 4-amino-4′-nitroazobenzene [Jeng et al., 1992c]	T_g = 115 °C. λ_{max} = 461 nm. n (533 nm) = 1.718. Corona poled at 115 °C for 1 hour. d_{33} (1.06 μm) = 34 pm/V. 70% of SHG activity remains stable after 20 days at ambient conditions.

TABLE 6.3 Experimental Results of Poled Polymer Studies—*(continued)*

Structure and Nomenclature (Ref)	Properties

Polyurethanes containing 4-amino-4′nitroazobenzene [Meyrueix et al., 1991a; 1991b]	Polymers prepared by melt condensation at 150 °C. $\chi_{333}(-\omega,\omega,0)$ values are measured at 830 nm after contact poling with 25 V/μm at 100 °C.
	
	  | | 1 | 2 | 3 |
	  |---|---|---|---|
	  | T_g (°C) | 74 | 94 | 116 |
	  | λ_{max} (nm) | 471 | 476 | |
	  | Mw (×10³) | 11 | 15 | 31 |
	  | n (830 nm) | 1.69 | 1.69 | 1.70 |
	  | χ_{333} (pm/V) | 14.8 | 10.3 | |

Polyurethanes containing 4-amino-4′nitroazobenzene [Chen et al., 1991]	Polymers prepared by condensation in dioxane solution. d_{33} is stabilized value measured at 1.064 μm after corona poling at about 130 °C. Activity is stable at ambient conditions.
	
	  | | 1 | 2 | 3 | 4 |
	  |---|---|---|---|---|
	  | T_g (°C) | 92 | 120 | 122 | 140 |
	  | λ_{max} (nm) | 460 | 470 | 455 | 463 |
	  | Dye (wt%) | 48 | 68 | 47 | 52 |
	  | Thickness (μm) | 0.04 | 0.05 | 0.22 | 0.28 |
	  | d_{33} (pm/V) | 108 | 250 | 119 | 223 |

Polyurethanes derived from 2,4-toluenediisocyanate/2-methyl-4-nitro-[N,N-bis(2-hydroxyethyl)]-aniline and other diols [Kitipichai et al., 1993]	In situ corona poling during melt polymerization at 120 °C. d_{33} values are measured at 1.064 μm; stabilized values after 180 days at 25 °C are given in *italics*.
	
	  | | 1 | 2 | 3 | 4 | 5 |
	  |---|---|---|---|---|---|
	  | T_g (°C) | 98 | 131 | 121 | 125 | 178 |
	  | λ_{max} (nm) | 373 | 472 | 431 | 518 | 388 |
	  | Mw (×10³) | 2.3 | 2.5 | 2.0 | 1.7 | 2.1 |
	  | d_{33} (pm/V) | 9 | 10 | 2.6 | 3.6 | 2 |
	  | | *0.5* | *5.5* | *6* | *0.9* | *2.5* |

Second-Order Nonlinear Optical Properties of Organic Materials

TABLE 6.3 Experimental Results of Poled Polymer Studies—(*continued*)

Structure and Nomenclature (Ref)	Properties

Main-chain polymers derived from (1) α-cyano-ester quinodimethanes, (2) *p*-oxy-α-cyanocinnamates, (3) *p*-thio-α-cyanocinnamates [Fuso, 1991; Hall et al., 1988; Green, 1987a; 1987b]

Polymers prepared by high temperature transesterification. Homopolymers are typically insoluble and high melting. Unit molecular nonlinearity $\mu\beta/n$ values are measured at 1.064 μm.

	1	2	3
T_g (°C)	none	<25	45
Dye (mol%)	50	67	100
Mw (×10^3)		70	
$\mu\beta/n$ (×10^{-48} esu)		140	

Main-chain homopolymers derived from (4-*N*-ethyl-*N*-(2-hydroxyethyl)amino-α-cyanocinnamate (1) and the corresponding acid (2) [Stenger–Smith et al., 1991; 1990]

Polymerization: (1) melt transesterification at 160 °C; (2) solution polycondensation at 25 °C.

	1	2
T_g (°C)	100–110	60–90
λ_{max} (nm)	442	442
Mw (×10^3)	27	5
Poling temp. (°C)	120	85–100
d_{33} (pm/V) at 1.064 μm	7	7

Main-chain polymer derived from 3-[(methyoxycarbonyl)methyl]-5-[4′-[*N*-ethyl-*N*-(2″-hydroxyethyl)amino] benzylidene]rhodanine [Francis et al., 1993b]

Polymers are prepared by transesterication in melt at 140 °C for 8 hours.
T_g = 63 °C.
M_w = 13 × 10^3.
λ_{max} = 473 nm.
n (790 nm) = 1.760.
Contact poled with 43 V/μm at 65 °C.
d_{33} (1.58 μm) = 7.3 pm/V.

4-amino-4′-alkylsulfonyltolane main-chain polymer [Zentel et al., 1993]

Polymers are corona poled during melt polymerization at 150 °C for a few hours.
T_g = 60 °C.
λ_{max} = 366 nm.
d_{33}(1.064 μm) = 12.5 pm/V.
Activity decays by 40% in 25 days at ambient conditions.

TABLE 6.3 Experimental Results of Poled Polymer Studies—(*continued*)

Structure and Nomenclature (Ref)	Properties

Structures 1, 2 & 3, 4: Main-chain polyesters derived from 6-hydroxyhexyloxyphenyl propenoic, azobenzoic, or ethenylbenzoic acids [S'heeren et al., 1993b]

Polymers are prepared by melt polycondensation at 220 °C, 1.3 mbar for 4 hours.
Films are of pale yellow in color.
d_{33} values are measured at 1.064 μm after corona poling at 10 °C above T_g.

	1	2	3	4
T_g (°C)	48	41	51	53
Dye (mol%)	30	43	63	34
Mw (×10³)		35	115	99
d_{33} (pm/V)		0.04	0.5	0.3

Main-chain accordionpolymers of α-cyano cinnamamides [Lindsay et al., 1992; Wang and Guan, 1992]

Polymer 2 was corona poled. Activity is found to be stable at room temperature.

	1	2
T_g (°C)	143	193
λ_{max} (nm)	≈425	
Mw (×10³)	55	
Poling temp. (°C)		215
d_{33} (pm/V) at 1.064 μm		3.5

Random main-chain polymers containing 4-N, N-dialkylamino-4'-hexylsulfonylazobenzene [Xu et al., 1993; 1992a]

Polymers are prepared by condensation in dioxane solution.
d_{33} values are measured at 1.064 μm after corona poling at about 120 °C.

	1	2	3
T_g (°C)	62	114	108
λ_{max} (nm)	440	436	443
Mw (×10³)	8	7	9
d_{33} (pm/V)	60	125	150
d_{33} (10 days) at ambient	12		140

Cross-Linked polymers

Thermal cross-linking is achieved by precuring at 100 °C for 3 hours and at 140 °C for 16 hours under a corona poling field.
λ_{max} = 410 nm.
n (633 nm) = 1.629.
d_{33} (1.06 μm) = 13.5 pm/V.

Second-Order Nonlinear Optical Properties of Organic Materials

TABLE 6.3 Experimental Results of Poled Polymer Studies—*(continued)*

Structure and Nomenclature (Ref)	Properties
Cross-Linked epoxy polymer from 4-nitro-1,2-phenylenediamine and bisphenol-A diglycidylether [Eich et al., 1989a]	No decay of SHG activity is observed for 36 mins at 80 °C.
Cross-linked epoxy polymer from N,N-(diglycidyl)-4-nitroaniline and N-(2-aminophenyl)-4-nitroaniline [Jungbauer et al., 1990]	Thermal cross-linking is achieved by precuring at 130 °C for 4 min and at 120 °C for 24 hours under a corona poling field. λ_{max} = 397 nm. High dye content: 63 wt%. n (633 nm) = 1.74. d_{33} (1.064 μm) = 50 pm/V. Stable activity is observed at 80% of initial value at 80 °C.
Cross-linked methacrylate polymer containing 4-N,N-dialkylamino-4'-N, N-dialkylaminosulfonylazobenzene [Wang et al., 1993; Allen et al., 1991]	Cross-linking is initiated with free radicals under a corona poling field. λ_{max} = 480 nm. r_{33} (633 nm) = 36 pm/V; r_{33} (1.3 μm) = 6.6 pm/V. No decay of activity is observed over 8 months at ambient conditions for A. Higher temporal stability found for B.
Cross-linked polymer from a reactive diamine derivative of 4-N, N-dimethylamino-4'-nitroazobenzene and oligomeric derivative of diglycidyl ether of bisphenol A [Hubbard, 1992]	Thermal cross-linking is achieved by precuring at 100 °C for 3 hours and at 130 °C for 2 hour under a corona poling field. $\rho^{\#}$ = 6 × 10^{20} cm^{-3} for B. d_{33} (1.064 μm) = 3(A)–6(B) pm/V. Temporal stability: τ_1 (25 °C) = 6(A)–4.1(B) days and τ_2 (25 °C) = 120(A)–300(B) days; τ_1 (85 °C) = 1.6(B) days and τ_2 (85 °C) = 120(B) days.

TABLE 6.3 Experimental Results of Poled Polymer Studies—*(continued)*

Structure and Nomenclature (Ref)	Properties
Triazine cross-linked polymer obtained from p(N,N-bis(4′-cyanatobenzyl)amino-p′-(2,2dicyanovinyl)azobenzene (Holland and Fang, 1992; Singer et al., 1991)	Polymers are prepared by polyacyclotrimerization at 150 °C under a poling field for several hours. $r_{33}(830\ nm) = 11$ pm/V. High stability is observed at 85 °C.
Isocyanate cross-linked polymer derived from tris-1-hexamethyleneisocyanate isocyanurate and 3-amino-5-[4′(N-ethyl-N-(2″-hydroxyethyl)amino)benzylidene]-rhodanine (Francis et al., 1993a)	Thermally cross-linked at 135 °C for 16 hours. $\lambda_{max} = 470$ nm. $n(1.3\ \mu m) = 1.611$. Doping level: 60 mol%. Contact poled with 100 V/μm at 135 °C. $d_{33}(1.58\ \mu m) = 6.9$ pm/V. $r_{33}(1.3\ \mu m) = 3.6$ pm/V. Activity decays by 30% in 150 days at 100 °C due to thermal decomposition.
Cross-Linked epoxy polymer from 1,2,7,8-diepoxyoctane and N-(3-hydroxy-4-nitrophenyl)-(S)-prolinoxy functionalized poly(p-hydroxystyrene) copolymer (Park et al., 1990)	Thermal cross-linking is achieved by precuring at 100 °C for 24 hours and at 180 °C for 1 hour under a corona poling field. High dye content: 16 wt%. diepoxide/phenol ratio: 0.5. $d_{33}(1.06\ \mu m) = 3$ pm/V. Temporal stability: $\tau_1(25\ °C) = 79$ days and $\tau_2(25\ °C) = 100$ days.

TABLE 6.3 Experimental Results of Poled Polymer Studies—(*continued*)

Structure and Nomenclature (Ref)	Properties

Cross-linked homo- and copolymers of isocyanatoethyl methacrylate, MMA, and dimethylaminoethyl methacrylate. (Cheng et al., 1992)

R =
A: $-N(\text{CH}_2\text{CH}_2-)_2-\text{C}_6\text{H}_4-\text{NO}_2$
B: $-N(\text{CH}_2\text{CH}_2-)_2-\text{C}_6\text{H}_4-\text{C}_6\text{H}_4-\text{SO}_2\text{C}_3\text{F}_7$
C: $-N(\text{CH}_2\text{CH}_2-)_2-\text{C}_6\text{H}_4-\text{CH}=\text{CH}-\text{C}_6\text{H}_4-\text{NO}_2$

Corona poled near T_g.
d_{33} values are measured at 1.064 μm.
d_{33} of A are stable at room temperature.
Activity decays by 10% in 100 days at 80°C for polymer B and C.
$\alpha(1.06\ \mu m) \approx -(1.5\text{–}3)$ dB/cm

(x, y, z)	$\rho^{\#}$ (10^{20}/cm^3)	d_{33} (pm/V)
A(1, 0, 0)	11	11
B(1, 1, 0)	6	10
C(1, 0, 0)	9	34
C(1, 0, 1)	6	39

1

2

Cross-linked side-chain polymers containing 4-*N,N*-dialkylamino-4'-alkylsulfonylazobenzene [Xu et al., 1992b; Shi et al., 1993]

Thermal cross-linking under corona poling field: (1) with initiators; (2) without initiator.

	1	2
λ_{max}(nm)	≈440	≈440
Poling temp. (°C)	115	160–180
d_{33}(pm/V) at 1.064 μm	30	50
d_{33}(100 °C 4 days)		47.5

Cross-linked epoxy polymers containing dialklyamino nitroazobenzene derivatives [Muller et al., 1993]

Cross-linked polymers are prepared with 10–20 wt% PMMA as a binder.
Films are characterized as having good to excellent optical quality.
r_{33} values are measured at 1.32 μm after corona poling at about 140 °C for 8 hours.

	1	2	3
T_g (°C)	165	150	140
Precure (°C)	none	140	140
Precure (hr)	none	4.5	2.5
r_{33} (pm/V)	1.6	0.9	3.5
r_{33} (80 °C for 30 min)	8.8	7.7	1.5

TABLE 6.3 Experimental Results of Poled Polymer Studies—(*continued*)

Structure and Nomenclature (Ref)	Properties
Cross-Linked polymer containing 4-amino-4′-nitroazobenzene (Yu et al., 1992)	Prepolymer containing ethynyl group is thermally cross-linked at 190 °C for 2 hours under a corona poling field. $\lambda_{max} \approx 470$ nm. $d_{33}(1.064$ μm$) = 20$ pm/V. Activity decays to 75% of initial value after 30 days at 90 °C.
Cross-Linked polyurethane containing 4-N,N-dialkylamino-4′-nitroazobenzene (Chen et al., 1992; Shi et al., 1992)	Oligomeric prepolymer is thermally cross-linked with triethanolamine at 160 °C for an hour under a corona poling field. $\lambda_{max} = 475$ nm. $n(633$ nm$) = 1.753$; $n(800$ nm$) = 1.692$. $d_{33}(1.06$ μm$) = 120$ pm/V. $r_{13}(633$ nm$) = 13$ pm/V; $r_{13}(800$ nm$) = 5$ pm/V. No decay of activity is observed at ambient for 4 months; activity stabilizes to 70% of initial value after 4 months at 90°C.
Cross-Linked random main-chain polymer containing 4-N,N-dialkyamino-4′-hexylsulfonylazobenzene and 3,3′-dianisidine diisocyanate (Xu, 1992; Ranon et al., 1993; Xu et al., 1993)	Thermally cross-linked under a corona poling field at 125 °C for 2 hours. Prepolymer $M_w = 8 \times 10^3$. $\lambda_{max} = 440$ nm. $d_{33}(1.064$ μm$) = 40$ pm/V. 90% activity remains stable at ambient conditions for >3 months.
Alkoxysilane derivative of 4-(4′nitrophenylazo)-phenylamine cross-linked with 1,1,1-tris(4-hydroyphenyl)ethane (Jeng et al., 1993)	Thermally cross-linking occurs at 200 °C. $T_g = 110$°C. $\lambda_{max} = 493$ nm. $n(533$ nm$) = 1.744$. Doping level: 50 mol%. Corona poled at 200 °C for 30 min. $d_{33}(1.06$ μm$) = 77$ pm/V. $d_{33}(24$ hrs, 105 °C$) = 62$ pm/V. No decay of activity is observed for 7 days at ambient conditions.

TABLE 6.3 Experimental Results of Poled Polymer Studies—*(continued)*

Structure and Nomenclature (Ref)	Properties
Alkoxysilane derivative of 4-(4'nitrophenylazo)-phenylamine cross-linked with phenylsiloxane polymer (Allied Signal Accuglass 204®) (Jeng et al., 1992a) $[(SiO)_{>1}(C_6H_5)_{<0.5}(OC_2H_5)_{<0.5}(OH)_{<0.5}]_n$	Thermal cross-linking occurs at 200 °C. λ_{max} = 493 nm. n(533 nm) = 1.537. Doping level: 0.1 g dye in 4 g of A204. Corona poled at 200 °C for 10 min. $d_{33}(1.06~\mu m)$ = 5.28 pm/V. $d_{33}(40~h, 100~°C)$ = 2.9 pm/V.
Alkoxysilane derivative of 4-(4'-nitrophenylazo)phenylamine/polyimide (Skybond®) composite (Jeng et al., 1992b)	T_g > 275 °C. λ_{max} = 466 nm. Doping level: 16 wt%. Excellent optical quality. Corona poled at 220 °C. $d_{33}(1.06~\mu m)$ = 13.7 pm/V. Stable $d_{33}(168~h, 120~°C)$ = 10 pm/V. Novelty: composite of polyimide and Si-O-Si network.
Interpenetrating network of epoxy and silicon-based cross-linked polymers (Marturunkakul et al., 1993)	Equal weight ratio of the two polymers is heated at 200 °C for 1 hour under a corona poling field. Films are of high optical quality. T_g = 176 °C. λ_{max} = 458 nm. n(533 nm) = 1.708. $d_{33}(1.064~\mu m)$ = 33 pm/V. No decay of activity is observed at 100 °C for 7 days. 50% decay is observed after 15 hours at 160°C.

TABLE 6.3 Experimental Results of Poled Polymer Studies—*(continued)*

Structure and Nomenclature (Ref)	Properties
Cross-linked acrylate polymer of 4'-N,N-bis(6-methacroyloxyhexyl)-amino-4-methylsulfonylstilbene (Robello et al., 1991b)	Liquid monomers are cross-linked by UV irradiation with initiator at room temperature. $T_g < 70$ °C. Good optical quality. Contact poled with 6.7 V/μm. $d_{33}(1.06\ \mu m) = 0.7$ pm/V. Activity decays to zero in several months at ambient conditions.
Cross-linked polymer of polyvinylcinnamate doped with 3-cinnamoyloxy-4-[4-(N,N-diethylamino)-2-cinnamoyloxy phenylazo] nitrobenzene (Mandal et al., 1991a; 1991b)	Cross-linking is induced by UV irradiation (2 mW/cm² at 254 nm for 3 to 10 min) at 70 °C. $T_g = 84$°C (before cross-linking). $\lambda_{max} = 520$ nm. Doping level: 10 wt%. $n(633\ nm) = 1.677$. Corona poled at 70 °C during UV irradiation. $d_{33}(1.064\ \mu m) = 11.5$ pm/V. $d_{33}(1.54\ \mu m) = 3.7$ pm/V. $r_{33}(633\ nm) = 9$ pm/V. No decay of activity is observed in 22 hours at ambient conditions
Cross-linked side-chain cinnamate or furylacrylate polymers doped with cinnamoyloxy or furylacryloyloxy functionalized 4-N,N-diethylamino 4'-azonitrobenzene (Muller et al., 1992b)	Photo-cross-linking with 1.8 mW/cm² at 312 nm at 75 °C for 4 (A) and 2 (B) h to achieve 60% reaction. B cannot be poled under irradiation and is poled with a cycle of poling (10 min at 25 °C) and cross-linking (20 min at −15 °C) for a total exposure time of 2 h. $r_{33}(1.32\ \mu m) = 0.6$ pm/V. Activity remains stable after 3 h at 80 °C.

TABLE 6.3 Experimental Results of Poled Polymer Studies—(continued)

Structure and Nomenclature (Ref)	Properties
Cross-linked side-chain cinnamate copolymers functionalized with 4-alkoxy-4'-biphenyl or 4-N,N-diethylamino 4'-azonitrobenzene [Kato et al., 1993]	Photo-cross-linking with UV light under corona poling field: (A) 1.5 min at 50 °C; (B) 10 min at room temperature. Both polymers were subsequently poled at 150 °C for 20 min. Activities are stable at room temperature. 	mol%	d_{33}(pm/V)	 A: 10, 6 33, 20 57, 30 B: 27, 150
Cross-Linked epoxy polymer containing 4-amino-4'-nitroazobenzene (Jeng et al., 1992c)	Prepolymer containing cinnamoyl group is cross-linked by UV irradiation at 3 mW/cm² for 10 min. λ_{max} = 461 nm. n(533 nm) = 1.718. Corona poled at 115 °C for 1 hour. d_{33}(1.064 μm) = 22 pm/V. 95% of SHG activity remains stable after 20 days at ambient conditions.
Cross-linked, random, main-chain polymer containing 4-N,N-dialkyamino-4'-hexylsulfonylazobenzene and cinnamoyl groups (Xu et al., 1993; Chen et al., 1991)	Photo-cross-linked under a corona poling field. Prepolymer M_w = 8.5 × 10³. Dye content = 53 wt%. λ_{max} = 443 nm. d_{33}(1.064 μm) = 150 pm/V. 90% activity remains stable at ambient conditions for 25 days.

The electro-optic coefficients of optimized poled polymers can be estimated with Eq. (6.43). Because there is no transparency constraint, some of the largest β_μ values found in Table 1 can be used for the estimate. Using similar number densities, local field factors, and order parameters as above, for n_ω = 1.6 and β_μ(1.9 μm) = 1000 × 10⁻³⁰ esu, a $r_{zzz} \approx 2.5 \times 10^{-6}$ esu (1000 pm/v) is obtained. This estimated value is over one order of magnitude higher than the best experimental results (see Table 3), indicating the potential for material optimization. This large potential nonlinearity also allows some flexibility in material engineering to meet other property requirements. For example, to maintain many of the desirable properties of polymers with high T_g, the chromophore contents may have to be limited. With highly nonlinear chromophores, given the low dielectric constant, device figures of merit $n^3 r_{zzz}/\epsilon$ comparable to the best of the inorganics is possible even with low number density.

Processibility

For side-chain polymers, homopolymers, containing pendant chromophores with large $\mu\beta_\mu$, generally, have higher nonlinearity than copolymers due to the higher number density. However, the intense absorption of the chromophores and their good electron-transferring abilities often render homopolymerization of their derivatives difficult either by radical initiation or photolysis. The highly polar nature of these molecules also results in poor solubility of the homopolymers making film formation from solution difficult or impossible. Highly cross-linked networks are desirable from the standpoint of improved temporal stability of the polar order. However, the cross-linked polymers are insoluble preventing film formation by spin casting. Although processing can be carried out at a prepolymer stage, the low viscosity of their solutions due to limited molecular weight and/or solubility, is also a limiting factor for film formation. Processing at the prepolymeric or cross-linking monomeric stage may also result in poor optical quality (see below) and low poling efficiency.

Optical Loss

The optical loss of a nonlinear optical polymer is influenced by many factors. Absorption loss due to O-H and C-H overtone absorption is present in all organics and can be considered as the background loss of the poled polymers. Much higher absorption loss can result from impurities in the polymeric or chromophoric portion of the polymer. The latter is generally much more important. Because the typical peak absorption coefficients of nonlinear optical chromophores are in excess of 100,000 cm^{-1}, even a small fraction (<0.001%) of isomeric impurity, which may have a shifted absorption peak, can lead to unacceptable absorption loss at an otherwise transparent region of the spectrum. In addition, because of the large bandwidth of CT transition, low loss propagation (about -1 dB/cm) is possible only at wavelengths 300–400 nm red-shifted from its peaks. This consideration may restrict the choice of chromophore for modulation applications of short-wavelength (around 830 nm) diode lasers.

Scattering often dominates the optical loss of a poled polymer. Scattering due to internal defects, such as voids, dust particles, and poling defects, can be eliminated by quality control, clean-room preparation, and reduced poling voltage. Surface scattering due to roughness can also be improved by controlling the polymer rheology and film forming conditions, such as choice of solvent and evaporation rate. Crystallinity and liquid crystallinity in polymers must be eliminated to reduce scattering loss. Copolymerization often frustrates the packing order and avoids crystallization. The liquid crystallinity of many side-chain polymers can be eliminated by reducing the flexibility of the chromophore-polymer linkages. For film forming approaches which involve mixing two or more components, such as cross-linking monomers, cross-linker with reactive prepolymer, and guest-host systems, scattering losses due to multiphasic behavior often dominate. Molecular weight control and altering the solubility of individual components can lead to single-phase materials at certain mixing concentrations. However, due to the often large mismatch in refractive indices between the different components, Rayleigh scattering due to concentration fluctuations can be significant even for single-phase materials. Density fluctuation because of poor transport during solid-state reactions for cross-linking networks can also result in high-loss materials. If the scattering centers are small compared to the wavelength of interest, Rayleigh scattering loss, which is proportional to λ^{-4}, becomes less of a problem for long-wavelength applications.

Temporal Stability of Nonlinearity

The general description for complex response in amorphous systems, including the poling and depoling dynamics of dipolar chromophores in a polymeric matrix, involves a distribution function of relaxation times (Ferry, 1980). This distribution is governed by the microscopic inhomogeneity of the polymer matrix and represents the distribution in barrier energy to orientational relaxation. For polymers near or below their glass transition temperatures, the width of this distribution is extremely broad, covering over five decades in time scales. Its peak position, shape, and, perhaps, width are temperature dependent, changing to faster time scales, narrow width, and higher symmetry, respectively, at higher temperature. Figure 6 shows the typical poling dynamic of a poled polymer with contact poling. It can be deduced that, at low temperatures, only a small fraction of the chromophores encounter orientational barriers which are sufficiently low to allow dipolar alignment within a few minutes of observation. The distribution function shifts to shorter time scales with each increment in temperature, increasing the fraction of chromophores which can orient during the observation time window. Above T_g, nearly all chromophores, in accordance with the Boltzmann factor in Eq. (6.29), respond to the poling field within the observation time, and the harmonic intensity saturates.

To create a quasi-static polar order, the poling field is switched off after the polymer is cooled to a temperature much below T_g, where the relaxation time is long. However, because the distribution function covers many time scales and its width often broadens at low temperature, some level of relaxation of the polar order cannot be prevented. Similar to poling dynamics, the polar order relaxes nonexponentially, again, dictated by the distribution of relaxation times. Environmental effects, such as the plasticizing effect of moisture, can influence this relaxation. Figure 7 shows the temporal decay of the nonlinear optical response of a typical poled polymer. The dynamic is well fitted with a stretch exponential function,

$$\chi^{(2)} \propto e^{-(t/\tau_0)^\delta}, \tag{6.78}$$

where τ_0 is an average relaxation time and δ is a width parameter for the distribution, with values $0 < \delta \leq 1$. A biexponential relaxation is sometimes used to approximate the depoling:

$$\chi^{(2)} \propto A e^{-t/\tau_f} + B e^{-t/\tau_s} \tag{6.79}$$

where τ_f and τ_s represent the relaxation of the fast and slow components. The extent of dipolar relaxation, within the life span of a poled-polymer device, depends on both τ_0 and δ. In increasing

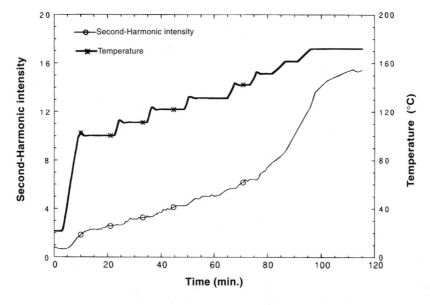

FIGURE 6.6 Contact poling of a typical side-chain nonlinear optical polymer.

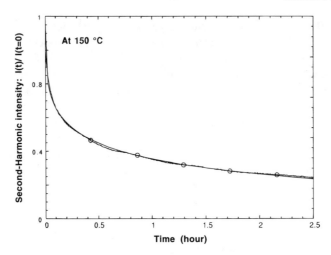

FIGURE 6.7 Temporal decay of poled nonlinearity at elevated temperature.

τ_0, it is desirable to keep the width of the distribution of relaxation times as narrow as possible, i.e., maximizing δ. For a narrow distribution, thermal activation and deactivation of the orientational mobility become more efficient because the majority of the chromophores have similar relaxation times.

In terms of material properties, the temporal stability of the polar order at a given application temperature can be improved by increasing its difference from the poling temperature, usually near T_g. Although there are other factors besides T_g which may lead to dipolar relaxation, in most poled-polymer studies, a high T_g has been found to be a minimum requirement. For many applications, a device must remain functional at temperatures as high as 120°C. Short excursions to higher temperatures (>250°C) may also be required for certain device processing steps. Therefore, acceptable temporal stability of the polar order is unlikely for polymers with T_g below 200°C.

The T_g of a polymer can be increased by reducing the flexibility in the polymer backbone. Phenyl and other cyclic substituents are known to increase T_g (see Table 3). The influence of T_g on the temporal stability of the polar order is additionally dependent on the coupling between the chromophore orientational mobility and the segmental motion of the polymer backbone. Several strategies have been used to increase this coupling. For side-chain polymers, the coupling can be improved by shortening the connecting linkages or using multiple linkages. The chromophores can also be incorporated into the polymer backbone as in the case of main-chain polymers. The over all rigidity of the polymeric matrix can be further enhanced by cross-linking. However, cross-linking increases the microscopic inhomogeneity in the polymer, resulting in a much broader relaxation distribution function which can reduce poling efficiency.

Polymers

Table 3 summarizes the reported properties of various poled polymers. They are classified according to the different routes of chromophoric incorporation. From the first studied guest-host composites, which have seen much renewed interest, to the highly cross-linked materials, which are under active development for the improved temporal stability of their polar order, relevant materials and nonlinear optical properties are tabulated. The reported results however may not indicate the best possible properties of a particular chemical approach, but serve to illustrate the diversity and flexibility of the poled-polymer approach. Crystalline, liquid crystalline, and functionalized sol-gel systems are not included in the review. In the remainder of our discussion, "temporal stability" refers to that of the poled nonlinearity.

Guest-Host Composites

The poled-polymer approach in an amorphous system was first demonstrated by dissolving 4-(N-ethyl-N-(2-hydroxyethyl)amino-4′-nitroazobenzene) (DR1) in poly(methylmethacrylate) (PMMA) (Singer et al., 1986). Characterization of this material was carried out with SHG. Although the number density was limited by solubility, a nonlinearity several times that of a commonly used inorganic material (KDP) was found. The quasi-static nature of the polar order was addressed in a novel experiment in which a ferroelectric host polymer, a copolymer of polyvinylidene fluoride (PVDF), was used in a room temperature poled system (Hill et al., 1987; Pantelis et al., 1988). The ferroelectric internal field in the host polymer was concluded to be responsible for a much improved orientational stability of the polar guest chromophores compared with other guest-host systems.

The guest-host approach was revisited when good temporal stability was reported for a high T_g polyimide polymer at a temperature as high as 200°C (Wu et al., 1991a). Guest chromophores were typically introduced into solutions of polyamic acids which was processible into thin films. Thermal and chemical imidization under a poling field at temperatures up to 360°C have resulted in impressive temporal stability at 200–300°C (Wu et al., 1991a; 1991b). A densification process during imidization at high temperature was alluded to as a factor contributing to the high temporal stability found in the polyimide guest-host system (Wu et al., 1991b). These early systems were very lightly doped, thus, leaving the high T_gs of the host polymers unchanged. At higher doping levels, the apparent T_gs of the composites were significantly depressed (Stahelin et al., 1992a; Ermer et al., 1992). The problem of a low poling field due to high ionic conductivity was also reported. The high poling temperature required in this approach also initiated a search for thermally stable chromophores.

Without the severe thermal stability requirement of using ultrahigh T_g polymers, cross-linked host polymers were investigated to improve the temporal stability of a guest-host system. In a composite of 4-N, N-dimethylamino-4′-nitrostilbene (DANS) and a highly cross-linked epoxy network, good temporal stability at room temperature was achieved (Hubbard et al., 1989). In a different approach, the water soluble nitrophenol was introduced into gelatin which served as a cross-linked host through hydrogen bonding (Ho et al., 1992). Although high nonlinearity was reported, temporal stability was poor. Using an inorganic glass network as a host matrix by way of sol-gel chemistry, DR1 was doped into a phenylsiloxane polymer (Jeng et al., 1992a). Poling was carried out during condensation reactions at 200°C. The reported temporal stability is however significantly lower than that of the polyimide approach at comparable poling temperature.

Although high temporal stability can be achieved with high T_g host polymers, problems such as low solubility, thermal degradation, sublimation, and the plasticizing effect of the guest molecules have prevented the attainment of both high stability and nonlinearity. Until chromophores with exceptional thermal stability and high nonlinearity can be developed, the guest-host approach is unlikely to yield materials with application potential. However, because many of the high T_g host polymers are well accepted by the electronic industry, significant advantages can be realized by the guest-host approach, including simplicity and ease of adaptation to current processing technology. At low number density, all other desirable properties of the host polymer can be preserved. These considerations have prompted much current chromophore research aiming toward the optimization of both nonlinearity and thermal stability.

Side-Chain Polymers

Functionalized polyalkyacrylates are the most studied polymer system for nonlinear optics. The good optical and dielectric properties of PMMA and the synthetic accessibility of its derivatives have, undoubtedly, contributed to the appeal of the acrylates. Early works on co- and homopolymers of functionalized polyacrylates demonstrated that large nonlinearity was possible due to the much higher number density compared with guest-host systems (Esselin et al., 1988; Singer et al., 1988; Hill et al., 1988, 1989). Polymers with methacroyloxyethoxy linkages were found to have T_g's typically between 100–140°C. Azobenzene and stilbene derivative functionalized polymers also confirmed the room temperature stability of their poled nonlinearity. Because of the smaller size of the side-chains and lower T_gs, benzene derivative functionalized polymers were found to have significantly lower temporal stability (Hayashi et al., 1992). The chromophore size dependence on orientation stability was further illustrated with a much longer three-ring system (Amano and Kaimo, 1990). High temporal stability was reported at 80°C. For polymers with the more flexible methacroyloxyhexyloxy linkage, the T_gs and temporal stabilities were generally reduced. Liquid crystallinity could also develop in the case of homopolymers with long linkages. Highly functionalized and insoluble acrylate polymers were shown to be processible at the prepolymer stage, where the reactive pre-polymer served to bind reactive chromophores. Isocyanate-hydroxy coupling chemistry was thermally activated to form the final polymer in the solid state (Cheng et al., 1992).

Exceptional temporal stability at 100°C of two acrylate polymers was reported recently. A coumaromethacrylate copolymer with isobornyl methacrylate was found to have a T_g of 170°C (Mortazavi et al., 1991). The higher T_g (about 35°C higher than that of a methyl methacrylate copolymer) was believed to result from the bulky isobornyl rings. A homopolymer of 2,5-diphenyl thiobarbituric acid derivatives was found to have a T_g of 194°C, nearly 70°C higher than that of a homopolymer with a 2,5-diethyl thiobarbituric acid group (Cheng and Tam, 1993). This represents the highest T_g reported for a side-chain acrylate polymer and required a poling temperature at 220°C. No thermal degradation or high ionic conductivity were observed at the poling temperature. Microcracks were found in films thicker than several microns due to brittleness. Copolymerization with diethyl thiobarbituric acid derivatives was found to improve the mechanical properties and eliminate the cracking problem. Because derivatives of the two accepting groups have similar nonlinearities, copolymerization did not lead to a reduction in nonlinearity. At 20 mol% of the diethyl derivative, T_g was found to decrease to 180°C.

Many other side-chain polymeric systems were investigated for nonlinear optics. The polyethylene-type backbone was studied early for ease of functionalization of poly(allylamine hydrochloride) (Eich et al., 1989b). A nitroaniline derivatized homopolymer yielded a high number density and nonlinearity with a T_g of 125°C. The polar order was found to be unstable at ambient conditions, probably due to the small size of the chromophore. Poled polymers with polystyrene-type backbones were obtained by the functionalization of chlormethylated polystyrene(PS) and poly(p-hydroxstyrene)(PHS). Good optical and dielectric properties, relatively high T_g, and ease of functionalization were quoted as motivations for these studies. Functionalized PSs have been obtained via polymer reaction (Ye et al., 1987) and polymerization of functionalized vinyl monomers (Hayashi et al., 1991). Low degree of functionalization (20–30%) due to limited solubility was reported for the former approach. With either method, materials with T_g of 100–120°C were obtained. These T_g values were not sufficiently high to insure temporal stability. For PHS, which has a high T_g of 155°C due to extensive hydrogen bonding, nearly complete functionalization can be obtained through polymer reaction at the hydroxyl group. However, as the functionalization level increases, the T_g decreases precipitously (from 155 to 96°C) due to the elimination of hydrogen bonds (Ye et al., 1989).

Poled polymers with polyvinylether-type backbones were also reported (S'heeren et al., 1992). Functionalization was carried out at the monomeric level with a wide range of phenyl and azobenzene derivatives. Only low T_g (40–60°C) materials were obtained. The reported nonlinear coefficients were obtained several days after poling and represent only the residual nonlinearity due to inevitably fast relaxation. A nitroaniline derivatized polyvinylalcohol copolymer was reported to have a T_g of 120°C and marginal room temperature stability (Sasaki, 1993). The conductive nature of polyvinylalcohol may also impact the poling of its derivatives. All of the side-chained polymers reviewed above, were derived from relatively low T_g parent polymers. Functionalization often resulted with materials with increased T_g due to the aromatic chromophoric contents.

Investigation of functionalized polymers derived from high T_g parents represented the next logical step. Several examples can be found in the literature. The high T_g (205–210°C) poly(2,6-dimethyl-1,4-phenylene oxide) has been functionalized with N-(4-nitrophenyl)-S-prolinol to high number density via a polymeric reaction (Dai et al., 1990). Functionalization up to 1.6 chromophores per monomeric unit was reported. A T_g of 173°C was found for the highly functionalized polymer. Contact poling resulted in reasonably high nonlinearity. The temporal stability of the polar order was reported to have a lifetime of about one year at room temperature. The surprisingly poor temporal stability for such a high T_g polymer could be due to the relatively small chromophore size. The high T_g of the polyimide backbone was also exploited in a promising approach (Lin et al., 1992). A nitroaniline-type chromophore with diamine functionalities was reacted with diacids to form a processible polymer. Thin films of this polymer were imidized at high temperatures (up to 240°C) under a poling field to form a functionalized polyimide with a T_g of 236°C. Good temporal stability of the nonlinearity was observed at 80°C.

Several high T_g functionalized polymers obtained from maleic anhydrides were reported recently. A maleic anhydride co-propylene copolymer with DR1 side-chains was found to have a T_g of 180°C (Bauer et al., 1993). The chromophore content was however quite low at 16 mol%. High T_g (170–180°C) as well as high functionalization (up to 96 mol%) were obtained for a series of azo dye derivatized maleic anhydride co-styrene and co-norbornadiene polymers (Ahlheim and Lehr, 1994). The high T_gs are likely due to the aromatic and cyclic groups in the backbone and the cyclic imide linkage to the chromophore. Although optical results have not been published, these polymers appear to be very promising.

The majority of device-related research has been carried out on side-chain acrylates. The chromophoric content of the acrylate polymers is determined by the target application. For SHG, methoxy nitrostilbene and nitroazobenzene derivatives have been reported. Sulfonyl derivatives of stilbene, tolane, and biphenyl were reported to be optimum for SHG of short-wavelength diode lasers in the blue region of the spectrum (Rikken et al., 1991, 1992; Seppen et al., 1991). For electro-optic modulation, stilbene, azobenzene, and more extended derivatives with nitro and dicyanovinyl acceptors yielded the highest nonlinearity. Harmonic generation in the red, green, and blue regions has been demonstrated with periodic poling (Khanarian et al., 1990), periodic photobleaching (Rikken et al., 1991), periodic stacking (Khanarian et al., 1993), modal dispersion (Rikken et al., 1993), Cherenkov (Sugihara et al., 1991), and noncollinear phase-matching (Otomo et al., 1993) schemes. None of these demonstrations have, however, shown sufficient efficiency for practical device. Electro-optic modulation up to 84 GHz (Leibfried et al., 1993) and sampling up to 460 GHz (Ferm, 1991) have been demonstrated, providing clear evidence of the ultrafast nature of the electro-optic response in organic materials. Modulators with attractive performance have been reported (Girton et al., 1991; Teng, 1992).

Main-Chain Polymers

Several main-chain polymers with simple functionalities, such as cyano, carbonyl, and amide groups, have been investigated for their nonlinear optical properties. Poly(vinylidenecyanide/vinylacetate) is well known for its piezoelectric properties because of a high concentration of polar cyano and carbonyl groups. When poled near its T_g at 180°C, nonlinear optical response was observed by SHG (Sato et al., 1987). At 1.06 μm, a d_{33} of 0.3 pm/V was found (Eich et al., 1988). However, when studied at 2.94 μm, an enormous value of 117 pm/V was reported (Azumai et al., 1990). The origin and validity of such a high coefficient are not clear. A large Kirkwood–Frohlich g factor, which describes the dipole orientation correlation, has been cited to explain the poling efficiency (Azumai et al., 1990). Nevertheless, even if the order parameter equals unity, the origin of the electronic nonlinearity remains unknown (the cyano group itself has very low hyperpolarizability). The discrepancy with the 1.06 μm result has also not been reconciled. Nevertheless, although few SHG studies of organic materials have been carried out with 3 μm radiation, certain vibronic resonance enhancement effects may not be ruled out.

When poled above their T_g, aromatic polyureas were found to exhibit a nonlinear optical response. Two materials were studied in some detail. A linear main-chain polyurea was found to have a relatively high T_g of 150°C and an absorption cut-off of 400 nm (Kajikawa et al., 1991; Nagamori et al., 1992). When poled at 180°C, a relatively stable nonlinearity of 1.7 pm/V was reported. For a polyurea containing phenyl side-chains, a lower T_g of 123°C and a short absorption cut-off of 308 nm were reported (Nalwa et al., 1993a, 1993b). A much higher d coefficient of 5.5 pm/V and good room temperature stability were also reported. The urea moiety is, undoubtedly, the source of the optical nonlinearity in these polymers. Their short absorption cut-offs and modest nonlinearity make them attractive candidates for SHG applications.

Several main-chain polymers prepared by condensation reactions between primary amine (or two secondary amines) functionalized chromophores and bifunctional epoxy linkers were reported (Teraoka et al., 1991; Gadret et al., 1991; Jungbauer et al., 1991; Twieg et al., 1992; Jeng et al., 1992c). Similarly, reaction between di-isocyanate linkers and diol functionalized chromophores also yielded main-chain polymers with polyurethane backbones (Meyrueix et al., 1991a, 1991b; Chen et al., 1991; Kitipichai et al., 1993). Most of the reported polymers employed chromophores whose amine or alcoholic functionalities were located at one end of the molecule. As a result, only one end of the chromophore (typically an amino or dialkylamino donor) was incorporated into the polymeric backbone. From the point of view of improved coupling between the chromophoric orientational motion and the backbone relaxation, this represented only a marginal improvement over the side-chain polymer approach because considerable backbone flexibility was allowed by the linkages. To impart rigidity to the polymer backbone, phenyl-containing epoxy or isocyanate linkers were used. Nevertheless, reported T_g values were typically between 80–130°C, comparable to values obtained in the side-chain polymer approach. Temporal stability was found only at ambient conditions, confirming expectations from structural and T_g considerations. The relatively bulky aromatic linkers also reduce the overall chromophoric content of the polymers. Exceptions are found in homopolymers obtained by placing the two epoxide groups directly on the chromophore (Twieg et al., 1992). The higher chromophoric content yielded much higher nonlinearity and T_g.

Main-chain polymers which contain chromophores in a head-to-tail arrangement within the polymer backbone have long been sought for nonlinear optics. The potential advantages include high temporal stability and a poling enhancement resulting from the backbone-correlated dipole orientation. Low T_g, low solubility, and high crystallinity of the main-chain polymers are among the material difficulties encountered in early studies (Fuso et al., 1991; Green et al., 1987a, 1987b; Hall, 1988). Some progress has been made recently yielding processable materials with T_gs ranging from 60–120°C. Thus far, no clear evidence of poling enhancement has been found. Results of temporal stability studies have been quite poor, similar to the experience with low T_g side-chain materials. Polar main-chain polymers with much higher T_g (167–187°C) were recently reported (Mitchell et al., 1992). Monomeric and oligomeric EFISH investigation of poling enhancement in these materials, in fact, found a decrease in poled nonlinearity. Unexpectedly low coefficients were also reported in many poled polymers. The reduced poling efficiency was explained by a high rotational viscosity of the main-chain.

The only encouraging results from the main-chain polymer approach were found in two examples in which a linear polar backbone was not sought, an "accordion" polymer in which the polymer backbone folded like an accordion with the chromophoric segments in head-to-tail arrangement (Lindsay et al., 1992). Despite rather low nonlinearity, a T_g of 143°C and room temperature temporal stability of the polar order were reported (Wang et al., 1992). The best result, thus, far was provided by a random main-chain polymer in which the chromophoric backbone consisted of a statistical mixture of head-to-head, tail-to-tail, and head-to-tail arrangements (Xu et al., 1992a, 1993). At 50% chromophoric content, a T_g of 114°C was found. Corona poling yielded a room temperature, stable d_{33} of 125 pm/V.

When the results from the random main-chain polymer study are compared to those obtained from an acrylate side-chain copolymer with similar T_g and chromophoric content, similar poling efficiency (within 80%) is shown (Xu et al., 1993). The temporal stabilities of the polar order are also similar in both polymers. This suggests that the degree of coupling between the polymer backbone motion and the chromophoric orientation are similar for both side-chain and main-chain polymers, despite their considerable structural differences. Therefore, there may be little incentive to tackle the synthetic challenge of the highly functionalized, processable, and high T_g, polar, main-chain materials.

Cross-Linked Polymer Networks

Cross-linking is a natural strategy aimed at stabilizing the order of parameter relaxation in poled polymers. With the exception of using ultrahigh T_g materials, such as polyimides, cross-linked networks have provided the highest temporal stability. For thermally activated cross-linking reactions, the curing temperature and time determine the rigidity of the network and, therefore, the temporal stability of the polar order. For photochemically activated cross-linking, the thermal treatment, which determines the transport of the reactants, and the photon dosage control the extent of cross-linking. The number, relative reactivities, and locations of functionalities on the chromophore can greatly influence the extent of cross-linking and the effectiveness of the polymer network in immobilizing chromophoric orientation. For polymerizable groups, two or more per monomer units are needed. If coupling chemistry is used, one of the monomeric pair must have at least two and the other three reactive sites. The large number of required reactive sites often lowers the chromophoric content in cross-linked networks. Intuitively, reactive sites distributed at both ends of the chromophore are more effective in immobilization. Poling must be carried out insitu during the cross-linking reaction. If a cross-linked network is synthesized from cross-linking monomers, a precuring step prior to film casting and poling is necessary to increase the viscosity and dielectric strength of the mixture. An appropriate polymeric binder can also be used. Preparation of a prepolymer with cross-linkable groups has been widely used to improve film forming and poling characteristics during cross-linking reactions.

For thermal cross-linking, numerous chemical routes have been demonstrated. They include the use of epoxy-amino coupling, radical-initiated thermal polymerization of multifunctionalized methacrylate monomers, epoxy-alcohol coupling, isocyanate-hydroxyl coupling, isocyanate-amino coupling, alcohol condensation of alkoxysilane, and ethynyl thermal intramolecular reaction. The most widely used are the epoxy and isocyanate coupling chemistry. A commercially available diepoxide (bisphenol-A) was used for the early work on cross-linked materials for nonlinear optics (Eich, 1989a). Reaction with a bisamino derivative of nitroaniline under a poling field yielded a polar cross-linked network with high temporal stability at 85°C. The nonlinear coefficients were greatly increased when the bulky bisphenol A was replaced with a nitroaniline derivative functionalized with two epoxide groups, thus, increasing the chromophoric content of the network (Jungbauer et al., 1990). The curing time in both materials was substantial (several hours) due to the low reactivity of aromatic amines. The importance of a highly cross-linked network in enhancing temporal stability was further illustrated by two comparative studies. When monomeric azo dyes functionalized with two or four methacrylate groups were polymerized, the resultant polymer of the bifunctional monomer behaved similarly to a side-chain polymer whereas that of the tetrafunctional monomer yielded much higher temporal stability (Allen et al., 1991). It was concluded that, at 60% completion of the reaction, only 10% of the bifunctional monomers were involved in network formation whereas 60% of the tetrafunctional monomers are cross-linked. Similar findings were made when an aliphatic amine functionalized azo dye was reacted with either a bisepoxide or a multifunctional oligomeric epoxy cross-linker (Hubbard et al., 1992).

For side-chain polymers whose backbones contain reactive sites, cross-linking was used to improve temporal stability with good results. An example was provided by the epoxy cross-linking of a polyhydroxy styrene copolymer (Park et al., 1990). Cross-linking of prepolymers containing reactive side-chains was also demonstrated with significant improvement of temporal stability (Xu et al., 1992b; Cheng et al., 1992; Muller et al., 1993). A reactive binder approach was reported in which a prepolymer-containing reactive isocyanate side-chains was shown to be a flexible route not only for side-chain polymers but also for cross-linked matrices through reaction with chromophores bearing multiple reactive alcohols (Cheng et al., 1992). Simple diols can also be used as passive cross-linkers. Cross-linking between isocyanate terminated main-chain oligomers and tri-functional alcohols as shown to be an alternative to precuring (Chen et al., 1992). The oligomeric system was reported to have good film forming properties and can be poled without further precure. Using an amino and alcohol bifunctional chromophore for the same purpose, the higher reactivities between amino and isocyanate groups were exploited to form a precured polymer which was processible into thin films (Francis et al., 1993a). Full curing under a poling field was accomplished by activating the isocyanate-hydroxy reaction at higher temperature. In many of these investigations, it was shown that anchoring the chromophores at both ends was advantageous. Cross-linking at the main chain was also illustrated with the intramolecular reaction of the ethynyl group (Yu et al., 1992) and amino-isocyanate chemistry (Ranon et al., 1993; Xu et al., 1993). In addition, alkoxysilane condensation chemistry was used to form a cross-linked network with tri-functional alcohol (Jeng et al., 1993). It was also illustrated that with only physical entanglement, as in the case of a interpenetrating network of phenoxy-silicon and epoxy polymers, good temporal stability of the polar order could be achieved (Marturunkakul et al., 1993).

Photo-cross-linking was explored early as an alternative route for cross-linked networks for nonlinear optics. It offers two potential advantages over thermal cross-linking. First, if a highly cross-linked network can be obtained under an electric field at low temperature, poling efficiency may be improved through both the Boltzmann factor and the possibility of a higher poling field. Secondly, under a poling electric field, the illuminated region of the polymer can be finely patterned, as in photolithography, creating not only a pattern in the refractive index but also a pattern in the nonlinear optical susceptibility through the subsequent dipolar relaxation in the uncross-linked regions.

The photo-chemistries of acryloxyethyl, cinnamoyl, and furylacrylate groups were used in various formats in the preparation of photo-cross-linked poled polymers. In an early demonstration, an acrylate functionalized chromophore was doped into a liquid trifunctional acryloxyethyl photo-cross-linker (Robello et al., 1991b). With an appropriate photosensitizer, the mixture was illuminated at room temperature under a poling field to form a cross-linked network with a quasi-static polar order. To increase the number density, a bisacrylate functionalized chromophore was prepared, and a liquid form of the chromophore was processed into thin film for photo-cross-linking (Robello et al., 1991b). Difficulties encountered in these early experiments included poor cross-linking due to dye absorption, high conductivity from an ionic sensitizer, photodegradation of the chromophore, and uneven cross-linking due to poor light penetration.

A much better result was reported when a biscinnamoyloxy functionalized chromophore was doped into a photo-cross-linkable polymer matrix (Mandal et al., 1991a; Mandal et al., 1991b). Photo-cross-linking with a few minutes of UV exposure under a poling field at T_g (70°C) resulted in both high nonlinearity and good temporal stability up to 60°C. This result was, however, disputed in a recent report which found that several hours of exposure were needed due to the low reactivity of the cinnamate group (Muller et al., 1992b). Poor dye incorporation due to the bulky nature of the cinnamate group was also reported. Improved reactivity was reported when furylacrylate was used (Muller et al., 1992b). An alkyl spacer was found to greatly improve the intermolecular cross-linking of these bulky groups. An alternating cycle of illumination and poling was found to be necessary because no corona poling field could be established during illumination (photo conductivity have been reported for this system (Li et al., 1991)). Nevertheless, good temporal stability of the polar order was found at temperatures as high as 130°C.

Results from the previous two studies appear to indicate that good temporal stability can be obtained at poling temperatures with photo-cross-linking. Other results, however, were less positive. In a different approach, the cinnamate group was functionalized into the side-chains of a chromophore-bearing copolymer (Kato et al., 1993). Although good temporal stability was reported at room temperature, substantial relaxation was found near the poling temperature. Cinnamoyl groups were also incorporated into the backbone of a side-chain epoxy polymer (Jeng et al., 1992c) and a random main-chain polymer (Xu et al., 1993; Chen et al., 1991). Photo-cross-linking of the main chains, in both cases, yielded dramatic improvement of their room temperature temporal stability.

6.7 Concluding Remarks

Impressive advances in both the fundamental understanding and property development of organic nonlinear optical materials have been made in the past two decades. For second-order materials, current research has clearly shifted from a scientific to a technological in emphasis. The challenges have also evolved from optimizing nonlinear optical responses to optimizing trade-offs with a wide range of physical as well as chemical properties. For molecular materials, ever more hyperpolarizable structures continue to be discovered. Although the high nonlinearity affords more flexibility in macroscopic organization, the real significance of such discoveries must be measured by the transparency, processibility, and environmental stability of these new molecules. Nature's limit in allowing the coexistence of hyperpolarizability and chemical stability in an organic molecule must be pursued. For single crystalline materials, early emphasis on the nitroaniline template has yielded a large number of candidate crystals for the harmonic generation of Nd:YAG lasers. Although these crystals offered much higher nonlinearity, their overall properties did not compete successfully with a wide array of inorganic alternatives. The number of inorganic crystals suitable for harmonic generation of short-wavelength diode laser is, however, much more limited. The organic approach, although constrained by an apparent transparency trade-off, may yield competitive crystals due to the much lower power levels of diode lasers. The organic approach is better adapted to electro-optic applications. With single crystals, many research opportunities remain, ranging from crystal discovery, their growth, and microfabrication, to the development of new applications which have not been explored by their inorganic counterparts. For polymers, the quasi-static nature of their nonlinearity continues to discourage applications. Although significant improvement has been made with high T_g polymers and cross-linked matrices, much critical evaluation remains. Plasticizing effects due to environmental factors may significantly degrade long-term performance. Because an important advantage of the polymer approach is its propensity for large-scale integration, demands on thermal and temporal dimensional stability are severe. These and other device requirements have not been fully accessed. In addition to exploratory material research in the diversity of the organic approach, candidate materials should be perfected and studied in detail.

References

Ahlheim, M., and Lehr, F. 1994 Electrooptically active polymers. Non-linear optical polymers prepared from maleic anhydride copolymers by polymer analogous reactions. *Macromol. Chem. Phys.* 195, 361–373.

Allen, S., Bone, D. J., Carter, N., Ryan, T. G., Sampson, R. B., Devonald, D. P., and Hutchings, M. G. 1991. In *Organic Materials for Non-Linear Optics*, Hann, R. A. and Bloor, D., eds., p.235. The Royal Society of Chemistry, London.

Allen, S., McLean, T. D., Gordon, P. F., Bothwell, B. D., Hursthouse, M. B., and Karaulov, S. A. 1988. A novel organic electro-optic crystal: 3-(1,1-dicyanoethenyl)-1-phenyl-4,5-dihydro-1H-pyrazole. *J. Appl. Phys.* 64(5):2583.

Amano, M., and Kaino, T. 1990. Second-order nonlinearity of a novel diazo-dye attached polymer. *J. Appl. Phys.* 68(12):6024.

Armstrong, J. A., Bloembergen, N., Ducuing, J., and Pershan, P. S. 1962. Interactions between lightwaves in a nonlinear dielectric. *Phys. Rev.* 127:1918.

Azumai, Y., Sato, H., and Seo, I. 1990. Second-harmonic generation of Er:YAG 2.94-μm laser radiation using an organic vinylidene cyanide/vinyl acetate thin film. *Optics Lett.* 15(17):932.

Bacquet, G., Bassoul, P., Cobellas, C., Simon, J., Thiebault, A., and Tournilhac, F. 1990. Highly polarizable zwitterions for nonlinear optics: Synthesis, structure and properties of phenoxide-pyridinium derivative. *Adv. Mater.* 2(6):311.

Bailey, R. T., Bourhill, G., Cruickshank, F. R., Pugh, D., Sherwood, J. N., and Simpson, G. S. 1993. The linear and nonlinear optical properties of the organic nonlinear material 4-nitro-4'-methylbenzylidene aniline. *J. Appl. Phys.* 73(4):1591.

Bailey, R. T., Bourhill, G., Cruickshank, F. R., Pugh, D., Sherwood, J. N., Simpson, G. S., and Varma, K. B. R. 1992. Linear electrooptic effect and temperature coefficient of birefringence in 4-nitro-4'-methylbenzylidiene aniline single crystals. *J. Appl. Phys.* 71(4):2012.

Bailey, R. T., Cruickshank, F. R., Guthrie, S. M. G., McArdle, B. J., Morrison, H., Pugh, D., Shepherd, E. A., Sherwood, J. N., and Yoon, C. S. 1989. Second-order optical nonlinearity and phase matching in 4-nitro-4'-methylbenzylidene aniline. *Mol. Cryst. Liq. Cryst.* 166:267.

Bailey, R. T., Cruickshank, F. R., Guthrie, S. M. G., McArdle, B. J., Morrison, H., Pugh, D., Shepherd, E. A., Sherwood, J. N., Yoon, C. S., Kashyap, R., Nayar, B. Y., and White, K. I. 1988a. Phase-matched optical second-harmonic generation in the organic crystal MBA-NP, (-)2-(α-methylbenzylamino)-5-nitropyridine. *J. Modern Optics.* 35(3):511.

Bailey, R. T., Cruickshank, F. R., Guthrie, S. M. G., McArdle, B. J., Morrison, H., Pugh, D., Shepherd, E. A., Sherwood, J. N., Yoon, C. S., Kashyap, R., Nayar, B. Y., and White, K. I. 1988b. The quality and performance of the organic nonlinear optical material (-)2-(α-methylbenzylamino)-5-nitropyridine (MBA-NP). *Optics Comm.* 65(3):229.

Barry, S. E., and Soane, D. S. 1991. Poling of polymeric film at ambient temperatures for second-harmonic generation. *Appl. Phys. Lett.* 58(11):1134.

Barzoukas, M., Blanchard-Desce, M., Josse, D., Lehn, J.-M., and Zyss, J. 1989a. Very large quadratic nonlinearities in solution of two push-pull polyene series: Effect of the conjugation length and of the end groups. *Chem. Phys.* 133:323.

Barzoukas, M., Josse, D., Fremaux, P., Zyss, J., Nicoud, J. F., and Morley, J. O. 1987. Quadratic nonlinear properties of N-(4-nitrophenyl)-L-prolinol and of a newly engineered molecular compound N-(4-nitrophenyl)-N-methylaminoacetonitrile: A comparative study. *J. Opt. Soc. Am. B.* 4(6):977.

Barzoukas, M., Josse, D., Zyss, J., Gordon, P., and Morley, J. O. 1989b. Quadratic hyperpolarizability of sulfur-containing systems. *Chem. Phys.* 139:359.

Bauer, S., Ren, W., Yilmaz, S., Wirges, W., Molzow, W.–D., Gerhard–Multhaupt, R., Oertel, U., Hanel, B., Haussler, L., Komber, H., and Lunkwitz, K. 1993. Nonlinear optical side-chain polymer with high thermal stability and its pyroelectric thermal analysis. *Appl. Phys. Lett.* 63(15):2018.

Bauerle, D., Betzler, K., Hesse, H., Kappan, S., and Loose, P. 1977. Phase-matched, second-harmonic generation in urea. *Phys. Status Solidi (A).* 42:119.

Baumert, J. C., Twieg, R. J., Bjorklund, G. C., Logan, J. A., and Dirk, C. W. 1987. Crystal growth and characterization of 4-(N, N-dimethylamino)-3-acetamidonitrobenzene, a new organic material for nonlinear optics. *Appl. Phys. Lett.* 51(19):1484.

Belikova, G. S., Golovey, M. P., Shigorin, V. D., and Shipulo, G. P. 1975. *Opt. Spectrosc. (USSR).* 38:441.

Betzler, K., Hesse, H., and Loose, P. 1978. Optical second-harmonic generation in organic crystals: urea and ammoniummaleate. *J. Mol. Struct.* 47:393.

Bierlein, J., Cheng, L. K., Wang, Y., and Tam, W. 1990. Linear and nonlinear optical properties of 3-methyl-4-methoxy-4'-nitrostilbene single crystals. *Appl. Phys. Lett.* 56(5):423.

Blanchard, P. M., and Mitchell, G. R. 1993. A comparison of photoinduced poling and thermal poling of azo-dye-doped polymer films for second-order nonlinear optical applications. *Appl. Phys. Lett.* 63(15):2038.

Bloembergen, N. 1965. *Nonlinear Optics.* W. A. Benjamin, New York.

Bosshard, C., Knopfle, G., Pretre, P., and Gunter, P. 1992. Second-order polarizabilities of nitropyridine derivatives determined with electric-field-induced second-harmonic generation and a solvatochromic method: A comparative study. *J. Appl. Phys.* 71(4):1594.

Bosshard, C., Sutter, K., Gunter, P., and Chapuis, G. 1989. Linear and nonlinear properties of 2-cyclooctylamino-5-nitropyridine. *J. Opt. Soc. Am. B.* 6(4):721.

Bosshard, C., Sutter, K., Schesser, R., and P., G. 1993. Electrooptic effects in molecular crystals. *J. Opt. Soc. Am. B.* 10(5):867.

Botrel, A., Illien, B., Rajczy, P., Ledoux, I., and Zyss, J. 1993. Intramolecular charge transfer in 5-phenyl-3H-1,2-dithiole-3-thione and 5-phenyl-3H-1,2,-dithole-3-one derivative molecules for quadratic nonlinear optics. *Theor. Chim. Acta.* 87:175.

Bottcher. 1973. *Theory of Electric Polarization.* Elsevier, New York.

Bourhill, G., Bredas, J. L., Cheng, L.–T., Marder, S. R., Meyers, F., Perry, J. W., and Tiemann, B. G. 1994. Experimental demonstration of the dependence of the first hyperpolarizability of donor-acceptor substituted polyenes on the ground-state polarization and bond length alternation. *Chem. Comm.*, accepted.

Buckley, A., Choe, E., DeMartino, R., Nelson, L. T., Stamatoff, G., Stuetz, D., and Yoon, H. 1986. *Proc. ACS Div. Polym. Mat: Sci. Eng.* 54:502.

Burland, D. M., Miller, R. D., Reiser, O., Twieg, R. J., and Walsh, C. A. 1992. The design, synthesis, and evaluation of chromopheres for second-harmonic generation in a polymer waveguide. *J. Appl. Phys.* 71(1):410.

Burland, D. M., Walsh, C. A., Kajzar, F., and Sentein, C. 1991. Comparison of hyperpolarizabilities obtained with different experimental methods and theoretical techniques. *J. Opt. Soc. Am. B.* 8(11):2269.

Cahill, P. A., Singer, K. D., and King, L. A. 1989. Anomalous dispersion phase-matched second-harmonic generation. *Opt. Lett.* 14(20):1137.

Canon. 1989. *Jpn. Patent Application.* Open No: 1-90424

Carenco, A., Jerphagnon, J., and Perigaud, A. 1977. Nonlinear optical properties of some m-disubstituted benzene derivatives. *J. Chem. Phys.* 66(8):3806.

Cassidy, C., Halbout, J. M., Donaldson, W., and Tang, C. L. 1979. Nonlinear optical properties of urea. *Opt. Comm.* 29(2):243.

Chemical, S. 1990. *Jpn. Patent Application.* Open No: 2-85828

Chemla, D.S., and Zyss, J. 1987. *Nonlinear Optical Properties of Organic Molecules and Crystals.* Vols. 1 and 2. Academic Press, New York.

Chen, M., Dalton, L. R., L., Y., Shi, Y., and Steier, W. H. 1992. Thermosetting polyurethanes with stable and large second-order optical nonlinearity. *Macromolecules.* 25:4032.

Chen, M., Yu, L., Dalton, L. R., Shi, Y., and Steier, W. H. 1991. New polymers with large and stable second-order nonlinear optical effects. *Macromolecules.* 24:5421.

Cheng, L.–T. 1994. Previously unpublished EFISH results. Some results to appear in upcoming publications. Collaborators: Donald, D, Ermer, S, Gilmour, S., Griffiths, J., Marder, T. B., Marder, S. R., Tam, W, and Tiemann, B. G.

Cheng, L.–T., and Tam, W. Polymeric compositions for nonlinear optics. *U. S. Patent* 5,272,218, 1993.

Cheng, L.–T., Tam, W., Feiring, A., and Rikken, G. L. J. A. 1990. Quadratic hyperpolarizabilities of fluorinated sulfonyl and carbonyl aromatics: Optimization of nonlinearity and transparency trade-off. *SPIE Proc.* 1337:203.

Cheng, L.-T., Tam, W., Marder, S. R., Stiegman, A. E., Rikken, G., and Spangler, C. W. 1991a. Experimental investigations of organic molecular nonlinear optical polarizabilities. 2. A study of conjugation dependences. *J. Phys. Chem.* 95(26):10643.

Cheng, L.-T., Tam, W., Stevenson, S., Meredith, G. R., Rikken, G., and Marder, S. R. 1991b. Experimental investigations of organic molecular nonlinear optical polarizabilities. 1. Methods and results on benzene and stilbene derivatives. *J. Phys. Chem.* 95(26)10631.

Cheng, L. T., Foss, R. P., Meredith, G. R., Tam, W., and Zumsteg, F. C. 1992. Preparation of polymeric films for NLO applications: Cross-Linked and side-chain polymers derived from homo- and copolymers containing reactive isocyanate groups. *Mol. Cryst. Liq. Cryst. Sci. Tech.-Sec. B: Nonlinear Optics.*

Clays, K., and Persoon, A. 1991. Hyper-Rayleigh scattering in solution. *Phys. Rev. Lett.* 66(23):2980.

Cockerham, M. P., Frazier, C. C., Guha, S., and Chauchard, E. A. 1991. Second-harmonic generation in derivatives and analogs of benzophenone and chalcone. *Appl. Phys. B.* 53:275.

Combellas, C., Petit, M. A., and Thiebault, A. 1992. New comb-like polymers with prospective nonlinear optical properties. *Makromol. Chem.* 193:2445.

Dai, D.-R., Marks, T. J., Yang, J., Lundquist, P. M., and Wong, G. K. 1990. Polyphenylene ether based thin-film nonlinear optical materials have high chromophore densities and alignment stability. *Macromolecules.* 23:1891.

Dao, P. T., Williams, D. J., McKenna, W. P., and Goppert–Berarducci, K. 1993. Constant current corona charging as a technique for poling organic nonlinear optical thin films and the effect of ambient gas. *J. Appl. Phys.* 73(5):2043.

Dewar, M. J. S. 1950. *J. Am. Chem. Soc..* 2329.

Dirk, C. W., Katz, H. E., and Schilling, M. L. 1990. Use of thiazole rings to enhance molecular second-order nonlinear optical susceptibility. *Chem. Mater.* 2:700.

Dulcic, A., Flytzanis, C., Tang, C. L., Pepin, D., Fetizon, M., and Hoppilliard, Y. 1981. Length dependence of the second-order optical nonlinearity in conjugated hydrocarbons. *J. Chem. Phys.* 74(3):1559-1563.

Dulcic, A., and Sauteret, C. 1978. The regularities observed in the second order hyperpolarizabilities of variously disubstituted benzenes. *J. Chem. Phys.* 69(8):3453.

Dulic, A., and Flytzanis, C. 1978. A new class of conjugated molecules with large second-order polarizability. *Opt. Comm.* 25(3):402.

Eich, M., Reck, B., Yoon, D. Y., Willson, C. G., and Bjorklund, G. C. 1989a. Novel second-order nonlinear optical polymers via chemical cross-linking-induced vitrification under electric field. *J. Appl. Phys.* 66(7):3241.

Eich, M., Sen, A., Looser, H., Bjorklund, G. C., Swalen, J. D., Twieg, R., and Yoon, D. Y. 1989b. Corona poling and real-time, second-harmonic generation study of a novel covalently functionalized amorphous nonlinear optical polymer. *J. Appl. Phys.* 66(6):2559.

Eich, M., Sen, A., Looser, H., Yoon, D. Y., Bjorklund, G. C., Twieg, R., and Swalen, J. D. 1988. In situ measurements of corona poling induced SHG in amorphous polymers. *SPIE Proc.* 971:128.

Ermer, S., Lovejoy, S., Leung, D., Spitzer, R., Hansen, G., and Stone, R. 1991. Synthesis and nonlinear optical activity of cumulenes. *SPIE Proc.* 1560:120–129.

Ermer, S., Valley, J. F., Lytel, R., Lipscomb, G. F., Van Eck, T. E., and Girton, D. G. 1992. DCM-polyimide system for triple-stack poled polymer electro-optic devices. *Appl. Phys. Lett.* 61(19):2272.

Esselin, S., Barny, P. L., Robin, P., Broussoux, D., Dubois, J. C., Raffy, J., and Pocholle, J. P. 1988. Second-harmonic generation in amorphous polymers. *SPIE Proc.* 971:120.

Falk, H., and Hemmer, D. 1992. On the chemistry of pyrrole pigments, LXXXVIII[1]: Nonlinear optical properties of linear oligopyrroles. *Monatshefte fur Chemie.* 123:779.

Ferm, P. M., Knapp, C. W., Wu, C., and Yardley, J. T. 1991. Femtosecond response of electro-optic poled polymers. *Appl. Phys. Lett.* 59(21):2651.

Ferry, J. D. 1980. *Viscoelastic Properties of Polymers.* John Wiley & Sons, New York.
Francis, C. V., White, K. M., Boyd, G. T., Moshrefzadeh, R. S., Mohapatra, S. K., Radcliffe, M. D., Trend, J. E., and Williams, R. C. 1993a. Isocyanate cross-linked polymers for nonlinear optics. 1. Polymers derived from 3-amino-5-[4′-(N-ethyl-N-(2″-hydroxyethyl)amino)benzylidene]-rhodanine. *Chem. Mater.* 5:506.
Francis, C. V., White, K. M., Newmark, R. A., and Stephens, M. G. 1993b. Processible main-chain optical polymer with a new acceptor group from melt polycondensation. *Macromolecules.* 26:4379.
Fuso, F., Padias, A. B., and Hall, H. K., Jr., 1991. Poly[(ω-hydroxyalky)thio-α-cyanocinnamates]. Linear polyesters with NLO-phores in the main chain. *Macromolecules.* 24:1710.
Gadret, G., Kajzar, F., and Raimond, P. 1991. Nonlinear optical properties of poled polymer. *SPIE Proc.* 1560:226.
Gallo, T. J., Kimura, T., Ura, S., Suhara, T., and Nishihara, H. 1993. Method for characterizing poled-polymer waveguides for electro-optic integrated-optical-circuit applications. *Opt. Lett.* 18(5):349.
Gilmour, S., Marder, S. R., Perry, J. W., and Cheng, L.–T. 1994a. Large second-order optical nonlinearities and enhanced thermal stabilities in extended thiophene-containing compounds. *Adv. Mater.*, accepted.
Gilmour, S., Marder, S. R., Tiemann, B. G., and Cheng, L.–T. 1993. Synthesis and first hyperpolarizabilities of acceptor-substituted β-apo-8′-carotenal derived compounds. *J. Chem. Soc. Chem. Comm.* 5:432–433.
Gilmour, S., Montgomery, R. A., Marder, S. R., Cheng, L.–T., Jen, A. K.–Y., Cai, Y., Perry, J. W., and Dalton, L. 1994b. Synthesis of diarylthiobarbituric acid chromophores with enhanced second-order optical nonlinearities and thermal stability. *J. of Chem. Soc.*, submitted.
Girton, D. G., Kwiatkowski, S. L., Lipscomb, G. F., and Lytel, R. S. 1991. 20 GHz electro-optic polymer Mach–Zehnder modulator. *Appl. Phys. Lett.* 58(16):1730.
Green, G. D., Hall, H. K., Jr., Mulvaney, J. E., Noonan, J., and Williams, D. J. 1987a. Donor-acceptor-containing quinodimethanes. Synthesis and copolyesterification of highly dipolar quinodimethanes. *Macromolecules.* 20:716.
Green, G. D., Weinschenk, J. I., Mulvaney, J. E., and Hall, H. K., Jr., 1987b. Synthesis of polyester containing a nonrandomly placed highly polar repeat unit. *Macromolecules.* 20:722.
Grossman, C. H., Wada, T., Yamada, A., Sasabe, H., and Garito, A. F. 1989. Maker fringe and phase-matched SHG studies of crystalline dicyano substituted benzene compound. In *Nonlinear Optics of Organics and Semiconductors*, Kobayashi, T., ed. Springer-Verlag, Berlin, Heidelberg.
Gunter, P., Bosshard, C., Sutter, K., Arend, H., Chapuis, G., Twieg, R. J., and Dobrowolski, D. 1987. 2-cyclooctylamino-5-nitropyridine, a new nonlinear optical crystal with orthorhombic sysmmetry. *Appl. Phys. Lett.* 50(9):486.
Halbout, J.–M., Blit, S., Donaldson, W., and Tang, C. L. 1979. Efficient phase-matched second-harmonic generation and sum-frequency mixing in urea. *IEEE J. of Quantum Electron.* QE-15(10):1176.
Hall, H. K., Jr. 1988. Exploratory synthesis of polymers possessing electrically conducting, liquid crystalline, piezoelectric and nonlinear optical activity. *J. Macromol. Sci.-Chem.* A25:729.
Hampsch, H. L., Torkelson, J. M., Bethke, S. J., and Grubb, S. G. 1990. Second-harmonic generation in corona-poled, doped polymer films as a function of corona processing. *J. Appl. Phys.* 67(2):1037.
Harada, A., Okazaki, Y., and Kamiyama, K. 1990. Generation of continuous wave, blue coherent light from a semiconductor laser using nonlinear optical fiber with organic core crystal. *CLEO Technical Digest.* 496.
Havinga, E. E., and Van Pelt, P. 1979. Intramolecular charge transfer studied by electrochromism of organic molecules in polymer matrices. *Mol. Cryst. Liq. Cryst.* 52:145.

Hayashi, A., Goto, Y., Nakayama, M., Kaluzynski, K., Sato, H., Kato, K., Kondo, K., Watanabe, T., and Miyata, S. 1992. Second-order nonlinear optical properties of poled polymers of vinyl chromophore monomers: Styrene, methacrylate, and vinyl benzoate derivatives having one aromatic ring push-pull chromophores. *Chem. Mater.* 4:555–562.

Hayashi, A., Goto, Y., Nakayama, M., Kaluzynski, K., Sato, H., Watanabe, T., and Miyata, S. 1991. Second harmonic generation of poled 4-[N-methyl-N-(4'-nitrophenyl)amino-methyl]styrene polymer film. *Chem. Mater.* 3:6.

Henderson, C. C., Cahill, P. A., Kowalczyk, T. C., and Singer, K. D. 1993. Thermal, optical, and nonlinear optical properties of tetrafluorinated donor-acceptor benzenes. *Chem. Mater.* 5(8):1059.

Herning, C. E. A. 1989. *Polymer Preprints, Japan.* 38(8):2607.

Hill, J. R., Dunn, P. L., Davies, G. J., Oliver, S. N., Pantelis, P., and Rush, J. D. 1987. Efficient frequency doubling in a poled PVDF copolymer guest/host composite. *Electron. Lett.* 23(13):700.

Hill, J. R., Pantelis, P., Abbasi, F., and Hodge, P. 1988. Demonstration of the linear electro-optic effect in a thermopoled polymer film. *J. Appl. Phys.* 64(5):2749.

Hill, J. R., Pantelis, P., Dunn, P. L., and Davies, G. J. 1989. Organic polymer films for second-order nonlinear applications. *SPIE Proc.* 1147:165.

Ho, E. S. S., Lizuka, K., Freundorfer, A. P., and Wah, C. K. L. 1991. Determination of electro-optical coefficients of 2-methyl-4-nitroaniline. *J. Appl. Phys.* 69(3):1173.

Ho, Z. Z., Chen, R. T., and Shih, R. 1992. Electro-optic phenomena in gelatin-based poled polymer. *Appl. Phys. Lett.* 61(1):4.

Holland, W. R., and Fang, T. 1992. A thermally stable cyanate polymer for nonlinear optical application. *Polym. Mater. Sci. Eng.* 66:500.

Horsthuis, W. H. G., and Krijnen, G. J. M. 1989. Simple measuring method for electro-optic coefficients in poled polymer waveguides. *Appl. Phys. Lett.* 55(7):616.

Hubbard, M. A., Marks, T. J., Lin, W., and Wong, G. K. 1992. Poled polymeric nonlinear optical materials. Enhanced second-harmonic generation temporal stability of epoxy-based matrices containing a difunctional chromophoric comonomer. *Chem. Mater.* 4:965–968.

Hubbard, M. A., Marks, T. J., Yang, J., and Wong, G. K. 1989. Poled polymeric nonlinear optical materials. Enhanced second-harmonic generation stability of cross-linkable matrix/chromophore ensembles. *Chem. Mater.* 1(2):167.

Huijts, R. A., and Hesselink, G. L. J. 1989a. Length dependence of the second-order polarizability in conjugated organic molecules. *Chem. Phys. Lett.* 156(2, 3):209.

Huijts, R. A., Jenneskens, L. W., van der Vorst, C. P. J. M., and Wreesmann, C. T. J. 1989b. Construction of the second-order optical nonlinearity in an organic side-chain polymer. *SPIE Proc.* 1127:165.

Hyuga, H., Goto, C., Okazaki, Y., Harada, A., Mitsumoto, S., Kamiyama, K., and Umegaki, S. 1993. An organic crystal 3,5-dimethyl-1-(4-nitrophenyl) pyrazole for compact visible light sources. *OSA Topical Meeting Compact Blue-Green Lasers Technical Digest.* 382.

Ikeda, H., Kawabe, Y., Sakai, T., and Kawasaki, K. 1989a. Nonlinear optical properties of retinal derivatives. *Chem. Lett.* 1989:1285.

Ikeda, H., Kawabe, Y., Sakai, T., and Kawasaki, K. 1989b. Second-harmonic generation in nonbenzenoid aromatics. *Chem. Phys. Lett.* 157(6):576–578.

Ikeda, H., Kawabe, Y., Sakai, T., and Kawasaki, K. 1989c. Second-order hyperpolarizabilities of barbituric acid derivatives. *Chem. Lett.* 1989:1803.

Ikeda, H., Sakai, T., and Kawasaki, K. 1991. Nonlinear optical properties of cyanine dyes. *Chem. Phys. Lett.* 179(5, 6):551.

Itoh, Y. 1989. *Mol. Cryst. Liq. Cryst.* 170:259.

Jen, A. K.-Y., Rao, V. P., Wong, K. Y., and Drost, K. J. 1993. Functionalized thiophenes: Second-Order nonlinear optical materials. *J. Chem. Soc., Chem. Comm.* 1993:90.

Jen, A. K.-Y., Rao, V. P., Wong, K. Y., Drost, K. J., and Mininni, R. M. 1992. Heteroaromatics: Exceptional materials for second-order nonlinear optical applications. *Mat. Res. Soc. Symp. Proc.* 247:59.

Jeng, R. J., Chen, Y. M., Chen, J. I., Kumar, J., and Tripathy, S. K. 1993. Phenoxysilicon polymer with stable second-order optical nonlinearity. *Macromolecules.* 26:2530.

Jeng, R. J., Chen, Y. M., Jain, A. K., Kumar, J., and Tripathy, S. K. 1992a. Second-Order optical nonlinearity on a modified sol-gel system at 100°C. *Chem. Mater.* 4:972.

Jeng, R. J., Chen, Y. M., Jain, A. K., Kumar, J., and Tripathy, S. K. 1992b. Stable second-order nonlinear optical polyimide/inorganic composite. *Chem. Mater.* 4:1141.

Jeng, R. J., Chen, Y. M., Mandal, B. K., Kumar, J., and Tripathy, S. K. 1992c. UV-curable, epoxy-based, second-order nonlinear optical material. *MRS Proc.* 247:111.

Josse, D., Dou, S. X., Zyss, J., Andreazza, P., and Perigaud, A. 1992. Near-infrared optical parametric oscillation in a N-(4-nitrophenyl)-L-prolinol molecular crystal. *Appl. Phys. Lett.* 61(2):121.

Jungbauer, D., Reck, B., Twieg, R., Yoon, D. Y., Willson, C. G., and Swalen, J. D. 1990. Highly efficient and stable nonlinear optical polymers via chemical cross-linking under electric field. *Appl. Phys. Lett.* 56(26):2610.

Jungbauer, D., Teraoka, I., Yoon, D. Y., Reck, B., Swalen, J. D., Twieg, R., and Willson, C. G. 1991. Second-order optical properties and relaxation characteristics of poled linear epoxy polymers with tolane chromophores. *J. Appl. Phys.* 69(12):8011.

Kajikawa, K., Nagamori, H., Takezoe, H., Fukuda, A., Ukishima, S., Takahashi, Y., Iijima, M., and Fukada, E. 1991. Optical second-harmonic generation from poled thin film of aromatic polyurea prepared by vapor deposition polymerization. *Jpn. J. Appl. Phys.* 30(10A):L1737.

Kajzar, F., Ledoux, I., and Zyss, J. 1987. Electric-field-induced optical second-harmonic generation in polydiacetylene solutions. *Phys. Rev. A.* 36(5):2210.

Karna, S. P., Zhang, Y., Samoc, M., Prasad, P. N., Reinhardt, B. A., and Dillard, A. G. 1993. Nonlinear optical properties of novel thiophene derivatives: Experimental and ab initio time-dependent coupled perturbed Hartree–Fock studies. *J. Chem. Phys.* 99(12):9984–9993.

Kato, M., Hashidate, S., Hirayama, T., Matsuda, H., Minami, N., Okada, S., and Nakanishi, H. 1993. Synthesis and properties of polymers having photo-cross-linkable moieties for second-order nonlinear optics. *J. Photopolym. Sci. Technol.* 6(2):211.

Kawamata, J., and Inoue, K. 1991. Large second-order nonlinearity of new organic materials of bis(benzylidene)cycloalkanone derivatives. *Jpn. J. Appl. Phys.* 30(8B):L1496.

Kawamata, J., Inoue, K., Kasatani, H., and Hikaru, T. 1992. New organic material of bis(benzylidene)cycloalkanone derivatives for efficient optical second-harmonic generation. *Jpn. J. Appl. Phys.* 31, Part 1(2A):254.

Kerkoc, P., Bosshard, C., Arend, H., and Gunter, P. 1989a. Growth and characterization of 4-(N, N-dimethylamino)-3-acetamidonitrobenzene single crystal cored fiber. *Appl. Phys. Lett.* 54(6):487.

Kerkoc, P., Zgonik, M., Sutter, K., Bosshard, C., and Gunter, P. 1989b. Optical and nonlinear optical properties of 4-(N, N-dimethylamino)-3-acetamidonitrobenzene single crystals. *Appl. Phys. Lett.* 54(21):2062.

Kerkoc, P., Zgonik, M., Sutter, K., Bosshard, C., and Gunter, P. 1990. 4-(N, N-dimethyamino)-3-acetamidonitrobenzene single crystals for nonlinear optical applications. *J. Opt. Soc. Am. B.* 7(3):313.

Khanarian, G., Mortazavi, M. A., and East, A. J. 1993. Phase-matched second-harmonic generation from free-standing periodically stacked polymer films. *Appl. Phys. Lett.* 63(11):1462.

Khanarian, G., Norwood, R. A., Haas, D., Feuer, B., and Karim, D. 1990. Phase-matched second-harmonic generation in a polymer waveguide. *Appl. Phys. Lett.* 57(10):977.

Kielich, S., Kozierowski, M., and Lalanne, J.-R. 1975. Second-Harmonic elastic light scattering by molecular liquid mixtures. *Le Journal De Physique.* 36:1015.

Kiguchi, M., Kato, M., and Taniguchi, Y. 1993. Measurement of second-harmonic generation from colored powders. *Appl. Phys. Lett.* 63(16):2165.

Kitaoka, Y., Sasaki, T., Nakai, S., and Goto, Y. 1991. New nonlinear optical crystal thienylchalcone and its harmonic generation properties. *Appl. Phys. Lett.* 59(1):19.

Kitaoka, Y., Sasaki, T., Nakai, S., Yokotani, A., Goto, Y., and Nakayama, M. 1990. Laser properties of new organic nonlinear optical crystal chalcone derivatives. *Appl. Phys. Lett.* 56(21):2074.

Kitipichai, P., Peruta, R. L., Korenbowski, G. M., and Wenk, G. E. 1993. In situ poling and synthesis of NLO chromophore-bearing polyurethanes for second-harmonic generation. *J. Poly. Sci. Part A: Polymer Chem.* 31:1365.

Knopfle, G., Bosshard, C., Schlesser, R., and Gunter, P. 1994. Optical, nonlinear optical, and electro-optical properties of 4'-nitrobenzylidene-3-acetamino-4-methoxyaniline (MNBA) crystals. *IEEE J. of Quant. Electron.*, to appear.

Kondo, K., Fukutome, M., Ohnishi, N., Aso, H., Kitaoka, Y., and Sasaki, T. 1991. Laser properties of nonlinear optical benzal barbituric acid crystals. *Jpn. J. Appl. Phys.* 30(12A):3419.

Kondo, K., Goda, H., Takemoto, K., Aso, H., Sasaki, T., Kawakami, K., Yoshida, H., and Yoshida, K. 1992. Micro- and macroscopic second-order nonlinear optical properties of fulvene compounds. *J. Mater. Chem.* 2(10):1097.

Kondo, T., Morita, R., Ogasawara, N., Umegaki, S., and Ito, R. 1989. A nonlinear optical organic crystal for waveguiding SHG devices: (-2)-(α-methylbenzylamino)-5-nitropyridine (MBANP). *Jpn. J. Appl. Phys.* 28(9):1622.

Kotler, Z., Hierle, R., Josse, D., Zyss, J., and Masse, R. 1992. Quadratic nonlinear-optical properties of a new transparent and highly efficient organic-inorganic crystal: 2-amino-5-nitropyridinium-dihydrogen phosphate (2A5NPDP). *J. Opt. Soc. Am. B.* 9(4):534.

Kurtz, S. K., and Perry, T. T. 1968. *J. Appl. Phys.* 39:3798.

Laidlaw, W. M., Denning, R. G., Verbiest, T., Chauchard, E., and Persoon, A. 1993. Large second-order optical polarizabilities in mixed valency metal complexes. *Nature.* 363:58.

Ledoux, I., Lepers, C., Perigaud, A., Badan, J., and Zyss, J. 1990. Linear and nonlinear optical properties of N-4-nitropheyl L-prolinol single crystals. *Optics Comm.* 80(2):149.

Leibfried, D., Schmidt–Kaler, F., Weitz, M., and Hansch, T. W. 1993. Phase-matched electrooptic modulator at 84 GHz for blue light: Theory, experimental test and application. *Appl. Phys. B.* 56:65.

Levine, B. F. 1976. Donor-acceptor charge transfer contributions to the second-order hyperpolarizability. *Chem. Phys. Lett.* 37(3):516.

Levine, B. F., and Bethea, C. G. 1975. Second and third order hyperpolarizability of organic molecules. *J. Chem. Phys.* 63(6):2666.

Levine, B. F., and Bethea, C. G. 1978a. Ultraviolet dispersion of the donor-acceptor charge transfer contribution to the second-order hyperpolarizability. *J. Chem. Phys.* 69(12):5240–5245.

Levine, B. F., Bethea, C. G., Thurmond, C. D., Lynch, R. T., and Bernstein, J. L. 1979. An organic crystal with an exceptionally large optical second-harmonic coefficient: 2-methyl-4-nitroaniline. *J. Appl. Phys.* 50(4):2523.

Levine, B. F., Bethea, C. G., Wasserman, E., and Leenders, L. 1978b. Solvent dependent hyperpolarizability of a merocyanine dye. *J. Chem. Phys.* 68(11):5042.

Levine, B. F., and Bethea, C. G. 1974. Molecular hyperpolarizability determination in solution and nonconjugated organic liquids. *Appl. Phys. Lett.* 24(9):445.

Li, D., Marks, T. J., and Ratner, M. A. 1992. Nonlinear optical phenomena in conjugated organic chromophores. Theoretical investigations via a π-electron formalism. *J. Phys. Chem.* 96:4325.

Li, L., Lee, J. Y., Yang, Y., Kumar, J., and Tripathy, S. K. 1991. Photoconductivity in a photo-cross-linkable second-order nonlinear optical polymer. *Appl. Phys. B.* 53:279.

Lin, J. T., Hubbard, M. A., Marks, T. J., Lin, W., and Wong, G. K. 1992. Poled polymeric nonlinear optical materials. Exceptional second-harmonic generation temporal stability of a chromophore-functionalized polyimide. *Chem. Mater.* 4:1148.

Lindsay, G. A., Stenger-Smith, J. D., Henry, R. A., Hoover, J. M., Nissan, R. A., and Wynne, K. J. 1992. Main-chain accordion polymers for nonlinear optics. *Macromolecules*, 25:6075.

Lipscomb, G. F., Garito, A. F., and Narang, R. S. 1981. An exceptionally large linear electro-optic effect in the organic solid MNA. *J. Chem. Phys.* 75(3):1509.

Mandal, B. K., Chen, Y. M., Lee, J. Y., Kumar, J., and Tripathy, S. K. 1991a. Cross-Linked stable second-order nonlinear optical polymer by photochemical reaction. *Appl. Phys. Lett.* 58(22):2459.

Mandal, B. K., Lee, J. Y., Zhu, X. F., Chen, Y. M., Prakeenavincha, E., Kumar, J., and Tripathy, S. K. 1991b. Thin film processing of NLO materials. V11 A new approach to design of cross-linked second-order nonlinear optical polymers. *Synthetic Metals.* 41–43:3143.

Marder, S. R., Beratan, D. N., and Cheng, L.-T. 1991. Approaches for optimizing the first electronic hyperpolarizability of conjugated organic molecules. *Science.* 252:103.

Marder, S. R., Cheng, L.-T., and Tiemann, B. G. 1992. The first molecular electronic hyperpolarizabilities of highly polarizable organic molecules: 2,6-di-tert-butylindoanilines. *J. Chem. Soc., Chem. Comm.* 1992:672.

Marder, S. R., Cheng, L.-T., Tiemann, B. G., Friedli, A. C., Blanchard-Desce, M., Perry, J. W., and Skindhoj, J. 1994. Large first hyperpolarizabilities in push-pull polyenes by tuning of the bond length alternation and aromaticity. *Science.* 263:511.

Marder, S. R., Gorman, C. B., Tiemann, B. G., and Cheng, L.-T. 1993. Stronger acceptors can diminish nonlinear optical response in simple donor-acceptor polyenes. *J. Am. Chem. Soc.* 115(7):3006.

Marder, S. R., Perry, J. W., and Schaffer, W. P. 1989. Synthesis of organic salts with large second-order optical nonlinearity. *Science.* 245:626.

Marturunkakul, S., Chen, J. I., Li, L., Jeng, R. J., Kumar, J., and Tripathy, S. K. 1993. An interpenetrating polymer network as a stable second-order nonlinear optical material. *Chem. Mater.* 5:592.

Mataga, N. K., T. 1970. Solvent effect on the electronic spectra. In *Molecular Interactions and Electronic Spectra.* Chapter 8. Marcel Dekker, New York.

Meredith, G. R., Buchalter, B., and Hanzlik, C. 1983. Third-Order optical susceptibility determination by third-harmonic generation. 1. *J. Chem. Phys.* 78:1533.

Meredith, G. R., Van Dusen, J. G., and Williams, D. J. 1982. *Macromolecules.* 15:1385.

Meyrueix, R., Lecomte, J. P., and Tapolsky, G. 1991a. Decay of the nonlinear susceptibility components in main-chain functionalized poled polymers. *SPIE Proc.* 1560:454.

Meyrueix, R., Lecomte, J. P., and Tapolsky, G. 1991b. Study of a novel class of second-order nonlinear optical (NLO) polyurethanes. In *Organic Molecules for Nonlinear Optics and Photonics*, Messier, J., ed. p. 161. Kluwer Academic Publishers, Netherlands.

Mignani, G., Kramer, A., Puccetti, G., Ledoux, I., Soula, G., Zyss, J., and Meyrueix, R. 1990a. A new class of silicon compounds with interesting nonlinear optical effects. *Organometallics.* 9(10):2640.

Mignani, G., Leising, F., Meyrueix, R., and Samson, H. 1990b. Synthesis of new thiophene compounds with large second-order optical nonlinearities. *Tetrahedron Lett.* 31(33):4743–4746.

Miller, R. D., Moylan, C. R., Reiser, O., and Walsh, C. A. 1993. Heterocyclic azole nonlinear optical chromopheres. 1. Donor-acceptor substituted pyrazole derivatives. *Chem. Mater.* 5(5):625.

Minemoto, H., Ozaki, Y., Sonoda, N., and Sasaki, T. 1993. Intracavity second-harmonic generation using a deuterated organic ionic crystal. *Appl. Phys. Lett.* 63(26):3565.

Mitchell, M. A., Hall, H. K., Jr., Mulvaney, J. E., Willand, C., Williams, D. J., and Hampsch, H. 1992. New high glass transition nonlinear optical polymers. *Polym. Prepr.* 33(1):1060.

Mitchell, M. A., Toimida, M., Padias, A. B., Hall, H. K., Jr., Lackritz, H. S., Robello, D. R., Willand, C. S., and Williams, D. J. 1993. Synthesis and investigation of the nonlinear optical properties of various p-aminophenyl sulfone oligomers. *Chem. Mater.* 5(7):1044–1051.

Morita, R., Ogasawara, N., Umegaki, S., and Ito, R. 1987. Refractive indices of 2-methyl-4-nitroaniline (MNA). *Jpn. J. Appl. Phys.* 26(10):L1711.

Morita, R., and Vidakovic, P. V. 1992. Angle and temperature tuning of phase-matched second-harmonic generation in N-(4-nitrophenyl)-N-methylaminoacetonitrile. *Appl. Phys. Lett.* 61(24):2854.

Morrell, J. A., Albrecht, A. C., Levin, K. H., and Tang, C. L. 1979. The electro-optic coefficients of urea. *J. Chem. Phys.* 71:5063.

Mortazavi, M. A., Knoesen, A., Kowel, S. T., Henry, R. A., and Hoover, J. M. 1991. Second-order nonlinear optical properties of poled coumaromethacrylate copolymers. *Appl. Phys. B.* 53:287.

Mortazavi, M. A., Knoesen, A., Kowel, S. T., Higgins, B. G., and Dienes, A. 1989. Second-harmonic generation and absorption studies of polymer-dye films oriented by corona-onset poling at elevated temperatures. *J. Opt. Soc. Am. B.* 6(4):733.

Moylan, C. R. 1992. Advances in characterization of molecular hyperpolarizabilities. *SPIE Proc.* 1775:198.

Moylan, C. R., Miller, R. D., Twieg, R. J., Betterton, K. M., Lee, V. Y., Matray, T. J., and Nguyen, C. 1993a. Synthesis and nonlinear optical properties of donor-acceptor substituted triaryl azole derivatives. *Chem. Mater.* 5(10):1499.

Moylan, C. R., Twieg, R. J., Lee, V. Y., Swanson, S. A., Betterton, K. M., and Miller, R. D. 1993b. Nonlinear optical chromophores with large hyperpolarizabilities and enhanced thermal stabilities. *J. Am. Chem. Soc.* 115(26):12599.

Muller, H., Nuyken, O., and Strohriegl, P. 1992a. A novel method for the preparation of polymethacrylates with nonlinear optically active side groups. *Makromol. Chem. Rapid Commu.* 13:125.

Muller, S., Chastaing, E., Barny, P. L., Robin, P., and Pelle, F. 1993. Quadratic nonlinear optical properties of thermally cross-linkable polymers. *Synth. Metals.* 54:139.

Muller, S., Le Barny, P., Chastaing, E., and Robin, P. 1992b. New photo-cross-linkable polymers for second-order nonlinear optics. *Mol. Eng.* 2:251.

Murayama, H., Muta, K., Takezoe, H., and Fukuda, A. 1992. Control of organic crystal orientation in glass capillary by modified Bridgman–Stockbarger method. *Jpn. J. Appl. Phys.* 31 Part 2(6A):L710.

Nagamori, H., Kajikawa, K., Takezoe, H., Fukuda, A., Ukishima, S., Iijima, M., Takahashi, Y., and Fukada, E. 1992. Poling dynamics and negligible relaxation in aromatic polyurea studied by in situ observation of second-harmonic generation. *Jpn. J. Appl. Phys.* 31:L553.

Nakatani, H., Hayashi, H., and Hidaka, T. 1992. Linear and nonlinear optical properties of 2-cyano-3-(2-methoxyphenyl)-2-propenoic acid methyl ester. *Jpn. J. Appl. Phys.* 31:1802.

Nalwa, H. S., Watanabe, T., Kakuta, A., Mukoh, A., and Miyata, S. 1993a. N-phenylated aromatic polyurea: A new nonlinear optical material exhibiting large second-harmonic generation and UV transparency. *Polymer.* 34(3):657.

Nalwa, H. S., Watanabe, T., Kakuta, A., Mukoh, A., and Miyata, S. 1993b. Second-Order nonlinear optical properties of an aromatic polyurea exhibiting optical transparency down to 300 nm. *Appl. Phys. Lett.* 62(25):3223.

Nogami, S., Nakano, H., Shirota, Y., Umegaki, S., Shimizu, Y., Uemiya, T., and Yasuda, N. 1989. Optical second-harmonic generation from a series of [Cyano(alkyloxycarbonyl)methylene]-2-ylidene-1,3-dithioles and their methyl and dimethyl derivatives. *Chem. Phys. Lett.* 155(3):338.

Ore, F. R., Hayden, L. M., Sauter, G. F., Pasillas, P. L., Hoover, J. M., Henry, R. A., and Lindsay, G. A. 1989. Electro-optic properties of new nonlinear side-chain polymers. *SPIE Proc.* 1147:26.

Otomo, A., Mittler–Neher, S., Bosshard, C., Stegeman, G. I., Horsthuis, W. H. G., and Mohlmann, G. R. 1993. Second harmonic generation by counterpropagating beams in 4-dimethylamino-4'-nitrostilbene side-chain polymer channel waveguides. *Appl. Phys. Lett.* 63(25):3405.

Oudar, J.-L. 1977a. Optical nonlinearities of conjugated molecules. Stilbene derivatives and highly polar aromatic compounds. *J. Chem. Phys.* 67(2):446.

Oudar, J.-L., and Chemla, D. S. 1977b. Hyperpolarizabilities of the nitroanilines and their relations to the excited state dipole moment. *J. Chem. Phys.* 66:2664.

Oudar, J. L., Chemla, D. S., and Batifol, E. 1977c. Optical nonlinearities of various substituted benzene molecules in the liquid state and comparison with solid state nonlinear susceptibilities. *J. Chem. Phys.* 67(4):1626.

Oudar, J. L., and Hierle, R. 1977d. An efficient organic crystal for nonlinear optics: Methyl-(2,4-dinitrophenyl)-aminopropanoate. *J. Appl. Phys.* 48(7):2699.

Oudar, J. L., and Le Person, H. 1975. Second-Order polarizabilities of some aromatic molecules. *Opt. Commun.* 15(2):258.

Oudar, J. L., and Zyss, J. 1982. Structrual dependence of nonlinear optical properties of methyl-(2,4-dinitrophenyl)-aminopropanoate crystals. *Phy. Rev. A.* 26(4):2016.

Paley, M. S., and Harris, J. M. 1989. A solvatochromic method for determining second-order polarizabilities of organic molecules. *J. Org. Chem.* 54:3774.

Pantelis, P., Hill, J. R., and Davies, G. J. 1988. Poled copoly(vinylidene fluoride-trifluoroethylene) as a host for guest nonlinear optical molecules. In *Nonlinear Optical and Electroactive Polymers*, Prasad, P. N. and Ulrich, D. R., eds. p. 229. Plenum Publishing, New York.

Park, J., Marks, T. J., Yang, J., and Wong, G. K. 1990. Chromophore-functionalized polymeric thin-film nonlinear optical materials. Effects of in situ cross-linking on second-harmonic generation temporal characteristics. *Chem. Mater.* 2:229.

Perry, J. W., Marder, S. R., Perry, K. J., Sleva, E. T., Yakymyshyn, C., Stewart, K. R., and Boden, E. P. 1991. Organic salts with large electro-optic coefficients. *SPIE Proc.* 1560:302.

Prasad, P. N., and Williams, D. J. 1991. *Introduction to Nonlinear Optical Effects in Molecules and Polymers.* John Wiley & Sons, New York.

Pu, L. S. 1992. New Materials: Cyclobutenediones for second-order nonlinear optics. *Nonlinear Optics.* 3:233.

Puccetti, G., Perigaud, A., Badan, J., Ledoux, I., and Zyss, J. 1993. 5-Nitrouracil: A transparent and efficient nonlinear organic crystal. *J. Opt. Soc. Am. B.* 10(4):733.

Ranon, P. M., Shi, Y., Steier, W. H., Xu, C., Wu, B., and Dalton, L. R. 1993. Efficient poling and thermal cross-linking of randomly bonded main-chain polymers for stable second-order nonlinearities. *Appl. Phys. Lett.* 62(21):2605.

Rao, V. P., Jen, A. K.-Y., Wong, K. Y., and Drost, K. J. 1993. Dramatically enhanced second-order nonlinear optical susceptibilities in tricyanovinylthiophene derivatives. *Chem. Comm.* 1118.

Rikken, G. L. J. A., Seppen, G. J. E., Nijhuis, S., and Meijer, E. W. 1991. Poled polymers for frequency doubling of diode laser. *Appl. Phys. Lett.* 58(5):435.

Rikken, G. L. J. A., Seppen, G. J. E., Staring, E. G. J., and Venhuizen, A. H. J. 1993. Efficient modal dispersion phase-matched frequency doubling in poled polymer waveguides. *Appl. Phys. Lett.* 62:2483.

Rikken, G. L. J. A., Seppen, G. J. E., Venhuizen, A. H. J., Nijhuis, S., and Staring, E. G. J. 1992. Frequency doubling of diode laser with poled polymer. *Philips J. Res.* 46:215.

Robello, D. R., Dao, P. T., Phelan, J., Revelli, J., Schildkraut, J. S., Scozzafava, M., Ulman, A., and Willand, C. S. 1992. Linear polymers for nonlinear optics. 2. Synthesis and electrooptical properties of polymers bearing pendent chromophores with methylsulfonyl electron-acceptor groups. *Chem. Mater.* 4:425.

Robello, D. R., Schildkraut, J. S., Armstrong, N. J., Penner, T. L., Kohler, W., and Willand, C. S. 1991a. *Polym. Prepr.* (Am. Chem. Soc., Div. Polym. Chem.). 32(3):78.

Robello, D. R., Willand, C. S., Scozzafava, M., Ulman, A., and Williams, D. J. 1991b. Nonlinear optical chromophores in photo-cross-linked matrices. *ACS Symp. Series.* 455:279.

Robinson, D. W., Abdel-Halim, H., Inoue, S., Kimura, M., and Cowan, D. O. 1989. Hyperpolarizabilities of 4-amino-4'-nitrodiphenyl sulfide and all of its chalcogen analogues. *J. Chem. Phys.* 90(7):3427.

Robinson, D. W., and Long, C. A. 1993. Quadratic hyperpolarizability of 4-(dimethylamino)benzonitrile in solvents of differing polarity. *J. Phys. Chem.* 97(29):7540.

Rosker, M. J., Cheng, K., and Tang, C. L. 1985. *IEEE J. Quantum Electron.* QE-21:1600.

S'heeren, G., Derhaeg, L., Verbiest, T., Samyn, C., and Persoons, A. 1993a. Nonlinear optical properties of polymers and thin polymer films. *Makromol. Chem., Makromol. Symp.* 69:193.

S'heeren, G., Persoons, A., Bolink, H., Heylen, M., Van Beylen, M., and Samyn, C. 1993b. Polymers containing nonlinear optical groups in the main chain. Second-harmonic generation in corona poled thin films. *Eur. Polym. J.* 29(7):981.

S'heeren, G., Persoons, A., Rondou, P., Van Beylen, M., and Samyn, C. 1993c. Synthesis of frequency doubling nonlinear optical polymers, functionalized with aminonitrostilbene dyes. Second harmonic generation in corona poled thin films. *Eur. Polym. J.* 29(7):975.

S'heeren, G., Vanermen, G., Samyn, C., Vanbeylen, M., and Persoons, A. 1992. Poly (vinyl ethers) as nonlinear optical materials, SHG in corona poled polymer films. *MRS Proc.* 247:129.

Sagawa, M., Kagawa, H., Kakuta, A., and Kaji, M. 1993. Generation of a violet phase-matched second-harmonic wave with a new organic single crystal, 8-(4'-acetylphenyl)-1,4-dioxa-8-azaspiro[4, 5] decane. *Appl. Phys. Lett.* 63(14):1877.

Sakai, K., Yoshikawa, N., Ohmi, T., Koike, T., Umegaki, S., Okada, S., Masaki, A., Matsuda, H., and Nakanishi, H. 1990. Crystal structure and nonlinear properties of stilbazolium derivatives having shortened cut-off wavelength. *SPIE Proceedings.* 1337:307.

Sasaki, K. 1993. Poled polymers for device applications. *J. Photopolym. Sci. Technol.* 6(2):221.

Sasaki, T., and Kitaoka, Y. 1990a. The growth and laser properties of organic crystal with extremely high order nonlinear optical coefficients. *Sen-i Gakkai Symp. Preprints.* A-15.

Sasaki, T., Yoshikawa, N., Ohmi, T., Koike, T., Umegaki, S., Okada, S., Masaki, A., Matsuda, H., and Nakanishi, H. 1990b. Crystal structure and nonlinear optical properties of stilbazolium derivatives having shortened cut-off wavelength. *SPIE Proc.* 1337:307.

Sato, H., Yamamoto, T., Seo, I., and Gamo, H. 1987. Second-Harmonic generation in amorphous vinylidene cyanide/vinyl acetate copolymer using a pulsed Nd:YAG laser. *Optics. Lett.* 12(8):579.

Sekkat, Z., and Dumont, M. 1992. Photoassisted poling of azo dye doped polymeric films at room temperature. *Appl. Phys. B.* 54:486.

Seppen, G. J. E., Rikken, G. L. J. A., Staring, E. G. J., Nijhuis, S., and Venhuizen, A. H. J. 1991. Linear optical properties of frequency doubling polymers. *Appl. Phys. B.* 53:282.

Shen, Y. R. 1984. *The Principles of Nonlinear Optics.* John Wiley & Sons, New York.

Shi, R. F., Wu, M. H., Yamada, S., Cai, Y. M., and Garito, A. F. 1993a. Dispersion of second-order optical properties of new high thermal stability guest chromophores. *Appl. Phys. Lett.* 63(9):1173.

Shi, Y., Ranon, P. M., Steier, W. H., Xu, C., Wu, B., and Dalton, L. R. 1993b. Improving the thermal stability by anchoring both ends of chromophores in the side-chain nonlinear optical polymers. *Appl. Phys. Lett.* 63(16):2168.

Shi, Y., Steier, W. H., Chen, M., Yu, L., and Dalton, L. R. 1992. Thermosetting nonlinear optical polymer: Polyurethane with disperse red 19 side groups. *Appl. Phys. Lett.* 60(21):2577.

Shuto, Y., Amano, M., and Kaino, T. 1991. Electrooptic light modulation and second-harmonic generation in novel diazo-dye substituted poled polymers. *IEEE Tran. Photonic Tech. Lett.* 3(11):1003.

Sigelle, M., and Hierle, R. 1981. Determination of the electrooptic coefficients of 3-methyl 4-nitropyridine 1-oxide by an interferometric phase-modulation technique. *J. Appl. Phys.* 52(6):4199.

Singer, K. D., E., S. J., King, L. A., Gordon, H. M., Katz, H. E., and Dirk, C. W. 1989. Second-Order nonlinear optical properties of donor- and acceptor-substituted aromatic compounds. *J. Opt. Soc. Am. B.* 6(7):1339.

Singer, K. D., and Garito, A. F. 1981. Measurements of molecular second order optical susceptibilities using dc induced second-harmonic generation. *J. Chem. Phys.* 75(7):3572–3580.

Singer, K. D., Kuzyk, M. G., Fang, T., Holland, W. R., and Cahill, P. A. 1991. Design consideration for multicomponent molecular-polymeric nonlinear optical materials. In *Organic Molecules for Nonlinear Optics and Photonics*, Messier, J., Kajzar, F. and Prasad, P., eds. p. 105. Kluwer Academic Publishers, Dordrecht, Boston, London.

Singer, K. D., Kuzyk, M. G., Holland, W. R., E., S. J., and Lalama, S. J. 1988. Electro-optic phase modulation and optical second harmonic generation in corona-poled polymer films. *Appl. Phys. Lett.* 53(19):1800.

Singer, K. D., Sohn, J. E., and Lalama, S. J. 1986. Second harmonic generation in poled polymer films. *Appl. Phys. Lett.* 49(5):248.

Stahelin, M., Burland, D. M., Ebert, M., Miller, R. D., Smith, B. A., Twieg, R. J., Volksen, W., and Walsh, C. A. 1992a. Reevaluation of the thermal stability of optically nonlinear polymeric guest-host systems. *Appl. Phys. Lett.* 61(14):1626.

Stahelin, M., Burland, D. M., and Rice, J. E. 1992b. Solvent dependence of the second order hyperpolarizability in p-nitroaniline. *Chem. Phys. Lett.* 191(3, 4):245.

Stenger–Smith, J. D., Fischer, J. W., Henry, R. A., Hoover, J. M., and Lindsay, G. A. 1990. Nonlinear optical polymer with chromophoric main chain. *Makromol. Chem., Rapid Commun.* 11:141.

Stenger–Smith, J. D., Fischer, J. W., Henry, R. A., Hoover, J. M., Nadler, M. P., Nissan, R. A., and Lindsay, G. A. 1991. Poly[(4-N-ethylene-N-ethylamino)-a-cyanocinnamate]: A nonlinear optical polymer with a chromophoric main chain. 1. Synthesis and spectral characterization. *J. Polym. Sci.: Part A: Polym. Chem.* 29:1623.

Stevenson, J. L. 1973. The linear electro-optic coefficients of meta-nitroaniline. *J. Phys. D: Appl. Phys.* 6:L13.

Stiegman, A. E., Craham, E., Perry, K. J., Khundkar, L. R., Cheng, L. T., and Perry, J. W. 1991. The electronic structure and second-order nonlinear optical properties of donor-acceptor acetylenes: A detailed investigation of structure-property relationships. *J. Am. Chem. Soc.* 113:7658.

Strohriegl, P. 1993. Esterification and amidation of polymeric acyl chlorides. A new route to polymethacrylates and polymethacrylamides with a variety of different side groups. *Makromol. Chem.* 194:363.

Sugihara, O., Kunioka, S., Nonaka, Y., Aizawa, R., Koike, Y., and Kinoshita, T. 1991. Second-Harmonic generation by Cerenkov-type phase matching in a poled polymer waveguide. *J. Appl. Phys.* 70(12):7249.

Sugiyama, Y., Suzuki, Y., Mitamura, S., and Nishiyama, T. 1993. Second-Order optical nonlinearities of substituted stilbenes and the related compounds containing a trifluoromethyl as the electron-withdrawing group. *Bull. Chem. Soc. Jpn.* 66:687.

Sutter, K., Bosshard, C., Ehrensperger, M., Gunter, P., and Twieg, R. J. 1988a. Nonlinear optical and electrooptical effects in 2-methyl-4-nitro-N-methylaniline (MNMA) crystals. *IEEE J. Quantum Electron.* 24(12):2362.

Sutter, K., Bosshard, C., Wang, W. S., Surmely, G., and Gunter, P. 1988b. Linear and nonlinear optical properties of 2-(N-prolinol)-5-nitropyridine. *Appl. Phys. Lett.* 53(19):1779.

Sutter, K., Hulliger, J., Knopfle, G., Saupper, N., and Gunter, P. 1991. Nonlinear optical properties of N-(4-nitro-2-pyridinyl)-phenylalanol (NNPA) single crystals. *SPIE Proceeding.* 1560:296.

Tao, X. T., Yuan, D. R., Zhang, Z., Jiang, M. H., and Shao, Z. S. 1992. Novel organic molecular second-harmonic generation crystal: 3-methoxy-4-hydroxy-benzaldehyde. *Appl. Phys. Lett.* 60(12):1415.

Teng, C. C. 1992. Travelling-wave polymeric optical intensity modulator with more than 40 GHz of 3-dB electrical bandwidth. *Appl. Phys. Lett.* 60(13):1538.

Teng, C. C., and Garito, A. F. 1983. Dispersion of the nonlinear second-order optical susceptibility of organic systems. *Phys. Rev. B.* 28(12):6766.

Teng, C. C., and Man, H. T. 1990. Simple reflection technique for measuring the electro-optic coefficient of poled polymers. *Appl. Phys. Lett.* 56(18):1734.

Teraoka, I., Jungbauer, D., Reck, B., Yoon, D. Y., Twieg, R., and Willson, C. G. 1991. Stablity of nonlinear optical characteristics and dielectric relaxation of poled amorphous polymers with main-chain chromophores. *J. Appl. Phys.* 69(4):2568.

Terhune, R. W., Maker, P. D., and Savage, C. M. 1965. Measurements of nonlinear light scattering. *Phys. Rev. Lett.* 14(17):681.

Tiemann, B. G., Cheng, L.-T., and Marder, S. R. 1993. The effect of varying ground-state aromaticity on the first molecular electronic hyperpolarizabilities of organic donor-acceptor molecules. *J. Chem. Soc., Chem. Comm.* 1993:735.

Tiemann, B. G., Marder, S. R., Perry, J. W., and Cheng, L.-T. 1990. Molecular and macroscopic second-order optical nonlinearity of substituted dinitrostilbenes and related compounds. *Chem. Mater.* 2(6):690.

Tomaru, S., Kawachi, M., and Kobayashi, M. 1984. Organic crystals growth for optical channel waveguides. *Optics. Comm.* 50(3):154.

Tomaru, S., Matsumoto, S., Kurihara, T., Suzuki, H., Ooba, N., and Kaino, T. 1991. Nonlinear optical properties of 2-adamantylamino-5-nitropyridine crystals. *Appl. Phys. Lett.* 58(23):2583.

Tsunbekawa, T., Gotoh, T., Mataki, H., Kondoh, T., Fukuda, S., and Iwamoto, M. 1990. New organic second-order nonlinear optical crystals of benzylidene-aniline derivative. *SPIE Proc.* 1337:272.

Twieg, R., M., E., Jungbauer, D., Lux, M., Reck, B., Swalen, J., Teraoka, I., Willson, C. G., Yoon, D. Y., and Zentel, R. 1992. Nonlinear optical epoxy polymers with polar tolan chromophores. *Mol. Cryst. Liq. Cryst.* 217:19.

Twieg, R. J., and Dirk, C. W. 1986. Molecular and crystal structure of the nonlinear optical material: 2-(N-prolinol)-5-nitropyridine (PNP). *J. Chem. Phys.* 85(6):3537.

Uemiya, T., Uenishi, N., Shimizu, Y., Yoneyama, T., and Nakatsu, K. 1990. Crystal growth and characterization of N-(5-nitro-2-pyridyl)-(s)-phenylaninol: A new organic material for nonlinear optics. *Mol. Cryst. Liq. Cryst.* 182A:51.

Uemiya, T., Uenishi, N., and Umegaki, S. 1993. Modified method of electric-field induced second-harmonic generation. *J. Appl. Phys.* 73(1):12.

Ukachi, T., Shigemoto, T., Komatsu, H., and Sugiyama, T. 1993. Crystal growth, and characterization of a new organic nonlinear optical material: L-N-(5-nitro-2-pyridyl) leucinol. *J. Opt. Soc. Am. B.* 10(8):1372.

Ulman, A., Willand, C. S., Kohler, W., Robello, D. R., Williams, D. J., and Handley, L. 1990. New sulfonyl-containing materials for nonlinear optics: Semiempirical calculations, synthesis, and properties. *J. Am. Chem. Soc.* 112(20):7083.

Verbiest, T., Clays, K., and Persoons, A. 1993. Determination of the hyperpolarizability of an octopolar molecular ion by hyper-Rayleigh scattering. *Opt. Lett.* 18(7):525.

Wang, C. H., and Guan, H. W. 1992. Electro-optics and second harmonic generation of nonlinear optical polymers. *SPIE Proc.* 1775:318.

Wang, C. H., Lloyd, A. D., Wherrett, B. S., Bone, D. J., Harvey, T. G., Ryan, T. G., and Carter, N. 1993. Electro-optic modulation using Fabry–Perot etalons containing a poled cross-linked polymer. *Opt. Mater.* 2:95.

Willetts, A., Rice, J. E., and Burland, D. M. 1992. Problems in the comparison of theoretical and experimental hyperpolarizabilities. *J. Chem. Phys.* 97(10):7590.

Wu, J. W., Binkley, E. S., Kenney, J. T., Lytel, R., and Garito, A. F. 1991a. Highly thermally stable electro-optic response in poled guest-host polyimide systems cured at 360°C. *Appl. Phys. Lett.* 69(10):7366.

Wu, J. W., Valley, J. F., Ermer, S., Binkley, E. S., Kenney, J. T., Lipscomb, G. F., and Lytel, R. 1991b. *Appl. Phys. Lett.* 58:225.

Wurthner, F., Effenberger, F., Wortmann, R., and Kramer, P. 1993. Second-Order polarizability of donor-acceptor substituted oligothiophenes: Substituent variation and conjugation length dependence. *Chem. Phys.* 173:305–314.

Xu, C., Wu, B., Dalton, L. R., Ranon, P. M., Shi, Y., and Steier, W. H. 1992a. New random main-chain, second-order nonlinear optical polymers. *Macromolecules.* 25:6716.

Xu, C., Wu, B., Dalton, L. R., Ranon, P. M., Shi, Y., and Steier, W. H. 1993. New cross-linkable polymers with second-order nonlinear optical chromophores in the main chain. *SPIE Proc.* 1852:198.

Xu, C., Wu, B., Dalton, L. R., Shi, Y., Ranon, P. M., and Steier, W. H. 1992b. Novel double-end cross-linkable chromophores for second-order nonlinear optical materials. *Macromolecules.* 25:6714.

Yakovlev, Y. O., and Poezzhalov, V. M. 1990. Nonlinear optical properties of p,p'-dihydroxydiphenyl sulfone crystals. *Sov. J. Quantum Electron.* 20(6):694.

Yang, G.–M., Bauer–Gogonea, S., Sessler, G. M., Bauer, S., Ren, W., Wirges, W., and Gerhard–Multhaupt, R. 1994. Selective poling of nonlinear optical polymer films by means of a monoenergetic electron beam. *Appl. Phys. Lett.* 64(1):22.

Yankelevich, D., Knoessen, A., A., E. C., and Kowel, S. 1991. Reflection-mode polymeric interference modulators. *SPIE Proc.* 1560:406.

Ye, C., Feng, Z., Wang, J., Shi, H., and Dong, H. 1992. Investigation of poled film of styrene copolymer with p-dicyanovinyl azo dye side chain. *Mol. Cryst. Liq, Cryst.* 218:165.

Ye, C., Marks, T. J., Yang, J., and Wong, G. K. 1987. Synthesis of molecular arrays with nonlinear optical properties, second-harmonic generation by covalently functionalized glassy polymer. *Macromolecules.* 20:2322.

Ye, C., Minami, N., Marks, T. J., Yang, J., and Wong, G. K. 1989. Synthetic approaches to stable and efficient frequency doubling materials. Second-order nonlinear properties of poled chromophore-functionalized glassy polymer. In *Nonlinear Optical Effects in Organic Polymers,* Messier, J., Kajzar, F., Prasad, P. and Ulrich, D., eds. p. 173. Kluwer Academic Publishers, Dordrecht, Boston, London.

Yitzchaik, S., Berkovic, G., and Krongauz, V. 1991. Charge injection asymmetry: A new route to strong optical nonlinearity in poled polymers. *J. Appl. Phys.* 70(7):3949.

Yoshimura, T. 1987. Characterization of the electro-optic effect in styrylpyridinium cyanine dye thin-film crystals by an ac modulation method. *J. Appl. Phys.* 62(5):2028.

Yu, L., Chan, W., Dikshit, S., Bao, Z., Shi, Y., and Steier, W. H. 1992. Thermally curable second-order nonlinear optical polymer. *Appl. Phys. Lett.* 60(14):1655.

Zentel, R., Baumann, H., Scharf, D., Eich, M., Schoenfeld, A., and Kremer, F. 1993. Second-order nonlinear optical main-chain polymer by polymerization of poled monomers. *Makromol. Chem. Rapid Commun.* 14:121.

Zhang, G. J., Kinoshita, T., and Sasaki, K. 1990. Second-harmonic generation of a new chalcone-type crystal. *Appl. Phys. Lett.* 57(3):221.

Zhang, N., Yuan, D. R., Tao, X. T., Xu, D., Shao, Z. S., Jiang, M. H., and Liu, M. G. 1993. Phase-matched second harmonic generation in new organic MHBA crystal. *Optics Comm.* 99:247.

Zhang, X.–C., Ma, X. F., Jin, Y., Lu, T.–M., Boden, E. P., Phelps, P. D., Stewart, K. R., and Yakymyshyn, C. P. 1992. Terahertz optical rectification from a nonlinear organic crystal. *Appl. Phys. Lett.* 61(26):3080.

Zyss, J., Chemla, D. S., and Nicoud, J. F. 1981. Demonstration of efficient nonlinear optical crystals with vanishing molecular dipole moment: Second-Harmonic generation in 3-methyl-4-nitropyridine-1-oxide. *J. Chem. Phys.* 74(9):4800.

Zyss, J., Nicoud, J. F., and Coquillary, M. 1984. Chirality and hydrogen bonding in molecular crystals for phase-matched second-harmonic generation: N-(4-nitrophenyl)-(L)-prolinol (NPP). *J. Chem. Phys.* 81(9):4160.

Zyss, J., and Oudar, J. L. 1982. Relations between microscopic and macroscopic lowest order optical nonlinearities of molecular crystals with one- or two-dimensional units. *Phys. Rev. A.* 26(4):2028.

Zyss, J., Van, T. C., Dhenaut, C., and Ledoux, I. 1993. Harmonic Rayleigh scattering from nonlinear octopolar molecular media: The case of crystal violet. *Chem. Phys.* 177:281.

PART B
PHOTONIC DEVICES AND OPTICS

SECTION IV
Devices

7
Optoelectronic Devices

Ilesanmi Adesida
University of Illinois

James J. Coleman
University of Illinois

7.1 Introduction 291
7.2 Semiconductor Diode Lasers 292
    Principles • Transverse Structure • I-V Characteristic • Light-Current Characteristic • Lateral Structure • Spatial Optical Profiles • Optical Spectrum • Modulation • Temperature Effects • Reliability
7.3 Photodetectors 301
    Principles • Quantum Efficiency • Responsivity • Bandwidth • Noise Equivalent Power • Photoconductors • PIN Photodiodes • Avalanche Photodiodes (APDs) • Metal-Semiconductor-Metal (MSM) Photodiode

7.1 Introduction

The advent of low-loss fibers has been a key to the recent emergence of optical methods as the backbone of long distance communications. The sheer amount of information that can be moved via optical communications is destined to have a huge impact on every aspect of human endeavor. The enormous scientific and economic potential of the foregoing fact is largely responsible for the dynamic pace of research and development activities in the areas of optoelectronic devices and optoelectronic integrated circuits (OEICs). These devices and circuits along with optical fibers constitute the basic components for the physical network needed to transmit information. This network has been dubbed the information superhighway because of the dense amount of traffic it is expected to support. Other areas in which optoelectronic devices have found applications are optical storage systems and entertainment systems, such as compact disks.

The basic optoelectronic devices which are usually made with semiconductor materials are heterojunction lasers, light emitting diodes (LEDs), and photodetectors. The integration of lasers or photodetectors with electronic devices, such as field effect transistors and/or bipolar transistors constitute OEICs. This integration can be of the hybrid or the monolithic form. Although silicon is the predominant material for microelectronics, it is not suitable for optoelectronic devices in which energy conversion from electrical to optical and vice versa is the fundamental mechanism of operation. Optoelectronic devices are made from compound semiconductors which have direct energy gaps. Energy conversion is more efficient in these materials. III-V compound semiconductors have been the traditional materials used for optoelectronic devices. However, the II-VI compounds have recently demonstrated the potential for optoelectronic devices operating at short wavelengths. The wide bandgap nitrides, which also have direct bandgaps, have demonstrated impressive results for lasers and photodetectors at short wavelengths. The properties of all these materials are discussed in other chapters of this handbook.

The intent of this chapter is to discuss the general principles and characteristics of semiconductor lasers and photodetectors. Although the types of lasers and photodetectors available are diverse, there is an underlying unity in the basic principles and characteristics so that figures of merit can be used to compare the various devices. Tranverse structure, spatial optical profiles, and modulation are some of the important laser characteristics that are discussed. Quantum efficiency, responsivity, bandwidth, and noise equivalent power are the important parameters used to characterize photodetectors. These are discussed and the relevant basic equations are provided. The different types of photodetectors—photoconductors, p-i-n photodiodes, and avalanche photodiodes, commonly used along with the relatively new metal-semiconductor-metal photodiodes, are included in this chapter.

7.2 Semiconductor Diode Lasers

Principles

The semiconductor diode laser is a two terminal optoelectronic device that can very efficiently convert electrical energy in the form of a current to optical energy in the form of a coherent, single-frequency beam of light. Thus, characterizing a laser diode requires understanding the electrical properties of the device, which are similar to those of a conventional semiconductor junction diode, and the optical properties of the device, which include the spatial and spectral properties of the emitted light. Injected current results in a large number of electron-hole pairs. These can recombine to yield photons with nearly unitary quantum efficiency. The photons propagate in a waveguide resonant cavity, establish optical gain, and, eventually, laser oscillation. A useful portion of the light is emitted from the laser cavity.

In concise terms, the basic processes are

a) injection of electron-holes pairs by means of an injected current,
b) efficient recombination of electron-hole pairs to form photons, and
c) generation of gain and oscillation from photons propagating in a resonant cavity.

Transverse Structure

The physical structure of a semiconductor laser generally consists of a number of epitaxial layers, typically five or more, engineered in thickness, composition, and doping to provide an electrically and optically efficient device. The schematic cross section of a typical $Al_xGa_{1-x}As$-GaAs semiconductor laser is shown in Fig. 7.1. Similar structures are used with other material systems. The p-n junction structure consists of two relatively thick confining layers, two thinner inner barrier layers, and an undoped active layer. The cap layer and substrate are heavily doped to facilitate low-resistance ohmic contacts. The energy band structure, consisting of the conduction band edge E_c and the valence band edge E_v, for this structure, called the separate confinement heterostruc-

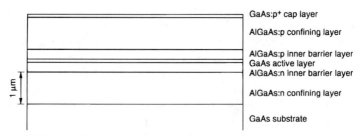

FIGURE 7.1 Schematic cross-section of a typical $Al_xGa_{1-x}As$-GaAs semiconductor laser diode.

Optoelectronic Devices

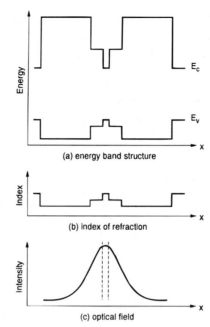

FIGURE 7.2 Energy band structure, index of refraction profile, and optical field profile for the separate confinement, heterostructural (SCH) laser.

ture (SCH), are shown in Fig. 7.2a, where x is measured into the wafer from the surface. The confining layers are chosen to have the widest bandgap energy, the active layer has the smallest bandgap energy, and the inner barrier layers have an intermediate bandgap energy. This structure provides for efficient confinement of injected electrons and holes in the active layer.

The index of refraction profile of the SCH structure, shown in Fig. 7.2b, provides effective optical confinement and determines the optical field profile of the laser shown in Fig. 7.2c. A key parameter in describing the transverse waveguide is the optical confinement factor Γ, defined as the fraction of the total optical field that overlaps the active layer. Another typical laser structure is the graded-index, separate confinement heterostructure (GRIN-SCH) laser, shown in Fig. 7.3.

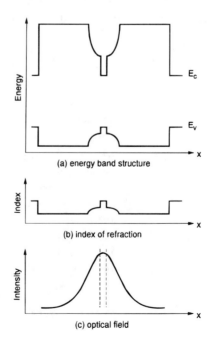

FIGURE 7.3 Energy band structure, index of refraction profile, and optical field profile for the graded index separate confinement, heterostructural (GRIN-SCH) laser.

In this structure, the inner barrier layers are replaced by a parabolically graded layer, similar to the graded-index optical fiber. There are many other laser structures, all of which provide electrical and optical confinement in a similar manner, but differ slightly in detail.

The quantum well laser differs only slightly from the structures described above. The chief difference is that the active layer is thin enough, usually less than 200 Å, so that quantum size effects are important. For these structures, the optical confinement factor Γ is considerably smaller, but the optical gain is larger, requiring much less current. In some cases, a number of closely coupled quantum wells are utilized in place of a single active layer.

I-V Characteristic

The i-V characteristic of a semiconductor laser is identical to that of any pn junction diode. A typical i-V characteristic is shown in Fig. 7.4. The electrical p-n junction is located in the active layer so the turn-on voltage, or contact potential V_o for the diode is given approximately by

$$V_o = qE_g, \tag{7.1}$$

where E_g is the active layer bandgap energy and q is the fundamental charge. Similarly, the emission wavelength λ of the laser is related to the active layer bandgap energy by

$$\lambda = \frac{hc}{E_g}, \tag{7.2}$$

where h is Planck's constant and c is the speed of light. For commonly used units of wavelength and energy, this gives

$$\lambda(\mu m) = \frac{1.239852}{E_g(eV)}. \tag{7.3}$$

The series resistance of the laser diode R_s is given by

$$\frac{1}{R_s} = \frac{dI}{dV}, \tag{7.4}$$

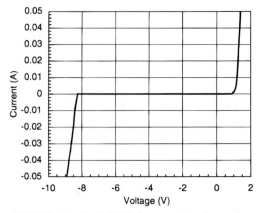

FIGURE 7.4 Typical i-V characteristic of a semiconductor laser.

where the slope is taken above the turn-on voltage. The largest contribution to series resistance is usually the contact resistance from the metal contacts. Typical and acceptable values for the series resistance are <1 Ω. Because of the increasing depletion width with reverse bias, the reverse breakdown voltage is determined by the doping, composition, and thickness of the active layer and the surrounding layers. Typical values are in the range of 8–10 V.

Light-Current Characteristic

The light-current characteristic for a typical laser diode is shown in Fig. 7.5. At low drive current levels, the power output is a relatively low-level, spontaneous emission. At laser threshold, indicated by a knee in the light-current characteristic, the gain in the resonant cavity equals the losses and laser action begins. Shown as a dashed line in Fig. 7.5 is a linear fit to the characteristic above the threshold. The intercept along the x axis is considered to be the laser threshold current whereas the slope is a measurement of the external, or differential, quantum efficiency η_{ext}. The threshold current is given by

$$I_{th} = wLJ_{th} \qquad (7.5)$$

where w is the width, L is the cavity length, and J_{th} is the threshold current density given by

$$J_{th}(A/cm^2) = \frac{J_o}{\eta} + \frac{1}{\eta\beta\Gamma}\left[\alpha_i + \frac{1}{2L}\ln\frac{1}{R_1R_2}\right], \qquad (7.6)$$

where η is the internal quantum efficiency (typically greater than 0.90), Γ is the confinement factor, J_o is the transparency current density, and β is the linear gain coefficient, which is defined as the differential of gain with respect to current density ($\gamma_i = \beta[J - J_o]$). R_1 and R_2 are the facet reflectivities. In conventional double heterostructure lasers, the transparency current density overwhelmingly dominates the threshold current density. For quantum well heterostructure lasers, a nonlinear gain coefficient, taken from $\gamma_i = J_o\beta \ln(J/J_o)$, is more appropriate, and results in a threshold current density given by

$$J_{th}(A/cm^2) = \frac{J_o}{\eta}\exp\left(\frac{1}{J_o\beta\Gamma}\left[\alpha_i + \frac{1}{2L}\ln\frac{1}{R_1R_2}\right]\right). \qquad (7.7)$$

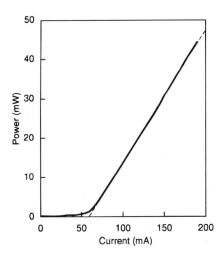

FIGURE 7.5 Light-current characteristic for a typical laser diode.

The very small confinement factor in thin, single quantum well, heterostructure lasers can be increased while maintaining the key elements of the quantum size effect by utilizing a multiple quantum well structure. For m wells, the total confinement factor, and, hence, modal gain, increases by approximately m times, with the penalty that the transparency current density also increases by a factor of m.

The threshold for the equations above is defined by setting the gain equal to the loss in the laser cavity. This loss consists of undesirable distributed loss α_i, such as residual optical absorption, and useful mirror loss α_m arising from light transmitted from the ends of the laser. The distributed loss is usually small in high-performance lasers ($\alpha_i = 3\text{--}15$ cm^{-1}) whereas the mirror loss, a function of cavity length and facet reflectivity, is given by

$$\alpha_m = \frac{1}{2L} \ln \frac{1}{R_1 R_2}. \tag{7.8}$$

The slope efficiency of the laser is given in terms of either a percent where

$$\eta_{\text{ext}} = \frac{1}{V} \frac{dP}{dI} \tag{7.9}$$

or simply in the units of W/A where

$$\eta_{\text{ext}} = \frac{dP}{dI}. \tag{7.10}$$

The external and internal quantum efficiencies are related by

$$\eta_{\text{ext}} = \eta_i \frac{\frac{1}{2L} \ln\left(\frac{1}{R_1 R_2}\right)}{\alpha_i + \frac{1}{2L} \ln\left(\frac{1}{R_1 R_2}\right)}. \tag{7.11}$$

The wall plug efficiency, at any value of drive current, is given by

$$\frac{P(I)}{I(V_o + IR_s)}. \tag{7.12}$$

The uncoated facets of a semiconductor laser have a natural facet reflectivity of $R_1 = R_2 = 0.30$. In most cases, it is desirable to have significant light emission from only a single facet, which can be obtained by providing dielectric optical coatings on the facets. High reflectivity (HR) facet coatings have a typical facet reflectivity of $R_1 = 0.90\text{--}0.95$, and antireflection (AR) facet coatings typically have a reflectivity of $R_2 = 0.10\text{--}0.05$. It is common to design the AR-HR coatings so that the product of R_1 and R_2 is ~ 0.09 as it is for uncoated facets.

Lateral Structure

Threshold current densities of optimized double heterostructure lasers are typically in the range of 500–750 A/cm^2 and, for quantum well heterostructure lasers, this value can be as small as 150 A/cm^2. The typical cavity length of a semiconductor laser is in the range of 300 to 1000 μm and cannot be significantly reduced without increasing the threshold current density. Reasonable drive circuitry and heat sinking require $I_{\text{th}} < 30$ mA, so that the width of the laser must be adjusted

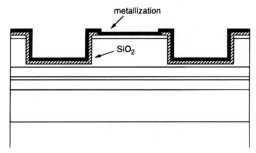

FIGURE 7.6 Cross-section of a ridge waveguide, index-guided laser structure.

accordingly. Ideally, the width is as small as possible. In addition to confining the current to a narrow stripe, it is also important to consider the beam shape. The transverse waveguide described above very strongly guides the optical wave and results in a transverse beam width of 1 μm or less at the facet. To minimize beam astigmatism, a similar strong narrow waveguide in the lateral direction is desirable.

An important example of a common, index-guided laser structure is the ridge waveguide laser shown in Fig. 7.6. After epitaxial growth of the structure, narrow mesa stripes are patterned by conventional lithography and etched with wet chemicals or by one of the dry etching methods near (<0.25 μm), but not through, the active region. The optical field outside the stripe is distorted by the proximity of the etched surface, oxide, and metallization resulting in a relatively large step in the lateral effective index. Because the index step is structural, the lateral waveguide formed is independent of the drive current and, if the stripe width and index step are appropriate with respect to the wavelength, fundamental lateral mode operation can be maintained over a wide range of drive current. In addition, the stripe width necessary for fundamental mode operation is usually only 1–3 μm, resulting in threshold currents of less than 20 mA for typical double heterostructure, ridge waveguide lasers.

Spatial Optical Profiles

The spatial near-field emission patterns of a semiconductor laser are defined by the transverse and lateral index of refraction profiles. As described above, the transverse index profile is generally characterized by a large index step and a thin active layer. In contrast, the lateral index step is usually much smaller with a wider guiding layer. Shown in Fig. 7.7 is the astigmatic near-field emission pattern at the facet for a typical laser diode. Shown for reference are the cross-sectional intensity profiles in the lateral and transverse directions with the index of refraction profiles as

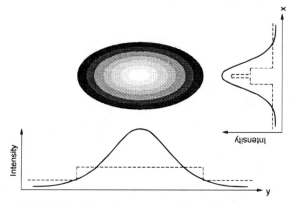

FIGURE 7.7 The astigmatic near-field emission pattern at the facet for a typical laser diode.

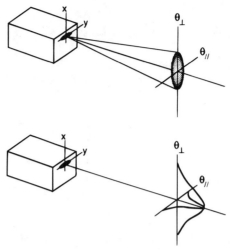

FIGURE 7.8 Schematic description of the Fourier transform relationship between the far-field emission pattern and the near-field emission pattern of a semiconductor diode laser.

dashed lines. The near-field beam width in each direction is characterized by a full width at half maximum (FWHM), typically in units of microns.

The far-field emission pattern of a semiconductor diode laser is, of course, the Fourier transform of the near-field emission pattern. The relationship is shown schematically in Fig. 7.8. The far-field beam width in each direction is characterized by a FWHM beam divergence, typically in units of degrees.

Optical Spectrum

The optical spectrum of a semiconductor laser diode is a convolution of the material gain spectrum of the active layer of the laser and the spectral response of the resonant cavity. The gain spectrum, shown in Fig. 7.9, results from the approximately parabolic band structure of semiconductor

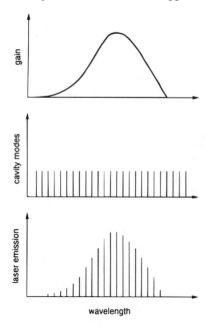

FIGURE 7.9 The gain spectrum, Fabry–Perot cavity mode spectrum, and the emission spectrum of a semiconductor laser.

Optoelectronic Devices

materials multiplied by a Fermi–Dirac occupancy probability function. This is a relatively broad function, usually many hundreds of angstroms wide. The cavity mode spectrum shown in Fig. 7.9 is for a typical Fabry–Perot resonator formed by cleaving and, perhaps, later coating plane, parallel, reflecting facets oriented along a fundamental crystal plane of the laser material. These cleaved facets yield a large number of strong resonances with a spacing given by

$$\Delta\lambda = \frac{\lambda^2}{2\pi L[n - \lambda \partial n/\partial \lambda]} \quad (7.13)$$

where L is the cavity length and $[n - \lambda \partial n/\partial \lambda]$ is the effective index of refraction, including dispersion. Typical values of $\Delta\lambda$ for normal cavity lengths are in the range of 3 Å. The product of material gain and cavity resonance is also shown in Fig. 7.9. At higher drive currents, a smaller number of longitudinal modes, or a single mode, often dominates the laser emission. With heating or additional drive current, however, mode hopping to nearby adjacent cavity modes can easily occur.

When the longitudinal mode stability of the Fabry–Perot resonator is insufficient for an application, greater stability against wavelength shift with temperature and mode hopping can be obtained by using a distributed feedback (DFB) or a distributed Bragg reflector (DBR) laser structure. These structures, processed with a wavelength-sensitive Bragg grating either distributed along the gain path (DFB) or at the end of the gain path (DBR), are characterized by a very narrow linewidth, single, resonant frequency. Shown in Fig. 7.10 is the longitudinal mode spectrum of a ridge waveguide DBR laser. The side mode suppression ratio for this family of structures is usually greater than 35 dB.

Modulation

The simplest and most direct method for modulating a semiconductor laser diode is modulation of the laser drive current. Direct current modulation is best described by rate equations for both carriers and photons. Solution of the rate equations gives the modulation frequency response,

$$\frac{\partial P}{\partial J} = \frac{-(1/qd)\Gamma A P_o}{\omega^2 - i\omega/\tau_s - i\omega A P_o - A P_o/\tau_p}, \quad (7.14)$$

where P is the photon density, J is the current density, A is a constant related to the gain coefficient, d is the active region thickness, τ_s is the spontaneous recombination lifetime, τ_p is the photon lifetime, and ω is the frequency. The modulation frequency response is shown for two output

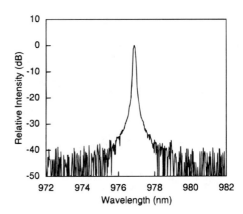

FIGURE 7.10 Longitudinal mode spectrum of a ridge waveguide, distributed Bragg reflector (DBR) laser.

power levels in Fig. 7.11. The frequency response is flat to frequencies approaching 1 GHz, rises to a peak value at some characteristic frequency, and, then, quickly rolls off. The peak in the frequency response is given approximately by

$$\omega_{max} = \left[\frac{AP_o}{\tau_p}\right]^{1/2}. \qquad (7.15)$$

Because A is related to the gain of the laser, higher frequency operation requires greater gain, greater photon density, and decreased photon lifetime.

Modulation of the laser drive current modulates the injected carrier density which, in turn, results in a modulation of the quasi-Fermi levels. Because the emission wavelength and, hence, frequency is related to the difference between the quasi-Fermi levels, modulation of the carrier density results in modulation of the emission frequency. Thus, amplitude modulation (AM) of the laser power output results in a corresponding frequency modulation (FM). The change and broadening of the emission frequency is called "chirp".

Temperature Effects

There are two key design parameters associated with variation of the operating junction temperature of a semiconductor laser diode. The first is the variation of the laser threshold current with temperature. This is usually quantified as a characteristic temperature T_o where the threshold current is given by

$$I_{th}(T) = I_o \exp(T/T_o) \qquad (7.16)$$

where I_o and T_o are only valid around a specific temperature, usually room temperature. Values of $T_o > 125$ K are observed for AlGaAs lasers whereas, for InGaAsP lasers, values in the range of 50–60 K are typical. The highest values are most desirable.

The second parameter is the change in wavelength with operating temperature $d\lambda/dT$. The wavelength for peak gain in most semiconductor materials varies with temperature in the range of 3–5 Å/°C. This is also the temperature variation of wavelength observed in Fabry–Perot lasers where there are a large number of closely spaced, cavity modes. Actually, of course, there is also mode hopping taking place. In DFB and DBR lasers, the mode spacing is much larger, and the inherent temperature dependence of the Bragg wavelength is much smaller, typically less than 0.8 Å/°C.

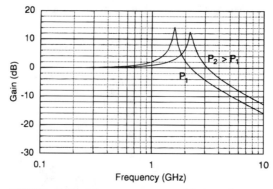

FIGURE 7.11 Direct current modulation frequency response of a semiconductor laser diode for two output power levels.

Reliability

The failure mechanisms for semiconductor lasers can be separated into two relatively broad categories. The first category is catastrophic failure associated with exceeding the maximum safe value of some operating parameter. Perhaps the most important example is catastrophic facet damage. When the optical power density at the laser facet reaches a certain value, catastrophic optical damage (COD) occurs. COD is a function of the shape of the near-field pattern, the drive current amplitude and pulse length, and the presence or absence of passivation on the facets. COD occurs rapidly and irreversibly at the laser's upper limit for power output.

The second general category of semiconductor laser diode failure is gradual degradation resulting from long term effects related to materials, such as defects or contaminants, or processing, such as handling damage. An estimate of the long term reliability and mean time between failure (MTBF) of a particular laser design is an important part of commercial laser development, especially for remote or space-based laser systems. The measurements utilized to establish reliability parameters include current constant power measurements, where device failure is defined as the laser output power falling to half its initial value, or constant output power current adjustment, where device failure is defined as the laser drive current rising to twice its initial value.

As semiconductor laser lifetimes have approached those of other solid state electronic devices, measured in tens of thousands of hours, accelerated life testing methods have become important. By measuring the failure rate or median lifetime as a function of temperature above room temperature (typically 30°C to 70°C), an activation energy E_a can be determined from the slope of the best fit line. This activation energy allows a preliminary analysis of the dominant failure mechanism, because certain failure mechanisms have characteristic activation energies, and extrapolation of a room temperature lifetime.

7.3 Photodetectors

Principles

A photodetector is a device that converts optical energy to electrical energy. The principal mechanism responsible for this transformation is photoconductivity. This property is exhibited by all semiconductors and it is the increase in conductivity brought about by the absorption of photons. The absorption of a photon results in the generation of an electron-hole pair. The electrons and holes separate to become mobile carriers which are transported through the semiconductor under the influence of an externally applied electric field. The transport of these carriers enhances the conductivity of the semiconductor.

In concise terms, the three basic processes involved in this conversion are

a) the absorption of photons and the resulting generation of carriers.
b) the transport of the generated carriers across the absorption or drift region under the influence of an applied field. Internal amplification of carriers can occur at this stage via various mechanisms. An example of a mechanism is impact ionization which occurs with the application of large electric fields.
c) collection of carriers constituting a photocurrent which flows through an external circuitry.

Perhaps, the simplest type of photodetector is the photoconductor which is simply a slab of an intrinsic semiconductor with two contacts. The electron-hole pairs, generated by absorption of photons in the material, are collected by oppositely biased contacts to constitute a photocurrent. Other types of photodetectors are photodiodes which are based on either the p-n junction or the metal-semiconductor junction (also called a Schottky-barrier). A p-i-n (or PIN) photodiode is a reverse-biased p-n junction with an intrinsic layer interposed between the p and n layers.

Because the depletion area is the only region supporting an electric field in a p-n junction, the intrinsic layer serves to increase the depletion layer width and therefore the photon absorption region of the device. The PIN photodiode is normally operated in a bias mode in which the device does not exhibit gain. Another type of photodetector based on a p-n junction is the avalanche photodiode (APD). APDs are operated at electric fields which are high enough to cause impact ionization and, thereby, generate more carriers, leading to the avalanche effect. The net effect of the avalanche process is a multiplication of carriers, resulting in gain for the output photocurrent of the device. As mentioned above, another important class of photodetectors, which have recently gained prominence, are the metal-semiconductor photodiodes which are made from metal-semiconductor junctions. The relative ease with which they can be fabricated has made them and the metal-semiconductor-metal (MSM) photodiodes attractive for some applications requiring monolithic integration.

As described above, there are some basic properties exhibited by all semiconductor photodetectors. These general properties can be quantified, to a certain extent, and have become figures of merit used in comparing photodetectors. The properties can be quantified in terms of quantum efficiency, responsivity, bandwidth, and noise equivalent power (NEP).

Quantum Efficiency

The quantum efficiency η of a photodetector is the number of electron-hole pairs generated per incident photon collected at the contacts. Quantum efficiency is determined by many factors. These include the fact that not all photons incident on the semiconductor will produce electron-hole pairs and that some of the photons may be reflected at the surface of the semiconductor. All these factors combine to reduce η. Quantum efficiency can therefore be given by

$$\eta = (1 - r)\zeta[1 - \exp(-\alpha L)] \tag{7.17}$$

where r is the optical power reflectance at the surface of the detector, ζ is the fraction of electron-hole pairs that actually contribute to the photocurrent, α is the absorption coefficient of the detector material in cm^{-1}, and L is the width of the detector's absorption region. By applying an antireflection coating on the detector's surface for the wavelength of operation, reflection can be reduced and, thereby, the factor $(1 - r)$ can be maximized. It is difficult to estimate the factor ζ because it depends on the quality of the materials. Carriers can be lost through recombination at the surface or in the bulk of the material which reduces the photocurrent. Modern epitaxial growth methods are now capable of producing high quality materials, and, therefore, for a practical estimation of quantum efficiency, ζ can be assumed as unity. The last factor $[1 - \exp(-\alpha L)]$ denotes the fraction of the incident optical power absorbed in bulk of the detector.

In terms of the quantities easily measured in the laboratory, quantum efficiency is given by

$$\eta = \frac{I_p/q}{P_i/h\nu} = \frac{h\nu}{q} \cdot \frac{I_p}{P_i} \tag{7.18}$$

where I_p is the detector photocurrent, P_i is the incident optical power, and $h\nu$ is the photon energy. The quantity η given above is known as the external quantum efficiency η_{ext}. The quantum efficiency is dependent on the absorption coefficient α which is a function of wavelength λ; therefore, η is dependent on λ. Fig. 7.12 shows the dependence of α on λ for some detector materials. Because only photons with energy greater than or equal to the bandgap energy E_g can be absorbed (i.e., $hc/\lambda \geq E_g$), the long-wavelength limit for a practical detector is the bandgap wavelength. Bandgap energies at 300 K for representative photodetector materials are displayed in Table 7.1. There is also a short-wavelength limit because α is very large at short wavelengths

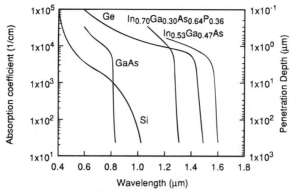

FIGURE 7.12 Wavelength dependence of the optical absorption coefficients of several semiconductor materials.

TABLE 7.1. Bandgap Energies (in eV) at 300K for Some Photodiode Materials

Material	Bandgap Energy (eV)
GaAs	1.42
GaSb	0.73
$GaAs_{0.88}Sb_{0.12}$	1.15
Ge	0.67
InAs	0.35
$In_{0.53}Ga_{0.47}As$	0.75
InP	1.35
Si	1.14

for most semiconductors, and, consequently, all of the incident photons are absorbed near the surface of the detector.

Responsivity

The responsivity $\mathcal{R}$ of a detector is the photocurrent in the device divided by the input optical power and is given by

$$\mathcal{R} = \frac{I_p}{P_i} = \frac{\eta q}{h\nu} = \eta \frac{\lambda(\mu m)}{1.24}. \tag{7.19}$$

The unit of responsivity is A/W. It is seen from this expression that, for a constant η, $\mathcal{R}$ should increase with λ. This is illustrated in Fig. 7.13. However, because α depends on λ, there is a region between the short- and long-wavelength limits over which $\mathcal{R}$ increases. For photodetectors which exhibit gain, the gain factor G can be accommodated in a more general equation for responsivity given by

$$\mathcal{R} = G\eta \frac{\lambda(\mu m)}{1.24}. \tag{7.20}$$

It is possible to degrade the responsivity of a detector by applying excessive incident optical power. The detector becomes saturated, thus, limiting its linear dynamic range, which is the range over which the relationship between the detector's output and the incident optical power is linear.

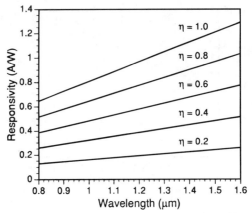

FIGURE 7.13 Responsivity versus wavelength for various external quantum efficiencies.

Bandwidth

The bandwidth B of a photodetector measures the shortest response time of the device. This property becomes very important when a photodetector is used in a data transmission circuit. The faster a detector can respond to a stream of optical pulses, the higher the density of the transmitted data can be. The response time of a photodetector is determined by three factors—transit time, diffusion time, and the device RC time constant.

Electron-hole pairs created by photons in the active region of a photodetector move in directions opposite to the contacts for collection under an applied electric field. The carriers move by drift and diffusion. If the electric field is sufficiently large, most of the carriers travel by drift, and they reach their scattering-limited or saturation velocity in the material. The velocity of holes is usually smaller than that of electrons, therefore, the time (i.e., transit time) it takes holes to drift across the active region of the detector limits the response time of the device. If electron-hole pairs are generated uniformly throughout the material, then, a severe transit time spread between electrons and holes can occur. Diffusion time limitations can occur only at low bias where the drift field is low. Because the diffusion process is slow, it can be a severe problem even though only a small number of carriers may be involved. A judicious design of the active area of the detectors and the application of an appropriate bias can make this limitation insignificant. The last factor is the resistance R and the capacitance C of the device and its associated circuitry. This composite RC network integrates the output current of the detector and, therefore, increases the response time. Different types of photodetectors are influenced by different combinations of these limitations which set their bandwidths. However, photodetectors of a given design and material do exhibit a constant gain-bandwidth product.

Noise Equivalent Power

Photodetectors are subject to several sources of noise that degrades their performance. The inherent randomness in the arrival of photons and the absorption of photons in the device serve as sources of noise. Various sources of current generation exist in all photodetectors. Some of these include current due to the incoming optical signal, current due to background radiation, and the dark current that is due to surface leakage, tunneling, and thermal generation of electron-hole pairs in or around the active region. All of these currents are generated randomly and contribute to shot noise. The amplification process that produces gain in some detectors is the avalanche mechanism. This is a random effect, and, therefore, there is a gain noise associated with such detectors. Another source of noise involves the random motion of carriers in resistive electrical materials at finite temperatures. There are parasitic resistances intrinsic to photodetectors and also resistances in circuits in which photodetectors are utilized. An example is a receiver circuit in which a detector serves as a source of input current to a preamplifier. The noise generated by these resistive elements is called thermal, or Johnson, or Nyquist noise. This noise is given by

$$\langle i_j^2 \rangle = \frac{4kT_{\text{eff}}B}{R_{\text{eff}}} \qquad (7.21)$$

where R_{eff} is the parallel combination of the detector and the preamplifier input resistances, B is the bandwidth, and T_{eff} is the effective temperature which is related to the noise figure NF of the amplifier:

$$T_{\text{eff}} = T(10^{\text{NF}/10} - 1) \qquad (7.22)$$

where T is the ambient temperature. It is, therefore, evident that, in the operation of a detector, the output signal must be above the noise level. The signal-to-noise ratio (SNR) is, therefore, an important characteristic in photodetectors, and it is related to sensitivity. The sensitivity of a photodetector is the minimum optical input power needed to achieve a given value of SNR. A measure of sensitivity is called noise equivalent power (NEP). NEP is the optical power (or photocurrent) required for the SNR to be unity over a 1-Hz bandwidth. Essentially, this measures when the photocurrent is exactly equal to the noise current. Thus, NEP measures the minimum detectable power in a photodetector. NEP depends on bandwidth, and, to find the optical power required to produce a SNR of unity for an entire measurement bandwidth, we have

$$P_i = \text{NEP}\sqrt{B}. \qquad (7.23)$$

Another figure of merit also useful for determining the ultimate detection limit is detectivity D^* given by

$$D^* = \frac{\sqrt{AB}}{\text{NEP}} \; (\text{cm} \cdot \text{Hz}^{1/2} \cdot \text{W}^{-1}) \qquad (7.24)$$

where A is the area of the photodetector on which light is incident. As with NEP, the reference bandwidth is taken as 1 Hz. D^* is usually expressed as D^* (λ, f, 1) where λ is the wavelength and f is the modulation frequency of the input optical signal. It must be noted that NEP and D^* are not equal to system sensitivity in actual applications because other noise sources, such as preamplifier noise, may dominate, especially in high speed (GHz) systems.

Photoconductors

Essentially, a photoconductor is a semiconductor material with two alloyed ohmic contacts. Its operation is based on the increase in conductivity of the semiconductor material due to the absorption of photons. The photogenerated electrons and holes are transported to the contacts under the influence of the electric field applied to the material by an external voltage source. The conductivity of the device is increased, and the signal is detected as an increased current flow under a constant-voltage bias. The conductivity of the material increases as a function of the photon flux. A practical form of photoconductors is shown in Fig. 7.14 consisting of interdigitated contacts on a semiconductor material. The interdigitated contacts are designed to minimize transit time and to maximize light transmission into the semiconductor material. To further improve quantum efficiency, an antireflection coating can be deposited on the device surface or light can be coupled in from the bottom if the substrate material has a wider bandgap than the absorption region. This is the case in Fig. 7.14 where InP is the substrate and, therefore, transparent to wavelengths longer than 0.92 μm at 300 K.

The quantum efficiency of a photoconductor is given by Eq. (7.17) and the responsivity is given by Eq. (7.20), where G is the internal photocurrent gain of the device. The gain is brought about by the fact that the photogenerated carriers contribute to current until they recombine. The gain is given by

$$G = \frac{\tau}{t_{tr}} \qquad (7.25)$$

where τ is the excess-carrier recombination lifetime and t_{tr} is the transit time of the majority carrier. The transit time is given by

$$t_{tr} = \frac{L}{v} \qquad (7.26)$$

where L is the channel length (distance between contacts) and v is the carrier velocity. It is seen from the these equations that, if the recombination lifetime is greater than the majority-carrier transit time, then many carriers will travel between the contacts before recombination takes place. This is photoconductor gain, and it can be below unity or well above unity depending on various factors including semiconductor material, size of the device, and the magnitude of the applied voltage.

The bandwidth of a photoconductor is given by

$$B = \frac{1}{2\pi\tau}. \qquad (7.27)$$

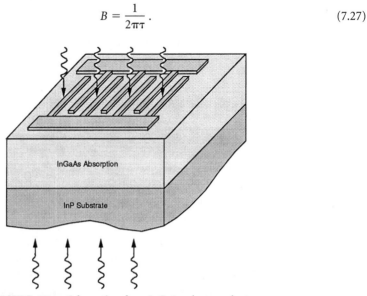

FIGURE 7.14 Schematic of an InGaAs photoconductor which is illuminated on the front and back sides.

It is seen that, whereas a long recombination time makes for a high gain, it also reduces bandwidth. So, a trade-off between gain and bandwidth exists for photoconductors. The gain-bandwidth product of a photoconductive detector is given by

$$GB = \frac{1}{2\pi t_{tr}} \qquad (7.28)$$

where t_{tr} is a constant for a given material and detector configuration.

The primary contributions to noise in photoconductive detectors are made by thermal or Johnson noise and the generation-recombination noise. The thermal noise given by Eq. (7.21) results from the random motion of carriers with average energy of kT contributing to the dark current of the device. The generation-recombination noise is due to fluctuations in the generation and recombination of carriers which, in turn, leads to fluctuations in the conductivity of the device. The generation-recombination noise is given by

$$\langle i_{G-R}^2 \rangle = \frac{4qI_oGB}{1 + \omega^2\tau^2} \qquad (7.29)$$

where I_o is steady-state output photocurrent and ω is the angular modulation frequency of the input optical signal.

To describe the overall noise performance of a photoconductor, the NEP is given by

$$\text{NEP} = \frac{8h\upsilon}{\eta}\left[1 + \frac{kT}{qG}(1 + \omega^2\tau^2)\frac{G_c}{I_o}\right] \qquad (7.30)$$

where G_c is the conductance of the channel. The dominant noise mechanism in a photoconductive detector is the thermal noise of the conducting channel. An increase in channel resistance is necessary to reduce thermal noise. If the thickness of the channel is reduced to increase the resistance, then quantum efficiency is reduced. To obtain the highest resistance and, hence, the lowest thermal noise achievable while maintaining high gain and quantum efficiency, it is necessary to utilize materials with the lowest carrier concentration. Thermal noise, generation-recombination noise, and dark current are high in semiconductors with smaller bandgaps. Although, photoconductors can have large gains, the gain may not be enough to surmount the inherent noise limitations to make them useful in many applications.

PIN Photodiodes

A p-i-n (PIN) photodetector is a p-n junction with an intrinsic (i) layer sandwiched between the p and n layers. In practice, the i-layer is either a p^- or n^- layer (i.e., lightly doped) which is inserted between the p^+ and n^+ layers. This structure is illustrated in Fig. 7.15 along with the energy band diagram under an applied bias. The PIN photodiode is operated in the reverse-bias mode, and, because the i-region has a low concentration of free carriers, it can be depleted with a minimum amount of voltage. Therefore, the depletion region extends through the entire i-region. When photons with energy greater than or equal to the bandgap energy are incident on the photodiode, electron-hole pairs are created. Carriers generated within a diffusion length of the depletion region diffuse into the i-region. These carriers along with all the carriers generated in the depletion region are transported by drift due to the applied reverse bias and are collected. The electric field in the depletion region is high and sufficiently uniform so that the carriers travel at saturation velocities. If all of the photogenerated carriers are collected, the quantum efficiency is given by equation (17) with L being the thickness of the region where the light is absorbed and $\zeta = 1$. L is usually assumed to be thickness of the i-region. As mentioned earlier,

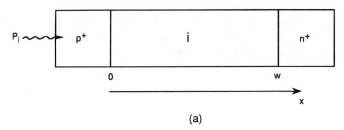

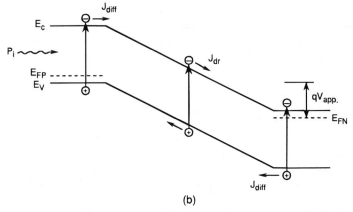

FIGURE 7.15 The p-i-n photodiode: (a) the layer structure and (b) the associated energy band diagram under an applied bias.

r can be made negligible by utilizing an antireflection coating to obtain good quantum efficiency. The internal quantum efficiencies $(1 - \exp(-\alpha L))$ for InGaAs and GaAs at various operating wavelengths are shown in Fig. 7.16. Because there is no internal optical gain associated with the PIN diode, the maximum internal quantum efficiency that can be expected is 100 %. The external quantum efficiency is given by Eq. (7.17) with $\zeta = 1$, and the responsivity is given by Eq. (7.20) with unity gain.

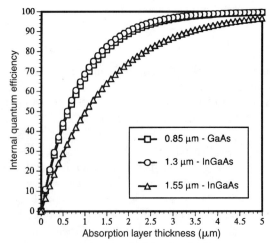

FIGURE 7.16 Internal quantum efficiency for GaAs and InGaAs at certain wavelengths and various absorption layer thicknesses.

Optoelectronic Devices

A schematic of a typical i-V characteristic of a PIN photodiode is shown in Fig. 7.17. This is the usual i-V characteristic of a p-n junction but with an added current $-i_p$ which is proportional to the incoming photon flux. The i-V relationship is given by

$$i = i_s[e^{eV/kT} - 1] - i_p. \tag{7.31}$$

The applied reverse bias (V_A) needed to deplete the i-region of thickness L is given by

$$L = \frac{\sqrt{2\varepsilon(V_o + V_A)}}{qN_c} \tag{7.32}$$

where ε is the dielectric constant, V_o is the built-in voltage, and N_C is the reduced carrier concentration, $N_C = N_A N_D / N_A + N_D$.

For high quantum efficiency or responsivity, it is necessary that $\alpha L \gg 1$. Of course, if L is large, the transit time of carriers across the i-region becomes large and device speed suffers. It is desirable to make α large, but it is a material property which cannot be changed. Therefore, there are many factors that must be considered to realize high-performance PIN photodiodes.

The response speed of PIN photodiodes can be limited by (a) the transit time of the photogenerated carriers across the depletion region, (b) the diffusion time of carriers generated outside the depletion region, and (c) the RC time constant with C consisting of junction capacitance and any other parasitic capacitances. The transit time is limited by the slower carriers, usually the holes. Another factor which can limit speed is charge trapping at heterojunctions. For the RC time constant, the junction capacitance of the photodiode is given by

$$C = \frac{\varepsilon A}{L} \tag{7.33}$$

where A is the cross-sectional area of the device. As seen here, the device capacitance can be minimized by increasing the thickness of the depletion region and by reducing the diameter of the device. Other parasitic capacitances associated with packaging can also be reduced. As mentioned previously, transit-time considerations suggest the use of a very thin i-layer to obtain high speed response. However, that increases parasitic junction capacitance which, in turn, reduces bandwidth.

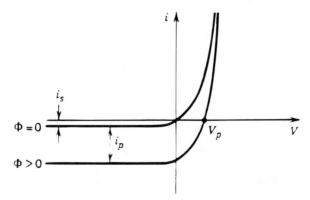

FIGURE 7.17 An i-V relationship of a typical photodiode. The two different curves represent the i-V relationship when the incident photon flux is equal to zero and greater than zero.

So, for a given detector area, an optimum i-layer thickness exists to obtain the highest speed possible. The transit-time limited frequency response of a PIN detector is given by

$$\frac{i(\omega)}{i(0)} = \frac{1}{(1 - e^{-\alpha L})} \left[\frac{e^{j\omega\tau_n - \alpha L} - 1}{j\omega\tau_n - \alpha L} - e^{-\alpha L} \frac{(e^{j\omega\tau_n} - 1)}{j\omega\tau_n} \right.$$
$$\left. + \frac{e^{j\omega\tau_p} - 1}{j\omega\tau_p} - e^{-\alpha L} \frac{(e^{j\omega\tau_p + \alpha L} - 1)}{j\omega\tau_p + \alpha L} \right] \quad (7.34)$$

where $i(\omega)$ is the detected current at an angular modulation frequency of ω, $i(0)$ is the dc current, L is the thickness of the i-layer, and $\tau_n = L/v_n$ and $\tau_p = L/v_p$ are the electron and hole transit times, and v_n and v_p are the electron and hole saturation velocities, respectively. For InGaAs, the saturation electron and hole velocities are 6.5×10^6 cm/s and 4.8×10^6 cm/s, respectively. Using this equation and the frequency response of a parallel RC network, the theoretical 3 dB bandwidth of InGaAs/InP PIN detectors of several diameters are plotted against i-layer thickness in Fig. 7.18. The results were calculated for 1.3 μm operation (i.e. $\alpha = 1.16$ μm^{-1}) and for a load resistance of 50 Ω. In practice, there are parasitic resistances (the shunt or junction resistance and the series resistance) and a stray capacitance that should be taken into account for higher accuracy. The transit-time limited response is also shown in the figure. The figure clearly demonstrates the compromises that are needed to design PIN photodiodes and also shows that devices with bandwidths over 60 GHz can be realized if the small diameter needed is not limiting for the particular application in hand.

An approximate expression for bandwidth for a detector with a very thin i-layer is

$$B = \frac{0.45v}{L} \quad (7.35)$$

for which $v_n = v_p = v$ has been assumed. Therefore, for a high speed detector where $\alpha L \ll 1$, the quantum efficiency from Eq. (7.17) is $(1 - r)\alpha L$, and consequently, the bandwidth-efficiency product is

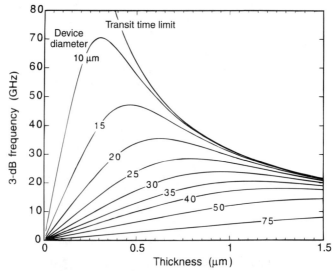

FIGURE 7.18 The theoretical frequency response of a p-i-n photodiode for various active layer thicknesses and dimensions.

Optoelectronic Devices

$$B \cdot \eta = 0.45\alpha v(1 - r). \tag{7.36}$$

Once again, the critical importance of a large α is clear. The large α for InGaAs at long wavelengths (1.3 μm and 1.55 μm) gives it an advantage over Si and Ge as a material of choice for PIN photodetectors for optical communication applications.

The sources of noise in a PIN photodiode are from (a) the current due to the photocurrent I_P, (b) the dark current due to generation in the depletion region I_D, and (c) the current due to background radiation I_B. Due to the random generation of these currents, they contribute to shot noise which can be expressed as

$$\langle i_s^2 \rangle = 2q(I_P + I_D + I_B)B \tag{7.37}$$

where B is the bandwidth. In addition, there is thermal or Johnson noise contributed with the shunt resistance of the diode and the input resistance of the following preamplifier stage. The Johnson noise for the PIN photodiode is given by Eq. (7.21). From these considerations, the NEP in units of watts for a PIN photodiode is given by

$$\text{NEP} = \frac{hv}{q\eta} \left[2q(I_P + I_D + I_B) + \frac{4kT}{R_{\text{eff}}} \right]^{1/2}. \tag{7.38}$$

It is seen that the key to increasing the sensitivity of the photodiode is to make R_{eff} and η as large as possible and make the I_B and I_D as small as possible. I_B, which is due to background radiation is usually very small, and I_D is also very small in a PIN photodiode because the device operates in the reverse-bias mode. The shot noise contributed by I_D is very small compared to that in photoconductors. Johnson noise is usually dominant in PIN detectors, but it can be minimized by optimizing the device and the circuit parameters. Therefore, PIN detectors are very useful for high speed, low noise applications such as encountered in optical communications.

In these applications which require operation at long wavelengths, photodiodes are made of InP-based materials, such as $In_{0.53}Ga_{0.47}As$ and InGaAsP. Heterojunctions of InP/InGaAs and InAlAs/InGaAs are needed for photodiodes to obtain higher breakdown and lower reverse leakage current than can be obtained using homojunctions. Practical heterostructures for PIN devices operating at infrared and long wavelengths are shown in Figs. 7.19a and b.

Avalanche Photodiodes (APDs)

An avalanche photodiode (APD) is essentially a p-n junction operated at high reverse-bias voltages close to the breakdown voltage. At such high voltages, photogenerated carriers in the depletion region gain kinetic energy from the induced electric field and travel at their saturation velocities in the host material. The carriers can acquire sufficient kinetic energies to undergo inelastic collisions with the lattice to create secondary electron-hole pairs. These secondary electron-hole pairs along with the primary carriers continue to drift and produce tertiary electron-hole pairs. To have an ionizing collision, a carrier must gain a threshold energy greater than the bandgap energy. Therefore, the critical field required to create ionizing collisions is material dependent and ranges from 10^4 to 10^5 V/cm. The process of creating carriers through ionizing collisions is called impact ionization, and the subsequent increase in the number of generated carriers is termed avalanche multiplication. This process is illustrated in the band diagram of Fig. 7.20 showing the direction of electron and hole injection via optical illumination. The avalanche multiplication process is an internal gain mechanism. This gain is typically much higher than that associated with a photoconductor. The quantum efficiency of an APD is given by Eq. (7.17) with L being the thickness of the absorption region.

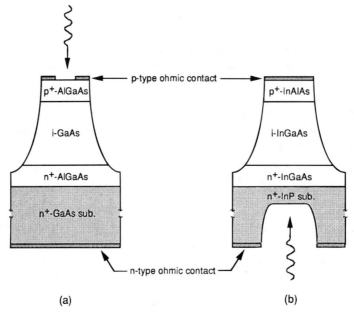

FIGURE 7.19 Schematics of p-i-n photodiodes utilizing (a) front side illumination and (b) back side illumination.

As described above, the avalanche process is characterized by impact ionization. The electron-hole pairs create drift in opposite directions, and, with sufficient energies, each carrier type can undergo impact ionization. The number of ionizing collisions per unit length is designated α_n and α_p, the ionization coefficients for holes and electrons, respectively. Among other things, α_n and α_p depend on materials and their band structures. Ionization coefficients are also dependent on the applied electric field, and an approximate relationship is given by

$$\alpha_n, \alpha_p = a \exp\left[-\frac{b}{E}\right] \quad (7.39)$$

where a and b are constants dependent on the type of material and doping density and E is the applied electric field. The field-dependent impact ionization coefficients for various semiconductor

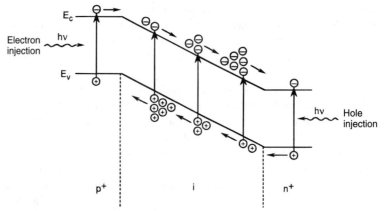

FIGURE 7.20 The energy band diagram of an APD illustrating the avalanche gain process.

materials have been determined and reported in the literature. In general, the two rates approach each other at fields higher than 3×10^5 V/cm for many semiconductors. Table 7.2 shows the values of a and b of Eq. (7.39) for various semiconductors.

The avalanche process produces a carrier multiplication factor M or an avalanche gain which is dependent on α_n and α_p. The responsivity of an APD is expressed by Eq. (7.20) where G is the gain and it is equivalent to the multiplication factor M. The low-frequency gain for electrons is given by

$$M = \left\{ 1 - \int_0^W \alpha_n \exp\left[-\int_0^x (\alpha_n - \alpha_p) dx' \right] dx \right\}^{-1} \quad (7.40)$$

where W is the width of the depletion region. When $\alpha_n \neq \alpha_p$ and the values of both ionization rates are independent of position as in p-i-n diodes, then,

$$M = \frac{(1 - \alpha_p/\alpha_n)\exp[\alpha_n W(1 - \alpha_p/\alpha_n)]}{1 - (\alpha_p/\alpha_n)\exp[\alpha_n W(1 - \alpha_p/\alpha_n)]}. \quad (7.41)$$

For $\alpha_n = \alpha_p = \alpha$, the multiplication factor M takes the form

$$M = \frac{1}{(1 - \alpha W)}. \quad (7.42)$$

It is observed that M goes to ∞ when $\alpha W = 1$ which signifies the condition for junction breakdown. Therefore, high values of M can be obtained for photodiodes biased near the breakdown voltage. A schematic of the avalanche gain in an APD is illustrated in Fig. 7.21 showing the sharp increase in gain as the reverse bias approaches the breakdown voltage. The critical field at which breakdown is initiated depends on impurity concentration in semiconductors and can be calculated from the one-sided abrupt p-n junction approximation.

The steady-state photocurrent in an APD in the presence of an avalanche gain is given by

$$I_p = \frac{q\eta P_i}{h\nu} M \quad (7.43)$$

TABLE 7.2 The Impact Ionization Coefficients for Various Semiconductors

Semiconductor Material	a(cm^{-1})		b(V/cm)		ref.
	Electrons	Holes	Electrons	Holes	
GaAs	$1.1 \cdot 10^7$	$5.5 \cdot 10^6$	$2.2 \cdot 10^6$	$2.2 \cdot 10^6$	1
GaAs	$3.82 \cdot 10^4$	$4.50 \cdot 10^4$	$3.80 \cdot 10^5$	$3.10 \cdot 10^5$	2
In$_{0.15}$Ga$_{0.63}$Al$_{0.22}$As	$3.19 \cdot 10^4$	$4.03 \cdot 10^4$	$4.03 \cdot 10^5$	$3.25 \cdot 10^5$	2
In$_{0.2}$Ga$_{0.8}$As	$5.9 \cdot 10^4$	$6.8 \cdot 10^4$	$3.0 \cdot 10^5$	$3.02 \cdot 10^5$	2
Si	$9.2 \cdot 10^5$	$2.4 \cdot 10^5$	$1.45 \cdot 10^6$	$1.64 \cdot 10^6$	3
InP	$5.5 \cdot 10^6$	$1.98 \cdot 10^6$	$3.10 \cdot 10^6$	$2.29 \cdot 10^6$	4
In$_{0.53}$Ga$_{0.47}$As	$2.27 \cdot 10^6$	$3.95 \cdot 10^6$	$1.13 \cdot 10^6$	$1.45 \cdot 10^6$	4
In$_{0.67}$Ga$_{0.33}$As$_{0.70}$P$_{0.30}$	$3.37 \cdot 10^6$	$2.94 \cdot 10^6$	$2.29 \cdot 10^6$	$2.40 \cdot 10^6$	4

[1] Ando, H., and Kanbe, H. Ionization coefficient measurement in GaAs by using multiplication noise characteristics, *Solid State Elec.*, 24, 629–634, 1981.

[2] Chen, Y. C., and Bhattacharya, P. K. Impact ionization coefficients for electrons and holes in strained In$_{0.2}$Ga$_{0.8}$As and In$_{0.15}$Ga$_{0.63}$Al$_{0.22}$As channels embedded in Al$_{0.3}$Ga$_{0.7}$As, *J. of Appl. Phys.*, 73(1), 465–467, 1993.

[3] Woods, M. H., Johnson, W. C., and Lampert, M. A. Use of a Schottky barrier to measure impact ionization coefficients in semiconductors, *Solid State Elec.* 16, 381–394, 1973.

[4] Osaka, F., Mikawa, T., and Kaneda, T. Impact ionization coefficients of electrons and holes in (100)-Oriented Ga$_{1-x}$In$_x$As$_y$P$_{1-y}$, *IEEE J. of Quan. Elec.*, 21(9), 1326–1338, 1985.

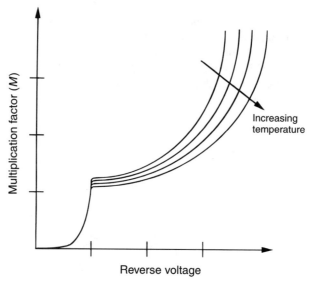

FIGURE 7.21 A representation of the multiplication factor as a function of reverse bias voltage and temperature for an APD.

where P_i is the incident optical power. For the case of an intensity-modulated incident optical signal, the root mean square (rms) signal with a modulation index m is expressed as

$$i_p = \frac{q\eta P_i m}{\sqrt{2}h\nu} M. \tag{7.44}$$

The avalanche process is a regenerative process and it can be time-consuming. It takes time to build up which implies that the higher the avalanche gain, the longer the avalanche process persists. This results in the presence of a large number of secondary carriers in the depletion region after the primary carriers have been collected. Therefore, the gain and bandwidth of an APD are inextricably linked together. This results in a finite gain-bandwidth product fixed for a given material and device structure. The gain-bandwidth product is given by

$$B \cdot M = \frac{1}{2\pi\tau_1} \tag{7.45}$$

where $\tau_1 = (\tau_n + \tau_p)/2$, τ_n is the electron transit time, and τ_p is the hole transit time. τ_n and τ_p are given by W/v_n and W/v_p where v_n and v_p are electron and hole saturation velocities, respectively. The equation above has only considered the transit-time effects on the response speed of a photodiode. Other usual effects are diffusion and RC limitations. Another effect particular to APDs is the time it takes carriers to complete the avalanche process, which is called the avalanche build-up time. Due to the randomness of the multiplication process, the build-up time is itself random. The dependence of the multiplication factor on frequency for $M_o > \alpha_n/\alpha_p$ is of the form

$$M(\omega) \approx \frac{M_o}{\sqrt{1 + \omega^2 M_o^2 \tau_t^2}} \tag{7.46}$$

where M_o is the dc value of the multiplication factor and t_t is an effective transit time through the avalanche region. A functional form that describes the effective transit time is

Optoelectronic Devices

$$\tau_t = N\tau_1(\alpha_n/\alpha_p) \tag{7.47}$$

where τ_1 is the real transit time through the avalanche region and N is a number slowly varying from $N = 1/3$ to 2 as α_p/α_n varies from 1 to 0.001.

Apart from the usual shot noise limitations in photodiodes, the randomness associated with the avalanche or multiplication process makes it a principal source of noise in APDs. These random fluctuations create a distribution of gain which causes excess noise in the device. For the condition where the avalanche is initiated by electrons (i.e., $\alpha_n > \alpha_p$), the excess noise factor is given by

$$F = M\left[1 - (1-k)\left(\frac{M-1}{M}\right)^2\right] \tag{7.48}$$

where k is the ratio of the ionization factors, i.e., α_p/α_n. An equivalent expression also exists for a hole-induced avalanche gain in which k is replaced by $k' = \alpha_n/\alpha_p$. There are two special cases, $k = 0$ and $k = 1$. The first case results in

$$F = 2 - \frac{1}{M} \tag{7.49}$$

whereas the second, denoting where both ionization coefficients are equal, results in

$$F = M. \tag{7.50}$$

The noise factor for various multiplication gains and ratios of ionization coefficient is shown in Fig. 7.22.

In an APD, the dark current and the current due to background radiation are all enhanced by the multiplication gain similar to that of the signal current in Eq. (7.43). Therefore the root mean shot noise is given by

$$\langle i_s^2 \rangle = 2q(I_P + I_D + I_B)FBM^2. \tag{7.51}$$

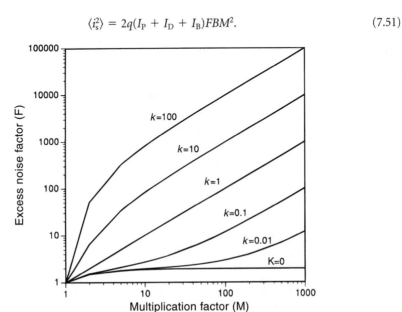

FIGURE 7.22 Excess noise factor as a function of multiplication gain and the ratio of ionization coefficients.

The Johnson or thermal noise for an APD is identical to that of a PIN photodiode. Therefore, the resulting NEP for an APD is

$$\text{NEP} = \frac{h\upsilon}{q\eta M}\left[2q(I_P + I_D + I_B)FM^2 + \frac{4kT}{R_{\text{eff}}}\right]^{1/2}. \quad (7.52)$$

It is observed that the NEP of an APD is almost identical to that of a PIN photodiode except for an extra gain factor M which appears in the denominator. It is evident, therefore, that the gain factor does reduce the NEP of an APD, making the device more sensitive than PIN photodiodes. Indeed, APDs have shown superior sensitivity compared to other photodiodes.

Important issues for consideration in designing and fabricating APDs concern dark current and high-speed performance. The contributions to the dark current of an APD include generation-recombination in the depletion region, tunneling of carriers across the bandgap, and leakage across junctions. Various methods have been developed to alleviate these problems. For example, guard rings and dielectric film deposition are used in Si APDs to minimize various forms of leakage currents. Modern APDs utilized for fiber-optic communications depend on InP related compounds. Lightwave operation at 1.55 µm relies on $In_{0.47}Ga_{0.53}As$ which has an energy bandgap of 0.75 eV. Large tunneling current is a problem if a material with such a low bandgap is used in homojunction devices. Therefore, a separate absorption and multiplication (SAM) structure, in which a low-field InGaAs region is utilized as the absorption region and a high-field InP region is the avalanching area, is now used for APDs. Figure 7.23 shows a typical InP/InGaAs SAM-APD heterostructure where the p^+-n^- junction is located in the high-bandgap (1.35 eV) InP material. In practice, a graded bandgap material (InGaAsP) is placed at the InP/InGaAs heterojunction to prevent a sharp energy bandgap discontinuity which can trap carriers and result in slow device response. This modification results in a structure called separate absorption-graded-multiplication APD (or SAGM-APD). These devices combine low dark current and high gain-bandwidth operations.

Other types of APDs have been designed to ensure preferential multiplication of one carrier type (i.e., k or k' approaches zero) over the other. These are the superlattice and the multiquantum-well APDs.

Metal-Semiconductor-Metal (MSM) Photodiode

A metal-semiconductor-metal (MSM) photodiode consists of two interdigitated electrodes which form back-to-back Schottky diodes on a semiconductor absorbing layer. The MSM is a planar

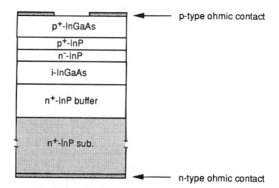

FIGURE 7.23 Schematic diagram of a separate avalanche and multiplication avalanche photodiode (SAM-APD).

Optoelectronic Devices

device which can be fabricated easily. This ease in fabrication along with the ease in integrating the MSM with conventional field-effect-transistor (FET) technology have provided the underlying impetus to gradually improve the MSM's efficiency, dark current, and bandwidth to the point where it can now compete with the PIN photodiode. The schematic of the MSM is shown in Fig. 7.24. The interdigitated electrodes are similar to those of the photoconductors of Fig. 14 but differ in the type of contacts made to the semiconductor. The contacts of photoconductors are ohmic. Photogenerated electrons and holes in MSMs are transported along the electric field lines to the oppositely biased contacts as shown in Fig. 7.24b. The electric field distribution is nonuniform due to the electrode geometry, with fields being strongest near the semiconductor's surface. Therefore, the applied bias must be sufficiently large to ensure the depletion of the entire absorbing region for the carriers to travel at their saturation velocities. The energy band diagram in Fig. 7.24c shows the flow of the generated electrons and holes and also shows how the Schottky barriers limit the injection of carriers from the metal contacts to the semiconductor.

The external quantum efficiency is basically given by Eq. (7.17), but taking the opaque interdigitated electrodes into account, we have

$$\eta = (1 - r)\frac{d}{d + w}\zeta[1 - \exp(-\alpha L)] \tag{7.53}$$

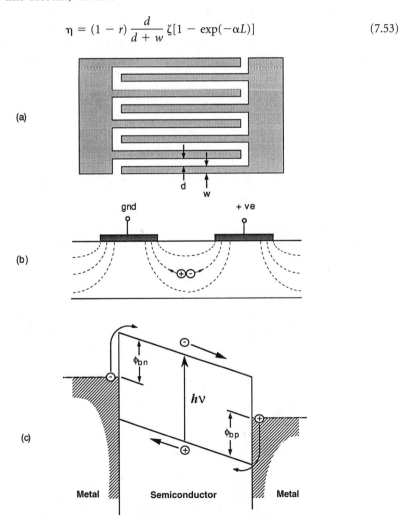

FIGURE 7.24 (a) Schematic of a MSM photodiode (b) cross-sectional view of the device displaying the direction carriers travel in response to an electric field, and (c) the associated energy band diagram.

where w is the electrode finger width and d is the finger spacing. It is seen that η is reduced by the opacity of the electrodes. The use of transparent Schottky contacts, such as indium tin oxide and cadmium tin oxide, to avoid the opacity of metal electrodes has been demonstrated. The shortcoming of this method is that carriers generated in the low-field region under the contacts degrade the bandwidth of the device. The responsivity of MSMs is given by Eq. (7.20). Ideally, the gain G should be unity. However, two possible gain mechanisms have been identified in MSMs. These are i) the gain due to avalanche multiplication in high-field region of the device, for example, around the sharp edges of the contacts and ii) the gain due to trapping of carriers. Trapping of carriers can occur at trapping centers on semiconductor surface (between fingers), on heterointerfaces, and at Schottky/semiconductor interfaces. These various gain mechanisms can have a detrimental influence on the dark current and high-speed response. The use of high quality epitaxial materials (low trap density), fabrication of high quality Schottky diodes (no interfacial oxides), and appropriate biasing of devices (avoid high fields at contacts) are some means of minimizing these effects.

The dark current of an epitaxial GaAs MSM is shown in Fig. 7.25. The photodiode has an area of 50 μm × 50 μm, a finger width of 1 μm, a finger spacing of 1 μm, and an absorbing layer thickness of 1 μm. A symmetrical trace of the above result should be obtained for negative voltages due to the back-to-back Schottky diode configuration. The dark current of an MSM is dominated by thermionic emission over the barrier as illustrated in Fig. 7.24. A simple one-dimensional model gives the flat band voltage V_{fb} at the forward-biased contact as

$$V_{fb} = \frac{qN_D d^2}{2\varepsilon} \tag{7.54}$$

where q is the electron charge, ε is the dielectric constant of the semiconductor, N_D is the residual doping in the semiconductor, and d is the electrode spacing. The maximum V_{fb} is limited by the avalanche breakdown voltage near the reverse-biased contact. At low bias voltage compared to V_{fb}, electron injection at the reverse-biased contact is the dominant factor in carrier transport. At higher bias, hole injection at the forward-biased contact become dominant. At higher voltages, there is a reach-through condition, when the edges of the depletion regions of the two Schottky

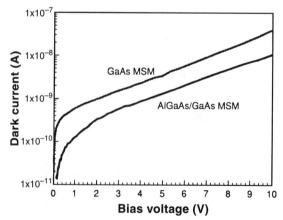

FIGURE 7.25 Comparison of the dark current of a 1 μm by 1 μm GaAs MSM photodiode with and without an AlGaAs cap layer (Seo, J. W., Metal-Semiconductor-Metal Photodetectors for Optoelectronic Receiver Applications, Ph.D. Thesis, University of Illinois at Urbana-Champaign, Urbana, IL, 1993, with permission.)

diodes merge. The total current density for a voltage higher than the reach-through voltage is given by

$$J_{\text{tot}} = A_n^* T^2 \exp\left[\frac{-q(\phi_{\text{bn}} - \Delta\phi_{\text{bn}})}{kT}\right]$$
$$+ A_p^* T^2 \exp\left[\frac{-q(\phi_{\text{bp}} - \Delta\phi_{\text{bp}})}{kT}\right] \quad (7.55)$$

where A_n^* and A_p^* are Richardson constants for electrons and holes, ϕ_{bn} and ϕ_{bp} are the barrier heights for electrons and holes, and, finally, $\Delta\phi_{\text{bn}}$ and $\Delta\phi_{\text{bp}}$ are the respective image force barrier lowerings. A_n^* and A_p^* are 8.15 and 74.4 A/cm^2/K^2 for GaAs, respectively. The dark current in Fig. 7.25 compares to that of a PIN photodiode. Improvements in MSM dark current have been shown to be possible by choosing an appropriate metal system for Schottky contacts and also by using surface epitaxial layers with higher bandgap than the absorbing layer. The latter example is demonstrated for an AlGaAs/GaAs MSM in Fig. 7.25 where the Schottky contacts were made to an Al$_{0.3}$Ga$_{0.7}$As layer grown on the GaAs absorbing layer.

The bandwidth of a MSM photodiode is determined by the carrier transit time and the RC time constant, and both depend on the physical dimensions of the device. The carrier transit time is proportional to the distance between the electrodes if the effects of velocity overshoot and fringing electric field are neglected. The transit time is limited by the slowest carriers (holes). Using a dc model, the resistance of a MSM photodiode is given by

$$R_{\text{tot}} = \frac{2R_o}{N} \quad (7.56)$$

where R_o is the resistance of a single electrode and N is the number of electrodes on each side of the two sets of interdigitated electrodes. The resistance of the electrodes is usually less than the 50 Ω load resistance. A major advantage of an interdigitated MSM is its low capacitance. For an undoped and infinitely thick semiconductor, the capacitance is given by

$$C = \frac{\kappa(k)}{\kappa(k')} \varepsilon_o (1 + \varepsilon_r) \frac{A}{w + d} \quad (7.57)$$

where ε_r is the relative dielectric constant of the semiconductor, A is the active area of the MSM, w and d are the finger width and spacing, respectively, and $\kappa(k)$ is the complete elliptic integral of the first kind given by

$$\kappa(k) = \int_0^{\pi/2} \frac{d\Phi}{\sqrt{1 - k^2 \sin^2\Phi}}, \quad (7.58)$$

$$k = \tan^2\left[\frac{\pi w}{4(w + d)}\right], \quad (7.59)$$

and

$$k' = \sqrt{1 - k^2}. \quad (7.60)$$

Using these expressions, it has been shown that the capacitance of a MSM is significantly lower than that of a PIN photodiode of corresponding size.

The transit-time limited response of the a surface-illuminated PIN photodiode can be adapted to the treatment of MSMs. The upper bound of the transit-time limited bandwidth of a MSM is given by Eq. (7.35) where the interdigitated spacing of the MSM is equal to the intrinsic layer thickness of the PIN. Because the effects of the carrier transit time and the RC time constant on the bandwidth are cumulative, then, the bandwidth of a MSM is given by

$$B = \left(\frac{d}{0.45v} + R_L C\right)^{-1} \quad (7.61)$$

where R_L is the MSM load resistance and $v_n = v_p = v$. Although, the finger width to finger spacing ratio is a design parameter for MSM, if a ratio of unity is assumed for practical reasons, then, Eq. (7.61) becomes

$$B = \left(\frac{d}{0.45v} + \frac{\pi K R_L}{d}\right)^{-1} \quad (7.62)$$

where K is a constant given by

$$K = \frac{K(k_o)}{K(k_o')} \varepsilon_0 (1 + \varepsilon_r) A \quad (7.63)$$

with

$$k_o = \tan^2\left(\frac{\pi}{8}\right). \quad (7.64)$$

The elliptic integrals in Eq. (7.63) are $K(k_o) = 1.58$ and $K(k_o') = 3.17$ for a device with $d = w$. Fig. 7.26 shows a plot of bandwidth as a function of finger spacing for an InGaAs MSM with the active area as a variable parameter. As shown in the curve, for small values of d the response is RC-limited while for larger values of d the response is transit-time limited.

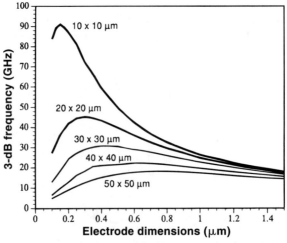

FIGURE 7.26 The theoretical frequency response of an InGaAs MSM photodiode as a function of electrode and active area dimensions for a device with equal electrode width and spacing.

It is possible to obtain the finger spacing d_{max}, which gives a maximum available bandwidth, by equating the derivative of Eq. (7.62) with respect to d to zero, yielding

$$d_{max} = \sqrt{0.45\pi\nu KR_L}. \qquad (7.65)$$

Therefore, for a device with a 50 × 50 μm^2 active area, an $In_{0.47}Ga_{0.53}As$ absorbing layer, and a 50 Ω load resistance, we obtain a d_{max} and B_{max} of 1 μm and 21.4 GHz, respectively. Figure 7.27 shows the pulse response waveforms of InAlAs/InGaAs MSMs with Ti/Au Schottky contacts at a bias of 10 V for various device finger widths/finger spacings and a 50 × 50 μm^2 active area. The full-width-at-half-maximums (FWHMs) of the response pulses for the devices are 42, 47, and 52 psec, respectively. Corresponding estimated bandwidths are 20, 14, and 11 GHz, respectively.

Various schemes have been adopted to improve the high speed response of MSMs. The use of thin absorbing layers and small finger spacing to finger width ratio will ensure very fast transit time but low quantum efficiency. The smaller ratio also increases the capacitance. Ion implantation in the absorbing region reduces transit time but also reduces sensitivity. This may contribute to excess noise and gain. In GaAs MSM design, the use of an AlGaAs barrier layer in the buffer region has been shown to improve response with a slight reduction in the dc responsivity. The barrier impedes slow photogenerated carriers from deep in the substrate, where the electric field is weak, from being collected by the electrodes. This leads to reduced transit time and a suppressed tail in the pulse response. To maximize charge collection and also reduce transit time, a graded layer is usually placed at the heterostructure interface to prevent a sharp, energy-band discontinuity.

Much like the PIN photodiode, the sources of shot noise in MSM are the photocurrent, the dark current, and the current due to background radiation. There is a also thermal noise component. The MSM photodiode has a lateral current flow with strong interactions with semiconductor interfaces like a FET device. Therefore, it is expected that MSM will exhibit a low-frequency, excess noise component with a $1/f\alpha$-like dependence. This excess $1/f^\alpha$ noise has been observed at low frequencies and high biases close to breakdown. Evidence shows that this excess noise is very dependent on the presence of traps. A significant excess noise was obtained at frequencies below 35 MHz for an InGaAs MSM which had a strained GaAs Schottky barrier enhancement layer. However, for

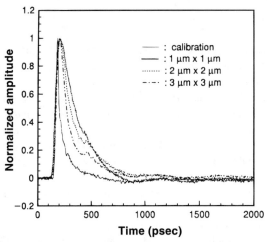

FIGURE 7.27 A typical time-domain pulse response of an InGaAs MSM for devices with different electrode dimensions (Seo, J. W., Metal-Semiconductor-Metal Photodetectors for Optoelectronic Receiver Applications, Ph.D. Thesis, University of Illinois at Urbana-Champaign, Urbana, IL, 1993, with permission.)

a high quality InGaAs MSM with an InAlAs enhancement layer, the device was shot-noise dominated at frequencies as low as 1 MHz. Presumably, the excess noise for the GaAs/InGaAs MSMs is due to the large number of traps present at the strained interface between GaAs and InGaAs. Therefore, for MSMs fabricated on high quality epitaxial layers, the dominant noise properties will be similar to those of PIN photodiodes.

Acknowledgments

The authors gratefully acknowledge the superb assistance of Walter Wohlmuth in preparing editing the photodetector aspects of this chapter. The work was supported by funding from the National Science Foundation (ECD 89-43166 and DMR 89-20538), the Advanced Research Projects Agency (MDA972-94-1-0004), and the Joint Services Electronics Program (N0014-90-J-1270).

References

Adachi, S. 1992. *Physical Properties of III-V Semiconductor Compounds, InP, InAs, GaAs, GaP, InGaAs, and InGaAsP*. John Wiley & Sons. New York.

Agrawal, G. P. and Dutta, N. K. 1993. *Semiconductor Lasers*, 2nd Ed. Van Nostrand Reinhold, New York.

Bhattacharya, P. 1993. *Properties of Lattice-Matched and Strained Indium Gallium Arsenide*. INSPEC, Institution of Electrical Engineers, London.

Bhattacharya, P. 1994. *Semiconductor OptoElectronic Devices*. Prentice-Hall, Englewood Cliffs, NJ.

Bowers, J. W., Burrus, C. A., and McCoy, R. J. 1985. InGaAs PIN photodetectors with modulation response to millimetre wavelengths, *Elec. Lett.*, 21(18):812.

Bowers, J. E. and Burrus, C. A. 1987. Ultrawide-Band long-wavelength p-i-n photodetectors, *J. of Lightwave Tech.*, 5(10):1339.

Brennan, K. F. 1988. Field and spatial geometry dependencies of the electron and hole ionization rates in GaAs/AlGaAs multiquantum well APD's, *IEEE Trans. on Elec. Dev.*, 35(5):634.

Butler, J. K. 1979. *Semiconductor Injection Lasers*. IEEE Press, New York.

Capasso, F. 1982. New ultra-low noise avalanche photodiode with separated electron and hole avalanche regions, *Elec. Lett.*, 18(1):12.

Capasso, F., Mohammed, K., and Cho, A. Y. 1985. Tunable barrier heights and band discontinuities via doping interface dipoles: An interface engineering technique and its device applications, *J. Vac. Sci. Tech. B*, 3(4):1245.

Cheo, P. K. 1989. *Handbook of Solid State Lasers*. Marcel Dekker, New York.

Chou, S. Y., Liu, Y., and Fischer, P. B. 1992. Terahertz GaAs metal-semiconductor-metal photodetectors with 25 nm finger spacing and finger width, *Appl. Phys. Lett.*, 61(4):477.

Chou, S. Y., Liu, Y., Khalil, W., Hsiang, T. Y., and Alexandrou, S. 1992. Ultrafast nanoscale metal-semiconductor-metal Photodetectors on bulk and low-temperature grown GaAs, *Appl. Phys. Lett.*, 61(7):819.

Coleman, J. J. 1992. *Selected Papers on Semiconductor Diode Lasers*. SPIE Optical Engineering Press, Bellingham, WA.

Dentan, M., and de Cremoux, B. 1990. Numerical simulation of the nonlinear response of a p-i-n photodiode under high illumination, *J. Lightwave Tech.*, 8(8):1137.

Dutta, N. K., Lopata, J., Berger, P. R., Wang, S. J., Smith, P. R., Sivco, D. L., and Cho, A. Y. 1993. 10 GHz bandwidth monolithic p-i-n modulation-doped field effect transistor photoreceiver, *Appl. Phys. Lett.*, 63(15):2115.

Ferry, D. K. 1985. *Gallium Arsenide Technology*. H. W. Sams, Indianapolis, IN.

Gooch, G. H. 1973. *Injection Electroluminescent Devices*. John Wiley & Sons, New York.

Hanatani, S., Nakamura, H., Tanaka, S., and Ido, T. 1994. Flip-Chip InAlAs/InGaAs superlattice avalanche photodiodes with back-illuminated structures, *Microwave and Optical Tech. Lett.*, 7(3):103.

Harder, Ch. S., van Zeghbroeck, B. J., Kesler, M. P., Meier, H. P., Vettiger, P., Webb, D. J., and Wolf, P., 1990. High-Speed GaAs/AlGaAs optoelectronic devices for computer applications, *IBM J. Res. and Dev.* 34(4):568.

He, L., Lin, Y., van Rheenen, A. D., van der Ziel, A., van der Ziel, J. P., and Young, A. 1990. Low-Frequency noise in small InGaAs/InP p-i-n diodes under different bias and illumination conditions, *J. Appl. Phys.*, 68(10):5200.

Hunsberger, R. G. 1984. *Integrated Optics: Theory and Technology*, 2nd Ed. Springer-Verlag, Berlin.

Jhee, Y. K., Campbell, J. C., Holden, W. S., Dentai, A. G., and Plourde, J. K. 1985. The effect of nonuniform gain on the multiplication noise of InP/InGaAsP/InGaAs avalanche photodiodes, *IEEE J. Quan. Elec.*, 21(12):1858.

Kahraman, G., Saleh, B. E. A., Sargeant, W. L., and Teich, M. C. 1992. Time and frequency response of avalanche photodiodes with arbitrary structure, *IEEE Trans. Elec. Dev.*, 39(3):553.

Kawaga, T., Kawamura, Y., and Iwamura, H. 1993. A wide-bandwidth low-noise InGaAsP-InAlAs superlattice avalanche photodiode with a flip-chip structure for wavelengths of 1.3 and 1.55 μm, *IEEE J. Quan. Elec.*, 29(5):1387.

Kim, J. H., Griem, H. T., Friedman, R. A., Chan, E. Y., and Ray, S. 1992. High-Performance back-illuminated InGaAs/InAlAs MSM photodetector with a record responsivity of 0.96 A/W, *IEEE Photonics Tech. Lett.*, 4(11):1241.

Klingenstein, M., Kuhl, J., Rosenzweig, J., Moglestue, C., and Axmann, A. 1991. Transit time limited response of GaAs metal-semiconductor-metal photodiodes, *Appl. Phys. Lett.*, 58(22):2503.

Kressel, H. 1982. *Topics in Applied Physics, Vol. 39, Semiconductor Devices for Optical Communication*. Springer-Verlag, Berlin.

Kressel, H. and Butler, J. K. 1977. *Semiconductor Lasers and Heterojunction LEDs*. Academic Press, New York.

Kuhl, D., Hieronymi, F., Böttcher, E. H., Wolf, T., Krost, A., and Bimberg, D. 1990. Very high-speed metal-semiconductor-metal InGaAs:Fe photodetectors with InP:Fe barrier enhancement layer grown by low pressure metalloorganic chemical vapour deposition, *Elec. Lett.*, 26(25):2107.

Kuhl, D., Hieronymi, F., Böttcher, E. H., Wolf, T., Bimberg, D., Kuhl, J., and Klingenstein, M. 1992. Influence of space charges on the impulse response of InGaAs metal-semiconductor-metal Photodetectors, *J. Lightwave Tech.*, 10(6):753.

Lim, Y. C., and Moore, R. A. 1968. Properties of alternately charged coplanar parallel strips by conformal mappings, *IEEE Trans. Elec. Dev.*, 15(3):173.

Lu, J., Surridge, R., Pakulski, G., van Driel, H., and Xu, J. M. 1993. Studies of high-speed metal-semiconductor-metal photodetector with a GaAs/AlGaAs/GaAs heterostructure, *IEEE Trans. Elec. Dev.*, 40(6):1087.

Lucovsky, G., Schwarz, R. F., and Emmons, R. B. 1963. Transit-Time considerations in p-i-n diodes, *J. Appl. Phys.*, 35(3):622.

Miller, S. E. and Kaminow, I. P. 1988. *Optical Fiber Telecommunications II*. Academic Press, New York.

Newman, D. H., and Ritchie, S. 1986. Sources and detectors for optical fiber communications applications: The first 20 years, *IEE Proc., Pt. J.*, 133(3):213.

Panish, M. B. and Casey, H. C. 1978. *Heterostructure Lasers, Part A: Fundamental Principles*. Academic Press, New York.

Panish, M. B. and Casey, H. C. 1978. *Heterostructure Lasers, Part B: Materials and Operating Characteristics*. Academic Press, New York.

Pankove, J. I. 1971. *Optical Processes in Semiconductors*. Dover, New York.

Parker, D. G. 1988. The theory, fabrication and assessment of ultra high speed photodiodes, *GEC J. Res.*, 6(2):106.

Pearsall, T. P. 1982. *GaInAsP Alloy Semiconductors*. John Wiley & Sons, New York.

Peredo, E., Decoster, D., Gouy, J. P., Vilcot, J. P., and Constant, M. 1994. Comparison of InGaAs/InP photodetectors for microwave applications, *Microwave Optical Tech. Lett.*, 7(7):332.

Peterson, R. L. 1987. Numerical study of currents and fields in a photoconductive device., *IEEE J. Quan. Elec.*, 23(7):1185.

Powers, J. P. 1993. *An Introduction to Fiber Optical Systems*. Richard D. Irwin and Aksen Associates, Homewood, IL.

Ridley, B. K. 1985. Factors Affecting Impact Ionisation in Multilayer Avalanche Photodiodes, *IEE Proc., Pt. J*, 132(3):177.

Sabella, R., and Merli, S. 1993. Analysis of InGaAs P-I-N photodiode frequency response, *IEEE J. Quan. Elec.*, 29(3):906.

Saleh, B. E. A. and Teich, M. C. 1991. *Fundamentals of Photonics*. John Wiley and Sons, New York.

Schichijo, H., and Hess, K. 1981. Band-Structure dependent transport and impact ionization in GaAs, *Phys. Rev. B*, 23(8):4197.

Schumacher, H., LeBlanc, H. P., Soole, J. B. D., and Bhat, R. 1988. An investigation of the optoelectronic response of GaAs/InGaAs MSM photodetectors, *IEEE Elec. Dev. Lett.*, 9(11):607.

Soole J. B. D., and Schumacher, H. 1991. InGaAs metal-semiconductor-metal photodetectors for long wavelength optical communications, *IEEE J. Quan. Elec.*, 27(3):737.

Soole, J. B. D. 1992. InGaAs MSM photodetectors for long wavelength fiber communications, *SPIE, High Speed Elec. OptoElectronics*, 1680:153.

Stillman, G. E., Robbins, V. M., and Tabatabaie, N. 1984. III-V Compound semiconductor devices: Optical detectors, *IEEE Trans. Elec. Dev.*, 31(11):1643.

Stillman, G. E. and Wolfe, C. M. 1977. Avalanche Photodiodes. In *Semiconductors and Semimetals*, R. K. Willardson and A. C. Beer, eds., pp. 291–393. Academic Press, New York.

Streetman, B. G. 1990. *Solid State Electronic Devices*, 3rd Ed. Prentice-Hall, Englewood Cliffs, NJ.

Thompson, G. H. B. 1980. *Physics of Semiconductor Laser Devices*. John Wiley & Sons, New York.

Tiwari, S., and Tischler, M. A. 1992. On the role of mobility and saturated velocity in the dynamic operation of p-i-n and metal-semiconductor-metal photodetectors, *Appl. Phys. Lett.*, 60(9):1135.

Tsang, W. T. 1985. *Semiconductors and Semimetals, Vol. 22, Lightwave Communications Technology, Part A Material Growth Technologies*. Academic Press, New York.

Tsang, W. T. 1985. *Semiconductors and Semimetals, Vol. 22, Lightwave Communications Technology, Part B Semiconductor Injection Lasers I*. Academic Press, New York.

Tsang, W. T. 1985. *Semiconductors and Semimetals, Vol. 22, Lightwave Communications Technology, Part C Semiconductor Injection Lasers II*. Academic Press, New York. NY.

Tucker, R. S., Taylor, A. J., Burrus, C. A., Eisenstein, G., and Wersenfeld, J. M. 1986. Coaxially mounted 67 GHz bandwidth InGaAs photodiode, *Elec. Lett.*, 22(17):917.

van Zeghbroeck, B. J., Patrick, W., Halbout, J-M., and Vettiger, P. 1988. 105-GHz bandwidth metal-semiconductor-metal photodiode, *IEEE Elec. Dev. Lett.*, 9(10):527.

Verdeyen, J. T. 1989. *Laser Electronics*, 2nd Ed. Prentice-Hall, Englewood Cliffs, NJ.

Wada, O., Nobuhara, H., Hamaguchi, H., Mikawa, T., Tackeuchi, A., and Fuji, T. 1989. Very high speed GaInAs metal-semiconductor-metal photodiode incorporating an AlInAs/GaInAs graded superlattice, *Appl. Phys. Lett.*, 54(1):16.

Waynant, R. W. and Ediger, M. N. 1994. *Electro-Optics Handbook*. McGraw-Hill, New York.

Wilson, J. and Hawkes, J. F. B. 1989. *Optoelectronics, An Introduction*, 2nd Ed. Prentice-Hall, Englewood Cliffs, NJ.

Wojtczuk, S. J., Ballantyne, J. M., Wanaga, S., and Chen, Y. K. 1987. Comparative study of easily integratable photodetectors, *J. Lightwave Tech.*, 5(10):1365.

Wright, D. R., Keir, A. M., Pryce, A. M., Birbeck, J. C. H., Heaton, J. M., Norcross, R. J., and Wright, P. J. 1988. Limits of electroabsorption in high purity GaAs and the optimisation of waveguide devices, *IEE Proc., Pt. J.*, 135(1):39.

Yariv, A. 1989. *Quantum Electronics*, 3rd Ed. John Wiley & Sons, New York.

Yi, M. B., Paslaski, J., Lu, L–T., Margalit, S., Yariv, A., Blauvelt, H., and Lau, K. 1985. InGaAsP p-i-n photodiodes for optical communication at the 1.3 μm wavelength, *J. Appl. Phys.*, 58(12):4730.

Zory, P. S. 1993. *Quantum Well Lasers*. Academic Press, New York.

8
Miniature Solid-State Lasers

	List of Symbols	326
8.1	Introduction	330
8.2	Fundamental Concepts	331
	Laser Components • Stimulated Emission • Models of Gain Media • Rate-Equation Model of Lasers	
8.3	Solid-State Gain Media	339
	Rare Earth Dopants • Transition-Metal Dopants • Properties of Select Host Crystals • Laser Parameters for Select Gain Media	
8.4	Cavity Design	350
	Issues in Laser Design • Fundamental-Transverse-Mode Operation • Pump Considerations • Polarization Control	
8.5	Spectral Control	357
	Single-Frequency Operation • Linewidth • Frequency Tuning	
8.6	Pulsed Operation	372
	Quasi-CW Operation • Q-Switched Operation • Gain-Switched Operation • Cavity-Dumped Operation • Mode-Locked Operation	

J. J. Zayhowski[*]
Massachusetts Institute of Technology/Lincoln Laboratory

J. Harrison
Schwartz Electro-Optics, Inc.

List of Symbols

$$2^* = \begin{cases} 1 & \text{for a four-level laser} \\ g_u(f_u + f_l) & \text{for a quasi-three-level laser} \\ 2 & \text{for a nondegenerate three-level laser} \end{cases}$$

$$2\dagger = \begin{cases} 1 & \text{for a traveling-wave laser} \\ 2 & \text{for a standing-wave laser} \end{cases}$$

$2\ddagger$ number of times the same cross-sectional area of the gain medium is traversed by light during one round trip within the laser cavity

$2\ddagger_s$ number of times the same cross-sectional area of the saturable absorber is traversed by light during one round trip within the laser cavity

3^* a number between 0 and 3, depending on the alignment of the electronic wave functions with the polarization of the optical field

A cross-sectional area of the optical field (m^2)

---

[*] Supported by the U.S. Department of the Air Force.

A_g cross-sectional area of the oscillating mode in the gain medium (m^2)
A_s cross-sectional area of the oscillating mode in the saturable absorber (m^2)
B probability per unit time that a photon will interact with a given inverted site (s^{-1})
B_p pump brightness (W m^{-2} st^{-1})
c speed of light in vacuum (m s^{-1})
c_{sh} specific heat (J kg^{-1} K^{-1})
C characterization of mechanical boundary conditions
C_{11} longitudinal elastic constant (Pa)
D_e energy-diffusion constant (m^2 s^{-1})
E electric-field amplitude (V m^{-1})
$E(t)$ electric-field amplitude function (V m^{-1})
E_l energy of the lower laser level relative to the ground level (m^{-1})
E_o output pulse energy (J)
E_s energy of a state (J)
f_g thermal occupation probability of states in the ground level
f_l thermal occupation probability of states in the lower laser level
f_s thermal occupation probability of a state within a manifold
f_u thermal occupation probability of states in the upper laser level
F_{eo} fraction of the optical length of the laser cavity filled with electrooptic material
F_g fraction of the optical length of the laser cavity filled with gain medium
$F_{N,th}$ fraction of the total population of active ions in the gain medium inverted at threshold
g gain coefficient (m^{-1})
g_0 unsaturated gain coefficient (m^{-1})
g_g degeneracy of states in the ground level
g_l degeneracy of states in the lower laser level
g_{rt} round-trip gain coefficient at the time a Q-switched pulse begins to develop
g_s degeneracy of a state
g_u degeneracy of states in the upper laser level
h Planck's constant (J s)
I circulating optical intensity (W m^{-2})
I_g circulating optical intensity in the gain medium (W m^{-2})
I_m circulating optical intensity of mode m (W m^{-2})
$I_m(z)$ total optical intensity of mode m at position z (W m^{-2})
I_o output optical intensity (W m^{-2})
I_s circulating optical intensity in the saturable absorber (W m^{-2})
I_{sat} circulating optical saturation intensity (W m^{-2})
k_B Boltzmann's constant (J K^{-1})
k_c thermal conductivity (W m^{-1} K^{-1})
k_m wave vector of mode m (m^{-1})
l length of the material (m)
l_g length of the gain medium (m)
$l_{op,rt}$ round-trip optical length of the laser cavity (m)
l_s length of the saturable absorber (m)
m mode identifier
m_l ratio of the optical length of the etalon cavity to the optical length of the gain cavity in a coupled-cavity Q-switched laser
Δm difference in longitudinal mode numbers
n refractive index of the material
n_g refractive index of the gain medium
N_0 effective population inversion immediately before Q switching
N_e population of the excited level

N_{eff}	effective population inversion
N_f	effective population inversion well after the peak of a Q-switched output pulse
N_g	population of the ground level
N_l	population of the lower laser level
N_{lm}	population of the lower laser manifold
N_m	population of a manifold
N_s	population of saturable absorber sites
N_t	population of active ions in the gain medium
N_{th}	effective population inversion at threshold
N_u	population of the upper laser level
$N_{u,\text{th}}$	population of the upper laser level at threshold
N_{um}	population of the upper laser manifold
$\dot{N}_p$	pump rate (s^{-1})
$\dot{N}_{u,\text{sp}}$	spontaneous decay rate of the upper-level population (s^{-1})
$\dot{N}_{u,\text{st}}$	stimulated decay rate of the upper-level population (s^{-1})
P_a	absorbed pump power (W)
P_i	absorbed pump power required to obtain inversion (W)
P_o	output power (W)
P_p	pump power (W)
P_{po}	peak output power (W)
P_{th}	pump power at threshold (W)
q	number of photons in the laser cavity
q_0	number of photons in the laser cavity at the time a Q-switched pulse begins to form
$\dot{q}_e$	photon emission rate (s^{-1})
$\dot{q}_l$	photon loss rate (s^{-1})
$\dot{q}_{\text{sp}}$	spontaneous-emission rate of photons (s^{-1})
$\dot{q}_{\text{st}}$	stimulated-emission rate of photons (s^{-1})
Q	"quality" of the optical cavity
Q_h	rate of heat deposition (W s^{-1})
r	radial coordinate (m)
r_0	characteristic thermally defined waist size (radius) of the oscillating mode (m)
r_{eo}	electrooptic coefficient (m V^{-1})
r_m	radius of the oscillating mode (m)
r_p	radius of the pump beam (m)
r_s	radial distance to the heat sink (m)
R_1	reflectivity of the interface between the gain cavity and the etalon cavity of a coupled-cavity Q-switched laser
R_2	reflectivity of the output coupler of the etalon cavity of a coupled-cavity Q-switched laser
R_o	reflectivity of the output coupler of a laser
R_W	ratio of the pump rate factor W_p to its threshold value
R_s	ratio of saturable to unsaturable cavity losses
S_p	pulse shape factor
t	time (s)
t_b	pulse build-up time (s)
t_{rt}	cavity round-trip time (s)
$t_{\text{rt,et}}$	round-trip time of the etalon cavity of a coupled-cavity Q-switched laser (s)
$t_{\text{rt,g}}$	round-trip time of the gain cavity of a coupled-cavity Q-switched laser (s)
t_{sw}	switching time (s)
t_w	pulse width (full width at half-maximum) (s)
T	temperature (K)

$T_{et,t=0}$	effective transmission of an etalon during the early formation of a Q-switched pulse
$T_{et,cw}$	cw transmission of an etalon
T_o	transmission of the output coupler of a laser
T_r	voltage rise time (s)
V_g	active volume of the gain medium (m³)
W_p	pump rate factor (s⁻¹)
$W_{p,th}$	pump rate factor at threshold (s⁻¹)
z	coordinate along the cavity axis (m)
α	absorption coefficient of the material (m⁻¹)
α_d	thermal diffusivity (m² s⁻¹)
α_e	thermal expansion coefficient (K⁻¹)
α_p	pump absorption coefficient (m⁻¹)
$\beta(1, 2)$	mode-discrimination factor
γ	Euler's constant
γ_o	output loss coefficient
γ_{rt}	round-trip loss coefficient
$\gamma_{rt,m}$	round-trip loss coefficient for mode m
$\gamma_{rt,p}$	round-trip parasitic loss coefficient
$\gamma_{rt,p,s}$	round-trip parasitic loss associated with the saturable absorber
$\gamma_{rt,s}$	round-trip saturable loss coefficient
Γ_{rt}	round-trip loss
$\Delta_{nl,T}$	fractional thermally induced change in optical length (K⁻¹)
ζ	single-mode inversion ratio
ζ_D	single-mode inversion ratio resulting from diffusion
ζ_{SH}	single-mode inversion ratio resulting from spatial hole burning
η	efficiency
η_a	area efficiency
η_e	pulse extraction efficiency
$\eta_{e,rep}$	extraction efficiency of a repetitively pulsed system
η_i	inversion efficiency
η_o	output-coupling efficiency
η_p	pump efficiency
η_q	quantum efficiency
η_s	slope efficiency
θ	angular coordinate
κ	heat-generating efficiency of the pump
λ	free-space wavelength (measured in vacuum) = c/ν (m)
λ_0	free-space wavelength of the gain peak (m)
λ_o	oscillating free-space wavelength (m)
$\Delta\lambda(1, 2)$	difference in the free-space wavelength of the first mode to oscillate and a potential second mode (m)
$\Delta\lambda_g$	full width at half-maximum of the gain peak (m)
ν	optical frequency (s⁻¹)
ν_0	optical center frequency (s⁻¹)
ν_m	optical frequency of mode m (s⁻¹)
ν_o	optical frequency of the oscillating mode (s⁻¹)
ν_p	optical frequency of the pump (s⁻¹)
$\delta\nu_G$	gaussian linewidth (s⁻¹)
$\delta\nu_L$	Lorentzian linewidth (s⁻¹)
$\Delta\nu$	frequency difference between adjacent longitudinal modes (s⁻¹)
$\Delta\nu_{fsr}$	free spectral range (s⁻¹)

$\Delta\nu_{osc}$ oscillating bandwidth (s^{-1})
$\rho(z)$ population of the upper laser manifold at position z (m^{-3})
ρ_0 unsaturated inversion density (m^{-3})
     population density at time $t = 0$ (m^{-3})
ρ_{cq} parameter used to characterize concentration quenching (at. %)
ρ_d doping concentration (at. %)
ρ_{eff} effective inversion density (m^{-3})
ρ_l density of sites in the lower laser level (m^{-3})
ρ_m mass density (kg m^{-3})
ρ_{th} threshold population inversion density (m^{-3})
ρ_u density of sites in upper laser level (m^{-3})
$\dot{\rho}_p(z)$ pump rate at position z ($m^{-3}\,s^{-1}$)
σ_a absorption cross-section (m^2)
σ_{eff} effective emission cross-section (m^2)
σ_g stimulated-emission cross-section (m^2)
σ_m effective stimulated-emission cross-section for mode m (m^2)
σ_r radiative cross-section of a transition (m^2)
σ_s cross-section of the saturable absorber (m^2)
τ spontaneous lifetime of the upper-state population (s)
τ_0 damping constant (s)
τ_c cavity lifetime (s)
τ_l spontaneous lifetime of the upper-state population in the limit of low doping (s)
τ_{nr} nonradiative lifetime of the upper-state population (s)
τ_p temporal separation between pulses (s)
τ_r radiative lifetime of the upper-state population (s)
$\phi(\omega)$ phase
ϕ_{et} phase length of an etalon
ϕ_m phase of mode m
$\Delta\phi$ phase difference between modes m and $m - 1$
ψ_{max} maximum angle between the mirrors of a Fabry–Perot cavity
$\langle\psi(1, 2)\rangle$ mode-correlation factor
ω frequency (s^{-1})
ω_0 characteristic frequency (s^{-1})
ω_{qs} quasi-static frequency (s^{-1})
ω_r relaxation frequency (s^{-1})

8.1 Introduction

This chapter presents fundamental concepts and formulas for designing and utilizing miniature solid-state lasers. Diode-pumped, miniature, monolithic, solid-state lasers offer an efficient, compact, and robust means of generating diffraction-limited, single-frequency radiation. In addition, their diminutive size results in high-speed tuning capabilities and short-pulsed operation unmatched by larger devices. Applications areas are as diverse as communications, spectroscopy, remote sensing, nonlinear optics, projection displays, and micromachining.

The text is structured to provide adequate background for those readers who are new to the area. Sections 8.2-8.4 present a broad introduction to laser parameters and modeling, gain media, and resonators. Section 8.5 discusses the interactions in the gain media that determine spectral purity and presents methods for controlling the frequency of a laser. The final technical section, Section 8.6, contains a thorough review of issues specific to various modes of pulsed operation. In particular, there is a detailed discussion of Q switching which presents analytic results of detailed temporal modeling.

Although the presentation in this chapter is tailored specifically toward miniature solid-state lasers, the concepts are generally applicable to other classes of lasers. A section entitled Further Information has been included at the end of the chapter to help direct the interested reader to references on other laser types.

8.2 Fundamental Concepts

Laser Components

A laser, generally, consists of three components: an active medium with energy levels that can be selectively populated, a pump to produce population inversion between some of these levels, and a resonant electromagnetic cavity that contains the active medium and provides feedback to maintain the coherence of the electromagnetic field. In a continuously operating laser, coherent radiation will build up in the cavity to the level required to balance stimulated emission and cavity losses. The system is then said to be lasing and radiation is emitted in a direction defined by the cavity.

Stimulated Emission

Absorption

Quantum theory shows us that matter exists only in certain allowed energy levels or states. In thermal equilibrium, lower energy states are preferentially occupied, with an occupation probability proportional to $\exp(-E_s/k_B T)$, where E_s is the energy of the state, T is absolute temperature, and k_B is Boltzmann's constant. When light interacts with matter, it is possible for one quantum of optical energy, a photon, to be absorbed while simultaneously exciting the material into a higher energy state. In this process, the energy difference between the states is equal to the energy of the absorbed photon, and energy is conserved.

Consider a simple material system with only two energy states, and assume that essentially all optically active sites within the material system are in the lowest energy state in thermal equilibrium. The probability that a randomly chosen photon in an optical field of cross-section A will be absorbed by a given absorption site with a radiative cross-section σ_r as it passes through a material is σ_r/A. If there are g_u identical (degenerate) high-energy states, the probability becomes $g_u \sigma_r/A$. The product $g_u \sigma_r$ is known as the absorption cross-section σ_a. When all of the photons in an optical field of intensity I and all of the absorption sites in a material of length dl are accounted for, the intensity of an optical field passing through the material changes by

$$dI = -I\rho_l \sigma_a dl, \tag{8.1}$$

where ρ_l is the density of absorption sites in the material. This equation has the solution

$$I(l) = I(0)\exp(-\alpha l), \tag{8.2}$$

where $I(0)$ is the intensity of the optical field as it enters the material at position $z = 0$ and $\alpha = \rho_l \sigma_a$ is the absorption coefficient of the material.

Population Inversion, Stimulated Emission, and Gain

In the material system discussed above, there is the possibility that some sites will be in an upper energy state. Such sites are referred to as inverted. In the presence of an optical field, a transition from an upper state to a lower state can be induced by the radiation, with the simultaneous emission of a photon in phase (coherent) with the stimulating radiation. This stimulated-emission process is the inverse of the absorption process.

The probability that a randomly chosen photon in an optical field of cross-section A will stimulate a given inverted site with a radiative cross-section σ_r as it passes through a material is σ_r/A. The radiative cross-section of the transition is proportional to the dipole strength of the transition and is the same for absorption and emission. If the low-energy state is degenerate, with a degeneracy of g_l, the probability of stimulated emission becomes $g_l\sigma_r/A$. The product $g_l\sigma_r$ is known as the emission or gain cross-section σ_g. When all of the photons in the optical field are accounted for and absorption and stimulated emission are included, Eq. 8.2 becomes

$$I(l) = I(0)\exp[(\sigma_g\rho_u - \sigma_a\rho_l)l], \tag{8.3}$$

where ρ_u is the density of inverted sites. If the material is forced out of thermal equilibrium (pumped) to a sufficient degree, so that $\sigma_g\rho_u > \sigma_a\rho_l$, stimulated emission occurs at a higher rate than absorption. This leads to the coherent growth or amplification of the optical field. The material is now said to have gain, with a gain coefficient

$$g = \left[\rho_u - \left(\frac{g_u}{g_l}\right)\rho_l\right]\sigma_g. \tag{8.4}$$

The term $[\rho_u - (g_u/g_l)\rho_l]$ is referred to as the effective inversion density ρ_{eff}, and $g = \rho_{eff}\sigma_g$. When $\sigma_g\rho_u = \sigma_a\rho_l$, there is no change in the intensity of an optical field as it passes through the material, and the material is said to be in a state of transparency.

Spontaneous Emission and Lifetime

In the absence of an optical field, materials with an inverted population will still evolve toward thermal equilibrium. Transitions from an upper state to a lower state may be facilitated through interactions with the lattice and accompanied by the emission of a phonon (quantum of lattice vibration) or the generation of heat. Alternatively, they can take place through the spontaneous emission of a photon.

The time associated with the spontaneous decay of the upper state is known as the spontaneous lifetime τ and is dominated by spontaneous optical transitions in many laser gain media. For dipole transitions, the radiative lifetime τ_r is related to the gain cross-section such that (Siegman, 1986)

$$\int \sigma_g(\nu)\, d\nu = \frac{3^*\lambda_0^2}{8\pi n_g^2\tau_r}, \tag{8.5}$$

where ν is frequency, λ_0 is the free-space wavelength of the gain peak, n_g is the refractive index of the gain medium, and 3^* is a number between 0 and 3, depending on the alignment of the electronic wave functions with the polarization of the optical field. The combination of large gain cross-section and large bandwidth implies short lifetime. In general, the spontaneous lifetime is related to the radiative lifetime τ_r and nonradiative lifetime τ_{nr} according to

$$\tau^{-1} = \tau_r^{-1} + \tau_{nr}^{-1}. \tag{8.6}$$

Bandwidth

In any material system, the energy levels have a finite spectral width. This results in a bandwidth for the optical transitions of the system. The bandwidth may be determined by effects which are common to all sites within the system, resulting in a homogeneously broadened transition, or

Miniature Solid-State Lasers

may be determined by local variations in material properties, leading to an inhomogeneously broadened transition. From the above discussion on Spontaneous Emission and Lifetime, the gain cross-section σ_g of an optical transition is related to the bandwidth of the transition and the radiative lifetime τ_r. A broad transition generally implies a small gain cross-section or a short spontaneous lifetime.

Models of Gain Media

Four-Level Gain Media

The energy-level structure of a laser plays an important role in obtaining inversion. Let us try to understand pumping and relaxation in an "ideal" four-level laser with the aid of Fig. 8.1. The pumping process, indicated by the upward arrow, is assumed to excite the system from the lowest energy level, denoted by g for ground level, to the highest level, denoted by e for excited level. Pumping might occur in a variety of ways, one of which could be through radiative excitation using light whose frequency coincides with the transition frequency between g and e. Level e is assumed to relax to the upper laser level u. The population of the upper laser level is radiatively transferred, either through spontaneous or stimulated emission, to the lower laser level l. Finally, the lower laser level can either relax to the ground level or absorb the laser radiation and repopulate the upper laser level.

Several conclusions concerning optimum operation can be made from this model. First, the relaxation rates from e to u and from l to g should be as rapid as possible to maintain the maximum population inversion between u and l. Second, the pumping rate between g and e should be sufficiently rapid to overcome the spontaneous emission from u to l. Third, the thermal-equilibrium population of l should be as small as possible. Fourth, decay of e to any level other than u should be as slow as possible (for optical pumping, e can decay radiatively to g) and the nonradiative decay of u should be slow. For radiative pumping, it is advantageous to have e be distinct from u and not to have rapid radiative decay of u to g.

Three-Level Gain Media

In a three-level laser, such as ruby, the lower laser level and the ground level are the same (Fig. 8.2). Therefore, a large fraction of the ground level must be depopulated to obtain population

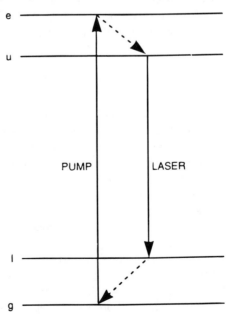

FIGURE 8.1 Schematic representation of a four-level system. Population is pumped from g to e and laser operation occurs on the transition between u and l.

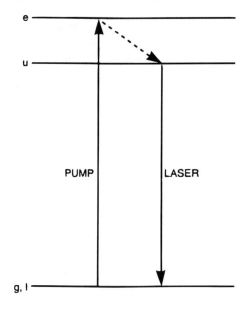

FIGURE 8.2 Schematic representation of a three-level system. Note that g and l are now the same level.

inversion. For this reason, pumping of a three-level laser requires an extremely high-intensity source.

Quasi-Three-Level Gain Media

Energy levels in ionic gain media are grouped into manifolds, as discussed in Section 8.3. Most diode-pumped systems that are referred to as "three-level" lasers have a lower laser level that is slightly above the ground level, typically within the same ground-state manifold (see Section 8.3). The term quasi-three-level is sometimes used to distinguish these systems from true three-level lasers (e.g., ruby). In quasi-three-level lasers, the lower laser level is partially occupied in thermal equilibrium and such lasers have properties that are intermediate between true three-level systems and four-level lasers.

Even though the populations of the different energy-level manifolds may be out of equilibrium, there is usually rapid thermalization within each manifold. The probability that a level within a manifold is occupied is given by $g_s f_s N_m$, where g_s is the degeneracy of states in the level, N_m is the total occupation of the manifold, and

$$f_s = \frac{\exp(-E_s/k_B T)}{\Sigma_m g_m \exp(-E_m/k_B T)}, \tag{8.7}$$

with the sum taken over all levels in the manifold. The population of the lower laser level N_l is, therefore, $g_l f_l N_{lm}$, where the subscript l denotes the lower laser level and lm denotes the lower manifold. Similarly, the population of the ground level N_g is $g_g f_g N_{lm}$. To make the model as general as possible, the upper laser level is assumed to be in an upper manifold with a population $N_u = g_u f_u N_{um}$.

The quasi-three-level model reduces to the three-level model when $g_g f_g = g_l f_l = 1$ and $g_u f_u = 1$. The four-level model is obtained when $g_g f_g = 1$, $g_l f_l = 0$, and $g_u f_u = 1$. In the discussion of Rate Equations given below, it is shown that the four-level model is also obtained without the restriction $g_u f_u = 1$ if the rate equations are written using an effective emission cross-section $\sigma_{\text{eff}} = g_u f_u \sigma_g$ instead of the spectroscopic gain cross-section σ_g. For parallel reasons, $g_u(f_u - f_l)\sigma_g$ is often referred to as the effective cross-section for quasi-three-level systems (see the discussion of Rate Equations below). When multiple closely spaced transitions overlap at a given frequency, the effective cross-section at that frequency is the sum of $g_u(f_u - f_l)\sigma_g$ over all transitions. One important example of this is the ~1.064-μm line in room-temperature Nd:YAG, where a pair of transitions centered at 1.06415 and 1.0644 μm both contribute to the effective cross-section.

Rate-Equation Model of Lasers

Rate Equations

Many of the properties of a laser can be determined from a rate-equation model for the population of the laser levels and the number of photons in the laser cavity. The rate equations provide a simple and intuitive, yet accurate, picture of the behavior of lasers. In the most simplified form, the increase in photon number within the laser cavity is balanced by the decrease in the population difference between the upper and lower laser levels. In addition, the population difference increases on account of pumping, whereas the photon number decreases due to intracavity losses (including absorption and scattering), diffraction of the beam out of the cavity, and transmission through the mirrors.

The rate-equation model can be derived as an approximation to the fundamental equations relating the electromagnetic field, the material polarization, and the quantum-state populations. The validity of the rate equations requires that the material polarization can be accurately approximated by assuming that it instantaneously follows the field; this is a situation which applies to most lasers. To describe the problem in terms of total population and total photon number within a laser cavity, it is necessary that the gain of the laser be small during one pass through the cavity, that the laser operate in a single longitudinal mode, and that the population and optical field be uniform within the cavity. (Many of these assumptions are reexamined in the discussion of Single-Frequency Operation in Section 8.5.)

As stated above, the number of photons q within the laser cavity is affected by two types of events, the emission of a photon by the gain medium ($\dot{q}_e$) and the escape of a photon from the cavity or absorption by unpumped transitions ($\dot{q}_l$). Photon emission can be either stimulated ($\dot{q}_{st}$) or spontaneous ($\dot{q}_{sp}$). Once a laser is above threshold, the stimulated-emission rate is much greater than the spontaneous rate into the oscillating mode and, to first order, spontaneous emission can be ignored. We will return to the issue of spontaneous emission in the next section when we discuss the build-up of a laser from noise.

The stimulated-emission rate is proportional to the number of photons within the cavity q, the effective population inversion N_{eff}, and the probability per unit time B that a given photon will interact with a given inverted site. The interaction probability B is the product of the probability that a photon will pass within the gain cross-section σ_g of a given inverted site as it completes a round trip within the laser cavity and the number of times the gain medium is traversed during the round trip, divided by the round-trip transit time. Mathematically, this reduces to $B = \sigma_g/A_g \times 2\ddagger/t_{rt}$, where A_g is the cross-sectional area of the laser mode, $2\ddagger$ is the number of times the gain medium is traversed (the notation $2\ddagger$ is used as a reminder that for the most common type of laser, a standing-wave laser, $2\ddagger = 2$), $t_{rt} = l_{op,rt}/c$ is the round-trip time of light in the laser cavity, $l_{op,rt} = \oint ndl$ is the round-trip optical length of the cavity, c is the speed of light in vacuum, n is the local refractive index, and the contour integral defining $l_{op,rt}$ is performed along the optical path. Therefore, the stimulated-emission rate $\dot{q}_{st} = 2\ddagger q N_{eff}\sigma_g/t_{rt}A_g$.

The escape of photons from the laser cavity and their absorption within the cavity are characterized by the cavity lifetime in the absence of any inversion, $\tau_c = t_{rt}/\gamma_{rt}$, where $\gamma_{rt} = -\ln(1 - \Gamma_{rt})$ is the round-trip loss coefficient and Γ_{rt} is the round-trip loss including transmission through the output coupler. The corresponding decrease in photon number $\dot{q}_l = q\gamma_{rt}/t_{rt}$. Thus, the total rate of change of the number of photons within the laser cavity is

$$\dot{q} = \dot{q}_{st} - \dot{q}_l = \frac{q}{t_{rt}}\left(\frac{2\ddagger N_{eff}\sigma_g}{A_g} - \gamma_{rt}\right). \tag{8.8}$$

We will now derive the rate equations for the population inversion of both a four-level and quasi-three-level laser (see the above discussion of Models of Gain Media). In both cases, the pump excites the active medium from the ground level to the excited level. It is assumed that the excited level quickly decays to the upper laser level (or manifold), so that the population of the excited level is nearly zero. Lasing occurs between the upper laser level and the lower laser level.

To derive the rate equation for the effective population inversion N_{eff}, we start by considering the population of the upper laser level N_u. The population of the upper level is affected by pumping $\dot{N}_p$, stimulated emission $\dot{N}_{u,st}$, and spontaneous decay $\dot{N}_{u,sp}$. For most pumping schemes, the pump rate is proportional to the number of ions in the ground level and can be written as $\dot{N}_p = W_p N_g$. The stimulated-emission process decreases the population of the upper laser level by one for every photon created, so that $\dot{N}_{u,st} = -\dot{q}_{st}$. Spontaneous decay is characterized by the spontaneous lifetime τ, corresponding to $\dot{N}_{u,sp} = -N_u/\tau$. Thus, the rate equation for the upper population level is

$$\dot{N}_u = \dot{N}_p + \dot{N}_{u,st} + \dot{N}_{u,sp} = W_p N_g - \frac{2\ddagger q N_{eff} \sigma_g}{t_{rt} A_g} - \frac{N_u}{\tau}. \tag{8.9}$$

In an ideal four-level laser there is a very rapid decay of the lower laser level to the ground level, so that N_l and $\dot{N}_l$ are approximately equal to zero and $N_{eff} \approx N_u$. Because the total number of active ions N_t is constant, $N_t \approx N_g + N_u$. Therefore, the rate equation for the effective population inversion of a four-level laser is

$$\dot{N}_{eff} = W_p(N_t - N_{eff}) - \frac{2\ddagger q N_{eff} \sigma_g}{t_{rt} A_g} - \frac{N_{eff}}{\tau}. \tag{8.10}$$

In a quasi-three-level laser, all of the population is in either the upper manifold or the lower manifold, and $N_t = N_{um} + N_{lm}$. The population of the upper laser level $N_u = g_u f_u N_{um}$, the population of the lower laser level $N_l = g_l f_l N_{lm}$, and the population of the ground level $N_g = g_g f_g N_{lm}$. As a result, the rate equation for the effective population inversion $N_{eff} = N_u - (g_u/g_l)N_l$ of a quasi-three-level laser reduces to

$$\dot{N}_{eff} = g_g f_g W_p (g_u f_u N_t - N_{eff}) - \frac{2\ddagger g_u (f_u + f_l) q N_{eff} \sigma_g}{t_{rt} A_g} - \frac{g_u f_l N_t + N_{eff}}{\tau}. \tag{8.11}$$

In many important laser systems, the lower laser level is essentially empty, and all of the ions are either in the ground level or the upper laser manifold. For such systems, Eqs. 8.8 and 8.11 can by written as

$$\dot{q} = \frac{q}{t_{rt}} \left(\frac{2\ddagger N_{um} \sigma_{eff}}{A_g} - \gamma_{rt} \right) \tag{8.12}$$

and

$$\dot{N}_{um} = W_p(N_t - N_{um}) - \frac{2\ddagger q N_{um} \sigma_{eff}}{t_{rt} A_g} - \frac{N_{um}}{\tau}, \tag{8.13}$$

where $\sigma_{eff} = g_u f_u \sigma_g$. These equations are the same as the rate equations for an ideal four-level laser except that the emission cross-section has been replaced by an effective emission cross-section and the population of the upper laser level has been replaced by the population of the ns manifold. The concept of an effective emission cross-section has been extended to quasi-three-level lasers; $\sigma_{\text{eff}} = g_u(f_u + f_l)\sigma_g$.

Build-Up from Noise

In the photon rate equations derived above, the term corresponding to spontaneous emission was left out. Laser action is initiated by spontaneous emission, or noise. As a result, these rate equations cannot account for the onset of lasing, as is seen by setting $q = 0$ at time $t = 0$. When spontaneous emission is properly taken into account, the photon rate equation becomes

$$\dot{q} = \frac{2\ddagger(qN_{\text{eff}} + 2\dagger N_u/2)\sigma_g}{A_g t_{\text{rt}}} - \frac{\gamma_{\text{rt}}}{t_{\text{rt}}}, \tag{8.14}$$

where

$$2\dagger = \begin{cases} 1 & \text{for a traveling-wave laser} \\ 2 & \text{for a standing-wave laser} \end{cases}.$$

The difference between a standing-wave laser and a travelling-wave laser can be understood by realizing that the intracavity optical field for a standing-wave laser is the sum of two counterpropagating traveling waves, each with its own one-half photon of noise. This one-half photon of noise stimulates optical transitions and initiates lasing.

Threshold

The threshold inversion for lasing is derived by requiring that the photon rate equation have a nontrivial solution in steady state, where $\dot{q} = 0$. This results in the condition $2\ddagger N_{\text{eff}}\sigma_g/A_g - \gamma_{\text{rt}} = 0$. Physically, the number of photons leaving the cavity must be balanced by the number of photons generated through stimulated emission. The threshold inversion, therefore, is given by

$$N_{\text{th}} = \frac{\gamma_{\text{rt}} A_g}{2\ddagger \sigma_g}. \tag{8.15}$$

The pump rate W_p required to reach threshold is obtained by setting $\dot{N}_{\text{eff}} = 0$, $\dot{q} = 0$, and $N_{\text{eff}} = N_{\text{th}}$. For a four-level laser

$$W_{p,\text{th}} = \frac{N_{\text{th}}}{(N_t - N_{\text{th}})\tau} \approx \frac{\gamma_{\text{rt}} A_g}{2\ddagger N_t \sigma_g \tau}, \tag{8.16}$$

where we have assumed that $N_{\text{th}} \ll N_t$. For a quasi-three-level laser,

$$W_{p,\text{th}} = \frac{g_u f_l N_t + N_{\text{th}}}{g_g f_g (g_u f_u N_t - N_{\text{th}})\tau}. \tag{8.17}$$

For the same value of τ, the threshold pump rate for a four-level laser is usually much smaller than the threshold pump rate for a quasi-three-level system. This is the basis of the superior performance of a four-level system over a quasi-three-level system in cw (continuous wave) operation.

Gain Saturation

The gain coefficient g of an active medium is defined as the fractional change in optical intensity per unit length as a light beam passes through. From the above discussions of Stimulated Emission and Rate Equations, it follows that $g = N_{\text{eff}} \sigma_g / V_g$, where V_g is the volume of active medium that interacts with the optical field. In the presence of a strong optical field, the population inversion is reduced and the gain is saturated. The rate equation for the population inversion (of a three- or four-level gain medium) in steady state can be rewritten in the form

$$g = \frac{g_0}{1 + I/I_{\text{sat}}}, \tag{8.18}$$

where g_0 is the unsaturated gain coefficient (the gain coefficient in the absence of an optical field), $I = q h \nu_o / t_{rt} A_g$ is the circulating optical intensity ($h\nu_o$ is the energy of one photon), and I_{sat} is the circulating saturation intensity. For a four-level laser,

$$g_0 = \frac{N_t \sigma_g W_p \tau}{V_g (W_p \tau + 1)} \tag{8.19}$$

and

$$I_{\text{sat}} = \frac{h\nu_o}{2 \ddagger \sigma_g \tau} (W_p \tau + 1). \tag{8.20}$$

It is often true that $W_p \tau \ll 1$, resulting in the simplification $g_0 = \dot{N}_p \sigma_g \tau / V_g$ and $I_{\text{sat}} = h\nu_o / 2\ddagger \sigma_g \tau$. The additional terms in Eqs. 8.19 and 8.20 correspond to bleaching (or saturating) of the pump transition by the pump. For a quasi-three-level laser,

$$g_0 = \frac{N_t \sigma_g (g_u f_u g_g f_g W_p \tau - g_u f_l)}{V_g (g_g f_g W_p \tau + 1)} \tag{8.21}$$

and

$$I_{\text{sat}} = \frac{h\nu_o (g_g f_g W_p \tau + 1)}{2 \ddagger g_u (f_u + f_l) \sigma_g \tau}. \tag{8.22}$$

Laser Efficiency

The steady-state photon rate equation ($\dot{q} = 0$) predicts that, above threshold ($q > 0$), the inversion density and the gain of a laser are clamped at their threshold values; the round-trip gain of the cavity is equal to the round-trip loss. With increased pumping, the gain remains fixed while the photon number and the output of the laser increase. For a four-level laser,

$$q = \frac{t_{rt}}{\gamma_{rt}} \left[W_p (N_t - N_{\text{th}}) - \frac{N_{\text{th}}}{\tau} \right]; \tag{8.23}$$

for a quasi-three-level laser,

$$q = \frac{t_{rt}}{g_u (f_u + f_l) \gamma_{rt}} \left[g_g f_g W_p (g_u f_u N_t - N_{\text{th}}) - \frac{g_u f_l N_t + N_{\text{th}}}{\tau} \right]. \tag{8.24}$$

The output intensity of a laser $I_o = q h \nu_o T_o / t_{rt} A$, where T_o is the transmission of the output coupler.

Because the photon number and the output intensity of the laser are linear functions of the pump rate, the efficiency of a laser is often discussed in terms of the slope efficiency η_s. Slope efficiency is defined as the ratio of the change in output power to the change in pump power of a laser once it has reached threshold, and is determined by five factors: the pump efficiency η_p, the area efficiency η_a, the inversion efficiency η_i, the quantum efficiency η_q, and the output-coupling efficiency η_o. Mathematically,

$$\eta_s = \eta_p \eta_a \eta_i \eta_q \eta_o. \tag{8.25}$$

The pump efficiency η_p is the ratio of the energy absorbed by the gain medium to the energy of the pump source. In an optically pumped laser, part of the incident optical energy may be reflected by the gain medium and part may be transmitted. Both of these effects decrease the pump efficiency. The area efficiency η_a is a measure of how well the pumped volume is used by the oscillating mode. If the cross-section of the pumped volume is much larger than the cross-section of the lasing mode, only a small portion of the pumped volume contributes to the gain of the system, and the area efficiency will be low. For a longitudinally pumped laser, the area efficiency can often be approximated by $\eta_a = 1/(r_p^2/r_m^2 + 1)$, where r_p is the radius of the pump beam and r_m is the radius of the oscillating mode. The inversion efficiency η_i is the ratio of the energy of a photon created during lasing ($h\nu_o$) to the energy required to invert one active site. In an optically pumped system, the inversion efficiency is the ratio of the energy of a photon at the oscillating frequency to the energy of an absorbed pump photon. The difference between the energies of the absorbed and emitted photons is referred to as the quantum defect. The quantum efficiency η_q is the fraction of the inverted sites that emit a photon into the oscillating mode. For an optically pumped system, 100% quantum efficiency implies that each absorbed pump photon results in one photon at the oscillating frequency. A laser operating with 100% quantum efficiency is said to have quantum-limited performance. Finally, the output-coupling efficiency η_o is the ratio of the output coupling to the total round-trip loss of the laser cavity.

The total efficiency η of a laser (power out divided by power in) is dependent on the slope efficiency and the laser threshold. For a constant slope efficiency,

$$\eta = \eta_s \left(1 - \frac{P_{th}}{P_p}\right), \tag{8.26}$$

where P_p is the total pump power and P_{th} is the pump power required to reach threshold.

8.3 Solid-State Gain Media

Rare-Earth Dopants

Crystals and glasses doped with rare earths are the most important subclass of solid-state gain media both from an applications perspective and in terms of the development of miniature lasers. Neodymium-doped yttrium aluminum garnet (Nd^{3+}:$Y_3Al_5O_{12}$ or Nd:YAG) has been the most widely used gain medium because of its unusual combination of favorable optical, thermal, and mechanical properties. Other rare-earth ions of particular interest are ytterbium (Yb^{3+}), thulium (Tm^{3+}), holmium (Ho^{3+}), and erbium (Er^{3+}). All of these have been operated as lasers both by conventional means (pumped by lamps) and as miniature lasers, including diode-pumped monolithic devices.

The optical transitions of interest for the five rare-earth ions discussed here are similar in that they correspond to transitions among an incomplete set of 4f electrons that are, generally, well shielded from the host crystal field by a complete xenon shell ($5s^2 5p^6$). The dopant ion is triply ionized, having given up a pair of 6s electrons and a single 4f electron. The transitions are characterized by sharp lines that vary little in energy from host to host. Transitions for the entire rare-earth series have been charted by Dieke and Crosswhite (Dieke and Crosswhite, 1963) in a form that is very convenient for reference. Figure 8.3 shows the positions of the energy manifolds of interest for diode pumping of Nd^{3+}, Yb^{3+}, Tm^{3+}, Ho^{3+}, and Er^{3+}. According to convention, the levels are designated by the spectroscopic notation $^{2S+1}L_J$, where the quantum number S (total spin) is an integer, J (total angular momentum) is either an integer or a half-integer, and L (total orbital angular momentum) is assigned a letter value (S, P, D, F, G, H, I, K, L, M, ... corresponding to 0, 1, 2, 3, 4, 5, 6, 7, 8, 9, ...). Each manifold consists of $2J + 1$ states. In manifolds with half-integer values of J, each state is at least doubly degenerate (there are at most $J + 1/2$ energetically distinct levels). The symmetry of the host crystal field determines any further degeneracy. Energy levels of atomic systems strongly affected by the host crystal field, such as the transition-metal dopants (discussed below), are designated by a different notation which reflects the symmetry of the modified electronic waveforms and the degeneracy of the levels.

Note that, although the interaction between the optically active electrons of the dopant ions and the host medium is weak, it is sufficient to provide electric-dipole strength to what are otherwise parity-forbidden dipole transitions. The resulting transitions are characterized by narrow linewidths and long lifetimes ($\sim 0.1 - 10$ ms). The former property leads to the generalization that rare-earth gain media are line tunable (the tuning range is a very small fraction of the oscillating frequency) as opposed to broadly tunable. The long lifetimes are particularly useful for pump-energy storage and the generation of high peak power via Q switching of the laser cavity.

Neodymium

The three most commonly used Nd lines correspond to transitions originating in the $^4F_{3/2}$ manifold. The highest gain occurs near 1.06 μm, terminating in the $^4I_{11/2}$ manifold. The 1.32-μm line ($^4F_{3/2} \rightarrow {}^4I_{13/2}$) has important applications in fiber networks (where the wavelength coincides with the propagation-loss minimum in silica fibers), although the effective cross-section σ_{eff} is roughly 1/5 of that at 1.06 μm. The 0.95-μm line terminates in the $^4I_{9/2}$ ground-state manifold and, typically, operates as a quasi-three-level system near room temperature.

Historically, Nd:YAG was developed for lamp-pumped systems in which pump absorption is achieved in a handful of discrete spectral bands located throughout the visible and near-infrared, where the broad lamp emission overlaps relatively narrow Nd transitions. In this process, the upper laser level is populated through the rapid, nonradiative relaxation of excited ions. One means of improving the net absorption is to co-dope the material with a broadband absorber that better matches the lamp spectrum while efficiently transferring the pump energy to the Nd system. The efficiency of the transfer process is sensitive to the crystal field and, although it is not effective in Nd:YAG, Cr^{3+} has been used as a sensitizing ion in low-field garnet hosts. Primary

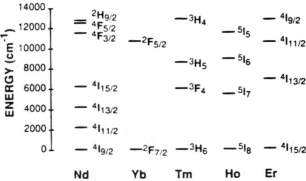

FIGURE 8.3 Partial energy-level diagrams for Nd, Yb, Tm, Ho, and Er.

among these are gadolinium scandium aluminum garnet ($Gd_3Sc_2Al_3O_{12}$ or GSGG) and yttrium scandium aluminum garnet ($Y_3Sc_2Al_3O_{12}$ or YSGG).

The optimal host material for Nd, in a given application, is determined by the detailed source requirements. The garnets have been extensively developed because of their excellent mechanical and thermal properties (e.g., hardness and thermal conductivity), and because garnet materials can be grown as relatively large boules. Among noncubic crystalline hosts, two uniaxial materials, yttrium lithium fluoride ($LiYF_4$ or YLF) and yttrium orthovanadate (YVO_4), are the most commonly used in applications involving miniature lasers.

YLF was originally developed as an alternative to YAG for systems in which thermo-optic distortions in Nd:YAG limited the performance at high average power. It is characterized by a negative bulk-lensing coefficient ($dn/dT < 0$) which offsets pump-induced changes in the optical path length due to thermal expansion. As a result, thermal lensing is greatly reduced relative to that in other Nd-doped materials, and YLF is often referred to as an athermal material. Furthermore, the large natural birefringence provides immunity to pump-induced birefringence. Another favorable property is the relatively low refractive index of YLF which increases parasitic thresholds in high-gain systems. Although Nd:YLF can now be grown in large boules of high optical quality, its relatively low fracture threshold continues to be a limitation both in high-average-power, lamp-pumped systems and in high-intensity, diode-pumped applications.

The combination of very high gain cross-section and short lifetime make $Nd:YVO_4$ a favorable gain medium in high-repetition-rate, Q-switched systems. In addition, it has received attention for diode-pumped systems required to operate over wide temperature ranges because the continuous range of useful absorption near 0.81 μm is quite broad (>10 nm). Historically, poor crystal quality has limited the use of YVO_4. However, with renewed interest in the material for diode-pumped lasers have come significant improvements in crystal growth, and good material is now generally available.

Apart from the basic optical, thermal, and mechanical considerations, the practical viability of a given crystalline host is critically dependent on the ability to synthesize large boules of high optical quality. A feature of diode-laser pumping is that the relatively high absorption coefficients achieved near 0.8 μm permit the use of smaller crystals. Figure 8.4 shows the absorption coefficient measured for Nd:YAG nominally doped at 1 at. % Nd, in the spectral region typically used for

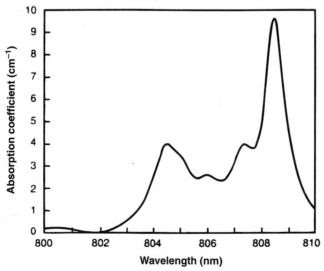

FIGURE 8.4 Absorption spectrum of Nd:YAG (1 at. %) in the region commonly used for diode pumping.

diode pumping. The peak absorption scales linearly with the Nd concentration, which is usually limited to ~1% to avoid a reduction in lifetime due to concentration quenching:

$$\tau \approx \frac{\tau_l}{1 + (\rho_d/\rho_{cq})^2}, \qquad (8.27)$$

where τ_l is the spontaneous lifetime of the upper-state population in the limit of low doping, ρ_d is the doping concentration, and ρ_{cq} is a parameter used to characterize the effects of concentration quenching. For Nd:YAG, $\rho_{cq} = 2.63$ at. %; similar values are obtained for other host crystals.

Concentration quenching stems from the presence of the $^4I_{15/2}$ manifold midway between the upper laser level and the ground state. With a sufficient density of Nd ions, energy transfer among ions enables a cross-relaxation process in which a single upper-level ion combines with a single ground-state ion to yield two additional ions in the $^4I_{15/2}$ manifold. One facet of the effort to develop miniature, diode-pumped Nd lasers has been the development of stoichiometric Nd materials in which concentration quenching is effectively inhibited even at very high Nd concentration. The inhibiting mechanism is believed to be the deviation from resonance of the intermediate $^4I_{15/2}$ levels. NdP_5O_{14} (NPP) and $LiNdP_4O_{12}$ (LNP) were early examples of stoichiometric Nd materials (Danielmeyer, 1975; Weber, 1975).

In applications, such as laser fusion, where extremely high peak powers are generated, very large gain elements are required to spread the incident beam over sufficient area to prevent irreversible damage to the optic. In such cases, a glass host is often used, in spite of the fact that the thermal characteristics are generally inferior to those of crystals. Also, unlike the case with crystalline hosts, Nd:glass is inhomogeneously broadened near room temperature due to local field variations inherent in disordered hosts. Tables of material characteristics and laser parameters for Nd:YAG, Nd:YLF, and Nd:YVO$_4$ are included below.

The miniature solid-state lasers in use today derive from early efforts to pump with GaAs light-emitting diodes and laser diodes (Newman, 1963; Keyes and Quist, 1964). With the advent of high-brightness GaAlAs lasers, it has become possible to operate miniature Nd lasers in end-pumped configurations with slope efficiencies (optical-to-optical) in excess of 50%.

Much of the work on monolithic Nd lasers has been devoted to the development of single-frequency sources for coherent applications in communications and remote sensing. One geometry that has proven successful is the nonplanar ring, illustrated in Fig. 8.5. In this device, the conventional approach toward eliminating spatial hole burning (see the discussion of Spatial Hole Burning in Section 8.5) has been implemented in a compact, multifaceted Nd gain element commonly referred to as a MISER (Monolithic Isolated Single-mode End-pumped Ring) (Kane and Byer, 1985; Kane et al., 1987). Unidirectional operation is achieved by a combination of the Faraday effect (induced by an applied magnetic field), the nonplanar geometry of the resonator, and the polarization dependence of the reflectivity of the coated output facet (Nilsson et al., 1989). To date, this approach has been limited to optically isotropic host materials [YAG, GGG (Day et al., 1989)]. The generation of nearly 1 W in a single-frequency, diffraction-limited beam has been reported from a diode-pumped, Nd:YAG MISER laser operating at 1.06 μm with a slope efficiency (output power vs. incident pump power) of 47% (Cheng et al., 1991). The same paper reported over 300 mW of output power with a slope efficiency of 60% for the same device pumped by a lower power, higher brightness diode laser.

A second monolithic geometry that has proven successful is the microchip laser, illustrated in Fig. 8.6. In a microchip laser, two faces of the gain element are polished flat and parallel and are dielectrically coated to form the cavity mirrors. The resonator is stabilized by pump-induced thermal guiding due to a combination of bulk lensing and end-face curvature (Zayhowski, 1991a; MacKinnon and Sinclair, 1992). In cubic crystals, unidirectional, transverse stress can be used to break the polarization degeneracy of the crystal and obtain linearly polarized output (Zayhowski and Mooradian, 1989b). Single-frequency operation well above threshold can be achieved if the

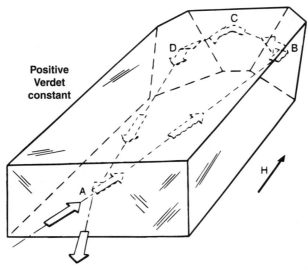

FIGURE 8.5 Illustration of a monolithic isolated single-mode end-pumped ring (MISER) laser. Polarization selection takes place at the curved, partially transmitting face (point A). At points B, C, and D, total internal reflection occurs. A magnetic field H is applied to establish unidirectional oscillation. Magnetic rotation takes place along segments AB and DA. The focused pump laser beam enters the crystal at point A, and the output beam emerges at the same point (Kane and Byer, 1985).

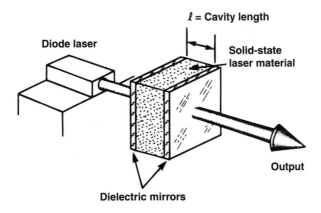

FIGURE 8.6 Illustration of a microchip laser.

longitudinal-mode spacing of the monolith is large compared to the gain bandwidth (Zayhowski and Mooradian, 1989a). For Nd:YAG at 1.06 μm, this dictates microchips with submillimeter crystal lengths. Due to the short absorption path, design optimization can include unusually high Nd concentrations in a trade-off between pump utilization (absorption) and radiation efficiency (concentration quenching).

The earliest work on microchip lasers concentrated on Nd-based stoichiometric gain media characterized by large absorption coefficients, and used an argon-ion laser as a pump source (Winzer et al., 1976; Winzer et al., 1978; and Krühler et al., 1979). Renewed interest in microchip lasers followed the first demonstration of a diode-laser-pumped device in Nd:YAG (Zayhowski and Mooradian, 1989a). Since then, diode-pumped, Nd microchip lasers oscillating at ~1.06 μm have been reported in many host materials, including YAG (Zayhowski and Mooradian, 1989a), NPP (Zayhowski and Mooradian, 1989a), LNP (Zayhowski and Mooradian, 1989c; Dixon et al., 1989), GSGG (Zayhowski and Mooradian, 1989a), YVO$_4$ (Taira et al., 1991), LaMgAl$_{11}$O$_{19}$ (Mermilliod et al., 1991), YCeAG (Gavrilovic et al., 1992), YLF (Leilabady et al., 1992), NdAl$_3$(BO$_3$)$_4$ (Amano, 1992), La$_2$O$_2$S (Zarrabi et al., 1993), and MgO:LiNbO$_3$ (MacKinnon et al., 1994). NdAl$_3$(BO$_3$)$_4$ microchip lasers have also been operated in a "self-frequency-doubled" mode, with output at 531 nm (Amano, 1992).

Ytterbium

The relevant energy-level structure of Yb^{3+} is unique among the rare-earth ions of interest in that it consists entirely of two manifolds. By pumping into the upper manifold with a source near 0.94 μm, it is possible to obtain quasi-three-level laser operation at 1.03 μm near room temperature. Due to the absence of higher lying manifolds, the efficiency of lamp-pumped systems is very poor. Renewed interest in Yb in recent years is the result of the development of strained-layer InGaAs diode lasers, and a large number of host crystals have been evaluated for use in diode-pumped Yb lasers (DeLoach et al., 1993). A room-temperature, diode-pumped microchip laser has been demonstrated in Yb:YAG (Fan, 1994), where the width of the ground-state manifold is close to 800 cm^{-1} ($\Delta\nu/c = 800$ cm^{-1}). The key laser parameters for Yb:YAG are included in Table 8.8.

Yb:YAG provides an excellent example of how high-brightness diode lasers have enabled useful applications of otherwise unfriendly laser media. Efficient operation of Yb:YAG requires a spectrally narrow pump source. Furthermore, that source must have adequate beam quality to allow efficient energy transfer to the laser mode while uniformly saturating the ground-state absorption throughout the gain region. Given the high efficiency of diode lasers and the low quantum defect inherent in the Yb^{3+} system, diode-pumped Yb:YAG has been proposed as an alternative to Nd:YAG in high-power systems. The small quantum defect provides an additional benefit in that the heat load imparted to the host is substantially lower at a given pump level than that of a comparable Nd laser (i.e., pumped near 0.8 μm for operation at 1.06 μm). Also, the long upper-state lifetime (~1 ms, compared to ~240 μs in Nd:YAG) reduces the pump requirements in low-repetition-rate Q-switched systems.

Thulium

The primary Tm laser transition is from the ^{3}F$_4$ manifold to the ground state (^{3}H$_6$), corresponding to laser action at ~2.0 μm. This wavelength is near the peak absorption of liquid water at 1.94 μm (especially important to medical applications) and is useful for eye-safe systems where Nd lasers are generally unsuitable. The key laser parameters for Tm:YAG at room temperature are listed in Table 8.9. In addition to YAG and other garnet hosts, YLF and YVO$_4$ have received considerable attention for Tm lasers.

The long lifetime of Tm is well suited to energy storage. Cr^{3+} sensitization has been employed to improve the efficiency of lamp-pumped Tm lasers in several garnet hosts including YAG (where a fortuitous overlap between the ^{2}E lines and the ^{3}H$_4$ manifold enables efficient energy transfer). Due to the low gain cross-section, however, damage often limits Q-switched operation.

For diode-pumped operation, GaAlAs lasers are employed at wavelengths just below 0.8 μm to pump the ^{3}H$_4$ manifold. Although the quantum defect is large, it is typically compensated for by a remarkable cross-relaxation process in which a single excited ^{3}H$_4$ ion combines with a single ground-state ion to produce two ions in the ^{3}F$_4$ upper laser level. The pump quantum efficiency of this process can approach 200% with adequate Tm concentration (~4 at. % in YAG). Note that this cross-relaxation process, sometimes referred to as self-quenching, is equivalent to concentration quenching in Nd. Reports of diode-pumped, monolithic Tm lasers include a Tm:YAG MISER laser (Kane and Kubo, 1991), and Tm:YAG (Storm, 1991) and Tm:YVO$_4$ (Zayhowski et al., 1995a) microchip lasers.

Selection of the Tm concentration involves a fundamental trade-off in the design of room-temperature, diode-pumped Tm lasers. High concentrations are required to efficiently absorb the

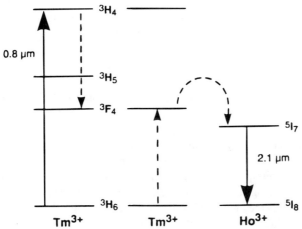

FIGURE 8.7 Energy transfer in diode-pumped materials co-doped with Tm and Ho.

pump power and to provide for efficient self-quenching. However, ground-state absorption increases in proportion to the Tm concentration. In practice, this problem has been overcome by employing Tm as a sensitizer for Ho, as described below.

Holmium

Table 8.10 includes the laser parameters for Ho:YAG. As a laser-active ion, Ho is similar to Tm. The relevant transition ($^5I_7$ to the $^5I_8$ ground state) is close in wavelength and lifetime to the $^3F_4 \rightarrow {}^3H_6$ Tm transition, although the effective gain cross-section is roughly five times larger. The key to efficient, low-threshold operation of Ho has been to use Tm as the pump-absorbing ion in co-doped crystals (e.g., Tm,Ho:YAG and Tm,Ho:YLF). This works because the upper laser level of Ho ($^5I_7$) lies just below the $^3F_4$ Tm level in energy. As a result, in thermal equilibrium the excited-ion population is shared according to the Boltzmann distribution between the Tm and Ho systems. Figure 8.7 illustrates the pump process. As long as the emission rate of the Ho laser is low compared to the equilibration rate, the Tm system acts as a reservoir for the Ho upper laser level. In Q-switched operation, the emission rate can be sufficiently high that only the energy stored in the Ho system is extracted in a single pulse. Typically, the concentration of Ho is an order of magnitude less than that of Tm and roughly half of the excited ions are in each system at room temperature.

Doubly sensitized material, Cr,Tm,Ho:YAG (CTH:YAG), is commonly used in lamp-pumped systems. In diode-pumped operation, upconversion can severely limit laser performance in Tm,Ho:YAG (Fan et al., 1988). The upconversion rate constant has been shown to be five times lower in Tm,Ho:YLF (Hansson et al., 1993), making it the preferred material in many cases. The list of diode-pumped monolithic Ho^{3+} lasers includes a Tm,Ho:YAG MISER laser (Kane and Kubo, 1991), and Tm,Ho:YAG (Storm and Rohrbach, 1989) and Tm,Ho:YLF (Harrison and Martinsen, 1994a) microchip lasers.

It is worth noting that the advent of room-temperature, mid-infrared diode lasers raises the prospect of directly pumping the $^3F_4$ Tm level in co-doped materials or resonantly pumping either thulium or holmium. An example of the latter, Ho:YAG pumped by a GaInAsSb/AlGaAsSb diode laser, was recently reported (Nabors et al., 1994).

Erbium

Er has transitions of interest near 2.9 and 1.5 μm. The longer wavelength has unique medical potential because it is close to a liquid-water absorption peak that is two orders of magnitude stronger than the peak near 2 μm. The proximity of the shorter wavelength to the dispersion minimum in silica fiber has led to considerable interest in Er lasers as sources for fiber-based communications systems. In addition, it is suitable for eye-safe remote-sensing applications. Diode-pumped microchip lasers have been reported at 2.8 μm in Er:YSGG (Harrison and Martinsen, 1994b) and 1.5 μm in Yb,Er:glass (Laporta et al., 1993).

For lamp-pumped operation, low-field garnets can be sensitized with Cr^{3+} for efficient energy transfer to the $^4I_{9/2}$ Er manifold. From there, the excited ions rapidly decay ($\tau < 1$ μs) to the $^4I_{11/2}$ manifold, the source of the 2.9-μm transition ($^4I_{11/2} \rightarrow {}^4I_{13/2}$). For diode-pumped operation, the $^4I_{9/2}$ level can be pumped directly near 0.97 μm. Based on the spontaneous lifetimes, the $^4I_{11/2}$ ($\tau = 1.5$ ms) to $^4I_{13/2}$ ($\tau = 5.1$ ms) transition appears to be a poor candidate for cw operation due to the apparent lower-level bottleneck. However, near-quantum-limited slope efficiency has been achieved in diode-pumped monolithic lasers (Dinerman et al., 1994) with several different host crystals. This surprising result is due to a fortuitous upconversion process ($^4I_{13/2} + {}^4I_{13/2} \rightarrow {}^4I_{15/2} + {}^4I_{9/2}$) that depletes the lower level while partially repumping the upper level ($^4I_{9/2} \rightarrow {}^4I_{11/2}$). In fact, by pumping directly into the $^4I_{11/2}$ manifold (resonant pumping), a slope efficiency consistent with a quantum efficiency in excess of 100% has been demonstrated in Er:GSGG (Stoneman and Esterowitz, 1992).

Yb^{3+} sensitization is used to enable direct pumping of the ground-state transition near 1.5 μm. CW operation is generally restricted to glass media due to two important characteristics. The first is that the efficiency of the upconversion process is significantly less in glasses than in crystalline hosts (for uncertain reasons). The second is that the lifetime of the $^4I_{11/2}$ manifold is generally much shorter in glasses (because nonradiative coupling is facilitated by the relatively low phonon frequencies), so that the effect of any upconversion is reduced.

Transition-Metal Dopants

The second major subclass of dopant materials is the transition-metal ions. Primary among these are chromium (Cr^{3+}, Cr^{4+}), titanium (Ti^{3+}), and cobalt (Co^{2+}). These are especially interesting in cases where the ion-host interaction results in a broadly tunable four-level system. Such lasers are powerful tools in applications, such as spectroscopy and remote sensing, where it is desirable to have a single source to investigate multiple subjects. In addition, the broad fluorescence spectra are well suited to the generation of subpicosecond pulses.

The transition metals are characterized by a number of 3d valence electrons in addition to a closed inner shell ($3s^23p^6$). In contrast to the rare-earth ions, the energy-level structure of the transition-metal ions is strongly influenced by the crystal field of the host. As a result, not all electronic levels have the same atomic spatial distribution. Broad tunability is possible when the optically active transition involves a pair of differently configured levels (Moulton, 1992). In such cases, the electronic transitions require phonon emission to reach the minimum vibrational energy within an electronic level. The atomic motion occurs slowly compared to photon emission, and optical transitions can occur corresponding to a broad range of vibrational states. Coupling between the electronic levels of the ions and the vibrational states of the host allows numerous single-photon/multiple-phonon transitions that satisfy the energy requirements of the system, resulting in vibronic broadening of the absorption and emission spectra.

The transition-metal lasers in use today represent a small portion of the systems that have been investigated, dating back to early work at Bell Laboratories (Johnson et al., 1963; Johnson et al., 1964; Johnson et al., 1966; and Johnson et al., 1967). Excited-state absorption and fluorescence quantum efficiency are often important factors in determining the viability of transition-metal-doped materials. In some cases, cryogenic operation is required to limit the nonradiative-transition rate. The laser performance of potential vibronic materials is difficult to predict accurately; each one requires careful experimental characterization. However, the promise of broadly tunable lasers, especially where there are gaps in the available spectrum, continues to spur active development of new materials.

Chromium

Cr^{3+} is the active ion in the ruby laser ($Cr:Al_2O_3$), the first laser system to be demonstrated (Maiman, 1960). It operates as a true three-level laser at 0.69 μm on a transition between the ^2E state and the $^4A_2$ ground state. The transition energy is relatively insensitive to the crystal field and the two levels have the same spatial distribution at equilibrium. Lamp pumping occurs via the vibronically broadened $^4A_2 \to {^4T_2}$ and $^4A_2 \to {^4T_1}$ transitions, with rapid, efficient decay to the upper laser level.

The position of the $^4T_2$ state is extremely sensitive to the host (its separation from the ground state is proportional to the strength of the crystal field). In low-field materials, it can have significant thermal population at room temperature due to the distribution of excitations among the $^4T_2$, ^2E, and intermediate $^2T_1$ levels, and may be the basis of a broadly tunable, four-level laser. Such is the case for alexandrite (Cr^{3+}-doped chrysoberyl or $Cr:BeAl_2O_4$) (Walling et al., 1985) which, though initially developed as a substitute for ruby, was the first vibronic laser based on Cr^{3+}. In $Cr:BeAl_2O_4$, the $^4T_2$ level ($\tau = 6.6$ μs) lies approximately 800 cm^{-1} above the ^2E level ($\tau = 1.5$ ms). If the emission rate remains low compared to the thermalization rate, the ^2E level can serve as an excitation reservoir for the near-infrared vibronic transition ($\sim 0.7 - 0.8$ μm). This leads to interesting thermal effects. For example, with increasing temperature, the effective gain cross-section increases, the effective upper-state lifetime decreases, and the gain peak shifts toward the red (partially due to thermal population of the ground-state manifold). In many cases, optimal laser performance on the vibronic transition is achieved well above room temperature.

Figure 8.8 shows a comparison of the relevant energy levels in generic high-field and low-field Cr^{3+}-doped materials. Although both might operate as three-level lasers on the $^2E \to {^4A_2}$ transition, only the low-field material promises operation as a vibronic laser. Note that, while it can be operated efficiently on the vibronic transition, $Cr:BeAl_2O_4$ is actually an intermediate-field material in that the ^2E state remains between the $^4T_2$ and $^4A_2$ states. Although vibronic operation has been demonstrated in many Cr^{3+}-doped materials, few have developed as practical sources in the near-infrared.

Recently a number of related hexafluoride crystals have shown unusual promise for tunable laser operation in the near-infrared (Payne et al., 1988; Payne et al., 1989; Smith et al., 1992; Chai et al., 1992; and Smith et al., 1993), particularly $Cr:LiCaAlF_6$ (Cr:LiCAF) and $Cr:LiSrAlF_6$ (Cr:LiSAF). Both are true Cr^{3+}-doped, low-field materials as depicted in Fig. 8.8, although, of

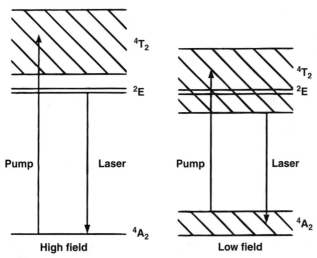

FIGURE 8.8 Partial energy-level diagrams for Cr^{3+} in high- and low-field hosts.

the two, Cr:LiCAF has a considerably higher field. As a result, Cr:LiSAF has a broader tuning range ($\sim$0.8 – 1.0 μm) but poorer mechanical and thermal properties.

Direct diode-laser pumping of the ^{2}E state is a promising means of achieving an efficient, broadly tunable, near-infrared, miniature laser. Materials, such as Cr:LiSAF, that can be doped to high levels are particularly attractive. As a first step, diode-pumped operation has been demonstrated in Cr:BeAl$_2$O$_4$ (Scheps et al., 1990), Cr:LiCAF (Scheps, 1991), Cr:LiSAF (Scheps et al., 1991), and Cr:LiSrGaF$_6$ (Scheps, 1992) using AlGaInP lasers operating near 0.67 μm. As a second step, a tuning range of 80 nm has been demonstrated with a diode-pumped Cr:LiSAF laser system consisting of a 0.5-mm-long microchip laser coupled to a dispersive external resonator (Zhang et al., 1992). As the power and brightness of available AlGaInP diode lasers continue to increase, so will the range of miniature lasers based on Cr^{3+}-doped materials. For example, the first diode-pumped, mode-locked Cr:LiSAF lasers were recently reported (Balembois et al., 1993; French et al., 1993).

When vibronic laser operation of Cr-doped fosterite (Cr:Mg$_2$SiO$_4$) was first reported in 1988 (Petricevic et al., 1988a) the active ion was believed to be Cr^{3+}. This material is of particular interest because it operates within the otherwise inaccessible $\sim$200-nm-wide tuning range centered near 1.25 μm. The fluorescence spectrum is now attributed to a vibronic transition of Cr^{4+} (Petricevic et al., 1988b; Verdun et al., 1988; and Moncorge et al., 1991). The broad absorption band is suitable for lamp pumping. It also allows pumping with a Nd:YAG laser at 1.06 μm, although the much stronger peak absorption near 0.75 μm suggests a means for diode-pumped operation. Currently, material quality and poor thermal properties remain important issues limiting the performance of Cr:Mg$_2$SiO$_4$ lasers. In addition, nonradiative decay impedes efficient cw operation at room temperature. However, at 77 K, pumped with a 12-W, 1.06-μm Nd:YAG laser, 2.8 W of cw output was obtained (Carrig and Pollock, 1993). Going from room temperature to 77 K, τ increased from 3.6 to 27 μs.

Titanium

Since its discovery as a vibronic laser material (Moulton, 1982), Ti^{3+}-doped sapphire (Ti:Al$_2$O$_3$) has proven to be a nearly ideal laser material (Moulton, 1986). Laser pumped near 0.5 μm (by an argon-ion laser, a frequency-doubled Nd laser, a copper-vapor laser, or a dye laser), it can be operated with nearly unitary quantum efficiency at room temperature either gain switched, cw, or mode locked. The peak gain near 0.8 μm is similar to that of Nd:YAG at 1.06 μm, and Ti:Al$_2$O$_3$ has been tuned (in both pulsed and cw operation) from below 0.7 μm to beyond 1.1 μm. The laser operates on a single-electron (3d^1) transition (^{2}E → ^{2}T$_2$); there are no higher lying levels to support excited-state absorption. Nonradiative decay only becomes significant above room temperature. The lifetime at room temperature (τ = 3.2 μs at 300 K) is only slightly below that at cryogenic temperatures (τ = 3.8 μs at 77 K). Furthermore, Al$_2$O$_3$ has excellent mechanical and thermal properties. To date, laser-pumped Ti:Al$_2$O$_3$ lasers have generated up to 2.0 J in pulsed operation (Muller et al., 1988) and 43 W cw (Erbert et al., 1991).

Early work on the development of Ti:Al$_2$O$_3$ lasers was plagued by a broad, featureless parasitic absorption that extended throughout the fluorescence band. Material was characterized in terms of the figure of merit, defined as the ratio of the peak absorption near 0.5 μm to the absorption at 0.8 μm. Figures of merit of a few tens were typical of early laser material. Later, the parasitic absorption was attributed to the presence of Ti^{4+}, which can be reduced by annealing (Aggarwal et al., 1988). Currently, the figure of merit of commercial material is commonly above 200, leading to negligible parasitic loss in most cases.

One drawback of Ti:Al$_2$O$_3$ is the difficulty of direct flashlamp pumping due to a short upper-state lifetime and limited overlap of the pump spectrum with typical lamp spectra. However, this has not prevented the introduction of commercial, lamp-pumped systems. Flashlamp-pumped pulse energies as high as 6.5 J have been reported (Brown and Fisher, 1993). The prospects for miniature, diode-pumped Ti:Al$_2$O$_3$ lasers are less clear. Direct pumping is likely to be achieved in the long term, assuming continued success in the development of green diode lasers. Currently, diode-pumped systems use diode-pumped, frequency-doubled Nd lasers to pump Ti:Al$_2$O$_3$. Lasers of this type have been demonstrated in both pulsed (Maker and Ferguson, 1990) and cw operation (Harrison et al., 1991).

Cobalt

Cryogenic laser operation of Co^{2+}-doped magnesium fluoride (Co:MgF$_2$) was first achieved in 1964 (Johnson et al., 1964), but it was not until 1988 that room temperature, pulsed operation was reported (Welford and Moulton, 1988). The laser transition ($^4T_2 \rightarrow {}^4T_1$) is relatively weak (Harrison et al., 1989) but extremely broad (Lovold et al., 1985). Tuning over the ranges 1.51 – 2.28 μm and 1.75 – 2.50 μm has been demonstrated at 77 K and 299 K, respectively (Moulton, 1992). The absorption peak near 1.3 μm is well suited to Nd-laser pumping. Although excited-state absorption is insignificant in Co:MgF$_2$, nonradiative decay is a major challenge in noncryogenic operation. The lifetime increases from 36 μs at room temperature to 1.3 ms at 77 K (Moulton, 1985). The material has good mechanical and thermal properties (in particular, low thermal lensing due to high thermal conductivity and low dn/dT), and boules can be grown with high optical quality.

Near-quantum-limited slope efficiency was demonstrated at room temperature by employing a low-loss resonator and pump pulses with sufficiently short rise times to achieve quasi-cw emission in a time short compared to τ (Welford and Moulton, 1988). Cryogenic temperatures are still required for cw operation. Currently, the prospects for direct diode pumping of Co:MgF$_2$ lasers are poor because there are few applications to drive the development of high-power diode lasers at 1.3 μm. A more realistic goal may be to utilize a diode-pumped Nd laser operating at 1.3 μm as a pump source in an all-solid-state system.

Properties of Select Host Crystals

Tables 8.1 – 8.4 list several important properties of four common host crystals: yttrium aluminum garnet (YAG), yttrium lithium fluoride (YLF), yttrium orthovanadate (YVO$_4$), and sapphire (Al$_2$O$_3$). The values listed are typical of the those reported in the literature, although, in many cases, there is a significant spread in the reported values. All values listed are for room temperature (300 K).

TABLE 8.1 Properties of Yttrium Aluminum Garnet (Y$_3$Al$_5$O$_{12}$)

Property	Symbol	Value	Units	Comments
Crystal symmetry				cubic
Lattice constant	a_0	12.01	Å	
Refractive index	n	1.818		1.064 μm
Temperature index variation	dn/dT	7.3×10^{-6}	K^{-1}	1.064 μm
Thermal expansion coefficient	α_e	7.5×10^{-6}	K^{-1}	
Thermal conductivity	k_c	0.13	W cm^{-1} K^{-1}	
Specific heat	c_{sh}	0.6	J g^{-1} K^{-1}	
Thermal diffusivity	α_d	0.046	cm^2 s^{-1}	
Mass density	ρ_m	4.55	g cm^{-3}	
Elastic constant	C_{11}	3.2×10^3	g cm^{-2}	
Poisson's ratio		0.25		
Hardness		1215	kg mm^{-2}	Knoop
		8–8.5		Moh's scale
Tensile strength		2.0	g cm^{-2}	
Melting point		1950	°C	

TABLE 8.2 Properties of Yttrium Lithium Fluoride (LiYF$_4$)

Property	Symbol	Value	Units	Comments
Crystal symmetry				tetragonal positive uniaxial
Lattice constant	a_0	5.242	Å	
	c_0	11.37	Å	
Refractive index	n_π	1.4704		E∥c, 1.053 μm
	n_σ	1.4482		E⊥c, 1.053 μm
Temperature index variation	dn/dT	-4.3×10^{-6}	K^{-1}	E∥c, 1.053 μm
		-2.0×10^{-6}	K^{-1}	E⊥c, 1.053 μm
Thermal expansion coefficient	α_e	8.0×10^{-6}	K^{-1}	E∥c
		1.3×10^{-6}	K^{-1}	E⊥c
Thermal conductivity	k_c	0.06	W cm^{-1} K^{-1}	
Specific heat	c_{sh}	0.79	J g^{-1} K^{-1}	
Thermal diffusivity	α_d	0.02	cm^2 s^{-1}	
Mass density	ρ_m	3.99	g cm^{-3}	
Elastic constant	C_{11}	7.6×10^2	g cm^{-2}	
Poisson's ratio		0.33		
Hardness		260–325	kg mm^{-2}	Knoop
		4–5		Moh's scale
Tensile strength		0.34	g cm^{-2}	
Melting point		825	°C	

TABLE 8.3 Properties of Yttrium Orthovanadate (YVO$_4$)

Property	Symbol	Value	Units	Comments
Crystal symmetry				tetragonal positive uniaxial
Lattice constant	a_0	7.119	Å	
	c_0	6.290	Å	
Refractive index	n_π	2.168		E∥c, 1.064 μm
	n_σ	1.958		E⊥c, 1.064 μm
Temperature index variation	dn/dT	2.9×10^{-6}	K^{-1}	E∥c, 632.8 nm
		8.5×10^{-6}	K^{-1}	E⊥c, 632.8 nm
Thermal expansion coefficient	α_e	11×10^{-6}	K^{-1}	E∥c
		4.4×10^{-6}	K^{-1}	E⊥c
Thermal conductivity	k_c	0.05	W cm^{-1} K^{-1}	
Specific heat	c_{sh}		J g^{-1} K^{-1}	
Thermal diffusivity	α_d		cm^2 s^{-1}	
Mass density	ρ_m	4.23	g cm^{-3}	
Elastic constant	C_{11}		g cm^{-2}	
Poisson's ratio				
Hardness		480	kg mm^{-2}	Knoop
		5.5		Moh's scale
Tensile strength			g cm^{-2}	
Melting point		1750–1940	°C	dependent on oxygen pressure

Laser Parameters for Select Gain Media

Tables 8.5 – 8.14 list important laser parameters for several gain media: Nd:YAG, Nd:YLF, Nd:YVO$_4$, Yb:YAG, Tm:YAG, Ho:YAG, Tm,Ho:YAG, Tm,Ho:YLF, Er:YAG, and Ti:Al$_2$O$_3$. The values listed are typical of the those reported in the literature, although, in many cases, there is a significant spread in the reported values. All values listed are for room temperature (300 K).

Miniature Solid-State Lasers 351

TABLE 8.4 Properties of Sapphire (Al_2O_3)

Property	Symbol	Value	Units	Comments
Crystal symmetry				rhombohedral positive uniaxial
Lattice constant	a_0	4.759	Å	
	c_0	12.99	Å	
Refractive index	n_π	1.755		E∥c, 694.3 nm
	n_σ	1.763		E⊥c, 694.3 nm
Temperature index variation	dn/dT	17×10^{-6}	K^{-1}	E∥c, 546.1 nm
		13×10^{-6}	K^{-1}	E⊥c, 546.1 nm
Thermal expansion coefficient	α_e	5.3×10^{-6}	K^{-1}	E∥c
		4.8×10^{-6}	K^{-1}	E⊥c
Thermal conductivity	k_c	0.42	$W\ cm^{-1}\ K^{-1}$	
Specific heat	c_{sh}	0.75	$J\ g^{-1}\ K^{-1}$	
Thermal diffusivity	α_d	0.14	$cm^2\ s^{-1}$	
Mass density	ρ_m	3.99	$g\ cm^{-3}$	
Elastic constant	C_{11}	3.5×10^3	$g\ cm^{-2}$	
Poisson's ratio		0.27		
Hardness		2100	$kg\ mm^{-2}$	Knoop
		9		Moh's scale
Tensile strength		5.5	$g\ cm^{-2}$	
Melting point		2040	°C	

TABLE 8.5 Laser Parameters for Neodymium-doped YAG (Nd^{3+}:$Y_3Al_5O_{12}$)

Property	Symbol	Value	Units	Comments
Nd^{3+} concentration (1.0 at. %)		1.39×10^{20}	cm^{-3}	
Quantum efficiency	η_q	1.0		
Wavelength at gain peak	λ_0	946	nm	
		1.064	μm	
		1.319	μm	
Linewidth	$\Delta\lambda_g$	0.8	nm	946 nm
		0.6	nm	1.064 μm
		0.6	nm	1.319 μm
Spontaneous lifetime	τ	230	μs	1.0 at %
Concentration-quenching parameter	ρ_{cq}	2.63	at. %	
Emission cross-section	σ_g	0.4×10^{-19}	cm^2	946 nm
		6.5×10^{-19}	cm^2	1.06415 μm
		1.2×10^{-19}	cm^2	1.0644 μm
		1.7×10^{-19}	cm^2	1.319 μm
Occupation probability	f_u	0.60		946 nm
	f_l	0.008		946 nm
	f_u	0.40		1.06415 μm
	f_l	0		1.06415 μm
	f_u	0.60		1.0644 μm
	f_l	0		1.0644 μm
	f_u	0.40		1.319 μm
	f_l	0		1.319 μm
Energy of lower level	E_l	857	cm^{-1}	946 nm
Effective emission cross-section	σ_{eff}	0.3×10^{-19}	cm^2	946 nm
		3.3×10^{-19}	cm^2	sum of 1.06415 and 1.0644 μm
		0.7×10^{-19}	cm^2	1.319 μm
Pump absorption coefficient	α_p	12	cm^{-1}	809 nm

TABLE 8.6 Laser Parameters for Neodymium-doped YLF (Nd^{3+}:$LiYF_4$)

Property	Symbol	Value	Units	Comments
Nd^{3+} concentration (1.0 at. %)		1.4×10^{20}	cm^{-3}	
Quantum efficiency	η_q	1.0		
Wavelength at gain peak	λ_0	1.047	μm	E∥c
		1.053	μm	E⊥c
Linewidth	$\Delta\lambda_g$	1.2	nm	E∥c, 1.047 μm
		1.4	nm	E⊥c, 1.053 μm
Spontaneous lifetime	τ	460	μs	1.0 at %
Concentration-quenching parameter	ρ_{cq}	2.77	at. %	
Emission cross-section	σ_g	4.4×10^{-19}	cm^2	E∥c, 1.047 μm
		2.2×10^{-19}	cm^2	E⊥c, 1.053 μm
Occupation probability	f_u	0.43		E∥c, 1.047 μm
	f_l	0		E∥c, 1.047 μm
	f_u	0.57		E⊥c, 1.053 μm
	f_l	0		E⊥c, 1.053 μm
Effective emission cross-section	σ_{eff}	1.9×10^{-19}	cm^2	E∥c, 1.047 μm
		1.2×10^{-19}	cm^2	E⊥c, 1.053 μm
Pump absorption coefficient	α_p	2.4	cm^{-1}	E∥c, 806 nm
		3.6	cm^{-1}	E∥c, 797 nm
		8.0	cm^{-1}	E⊥c, 797 nm
		11	cm^{-1}	E⊥c, 792 nm

TABLE 8.7 Laser Parameters for Neodymium-doped YVO_4 (Nd^{3+}:YVO_4)

Property	Symbol	Value	Units	Comments
Nd^{3+} concentration (1.0 at. %)		1.252×10^{20}	cm^{-3}	
Quantum efficiency	η_q	1.0		
Wavelength at gain peak	λ_0	1.064	μm	E∥c
Linewidth	$\Delta\lambda_g$	0.8	nm	
Spontaneous lifetime	τ	100	μs	1.0 at %
Concentration-quenching parameter	ρ_{cq}	2.76	at. %	
Emission cross-section	σ_g	30.0×10^{-19}	cm^2	
Occupation probability	f_u	0.52		
	f_l	0		
Effective emission cross-section	σ_{eff}	16×10^{-19}	cm^2	
Pump absorption coefficient	α_p	41	cm^{-1}	E∥c, 808.5 nm
		11	cm^{-1}	E⊥c, 808.5 nm

TABLE 8.8 Laser Parameters for Ytterbium-doped YAG (Yb^{3+}:$Y_3Al_5O_{12}$)

Property	Symbol	Value	Units	Comments
Yb^{3+} concentration (5.5 at. %)		7.6×10^{20}	cm^{-3}	
Quantum efficiency	η_q	1.0		
Wavelength at gain peak	λ_0	1.030	μm	
Linewidth	$\Delta\lambda_g$	9	nm	
Spontaneous lifetime	τ	0.95	ms	
Emission cross-section	σ_g	3.1×10^{-20}	cm^2	
Occupation probability	f_u	0.7		
	f_l	0.047		
Energy of lower level	E_l	612	cm^{-1}	
Effective emission cross-section	σ_{eff}	2.3×10^{-20}	cm^2	
Pump absorption coefficient	α_p	4.9	cm^{-1}	968 nm
		5.1	cm^{-1}	941 nm

TABLE 8.9 Laser Parameters for Thulium-doped YAG (Tm^{3+}:$Y_3Al_5O_{12}$)

Property	Symbol	Value	Units	Comments
Tm^{3+} concentration (6.0 at. %)		8.3×10^{20}	cm^{-3}	
Quantum efficiency	η_q	1.8		
Wavelength at gain peak	λ_0	2.013	μm	
Spontaneous lifetime	τ	10	ms	
Emission cross-section	σ_g	3.6×10^{-21}	cm^2	
Occupation probability	f_u	0.46		
	f_l	0.017		
Energy of lower level	E_l	588	cm^{-1}	
Effective emission cross-section	σ_{eff}	1.7×10^{-21}	cm^2	
Pump absorption coefficient	α_p	6.2	cm^{-1}	786 nm
		5.4	cm^{-1}	781 nm

TABLE 8.10 Laser Parameters for Holmium-doped YAG (Ho^{3+}:$Y_3Al_5O_{12}$)

Property	Symbol	Value	Units	Comments
Ho^{3+} concentration (4.0 at. %)		5.6×10^{20}	cm^{-3}	
Quantum efficiency	η_q	1.0		
Wavelength at gain peak	λ_0	2.092	μm	
Spontaneous lifetime	τ	7	ms	
Emission cross-section	σ_g	1×10^{-19}	cm^2	
Occupation probability	f_u	0.10		
	f_l	0.016		
Energy of lower level	E_l	458	cm^{-1}	
Effective emission cross-section	σ_{eff}	1×10^{-20}	cm^2	
Pump absorption coefficient	α_p	5.8	cm^{-1}	1.907 μm

TABLE 8.11 Laser Parameters for Thulium-Holmium YAG (Tm^{3+},Ho^{3+}:$Y_3Al_5O_{12}$)

Property	Symbol	Value	Units	Comments
Tm^{3+} concentration (6.0 at. %)		8.3×10^{20}	cm^{-3}	
Ho^{3+} concentration (0.4 at. %)		5.6×10^{19}	cm^{-3}	
Quantum efficiency	η_q	1.8		
Wavelength at gain peak	λ_0	2.092	μm	
Spontaneous lifetime	τ	8	ms	
Effective lifetime		1 − 3	ms	due to upconversion
Emission cross-section	σ_g	1×10^{-19}	cm^2	
Occupation probability	f_u	0.10		
	f_l	0.016		
Energy of lower level	E_l	458	cm^{-1}	
Effective emission cross-section	σ_{eff}	1×10^{-20}	cm^2	
Fraction of excitation in Ho^{3+}		0.58		
Pump absorption coefficient	α_p	6.2	cm^{-1}	786 nm
		5.4	cm^{-1}	781 nm

8.4 Cavity Design

Issues in Laser Design

The design of a laser is dictated by many interdependent factors, including the requirements placed on the output beam (wavelength, spectral purity, tunability, divergence, polarization, power, power stability), the operating environment (temperature, humidity, vibration, acceleration, externally applied forces), and practical considerations (size, cost, available power, pump-source characteristics). There is an increasingly large number of gain media, cavity designs, and pump

TABLE 8.12 Laser Parameters for Thulium-Holmium YLF (Tm^{3+},Ho^{3+}:$LiYF_4$)

Property	Symbol	Value	Units	Comments
Tm^{3+} concentration (6.0 at. %)		8.3×10^{20}	cm^{-3}	
Ho^{3+} concentration (0.4 at. %)		5.6×10^{19}	cm^{-3}	
Quantum efficiency	η_q	1.8		
Wavelength at gain peak	λ_0	2.067	μm	E∥c
Spontaneous lifetime	τ	15	ms	
Effective lifetime		~5	ms	due to upconversion
Emission cross-section	σ_g	1.4×10^{-19}	cm^2	
Occupation probability	f_u	0.13		
	f_l	0.032		
Energy of lower level	E_l	315	cm^{-1}	
Effective emission cross-section	σ_{eff}	2.3×10^{-20}	cm^2	
Fraction of excitation in Ho^{3+}		0.56		
Pump absorption coefficient	α_p	2.7	cm^{-1}	E⊥c, 795 nm
		4.3	cm^{-1}	E∥c, 792 nm
		6.1	cm^{-1}	E∥c, 781 nm

TABLE 8.13 Laser Parameters for Erbium-doped YAG (Er^{3+}:$Y_3Al_5O_{12}$)

Property	Symbol	Value	Units	Comments
Er^{3+} concentration (30 at. %)		4.2×10^{21}	cm^{-3}	
Quantum efficiency	η_q	1.6		
Wavelength at gain peak	λ_0	2.937	μm	
Linewidth	$\Delta\lambda_g$	8	nm	
Spontaneous lifetime	τ	100	μs	
Emission cross-section	σ_g	2.7×10^{-20}	cm^2	
Occupation probability	f_u	0.21		
	f_l	0		
Effective emssion cross-section	σ_{eff}	5.6×10^{-21}	cm^2	
Pump absorption coefficient	α_p	6.2	cm^{-1}	974 nm
		13	cm^{-1}	965 nm
		11	cm^{-1}	960 nm

TABLE 8.14 Laser Parameters for Titanium-doped Sapphire (Ti^{3+}:Al_2O_3)

Property	Symbol	Value	Units	Comments
Ti^{3+} concentration (0.09 wt. %)		3×10^{19}	cm^{-3}	
Quantum efficiency	η_q	1.0		
Wavelength at gain peak	λ_0	790	nm	
Linewidth	$\Delta\lambda_g$	300	nm	
Spontaneous lifetime	τ	3.15	μs	
Emission cross-section	σ_g	3.5×10^{-19}	cm^2	E∥c, 790 nm
Pump absorption coefficient	α_p	2.2	cm^{-1}	E∥c, 532 nm
		2.9	cm^{-1}	E∥c, 490 nm

configurations that have been employed in lasers, and several texts have been written on the subject of laser design. No one design is well suited to all applications; every laser is optimized for operation at one point in the multidimensional parameter space outlined above.

A very important issue in the design of many lasers is the extraction of heat from the gain medium. In the process of pumping the gain medium, heat is generated. As the temperature of the gain medium changes, so do its physical length and refractive index. Each of these contributes to changes in the optical length and resonant frequencies of the laser cavity. (These changes are especially significant in miniature lasers where the active volume constitutes much of the resonator.)

Nonuniform heating results in a nonuniform refractive index and internal stress. Index gradients lead to thermal lensing, which changes the confocal parameters of the laser cavity and can destabilize an otherwise stable cavity, or vice versa. Internal stress leads to stress birefringence and, eventually, stress-induced fracture.

Other issues that must be considered in high-power lasers are nonlinear optical effects and optical damage. The electrical field within the optical beam of a high-power laser can be large enough to damage optical components. This is particularly important in high-peak-power pulsed lasers. At optical intensities below the optical damage level, deleterious nonlinear optical interactions can still degrade the performance of the laser and even destroy the device. One example is stimulated Brillouin (acoustic wave) scattering in fiber lasers. In this case, nonlinear interactions create acoustic waves which can shatter the ends of the fiber.

Fundamental-Transverse-Mode Operation

Conventional Cavity Designs

One common feature of many applications for miniature lasers is that mode quality is at least as important as total power. As a result, miniature lasers are usually designed to operate in the fundamental transverse mode (see the discussion of Pump Considerations below). Most of the results and formulas presented in this chapter are derived for fundamental-mode operation and may require some modification for lasers operating in multiple transverse modes.

The transverse modes of a laser are determined by the cavity design and the pump deposition profile. The cavities for many miniature solid-state lasers are small versions of larger devices, designed according to the same principles (Hall and Jackson, 1989). However, the use of longitudinal pumping allows additional possibilities for obtaining stable cavity modes.

Fabry–Perot Cavities

Thermal Guiding. Efficient lasers can be produced using small, longitudinally pumped, standing-wave laser cavities defined by two plane mirrors. The planar uniformity of such a cavity is broken by the pump beam, which deposits heat as it pumps the crystal. The heat diffuses away from the pump beam, generally resulting in a radially symmetric temperature distribution. In materials with a positive change in refractive index n with temperature T ($dn/dT > 0$), such as Nd:YAG, this results in a thermal waveguide. In addition, when the cavity mirrors are deposited on the gain medium, there is some thermally induced curvature of the mirrors as the warmer sections of the gain medium expand or contract. In materials with a positive thermal expansion coefficient α_e this effect also contributes to the stabilization of the transverse mode. In some materials, such as Nd:YLF, this term dominates and can lead to stable transverse-mode operation in an otherwise flat-flat cavity, despite a negative dn/dT. Another effect is strain-induced variation of the refractive index caused by nonuniform heating and expansion of the gain medium. This effect tends to be less important than the others in determining the transverse mode characteristics of the cavity, although it can cause significant local birefringence. It will be ignored in this section.

When both index change and thermal expansion are considered, the variation in the optical length of a material nl (where l is the physical length of the material) as a function of temperature is given by

$$\frac{\delta(nl)}{nl} = \left(C\alpha_e + \frac{1}{n}\frac{dn}{dT}\right)\delta T, \tag{8.28}$$

where C is a number between 0 and 1, depending on whether the material is constrained ($C = 0$) or free to expand ($C = 1$). For the case of nonuniform heating, the thermal expansion of the warmer sections of the material will be constrained by the cooler regions, and C may be a function of the thermal gradients. If the cavity length is short compared to the confocal parameter, as in the case of microchip lasers, the total change in optical length as a function of transverse cavity position can be modeled as a simple lens between the two flat mirrors or as an axially uniform waveguide with a radially varying index.

To simplify the analysis, we will assume that C is independent of the position within the laser cavity, so that $\delta(nl) = nl\Delta_{nl,T}\delta T$, where $\Delta_{nl,T} = C\alpha_e + dn/ndT$ is a constant. For a monolithic, longitudinally pumped, short cavity with radial heat flow, the radius of the oscillating mode r_m is given by (Zayhowski, 1991a)

$$r_m^2 = \lambda_o \left[\frac{(r_0^2 + r_p^2)lk_c}{\pi \kappa P_a n \Delta_{nl,T}}\right]^{1/2}, \qquad (8.29)$$

where λ_o is the oscillating wavelength, l is the cavity length, k_c is the thermal conductivity of the gain medium, κ is the heat-generating efficiency of the gain medium, P_a is the absorbed pump power, r_p is the average radius of the pump beam within the laser cavity, and r_0 determines the waist size of the oscillating mode for extremely small pump radii. Once $\Delta_{nl,T}$ and r_0 are determined, this equation does an excellent job of describing the pump-power dependence of the oscillating-mode radius.

For a thermally guided Fabry–Perot laser cavity to create a symmetric fundamental transverse mode, parallelism between the cavity mirrors is critical. The maximum angle ψ_{max} that can be tolerated between the mirrors (for symmetric fundamental-mode operation) is given by (Zayhowski, 1991a)

$$\psi_{max} = 1.8 \frac{\kappa P_a \Delta_{nl,T}}{4\pi k_c r_p}. \qquad (8.30)$$

A linear thermal gradient across the oscillating mode will also contribute to the effective wedge between the two mirrors, through both the thermal expansion and the temperature-induced change in index.

Aperture Guiding. In three-level or quasi-three-level lasers, there can be significant absorption of the oscillating radiation in unpumped regions of the gain medium. For longitudinally pumped devices, this creates a radially dependent loss (aperture) which can restrict the transverse dimensions of the lasing mode, resulting in smaller mode radii than predicted by Eq. 8.29. Aperture guiding may be important in Yb:YAG microchip lasers (Fan, 1994).

In principle, aperture guiding (gain guiding) can exist even in ideal four-level lasers with no absorption of the oscillating radiation. The absence of gain in the unpumped regions of the gain medium is sufficient to define a stable transverse mode (Kogelnik, 1965). This effect is usually insignificant compared to thermal guiding.

Pump Considerations

To ensure oscillation in the fundamental transverse mode, it is necessary for the fundamental mode to use most of the gain available to the laser. If the radius of the fundamental mode is much less than the radius of the pumped region of the gain medium, higher order transverse modes will oscillate.

For longitudinally pumped lasers, the radius r_m of the oscillating mode in the gain medium and the desired round-trip gain $g_{rt} = 2\ddagger gl_g$ (where l_g is the length of the gain medium) together determine the required brightness of the pump source B_p, where brightness is defined as intensity per unit solid angle. For an incoherent pump source, the required brightness is given by (Fan and Sanchez, 1990)

$$B_p > \frac{4}{P_p}\left(\frac{h\nu_p g_{rt} l_g}{2\ddagger n_g \sigma_{eff}\tau\eta_p}\right)^2, \qquad (8.31)$$

where P_p is the total pump power, $h\nu_p$ is the energy of a pump photon, and η_p is the fraction of the pump power absorbed. For a diffraction-limited pump source, the brightness requirement is reduced by a factor of four (Fan and Sanchez, 1990):

$$B_p > \frac{1}{P_p}\left(\frac{h\nu_p g_{rt} l_g}{2\ddagger n_g \sigma_{eff}\tau\eta_p}\right)^2. \qquad (8.32)$$

Single-transverse-mode operation is obtained by focusing the pump light within the volume of the fundamental oscillating mode.

Polarization Control

It is often desirable for a laser to oscillate in a single linear polarization. For lasers with isotropic gain media, this can be ensured by including a polarizing element, such as a Brewster plate, within the cavity. In single-frequency devices, the presence of any tilted optical element may be sufficient to polarize a laser because the reflection coefficient of surfaces is often different for s- and p-polarized waves; only a small amount of modal discrimination is needed to select one of two frequency-degenerate polarizations (see the discussion of Single-Frequency Operation in Section 8.5). If the design of a cavity does not favor a given polarization, the polarization degeneracy of an isotropic gain medium can often be removed by applying a uniaxial transverse stress.

If there is very little polarization selectivity within a laser cavity, feedback from external surfaces can determine the polarization of the oscillating mode. Although this effect is usually undesirable, it has been used to controllably switch the polarization of microchip lasers at rates up to 100 kHz (Zayhowski, 1991b).

8.5 Spectral Control

Single-Frequency Operation

Multimode Operation

The frequency spacing between adjacent longitudinal modes of a cavity (also known as the free spectral range) is given by

$$\Delta\nu_{fsr} = \frac{c}{l_{op,rt}} = \frac{1}{t_{rt}}. \qquad (8.33)$$

For most solid-state lasers, this is much less than the gain bandwidth. For example, the free spectral range of a 10-cm-long empty cavity is 1.5 GHz; the gain bandwidth of commonly used solid-state laser media is greater than 100 GHz. As a result, lasers tend to oscillate at several frequencies simultaneously.

Although the above statement is true, the reasons are more subtle than they may initially seem. In the early days of lasers, it was believed that lasers with homogeneously broadened gain spectra should operate in a single longitudinal mode. The reasoning behind this can be understood from the rate equations. If the optical intensity within the laser cavity is assumed to be uniform, the steady-state solution to the photon rate equation fixes the inversion density at its threshold value. The first cavity mode to lase (the one with the highest net gain) clamps the inversion density and no other mode can reach threshold. The flaw in this reasoning lies in the assumption of uniform optical intensity. Lasers with both homogeneously and inhomogeneously broadened gain media tend to oscillate in several longitudinal modes as a result of spatial and spectral hole burning.

Spatial Hole Burning and Energy Diffusion

Spatial hole burning (Tang et al., 1963) and energy diffusion are responsible for shaping the population inversion in standing-wave laser cavities. The spatial distribution of the population inversion determines the gain of the different cavity modes and, therefore, which mode or modes will oscillate.

Spatial Hole Burning. In standing-wave laser cavities, the coherent superposition of the optical fields traveling in opposite directions within the cavity results in a sinusoidal intensity distribution. At the maxima of the intensity distribution, there is strong gain saturation, and the population inversion is depleted. However, at nulls in the optical field, the oscillating mode is unable to deplete the inversion. As a result, the inversion density is no longer uniform, but has "holes" at the positions corresponding to the peaks in the optical intensity. This phenomenon is known as spatial hole burning. The gain at the nulls in the optical field will continue to increase as the gain medium is pumped harder. Because other cavity modes have a different spatial profile than the first mode and can use the population inversion at these positions, this can lead to multimode operation. These ideas are illustrated in Fig. 8.9.

If we make the assumption that the round-trip loss of the cavity is relatively low, so that all of the maxima are of nearly equal intensity, the intensity of mode m within the gain medium can be written as

$$I_m(z) = 4I_m \sin^2(k_m z + \phi_m), \qquad (8.34)$$

where I_m is the circulating intensity of mode m, k_m and ϕ_m are determined by the oscillating

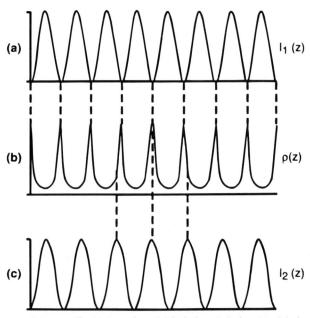

FIGURE 8.9 Illustration of spatial hole burning showing (a) the intensity profile of the first longitudinal mode to lase $I_1(z)$, (b) the population inversion in the presence of the first oscillating mode $\rho(z)$, and (c) the intensity profile of the second longitudinal mode to lase $I_2(z)$.

Miniature Solid-State Lasers

frequency and the geometry of the cavity, and z is the dimension along the cavity axis. For a cavity with a homogeneous four-level gain medium located between $z = 0$ and $z = l_g$, and assuming that energy diffusion occurs slowly compared with the stimulated relaxation of the inversion density (the effects of energy diffusion will be discussed shortly), we can write the cw rate equations for each of the oscillating modes as

$$\sigma_m \int_0^{l_g} \rho(z) I_m(z) \, dz = \gamma_{rt,m} I_m, \tag{8.35}$$

and the rate equation for the upper-manifold population density $\rho(z)$ as

$$\dot{\rho}_p(z) = \sum_m \frac{\sigma_m \rho(z) I_m(z)}{h\nu_m} + \frac{\rho(z)}{\tau}, \tag{8.36}$$

where σ_m is the value of the effective emission cross-section at the frequency ν_m corresponding to mode m, $\gamma_{rt,m}$ is the round-trip loss coefficient for mode m, and bleaching of the pump transition is neglected under the assumption that the pump rate $\dot{\rho}_p(z)$ is independent of the inversion density. (For a spatially uniform optical intensity and inversion density, Eqs. 8.35 and 8.36 reduce to Eqs. 8.12 and 8.13.)

For single-mode operation (oscillating in mode $m = 1$), we can solve Eqs. 8.35 and 8.36 for the intensity I_1 (Casperson, 1980):

$$\frac{4I_1}{I_{sat,1}} = \frac{2\langle\rho_0\rangle}{\rho_{th}} - \left(\frac{2\langle\rho_0\rangle}{\rho_{th}} + \frac{1}{4}\right)^{1/2} - \frac{1}{2}, \tag{8.37}$$

where

$$\langle\rho_0\rangle \equiv \frac{1}{l_g} \int_0^{l_g} \rho_0(z) \, dz \tag{8.38}$$

is the average unsaturated inversion density within the gain medium (which is proportional to the absorbed pump power), $\rho_{th} = \gamma_{rt,1}/2\sigma_1 l_g$ is the threshold inversion density, $I_{sat,1} = h\nu_1/\sigma_1\tau$, and it has been assumed that the active portion of the gain medium is much longer than the wavelength of oscillation. Because the output intensity of the laser is $T_o I_1$, where T_o is the transmission of the output coupler, Eq. 8.37 shows that, in contrast to the results predicted in the discussion of the Rate-Equation Model of Lasers in Section 8.2, for a single-mode standing-wave laser, the output power is not a linear function of the pump power. The neglect of spatial hole burning (the use of Eqs. 8.12 and 8.13) leads to an overestimate of the laser intensity; for a single-frequency laser, the laser intensity is overestimated by a factor of 1.5 near threshold.

The condition for single-mode operation reduces to

$$\int_0^{l_g} \rho(z) \left[\frac{I_1(z)}{I_1} - \beta(1, 2) \frac{I_2(z)}{I_2}\right] dz < 0, \tag{8.39}$$

where the discrimination factor $\beta(1, 2)$ is given by

$$\beta(1, 2) \equiv \frac{\sigma_1 \gamma_{rt,2}}{\sigma_2 \gamma_{rt,1}}, \tag{8.40}$$

and the subscript 2 denotes the second mode to oscillate. For a Lorentzian gain profile, if we assume that one cavity mode falls exactly at line center and that the cavity losses are the same for all potential oscillating modes,

$$\beta(1, 2) = 1 + \left[\frac{2\Delta\lambda(1, 2)}{\Delta\lambda_g}\right]^2, \quad (8.41)$$

where $\Delta\lambda_g$ is the full width at half-maximum of the gain peak, $\Delta\lambda(1, 2) = \Delta m \lambda_0^2/ct_{rt}$ is the difference in wavelength between the first mode to oscillate and a potential second mode, Δm is the difference in the longitudinal mode numbers of the first and second modes (the difference in the number of intensity peaks along the cavity axis), and λ_0 is the wavelength of the gain peak. (The Lorentzian-lineshape assumption must be used with care in solid-state lasers. Even if a given transition has a Lorentzian spectrum, the second mode to oscillate may correspond to a different transition.)

The solution to Eq. 8.39 is (Zayhowski, 1990a)

$$\langle\rho_0\rangle < \zeta_{SH}(1, 2)\rho_{th}, \quad (8.42)$$

where $\zeta_{SH}(1, 2)$ is the ratio of the maximum single-mode inversion density to the threshold inversion density and is a function of the cavity parameters, the gain cross-sections at the frequencies of modes 1 and 2, and the phase relationship between modes 1 and 2:

$$\zeta_{SH}(1, 2) \equiv \left(\frac{\beta(1, 2) - 1}{1 - \langle\psi(1, 2)\rangle} + 1\right)\left(\frac{2[\beta(1, 2) - 1]}{1 - \langle\psi(1, 2)\rangle} + 1\right). \quad (8.43)$$

Here, the mode correlation factor $\langle\psi(1, 2)\rangle$ is the weighted spatial average of the cosine of the phase difference between the standing-wave intensity patterns generated by modes 1 and 2:

$$\langle\psi(1, 2)\rangle \equiv \frac{1}{l_g\langle\rho_0\rangle} \int_0^{l_g} \rho_0(z) \cos[2(k_1 - k_2)z + 2(\phi_1 - \phi_2)] \, dz. \quad (8.44)$$

The correlation factor $\langle\psi(1, 2)\rangle$ can be solved in closed form for many common laser configurations, including lasers with uniformly pumped gain media and longitudinally pumped lasers. For a longitudinally pumped gain medium positioned adjacent to the end mirror of the cavity through which it is pumped,

$$\langle\psi(1, 2)\rangle = \frac{1 - \exp(-\alpha_p l_g)\cos[2(k_1 - k_2)l_g] - [2(k_1 - k_2)/\alpha_p]\sin[2(k_1 - k_2)l_g]}{[1 - \exp(-\alpha_p l_g)][1 + 4(k_1 - k_2)^2/\alpha_p^2]}, \quad (8.45)$$

where α_p is the absorption coefficient of the gain medium at the pump wavelength. When the active region fills the entire cavity (as is common in microchip lasers),

$$\langle\psi(1, 2)\rangle = \frac{1}{1 + [2\pi\Delta m/(\alpha_p l_g)]^2}. \quad (8.46)$$

For a uniformly pumped gain medium, Eqs. 8.45 and 8.46 reduce to

Miniature Solid-State Lasers

$$\langle\psi(1, 2)\rangle = \frac{\sin(2\pi\Delta m F_g)}{2\pi\Delta m F_g}, \quad (8.47)$$

where F_g is the fraction of the cavity's optical length occupied by the gain medium, and

$$\langle\psi(1, 2)\rangle = 0, \quad (8.48)$$

respectively.

$\zeta_{SH}(1, 2)$ is plotted as a function of $\beta(1, 2)$, for $\langle\psi(1, 2)\rangle = \{1, 0, -1\}$ in Fig. 8.10. $\langle\psi(1, 2)\rangle = 1$ corresponds to perfect correlation between modes 1 and 2. In this case, the first mode to lase efficiently depletes the gain for the second mode, and the second mode will never reach threshold. Two orthogonally polarized modes of a two-mirror cavity with the same wave vector k are one example of perfect correlation. $\langle\psi(1, 2)\rangle = -1$ corresponds to the case of anticorrelated modes, where the maxima of one mode line up with the minima (nulls) of the other. This is the most favorable situation for the onset of a second lasing mode and occurs when a thin gain medium is located in the center of a standing-wave cavity.

Energy Diffusion. In the presence of spatial hole burning, energy diffusion moves some of the inverted population away from the nulls in the optical field toward the maxima, where it can be effectively depleted. As a result, energy diffusion allows single-mode operation of lasers at power levels well above the multimode threshold determined by Eq. 8.42. In the presence of energy diffusion, Eq. 8.36 takes the form

$$\dot{\rho}_p(z) = \sum_m \frac{\sigma_m \rho(z) I_m(z)}{h\nu_m} + \frac{\rho(z)}{\tau} - D_e \frac{d^2\rho(z)}{dz^2}, \quad (8.49)$$

where D_e is the energy diffusion constant. From this equation, a conservative estimate for the

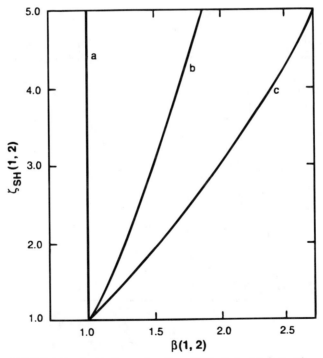

FIGURE 8.10 $\zeta_{SH}(1, 2)$ as a function of $\beta(1, 2)$ for (a) $\langle\psi(1, 2)\rangle = 1$, (b) $\langle\psi(1, 2)\rangle = 0$, and (c) $\langle\psi(1, 2)\rangle = -1$.

onset of lasing in a second mode can be obtained. In the presence of energy diffusion, single-mode operation is maintained as long as (Zayhowski, 1990c)

$$\langle \rho_0 \rangle < \zeta_D(1, 2)\rho_{th}, \tag{8.50}$$

where

$$\zeta_D(1, 2) \equiv (4k_1^2 D_e \tau - 1)\left(\frac{\beta(1, 2) - 1}{1 - \langle \psi(1, 2) \rangle}\right) + 4k_1^2 D_e \tau \left(\frac{[\beta(1, 2) - 1]}{1 - \langle \psi(1, 2) \rangle}\right)^2. \tag{8.51}$$

Expressions 8.42 and 8.50 are complementary. Expression 8.42 accounts for spatial hole burning and neglects the effects of energy diffusion. Expression 8.50 accounts for energy diffusion in the presence of strong spatial hole burning. Whichever predicts a larger single-mode inversion ratio will provide a better estimate of the maximum single-mode value of $\langle \rho_0 \rangle/\rho_{th}$. It is, therefore, reasonable to define a function $\zeta(1, 2)$ which is the larger of $\zeta_{SH}(1, 2)$ and $\zeta_D(1, 2)$. The inversion ratio $\zeta(1, 2)$ is shown as a function of $[\beta(1, 2) - 1]/[1 - \langle \psi(1, 2) \rangle]$ for several values of $4k_1^2 D_e \tau$ in Fig. 8.11.

Spectral Hole Burning and Cross-Relaxation

Homogeneous gain broadening occurs when each optically active site sees exactly the same environment and, therefore, the excited states of each site have the same energy distribution. This is often the case in crystalline solid-state gain media at room temperature. In other materials (e.g., glasses), each site sees a slightly different environment. The result is a collection of homogenous lines with slightly different center frequencies. The ensemble effect is called inhomogeneous gain broadening.

Spectral Hole Burning. In an inhomogeneous system, only those excitations with gain at the lasing frequency are able to participate efficiently in the stimulated-emission process. As a result,

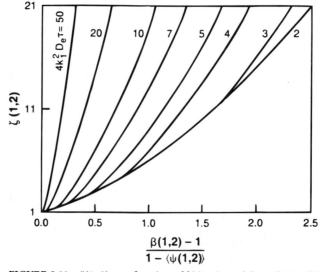

FIGURE 8.11 $\zeta(1, 2)$ as a function of $[\beta(1, 2) - 1]/[1 - \langle \psi(1, 2) \rangle]$ for several values of $4k_1^2 D_e \tau$.

only those excitations become depleted, producing a gain spectrum which has a dip at the lasing frequency. This is known as spectral hole burning and is illustrated in Fig. 8.12.

For weak hole burning, the width of the spectral hole is typically between one and two times the homogeneous linewidth, depending on the homogeneous line shape. As the intracavity optical field increases, it can draw energy from an increasingly large inversion density because many of the homogeneous lines associated with different center frequencies will have spectral tails extending to the oscillating frequency. As a result, the width of the spectral hole increases with increasing optical intensity, and the inhomogeneous gain saturates more slowly than any of the individual homogeneous lines. For an inhomogeneous gaussian ensemble of homogeneous Lorentzian lines (Siegman, 1986), the width of the spectral hole is twice the Lorentzian linewidth at low optical intensities and increases as $\sqrt{I/I_{sat}}$; the gain at the oscillating frequency saturates as $1/\sqrt{1 + I/I_{sat}}$ rather than as $1/(1 + I/I_{sat})$. As in the case of spatial hole burning, the output power of the laser is a superlinear function of pump power near threshold.

Cross-Relaxation. Excitations which do not contribute to the lasing process for the first oscillating mode may contribute to the onset of lasing for other modes. If the separation between the potential oscillating modes is greater than the width of the spectral holes, the inversion densities seen by the modes are nearly independent. The condition for single-mode operation is, therefore,

$$\langle \rho_0 \rangle < \beta(1, 2)\rho_{th}. \tag{8.52}$$

At higher inversion densities (pump powers), multimode behavior results.

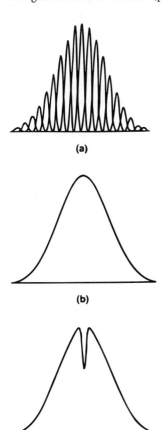

FIGURE 8.12 Illustration of spectral hole burning showing (a) several closely spaced homogeneously broadened spectra, (b) the inhomogeneously broadened spectrum resulting from the sum of several homogeneously broadened spectra, and (c) the inhomogeneously broadened spectrum with a spectral hole burned in the center due to the saturation of one of the homogeneously broadened components.

In many material systems there is some cross-relaxation (or spectral diffusion) of the individual homogenous lines in an inhomogeneous gain spectrum. This is usually a small but significant effect and is directly analogous to energy diffusion in the case of spatial hole burning. In the limit of very fast cross-relaxation, the inhomogeneous line becomes effectively homogeneous for all practical purposes. The mechanisms responsible for cross-relaxation in solids include dipole interactions between neighboring sites, radiation trapping, and phonon coupling.

Single-Frequency Lasers

There are many techniques for obtaining single-frequency operation of a laser. Several of these involve introducing a dispersive element into the cavity to provide a frequency-dependent loss, thereby increasing the modal discrimination factor $\beta(1, 2)$ enough to select an individual longitudinal cavity mode. Examples of such elements are prisms, gratings, Fabry–Perot etalons, and combinations of birefringent filters and polarizing elements. In some cases, a cavity may require multiple dispersive elements to obtain enough frequency selectivity to ensure single-frequency operation.

Alternatively, single-mode operation can be obtained in homogeneously broadened gain media by designing the laser cavity to eliminate or reduce the effects of spatial hole burning, resulting in a large inversion ratio $\zeta(1, 2)$ (see the above discussion of Spatial Hole Burning and Energy Diffusion). This approach usually requires interferometric control of the cavity length because it is desirable to have the oscillating mode located at the center of the gain peak so that the discrimination factor $\beta(1, 2)$ is as large as possible [if two modes straddle the gain peak $\beta(1, 2) \approx 0$ and $\zeta(1, 2) \approx 1$]. Nevertheless, there are several practical devices available which rely on this concept.

Spatial hole burning is almost completely eliminated in traveling-wave lasers, such as unidirectional ring lasers (Clobes and Brienza, 1972; Kane and Byer, 1985). (A small amount of hole burning is still present if the gain medium is located near a mirror, where fields propagating in different directions overlap.) As a result, there is a uniform intracavity optical intensity and the first mode to oscillate clamps the gain so that no other mode can reach threshold. A variation of the same idea, the "twisted-mode" laser, involves placing a quarter-wave plate on either side of the gain medium in a two-mirror cavity (Evtuhov and Siegman, 1965; Wallmeroth, 1990). In such a laser, the optical fields traveling in opposite directions in the gain medium are orthogonally polarized and do not interact coherently. The optical intensity within the gain medium is uniform, and there is no spatial hole burning.

Spatial hole burning is also reduced in standing-wave cavities with a large output coupling. In this case, the standing-wave field of the cavity is modified by the asymmetry between oppositely traveling waves. The wave traveling toward the output coupler is more intense than the wave traveling away from the output coupler. The more the ratio of the intensities of these two traveling waves deviates from unity, the smaller the contrast between the maximum and minimum axial intensity and the smaller the effect of spatial hole burning. There is, therefore, more uniform gain saturation.

Standing-wave lasers will also operate in a single longitudinal mode at higher powers when the active region of the gain medium is strongly localized near one end mirror of the cavity than when it is located near the center. This is due to the pinning of the antinodes of all cavity modes at the mirror, resulting in a strong correlation between modes ($\langle\psi(1, 2)\rangle \approx 1$). Localization of the active region may be the result of positioning a thin gain element at one of the mirrors. Alternatively, it may result from longitudinally pumping a gain medium which strongly absorbs the pump radiation, even though the gain medium itself may be large (Kintz and Baer, 1990). Examples of appropriate gain media include Nd:YVO$_4$ and stoichiometric materials. The best results are obtained by using a short gain element with a large absorption coefficient placed adjacent to one of the cavity end mirrors.

Single-mode operation has also been obtained by reducing the length of the cavity so that the longitudinal mode spacing is comparable to, or greater than, the gain bandwidth (Zayhowski and Mooradian, 1989a). For very short cavity lengths, $\zeta(1, 2)$ is a strong function of cavity length. Nd:YAG microchip lasers have operated in a single longitudinal mode at pump powers in excess of 22 times threshold; Nd:YVO$_4$ microchip lasers have produced 240 mW of single-frequency output.

A saturable absorber in a standing-wave laser cavity can also help to discriminate against multifrequency oscillation. As the oscillating mode burns holes in the gain, it also burns holes in the loss. This approach can, however, severely reduce the efficiency of the device.

In some gain media there is a large amount of energy diffusion. As discussed above, energy diffusion mitigates the effects of spatial hole burning and can result in single-mode oscillation. Energy diffusion is expected to be most important in gain media with a high density of excited states, such as stoichiometric materials, but still plays a significant role in more dilute materials. In semiconductor materials, where the excited states (electrons and holes) themselves diffuse, energy diffusion can completely eliminate the effects of spatial hole burning.

Linewidth

One contribution to the finite spectral width of all lasers is the coupling of spontaneous emission (or noise) to the oscillating mode (Schawlow and Townes, 1958; Lax, 1966), which results in a Lorentzian power spectrum. For many ultrastable lasers, this contribution alone determines the fundamental linewidth, and it is common practice to obtain the fundamental linewidth of such a laser by fitting the tails of the measured power spectrum to a Lorentzian curve. In very small lasers, there is a second important contribution to the fundamental linewidth — the thermal fluctuations of the cavity length at a constant temperature (Jaseja et al., 1963). This contribution is expected to result in a gaussian power spectrum and, unlike the Lorentzian contribution discussed above, is independent of the output power of the device. In lasers with a very small mode volume, such as microchip lasers, the contribution due to thermal fluctuations is much larger than that due to spontaneous emission (Zayhowski, 1990b). However, because a gaussian curve decays more quickly than a Lorentzian curve, the tails of the power spectrum still correspond to the Lorentzian contribution.

The output of a laser contains both stimulated and spontaneous emission. Because the phase of the spontaneous emission is not correlated with the phase of the stimulated emission, this leads to a random variation in the net phase of the optical field. As a result, the fundamental linewidth of a laser has a full width at half-maximum of

$$\delta \nu_L = \frac{1}{2\pi} \frac{\dot{q}_{sp}}{2q}, \qquad (8.53)$$

where the 2 in the denominator of $\dot{q}_{sp}/2q$ results from averaging over all possible phases of the spontaneous emission relative to the stimulated emission and the prefactor $1/2\pi$ converts phase uncertainty to frequency width. If we approximate the optical field within the laser cavity as uniform (i.e., neglect spatial hole burning), the spontaneous-emission rate is clamped at the threshold-emission rate (see the discussion of the Rate Equation Model of Lasers in Section 8.2):

$$\dot{q}_{sp} = \frac{2 \dagger N_{u,th}}{2 N_{th}} \tau_c, \qquad (8.54)$$

where $N_{u,th}$ is the population of the upper laser level at threshold. The resulting Lorentzian linewidth is given by

$$\delta \nu_L = \frac{2\dagger N_{u,th}}{2N_{th}} \frac{h\nu_0}{4\pi P_o} \frac{\gamma_{rt}\gamma_o}{t_{rt}^2}, \qquad (8.55)$$

where h is Planck's constant, ν_0 is the center frequency of the laser, P_o is the output power, $\gamma_{rt} = \gamma_{rt,p} + \gamma_o$ is the round-trip loss coefficient, $\gamma_{rt,p}$ is the parasitic round-trip cavity loss coefficient (not including transmission through the output coupler), $\gamma_o = -\ln(R_o)$ is the output coupling loss coefficient, and R_o is the reflectivity of the output coupler.

The spectral-broadening effects of thermal fluctuations in cavity length are easily calculated using the principle of equipartition of energy. This principle, derived from classical mechanics, states that whenever the energy of a system may be written as a sum of independent terms, each of which is quadratic in the variable representing the associated degree of freedom, then, in equilibrium at absolute temperature T, each of the terms (that is, each degree of freedom) contributes $k_B T/2$, where k_B is Boltzmann's constant, to the energy of the system. In a monolithic standing-wave cavity this leads to the expression (Zayhowski, 1990b)

$$C_{11} \left\langle \left(\frac{\delta l}{l}\right)^2 \right\rangle V_g = k_B T, \qquad (8.56)$$

where C_{11} is the longitudinal elastic constant of the gain medium, δl is the change in the cavity length l, and the brackets indicate averaging over time. This results in a gaussian contribution to the fundamental linewidth, with a full width at half-maximum given by

$$\delta \nu_G = \nu_0 \left[\frac{8 k_B T \ln(2)}{C_{11} V_g}\right]^{1/2}. \qquad (8.57)$$

For microchip lasers, the value of $\delta \nu_G$ varies with pump power and pump-beam diameter, but is typically between 5 and 7 kHz, compared to a Lorentzian contribution to the linewidth, $\delta \nu_L$, of only a few hertz (Zayhowski, 1990b).

Other factors which contribute to the linewidth of a laser include fluctuations of the pump power, optical feedback into the laser cavity, mechanical vibrations, and temperature variations. These contributions are less fundamental, however, and can be controlled. In addition, they tend to occur on a longer time scale than spontaneous emission and thermal fluctuations. Although they are often not important factors in attempts to measure the fundamental linewidth of a laser, they are important contributions to the frequency fluctuations of lasers on time scales of practical interest.

All fluctuations in the frequency of a laser, including those resulting from spontaneous emission and thermal fluctuations, can be actively compensated for and a laser can be locked to an external reference with extreme precision (Hough et al., 1984; Man and Brillet, 1984; and Day et al., 1990). Note that reducing the frequency fluctuations below the fundamental limits may entail an increase in the intensity noise (Ohtsu and Kotajima, 1985).

Frequency Tuning

Mode Selection

Frequency tuning of a laser can occur in one of two ways. If the longitudinal mode spacing of the laser cavity is much less than the gain bandwidth, the cavity is capable of supporting several modes, each at a different frequency. A single frequency can be selected through the insertion of one or more frequency-selective dispersive elements (see the above discussion of Single-Frequency Lasers). In all of the examples listed there, a small repositioning of the frequency-selective element results in the selection of a new longitudinal mode (and, hence, a new operating frequency). The frequency-selective element is used to select one of the several cavity modes, and discrete tuning is obtained.

Rapid tuning can be obtained through the use of electrooptic components. However, if the frequency selectivity of the tuning element is weak, cavity dynamics rather than the response time of the electrooptic element determine how quickly the laser will switch frequencies. This is often the case in single-frequency unidirectional ring lasers. In such lasers, the additional loss introduced at the original oscillating frequency to switch modes may be small, and the original mode will continue to oscillate, depleting the population inversion. Therefore, the net gain at the newly selected frequency is initially small, and radiation in the new mode builds up slowly. Eventually the old mode decays, releasing the gain to the new mode, which, then, rapidly reaches its cw intensity. Under these conditions, the switching time is approximately given by the expression (Schulz, 1990)

$$t_{sw} = \frac{\tau_c}{\delta\gamma_{rt}} = \frac{t_{rt}}{\gamma_{rt}\delta\gamma_{rt}}, \tag{8.58}$$

where $\delta\gamma_{rt}$ is the change in the round-trip cavity loss coefficient induced by the frequency-selective element for the original mode. In some cases it requires less time to turn a laser off and restart it in the new mode than to switch modes while it is oscillating.

Mode Shifting

The other way a laser can be tuned is to change the frequency of a given cavity mode by changing the optical length of the cavity:

$$\frac{\delta\nu_o}{\nu_o} = -\frac{\delta(l_{op,rt})}{l_{op,rt}} = -\frac{\delta(t_{rt})}{t_{rt}}. \tag{8.59}$$

Because the cavity's optical length can be changed continuously, this leads to continuous tuning. This type of tuning is limited by the free spectral range of the cavity (except under transient conditions). Once the cavity modes are shifted by a full free spectral range, an adjacent cavity mode is positioned at the frequency where the initial mode started. For the same reasons that the initial mode was originally favored, the adjacent mode is now favored, and the laser will have a tendency to mode hop (the longitudinal mode number will change by 1), returning to the original frequency.

The optical length of the elements in a laser cavity can be changed by a variety of techniques including thermal tuning, stress tuning, and electrooptic tuning. Pump-power modulation represents a special case of thermal tuning. Each of these techniques allows continuous frequency modulation (FM) of a single longitudinal cavity mode and is discussed below. To calculate the total change in the optical length of the laser cavity, one simply sums the changes in the optical lengths of each of the intracavity elements, keeping in mind that the total change in the physical length of the cavity may be constrained by the mechanical support structure of the device. For example, the expansion of the gain medium may be partially offset by the shortening of an adjacent air gap.

Thermal Tuning. Changing the temperature of an element in a laser cavity is often the simplest way to tune the device. The response time of the cavity, however, is limited by the thermal diffusion time, which is usually relatively long (typically several milliseconds or more). The temperature-induced change in the optical length of an unconstrained piece of material is given by

$$\frac{\delta(nl)}{nl} = \left(\alpha_e + \frac{1}{n}\frac{dn}{dT}\right)\delta T, \tag{8.60}$$

where α_e is the thermal expansion coefficient, n is the refractive index, l is the length of the material, and δT is the change in temperature.

Stress Tuning. The cavity modes of a resonator will also tune as elements within the cavity are squeezed. For squeezing transverse to the optic axis, the main effect is usually an elongation of the material along the optical axis of the resonator (Owyoung and Esherick, 1987). Superimposed on this effect is the stress-optic effect. In crystals with cubic symmetry, the stress-optic effect can split the frequency degeneracy of orthogonally polarized optical modes because the stress-optic coefficients are different for light polarized parallel to and perpendicular to the applied stress. For squeezing along the optic axis of the cavity, there is a compression of the element and the frequency degeneracy of orthogonally polarized modes remains unchanged. By using a piezoelectric transducer to squeeze a monolithic microchip laser, tuning has been obtained at modulation frequencies up to 20 MHz, although linear response was obtained only at modulation frequencies up to 80 kHz (Zayhowski and Mooradian, 1989b). The linear tuning response in that experiment was 300 kHz/V. Much larger tuning responses can be obtained by piezoelectrically driving a separate cavity mirror.

Electrooptic Tuning. For applications including frequency-modulated optical communications and chirped coherent laser radar, extremely high rates of tuning are required. These rates can only be achieved electrooptically.

The frequency response of a laser whose optical length is varied is well understood. When a linear voltage ramp is applied, the frequency of the laser undergoes a series of steps whose spacing in time is the cavity round-trip time t_{rt} (Genack and Brewer, 1978). When the rise time T_r of the voltage is long compared to the cavity round-trip time and, if we assume that the change in the optical length of the cavity tracks the applied voltage, the frequency has an approximately linear chirp with a fractional deviation from linearity of $t_{rt}/2T_r$. (This deviation from linearity can be corrected by using a voltage waveform that is quadratic during the first cavity round-trip time and linear thereafter.) Also, because two steps are required to define a modulation frequency (one step up and one step down), the maximum response frequency of the cavity is $1/2t_{rt}$.

The sensitivity of electrooptic voltage-to-frequency conversion increases linearly with the percentage of the cavity length occupied by the electrooptic element. If we ignore piezoelectric and electrostrictive effects,

$$\frac{\delta(\nu_o)}{\nu_o} = -F_{eo}n^2 r_{eo}E, \tag{8.61}$$

where F_{eo} is the fraction of the cavity's round-trip optical length occupied by the electrooptic material, r_{eo} is the appropriate electrooptic coefficient, and E is the magnitude of the applied electric field. For high-sensitivity tuning, it is desirable to fill the cavity with as large a fraction of electrooptic material as possible. However, it is often still important to keep the total cavity length as small as possible, for the reasons discussed above. In addition, as the length of the electrooptic crystal increases, the capacitance between the electrodes on the crystal rises (for fields applied transverse to the crystal length), resulting in higher energy requirements and slower electrical response.

Most electrooptic crystals are piezoelectrically active. As a result, when a voltage is applied to the crystal, both stress (and the concomitant strain) and refractive index are modulated. Although the effects of stress may normally be small compared to the index modulation, a free-standing crystal can act as a high-Q acoustic cavity. At the resonant frequencies, the piezoelectric effect can cause a greatly enhanced FM response of the laser. Depending on the dimensions of the crystal, these resonances can fall between a few kilohertz and several megahertz. To eliminate these resonances, the electrooptic material can be bonded to materials with a similar acoustic impedance to transmit the electrically excited acoustic waves out of the crystal (Schulz and Henion, 1991a).

For the reasons discussed above, the high-speed tuning of a laser is most linear when the shortest possible laser cavity is used. Short cavity length offers an advantage in tuning range as well, because the amount of frequency tuning that can be obtained without mode hopping is limited by the free spectral range of the laser (Eq. 8.33). As a result, the short length of composite-cavity, electro-optically tuned microchip lasers makes them particularly attractive as solid-state lasers capable of high sensitivity, a large continuous tuning range, and a high tuning rate (Zayhowski et al., 1993). Such devices have been continuously tuned over a 30-GHz range with a tuning sensitivity of ~14 MHz/V. The tuning response was relatively flat from dc to 1.3 GHz. Even higher rates of modulation should be possible.

Pump-Power Modulation. Changes in pump power typically induce frequency changes in the output of solid-state lasers. As the pump power increases, more thermal energy is deposited in the gain medium, raising its temperature and changing both the refractive index and length. Because frequency tuning via pump-power modulation relies on thermal effects, it is often thought to be too slow for many applications. In addition, modulating the pump power has the undesirable effect of changing the amplitude of the laser output. However, for very small lasers, significant frequency modulation, at relatively high modulation rates, can be obtained with little associated amplitude modulation. This technique has been used to phase-lock two lasers (Keszenheimer et al., 1992). When it can be used, pump-power modulation has advantages over other frequency modulation techniques because it requires very little power, no high-voltage electronics, no special mechanical fixturing, and no additional intracavity elements. Another important reason to understand the frequency response of a laser to changes in pump power is that pump-power fluctuations are often responsible for a large portion of the frequency fluctuations of the laser output.

The following results are obtained for a small, longitudinally pumped, monolithic solid-state laser. Both the pump beam and oscillating mode are assumed to be gaussian, with constant amplitude and diameter as they propagate along the cavity axis. Because the thermal conductivity of most solid-state gain media is much greater than that of either dielectric coatings or air, it is assumed that all heat flow is directed in the plane normal to the cavity axis. To further simplify the analysis, the transverse dimensions of the gain medium are initially modeled as infinite. A discussion of the effect of heat sinking the gain medium at a finite distance from the cavity axis is included later in this section.

The variation in the optical length of the laser cavity as a function of temperature is given by Eq. 8.28. For the case of nonuniform heating, the thermal expansion of the warmer sections of the crystal will be constrained by the cooler regions, and Eq. 8.28 will not be completely accurate (see the discussion of Fabry-Perot Cavities in Section 8.4). In addition, there will be internal stress which could modify the refractive index through the stress-optic coefficient. These effects will usually represent only small corrections to Eq. 8.28 and are neglected in the following analysis.

With the approximations stated above, the crystal temperature averaged over the volume of a laser mode with waist radius r_m in a cavity of length l is given by

$$T_{\text{ave}} = \frac{2}{\pi r_m^2 l} \int_0^l \int_0^{2\pi} \int_0^{\infty} T \exp\left(-\frac{2r^2}{r_m^2}\right) r \, dr \, d\theta \, dz. \tag{8.62}$$

The temperature distribution T within the gain medium satisfies the equation

$$\frac{dT}{dt} = \alpha_d \nabla^2 T + \frac{Q_h}{\rho_m c_{sh}},\tag{8.63}$$

where c_{sh} is the specific heat, ρ_m is the mass density, $\alpha_d = k_c/\rho_m c_{sh}$ is the thermal diffusivity, k_c is the thermal conductivity, and Q_h is the rate of heat deposition per unit volume. The spectral components of the rate of heat deposition $Q_h(\omega, r)$ are proportional to the pump power $P_p(\omega)$ at frequency ω:

$$Q_h(\omega, r)\sin(\omega t) = \frac{2\kappa P_p(\omega)}{\pi r_m^2 l} \exp\left(-\frac{2r^2}{r_p^2}\right)\sin(\omega t),\tag{8.64}$$

where κ is the heat-generating efficiency of the pump and r_p is the radius of the pump beam.

The frequency response of the laser to pump-power modulation, obtained from the above equations, is given by (Zayhowski and Keszenheimer, 1992)

$$\left|\frac{dv_o(\omega)}{dP_p(\omega)}\right|\sin[\omega t + \phi(\omega)] = \frac{v_o \kappa}{4\pi k_c l}\left(C\alpha_e + \frac{1}{n}\frac{dn}{dT}\right)$$

$$\times \left\{\left[\frac{\pi}{2} - \text{Si}\left(\frac{\omega}{\omega_0}\right)\right]\cos\left[\omega\left(t + \frac{1}{\omega_0}\right)\right] + \text{Ci}\left(\frac{\omega}{\omega_0}\right)\sin\left[\omega\left(t + \frac{1}{\omega_0}\right)\right]\right\},\tag{8.65}$$

where

$$\omega_0 = \frac{8\alpha_d}{r_m^2 + r_p^2},\tag{8.66}$$

and Si and Ci are the sine and cosine integrals. The magnitude and phase of this frequency response function are shown in Fig. 8.13. Asymptotically,

$$\lim_{\omega \gg \omega_0}\left|\frac{dv_o(\omega)}{dP_p(\omega)}\right| = \frac{v_o \kappa}{4\pi k_c l}\left(C\alpha_e + \frac{1}{n}\frac{dn}{dT}\right)\frac{\omega_0}{\omega},\tag{8.67}$$

$$\lim_{\omega \gg \omega_0} \phi(\omega) = \frac{\pi}{2},\tag{8.68}$$

$$\lim_{\omega \ll \omega_0}\left|\frac{dv_o(\omega)}{dP_p(\omega)}\right| = \frac{v_o \kappa}{4\pi k_c l}\left(C\alpha_e + \frac{1}{n}\frac{dn}{dT}\right)\left|\gamma + \ln\left(\frac{\omega}{\omega_0}\right)\right|,\tag{8.69}$$

$$\lim_{\omega \ll \omega_0} \phi(\omega) = \pi,\tag{8.70}$$

where $\gamma \approx 0.577$ is Euler's constant.

The logarithmic divergence of the response function for small ω is a result of the approximation that the gain medium has infinite transverse dimensions. In the quasi-static limit ($\omega \leq \omega_{qs}$, with ω_{qs} to be defined below), with a radially symmetric heat sink of infinite capacity at a distance r_s from the cavity axis, the change in laser frequency is expressed by (Zayhowski and Keszenheimer, 1992)

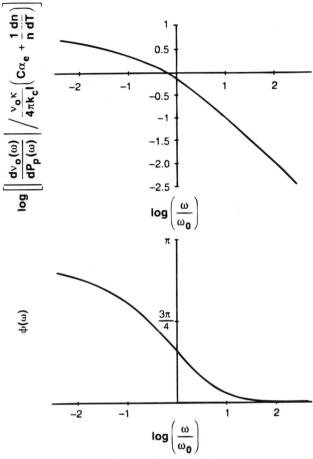

FIGURE 8.13 Magnitude and phase of the frequency response of a miniature solid-state laser to pump-power modulation.

$$\frac{dv_o(\omega_{qs})}{dP_p(\omega_{qs})} = -\frac{v_o\kappa}{4\pi k_c l}\left(C\alpha_e + \frac{1}{n}\frac{dn}{dT}\right)$$

$$\times \left\{\gamma + \ln\left(\frac{2r_s^2}{r_m^2 + r_p^2}\right) + \text{Ei}\left[-\frac{2r_s^2(r_m^2 + r_p^2)}{r_m^2 r_p^2}\right] - \text{Ei}\left(-\frac{2r_s^2}{r_m^2}\right) - \text{Ei}\left(-\frac{2r_s^2}{r_p^2}\right)\right\}, \quad (8.71)$$

where Ei is the exponential integral. This function is illustrated in Fig. 8.14. Asymptotically,

$$\lim_{r_s \gg r_m, r_p} \frac{dv_o(\omega_{qs})}{dP_p(\omega_{qs})} = -\frac{v_o\kappa}{4\pi k_c l}\left(C\alpha_e + \frac{1}{n}\frac{dn}{dT}\right)\left\{\gamma + \ln\left(\frac{2r_s^2}{r_m^2 + r_p^2}\right)\right\}. \quad (8.72)$$

In this limit and in the limit $\omega_{qs} \ll \omega_0$, Eq. 8.65 is equivalent to Eq. 8.71 if

$$\omega_{qs} = \frac{4\alpha_d}{r_s^2}\exp(-2\gamma). \quad (8.73)$$

Equation 8.73 can be interpreted as defining the quasi-static frequency limit. For modulation frequencies below ω_{qs}, the response function is given approximately by Eq. 8.71; at higher modulation frequencies Eq. 8.65 becomes more accurate.

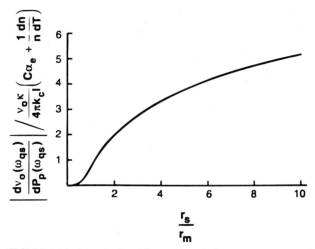

FIGURE 8.14 Magnitude of the quasi-static frequency response of a miniature solid-state laser to pump-power modulation as a function of r_s for $r_m = r_p$.

A very important aspect of the frequency response of a laser to pump-power modulation is its dependence on the dimensions of the laser. The characteristic frequencies of the system ω_0 and ω_{qs} increase as $1/(r_m^2 + r_p^2)$ and $1/r_s^2$, respectively. For the highest frequency response, all of the transverse dimensions should be kept as small as possible. The magnitude of the response function is inversely proportional to the cavity length. This also means that, for a laser with a given slope efficiency, the ratio of the frequency modulation to amplitude modulation, [FM(ω)/AM(ω)], for a given pump modulation, increases as the inverse of the cavity length. For large FM(ω)/AM(ω), the cavity length should be kept as short as possible. With their short cavity lengths (typically $l < 1$ mm) and small transverse dimensions ($r_m, r_p \approx 0.1$ mm, $r_s \approx 0.5$ mm), microchip lasers are well suited for frequency tuning via pump-power modulation.

8.6 Pulsed Operation

Quasi-CW Operation

Quasi-CW operation (also called long-pulse or normal-mode operation) refers to a pulsed laser with a pulse duration long enough for all relevant parameters within the system to approach their equilibrium values. Although the behavior of the system is cw-like at the end of the pulse, it is, in general, quite different at the beginning of the pulse.

Consider a laser with a step-function pump source. The pump may quickly create a population inversion. It will take some time, however, for an oscillating mode to build up from spontaneous emission. During this time, the inversion density may greatly exceed the cw threshold value. The large inversion density can, eventually, result in an optical intensity well in excess of the cw value. This, in turn, will drive the inversion density below threshold, substantially reducing the laser intensity. The entire process may then start again. For a single-mode laser, this often leads to regular spiking at the beginning of the pulse. The process is damped, however, and with time the intensity of the spiking decreases. Eventually, spiking gives way to damped oscillations (known as relaxation oscillations) in the optical intensity and, finally, cw-like behavior.

In high-power lasers, the initial spike can be sufficiently intense to damage the laser. One technique used to reduce the intensity of this spike and damp the relaxation oscillations is to include a nonlinear crystal within the cavity (Statz et al., 1965; Kennedy and Barry, 1974; and

Jeys, 1991). Nonlinear frequency generation in the crystal limits the initial spike by serving as an intensity-dependent loss.

Figure 8.15 shows computer solutions to the rate equations for a quasi-cw laser. In these simulations, the spiking was heavily damped and cw-like behavior was quickly obtained. In a multimode laser, the interaction between modes can lead to mode hopping, mode beating, and prolonged, highly irregular spiking.

Relaxation Oscillations

Relaxation oscillations occur whenever the population inversion of a laser is disturbed from its equilibrium value. They are a result of the coupling between the population inversion and the photon density within the laser cavity, as described above. For most solid-state lasers, the relaxation oscillations are underdamped and many oscillations occur before steady-state is reached.

The relaxation-oscillation frequency ω_r and damping time τ_0 can be derived by linearizing the rate equations (Eqs. 8.8, 8.10, and 8.11) about the steady-state operating point. If we assume $\tau_0 > 1/\omega_r$, the results for a single-mode four-level laser are

$$\tau_0 = \frac{2\tau(1 - F_{N,\text{th}})}{R_W} \tag{8.74}$$

and

$$\omega_r = \left[\frac{(R_W - 1)\gamma_{rt}}{\tau t_{rt}} - \frac{1}{\tau_0^2}\right]^{1/2}, \tag{8.75}$$

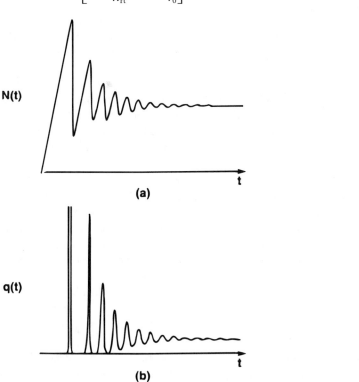

FIGURE 8.15 Computer solutions to the rate equations for a laser with a step-function pump source showing (a) the population inversion $N(t)$ and (b) the photon number $q(t)$.

where $R_W = W_p/W_{p,th}$ is the number of times the pump rate factor exceeds its threshold value and $F_{N,th} = N_{th}/N_t$ is the ratio of the effective population inversion at threshold to the total population of active sites in the active region of the gain medium. Because $F_{N,th} \ll 1$ for most four-level lasers, this term is often dropped from the numerator of Eq. 8.74. The oscillatory behavior of a quasi-three-level laser is similar to that of a four-level laser, with

$$\tau_0 = \frac{2\tau F_{N,th}(g_u f_u - F_{N,th})}{g_u f_u R_W(g_u f_l + F_{N,th}) - g_u f_l(g_u f_u - F_{N,th})} \quad (8.76)$$

and

$$\omega_r = \left[\frac{(R_W - 1)(g_u f_l + F_{N,th})\gamma_{rt}}{\tau t_{rt} F_{N,th}} - \frac{1}{\tau_0^2} \right]^{1/2}. \quad (8.77)$$

Note that, if $\tau_0 < 1/\omega_r$, the oscillations are overdamped and spiking will not occur. Although this condition is not satisfied in solid-state lasers, it is common in gas lasers because of their relatively long upper-state lifetimes.

Because oscillations in the intensity of a laser generate spectral sidebands, the frequency spectrum of a free-running laser typically shows a pronounced feature at the relaxation frequency. This resonance can be reduced through the use of electronic feedback to the pump (Kane, 1990).

Amplitude Modulation

The output power of a laser can be controlled by changing the pump power, the output coupling, or the intracavity loss. This type of amplitude modulation is usually limited to frequencies below the frequency of the relaxation oscillations. The relaxation frequency characterizes the response time of the cavity. There is resonant enhancement of the modulation response near the relaxation frequency; above the relaxation frequency, the response rolls off.

Methods used for direct amplitude modulation of a laser may have the side effect of introducing frequency modulation as well. For example, changing the pump power affects the thermal load on the gain medium and, therefore, the temperature. This, in turn, affects the refractive index, changing the optical length of the cavity and the oscillating frequency. For AM applications where frequency stability is critical, it is often better to modulate the laser power external to the cavity.

Q-Switched Operation

Q switching is a means of generating short, high-peak-power pulses with relatively low-power pump sources. In a Q-switched laser, the loss of the cavity is maintained at a high level until a large population inversion is achieved. At such time, the loss is rapidly decreased so that the inversion is well above its new threshold value, resulting in a short, high-power output pulse. Q switching relies on the fact that the lifetime of the population inversion is much longer than the output pulse width. The gain medium is therefore able to store energy, which can be quickly released in the form of an output pulse. The cavity loss is used to control the performance of the laser.

Q-Switched Rate Equations

The standard tools for analyzing the performance of a Q-switched laser are the rate equations (Wagner and Lengyel, 1963). To model the Q-switched operation of a laser, we will start with the rate equations derived in Section 8.2 and assume uniform cw pumping of the gain medium. Because the Q-switched output pulses from the laser are much shorter than both the spontaneous lifetime and the pump period (time between output pulses), spontaneous relaxation and pumping can be safely neglected during the development of the output pulse. (As shown in the discussions of Build-Up from Noise in Section 8.2 and Pulse Build-Up Time below, spontaneous emission is still important in initiating the lasing process.) This reduces the rate equations for a single-mode laser to

$$\frac{\dot{q} t_{rt}}{\gamma_{rt}} = q\left(\frac{N_{eff}}{N_{th}} - 1\right) \tag{8.78}$$

and

$$\frac{\dot{N}_{eff} t_{rt}}{\gamma_{rt}} = -\frac{2^* q N_{eff}}{N_{th}}, \tag{8.79}$$

where

$$2^* = \begin{cases} 1 & \text{for a four-level laser} \\ g_u(f_u + f_l) & \text{for a quasi-three-level laser} \\ 2 & \text{for a nondegenerate three-level laser} \end{cases}$$

The value of 2^* may depend on the length of the output pulse from the Q-swithched device; if the lower level of a four-level system has a decay time longer than the output pulse, the system will behave like a quasi-three-level system under Q-switched operation (Fan, 1988).

These rate equations will tend to overestimate the efficiency of standing-wave Q-switched lasers because they do not include the effects of spatial and spectral hole burning (see the discussions of Spatial Hole Burning and Energy Diffusion, and Spectral Hole Burning and Cross-Relaxation in Section 8.5). In addition, many Q-switched lasers operate with a large round-trip gain, so that the optical field within the cavity is not uniform. Although this mitigates the effects of spatial hole burning, it violates one of the assumptions used to reduce the rate equations to such a simple form. The net result is to overestimate the peak power and energy efficiency obtained from the device, and to distort the pulse shape (Stone, 1992). These effects are often small, however, and Eqs. 8.78 and 8.79 are extremely powerful tools.

For most of this section, we will assume that the loss of the laser cavity is rapidly switched at time $t = 0$ such that, for $t < 0$, the effective inversion N_{eff} within the cavity is below its threshold value and, for $t > 0$, it is above the new threshold inversion N_{th}. With these assumptions, we can derive expressions for the maximum peak power, maximum pulse energy, and minimum pulse width of a single Q-switched output pulse, along with the corresponding cavity output couplings. The maximum power efficiency of a repetitively Q-switched laser and the corresponding output coupling are also obtained. The instantaneous switching approximation will be examined in the discussion of Pulse Build-Up Time below.

Maximum Peak Power

To determine the maximum peak power that can be achieved from a Q-switched laser, we start by dividing Eq. 8.78 by 8.79 and integrating with respect to inversion density, to obtain

$$2^*(q - q_0) = N_0 - N_{eff} - N_{th} \ln\left(\frac{N_0}{N_{eff}}\right), \tag{8.80}$$

where q_0 and N_0 are the number of photons in the cavity and the total population inversion at the time the output pulse begins to develop. Using the fact that the inversion density is equal to its threshold value at the peak of the pulse ($\dot{q} = 0$) and neglecting the initial number of photons in the cavity, we obtain a peak output power

$$P_{po} = \frac{h\nu_o N_{th}}{2^* t_{rt}} \left[\frac{N_0}{N_{th}} - 1 - \ln\left(\frac{N_0}{N_{th}}\right) \right] \gamma_o. \qquad (8.81)$$

This is maximized when

$$\frac{\gamma_{rt} + \gamma_o}{\gamma_{rt}} \ln\left(\frac{N_0}{N_{th}}\right) = \frac{N_0}{N_{th}} - 1. \qquad (8.82)$$

P_{po} is plotted as a function of $1/\gamma_o$ in Fig. 8.16 for the case of negligible parasitic loss ($\gamma_o = \gamma_{rt}$). In this case, the maximum peak power (Zayhowski and Kelley, 1991)

$$P_{po,max} = \frac{0.102 N_0^2 h\nu_o \gamma_{rt}}{2^* N_{th} t_{rt}}, \qquad (8.83)$$

and is obtained for

$$\gamma_o = 0.28 g_{rt}, \qquad (8.84)$$

where $g_{rt} = 2\ddagger N_0 \sigma_g l_g / V_g$ is the round-trip gain coefficient at the time the pulse begins to develop. The presence of parasitic loss increases the optimal value of γ_o.

Maximum Pulse Energy

The quantum extraction efficiency η_e of a pulse is determined by the change in the population inversion density during the pulse:

$$\eta_e = \frac{N_0 - N_f}{N_0}, \qquad (8.85)$$

where N_f is the population inversion well after the peak of the output pulse. By combining Eqs. 8.85 and 8.80, we obtain the implicit relationship

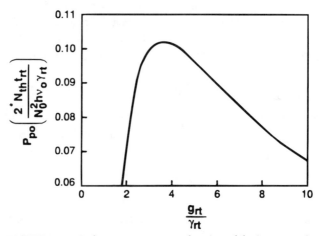

FIGURE 8.16 Peak output power as a function of the inverse cavity output coupling for a cavity with negligible parasitic loss.

$$\frac{N_0}{N_{th}} = -\ln\left(\frac{1-\eta_e}{\eta_e}\right). \tag{8.86}$$

The energy in the output pulse E_o is proportional to the extraction efficiency and is given by

$$E_o = \frac{\eta_e N_0 h\nu_o \gamma_o}{2^*\gamma_{rt}}. \tag{8.87}$$

The maximum value of E_o (Degnan, 1989)

$$E_{o,max} = \frac{N_0 h\nu_o \gamma_{rt,p}}{2^* g_{rt}}\left[\frac{g_{rt}}{\gamma_{rt,p}} - 1 - \ln\left(\frac{g_{rt}}{\gamma_{rt,p}}\right)\right], \tag{8.88}$$

and is obtained for

$$\gamma_o = -\gamma_{rt,p}\left[\frac{(g_{rt}/\gamma_{rt,p}) - 1 - \ln(g_{rt}/\gamma_{rt,p})}{\ln(g_{rt}/\gamma_{rt,p})}\right]. \tag{8.89}$$

Minimum Pulse Width

The width of a pulse t_w (full width at half-maximum) can be determined from the energy in the pulse E_o, the peak power of the pulse P_{po}, and the shape of the pulse:

$$t_w = \frac{S_p E_o}{P_{po}} = \frac{S_p \eta_e N_0 t_{rt}}{N_{th}[N_0/N_{th} - 1 - \ln(N_0/N_{th})]\gamma_{rt}}, \tag{8.90}$$

where S_p is a number that characterizes the pulse shape. Single-mode Q-switched pulses have a pulse shape factor (S_p) of ~0.86 (Zayhowski and Kelley, 1991), which is midway between the pulse shape factor for sech- and sech²-shaped pulses ($S_p = 0.84$ and 0.88, respectively). The rising edge of the pulse is exponential with a rise time determined by the gain of the cavity; the exponential decay of the trailing edge is determined by the cavity lifetime.

The exact solution to the rate equations for the pulse width is shown as a function of the inverse cavity loss $1/\gamma_{rt}$ in Fig. 8.17. The minimum pulse width (Zayhowski and Kelley, 1991)

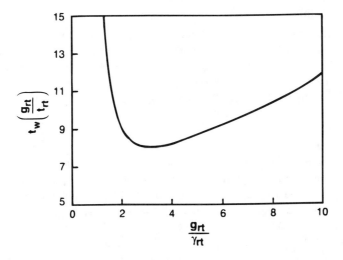

FIGURE 8.17 Pulse width as a function of the inverse cavity loss.

$$t_{w,\min} = \frac{8.1 t_{rt}}{g_{rt}}, \qquad (8.91)$$

is obtained for

$$\gamma_o = 0.32 g_{rt} - \gamma_{rt,p}. \qquad (8.92)$$

This result is independent of whether the laser is a three-level or four-level system.

Maximum Power Efficiency of a Repetitively Q-Switched Laser

For a repetitive train of identical pulses, the inversion density not used by one pulse may contribute to the next pulse, improving the net efficiency of the system. With the assumption of continuous pumping, the extraction efficiency of a repetitive system becomes

$$\eta_{e,\text{rep}} = \frac{\eta_e [1 - \exp(-\tau_p/\tau)]\tau}{[1 - (1 - \eta_e)\exp(-\tau_p/\tau)]\tau_p}, \qquad (8.93)$$

where τ_p is the time between Q-switched pulses. The total efficiency of the laser is

$$\eta = \eta_p \eta_a \eta_i \eta_{e,\text{rep}} \eta_o, \qquad (8.94)$$

where $\eta_o = \gamma_o/\gamma_{rt}$, and is maximized when (Zayhowski and Kelley, 1991)

$$\gamma_o \left[\frac{1 - \exp(-\tau_p/\tau)}{1 - (1 - \eta_e)\exp(-\tau_p/\tau)} \right] = -\gamma_{rt,p} \left[\frac{\eta_e}{(1 - \eta_e)\ln(1 - \eta_e)} + 1 \right]. \qquad (8.95)$$

In the absence of parasitic loss, the maximum energy efficiency is obtained by making the output coupling as small as possible to maximize the single-pulse extraction efficiency and minimize the effects of spontaneous emission. On the other hand, if the effect of spontaneous emission is unimportant (e.g., when the interpulse spacing is small compared to the spontaneous lifetime), the output coupling should be made as large as possible to minimize the parasitic loss.

Pulse Build-Up Time

All of the above results for the Q-switched operation of lasers were obtained on the assumption that the Q of the laser cavity is changed instantaneously. In practice, changing the Q often requires the switching of high voltages or mechanical motion of components, and is difficult to do rapidly. As a result, the instantaneous Q-switching approximation may not be valid. Fortunately, the magnitude and shape of the Q-switched output pulse are determined by the cavity Q at the time of the output pulse (within t_w of the peak) and are insensitive to the details of the way in which the Q is switched. The results obtained above are, therefore, valid as long as the cavity Q does not change significantly during the output pulse. Because the pulse build-up time is usually much greater than the pulse width, this is often the case. To illustrate this, it is helpful to calculate the pulse build-up times for the cases of instantaneous Q switching and gradual Q switching.

Because the optical intensity is small during most of the pulse build-up time, the inversion density can be approximated as constant during the pulse development. This allows us to solve Eq. 8.78 analytically to obtain an approximate expression for the pulse build-up time. For a laser cavity that is sufficiently lossy between output pulses, the pulse must build up from spontaneous emission. If the cavity Q is changed instantaneously, the pulse build-up time for a minimum-width pulse is given by (Zayhowski and Kelley, 1991)

$$t_b \approx \frac{\ln(0.63N_0)t_{rt}}{2.1\gamma_{rt}}, \tag{8.96}$$

which is typically an order of magnitude greater than the pulse width. If the value of the cavity loss changes linearly with time, lowering the threshold inversion N_{th} from an initial value of slightly greater than N_{eff} to a value of $0.32N_{eff}$ just before the peak of the output pulse, the pulse build-up time can be increased by a factor of ~ 3.3. The switching time is further increased if the initial value of N_{th} is much greater than N_{eff} or if N_{th} never reaches $0.32N_{eff}$. Also, the change in cavity Q can continue beyond the time when the output pulse is generated, as long as it is reversed early enough to prevent the generation of a second pulse.

Active Q Switching

Intracavity Q Switches. Conventional Q switches are located within the laser cavity and control the amount of cavity loss. The first devices used to Q switch a laser were mechanical switches, such as rotating mirrors. Their relatively slow speed has led to their replacement by acousto-optic and electrooptic Q switches, except in cases where optical damage limits the use of the alternative technologies.

There are two contributions to the switching time of an acousto-optic Q switch. First, the transducer is usually part of a resonator with a finite response time. The second component is the time required for the acoustic excitation to propagate across the diameter of the optical beam. Recently, acousto-optic TeO_2 Q switches have been reported that operate at frequencies of 108 MHz, generating switching speeds of 18 ns, limited by the transit time of the acoustic signal across the oscillating mode (~ 100-μm diameter). Using these switches, Q-switched pulses as short as 600 ps have been generated (Plaessmann et al., 1994).

Electrooptic Q switches are capable of faster switching than acousto-optic devices because the speed of light is faster than the speed of sound. However, they usually require high-voltage electronics. When used in a coupled-cavity configuration, electooptic Q switches have produced the shortest Q-switched pulses yet obtained from an actively Q-switched solid-state laser (Zayhowski and Dill, 1992). (Shorter Q-switched pulses have been obtained from passively Q-switched devices, as discussed below.)

Coupled-Cavity Q Switches. Coupled-cavity electrooptically Q-switched microchip lasers (illustrated in Fig. 8.18) have, to date, demonstrated the shortest Q-switched pulses obtained from an actively Q-switched solid-state laser. From a Nd:YAG device, pulse widths of 270 ps have been

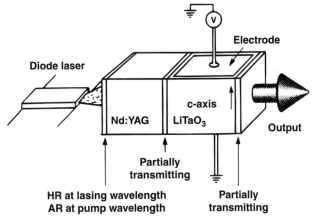

FIGURE 8.18 Illustration of one embodiment of an coupled-cavity electrooptically Q-switched microchip laser (HR, highly reflecting; AR, antireflecting).

obtained, and repetition rates of 500 kHz were demonstrated (Zayhowski and Dill, 1992). From a Nd:YVO$_4$ laser, pulse widths of 115 ps and repetition rates of 2.25 MHz were obtained (Zayhowski and Dill, 1995). Coupled-cavity electrooptically Q-switched microchip lasers have also produced the highest peak power (>90 kW) of any laser pumped by a single (1-W) laser diode (Zayhowski and Dill, 1995).

The principle behind the operation of the coupled-cavity Q-switched laser is that the etalon formed by the electrooptic element serves as a variable-reflectivity output coupler for the gain cavity (defined by the two mirrors adjacent to the gain medium). The potential lasing modes of the device are determined primarily by the gain cavity. In the low-Q state, the variable etalon must have a high transmission for all potential lasing modes so that none can reach threshold. In the high-Q state, the reflectivity of the etalon is high for the desired mode, and a Q-switched output pulse develops. To ensure that all potential modes of the gain cavity can be simultaneously suppressed, the optical length of the variable etalon must be nearly an integral multiple of the optical length of the gain cavity. The higher the Q of the etalon, the tighter the tolerance on length. For an isotropic gain medium, such as Nd:YAG, the length tolerance imposes a restriction on the birefringence of the electrooptic material, because oscillation must be suppressed for modes of both polarizations.

Before Eq. 8.92 can be used with the coupled-cavity microchip laser, we must have an expression for the effective output coupling. Similarly, before Eq. 8.91 can be used to determine the minimum pulse width of a coupled-cavity microchip laser, we must have an expression for the effective round-trip time of the laser cavity.

The transmission of the electrooptic etalon is dynamically dependent on the rate of formation and decay of the pulse. For example, if the gain cavity and the etalon have the same optical length, the light within the gain cavity is amplified by the round-trip gain of the cavity before it is recombined with light that is reflected from the far mirror of the etalon. As a result, the effective transmission of a lossless etalon, as seen at the interface between the gain medium and the etalon during the early formation of the pulse, is given by

$$T_{et,t=0} = \frac{[1-R_1][1-R_2\exp(-m_l g_{rt})]}{1+R_1R_2\exp(-m_l g_{rt})-2\cos(\phi_{et})\sqrt{R_1R_2}\exp(-m_l g_{rt}/2)}, \quad (8.97)$$

where R_1 is the reflectivity of the interface between the gain medium and the electrooptic material, R_2 is the reflectivity of the output mirror on the etalon, $m_l = t_{rt,et}/t_{rt,g}$ is the ratio of the round-trip time of light in the etalon ($t_{rt,et}$) to the round-trip time of light in the gain cavity ($t_{rt,g}$), and $\phi_{et} = 2\pi(t_{rt,et} - t_{rt,g})\nu_o$. To hold off lasing, we must be able to satisfy the relationship

$$T_{et,t=0} \geq 1 - \exp(\gamma_{rt,p} - g_{rt}). \quad (8.98)$$

Near the peak of the output pulse, the gain of the laser is saturated and the effective transmission of the etalon approaches its cw value

$$T_{et,cw} = \frac{(1-R_1)(1-R_2)}{1+R_1R_2-2\cos(\phi_{et})\sqrt{R_1R_2}}. \quad (8.99)$$

Because the pulse width of the laser is determined by the transmission of the etalon near the peak of the pulse, it is this value of transmission that should be used as the output coupling Γ_o [$\gamma_o = -\ln(1-\Gamma_o)$] in Eq. 8.92.

The transit time of light in the etalon affects the behavior of the coupled-cavity Q-switched laser in a second way. After leaving the gain cavity, light must escape the etalon, which has an associated decay time. To minimize this effect and reduce the pulse width, it is desirable to keep

Miniature Solid-State Lasers

R_2 as small as possible. The minimum pulse width is therefore obtained by selecting values of R_1 and R_2 that satisfy the equations

$$\frac{[1 - R_1][1 - R_2 \exp(-m_l g_{rt})]}{1 + R_1 R_2 \exp(-m_l g_{rt}) - 2 \sin(2\pi\nu_o \delta_1)\sqrt{R_1 R_2} \exp(-m_l g_{rt}/2)} = 1 - \exp(\gamma_{rt,p} - g_{rt}) \quad (8.100)$$

and

$$\frac{(1 - R_1)(1 - R_2)}{1 + R_1 R_2 + 2 \sin(2\pi\nu_o \delta_2)\sqrt{R_1 R_2}} = 1 - \exp(\gamma_{rt,p} - 0.32 g_{rt}), \quad (8.101)$$

where $\delta_1 + \delta_2 = \delta(t_{rt,et}) \leq 1/2\nu_o$ is the total change in the round-trip time of the electrooptic etalon induced during Q switching and the amount of change associated with δ_1 and δ_2 is determined by the exact length of the etalon before Q switching.

For low-gain operation ($g_{rt} \ll 1$), the solution to Eqs. 8.100 and 8.101 yields a high value for R_1 and a low value for R_2, and the effective round-trip time of a properly optimized coupled-cavity laser is approximately the round-trip time of the gain cavity. In very high-gain systems, R_2 becomes larger than R_1, and the effective round-trip time of the coupled cavity approaches the sum of the round-trip times of the gain cavity and the etalon. A simple expression which satisfies these asymptotic constraints and has a reasonable fit to numerical simulations is

$$t_{rt} = t_{rt,g}\left[1 + \frac{m_l(\gamma_{rt,p} - 0.32 g_{rt})}{\ln(R_2)}\right]. \quad (8.102)$$

Figure 8.19 shows the resulting normalized effective round-trip time of an optimized coupled cavity laser with $\delta_1 = \delta_2 = 1/4\nu_o$ as a function of gain for $m_l = 1$ and 2, for $\gamma_{rt,p} = 0$. Over a large part of the potential operating range of the coupled-cavity Q-switched laser, the effective cavity length is the length of the gain cavity; the length of the Q switch does not contribute significantly.

Passive Q Switching

Passive Q switching can be obtained through the use of an intracavity saturable absorber. In a cavity containing a saturable absorber, the total round-trip cavity loss γ_{rt} consists of three components: an output-coupling loss γ_o, a parasitic loss $\gamma_{rt,p}$, and a saturable loss $\gamma_{rt,s}$,

$$\gamma_{rt} = \gamma_o + \gamma_{rt,p} + \gamma_{rt,s}. \quad (8.103)$$

The rate-equation analysis of Q switching due to a saturable absorber (Szabo and Stein, 1965) shows that the saturable absorber acts as an ideal instantaneous Q switch if

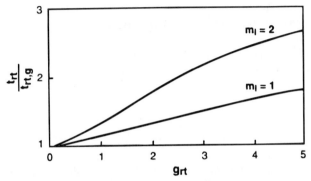

FIGURE 8.19 Effective cavity round-trip time versus gain for an optimized coupled-cavity Q-switched laser with $m_1 = 1$ and $m_1 = 2$.

$$I_g \sigma_g < 10 I_s \sigma_s, \tag{8.104}$$

where I_g is the optical intensity in the gain medium, σ_g is the stimulated-emission cross-section in the gain medium, I_s is the optical intensity in the saturable absorber, $\sigma_s = \gamma_{rt,s} A_s / 2_s^{\ddagger} N_s$ is the cross-section of the saturable absorber, A_s is the cross-sectional area of the oscillating mode in the saturable absorber, $2_s^{\ddagger}$ is the number of times the same cross-sectional area of the saturable absorber is traversed by light during one round trip within the laser cavity, and N_s is the number of intracavity saturable absorber sites. In this case, the intracavity loss is given by Eq. 8.103 until the laser reaches threshold, at which time it immediately drops to

$$\gamma_{rt} = \gamma_o + \gamma_{rt,p}. \tag{8.105}$$

Under these ideal conditions, the equations developed for active Q switching are applicable. As long as $I_g \sigma_g < 3 I_s \sigma_s$, the minimum pulse width obtainable should be within a factor of two of that predicted for ideal conditions.

For the condition described by Eq. 8.104, the saturable absorber will prevent the onset of lasing until the inversion within the cavity reaches a value of

$$N_{th} = \frac{(\gamma_o + \gamma_{rt,p} + \gamma_{rt,s}) A_g}{2 \ddagger \sigma_g}. \tag{8.106}$$

The onset of lasing, at this point, produces a high intracavity optical field which quickly saturates the saturable component of the loss, increasing the cavity Q and resulting in a Q-switched output pulse with a pulse width

$$t_w = \frac{S_p t_{rt}}{\gamma_{rt,s}} \left[\frac{R_s (1 + R_s) \eta_e}{R_s - \ln(1 + R_s)} \right], \tag{8.107}$$

where $R_s = \gamma_{rt,s} / (\gamma_{rt,p} + \gamma_o)$ is the ratio of saturable to unsaturable cavity losses. For a given amount of saturable loss, the pulse width asymptotically approaches its minimum value of $t_w = 4 S_p t_{rt} / \gamma_{rt,s}$ in the limit of large unsaturable losses ($\gamma_{rt,p} + \gamma_o \gg \gamma_{rt,s}$). However, this minimum pulse width is obtained at the expense of high threshold and low efficiency. In the opposite limit ($\gamma_{rt,p} + \gamma_o \ll \gamma_{rt,s}$), the pulse width asymptotically approaches $t_w = S_p t_{rt} / (\gamma_{rt,p} + \gamma_o)$, the threshold of the laser is reduced, and the extraction efficiency is high. The best compromise between pulse width, threshold, and efficiency will depend on the application of the laser. At low pulse repetition rates ($\tau_p > \tau$), passively Q-switched lasers have excellent pulse-to-pulse amplitude and pulse-width stability.

For a cw-pumped passively Q-switched laser, the repetition rate of the pulses increases as the pump power is increased above threshold. At high pulse repetition rates ($\tau_p \ll \tau$), there is a tendency for the pulse train to bifurcate into alternating strong and weak pulses, with a typical amplitude ratio of ~0.9 (Zayhowski and Dill, 1994). The timing interval between the pulses varies in accordance with the amplitudes; the period preceding a weak pulse is shorter than the period preceding a strong pulse. The average interpulse period is given by

$$\tau_p = \frac{\tau (P_{th} - P_i)}{P_a - P_i}, \tag{8.108}$$

where P_a is the total amount of pump power absorbed within the lasing mode volume, P_i is the absorbed pump power required to obtain inversion ($N_{eff} > 0$), and P_{th} is the absorbed pump power necessary to reach threshold (Eq. 8.106). At still higher pulse repetition rates (higher pump powers), the pulse train has a tendency to subdivide further, into a train containing pulses of three or more amplitudes which replicate themselves periodically. Although the tendency is toward an increasing number of pulse amplitudes with increasing repetition rate, this trend is not always observed and depends on the exact pump conditions.

In some solid-state saturable absorbers, the contrast ratio $\gamma_{rt,s}/\gamma_{rt,p,s}$, where $\gamma_{rt,p,s}$ is the parasitic loss associated with the saturable absorber, is limited by excited-state absorption or other intrinsic material properties, and may be small. This can limit the efficiency of a passively Q-switched device.

Diode-pumped, passively Q-switched monolithic lasers using Cr^{4+},Nd^{3+}:YAG as both the gain medium and saturable absorber (Cr^{4+} acts as a saturable absorber at 1.064 μm in YAG) have demonstrated output pulses with durations between 3.0 ns (Zhou et al., 1993) and 290 ps (Wang et al., 1995). Pumped with a 1-W fiber-coupled diode laser, passively Q-switched composite-cavity microchip lasers, containing Nd^{3+}:YAG and Cr^{4+}:YAG, have demonstrated pulses as short as 200 ps with peak powers in excess of 30 kW (Zayhowski and Dill, 1994; Zayhowski et al., 1995b). By using a 12-W fiber-coupled diode array as the pump, 130-kW peak powers have been obtained in 1-ns pulses (Zayhowski, 1997a). A variety of miniature nonlinear optical devices, including harmonic generators (Zayhowski, 1996a; Zayhowski, 1996b), parametric amplifiers (Zayhowski, 1997a), parametric oscillators (Zayhowski, 1997b), and broad-band fiber-based Raman amplifiers (Zayhowski, 1997c; Agrawal, 1995) have been used to frequency convert the output of these lasers, accessing the entire spectrum from 5 μm to 200 nm in extremely compact optical systems. Pulses as short as 180 ps have been obtained by Q switching Nd:LaSc$_3$(BO$_3$)$_4$ microchip lasers with semiconductor antiresonant Fabry-Perot saturable absorbers (Barun et al., 1996). More recently, semiconductor saturable absorbers have been used to generate 56-ps Q-switched pulses from Nd:YVO$_4$ microchip lasers (Braun et al., 1997). Although the devices based on semiconductor saturable absorbers operate at lower peak powers and lower energies than the Cr^{4+}:YAG-based lasers, they have been pulsed at repetition rates up to 7 MHz. All of the passively Q-switched lasers mentioned above operate on a ~1.0-μm Nd line. Passively Q-switched microchip lasers operating on the ~1.3-μm (Keller, 1996) and ~946-nm (Zayhowski, 1997c) Nd lines have also been demonstrated.

Gain-Switched Operation

Gain switching is another way to obtain short, high-peak-power pulses from a laser. The idea is to rapidly increase the pump power so that the population inversion of the laser is well in excess of the threshold value by the time the first output spike develops. The optical pulse then drives the population inversion below its threshold value, as described in the above discussion of Quasi-CW Operation. The pump power is subsequently reduced so that the inversion remains below threshold and a single pulse is obtained.

The equations that describe a Q-switched pulse also describe a gain-switched pulse. The difference is that the operation of the laser is controlled by gain rather than loss. Once we know the inversion density at the beginning of the output pulse, all of the formalism developed in the above discussion of Q-Switched Operation applies. Essentially, given the temporal pump profile, we must calculate the pulse build-up time and integrate the pump rate up to that time.

Gain-switched Nd:YAG microchip lasers, pumped by 36-ns-long pulses from a Ti:Al$_2$O$_3$ laser, have produced pulses as short as 760 ps (Zayhowski et al., 1989). Diode-pumped, 1.3-μm Nd:YAG microchip lasers have been gain switched to produce a 100-kHz train of pulses with durations of 170 ns and peak powers of 1.8 W (Zayhowski, 1990b). In this case, the lasers were resonantly pumped at their relaxation frequency to minimize both the pulse width and the fluctuations in pulse amplitude.

Cavity-Dumped Operation

In principle, cavity dumping allows the energy in a laser cavity to be output in a time comparable to the cavity round-trip time. The concept is to rapidly introduce (within a cavity round-trip

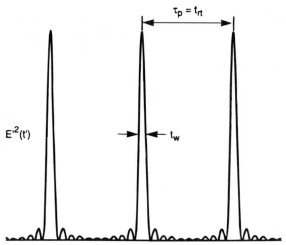

FIGURE 8.20 Train of mode-locked pulses made up of 11 modes of equal intensity.

time) a large output coupling (~ 100%) into a cavity which previously had no output coupling. This works well for large laser cavities. As the length of a laser cavity becomes small, however, the duration of a cavity-dumped pulse becomes limited by the switching time of the output-coupling mechanism (see the above discussion of Active Q Switching). For small laser cavities, Q switching is more efficient and produces shorter pulses. Methods of cavity dumping include the use of an electrooptic Pockels cell and polarizing beam splitter, and acousto-optic devices.

Mode-Locked Operation

Mode locking refers to the situation where the phases of several cavity modes are fixed (or locked) with respect to each other such that the electric fields add coherently and constructively for a short period of time. This allows the generation of a train of high-intensity, ultrashort pulses. To understand this, consider the case of $2n + 1$ equally spaced longitudinal modes oscillating with the same amplitude E_0. Assume that the phases ϕ_m of the modes are locked according to $\phi_m - \phi_{m-1} = \Delta\phi$, where $\Delta\phi$ is a constant. The total electric field is the sum of all of these modes:

$$E(t) = \sum_{m=-n}^{n} E_0 \exp\{2\pi i[(\nu_0 - m\Delta\nu)t + m\Delta\phi]\} = E'(t')\exp(2\pi i\nu_0 t), \qquad (8.109)$$

where ν_0 is the frequency of the center mode, $\Delta\nu$ is the frequency difference between two adjacent modes, $t' = t + \Delta\phi/\Delta\nu$, and

$$E'(t') = E_0 \frac{\sin[\pi(2n + 1)\Delta\nu t']}{\sin[\pi\Delta\nu t']}. \qquad (8.110)$$

Equation 8.109 shows that $E(t)$ can be represented in terms of a wave with a carrier frequency ν_0 whose amplitude is time dependent. The intensity of this wave is given by $E'^2(t')$, which consists of a train of pulses whose peak intensity is $E'^2(0) = (2n + 1)^2 E_0^2$, pulse width is $t_w = 1/(2n + 1)\Delta\nu$, and separation between pulses is $\tau_p = 1/\Delta\nu$. Because the total oscillating bandwidth $\Delta\nu_{osc}$ is given by $(2n + 1)\Delta\nu$ and the frequency separation $\Delta\nu$ is $1/t_{rt}$, the mode-locked laser just described produces a train of output pulses with a pulse width given by the inverse of the oscillating bandwidth and a separation between pulses equal to the cavity round-trip time. Figure 8.20 shows $E'^2(t')$ for $n = 5$. A miniature Nd:YLF laser has been mode locked to produce a train of 13-ps pulses at a 5-GHz repetition rate (Schulz and Henion, 1991b). Semiconductor saturable absorbers have been used to obtain mode-locked pulses as short as 10 fs (Keller, 1996).

In the above example, the phases of all of the modes were locked so that the output pulse had the minimum possible duration. Such a pulse is referred to as transform limited, because the temporal profile is the Fourier transform of the spectral profile. This need not be the case — it is possible to obtain longer pulses (but not shorter pulses). Also, it is worth noting that, unlike the other pulsed schemes described in this section, a mode-locked laser is a cw device, and there is phase coherence between pulses.

The discussion above tells us what mode locking is, but sheds little light on how or why it occurs. Mode locking can be broken into two categories, AM mode locking (produced with an amplitude modulator within the laser cavity) and FM mode locking (produced with a phase or frequency modulator within the laser cavity) and can be obtained actively or passively. In general, mode locking will occur if the net gain for a mode-locked train of pulses is greater than the net gain for any other combination of cavity modes.

AM Mode Locking

In AM mode locking, the loss of some element in the laser cavity is modulated at the round-trip cavity frequency. This modulates the amplitude of the optical field and generates sidebands that are resonant cavity modes. With the proper phase relationship between these modes, most of the light is incident on the loss element during its minimum loss. This combination of modes sees a lower loss than any other combination of modes and is, therefore, favored. The same result occurs if the gain of the cavity is modulated. Gain modulation through modulation of the pump source is known as synchronous pumping.

In an ideal mode-locked system, the relative phases of the cavity modes remain constant after a complete round trip in the laser cavity. In such a system, once the phase relationship between the modes is established, it will persist even in the absence of amplitude modulation. One effect that can destroy the phase relationship and, therefore, disrupt mode locking, is dispersion. Another is spontaneous emission (noise), because the phase of the noise is uncorrelated to the phase of the oscillating modes. Mode locking occurs when there is sufficient loss (or gain) modulation to overcome these effects.

The duration of the mode-locked pulse and the net intracavity gain are strongly dependent on the intracavity dispersion. Excess dispersion will increase the duration of the pulses. Typically, such pulses are frequency chirped, with a different optical frequency at the beginning of the pulse than at the end. To minimize the round-trip dispersion of the cavity, prisms are often used to compensate for the dispersion of other intracavity elements, such as the gain medium.

Passive mode locking can occur when a laser cavity contains a nonlinear optical element, such as a saturable absorber. In this case, the more intense the light incident on the saturable absorber, the less the total absorption. The total loss of the cavity is, therefore, minimized by putting all of the energy into short pulses. This is, essentially, self-induced AM mode locking. To work most effectively, the recovery time of the saturable absorber should be longer than the pulse duration, but shorter than the round-trip time of the cavity. Also, the round-trip intracavity dispersion must be small enough so that the superposition of the modes after one round trip produces a pulse of sufficient peak power to maintain the required nonlinear steady state. A similar effect is obtained with the combination of a Kerr lens (often formed within the gain medium itself) and an aperture within the cavity. Other techniques include the use of interferometric elements containing nonlinear media.

Passive mode locking must be initiated by the presence of a pulse within the cavity. If the optical intensity within the cavity is uniform in time, there is no loss or gain element that is modulated at the round-trip cavity frequency to induce mode locking. Noise, however, is capable of introducing a small amplitude modulation on the optical field. In some lasers, this small modulation is sufficient to start the mode-locking process. Such lasers are referred to as self-starting. In other systems, a pulse (or AM modulation) must be intentionally introduced into the cavity to start the mode-locking process. Once started, however, mode locking may persist indefinitely.

FM Mode Locking

In FM mode locking, the optical length of the laser cavity (physical length or refractive index) is modulated at the round-trip cavity frequency. As in the case of AM mode locking, the modulation generates sidebands that are resonant cavity modes. For simplicity, consider the case where one of the cavity mirrors is moving sinusoidally along the direction of the cavity axis. Light incident on a moving mirror will be Doppler shifted. As a result, it will not reproduce itself after one round trip and will not produce a coherent oscillating mode. Light incident on the mirror at its turning points (maximum or minimum cavity length) is reflected by a stationary mirror and does not experience a Doppler shift. Modulation of the refractive index at some point in the cavity has the same effect. Mode locking will occur when there is sufficient modulation to overcome the effects of dispersion and noise. The duration of the mode-locked pulse and the net intracavity gain are (as in the case of AM mode locking) strongly dependent on intracavity dispersion.

References

Aggarwal, R. L., Sanchez, A., Stuppi, M. M., Fahey, R. E., Strauss, A. J., Rapoport, W. R., and Khattak, C. P. 1988. Residual infrared absorption in as-grown and annealed crystals of Ti:Al$_2$O$_3$, *IEEE J. Quantum Electron.* **24**, 1003.

Agrawal, G. P. 1995. *Nonlinear Fiber Optics*, 2nd ed. Academic Press, San Diego, Ch. 8.

Amano, S. 1992. Microchip NYAB green laser, *Rev. Laser Eng.* **20**, 723 (in Japanese).

Balembois, F., Georges, P., and Brun, A. 1993. Quasi-continuous-wave and actively mode-locked diode-pumped Cr^{3+}:LiSrAlF$_6$ laser, *Opt. Lett.* **18**, 1730.

Braun, B., Kärtner, F. X., Keller, U., Meyn, J.-P., and Huber, G. 1996. Passively Q-switched 180-ps Nd:LaSc$_3$(BO$_3$)$_4$ microchip laser, *Opt. Lett.* **24**, 405.

Braun, B., Kärtner, F. X., Zhang, G., Moser, M., and Keller, U. 1997. 56-ps passively Q-switched diode-pumped microchip laser, *Opt. Lett.* **22**, 381.

Brown, A. J. W., and Fisher, C. H. 1993. A 6.5-J flashlamp-pumped Ti:Al$_2$O$_3$ laser, *IEEE J. Quantum Electron.* **29**, 2513.

Carrig, T. J., and Pollock, C. R. 1993. Performance of a continuous-wave fosterite laser with krypton ion, Ti:sapphire, and Nd:YAG pump lasers, *IEEE J. Quantum Electron.* **29**, 2835.

Casperson, L. W. 1980. Laser power calculations: Sources of error, *Appl. Opt.* **19**, 422.

Chai, B. H. T., Lefaucheur, J.-L., Stalder, M., and Bass, M. 1992. Cr:LiSr$_{0.8}$Ca$_{0.2}$AlF$_6$ tunable laser, *Opt. Lett.* **17**, 1584.

Cheng, E. A. P., and Kane, T. J. 1991. High-power single-mode diode-pumped Nd:YAG laser using a monolithic nonplanar ring resonator, *Opt. Lett.* **16**, 478.

Clobes, A. R., and Brienza, M. J. 1972. Single-frequency traveling-wave Nd:YAG laser, *Appl. Phys. Lett.* **21**, 265.

Danielmeyer, H. G. 1975. Stoichiometric laser materials, in *Festkörperprobleme* (*Advances in Solid State Physics*), Queisser, H.-J., ed. Pergamon/Vieweg, Braunschweig, **XV**, 253.

Danilov, A. A., Denisov, A. L., Zharikov, E. V., Zagumennyi, A. I., Lutts, G. B., Nikol'skii, M. Yu., Tsvetkov, V. B., and Shcherbakov, I. A. 1988. Laser utilizing gadolinium scandium aluminum garnet activated with chromium and neodymium, *Sov. J. Quantum Electron.* **18**, 1097.

Day, T., Nilsson, A. C., Fejer, M. M., Farinas, A. D., Gustafson, E. K., Nabors, C. D., and Byer, R. L. 1989. 30 Hz-linewidth, diode-laser-pumped, Nd:GGG nonplanar ring oscillators by active frequency stabilization, *Electron. Lett.* **25**, 810.

Day, T., Gustafson, E. K., and Byer, R. L. 1990. Active frequency stabilization of a 1.062-μm, Nd:GGG, diode-laser-pumped nonlinear ring oscillator to less than 3 Hz of relative linewidth, *Opt. Lett.* **15**, 221.

Degnan, J. J. 1989. Theory of the optimally coupled Q-switched laser, *IEEE J. Quantum Electron.* **25**, 214.

DeLoach, L. D., Payne, S. A., Chase, L. L., Smith, L. K., Kway, W. L., and Krupke, W. F. 1993. Evaluation of absorption and emission properties of Yb^{3+} doped crystals for laser application, *IEEE J. Quantum Electron.* **29**, 1179.

Dieke, G. H., and Crosswhite, H. M. 1963. The spectra of the doubly and triply ionized rare earths, *Appl. Opt.* **2**, 675.

Dinerman, B. J., Harrison, J., and Moulton, P. F., 1994. CW and pulsed operation at 3 μm in Er^{3+}-doped crystals, in *OSA Proc. Advanced Solid State Lasers*, Fan, T. Y., and Chai, B. H. T., eds. Optical Society of America, Washington, D.C., **20**, 168.

Dixon, G. J., Lingvay, L. S., and Jarman, R. H. 1989. Properties of close-coupled, monolithic lithium neodymium tetraphosphate laser, in *SPIE Growth, characterization, and applications of laser host and nonlinear crystals* **1104**, 107.

Erbert, G., Bass, I., Hackel, R., Jemkins, S., Kanz, K., and Paisner, J. 1991. 43-W, cw Ti:sapphire laser, *Conf. Lasers and Electro-Optics Tech. Dig.*, **11**, 390.

Evtuhov, V., and Siegman, A. E. 1965. A twisted-mode technique for obtaining axially uniform energy density in a laser cavity, *Appl. Opt.* **4**, 142.

Fan, T. Y. 1988. Effect of finite lower level lifetime on Q-switched lasers, *IEEE J. Quantum Electron.* **24**, 2345.

Fan, T. Y. 1994. Aperture guiding in quasi-three-level lasers, *Opt. Lett.* **19**, 554.

Fan, T. Y., and Byer, R. L. 1988. Diode laser-pumped solid-state lasers, *IEEE J. Quantum Electron.* **24**, 895.

Fan, T. Y., and Sanchez, A. 1990. Pump source requirements for end-pumped lasers, *IEEE J. Quantum Electron.* **26**, 311.

Fan, T. Y., Huber, G., Byer, R. L., and Mitzscherlich, P. 1988. Spectroscopy and diode laser-pumped operation of Tm,Ho:YAG, *IEEE J. Quantum Electron.* **24**, 924.

French, P. M. W., Mellish, R., Taylor, J. R., Delfyett, P. J., and Florez, L. T. 1993. Mode-locked all-solid-state diode-pumped Cr:LiSAF laser, *Opt. Lett.* **18**, 1934.

Genack, A. Z., and Brewer, R. G. 1978. Optical coherent transitions by laser frequency switching, *Phys. Rev. A* **17**, 1463.

Gavrilovic, P., O'Neill, M. S., Meehan, K., Zarrabi, J. H., and Singh, S. 1992. Temperature-tunable, single frequency microcavity lasers fabricated from flux-grown YCeAG:Nd, *Appl. Phys. Lett.* **60**, 1652.

Hall, D. R., and Jackson, P. E., eds. 1989. *The Physics and Technology of Laser Resonators.* Adam Hilger, Bristol and New York.

Hansson, G., Callenås, A., and Nelsson, C. 1993. Upconversion studies in laser pumped Tm,Ho:YLiF$_4$, in *OSA Proc. Advanced Solid State Lasers*, Pinto, A., and Fan, T. Y., eds. Optical Society of America, Washington, D.C., **15**, 446.

Harrison, J., and Martinsen, R. J. 1994a. Thermal modeling for mode-size estimation in microlasers with application to linear arrays in Nd:YAG and Tm,Ho:YLF, *IEEE J. Quantum Electron.* **30**, 2628.

Harrison, J., and Martinsen, R. J. 1994b. Operation of linear microlaser arrays near 1 μm, 2 μm, and 3 μm, in *OSA Proc. Advanced Solid State Lasers*, Fan, T. Y., and Chai, B. H. T., eds. Optical Society of America, Washington, D.C., **20**, 272.

Harrison, J., Welford, D., and Moulton, P. F. 1989. Threshold analysis of pulsed lasers with application to a room-temperature Co:MgF$_2$ laser, *IEEE J. Quantum Electron.* **25**, 1708.

Harrison, J., Finch, A., Rines, D. M., Rines, G. A., and Moulton, P. F. 1991. Low-threshold, cw, all-solid-state Ti:Al$_2$O$_3$ laser, *Opt. Lett.* **16**, 581.

Hough, J., Hils, D., Rayman, M. D., Ma, L.–S., Hollberg, L., and Hall, J. L. 1984. Dye-laser frequency stabilization using optical resonators, *Appl. Phys. B* **33**, 179.

Jaseja, T. S., Javan, A., and Townes, C. H. 1963. Frequency stability of He-Ne masers and measurements of length, *Phys. Rev. Lett.* **10**, 165.

Jeys, T. H. 1991. Suppression of laser spiking by intracavity harmonic generation, *Appl. Opt.* **30**, 1011.

Johnson, L. F., Dietz, R. E., and Guggenheim, H. J. 1963. Optical maser oscillation from Ni^{2+} in MgF_2 involving simultaneous emission of phonons, *Phys. Rev. Lett.* **11**, 318.

Johnson, L. F., Dietz, R. E., and Guggenheim, H. J. 1964. Spontaneous and stimulated emission from Co^{2+} ions in MgF_2 and ZnF_2, *Appl. Phys. Lett.* **5**, 21.

Johnson, L. F., Guggenheim, H. J., and Thomas, R. A. 1966. Phonon-terminated optical masers, *Phys. Rev.* **149**, 179.

Johnson, L. F., Guggenheim, H. J., and Thomas, R. A. 1967. Phonon-terminated coherent emission from V^{2+} ions in MgF_2, *J. Appl. Phys.* **38**, 4837.

Kane, T. J. 1990. Intensity noise in diode-pumped single-frequency Nd:YAG lasers and its control by electronic feedback, *IEEE Photon. Technol. Lett.* **2**, 244.

Kane, T. J., and Byer, R. L. 1985. Monolithic, unidirectional single-mode Nd:YAG ring laser, *Opt. Lett.* **10**, 65.

Kane, T. J., and Kubo, T. S. 1991. Diode-pumped single-frequency lasers and Q-switched lasers using Tm:YAG and Tm,Ho:YAG, in *OSA Proc. Advanced Solid State Lasers*, Jenssen, H. P., and Dubé, G., eds. Optical Society of America, Washington, D.C., **6**, 136.

Kane, T. J., Nilsson, A. C., and Byer, R. L. 1987. Frequency stability and offset locking of a laser-diode-pumped Nd:YAG monolithic nonplanar ring oscillator, *Opt. Lett.* **12**, 175.

Keller, U. 1996. Modelocked and Q-switched solid-state lasers using semiconductor saturable absorbers, *Conf. Proc. of 9th Annual Meeting IEEE Lasers and Electro-Optics Society*, **1**, 50.

Kennedy, C. J., and Barry, J. D. 1974. Stability of an intracavity frequency-doubled Nd:YAG laser, *IEEE J. Quantum Electron.* **QE-10**, 596.

Keszenheimer, J. A., Balboni, E. J., and Zayhowski, J. J. 1992. Phase locking of 1.32-µm microchip lasers through the use of pump-diode modulation, *Opt. Lett.* **17**, 649.

Keyes, R. J., and Quist, T. M. 1964. Injection luminescent pumping of $CaF_2:U^{3+}$ with GaAs diode lasers, *Appl. Phys. Lett.* **4**, 50.

Kintz, G. J., and Baer, T. 1990. Single-frequency operation in solid state laser materials with short absorption depths, *IEEE J. Quantum Electron.* **26**, 1457.

Kogelnik, H. 1965. On the propagation of Gaussian beams of light through lenslike media including those with a loss or gain variation, *Appl. Opt.* **4**, 1562.

Krühler, W. W., Plättner, R. D., and Stetter, W. 1979. CW oscillation at 1.05 and 1.32 µm of $LiNd(PO_3)_4$ lasers in external resonators and in resonators with directly applied mirrors, *IEEE J. Quantum Electron.* **QE-14**, 840.

Kubo, T. S., and Kane, T. J. 1992. Diode-pumped lasers at five eye-safe wavelengths, *IEEE J. Quantum Electron.* **28**, 1033.

Laporta, P., Taccheo, S., Longhi, S., and Svelto, O. 1993. Diode-pumped microchip Er-Yb laser, *Opt. Lett.* **18**, 1232.

Lax, M. 1966. Quantum noise V: Phase noise in a homogeneously broadened maser, in *Physics of Quantum Electronics*, Kelley, P. K., Lax, B., and Tannenwald, P. E., eds. McGraw-Hill, New York, p. 735.

Leilabady, P. A., Anthon, D. W., and Gullicksen, P. O. 1992. Single-frequency Nd:YLF cube lasers pumped by laser diode arrays, in *Conf. Lasers and Electro-Optics Tech. Dig.*, **12**, 54.

Lovold, S., Moulton, P. F., Killinger, D. K., and Menyuk, N. 1985. Frequency tuning characteristics of a Q-switched $Co:MgF_2$ laser, *IEEE J. Quantum Electron.* **QE-21**, 202.

MacKinnon, N., and Sinclair, B. D. 1992. Pump power induced cavity stability in lithium neodymium tetraphosphate (LNP) microchip lasers, *Opt. Comm.* **94**, 281.

MacKinnon, N., Norrie, C. J., and Sinclair, B. D. 1994. Laser-diode-pumped, electro-optically tunable Nd:MgO:LiNbO$_3$ microchip laser, *J. Opt. Soc. Am. B* **11**, 519.

Maiman, T. H. 1960. Stimulated optical radiation in ruby masers, *Nature* **187**, 493.

Maker, G. T., and Ferguson, A. I. 1990. Ti:sapphire laser pumped by a frequency-doubled diode-pumped Nd:YLF laser, *Opt. Lett.* **15**, 375.

Man, C. N., and Brillet, A. 1984. Injection locking of argon-ion lasers, *Opt. Lett.* **9**, 333.

Mermilliod, N., François, B., and Wyon, Ch. 1991. LaMgAl$_{11}$O$_{19}$:Nd microchip laser, *Appl. Phys. Lett.* **59**, 3520.

Moncorge, R., Cormier, G., Simkin, D. J., and Capobianco, J. A. 1991. Fluorescence analysis of chromium-doped fosterite (Mg$_2$SiO$_4$), *IEEE J. Quantum Electron.* **27**, 114.

Moulton, P. F. 1982. Ti:Al$_2$O$_3$—a new tunable solid state laser, *M.I.T. Lincoln Lab. Solid State Res. Rep.*, DTIC AD-A124305/4, 15.

Moulton, P. F. 1985. An investigation of the Co:MgF$_2$ laser system, *IEEE J. Quantum Electron.* **QE-21**, 1582.

Moulton, P. F. 1986. Spectroscopic and laser characteristics of Ti:Al$_2$O$_3$, *J. Opt. Soc. Am. B*, **3**, 125.

Moulton, P. F. 1992. Tunable solid-state lasers, *Proc. IEEE* **80**, 348.

Muller III, C. H., Lowenthal, D. D., Kangas, K. W., Hamil, R. A., and Tisone, G. C. 1988. 2.0-J Ti:sapphire laser oscillator, *Opt. Lett.* **13**, 380.

Nabors, C. D., Ochoa, J., Fan, T. Y., Sanchez, A., Choi, H., and Turner, G. 1994. 1.9-μm-diode-laser-pumped, 2.1-μm Ho:YAG laser, in *Conf. Lasers and Electro-Optics Tech. Dig.*, **8**, 172.

Nilsson, A., Gustafson, E. K., and Byer, R. L. 1989. Eigenpolarization theory of monolithic nonplanar ring oscillators, *IEEE J. Quantum Electron.* **25**, 767.

Newman, R. 1963. Excitation of Nd^{3+} fluorescence in CaWO$_4$ by recombination radiation in GaAs, *J. Appl. Phys.* **34**, 437.

Ohtsu, M., and Kotajima, S. 1985. Linewidth reduction of a semiconductor laser by electrical feedback, *IEEE J. Quantum Electron.* **QE-21**, 1905.

Owyoung, A., and Esherick, P. 1987. Stress-induced tuning of a diode-laser-excited monolithic Nd:YAG laser, *Opt. Lett.* **12**, 999.

Payne, S. A., Chase, L. L., Newkirk, H. W., Smith, L. K., and Krupke, W. F. 1988. LiCaAlF$_6$:Cr^{3+}: A promising new solid-state laser material, *IEEE J. Quantum Electron.* **24**, 2243.

Payne, S. A., Chase, L. L., Smith, L. K., Kway, W. L., and Newkirk, H. W. 1989. Laser performance of LiSrAlF$_6$:Cr^{3+}, *J. Appl. Phys.* **66**, 1051.

Petricevic, V., Gayen, S. K., and Alfano, R. R. 1988a. Laser action in chromium-activated fosterite, *Appl. Phys. Lett.* **52**, 1040.

Petricevic, V., Gayen, S. K., and Alfano, R. R. 1988b. Laser action in chromium-activated fosterite for near-infrared excitation: Is Cr^{4+} the lasing ion? *Appl. Phys. Lett.* **53**, 2590.

Plaessmann, H., Yamada, K. S., Rich, C. E., and Grossman, W. M. 1994. Subnanosecond pulse generation from diode-pumped acousto-optically Q-switched solid-state lasers, *Appl. Opt.* **32**, 6616.

Schawlow, A. L., and Townes, C. H. 1958. Infrared and optical masers, *Phys. Rev.* **12**, 1940.

Scheps, R. 1991. Cr:LiCaAlF$_6$ laser pumped by visible laser diodes, *IEEE J. Quantum Electron.* **27**, 1968.

Scheps, R. 1992. Laser-diode-pumped Cr:LiSrGaF$_6$ laser, *IEEE Photon. Technol. Lett.* **4**, 548.

Scheps, R., Gately, B. M., Myers, J. F., Krasinski, J. S., and Heller, D. F. 1990. Alexandrite laser pumped by semiconductor lasers, *Appl. Phys. Lett.* **56**, 2288.

Scheps, R., Myers, J. F., Serreze, H. B., Rosenberg, A., Morris, R. C., and Long, M. 1991. Diode-pumped Cr:LiSrAlF$_6$ laser, *Opt. Lett.* **16**, 820.

Schulz, P. A. 1990. Fast electro-optic wavelength selection and frequency modulation in solid state lasers, *Lincoln Lab. J.* **3**, 463.

Schulz, P. A., and Henion, S. R. 1991a. Frequency-modulated Nd:YAG laser, *Opt. Lett.* **16**, 578.

Schulz, P. A., and Henion, S. R. 1991b. 5-GHz mode locking of a Nd:YLF laser, *Opt. Lett.* **16**, 1502.

Siegman, A. J. 1986. *Lasers*. University Science Books, Mill Valley, CA.

Smith, L. K., Payne, S. A., Kway, W. L., Chase, L. L., and Chai, B. H. T. 1992. Investigation of the laser properties of Cr^{3+}:LiSrGaF$_6$, *IEEE J. Quantum Electron.* **28**, 2612.

Smith, L. K., Payne, S. A., Krupke, W. F., DeLoach, L. D., Morris, R., O'Dell, E. W., and Nelson, D. J. 1993. Laser emission from the transition-metal compound LiSrCrF$_6$, *Opt. Lett.* **18**, 200.

Statz, H., DeMars, G. A., and Wilson, D. T. 1965. Problem of spike elimination in lasers, *J. Appl. Phys.* **36**, 1510.

Stone, D. H. 1992. Effects of axial nonuniformity in modeling Q-switched lasers, *IEEE J. Quantum Electron.* **28**, 1970.

Stoneman, R. C., and Esterowitz, L. 1992. Efficient resonantly pumped 2.8-μm Er^{3+}:GSGG laser, *Opt. Lett.* **17**, 816.

Storm, M. E. 1991. Spectral performance of monolithic holmium and thulium lasers, in *OSA Proc. Advanced Solid State Lasers*, Shand, M. L., and Jenssen, H. P., eds. Optical Society of America, Washington, D.C. **5**, 186.

Storm, M. E., and Rohrbach, W. W. 1989. Single-longitudinal-mode lasing of Ho:Tm:YAG at 2.091 μm, *Appl. Opt.* **28**, 4965.

Szabo, A., and Stein, R. A. 1965. Theory of laser giant pulsing by saturable absorber, *J. Appl. Phys.* **36**, 1562.

Taira, T., Mukai, A., Nozawa, Y., and Kobayashi, T. 1991. Single-mode oscillation of laser-diode-pumped Nd:YVO$_4$ microchip lasers, *Opt. Lett.* **24**, 1955.

Tang, C. L., Statz, H., and DeMars, G. 1963. Spectral output and spiking behavior of solid-state lasers, *J. Appl. Phys.* **34**, 2289.

Verdun, H. R., Thomas, L. M., Andrauskas, D. M., and McCollum, T. 1988. Chromium-doped fosterite laser pumped with 1.06 μm radiation, *Appl. Phys. Lett.* **53**, 2593.

Wagner, W. G., and Lengyel, B. A. 1963. Evolution of the giant pulse in a laser, *J. Appl. Phys.* **34**, 2040.

Walling, J. C., Heller, D. F., Samelson, H., Harter, D. J., Pete, J. A., and Morris, R. C. 1985. Tunable alexandrite lasers: Development and performance, *IEEE J. Quantum Electron.* **QE-21**, 1568.

Wallmeroth, K. 1990. Monolithic integrated Nd:YAG laser, *Opt. Lett.* **15**, 903.

Wang, P., Zhou, S.-H., Lee, K. K., and Chen, Y. C. 1995. Picosecond laser pulse generation in a monolithic self-Q-switched solid-state laser, *Opt. Commun.* **114**, 439.

Weber, H. P. 1975. Review of Nd-pentaphosphate lasers, *Opt. Quantum Electron.* **7**, 431.

Welford, D., and Moulton, P. F. 1988. Room temperature operation of a Co:MgF$_2$ laser, *Opt. Lett.* **13**, 975.

Winzer, G., Möckel, P. G., Oberbacher, R., and Vité, L. 1976. Laser emission from polished NdP$_5$O$_{14}$ crystals with directly applied mirrors, *Appl. Phys.* **11**, 121.

Winzer, G., Möckel, P. G., and Krühler, W. W. 1978. Laser emission from miniature NdAl$_3$(BO$_3$)$_4$ crystals with directly applied mirrors, *IEEE J. Quantum Electron.* **QE-14**, 840.

Zayhowski, J. J. 1990a. Limits imposed by spatial hole burning on the single-mode operation of standing-wave laser cavities, *Opt. Lett.* **15**, 431.

Zayhowski, J. J. 1990b. Microchip lasers, *Lincoln Lab. J.* **3**, 427.

Zayhowski, J. J. 1990c. The effects of spatial hole burning and energy diffusion on the single-mode operation of standing-wave lasers cavities, *IEEE J. Quantum Electron.* **26**, 2052.

Zayhowski, J. J. 1991a. Thermal guiding in microchip lasers, in *OSA Proc. Advanced Solid State Lasers*, Jenssen, H. P., and Dubé, G., eds. Optical Society of America, Washington, D.C., **6**, 9.

Zayhowski, J. J. 1991b. Polarization-switchable microchip lasers, *Appl. Phys. Lett.* **58**, 2746.

Zayhowski, J. J. 1996a. Microchip lasers create light in small places, *Laser Focus World*. Penwell Publishing Co., Tulsa, Oklahoma, April 1996, p. 73.

Zayhowski, J. J. 1996b. Ultraviolet generation with passively Q-switched microchip lasers, *Opt. Lett.* **21**, 588; Ultraviolet generation with passively Q-switched microchip lasers: errata, *Opt. Lett.* **21**, 1618.

Zayhowski, J. J. 1997a. Periodically poled lithium niobate optical parametric amplifiers pumped by high-power passively Q-switched microchip lasers, *Opt. Lett.* **22**, 169.

Zayhowski, J. J. 1997b. Microchip optical parametric oscillators, to be published.

Zayhowski, J. J. 1997c. Unpublished.

Zayhowski, J. J., and Dill III, C. 1992. Diode-pumped microchip lasers electro-optically Q switched at high pulse repetition rates, *Opt. Lett.* **17**, 1201.

Zayhowski, J. J., and Dill III, C. 1994. Diode-pumped passively Q-switched picosecond microchip lasers, *Opt. Lett.* **19**, 1427.

Zayhowski, J. J., and Dill III, C. 1995. Coupled-cavity electro-optically Q-switched Nd:YVO$_4$ microchip lasers, *Opt. Lett.* **20**, 716.

Zayhowski, J. J., Harrison, J., Dill III, C., and Ochoa, J. 1995a. Diode-pumped Tm:YVO$_4$ microchip laser, *Appl. Opt.*, **34**, 435.

Zayhowski, J. J., and Kelley, P. L. 1991. Optimization of Q-switched laser, *IEEE J. Quantum Electron.* **27**, 2220; Corrections to Optimization of Q-switched laser, *IEEE J. Quantum Electron.* **29**, 1239.

Zayhowski, J. J., and Keszenheimer, J. A. 1992. Frequency tuning of microchip lasers using pump-power modulation, *IEEE J. Quantum Electron.* **28**, 1118.

Zayhowski, J. J., and Mooradian, A. 1989a. Single-frequency microchip Nd:YAG lasers, *Opt. Lett.* **14**, 24.

Zayhowski, J. J., and Mooradian, A. 1989b. Frequency-modulated Nd:YAG microchip lasers, *Opt. Lett.* **14**, 618.

Zayhowski, J. J., and Mooradian, A. 1989c. Microchip lasers, in *OSA Proc. Advanced Solid State Lasers*, Shand, M. L., and Jenssen, H. P., eds. Optical Society of America, Washington, D.C., **5**, 288.

Zayhowski, J. J., Ochoa, J., and Dill III, C. 1995b. UV generation with passively Q-switched picosecond microchip lasers, in *Conf. Lasers and Electro-Optics Tech. Dig.*, **15**, 139.

Zayhowski, J. J., Ochoa, J., and Mooradian, A. 1989. Gain-switched pulsed operation of microchip lasers, *Opt. Lett.* **14**, 1318.

Zayhowski, J. J., Schulz, P. A., Dill III, C., and Henion, S. R. 1993. Diode-pumped composite-cavity electrooptically tuned microchip laser, *IEEE Photon. Technol. Lett.* **5**, 1153.

Zarrabi, J. H., Gavrilovic, P., Williams, J. E., O'Neill, M. S., and Singh, S. 1993. Single-frequency, diode-pumped, neodymium-doped lanthanum oxysulfide microchip laser, in *Conf. Lasers and Electro-Optics Tech. Dig.*, **11**, 588.

Zhang, Q., Dixon, G. J., Chai, B. H. T., and Kean, P. N. 1992. Electronically tuned diode-laser-pumped Cr:LiSrAlF$_6$ laser, *Opt. Lett.* **17**, 43.

Zhou, S., Lee, K. K., Chen, Y. C., and Li, S. 1993. Monolithic self-Q-switched Cr,Nd:YAG laser, *Opt. Lett.* **18**, 511.

Further Information

Popular discussions of lasers and their applications and history are given in:

J. Hecht, *Laser Pioneers*, revised ed. Academic Press, San Diego, 1985.

J. Hecht and R. Teresi, *Laser, Supertool of the 1980s*. Ticknor and Fields, New Haven and New York, 1982.

A general discussion of laser science and technology, accessible to anyone with a background of electromagnetic theory, basic quantum mechanics, and calculus, can be found in any of the following:

O. Svelto, *Principles of Lasers*, 3rd ed. Plenum Press, New York and London, 1989.

A. Yariv, *Quantum Electronics*, 3rd ed. John Wiley and Sons, New York, 1988.

A. J. Siegman, *Lasers*. University Science Books, Mill Valley, CA, 1986.

K. Shimoda, *Introduction to Laser Physics*, 2nd ed. Springer-Verlag, New York, 1986.
D. A. Eastham, *Atomic Physics of Lasers*. Taylor & Francis, London and Philadelphia, 1986.
I. P. Kaminow and A. E. Siegman, *Laser Devices and Applications*. IEEE Press, New York, 1973.

Detailed theoretical discussion of lasers is contained in:
H. Haken, *Laser Theory*. Springer-Verlag, New York, 1984.
M. Sargent III, M. O. Scully, and W. E. Lamb, Jr., *Laser Physics*. Addison–Wesley, Reading, MA, 1974.

The underlying physics of laser resonators and its application to the design of many types of lasers in use today is the topic of:
The Physics and Technology of Laser Resonators, D. R. Hall and P. E. Jackson, eds. Adam Hilger, Bristol and New York, 1989.

Specific types of lasers are discussed in:
W. Koechner, *Solid-State Laser Engineering*, 3rd ed. Springer-Verlag, New York, 1992.
Tunable Lasers, L. F. Mollenauer and J. C. White, eds. Springer-Verlag, New York, 1988.
D. C. Brown, *High-Peak-Power Nd:Glass Laser Systems*. Springer-Verlag, New York, 1981.
A. A. Kaminskii, *Laser Crystals*. Springer-Verlag, New York, 1981.
F. J. Duarte and L. W. Hillman, *Dye Laser Principles with Applications*. Academic Press, San Diego, 1990.
Dye Lasers, F. P. Schäfer, ed. Springer-Verlag, New York, 1973.
C. S. Willett, *Introduction to Gas Lasers: Population Inversion Mechanisms*. Pergamon Press, Oxford, 1974.
C. G. B. Garrett, *Gas Lasers*. McGraw-Hill, New York, 1967.
H. C. Casey, Jr. and M. B. Panish, *Heterostructure Lasers*, Academic Press, New York, Parts A and B, 1978.

A fairly comprehensive review of recent work in diode-pumped solid-state lasers is provided by the two articles:
D. W. Hughes and J. R. M. Barr, Laser diode pumped solid-state lasers, *J. Phys. D: Appl. Phys.* **25**, 563 (1992).
T. Y. Fan and R. L. Byer, Diode laser pumped solid-state lasers, *IEEE J. Quantum Electron.* **24**, 895 (1988).

A concise source of data in tabular and graphical form is provided for workers in the areas of laser research and development in the series:
Handbook of Laser Science and Technology, M. J. Weber, ed. CRC Press, Boca Raton, FL, Vols. I and II, 1982; Vols. III and IV, 1986; Vol. V, 1987; Supp. 1, 1991.

9
Optical Modulators

John N. Lee
U.S. Naval Research Laboratory

9.1 Introduction 393
General Modulator Considerations • Optically Addressed Modulators • Electrically Addressed Modulators

9.2 Electro-Optic Devices 398
Electro-Optic Effect and Materials Considerations • Electro-Optic Modulators

9.3 Acousto-Optic Devices 411
Acousto-Optic Effect and Materials Considerations • Acousto-Optic Modulators

9.4 Magneto-Optic Devices 418
Faraday-Rotational Materials and Devices • Magnetostatic Wave Interactions and Devices

9.5 Mechanical and Micromechanical Devices 421
Mechanical Devices • Membrane Devices • Silicon Mechanical Devices

9.6 Thermal Devices 424
Thermoplastic and Oil Film Devices • Phase-Transition Devices

9.7 Specialized and Emerging Devices 425
Direct modulation of Light Sources • Biological Materials

9.8 Summary 428

9.1 Introduction

Optical modulators are devices that alter the temporal and spatial character of a light beam. Such devices are required in virtually all areas of optical technology, particularly, optical communication, storage, signal and image processing, and optical sensing. Other chapters in this handbook provide in-depth description of application areas such as optical communications (Chap. 16), optical data storage (Chap. 17), and optical signal processing (Chap. 14). Many of the technologies and processes used in the construction of optical modulators are important aspects of these other chapters, especially, the chapter on integrated optics (Chap. 11). This chapter will provide a systematic description and categorization of modulators, describing their basic operating principles, current device-performance levels, fundamental limitations, and the present status of specific devices.

A large variety of modulation devices have been investigated (Lee and Fisher, 1987; Yariv, 1989). But, generally, there is no single "best" modulator; the choice of device is highly dependent on the specific application. By categorizing the variety of modulators and their performance parameters, we hope users will gain an understanding of the trade-offs in selecting one device over another. The first broad categorization we provide here is according to modulation mechanism, with a second according to whether the optical modulator is electrically or optically addressed.

We feel that identification of modulation mechanism roughly identifies and bounds a number of parameters important to performance and implementation, e.g., whether the requisite speed, frame rate, and resolution are achievable and whether the device size, power consumption, and mechanical and electrical circuit configuration are compatible with the application. The major categories of modulation mechanisms we use are:

—Electro-optic
—Acousto-optic
—Magneto-optic
—Mechanical and Micromechanical
—Thermal

We also briefly cover some specialized and emerging techniques, such as directly modulated light sources and biological materials.

General Modulator Considerations

The input light to a modulator can be described very generally as a spatio-temporal amplitude and phase function

$$A(x, y, t)\exp\{-j([\Omega t + k_1 x + k_2 y + \phi(x, y, t)]\} \tag{9.1}$$

with the amplitude function $A(x, y, t)$ and the phase information represented by the expression in brackets. Optical modulators, then, generally modify the phase, polarization, and amplitude or intensity of the input light. An optical switch would be the simplest example of amplitude modulation. Optical phase modulation is often desired, and such modulation can be converted into an amplitude modulation by techniques discussed below. Large modulations in phase can also be interpreted as optical-frequency modulation. Finally, modulation of optical wavelengths, i.e., extremely large changes in frequency Ω (~terahertz), can also be desired. Usually it is desirable to modulate only one optical parameter at a time; simultaneous changes in the other parameters must, then, be avoided in an acceptable modulator.

The simplest type of modulator is the temporal-only or point modulator, where the modulation signal is identical within any cross-section perpendicular to the propagation of the light beam. Major objectives for temporal modulators are

—ultrahigh modulation bandwidth,
—linearity of modulation, and
—high efficiency.

Other issues, such as power-handling capability, size, cost, and convenience can also be important. A more general modulation device performs both spatial and temporal modulation; such a device is usually called a spatial light modulator (SLM). With spatial light modulators, there is the additional objective of maximizing, simultaneously, the frame rate and, within a given frame, the spatial resolution (or equivalently, the number of pixels or the space-bandwidth product).

The linear operating range is often limited in modulators. For example, there is often a fundamental sine-squared transfer function, such as might be obtained with crossed polarizers. Quasi-linear operation with a sine-squared device is obtained either by operating in the small-signal regime or by operating around the quadrature ($\pi/4$) point. Predistortion of the input signals might be performed to give a linear output relative to the original input, but this is often

Optical Modulators

difficult in practice, especially, for spatial light modulators. Deviations from linearity will lead to undesired harmonic responses and third-order mixing terms in the output when the modulator is operated at high drive level, over large bandwidth, and with a multicomponent drive signal. These undesirable outputs reduce the usable dynamic range of the modulator and can be quantitatively modeled from Taylor series expansions of the modulator response (Hecht, 1976; Korotsky and DeRidder, 1990; Johnson and Roussell, 1988). Dynamic range is also affected by the efficiency of the modulator. High efficiency usually results in a larger quasi-linear response range.

Fundamental noise limits on dynamic range should always be kept in mind in the application of optical modulators. These can be either thermal noise or quantum noise. Thermal noise limits arise in all electrical components of the modulator, and is given by kTB where k is the Boltzmann constant, T is the absolute temperature, and B is the temporal bandwidth of the drive signal. In the optical domain, photon shot-noise is the fundamental limit and often dominates over thermal noise. The discrete-photon nature of light give a quantum noise limit of hf, where h is Planck's constant and f is the frequency of the light. As an example, at $T = 300$ K thermal noise kT is 4.14×10^{-21} J. For an optical beam at 850-nm wavelength, the photon noise limit is 2.34×10^{-19} J. Thus, for equal amounts of power and the same detection bandwidth, the noise in the optical beam is 18 dB higher.

Spatial light modulation requires additional considerations. Such modulation may be either one-dimensional (1-D) or two-dimensional (2-D). The SLM may be either optically or electrically addressed, as will be discussed below in secs. 1.2 and 1.3. Figure 9.1 illustrates an optically addressed 2-D SLM. In an electrically addressed SLM, the write beam is replaced by electrical signals, such as in matrix-addressed displays and electron-beam-addressed cathode ray tubes. The modulation may be either spatially continuous or within discrete pixels. The performance characteristics of an SLM are strongly related to the spatial dimensions of the data display, e.g., pixel pitch, size, and shape (Turner et al., 1995). If the pixel pitch is d mm, then, the maximum spatial frequency (line pairs/mm) is given by

$$f_n = 1/(2d). \tag{9.2}$$

The information capacity of an SLM of area A, also known as the space-bandwidth product, is given by

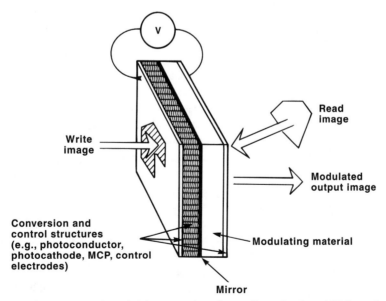

FIGURE 9.1 Generic sandwich construction of a two-dimensional spatial light modulator with optical addressing.

$$N^2 = 4A(f_{\max})^2. \tag{9.3}$$

For applications where the spatial spectrum of the 2-D data is important, a pixelated modulator introduces an envelope function $\mathrm{sinc}(f_x a)\mathrm{sinc}(f_y b)$ onto the spectrum, where a and b are the dimensions of a rectangular pixel, and f_x and f_y are the spatial frequencies in the x and y directions.

For optical processing applications the purpose of an SLM is not only to perform light modulation, but also to provide three-terminal device operation somewhat analogous to that of a transistor, but controlling spatio-temporal optical beams instead of temporal electrical signals. Essentially, the three ports of the three-terminal SLM are the input optical beam (input signal), the readout optical beam (second signal input port and power supply for the output), and the final output beam (output signal). A three-terminal device isolates the input from the output and provides signal gain and storage. In addition, such devices can add, subtract, and multiply arrays of data in parallel and perform linear and nonlinear transformations. If the SLM is optically addressed, it is also possible to perform photodetection and optical conversions, such as from incoherent to coherent light and from one wavelength to another.

The definition of an adequate SLM can vary greatly. The major parameters are resolution, frame rate, and dynamic range. At one extreme is a goal of very high resolution (>million pixels); at another extreme is a goal for extremely high speed (10 ns or less) but at much lower resolution. It is usually not necessary to achieve, simultaneously, the two extreme figures. It can be expected that a variety of SLMs should be developed, each optimal for particular needs. A measure of comparison among the various SLM devices is to look at the product of frame resolution and frame rate, which produces a figure of merit in units of pixels/second. A figure as high as 10^{12} pixels/s can arise in demanding applications. The noise and dynamic range performance for an SLM is affected by a wide range of factors, in addition to the general ones given above. These include spatial uniformity of response, cross-talk mechanisms, such as electrical cross-talk and optical scatter (Brown et al., 1989), and coherent effects, such as speckle. Some analyses have been done for specific SLMs by modeling them as components in a numerical linear system (Taylor and Casasent, 1986; Batsell et al., 1990).

Optically Addressed Modulators

Optical addressing is of concern, primarily, for spatial light modulators. Many of the earliest optically addressed SLMs adopted the basic sandwich structure of Fig. 9.1. A bias voltage applied to the sandwich is shunted within the illuminated regions of the photosensitive layer. In the reflective configuration illustrated, there is a mirror and light-blocking layer at the center of the sandwich, which allows the written input information to be read out by reflection from the modulating-material side of the SLM. An optically addressed SLM, generally, requires a photosensor integrated with the optical material to maximize speed and to reduce drive power and voltages. In some cases, the modulating material can also function as the photosensor (as will be discussed below), but, in most cases, one deposits a photosensitive layer onto the modulating material, using exactly the geometry shown in Fig. 9.1. Photoconductors, such as CdS, PVK:TNF, amorphous Se, CdS, ZnS, and ZnCdS have been deposited onto various modulating materials. The performance of these photoconductors can limit the device performance. For example, photoconductors, such as amorphous Se, exhibit carrier trapping effects that limit frame rates to <1000/s (Armitage et al., 1985). More recent SLM devices have tended to use amorphous silicon (a-Si) or polysilicon thin film transistors (Ashley et al., 1988) and photodiodes (Armitage et al., 1985).

Electrically Addressed Modulators

There are several electrical addressing mechanisms to consider for SLMs.

Electron-Beam

An electron beam, such as in a cathode ray tube, may be scanned across the modulating material or control structures, resulting in an activating electric field. It is also possible for a light beam to be scanned across a photosensitive material to produce electrons that impinge on the modulating material; this case should also be considered electrical addressing, because the scanning action must be electrically controlled.

Electrode Matrix

The electrode matrix can be either one- or two-dimensional. For 2-D SLMs, a rectangular grid of electrodes is placed on the modulating material, and circuits are placed at the ends of row and column electrodes. The orthogonal lines are usually in the same plane, but it is possible to place them on opposite faces of the material. Pixels are addressed at the intersections of the grid lines. A simple "passive matrix" of electrodes suffers from the need to distinguish between a pixel, where both the row and the column electrode are activated (true signal), and a pixel, where only one of the electrodes is activated (half-select); this, generally, requires the material to have a nonlinear response.

Circuitry can also be placed at the grid intersections. Use of arrays of thin-film transistors has been developed in the display industry, i.e., "active matrix" displays, especially for liquid crystal displays and televisions (See Chap. 18). Active matrix addressing eliminates the half-select problem. The transistors are generally fabricated from a-Si:H deposited on a glass substrate; polycrystalline silicon and, less commonly, CdSe are also used. The transistor, then, connects to a transparent electrode over the liquid crystal region.

In principle, it is possible to place more circuitry at the grid intersections using microchip fabrication techniques. This would increase the functionality of an SLM, especially, to perform nonlinear functions, such as logarithm, square root, inversion, hardclipping, switching, and general arithmetic functions, all of which are difficult to perform using only the physical properties of the modulating materials.

Traveling Wave Addressing

Spatial information can be introduced into an optical modulator in the form of a traveling wave. The prime example of this type of addressing is the acousto-optic device to be discussed in sec. 9.3. The spatial information is transitory; the duration depends on the velocity of the wave in the medium.

It is also possible to use a traveling wave to perform the addressing functions in an SLM. Surface acoustic waves in a piezoelectric medium have been primarily employed, because there is a strong acousto-electric field accompanying the wave at the surface of the medium that can be used for activation purposes. Thus, an acoustic impulse can be used to probe or activate a thin-film optical medium (which might also be the piezoelectric acoustic medium).

In the following sections, the various modulators will be grouped according their modulation mechanism. For each group, we shall discuss, first, the modulation mechanism and available materials. Devices are, then, tabulated and described. Within each group, devices are identified as either electrically or optically addressed. Performance numbers can be expected to change with time if a technology is not mature. Hence, the text that corresponds to the tabulations will indicate whether technological limits have been reached and what those limits might be. Also, within the tabulations, we indicate the availability of the listed performance as being commercial, laboratory prototype, laboratory experiment, etc.

9.2 Electro-Optic Devices

Electro-Optic Effect and Materials Considerations

The classic electro-optic effect is an alteration of the optical refractive index due to application of an electric field. This effect is employed in materials without a center of symmetry, as can be identified by crystal class (Koster et al., 1963), and as in an electrically poled semiamorphous material. The application of an external electric field to such noncentrosymmetric materials changes the inverse of the refractive index n according to

$$\Delta(1/n^2)_i = \sum_{j=1}^{3} r_{ij}E_j,$$

$$i = 1, 2, \ldots, 6$$
$$= x, y, z, yz, xz, xy \qquad (9.4)$$

where r_{ij} are called the electro-optic coefficients, and the index i corresponds to the coordinates of the general "index ellipsoid" that specifies light velocity and polarization with direction when an electric field is applied (Yariv, 1989).

Figure 9.2 shows how a change in one of the refractive indices of a birefringent material can be used to produce optical intensity modulation. The birefringent electro-optic material is placed between crossed polarizers as illustrated in Fig. 9.2. A voltage (V) applied to electrodes deposited on the birefringent material produces an electric field within the material. In Fig. 9.2, this field is along the direction of light propagation, although the field may also be transverse to the light as described by Eq. (9.4). The field changes the extraordinary and ordinary indices of refraction (n_e and n_o, respectively) by different amounts. Hence, linearly polarized incident light will exit as elliptically polarized light. Crossed polarizers, then, transmit a fraction of the light intensity, according to a sine-squared function of applied voltage,

$$I_o/I_{in} = \sin^2[(\pi/2)(V/V_\pi)] \qquad (9.5)$$

where the half-wave voltage V_π is as defined following Eq. (9.9) below. Note that Eq. (9.5) represents a nonlinear transfer. The sine-squared function occurs often in optical modulation, where the modulated and unmodulated light beams comprise a two-mode coupled system (Yariv, 1989; also, see Chap. 11).

The origin of the electro-optic effect is the macroscopic polarizability of a material due to an external field E. The polarizability of any medium is obtained by summing over all molecular sites k and is given by

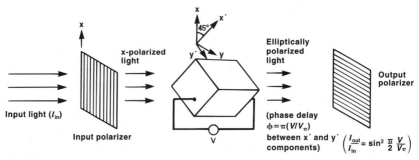

FIGURE 9.2 Intensity modulation with a birefringent electro-optic material between crossed polarizers.

Optical Modulators

$$P = \sum P_k = P_o + \chi^{(1)}E_{opt} + \chi^{(2)}E_{opt}E + RE_{opt}EE + \text{higher orders} \tag{9.6}$$

where E_{opt} is the optical field. The quantity $\chi^{(2)}$ in Eq. (9.5) is a measure of the *linear electro-optic* effect, often referred to as the Pockels effect. $\chi^{(2)}$ is related to r_{ij} by

$$(\epsilon^2/\epsilon_o)r_{ij} = \chi^{(2)}$$

The fourth term in Eq. (9.6), with coefficient R, represents a quadratic dependence on the applied field, often referred to as the Kerr effect. Other triple-field terms are possible if E_{opt} replaces either or both E factors, giving rise to two-beam mixing (or second-harmonic optical generation) and three-beam mixing.

The most familiar and available electro-optic materials are, generally, crystalline inorganics. Lithium niobate (LiNbO$_3$) and lanthanum-modified lead zirconate titanate (Pb$_{1-x}$La$_x$(Zr$_y$Ti$_{1-y}$)$_{1-x/4}$O$_3$, or PLZT) exemplify materials with linear and quadratic refractive-index changes with field, respectively. Table 9.1 lists the most widely used inorganics.

Organic electro-optic materials have recently been developed which try to exploit intrinsic capabilities of certain organic molecules and long-chain polymers to provide orders of magnitude larger effects at lower drive powers than traditional inorganics (Singer et al., 1988; Garito et al., 1984; Singer and Garito, 1981; see Chap. 6). However, a variety of other effects can be included under the electro-optic category. These include

—the photorefractive effect(see Chap. 13), which is a manifestation of the classic electro-optic effect on a microscopic scale (Pepper et al., 1990),
—alteration of optical properties of liquid-crystal materials (both twisted nematic and ferroelectric, e.g., see Chap. 4) (Goodby and Patel, 1986; Blinov, 1983; DeJeu, 1980), and
—alteration of material absorptivity (the imaginary part of the complex index of refraction) in materials, such as GaAs/GaAlAs-layer quantum wells (Miller et al., 1984, see Chap. 1).

Inorganic Materials

Linear Electro-Optic Materials. The basics of classic linear electro-optic modulation in inorganic crystals can be illustrated by considering a uniaxial crystal with the E-field along the z axis ($i = j = 3$). Differentiating Eq. (9.4) gives

$$\Delta n_3 = -(n^3/2)r_{33}E_3 \tag{9.7}$$

Inserting Eq. (9.7) into the phase term of the optical wave

$$E = A \exp\{j(\omega t - \omega n_o z/c)\} \tag{9.8}$$

gives a phase change of

$$\Delta\phi = -(\omega/c)(n^3/2)r_{33}V \tag{9.9}$$

where $V = EZ$ is the applied voltage across a sample of thickness Z. If $\Delta\phi = \pi$, then, $V = V_\pi$ is known as the half-wave voltage. For a two-dimensional SLM, minimum thickness Z is desired to maximize number of pixels (or space-bandwidth), because the minimum pixel dimension is roughly equal to Z due to the fringing electric fields in small spatial features. Thus, a 100-μm thick sample requires $\Delta n = 0.005$ to produce an optical path difference $\Delta nZ = \lambda/2$, i.e., $\Delta\phi = \pi$, at $\lambda = 1000$ nm.

TABLE 9.1 Selected Inorganic Electro-Optic Materials and Properties at 300 K

Linear Electro-optic Materials				
Material	Electro-optic Coefficient	Index of Refraction	$n_o^3 r$ (10^{-12} m/V)	ϵ/ϵ_o
$\lambda = 550$ nm	Linear coeff. (10^{-12} m/V)			
KD*P (KD_2PO_4)	$r_{63} = 23.6$	$n_o = 1.51$ $n_e = 1.47$	80	‖c-50 ⊥c-20
KDP (KH_2PO_4)	$r_{41} = 8.6$ $r_{63} = 10.6$	$n_o = 1.51$ $n_e = 1.47$	(r_{41})-29 (r_{63})-34	‖c-45
$LiNbO_3$	$r_{33} = 30.8$ $r_{13} = 8.6$ $r_{22} = 3.4$ $r_{42} = 28$	$n_o = 2.29$ $n_e = 2.20$	(r_{33})-328 (r_{22})-37	⊥c-98 ‖c-50
$LiTaO_3$	$r_{33} = 30.3$ $r_{13} = 5.7$	$n_o = 2.175$ $n_e = 2.18$	(r_{33})-314	‖c-43
$BaTiO_3$	$r_{33} = 23$ $r_{13} = 8$ $r_{42} = 1640$	$n_o = 2.437$ $n_e = 2.365$	(r_{33})-334	⊥c-4300 ‖c-106
$Sr_xBa_{1-x}NbO_3$ (SBN)	$r_{33} = 450$ for x = 0.75	$n_o = 2.346$ $n_e = 2.310$	(r_{33})-5800	⊥c-3000
$KTa_xNb_{1-x}O_3$ (KTN)	$r_{33} = 1400$ $r_{42} = 3000$	2.25	(r_{33})-16000 (r_{42})-34000	⊥c-3000
Quartz (SiO_2)	$r_{41} = 0.2$ $r_{63} = 0.93$	$n_o = 1.54$ $n_e = 1.55$	(r_{41})-0.7 (r_{63})-3.4	⊥c-4.3 ‖c-4.3
GaP	$r_{41} = 0.97$	$n_o = 3.31$	(r_{41})-29	
ZnTe	$r_{41} = 3.9$	$n_o = 2.79$	(r_{41})-77	7.3
CdTe	$r_{41} = 6.8$	$n_o = 2.60$	(r_{41})-120	
GaAs (10600 nm)	$r_{41} = 1.6$	$n_o = 3.34$	(r_{41})-59	11.5

Quadratic Electro-optic Materials				
$\lambda = 550$ nm	Quad. Coeff. (10^{-16} m^2/V^2)R	Index of Refraction	$n_o^3 R$ (10^{-16} m^2/V^2)	ϵ/ϵ_o
$KTa_xNb_{1-x}O_3$ (KTN)	> 0.8	2.25	9	3000
$Pb_{1-x}La_x(Zr_y Ti_{(1-y)(1-x/4)})O_3$ (PLZT)	3.8, for x = 0.07, y = 0.65	2.55	63	4500

Source: (Kaminow and Turner, 1971; Yariv, 1989)

Regardless of whether an electro-optic material is used in an optically or electrically addressed device (see Sec. 9.1 and Electro-optic Modulators, below, for a discussion of such SLMs), an *RC* time constant can be the limit on the device speed. We can model the electro-optic crystal as a capacitor of area *A*, thickness *Z*, and capacitance *C*. The half-wave charge for this capacitor, Q_π is related to the half-wave voltage V_π by

$$Q_\pi = CV_\pi = (\epsilon A V_\pi)/(4\pi \epsilon_o Z) \tag{9.10}$$

where ϵ is the dc dielectric constant, and ϵ_o is the permittivity of free space. It is instructive to consider the response time for an electrically addressed device where *R* corresponds to an external load resistor R_L. (The optically addressed case can be described by equivalent circuit models). Because $R_L = V/I$, we obtain the dependence of the device response τ on ϵ, V_π and *I* the current in the drive circuit.

Optical Modulators 401

$$\tau = (\epsilon A V_\pi)/(4\pi\epsilon_o Z I) \tag{9.11}$$

In particular, note that the speed of response scales inversely with ϵ.

The drive-power requirements for high-frequency modulation in the electrically addressed case can be derived using Eq. (9.4). The above electro-optic crystal capacitor operating into a load resistor R_L, requires a power to modulate over the bandwidth $\Delta f = 1/(R_L C)$ of

$$P = \left(\frac{V^2}{2R_L}\right) = \frac{(\Delta\phi)^2 \lambda^2 A\epsilon(\Delta f)}{4\pi Z n^6 r^2}. \tag{9.12}$$

Note, in particular, the importance of material properties n and ϵ. The important objective of minimizing P requires maximizing the quantity

$$F = n^3 r/\epsilon. \tag{9.13}$$

F, therefore, represents an important figure of merit for material evaluation. Using a material having small ϵ will not only maximize F, but will also shorten the time response according to Eq. (9.11). Thus, the quantities n, ϵ and $n^3 r$ are listed in Table 9.1.

In Table 9.1 it is worth noting that one of the strongest crystalline electro-optic materials is LiNbO$_3$(r_{33} = 30.8 × 10^{-12} m/V), but this material, with n = 2.29 and ϵ = 98, requires relatively high drive powers. Using Eq. (9.7), the half-wave index change Δn = 0.005, in a 100-μm thick sample of LiNbO$_3$, requires E = 2.5 × 10^7 V/m, or a voltage of 2500V across the sample. SLMs developed using classic electro-optic materials, such as LiNbO$_3$ and potassium dihydrogen phosphate (KDP) or its deuterated version KD*P, show limitations on frame rate from the considerations of Eqs. (9.11) and (9.12), especially when high spatial resolution is desired, i.e., use of very thin samples with large capacitance. If ϵ can be decreased and if electro-optic materials can be produced in thin films, both resolution and speed should be increased. Thin-film deposition of electro-optic materials on a variety of substrates would also allow easier integration with electronic circuits. Efforts to deposit LiNbO$_3$ have continued over the past decade, and progress has been made in major problem areas, such as the tendency for the films to be polycrystalline and for the electro-optic coefficients to be lower than for the bulk materials (Griffel et al., 1989). However, significant work remains to be done at this time.

In the past decade, new inorganics have been developed with high electro-optic coefficients. Most notable among these are tungsten-bronze compounds, e.g., strontium barium niobate, Sr$_x$Ba$_{1-x}$NbO$_3$ or SBN (Neurgaonkar and Cory, 1986), and KTa$_x$Nb$_{1-x}$O$_3$ (KTN) which can be grown in thin-film form on silicon and GaAs.

Quadratic Electro-Optic Materials. A major class of solid-state inorganics that exhibits a strong quadratic effect is lanthanum-modified lead zirconate titanate (Pb$_{1-x}$La$_x$(Zr$_y$Ti$_{1-y}$)$_{1-x/4}$O$_3$ or PLZT) material. Commercially available samples are ceramics composed of hot-pressed microcrystallites. As an example of the strength of the quadratic electro-optic effect, commercial PLZT with x = 0.07, and y = 0.65 has R = 3.8 × 10^{-16} m^2/V^2. Comparing PLZT to a typical linear material, about 100 volts are required for the quadratic material to produce the same effect as 1 volt in the linear material, but if the linear material requires 1000 volts, e.g., such as across a platelet of LiNbO$_3$, the voltage requirement for the quadratic material is comparable (3000 volts). However, the response time of hot-pressed PLZT material is limited to about 50 μs typically, because of its microcrystalline nature, and sets the ultimate limit to frame rate in devices using this material.

As with linear electro-optic materials, it is desirable to produce thin-film, single-crystal quadratic materials. There has been significant progress in thin-film deposition of PLZT and other quadratic electro-optic materials. Single-crystal, micrometer-thick PZT and PLZT films have been produced on crystalline and glass substrates with quadratic behavior consistent with expected bulk behavior (Preston and Haertling, 1992). $KTa_xNb_{1-x}O_3$ (KTN) also exhibits a quadratic effect and, as noted above, has also been grown in thin films. Rf-sputtered films of KTN have been grown on substrates of silicon and GaAs with quadratic coefficients $R > 0.8 \times 10^{-16} m^2/V^2$.

Organic Materials

Organic materials, generally, have lower ϵ than inorganics and, thus, have the potential to lower modulator drive power or to provide higher bandwidths. Optical organic materials are covered in detail in Chapter 6. For optical modulation, these materials possess large polarizabilities arising from large changes between ground and excited states in the electron wave functions of the molecular π bonds, resulting in large second-order molecular susceptibilities. Organic crystals have been investigated extensively (Chemla and Zyss, 1987). The most developed of these is 2-methyl-4-nitroaniline (MNA). Values of $\chi^{(2)}$ of 67 pm/V have been measured for MNA, which compares favorably with $LiNbO_3$. However, this falls considerably short of the potential performance derived from intrinsic hyperpolarizabilities for organic dye molecules, which predict 50 times larger electro-optic coefficients than for MNA (Marowsky et al., 1988).

Major issues, now, with organic crystals include fabrication of large samples with good optical quality, fabrication of thin-film samples (as discussed previously for the inorganics), and material lifetime. Hence, effort has also been directed toward polymer systems. One of the advantages of polymers is the ability to be deposited as thin films on a variety of substrates by spinning techniques. Active species, or chromophores, such as dye molecules are introduced into a polymer host. Early research was performed with guest-host systems, such as red dye molecules like hemicyanine in PMMA (Singer et al., 1988). Material is poled with a strong electric field, such as with a corona discharge, that avoids problems with electric-field breakdown in the material. Poling is performed at a temperature above the glass transition temperature of the polymer to orient all of the electric dipoles, and the electric field remains on as the sample is slowly cooled below the glass transition temperature. Retention of the poled character is an intrinsic problem; the poled state is not one of thermodynamic equilibrium, as the natural tendency is for neighboring dipoles to orient antiparallel. Stability in the guest-host systems has been found to be especially problematic, with less than half of the poling remaining after a few days. More recently, the chromophores have been attached as sidegroups along the length of a stiff polymer chain, such as those in polystyrene, to produce a so-called *copolymer*. Retention of about 90% of the poling after thirty days was demonstrated early (Singer and Garito, 1981). Further progress in lifetime has been achieved through cross-linking polymer chains and use of polymers with glass transition temperatures well above room temperature (Mandal et al., 1991; Stahelin et al., 1992). However, additional work remains to be done at this time to insure lifetimes of many years under conditions of repeated use, elevated temperature, and environmental stress. Most present electro-optic organic polymers have electro-optic performance somewhat less than that of $LiNbO_3$.

Liquid Crystals

For a given applied voltage, liquid crystals produce the largest electro-optic effects of any present material. Liquid crystals are long-chain polymers possessing phases that are uniaxial. These phases include the uniaxial smectic and nematic phases, and the technologically important twisted-nematic phase, which exhibit very strong birefringence. The change in birefringence of twisted-nematic material can be exploited using the crossed-polarizer configuration, shown in Fig. 9.2, to convert polarization changes in the transmitted light into optical intensity modulation. Twisted-nematic materials are widely employed for commercial displays, such as computer screens and hand-held televisions. We will discuss various liquid crystal modulators in Sec. 2.2.2. (See Chap. 18 for a discussion of liquid crystal displays.) Liquid crystal materials can also be made very sensitive to temperature changes. Devices that employ these changes are discussed in sec. 9.6 as thermal devices.

Typically, $\Delta n = n_e - n_o \sim 0.2$–$0.4$ is possible in liquid crystals. Δn generally increases with decreasing wavelength and temperature (Blinov, 1983; Dejeu, 1980). Application of relatively small electric fields, corresponding to a few volts, can change the direction of the axis for the liquid crystal chains, with correspondingly large refractive-index changes. An alternative interpretation is that the field rotates the index ellipsoid. Values of $\chi^{(2)}$ for liquid crystals can be very large, on the order of 10^{-6} m/V. By comparison, values for inorganics are in the range of 10^{-12} m/V. The large field-induced, refractive-index changes in liquid crystals are due to physical movement of the molecular chains. Larger electric fields will speed molecular reorientation somewhat. The electric field induces electric dipoles in the twisted-nematic material. The dipolar field interaction, then, causes the molecular chains to reorient along the field direction. Larger fields increase the dipolar fields and, hence, the speed of response, up to limits due to viscosity, but reversion to the original state must occur naturally at a characteristic relaxation rate that is typically milliseconds. The fastest nematic response observed has been a 1-ms rise time employing a voltage undershoot effect in the nematic material (Sayyah et al., 1992). Hence, there has been great interest recently in ferroelectric liquid crystal (FLC) materials (Clark and Lagerwall, 1980).

Unlike the twisted-nematic material where the dipolar field is induced, the FLC molecules possess permanent electric dipoles, allowing rapid switching between different optical polarization states. Intrinsic speeds at a few volts are typically on the order of a few microseconds, again limited by material viscosity. FLC materials are covered in depth in Chapter 4 of this handbook. Two types of material have been employed–smectic C and smectic A. The surface-stabilized Smectic C FLC material constrains the molecular director (the average direction of the elongated liquid-crystal molecules) to only one particular value of tilt angle from the axis of symmetry. Material with a tilt angle of 22.5° has been extensively employed in actual devices. This tilt angle defines a polarization axis. The material is used as a switchable, half-wave plate between crossed polarizers. An applied field switches the optic axis of the half-wave plate (or equivalently the molecular director of the FLC) through 45°, i.e., twice the tilt angle, so that light polarized along the original axis has its polarization rotated by 90°. When placed between crossed polarizers, optical transmission is switched between on and off states. The contrast ratio depends on how perfectly the FLC behaves as a half-wave plate. This tends to be a more complex manufacturing problem for the FLC material than for inorganic electro-optic materials. For example, the thickness and composition of the FLC film must be carefully controlled or else plane-polarized input light gives elliptically polarized output light.

The smectic C material, operated as a half-wave plate, provides only two- or three-state (binary or ternary) intensity modulation, although devices can be designed to use either time- or space-multiplexing to provide the equivalent of grey scale (See Electro-optic Modulators, below). The smectic A material allows a continuous variation of the optic-axis tilt angle with applied voltage (Anderson et al., 1988), so that devices with analog intensity modulation are possible using this material in conjunction with a polarizer. Increased dynamic range (number of gray levels) and temperature sensitivity are present areas of development for smectic A devices. To increase the dynamic range, the maximum tilt angle of smectic A materials needs to be increased. Presently, tilt angles of 18° have been achieved. However, the goal is a tilt angle of 22.5° because this angle would allow full on-off intensity swing in a single film when placed between crossed polarizers. The induced tilt angle is a strong function of temperature, so that device temperature might have to be controlled to have reproducible grey levels. Finally, although a large number of analog gray-scale levels is desirable, it is also desirable to keep activation voltages below 5 volts. Such voltages may limit the number of gray levels. Because smectic A materials are not yet fully optimized, the smectic C devices are much more available now.

Photorefractive Materials

The discussion, so far, of inorganic, organic, and liquid crystal electro-optic materials relates to optical changes induced in bulk samples, such as via electrodes attached to the sample. It is also possible to consider the classic linear electro-optic effect on a microscopic scale. Charge (electrons or holes) is produced at illuminated regions of certain photoconductive materials. Electric fields, then, cause the charge to be transported to nonilluminated regions where they become trapped. The resultant charge pattern produces a corresponding spatial electric-field pattern that alters the refractive index of the material (Pepper et al., 1990), and these refractive-index changes are optically read out, resulting in a very simple SLM structure. Use of such *photorefractive* materials in optically addressed SLMs is particularly attractive, because the material is both the photosensor and the modulator. Optical control of other optical beams occurs via interaction at the atomic and molecular level. Photorefraction is discussed in detail in Chapter 13 of this handbook.

The first materials, where the photorefractive effect was observed, were $LiNbO_3$ and $LiTaO_3$ (Chen, 1969). The effect was observed in both the pure and doped materials. The charge transport mechanism in the pure material results from electron charge cloud distortions in the vicinity of the ions. For material doped with iron, it was found that photoconductivity causes charge transfer between Fe^{2+} and Fe^{3+} impurity sites. Since these early findings, other ferroelectrics exhibiting photorefraction have been extensively studied. The most popular of these are $BaTiO_3$ (White and Yariv, 1984) and the tungsten-bronze compounds, e.g., strontium barium niobate, $Sr_xBa_{1-x}NbO_3$ (Rakuljic et al., 1988; Neurgaonkar and Cory, 1986).

Very efficient photorefractive materials have been developed from photoconductors with high charge mobility. The photoconductor that has undergone the most study is bismuth silicon oxide ($Bi_{10}SiO_{12}$ or BSO)(White and Yariv, 1984). The drift velocity of the charge can be increased by applying large electric fields, so that charge separation is maximized and response times of <1 ms are possible.

An important consideration for photorefractives is wavelength of activation. The bandgap for BSO is relatively large, so that visible wavelengths are required to induced photoconduction. It is desirable to have materials responsive in the near-IR region illustrated by semiconductor-laser wavelengths becoming more common in optical processing systems (Lee and Fisher, 1987). Doped semiconductors have, therefore, been investigated as photorefractives. CdTe:V has been demonstrated at 1500 nm (Partovi et al., 1990), and GaAs:Cr (Albanese et al., 1986) and InP:Fe (Valley et al., 1988) have been demonstrated at 1060 nm.

Readout of photorefractives must be done at a longer wavelength than for activation to avoid further charge generation that would erase the stored information.

III-V Compounds and Multiple Quantum Well (MQW) Materials

Application of electric fields to semiconductors in the III-V class can strongly alter the phase and amplitude of transmitted light. In particular, it is possible to obtain strong electroabsorption, i.e., change in absorption coefficient κ with applied field, using quantum-well structures. III-V semiconductors, of great importance both in electronics and photonics, are discussed in detail in Chapter 1 of this handbook. They include basic compounds formed from group III and V elements of the Periodic Table, such as GaAs, GaP, and InP, and also the ternary compounds, like GaAlAs, and quaternary compounds, like InGaAsP. The percentages of the various group III and V constituents can be adjusted to vary the bandgap and refractive index of the material, and the varied compounds can be layered to produce LEDs and lasers. Because the bandgaps of these materials can be altered, optical devices can be made for various wavelengths. For example, devices with GaAlAs/GaAs layers generally operate at 850 nm, whereas InGaAsP/InP devices can be made for operation at either the 1300 or 1500-nm region to match wavelengths of interest for optical fiber systems.

Heretofore, the optical index of refraction has been considered a real quantity. If the index is defined as a complex quantity, the optical field represented by Eq. (9.6) can be rewritten as

$$E = A \exp\{j[\omega t - \omega n_o(1 - j\kappa)z/c]\}. \tag{9.14}$$

If the refractive index is redefined as

$$n' = n_o(1 - j\kappa), \tag{9.15}$$

the complex component corresponds to absorption in the medium. Hence, an absorption change induced by an applied field should also be considered an electro-optic effect under this generalized definition. Further, it is important to remember that a change in the real part of the refractive index accompanies the change in absorption coefficient. Changes in the real and imaginary components of n are related by the well-known Kramers–Kronig relationship. Hence, phase modulators are also possible, often operating off the peak absorption point but where significant phase change is still possible.

Until recently, it has been difficult to obtain large changes in absorption in bulk GaAs through mechanisms such as the Franz–Keldysh effect, a red shift of the band edge to longer wavelengths with an applied electric field. However, with multiple quantum-well structures, orders of magnitude improvement have been achieved (Miller et al., 1984). Quantum wells are very thin ($\sim$10-nm) layers of various III–V materials, grown by methods such as molecular or chemical beam epitaxy. For multiple quantum-well (MQW) structures there are two or more types of doped or undoped layers that are stacked in some alternating scheme. Excitons in the lower bandgap layers are quantum-confined, so that the binding energy between electrons and holes is increased. As a result, when electric fields are applied, the excitons do not become ionized, and the exciton absorption line remains sharp. Band shifting stills occurs with application of the field, and large changes in absorption are possible. The absorption change is maximal at an optical wavelength that depends on the material composition. As an example, for a multiple quantum-well stack of alternating GaAlAs and GaAs layers optimized at 850 nm wavelength, the absorptivity α changes from 6×10^3 to 8×10^3 cm^{-1} (Aull et al., 1988). Thus, even with the larger effects with quantum wells, the values of α still require relatively thin layers to minimize the residual absorption. As a result, contrast enhancement is often needed. One enhancement technique is to fabricate the MQW material into an asymmetric Fabry–Perot, although this can limit spatial resolution and narrows the wavelength band of operation. For optical addressing at high resolution and frame rate, it is possible to use a two-beam coupling configuration between reference and signal beams to create transient spatial gratings within the material (Bowman et al., 1994). These gratings, then, diffract a third read-out beam.

A potential attraction of modulators fabricated with III–V materials is the hope that light sources, modulators, and drive circuits can all be fabricated from these materials and integrated on a single chip. This integration has not yet been totally achieved, because the fabrication processes for the three types of devices are either not entirely compatible or the entire fabrication process for an integrated chip is complex compared to current capabilities.

Electro-Optic Modulators

Temporal Modulators

Major objectives in temporal electro-optic modulators are ultrahigh modulation bandwidth, linearity, and high sensitivity (or low drive power). The primary technique for achieving high bandwidth with good depth of modulation without inordinately high drive power is to use integrated- or guided-optical-wave devices. Integrated optic techniques and devices are described in detail in Chapter 11 by Alferness. Bulk modulators consisting of discrete samples of electro-optic material in the configuration of Fig. 9.2 (to obtain amplitude modulation) are possible, but only with high drive power and voltages, as determined by Eq. (9.12). In particular, note that the cross-sectional area A for bulk samples is, generally, much larger than optical waveguide thicknesses which are a few micrometers maximum. Efficient phase modulation in integrated-optic devices results from confinement of the optical wave and from matching optical velocities with electrical-signal velocities in transmission-line electrodes along a long section of guide (~cm). Construction of two-arm waveguide interferometers, such as a Mach–Zehnder interferometer with one modulated arm, converts the phase change produced by the electro-optic effect (see Eq. (9.9)) into an amplitude-modulated output. (See Chap. 11 for further description of such interferometers.) Performance in excess of 50 GHz has been shown in LiNbO$_3$ devices (Dolfi and Ranganath, 1992; Seino et al., 1988). Use of organic electro-optic materials, which have comparatively lower values of ϵ, can theoretically result in devices with lower drive power or performance at higher bandwidths, up to 100 GHz.

Temporal modulation can be obtained with integrated-optic devices in III-V materials employing either phase modulation (Wakita et al., 1992) or electroabsorption (Suzuki et al., 1992; Devaux et al., 1994) in multiple quantum wells. Direct intensity modulation bandwidth as high as 18 GHz has been demonstrated with electroabsorption in InGaAsP, and the use of a semiconductor amplifier within the device allows low-loss operation (Devaux et al., 1994). Greater than 20 GHz phase modulation has been obtained in InGaAlAs/InAlAs modulators (Wakita et al., 1992).

Using either a single Mach–Zehnder integrated-optic or a bulk phase modulator in the configuration of Fig. 9.2 produces an optical intensity output whose sine-square varies with drive signal, as in Eq. (4.5). Quasi-linear operation is obtained either by operating in the small-signal regime or by operating around the quadrature ($\pi/4$) point. Predistortion of the input signals might be performed to give a linear output relative to the original input (Childs and O'Byrne, 1990), but this is often difficult in practice. Operating within the quasi-linear regions has additional negative consequences: for a broadband, multiple-component, input signal the sine-squared transfer function produces spurious in-band intermodulation-product signals. The usable dynamic range is, therefore, reduced. The production of spurious third-order signals can be seen from a Taylor series expansion of the modulated quadrature-point signal

$$1 \pm \sin\{(\pi V(t)/V_\pi)\}$$

as

$$1 \pm (\pi V(t)/V_\pi) \pm [(1/6)(\pi V(t)/V\pi)^3 \pm \text{higher order terms}].$$

The nonlinear third-power term will cause a triple-product mixing of the frequencies of two different signal components. Linearized modulators that greatly reduce third-order distortion have been demonstrated using multiple Mach-Zehnder devices, either in parallel or in series (Korotsky and DeRidder, 1990; Johnson and Roussell, 1988). The third-order term in the Taylor series can be canceled to more than 20 dB by carefully controlling the splitting of the optical power and the RF drive signal to each of the Mach–Zehnder devices (Skeie and Johnson, 1991).

Spatial Light Modulator Devices

In this section we will tabulate the most important of the electro-optic spatial light modulators (SLMs) and provide some descriptions of optically and electrically addressed SLM devices to guide the reader through Table 9.2.

Table 9.2 summarizes the performance of the major electro-optic SLMs. The performance parameters are defined as follows:

—Space-bandwidth/resolution is that at 50% MTF.
—Frame rate is for the space-bandwidth/resolution indicated.

The organization of the Table is as follows:

Optical Modulators

TABLE 9.2 Selected Electro-Optic Spatial Light Modulators and Performance

Device/Name	Address Method	Frame Rate (Hz)	Space or Time Bandwidth/Resolution (pixels/mm)	Remarks
KD*P (KD$_2$PO$_4$)/ Titus	Electrical e-beam	30	20/mm	Limited commercial availability
KD*P (KD$_2$PO$_4$)/ Phototitus	Optical Se film Si diodes	500 500	15/mm 10/mm	Limited commercial availability
Bismuth silicon oxide (BSO) or bismuth germanium oxide (BGO)/ PROM, PRIZ	Optical BSO, BGO	500	10/mm	Limited commercial availability
LiNbO$_3$/ MSLM (Hamamatsu, Optron)	Electrical e-beam (microchannel plate)	100	10/mm	Long storage times due to large DC dielectric constant (days to months); commercially available
LiNbO$_3$/ MSLM (Hamamatsu, Optron)	Optical photocathode + microchannel plate	100	10/mm	very high optical sensitivity: 10 pJ/cm^2; Long storage times: days to months; commercially available
LiNbO$_3$/ TIR (Xerox)	Electrical Si Chip	10,000	20/mm	commercially available 1-D modulator
PLZT (Univ. CA, San Diego)	Electrical/Optical Si transistor/ phototransistor	10,000	10/mm	Laboratory prototype; analog modulation
Twisted nematic liquid crystal/ LCTV	Electrical thin-film transistor arrays	30	typically 640 × 480 pixels; > 1000 × 1000 displays possible	Widely available and inexpensive
Twisted nematic liquid crystal/ LCLV (Hughes, Hamamatsu, Control Optics, Thomson CSF)	Optical CdS/CdSe BSO Si diode array amorph. Si	30	30/mm	Instrinsic decay time limits frame rate; Commercially available
Ferroelectric Liquid Crystal/ (Displaytech, Boulder Nonlinear Systems)	Electrical	10,000	256 × 256 space bandwidth	commercially available; binary Modulation/ analog with space/ time multiplexing

TABLE 9.2 Selected Electro-Optic Spatial Light Modulators and Performance—(*continued*)

Device/Name	Address Method	Frame Rate (Hz)	Space or Time Bandwidth/Resolution (pixels/mm)	Remarks
Ferroelectric Liquid Crystal/ (Rockwell, Hamamatsu, Control Optics Univ. Colorado, Displaytech, CTL Smectic)	Optical	10,000	32/mm	Commercially available; binary Modulation.
GaAs, Multiple Quantum Well (NRL)	Optical	6×10^5	160/mm	Laboratory Prototype
GaAs, Multiple Quantum Well (Martin-Marietta, MIT Lincoln Labs)	Electrical Si chip	10,000	64×64	Laboratory Prototype
GaAs, Multiple Quantum Well/ SEED array (Bell Labs)	Optical	10^6	16×16	Limited commercial availability. High power required 100 MHz capability
GaAs, Multiple Quantum Well	Electrical surface acoustic wave (SAW)	10^5	1000 spots time-bandwidth	1-D device Laboratory Prototype
Photorefractive LiNbO$_3$	Electrical readout: SAW optical writing	10^5	1000 spots time-bandwidth	1-D device Laboratory Prototype

—The SLM is identified first by modulating material.
—If there is a common name for the modulator, it is, then, given.
—If the device is commercially available, then, several current manufacturers are listed in parentheses.
—The SLM is, then, identified as either electrically or optically addressed.
—If more than one method exists for addressing, then, performance figures are provided for each.
—Unique aspects of particular SLMs are provided under "Remarks". Availability and development status are also given.

Optically Addressed Devices. We will now discuss the various optically addressed devices that employ electro-optic modulating materials.

Electro-optic inorganic crystals are employed in various devices, such as KD*P in the Phototitus (Donjon et al., 1973; Casasent, 1978a; Armitage et al., 1985) and LiNbO$_3$ in the microchannel-plate spatial light modulator (MSLM) (Schwartz et al., 1985). These devices use photoconductors and photocathodes, respectively, as the light-sensitive element. The microchannel plate in the MSLM multiplies the photoelectrons from a photocathode, allowing an alternate means of signal gain (other than a stronger read-out beam).

Twisted-nematic, liquid-crystal devices are currently the most available of all SLMs. The twisted-nematic, liquid-crystal light valve (LCLV) SLM, first developed more than a decade ago (Bleha et al., 1978; Augborg et al., 1982), is an optically addressed device using a layer of CdS or CdSe as the photosensitive layer. Resolution and sensitivity depend on film quality. Alternatively, an array of silicon Schottky diodes has been used for an optically addressed LCLV (Sayyah et al., 1991) where resolution is limited by the Schottky diode periodicity.

Binary-only ferroelectric devices have been optically addressed with a photosensitive layer of hydrogenated amorphous silicon (a-Si:H) over the ferroelectric liquid crystal (Jared and Johnson, 1991; Roe and Schehrer, 1993).

Ferroelectric ceramics, such as in the PLZT devices (Land, 1978) can be optically addressed by overlaying the material with an array of phototransistors. Islands of recrystallized polysilicon are deposited on a PLZT substrate. Each Si island contains a phototransistor which controls the potential across the surface of an adjacent region of the PLZT (Lee et al., 1986). A read beam transmitted through this PLZT region and on through the substrate is, thus, modulated by the transverse electro-optic effect. Alternatively, reflective read-out can be obtained by adding a mirror and using a two-pass read-out beam configuration, and PLZT islands can be deposited on a silicon chip (Jin et al., 1994).

The electro-optic modulating material itself can also act as the photosensor. The Pockels read-out optical modulator (PROM) (Horowitz and Corbett, 1978; Owechko and Tanguay, 1982) and the PRIZ (Petrov, 1981) device use photoconductors, either BSO ($Bi_{12}SiO_{20}$) or BGO ($Bi_{12}GeO_{20}$). In the PROM, the modulating material is sandwiched between two transparent electrodes to which an external field bias voltage is applied. In the PRIZ, the fields are transverse across the writing face. This configuration was shown to be useful for edge enhancement of images and to have better resolution than the PROM configuration. The bias voltage causes drifting of the photocharge produced by an optical writing beam with wavelength shorter than the material bandgap. Thus, a spatial refractive-index pattern corresponding to the input pattern is produced by the electro-optic effect, i.e., photorefraction. To avoid producing more photocharge during readout of the device, a beam at a wavelength longer than that corresponding to the bandgap is used. Hence, it is also possible to use such devices as wavelength converters. Other photorefractive materials that have been employed include $BaTiO_3$, KTN, SBN and iron-doped $LiNbO_3$.

Optical activation of electroabsorption changes in III-V MQW structures can be obtained via the photovoltages from a photosensor array. Optical readout of the MQW material is often enhanced by using the material within an asymmetric Fabry–Perot cavity to maximize the electroabsorption. However, this results in a very narrow operating optical bandwidth, and the spatial extent of the individual Fabry–Perot cavities limits device resolution. Another means of optical addressing and readout is to use two-beam coupling between a reference and signal beam to create transient spatial gratings in the MQW material that, then, diffract a third laser beam (Bowman et al., 1994). High spatial resolution (<6 μm) and diffraction efficiency ($>1\%$) is obtained by proton implantation in buffering surface layers to prevent transverse movement of charge, whereas speed is unaffected because the nonimplanted, optically active layer retains high charge mobility across the layer. Writing/erasure at rates $>600,000$ frames/s have been shown possible with a 1-μm thick sample.

A major application of III-V quantum-well materials for SLMs has been in constructing arrays of self-electro-optic devices (SEEDs) which can provide nonlinearities needed for optical logic or threshold functions. The construction of a single SEED is shown in Fig. 9.3a. A MQW structure is sandwiched between layers of bulk GaAs. The interfaces between the MQW section and the bulk form p-n junctions. The SEED, thus, acts as a photodetector of the incident light. However, the absorption of light produces a change in the bias voltage on the MQW, altering the optical transmission of the device. The transfer function for incident intensity versus output intensity has a bistable character, as shown in Fig. 9.3b.

We note that a major attraction of the MQW type of SLM is the potential speed of response. The band-shifting mechanisms occur on a picosecond time frame. However, the speed of actual devices depends on the optical power levels, and the achievable speeds are, thus, limited by the

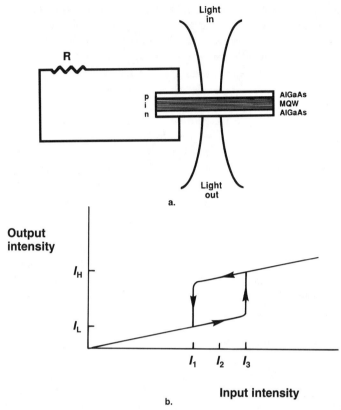

FIGURE 9.3 Construction of a) a self-electro-optic device (SEED) for b) bistable intensity modulation.

capacity for the device to dissipate energy. For example, if a SEED device can dissipate a maximum of 1W/cm^2, a maximum speed of 100 MHz results (Yu and Forrest, 1993), independent of various device constructs, such as symmetric SEEDs (S-SEEDs), whose output is proportional to the difference in optical power between two beams.

Electrically Addressed Devices. Electrically-addressed SLMs can employ the same electro-optic materials as the optically addressed SLMs. Any material-dictated performance properties will be common to the two.

Electron-Beam Devices. A scanning electron beam, as in a cathode ray tube, is used with electro-optic crystals, such as KD*P in the Titus (Groth and Marie, 1970; Casasent, 1978b) and LiNbO$_3$ in the MSLM (Schwartz et al., 1985). In the latter, faster writing is achieved with a microchannel plate that amplifies the electron beam (rather than photoelectrons as in the optically-addressed MSLM).

Electrode Matrix Devices. An electro-optic modulator with a 1-D array of electrodes is the total internal reflection (TIR) modulator. A VLSI silicon ship, with as many as 5000 electrodes with driver circuitry, is overlaid onto an electro-optic plate of LiNbO$_3$ (Johnson et al., 1983).

Representative of the 2-D active and passive matrix-addressed devices are the various liquid-crystal televisions (LCTVs) (Blechman, 1986) and displays. LCTVs have been modified to perform as inexpensive SLMs (McEwan et al., 1985; Liu et al., 1985). However, it is important to realize that display devices are not specifically built for use with coherent light. For example, their flatness

Optical Modulators 411

and cosmetic qualities may not be adequate for maintaining optical amplitude and phase. A radius of curvature of less than 10 m will reduce the peak of a correlation function (relative to its sidelobes) by more than 10% (Turner et al., 1994). Further, the large size of the pixels in liquid crystal displays can result in poor diffraction efficiency, as discussed in Sec. 9.1.

If large electrode matrices are employed, it is possible to use spatial multiplexing to obtain effective grey-scale display from materials that provide only binary response. For example, a 4 × 4 "super pixel" would provide 16 grey levels. If high-speed response is available from both the electrode matrix and the material, e.g., ferroelectric liquid crystals (compared to twisted nematic), time multiplexing is also possible to provide effective grey scale for applications that temporally integrate light signals, i.e., light is switched on for varying times up to the integration period. The maximum number of grey levels is the ratio of the integration period to the device response time.

In the 2-D active matrix devices, placing more circuitry at the grid intersections, as described in Sec. 9.1, makes possible an SLM with "smart pixels". Such SLMs are optically addressable if the circuitry includes a photodetector with drive amplifier for modulation, such as with the PLZT device mentioned above (Lee et al., 1986).

Traveling Wave Addressing. Acoustic-wave devices have been used for fast serial addressing of electrical signal-storage devices (Cafarella, 1978), and of a photorefractive optical memory device (Lee, 1983). More recently, addressing of a multiple, quantum-well, optical modulator has been demonstrated using surface acoustic waves (SAWs) (Jain and Bhattacharjee, 1989). The device material is essentially the p-i-n GaAs/GaAlAs structure of Fig. 9.3a. Because GaAs is a material that is both piezoelectric and electro-optic, it can be addressed with the electric fields that accompany the SAW, i.e., the acoustoelectric field. A SAW is propagated in one of the bulk GaAs layers, resulting in electroabsorption modulation of the MQW layer. Limitations of such acoustic-wave devices are that they are intrinsically 1-D and information continuously flows through the device at the velocity of sound. These limitations are the same as those discussed in sec. 9.3 below for acousto-optic devices.

9.3 Acousto-Optic Devices

Acousto-Optic Effect and Materials Considerations

Acousto-Optic devices use the interaction between light and refractive-index changes induced by an ultrasonic wave (Uchida and Niizeki, 1973). Such devices have been very effective as temporal point modulators and one-dimensional (1-D) spatial light modulators for optical processing systems (Berg and Lee, 1983).

In acousto-optic materials, the ultrasonic wave induces refractive-index changes (Δn) via the photoelastic effect. These index changes have the periodicity, amplitude, and phase modulation of the acoustic wave, and they act as a phase grating to diffract the incident optical beam. As in any grating, there are, generally, multiple diffraction orders, but the generation of multiple orders is usually not technologically desirable. This Raman–Nath regime (Lee and VanderLugt, 1989) arises when the grating is weak (Δn small) and the interaction distance short. Hence, the interaction length is increased. With a sufficiently thick diffraction grating (thickness Z) at a particular incident angle, only one diffraction order persists, with destructive interference occurring at all other diffraction angles. In an isotropic medium, maximum diffraction efficiency is obtained when the incident light direction is at an angle θ_B to the acoustic wavefront. The diffracted light is also at θ_B, i.e., deflected by an angle $2\theta_B$. (The case of anisotropic media will be discussed.) This mode of diffraction is called the Bragg regime, analogous to the Bragg effect occurring with X-ray diffraction in crystal lattices. θ_B is called the Bragg angle and, for the isotropic case, is given by (Berg and Lee, 1983)

$$\sin \theta_B = \lambda/(2\Lambda) = \lambda f/2v \qquad (9.16)$$

where λ is the optical wavelength, Λ is the acoustic carrier wavelength, v is the acoustic velocity and f is the acoustic frequency. Bragg diffraction is, therefore, the basis of all acousto-optic light devices.

The basic construction of an acousto-optic device is shown in Fig. 9.4. The device is electrically addressed by a temporal signal. A piezoelectric material is used to construct an electrical-to-acoustic transducer that is attached to the acousto-optic medium. The transducer is driven by signal $r(t)$, giving the acoustic space/time signal $r(t - x/v)$. The thinner the transducer, the higher the operating frequency (Young and Yao, 1981). Present mechanical thinning techniques allow devices that operate at RF frequencies up to 4 GHz. However, acoustic attenuation (discussed below) may limit frequency before this point.

The medium is illuminated by a light beam of wavelength λ that may be modulated in both space and time. The thickness of the acoustic signal Z is usually made sufficiently large so that one operates in the Bragg regime. Quantitatively, we characterize the transition from multiple-order diffraction (Raman–Nath regime) to the Bragg regime by the expression (Klein and Cook, 1967)

$$Q = (2\pi\lambda Z)/n_o \Lambda^2 \qquad (9.17)$$

For $Q > 7$, the intensity of the first diffraction order exceeds 90% of the maximum.

Acousto-optic modulators can also be fabricated using thin-film optical waveguide structures (Berg and Lee, 1983). Then, surface acoustic waves (SAWs) are employed. A SAW transducer is deposited on top of the optical waveguide (see Chap. 11). Materials that are both piezoelectric and photoelastic are favored, because this results in the simplest construction. Hence, materials, such as $LiNbO_3$ and $LiTaO_3$, are popular for acousto-optic-SAW devices. Limitations on minimum photolithography linewidth in the SAW transducer structure presently restrict these devices to about 1 GHz operating frequency.

In addition to a deflection of the beam in the acousto-optic interaction, there is also a multiplication of the data on the light carrier by that on the acoustic wave. The multiplicative operation of an acousto-optic cell is illustrated in Fig. 9.4, where an electrical signal $r(t)$ is transducted into an acoustic signal which then diffracts an spatio-temporal optical signal $s(x, t)$ at wavelength λ. The light amplitude and phase are multiplied by those of the acoustic wave.

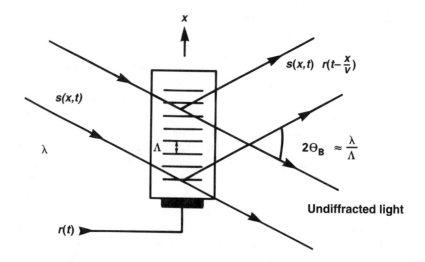

FIGURE 9.4 Multiplication operation with an acousto-optic cell.

Further, the frequency of the light is shifted by the frequency of the acoustic wave in a direction so as to conserve momentum in the phonon-photon interaction, or equivalently, to obey the Doppler-shift relationship. The diffracted light, therefore, represents single-sideband modulated information. It is possible to have either Doppler upshift or downshift in the diffracted light. The upshift case is shown in Fig. 9.4, whereas the downshift case occurs for $-\theta_B$ angle of incidence. For a given drive signal, the data on the diffracted light in these two cases are complex conjugates of each other. The drive signal $r(t)$ must be a real signal, and it must be on a sinusoidal carrier, because acousto-optic cells cannot support low temporal or spatial frequencies (Lee and VanderLugt, 1989). However, $r(t)$ can represent complex quantities (Berg and Lee, 1983, Chap. 11); for example, $r(t)$ can be made the sum of two real components whose carriers are in quadrature.

Aside from the traveling-wave character of the index changes, the optical modulation is entirely analogous to the method employed on many SLMs using data encoded on a spatial carrier. The resolution in an acousto-optic modulator is usually given by the time-bandwidth product of the device. The product of the time aperture (T) and the bandwidth (B) of the acousto-optic cell gives the number of deflection spots possible in an acousto-optic deflector. A diffraction-limited spot has an angular extent given by

$$\Delta\theta = \lambda/D = \lambda/(vT) \tag{9.18}$$

but the total angular capability of the device, from Eq. (9.16), is given by

$$\lambda\Delta f/v = \lambda B/v. \tag{9.19}$$

The ratio is the number of spots, or TB.

Material parameters strongly affect achievable device diffraction efficiency, bandwidth, and/or aperture size. Most good acousto-optic materials have been identified by noting the strong dependence of various figures of merit on the index of refraction n and acoustic velocity v. The popular figure of merit M_2 is obtained from the expression for diffraction efficiency (Dixon, 1967b),

$$I_o/I_{in} = C_o \sin^2\{\sqrt{M_2(\pi^2/2\lambda^2)(L/H)C_{RF}P_{RF}}\} \tag{9.20}$$

where C_o and C_{RF} are constants relating to gain/loss in the optical and RF systems, L and H are the acoustic transducer dimensions along and perpendicular to the light, respectively. The sine-squared relationship again arises from a coupled mode formalism, as described in Sec. 9.2 for electro-optic devices (Yariv, 1989; also see Chap. 11). The constant M_2 is related to material properties by (Uchida and Niizeki, 1973)

$$M_2 = (n^6 p^2)/(\rho v^3) \tag{9.21}$$

where p is the photoelastic constant and ρ is the material density.

Table 9.3 summarizes properties and recent device performance for the most important existing and emerging acousto-optic materials. It is worth noting that it is not always desirable to focus on the optimum for any one single parameter. For example, LiNbO$_3$ and TeO$_2$ are popular in large measure due to their availability, in spite of their being nonoptimum in figure of merit and attenuation, respectively. Availability of material means that good, homogeneous optical transmission, low defect levels (to minimize optical loss), and low optical scatter are routinely obtained (Brown et al., 1989). Note that, in Table 9.3, values of M_2 are specified at particular wavelengths at room temperature. Eq. (9.21) shows that variations with wavelength and temperature in quantities, such as n, can have a large effect on M_2.

TABLE 9.3 Selected Acousto-Optic Materials and Properties at Room Temperature

Material	Index of Refraction for Optical Mode	Acoustic Mode and Velocity (10^5 cm/sec)		Figure of Merit, M_2 (10^{-18} s³/g)	Acoustic Attenuation at 500 MHz (dB/μsec)
$\lambda = 633$ nm					
Fused silica (SiO2)	1.46-⊥	longitudinal	5.96	1.56	1.8
Dense flint glass (SF-59)	1.95-∥	long.	3.26	19	
LiNbO$_3$	2.20-∥	long.[100]	6.57	7.0	0.03
		shear[100]	3.60	13.0	
TeO$_2$	2.27-⊥	long.[001]	4.26	34.5	1.6
	2.27-∥	shear[110]	0.617	793	73.
GaP	3.31-∥	long.[110]	6.32	44.6	0.6
	3.31-arb.	shear[100]	4.13	24.1	
TiO$_2$	2.58-[010]	long.[100]	8.03	3.9	0.11
PbMoO$_4$	2.39-⊥	long.[001]	3.66	36.1	1.4
	2.26-∥			36.3	
H$_2$O	1.33-arb.	long.	1.5	126	90.
$\lambda = 1150$ nm					
As$_2$S$_3$ glass	2.46-∥	long.	2.6	347	1.7
Tl$_3$AsS$_4$	2.63-∥	long.[001]	2.15	510	0.3
$\lambda = 10,600$ nm					
Ge	4.00-∥	long.[111]	5.50	840	1.4
	4.00-arb.	shear[100]	4.13	290	0.7
Te	4.80-∥	long.[100]	2.2	4400	6.

Notes: —M_2 varies with wavelength and temperature, due to dependency of the parameters in Eq. (9.21) on such.
—∥ and ⊥ indicate optical polarization directions with respect to acoustic-wave propagation direction
Source: (Uchida and Ohmachi, 1969; Dixon, 1967 a,b; Coquin et al., 1971; Singh and Todd, 1990; Blistanov et al., 1982)

To find the maximal values of M_2 for a given material requires detailed calculations. The stiffness tensor of a material is used in solving the Christoffel matrix (Dieulesant and Royer, 1980) giving acoustic-mode velocities and polarizations. The strain tensor associated with these modes is used to calculate a strain-perturbed index ellipsoid which in turn is related to an optical polarization direction to give an effective photoelastic constant and, hence, the resultant M_2. Even though there are only a few good acousto-optic materials, calculations are needed for each to identify optimum acoustic modes and directions for particular applications. Hence, Table 9.3 provides data on different acoustic modes for a given material.

Also listed in Table 9.3 is acoustic attenuation, another material parameter of importance to acousto-optic device design. Attenuation of the acoustic wave limits the center frequency, bandwidth of operation, and the aperture size. As noted above, the number of spots in a deflector is a product of the time aperture and the bandwidth of the deflector. Further, high attenuation can result in heating of the acoustic material, which, in turn, can lead to effects, such as change in the acoustic velocity (and, hence, change in M_2) and defocusing of the optical beam. Attenuation is shown in Table 9.3 in units of dB per microsecond of acoustic propagation time. Although the basic mechanisms of acoustic attenuation are known (Uchida and Niizeki, 1973; Woodruff and Ehrenreich, 1961), observed values depend on the quality of material fabrication. Theoretically, the dependence of attenuation on frequency (f) is expected to be f^2 (Woodruff and Ehrenreich, 1961):

$$\alpha = (f^2 T)/(v^4) \tag{9.22}$$

where T is the absolute temperature. Hence, the acoustic attenuation coefficients in Table 9.3 are normalized to the value at 500 MHz. Deviations from quadratic are often observed in practice (Chang and Lee, 1983). This is usually ascribed to variation in material quality. Materials, such as LiNbO$_3$ and rutile (TiO$_2$), have the lowest attenuation coefficients (tenths of a dB per cm at 1 GHz acoustic frequency). LiNbO$_3$ has benefited from intense materials-growth efforts to improve the quality, but the M_2 of LiNbO$_3$ is still not particularly high compared to other materials, such as TeO$_2$ (7 vs. 793). But acoustic attenuation limits the usable bandwidth of TeO$_2$ Bragg cells to about 50 MHz. However, the throughput limit for an acousto-optic system is determined by the *product* of the bandwidth and the time-bandwidth (Lee and VanderLugt, 1989). The extraordinarily slow acoustic velocity of TeO$_2$, thus, allows for large time-bandwidth product devices (>5000). It is, therefore, of interest to search for acousto-optic materials with even slower velocities than TeO$_2$ for larger time-bandwidth capability. There has been recent promising crystal-growth work with the mercurous halide compounds Hg$_2$Cl$_2$ and Hg$_2$Br$_2$ (Goutzoulis and Gottlieb, 1988), which have about half the acoustic velocity of TeO$_2$.

The number of possible acoustic modes covered by Table 9.3 is somewhat limited due to space limitations. Extensive compilations of the acoustic and acousto-optic properties of materials can be found elsewhere for those with unique requirements (Blistanov et al., 1982; Uchida and Niizeki, 1973).

Acousto-Optic Modulators

Acousto-optic modulators are widely available for temporal and spatial modulation. The 1-D acousto-optic SLMs (or Bragg cells) have, perhaps, undergone the most successful development of any SLM. Primary issues in such devices are bandwidth and diffraction efficiency, and linearity. Diffraction efficiency is given by Eq. (9.20), and has been discussed in Sec. 3.1 primarily from a materials perspective. However, the nonlinear sine-squared response of acousto-optic devices in Eq. (9.20) places limitations on wide-bandwidth, or multiple-frequency performance due to generation of harmonic distortion and in-band frequency components. As in the case of electro-optic modulators described in sec. 2.2.1, one can consider quasi-linear operation about the quadrature point, but this does not totally eliminate the production of such intermodulation products. Quadrature point operation has the additional difficulty of requiring much more electrical drive power for the modulator. The consequent requirement for high-power distortion-free amplifiers is difficult to satisfy for high-bandwidth operation. For the multifrequency, small-signal case, the usable dynamic range is the ratio of the intensity of intermodulation products to the signal intensity at the true frequency. To keep the intermodulation-free dynamic range under control, it is necessary to limit the diffraction efficiency per frequency to about 1% (Hecht, 1976).

Temporal Modulators

For temporal-only modulation, we can visualize the acousto-optic cell in Fig. 9.4 to have zero length so that the space/time signal $f(t - x/v)$ degenerates to $f(t)$, giving a "point" modulator. The modulation bandwidth is inversely proportional to the transit time of the acoustic wave through the incident light beam, and diffraction efficiency is highest near the transducer for the acousto-optic cell. Hence, the cell is illuminated with a focused cone of light near the transducer. The diffracted light is a similar cone modulated by $f(t)$. Bandwidths of up to 1000 MHz are possible. However, trade-offs exist. If diffraction efficiency is maximized by increasing Q (Eq. 9.17), the output cone becomes elliptical because of optical-acoustic wave mismatching (Young and Yao, 1981).

The acousto-optic modulator has several unique features:

— Although such modulators are generally used in conjunction with laser beams, it is also possible to modulate, simultaneously, several wavelengths or even white light.
— The modulator also shifts the frequency of the optical beam, which is desired in some applications.
— Although an alternative is to directly modulate the current in light sources, such as laser diodes and light emitting diodes, sometimes the acousto-optic modulator provides better coherence in the modulated beam.

Spatial Modulators

One-Dimensional Devices. 1-D Bragg cells perform optical deflection, and they fill particular but important roles in optical processing. However, because they are 1-D, a single acousto-optic cell exploits only a fraction of the capability of optics for processing. Although optical capability can be increased by using multiple cells operating in parallel, such devices still handle only 1-D data, e.g., temporal data. Further, there is no information-storage capability, because the information is on an acoustic wave flowing continuously through the cell. Nonetheless, development of 1-D optical processors has been pursued for real-time, high-bandwidth, time-series data processing, and such processors are generally more advanced than true 2-D processors. Reasons for this include achievement of very high throughput with 1-D light modulators, and slower development of 2-D SLMs. For example, with acousto-optic devices it is possible to input data at several gigasamples per second. Further, it is possible to use acousto-optic devices to perform some 2-D processing operations, e.g., ambiguity functions and folded spectra (see Chap. 14). These are operations that can be factored into multiple 1-D terms and entered into orthogonally situated acousto-optic cells. Performance parameters that have been achieved in present acousto-optic cells are listed in Table 9.3.

For some advanced Bragg cells constructed of birefringent material, low values of M_2, i.e., low efficiency, have been overcome by employing anisotropic Bragg diffraction. With the isotropic case described by Eq. (9.16), there is a strong trade-off between the cell bandwidth and the diffraction efficiency due to phase mismatch (Dixon, 1967a), i.e., momentum is not conserved over the entire interaction bandwidth, because the optical momentum (K) vectors are not perfectly connected by the acoustic K-vector ($K_{acoustic}$) except at one acoustic frequency. Phase matching is achieved over a wide bandwidth, but with reduced diffraction efficiency, by using acoustic transducers with very small lateral dimensions. Acoustic diffraction (due to the finite extent of the transducer) distributes the acoustic energy over a wide range of angles. In birefringent materials and over specific frequency ranges, the loss of diffraction efficiency can be minimized and large bandwidths obtained, with the technique of tangential phase matching (Dixon, 1967a) which acousto-optically couples oppositely polarized diffracted and undiffracted modes. This technique is illustrated in Fig. 9.5. The light vectors K_{in} and K_{out} each terminate on one of two optical-velocity (or refractive-index) surfaces, i.e., for ordinary or extraordinary polarized light. The

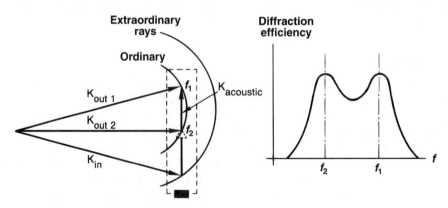

FIGURE 9.5 Anisotropic acoustooptic Bragg diffraction

Optical Modulators

incident and diffracted light directions are generally not symmetrically situated with respect to $K_{acoustic}$, as in the isotropic case, because the optical-velocity surfaces can be ellipsoidal rather than spheroidal as in the isotropic case. $K_{acoustic}$ connects optical K vectors, as desired, except that it is almost tangent to one optical-velocity surface. Only small changes in the angular direction of the acoustic wave vector with frequency are required to maintain near-conservation of momentum. Hence, the need for acoustic diffraction is minimal, and it is possible to increase the length of the transducer along the direction of the optical beam, thereby, recovering diffraction efficiency. As the acoustic frequency increases, for a fixed direction for $K_{acoustic}$, exact phase matching occurs at two points, f_2 and f_1, giving the wideband double-peaked band frequency response of Fig. 9.5. The degree to which this can be achieved depends on the shape of the optical velocity surfaces of the birefringent material and, hence, on the crystal "cut" or the acoustic-wave direction. From the definition of M_2, we want a crystal cut where the photoelastic constant p and the optical index n are maximized and the acoustic velocity v is minimized. Using such an optimization procedure, shear mode $LiNbO_3$ cells have been demonstrated with diffraction efficiencies of 44%/RF watt over the 2–3 GHz band and 12%/RF watt over the 2–4 GHz band at 633 nm wavelength (Chang and Lee, 1983).

Two-Dimensional Devices. Acousto-optic devices can be extended to two-dimensional architectures by either i) constructing an array of 1-D devices in a single crystal or by ii) constructing a device with acoustic transducers on orthogonal edges of a crystal. In either of these approaches, a large, high quality crystal is required. For multichannel acousto-optic devices, a large number of channels is desirable, but, in the absence of electrical crosstalk, the major fundamental limitation is acoustic diffraction. Acoustic diffraction depends not only on the dimensions of the transducer, but also on material. Ideally, the near-field acoustic wavefront is planar over the aperture of the transducer, and diffraction effects are observed in the far field. Hence, it is desirable to maintain a near-field condition over as long a propagation distance as possible. In an isotropic medium the transition from near to far field occurs at a distance from the transducer of approximately

$$D = H^2/8\Lambda \qquad (9.23)$$

where H is the vertical dimension of the transducer and Λ is the acoustic wavelength. However, for an anisotropic medium, D can be increased by an additional factor of $\chi = (1 - 2b)^{-1}$. The quantity b is the coefficient of the θ^2 term in a power series representation of the acoustic slowness surface (Cohen, 1967)

$$K_a(\theta_a) = K_a(1 + b\theta_a^2 + d\theta_a^4 + \ldots\ldots\ldots) \qquad (9.24)$$

where θ_a is the acoustic-wave direction relative to the normal to the transducer. For shear-mode TeO_2, the quantity χ is 0.02 so that the acoustic spreading is very rapid. However, for the fast shear mode propagating along [110] in GaP, χ is 38, and the acoustic spreading is very slow (Hecht and Petrie, 1980). Because of the severe acoustic spreading in shear-mode TeO_2, multichannel devices have been constructed only of longitudinal-mode TeO_2 where the spreading factor χ is 2. A 32-channel TeO_2 device at a center frequency of 250 MHz has been demonstrated (VanderLugt et al., 1983; Amano and Roos, 1987).

The second 2-D approach uses orthogonally propagating acoustic waves. Using two separate 1-D cells, orthogonally situated, and anamorphic optics between the cells to pass light from one to the other, one can perform 2-D beam steering and 2-D processing (Turpin, 1981). A much more compact alternative to individual 1-D devices is a single, large, square crystal with transducers along two edge facets and light propagating perpendicular to the square aperture. This approach is feasible only in cubic materials. TeO_2 satisfies this criterion, and shear-mode devices have been demonstrated. However, if the two acoustic waves propagate within the same volume of crystal, nonlinear mixing of the signals will occur at lower power levels. This is particularly true with

TeO$_2$ which exhibits acoustic nonlinearities at relatively low power levels. Thus, devices are made with the acoustic transducers offset from each other along the optical-beam direction. The transducer dimension along the optical path and the offset distance must be minimized to prevent loss of resolution or to avoid use of optics with a large depth of focus.

Optical Wavelength Control and Filtering

Many optical modulators are sensitive to the wavelength of operation. However, acousto-optic modulators are unique in that, without a large degree of modification, they can be used as wavelength-selective elements. Specific wavelengths can be selected from a broadband input spectrum, by driving an acousto-optic device with an appropriate set of RF frequencies. This filtering action is notable in that a multiplicity of wavelengths may be selected simultaneously, unlike optical glass and interference filters and resonant tunable structures, such as Fabry–Perot cavities. This multiple-wavelength capability allows applications, such as color correction in displays (Young and Belfatto, 1993), and wavelength and spectral modulation as an added dimension in information processing, as in wavelength division multiplexed, fiber optic systems (Smith et al., 1990).

Acousto-optic tunable filters generally operate in the anisotropic diffraction mode (see above) to maximize the distance over which optical and acoustic waves are phase matched (Chang, 1983), because the longer the resonance interaction length, the higher the spectral resolution. Therefore, the RF frequencies required for specific wavelengths is governed by phase-matching conditions for anisotropic diffraction (Cohen, 1967; Dixon, 1967a). Both bulk and integrated-optic devices are possible. The latter are integrable with optical fiber and utilize the polarization anisotropy in thin-film optical waveguides in addition to material anisotropy (Smith et al., 1990)

9.4 Magneto-Optic Devices

If plane-polarized light passes through an optically transparent magneto-optic material, the Faraday effect rotates the plane of polarization of the light as a function of magnetic field strength H and sample thickness d along the light-propagation direction according to

$$\vartheta = VHd \tag{9.25}$$

where V is the Verdet constant, defined as the rotation per unit field and path length. The polarization change can be employed for either phase or amplitude optical modulation using polarizers such as described for electro-optic modulation (Fig. 9.2). ϑ depends only on the direction of H and the sign of V, not on the light propagation direction, making the effect useful for isolation against back-reflected optical beams. The polarization rotation arises from a difference in the refractive indices for right-handed and left-handed circularly polarized light. It can be shown that the normal modes of a light wave in a magnetized medium are circularly polarized (Jenkins and White, 1976). Faraday rotation can occur in materials with or without a permanent internal magnetization. An example of the former case is terbium gallium garnet, with $V = .002$ deg/(cm-Oersted) at a wavelength of 1064 nm. The latter case involves materials with paramagnetic or diamagnetic atomic species, such as doped glasses. A representative paramagnetic glass is terbium-doped borosilicate glass with $V = 0.0058$ deg/(cm-Oersted) at a wavelength of 531 nm (Butler and Venturini, 1987).

For materials with internal magnetization, an externally applied field is necessary only to saturate the magnetic state, and the polarization rotation in the magnetically saturated state is described by the Faraday rotation ϕ_F with units of degrees/cm. Ferrite garnets are, perhaps, the best materials with strong Faraday rotation and include yttrium iron garnet (YIG) with $\phi_F = 280$ deg/cm at a wavelength of 1000 nm (Dillon, 1978). Bismuth-doped YIG films exhibit ~10 times larger values of ϕ_F. Values of ϕ_F are strongly dependent on wavelength, ϕ_F usually increases steeply at shorter wavelengths, but optical absorption also tends to increase commensurately. Hence, a figure of merit is often the ratio of ϕ_F to absorption.

A second method of employing magneto-optics is to use magnetostatic waves for light modulation. 1-D spatial light modulators have been constructed using magnetostatic traveling waves analogously to 1-D acousto-optic devices (Fisher et al., 1982). Magnetostatic waves (MSWs) are slow-velocity electromagnetic waves that propagate in magnetic ferrites (Damon and Eshbach, 1961), such as thin-film YIG, under application of a bias magnetic field. MSW-optical devices are electrically addressed similarly to surface-acoustic-wave acousto-optic devices (see Sec. 3.2). The electrical-to-MSW transducer is usually a simple microwave stripline antenna on the surface of the film. The major differences with acousto-optics are that modulation can be performed at high microwave frequencies, up to 30 GHz, the frequency of operation and wave velocity are adjustable with the bias field, and the devices must be thin film (see below).

The performance of both the Faraday-rotation and MSW spatial light modulator devices is summarized below and in Table 9.4.

Faraday-Rotational Materials and Devices

Both temporal and spatial light modulators employing Faraday rotation have been constructed. Faraday rotation is used in a crossed-polarizer configuration like that of Fig. 9.2 to produce intensity modulation, with the magneto-optic device replacing the electro-optic device in the figure. A temporal modulator is generally constructed of garnets, such as TGG, YIG, and bismuth-doped YIG. An external magnetic field allows control of the amount of polarization rotation. 2-D spatial light modulators have been constructed by fabricating an array of Faraday-rotational elements from films of garnet material, such as bismuth-doped YIG (Ross et al., 1983). To date only electrical matrix-addressed SLM devices have been constructed; the addressing issues are as discussed in Sec. 9.1.

In the electrically addressed 2-D SLMs, thin-film material is pixelated into mesas, with the surrounding channels containing the addressing electrode. Application of current to the sur-

TABLE 9.4 Selected Magneto-Optic Spatial Light Modulators and Performance

Device/Name	Address Method	Frame Rate (Hz)	Space-Bandwidth/ Resolution (pixels/mm)	Remarks
YIG:Bi, Faraday rotation/ (LIGHT-MOD, Litton SIGHT-MOD, Semetex, Faroptic LISA, Phillips)	Electrode matrix	1000	512 × 512	Commercially available; permanent storage
YIG/YIG:Bi, Magnetostatic wave/ (NRL, Univ. CA Irvine, Westinghouse)	Microwave stripline antenna	100 ns	100 time-bandwidth	1-D traveling-wave, up to 30 GHz carrier input. 4–12%/watt diffraction efficiency demonstrated in laboratory prototype

rounding electrodes transforms a mesa into a single magnetic domain in one of two opposite magnetic orientations. The two orientations produce equal but opposite Faraday rotation in the polarization of a transmitted read-out beam, while an unaddressed mesa is observed to maintain the input polarization. If the magneto-optic material is placed between crossed polarizers, as in Fig. 9.2, either two or three discrete amplitude modulation levels can be obtained. In the former, the rotated polarization state due to one orientation, say the clockwise-rotated state, is aligned with the input polarization, and every pixel is addressed. This mode maximizes contrast. For the latter, or ternary-state case, the unrotated state is also used, and its amplitude transmittance is sensed. Using an optical arrangement that senses the direction of polarization rotation, polarization- or phase-encoding is possible, because the optical phases for the two orientations are 180° from each other (Ross et al., 1983).

Operation of the device requires that one corner of each mesa be modified to allow a pixel to be activated, but without activating any of the other three neighboring pixels. In the commercially available LIGHTMOD/SIGHTMOD devices, one corner is ion-implanted to lower the threshold for domain switching. The conjunction of currents in both crossing matrix electrodes at the ion-implanted corner produces high magnetic fields which induce reversal of the magnetic domain of the pixel. In the LISA device, a small resistive element at a corner thermally induces domain flipping.

One of the advantages of the magneto-optic SLM is its intrinsic long-term storage capability, because domain orientation remains constant in the absence of an addressing signal. Therefore the magneto-optic SLM does not require periodic frame updating, and random pixel-by-pixel addressing is possible. Random pixel addressing can also reduce average power consumption. The required currents for switching can lead to the perception that magnetic devices have high power consumption. To reduce switching current requirement, a thinner mesa with higher Faraday rotation can be used. An added benefit is that larger arrays at a given power level are possible. Hence, bismuth-doped films have been employed in more recent devices. Faraday rotation increases with bismuth content. However, lattice constants increase with doping. To maintain lattice matching between the grown film and the substrate, therefore, requires so-called large-lattice-constant substrates of doped gallium gadolinium garnet (GGG).

Magnetostatic Wave Interactions and Devices

Optical modulation with MSWs is very similar to that shown for acousto-optics in Fig. 9.4. The MSWs induce refractive index changes in the optical media. Conservation of momentum results in deflection of the incident light and a frequency shift by the MSW frequency, but there are major differences between MSW-optical and acousto-optic device parameters.

- —MSW velocities are roughly 10 times higher than typical acoustic velocities. Thus, MSW delay lines are several hundreds of nanoseconds in length, compared to microseconds for acoustic-optic delay lines.
- —MSW center frequencies are in the microwave region, from 1–30 GHz, and can be tuned with an applied external magnetic field. By contrast, acousto-optic devices presently operate at no more than 4 GHz due to acoustic attenuation. Although modulation at these high microwave frequencies is also possible with laser diodes and with traveling-wave, electro-optic modulators, these alternatives provide only temporal, not spatial, light modulation. MSWs, therefore, allow direct light modulation at microwave frequencies with potentially large time-bandwidth products.
- —MSWs must generally be propagated in thin-film geometries. In bulk samples of ferrite, shape-factor demagnetizing fields within the sample are generally not spatially constant, leading to spatial inhomogeneities in MSW energy density and nonconstant wave propagation. However, a thin, infinite sheet has constant shape-factor demagnetization. Thus, these optical modulators are best employed in a guided-wave configuration. For example, a YIG

film is grown on a lattice-matched substrate of gallium gadolinium garnet (GGG) and serves as both a magneto-optic medium and an optical waveguide. Highest quality growth is usually with liquid-phase epitaxy techniques.

—MSWs are excited with a stripline antenna, as opposed to either a bulk piezoelectric plate or SAW transducer for acousto-optic devices.

Optical interactions with MSWs were first demonstrated in epitaxial YIG films (Fisher et al., 1982). Subsequently, devices for spectrum analysis and correlation (Fisher, 1985; Tsai et al., 1985) have been demonstrated. Diffraction efficiencies of about 4%/watt were observed at an optical wavelength of 1300 nm, and bandwidths of 30 MHz in a homogeneous bias field and 350 MHz in gradient fields (Wey et al., 1986) have been achieved. The gradient field approach, which biases different spatial regions of the sample to different center frequencies, has potential for bandwidths as high as 1 GHz but with lower diffraction efficiency than the homogeneous case.

Higher diffraction efficiencies are desired, and bismuth-doped iron-garnet films such as Bi:YIG, have been investigated for this purpose (Tamada et al., 1988; Butler et al., 1992). The theory for the optical-MSW interaction shows that the efficiency scales as the square of ϕ_F (Fisher et al., 1982; Fisher, 1985). Bismuth-doped films have an order of magnitude higher Faraday rotation constants than pure YIG, giving the potential to provide 100 times higher diffraction efficiencies. Experimentally, a diffraction efficiency of 12% over a tuning range of 3.7–12 GHz and a bandwidth of 150 MHz using Bi:YIG with $\phi_F = -2313$ deg/cm has been reported (Young and Tsai, 1989). Because optical transparency, MSW attenuation, and Faraday rotation must all be simultaneously optimized, to date there has been limited use of these doped materials in optical-MSW devices.

9.5 Mechanical and Micromechanical Devices

Mechanical devices are, perhaps, the oldest method for light modulation. These include high-speed shutters, such as found on cameras, rotating slotted-disk optical choppers, and vibrating resonant structures, such as tuning forks (Woodruff, 1991). Such mechanical devices are generally limited to temporal modulation only. Micromechanical mechanisms have also been employed to temporally and spatially modulate light via physical movement of very thin (<0.1 μm) membranes of polymer materials or silicon structures (e.g., cantilevered beams, torsion bars, optical waveguides, etc.) (Somers, 1972; Fisher et al., 1986; Ohkawa et al., 1989). These modulators are summarized below and in Table 9.5.

Mechanical Devices

Temporal modulation is achieved with mechanical shutters and choppers. A shutter allows both the frequency and the duty cycle to be varied, whereas these parameters are fixed in a chopper. For example, a tuning-fork chopper provides a sinusoidal intensity modulation, and a rotary slotted-plate chopper gives a triangular or trapezoidal modulation depending on slot size relative to the beam. The maximum frequencies for shutters and choppers are generally in the 1–10 kilohertz regime, limited by inertia. Advantages of mechanical systems include independence of modulation on wavelength and the ability to handle high optical power levels. In practice, the opaque portions of shutters and choppers must completely absorb or redirect the optical beam without scattering light in the modulated beam direction. A device which combines most of the desired features in mechanical modulation is the magnetic flexure shutter, consisting of a single flexing, optically reflective blade at a near-grazing angle to the optical beam direction. It offers high speed (>10 kHz), reliability, and high optical power-handling capability (up to 100 megawatts) with little optical backscatter (10^{-6} of incident) (Woodruff, 1991).

TABLE 9.5 Selected Micromechanical Spatial Light Modulators and Performance

Device/Name	Address Method	Frame Rate (Hz)	Space-Bandwidth/ Resolution (pixels/mm)	Remarks
Deformable membrane/ Membrane light modulator—MLM (Perkin-Elmer/ Hughes)	Optical/ Si photodiodes Electrical	100 100	10/mm 100 × 100	Limited availability
Deformable elastomer/ RUTICON (Xerox, Harris, IBM)	Optical/ amorphous Se Electrical/ e-beam	100 25	100/mm 15/mm	Limited availability
Deformable membrane device—DMD (Texas Instruments)	Optical/Si phototransistors Electrical/ Si CCDs/ transistors	500 500	128 × 128 128 × 128	Limited availability: superseded by cantilevered beam device
Cantilevered micromechanical beams—SiO$_2$ or metal film—DMD (Texas Instruments, TRW)	Electrical/ silicon circuits	500 30	256 × 256 640 × 480	Commercially available
Thermoplastic (Newport, Honeywell, Fujinon, Kalle-Hoechst)	Optical/ PVK:TNF	10	1000/mm	Commercially available; limited lifetime
Photoemitter membrane light modulator— PEMLM (NRL)	Optical/ photocathode + microchannel plate	1000	80/mm	Laboratory prototype; very high optical sensitivity: 10 pJ/ cm^2; storage time: days
Membrane light modulator (Optron system, Inc.)	Electrical/ e-beam + charge transfer plate	30	256 × 256	Commercially available

Membrane Devices

Membranes modulate the optical phase via mechanical motions as the light reflects off either the membrane or a deposited, reflective, metallic layer. The framing speed is potentially very high, because the required membrane motion is only a fraction of an optical wavelength and the membrane element can be very small. A few micrometers of spatial resolution has been demonstrated (Fisher et al., 1986). Membrane devices can be either optically or electrically addressed. Addressing issues are as discussed in Sec. 9.1.

Membranes must be deposited on some substrate that performs the conversion and control function indicated in Fig. 9.1. Membrane devices that have undergone the most development are the membrane light modulator (MLM) (Preston, 1969; Reizman, 1969), where a thin conducting membrane is stretched over an array of insulating holes, the deformable membrane device or DMD (Pape, 1985), where an insulating membrane is laid on silicon circuits, and the photoemitter membrane light modulator or the PEMLM (Somers, 1972; Fisher et al., 1986), where an insulating membrane is laid onto a microchannel plate. Fig. 9.6 shows the construction of the PEMLM. A

Optical Modulators

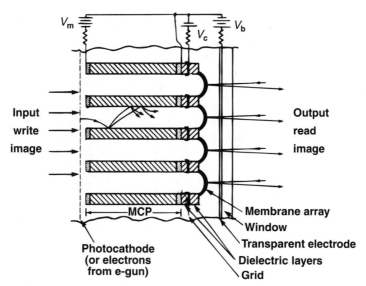

FIGURE 9.6 Construction of the photoemitter membrane light modulator (PEMLM).

photocathode on the left side of Fig. 9.6 produces photoelectrons in response to an optical input beam, the photoelectrons are amplified by the microchannel plate, and the amplified beam deposits charge onto a membrane. The charged membrane acts against a transparent electrode through which the read-out beam passes. The membrane is coated on the read-out side with a material like indium film which is reflective without being electrically conductive (Fisher et al., 1986). An electrically addressed variant of the PEMLM uses a charge-transfer plate, a microchannel-plate structure whose pores have been filled with metal (Optron Systems, Inc., 1994). An electron beam then addresses one side of the charge-transfer plate. Advantages are that the modulating membrane material can be in air and is, therefore, easily accessed and repaired, and there is no need for a transparent electrode on the read-out side. However, the membrane acts against the metal in the pores and can only be pulled into the pores, somewhat limiting the phase dynamic range.

The ultimate speed of a membrane device is determined by the resonance frequency of the membrane, which, for a circular geometry, is given by (Fisher et al., 1984)

$$f_o = 0.38R\sqrt{\tau/M} \qquad (9.26)$$

where τ is the tension, R the radius, and M the mass density of the membrane. The membrane material used in the initial PEMLM and DMD devices was nitrocellulose. For this material, a 40-nm thick circular membrane with a 30-nm reflective indium coating and $R = 12$ μm, $M = 2.85 \times 10^{-4}$ kg/m^2, and $\tau = 0.8$ N-m, gives a 1.66-MHz frequency response. However, the framing rate of a membrane device is invariably much less than this speed. Limitations are usually first encountered in the drive current required to charge the membrane for a given membrane deflection Z. For the DMD, the limit is the current for the silicon circuit, whereas, for the PEMLM, it is the amplified current from the microchannel plate. Membrane deflection amplitude is given by (Somers, 1972)

$$Z = (E^2 R^2 \epsilon_o)/(8\tau) \qquad (9.27)$$

where E is the electrostatic field at the surface of the membrane (V/m), and ϵ_o is the dielectric constant of free space. For example, for a PEMLM device with a nitrocellulose membrane, to produce a deflection of $Z = \lambda/4$ (full-contrast reversal), the drive-current requirements limit the frame rate to about 1.5 kHz (Fisher et al., 1986). However, examination of Eq. (9.27) shows that the speed can be optimized by reducing the required value for E, which, for a given Z, requires lowering the membrane tension. τ can be lowered by using a polymer material which has lower surface energy than nitrocellulose. A multilayer membrane structure of parylene on PTFEP (Poly(Bis(Triflouroethoxy)Phosphazene), a flouropolymer), when deposited onto SiO_x (the surface material of the microchannel plate), has been shown to have 10 kHz capability, while also exhibiting a long lifetime compared to nitrocellulose and stable mechanical behavior after break-in cycling (Rolsma and Lee, 1990). The parylene has high mechanical strength that holds up at elevated temperatures (>50 C), but the underlying PTFEP film has a low surface energy which reduces the adhesion forces to the SiO_x substrate and leads to lower tension in the parylene film. The PTFEP does not have sufficiently robust mechanical characteristics to allow it to be used alone.

The contrast ratio is limited by device construction. The fundamental thermal noise contribution to the deflection Z is much less than the $\lambda/4$ required for full contrast reversal.

Silicon Mechanical Devices

For temporal modulation only, guided-wave Mach–Zehnder interferometer devices have been demonstrated with either glass (Ohkawa et al., 1989) or a silicon dioxide/silicon nitride stack (Vadekar et al., 1994) deposited on silicon. These devices have been used for optical sensing of acoustic signals. One arm of the interferometer is subjected to an applied pressure, resulting in a change in optical path length and, hence, a phase shift with respect to the reference arm. Mode attenuation due to bending has been shown to be negligible, and efficiency was increased by acoustic impedance matching to the applied source (Vadekar et al., 1994).

For spatial modulation, the DMD using polymer membranes on silicon circuits has the disadvantage of involving two very different fabrication processes (Pape, 1985), and there has been difficulty in obtaining defect-free devices larger than 256 × 256. This has led to investigation of an all-silicon modulator (Brooks, 1985). Still designated the DMD, it uses optically reflective micromechanical silicon structures, such as cantilevered beams (Brooks, 1985) and torsion beams fabricated out of silicon by an etching process (Hornbeck, 1989). Again, a silicon circuit below the micromechanical structure provides addressing and activation voltages that deflect the structure. The deflection changes the angular position of the reflected light beam, producing intensity modulation in the read-out system. Recent efforts have been focused on larger 2-D arrays, e.g., 640 × 480 pixels for display purposes. Addressing of such arrays at a high frame rate is a problem. Electrical addressing requires a number of parallel high-speed lines, and optical addressing requires integrating photodetectors as part of a large optoelectronic integrated circuit. Other optical and electrical addressing issues are as discussed in Sec. 9.1, generally involving VLSI fabrication issues rather than materials issues.

9.6 Thermal Devices

Strong thermal variations in some material properties have been exploited for modulation. The primary ones are plastics and oil films, that deform when heated, and materials that undergo phase transitions. The most well-known of these is the phase transition of liquid crystals from the smectic to the nematic phase. Another that has been used is the metal-to-insulator phase transition in vanadium dioxide (VO_2).

Both electrically and optically addressed devices have been demonstrated, and addressing issues are as discussed in Sec. 9.1.

Thermoplastic and Oil Film Devices

Plastics and oil films can be deformed by heating. Local heating can be employed for spatial light modulation. In both types of devices, heating has been performed electrically with a scanning electron beam (Turpin, 1974, Schneeburger et al., 1979). Optical addressing has been achieved in the oil-film device by a light-absorbing plastic substrate (Schneeburger et al., 1979). Deformations on the order of one optical wavelength phase-modulate the read-out beam, as in the membrane devices discussed in Sec. 9.5.

Thermoplastics have been employed in optically addressed devices in a sandwiched geometry almost exactly like that shown in Fig. 9.1. The thermoplastic material is sandwiched between a metallic electrode layer and a photosensor layer (Colburn and Chang, 1978). Materials used for the photosensor layer have included PVK:TNF, amorphous Se, CdS, ZnS, and ZnCdS. Electrically addressed devices using thermoplastics and other elastomers, such as those for the Ruticon devices (Ralston and McDaniel, 1979), have also been constructed using an ionized gas or conducting liquid rather than a metallic electrode to apply voltage. The various electrode configurations have overcome some, but not all of the problems with limited lifetime, especially in thermoplastics. Thus, there has been relatively little recent work (Colburn and Chang, 1978; Ralston and McDaniel, 1979).

Phase-Transition Devices

A highly ordered smectic phase of liquid crystals is possible at room temperature. Heating of the material, with an electron/optical beam or a pixel heating element, changes it into the nematic phase. Upon cooling, the material transforms into a disordered smectic phase which strongly scatters light. Schlieren read-out optics produce the intensity-modulated output. Grey scale is obtained by partial writing/erasure. The nematic-to-smectic mechanism has been employed for very large high definition displays (2200 × 3400) using near-IR semiconductor lasers (Kahn et al., 1987). However, the response speed is slow—on the order of a second, and the spatial temperature uniformity of the device must usually be closely controlled.

VO_2 exhibits a phase transition between a highly reflective metallic phase and a low-reflectance insulating phase (Eden, 1979). Generally, the VO_2 phase-transition devices provide only binary intensity modulation, with typically a 10:1 contrast ratio. However, the temperature of the material must be held close to the transition temperature to minimize the required switching energy and time. The response time of the material can be reasonably fast, a few hundred nanoseconds (Eden, 1979), for temporal switching applications. For spatial light modulation, it is usually the addressing mechanisms that limit the frame rate.

9.7 Specialized and Emerging Devices

Direct Modulation of Light Sources

Modulation can be achieved by directly controlling the electrical power to an optical source. The primary, directly modulated light sources are semiconductor laser diodes and LEDs because of their small size (μm-sized emission facet) and their high electrical-to-optical conversion efficiency. Modulation is achieved by controlling the drive current in these devices. With laser diodes the electrical-to-optical conversion efficiency can be close to unity. Because the most popular laser diodes and LEDs are fabricated with III-V compounds, such as GaAlAs and InGaAsP, there is the hope that such light sources can be integrated on a single chip together with high-speed drive circuitry fabricated with these same III-V materials.

Temporal Modulation

Temporal modulation of laser diodes and LEDs is quite common and capable of very high bandwidth (tens of GHz), but is presently used, primarily, for only a single channel. Achievable

TABLE 9.6 Selected Thermal Spatial Light Modulators and Performance

Device/Name	Address Method	Frame Rate (Hz)	Space-Bandwidth/ Resolution (pixels/mm)	Remarks
Deformable elastomer/ RUTICON (Xerox, Harris, IBM)	Optical amorph.Se Electrical e-beam	100 25	100/mm 15/mm	Limited availability
Thermoplastic (Newport, Honeywell, Fujinon, Kalle-Hoechst)	Optical PVK:TNF	10	1000/mm	Commercially available; limited lifetime
Vanadium Dioxide— VO_2 metal-to-insulator phase transition (LTV)	Optical Electrical		150/mm 20/mm	Limited availability; high writing power needed
Oil film-(Talaria-GE/ Martin-Marietta, Eido- phor-Greytag)	Electrical e-beam	30	100/mm (1000 × 1000)	Limited commercial availability
Oil film	Optical heat absorp. in substrate	5	4/mm	Limited commercial availability
Smectic-to-nematic liquid crystal phase transition (Greyhawk Systems)	Optical IR laser heating	0.1	16/mm (2200 × 3400)	Commercial display product

Refs. See text

optical power, dynamic range and linearity of modulation are major issues. Good linearity of modulation is possible with LEDs, but the optical power output is low, typically <1 mW. Power from presently available single-element, single-mode, laser diodes is >100 mW, but modulation over a large dynamic range is still a challenge. In addition, the laser drive current must exceed a threshold before lasing with a linear output vs. drive current. The ideal behavior of laser diode output vs. input is shown in Fig. 9.7. The threshold current has been reduced using quantum-well structures in the active gain region (Lau et al., 1988).

High-bandwidth modulation of a laser diode can cause modal instability, with attendant intensity fluctuations, i.e., even if a laser is single-mode, the intensity can abruptly jump if the modulation amplitude puts the laser near a point where mode competition and mode hopping occur. In practice, one must also be careful to avoid backreflection into the laser cavity, because this will contribute to mode instability. Another serious problem is shifting or chirping of the optical frequency with modulation. Chirping arises from changes in the carrier density (N) during modulation such that the later portions of a driving signal see a different refractive index n (which is proportional to N) than the earlier portions. The linearity and optical frequency stability of laser diodes has been improved with the use of distributed feedback (DFB). These lasers have a periodic grating structure within the cavity that continuously Bragg-scatters a portion of the light backwards. This limits the number of longitudinal modes which can propagate, leading to better mode stability, higher spectral purity, and suppression of chirping during modulation (Nakamura et al., 1973). Present laser diodes can be modulated at bandwidths greater than 20 GHz. Quantum-well laser structures can, theoretically, reach the 30-GHz regime (Uomi et al., 1985), based on limits due to relaxation oscillations of charge carrier density in the laser cavity. The dynamic range at high modulation bandwidths can be increased by using an external-cavity

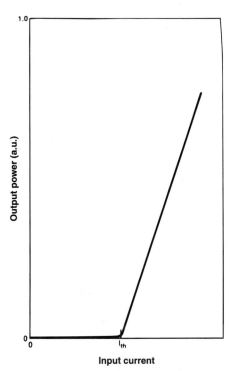

FIGURE 9.7 Typical dependence of optical output on electrical drive current for a laser diode.

configuration where a well-behaved, low-power laser is used to control the behavior of a high-power laser.

Spatial Modulation

Arrays of either incoherent light sources or laser diodes with individual element addressing can be employed for spatial light modulation. The addressing issues are as discussed in Sec. 9.1.

One approach is to use incoherent light by adapting existing luminous displays (Apt, 1985; Blechman, 1986). This approach, like the case described earlier for SLMs based on transmissive liquid-crystal televisions, takes advantage of the vast amount of technological development put into realizing large high-resolution displays. A variety of emissive display technologies are available, including cathode-ray tubes, gas plasma displays, electroluminescent panels, and LED displays. However, the major problems with this approach are i) luminous intensity is often insufficient for optical processing, ii) with incoherent light, phase cannot be processed without using some encoding scheme that sacrifices a parameter, such as resolution or dynamic range, iii) only electrical addressing is available for these devices, so that a single device cannot perform image multiplication, and iv) the cosmetic quality and flatness are often not sufficient for SLM applications.

The coherent-light approach, using an array of semiconductor laser diodes, can be implemented with either electrical or optical addressing. For example, a laser amplifies an input optical at the cavity wavelength. An approach that has excited much interest is use of a large number of individually addressable semiconductor lasers on a single wafer. In principle, such an array can be either optically or electrically addressed, but the latter is presently the rule. Conventional edge-emitting (in-plane) lasers can be sliced from a wafer to produce a line array of sources. Stacking such line arrays to produce a large 2-D array has been achieved, but mainly as a high-power optical source; it is difficult and costly to provide addressing. A 2-D array of surface-emitting lasers integrated on a single wafer can be achieved in either of two ways. The first is to combine a 45° mirror structure with each laser in an edge-emitting array to redirect the beam upwards (Liau et al., 1984) The second approach is to construct the array using vertical-cavity, surface-emitting lasers (VCSELs)(Chang–Hasnain et al., 1991). Construction of such integrated laser arrays involves the III-V compounds discussed above, but with the more-complex structures involved in laser fabrication. Several materials-related issues arise. High electrical-to-optical efficiency is required for the lasers to keep within power dissipation capabilities of a wafer. Use of a quantum-well structure in the construction of the semiconductor lasers is imperative to minimize the lasing threshold and maximize the electrical-to-optical efficiency (Lau et al., 1988). For the VCSEL approach, present devices also tend to exhibit high resistance (Tai et al., 1990), primarily due to the mirror layers for the laser cavity. This further increases the heating of the wafer.

Biological Materials

Biological compounds, often embedded in membranes, constitute a large subset of organics and are largely unexplored for optical modulation. The compound of most recent interest for optical modulation is bacteriorhodopsin (Oesterhelt et al., 1971; Hampp et al., 1992). Bacteriorhodopsin consists of a chain of amino acids to which a chromophoric group is bound. To make an SLM, the material is placed in a 2-D matrix embedded in a cell membrane. Light activation produces a gradient in the hydrogen ion (or proton) concentration across the membrane. This gradient is normally needed for synthesis of other biological compounds. Light at about 570 nm excites the bacteriorhodopsin from the so-called B state to the long-lived intermediate M state which has a lifetime of about 10 ms. The M state would naturally relax back to the B state, but 412 nm light can also initiate the transition. Hence, optical beams at 570 and 412 nm form the writing and read beams, respectively, of the SLM. Although bacteriorhodopsin-based devices are presently still in the developmental stage, they indicate the potential for biological materials in SLM devices.

9.8 Summary

Many materials and devices have been explored for optical modulation. Some are quite mature, such as mechanical devices and many inorganic electro-optic and acousto-optic materials. In other cases, the materials are well developed, but significant device development still remains. Finally, in some cases, both materials and devices are immature. In general, temporal and 1-D modulators, such as acousto-optic modulators, are generally much more mature than 2-D, spatial light modulators.

References

Albanese, G., Kumar, J. and Steier, W. H. 1986. Investigation of the photorefractive behavior of chrome-doped GaAs by using two-beam coupling, *Opt. Lett.* 11, 650.

Amano, M., and Roos, E. 1987. 32-channel acousto-optic Bragg cell for optical computing, in *Acoustooptic, Electrooptic, and Magnetooptic Devices and Applications*, Proc. SPIE 753, 37.

Anderson, G., Dahl, I., Kuczynski, W., Lagerwall, W. T., Sharp, K., and Stebler, B. 1988. The soft-mode ferroelectric effect, *Ferroelectrics* 84, 285.

Anderson, G. W., Webb, D., Spezio, A. E., and Lee, J. N. 1991. Advanced channelization devices for rf, microwave, and millimeterwave applications, *Proc. IEEE* 79, 355.

Apt, C. M. 1985. Perfecting the picture, *IEEE Spectrum* 22 (7), 60.

Armitage, D., Anderson, W. W., and Karr, T. J. 1985. High-speed spatial light modulator, *IEEE J. Quan. Electron.* QE-21, 1241.

Ashley, P. R., Davis, J. H., and Oh, T. K. 1988. Liquid crystal spatial light modulator with a transmissive amorphous silicon photoconductor, *Appl. Opt.* 27, 1797.

Augborg, P., Huignard, J. P., Hareng, M., and Mullen, R. A. 1982. Liquid crystal light valve using bulk monocrystalline $Bi_{12}SiO_{20}$ as the photoconductive material, *Appl. Opt.* 21, 3706.

Aull, B. F., Kirby, B. N., Goodhue, W. D., Burke, B. E. 1988. Multiple-quantum-well CCD spatial light modulators, *Spatial LIght Modulators and Applications*, OSA 1988 Technical Digest Series, Optical Society of America, Washington, D.C., Vol. 8.

Batsell, S. G., Jong, T. L., Walkup, J. F., and Krile, T. F. 1990, Noise limitations in optical linear algebra processors, *Appl. Opt.* 29, 2084.

Berg, N. J., and Lee, J. N. eds. 1983. *Acousto-Optic Signal Processing: Theory and Implementation*, Marcel Dekker, New York.

Blechman, F. 1986. Pocket Television Receivers, *Radio Electronics* 57, Pt. 1, July, 39 and Pt. 2, Aug., 47.

Bleha, W. P., Lipton, L. T., Weiner–Avnear, E., Grinberg, J., Rief, P. G., Casasent, D., Brown, H. B., and Markevitch, B. V. 1978. Application of the liquid crystal light valve to real-time optical data processing, *Opt. Eng.* 17, 371.

Blinov, L. M. 1983. *Electro-optical and Magneto-optical Properties of Liquid Crystals*, John Wiley & Sons, New York.

Blistanov, A. A., Bondarenko, V. S., Perelomova, N. V., Strizhevskaya, F. N., Chkalova, V. V., and Shaskol'skaya, M. P. 1982. *Acoustic Crystals*, Publishing House Nauka, Moscow.

Bowman, S. R., Rabinovich, W. S., Kyono, C. S., Katzer, D. S., and Ikossi–Anastasiou, K. 1994. High resolution spatial light modulators using GaAs/AlGaAs multiple quantum wells, *Appl. Phys. Lett.* 65, 956.

Brooks, R. E. 1985. Micromechanical light modulator on silicon, *Opt. Eng.* 24, 101.

Brown, R. B., Craig, A. E., and Lee, J. N. 1989. Predictions of stray light modeling on the ultimate performance of acousto-optic processors, *Opt. Eng.*, 28, 1299.

Butler, J. C., Kramer, J. J., Lee, J. N., Ings, J. B. and Belt, R. F. 1992. Optical and Electrical Characterization of Magnesium-Doped Bismuth-Substituted Lutetium Iron Garnet Thin Films, *J. Appl. Phys.* 71, 924.

Butler, M. A. and Venturini, E. L. 1987. High frequency Faraday rotation in FR-5 glass, *Appl. Opt.* 26, 1581.

Cafarella, J. H. 1978. Acoustoelectrical signal-processing devices with charge storage, Proc. 1978 IEEE Ultrasonics Symp., IEEE Cat. No. 78CH1344-1SU, 767.

Casasent, D., 1978a. Photo DKDP Light Valve: A Review, *Opt. Eng.* 17, 365.

Casasent, D., 1978b. E-beam DKDP Light Valves, *Opt. Eng.* 17, 344.

Chang, I. C. 1983. Acousto-Optic tunable filters. In *Acousto-Optic Signal Processing: Theory and Implementation*, Berg, N. J., and Lee, J. N. eds., Marcel Dekker, New York, p. 139.

Chang, I. C., and Lee, S. 1983. Efficient wideband acousto-optic cells, Proc. 1983 IEEE Ultrasonics Symp., IEEE No. 83CH 1947-1, 427.

Chang–Hasnain, C. J., Maeda, M. W., Harbison, J. P., Florez, L. T. and Lin, C. 1991. Monolithic multiple wavelength surface emitting laser arrays, *J. Lightwave Technol.* 9, 1665.

Chemla, D. S., and Zyss, J., eds. 1987. *Nonlinear Optical Properties of Organic Molecules and Crystals*. Academic Press, London.

Chen, F. S., 1969. Optically induced changes of refractive indices in $LiNbO_3$ and $LiTaO_3$, *J. Appl. Phys.* 40, 3389.

Childs, R. B., and O'Byrne, V. A. 1990. Multichannel AM video transmission using a high-power Nd:YAG laser and linearized external modulator, *IEEE J. Select. Areas Comm.* 8.

Clark, N. A., and Lagerwall, S. 1980. Submicrosecond bistable electro-optic switching in liquid crystal, *Appl. Phys. Lett.* 36, 899.

Cohen, M. G. 1967. Optical study of ultrasonic diffraction and focusing in anisotropic media, *J. Appl. Phys.* 38, 3821.

Colburn, W. S., and Chang, B. J. 1978. Photoconductor-thermoplastic image transducer, *Opt. Eng.* 17, 334.

Coquin, G. A., Pinnow, D. A., and Warner, A. W. 1971. Physical properties of lead molybdate relevant to acousto-optic device applications, *J. Appl. Phys.* 42, 2162.

Damon, R. W., and Eshbach, J. R. 1961. Magnetostatic modes of a ferromagnet slab, *J. Phys. Chem. Solids* 19, 308.

DeJeu, W. H., 1980. *Physical Properties of Liquid Crystalline Materials*, Gordon and Breach, New York, Chapters 4–5.

Devaux, F., Muller, S., Ougazzaden, A., Mircea, A., Ramdane, A., Krauz, P., Semo, J., Huet, F., Carre, M., and Carenco, A. 1994. Zero-loss multiple-quantum-well electroabsorption modulator with very low chirp, *Appl. Phys. Lett.* 64, 954.

Dieulesaint, E., and Royer, D. 1980. *Elastic Waves in Solids*. John Wiley & Sons, Chichester

Dillon, J. F. 1978. Magneto-optical properties of magnetic garnets, in *Physics of Magnetic Garnets*, A. Paoletti, ed. North-Holland, New York, p. 379.

Dixon, R. W., 1967a. Acoustic diffraction of light in anisotropic media, *IEEE J. Quantum Electron.* QA-3, 85.

Dixon, R. W. 1967b. Photoelastic properties of selected materials and their relevance for applications to acoustic light modulators and scanners, *J. Appl. Phys.* 38, 5149.

Dolfi, D. W., and Ranganath, T. R. 1992. 50 GHz velocity-matched broad wavelength $LiNbO_3$ modulator with multimode active sections, *Electron. Lett.* 28, 1197.

Donjon, J., Dumont, F., Grenot, M., Hazan, J. P., Marie, G., and Pergrale, J. 1973. A Pockels-effect light valve: Phototitus. Applications to optical image processing, *IEEE Trans. Elec. Dev.* ED-20, 1037.

Eden, D. D. 1979. Some applications involving the semiconductor-to-metal phase transition in VO_2, *Optical Processing Systems Proc. SPIE* 185, 97.

Fisher, A. D. 1985. Optical signal Processing with Magnetostatic Waves, *Circuits, Systems, Signal Processing* 4, 265.

Fisher, A. D., Lee, J. N., Gaynor, E. S., and Tveten, A. B. 1982. Optical guided-wave interactions with magnetostatic waves at microwave frequencies, *Appl. Phys. Lett.* 41, 779.

Fisher, A. D., Ling, L. C., and Lee, J. N. 1984. A high performance photo-emitter membrane spatial light modulator, in *Spatial Light Modulators and Applications*, U. Efron, ed., Proc. SPIE 465, 36.

Fisher, A. D., Ling, L. C., Lee, J. N., and Fukuda, R. C. 1986. The photo-emitter membrane light modulator, *Opt. Eng.* 25, 271.

Garito, A. F., Teng, C. C., Wong, K. Y. and Zammani' Khamiri, O. 1984. Molecular optics: Nonlinear optical processes in organic and polymer crystals, *Mol. Cryst. Liq. Cryst.* 106, 219.

Goodby, J. W., and Patel, J. S. 1986. Properties of Ferroelectric Crystals, *Liquid Crystals and Spatial Light Modulators*, Proc. SPIE, 52

Goutzoulis, A. P., and Gottlieb, M. S. 1988. Design and performance of optical activity based Hg_2Cl_2 Bragg cells, *Advances in Optical information Processing III*, Proc. SPIE 936, 119.

Griffel, G., Ruschin, S., and Croitrou, N. 1989. Linear electro-optic effect in sputtered polycrystalline $LiNbO_3$ films, *Appl. Phys. Lett.* 54, 1385.

Groth, G., and Marie, G. 1970. Information input in an optical pattern recognition system using a relay tube based on the Pockels effect, *Opt. Comm.* 2, 133.

Hampp, N., Thoma, R., Oesterhelt, D., and Brauchle, C. 1992. Biological photochrome bacteriorhodopsin and its genetic variant ASP96→Asn as media for optical pattern recognition, *Appl. Opt.* 31, 1834.

Hecht, D. L. 1976. Spectrum analysis using acoustooptic devices, in *Acoustooptics*, Proc. SPIE 90, 148.

Hecht, D. L., and Petrie, G. W. 1980. Acousto-optic diffraction from acoustic anisotropic shear modes in gallium phosphide, *Proc. 1980 IEEE Ultrasonics Symp., IEEE No. 80CH1602-2*, 474.

Hornbeck, L. J. 1989. Deformable-mirror spatial light modulators, in *Spatial Light Modulators and Applications III*, Proc. SPIE 1150, 86.

Horowitz, B. A., and Corbett, F. J. 1978. The PROM - Theory and application of the Pockels readout optical modulator, *Opt. Eng.* 17, 353.

Jain, F. C. and Bhattacharjee, K. K. 1989. Multiple quantum well optical modulator structures using surface acoustic wave induced stark effect, *IEEE Photon. Technol. Lett.* 1, 307.

Jared, D. A., and Johnson, K. M. 1991. Optically addressed thresholding very-large-scale-integration liquid crystal spatial light modulator, *Opt. Lett.* 16, 967.

Jenkins, F. A. and White, H. E. 1976. *Fundamentals of Optics*, 4th Ed. McGraw-Hill, New York.

Jin, M. S., Wang, J. H., Ozguz, V., and Lee, S. H. 1994. Monolithic integration of a silicon driver circuit onto a lead lanthanum zirconate titanate substrate for smart spatial light modulator fabrication, *Appl. Opt.* 33, 2842.

Johnson, L. M., and Rousell, H. V. 1988. Reduction of intermodulation distortion in interferometric optical modulators, *Opt. Lett.* 13, 928.

Johnson. R. V., Hecht, D. L., Sprague, R. A., Flores, L. N., Steinmetz D. L., and Turner, W. D. 1983. Characteristics of the linear array total internal reflection (TIR) electrooptic spatial light modulator for optical information processing, *Opt. Eng.* 22, 665.

Kahn, F. J., Kendrick, P. N., Leff, J., Livoni, L. J., Loucks, B. E., and Stepner, D. 1987, A paperless plotter display system using a laser smectic liquid-crystal light valve, 1987 SID International Symposium, Digest of Technical Papers, paper 14.1, p. 254.

Kaminow, I. P., and Turner, E. 1971. Linear electrooptic materials, in *Handbook of Lasers*. Chemical Rubber Co., Cleveland, Chapter 15.

Klein, W. R., and Cook, B. D. 1967. Unified approach to ultrasonic light diffraction, *IEEE Trans. Sonics Ultrason.* SU-14, 123.

Korotsky, S. K., and DeRidder, R. M. 1990. Dual parallel modulation schemes for low-distortion analog optical transmission, *IEEE J. Select. Areas Commun.* 8, 1377.

Koster, G. F., Dimmock, J. O., Wheeler, R. G., and Statz, H. 1963. *Properties of the Thirty-Two Point Groups*, MIT Press, Cambridge, MA.

Land, C. E. 1978. Optical information storage and spatial light modulation in PLZT ceramics, *Opt. Eng.*, 17, 317.

Lau, K. Y., Derry, P. L., and Yariv, A. 1988. Ultimate limit in low threshold quantum well GaAlAs semiconductor lasers, *Appl. Phys. Lett* 52, 88.

Lee, S. H., Esner, S. C., Title, M. A., and Drabik, T. J. 1986. Two-dimensional silicon/PLZT spatial light modulators: Design considerations and technology, *Opt. Eng.* 25, 250.

Lee, J. N. 1983. Signal processing using an acousto-optic memory device, in *Acousto-optic Signal Processing: Theory and Implementation*, N. J. Berg and J. N. Lee, eds. Marcel Dekker, New York, p. 203.

Lee, J. N., and Fisher, A. D. 1987. Device developments for optical information processing, *Adv. Electron. Elec. Phys.* 69, 115.

Lee, J. N., and VanderLugt, A. 1989. Acousto-optic Signal Processing and Computing, *Proc. IEEE* 7, 1528.

Liau, Z. L., Walpole, J. N., and Tsang, D. Z. 1984. Mass-transported GaInAsP/InP buried-heterostructure lasers and integrated mirrors, Technical Digest 7th Topical Meeting on Integrated and Guided-Wave Optics, paper TuC5, Optical Society of America, Washington, D.C.

Liu, H. K., Davis, J. A., and Lilly, R. A. 1985. Optical data processing properties of a liquid-crystal television light modulator, *Opt. Lett.* 10, 635.

Mandal, B. K., Chen, Y. M., Lee, J. Y., Kumar, J., and Tripathy, S. 1991. Cross-linked stable second-order nonlinear optical polymer by photochemical reaction, *Appl. Phys. Lett.* 58, 2459.

Marowsky, G., Chi, L. F., Mobius, D., Steinhoff, R., Shen, Y. R., Dorsch, D., and Rieger, B. 1988. Nonlinear optical properties of hemicyanine monolayers and the protonation effect, *Chem. Phys. Lett.* 147, 420.

McEwan, J. A., Fisher, A. D., Rolsma, P. B., and Lee, J. N. 1985. Optical processing characteristics of a low cost liquid crystal display, *J. Opt. Soc. Amer.* A 2 (13), 8.

Miller, D. A. B., Chemla, D. S., Damen, T. C., Gossard, A. C., Wiegmann, W., Wood, T. H., and Burrus, C. A. 1984. Novel hybrid optically bistable switch: The quantum well self-electro-optic effect device, *Appl. Phys. Lett.* 45, 13.

Nakamura, M., Yen, H. W., Yariv, A., Garmire, E., Somekh, S. 1973. Laser oscillation in epitaxial GaAs waveguides with corrugation feedback, *Appl. Phys. Lett.* 23, 224.

Neurgaonkar, R. R., and Cory, W. K. 1986. Progress in photorefractive tungsten bronze crystals, *J. Opt. Soc. Am.* B, 3, 274.

Oesterhelt, D., and Stoeckenius, W. 1971. Rhodopsin-like protein from the purple membrane of Halobacterium halobium, *Nature* 233, 149.

Ohkawa, M., Izutsu, M., and Sueta, T. 1989. Integrated optic pressure sensor on silicon substrate, *Appl. Opt.* 28, 5153.

Optron Systems, Inc., *Wire-, Optically- and Electron-Beam-Addressed Membrane Light Modulator, Product Descriptions*, Bedford, MA, 1994.

Owechko, Y. and Tanguay, A. R. 1982. Effects of operating mode on electrooptic spatial light modulator resolution and sensitivity, *Opt. Lett.* 7, 587.

Pape, D. R. 1985. Optically addressed membrane spatial light modulator, *Opt. Eng.* 24, 107.

Partovi, A., Millerd, J., Garmire, E. M., Ziari, M., Steier, W. H., Trivedi, S. B., and Klein, M. B. 1990. Photorefractivity at 1.5 μm in CdTe:V, *Appl. Phys. Lett.* 57, 846.

Pepper, D. M., Feinberg, J., and Kukhtarev, N. V. 1990. The photorefractive effect, *Scientific American*, Oct. 1990, 62.

Petrov, M. P. 1981. Electrooptic photosensitive media for image recording and processing, *Current Trends in Optics*, F. T. Arecchi and F. R. Aussenegg, eds., p. 161, Taylor Francis, London.

Preston, K. P. 1969. The membrane light modulator and its application to optical computers, *Optica Acta* 16, 579.

Preston, K. D., and Haertling, G. H. 1992. Comparison of electro-optic lead-lanthanum zirconate titanate films on crystalline and glass substrates, *Appl. Phys. Lett.* 60, 2831.

Rakuljic, G. A., Sayano, K., Agranat, A., Yariv, A., and Neurgaonkar, R. R. 1988. Photorefractive properties of Ce- and Ca-doped $Sr_{0.6}Ba_{0.4}NbO_3$, *Appl. Phys. Lett.* 53, 1465.

Ralston, L. M., and McDaniel, R. V. 1979. Experimental evaluation of an elastomer storage device, *Optical Processing Systems* Proc. SPIE 185, 86.

Reizman, F. 1969. *Proc. Electrooptic System Design*, 225.

Roe, M. G., and Schehrer, K. L. 1993. High-speed and high-contrast operation of ferroelectric liquid crystal optically addressed spatial light modulators, *Opt. Eng.* 32 (7), 1662.

Rolsma, P. B., and Lee, J. N. 1990. Polymer membrane properties and structures for membrane light modulator applications, *Opt. Lett* 15, 712.

Ross, W. E., Psaltis, D., and Anderson, R. H. 1983. Two-dimensional magneto-optic spatial light modulator for signal processing, *Opt. Eng.* 22, 485.

Sayyah, K., Efron, U., Forber, R. A., Goodwin, N. W., and Reif, P. G. 1991. Schottky diode silicon liquid crystal light valve, in *Liquid-Crystal Devices and Materials*, Proc. SPIE *1455*, 249.

Sayyah, K., Wu, C. S., Wu, S. T., and Efron, U. 1992. Anomalous liquid crystal undershoot effect resulting in a nematic liquid crystal-based spatial light modulator with one millisecond response time, *Appl. Phys. Lett.* 61, 883.

Schneeberger, B., Laeri, F., Tschudi, T., and Mast, F. 1979. Real-time spatial light modulator, *Opt. Comm.* 31, 13.

Schwartz, A., Wang, X.-Y., and Warde, C. 1985. Electron-beam-addressed microchannel spatial light modulator, *Opt. Eng.* 24, 119.

Seino, M., Mekada, N., Namiki, T., and Nakajima, H. 1988. 33-GHz-cm Broadband $Ti:LiNbO_3$ Mach–Zehnder modulator, ECOC '88 Conference, Digest of Paper Summaries, paper ThB22-5, 433.

Singer, K. D. and Garito, A. F. 1981. Measurements of molecular second order optical susceptibilities using dc induced second harmonic generation, *J. Chem. Phys.*, 75(7), 3572.

Singer, K. D., Kuzyk, M. G., Holland, W. R., Sohn, J. E., Lalama, S. J., Comizzoli, R. B., Katz, H. E. and Schilling, M. L. 1988. Electro-optic phase modulation and optical second-harmonic generation in corona-poled polymer films, *Appl. Phys. Lett.* 53, 1800.

Singh, N. B., and Todd, D. J. 1990. *Growth and Characterization of Acousto-optic Materials*, Trans Tech Publications, Switzerland

Skeie, H., and Johnson, R. V. 1991. Linearization of electro-optic modulators by a cascade coupling of phase modulating electrodes, *Integrated Optical Circuits*, Proc. SPIE 1583, 153.

Smith, D. A., Baran, J. E., Cheung, K. W., and Johnson, J. J. 1990, Polarization-Independent acoustically tunable filter, *Appl. Phys. Lett.* 60, 1538.

Somers, L. E. 1972. The photoemitter-membrane light modulator image transducer, *Adv. Elec. Electron Physics* 33A, 493.

Stahelin, M., Burland, D. M., Ebert, M., Miller, R. D., Smith, B. A., Tweig, R. J., Volksen, W., and Walsh, C. A. 1992. Reevaluation of the thermal stability of optically nonlinear polymeric guest-host systems, *Appl. Phys. Lett.* 61, 1626.

Suzuki, M., Tanaka, H., and Matsushima, Y. 1992. InGaAsP electroabsorption modulator for high-bit-rate EDFA system, *Photon. Technol. Lett.* 4, 586.

Tai, K., Yang, L., Wang, Y. H., Wynn, J. D., and Cho, A. Y. 1990. Drastic reduction of series resistence in doped semiconductor distributed Bragg reflectors for surface emittting lasers, *Appl. Phys. Lett.* 56, 1942.

Tamada, H., Kaneko, M., and Okamoto, T. 1988. TM-TE optical-mode conversion induced by a transversely propagating magnetostatic wave in $(BiLu)_3Fe_5O_{12}$ film, *J. Appl. Phys.* 64, 554.

Taylor, B. K. and Casasent, D. P. 1986, Error-source effects in a high accuracy optical finite-element processor, *Appl. Opt.* 25, 966.

Tsai, C. S., Young, D., Chen, W., Adkins, L., Lee, C. C., and Glass, H. 1985. Noncollinear coplanar magneto-optic interaction of guided optical waves and magnetostatic surface waves in yttrium iron garnet-gadolinium gallium garnet waveguide, *Appl Phys. Lett.* 47, 651.

Turner, R. M., Johnson, K. M., and Serati, S. 1995. High-speed compact optical correlator design and implementation, in *Design Issues in Optical Processing*, J. N. Lee, ed. Cambridge Press, Cambridge, Chap. 5, pp. 169–219.

Turpin, T. Real-time input transducer for coherent optical processing, 1974. Digest 1974 Int. Optical Computing Conference, p. 34.

Turpin, T. M. 1981. Spectrum analysis using optical processing, *Proc. IEEE* 69, 79.

Uchida, N. and Niizeki, N. 1973. Acousto-optic deflection materials and techniques, *Proc. IEEE* 61, 1073.

Uchida, N. and Ohmachi, Y. 1969. Elastic and photoeleastic properties of of TeO_2 single crystal, *J. Appl. Phys.* 42, 4692.

Uomi, K., Mishima, T., and Chinone, N. 1985. Ultra-high relaxation oscillation frequency (up to 30GHz) of highly p-doped GaAlAs multiquantum well lasers, *Appl. Phys. Lett.* 51, 78.

Vadekar, A., Nathan, A. and Huang, W. P. 1994. Analysis and design of an integrated silicon ARROW Mach–Zehnder micromechanical interferometer, *J. Lightwave Technol.* 12, 157.

Valley, G. C., McCahon, S. W., and Klein, M. B. 1988. Photorefractive measurement of photoionization and recombination cross-sections in InP:Fe, *J. Appl. Phys.* 64, 6684.

VanderLugt, A., Moore, G. S., and Mathe, S. S. 1983. Multichannel Bragg cell compensation for acoustic spreading, *Appl. Opt.* 22, 3906.

Wakita, K., Kotaka, I., and Asai, H. 1992. High-Speed InGaAlAs/InAlAs multiple quantum well electrooptic phase modulators with bandwidths in excess of 20 GHz, *Photon. Technol. Lett.* 4, 29.

Wey, A. C. T., Tuan, H. S., Parekh, J. P., Craig, A. E., and Lee, J. N. 1986, Inhomogeneous field MSFVW-optical interaction, *Integrated Optical Circuit Engineering IV*, Proc. SPIE 704, 51.

White, J. O., and Yariv, A. 1984. Photorefractive crystals as optical devices, elements, and processors, in *Solid State Optical Control Devices*, Proc. SPIE 464, 7.

Woodruff, D. C. 1991. New shutter designs meet challenge of laser systems, *Laser Focus World* 27, 129.

Woodruff, T. O., and Ehrenreich, H. 1961. Absorption of sound in insulators, *Phys. Rev.* 123, 1553.

Yariv, A. 1989. *Quantum Electronics*, 3rd Ed. John Wiley & Sons, New York, Chapters 14 and 16.

Young, D., and Tsai, C. S. 1989. X-band magneto-optic Bragg cells using bismuth-doped yttrium iron garnet waveguides, *Appl. Phys. Lett.* 55, 2242.

Young, E. H., and Belfatto, R. V. 1993. Polychromatic acousto-optic modulators let users tailor output wavelengths, in *Laser Focus World*, Nov. 1993, 179.

Young, E. H. and Yao, S. K. 1981. Design considerations for acousto-optic devices, *Proc. IEEE* 69, 54.

Yu, S., and Forrest, S. 1993. Implementations of smart pixels for optoelectronic processors and interconnection systems: I. Optoelectronic gate technology, and II. SEED-based technology and comparison with optoelectronic gates, *J. Lightwave Technol.* 11, 1659.

10
Optical Fibers

D. A. Nolan
Corning Incorporated

10.1	Introduction	435
10.2	Optical Propagation	436
10.3	Multimode Fibers	441
10.4	Attenuation	443
10.5	Single-Mode Propagation	443
10.6	Dispersion-Shifted Fiber	445
10.7	Fiber Fabrication	445
10.8	Polarization	446
10.9	Optical Fiber Amplification	449
10.10	Optical Nonlinearities in Fibers	451
10.11	Solitons in Optical Fiber	452

10.1 Introduction

Optical communication technology continues to expand after many years of rapid growth. It was over twenty years ago that Corning Incorporated (Kapron et al., 1970) demonstrated that the attenuation in doped silica fiber was low enough that long-distance communication through optical fibers would be advantageous over copper. Soon thereafter, fiber-optic systems proved feasible. In the 70's, the first commercial systems used multimode fibers to better capture the available light and because the state of the art of the associate components was better suited for a large-core fiber. In the 80's, long-haul transmission proved, beyond a doubt, that fiber-optic transmission was a commercial reality. The use of single-mode fiber with laser diodes enabled transmission rates to reach the gigabit-per-second rate over tens of kilometers. Almost immediately, single-mode fiber replaced multimode fiber in trunk line applications. Multimode fiber, however, found important applications in local area networks.

At this point, research activities in the area of new optical fibers and applications of such fibers have continued to expand at a rapid pace. Areas of research, for example, have included new index profiles for transmission in the 1550 nm window, rare earth-doped fibers for optical amplifiers and fiber lasers, the demonstration of soliton transmission in fiber, the demonstration of a number of important nonlinear effects in fibers, and the development of numerical techniques to simulate optical propagation. Meanwhile, process research has led to rapid reduction in the price of fiber. Commercially, the deployment of fiber has expanded beyond the trunk lines to feeder lines, to major business users and, in a number of cases, to the home. All of these activities have progressed in parallel with the development of new active and passive components and advances in communication systems and networks.

In this chapter, we will review the fundamental properties of optical fibers. This will include the material and fabrication properties and optical propagation fundamentals. Propagation in the optical fibers used in today's communication systems is best described using the concept of guided modes. In Section 10.2, the concept of modes of propagation will be introduced. Multimode (section 10.3) and single-mode fiber will be discussed with particular attention being paid to single-mode fiber. This will include single-mode fiber for light transmission at 1.31 μms and 1.55 μms. Index profiles suitable for transmission in the 1.55 μm window, known as dispersion-shifted fibers, are receiving much attention today because the intrinsic loss of the fiber is low in this window and, more recently, because erbium-doped, fiber-optic amplifiers are expected to change the transmission architecture. These amplifiers are effective only for light transmission in the 1.55 μm window. In section 10.4, we will discuss the causes of attenuation in optical fiber. Attenuation of the optical signal in today's fiber approaches the theoretical minimum due to Rayleigh scattering. In Section 10.5, the important single-mode parameters will be introduced and in section 10.6, the dispersion-shifted profiles are shown capable of low-dispersion propagation at 1.55 μm.

In section 10.7, we will discuss the optical fiber fabrication process. This includes techniques for fabricating the fiber blank and drawing the blank into fiber. Today's blank manufacturing processes include the outside vapor deposition process, in which soot particles are collected on a bait rod and then sintered to glass at elevated temperatures, the chemical deposition process, in which glass particles are formed inside a glass tube and deposited onto the wall, and the axial deposition process where the soot stream of the outside deposition process is deposited axially. Important to the reliability is the organic coating put on the fiber during the fiber-drawing process. The finished fiber is cabled after manufacturing to be ready for deployment.

In section 10.8, we will discuss more advanced topics related to the propagation of electromagnetic waves in a fiber. This will include the subject of modes of polarization. Polarization-mode dispersion (PMD) has become an important issue for ultralong-distance communication and for analog transmission for video transport. For these systems, polarization-mode dispersion must be kept to a minimum value. A significant amount of research has been recently published addressing the concept requirements and measurement of PMD. On the other hand, polarization-maintaining fiber enables one to retain a given polarization through the fiber. These fibers are used in sensors, such as the fiber-optic gyroscope.

The fiber-optic amplifier, section 10.9, represents a major advance in fiber-optic technology. Commercialization of this device is in the initial stages. Amplification of the signal enables one to communicate over ultralong distances without the need for expensive electronic regeneration. This is expected, in part, to enable two important technologies: wavelength division multiplexing and new applications in nonlinear optics, such as soliton pulse propagation. The use of amplifiers in optical communication systems is not straightforward from a system's point of view because the optical nonlinearities that develop at these higher powers can significantly affect signal quality, section 10.10. Research into new applications of fiber optics via amplifiers will, no doubt, continue to expand. The promise of optical communication using many wavelengths and exploiting nonlinearities, such as soliton propagation, section 10.11, is indeed exciting.

10.2 Optical Propagation

The phenomenon of total internal reflection first reported by Tyndall (1884), enables us to understand how one can transmit light over many kilometers through a fiber. Geometrical optics can be used to describe the repeated reflection of light through the guide. The description is useful for analyzing optical propagation through multimode fiber and can be used, to a limited extent, to understand single-mode propagation. Figure 10.1 shows the variation of the index squared with radius for gradient-index multimode and step-index, single-mode fibers. For a step-index fiber, light undergoes total internal reflection if the angle made by the ray is less than the critical angle. The critical angle θ_c is given (assuming launch from air into fiber) by

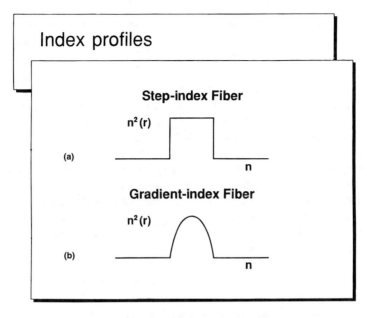

FIGURE 10.1 Index profiles a) step index; b) parabolic index.

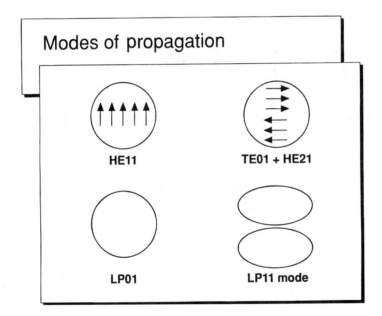

FIGURE 10.2 Schematic of the LP modes.

$$\sin(\theta_c) = (n_{co}^2 - n_{cl}^2)^{1/2}. \qquad (10.1)$$

Therefore, light propagating at angles less than the critical angle (θ_c) is guided and is described with an acceptance angle called the numerical aperture (NA), defined by

$$NA = \sin(\theta_c). \qquad (10.2)$$

Light is, of course, refracted when input to the guide as described by Snell's law.

$$\sin(\varphi_i) = n_{co} \sin(\varphi_r) \qquad (10.3)$$

The index of the core n_{co} is slightly higher than that of the cladding n_{cl} which, for silica, is 1.457 in the visible. Rays incident with different angles will propagate along the fiber with different speeds. This, of course, results in a distortion of the signal or pulse and, therefore, a decrease in the possible signal rate or temporal bandwidth. To circumvent this problem, two approaches have been considered. The first is the use of graded-index multimode fibers. In a step-index, multimode fiber, the rays of higher propagation angle are required to traverse along a greater optical path. The optical path can be affected by altering the index with radius. In a graded-index multimode fiber, the use of a parabolic index profile (Gloge and Marceteli, 1973) effectively balances the optical path of the higher angle rays with the optical path of the rays making a smaller traverse angle with the axis of the fiber. Later, Olshansky and Keck (1975) showed that a more optimum profile is of the form

$$n^2(r) = n_{co}^2(0) - 2\Delta(r/a)^\alpha \qquad (10.4a)$$

where a is the core diameter, the exponent α depends on the wavelength of operation and Δ is determined from the core and clad indices by

$$\Delta = (n_{co}^2(0) - n_{cl}^2)/n_{co}^2(0). \qquad (10.4b)$$

The exponent α is a function of the index change with wavelength or index dispersion.

The use of single-mode fiber eliminates the need to balance the optical paths of different rays because only one or mode propagates through the fiber. Analysis of pulse transmission through fibers is most effective using the concept of modal propagation rather than a ray-tracing analysis. Here, the scalar wave equation is used to determine the modes of propagation within the fiber. The scalar wave equation is derived from Maxwell's equations.

$$\overline{\nabla} \times \overline{E} = -\partial \overline{B}/\partial t, \qquad (10.5)$$

$$\overline{\nabla} \times \overline{H} = \partial \overline{D}/\partial t, \qquad (10.6)$$

$$\overline{\nabla} \cdot \overline{D} = 0, \qquad (10.7)$$

and

$$\overline{\nabla} \cdot \overline{B} = 0. \qquad (10.8)$$

Here, $\overline{E}$ and $\overline{H}$ are the field vectors. $\overline{D}$ and $\overline{B}$ are the respective flux densities and can be expressed as

$$\overline{D} = \epsilon_0 \overline{E} + \overline{P} \qquad (10.9)$$

and

$$\overline{B} = \mu_0 \overline{H} + \overline{M}. \qquad (10.10)$$

In Eqs. 10.9 and 10.10, P and M are the induced electric and magnetic polarizations, and ϵ_0 and μ_0 are the vacuum permitivity and permeability. The scalar wave equation is obtained by taking the curl of Eq. 10.5

Optical Fibers

$$\bar{\nabla} \times \bar{\nabla} \times \bar{E} = -\partial \bar{\nabla} \times \bar{B}/\partial t \tag{10.11}$$

using

$$\bar{\nabla} \times \bar{B} = \bar{\nabla} \times [\mu_0 \bar{H} + \bar{M}]. \tag{10.12}$$

Because glass fibers are an insulator, we set

$$\bar{\nabla} \times \bar{M} = 0 \tag{10.13}$$

and, then,

$$\bar{\nabla} \times \bar{\nabla} \times \bar{E} = -\partial \bar{\nabla} \times \mu_0 \bar{H}/\partial t$$
$$= \mu \partial/\partial t \partial \bar{D}/\partial t. \tag{10.14}$$

The vector wave equation can now be written

$$\bar{\nabla} \times \bar{\nabla} \times \bar{E} = -\mu_0 \epsilon_0 \partial^2 \bar{E}/\partial t^2 - \mu_0 \partial^2 \bar{P}/\partial t^2. \tag{10.15}$$

Using the expression

$$\bar{\nabla} \times \bar{\nabla} \times \bar{E} = \bar{\nabla}(\bar{\nabla} \cdot \bar{E}) - \nabla^2 \bar{E} \tag{10.16}$$

and because glass fibers are an insulating material,

$$\bar{\nabla} \cdot \bar{E} = 0, \tag{10.17}$$

and the vector wave equation becomes

$$\bar{\nabla}(\bar{\nabla} \cdot \bar{E}) = -\mu_0 \epsilon_0 \partial^2 \bar{E}/\partial t^2 - \mu_0 \partial^2 \bar{P}/\partial t^2. \tag{10.18}$$

The susceptibility relates the electric field with the polarization according to

$$\epsilon \bar{E} = \epsilon_0 \bar{E} + \bar{P} \tag{10.19}$$

and because the speed of light in a vacuum is related to the permitivity and permeability

$$c = \frac{1}{\sqrt{\epsilon_0 \mu_0}}, \tag{10.20}$$

the vector wave equation becomes

$$\nabla^2 \bar{E} = -\mu_0 \epsilon \frac{\partial^2 \bar{E}}{\partial t^2}. \tag{10.21}$$

In the case of optical fibers, the dielectric constant is the index squared times ϵ_0 so that the vector wave equation is

$$\nabla^2 \bar{E} = -\mu_0 \epsilon_0 n^2 \frac{\partial^2 \bar{E}}{\partial t^2}$$
$$= -\frac{n^2}{c^2} \frac{\partial^2 \bar{E}}{\partial t^2}. \tag{10.22}$$

We can separate the spatial (r) and temporal (t) variables by assuming

$$\overline{E}(\overline{r}, t) = \overline{E}(\overline{r})e^{i\omega t} \tag{10.23}$$

so that

$$\nabla^2 \overline{E}(r) = -n^2 \frac{\omega^2}{c^2} \overline{E}(r)$$

$$= -n^2 k_0^2 \overline{E}(r) \tag{10.24}$$

where the free-space wave number is given by

$$k_0 = \frac{2\pi}{\lambda}. \tag{10.25}$$

In the scalar wave approximation, the radial and azimuth components of the electric and magnetic fields can be determined from the axial components. For this reason, we need solve only Eq. 10.23 for the axial component.

$$\frac{\partial^2 E_z}{\partial r^2} + \frac{1}{r}\frac{\partial E_z}{\partial \tau} + \frac{1}{r^2}\frac{\partial^2 E_z}{\partial z^2} + n^2 k_0^2 E_z = 0. \tag{10.26}$$

We assume that

$$E_z(r, \theta, z) = \Psi(r)e^{i\beta z} \tag{10.27a}$$

and

$$\frac{\partial^2 \Psi(r)}{\partial r^2} + \frac{1}{r}\frac{\partial \Psi(r)}{\partial r} + \left(n^2(r)k_0^2 - \frac{l^2}{r^2} - \beta^2\right)\Psi(r) = 0. \tag{10.27b}$$

The term β is referred to the propagation constant, and its value is bounded by

$$\frac{2\pi}{\lambda} n_{co} > \beta > \frac{2\pi}{\lambda} n_{cl} \tag{10.28}$$

where n_{co} is the maximum index of the core and n_{cl} is the index of the cladding.

At this point, it is useful to analyze Eq. 10.27 for different fiber-index distributions. These solutions enable one to determine the modes of propagation for a given index profile. Profiles usually considered (Snyder and Love, 1983) include step, parabolic and gaussian distributions of the core index.

The step-index profile can be described using Eq. 10.4 with $\alpha = \infty$. The number of bound modes depends on the numerical aperture. Eq. 10.2, or Δ, Eq. 4b. The solutions to Eq. 10.27 for the step-index profile are the well-known Bessel functions. So that the fields are finite at $r = 0$ and ∞,

$$\Psi_m = N^{1/2} J_l(u)_{r \leq a} \tag{10.29a}$$

$$= N^{1/2} K_l(w)_{r \geq a}. \tag{10.29b}$$

Here, N is a normalization factor,

$$u^2 = (n_{co}^2 k_0^2 - \beta^2)a^2, \tag{10.30}$$

$$w^2 = (\beta^2 - n_{cl}^2 k_0^2)a^2, \tag{10.31}$$

and

$$v^2 = u^2 + w^2. \tag{10.32}$$

A set of eigenvalue equations determines the propagation constant β. The eigenvalue equations are determined by requiring that both the tangential electric and magnetic fields are continuous at the core-clad boundary. Both the tangential electric and magnetic fields can be derived from the axial components (Agrawal, 1992). However, Gloge (1971) has shown that the longitudinal electric and magnetic field components are essentially zero for small Δ. In this case, one characteristic equation can be used for all modes:

$$u[J_{l-1}(u)/J_l(u)] = -w\{[k_{l-1}(w)/k_l(w)]\}. \tag{10.33}$$

This expression is termed the linearly polarized (LP) mode approximation. For $l \geq 1$, each solution comprises four modes: the two polarizations and the orientation of the lobes associated with l. The fields are of the form

$$\Psi_l = \frac{1}{2}[J_l(ur/a)/J_l(u)][\cos(l+1)0 + \cos(l-1)\phi]r \leq a$$

and

$$\frac{1}{2}[k_l(wr/a)/k_l(w)][\cos(l+1)0 + \cos(l-1)\phi]r \geq a. \tag{10.34}$$

The linearly polarized mode approximation is very useful for analyzing the key attributes of dispersion in both multimode and single-mode fibers. It is also useful, to a limited extent, in analyzing the propagation of polarized modes within fibers.

10.3 Multimode Fibers

As mentioned above, gradient-index cores are used to minimize mode propagation differences within fibers. The scalar wave equation can be used to determine the mode-delay differences as a function of the index profile. For gradient-index profiles the parameters of Eq. 4 are useful for determining the number of modes within the fiber and the mode-delay differences. This parametric description of the index is used in Eq. 10.27 to determine the number of guided modes. The solutions to Eq. 10.27 with $\alpha = 2$ are the Laguerre–Gauss functions:

$$n^2(r) = n_{co}^2(1 - g^2 r^2). \tag{10.35}$$

with (Vassallo, 1993)

$$g^2 \equiv 2\Delta/a^2.$$

The solutions are

$$\emptyset_{nr}(r) = e^{-kn_{co}^2 gr^2/2}(\sqrt{kn_{co}}\,r)^\eta L_\mu^\eta(kn_{co}gr^2). \tag{10.36}$$

The propagation constants $\beta_{n,m}$ are determined from the eigenvalues

$$\beta_{m,n}^2 = k^2 n_{co}^2 - 2(\mu + 2\eta + 1)kn_{co}g. \tag{10.37}$$

The associated propagation delay time is given by

$$\tau_{\mu,\eta} = \frac{L}{c}\frac{\partial \beta}{\partial k}. \tag{10.38}$$

Solving Eq. 10.36 for $\tau_{\mu,\eta}$, it has been shown that $\tau_{\mu,\eta}$ is independent of μ and η to the first order in Δ and, therefore, the parabolic profile optimizes the index for a temporal bandwidth. However, Olshansky and Keck (1975) have shown that, because the index of glass is a function of wavelength, the parabolic profile is, in fact, not optimized for bandwidth. They used the exponent α to characterize the radial dependence of the index and, using the WKB approximation, showed that, for minimum pulse distortion,

$$\alpha = 2 + y - \Delta\frac{(4 + y)(3 + y)}{5 + 2y}. \tag{10.39a}$$

where

$$y = \frac{-2n_1}{n_1 - \frac{dn_{co}}{d\lambda}}\frac{\lambda}{\Delta}\frac{d\Delta}{d\lambda}.$$

As can be seen from Eq. 10.38, the temporal bandwidth depends on the value of Δ. In practice, bandwidths as high as a Ghz/km have been achieved with Δs of 2%.

The WKB method is used to solve equations of the form

$$\frac{d^2 f}{dx^2} = P(x)f(x) = 0. \tag{10.40}$$

For circular fibers, the transformation

$$r = e^\xi \tag{10.41}$$

is used to express the radial wave equation in the form of Eq. 10.27, after which one obtains an equation for the propagation constant

$$\int_{r_1}^{r_2}\left[k^2 n^2(r) - \beta_{m,n}^2 - \frac{m^2}{r^2}\right]^{1/2} dr = \frac{\pi}{2} + N\pi. \tag{10.42}$$

Here, r_1 and r_2 are the turning points (where the bracket expression is zero) and, for $m = 0$, r_1 is chosen as zero. The WKB is most useful for fibers with a large number of modes, i.e., core

Optical Fibers

radius much longer than the wavelength of the propagating light. For more information on the application of the WKB method to circular guides, see Morse and Feschbach, 1953.

10.4 Attenuation

In addition to the requirement of minimum pulse broadening during pulse propagation, minimum loss in optical power is also required. The optical fiber fabrication process minimize the introduction of common transition-metal impurities, such as iron, copper, cobalt, etc., into the glass. The important source of attenuation in fiber can be classified as due to molecular vibrational resonance, vibrational resonances of the OH ion, and Rayleigh scattering. The molecular vibrational resonance in optical fiber glasses occurs in the infrared region beyond 5 μms, but these bands have tails that extend into the region at which optical signals propagate (0.8 to 1.6 μms). Although small, these bands contribute a loss to optical propagation on the order of 0.5 dB/km and less. Vibrational resonances in fibers can also result from the inclusion of OH ions in the silica matrix. An important vibrational resonance of the OH ion occurs at 2.71 μms. There are overtones and combination bands at 0.95, 1.25, and 1.38 μms due to the 2.71-μm resonance (Agrawal, 1992). Optical fiber fabrication processes limit the amount of OH ions through proper doping procedures, and it is possible to fabricate fibers with OH ion inclusions on the order of parts per billion. The windows of transmission of 0.8, 1.3 and 1.55 μm are chosen so that losses due to OH ion vibrational resonances are limited.

Rayleigh scattering of light is strongly wavelength-dependent and varies inversely with the fourth power of the wavelength. This source of optical loss results from concentration fluctuations at high temperatures that are frozen in place as the glass cools through the transformation region. One can calculate an effective loss coefficient due to Rayleigh scattering by the expression

$$\alpha_R = \frac{8\pi^3}{3\lambda^4}(n^2 - 1)KTB \qquad (10.43)$$

Here, B represents the isothermal compressibility of the glass, T is the fictive temperature and K is Boltzman's constant. At 1550 nms, losses resulting from Rayleigh scattering are on the order of 0.13 dB/km. This window is minimum with respect to optical attenuation as it is optimized with respect to the losses resulting from Rayleigh scattering (which increase with lower wavelength) and those from molecular vibrational resonances (which increase at higher wavelengths). Losses in this window are as low as 0.2 dB/km, whereas in the 1310 nm window, they are on the order of 0.35 dB/km (Fig. 10.3).

10.5 Single-Mode Propagation

The gradient-index profile is especially useful for minimizing temporal pulse broadening in multimode fibers. In single-mode fibers, this balance is, of course, not required because the information is carried by one optical mode. Pulse broadening does result, however, from both intramode dispersion and polarization-mode dispersion. Polarization-mode dispersion will be discussed in Section 10.8. Intramode dispersion results, primarily, from the fact that the index varies with wavelength (material dispersion), and the confinement of the light within the guide varies with wavelength. Light of high wavelength occupies more of the cladding and, as such, experiences a slight shift in time delay (profile dispersion). The material dispersion results from the fact that the index varies with wavelength and in germania-doped silica is a minimum near 1.3 μms. The wavelength at which the minimum occurs is a function of the dopant level. The wavelength dependence of the index is generally described using the Sellmeier coefficients

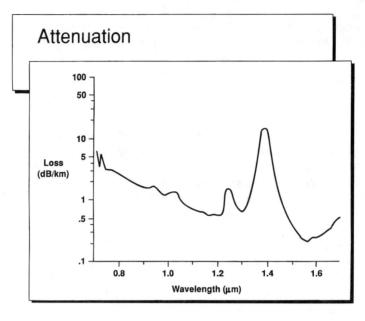

FIGURE 10.3 Absoption, including OH, with wavelength.

$$n^2 - 1 = \sum_{j=1}^{N} \frac{A_j \lambda^2}{\lambda^2 - l_j^2}. \quad (10.44)$$

The parameters A_j and l_j are obtained by fitting the index data and depend on the glass-dopant levels. These parameters represent the oscillator strength and resonance wavelength respectively.

Pulse broadening in single-mode fibers is characterized by the term dispersion. Dispersion represents the change in time delay with wavelength and is measured in psecs/(nm · km)

$$D = \frac{\partial}{\partial \lambda}\left[\frac{\partial \beta}{\partial k}\right]\frac{L}{c}. \quad (10.45)$$

The dispersion can be further expressed as

$$D = \frac{\partial}{\partial \lambda}\left[\frac{\partial \lambda}{\partial k} \cdot \frac{\partial \beta}{\partial \lambda}\right]\frac{L}{c}$$

$$= -\frac{\lambda}{\pi}\left[\frac{\partial \beta}{\partial \lambda} + \frac{\lambda}{2}\frac{\partial^2 \beta}{\partial \lambda^2}\right]\frac{L}{c}. \quad (10.46)$$

In step-index, single-mode fibers, the dispersion varies nearly linearly from approximately -1 psecs/(nm · km) at 1280 nm to 15 psecs/(nm · km) at 1550 nm. The propagation constant β is determined from the eigenvalue equation, Eq. 10.32, for a step-index guide.

Another important parameter for single-mode guides is the spot size. The field distribution calculated using the eigenfunction can be approximated with the expression

$$\Psi(r) = \exp\left(-\frac{1}{2}\left(\frac{r}{r_0}\right)^2\right). \quad (10.47)$$

The spot size for standard single-mode fibers or mode-field diameter is on the order of 9 μms.

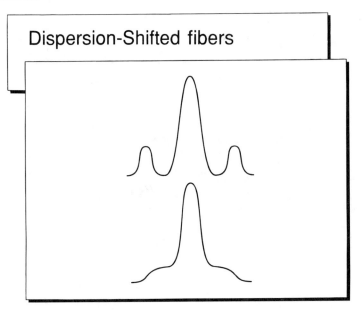

FIGURE 10.4 Dispersion-Shifted profiles.

For a given index profile, the spot size increases with wavelength on the order of 1 μm per 200 nms. Larger spot size enables simpler fiber-to-fiber connection and splicing, but bending losses, both from macro and micro bending, increase with increasing spot size.

10.6 Dispersion-Shifted Fiber

As mentioned above, the attenuation of silica-based fibers is a minimum near 1.55 μms. For this reason, there is considerable interest in transmitting information at this wavelength. Also, with the recent availability of erbium-doped fiber amplifiers at 1.55 μms, this window of operation is most advantageous for long distance communication. Step-index, single-mode fibers, however, exhibit considerable pulse broadening in this window, approximately 18 psecs/(nm · km). Figure 10.4 shows the segmented core profile of a dispersion-shifted fiber (Bhagavatula et al., 1984) and that of the platform profile. The segmented core profile was the first dispersion-shifted profile capable of shifting the λ_0 while maintaining other important parameters.

Solution of the radial equation, Eq. 10.27, using the index profile shown, shows that the dispersion minimum can be easily shifted to the 1.55 μm window, while at the same time maintaining other important properties of the fiber (Bhagavatula et al., 1984). Other important parameters include both bend loss, spot size (r_0), and cutoff wavelength (λ_c). The cutoff wavelength refers to the wavelength below which multimode propagation exists. These dispersion-shifted profiles are now finding application in transoceanic communication and the trunk lines. The use of dispersion-shifted fibers with erbium-doped amplifiers requires a minimum dispersion, on the order of 0.2 psec(nm · km) because optical nonlinearities distort the signal at the zero dispersion wavelength λ_0. Of particular concern is four-wave mixing. Later, we will summarize the important nonlinearities and how they affect signal distortion.

10.7 Fiber Fabrication

Modern fiber-optic processes enable the fabrication of low-cost fibers of excellent quality. In all glass-forming processes, silica tetrachloride and the dopants, such as germanium tetrachloride,

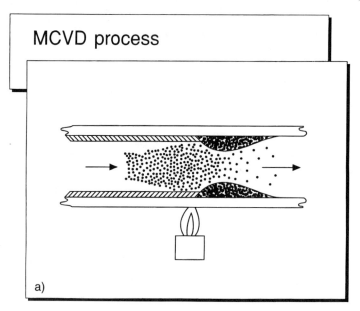

FIGURE 10.5 Three blank fabrication processes: a) MCVD; b) OVD; c) VAD.

are delivered to the reaction region as vapors, where silica and germania are formed. The byproduct is chlorine gas, Cl_2. In addition to glass blank fabrication, high-speed draws (greater than one meter per second) are used to draw and coat the fiber with an organic material that protects the fiber from handling and from the environment.

There are basically three methods used to form the glass blank. The modified chemical vapor deposition process (MCVD) (Nagel et al., 1985), the outside vapor deposition process (OVD) (Morrow et al., 1985) and the vapor-axial deposition process (VAD) (Nilzeki et al., 1985). In the MCVD process, successive layers of SiO_2 and dopants, which include germania, phosphorous, and fluorine, are deposited on the inside of a fused silica tube by mixing the chloride vapors and oxygen at a temperature on the order of 1800 °C. The temperature is maintained using a burner which traverses the outside of the tube. After the dopants are deposited, the temperature of the tube is raised, via the burner, to collapse the tube. In the OVD process, the core and cladding layers are deposited on a rotating mandrel by the flame hydrolysis process. Upon completing the deposition process, the mandrel is removed from the preform, and the preform is sintered in a furnace to form the fiber blank. The VAD process is also a flame hydrolysis process, but, in this technique, the soot is deposited axially. Figure 10.5 is a schematic of the three blank fabrication processes.

In the drawing process, the blank is fed from above into the drawing portion of the furnace while being drawn from the bottom using tractors. The fiber is then wound onto a drum while being monitored for tensile strength. The temperature during draw is on the order of 2000°C. After exiting the furnace, the fiber is coated with a UV-curable coating before winding on the drum.

10.8 Polarization

Single-mode fibers capable of maintaining an input linear polarization are known as polarization-preserving fibers. These fibers employ either an elliptical core or stress rods placed 180° from each other and outside the core. The elliptical core and/or stress rods introduce a birefringence that removes the degeneracy of the orthogonally polarized modes. The birefringence is defined as the difference between the effective indices of the two orthogonal modes and is on the order

Optical Fibers

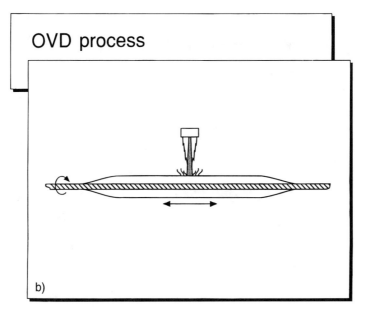

FIGURE 10.5 Continued.

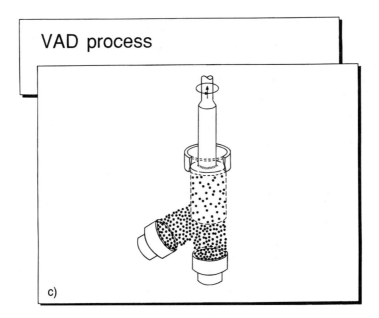

FIGURE 10.5 Continued.

of 10^{-4}. Figure 10.6 shows the index profile for stress rod and elliptical core polarization-maintaining fibers. Polarization-preserving fibers are not used at present in telecommunication systems but rather find applications in the area of fiber-optic sensors. As an example, fiber-optic gyroscopes use polarization-preserving fiber to prevent coupling from one polarization to the other in the gyroscope coil. Such coupling leads to inaccurate sensing and is eliminated by using only one polarization. An important parameter for these fibers is the beat length, defined as the wavelength divided by the birefringence (δn). This parameter describes the beating which results from the fact that the propagation speed along the fast axis is considerably larger than that along

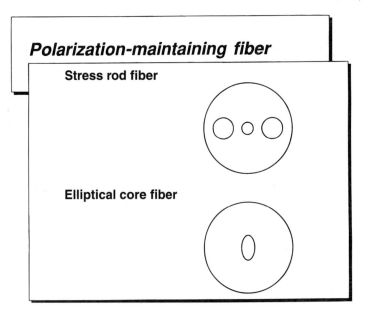

FIGURE 10.6 Polarization-Maintaining fibers, index profiles.

the slow axis. Highly polarized fiber exhibits beat lengths on the order of less than a centimeter. Standard fibers have very small birefringence and exhibit beat lengths greater than ten meters.

The slight differences in propagation speeds for the two polarization modes in standard single-mode fiber lead to polarization-mode dispersion (PMD) (Rashleigh, 1983). Polarization-mode dispersion for standard fiber is, typically, significantly less than 1 psec/sqrt(km). The inverse square root length dependence results from the fact that the polarization randomly couples as the light propagates through the fiber. Transoceanic systems using optical amplifiers eliminate the need for expensive electronic regenerators. Without signal regeneration, polarization-mode dispersion can become significant over hundreds of kilometers. Also, CATV analog transmission systems can be sensitive to PMD when component polarization-dependent loss and significant laser chirping occur. In this situation, composite second-order distortion (CSO) (Peterman, 1981) can significantly degrade the signal.

As mentioned, small-core eccentricities, on the order of a few percent, can be the source of PMD. Significant literature exists on the effect of core ellipticity on polarization dispersion for step-index fibers. However, these perturbation models predict that the polarization dispersion increases linearly with length. The square root length dependence that is experimentally observed results from random coupling of the polarization modes. In step-index fibers, polarization dispersion (Kapron et al., 1972) results from both form birefringence (Snyder and Love, 1983) and stress birefringence (Rashleigh, 1983), both of which depend on core ellipticity and are approximated in the low ellipticity limit as

$$\Delta\beta(\text{FORM}) = \frac{E(2\Delta)^{3/2}}{\rho} \frac{4}{V^3} \frac{(\ln v)^3}{1 + \ln v} \qquad (10.48)$$

and

$$\Delta\beta(\text{Stress}) = \left(1 - \frac{u^2}{v^2}\right) \frac{C_s}{1 - \eta_\rho} \Delta\alpha\Delta T. \qquad (10.49)$$

Optical Fibers

In the above equations, u and v are defined in Eqs. 10.30–10.32, and e is the eccentricity which is defined as

$$e = \frac{A - B}{A + B} \tag{10.50}$$

and

$$E = 1 - \frac{B^2}{A^2},$$

where A and B are the major and minor diameters. The stress birefringence depends on the expansion differences of the core and cladding, $\Delta\alpha$, which, in turn, depend on the dopant concentrations. The stress birefringence also depends on the fictive temperature T which is used to approximate a point below which the glass structure cannot change on the same time scale as the cooling rate, a stress optic coefficient C_s, and Poisson's ratio h_r.

The measured value of PMD can be sensitive to fiber deployment. This is because the amount of mode coupling varies with deployment and with the environment. The effect of mode coupling on PMD is modeled using the statistics of random occurrences (Matera and Someda, 1992). This analysis explains the square root length dependence of PMD and the distribution of PMD values with changes in the environment and the fact that repeated measurements of PMD show that the observed values obey a Maxwellian process:

$$P(\Delta\tau) = \frac{2\tau}{\sqrt{2\pi q^3}} e^{-(\Delta\tau^2/2q^2)} \tag{10.51}$$

In Eq. 10.51, q^2 is the variance. Another aspect of the random coupling of the polarization modes is a significant wavelength dependence of PMD (Matera and Someda, 1992; Poole and Wagner, 1986).

One final and important aspect of polarization modes in fibers is the evolution of the polarization states as light propagates through a fiber. Light linearly polarized as it enters the fiber not only evolves into circularly or elliptically polarized light, as it propagates, but significant mode coupling occurs. This complicates the analysis and measurement (Hoffner, 1993), of PMD. Recently, techniques to minimize the effects of mode coupling on the stability of PMD have been described by Judy (1994).

10.9 Optical Fiber Amplification

An important aspect of optical fiber research is rare-earth doping for amplification and lasing. Amplification in optical fiber has not only renewed interest in the materials and propagative aspects of fiber research, but it has also significantly affected the systems aspects of optical communication. Doping of fibers for optical amplification has been under study since the 1960s (Koester and Snitzer, 1964) when neodymium was used as a dopant. Interest in rare-earth doping was renewed in the 1980s when research scientists at the University of Southampton (Meavs et al., 1987) showed amplification in the 1.55 micron region, which coincidentally is the low-loss transmission window. The fact that erbium-doping enables amplification in the transmission window of lowest loss has attracted much interest in this aspect of optical fiber research.

In the erbium-doped fiber amplifier (in its simplest form), Fig. 10.7, an erbium-doped fiber with lengths on the order of meters and dopant levels on the order of 2 ppm, is spliced to a wavelength-dependent, fiber-optic coupler. The coupler enables one to continuously pump the erbium-doped fiber with light emitted from a high-power semiconductor laser diode at 980 or 1480 nm. Filters and optical isolators are often included to minimize spontaneous emission noise and reflections. The pump light is used to excite ions from the ground state to an excited state. Signal light entering the fiber initiates stimulated emission and is coherently amplified. Years of research at many laboratories has led to the development of erbium amplifiers. Such technical issues as wavelength dependence of gain, gain saturation, polarization dependence and spontaneous emission, among others, have been carefully studied. Spontaneous emission occurs when ions in the excited state spontaneously relax to the ground state contributing to noise. This phenomenon in itself significantly affects the signal-to-noise ratio of an amplifier-based communication system. Another important parameter of the optical amplifier is the concentration of erbium ions. An optimum concentration of erbium ions avoids ion clustering which alters the excited states and results in elevating one ion to a higher state and emission to the ground of neighboring ions.

Research in the areas of rare-earth doped optical fibers is far from complete. Issues, such as multiple wavelength amplification, need to be further addressed. Research into the possibilities of using erbium-doped fibers in a lasing configuration is useful for picosecond pulse sources. Another aspect to rare-earth doping is the interest in amplification at 1.3 μms. Amplification in the low-loss transmission windows requires dopants other than erbium (praseodymium co-doped with neodymium, for example), and also requires the use of a fluoride glass host (Ohishi et al., 1991). In a silica glass, the phonon vibrational spectrum affects the amplification process. In a fluoride-based glass, however, the phonon edge is shifted to higher frequencies, thereby, enabling reasonable amplification in the 1.3 μm window. Other issues remain, however, including the amount of gain and the wavelength dependence of the gain.

Fluoride fibers are melted at temperatures far below that of silica and, therefore, cannot be fused to silica fibers. The index is not well matched to silica and, therefore, leads to strong back reflections. More importantly, the fabrication technology of fluoride-based fibers is less advanced than that of silica. Nonetheless, interest in amplification at 1.3 μms remains because there is a

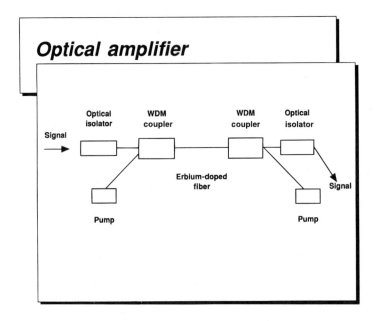

FIGURE 10.7 Schematic (simple) of the optical amplifier.

huge installed base of optical fiber that is optimized for minimum pulse distortion at 1.3 μms rather than 1.55 μms.

10.10 Optical Nonlinearities in Fibers

Nonlinear effects in silica fibers have been considered unimportant. However, the use of optical amplifiers and the promise of dense wavelength-division multiplexing imposes a number of limitations (Chraplyvy, 1990) on the ultimate bandwidth of optical communication. Optical nonlinearities will be discussed in detail in other chapters, but it is important to mention the limitation that these nonlinearities impose on communication bit rates. These nonlinearities are stimulated Brillouin scattering, four-wave mixing, cross-phase modulation, and stimulated Raman scattering.

Multiple wavelength mixing was first observed in optical fibers by Hill et al. (1978). The output power (Inoue, 1992) of light generated at a fourth wavelength can be written as

$$P_F(L) = \frac{1024\pi^6}{n^4\lambda^2 c^2}(D_x)^2 \frac{P_i(0)P_j(0)P_k(0)}{A_{eff}^2} e^{-\alpha L} \frac{(1-e^{-\alpha L})^2}{\alpha^2} \eta \tag{10.52}$$

In Eq. 10.52, i, j and k represent the three input wavelengths at $Z = 0$, P_F is the power at wavelength F and at $Z = L$. A_{eff} is the effective area of the guide, α is the loss, and D_x is a degeneracy factor. The efficiency factor η is given by

$$\eta = \frac{\alpha^2}{\alpha^2(\Delta\beta)^2}\left[1 + \frac{4e^{-\alpha L}\sin^2(\Delta\beta L/2)}{(1-e^{-\alpha L})^2}\right] \tag{10.53}$$

where

$$\Delta\beta = \beta(f_i) + \beta(f_j) - \beta(f_k) - \beta(f_F). \tag{10.54}$$

In Eq. 10.54, $\Delta\beta$ is the difference in propagation constants, which depends on the dispersion of the fiber. Numerical and systems studies have shown the importance of using a finite amount of dispersion to avoid generating new wavelengths while depleting the signal wavelength. Finite dispersion is also important at single channel operation over very long lengths because multiple wavelength mixing occurs through the wavelengths generated with amplifiers by amplified spontaneous emission (Marcuse, 1991). A major impact of four-wave mixing is that it forces designers to operate their signal wavelength away from a dispersion zero. Over long distances, this require chromatic or dispersion compensation.

Brillouin scattering is the interaction of light with acoustical vibrations in fiber. The signal or carrier wave is shifted to longer wavelengths (Stokes shifted) with the simultaneous emission of an acoustical phonon. The amount of power generated can be characterized with an exponential gain coefficient, g_B (Stolen, 1979):

$$g_B = 4 \times 10^{-9} \text{ cm/W.} \tag{10.55}$$

The gain coefficient enables one to calculate the amount of stimulated Brillouin scattering with length as a function of incident light power. The wavelengths generated are separated from the carrier by less than one thousandth of a nanometer. This slight shift in wavelength would not be expected to cause problems, except for the fact that the generated light is scattered backwards, depleting the carrier signal and, at times, affecting the transmission laser. It is estimated that

Brillouin scattering becomes an issue when the power in an optical fiber is on the order of a milliwatt (Chraplyvy, 1990).

Raman scattering results from an interaction of the incident light with molecular vibrations in the fiber. As with Brillouin scattering, Raman scattering can be characterized with a gain coefficient (Stolen and Ippen, 1973)

$$g_R = 7 \times 16^{-12} \text{ cm/W}. \tag{10.56}$$

Comparison of Eqs. 10.55 and 10.56 shows that the threshold for Raman scattering occurs at power levels three orders of magnitude higher than that for Brillouin scattering. Raman scattering can be a problem, however, because the wavelength or bandwidth of interaction is greater than a hundred nanometers. Raman scattering can cause serious cross-talk for multiple channel systems of significant power and will ultimately limit the information transmission capacity of optical fiber systems. It is estimated that Raman scattering becomes an issue (Chraplyvy, 1990) when the total power in an optical fiber is on the order of 1 watt.

Self- and cross-phase modulation refers to the fact that the index of glass is intensity-dependent (Islam, 1992):

$$n = n_0 + n_2 I \tag{10.57}$$

where n_0 represents the linear index and the second terms includes the intensity-dependent refractive index n_2 and the optical intensity I.

In silica, n_2 is on the order of 3×10^{-16} cm^2/W and 6×10^{-16} cm^2/W for self- and cross-phase modulation, respectively. Both cross- and self-modulation affect the phase and, hence, the arrival time of a pulse. Changes in power and the modulation of power with other carriers limit the amount of power in fibers. It is estimated (Chraplyvy, 1990) that self- and phase-modulation become an issue when the power in a fiber is on the order of 10 milliwatts.

10.11 Solitons in Optical Fiber

In this section, we will briefly introduce the concept of soliton pulses in optical fiber. This is a fascinating research topic, and it is expected that the use of soliton pulses in optical fibers will enable transmission over transoceanic distances at the highest of possible bit rates.

Soliton propagation in fiber is a nonlinear phenomenon, and such pulses are sensitive to the amount of optical power. Attenuation severely limits the distances over which a soliton pulse can travel without significant distortion. The development of the erbium-doped optical amplifier has spurred research activity toward commercializing transoceanic soliton systems. The coincidence of minimum fiber attenuation in the 1550 nm telecommunication window and the strong gain in this same window with the erbium amplifier is, indeed, encouraging. In fact, by amplifying the signal every 30 kms, soliton pulses can travel thousands of kilometers without distortion. One can also use multiple wavelengths (WDM) without the imposition of a number of the nonlinearities mentioned in the previous section. For these reasons, it is expected that, ultimately, soliton transmission will be the technique of choice for transoceanic communication. However, more research is required before the potential of soliton communication can be fully realized.

Solitons occur in nature in many different ways, optical pulses being one of them. Scott Russel is credited with first observing and recording solitary waves in a barge canal in Great Britain in 1938. He followed the "large solitary elevation" on horseback and noted that it did not change in form or speed for miles. He derived equations describing the velocity of the waves and reported on his work at the Liverpool meeting of the British Association for the Advancement of Science. In 1967, a group of mathematical physicists (Gandner et al., 1967) from Princeton University

Optical Fibers

solved the so-called Korteweg–de Vries equations describing the nonlinear hydrodynamic wave. The Russian pair, Zahkharov and Shabat (1971) considered the nonlinear propagation of optical waves in a two-dimensional medium. They showed that the nonlinear Schroedinger equation could be solved using the inverse scattering theorem and pointed out the relationship between spatial dispersion and optical intensity.

Hasegawa and Tappert (1973) showed, theoretically, that temporal solitons should exist in fiber and pointed out the possibility of using them for optical communication. Molleneauer, Stolen and Gordon (1980) were the first to observe solitons experimentally in fiber. They built a "color center" laser that enabled them to generate and launch narrow temporal pulses of power levels significant enough to develop into soliton pulses. This experimental observation can be considered to mark the beginning of a new technology aimed at enabling dispersionless transmission over transoceanic distances. There are, however, many obstacles yet to be overcome before this technology can be fully utilized.

Nonlinear pulse propagation in fiber is described with the so-called nonlinear Schroedinger equation

$$-i \frac{\partial u}{\partial z} = \frac{\lambda^2 D}{4\pi c} \frac{\partial^2 u}{\partial t^2} + \frac{2\pi n_2}{\lambda A_{\text{eff}}} |u|^2 u. \tag{10.58}$$

Here, D is the dispersion and u is the pulse amplitude which varies both spatially and temporally. The soliton pulse has both a temporal and spectral width, both of which are described with hyperbolic secant functions (Islam, 1992):

$$u(t) = \sec h[t] \cdot e^{iz/2} \tag{10.59}$$

and

$$\bar{u}(w) = \frac{1}{2} \sec h \left[(w - w_0) \frac{\pi t_c}{2} \right]. \tag{10.60}$$

In Eq. 10.58, t_c is the temporal width and w_0 the center frequency. An important parameter is the power P_c at which the nonlinearity and dispersion balance (Islam, 1992):

$$P_c = \frac{\lambda A_{\text{eff}}}{2\pi n_2 Z_c} \tag{10.61}$$

where

$$Z_c = \frac{2\pi c}{\lambda^2 D} t_c^2. \tag{10.62}$$

The parameter Z_c characterizes the distance at which the pulse begins to spread. Another important parameter is the soliton period, characterized by Z_c times $\pi/2$.

At this point, we will terminate our introduction of soliton pulses in optical fibers. We should mention, however, that current research interest is in the areas of minimizing pulse jitter (Gordon–Haus jitter) using frequency filters (Mollenauer et al., 1992), soliton-soliton interactions, and the use of wavelength division multiplexing (WDM) with solitons.

The possibility of switching light at ultrafast speeds using solitons is also a topic for research. Soliton pulses have properties that make them attractive in this regard.

Defining Terms

Attenuation: Propagation loss of an optical input or signal. Quantified with an exponential coefficient and measured in dB/km.
Birefringence: Effective index difference of the two polarization modes in single-mode fiber.
Brillouin scattering: Scattering of light by an acoustical phonon. In an optical fiber, the light predominantly scatters in the background direction.
Critical angle: The angle below which total internal reflection occurs, dependent on the index difference between the propagation and the surrounding medium.
Dispersion-Shifted fiber: Optical fiber in which the wavelength of minimum dispersion is shifted further to the infrared by an index profile design.
Fiber blank: Core and clad glass composite which is drawn into fiber.
Fiber draw: Equipment used to draw fiber from the blank to 125 μms and less. Simultaneously, the fiber is coated with an organic coating to protect the glass from the environment.
Fiber-Optic amplifier: Device composed of rare-earth-doped fiber and components, such as couplers and isolators. The device is capable of amplifying an input signal by over three orders of magnitude.
Gradient-Index profile: Fiber-optic index profile where the core index is graded with core radius. Typically, it is parabolic with radius.
Maxwell's equations: Differential equations describing the relationships of the electric and magnetic field vectors of light.
MCVD process: Optical blank making process, where gases are input inside a rotating tube, react and, under a thermophoresis process, are deposited onto the wall as glass.
Multimode fibers: Optical fiber capable of guiding more than one mode. Typically hundreds of modes are guided.
Nonlinear index: The index component dependent on optical power. In silica based fiber, this parameter becomes above about one milliwatt.
Outside vapor deposition process: Optical blank making process, where glass soot particles are deposited onto a rotating mandrel and, then, consolidated to glass.
Polarization: In optical fiber, the vector components of the propagating light. Includes both linear and circular components.
Raman scattering: Scattering of light at high optical powers, resulting from an interaction of light with the molecular vibrations of the glass.
Rayleigh scattering: Scattering of light from particles with differences in index. Affects the fundamental limit of attenuation in an optical fiber.
Single-Mode fiber: Fiber capable of propagating only one mode at the wavelength of interest, but includes modes of both polarizations.
Self-Phase modulation: Results from the fact that the index of light is intensity-dependent. Power-dependent changes in index affect the phase delay of the signal.
Step-Index profile: The fiber-optic index profile where the core and clad index are constant with radius, the clad glass being less in index than the core.

References

Agrawal, G. P., *Fiber-Optic Communication Systems*, John Wiley & Sons, 1992.
Bhagavatula, V. A., Spotz, M. S., and Love, W. F., 1984. Dispersion Shifted Segmented Core Single-Mode Fibers, *Optics Lett.* 9, 186.
Born, M., and Wolf, E., *Principles of Optics*, Pergamon Press, 1980.
Chraplyvy, A. R. 1984. Optical power limits in multichannel WDM systems due to stimulated Brillouin scattering, *Electron. Lett.*, 20, 58.

Chraplyvy, A. R. 1990. Limitations on lightwave communications imposed by optical-fiber nonlinearities, *J. Lightwave Technol.*, 8, 1548.

Fleming, J. W., 1978. Material Dispersion in Light Guide Glasses, *Electron. Lett.* 14, 326.

Gardner, C. S., Greene, J. M., Krusal, M. D., and Miura, R. M. 1967. Method for solving the Korteweg–de Vries equation, *Phys. Rev. Lett.*, 19, 1095.

Gloge, D. 1971. Weakly guiding fibers, *Appl. Opt.* 10, 2252.

Gloge, D., and Marcateli, E. A. J. 1973. Multimode theory of graded-core fibers, *Bell Syst. Tech. J.*, 52, #9, 1563.

Hasegawa, A., and Tappert, F. D. 1973. Theory of stationary nonlinear optical pulses in dispersive dielectric fiber, 1, Anomalous dispersion, *Appl. Phys. Lett.*, 23, 142.

Heffner, B. L. 1993. Accurate automated measurement of differential group delay dispersion and principal state variation using Jones matrix eigenanalysis, *IEEE Photon. Tech. Lett.*, 5, 814.

Hill, K. O., Johnson, D. C., Kawasaki, B. S., and MacDonald, R. I. 1978. CW three-wave mixing in single-mode optical fibers, *J. Appl. Phys.*, 49, 5098.

Inoue, K. 1992. Four-Wave Mixing in an Optical Fiber in the Zero-Dispersion Wavelength Region, *J. Lightwave Techno.*, 10, 1553.

Islam, M. N., *Ultrafast Fiber Switching Devices*, Cambridge U. Press, 1992.

Judy, A. F., Improved PMD stability in optical fibers and cables, International Wire and Cable Symposium Proc., 1994.

Kapron, F. P., Borrelli, N. F., and Keck, D. B. 1972. Birefringence in dielectric optical waveguides, *IEEE J. Quantum Electron.* 8, 222.

Kapron, K. P., Keck, D. B., and Maurer, R. D. 1970. Radiation losses in optical waveguides, *Appl. Phys. Lett.* 17, 423.

Keck, D. B., Optical Fiber Waveguides, in *Fundamentals of Optical Fiber Communications*, Barnoski, M. K., ed., 1981.

Koester, C. J., and Snitzer, E. A. 1964. Amplification in a fiber laser, *Appl. Opt.*, 3, 1182.

Marcuse, D. 1991. Single channel operation in very long nonlinear fibers with optical amplifiers at zero dispersion, *J. Lightwave Tech.*, 9, 356.

Matera, F., and Someda, C. G., *Random Birefringence and Polarization Dispersion in Anisotropic and Nonlinear Optical Waveguides*, Someda and Stegeman, eds., Elsevier 1992.

Mears, R. J., Reekie, L., Jauncey, I. M., and Payne, D. N. 1987. *Electron. Lett.*, 23, 1026.

Mollenauer, L. F., Stolen, R. H., and Gordon, J. P. 1980. Experimental observation of picosecond pulse narrowing and solitons in optical fibers, *Phys. Rev. Lett.*, 45, 1095.

Mollenauer, L. F., Gordon, J. P., and Islam, M. N. 1986. Soliton propagation in long fiber with periodically compensated loss, *IEEE J. Quantum Electron.*, QE-22, 157.

Mollenauer, L. F., Gordon, J. P., and Evangelides, S. G. 1992. The sliding-frequency guiding filter: An improved form of soliton jitter control, *Optics Lett.*, 17, 22, 1575.

Morrow, A. J., Sarkar, A., and Schultz, P. C., *Optical Fiber Communications*, Li, T., ed., Academic Press, 1985, Vol. 1, Chapter 1.

Morse, P. M., and Feschbach, H., *Methods of Theoretical Physics*, McGraw-Hill, New York, 1953.

Nagel, S. R., MacChesney, J. B., and Waller, K. L., Optical Fiber Communications, Li, T., ed., Academic Press, 1985, Vol. 1, Chapter 1.

Niizeki, N., Inagaki, N., and Edahiro, T., *Optical Fiber Communications*, Li, T., ed., Academic Press, 1985, Vol. 1, Chapter 3.

Ohishi, Y., Kanamori, T., Kitagawa, T., Tagahashi, S., Snitzer, E., and Sigel, G. H., Jr., Pr^{3+}-Doped fluoride fiber amplifier operating at 1.31 μm, Optical Fiber Conf., San Diego, Calif. (PD-2), 1991.

Olshansky, R., and Keck, D. B., Material effects on minimizing pulse broadening, Optical Fiber Transmission Conf., OSA, TUC5, 1975.

Petermann, K. 1981. Nonlinear transmission behavior of a single-mode fiber transmission line with polarization coupling, *J. Opt. Comm.*, 2, 59.

Poole, C. D., and Wagner, R. E. 1986. Phenomenological approach to polarization dispersion in long single-mode fibers, *Electron. Lett.*, 22, 1029.

Rashleigh, S. C. 1983. Origins and control of polarization effects in single-mode fibers, *J. Lightwave Technol.*, 1, 312.

Snyder, A. W., and Love, J. P., *Optical Waveguide Theory*, Chapman and Hall, 1983.

Stolen, R. H., Nonlinear Properties of Optical Fibers, in *Optical Fiber Telecommunications*, Miller, S. E., and Chynoweth, A. G., eds., Academic, New York, 1979.

Stolen, R. H., and BjorKholm, J. E. 1982. Parametric amplification and frequency conversion in optical fibers, *IEEE J. Quantum Electron.*, 18, 1062.

Stolen, R. H., and Ippen, E. P. 1973. Raman gain in glass optical waveguides, *Appl. Phys. Lett.*, 22, 276.

Tyndall, J., 1884. *Proc. Roy. Inst.*, 1, 466.

Vassallo, C., *Optical Waveguide Concepts*, Elsevier, 1993.

Zahkharov, V. E., and Shabat, A. B. 1971. *Zh. Eksp. i Teor. Fiz.* 61, 118.

SECTION V
Optics

11

Binary Optics

Michael W. Farn
Massachusetts Institute of Technology/Lincoln Laboratory

11.1 Introduction 459
A Working Definition • Application Areas • Chapter Overview
11.2 Designs Based on Ray Tracing 463
Fundamentals • Phase Profiles of Common Elements • Design of Monochromatic Systems • Design of Wideband Systems • Other Applications • Optical Performance
11.3 Designs Based on Scalar Diffraction Theory 479
Fundamentals • Designs Utilizing Multiple Diffraction Orders • Beam Shapers and Homogenizers • Other Applications
11.4 Designs Based on Maxwell's Equations 488
Fundamentals • Applications
11.5 Fabrication Using Multiple Masks 490
Mask Generation • Photolithography • Micromachining Techniques
11.6 Alternate Fabrication Methods 495
Other Micromachining Approaches • Replication Methods

11.1 Introduction

A Working Definition

What is binary optics? There is no strict definition because, as a result of the historical development of the field, binary optics means different things to different people. The first element, which was referred to by the term "binary optics," was formally presented in 1980 (Veldkamp, 1980). In this laser radar application, the binary optic device shown in Fig. 11.1 split a single laser beam into multiple beams, each of which was used as a local oscillator for photomixing. The design of the element was based on a combination of communications theory and scalar diffraction theory, the fabrication was achieved using computer-generated data and VLSI fabrication techniques, and the resulting element was truly a binary surface-relief structure. Since this original element, however, the field has expanded in many directions. Design methods currently range from classical lens design to the numerical solution of Maxwell's equations, application areas run the gamut from the entertainment industry to medical products to military missions, and fabrication technologies can be as diverse as diamond turning and ion milling. In addition, modern-day binary optic devices are usually multilevel or continuous surface reliefs, rather than strictly binary, as in the original application. The "binary" in the name, however, is still relevant because a binary coding technique is used to extend the original fabrication method to create the current multilevel profiles.

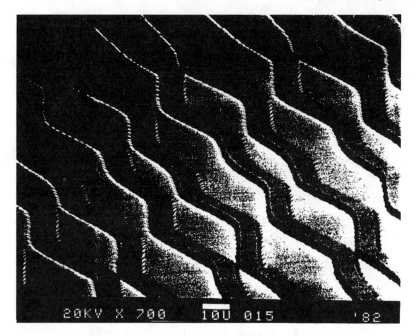

FIGURE 11.1 The original binary optics element (Veldkamp, 1980).

As a result of this diversification, it is difficult to come up with an accurate, concise definition for the field. However, many binary optic elements are characterized by the following:

- VLSI-based fabrication - Binary optics are usually manufactured using semiconductor fabrication technology, in particular photolithography and micromachining, and the entire fabrication goal is to micromachine a desired surface profile to optical tolerances. Historically, the fabrication techniques are what have distinguished binary optics from other technologies.
- Surface relief - The optical properties of binary optics are solely the result of the surface profile of the element. Volume effects, such as those in thick holograms, graded-index optics or photorefractives, are nonexistent. Because the surface profile contains the entire functionality of the device and the profiles are typically thin (on the order of several microns deep), binary optics may be inexpensively replicated by method, such as embossing.
- Free-space, passive optics - Binary optic components are passive optical elements to be used in free-space applications. Passivity is a result of the surface-relief characteristic (i.e., no transistors, electro-optic effects, etc.), while the exclusion of integrated optics is more for historical reasons than technical ones.

Taken together, these characteristics provide a working definition of binary optics. The definition is "working" in the sense that, even though it is not entirely accurate today and will certainly change tomorrow, it still embodies the crux of the field.

Application Areas

Why or when might one use binary optics? In the majority of successful applications, binary optic elements are used because they offer one of the following two advantages:

- Novel functionality - Most binary optic devices rely on diffraction for the manipulation of light, whereas conventional optical elements rely on refraction or reflection. The fundamental difference between diffraction, refraction, and reflection allows binary optic components to perform functions which could not be achieved otherwise, one example being lenses with simultaneous multiple focal lengths. In addition, the unique fabrication process also allows the manufacture of components which could not be made by other technologies. For example, the ability to make arrays of microoptics, each of which can be of a different shape and perform a different function, is unique to binary optics.

- Increased performance - For existing systems, the use of binary optics may result in increased performance. Typically, this higher performance is a wider field of view, higher numerical aperture, lower mass, lower volume, elimination of exotic materials, or lower cost.

Table 11.1 lists some current applications for binary optics. There are two points of interest in Table 11.1. First, the applications represent a diverse spectrum, ranging from commercial to military, from simple to complex, and from commercially available to still in development. Second, in the majority of applications, binary optic elements are not the end-product. Rather, the binary optic component is used to increase the performance of a larger system. In general, it is true that binary optics is an enabling technology.

Chapter Overview

As shown in Fig. 11.2, there are four basic tasks in a typical binary optic project cycle. The first is the optical design of the element, the end result being an optical prescription. The second is the translation of this optical prescription into a fabrication prescription. The third task is the actual fabrication of the element, and the final task is testing of the completed element. This chapter will focus on the two dominant tasks - optical design and fabrication, with more emphasis placed on optical design.

Optical Design

Optical design can be further categorized according to the optical theory used to model the element: geometrical optics, scalar diffraction theory, or vector diffraction theory. Each of these methods will be discussed in detail in sections 11.2 to 11.4, respectively. The following is a comparison of the three theories.

One way to characterize the three optical design methods is to consider how the binary optic element is modeled and which optical effects are neglected. In the geometrical optics approach, the binary optic model is based on the grating, and rays are traced through the element using the grating equation. This model is the simplest but neglects diffraction effects except at the exit pupil, reflections at material interfaces, polarization effects, and the effect of the finite thickness of the element. In the scalar diffraction approach, binary optic devices are modeled as infinitely

TABLE 11.1 Examples of Binary Optics Applications

Application Area	Example Application
Medical	Bifocal intraocular lens
Consumer	Optical heads for compact disks
	Increased sensitivity of CCD cameras
	Stereovision goggles
Industrial	Beam shaping for lasers
	Structured light for machine vision
Optics	Null testing of aspherics
	Hartman-shack wavefront sensors
	Grating beam splitters
	Microoptics for semiconductor lasers
Military	Aberration correction of wideband imaging systems
	Hardening of detectors arrays

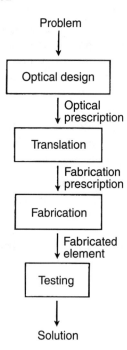

FIGURE 11.2 A typical binary optics project cycle.

thin phase plates and wavefronts are propagated from one surface to another by the scalar diffraction integrals. In this model, most of the previously neglected diffraction effects are accounted for, but secondary reflections, polarization effects, and the thickness of the element are still neglected. Finally, the vector approach is the most complete, as it is a numerical solution to Maxwell's equations, and all effects predicted by Maxwell's equations, including those mentioned above, are included.

The underlying assumptions determine which applications can be adequately modeled by a given theory. Typical geometrical optic applications include the design of almost all imaging systems, which run the gamut from purely diffractive systems to diffractive modifications of conventional wideband systems, laser systems, and microoptics which are large enough that diffraction from them can be neglected. In short, if an application would normally be designed using geometrical optics, then ray tracing will probably still be adequate to model a version which includes binary optics. The major difference between geometrical optics and scalar theory is that the scalar diffraction integrals include diffraction effects not easily handled by geometrical optics. Therefore, applications which rely specifically on these effects must be modeled by scalar theory. Examples include the design of laser cavities, microoptics which are small enough to introduce their own diffraction effects, diffusers, beam shaping optics for lasers, and grating beam splitters (i.e., gratings which split one incoming beam into many outgoing diffraction orders). Finally, vector theory is required when the assumptions implied by scalar theory are no longer valid. In many applications in this regime, the binary optic element is a sort of artificial crystal and the device operates by using the resonances or lack of resonances in the crystal. Examples are antireflection skins, artificially birefringent materials (i.e., form birefringence), artificial index materials, and wavelength filters.

Because each of these theories is quite different from the others, it is also instructive to examine the available tools and users in each case. As one might expect, the geometrical optic model is the most mature, and much of the commercially available lens design software can ray trace diffractive optical elements. However, although codes can accurately predict the direction of different diffraction orders, the calculation of diffraction efficiencies is often inadequate and must be supplemented by other means. Because many of the geometrical optic applications for diffractive

Binary Optics

optics can be considered classical lens design problems, a lens design background is most appropriate for this type of application. In the scalar and vector regimes, there is not much, if any, general purpose software. As a result, in the scalar diffraction case, the user will code the well-accepted scalar diffraction integrals in the language and on the machine of his choice. To do this, a background in Fourier optics and numerical methods would be most useful. In the vector diffraction case, the designer will also have to code his own tools, but the case differs from the scalar situation in the following respects. First, the vector case is more numerically intense so that the choice of machines is limited and there is a stiffer penalty for inefficient programming. Second, the numerical implementations of the theories are not as mature or as well accepted as in the scalar case. As a result, the designer may have to do some research in addition to any programming. For this type of work, a background in differential equations and numerical methods is useful.

Finally, the three theories are sometimes correlated with the size of the features on the binary optic elements. In cases where the feature is much larger than the wavelength, geometrical optics is used. If the features are on the order of a wavelength or smaller, the vector theory is used, and, in the intermediate case, the scalar theory is appropriate. Although this generalization is not entirely accurate, it does give one a sense of the differences among the approaches.

Fabrication

Fabrication can be categorized into the traditional binary optic approach, which uses multiple photomasks, and other fabrication processes. These are discussed in sections 11.5 and 11.6, respectively.

In the traditional binary optic process, a continuous surface relief is approximated by a staircase profile, and the staircase profile is created by a process utilizing a set of photolithographic masks. The first mask results in a surface profile with two steps, as in the example of Fig. 11.1, and each succeeding mask then doubles the number of steps. Thus, the use of N masks results in a staircase surface profile with 2^N steps. In more detail, the process begins with a translation of the optical prescription to the fabrication prescription, which is the required set of lithographic masks. This conversion is theoretically straightforward but can often result in large amounts of data, and so the major issue is not how to do the conversion but how to do it efficiently, in terms of computation time and data manipulation. The masks are usually fabricated by an electron beam, and the mask set is then used to fabricate the binary optic element. The fabrication of the binary optic device can be broken down into two major steps: photolithography and micromachining. In the photolithography step, the mask set is used to pattern a photoresist, which then acts as a protective layer for part of the underlying optical substrate. The unprotected part of the substrate is subsequently removed by the micromachining step, typically by reactive ion etching or ion milling.

The alternative fabrication processes attempt to generate continuous profiles analogous to numerically controlled milling, except that they occur on a microscopic scale. Examples include diamond turning, laser ablation, and the analog exposure of resist by laser scanning.

Finally, once a master is manufactured, copies may be mass produced by processes such as embossing or casting, as described in section 11.6.2.

11.2 Designs Based on Ray Tracing

In applications for which ray tracing is adequate, binary optic elements are modeled as a generalization of the linear grating. This section first presents the theory underlying this model, including the generalized grating model and the two common methods for ray tracing binary optics: the grating equation and the Sweatt model. Next, tabulates the phase profiles for some common elements are tabulated.

Finally, this section consider the classes of applications which can be designed using geometrical optics: the correction of monochromatic systems, the correction of wideband systems, and miscellaneous applications. In the monochromatic case, the binary optic element can have significant optical power and is often a direct substitute for a refractive counterpart. In the broadband case, the binary optic structure usually has only enough optical power to balance the chromatic aberrations and/or it may also perform some residual correction of the monochromatic aberrations, similar in function to a Schmidt corrector. In the miscellaneous category, the three most interesting applications are the correction of chromatic aberrations in purely diffractive systems, the thermal correction of systems, and the use of binary optics as microoptics. In geometrical optic designs, binary optics typically offers the following advantages:

- Increased design freedom in correcting aberrations, resulting in better system performance.
- Reduction in overall volume, mass, number of elements and/or number of materials.
- The generation of arbitrary lens shapes and phase profiles, thereby allowing functionality which could not be realized otherwise.

The final section under the geometrical optics heading considers the optical performance of binary optics, which can be broken down into wavefront quality, which usually is not an area of concern, and diffraction efficiency, which almost always is an area of major concern.

Fundamentals

The Generalized Grating Model

The geometrical optic model of binary optics is based on the premise that, in a sufficiently local area, a binary optic structure behaves like a linear grating. In particular, a binary optic component with phase profile $\phi(x, y)$, where z is taken as the optical axis and the element is assumed to lie on a planar substrate, may be locally modelled as a grating of spatial frequency

$$\mathbf{f}(x, y) = \frac{1}{2\pi} \nabla \phi \tag{11.1}$$

where ∇ is the gradient operator. The direction of the vector $\mathbf{f}$ indicates the orientation of the local grating, and the magnitude determines the period, which is explicitly given by

$$T(x, y) = 2\pi/|\nabla \phi| \tag{11.2}$$

where $|\ |$ is the vector magnitude. Using this concept of a local grating, results which are derived for linear gratings can be generalized to the case of a binary optic element of arbitrary phase profile. Accordingly, much of what follows is derived for the case of linear gratings with the intent that it will be generalized in such a manner. For example, efficiency models for the linear grating are presented later. By using the local grating concept, these linear grating estimates can be used to predict the local efficiency of a binary optic element, which, in turn, can then be combined to derive an average efficiency for the entire element.

To define a binary optic component, one must specify the phase function $\phi(x, y)$, the design wavelength λ_0, and the surface on which the component lies (i.e., the substrate shape). If the phase function is expressed in units of radians, then the design wavelength is not required for ray tracing, but it is still required to determine the blaze depth of the grating. For convenience, this section assumes that the phase function is expressed in radians and that the binary optic element lies on a planar substrate, with (x, y) being the coordinates in the plane and z the coordinate of the optical axis. In the case where the substrate is not planar (e.g., a binary optic structure on one surface of a non plano lens), the following results are still applicable if z is taken as the coordinate normal to the surface and (x, y) are taken as orthogonal coordinates tangent to the surface, that is, the curved surface is locally approximated as a plane.

Binary Optics

The phase function can be defined in any manner as long as it is defined at each point on the surface. Commonly used methods are explicit analytical expression for simple elements, decomposition into polynomials in (x, y), decomposition into polynomials in the radial coordinate $r = \sqrt{x^2 + y^2}$ for circularly symmetric cases, decompositions using other basis functions (e.g., Zernicke polynomials), and interpolation of look-up tables. Of these, the polynomial representations are the most commonly used in ray-tracing software. In addition, many ray-tracing codes also include a spherical term to allow for the convenient representation of optically generated diffractive optics. For binary optics, the spherical term is usually set to zero, and the polynomial decomposition alone is used to describe the profile. The binary optic element is then optimized by optimizing the polynomial coefficients.

The Grating Equation

A binary optics element with phase $\phi(x, y)$ can be ray traced by using the local grating concept and applying the grating equation. In the general case, the vector form of the grating equation is

$$n_2(\mathbf{n} \times \mathbf{r}_2) = n_1(\mathbf{n} \times \mathbf{r}_1) + \frac{m\lambda}{2\pi}(\mathbf{n} \times \nabla\phi) \tag{11.3}$$

where $\mathbf{r}_1$ is the unit vector describing the direction of the incident ray, n_1 is the index of the medium on the incident side, $\mathbf{r}_2$ and n_2 are the corresponding quantities for the diffracted ray, $\mathbf{n}$ is the direction of the surface normal, m and λ are the diffraction order and wavelength of interest, and $\times$ is the vector cross-product (Welford, 1986). If the substrate surface is planar, then the grating equation reduces to

$$n_2 L_2 = n_1 L_1 + \frac{m\lambda}{2\pi} \frac{\partial \phi}{\partial x} \tag{11.4}$$

and

$$n_2 M_2 = n_1 M_1 + \frac{m\lambda}{2\pi} \frac{\partial \phi}{\partial y} \tag{11.5}$$

where the ray directions are defined by the direction cosines (L, M, N) corresponding to the coordinates (x, y, z). In geometrical designs, the element is usually used in the first order ($m = 1$).

The Sweatt Model

In the Sweatt model, the binary optic element is replaced by an equivalent refractive element (Sweatt, 1979). The Sweatt model is important because it allows results derived for refractive optics to be directly applied to binary optics and because it allows the modeling of binary optics on software which does not support the grating equation. The refractive equivalent of a binary optic element with phase $\phi(x, y)$ is defined by a thickness profile and refractive index of

$$t(x, y) = \frac{\lambda_0}{n_0 - 1} \frac{\phi(x, y)}{2\pi} + t_0 \tag{11.6}$$

and

$$n(m, \lambda) - 1 = \frac{m\lambda}{\lambda_0}(n_0 - 1), \tag{11.7}$$

where t_0 is a constant chosen to make the refractive element as thin as possible, but still having a positive thickness, and n_0 is the index of the material at the design wavelength λ_0. As n_0 approaches infinity, the Sweatt model approaches the grating equation, and, in practice, values of $n_0 = 10\,000$ are sufficiently high for accurate results (Farn, 1992). Note that the effective index of refraction is a function of both the wavelength and the diffraction order of the ray being traced.

In the special case of a binary optic lens described by Eq. 11.10, the more accurate Sweatt lens can be used, as illustrated in Fig. 11.3 (Sweatt, 1977). In this special case, the first surface collimates the incoming rays and the second surface, then, focuses them to the image point, thus, perfectly imaging the object point (x_1, y_1, z_1) to the image point (x_2, y_2, z_2) just as the diffractive element does. The shapes of these surfaces are hyperbolic with curvatures of

$$c_1 = 1/[(1 - n_0/n_1)z_1] \tag{11.8}$$

and

$$c_2 = 1/[(1 - n_0/n_2)z_2] \tag{11.9}$$

and eccentricities of (n_0/n_1) and (n_0/n_2), with the axis of each surface passing through the respective point source. The refractive index of Eq. 11.7 is still valid.

Phase Profiles of Common Elements

A lens used to image point (x_1, y_1, z_1) to point (x_2, y_2, z_2) at wavelength λ_0 in the m_0th diffraction order has a phase profile of

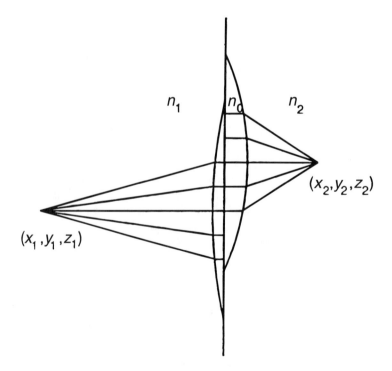

FIGURE 11.3 The Sweatt lens.

$$\phi(x, y) = \frac{k_0}{m_0} [n_1 z_1(\sqrt{(x - x_1)^2/z_1^2 + (y - y_1)^2/z_1^2 + 1} - 1)$$
$$- n_2 z_2(\sqrt{(x - x_2)^2/z_2^2 + (y - y_2)^2/z_2^2 + 1} - 1)] \qquad (11.10)$$

where $k_0 = 2\pi/\lambda_0$, n_1 and n_2 are the indices on the object and image side, and the binary optic lens is planar and located at $z = 0$. The paraxial focal length in air of the m_0th order at wavelength λ_0 is given by the lens law,

$$1/f_0 = n_2/z_2 - n_1/z_1. \qquad (11.11)$$

The paraxial phase profile of this same lens is given by

$$\phi(r) = \frac{-k_0 r^2}{2 m_0 f_0}, \qquad (11.12)$$

and the local period, based upon the paraxial phase, is expressed as

$$T(r) = m_0 \lambda_0 f_0 / r. \qquad (11.13)$$

The exact local period based on Eq. 11.10 usually is not used because the paraxial estimate is accurate enough for most purposes and is much simpler to evaluate. If the lens is used in order m at wavelength λ, then the paraxial focal length in air will be shifted to

$$f = \frac{m_0 \lambda_0}{m \lambda} f_0. \qquad (11.14)$$

If the m_0th order of a grating deflects a plane wave of wavelength λ_0 and with direction cosines (L_1, M_1, N_1) to the direction (L_2, M_2, N_2), then, the grating is described by

$$\phi(x, y) = \frac{k_0}{m_0} [x(n_2 L_2 - n_1 L_1) + y(n_2 M_2 - n_1 M_1)], \qquad (11.15)$$

and the local period is constant at

$$T = m_0 \lambda_0 / \sqrt{(n_2 L_2 - n_1 L_1)^2 + (n_2 M_2 - n_1 M_1)^2}. \qquad (11.16)$$

If the grating is used in order m at wavelength λ, then, the diffracted direction cosines will be

$$L = \left[n_1 L_1 + \frac{m\lambda}{m_0 \lambda_0} (n_2 L_2 - n_1 L_1) \right] / n_2 \qquad (11.17)$$

and

$$M = \left[n_1 M_1 + \frac{m\lambda}{m_0 \lambda_0} (n_2 M_2 - n_1 M_1) \right] / n_2. \qquad (11.18)$$

An axicon, the m_0th order of which diffracts a normally incident plane wave of wavelength λ_0 into the radial direction with direction cosine L_2, is described by

$$\phi(r) = \frac{k_0}{m_0}[r(n_2L_2)] \tag{11.19}$$

with the radial period given by

$$T = m_0\lambda_0/|n_2L_2|. \tag{11.20}$$

The radial direction cosine at order m and wavelength λ is given by

$$L = \frac{m\lambda}{m_0\lambda_0}L_2. \tag{11.21}$$

A grating is obviously periodic. However, paraxial lenses are also periodic, but in r^2, and axicons are periodic in r. Nonparaxial lenses are also periodic, but with respect to a variable which does not have a simple form. This periodicity allows results derived for gratings to be applied to lenses and axicons.

Design of Monochromatic Systems

The designs presented in this section will generally not be useful for wideband systems due to the inherent chromatic sensitivity of diffractive optics. One significant exception is systems where the wavelengths of interest have a harmonic relationship (e.g., a system containing light and its frequency-doubled harmonic). This special case takes advantage of the fact that the ray trace of the first order of λ_0 is identical to that of the mth order of λ_0/m and, in general, all diffractive ray traces for which $m\lambda$ is constant are identical.

The following is based on (Buralli and Morris, 1991) and (Buralli and Morris, 1989). Consider the third-order design of a diffractive singlet which is to be used monochromatically at wavelength λ_0. Let the rotationally symmetric system consist of a diffractive singlet on a substrate of curvature c and a stop located at $\bar{l}$, as shown in Fig. 11.4. Let the object and image be located in air at l_1 and l_2 with u_1 and u_2 as the angles of the paraxial marginal rays and y as the paraxial marginal ray height. In addition, let the object have height $\bar{y}_1$ and the paraxial chief ray height $\bar{y}$. Given these constraints, the binary optic singlet will have the form

$$\phi(r) = \frac{-k_0 r^2}{2m_0 f_0} + 2\pi G r^4 \tag{11.22}$$

where f_0 is the focal length defined in Eq. 11.11, m_0 and k_0 are the diffraction order and wave number of interest and G is the fourth-order phase coefficient. Higher orders are not of interest

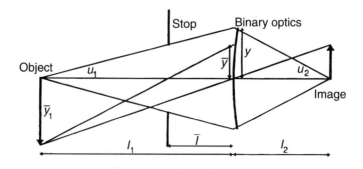

FIGURE 11.4 Paraxial quantities for the diffractive singlet problem.

in a third-order analysis, and odd orders would result in a cusp at the origin. Now introduce the following intermediate quantities: a bending parameter $B = 2cf_0$, a conjugate parameter $T = (u_1 + u_2)/(u_1 - u_2)$, a stop shift parameter $\epsilon = \bar{y}/y$, and the Lagrange invariant $H = u_1\bar{y}_1$.

As a function of the normalized object coordinate h and normalized pupil coordinates ρ and θ, the aberrations of this system are then given by

$$W(h, \rho, \theta) = \frac{1}{8} S_1 \rho^4 + \frac{1}{2} S_2 h \rho^3 \cos\theta + \frac{1}{2} S_3 h^2 \rho^2 \cos^2\theta$$

$$+ \frac{1}{4}(S_3 + S_4) h^2 \rho^2 + \frac{1}{2} S_5 h^3 \rho \cos\theta, \qquad (11.23)$$

where

$$S_1 = \frac{y^4}{4f_0^3}(1 + B^2 + 4BT + 3T^2) - 8\lambda_0 m_0 G y^4, \qquad (11.24)$$

$$S_2 = \frac{-y^2 H}{2f_0^2}(B + 2T) + \frac{\epsilon y^4}{4f_0^3}(1 + B^2 + 4BT + 3T^2)$$

$$- 8\epsilon\lambda_0 m_0 G y^4, \qquad (11.25)$$

$$S_3 = \frac{H^2}{f_0} - \frac{2\epsilon y^2 H}{2f_0^2}(B + 2T) \qquad (11.26)$$

$$+ \frac{\epsilon^2 y^4}{4f_0^3}(1 + B^2 + 4BT + 3T^2) - 8\epsilon^2 \lambda_0 m_0 G y^4, \qquad (11.27)$$

$$S_4 = 0,$$

and

$$S_5 = \frac{3\epsilon H^2}{f_0} - \frac{3\epsilon^2 y^2 H}{2f_0^2}(B + 2T)$$

$$+ \frac{\epsilon^3 y^4}{4f_0^3}(1 + B^2 + 4BT + 3T^2) - 8\epsilon^3 \lambda_0 m_0 G y^4. \qquad (11.28)$$

The quantities S_1 to S_5 represent spherical aberration, coma, astigmatism, Petzval field curvature and distortion. The field curvature is identically zero and the remaining four aberrations may be controlled by the three degrees of freedom: the stop position, the substrate curvature, and the fourth-order phase coefficient. By choosing the phase coefficient

$$G = (1 - T^2)/(32 m_0 \lambda_0 f_0^3) \qquad (11.29)$$

and the stop position so that

$$\epsilon = 2Hf_0/[y^2(B + 2T)], \qquad (11.30)$$

the coma and astigmatism terms can be reduced to zero. Designs based on this concept are discussed in further detail in (Buralli and Morris, 1991) and (Buralli and Morris, 1989), whereas (Buralli and Morris, 1992a) extends this approach to the design of two- and three-element telescopes.

Other problems of similar complexity are considered in (Welford, 1986), including the aberrations introduced when a perfectly corrected system is used at a different wavelength or at different conjugates and a discussion of special imaging conditions (e.g., aplanatic imaging).

Another class of monochromatic applications is the use of binary optics with lasers. In these applications, a binary optic lens or a diffractive/refractive combination is used as a direct replacement of a refractive asphere. Because the diffractive and refractive optical designs perform similar optical functions, the choice of element depends on other factors such as cost, durability, or ease of manufacture (Gruhlke et al., 1992).

Design of Wideband Systems

In wideband systems, binary optic elements are used to correct chromatic or residual monochromatic aberrations. Because their strong wavelength dependence precludes their use with any significant optical power, they are usually combined with conventional elements in so-called hybrid systems and the design of these systems can be described by conventional lens design principles by taking advantage of the Sweatt model.

Chromatic Correction

One of the most mature applications of binary optics is the chromatic correction of wideband imaging systems. (Anderson et al., 1993 and Chen, 1992) describe some recent examples and the following is based on (Swanson, 1989) and (Stone and George, 1988), and primarily taken from (OSA, 1994), with permission. The chromatic behavior of binary optics can be understood by using the Sweatt model, which states that a binary optic lens behaves like an ultrahigh index refractive lens with an index which varies linearly with wavelength (let $n_0 \to \infty$ in Eq. 11.7). Accordingly, it can be used to correct the primary chromatic aberration of a conventional refractive lens but cannot correct the secondary spectrum. For the design of achromats and apochromats, an effective Abbe number and partial dispersion can be calculated. For example, using the C, d, and F lines, the Abbe number is defined as $V = [n(\lambda_d) - 1]/[n(\lambda_F) - n(\lambda_C)]$. Substituting Eq. 11.7 and letting $n_0 \to \infty$, then, yields

$$V = \lambda_d/(\lambda_F - \lambda_C) = -3.45. \qquad (11.31)$$

Similarly, the effective partial dispersion, using the g and F lines, is

$$P = (\lambda_g - \lambda_F)/(\lambda_F - \lambda_C) = .296. \qquad (11.32)$$

By using these effective values, the conventional procedure for designing achromats and apochromats can be extended to designs in which one element is a binary optic lens.

Unlike all other materials, a binary optic lens has a negative Abbe number. Thus, an achromatic doublet can be formed by combining a refractive lens and a binary optic lens, both with positive power, thus significantly reducing the required lens curvatures and allowing for larger apertures. Similarly, the unusual Abbe number and partial dispersion also allow the design of apochromats with reduced lens curvatures and larger apertures.

Correction of Monochromatic Aberrations

The correction of aberrations in monochromatic systems has been discussed previously. In wideband systems, binary optic components are limited to the correction of residual aberrations, as in a Schmidt corrector. As a simple example (Swanson, 1989), taken from (OSA, 1994) with permission, consider a refractive system which suffers from third-order spherical aberration given by

$$W(\rho) = \frac{1}{8} S_1 \rho^4, \quad (11.33)$$

where the variables are as defined for Eq. 11.23. This is equivalent to a phase aberration of $kW(\rho)$, where $k = 2\pi/\lambda$, and a binary optic corrector with phase

$$\phi_C(\rho) = -k_0 \frac{1}{8} S_1 \rho^4 \quad (11.34)$$

will completely correct the aberration at wavelength λ_0. However, at other wavelengths, there will still be a residual spherochromatism of

$$kW(\rho) + \phi_C(\rho) = k \frac{1}{8} [(1 - \lambda/\lambda_0) S_1] \rho^4 \quad (11.35)$$

because the change in wavelength affects the phase of the original aberration, but it does not affect the phase produced by the binary optic corrector.

Other Applications

Other applications include the chromatic correction of purely diffractive systems, the use of binary optics for thermal correction, and the use of binary optics as microoptics.

It is often pointed out that an entirely diffractive imaging system with positive power cannot be corrected at two wavelengths (Buralli and Rogers, 1989). This statement, while true, implicitly assumes that the diffractive elements are lenslike, the system is imaging in the conventional sense, and only a single diffraction order is utilized. If one or more of these assumptions is relaxed, then the original statement no longer holds, as shown by the following designs. In Fig. 11.5a,

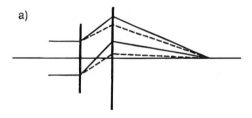

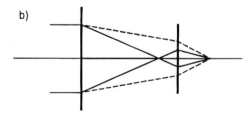

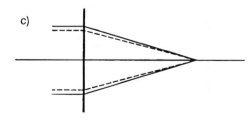

FIGURE 11.5 Unconventional diffractive designs which are corrected at two wavelengths. The shorter wavelength is shown by the dashed line. a) use of nonlenses; b) unconventional imaging; c) use of multiple orders.

the chromatic behavior of one diffractive element compensates for the other, thus, bringing two wavelengths to the same focus (Farn and Goodman, 1991). A degenerate case of this design occurs when the two elements are linear gratings of the same period, but blazed for opposite orders. For example, the first grating may be blazed for the +1 order and the second for the −1 order. In this case, all rays which are parallel upon entering the system are also parallel when they exit the system, regardless of wavelength. Figure 11.5b shows a design based on the one-glass achromat (Levi, 1980). The intermediate focus of the longer wavelength insures that the two wavelengths, while focusing at the same point, will have magnifications which are opposite in sign. Finally, as mentioned previously, wavelength/diffraction order combinations which have the same $m\lambda$ product will also ray trace the same, as depicted in Fig. 11.5c.

The approach to thermal correction is analogous to that of chromatic correction in that a thermal Abbe number for refractive and diffractive elements can be derived and the thermal variations then balanced (Londono et al., 1993). Physically, a change in temperature results in an index change and a dimensional change. For refractive elements, both of these effects will affect a ray trace, whereas, for diffractive elements, only the dimensional change (i.e., change in grating periodicity) changes the ray trace.

As a result of the inherent accuracy of the fabrication process, binary optic technology is well suited for the fabrication of microoptics and microoptic arrays. As with larger binary optics components, the phase profile and aperture of each microoptic are not limited to any specific shapes and the arraying of these microoptics is also not limited to any specific grid patterns. One extreme possibility would be arrays of arbitrarily shaped microoptics on an irregular grid, with each microoptic optimized for a different function. The other extreme would be a regular array of identical microoptics fully covering a large area. As an additional benefit, binary optic technology allows variations across the entire array to be held to optical tolerances, whereas, in most other microoptics technologies, even if each microoptic can meet such tolerances, variations across the array will not.

Although binary optics is usually synonymous with diffractive optics, microoptics is one area where this is not the case. Binary optic structures are constrained by the fact that the overall height of the surface profile is limited, typically, to no more than 3 μm in normal cases, but up to 20 μm in extreme cases. For large optics, this height limitation precludes the fabrication of refractive optics, but this is not so for microoptics. Consider a spherical microlens of curvature c and aperture half-width of r_{max}, where r_{max} is the radius of the circle which would circumscribe the aperture. The height of this spherical cap would be given by

$$h_{max} = cr_{max}^2/[1 + \sqrt{1 - (cr_{max})^2}] \tag{11.36}$$

which can be less than the height limitation in many cases.

Microoptics are often produced as arrays, and Table 11.2 gives the radii of several circles used to characterize arrays. In the table, min{x, y} is the lesser of x and y whereas max{x, y} is the greater of the two quantities. The inscribed circle is the maximum size round lens which could be placed on a given grid pattern, the equal area circle is used to define an equivalent F/# for

TABLE 11.2 Radii of Equivalent Circles for Common Grid Shapes

Tile shape	Inscribed circle	Equal Area Circle	Circumscribed Circle
rectangle $2a \times 2b$	min{a, b}	$\sqrt{4ab/\pi}$	$\sqrt{a^2 + b^2}$
hexagon	min{$c, c(a+b)/\sqrt{b^2+c^2}$}	$\sqrt{2(2a+b)c/\pi}$	max{$(a+b), \sqrt{a^2+c^2}$}

lenses which fill 100% of the grid pattern, as is typically the case for binary optics devices, and the radius of the circumscribed circle is required to calculate many fabrication parameters.

In their application, microoptics are often combined with sources, commonly laser diodes or detectors. Accordingly, two significant applications of microoptics use them to shape the beam from laser diodes or to concentrate light onto a detector.

Optical Performance

The optical performance of a binary optic element can be described by its wave-front quality and diffraction efficiency. Wave front quality is usually not an issue because the diffractive structure usually introduces an insignificant amount of wave front error. The diffraction efficiency, however, is usually a major issue because the efficiency of even a perfectly fabricated device can be significantly less than 100%. In particular, one must not only be concerned with how much light is diffracted into the order of interest but also with how the remaining light, which is primarily diffracted into other orders, will affect the system performance (Buralli and Morris, 1992b). In the following treatment, these issues will be discussed in the context of the linear grating, with the ideas being extended to the general case via the local grating concept introduced previously.

Diffraction Efficiency

The diffraction efficiency of a linear grating is a function of the wavelength, polarization and incident angle of the light, the surface profile and period of the grating, and of the indices of the materials involved. The rigorous solution of Maxwell's equations can provide accurate solutions to this problem (Moharam and Gaylord, 1985) and should be used when the scalar theory is not adequate, but it is computationally intensive and the results can be nonintuitive. Therefore, it will not be discussed further. Instead, this section focuses on the scalar theory model, which is a simplified solution to this problem, but still quite useful as long as its limitations are kept in mind. In addition, this section assumes that the grating is to be blazed for the $+1$ order, which is taken as diffracting rays in the $+x$ direction, because most geometrical designs are designed for use in this order.

Scalar Theory Predictions. The following is based on (Swanson, 1991). Consider the problem of Fig. 11.6a. The grating has period T in the x direction and height h, light of wavelength λ is incident with direction cosines (L_1, M_1, N_1), and the two media have indices n_1 and n_2. In the scalar theory approach, the grating is modeled as the infinitely thin phase screen of Fig. 11.6c. The phase delay is assumed to be linear in x and the only unknown is the optical path difference P between rays which traverse each end of one period. As the period approaches infinity, this path difference can be estimated by considering two rays which traverse a step of height h, as shown in Fig. 11.6b. The optical path difference is given by $P = s_1 n_1 - s_2 n_2$. Solving Snell's law and substituting then yields

$$P = h(n_1 N_1 - n_2 N_2), \tag{11.37}$$

where

$$N_2 = \sqrt{1 - (L_2^2 + M_2^2)},$$

$$L_2 = n_1 L_1 / n_2, \tag{11.38}$$

$$M_2 = n_1 M_1 / n_2, \tag{11.39}$$

and where (L_2, M_2, N_2) is the direction of the ray refracted by the step of Fig. 11.6b but not the direction of the diffracted ray in Fig. 11.6a.

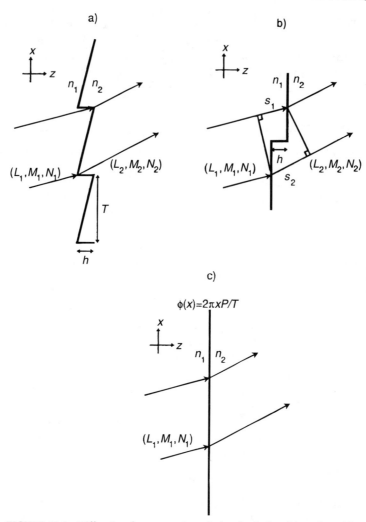

FIGURE 11.6 Diffraction from a grating. a) the physical problem, h positive as shown; b) optical path difference introduced by a step of height h; c) the resulting scalar phase screen model.

Based on this path difference, the grating can be modeled as a phase screen with the phase of one period given by

$$\phi(x) = kPx/T, \tag{11.41}$$

where $k = 2\pi/\lambda$. Based on the approach of Eq. 11.63–11.65, the Fourier series decomposition of this function predicts a diffraction efficiency of

$$\eta(m) = \text{sinc}^2(P/\lambda - m) \tag{11.42}$$

for the mth order, where $\text{sinc}(x) = \sin(\pi x)/(\pi x)$. This equation predicts a first order diffraction efficiency of 100% if $P = \lambda$. For normally incident light at the design wavelength λ_0, combining this condition with Eq. 11.37 yields a grating height of

$$h_0 = \lambda_0/(n_1 - n_2). \tag{11.43}$$

Binary Optics

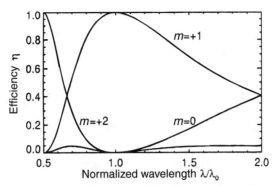

FIGURE 11.7 Scalar diffraction efficiency of a grating blazed for the $m = +1$ order, material dispersion neglected. Parasitic orders $m = 0$ and $m = +2$ are also shown.

Although this blaze height has been derived assuming normally incident light and large grating periods, it is accurate enough to be used as the actual blaze height in most cases and can be used, at least, as an approximate estimate in all cases. The following section presents some ad hoc modifications which will further increase the accuracy of the blaze height estimate.

Similar remarks also apply to the efficiency estimate of Eq. 11.42. In particular, in cases where Eq. 11.43 is inaccurate, the efficiency equation, as a result of its dependence on Eq. 11.37, will also be inaccurate. However, even in these cases, the efficiency equation is still useful because it still shows the dependence of efficiency on incidence angle and wavelength. In general, the effect of incidence angle is negligible except for large angles (e.g., greater than 30°). Of greater concern is the significant efficiency loss which occurs with changing wavelength. As shown in Fig. 11.7, light which is not diffracted into the $+1$ order is primarily diffracted into the 0 order at longer wavelengths and into the $+2$ order at shorter wavelengths. For wideband systems, this efficiency loss can be significant and will affect the choice of the design wavelength. For example, if a system has a bandwidth of $\lambda_{min} < \lambda < \lambda_{max}$, then, choosing the design wavelength such that $1/\lambda_0$ is the mean of $1/\lambda_{min}$ and $1/\lambda_{max}$, will insure that the efficiencies at the extreme wavelengths are equal. This choice of design wavelength, which results in the efficiency curves of Fig. 11.8, maximizes the minimum efficiency but does not maximize the average efficiency.

The binary optic fabrication process usually approximates a true blaze by an N-level staircase. In this case, if each step has height h/N, then, each step will introduce an optical path difference of P/N, where P is given by Eq. 11.37. Note that h is the physical height of the corresponding blazed grating and is not the physical height of the staircase approximation, which is given by

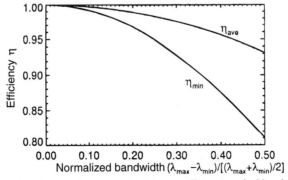

FIGURE 11.8 Average and minimum efficiencies of a blazed grating used in broadband illumination.

TABLE 11.3 Diffraction Efficiencies of an N-Level Device at the Design Wavelength

Number of Levels N	$\eta(1)$	$\eta(1 - N)$	$\eta(1 + N)$
∞	1.000	0.000	0.000
16	0.987	0.004	0.003
12	0.977	0.008	0.006
8	0.950	0.019	0.012
6	0.912	0.036	0.019
4	0.811	0.090	0.032
3	0.684	0.171	0.043
2	0.405	0.405	0.045

$(N - 1)h/N$. Following the same approach as previously, then results in an efficiency (Dammann, 1979) of

$$\eta(m) = \text{sinc}^2(m/N)\text{sinc}^2(P/\lambda - m)/\text{sinc}^2[(P/\lambda - m)/N]. \quad (11.44)$$

The quotient in the second and third terms is always less than one, and so, the maximum efficiency is given by the first term and occurs under the same conditions as in the blazed case (i.e., a height of h_0 or step height of h_0/N for normal illumination). For binary optic elements used in the first order, Eq. 11.44 yields the often-quoted efficiency estimate of $\text{sinc}^2(1/N)$ for an N-level grating, as given in the second column of Table 11.3.

The overall efficiency response of a staircase grating may be thought of as follows. With respect to wavelength and incidence angle, the first-order efficiency is similar in shape to the blazed case, but reduced by approximately $\text{sinc}^2(1/N)$, with the lost energy being diffracted into orders which are a multiple of N orders away from the first order, as shown in the third and fourth columns of Table 11.3. For example, for an eight-level device, the first-order efficiency is reduced by about 5%, with that 5% going primarily into the -7th, $+9$th and higher orders.

In the reflective case of Fig. 11.9, the optical path difference introduced by a step of height h is given by

$$P = h(2n_1N_1), \quad (11.45)$$

where n_1 is the index of the surrounding medium and (L_1, M_1, N_1) are the direction cosines of the incident ray. The corresponding optimal height for normal incidence is given by

$$h_0 = \lambda_0/(2n_1). \quad (11.46)$$

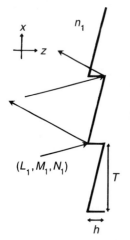

FIGURE 11.9 Diffraction from a reflective grating, h positive as shown.

Binary Optics

The previous results of this section are still valid for the reflective case except that Eq. 11.45 and 11.46 should be used in place of Eq. 11.37 and 11.43.

The scalar theory is more accurate for gratings which are thin and have large periods. This concept can be quantified by using the dimensionless parameter

$$Q = 2\pi\lambda h/(\min\{n_1 N_1, n_2 N_2\} T^2), \tag{11.47}$$

where $\min\{x, y\}$ is the lesser of x and y. If $Q \ll 1$, then, the scalar theory is applicable (Magnusson and Gaylord, 1978). As a rule of thumb, the scalar theory begins to lose validity when the grating period falls below ten wavelengths (i.e., a grating with a period less than $10\lambda_0$ or a lens faster than $F/5$), and lower efficiencies can be expected in these cases (Swanson, 1991; Gremaux and Gallagher, 1993). In addition to neglecting the influence of grating period and thickness, the scalar theory also neglects polarization effects and Fresnel reflections. Accordingly, if any of these effects are important, then the scalar theory model will be inadequate.

Scalar Theory Modifications. The most significant shortcoming of the scalar theory is that it does not predict the efficiency loss due to small grating periods and finite grating heights. This efficiency loss can be estimated by multiplying the scalar efficiency prediction by the factor (Swanson, 1991)

$$F = [1 - h\lambda/(T^2\sqrt{\min\{n_1, n_2\}^2 - (\lambda/T)^2})]^2. \tag{11.48}$$

This multiplier accounts for the degradation in an ad hoc manner and should only be used to predict general behavior. In most cases, it is not accurate enough to replace a rigorous solution to Maxwell's equations.

A second shortcoming of the scalar theory is that, for small periods and low index materials, the optimal blaze heights predicted by Eq. 11.43 and 11.46 can be off by a significant amount. However, conventional blaze height arguments can be used to modify these estimates in an ad hoc manner (Swanson, 1991). For example, consider the grating of Fig. 11.6a. The conventional argument states that rays which are diffracted by the grating and those which are refracted off the grating facet should have the same direction cosines. Applying both the grating equation and Snell's law to this situation and solving, then, yields a blaze height of

$$h_0 = \lambda_0/(n_1 N_1 - n_2 N_2), \tag{11.49}$$

where

$$N_2 = \sqrt{1 - (L_2 + M_2^2)}, \tag{11.50}$$

$$L_2 = (n_1 L_1 + \lambda_0/T)/n_2, \tag{11.51}$$

$$M_2 = n_1 M_1/n_2, \tag{11.52}$$

and (L, M, N) are the direction cosines of the incident and diffracted rays shown in Fig. 11.6a. Note that this equation reduces to Eq. 11.43 for the case of normal illumination as T approaches infinity. Similarly, the blaze depth for the reflecting case is given by

$$h_0 = \lambda_0/[n_1(N_1 + N_2)], \tag{11.53}$$

where

$$N_2 = \sqrt{1 - (L_2^2 + M_2^2)}, \quad (11.54)$$

$$L_2 = L_1 + \lambda_0/(n_1 T), \quad (11.55)$$

and

$$M_2 = M_1. \quad (11.56)$$

In addition to the ad hoc modifications given above, there have been some initial efforts to numerically optimize the grating profiles using vector theory, but no simple algorithms have yet been devised.

Manufacturing Tolerances

Manufacturing errors can affect both the wave front quality and the diffraction efficiency of a binary optic element. For example, if the periods of a grating are slightly aperiodic, but still ideally blazed, then the aperiodicity will introduce a wave-front error but will not significantly degrade the diffraction efficiency. On the other hand, if the grating is perfectly periodic, but the blaze shape is slightly in error, then the first order will be a perfect plane wave, but of lower efficiency. Of these effects, the efficiency issue usually overshadows the wave front issue. This section presents simple models for these effects, based on the scalar theory.

Errors in the blaze profile can be classified as either height or lateral errors, and both errors will result in lower diffraction efficiency (Cox et al., 1991; Ferstl et al., 1992; Emerton, 1985). In the case of height errors, if the actual grating height is in error by δ, as shown in Fig. 11.10a, then, by Eq. 11.42, the efficiency will be reduced by a multiplicative factor of

$$F = \text{sinc}^2(\delta/h_0), \quad (11.57)$$

where h_0 is the ideal blaze height. In practice, this error is not a major concern because most processes can be controlled to $\delta/h_0 < 0.05$ resulting in $F > 0.99$. Lateral errors, however, are of great concern. Lateral errors create portions of the grating which do not contribute to the overall efficiency, and these portions can be modeled as opaque regions, indicated by the black regions in Fig. 10b. For example, if the grating is diamond-turned, then, the finite size of the tool tip will result in incorrect blazing over part of the grating. Similarly, for binary optic components

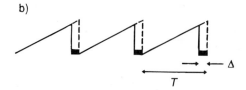

FIGURE 11.10 Models used to estimate the loss of diffraction efficiency due to a) height errors and b) lateral errors. The dashed line is the ideal profile.

Binary Optics

fabricated by multiple mask techniques, misalignments between mask layers will result in areas which deviate substantially from the desired profile, and these areas will not contribute significantly to the first-order efficiency of the grating. If the total length of such areas is Δ, then, the efficiency will be reduced by the fill factor squared:

$$F = (1 - \Delta/T)^2. \tag{11.58}$$

Wave-front errors can be calculated by comparing an ideal wave front with that produced by a grating with fabrication errors. If the error in grating height varies uniformly from $-\delta$ to $+\delta$ as shown in Fig. 11.11a, then, the RMS wave-front error in waves is approximately

$$\text{RMS} = \frac{1}{3} (\delta/h_0). \tag{11.59}$$

If the uncertainty in the position of each grating transition is uniformly distributed from $-\Delta$ to $+\Delta$ as shown in Fig. 11.11b, then, the RMS wave-front error will be approximately

$$\text{RMS} = \frac{1}{2} (\Delta/T). \tag{11.60}$$

Choosing worst case numbers of $\delta/h_0 = 0.05$, $T = 4\mu m$, and $\Delta = 0.2\mu m$ yields worst case RMS wave-front errors of 1/60 and 1/40 wave, respectively. Accordingly, wave-front quality is usually insignificantly affected by fabrication errors, but is limited by the quality of the substrate.

11.3 Designs Based on Scalar Diffraction Theory

Designs based on scalar diffraction theory take advantage of the wave nature of light. In particular, the binary optic device adjusts the phase of an incoming wave so that the wave then propagates in some desired manner. For example, the adjusted wave may produce a desired intensity distribution at a point downstream of the binary optic element. The difference between scalar and geometrical designs is that scalar designs rely specifically on diffraction effects which are largely neglected by the geometrical theory.

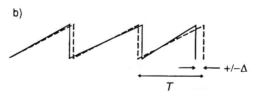

FIGURE 11.11 Models used to estimate wavefront errors due to a) height errors and b) lateral errors. The dashed line is the ideal profile.

This section first introduces the foundations for the numerical implementation of the scalar theory, followed by some common applications.

Fundamentals

In the scalar regime, a binary optic component with height profile $h(x, y)$ is modeled as a phase screen introducing a phase delay of

$$\phi(x, y) = k(n_1 - n_2)h(x, y), \tag{11.61}$$

where n_1 and n_2 are the indices of the two media involved. This equation is based on the same assumptions as Eq. 11.43. However, there is also a significant difference between the phase profile, as used in the scalar theory, and that used in the geometrical optics case. In particular, note that the phase profile given above is a function of the wavelength, whereas, in the geometrical optic case, the phase profile is independent of wavelength (e.g., see the example of section 2.4.2). The reason is that in the geometrical optic case, the phase profile is that of only the first-order diffracted wave, whereas, in the scalar case, the phase profile is that resulting from the combination of all orders. In general, as the wavelength changes, the phase profile of each order will remain the same, but the strengths of the diffraction orders will change so that the overall phase profile will also change. In addition, in the geometrical optic case, the phase profile will generally be continuous, spanning many multiples of 2π, whereas the phase profile in the scalar case spans only approximately 2π radians.

In other words, if the binary optics introduces a phase $\phi_0(x, y)$ at the design wavelength λ_0 and this phase accounts for all diffraction orders, then, the phase introduced at wavelength λ, including all diffraction orders, will be given by

$$\phi(x, y) = \frac{\lambda_0}{\lambda} [\phi_0(x, y) \bmod(2\pi)], \tag{11.62}$$

where $x \bmod y$ is the remainder of x divided by y. Contrast this with the geometrical optic case in which $\phi(x, y) = \phi_0(x, y)$ for all wavelengths.

The propagation of wave fronts in the scalar case is usually modeled by the Rayleigh–Sommerfeld, Fresnel, or angular spectrum approach, whichever is appropriate. Because these theories are described in length elsewhere, this section concentrates on special cases which occur for binary optics and numerical methods for implementing the integrals.

Gratings

As mentioned previously, the case of the linear grating is important because other problems can be solved by generalization or analogy. In the scalar theory, a diffraction grating of period T and phase $\phi(x)$ is decomposed into its Fourier coefficients $C(m)$, with

$$C(m) = \frac{1}{T} \int_0^T \exp[j\phi(x)]\exp(-j2\pi mx/T)\, dx, \tag{11.63}$$

$$\exp[j\phi(x)] = \sum_{m=-\infty}^{\infty} C(m)\exp(j2\pi mx/T), \tag{11.64}$$

and the relative intensity or efficiency of the mth diffracted order of the grating given by

$$\eta(m) = |C(m)|^2. \tag{11.65}$$

For propagation, each order is typically propagated separately, using the angular spectrum approach, and the resulting orders are then recombined using Eq. 11.64 to give the field at the plane of interest. In the angular spectrum approach, a plane wave with direction cosines (L, M, N), propagating over a distance z in air, experiences a transfer function of

$$H(L, M, N) = \exp(jkzN). \tag{11.66}$$

If the incident field is not a plane wave, it can be decomposed into plane waves by the fast Fourier transform (FFT). If the FFT is used for this and the field is sampled at increments of Δx, then the resulting N-point FFT will be sampled in increments of

$$\Delta L = \lambda/(N\Delta x). \tag{11.67}$$

If the medium is not air, then the previous two equations are still valid, provided that the wavelength in the medium is used (i.e., replace λ by λ/n).

Several common grating shapes are shown in Fig. 11.12 and their Fourier coefficients are given below. The binary optics fabrication process results in piecewise flat elements and, in the general N-step case of Fig. 11.12a, the Fourier coefficients take the form

$$C(m) = \sum_{n=0}^{N-1} \exp(j\phi_n)\delta_n \exp(-j2\pi m\Delta_n)\mathrm{sinc}(m\delta_n) \tag{11.68}$$

where

$$\delta_n = (x_{n+1} - x_n)/T, \tag{11.69}$$

$$\Delta_n = (x_{n+1} + x_n)/(2T), \tag{11.70}$$

ϕ_n are the phases of each step, and x_n are the transition points between steps. If each step of the piecewise flat profile has width Δx, as shown in Fig. 11.12b, then, Eq. 11.68 reduces to

$$C(m) = \exp(-j\pi m/N)\mathrm{sinc}(m/N)\mathrm{FFT}[\exp(j\phi_n)] \tag{11.71}$$

where

$$\mathrm{FFT}[x_n] = \frac{1}{N}\sum_n x_n \exp(-j2\pi mn/N). \tag{11.72}$$

If the phase is also incremented in steps of $\Delta\phi$, as in Fig. 11.12c, then, Eq. 11.71 further reduces to (Dammann, 1979)

$$C(m) = \exp\{j\pi[N-1]\alpha - m/N]\}\mathrm{sinc}(m/N)$$
$$\times \mathrm{sinc}(N\alpha)/\mathrm{sinc}\,\alpha, \tag{11.73}$$

where

$$\alpha = \Delta\phi/(2\pi) - m/N. \tag{11.74}$$

The remaining three profiles in Fig. 11.12 are common continuous profiles. Figure 11.12d is a linear ramp, with coefficients

$$C(m) = \exp\{j\pi[\phi_0/(2\pi) - m]\}\mathrm{sinc}[\phi_0/(2\pi) - m], \tag{11.75}$$

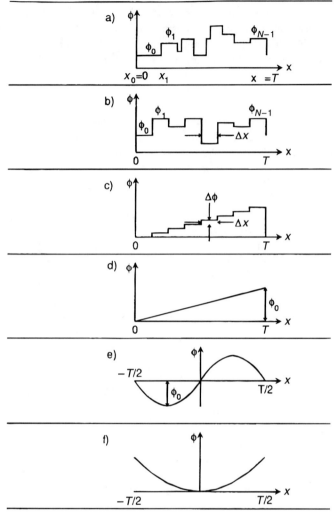

FIGURE 11.12 Common grating profiles: a) piece-wise flat; b) equal-width steps c) staircase; d) linear ramp; e) paraxial lens array (i.e., parabolic); f) sinusoidal.

whereas Fig. 11.12e is a sinusoidal phase grating with

$$C(m) = J_m(\phi_0), \quad (11.76)$$

where J_m is the mth order Bessel function of the first kind. Finally, a microlens array is also a grating of parabolic phase profile, as shown in Fig. 11.12f. In particular, an array consisting of paraxial lenses of first-order focal length f_0 at wavelength λ_0 will have diffraction orders described by

$$C(m) = \exp[j(\pi/2)(m/\alpha)^2]\text{conj}[\text{Fr}(m/\alpha + \alpha) - \text{Fr}(m/\alpha - \alpha)] \quad (11.77)$$

where

$$\alpha = T/\sqrt{2\lambda_0 f_0}, \quad (11.78)$$

$$\text{Fr}(x) = \int_0^x \exp(j\pi t^2/2)\, dt, \quad (11.79)$$

and conj [ ] denotes the complex conjugate.

Binary Optics

The case of hexagonal arrays can be evaluated directly (Roberts et al., 1992) or by taking advantage of the hexagonal pattern's rectangular periodicity, as indicated by the dashed line in Fig. 11.13a. The resulting rectangular diffraction pattern is shown in Fig. 11.13b by the dashed lines, with the corresponding hexagonal orders marked by x's. The nonhexagonal orders are identically zero due to the symmetry in the original rectangular period.

Scalar Diffraction Integrals

A detailed discussion of the numerical methods used to evaluate the scalar diffraction integrals (Barakat, 1980) is beyond the scope of this section, which concentrates, instead, on the simplest numerical implementations.

Consider the case in which an incident field $U_1(x, y)$ propagates over a distance z and an FFT implementation of the Fresnel diffraction integral is used to calculate the resulting field $U_2(x, y)$. First, the integrand is sampled in increments of Δx_1 and Δy_1 to yield the series

$$f(m_1, n_1) = \exp\{jk[(m_1\Delta x_1)^2 + (n_1\Delta y_1)^2]/(2z)\}U_1(m_1\Delta x_1, n_1\Delta y_1), \tag{11.80}$$

which consists of $M \times N$ samples, where the number of samples includes any zero padding. The integral can then be approximated by

$$g(m_2, n_2) = \text{FFT}[f(m_1, n_1)], \tag{11.81}$$

where the FFT is the two-dimensional version of Eq. 11.72, and m_1, m_2, n_1, and n_2 are index variables. Samples of the desired field are, then, approximated by

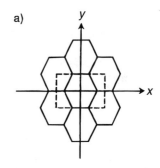

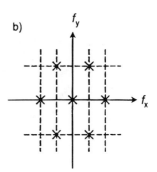

FIGURE 11.13 Hexagonal arrays: a) The hexagonal array is also a rectangular array, with a basic cell defined by the dashed rectangle. b) The corresponding diffraction orders.

$$U_2(m_2\Delta x_2, n_2\Delta y_2) \doteq \frac{\exp(jkz)}{j\lambda z} \exp\{jk[(m_2\Delta x_2)^2 + (n_2\Delta y_2)^2]/(2z)\}$$

$$\times M\Delta x_1 N\Delta y_1 g(m_2, n_2) \tag{11.82}$$

where

$$\Delta x_2 = \lambda z/(M\Delta x_1) \tag{11.83}$$

and

$$\Delta y_2 = \lambda z/(N\Delta y_1). \tag{11.84}$$

If the medium is not air, then the wavelength in the medium should be used. In an analytical approach, the Fraunhofer approximation is easier to evaluate because the quadratic phase factors are eliminated. However, in a numerical implementation, this advantage is lost, so the Fraunhofer approximation is rarely used.

In the circularly symmetric case, the quasi-fast Hankel transform may be used (Siegman, 1977). In this case, the integrand is first represented by the discrete function

$$f(n) = \begin{cases} r_1(n)\exp[jkr_1^2(n)/(2z)]U_1[r_1(n)] & n \in [0, N-1] \\ 0 & n \in [N, 2N-1] \end{cases} \tag{11.85}$$

where

$$r_1(n) = r_{01}\exp(\alpha n) \tag{11.86}$$

which consists of $2N$ samples, the last N of which are zero. The variables r_{01} and α determine the sampling of the incident field, and the continuous form, which is integrated from zero to infinity, will be approximated by a summation over the range $[r_{01}, r_{01}\exp(\alpha N - 1)]$. Next, the transform kernel is represented by the $2N$-point sequence

$$h(n) = 2\pi\alpha \frac{r_{01}r_{02}\exp(\alpha n)}{\lambda z} J_0\left[2\pi \frac{r_{01}r_{02}\exp(\alpha n)}{\lambda z}\right], \tag{11.87}$$

where J_0 is the zero-order Bessel function and r_{02} determines the sampling of the field U_2. The diffraction integral is approximated by the correlation of these two sequences, which may be implemented by $2N$-point FFT's, yielding

$$g(n) = 2N\text{FFT}\{\text{FFT}[f(n)] \cdot \text{IFFT}[h(n)]\} \tag{11.88}$$

where the FFT is as defined in Eq. 11.72 and the corresponding inverse IFFT uses the conjugate kernel and drops the $1/N$ factor (actually $1/(2N)$ using the current notation). The resulting field is then given by

$$U_2[r_2(n)] \doteq \frac{\exp(jkz)}{j\lambda z} \exp[jkr_2^2(n)/(2z)]$$

$$\times \lambda z g(n)/r_2(n) \quad \text{for} \quad n \in [0, N-1] \tag{11.89}$$

where

$$r_2(n) = r_{02}\exp(\alpha n) \tag{11.90}$$

and the last N points in $g(n)$ are discarded.

The use of the FFT in both of these methods increases the speed of the computation. However, it sometimes unnecessarily constrains the sampling of both U_1 and U_2. For example, perhaps U_2 need only be sampled at one point, as is the case for a Strehl calculation, or perhaps the sample points are required on a differently shaped grid. In these cases, other means may be more efficient to evaluate the diffraction integrals (Roose et al., 1993).

Designs Utilizing Multiple Diffraction Orders

In most geometrical designs, the binary optic component is designed to operate in a single order, and the grating profile is chosen to be blazed for that order. In many scalar designs, it is desirable to use more than one order, and the grating profile must be redesigned to achieve this purpose.

Grating Beam Splitters

Grating beam splitters are gratings in which the diffraction efficiencies are designed to follow a given weighting. A common example is the 1:K fanout problem in which an incident plane wave is divided into K diffraction orders of equal intensity. Grating beam splitters can be classified as either continuous phase or binary phase and their performance is evaluated by how closely the actual diffraction efficiencies match the desired weighting and by the total efficiency of the orders of interest. Fanouts of 1:100 with efficiencies over 80% and nonuniformities of less than 10% are typical figures. The continuous designs usually have higher total efficiencies, whereas the binary designs usually are easier to fabricate. Table 11.4 shows upper bounds for the total efficiency in the 1:K fanout case (Krackhardt et al., 1992). It should be noted that these bounds are not the tightest possible and that splitting into an odd number of diffraction orders is usually more efficient than splitting into an even number of orders. Table 11.5 gives some common simple designs.

The design of continuous phase gratings is analogous to the phase retrieval problem, and the same methods are used to solve both problems (Wong and Swanson, 1993). The resulting phase profiles are "wiggly" in nature, suggesting the superposition of weighted sinusoids. Due to the large number of degrees of freedom, there are no constrains as to what types of efficiency weightings may be attained.

In contrast, the efficiency weightings of a binary or Dammann grating, which is constrained to take on one of two possible phases (typically 0 and π), are constrained to have the following properties. First, the efficiencies of orders m and $-m$ must always be equal. Second, the grating is fully defined by the two phases and the locations of the transitions between the phases. Furthermore, the transition points alone define the relative weightings of all nonzero orders, that

TABLE 11.4 Upper Bounds on the Diffraction Efficiency of 1:K Grating Beam Splitters

K	Continuous designs	Binary Designs
2	0.81	0.81
3	0.94	0.94
5	0.96	0.87
7	0.98	0.90
9	0.99	0.88
11	0.98	0.89
13	0.99	0.87
>13	0.97–0.99	0.87–0.88

TABLE 11.5 Grating Multiplexers of Period T, $0 < x < T$

Phase Expression	$\eta(-1)$	$\eta(0)$	$\eta(1)$	Remarks
$\phi(x) = \begin{cases} 0 & 0 < x < T/2 \\ \pi & T/2 < x < T \end{cases}$	0.41	0.00	0.41	Binary 1:2 splitter
$\phi(x) = \begin{cases} 0 & 0 < x < T/2 \\ 2.01 & T/2 < x < T \end{cases}$	0.29	0.29	0.29	Binary 1:3 splitter
$\phi(x) = \pi x/T$	0.05	0.41	0.41	Continous 1:2 splitter
$\phi(x) = \arctan[2.657 \cos(2\pi x/T)]$	0.31	0.31	0.31	Continuous 1:3 splitter

Source: Handbook of Optics. 1994. Optical Society of America.
McGraw-Hill, New York. With permission.

is, the ratio $\eta(m)/\eta(n)$, for all m, $n \neq 0$ is determined solely by the transition points and does not change no matter which two phase values are chosen. Third, the two phases determine only how much energy is in the zeroth order compared to the nonzero orders. As a result, it is common first to determine the transition points such that the nonzero orders have the correct weightings and then use the choice of phase to balance the zero order. The design of binary gratings is different from that of continuous gratings, and the most common approaches are based on Dammann's method or search methods (Vasara et al., 1992). In addition, tables of binary designs have been compiled (Killat et al., 1982; Krackhardt, 1989).

Multifocal Lenses

The concepts used to design gratings with multiple orders can be directly extended to lenses and axicons by taking advantage of the periodicities mentioned previously, and the multiple diffraction orders will appear as multiple foci (King, 1985).

As an example, taken from (OSA, 1994) with permission, consider the paraxial design of a bifocal lens, as is used in intraocular implants. Half the light should see a lens of focal length f_0, whereas the other half should see no lens. A diffractive lens of focal length f_0, but with the light split evenly between the 0 and +1 orders, will achieve this function. The phase profile of a single focus lens is given by $\phi(r) = -k_0 r^2/(2f_0)$. This phase, with the 2π ambiguity removed, is plotted in Fig. 11.14a as a function of r and in Fig. 11.14b as a function of r^2, where the periodicity in r^2 is evident. To split the light between the 0 and +1 orders, the blaze of Fig. 11.14b is replaced by the 1:2 continuous splitter of Table 11.5, resulting in Fig. 11.14c. This is the final design, and the phase profile is displayed in Fig. 11.14d as a function of r.

Beam Shapers and Homogenizers

Another major application area is the reshaping and/or homogenization of laser beams, and these designs can be categorized as either deterministic or statistical, depending on whether or not the characteristics of the incident laser beam are known and stable.

If the beam is well behaved, then the design problem is greatly simplified and beam shaping can typically be achieved with efficiencies of 90% or greater. If the required output beam is large enough so that diffraction effects are negligible, as shown in Fig. 11.15a, then the reshaper can be designed by ray tracing. The crux of this approach is that there is a coordinate transform which maps the power of the laser beam from its original distribution to the desired one, and the binary optic element is then used to implement this transform (Hossfeld et al., 1991; Bryngdahl, 1974). The ray-tracing approach is a simple and effective one, but it can sometimes result in unpredicted interference effects, similar in nature to the spot of Arago. An extreme example of this is shown in Fig. 11.15b. Here, the desired flat-top distribution is nearly diffraction-limited and the ray-tracing approach is inadequate due to diffraction effects. Accordingly, the design,

Binary Optics

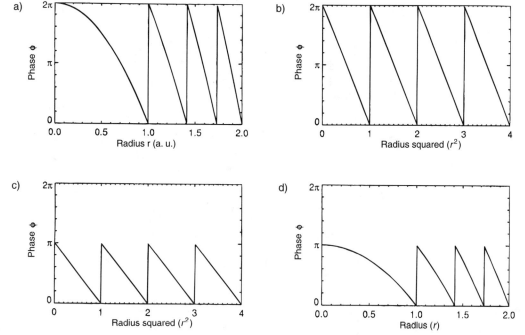

FIGURE 11.14 Designing a bifocal lens: a) lens with a single focus; b) same as a, but showing periodicity in r^2; c) substitution of a beam-splitting design; d) Same as c, but as a function of r.

which is simply a flat plate with a phase-shifted central portion, is accomplished by using an approach similar to the ones used in phase retrieval and apodization problems (Cordingley, 1993).

If the beam shape cannot be predicted a priori, then a design which will be robust to fluctuations in the beam is required. Most of these approaches depend on dividing the full aperture of the beam into subapertures, each of which is imaged into the desired output shape. For example, in Fig. 11.15c, each fly's eye lens independently images its subaperture onto the full output beam, thus averaging out any local nonuniformities in the incident beam (Deng, 1986). In these approaches, it is important to consider whether the beamlets from each subaperture add coherently or incoherently. Continuing the example of Fig. 11.15c, if the incident beam is spatially coherent, then the fly's eye array will behave like a grating and the beamlets will interfere to produce sharp diffraction orders rather than the smooth, flat top produced by a spatially incoherent beam. In the case of coherent beams, such a design will still work if the statistical fluctuations in the beam are enough to average out any interference patterns. In an extreme case of this approach, the subapertures are reduced in size until diffraction alone is sufficient to map the subaperture onto the full output beam, as shown in Fig. 11.15d. Furthermore, each subaperture is randomly phased with respect to the others to prevent any regular interference patterns, and, hence, the resulting profile is the diffraction pattern of the subaperture modulated by a speckle pattern (Kato et al., 1984; Kurtz, 1972).

Other Applications

Other applications include components for optical interconnects (Walker and Jahns, 1992), diffractive switches for laser beam steering (Goltsos and Holz, 1990), the coherent combination of lasers (Leger et al., 1988), and components for optical disks (Ono et al., 1989), among others.

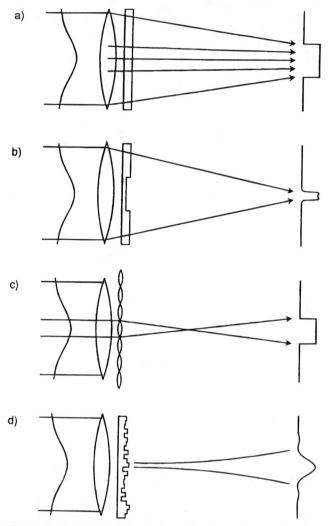

FIGURE 11.15 Methods for reshaping a laser beam a) Ray tracing allows the redistribution of energy. b) In this case, diffraction must be accounted for. c) Fly's eye approach d) Random phase plate.

11.4 Designs Based on Maxwell's Equations

There are two common situations which lead to designs in the vector regime. In the first, the design is based on a more approximate theory or, perhaps, just an intuitive idea, but the resulting surface relief has such fine features that a vector approach is required. In the second, the device is specifically designed to take advantage of a phenomenon, such as form birefringence, which is neglected by more approximate approaches.

Fundamentals

There are a limited number of approaches for attacking problems in the vector regime. The brute force numerical solution of Maxwell's equations is technically possible but limited in usefulness by practical issues. For example, the finite-difference, time-domain approach requires a grid sampling of approximately ten samples per wavelength. In the visible, a cube just 10 μm on a side would require over a million nodes.

Binary Optics

Of more practical interest is the solution of Maxwell's equations for a plane wave incident on a periodic structure. The two major approaches to solving this problem are the space harmonic approach and the modal approach (Moharam and Gaylord, 1985). In the modal approach, the field in the grating region is represented as a superposition of the eigenmodes of the grating. This approach is more numerically stable and converges more quickly than the space harmonic approach. However, it also requires that the eigenmodes of the grating profile be known and, unfortunately, eigenmodes have only been determined for the simplest cases, such as the square wave grating (Sheng et al., 1982). Accordingly, the space harmonic approach is more widely used. In this approach, the field in the grating region is represented as a set of plane waves which couple energy to each other as they propagate through the grating region. No matter which approach is used, the vector nature of the problem makes the optical design of gratings both nonintuitive and computationally intensive (Sheridan and Sheppard, 1990).

If the grating period is much smaller than the wavelength, then it can be argued that the light will not respond to the individual index variations of the grating, but will, instead, respond to some average index (i.e., an effective index). Accordingly, this approach is known as effective medium theory. For the square wave grating shown in Fig. 11.16, the effective indices are the solutions to the equations

$$\sqrt{n_1^2 - n_O^2}\tan[\pi T(1-f)\sqrt{n_1^2 - n_O^2}/\lambda] + \sqrt{n_2^2 - n_O^2}\tan[\pi Tf\sqrt{n_2^2 - n_O^2}/\lambda] = 0 \quad (11.91)$$

and

$$(\sqrt{n_1^2 - n_E^2}/n_1^2)\tan[\pi T(1-f)\sqrt{n_1^2 - n_E^2}/\lambda] + (\sqrt{n_2^2 - n_E^2}/n_2^2)\tan[\pi Tf\sqrt{n_2^2 - n_E^2}/\lambda] = 0, \quad (11.92)$$

where n_O is the case where the electric field is parallel to the grating grooves, n_E is the case where the electric field cuts across the grating grooves, and f is the fraction of the grating with index n_2 (Rytov, 1956). An approximate solution to these equations is given (Raguin and Morris, 1993) by

$$n_{O2}^2 = n_{O0}^2 + \frac{1}{3}[\pi Tf(1-f)/\lambda]^2(n_2^2 - n_1^2)^2 \quad (11.93)$$

and

$$n_{E2}^2 = n_{E0}^2 + \frac{1}{3}[\pi Tf(1-f)/\lambda]^2(1/n_2^2 - 1/n_1^2)^2 n_{E0}^6 n_{O0}^2 \quad (11.94)$$

where n_{O0} and n_{E0} are the zero order solutions

$$n_{O0}^2 = (1-f)n_1^2 + fn_2^2 \quad (11.95)$$

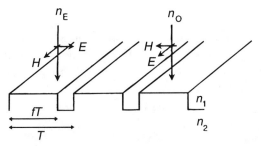

FIGURE 11.16 A grating of small period behaves like a medium of index n_E or n_O depending on polarization.

and

$$1/n_{E0}^2 = (1 - f)/n_1^2 + f/n_2^2. \tag{11.96}$$

Applications

The effective index, which ranges from n_1 to n_2, is determined by the fill factor f, and a number of applications are based on this concept. In antireflection applications, the fill factor is gradually increased in the direction of light propagation, thus, producing a smooth transition from one index to the other (Raguin and Morris, 1993). The resultant structures are similar to the spikes seen in anechoic chambers. If the fill factor is modulated in the transverse direction, the resulting binary structure will be the equivalent of a graded-index structure, similar to the production of gray scale by halftoning (Haidner et al., 1992). As shown in the above equations, the effective index depends on the polarization of the incoming light. Therefore, the structure shown in Fig. 11.16 is strongly birefringent because, in essence, it is an artificial crystal (Cescato et al., 1990). Loosely extending this crystal concept, it is possible to develop spectral filters (passband or stopband filters) by engineering the device to have resonances at certain wavelengths. Depending on the exact nature of the resonance, the filter can be either fairly broad (Rhoads et al., 1982) or extremely narrow (Wang and Magnusson, 1992).

11.5 Fabrication Using Multiple Masks

The traditional binary optics fabrication process is shown in Fig. 11.17. In this process, a set of photolithographic masks is used to produce a staircase approximation to the desired thickness profile. The process begins with an optical substrate, which is coated with photoresist. The resist is then exposed using the first mask of the mask set, as shown in Fig. 11.17a, and subsequent development removes resist either in the exposed areas for positive resists or the unexposed areas for negative resists. The substrate areas, which are no longer protected by resist, are then micromachined, as shown in Fig. 11.17b, producing the profile depicted by the dashed lines. Usually material is removed, but, in some processes, material is added. After micromachining, the optical substrate will have a surface profile with two levels, and repetition of this process, as shown for the second mask in parts c and d of Fig. 11.17, yields a staircase profile with 2^N levels if N masks are used. This process consists of three major steps: generating the mask set, transferring the pattern from the mask to the substrate via photolithography, and micromachining the patterned substrate. The mask generation step is described in some detail because optical designers commonly participate in this step. However, photolithography and micromachining are best handled by personnel specifically trained for these tasks, rather than an optical designer, so the descriptions given here are superficial.

Mask Generation

In the mask generation stage, the optical prescription of the binary optic element, which consists of a phase profile $\phi(x, y)$ and a design wavelength λ_0, is converted to a set of N photolithographic masks. This begins by converting the phase profile to a desired height profile, usually by the equation

$$h(x, y) = \frac{\lambda_0}{2\pi(n_1 - n_2)} (\phi \bmod 2\pi) \tag{11.97}$$

which is based on the same assumptions as Eq. 11.43. The desired profile, which varies from 0

Binary Optics

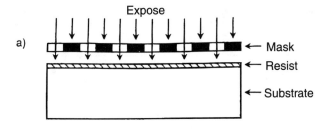

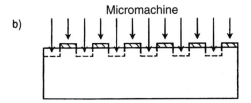

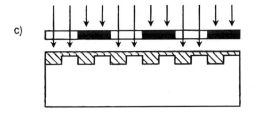

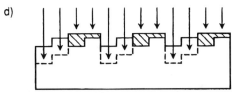

FIGURE 11.17 Binary optics fabrication based on multiple masks; a) exposure of resist using the first mask; b) micromachining the exposed areas of the substrate; c) and d) repetition of a and b using the second mask.

to $h_0 = \lambda_0/(n_1 - n_2)$, will be approximated by a staircase profile, with the transition points of the staircase occuring along constant thickness contours spaced in increments of h_0/N. If Eq. 11.97 is used, these contours will also be contours of constant phase. Table 11.6 defines the masks required to achieve this staircase approximation. The layer one mask is required for a one-mask process, layers one and two for a two-mask process, etc.

The contours can be calculated in several ways, depending on the nature of the thickness profile (Logue and Chisholm, 1989). If the thickness profile is described by a simple analytical expression, as is the case for most of the elements described in this chapter, then the contours may be determined analytically. If the phase profile is piecewise flat (e.g., the element consists of a checkerboard pattern with each square having a different phase), then the contours may also be determined analytically, but the resulting contour will consist of a large number of disjoint curves, in contrast to the previous case (e.g., a Fresnel lens) where the contour is usually one simple curve. Also note that, in these two cases, the resulting contours will be exact, which is not true in the following two cases. If the binary optic component is small in size or the phase

TABLE 11.6 Processing Steps for Binary Optics

Mask Layer	Etch Region, defined by $h(x, y)$	Etch Region, defined by $\phi(x, y)$	Etch Depth
1	$0 < h \bmod (h_0) < h_0/2$	$0 < \phi \bmod 2\pi < \pi$	$h_0/2$
2	$0 < h \bmod (h_0/2) < h_0/4$	$0 < \phi \bmod \pi < \pi/2$	$h_0/4$
3	$0 < h \bmod (h_0/4) < h_0/8$	$0 < \phi \bmod \pi/2 < \pi/4$	$h_0/8$
4	$0 < h \bmod (h_0/8) < h_0/16$	$0 < \phi \bmod \pi/4 < \pi/8$	$h_0/16$

Source: Handbook of Optics. 1994. Optical Society of America. McGraw-Hill, New York. With permission.

contours are nearly straight, then, the contours may be estimated by calculating the thickness at every point on a regular grid and interpolating the location of the contours. The limitation to this approach is that the grid must be dense enough to reduce the interpolation error to the required level, but it must also be large enough to cover the entire element. For many elements, this approach is not feasible because of limited computer speed and/or storage. Also note that this approach is not the same as sampling the thickness profile on a regular grid and then pretending that the element is piecewise flat with the thickness of each grid cell given by the sampled value. In particular, the piecewise flat assumption should not be used unless the thickness profile truly is piecewise flat. Finally, if the phase profile is continuous, then the contours can be traced one at a time, using a method similar to those used for tracing fringes on an interferogram (Krishnaswamy, 1991). In this approach, the contour may be locally modeled as a circle and the normal and tangent directions to the contour, as well as the radius of the osculating circle, are usually required. These are depicted in Fig. 11.18 and given by

$$\hat{n} = \nabla\phi/|\nabla\phi|, \tag{11.98}$$

$$\hat{t} = \hat{n} \times \hat{z}, \tag{11.99}$$

and

$$1/R = [(\phi_y\phi_{yx} - \phi_{yy}\phi_x)^2 + (\phi_{xy}\phi_x - \phi_{xx}\phi_y)^2]^{1/2}/|\nabla\phi|^2 \tag{11.100}$$

where $\hat{n}$ and $\hat{t}$ are unit vectors normal and tangential to the contour but in the plane of the

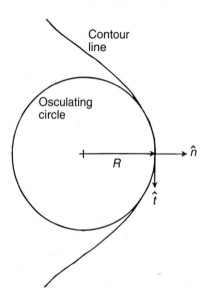

FIGURE 11.18 The contour line may be locally modelled as a circle of radius R.

Binary Optics

osculating circle, $\hat{z}$ is the unit vector out of the plane, and the notation ϕ_x denotes the partial derivative of ϕ with respect to x, etc.

In addition to mathematically determining the contours, the contours must also be represented in a manner compatible with the mask-making equipment. Common formats include GDSII and CIF (Rubin, 1987; Mead, 1980). Besides syntax rules, all formats also have rules regarding which shapes can be represented. Of particular note for binary optics, curved contours almost always must be approximated by a series of line segments with the end points of those line segments lying on a predefined grid. Because the grid spacing can, typically, range from 0.05 μm to 0.5 μm, snapping of coordinates to the grid usually has an insignificant effect if the grid spacing is chosen correctly. However, the approximation of curves by line segments cannot be neglected, and line segments are usually chosen so that the deviation between the actual curve and the line segment matches the grid resolution. In particular, if a circle of radius R is to be approximated with a maximum deviation of δ, then the line segment may be defined by

$$\theta_1 = 2 \arccos(1 - \delta/R_1) \doteq \sqrt{8\delta/R_1}, \tag{11.101}$$

$$R_1 = R, \tag{11.102}$$

or

$$\theta_2 = 2 \arccos(1 - 2\delta/R_2) \doteq \sqrt{16\delta/R_2}, \tag{11.103}$$

$$R_2 = R + \delta, \tag{11.104}$$

as shown in Fig. 11.19. The aggressive approach results in less data. However, it is more difficult to implement, especially for the case of arbitrarily shaped contours.

If the features on the photomask are coarse (typically greater than 10 μm), then it is possible to produce the photomask optically by photoreduction. However, in most cases, the photomasks

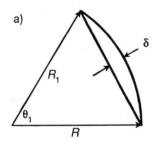

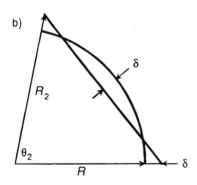

FIGURE 11.19 Approximating a circle of radius R by a line segment; a) conservative approach; b) aggressive approach.

are written by electron beams, most of which use the MEBES format as input. Strictly speaking, the formats mentioned previously cannot be used to run a MEBES machine. However, they can be translated or fractured to MEBES with an insignificant loss in fidelity, so that the distinction is transparent to the binary optic designer. Most commercial mask vendors can routinely write photomasks of sizes up to 7 inches square with a minimum feature size, or critical dimension (CD), of 0.8 μm in the case of general patterns. If the pattern is an array of equal width lines and spaces, then the achievable CD may be as small as 0.3 μm. The CD of a binary optic element is given by the minimum local period, as defined in Eq. 11.2, divided by the number of levels in the staircase profile.

In writing the photomask, the MEBES machine effectively represents the entire mask as an array of pixels, which typically range in size from 0.05 to 0.5 μm, and the electron beam is then scanned across the field in a raster fashion, exposing the desired pixels. As a result of this pixelization, curves and gently sloping lines will have noticeably jagged edges when viewed under the microscope. Fortunately, these artifacts usually do not significantly affect the optical performance of the element. In addition, the electron beam is usually scanned by electrostatic deflection in one direction while the mask is mechanically translated in the other direction. This difference in physical mechanisms results in a preferred direction for the machine, and masks for asymetric patterns (e.g., linear gratings) should take advantage of this.

Although most of this section has dealt with feature generation for the binary optic element, more time is usually spent generating the remainder of the mask, which typically includes alignment marks, control features, and labels. Alignment marks are required for aligning the mask to the substrate, aligning the mask or substrate to any jigs, and for aligning the finished piece in the optical system. Control features are diagnostic areas which are used to monitor the fabrication process. For example, the depth of a micromachining step can be checked by witness etch areas, and the fidelity of the photolithographic process can be checked by line and space patterns. Finally, labels allow both the fabricator and the optical designer to easily determine the device name, orientation, mask layer number, etc.

Photolithography

The photolithography step transfers the pattern from the mask to the photoresist covering the substrate. The most common methods are contact, proximity, and projection lithography. In contact and proximity printing, the mask is either in direct contact with or in close proximity to the resist, and the pattern is produced by shadow casting. Contact photolithography results in higher resolution and better linewidth fidelity because there are no diffraction effects, whereas proximity printing reduces the chance of physically damaging the resist layer because the mask and resist are separated by 5–50 μm. Both methods require 1:1 masks and, in comparison to projection printing, they offer lower equipment costs and more flexibility in substrate sizes and materials. In projection photolithography, the mask and substrate are separated, and the mask is then imaged onto the resist-covered substrate, with a demagnification ranging from 1 to 20 times. Projection printers are good for high volume manufacturing, such as the production of microprocessors, but, for binary optics, they have limited usefulness because they are expensive, the required alignment accuracies are difficult to achieve, and they have limited fields of view, typically on the order of 2 cm square.

After exposure, the photoresist is developed. For positive photoresists, the exposed areas are removed by the developer, whereas, for negative resists, the unexposed areas are removed. The underlying substrate in these areas can then be micromachined while the remainder of the substrate is protected by the remaining resist.

Because photolithography is easiest on flat surfaces, the masks are usually processed in order of shallowest etch depth to deepest (i.e., from layer 4 to layer 1 of Table 11.6). Occasionally, a project requires a surface relief which is deeper than can be handled by normal resists, which are typically 0.5–2.0 μm thick. In these cases, more complex processes are required (Stern and Medeiros, 1993).

Micromachining Techniques

In the micromachining step, material is removed from the unprotected areas of the substrate. The two most common micromachining methods are reactive ion etching (RIE) and ion milling. In RIE, a plasma is formed at the substrate surface and particles are directed at the surface (Stern et. al., 1991a). Substrate material is removed by chemical interactions between the plasma and the substrate and also by mechanical abrasion. Because of the chemical component, the etch attacks different materials at different rates (i.e., the etch is selective), and a different process must be developed for each material. In addition, although fused silica may be etched successfully in this manner, glasses with impurities usually cannot because the impurities react adversely with the chemicals. In contrast, ion milling is similar to sandblasting. The substrate surface is bombarded by inert ions (usually Ar) and material is removed solely by mechanical means (Stern, et. al., 1991b). Because the process does not involve any chemistry, the process can be used for any material but it is usually less selective and also slower than RIE.

Finally, several other micromachining techniques are worth mentioning. In wet chemical etching, the substrate is immersed in a solution which chemically removes material (Lee, 1979). This method can be fast, but it also has lower fidelity because the chemical reaction proceeds in all directions (i.e., the etch is isotropic) and the photoresist provides protection only in one direction. Crystallographic etching is fast but not of general utility, because the etching must follow crystallographic planes. Binary optic structures have also been fabricated by depositing material, rather than removing it, but this approach is more difficult to control (Jahns and Walker, 1990).

11.6 Alternate Fabrication Methods

Fabrication based on multiple masks is not the only method for manufacturing binary optics. This section describes some of the other promising approaches as well as the most common replication methods.

Other Micromachining Approaches

Most other fabrication approaches rely on directly micromachining the required surface relief, either in the final substrate material or in an intermediate material, such as a photoresist. Almost all of these processes can be described as a tool traversing a surface, removing material in a specific manner due to the tool shape and tool travel path. The problem of calculating the correct tool path, given the tool shape and desired final profile, is essentially a deconvolution problem which may be solved by an iterative method, such as the van Cittert deconvolution algorithm (Crilly, 1991). The feature which distinguishes one fabrication method from another is the tool type, which is discussed below.

Compared to the traditional binary optic fabrication, these micromachining approaches have the advantage of fewer processing steps, and, in particular, layer to layer alignments are no longer required. However, the disadvantage is that the required processes are more difficult to control (e.g., compare the on-off exposure of photoresist used in the traditional approach to the analog exposure required in some of the approaches below) and also take much longer to complete because the tool must traverse the entire surface sequentially.

Of the approaches which directly figure the final surface, diamond turning is the most commercially mature, especially for infrared optics (Riedl and McCann, 1991). In this technology, the substrate is mounted on a rotating spindle and then cut by a diamond tool which moves radially. The process is similar to a lathe and a basic trade-off occurs in selecting the tool size. Larger tools remove more material, thus reducing cutting time and tool wear, but smaller tools resolve finer features.

Ion milling, described previously, can also be used to directly figure a surface. In this method, a stream of ions is focused to a spot, which can be as small as 50 nm, and this is then used as the tool (DeFreez et al., 1989). Compared to most other tools, which are on the order of 1 μm or larger, ion milling can resolve much finer features, but the small tool size also results in longer processing times and limited element sizes.

Two other promising techniques for directly figuring substrates are the direct ablation of materials by lasers (Veiko et al., 1993), primarily excimer lasers, and laser-assisted chemical etching of materials (Gratrix and Zarowin, 1991).

An alternative to micromachining the surface in a single step is, to first produce a latent image in the material via an exposure step and to then remove the material in a subsequent development step. For example, if a photoresist is used, the resist may be exposed in an analog fashion by scanning a laser across the surface (Gale et al., 1992) or by using a gray-scale mask (Andersson et al., 1990). In the development step, differing amounts of resist are removed due to the variable exposure, and the remaining resist has the desired surface profile. This approach can also be used with electron-beam resists rather than photoresists (Fujita et al., 1982), with the advantage that electron-beams, like ion beams, can be configured to write with much smaller spot sizes. A general disadvantage to this approach is that the original material must be some sort of resist and a second, nontrivial step is required to transfer the surface profile from the resist to the actual substrate material.

Replication Methods

In many cases, the most economical way to produce large quantities of a binary optic element is to fabricate the first element by one of the previously mentioned techniques and then to replicate it by a method, such as embossing, casting, or molding. The first step, in almost all cases, is to produce a metal master which is a negative of the desired surface profile. For example, this may be achieved by coating the original with a thin conductive layer, immersing the piece in a nickel bath, and then building up a thicker nickel layer via electroplating. The nickel layer can then be separated from the original and used as a metal master. If required, the process may be repeated to produce multiple masters or to reverse the profile.

In embossing, the metal master is pressed into plastic which has been temporarily softened (Cowan and Slafer, 1985). Replicas can be formed very inexpensively by wrapping the metal master around a drum and then feeding in the plastic on long sheets, stamping a replica on each revolution of the drum. This method is economical but is limited to shallow, uncomplicated surface reliefs with low aspect ratios. In an alternate embossing process, the metal master is pressed into a photopolymer, the embossing is fixed by exposure to ultraviolet light, and then the master is removed (Shvartsman, 1993). This process has been used to reproduce complicated binary optics elements with high fidelity and good dimensional stability.

The remaining replication processes require that the metal master be combined with a second piece, commonly a flat, to form a mold. In conventional injection molding, plastic is heated and then injected into the mold. After the assembly cools, the mold is separated and the replica released (Goto et al., 1989). Alternatively, the injected resin may be cured by exposure to ultraviolet light, thus avoiding the long cooling stage (Tanigami et al., 1989). Sol-gels are processed in a similar fashion, but the resulting material is a glass rather than a polymer (Nogues and Howell, 1992). The potential drawbacks to these processes are that only one piece can be formed at a time and shrinkage during cooling or curing can be significant.

Defining Terms

Achromat: Lens corrected at two wavelengths. First-order design is based on the Abbe number.
Angular spectrum approach: Scalar diffraction approach used to propagate fields in the spatial frequency domain.
Apochromat: Lens corrected at three wavelengths. First-order design is based on the Abbe number and partial dispersion.
Blaze height: The grating height required to maximize the efficiency in a desired diffraction order for a given wavelength.
Dammann grating: Grating beam splitter which is binary (i.e., constrained to two phases).
Diffraction efficiency: Energy in a specific diffraction order of a binary optic element, relative to the total incident energy.
Effective medium theory: Model which replaces a fine-featured structure with a medium possessing the "average" properties of the structure.
Fly's eye lens: Type of microlens array.
Form birefringence: Birefringence which is the result of a man-made periodic structure which has a periodicity much less than a wavelength.
Fresnel approximation: Scalar diffraction integral used to propagate fields in the space domain.
Generalized grating model: An approach which locally models a diffractive optic by a linear grating.
Grating beam splitter: Grating with diffraction efficiencies which are designed to follow a specific weighting.
Grating equation: Equation which governs the ray tracing of binary optics.
Hybrid system: System which combines refractive/reflective elements with binary optic elements.
Micromachining: VLSI-based fabrication process which produces a surface relief on a microscopic scale. Common processes include reactive ion etching, ion milling, and wet chemical etching.
Phase screen: Approach which models binary optic elements as instantaneously introducing a phase delay.
Photolithography: Process for transferring a pattern from a photomask to a photoresist. Common types of photolithography include contact, proximity, and projection photolithography.
Photomask: Chrome on glass mask, which is usually written by an electron beam and used to pattern photoresist.
Rigorous coupled wave analysis: Method for solving Maxwell's equations for plane wave diffraction from a grating.
Sweatt lens: Modification of the Sweatt model, in which the binary optic component is a lens.
Sweatt model: Model for binary optics which replaces the binary optic element by an equivalent refractive element.

References

Anderson, J. W. 1993. Thermal weapon sight tws an/pas-13 diffractive optics designed for producibility. *Conf. Binary Optics*, NASA Conference Publication 3227: 303.

Andersson, H. 1990. Single photomask, multilevel kinoform in quartz and photoresist: Manufacture and evaluation. *Appl. Opt.* 29: 4259.

Barakat, R. Calculation of integrals encountered in optical Diffraction Theory. In *The Computer in Optical Research Methods and Applications*, Frieden, B. R., ed., Springer-Verlag, New York, 1980, p. 35–80.

Bryngdahl, O. 1974. Geometrical transforms in optics, *J. Opt. Soc. Am.* 64: 1092.

Buralli, D. A., and Morris, G. M. 1989. Design of a wide field diffractive landscape lens. *Appl. Opt.* 28: 3950.

Buralli, D. A., and Morris, G. M. 1991. Design of diffractive singlets for monochromatic imaging, *Appl. Opt.* 30: 2151.

Buralli, D. A., and Morris, G. M. 1992a. Design of two- and three-element diffractive Keplerian telescopes. *Appl. Opt.* 31: 38.

Buralli, D. A., and Morris, G. M. 1992b. Effects of diffraction efficiency on the modulation transfer function of diffractive lenses. *Appl. Opt.* 31: 4389.

Buralli, D. A., and Rogers, J. R. 1989. Some fundamental limitations of achromatic holographic systems. *J. Opt. Soc. Am.* A6: 1863.

Cescato, L. H. 1990. Holographic quarterwave plates. *Appl. Opt.* 29: 3286.

Chen, C. W. 1992. Application of diffractive optical elements in visible and infrared optical systems. *Proc. Soc. Photo-Opt. Instrum. Eng.* CR41: 158.

Cordingley, J. 1993. Application of a binary diffractive optic for beam shaping in semiconductor processing by lasers. *Appl. Opt.* 32: 2538.

Cowan, J. J., and Slafer, W. D. 1985. Holographic embossing at Polaroid: The Polaform process. *Proc. Soc. Photo-Opt. Instrum. Eng.* 600: 49.

Cox, J. A. 1991. Process error limitations on binary optics performance. *Proc. Soc. Photo-Opt. Instrum. Eng.* 1555: 80.

Crilly, P. B. 1991. A quantitative evaluation of various iterative deconvolution algorithms. *IEEE Trans. on Instrum. and Meas.* 40: 558.

Dammann, H. 1979. Spectral characteristics of stepped-phase gratings, *Optik* 53: 409.

DeFreez, R. K. 1989. Focused ion-beam micromachined diode laser mirrors, *Proc. Soc. Photo-Opt. Instrum. Eng.* 1043: 25.

Deng, X. 1986. Uniform illumination of large targets using a lens array. *Appl. Opt.* 25: 377.

Emerton, N. 1985. Manufacturing Tolerances for Blazed Diffractive Optical Elements. *Acta Polytechnica Scandinavica: Applied Physics* Ph-149: 308.

Farn, M. W. 1992. Quantitative comparison of the general Sweatt model and the grating equation. *Appl. Opt.* 31: 5312.

Farn, M. W., and Goodman, J. W. 1991. Diffractive doublets corrected at two wavelengths. *J. Opt. Soc. Am.* A8: 860.

Ferstl, M. 1992. Blazed Fresnel zone lenses approximated by discrete step profiles: Effects of fabrication errors. *Proc. Soc. Photo-Opt. Instrum. Eng.* 1732, 89.

Fujita, T. 1982. Blazed gratings and Fresnel lenses fabricated by electron-beam lithography. *Opt. Lett.* 7: 578.

Gale, M. T. 1992. Fabrication of kinoform structures for optical computing. *Appl. Opt.* 31: 5712.

Goltsos, W., and Holz, M. 1990. Agile beam steering using binary optics microlens arrays. œ29: 1392.

Goto, K. 1989. Plastic grating collimating lens. *Proc. Soc. Photo-Opt. Instrum. Eng.* 1139, 169.

Gratrix, E. J., and Zarowin, C. B. 1991. Fabrication of microlenses by laser assisted chemical etching. *Proc. Soc. Photo-Opt. Instrum. Eng.* 1544: 238.

Gremaux, D. A., and Gallagher, N. C. 1993. Limits of scalar diffraction theory for conducting gratings. *Appl. Opt.* 32: 1948.

Gruhlke, R. 1992. Laser energy distributions produced by the use of diffractive optical elements. *Proc. Soc. Photo-Opt. Instrum. Eng.* 1834: 152.

Haidner, H. 1992. Zero-order gratings used as an artificial distributed index medium. *Optik* 89: 107.

Hossfeld, J. 1991. Rectangular focus spots with uniform intensity profile formed by computer generated holograms. *Proc. Soc. Photo-Opt. Instrum. Eng.* 1574: 159.

Jahns, J., and Walker, S. 1990. Two-Dimensional array of diffractive microlenses fabricated by thin film deposition. *Appl. Opt.* 29: 931.

Kato, Y. 1984. Random phasing of high-power lasers for uniform target acceleration and plasma-instability suppression. *Phys. Rev. Lett.* 53: 1057.

Killat, U. 1982. Binary phase gratings for star couplers with high splitting ratios. *Fiber and Integrated Optics* 4: 159.

King, P. R. 1985. Design of diffractive surface relief lenses with more than one focus. *Acta Polytechnica Scandinavica: Applied Physics* Ph-149: 312.

Krackhardt, U. Binaere Phasengitter als Vielfach-Strahlteiler. *Diplomarbeit, Universitaet Erlangen-Nuernberg*, Erlangen, Germany 1989.

Krackhardt, U. 1992. Upper bound on the diffraction efficiency of phase-only fanout elements. *Appl. Opt.* 31: 27.

Krishnaswamy, S. 1991. Algorithm for computer tracing of interference fringes. *Appl. Opt.* 30: 1624.

Kurtz, C. N. 1972. Transmittance characteristics of surface diffusers and the design of nearly band-limited binary diffusers. *J. Opt. Soc. Am.* 62: 982.

Lee, W. H. 1979. High efficiency multiple beam gratings. *Appl. Opt.* 18: 2152.

Leger, J. R. 1988. Coherent laser beam addition: An application of binary-optics technology. *The Lincoln Lab J.* 1: 225.

Levi, L. 1980. A one-glass achromatic doublet. *Appl. Opt.* 27: 1491.

Logue, J., and Chisholm, M. L. 1989. General approaches to mask design for binary optics. *Proc. Soc. Photo-Opt. Instrum. Eng.* 1052: 19.

Londono, C. 1993. Athermalization of a single-component lens with diffractive optics. *Appl. Opt.* 32: 2295.

Magnusson, R., and Gaylord, T. K. 1978. Diffraction regimes of transmission gratings. *J. Opt. Soc. Am.* 68: 809.

Mead, C. A. *Introduction to VLSI Systems*. Addison-Wesley, Reading, MA., 1980.

Moharam, M. G., and Gaylord, T. K. 1982. Diffraction analysis of dielectric surface-relief gratings. *J. Opt. Soc. Am.* 72: 1385.

Nogues, J. L. R., and Howell, R. L. 1992. Fabrication of pure silica micro-optics by sol-gel processing. *Proc. Soc. Photo-Opt. Instrum. Eng.* 1751: 214.

Ono, Y. 1989. Computer generated holographic optical elements for optical disk memory read/write heads. *Proc. Soc. Photo-Opt. Instrum. Eng.* 1052: 150.

Optical Society of America, *Handbook of Optics*. McGraw-Hill, New York, 1994.

Raguin, D. H., and Morris, G. M. 1993. Antireflection structured surfaces for the infrared spectral region. *Appl. Opt.* 32: 1154.

Rhoads, C. M. 1982. Mid-infrared filters using conducting elements. *Appl. Opt.* 21: 2814.

Riedl, M. J. and McCann, J. T. 1991. Analysis and performance limits of diamond turned diffractive lenses for the 3–5 and 8–12 micrometer regions. *Proc. Soc. Photo-Opt. Instrum. Eng.* CR38: 153.

Roberts, N. C. 1992. Binary phase gratings for hexagonal array generation. *Opt. Comm.* 94: 501.

Roose, S. 1993. An efficient interpolation algorithm for Fourier and diffractive optics. *Opt. Comm.* 97: 312.

Rubin, S. M., *Computer Aids for VLSI Design*. Addison-Wesley, Reading, MA., 1987.

Rytov, S. M. 1956. Electromagnetic properties of a finely stratified medium. *Soviet Physics JETP* 2: 466.

Sheng, P. 1982. Exact eigenfunctions for square-wave gratings: application to diffraction and surface-plasmon calculations. *Phys. Rev. B* 26: 2907.

Sheridan, J. T., and Sheppard, C. J. R. 1990. An examination of the theories for the calculation of diffraction by square wave gratings: 3. Approximate theories. *Optik* 85: 135.

Shvartsman, F. P. 1993. Replication of diffractive optics. *Proc. Soc. Photo-Opt. Instrum. Eng.* CR49, in press.

Siegman, A. E. 1977. Quasi Fast Hankel transform. *Opt. Lett.* 1: 13.

Stern, M. B., and Medeiros, S. S. 1992. Deep three-dimensional microstructure fabrication for infrared binary optics. *J. Vac. Sci. and Tech. B* 10: 2520.

Stern, M. B. 1991a. Fabricating binary optics: process variables critical to optical efficiency. *J. Vac. Sci. and Tech. B* 9: 3117.

Stern, M. B. 1991b. Binary optics microlens arrays in CdTe. *Mat. Res. Soc. Symp. Proc.* 216: 107.

Stone, T., and George, N. 1988. Hybrid diffractive-refractive lenses and achromats. *Appl. Opt.* 27: 2960.

Swanson, G. J. 1989. Binary optics technology: The theory and design of multilevel diffractive optical elements, M.I.T. Lincoln Laboratory Technical Report 854, NTIS Publ. AD-A213-404.

Swanson, G. J. 1991. Binary Optics Technology: Theoretical Limits on the Diffraction Efficiency of Multilevel Diffractive Optical Elements, M.I.T. Lincoln Laboratory Technical Report 914.

Sweatt, W. C. 1977. Describing holographic optical elements as lenses. *J. Opt. Soc. Am.* 67: 803.

Sweatt, W. C. 1979. Mathematical equivalence between a holographic optical element and an ultrahigh index lens. *J. Opt. Soc. Am.* 69: 486.

Tanigami, M. 1989. Low-wavefront aberration and high-temperature stability molded micro Fresnel lens. *IEEE Photonics Tech. Lett.* 1: 384.

Vasara, A. 1992. Binary surface-relief gratings for array illumination in digital optics. *Appl. Opt.* 31: 3320.

Veiko, V. P. 1993. Laser technologies for micro-optics fabrication. *Proc. Soc. Photo-Opt. Instrum. Eng.* 1874: 93.

Veldkamp, W. B. 1980. Developments in laser-beam control with holographic diffraction gratings. *Proc. Soc. Photo-Opt. Instrum. Eng.* 255: 136.

Walker, S. J., and Jahns, J. 1992. Optical clock distribution using integrated free-space optics. *Opt. Comm.* 90: 359.

Wang, S. S., and Magnusson, R. 1993. Theory and applications of guided-mode resonance filters. *Appl. Opt.* 32: 2606.

Welford, W. T. *Aberrations of Optical Systems.* Adam Hilger, Boston, 1986.

Wong, V. V., and Swanson, G. J. 1993. Design and fabrication of a Gaussian fan-out optical interconnect. *Appl. Opt.* 32: 2502.

For Further Information

The conferences and special journal issues sponsored by the many professional societies give an accurate picture of current applications and research in binary optics. Volume 9 of the 1992 OSA Technical Digest series is a topical meeting on diffractive optics, and many of these papers also occur in the May 10, 1993 feature issue of *Applied Optics*. The SPIE sponsors two conference series which are of interest. Volumes 883 (1988), 1052 (1989), 1211 (1990), 1555 (1991) and 2152 (1994) are part of the *Computer and Optically Generated Holographic Optics* series. Volumes 1544 (1991), 1751 (1992) and 1992 (1993) make up the *Miniature and Microoptics* series.

In the area of optics, Koronkevich's "Computer synthesis of diffraction optical elements" in *Optical Processing and Computing*, Arsenault et al., ed., is an early description of many of the principles currently used in binary optics. In addition, *Computer-Generated Holograms and Diffractive Optics* and *Holographic and Diffractive Lenses and Mirrors*, volumes 33 and 34 of SPIE's Milestone Series, contain many of the original papers which are pertinent to binary optics today.

For fabrication, Sze's *VLSI Technology* is a general introduction to VLSI fabrication techniques, whereas *Lithography for VLSI*, Einspruch and Watts, eds., and *Plasma Processing for VLSI*, Einspruch and Brown, ed., concentrate specifically on the two areas most pertinent to binary optics. These two books are volumes 16 and 8 of Academic Press' VLSI Electronics Series. Finally, *Holography Market Place*, Kluepfel and Ross, eds., describes some of the currently available mass replication techniques.

Occasionally, a less in-depth treatment is desirable. Veldkamp and McHugh's "Binary optics" in the May 1992 issue of *Sci. Am.* and Lee's "Recent advances in computer generated hologram applications" in the July 1990 issue of *Opt. and Phot. News* are both short articles which discuss the technical issues and applications of the field. Farn and Veldkamp's "Binary optics: Trends and limitations" in *Conference on Binary Optics*, NASA Conference Publication 3227 (1993), discusses nontechnical issues as well.

12
Gradient-Index Glass Waveguide Devices

S. Iraj Najafi
École Polytechnique, Montreal Quebec, Canada

Seppo Honkanen
Optonex Ltd., Espoo, Finland

12.1 Introduction 502
12.2 Ion-Exchanged Glass Waveguides 503
  Buried Waveguide Fabrication • Ion-Exchanged Waveguides with Ionic Masking • Double Ion Exchange Process
12.3 Theoretical Analysis 506
  Single Ion Exchange • Double Ion Exchange
12.4 Losses in Ion-Exchanged Waveguides 509
12.5 Devices 512
  Mach–Zehnder Interferometers • Ring Resonator • Waveguides with Gratings • Computer-Generated Waveguide Holograms • Rare-Earth-Doped Waveguides • Semiconductor, Quantum Dot Glass Waveguides

12.1 Introduction

Optical materials containing a distribution of refractive indices are called gradient-index (GRIN) materials. Typically, GRIN materials are used in miniature optical systems (Houde–Walter, 1988) and in microoptic devices for fiber optics (Tomlinson, 1988). Good examples of GRIN components are GRIN-rod lenses (Sakomoto, 1992) and planar microlenses (Intani et al., 1992). The most common fabrication method for GRIN materials and components is ion exchange in glass (Doremus, 1964).

GRIN materials and components are used also in integrated optics, which is a technology utilizing planar optical waveguides. In integrated optics, GRIN materials are used in two distinct ways: 1) The waveguide itself has a gradient-index profile. This is very common because many of the important waveguide fabrication processes, such as Ti diffusion in $LiNbO_3$ and ion exchange in glass, result in gradient-index waveguiding layers. 2) The operation of an integrated optic device can be based on effective index gradients within the device. Both transverse and longitudinal effective index gradients are utilized. Good examples of devices utilizing transverse effective index gradients are integrated optical lenses (Kenan, 1988), such as Luneburg and Fresnel lenses. Here, in section 5.4, we describe new integrated optic devices, computer-generated waveguide holograms, whose operation is also based on transverse effective index gradients. Several integrated optic devices utilize longitudinal effective index gradients. Operation of these devices is, typically, based on slowly (adiabatically) tapering waveguide sections. Many of the devices described in section 12.5 use these longitudinal effective index gradients.

Recently, gradient-index glass waveguides and devices have become one of the most important areas of the diverse field of gradient-index optics mostly due to the rapidly growing need for high performance and inexpensive passive components for optical communication, such as achromatic power dividers and wavelength selective couplers, Keck et al., 1989; Stern, 1991. As in the case of micro-optic GRIN components, ion exchange in glass is, definitely, the most important fabrication method for gradient-index glass waveguide devices. Therefore, in this chapter, we concentrate on GRIN glass waveguides and devices produced by ion exchange processes.

Ion exchange in glass is a cationic diffusion process that takes place in multicomponent oxide glasses. These glasses can be classified according to the bond strength between their cations and the oxygen atoms. Oxides with strong bonds (e.g., SiO_2, B_2O_3, GeO_2, P_2O_5) are called "network formers". Glasses with large amount of network formers have higher transition temperatures. "Network intermediates," such as ZnO and PbO, contribute to the strength of the network but cannot form glasses by themselves. "Network modifiers" (e.g., Na_2O, CaO, K_2O) have looser bonds to the remainder of the network. Under certain conditions (elevated temperature, applied electric field), it is possible to replace some of the network modifiers in glass by other ions with the same valence and chemical properties. Historically, large ions have been used to replace smaller ions in glass to strengthen its surface (Kistler, 1962). At the exchange temperature, the glass volume can adjust to accommodate the larger ions. As the glass cools down, large compressive surface stresses develop resulting in stronger glasses due to microcrack formation.

The ion exchange process also modifies the electrical and optical properties of the glass. The new ions have different polarizability and size. The ion exchange process has been employed to produce optical waveguides by increasing the refractive index in selected parts of a glass substrate (Giallorenzi et al., 1973; Findakly, 1985; Ramaswany and Srivastava, 1988; Ross, 1989; Najafi, 1992a; Izawa and Nakagome, 1972). Ion exchange in glass is carried out at elevated temperatures (200 °C to 600 °C). Two basic forces drive the exchange: thermal agitation and nonzero mobility of certain ions and a potential difference set up across the glass, causing an ion current flow. The thermal force can be used alone, in many instances, to fabricate optical waveguides. An electric field is usually used to make larger waveguides or to bury the waveguide under the surface of the substrate.

In this chapter, we first (in section 12.2) describe the different types of ion exchange processes and ion-exchanged glass waveguides. Section 12.3 presents a theoretical analysis for single and double ion exchange processes. In section 12.4, the propagation losses in different types of ion-exchanged glass waveguides are compared. Finally, section 12.5 presents examples of devices produced by ion-exchanged GRIN waveguides. Here, the focus is on very recent developments in the field. The earlier works are reviewed in Najafi (1992a) and Ross (1989).

12.2 Ion-Exchanged Glass Waveguides

Two basic processes are used to fabricate ion-exchanged glass waveguides: ion exchange from molten salt and the silver-film technique. Figure 12.1 schematically shows these two fabrication processes. In the molten salt method, a glass substrate is immersed in a molten salt, at 300 °C to 500 °C, containing an ion, such as Ag^+, K^+, Cs^+, Tl^+. An ion in glass (usually Na^+) is exchanged with an ion in the molten salt resulting in an increased refractive index on the glass surface. In the silver-film technique, a thin silver film is deposited on the glass substrate. The sample is heated to 150 °C to 300 °C, and an electric field (1 to 100 volts) is applied across it. Then, the silver ions are introduced into the glass by an electrochemical reaction in which silver is oxidized: $Ag \rightarrow Ag^+ + e^-$. A voltage of 1 volt is enough to overcome the chemical potential barrier between the silver film, the glass, and diffuse silver ions inside the substrate. A detail explanation of ion-exchanged waveguide fabrication from molten salt and silver film is provided in Najafi (1992a).

The increase in the glass refractive index Δn due to ion exchange depends, primarily, on the substrate and diffusing ions. The diffusion depth $d = 2\sqrt{Dt}$ is a function of the diffusion coefficient

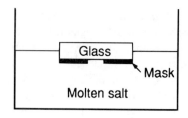
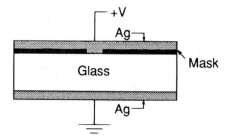

FIGURE 12.1 Schematics of ion exchange waveguide fabrication using molten salt (top) and silver film (bottom).

D, the diffusion time t and the temperature T. Potassium ion exchange results in a relatively small index increase. Δn as small as 0.003 has been reported. A large Δn (up to 0.09) can be achieved by silver and thallium ion exchange. In earlier works, ordinary microscope slides were used to make ion-exchanged waveguides (Najafi, 1992a). The composition of these glasses may vary from batch to batch which makes them unsuitable for fabricating reproducible waveguides. Recently, Corning 0211 glass has been utilized for ion exchange waveguide fabrication. It has been shown that very low-loss, reproducible glass integrated optical circuits can be produced from this glass (Najafi, 1992a). Table 12.1 gives the composition and the annealing temperature of Corning 0211. Different ion exchanges were carried out in this glass, and Δn and D for these processes were determined. Table 12.2 summarizes the fabrication conditions and the values obtained for Δn and D.

TABLE 12.1 Constituents of Corning 0211 Glass. Its Annealing Temperature Is 550 °C

SiO_2	65 wt %
Na_2O	7
K_2O	7
B_2O_3	9
Al_2O_3	2
ZnO	7
TiO_2	3

TABLE 12.2 Fabrication Conditions, Maximum Index Change Δn, and Diffusion Coefficient D for Corning 0211 Glass

Melt	Temperature (°C)	$D(\mu m^2/min)$	Δn
$AgNO_3$	300	0.028	0.06
$CsNO_3$	540	0.001	0.025
KNO_3	400	0.14	0.005
$KNO_3:NaNO_3$ (100:7.5 Wt%)	400	0.075	0.004
$KNO_3:NaNO_3$ (100:15 Wt%)	400	0.07	0.003
$KNO_3:TlNO_3$	480–490	0.02–0.03	0.04–0.09
Silver film ~ 343°C @ current density = 114 $\mu A/cm^2$ ($D \sim 0.12~\mu m^2/min$)			0.05

Gradient-Index Glass Waveguide Devices

The ion exchange process is used in conjunction with the standard photolithographic process to make channel waveguides. To do so, openings are prepared in a metallic or dielectric mask on the surface of the glass substrate and ion exchange is carried out through these openings to obtain channel waveguides.

In addition to the one-step process described above, different variations of the ion exchange process have also been employed to make glass waveguides with improved and more interesting optical characteristics. Most of these processes require two steps of ion exchange and, consequently, are referred to as "two-step" processes. Following, we describe these processes.

Buried Waveguide Fabrication

To fabricate a buried waveguide, first, a surface waveguide is made by an ion exchange process using the molten salt or silver-film techniques (Izawa and Nakagome, 1972; Ramaswamy and Najafi, 1986; Seki et al., 1988; Tervonen et al., 1991). Then, the sample is placed in a molten salt which does not contain the diffusing ions used in the first step. Sodium nitrate is usually utilized. Some of the ions diffuse back in the molten salt and some move farther inside the waveguide. Consequently, the waveguide becomes buried. By adequate selection of fabrication parameters, waveguides with nearly circular mode profiles can be achieved. An electric field can be applied across the substrate to bury the waveguide deeply inside the substrate. However, the application of an electric field considerably complicates the fabrication process.

Ion-Exchanged Waveguides With Ionic Masking

The fabrication of channel waveguides by a double-ion exchange with an ionic mask is sketched schematically in Fig. 12.2 (Honkanen et al., 1992). In the first process step, the substrate is masked, for example, by aluminum stripes, which define the waveguides required, and is immersed in molten KNO_3. The mask is, then, removed and the substrate is immersed in molten $AgNO_3$ at a much lower temperature. At such a low temperature, the exchanged potassium ions have very low mobility, and the silver ions can enter the glass almost exclusively in the previously masked

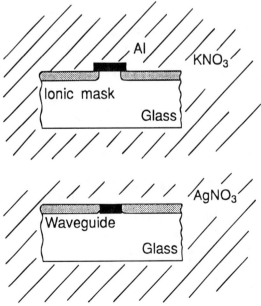

FIGURE 12.2 Silver ion-exchanged waveguide fabrication by potassium ionic masking.

regions. Because the index increase due to silver ions is much greater than the index increase due to potassium ions, the silver ion-exchanged regions serve as channel waveguides.

In channel waveguide fabrication, the ion exchange process with ionic masking has potential advantages compared to the conventional silver ion exchange processes. The width of the channel waveguides is limited by the ionic mask more effectively than by a conventional surface mask, which does not slow the lateral diffusion of silver ions. Also, the ionic masking process yields very low waveguide losses, because the use of a metallic mask in connection with silver ion exchange is avoided. A comparative study of losses in glass waveguides (including waveguides with ionic masking) has been performed and will be discussed in section 12.4.

Double Ion Exchange Process

Potassium and silver ion exchange is usually employed to fabricate waveguides by a double ion exchange process (Li et al., 1991). First, potassium ion exchange is carried out in pure molten potassium nitrate. Then, silver ion exchange is performed in pure, molten silver nitrate. Potassium-silver double-ion-exchanged waveguides have the advantages of both silver and potassium ion-exchanged waveguides made by a single-step process. They have the tight mode confinement of silver (due to a large Δn) and the low propagation losses of potassium ion-exchanged waveguides. The losses in these waveguides are discussed in section 12.4.

The potassium-silver ion exchange process has been utilized to produce dual-core waveguides (Li et al., 1991). These waveguides have two distinct guidance regions created by the potassium and silver ions. Both regions can guide light which allows fabrication of more sophisticated waveguides. This process has been used in fabricating devices, such as computer-generated waveguide holograms, waveguides with gratings, power dividers, wavelength multi/demultiplexers and ring resonators, which will be discussed in section 12.5.

12.3. Theoretical Analysis

Single Ion Exchange

There are two basic categories of ion exchange processes. In thermal diffusion processes, the dominating transport mechanism for ions in the glass is diffusion. In field-assisted processes, transport by migration in the electric field, caused by the external field, prevails.

The relative concentration of diffused ions C in fabricating slab waveguides can be obtained from the one-dimensional equation (Tervonen, 1992),

$$\frac{\partial C}{\partial t} = \frac{\partial}{\partial x}\left[\frac{D\frac{\partial C}{\partial x}}{C(M-1)+1}\right] - \frac{MJ_0\frac{\partial C}{\partial x}}{[C(M-1)+1]^2}, \qquad (12.1)$$

where M is the ratio of self-diffusion constants for the two ions involved in the exchange and J_0 is the flux proportional to the electric current density with the units of speed. The first term on the right includes the effect of diffusion and, the second term, the effect of migration due to the electric current flow through the substrate.

For thermal ion exchange, which is the most commonly used waveguide fabrication process, a simple diffusion equation with concentration-independent rates of diffusion and migration is obtained:

$$\frac{\partial C}{\partial t} = D^2 \frac{\partial^2 C}{\partial x^2}. \qquad (12.2)$$

Gradient-Index Glass Waveguide Devices

Here the two ions have the same mobilities $M = 1$. For a simple thermal diffusion, the equation has the analytic form,

$$C(x, t) = C_0 \, \text{erfc}\left(\frac{x}{2\sqrt{Dt}}\right), \tag{12.3}$$

where C_0 is a constant and the complementary error function is given by

$$\text{erfc}(z) = \frac{2}{\sqrt{\pi}} \int_z^\infty e^{-t^2} \, dt. \tag{12.4}$$

Figure 12.3 illustrates the calculation of the index profile in a silver ion-exchanged glass sample. The index profile of the sample is also determined by scanning electron microscopy of a sample made by ion exchange in molten silver nitrate.

Equation 12.1 is used to study slab waveguide fabrication by the field-assisted migration process. In a typical process, the boundary condition at the surface is $C(0, t) = 1$ and $M < 1$, so that the dopant ions migrating into the glass have smaller mobility than the original ions. Through the balance of diffusion and migration, this leads to the formation of the stationary-shape profile penetrating deeper into the substrate with speed J_0. If $M = 1$ (i.e., the two ions have equal mobilities), a stationary profile will not develop. In this case, the solution is given by (Tervonen, 1992)

$$C = 0.5\left[\text{erfc}\left(\frac{x - J_0 t}{2\sqrt{Dt}}\right) + \exp\left(\frac{J_0 x}{D}\right) \text{erfc}\left(\frac{x + J_0 t}{2\sqrt{Dt}}\right)\right] \tag{12.5}$$

For deep penetration, the first term dominates, and the second one can be ignored. The profile is a complementary error type, shifted to depth $J_0 t$.

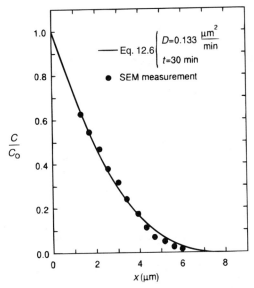

FIGURE 12.3 Theoretical prediction and experimental comparison of ion concentration in a silver ion-exchanged waveguide by a purely thermal process using molten silver nitrate (Ramaswamy and Najafi, 1986).

For more general solutions, numerical methods should be used to solve Eq. 12.1. Finite-difference methods have proven to be the most powerful for this purpose (Tervonen, 1992). These are needed to calculate anything but the stationary profile, when the mobilities of the two ions are unequal.

For channel waveguides, ion exchange is limited to a strip region on the glass surface. Two-dimensional concentration distributions are obtained. No accurate analytic solutions exist for channel waveguide concentration distributions. Finite-difference methods in two dimensions should be used (Tervonen, 1992). Material parameters, proper boundary conditions, and electric field distribution in the glass define the process. Software to calculate ion concentration in glass waveguides by the finite-difference method, adapted to small computers, is commercially available (Optonex).

Modeling of optical propagation in slab waveguides is based on the widely used WKB approximation, which is utilized to calculate the mode propagation constant for a known waveguide profile (Tervonen, 1992). For channel waveguides, finite-difference (Tervonen, 1992) and finite-element methods (Lamouche and Najafi, 1990) have been employed. The finite-element methods are more accurate but also more complicated than finite-difference methods. Scalar, finite-difference methods are less complicated, faster, and accurate enough in ion-exchanged glass waveguides.

To analyze channel, ion-exchanged waveguides, the concentration distribution of the waveguide is calculated by the finite-difference method and transferred to the refractive index profile. A linear relationship between the concentration distribution and the refractive index profile is assumed and the proportionality factor is experimentally determined. This factor depends on the glass composition and wavelength. Then, the scalar wave equation is solved by the finite-difference method to determine the propagation constant and the mode profile of the guided light (Tervonen, 1992; Optonex).

Double-Ion Exchange

The index profiles of double-ion-exchanged waveguides have two distinct regions. For example, in potassium-silver, double-ion-exchanged waveguides, the refractive index of the waveguide close to its surface is similar to that of single, silver ion-exchanged waveguide, but farther inside resembles that of the single, potassium ion-exchanged waveguide. Due to the multi-ion exchange process involved in fabricating double-ion-exchanged waveguides, it is quite complex to obtain an analytical solution for the refractive index or even calculate it numerically. An empirical solution is much simpler. Different mathematical expressions have been tried to determine the form of the index profile in potassium-silver double-ion-exchanged waveguides (Auger and Najafi, 1994a). To do so, the refractive index of the double-ion-exchanged waveguide is represented by

$$n(x) = \Delta n_{Ag} f\left(\frac{x}{d_{Ag}}\right) + \Delta n_k \, \text{erfc}\left(\frac{x}{d_k}\right) + n_b \quad (12.6)$$

where Δn_{Ag} and Δn_k are the maximum index changes due to silver and potassium ion exchange, d_{Ag} and d_k are the depth of the silver and potassium regions, n_b is the substrate refractive index, and $f(x/d_{Ag})$ is a normalization function. Different expressions are assumed for this function and Eq. 12.6 is fitted to the experimentally measured refractive index. Fig. 12.4 depicts a typical example of the fit. The waveguide is made by an initial K^+ exchange (at 400 °C for 15 hrs) followed by an Ag^+ exchange (at 300 °C for 15 hrs). The experimental index profile is determined by the WKB method (Najafi, 1992a).

The truncated quadratic function

$$f = 1 - \frac{x}{d} - b\left(\frac{x}{d}\right)^2 \quad (12.7)$$

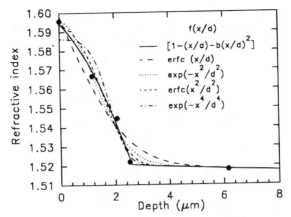

FIGURE 12.4 Fit of Eq. 12.6 using different trial functions $f(x/d_{Ag})$ to obtain experimental data for a potassium-silver double-ion-exchanged waveguide (Auger and Najafi, 1994a). Exchange time is 15 hours for both K$^+$ and Ag$^+$. Exchange temperatures: 400 °C (K$^+$) and 300 °C (Ag$^+$).

provides the best fit of Eq. 12.6 to the experimentally measured refractive index. The constant b is determined to be 1.7 ± 0.5 for the substrate used in the experiment (BK7).

Figure 12.5 compares the theoretically calculated effective indices with the experimentally measured ones. In addition to very good agreement between the theoretical predictions and the experimental ones, it is interesting to note a discontinuity in the slope of the curves for each guided mode. For a given mode, this discontinuity represents the transition between confinement of the light in the silver and potassium regions and in the silver region only.

X-ray photoelectron spectroscopy (XPS) and birefringence measurement in the waveguides have also been used to test the validity of the proposed model for the potassium-silver double-ion-exchanged waveguides (Auger and Najafi, 1994a). The proposed model satisfactorily predicts the index profile and optical behavior of these waveguides. This model is generalized for the channel waveguide (Auger and Najafi, 1994b). It is employed in conjunction with Beam Propagation Method (BPM) to study double-ion exchanged directional couplers (Auger and Najafi, 1994b). An effort is underway to generalize this model for channel waveguides.

12.4 Losses in Ion-Exchanged Waveguides

Potassium and silver ion exchanges have been the most popular processes for producing glass waveguides. Single, potassium ion-exchanged waveguides have low propagation losses and low mode mismatch losses with optical fibers. However, these waveguides have rather high bending losses due to their small Δn. Single, silver ion-exchanged waveguides have a tight mode confinement. By just postannealing the waveguides, their mode profiles are enlarged and mode mismatch losses below 0.2 dB with optical fibers are achieved (Honkanen et al., 1992; Najafi, 1992a). Waveguides made by silver ion exchange through a metallic mask have high propagation losses due to silver metal colloids formed during the exchange process. However, by using a dielectric mask, losses below 0.2 dB/cm are achieved (Najafi, 1992a). The lowest coupling and propagation losses are obtained by employing buried waveguides (see section 2.1) and utilizing substrate glasses developed specially for ion exchange. Coupling losses below 0.1 dB and propagation losses below 0.05 dB/cm have been reported (McCourt, 1993). However, the losses obtained by utilizing nonburied waveguides and standard multipurpose substrate glasses are low enough for most glass waveguide applications.

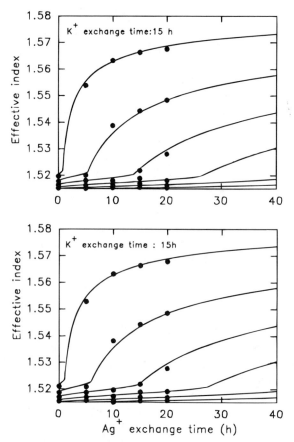

FIGURE 12.5 Variation of effective index with silver ion-exchange time. $\lambda = 0.6328$ μm. The K^+ exhange time is 15 hours (Auger and Najafi, 1994). TE modes (top). TM modes (bottom).

A systematic study has been carried out to determine the propagation losses in nonburied potassium and silver ion-exchanged waveguides. The fabrication processes and parameters of the waveguides studied are given in Table 12.3. All of the waveguides were fabricated from 0.5-mm thick, multipurpose, glass substrate (Corning 0211). About 100-nm thick aluminum film was deposited on the substrate, and mask openings were defined by conventional photolithography.

Samples S1 and S2 were fabricated using the following double-ion exchange process. First, substrate glasses with aluminum mask openings were immersed in molten potassium nitrate at 400 °C for 2.5 h. Then, the samples with the same mask were immersed in a molten silver nitrate bath at 300 °C for 6 h and 3.5 h, respectively. This process produces dual-core waveguides, because there are two distinct regions corresponding to the two types of ion exchange processes. The potassium ion-exchanged region is much deeper than the silver ion exchange region because the effective diffusion coefficient for potassium ions in the first step is much larger than for silver ions in the second ion exchange process (Li et al., 1991).

Sample S3 was made by the same double-ion exchange process as samples S1 and S2, but using a much shorter potassium ion exchange time (20 min). This results in similar depths for potassium and silver diffusion, and the waveguide has only one guiding core.

To better understand the reasons for losses (or loss reduction) in double-ion-exchanged waveguides, a sample S4 was fabricated by the following process. In the first step, a slab with a potassium ion enriched layer was produced by diffusion into an unmasked substrate in molten potassium

TABLE 12.3 Fabrication Parameters and Loss Measurement Results of the Waveguides Studied

Sample	Fabrication Process and Parameters	Mask Width (μm)	Propagation Losses	Reference
S1	Dual-core channel waveguide (Al mask) K^+: 400 °C, 2.5 h; Ag^+: 300 °C, 6 h.	4	0.30 dB/cm ($\pm$0.15 dB)	Wang et al., 1993
S2	Dual-core channel waveguide (Al mask) K^+: 400 °C, 2.5 h; Ag^+: 300 °C, 3.5 h.	4	0.24 dB/cm ($\pm$0.15 dB)	Wang et al., 1993
S3	Double-ion-exchanged channel waveguide (Al mask) K^+: 400 °C, 20 min; Ag^+: 300°C, 2 h.	4	0.22 dB/cm ($\pm$0.10 dB)	Wang et al., 1993
S4	Slab layer: K^+: 400 °C, 20 min; Channel waveguide: Ag^+: 300 °C, 5 h.	4	0.60 dB/cm ($\pm$0.20 dB)	Wang et al., 1993
S5	Channel waveguide: (Al mask) Ag^+: 300 °C, 2 h.	4	1.30 dB/cm ($\pm$0.20 dB)	Wang et al., 1993
S6	Ionic mask: K^+: 400 °C, 20 min; Channel waveguide: Ag^+: 300 °C, 2 h.	5	0.10 dB/cm ($\pm$0.10 dB)	Wang et al., 1993
S7	Channel waveguide: (Al mask) K^+: 400 °C, 2.5 h.	4	0.2 dB/cm ($\pm$0.1 dB)	Honkanen et al., 1993

nitrate at 400 °C for 20 min. Then, an aluminum film was deposited and mask openings defined by photolithography. Silver ion exchange was carried out in a silver nitrate bath at 300 °C for 5 h through the aluminum mask openings.

For comparison, silver ion-exchanged waveguides using metallic (sample S5) and dielectric (sample S6) masks were made. Sample S5 was fabricated by a one-step silver ion exchange through an opening in an aluminum mask by immersing the substrate in a silver nitrate bath at 300 °C for 2 h. Sample S6 was fabricated by silver ion exchange with ionic masking. First, a 5-μm-wide aluminum stripe was patterned by photolithography to define the channel waveguide required. Then, the sample was immersed in a potassium nitrate bath at 400 °C for 20 min to form the dielectric ionic mask. The aluminum stripe was removed, then, and the sample was immersed in a silver nitrate bath at 300 °C for 2 h. During this process step, diffusion of silver ions in the potassium ion-exchanged region is slightly slower than in the previously masked channel region. Because the index increase due to silver ions is much greater than due to potassium ions, the pure, silver ion-exchanged region serves as a channel waveguide. Finally, a potassium ion-exchanged waveguide (sample S7) was fabricated using a metallic mask. Ion exchange was carried out at 400 °C for 2.5 h.

Accurate loss measurements of the waveguides fabricated by the different processes and parameters were performed at the 1.296 μm wavelength as explained in Wang et al. (1993). Table 12.3 summarizes the measurements results.

The propagation losses of double-ion-exchanged waveguides S1, S2, and S3 are considerably lower compared to the one-step silver ion-exchanged waveguide S5 and only slightly higher than the losses of the waveguides S6 and S7. We believe that the main reason for this loss reduction is the high potassium ion concentration, and consequently, low sodium ion concentration, below the edges of the aluminum mask due to the first (potassium) ion exchange. Because the potassium ions are not very mobile during the silver ion exchange step, the possibility for silver ions to reduce to atoms below the mask edges is decreased. In the case of sample S4, which was made with silver ion exchange through the potassium ion-exchanged slab layer, we initially expected to get even lower losses than with the other double-ion-exchanged waveguides (S1, S2, and S3) because, in S4, the potassium ion concentration is high below the whole mask region, not only below the mask edges. However, the losses in S4 are higher than in the other double-ion-exchanged waveguides. We believe that the surface of the aluminum mask in samples S1, S2, and S3 is at least partly oxidized during the first (potassium) ion exchange, which also reduced the silver precipitation during the second (silver) ion exchange. With sample S4, the aluminum mask is deposited after the potassium ion exchange.

12.5 Devices

Glass integrated optic devices can be used to make sophisticated optical circuits for practical applications. Figure 12.6 illustrates a complex glass integrated optical circuit for optical communication systems. It can amplify the signals at 1.3 μm and 1.55 μm wavelengths, combine these two signals, and divide them into eight parts. Most of the devices to make this circuit are developed and will be discussed in this section. In addition, we will review some other recent devices that have been demonstrated and discuss their applications.

Mach–Zehnder Interferometers

Mach–Zehnder interferometer configurations are employed to make interesting glass integrated optic devices such as wavelength filters (Najafi, 1992a; Najafi et al., 1992b; Lefebvre et al., 1993), wavelength multiplexers (Wang et al., 1992a; Zhang et al. 1993; Zhang et al., 1994b) and wavelength demultiplexers (Najafi, 1992a; Wang et al., 1992a; Zhang et al., 1993; Tervonen et al., 1993; Zhang et al., accepted; Zhang et al., 1994a).

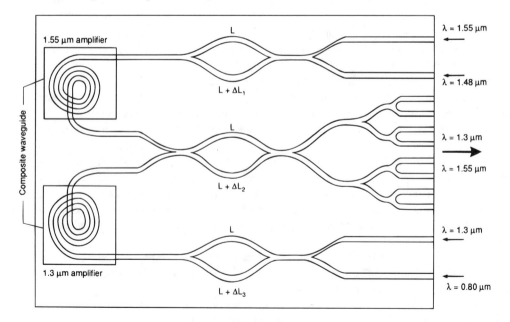

FIGURE 12.6 A glass integrated optical circuit for optical communication systems.

Three- and four-port Mach–Zehnder interferometers have attracted a lot of attention. Three-port interferometers to combine or separate the following pairs of wavelengths have been demonstrated: 1.3 μm/1.55 μm (Najafi, 1992a; Zhang et al., 1993), 0.98 μm/1.55 μm (Wang et al. 1992a), 0.81 μm/1.3 μm (Zhang et al., 1994a), and 1.48 μm/1.55 μm (Zhang et al., 1994b); Zhang et al., 1994a). Fig. 12.7 depicts, schematically, a three-port Mach–Zehnder interferometer. It is composed of an adiabatic, asymmetric Y-branch and two symmetric Y-branches, connected by two waveguides with different lengths. This path-length difference in the actual interferometer part of the device determines its wavelength-dependent operation. When the device is used as a demultiplexer, light is coupled to the single input waveguide. Light is, then, split equally by the first symmetric Y-branch into the two interferometer arms and combined by the second symmetric Y-branch. Due to the path-length difference, the phase difference between light from the two interferometer arms depends on wavelength. For proper operation, the phase differences have to be zero and π at the two wavelengths to be demultiplexed. At the symmetric/asymmetric Y-branch junction, light (shorter wavelength) with a zero phase difference excites the fundamental mode and is coupled to the wider arm of the asymmetric, adiabatic Y-branch, and light (longer wavelength) with a phase difference π excites the antisymmetric mode of the junction and is coupled to the narrower arm of the asymmetric, adiabatic Y-branch. As a multiplexer, the device operates similarly but in the opposite direction.

Because the behavior of the Y-branches depends very little on the wavelength, the spectral transmission properties of the WDM-device are determined only by the optical path-length difference between the two interferometer arms. The path-length difference is realized by using different lengths of curved and straight waveguides in the interferometer arms. In designing these devices, the shift of the mode field (Wang et al., 1992b) in the curved waveguides also has to be taken into account. The shift of mode fields in curved waveguides and, therefore, also the spectral transmission properties of the WDM device depend on fabrication conditions. This dependence is smaller if shorter path-length differences and, larger bend radii in curved waveguides are used. Device and fabrication parameters for the three-port Mach–Zehnder interferometers are summarized in Table 12.4. The devices are made by the potassium-silver double-ion exchange process. Optical characteristics of the fabricated devices (extinction ratios and propagation losses) are summarized in Table 12.5. The extinction ratio is defined as the ratio of the output from the two ports at a wavelength (when device is used as a demultiplexer). The propagation losses are determined as explained in Wang et al. (1993).

The four-port Mach–Zehnder interferometer (Wang et al., 1992a) (see Fig. 12.8) is designed for wavelength multi/demultiplexer application in Er-doped fiber amplifiers and ring lasers (at 1.55 μm wavelength region) which are pumped at $\lambda = 0.98$ μm. It consists of two identical hybrid couplers. The mask width for the narrower ports of the coupler (ports II and IV) is 2.5 μm and, for ports I and IV, is 5.5 μm. The two arms of the interferometer have the same width (4 μm) with a path-length difference of $\Delta L = 0.981$ μm. The branching angles are 3 mrad. The

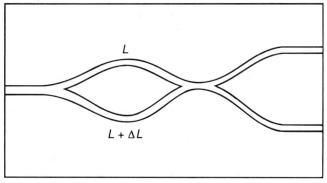

FIGURE 12.7 Schematics of a three-port Mach–Zehnder interferometer.

TABLE 12.4 Device Parameters of Three-Port Mach–Zehnder Interferometer WDMs

Operation Wavelengths (μm)	Waveguide Width (μm)		Other parts	Path Length Difference (μm)	Bend Radius (mrad)	Branching Angle (mrad)	Offsets (μm)	Reference
	Narrowport	Wideport						
1.3/1.55	2.5	5.5	4	2.565	50	3	0.25*, 0.50**	1, 2, 3
1.48/1.55	2.5	5.5	4	10.75	30	3	0.7*, 1.4**	1, 2, 3
0.807/1.3	2.5	5.5	4	3.45	30	3	0.5*, 1.0**	3

Note: All of the devices are made using the following fabrication parameters: first, potassium ion exchange in pure KNO_3 at 400 °C for 165 min., second, silver ion exchange in pure $AgNO_3$ at 300 °C for 210 min., and third, postannealing at 300 °C for 68 min. Glass substrate: Corning 0211.

*Curved and straight waveguides are joined.
**Two curved waveguides are joined.
1. Zhang et al., 1994b.
2. Zhang et al., 1996.
3. Zhang et al., 1994a.

TABLE 12.5 Optical Characteristics of Three-Port Mach–Zehnder Interferometer WDMs

Device	Test Wavelength (μm)	Facet to facet loss (dB)	Excess loss (dB)	Extinction Ratio (dB)	References
1.3/1.55*	1.273/1.557	2.65	1.65	13**/12**	1, 3, 4
1.48/1.55	–/1.557	2.49	1.49	–/19**	2, 3, 4
0.807/1.3	–/1.273	3.00	2.50	–/7**	4

*The multiplexer is followed by a 1/8 splitter.
**Actual extinction ratio is higher because the test wavelength does not match the designed operation wavelength.
1. Zhang et al., 1993.
2. Zhang et al., 1994b.
3. Zhang et al., 1996.
4. Zhang et al., 1994a.

bend radius in the Mach–Zehnder interferometer is 17.5 mm and 0.75-μm lateral offsets are used when the straight and curved waveguides are connected. The device operates as follows. The local antisymmetric mode around $\lambda = 1$ μm is divided by the symmetric Y-branch of the first coupler into two parts, and they undergo an additional phase difference π between them after they propagate through the different path lengths. They recombine in the second symmetric Y-branch with a total phase difference of 2π and excite the local fundamental mode in the second hybrid coupler, which, in turn, is cross-coupled to the wider output arm (port III) by the second nonsymmetric Y-branch. The two parts of the local antisymmetric mode for the wavelength around $\lambda = 1.5$ μm undergo an additional 2π phase difference between them after traveling through the two arms of the interferometer. They reach the second symmetric Y-branch with a total phase difference of 3π, and recombine to the first higher order local antisymmetric mode in the second hybrid coupler, which is coupled to the narrower output arm (port IV).

The device is made by potassium-silver double-ion exchange process with a Corning 0211 glass substrate. First, potassium ion exchange is carried out in pure, molten potassium nitrate at 400 °C for 165 min. Then, silver ion exchange is performed using a pure, molten silver nitrate bath at 300 °C for 210 min. Figure 12.8 shows the optical response of the fabricated device. This device is employed in the fabrication of a ring resonator which is explained in the next section.

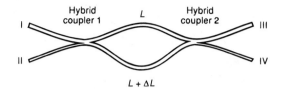

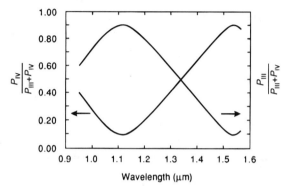

FIGURE 12.8 Four-port Mach–Zehnder interferometer (top). Measured transmission spectra from the two output ports (bottom) (Wang et al., 1992).

Ring Resonator

Integrated optical ring resonators can be used as tunable wavelength filters (or optical frequency filters) and in various sensors, such as rotation and temperature sensors. It can also be utilized to accurately measure propagation losses of low-loss waveguides.

A ring resonator is constructed (Wang et al., 1992b) by connecting ports I and III of the four-port Mach–Zehnder interferometer discussed in section 5.1 (see Fig. 12.8). The ring is composed of two half-circles with radii of 15 mm and 20 mm and a total path length of ~12 cm. The device is produced by the double-ion exchange process with Corning 0211 glass substrate. First, potassium ion exchange is carried out at 400 °C for 140 min. (pure KNO_3). Then, silver ion exchange is performed at 300 °C for 300 min. (pure $AgNO_3$).

The fabricated ring resonator is tested at $\lambda = 1.523$ μm. To change the optical path length of the ring, it is locally heated. Then, the heater is removed, and resonant peaks are periodically scanned through while the sample cools down. Figure 12.9 shows the measured response of the ring resonator. It has a finesse of 5 and propagation losses of 0.17 dB/cm. It is being employed in conjunction with Er-doped sol-gel films to produce a ring laser.

Waveguides with Gratings

Glass waveguides with gratings are wavelength- and direction-selective. They can be used, for example, in fabricating narrowband filters/WDMs/mirrors, and for three-dimensional integration (Najafi and Fallahi, 1994c; Li and Najafi, 1993; Tewonen et al., 1994; Najafi, 1994a; Najafi, unpublished). Significant progress has been achieved in modeling and fabricating glass waveguides with linear and circular gratings (Najafi, 1992a; Najafi and Fallahi, 1994c; Li and Najafi, 1993; Najafi, 1994a). Following, we describe three integrated optical devices with gratings.

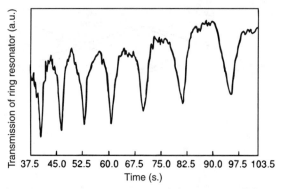

FIGURE 12.9 Resonance curve of ring resonator (Wang et al., 1992).

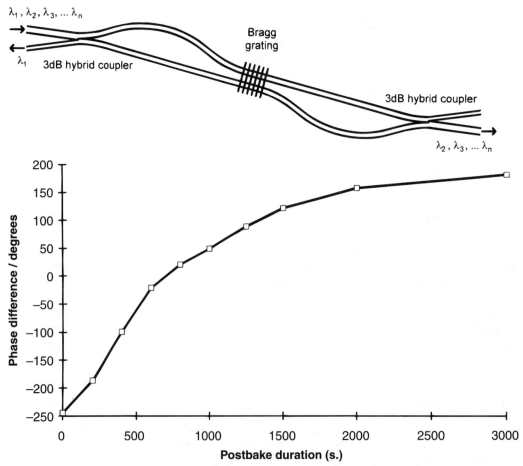

FIGURE 12.10 Schematics of the ion-exchanged, narrowband, add-drop, WDM device (top). Phase difference between light in the two arms of the Mach–Zehnder interferometer before the Bragg grating as a function of post annealing time (bottom) (Tervonen et al., 1994).

An interesting narrowband, add-drop wavelength multi/demultiplexer (WDM) based on a combination of a Bragg grating and a symmetrical Mach–Zehnder interferometer with 3 dB directional couplers at the input and output of the device is described in Najafi (1992a). Recently, a similar, narrowband, add-drop WDM device has been proposed (Tervonen et al., 1994), in which the directional couplers are replaced by four-port hybrid couplers already discussed in section 5.1. In the new add-drop WDM device, the total lengths of the two arms of the Mach–Zehnder interferometer are the same, but they have a rather long path-length difference before and after the Bragg grating as sketched in Fig. 12.10. The device operates as follows: non-Bragg-resonant-wavelength light (λ_2, λ_3, λ_4, λ_5) coupled to the input port is split equally, with no phase difference, in the first hybrid coupler to the two arms of the interferometer. Because the total optical path length in the two interferometer arms is exactly the same, light from the two arms arrives at the second hybrid coupler without any phase difference and is coupled to the wider output port of the second hybrid coupler; Bragg resonant light (λ_1) is reflected by the Bragg grating and, for proper operation, the total phase difference between light from the two arms as they recombine at the first hybrid coupler is π, and light is coupled to the narrower port of the first hybrid coupler (Tervonen et al., 1994). The advantage of the device is that the phase difference can be tuned simply by thermally annealing the device, because the path-length difference before the grating is realized by using a curved waveguide in one of the arms enabling thermal tuning (Zhang et al., accepted (b); Tervonen et al., 1994). Fig. 12.10 shows the calculated phase difference accumulated before the grating as a function of postannealing time at 343°C. The calculation is performed for a silver ion-exchanged waveguide in Corning 0211 substrate glass and for the 1.55 μm wavelength. The geometrical path-length difference before the grating is about 70 μm. The possibility for tuning is very important, because, otherwise, even very small errors in alignment of the grating would deteriorate the behavior of the device. Here, tuning is possible, because the behavior of the hybrid couplers is not sensitive to thermal annealing as is the case with 3-dB directional couplers (Tervonen et al., 1993).

An alternative glass WDM device(Najafi, unpublished) is shown in Fig. 12.11. Four channel waveguides, which form the four input/output ports of the device, are connected to each other by four tapered waveguides. In the center of the device, a 45° linear grating with period $\Lambda = \lambda/\sqrt{2}N$, N being the effective index of the guided mode, is constructed. The input light at λ_1 injected through input ports 1 or 2 is reflected normal to the input light and routed towards the output ports 3 or 4, respectively. The tapered waveguides minimize the losses. The grating can reflect the input light partially or totally. This depends mainly on the ratio of grating depth to waveguide depth (Najafi, 1992a). Potassium ion-exchanged waveguides are suitable for partial reflection. Silver ion-exchanged waveguides can be employed for total reflection. This WDM device is compact, and has very low losses. It is suitable for making WDM arrays for multi/demultiplexing in systems where more than two closely spaced signal wavelengths are used. It can also be used as a mirror in integrated optical circuits.

Recently, waveguides with circular gratings have attracted attention (Najafi, 1994a). Electron-beam and focused-ion-beam lithography are employed to write circular gratings on a PMMA resist deposited on the sample. The gratings are, then, transferred to the waveguide by reactive ion etch. Figure 12.12 displays the microphotograph of a circular grating etched in a glass waveguide. The grating period and depth are 0.672 μm and 0.1 μm, respectively. Figure 12.13 illustrates a circular grating waveguide device and an example of its application in three-dimensional integration. The device is called a "tap power divider". It consists of a central circular waveguide connected to four tapered waveguides, and a circular grating concentric with the central waveguide. The input light is directed perpendicularly onto the surface of the grating. The grating couples the light into the central waveguide. Light propagates in the central waveguide and distributes into the tapered waveguides. The output light intensity from the ports depends, primarily, on the coupling efficiency of the grating, the central waveguide radius, and the input

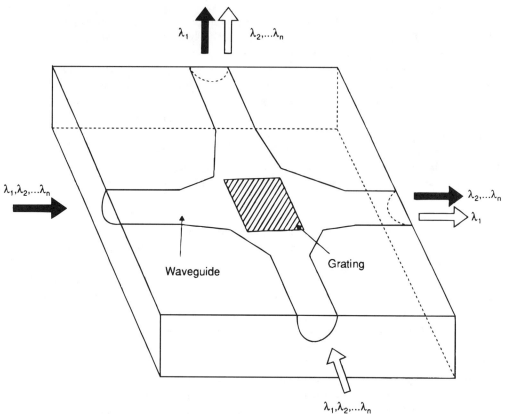

FIGURE 12.11 Ion-exchanged, narrowband, 90° reflective, WDM device (Najafi and Kavehrad, unpublished).

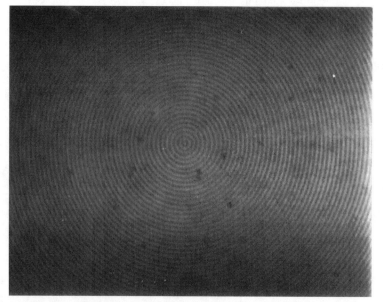

FIGURE 12.12 Microphotograph of a circular grating etched in a glass waveguide. Grating period and depth are 0.672 and 0.1 μm, respectively.

Gradient-Index Glass Waveguide Devices

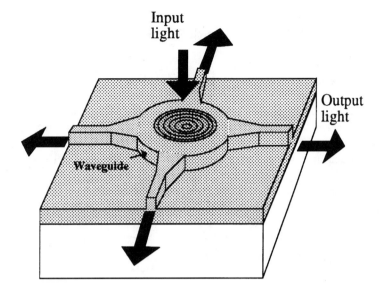

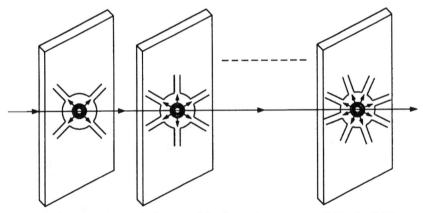

FIGURE 12.13 Schematic diagram of circular grating tap-power divider (top). 3D-integration by tap-power divider (bottom) (Najafi, 1994; Najafi et al., 1993).

width of the tapered waveguide. To tap only a small fraction of the signal light, the grating should be shallow and/or the grating period should be slightly different from the ideal coupling condition at the signal light wavelength. The device is designed to operate at the 514-nm wavelength. A slab waveguide is made by potassium ion exchange at 400°C for 15 min in a Corning 0211 glass substrate, resulting in a single-mode waveguide at the 514-nm wavelength. The sample is coated with a 0.15μm thick SiO_2 layer to prevent contaminating the reactive ion etch (RIE) equipment with the sodium in Corning 0211 glass. The sample is, then, coated with a 300 Å aluminum layer and a PMMA photoresist. The circular grating with a 0.36-μm period and 300-μm diameter is written in PMMA by focused ion beam lithography. The grating is etched by RIE in the aluminum layer and, then, in glass. The four-port waveguide is patterned by the liftoff technique and RIE. The light (at 514 nm) directed onto the grating is partially coupled into the central waveguide and transmitted through the output ports. The output from each port is about 0.1%

of the incident light. This amount can be varied easily by changing the grating period and depth. This structure can also be used to make multiport lasers.

Computer-Generated Waveguide Holograms

The Potassium-silver double-ion exchange process is used to make Dammann computer-generated waveguide hologram (CGWH) (J. Saarinen et al., 1992) for 1/8 beam splitting. The diffraction efficiencies of eight orders $\{-7, -5, -1, +1, +3, +5, +7\}$ are equalized and maximized using Fourier optics and nonlinear parametric optimization. The thickness of the grating is chosen to modulate the phase of the wave front passing through the CGWH by π rad. Figure 12.14 depicts a microphotograph of the fabricated double-ion-exchanged grating. First, a potassium ion-exchanged waveguide is made at 400 °C in a Corning 0211 glass substrate. Ion exchange is carried out for 45 min. Then, the grating is processed by silver ion exchange at 300°C for 70 min. The device is tested using $\lambda = 0.6328$-μm HeNe laser light. Figure 12.14 shows a photograph of the spot array generated by the waveguide grating. The bright, undiffracted, zeroth order is due to mode mismatch of the guided light in the potassium-silver double and potassium ion-exchanged regions.

Aiming at a higher diffraction efficiency, kinoform gratings are then employed to make CGWH 1/8 beam splitters (Saavinen et al., 1994b). This is a Fourier transform type phase grating consisting of 50 rectangular cells with equal width within each period. The phase level (the lengths of cells) are quantized to 119 levels per phase delay of 2π rad ($\sim$0.25 μm per quantization step). Figure 12.15 illustrates the profile of the designed kinoform grating (diffraction orders $-4, -3, -1, -1, 1, 2, 3, 4$). The ion exchange parameters are selected to achieve the designed phase modulation assuming that the modulated region produces a phase delay of

$$\phi(x) = \frac{2\pi}{\lambda_0} \int_0^h \Delta h_{\text{eff}}(x) \, dz$$

where λ_0 is the wavelength in vacuum and h is the modulated region thickness. The wave front is propagated in the z direction. An in-depth study of the effect of fabrication parameters on waveguide properties in potassium-silver double-ion-exchanged was carried out to optimize the optical characteristics of the CGWH (Saavinen et al., 1994b). Based on this study, new further optimized CGWH's were designed and their optical characterization is underway and will be reported in the near future (Saavinen et al., 1994a).

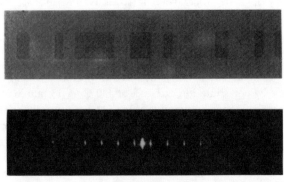

FIGURE 12.14 Microphotographs of double-ion-exchanged Dammann gratings (top). Photograph of spot array generated by grating (bottom) (Saarinen et al., 1992).

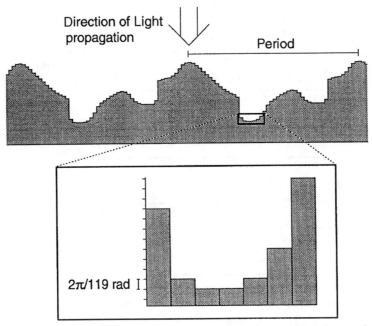

FIGURE 12.15 Phase profile of a computer-generated waveguide hologram for 1/8 beam splitting. The inset shows the quantization of phase level (Saarinen et al., 1994).

Rare Earth-Doped Waveguides

Glass waveguides doped with rare earths are used to make optical amplifiers and lasers (Najafi, 1992a; Najafi, 1991). To achieve laser oscillation, straight channel waveguide cavities with dielectric and Bragg grating mirrors (Najafi and Fallahi, 1994c; Najafi, 1989) and ring resonators (Wang et al., 1992b) are proposed. Two different waveguide configurations are employed to produce rare-earth-doped waveguides; ion exchange in a rare-earth-doped glass substrate and a composite waveguide structure. In the first configuration, ion exchange is simply carried out in a neodymium-doped glass substrates to produce amplifiers and lasers around 1.06 μm and 1.3 μm wavelengths (Sanford et al., 1990; Aoki et al., 1990a; Aoki et al., 1990b; Miliou et al., 1993). In particular, a waveguide with 15-dB amplification at 1.06 μm wavelength was demonstrated (Miliou et al., 1993). Although the first rare-earth-doped waveguide was reported in 1974 (Saruwatari and Izama, 1974), there was little effort until 1988 (Najafi, 1988) in this field. Since then, there has been significant progress in fabricating of neodymium- and erbium-doped waveguides, amplifiers, and lasers. However, this field is in its early stages, and much research and development is needed to produce devices for practical applications.

An important issue in fabricating rare-earth-doped waveguide devices, using the first configuration, is the production of suitable substrates for both light amplification and ion exchange. The ion exchange process considerably modifies the optical property of the substrate (in particular, its photoluminescence lifetime). This effect must be taken into account in preparing the substrate glass to obtain devices with optimum performance.

The composite structure (Honkanen et al., 1992a) (see Fig. 12.16) eliminates the above problem. In this technique, an ion-exchange waveguide is made in an undoped glass. Then, a rare-earth-doped glass is pressed against the fabricated waveguide. With adequate selection of the refractive indices of the two glasses and fabrication parameters, a significant amount (50 percent or so) of light can be guided in the doped glass (Najafi and Honkanen, 1992).

The rare-earth-doped composite glass waveguide amplifiers are one of the main devices employed in the integrated optical circuit in Fig. 12.6. This waveguides can also be used in

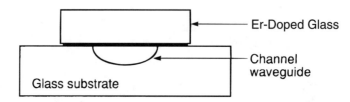

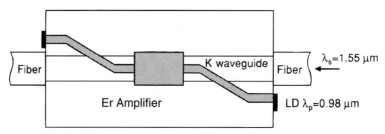

FIGURE 12.16 Schematic diagram of cross-section of composite rare-earth-doped glass waveguide (top). An Er-doped amplifier using a composite waveguide (bottom).

conjunction with Y-branch multiplexers (Honkanen et al., 1992b) to make compact, individual, glass, waveguide amplifiers. Figure 12.16 shows an example of such amplifier in which two laser diodes ($\lambda = 0.98$ μm) are butt-joined to the silver ion-exchanged waveguides. The signal from the fiber is coupled into the potassium ion-exchanged waveguide. The two wavelengths are multi/demultiplexed by the asymmetric Y-branches, and amplification is performed by the composite waveguides. In this device, the coupling losses are minimized because the potassium ion-exchanged waveguide are well matched to optical fibers, and, with silver ion-exchanged waveguides, efficient coupling to laser diodes is achieved.

Rare-earth-doped waveguides also provide the possibility of making amplifiers and lasers at the visible by upconversion. Green emission is observed in an erbium-doped glass waveguide (Ohtsuki et al., 1994; Francois et al., 1994). A phosphate glass, suitable for ion exchange waveguide fabrication, with a high Er-concentration has been developed. The waveguide was made by the silver-film ion exchange technique. The details of the glass and waveguide fabrication processes are explained in Honkanen et al., (1991). Table 12.6 gives the composition of the glass used to produce the waveguide. For optical characterization, pump light around the 800-nm wavelength was used and upconversion fluorescence was measured by spectrum analyzer. Figure 12.17 displays the upconversion fluorescence spectrum and a photograph of the green light emission in the waveguide (Ohtsuki et al., 1994; Francois et al., 1994). The emission peaks around 530 nm and 550 nm correspond to $^2H_{11/2} \rightarrow {}^4I_{15/2}$ and $^4I_{3/2} \rightarrow {}^4S_{15/2}$ transitions, respectively. A bright green

TABLE 12.6 Er-Doped Phosphate Glass Composition

Component	Weight percent
P_2O_5	62.4
Li_2O	5.1
Na_2O	4.4
ZnO	12.9
Al_2O_3	4.5
ErF_3	10.1
Others	0.6

Gradient-Index Glass Waveguide Devices

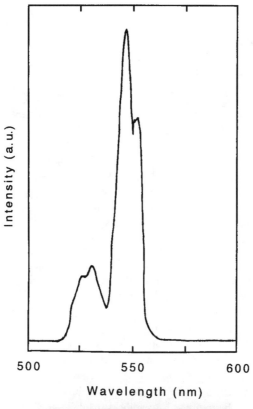

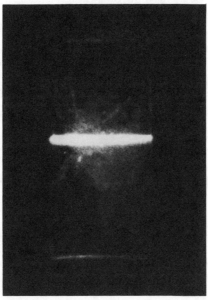

FIGURE 12.17 Upconversion fluorescence spectrum of bulk Er-doped phosphate glass (top). Photograph of the green light generated by upconversion in the waveguide (bottom).

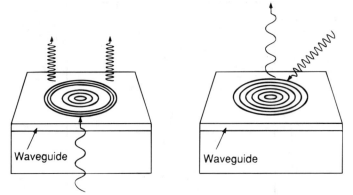

FIGURE 12.18 Circular cavity rare-earth-doped light emitting sources (Najafi and Fallahi, 1994; Najafi, 1994).

light is emitted in all directions when the pump is coupled in the waveguide. This implied that the emission is spontaneous and not stimulated. A part of this light is captured and guided by the waveguide. The measured transmitted spectrum of the waveguide is similar to that of the bulk glass sample. The observed green light emission suggested the possibility of diode-laser-pumped, small-size, short-wavelength, visible lasers.

Recently, there has been a new trend in fabricating laser cavities. Circular grating Bragg reflectors are employed to form laser cavities, and semiconductor lasers are successfully produced (Najafi, 1994). These lasers have low divergence and a relatively symmetric beam emission. Similar structures can be utilized to make rare-earth-doped light emitting sources in the visible and infrared. Such sources can be pumped by readily available laser diodes. Figure 12.18 shows two possible configurations for rare-earth-doped circular grating lasers at the visible and infrared.

The fabrication of circular grating, however, requires, costly equipment. It takes a rather long time to make each grating, and the gratings are produced individually, resulting in high final costs. An alternative to obtain a symmetric beam laser is to employ variable depth or width gratings. Figure 12.19 displays, schematically, a symmetric-beam, surface-emitting, rare-earth-doped, glass waveguide laser (Najafi, 1994b). The laser is pumped through the dielectric mirror end. The first grating, which forms the laser cavity with the dielectric mirror, allows a few percent

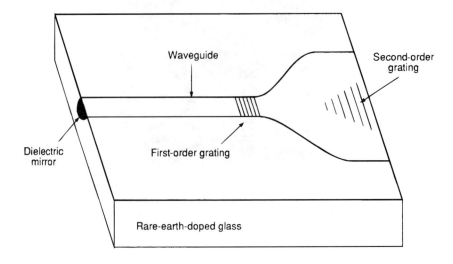

FIGURE 12.19 Symmetric beam, surface emitting laser (Najafi, 1994).

Gradient-Index Glass Waveguide Devices

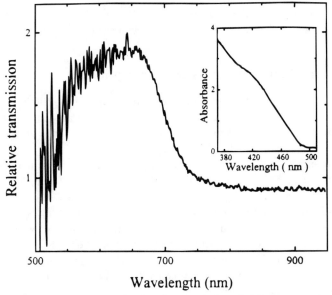

FIGURE 12.20 Measured spectral transmission of CdS quantum dot waveguide (Lee et al., submitted). The inset is the absorption spectrum of the quantum dot, glass substrate.

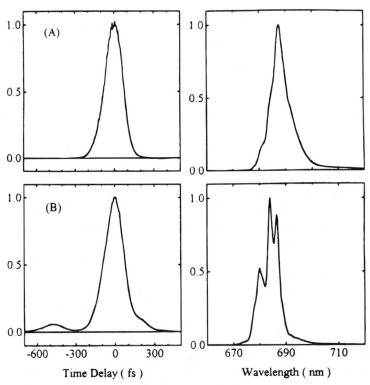

FIGURE 12.21 The cross-correlation and spectrum of the femtosecond pulse after propagating 8-mm long, CdS, quantum dot waveguide (Lee et al., submitted). 110 fs input pulse profile and its spectrum (top). The output pulse shape and spectrum with 12 GW/cm^2 input intensity (bottom).

of the emitted light to go into the adiabatic waveguide. The dimensions of the variable width grating and tapered waveguide are selected to achieve a symmetric diffracted beam of the laser light.

Semiconductor, Quantum Dot Glass Waveguides

Semiconductor, quantum dot waveguides are interesting because the carrier motion is restricted in all three dimensions. The presence of quantum-confined transitions in the dots leads to narrowing of the optical spectra (gain and absorption) (Peyghambarian et al., 1993), making them potentially attractive for developing superior devices.

Potassium ion exchange is used to make channel waveguides in a cadmium sulfide (CdS), quantum dot, glass sample (Lee et al., submitted). A special glass containing 15% CdS quantum dots is prepared by sol-gel technology. Eight-mm long waveguides are made by ion exchange at 400 °C for 16.5 hours in a pure, molten potassium nitrate. Figure 12.20 shows the measured transmission spectrum of the waveguide. The transmission drop for short wavelengths is due to the long tail absorption of the CdS quantum dots. The sudden drop in transmission near 700 nm corresponds to the cutoff frequency of the waveguide. The inset of Fig. 12.20 displays the absorption spectrum of the quantum dot sample itself. The lowest quantum-confined transition appears as a shoulder in the spectrum.

The femtosecond pulse propagation study was performed using the cross-correlation technique (Peyghamberian, 1994). The test beam was selected at 687 nm for the pulse to be in the high transition state. The full width at half maximum of the test beam was 100 fs. The time and spectral profiles of the input and output pulses of the waveguide are shown in Fig. 12.21. For an intensity of 12 GW/cm^2, the pulse develops into three peaks after it propagates through the waveguide with broadened and modulated spectrum. The pulse breakup is due to either the result of coherent effects or launching of a soliton in the quantum dot waveguide. Further investigation is needed to understand fully the origin of the pulse breakup.

References

Aoki, H., Maruyama, O., and Asahara, Y. 1990a. Glass waveguide laser. *IEEE Photon. Tech. Lett.* 2 (7), 459.

Aoki, H., Maruyama, O., and Asahara, Y. 1990b. Glass waveguide laser operated around 1.3 μm. *Electron. Lett.* 26 (22), 1910.

Auger, P., and Najafi, S. I. 1994a. Potassium and silver double-ion-exchanged slab glass waveguides: characterization and modeling, to appear in June issue of *Appl. Opt.*

Auger, P., and Najafi, S. I. 1994b. New method to design directional coupler dual wavelength multi/demultiplexer with bends at both extremities, *Optics Comm.* 3, 43. Also, see Najafi, 1994b.

Doremus, R. H. 1964. Exchange and diffusion of ions in glass. *J. Phys. Chem.* 68, 2212.

Findakly, T. 1985. Glass waveguides by ion exchange: A review. *Opt. Eng.* 24, 244.

François, V., Najafi, S. I., Lefebvre, P., Lafrenière, S., Ohtsuki, T., Peyghambarian, N., Honkanen, S., Fallahi, M., and Orcel, G. Progress towards realization of rare-earth-doped glass integrated optic lasers, Intl. Symp. on Integrated Optics: Conf. on Nanofabrication Technologies and Device Integration, Lindau (Germany), April 1994.

Giallorenzi, T. G., West, E. J., Kirk, R., Gunther, R., and Andrews, R. A. 1973. Optical waveguides formed by thermal migration of ions in glass. *Appl. Opt.* 12, 1240.

Honkanen, S., Najafi, S. I., Poyhonen, P., Orcel, G., Wang, W. J., and Chrostowski, J. 1991. Silver-film ion-exchanged single-mode waveguides in Er-doped phosphate glass. *Electron. Lett.* 27 (23), 2167.

Honkanen, S., Najafi, S. I., and Wang, W. J. 1992a. Composite rare-earth-doped glass waveguides. *Electron. Lett.* 28 (8), 746.

Honkanen, S., Najafi, S. I., Wang, W. J., Lefebvre, P., and Tervonen, A. Integrated optical devices in glass by ionic masking, SPIE OE/Fiber Symp., Conf. on Integrated Optical Circuits, Sept. 1992b. Also, Honkanen, S., Li, M. J., Wang, W. J., and Najafi, S. I. Ion-exchange process for advance glass waveguides, Proc. 1st Intl. Workshop on Photonic Networks, Components and Applications. Montebello, Oct. 1990, World Scientific, New Jersey, 1991.

Honkanen, S., Najafi, S. I., Wang, W. J., Lefebvre, P., and Li, M. J. 1992c. Single-mode glass channel waveguides by ion exchange with ionic masking. *Optics Comm.* 94, 54.

Honkanen, S., Poyhonen, P., Tervonen, A., and Najafi, S. I. 1993. Waveguide coupler for potassium- and silver-ion-exchanged waveguides in glass. *Appl. Opt.* 32 (12), 2109.

Houde–Walter, S. 1988. Recent progress in gradient index optics. *SPIE Proc.* 935, 2.

Intani, D., Baba, T., and Iga, K. 1992. Planar microlens relay optics utilizing laterla focusing. *Appl. Opt.* 31, 5255.

Izawa, T., and Nakagome, H. 1972. Optical waveguide formed by electrically induced migration of ions in glass plate. *Appl. Phys. Lett.* 21, 584.

Keck, D. B., Morrow, A. J., Nolan, D. A., and Thompson, D. A. 1989. Passive components in the subscriber loop. *J. Lightwave Technol.* LT-7, 1623.

Kenan, R. P. 1988. Gradient-index devices for integrated optics. *SPIE Proc.* 935, 105.

Kistler, S. S. 1962. Stresses in glass produced by nonuniform exchange of monovalent ions. *J. Am. Ceram. Soc.* 45, 59.

Lamouche, G., and Najafi, S. I. 1990. Scalar finite-element evaluation of cut-off wavelength in glass waveguides and comparison with experiment. *Can. J. Phys.* 68, 1251. Also, 1991. Accurate analysis of ordinary and grating-assisted ion-exchanged glass waveguides. *Opt. Eng.* 30 (9), 1365.

Lee, S. G., Kang, K. I., Guerreiro, P., Wright, E., Peyghambarian, N., Najafi, S. I., Zhang, G., Li, C. Y., Takada, T., and Mackenzie, J. Fabrication, optical characterization and femtosecond pulse propagation in a semiconductor quantum waveguide, submitted for publication.

Lefebvre, P., Honkanen, S., Najafi, S. I., and Tervonen, A. 1993. Nonsymmetrical potassium ion-exchanged Mach–Zehnder interferometers in glass. *Opt. Comm.* 96, 36.

Li, M. J., Honkanen, S., Wang, W. J., Leonelli, R., Albert, J., and Najafi, S. I. 1991. Potassium and silver ion-exchanged dual-core glass waveguides with gratings. *Appl. Phys. Lett.* 58 (23), 2607.

Li, M. J., and Najafi, S. I. 1993. Polarization dependence of grating-assisted waveguide Bragg reflectors. *Appl. Opt.* 32 (24), 4517.

McCourt, M. Status of glass and silicon-based technologies for passive components. Proc. European Conf. on Integrated Optics, Neuchâtel, April 1993, pp. 9-1 to 9-4.

Miliou, A. N., Cao, X. F., Srivastava, R., and Ramaswamy, R. V. 1993. 15 dB amplification at 1.06 μm in ion-exchanged silicate glass waveguides. *IEEE Photon. Tech. Lett.* 4 (4), 416.

Najafi, S. I. Ion-exchanged rare-earth-doped waveguides, Intl. Conf. in Optical Science and Eng.: Glasses for Optoelectronics, Paris 1989, paper #24.

Najafi, S. I. Rare-earth-doped waveguides for integrated optics. COST 217 Intl. Workshop on Active Fibers, Helsinki, June 1991.

Najafi, S. I. *Introduction to Glass Integrated Optics*, Artech House, Boston, 1992a.

Najafi, S. I., Lefebvre, P., Albert, J., Honkanen, S., Vahid-Shahidi, A., and Wang, W. J. 1992b. Ion-exchanged Mach–Zehnder interferometers in glass. *Appl. Opt.* 31 (18), 3381.

Najafi, S. I., and Honkanen, S., Ion-exchanged glass waveguides for active applications, Electrochem. Soc. Annu. Meet., Toronto, Oct. 1992c.

Najafi, S. I., Fallahi, M., Lefebvre, P., Wu, C., and Templeton, I. 1993. Integrated optical circular grating tap-power-divider. *Electron. Lett.* 29 (16), 1417.

Najafi, S. I. Circular gratings and applications in integrated optics. Conf. Nonlinear Optics for High-Speed Electronics, Los Angeles, Jan. 1994. *SPIE Proc.* #2145, 1994a.

Najafi, S. I. Recent progress in glass integrated optical circuits, Conf. on Integrated Optics and Microstructures, SPIE Annu. Meet., San Diego, July 1994b.

Najafi, S. I., and Fallahi, M. 1994c. Circular grating surface emitting lasers, Int. Symp. on Integrated Optics: Conf. on Nanofabrication Technologies and Device Integration, Lindau (Germany), April 1994.

Najafi, S. I., unpublished. Also, see Najafi, 1994b.

Ohtsuki, T., Peyghambarian, N., Najafi, S. I., François, V., Honkanen, S., and Orcel, G. Upconversion in ion-exchanged Er-doped phosphate glass waveguides, Intl. Conf. on Applications of Photonic Technology, Toronto, June 1994.

Ionex by Optonex, Espoo (Finland).

Peyghambarian, N., Koch, S., and Mysyrowicz, A. *Introduction to Semiconductor Optics*. Prentice Hall, Englewood Cliffs NJ, 1993.

Peyghambarian, N., Quantum dot glass integrated optical devices, SPIE Critical Review Conf. Integrated Optics and Optical Fiber Devices, Glass Integrated Optics and Optical Fiber Devices, Najafi, S. I., ed., 1994.

Ramaswamy, R. V., and Najafi, S. I. 1986. Planar, buried, ion-exchanged glass waveguides: Diffusion characteristics. *IEEE J. Quantum Electron.* QE-22 (6), 883.

Ramaswamy, R. V., and Srivastava, R. 1988. Ion-exchanged glass waveguides: A review. *J. Lightwave Technol.* 6, 984.

Ross, L. 1989. Integrated optical components in substrate glasses. *Glastech. Ber.* 62, 285.

Saruwatari, N., and Izawa, T. 1974. Nd-glass laser with three-dimensional optical waveguide. *Appl. Phys. Lett.* 24, 603.

Sanford, N. A., Malone, K. J., and Larson, D. R. 1990. Integrated-optic laser fabricated by field-assisted ion exchange in neodymium-doped soda-lime-silicate glass. *Opt. Lett.* 15 (7), 366.

Saarinen, J., Huttunen, J., Honkanen, S., Najafi, S. I., and Turunen, J. 1992. Computer-generated waveguide holograms by double ion exchange process in glass. *Electron. Lett.* 28 (9), 876.

Saarinen, J., Honkanen, S., Zhang, G., and Najafi, S. I. 1994a. Modified double ion exchange method for fabrication of phase gratings in glass waveguides, Integrated Optics Symp., Conf. Nanofabrication Technologies and Device Integration, Lindau, Germany, April 13–14, 1994a.

Saarinen, J., Honkanen, S., Najafi, S. I., and Huttunen, J. 1994b. Double ion exchange process in glass for the fabrication of computer-generated waveguide holograms, to appear in June issue of *Appl. Opt.*

Sakamoto, T. 1992. Coupling characteristic analysis of single-mode and multimode optical-fiber connectors using gradient-index-rod lenses. *Appl. Opt.* 31, 5184.

Seki, M., Hashizume, H., Sugawara, R. 1988. Two-step purely thermal ion exchange technique for single-mode waveguide devices in glass. *Electron. Lett.* 24, 1258.

Stern, J. R. Prospects for passive optical networks, Proc. 17th European Conf. Optical Communication ECOC'91 and 8th Int. Conf. on Integrated Optics and Optical Fiber Communication IOOC'91, Paris, Sept. 9–12, 1991, Publi tregor, Lannion, p. 134.

Tervonen, A., Pöyhönen, P., Honkanen, S., Tahkokorpi, M., and Tammela, S. 1991. Examination of two-step fabrication methods for single-mode fiber compatible ion-exchanged waveguides. *Appl. Opt.* 30, (3), 338.

Tervonen, A. Theoretical analysis of ion-exchanged glass waveguides. In *Introduction to glass integrated optics*, S. I. Najafi, ed., Artech House, Boston, 1992, Chapter 4.

Tervonen, A., Honkanen, S., and Najafi, S. I. 1993. Analysis of symmetric directional couplers and asymmetric Mach–Zehnder interferometers as 1.30- and 1.55-μm dual-wavelength demultiplexers/multiplexers. *Opt. Eng.* 32 (9), 2083.

Tervonen, A., Honkanen, S., and Najafi, S. I. New ion-exchanged narrow-band add-drop WDM-device: design considerations, Int. Conf. on Applications of Photonic Technology, Toronto, June 1994.

Tomlinson, W. J. 1988. Optical components for fiber systems. *SPIE Proc.* 935, 78.

Wang, W. J., Honkanen, S., Najafi, S. I., and Tervonen, A. 1992a. Four-port guided-wave nonsymmetric Mach–Zehnder interferometer. *Appl. Phys. Lett.* 61 (2), 150.

Wang, W. J., Honkanen, S., Najafi, S. I., and Tervonen, A. 1992b. New integrated optical ring resonator in glass. *Electron Lett.* 28 (21), 1967.

Wang, W. J., Honkanen, S., Najafi, S. I., and Tervonen, A. 1993. Loss characteristics of potassium and silver double ion-exchanged glass waveguides. *Appl. Phys.* 74 (3), 1529.

Zhang, G., Honkanen, S., Najafi, S. I., and Tervonen, A. 1993. Integrated 1.3 μm/1.55 μm wavelength multiplexer and 1/8 splitter by ion exchange in glass. *Electron. Lett.* 29 (12), 1064.

Zhang, G., Honkanen, S., Najafi, S. I., and Tervonen, A. 1994a. Glass integrated optical circuit for 1.3/1.55 μm optical communication circuit, Integrated Optics Symp., Conf. on Nanofabrication Technologies and Device Integration, Lindau, Germany, April 13–14.

Zhang, G., Honkanen, S., Tervonen, A., and Najafi, S. I. 1994b. Glass integrated optics circuit for 1.48/1.55 μm and 1.3/1.55 μm wavelength division multiplexing and 1/8 splitting, *Applied Optics*, 33, 3371.

Zhang, G., Honkanen, S., Tervonen, A., Najafi, S. I., and Katila, P. 1996. Ion-exchanged glass waveguide Mach–Zehnder interferometers for wavelength multi/demultiplexing and the effect of thermal post-annealing on spectral transmission, accepted for publication in *Pure and Applied Optics*.

13

Design Methodology For Guided-Wave Photonic Devices

G. L. Yip
McGill University, Montreal, Quebec, Canada

List of Symbols 530
13.1 Introduction 532
13.2 The Effective-Index Method 532
  Two-Dimensional, Step-Index Waveguides • Two-Dimensional, Graded-Index Waveguides
13.3 Experimental Characterization and Theoretical Modeling of Slab Waveguides 537
  Waveguides by Ion Exchange in Glass • Waveguides by Field-Assisted Ion Exchange in Glass • $LiNbO_3$ Waveguides by Ti Diffusion and Proton Exchange
13.4 Characterizing and Modeling Channel Guides 545
  Ion-Exchanged Channel Waveguides • Effect of a Dielectric Cladding • $LiNbO_3$ Waveguides by Ti Diffusion and Proton Exchange • Electro-Optical Index Change in Ti:$LiNbO_3$ Waveguides • Electrode Analysis and Design
13.5 The Beam Propagation Method 559
13.6 Illustrative Examples: Design and Fabrication of Some Guided-Wave Photonic Devices 561
  A Passive Three-Branch Power Divider by K^+-Ion Exchange in Glass • A Directional Coupler Power Divider by Two-Step K^+-Ion Exchange in Glass • A Widened X-Branch Wavelength Demultiplexer at 1.3 μm and 1.55 μm by K^+-Ion Exchange in Glass • An Electro-Optic Ti-Diffused $LiNbO_3$ Ridge Waveguide Mode-Confinement Modulator • Design and Fabrication of a Y-Branch TE-TM Mode Splitter in $LiNbO_3$ by Proton Exchange and Ti Diffusion
13.7 Conclusions 581

List of Symbols

x depth in the waveguide
N effective guide index
β_p propagation constant for the depth mode
k_o wave number in free space

κ	transverse propagation constant in the slab waveguide
γ_s	decay constant in the substrate
γ_c	decay constant in the cover
n_f	film index
n_c	cover index
n_b	substrate or bulk index
Φ_s	phase shift from the film-substrate interface
Φ_c	phase shift from the film-cover interface
b_E	normalized guide index for TE modes
a_E	asymmetric measure for TE modes
V	normalized frequency of an optical waveguide
m	mode order
b_m	normalized guide index for TM modes
a_M	asymmetric measure for TM modes
$\mathbf{E}$	electric field
$\mathbf{H}$	magnetic field
$n(x, y)$	local refractive index of an inhomogeneous medium
ω	angular frequency
Ψ	transverse field component
N_{eff}	same as N
β	propagation constant in the z direction
K^+	potassium ion
d	diffusion depth
D_e	effective diffusion coefficient
t	diffusion time
T	temperature
C_1, C_2	arbitrary constants associated with D_e
x	normalized waveguide depth
x_t	normalized turning point
F_e	effective field-dependent coefficient
K_o	a constant associated with field-assisted ion-exchange
μ_k	mobility of potassium ion in glass
E_a^t	total applied field
E_a	applied field
E_o	a constant field
n_s	surface index
Δn_s	change in the surface index over the substrate index
n_p	prism index
λ	wavelength
$c(x, t)$	ion concentration density
ρ	atomic density
Ti	titanium
Tl^+	thallium ion
He-Ne	helium-neon
τ	thickness of Ti film
D_x, D_y	diffusion coefficient along the x or y direction
D_o	a constant associated with D_x or D_y
ϵ_a	thermal activation energy
k_B	Boltzmann constant
σ_o	constant associated with a Gaussian function
σ_1	constant associated with the Hermite-Gaussian function

Na^+ sodium ion
w channel guide width
k_1, k_2 transverse propagation constants
γ_3, γ_4 transverse decay constants
$\epsilon_i, i = x, y, z$ relative permittivity along the i direction
$D_i, i = x, y, z$ electric displacement vector along the i direction
r_{ij} electro-optic or Pockels coefficients
n_e extraordinary index
n_o ordinary index
ϕ electric potential
ν Fourier transform variable

Other symbols used are as defined in the text.

13.1 Introduction

Since the early 1970s, there has been an explosion of research activities in realizing planar integrated waveguide devices fabricated by the technologies of ion-exchange in glass (Izawa and Nakagome, 1972; Giallorenzi et al., 1973), Ti diffusion into $LiNbO_3$ crystals (Schmidt and Kaminow, 1974) and proton exchange in $LiNbO_3$ substrates (Jackel et al., 1982). Early efforts involving diffusion processes were directed toward conceptualization and experimental realization of such devices, which have great signal processing capabilities in optical fiber communication and sensor applications. Some of these capabilities include power division, wavelength division multiplexing/demultiplexing, switching, modulation, polarization splitting, and so on. However, to obtain high performance in these devices, it is important, first, to develop accurate designs. For this purpose, detailed and accurate information on the characteristics of slab and channel guides in respective substrates must be known in relation to their fabrication conditions because slabs and channels often form the basic units in the more complicated structures of waveguide devices. Therefore, "design methodology" does not merely involve a mathematical or computational process. It is more important, first, to establish an accurate refractive-index model for the waveguide, fabricated from the results of experimental characterization, before mathematical and computer-aided design procedures can become effective. It goes without saying, however, that clever mathematics and efficient, accurate numerical algorithms would help provide accurate designs.

In this chapter, we will first explain in section 13.2 the popularly used effective-index method (EIM) for analyzing waveguide properties and then start with the characterizations of slab waveguides in section 13.3 and of channel waveguides in section 13.4. The characterization of channel guides is tackled via that of the slab guides. Section 13.5 presents the beam propagation method (BPM) as a very useful design tool. All of the materials presented in sections 13.2 through 13.5 are, then, used in the illustrative examples discussed in section 13.6, where both the design and measured device performance parameters are compared and discussed. Finally, we conclude our chapter in section 13.7. It is hoped that the methodology presented here will help researchers and engineers in university, industry, and government laboratories to achieve design-optimized guided-wave devices.

13.2 The Effective Index Method:

Two-Dimensional, Step-Index Waveguides

A simple technique, which can be used to analyze and provide good predictions for the propagation characteristics of a channel guide, is the well-known "Effective Index Method" (EIM) (Krox and Toulios, 1970). Assuming a uniform index in the direction of propagation along the z axis, the

Design Methodology For Guided-Wave Photonic Devices

method reduces a 3-D problem to a 2-D one in the transverse plane. In the step-index case, the solution of the 2-D problem starts by taking the equivalence of the rectangular waveguide to the problem of two slab waveguides, one in the x direction (the depth problem) and the other in the y direction (the lateral problem) as illustrated in Fig. 13.1.

Let us consider the asymmetric waveguide structure in Fig. 13.1b and define the effective guide index $N = \beta_d/k_0$ for the depth mode, where β_d is the related propagation constant and k_0 the wave number in free space. The effective index N should satisfy the following dispersion relationship (Tamir, 1988)

$$Kh = m\pi + \Phi_s + \Phi_c, \quad m = 0, 1, 2 \cdots \quad (13.1)$$

where m is the mode order and Φ_s and Φ_c are the phase shifts related to the total internal reflection of light from the film-substrate and film-cover interfaces, respectively. The propagation constant K and the decay constants γ_s and γ_c in the substrate and cover, respectively, along the x direction are given by

$$K = k_0\sqrt{n_f^2 - N^2},$$

$$\gamma_s = k_0\sqrt{N^2 - n_b^2},$$

and

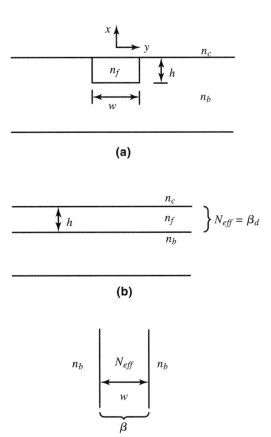

FIGURE 13.1 The Effective Index Method (EIM).

$$\gamma_c = k_0\sqrt{N^2 - n_c^2}. \qquad (13.2)$$

Usually, $n_f > N > n_c$.

For TE modes with field components E_y, H_x, and H_z, it is convenient to define a normalized guide index b_E, an asymmetric measure a_E, and a normalized frequency V as follows:

$$b_E = (N^2 - n_b^2)/(n_f^2 - n_b^2),$$
$$a_E = (n_b^2 - n_c^2)/(n_f^2 - n_b^2), \qquad (13.3)$$

and

$$V = k_0 h\sqrt{n_f^2 - n_b^2},$$

and the expression for the phase shifts Φ_s and Φ_c are

$$\tan \Phi_s = \gamma_s/K, \qquad (13.4)$$

and

$$\tan \Phi_c = \gamma_c/K$$

Using Eqs. (13.2)–(13.4), the dispersion relationship, Eq. (13.1) can be written in the normalized form

$$V\sqrt{1 - b_E} = m\pi + \tan^{-1}\sqrt{b_E/(1 - b_E)} + \tan^{-1}\sqrt{(b_E + a_E)/(1 - b_E)}. \qquad (13.5)$$

Given the parameters V and a_E, Eq. (13.5) can be solved via a root-search technique to yield b_E. For the TM modes with field components E_x, H_y, and E_z, it is necessary to define separately the normalized guide index b_M and the asymmetry measure a_M as (Nishihava et al., 1989)

$$b_M = [(N^2 - n_b^2)/(n_f^2 - n_b^2)][n_f^2/(n_b^2 q_s)], \qquad (13.6)$$

where

$$q_s = \frac{N^2}{n_f^2} + \frac{N^2}{n_b^2} - 1$$

and

$$a_M = \left(\frac{n_f}{n_c}\right)^4 \frac{(n_b^2 - n_c^2)}{(n_f^2 - n_b^2)} \qquad (13.7)$$

Using Eqs. (13.6)–(13.7), the corresponding dispersion relationship for the TM modes in the normalized form is

$$V\left[\sqrt{q_s}\frac{n_f}{n_b}\right]\sqrt{1 - b_M} = m\pi + \tan^{-1}\sqrt{\frac{b_M}{1 - b_M}} + \tan^{-1}\sqrt{\frac{b_M + a_M(1 - b_M d_f)}{1 - b_M}},$$

$$m = 0, 1, 2, \qquad (13.8)$$

where

$$d_f = \left(1 - \frac{n_b^2}{n_f^2}\right)\left(1 - \frac{n_c^2}{n_f^2}\right)$$

Now, let us consider the lateral problem as depicted in Fig. 13.1. If we have examined the TE mode (E_y, H_x, and H_z) in the above depth problem, the guided mode of interest in the lateral problem should be regarded as the TM mode because E_y is polarized along the y direction and, hence, perpendicular to the waveguide boundaries. The EIM method uses the propagation characteristics of the equivalent slab guide in Fig. 13.1c to yield those of the original channel guide in the y direction. This can be achieved by simply substituting the effective index $N_{eff} = N$ and the film and substrate indices, n_f and n_b in Eq. 13.8, the dispersion equation for the TM mode.

Two-Dimensional, Graded-Index Waveguides

The method outlined above can be extended to analyze 3-D, graded-index channel guides, whose index variations are confined to the transverse directions x and y. The index in the direction of the propagation z is taken as uniform.

The procedure of analysis is as indicated in Figs. 13.2a–13.2e. Because the calculation of the effective index distribution and the propagation constant of the channel guide necessitates the solutions of some wave equations, the pertinent wave equations are given below. Applying Maxwell's equations to an inhomogeneous medium with a refractive index given by $n(x, y)$, we can get the following vector wave equations for the electric field **E** and magnetic field **H**:

$$\nabla^2 \mathbf{E} + \left[\frac{\mathbf{E} \cdot \nabla n^2}{n^2}\right] + k_0^2 n^2 \mathbf{E} = 0 \qquad (13.9)$$

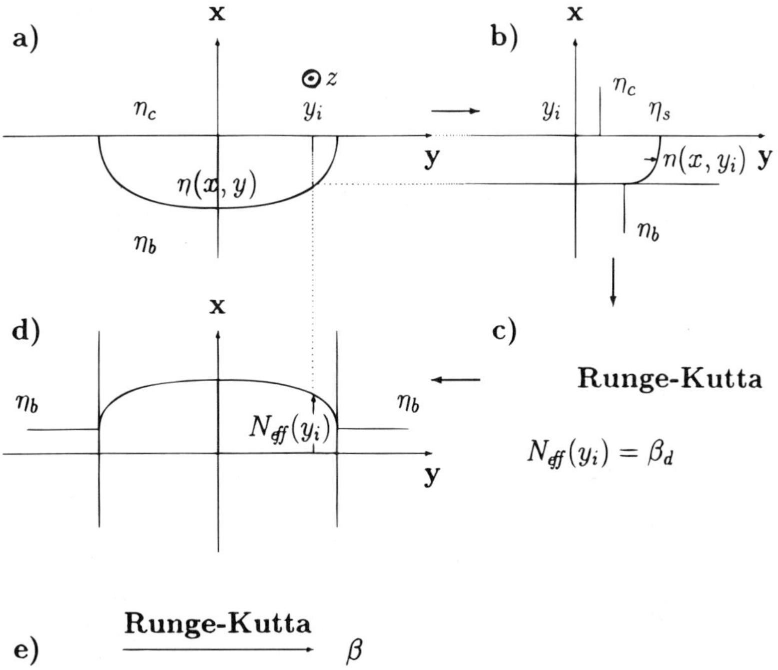

FIGURE 13.2

and

$$\nabla^2 \mathbf{H} + \left[\frac{\nabla n^2}{n^2} \times (\nabla \times \mathbf{H})\right] + k_0^2 n^2 \mathbf{H} = 0. \tag{13.10}$$

With a uniform index in the z direction, vector wave equations just for the transverse field components can be separated from Eqs. (13.9) and (13.10) as

$$\nabla_t^2 \mathbf{E}_t + \nabla_t \left\{ \mathbf{E}_t \cdot \left[\frac{\nabla_t n^2(x, y)}{n^2(x, y)}\right] \right\} + [k_0^2 n^2(x, y) - \beta^2] \mathbf{E}_t = 0 \tag{13.11}$$

and

$$\nabla_t^2 \mathbf{H}_t + \frac{\nabla_t n^2(x, y)}{n^2(x, y)} \times (\nabla_t \times \mathbf{H}_t) + [k_0^2 n^2(x, y) - \beta^2] \mathbf{H}_t = 0, \tag{13.12}$$

where it is assumed that the traveling wave term is of the form $\exp j(\omega t - \beta z)$ and

$$\nabla_t^2 = \frac{\partial^2}{\partial x^2} + \frac{\partial^2}{\partial y^2}.$$

For diffused waveguides, whose index variations in the transverse directions (x, y) are slow, the term $\nabla_t n^2(x, y)$ in Eqs. (13.11) and (13.12) can sometimes be neglected. Under this so-called scalar wave approximation, scalar wave equations for the transverse field components are

$$\nabla_t^2 \Psi + [k_0^2 n^2(x, y) - \beta^2] \Psi = 0, \tag{13.13}$$

where $\Psi = E_y(x, y)$ for the TE modes and $\Psi = H_y(x, y)$ for the TM modes. Using the separation of variables, we look for the solutions of the form $\Psi(x, y) = F(x, y) \cdot G(y)$. Substituting this in Equation (13.13) and assuming that the variations of $F(x, y)$ with y can be neglected (which has been numerically verified with practical problems), $F(x, y)$ and $G(y)$ can be shown to satisfy the following equations respectively:

$$\frac{\partial^2 F}{\partial x^2} + k_0^2 [n^2(x, y) - N_{\text{eff}}^2(y)] F = 0 \tag{13.14}$$

for the depth problem, and

$$\frac{\partial^2 G}{\partial y^2} + [k_0^2 N_{\text{eff}}^2(y) - \beta^2] G = 0 \tag{13.15}$$

for the lateral problem.

Starting from the channel guide with a given index distribution $n(x, y)$ as shown in Fig. 13.2a, we fix a specific position at y_i along the y-direction and calculate the local effective index $N_{\text{eff}}(x, y_i)(=\beta_d)$ for the depth mode by solving equation (13.14). This is equivalent to solving the wave equation for a one-dimensional slab guide in the y direction. If we carry out this process for a sufficient number of positions y_i's, we will be able to construct a lateral effective index profile $N_{\text{eff}}(x, y)$ as shown in Figure (13.2d) for the lateral problem in order to get β, the propagation constant for the original channel guide. This second step is again equivalent to solving the wave equation for a slab guide in the x direction.

Design Methodology For Guided-Wave Photonic Devices

For inhomogeneous slab guides occurring in practical situations, exact analytical field solutions are rare. We have to resort to numerical methods. Among an almost innumerable number of numerical methods, our laboratory has had experience with two very accurate and reliable techniques, namely the Raleigh–Ritz variational procedure and the Runge–Kutta method. We will encounter them in Section 13.4.

13.3 Experimental Characterization and Theoretical Modeling of Slab Waveguides

Waveguides by Ion Exchange in Glass

Due to space limitations, the characterization of slab waveguides by K^+-ion exchange in soda-lime glass substrates is used for illustration. For $AgNO_3$ melt, Stewart et al. (1977) found a linear relationship between d and $t^{1/2}$ given by

$$d = \sqrt{D_e t} \tag{13.16}$$

where d is the diffusion depth, t the diffusion time in seconds and D_e has been defined as the effective diffusion constant. Further, they also found that the dependence of D_e on the inverse temperature $1/T$ was given by the exponential relationship

$$D_e = C_1 \exp(-C_2/T) \; m^2 sec^{-1}. \tag{13.17}$$

Hence, from Eqs. (13.16) and (13.17)

$$d = (60C_1)^{1/2} \times 10^6 t^{1/2} \exp(-C_2/2T) \; \mu m \tag{13.18}$$

where t is in minutes, T in degrees Kelvin, and C_1 and C_2 are constants. It was intuitively thought (Yip and Albert, 1985) that, for ion-exchanged waveguides employing soda-lime glass substrates and a potassium nitrate (KNO_3) melt, relationships similar to Eqs. (13.16)–(13.18) should hold. Our task was, therefore, to determine C_1 and C_2 experimentally for KNO_3. The importance of establishing such relationships for KNO_3 is that, given (T, t) of such a waveguide, one can establish the waveguide's important parameters quickly without the need for further measurement.

To fabricate the diffused planar waveguides, the soda-lime glass microscope slides were immersed in molten KNO_3 at different temperatures and for different durations. The melt temperatures ranged from 360 to 440°C, and the diffusion times ranged from 30 minutes to 24 hours. The effective indices of the guided modes in each sample were obtained by measuring the synchronous angles with a prism coupler (flint glass, $n_p = 1.785$) (Tien and Ulrich, 1970). An accuracy of $\pm 1'$ was obtained in the measured synchronous angle θ_j, corresponding to an accuracy of $\pm 1 \times 10^{-4}$ in the measured effective index n_e. The measurements were initially performed with a He-Ne laser at 0.63 μm.

Assuming a gaussian index distribution in the diffused waveguide

$$n(x) = n_b + \Delta n_s \exp(-x^2), \tag{13.19}$$

where $\Delta n_s = n_s - n_b$, $x = x/d$, n_b is the substrate index n_s is the surface index and d the effective guide depth such that $n(d) = n_b + \Delta n_s/e$, the well-known WKB dispersion relationship for guided modes [Equation (2.1) of Stewart et al., 1977] was used. This relationship can be expressed as

$$d = \frac{(m + 3/4)\pi}{k_0 \int_0^{x_t} [n^2(x) - N_e^2]^{1/2} \, dx}, \quad m = 0, 1, 2 \cdots \quad (13.20)$$

where x_t is the normalized turning point and N_e is the effective mode index. The phase shift on reflection at the glass–air interface has been taken to be $-\pi$, here, for both TE and TM modes. Equation (13.20) contains two unknowns, n_s and d. Hence, given any pair of measured mode indices for a waveguide, one can eliminate d by using Equation (13.20) and determine n_e by a root-search technique in the resulting equation. Then, n_s can be substituted in Equation (13.20) to get d. This procedure can be repeatedly applied to all the possible pairs of the mode indices for a specific waveguide sample, and the average values of its n_s and d can be determined.

Our measurement data has confirmed the previous observation that the surface index value n_s is affected only by the diffusion temperature and is independent of the diffusion time (Chartier et al., 1978). In Figure 13.3, the theoretical dispersion curves computed from Equation (13.20) for the TE modes in the samples prepared at 400°C are presented together with experimentally measured data. The agreement seems good. The difference between the measured mode index and the theoretical index was calculated for each of the 167 measured mode indices, and the average of this difference was found to be $|n_e \text{ (meas.)} - n_e \text{ (theo.)}|_\text{av.} = (1.6 \pm 1.2) \times 10^{-4}$ with the largest single deviation being 6×10^{-4}. Similar, though not identical, results for the TM modes are not presented here for brevity.

The measured data of d versus $t^{1/2}$ are presented in Fig. 13.4 for the TE modes also showing a linear relationship between d and $t^{1/2}$ for KNO_3 as for $AgNO_3$ (Stewart et al., 1977). Table 13.1 summarizes the results of our measurements and values of D_e obtained from Eq. 13.16 and Fig. 13.4. The results in Table 13.1 were, then, used to plot $\log_e D_e$ versus $(1/T)$ in Fig. 13.5,, where, with the help of a linear-regression method, $C_1^{TE} = 7.82 \times 10^{-6}$ m²/sec, $C_2^{TE} = 1.489 \times 10^4$ K and $C_1^{TM} = 1.604 \times 10^{-5}$ m²/sec, $C_2^{TM} = 1.54 \times 10^4$ K respectively, were determined. To characterize a slab waveguide, with (T, t) given, its effective guide depth can be obtained by using Eq. (13.18). The value of n_s can be obtained by referring to Table 13.1 using interpolation, if necessary. Anisotropy in ion-exchanged wave guides, as suggested by the differences in Δn_s^{TE} and Δn_s^{TM} was also observed previously (Izawa and Nakagome, 1972).

Because the practical significance of the ion-exchanged waveguide devices now becomes more and more evident in single-mode wideband fiber communication systems operating at 1.31 and 1.55 μm, the diffused slab waveguides made by the K^+-ion exchange were also characterized

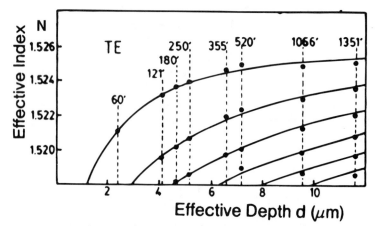

FIGURE 13.3 Theoretical dispersion curves. SPIE (*Optics Letters*). (*Source*: Yip, G. L. and Albert, J. 1985. Characterization of planar optical waveguides by K^+-ion exchange in glass. *Opt. Lett.* 10:151–153. With permission.)

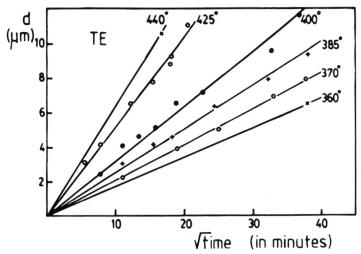

FIGURE 13.4 Effective guide depth. *SPIE (Optics Letters)*. (*Source*: Yip, G. L. and Albert, J. 1985. Characterization of planar optical waveguides by K⁺-ion exchange in glass. *Opt. Lett.* 10:151–153. With permission.)

TABLE 13.1 Measured Surface Index Change

Temp. (°C)	Δn_s^{TE} ($\times 10^{-3}$)	Δn_s^{TM} ($\times 10^{-3}$)	D_e^{TE} (m²/sec) ($\times 10^{-16}$)	D_e^{TM} (m²/sec) ($\times 10^{-16}$)
360	9.2 ± 0.3	11.3 ± 0.3	4.91	4.53
370	9.1 ± 0.5	10.7 ± 0.4	7.18	6.50
385	8.8 ± 0.2	10.8 ± 0.4	10.82	10.61
400	8.7 ± 0.3	10.5 ± 0.5	17.14	18.02
425	8.7 ± 0.3	10.4 ± 0.4	45.21	42.16
440	8.4 ± 0.5	10.0 ± 0.6	66.78	68.81

[a] All of the readings must be multiplied by the factor indicated to yield the actual values.

SPIE (Optics Letters). (*Source*: Yip, G. L. and Albert, J. 1985. Characterization of planar optical waveguides by K⁺-ion exchange in glass. *Opt. Lett.* 10:151–153. With permission.

experimentally at these waveguides by the same prism-coupler technique. Figure 13.6 shows the measurement setup, where a He-Ne laser beam at 0.63 μm was used to help with alignment because the He-Ne laser beam is invisible at 1.152 and 1.523 μm.

The refractive indices of the prism coupler (flint glass), at 1.152 and 1.523 μm, were found by interpolating, from data in Optics Guide 4 (1988), to be n_p = 1.7523 and 1.74485, respectively. The corresponding indices for the substrate glass n_b = 1.5030 ± 1 × 10^{-4} and 1.4984 ± 1 × 10^{-4}, respectively, were determined by employing the multilayer Brewster angle measurement method (Xiang and Yip, 1992). Because a He-Ne laser at 1.31 μm is unavailable, the data at this wavelength cannot be obtained experimentally but can be calculated by interpolation of the data at 1.152 and 1.523 μm.

Some of our measured results are summarized in Tables 13.2 and 13.3, and results for other diffusion times show that the gaussian profile yields the best fit to the measured mode indices, whereas the exponential profile fits worst.

Characterization of ion-exchanged waveguides with diluted KNO_3 was also reported by Kishi-oka (1991).

Waveguides by Field-Assisted Ion Exchange in Glass

Field-assisted ion exchange has the advantage of being a much faster process capable of producing various index profiles (Chartier et al., 1978), including the step-index profile (Izawa and Nagakome,

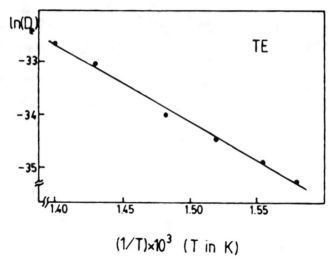

FIGURE 13.5 The relation between $\log_e$. *SPIE (Optics Letters)*. (*Source*: Yip, G. L. and Albert, J. 1985. Characterization of planar optical waveguides by K$^+$-ion exchange in glass. *Opt. Lett.* 10:151–153. With permission.)

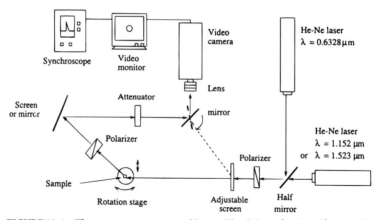

FIGURE 13.6 The measurement set-up. (*Source*: Yip, G. L. et al., 1991. Characterization of planar optical waveguides by K$^+$-ion exchange in glass at 1.152 and 1.523 μm. *SPIE Proc. Integrated Optical Circuits*, K. K. Wong, ed. Vol. 1583, 3–4, Sept., pp. 14–18, Boston, MA. With permission.)

TABLE 13.2 Measured Surface Index Change

λ (μm)	Δn_s^{TE} ($\times 10^{-3}$)	Δn_s^{TM} ($\times 10^{-3}$)	D_e^{TE} (m²/sec) ($\times 10^{-16}$)	D_e^{TM} (m²/sec) ($\times 10^{-16}$)
1.152	8.56 ± 0.11	10.05 ± 0.36	10.33	10.04
1.523	8.57 ± 0.26	10.34 ± 0.24	11.53	11.27

(*Source*: Yip, G. L. et al., 1991. Characterization of planar optical waveguides by K$^+$-ion exchange in glass at 1.152 and 1.523 μm. *SPIE Proc. Integrated Optical Circuits*, K. K. Wong, ed. Vol. 1583, 3–4, Sept., pp. 14–18, Boston, MA. With permission.)

TABLE 13.3 Comparisons of Measured TM-Mode Indices

m	N (Meas.)	N (Theor.)			Error $(\times 10^{-4})^d$		
		Gaussian[a]	exp.[b]	erfc[c]	Δ^a	Δ^b	Δ^c
0	1.510577	1.510490	1.512143	1.511102	0.87	15.66	5.25
1	1.507414	1.50746	1.507863	1.507611	0.43	4.49	1.97
2	1.505037	1.505152	1.505395	1.505230	0.15	3.58	1.93
3	1.503524	1.503417	1.503949	1.503675	1.07	4.25	1.51

[a] $n_s = 1.5116$, $d = 11.4$ μm.
[b] $n_s = 1.5198$, $d = 7.27$ μm.
[c] $n_s = 1.5161$, $d = 14.85$ μm.
[d] $\Delta = |N \text{ (Meas.)} - N \text{ (Theor.)}|$.

(*Source*: Yip, G. L. et al., 1991. Characterization of planar optical waveguides by K$^+$-ion exchange in glass at 1.152 and 1.523 [mu]m. *SPIE Proc. Integrated Optical Circuits*, K. K. Wong, ed. Vol. 1583, 3–4, Sept., pp. 14–18, Boston, MA. With permission.)

1972; Abou-el-Liel and Leonberger, 1988). It is also a simple and attractive technique in producing deep buried waveguides through a two-step process (Chartier et al., 1980), important for index-profile tailoring to achieve efficient coupling with optical fibers. Fig 13.7 gives the schematic of a typical experimental setup (Yip et al., 1990).

For Ag$^+$ ions, it was found (Ramaswamy and Najafi, 1986) that a linear combination of the diffusive and electromigrative terms was more appropriate for expressing the guide depth. For K$^+$ ions, a similar approach can be adopted. Hence,

$$d = \sqrt{D_e t} + F_e t + K_0 \tag{13.21}$$

where F_e is defined as the effective field-dependent coefficient and K_0 is a constant. The electromigrative coefficient F_e is linearly proportional to the total applied field E_a^t so that

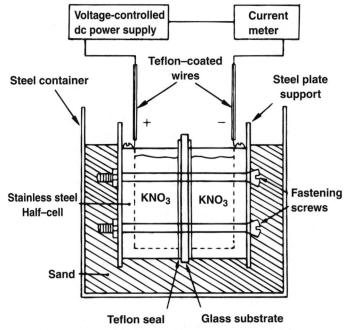

FIGURE 13.7 Schematic of the experimental apparatus. (*Source*: Yip, G. L., Noutsios, P. C., and Kishioka, K. 1990. Characteristics of optical waveguides made by electric-field-assisted K$^+$-ion exchange. *Opt. Lett.* 15:789–791. With permission.)

$$F_e = \mu_k E_a^t \tag{13.22}$$

where μ_k is a constant defined as the K$^+$ ion mobility in glass. Relationships in Eqns. (13.21) and (13.22) again yield the guide depth, given the fabrication conditions (E_a^t, t).

For fabrication details of planar waveguides by field-assisted ion exchange in our scheme, the readers are referred to (Yip et al., 1990). To avoid cracking the glass substrates immediately after the ion-exchange process, the samples were kept in the furnace for an additional ten minutes before being exposed for cooling in air. This extra time contributes to the K_0 value in Eq. 13.21. Another practical point to note is the observed existence, by several research groups, of a battery effect manifested as a potential drop across the electrodes even after the removal of the applied voltage. This potential drop (1.3V in our case) can be represented by a constant field E_0 between the two electrodes so that $E_a^t = E_a - E_0$, where E_a refers to the applied field.

Like the case of purely thermal diffusion, the surface index n_s, here, again depends only on the temperature and is independent of the diffusion time. The refractive index (concentration) profile of constant, applied-current (Abou-el-Liel and Leonberger, 1988) or electric-field-assisted (Miliou et al., 1989) K$^+$-ion exchanged planar waveguides has been found to be step like, and, hence, theoretical dispersion curves can be plotted with this step model. Although this step profile may be suitable for the case of either high current or electric field, it may not yield accurate results in our experiments using lower fields. For the more general case, we have proposed a modified Fermi (MF) index profile (Chen and Yang, 1985), which was previously used to model the Tl$^+$-ion-exchanged glass waveguides, namely,

$$n(x) = n_b + \Delta n_s \left\{ 1 - \exp\left(-\frac{d}{a}\right) + \exp\left(\frac{x-d}{a}\right) \right\}^{-1} \tag{13.23}$$

where a is a fitting parameter associated with the profile shape. The effective guide depth is defined as $n(d) = n_b + \Delta n_s[2 - \exp(-d/a)]$. The value of d/a in our experiments varies from 4 to 23 so that the effective guide depth occurs at almost the half-point of the profile.

With a set of fabrication conditions (T, E_a^t, t) and using Eqs. (13.20) and (13.23), given any pair of measured mode indices of a waveguide sample, one can determine the set of parameters (Δn_s, d, a) by a root-search technique similar to the way just discussed in the previous section. This procedure is normally repeated for all the possible pairs of the mode indices for a specific waveguide sample, and a is chosen so as to yield the minimum deviations of Δn_s and d from their average values.

In Fig. 13.8, the TE mode dispersion curves, N_e vs. d, were plotted, using a step-index profile, in dashed curves with the measured mode indices for a sample prepared at $T = 385°C$ and $E_a = 21.1$ V/mm. In comparison, the solid dispersion curves, using the MF profile and the WKB dispersion relationship yield better agreement with the measured data. In Fig. 13.9, the plots of d versus t clearly established a linear relationship between d and t for all of the applied fields. By using a least-squares fit, values of F_e and K_0 can be obtained for each E_a. In Fig. 13.10, values of F_e are plotted against E_a^t, where the use of a linear regression yields the constant μ_k(TE) = 21.33 μm^2/V min and μ_k(TM) = 20.52 μm^2/V min. The relevant coefficients and index changes determined at different applied fields are summarized in Tables 13.4 and 13.5. These results are believed to cover most of the practical conditions under which passive devices can be fabricated at $T = 385°C$. In Table 13.4, each value of a entered is the average over all the values of a obtained through optimization for all the samples prepared under each specified field.

LiNbO$_3$ Waveguides by Ti Diffusion and Proton Exchange

Similarly, although more complex in detail, LiNbO$_3$ waveguides by Ti diffusion and proton exchange (P.E.) can also be characterized. For brevity, only Ti-diffused waveguides are discussed here.

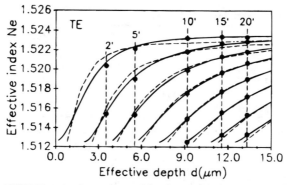

FIGURE 13.8 Comparison of the theoretical TE. (*Source*: Yip, G. L., Noutsios, P. C., and Kishioka, K. 1990. Characteristics of optical waveguides made by electric-field-assisted K^+-ion exchange. *Opt. Lett.* 15:789–791. With permission.)

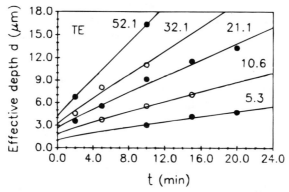

FIGURE 13.9 Effective guide depths versus. (*Source*: Yip, G. L., Noutsios, P. C., and Kishioka, K. 1990. Characteristics of optical waveguides made by electric-field-assisted K^+-ion exchange. *Opt. Lett.* 15:789–791. With permission.)

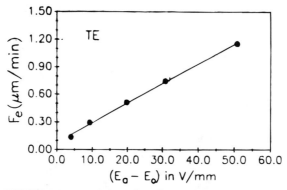

FIGURE 13.10 Variation of coefficient F. (*Source*: Yip, G. L., Noutsios, P. C., and Kishioka, K. 1990. Characteristics of optical waveguides made by electric-field-assisted K^+-ion exchange. *Opt. Lett.* 15:789–791. With permission.)

TABLE 13.4 Measured Surface Index at 385 °C

E_a (V/mm)	Δn_s(TE) ($\times 10^{-3}$)	Δn_s(TM) ($\times 10^{-3}$)	a(TE) (μm)	a(TM) (μm)
5.3	10.8 ± 0.0	13.7 ± 0.3	0.53	0.64
10.6	11.2 ± 0.2	13.3 ± 0.1	0.69	0.52
21.1	11.1 ± 0.1	13.4 ± 0.2	0.65	0.57
32.1	11.2 ± 0.1	13.6 ± 0.2	0.61	0.60
52.1	11.1 ± 0.0	13.5 ± 0.2	0.78	0.79

(*Source*: Yip, G. L., Noutsios, P. C., and Kishioka, K. 1990. Characteristics of optical waveguides made by electric-field-assisted K$^+$-ion exchange. *Opt. Lett.* 15:789–791. With permission.)

TABLE 13.5 Diffusion Coefficients and Depth Parameters for Various Applied Fields at 385 °C

E_a (V/mm)	F_e(TE) (μm/min)	F_e(TM) (μm/min)	K_0(TE) (μm)	K_0(TM) (μm)
5.3	0.14	0.14	0.96	0.81
10.6	0.29	0.29	1.74	1.76
21.1	0.51	0.50	2.56	2.72
32.1	0.74	0.76	3.09	3.01
52.1	1.15	1.11	4.08	4.30

(*Source*: Yip, G. L., Noutsios, P. C., and Kishioka, K. 1990. Characteristics of optical waveguides made by electric-field-assisted K$^+$-ion exchange. *Opt. Lett.* 15:789–791. With permission.)

For lengthy diffusion times compared to the time required for the titanium film to completely enter the crystal, an appropriate analytical concentration profile for the diffused Ti ions is given by Schmidt and Kaminow (1974) and Burns et al. (1979):

$$c(x, t) = \left(\frac{2}{\sqrt{\pi}}\right)\left(\frac{\alpha \tau}{d_x}\right)\exp\left(-\frac{x^2}{d_x^2}\right), \quad (13.24)$$

where α is the atomic density of the Ti film and τ is its thickness. d_x is the diffusion depth, given by

$$d_x = 2\sqrt{D_x t} \quad (13.25)$$

and D_x, the diffusion coefficient, is given by

$$D_x = D_0 \exp\left(-\frac{\epsilon_a}{k_B T}\right) \quad (13.26)$$

where D_0 is a constant, ϵ_a is the thermal activation energy, k_B is the Boltzmann constant, and T is the temperature.

As before, if the index change is assumed to be proportional to the concentration profile,

$$\Delta n(x, t) = Kc(x, t) = \Delta n_s \exp\left\{-\left(\frac{x}{d}\right)^2\right\}. \quad (13.27)$$

Hence,

$$\Delta n_s = \left\{\frac{2}{\sqrt{\pi}} \alpha \frac{dn}{dc}\right\} \frac{\tau}{d_x} \quad (13.28)$$

Design Methodology For Guided-Wave Photonic Devices

where the surface index change $\Delta n_s = n_s - n_b$, n_s is the surface index, n_b the bulk index, and $dn_e/dc = 1.6 \times 10^{-23}$ cm^3 for Ti.

The functional dependence of $c(x, t)$, d_x, D, and $\Delta n(x, t)$ on the various parameters as expressed in Eqs. (13.24)–(13.27) have been verified experimentally by several research groups employing different measurement techniques, such as mode index measurements (Naitoh et al., 1977), the x-ray microanalyzer method (XMA) (Minakata et al., 1978), and the secondary-ion mass spectrometry (SIMS) techniques (Burns et al., 1979). Our interest will be focused on the experimental data obtained via the prism-coupler method as discussed in the previous section. Naitoh et al. (1977) have shown that the experimental linear relationships between the various parameters as plotted in Figs. 13.11–13.13 qualitatively agree with their relationships expressed in Eqs. (13.24)–(13.27) for the TE modes propagating in a y-cut Ti:LiNbO$_3$ waveguide. Ctyroky et al., 1984 observed, through their measurements of mode indices in y-cut and z-cut LiNbO$_3$ substrates that no significant differences have been found between these two kinds of substrates.

13.4 Characterizing and Modeling Channel Guides

Ion-Exchanged Channel Waveguides

For a channel guide the index distribution in the depth direction is still modeled by a gaussian distribution. In the lateral direction, a simple way is to use the step-index model based on the EIM, where the channel width is defined by the photomask aperture for the channel guide. This model offers reasonable accuracy as illustrated in the problem of a three-branch power divider to be presented in Section 6.1.

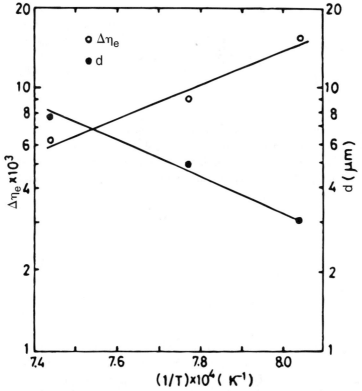

FIGURE 13.11 Variations of the index change. (*Source*: Naitoh, H., Nunoshita, M., and Nakayama, T. 1977. Mode control of Ti-diffused LiNbO$_3$ slab optical waveguides. *Appl. Pot.* 16(9)L2546–2549. With permission.)

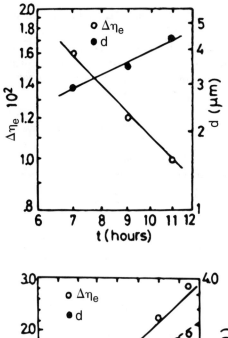

FIGURE 13.12 Variations of the index. (*Source*: Naitoh, H., Nunoshita, M., and Nakayama, T. 1977. Mode control of Ti-diffused LiNbO$_3$ slab optical waveguides. *Appl. Pot.* 16(9)L2546–2549. With permission.)

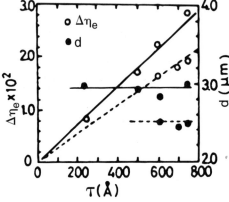

FIGURE 13.13 Variations of the index. (*Source*: Naitoh, H., Nunoshita, M., and Nakayama, T. 1977. Mode control of Ti-diffused LiNbO$_3$ slab optical waveguides. *Appl. Pot.* 16(9)L2546–2549. With permission.)

A more accurate model can be constructed as outlined in Section 2.2. To do this, the 2-D refractive index distribution $n(x, y)$ must first be established. The index $n(x, y)$ is assumed to be proportional to the normalized concentration $C(x, y)$ of incoming ions diffused into the glass as indicated in Eq. (13.28), and C satisfies the diffusion equation (13.29) (Crank, 1975):

$$n(x, y) = n_b + \Delta n_s C(x, y), \qquad (13.28)$$

$$\frac{\partial c}{\partial t} = \frac{\partial}{\partial x}\left[\frac{D_1}{1 - \alpha c}\frac{\partial C}{\partial x}\right] + \frac{\partial}{\partial y}\left[\frac{D_1}{1 - \alpha c}\frac{\partial c}{\partial y}\right], \qquad (13.29)$$

with $C = C_1/C_0$ and $\alpha = 1 - D_1/D_0$, where C_1 and D_1 are the concentration and self-diffusion coefficients of the incoming ions and C_0 and D_0 are those of the outgoing ions in the glass prior to the exchange. The values of D_1 for K$^+$-ions at 0.63 μm, for example, can be obtained from our previous characterizations of the planar K$^+$-ion exchange in soda-lime glass (Yip and Albert, 1985) and at 1.33 and 1.55 μm from Yip et al. (1991). α has been determined to be 0.998 (Doremus, 1969).

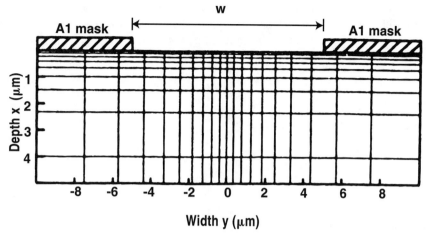

FIGURE 13.14 A nonuniform computation. (*Source*: Albert, J. and Yip, G. L. 1988. Wide single-mode channels and directional coupler by two-step ion-exchange in glass. *IEEE J. Lightwave Technol.* 6:552–563. With permission.)

To increase accuracy and save computer time, a nonuniform grid, finer in the region near (0,0) and coarser away from the origin should be used as shown in Fig. 13.14.

As an illustrative example, an explicit three-level, finite-difference scheme was used (Gerald, 1970) to solve Eq. 13.29 for a two-step ion-exchange process (Yip and Finak, 1984) (developed in our laboratory to produce a wide, weakly guiding, single-mode channel 10-μm wide to be compatible for butt-coupling with a single-mode fiber). This process consists of an initial exchange (time t_1) through a finite opening in an aluminum mask, to define a guiding channel, and a second exchange (time t_2), after the removal of the Al mask, over the whole plane of the substrate to adjust the lateral waveguiding properties of the channels by modifying the effective index of the surrounding areas. Figure 13.15 shows typical concentration contours for $T = 385°C$, $t_1 = 17$ min., and $t_1 + t_2 = 1$ hr.

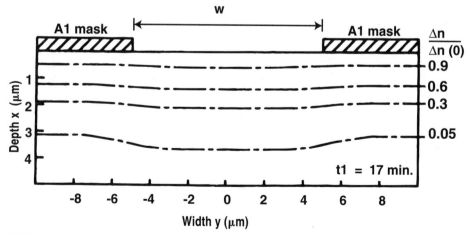

FIGURE 13.15 An example of concentration. (*Source*: Albert, J. and Yip, G. L. 1988. Wide single-mode channels and directional coupler by two-step ion-exchange in glass. *IEEE J. Lightwave Technol.* 6:552–563. With permission.)

The method used to solve Eq. (13.14) for this example, was the Raleigh–Ritz variational procedure with Hermite–Gaussian basis functions $\phi_i(x)$ (Taylor, 1976; Albert and Yip, 1988). The eigenvalues $N_{eff}(y)$ of Eq. (13.14) can be obtained by finding the field $\Psi(x)$, where

$$\Psi(x) = \sum_{i=1}^{M} C_i \phi_i(x) \qquad (13.30)$$

which minimizes the functional,

$$N_{eff}^2(y) = \max_{\psi} \frac{\int_{-\infty}^{\infty} dx [n^2(x,y)\Psi^2 (\nabla \Psi)^2 / k_0^2]}{\int_{-\infty}^{\infty} dx \Psi^2}. \qquad (13.31)$$

It was found that $N_{eff}(y)$ is correct to one part in 10^5 for $M = 21$. Figure 13.16 shows some sample plots for $N_{eff}(y)$ for $t_1 = 9$ min. and 17 min, $t_1 + t_2$ being kept at 1 hr.

The sets of values of $N_{eff}(y)$ numerically obtained for each sample with a fixed t_1 can be fitted, via a least-squares method, with the following two error functions (erf) to provide a good model for the lateral effective index distribution:

$$N_{eff}(y) = N(\infty) + \frac{[N(0) - N(\infty)]}{2}$$

$$\times \left[\mathrm{erf}\frac{(y + w/2)}{H} - \mathrm{erf}\frac{(y - w/2)}{H} \right] \qquad (13.32)$$

where H is a fitting parameter $N(0) = N_{eff}(y = 0)$, $N(\infty) = N_{eff}(y = \infty)$ and w is the aperture width in the photomask. Here, H can be approximated by $H = (D_y t_2)^{1/2}$ with $D_y = D_e$. The good agreement between the fitting function given by Eq. (13.32) and the numerical results can be readily seen in Fig. 13.16. *The channel guide formed by a single ion-exchange process can be treated as a special case of the two-step process, where t_2, becomes zero.* Thus the combination of Eq. (13.32)

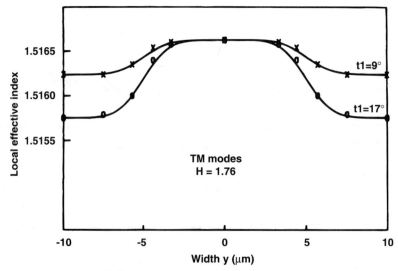

FIGURE 13.16 Local effective indices. (*Source*: Albert, J. and Yip, G. L. 1988. Wide single-mode channels and directional coupler by two-step ion-exchange in glass. *IEEE J. Lightwave Technol.* 6:552–563. With permission.)

and $H = (D_e t_2)^{1/2}$ yields a model for the lateral effective index profile of two-step, ion-exchanged waveguides, which depends exclusively on data derived from planar waveguide characterizations (i.e., $N(0)$, $N(\infty)$, and D_e). Of course, for the more general cases, $t_1 + t_2$ will not be kept at one hour, but would vary.

Having established the effective index model $N_{\text{eff}}(y)$, the propagation characteristics of the channel guide (i.e., the lateral problem) can be studied by solving Eq. (13.15) for β and $G(y)$ by either the WKB method or a single-function, variational method (Sharma et al., 1980) with good agreement between the two. For the latter, we use a single function for the first two modes, namely, a Gaussian for the fundamental mode

$$G_0(y) = e^{-\sigma_0 y^2} \tag{13.33}$$

and the Hermite–Gaussian of order 1 for the second mode,

$$G_1(y) = \sqrt{\sigma_1}\, y\, e^{-\sigma_1 y^2}. \tag{13.34}$$

Some dispersion curves of the channel guides are presented in Fig. (13.17), where

$$b = \frac{(\beta/k_0)^2 - N^2(\infty)}{N^2(0) - N^2(\infty)},$$

and

$$V = k_0 w \sqrt{N^2(0) - N^2(\infty)} \tag{13.35}$$

In Figure 13.2 in Section 2.2, it was indicated that the numerical Runge–Kutta (R-K) method can be used to calculate the effective index in the depth problem and, hence, to construct the lateral effective-index distribution for a channel guide as shown in Fig. 13.2b–d. Finally, it can also be used to calculate the channel guide's propagation constant as shown in Fig. 13.2d–e. The R-K method will not be outlined (Ralston and Wilf, 1960; Gill, 1951) but its use will be illustrated through two specific examples.

The constant concentration contours of two single-channel guides, with a width $w = 6$ μm and $w = 12$ μm, respectively, and produced by a single K^+-ion exchange, are shown in Figs. 13.18a and b with their fabrication conditions specified. Using Eq. (13.29), the 2-D index distribution $n(x, y)$ can be calculated. Following the procedures indicated in Figs. 13.2a and b, $N_{\text{eff}}(y_i)$ can be computed at discrete points, as shown in Figs. 13.19a and b, by the R-K method for the TM modes at 1.523 μm.

Under the same fabrication conditions, the diffusion problem of two parallel channel guides, for example, used in a directional coupler, can also be simulated for different waveguide spacings s (separation distance) and p (distance between the centers of the guides and $s = p - w$) as shown in Figs. 13.20a and b. The corresponding effective-index distributions $N_{\text{eff}}(y_i)$ are presented in Figs. 13.21a and b for the TM modes at 1.523 μm. It can be seen that, in the special use of zero separation distance ($s = 0$), the two guides merge into one with a width of $w = 12$ μm. The effective index $N_{\text{eff}}(y_i)$ of this composite guide is, then, identical to that shown in Figure 13.19b. In the case of very large separation distances ($s > 12$ μm, say), the two guides have less and less interaction between them, and the effective index $N_{\text{eff}}(y_i)$ looks more and more like that for each individual guide, centered at $y = p/2$ and $-p/2$ as shown in Figure 13.21b.

In Figs. 13.19 and 13.21, the values of N_{eff} resulting from the exact numerical R-K solution, the erf fit, and the step-index approximation are compared for the TM modes. The results under the step-index approximation have been obtained by neglecting side-diffusion ($\partial/\partial y = 0$) in Eq.

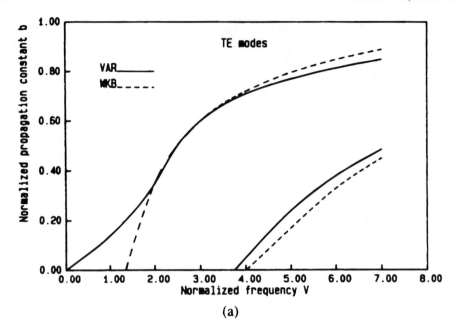

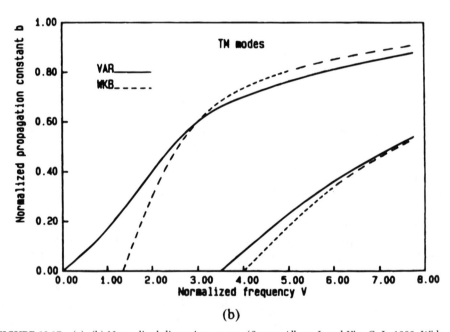

FIGURE 13.17 (a), (b) Normalized dispersion curves. (*Source*: Albert, J. and Yip, G. L. 1988. Wide single-mode channels and directional coupler by two-step ion-exchange in glass. *IEEE J. Lightwave Technol.* 6:552–563. With permission.)

13.29 and by assuming $N_{\text{eff}}(y) = N_{\text{eff}}(0)$ within the width of the channel guide and $N_{\text{eff}}(y) = n_b$ outside. The erf fit is very good for the middle of the guide, but the agreement deteriorates near the guide's two edges, where, however, the contribution of the index values to the guide's propagation constant becomes less important. The step-index model appears to be a poorer one, especially for narrower guides seen in Fig. 13.19a, where its $N_{\text{eff}}(0)$ is significantly higher than that obtained from the exact solution. For a two-guide structure with $0 < s < 12$ μm, the erf fit for $N_{\text{eff}}(y)$ is

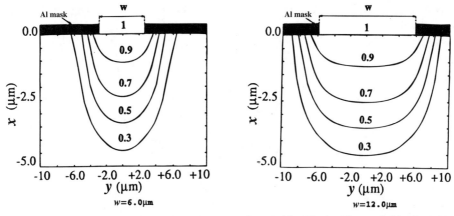

FIGURE 13.18 Constant concentration contours. a,b L. Babin, Thesis. (*Source*: Babin, L. 1993. Optimization and fabrication of a widened x-branch optical demultiplexer in glass. M. Eng. thesis. Dept. of Electrical Engineering, McGill University, Montreal, Quebec, Canada. With permission.)

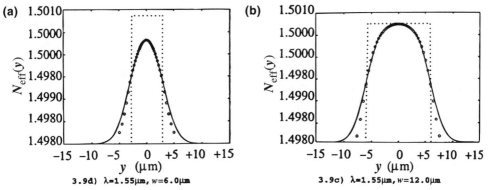

FIGURE 13.19 Discrete values of the local effective. a, Babin, Thesis. (*Source*: Babin, L. 1993. Optimization and fabrication of a widened x-branch optical demultiplexer in glass. M. Eng. thesis. Dept. of Electrical Engineering, McGill University, Montreal, Quebec, Canada. With permission.)

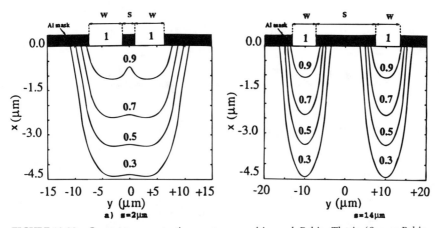

FIGURE 13.20 Constant concentration contours resulting. a, b Babin, Thesis. (*Source*: Babin, L. 1993. Optimization and fabrication of a widened x-branch optical demultiplexer in glass. M. Eng. thesis. Dept. of Electrical Engineering, McGill University, Montreal, Quebec, Canada. With permission.)

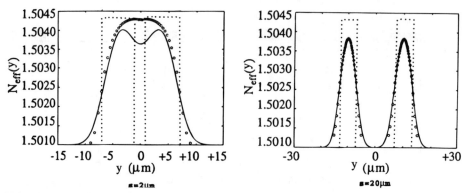

FIGURE 13.21 Discrete values of the local effective index. a, b Babin, Thesis. (*Source*: Babin, L. 1993. Optimization and fabrication of a widened x-branch optical demultiplexer in glass. M. Eng. thesis. Dept. of Electrical Engineering, McGill University, Montreal, Quebec, Canada. With permission.)

typically as presented in Fig. 13.21a, where it is seen to deviate somewhat from that by the R-K method. This should not be surprising because the erf fit attempted, here, is based on the single channel fit represented by Equation 13.32. However, in spite of this deviation, the resulting propagation constant β calculated will still be of sufficient accuracy as discussed in the following paragraph.

In Table 13.6, the effective index $N_{eff} = \beta/k_0$, of the first-order and second-order TM modes for the two different channel guides, calculated by the EIM, employing the lateral effective index $N_{eff}(y)$ established via the R-K method, the erf and the step-index model, respectively, are presented and compared. The errors are relative to the exact numerical results. The erf model is seen to yield closer results. The errors in β are of the order of $(5 - 9) \times 10^{-5}$.

Using the EIM and step-index or erf fit model for $N_{eff}(y)$, extensive, normalized, universal dispersion curves can be plotted as shown in Fig. 13.22 for channel guides built by a single K$^+$-Na$^+$ ion exchange. For this, four normalized waveguide parameters should be defined (Hocker and Burns, 1977) as follows:

$$b_0 = (N_p^2 - n_b^2)/(n_s^2 - n_b^2),$$

$$V_0 = k_0 d \sqrt{(n_s^2 - n_b^2)},$$

$$b' = \left[\left(\frac{\beta}{k_0}\right)^2 - n_b^2\right] \Big/ (N_p^2 - n_b^2), \tag{13.36}$$

TABLE 13.6 Calculated Effective Index

	R-K Numerical Solution	'erf fit'		Step-Index Model	
		N_{eff}	Error ($\times 10^{-5}$)	N_{eff}	Error ($\times 10^{-5}$)
1st order mode	1.503608	1.503666	+5.8	1.503800	+19.2
2nd order mode	1.502075	1.502148	+7.3	1.502309	+23.4
	w = 12.0 μm, t_d = 270 min, λ = 1.31 μm, TM modes				
1st order mode	1.499946	1.499955	+4.9	1.500110	+16.4
2nd order mode	1.498351	1.498441	+9.0	1.498486	+13.5
	w = 12.0 μm, t_d = 270 min, λ = 1.55 μm, TM modes				
1st order mode	1.502705	1.502611	−9.4	1.503043	+43.2
	w = 6.0 μm, t_d = 270 min, λ = 1.31 μm, TM modes				
1st order mode	1.499008	1.499107	+9.9	1.499361	+35.2
	w = 6.0 μm, t_d = 270 min, λ = 1.55 μm, TM modes				

(*Source*: Babin, L. 1993. Optimization and fabrication of a widened x-branch optical demultiplexer in glass. M. Eng. thesis. Dept. of Electrical Engineering, McGill University, Montreal, Quebec, Canada. With permission.)

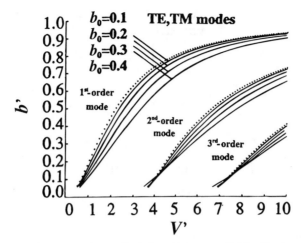

FIGURE 13.22 Normalized dispersion curve. Babin, Thesis. (*Source*: Babin, L. 1993. Optimization and fabrication of a widened x-branch optical demultiplexer in glass. M. Eng. thesis. Dept. of Electrical Engineering, McGill University, Montreal, Quebec, Canada. With permission.)

and

$$V' = k_0 w \sqrt{(N_p^2 - n_b^2)} = v_0 b_0^{1/2} \frac{w}{d}$$

where N_p is the effective index of a slab waveguide ion exchanged for a time t. All of the other parameters have been defined previously.

For weakly guiding K^+-Na^+ ion-exchanged guides, $n_s \simeq n_b$ and, hence, the universal curves $b' - V'$ are roughly the same for both polarizations. However, channel guides are still polarization-dependent because values of N_p differ for the TE and TM modes. Within the step-index model, the $b' - V'$ curves are b_0-independent, but are b_0-dependent within the erf model, which includes side-diffusion effects. The parameters N_p and b_0 strongly depend on the diffusion time t, increasing with t. For short diffusion times, side-diffusion effects are limited and do not contribute much difference in the propagation constants of the channel guides. However, for 200 min. $< t <$ 400 min. (used in a design example in Section 6.3) side-diffusion effects could contribute significant differences to the propagation constants of the channel guide modes.

It is important to point out, here, that this section has shown how the channel guides can be modeled and their propagation characteristics analyzed with the EIM, *using experimental data from slab guide characterizations only*. The accuracies in the lateral-index profile models will be tested in the design and implementation of some channel guide devices to be presented in section 13.6.

Effect of a Dielectric Cladding

In many situations, a dielectric cladding is deposited on a slab or channel guide to enhance the guide's effective index for light beam processing. In $LiNbO_3$ electro-optical waveguide devices, a dielectric cladding on a channel guide is usually deposited to serve as a buffer layer between the guide and an electrode to reduce device losses. Hence, it is important to know the change in the guide's effective index due to dielectric cladding in the device design.

Assuming a Gaussian index distribution in the diffused slab or channel guide, an equivalent homogeneous slab model is chosen with the criteria that the effective indices N_{eff} in both waveguides

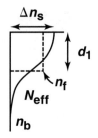

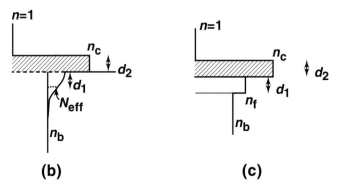

FIGURE 13.23 Modeling of a slab. a, b, c M. Belanger, Doctoral Theses. (*Source*: Belanger, M. 1986. Theoretical and experimental studies on active and passive 3-branch waveguides and their derivatives. Ph.D. thesis. Dept. of Electrical Engineering, McGill University, Montreal, Quebec, Canada. With permission.)

are equal and that the thickness of the homogeneous model is given by the penetration depth $d_1 = x_t$ (also called the turning point). The value of N_{eff} for an inhomogeneous slab guide can be calculated from the WKB dispersion relationship, Eq. (13.20), which can be rewritten in terms of the normalized waveguide parameters v_1 and b_1 as

$$V_1 \int_0^{\bar{x}_t} [f(\bar{x}) - b_1]^{1/2} \, d\bar{x} = (m + 3/4)\pi; \qquad m = 0, 1, 2 \cdots \qquad (13.37)$$

where $f(\bar{x}) = \exp(-\bar{x}^2)$ and $v_1 = v_0$, $b_1 = b_0$ as defined in Equation (13.36).

Then, N_{eff} can be substituted in the dispersion Eq. (13.5) for the TE modes or Eq. (13.8) for TM modes in an equivalent homogeneous slab waveguide to obtain its equivalent n_f, where $n_c = 1$ for free space.

A dispersion relationship for this equivalent homogeneous slab waveguide of thickness d_1, covered on top by a uniform cladding layer of thickness d_2 and index n_c, and then free space, and bounded below by an infinite medium of index n_b, can, then, be derived, in the case of the TM modes, as given in Equation (13.38):

$$k_2\left[1 - \frac{k_2}{\gamma_3}\tan k_2 d_2\right]\left[\tan k_1 d_1 + \frac{k_1}{\gamma_4}\right] - k_1\left[1 - \frac{k_1}{\gamma_4}\tan k_1 d_1\right]\left[\tan k_2 d_2 + \frac{k_2}{\gamma_3}\right] = 0 \qquad (13.38)$$

where

$$k_1 = (k_0^2 n_f^2 - \beta^2)^{1/2},$$
$$k_2 = (k_0^2 n_c^2 - \beta^2)^{1/2}$$
$$\gamma_3 = (\beta^2 - k_0^2)^{1/2},$$
$$\gamma_4 = (\beta^2 - k_0^2 n_b^2)^{1/2}, \tag{13.39}$$

and β is the propagation constant of the fundamental mode in the equivalent homogeneous slab guide with a film index n_f. Equation (13.38) can then be used to compute the new values of $N_{eff} = \beta/k_0$, and, hence, the index change ΔN caused by the cladding (n_c).

LiNbO$_3$ Waveguides by Ti Diffusion and Proton Exchange

Like the ion-exchanged channel guides in glass, the index distribution along the depth direction can be modeled by a gaussian function. In the lateral direction, the simpler step-index model, which neglects side diffusion, can yield reasonably accurate design predictions. A more accurate model, which takes into consideration the effects of side diffusion, is, again, given by the combination of two error functions. Hence, the general index distribution for a two-dimensional channel guide can be represented by (Burns et al., 1979; Ctyroky et al., 1984; Fukuma et al., 1978)

$$\Delta n(x, y) = \Delta n_s \exp\left(-\left(\frac{x}{d}\right)^2\right)\left\{\text{erf}\left(\frac{W/2 - y}{d_y}\right) + \text{erf}\left(\frac{W/2 + y}{d_y}\right)\right\} \tag{13.40a}$$

where Δn_s is the surface index change at the center of the channel guide, and

$$d_x = 2\sqrt{D_x t},$$
$$d_y = 2\sqrt{D_y t}, \tag{13.40b}$$

where D_x and D_y are the diffusion coefficients along the depth and lateral direction and d_x and d_y are the corresponding diffusion lengths.

Electro-Optical Index Change in Ti:LiNbO$_3$ Waveguides

Due to the excellent electro-optical properties and low propagation loss of LiNbO$_3$ crystals, electro-optical devices, such as modulators switches, polarization splitters, etc. have been designed and implemented in LiNbO$_3$ substrates (Alferness, 1988). Fundamentals of crystal optics and applications to electro-optical devices are discussed in Kaminow (1974). A very brief account of the index change due to voltage-induced electro-optical effects through an electrode system on the crystal surface will be given here.

LiNbO$_3$ is a negative, uniaxial crystal with a dielectric tensor given by (Alferness, 1988)

$$\bar{\epsilon} = \begin{bmatrix} \epsilon_x & 0 & 0 \\ 0 & \epsilon_y & 0 \\ 0 & 0 & \epsilon_z \end{bmatrix} = \begin{bmatrix} n_x^2 & 0 & 0 \\ 0 & n_y^2 & 0 \\ 0 & 0 & n_z^2 \end{bmatrix} \tag{13.41}$$

where $\epsilon_x = \epsilon_y \neq \epsilon_z$, $n_x = n_y = n_0$ and $n_z = n_e$ ($<n_0$, hence the name negative). For waves with an electric field polarized along either the x or y axes, the propagating waves "see" only the indices

n_x or n_y. Such waves are called ordinary waves and x, y correspond to the ordinary axis. For waves with an electric field polarized along the z axis, they "see" only the index n_z and propagate with a different phase velocity. They are, hence, called extraordinary waves and the z axis in this case corresponds to the optical or c axis.

Electromagnetic wave propagation in a crystal is characterized by the so-called indicatrix or index ellipsoid. The orientation of this ellipsoid model is related to the crystal axes, which correspond to the ellipsoid's principal axes. The semiaxes of the ellipsoid along the x, y and z directions give the principal refractive indices n_x, n_y, and n_z, respectively. Index changes via electro-optic effects can, therefore, be described in terms of changes in the semiaxes of the index ellipsoid. The index ellipsoid is given by (Kaminow, 1974; Nishihara et al., 1989)

$$\frac{x^2}{n_x^2} + \frac{y^2}{n_y^2} + \frac{z^2}{n_z^2} = 1 \tag{13.42}$$

where $i = D_i/\sqrt{2\epsilon_0 \omega_e}$, $i = x, y, z$. D_i is the electric displacement vector, ϵ_0 the free space permittivity and ω_e the stored electric energy density in the crystal.

Active integrated optical devices utilize the electro-optic index change $\delta n_{e.o.} = \delta n$ induced in certain optical crystals by an applied electric field. The induced index change causes the index ellipsoid to deform. This induced index change $\Delta(n^{-2})$ is given by (Kaminow, 1974)

$$\Delta(n^{-2}) = rE + R^2 E^2. \tag{13.43}$$

The first term varies linearly with the applied electric field and is known as the Pockels effect. The second term varies with the square of the applied field and is known as the Kerr effect. Our subsequent discussions will be confined to the linear Pockels effect. The induced electro-optic index change alters the shape, size, and orientation of the ellipsoid. The general expression for the deformed index ellipsoid is given by

$$a_{11}x^2 + a_{22}y^2 + a_{33}z^2 + 2a_{23}yz + 2a_{31}zx + 2a_{12}xy = 1, \tag{13.44}$$

where x, y, z, generally, no longer form the principal axes as in Eq. (13.42). The constants a_i, in Eq. (13.44) are related to the three components of the applied electric field E through the following 6×3 electro-optic tensor:

$$\begin{bmatrix} a_{11} - n_x^{-2} \\ a_{22} - n_y^{-2} \\ a_{33} - n_z^{-2} \\ a_{23} \\ a_{31} \\ a_{12} \end{bmatrix} = \begin{bmatrix} r_{11} & r_{12} & r_{13} \\ r_{21} & r_{22} & r_{23} \\ r_{31} & r_{32} & r_{33} \\ r_{41} & r_{42} & r_{43} \\ r_{51} & r_{52} & r_{53} \\ r_{61} & r_{62} & r_{63} \end{bmatrix} \cdot \begin{bmatrix} E_x \\ E_y \\ E_z \end{bmatrix} \tag{13.45}$$

where r_{ij} are called electro-optic or Pockels coefficients, and n_x, n_y, n_z are the refractive indices in the respective directions. When the applied E-field is zero, Equation (13.44) reduces to (13.42).

For LiNbO$_3$, the $[r_{ij}]$ tensor is given by (Alferness, 1988)

$$\begin{bmatrix} 0 & -r_{22} & r_{13} \\ 0 & r_{22} & 0 \\ 0 & 0 & r_{33} \\ 0 & r_{51} & 0 \\ r_{51} & 0 & 0 \\ -r_{22} & 0 & 0 \end{bmatrix} \tag{13.46}$$

where $r_{13} = 8.6\times$, $r_{33} = 30.8\times$, $r_{22} = 3.4\times$ and $r_{51} = 28 \times 10^{-12}$ m/V respectively.

Design Methodology For Guided-Wave Photonic Devices

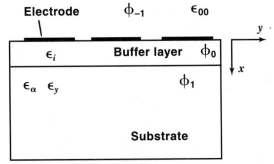

FIGURE 13.24 Electrode configuration. (*Source*: Sekerka-Bajbus, M. A. 1989. The BPM design optimization and implementation of 3-branch waveguide devices. M. Eng. thesis. Dept. of Electrical Engineering, McGill University, Montreal, Quebec, Canada. With permission.)

When $n_x = n_y = n_0$, $n_z = n_e$, and $E_x = E_y = 0$, through Eqs. (13.45) and (13.46), Eq. (13.44) becomes

$$(n_0^{-2} + r_{13}E_z)x^2 + (n_0^{-2} + r_{13}E_z)y^2 + (n_e^{-2} + r_{33}E_z)z^2 = 1. \tag{13.47}$$

Because the electro-optically induced index changes are usually very small, i.e., $|r_{13}n_0^2 E_z| \ll 1$ and $|r_{33}n_e^2 E_z| \ll 1$, Eq. (13.47) can be rewritten as

$$\frac{x^2}{\left(n_0 - \frac{n_0^3}{2}r_{13}E_z\right)^2} + \frac{y^2}{\left(n_0 - \frac{n_0^3}{2}r_{13}E_z\right)^2} + \frac{z^2}{\left(n_e - \frac{n_e^3}{2}r_{33}E_z\right)^2} = 1 \tag{13.48}$$

The refractive index changes due to an applied field E_z are, hence, respectively,

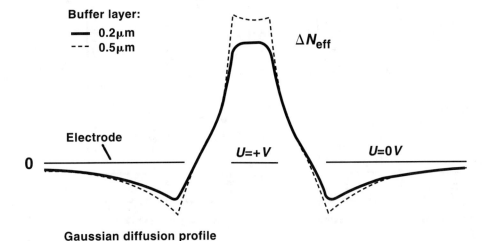

FIGURE 13.25 Electro-optically induced change. (*Source*: Sekerka-Bajbus, M. A. 1989. The BPM design optimization and implementation of 3-branch waveguide devices. M. Eng. thesis. Dept. of Electrical Engineering, McGill University, Montreal, Quebec, Canada. With permission.)

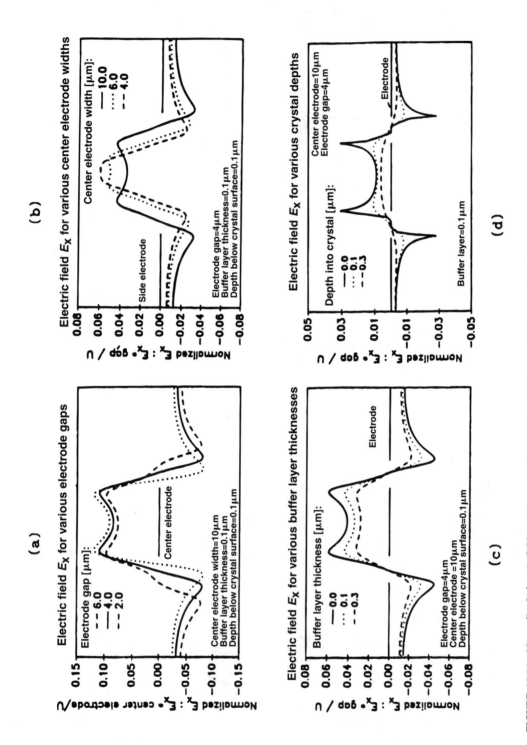

FIGURE 13.26 Normalized electric field E_x. (Source: Sekerka-Bajbus, M. A. 1989. The BPM design optimization and implementation of 3-branch waveguide devices. M. Eng. thesis. Dept. of Electrical Engineering, McGill University, Montreal, Quebec, Canada. With permission.)

$$\Delta n_x = \Delta n_y = -\frac{1}{2} r_{13} n_0^3 E_z$$

and

$$\Delta n_z = -\frac{1}{2} r_{33} n_e^3 E_z. \tag{13.49}$$

For LiNbO$_3$, $n_0 = 2.286$ and $n_e = 2.2$ at 0.633 μm.

Electrode Analysis and Design

For active electro-optic devices, an electrode system is always needed on a crystal surface to effect electro-optic control of guided waves, for example, in switching and modulation and also for fine tuning refractive indices in slab and channel guide devices, such as directional couplers and Mach–Zender interferometers, etc. A low-voltage and low-power loss to achieve the desired device functions are important design considerations.

The electric field distribution in the substrate can often be obtained analytically by conformal mapping for electrodes deposited directly on the LiNbO$_3$ surface (Vandenbulcke and Lagasse, 1974; Ramer, 1982). However, metal electrodes would attenuate optical waves, and, in practice, they are normally separated from the crystal surface by a thin buffer layer with a lower index to reduce the attenuation of optical waves. The presence of a buffer layer, nevertheless, causes a weakening of the applied electric field strength for the intended electro-optic effects and also makes the conformal mapping solution of the field distribution difficult, if not impossible. Fortunately, a semianalytical and numerical solution of the potential problem with a buffer layer is now available (Thylén and Gravestrand, 1986) and will be employed here.

A three-electrode system is used to illustrate this method, as shown in Fig. 13.24, where the space is divided into three regions marked by $-1(x < 0)$, $0(0 < x < d_0)$, and $1(x > d_0)$, respectively. The potential (due to the electrodes) to be solved must satisfy Laplace's equation:

$$\epsilon_x \frac{\partial^2 \phi}{\partial x^2} + \epsilon_y \frac{\partial^2 \phi}{\partial y^2} = 0 \tag{13.50}$$

with the following boundary conditions:

$$\begin{cases} \phi(x \to \infty, y) = 0, \\ \text{at the electrodes, } \phi(0, y) = V, \text{ the applied voltage, and} \\ \text{between the electrodes, the normal derivative } \frac{\partial \phi}{\partial n} \text{ is zero.} \end{cases} \tag{13.51a}$$

At the buffer-substrate interface, there should be

$$\begin{cases} \text{continuity of potential, } \phi_1(d_0, y) = \phi_0(d_0, y), \text{ and} \\ \text{continuity of } D_x, \epsilon_x \frac{\partial \phi_1}{\partial x}(x, y)\big|_{x=d_0} = \epsilon_i \frac{\partial \phi_0}{\partial x}(x, y)\big|_{x=d_0}. \end{cases} \tag{13.51b}$$

The semianalytical method of solution uses the Fourier transform technique. Let $\phi(x, y)$ be the original field and $\Phi(x, \nu)$ its Fourier transform in the y direction, where ν is the Fourier transform variable. Fourier transforming Eq. (13.50), one gets (Thylén and Gravestrand, 1986)

$$\phi_1(x, y) = \int_{-\infty}^{\infty} dv \, \Phi_1(d_0, v) e^{j2\pi vy} e^{-2\pi|v|(x-d_0)\sqrt{\epsilon_y/\epsilon_z}} \qquad (13.52)$$

throughout the substrate, provided $\Phi_1(d_0, v)$ is known. $\Phi_1(d_0, v)$ is related to $\Phi_0(0, v)$ by

$$\Phi_1(d_0, v) = \Phi_0(0, v)/[\cosh(2\pi|v|d_0) + \sinh(2\pi|v|d_0 \alpha_1)] \qquad (13.53)$$

where $\alpha_1 = (\epsilon_x \epsilon_y)^{1/2}/\epsilon_i$

Hence, $\Phi_0(0, v)$ must be known, but $\Phi_0(0, v)$ (Thylén and Gravestrand, 1986) must satisfy

$$\int_{-\infty}^{\infty} dv \, |v| \Phi_0(0, v) e^{j2\pi vy} \left[\epsilon_i \frac{\sinh(2\pi|v|d_0) + \alpha_1 \cosh(2\pi|v|d_0)}{\cosh(2\pi|v|d_0) + \alpha_1 \sinh(2\pi|v|d_0)} + \epsilon_\infty \right] = 0, \qquad (13.54)$$

where $\Phi_0(0, v)$ can be solved iteratively.

The numerical procedure in the electrode analysis and design can be summarized as follows:

1. Initially, guess of the surface potential $\phi_0(0, y)$.
2. FFT this potential to get $\Phi_0(0, v)$.
3. Multiply by $-2\pi|v|$, and use an inverse FFT to obtain the corresponding electric displacement D_x at the surface $x = 0$. $D_x = \epsilon_0(\partial \phi/\partial x)(0, y)$.
4. Modify D_x by setting all values *between* the electrodes to zero, i.e., $D_x = 0$ at $x = 0$.
5. FFT the resulting potential and divide by $-2\pi|v|$.
6. Inverse FFT to obtain the modified surface potential $\phi(0, y)$. Set the potential on the electrodes to the applied voltage, and use an interpolation scheme to make the potential continuous between the electrodes.
7. Repeat the iteration from step 2 until the potential and the electric displacement D_x satisfy their boundary conditions.

The above procedure yields numerical results which converge very fast even for very rough initial guesses. Once the surface potential for a specific electrode configuration has been determined, the potential and field distributions can be calculated, using the analytical expressions given in Eqs. (13.52)–(13.54). The corresponding index change, induced electro-optically, at any point, can also be determined. Consequently, the initial gaussian index profile is perturbed by the electro-optic effect:

$$n(x) = n_s + \Delta n_s \exp(-x^2/d_x^2) + \delta n(x)_{e.o.} . \qquad (13.55)$$

For the three-electrode configuration shown in Fig. 13.24, the resulting effective index change is shown in Fig. 13.25 for a voltage V applied to the central electrode while the potentials at the two outer electrodes are maintained at zero. The normalized electric field E_x for various electrode parameters is shown in Figs. 13.26a–d. Reducing the electrode gap, the central electrode width, and the buffer layer thickness increases the electric field strength.

13.5 The Beam Propagation Method

Fleck et al. (1976) proposed a new numerical method for solving the scalar Helmholtz equation. The usefulness of this technique, which was originally developed to trace laser beam propagation in the atmosphere, has been successfully demonstrated in the problems involving beam propagation in optical fibers later (Feit and Fleck, 1978; and 1980). It has since been known as the beam propagation method (BPM) and permits a combined treatment of both the guided and radiation modes in weakly guiding waveguides. More recently, this method has also been used in analyzing and designing integrated optical waveguide devices (Feit et al., 1983).

There are, two popular formulations of the BPM. Both can be implemented with the numerical Fast Fourier Transform (FFT). The more commonly used BPM formulation is based on the parabolic or Fresnel approximation of the scalar Helmholtz equation and is applicable to problems where the optical fields vary slowly along the propagation direction over distances of the order of a wavelength. When the Fresnel approximation is not valid, another more accurate form of the BPM based on the full scalar Helmholtz equation can be derived. For most problems, however, both formulations would lead to about the same numerical results. These schemes are referred to as FFT-BPM.

One drawback in the BPM technique is that, if the electric field propagates to the edge of the computational window in the transverse (x, y) plane, it will be folded back to the opposite side of the window and superposed on the propagation field during succeeding steps, thus, causing high-frequency numerical stability of the solution. This was pointed out by Saijonmaa and Yevick (1983) in their study of losses in bent optical waveguide fibers. They suggested that this can be avoided by absorbing the field at the edge of the grid, either by setting the field to zero at a few grid points close to the edge of the window or by inserting a large imaginary component in the refractive index at these points, thus, simulating the effect of a lossy cladding. The absorber function must be set up to ensure that the electric field is absorbed gradually near the window boundaries.

Besides the conventional FFT-BPM, the beam propagation method using finite differences, called FD-BPM, to solve the scalar paraxial wave equation has also been developed (Chung and Dagli, 1990) and compared to FFT-BPM (Chung and Dagli, 1990; Scarmozzino and Osgood, 1991).

The aforementioned schemes, however, can solve only the scalar wave equation under paraxial approximation. This may be adequate for some weakly guiding devices but not for those which are polarization-sensitive or those with abrupt index changes (Liu and Li, 1991): In the latter cases, one cannot neglect the vectorial nature of the guided waves in the analysis and simulation. Recently, a vector beam propagation method based on a finite difference scheme, called FD-VBPM, has been developed, accompanied by a detailed analysis and assessment performed for two-dimensional (2-D) waveguide structures (Huang et al., 1992).

Instead of the conventional absorbing boundary condition, a numerical, *transparent* boundary condition (TBC) developed more recently (Hadley, 1991), which allows the passage of the traveling waves through the edges of the computation window, can be recommended for the FD-VBPM scheme.

Unlike the absorber boundary condition, the algorithm for the TBC does not contain a parameter to be adjusted for each specific problem and is, therefore, problem-independent. It adapts naturally to a finite-difference scheme and has been shown to be accurate and robust for both 2-D and 3-D problems.

The reference index n_0 should be chosen so that the Fresnel or paraxial approximation is satisfied. If the variation of the refractive index over the transverse cross-section of the waveguide structure is small, the refractive index of the cladding can be chosen to be n_0. Otherwise, the reference index n_0 has to be chosen so as to minimize the variation of the transverse field components along the z direction. For single-mode waveguides, the propagation constant β of the fundamental mode can be chosen to be the reference index i.e., $n_0 = \beta/k_0$. For dual-mode waveguides, such as directional couplers and two-mode interference (TMI) cross-channel waveguides, n_0 can be chosen to be given by $n_0 = (\beta_e + \beta_0)/2$ where β_e and β_0 are the propagation constants of the even and odd mode of the structure, respectively.

Due to the page limitation, mathematical details of the various BPM schemes are not given here but have been fully documented in the references cited above.

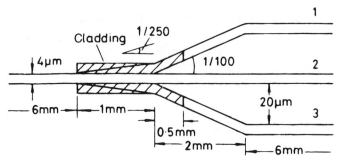

FIGURE 13.27 Configuration of a three-branch power divider. (*Source*: Yip, G. L. and Serkerka-Bajbus, M. A. 1988. Design of symmetric and asymmetric passive 3-branch power dividers by beam propagation method. *Electron. Lett.* 25:1584–1586. With permission.)

13.6 Illustrative Examples: Design and Fabrication of Some Guided-Wave Photonic Devices

The characterization and modeling procedures described in the previous sections have been employed in the theoretical design and experimental realization of several guided-wave photonic devices. In this section, we will present some of these devices as illustrative examples. Wherever possible, design and experimental results are both presented and comparisons made.

A Passive Three-Branch Power Divider by K$^+$-Ion Exchange in Glass

This device consists of an input channel guide followed by a widening taper which divides the incoming optical power into three output branches. The geometry and dimension of this device are given in Fig. 13.27. A layer of dielectric cladding can be deposited onto the relevant regions to increase the index in the taper and branch regions of arm 1 and 3, to channel more power into the outer arms 1 and 3 for equal power division. The design must be pursued with predetermined

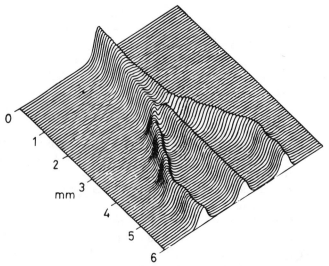

FIGURE 13.28 Optical field distribution. (*Source*: Yip, G. L. and Serkerka-Bajbus, M. A. 1988. Design of symmetric and asymmetric passive 3-branch power dividers by beam propagation method. *Electron. Lett.* 25:1584–1586. With permission.)

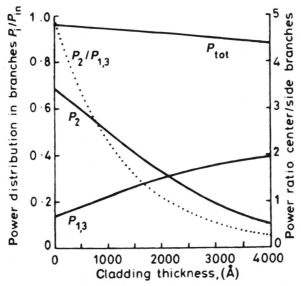

FIGURE 13.29 Power distribution in individual branches. *Electron. Lett.* (*Source*: Yip, G. L. and Serkerka-Bajbus, M. A. 1988. Design of symmetric and asymmetric passive 3-branch power dividers by beam propagation method. *Electron. Lett.* 25:1584–1586. With permission.)

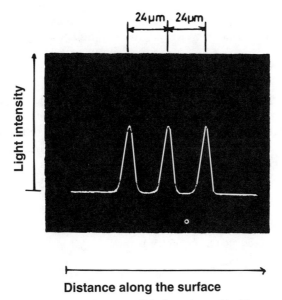

FIGURE 13.30 The output light intensity profile. (*Source*: Haruna, M., Belanger, M.n and Yip, G. L. 1985. Passive 3-branch optical power dividers by K$^+$-ion exchange in glass. *Electron. Lett.* 21:535–536. With permission.)

fabrication conditions, which can be optimized for the design outcome in an iterative manner. The channel guides were fabricated by K$^+$-ion exchange in a soda-lime glass substrate (n_b = 1.512) through an Al mask immersed in molten KNO$_3$ at 370 °C for one hour. This yielded single-mode channel guides with d = 1.53 μm and Δn_s = 0.010. Also, a layer of cladding, using Corning glass 7059 with $n_c \approx 1.544$ was Rf-sputter deposited. With these fabrication data, the effective index of the diffused channel guides can be calculated by solving the dispersion relationship for an equivalent homogeneous slab waveguide. The increase in the effective index in the waveguide regions under the cladding is obtained by solving the dispersion relationship for the guided modes in the homogeneous slab guide, covered on top by a uniform cladding of thickness $d_{2,i}$ (i = 1, 3) and index n_c, then, free space, and bounded below by an infinite medium of index n_b (substrate) as discussed in Section 4.2. In the lateral direction, the simplest step-index model was chosen for the channel guide. The optical field distribution in the device was calculated by the FFT beam propagation method (FFT-BPM) as presented in Section 5.1 and Yip and Sekerka–Bajbus (1988) and shown in Fig. 13.28 for the case of equal power division. The power distribution in the individual branches, as a function of the cladding thickness, is given in Fig. 13.29. The designed device has been fabricated and measured (Haruna et al., 1985). Fig. 13.30 shows the measured output light intensity profile by a TV camera for the case of equal power division. The theoretical cladding thickness required for equal power division can be estimated from Fig. 13.29 to be 0.215 μm. This compares favorably with the measured result of 0.2 μm (Haruna et al., 1985).

A Directional Coupler Power Divider by Two-Step K⁺-Ion Exchange in Glass

The characterization and modeling of a channel guide by a two-step, ion-exchange process as discussed in Section 4.1 were used in realizing a directional coupler power divider (Albert and Yip, 1988) as shown in Fig. 13.31.

A two-step ion-exchange process (Yip and Finak, 1984) was conceived and adopted for realizing a single-mode channel guide because it has the advantage of relaxing the channel width to 10 μm instead of 4–6 μm to render the channel guide more compatible for coupling to a single-mode fiber, whose core diameter is typically 8–10 μm. This fabrication procedure is outlined in Fig. 13.32.

The approach adopted to analyze and design the directional coupler is to calculate the normal modes of the structure consisting of two parallel channels. Then, by projecting the optical field incident in one branch onto the symmetric and antisymmetric modes of the coupler, the output characteristics can be obtained by propagating the two modes, thus, launched with their respective propagation constants β_e and β_0, respectively, down the length L of the coupling region. The interference between the two normal modes causes the power to oscillate between the two channels until their separation increases to a value where coupling ceases. The normal modes can be found by solving Eq. (13.14) with the following index profile $N_{eff}(y)$, using Eq. (13.32):

$$N_{eff}(y) = N(y + w/2) + N(y - w/2) - N(\infty) \qquad (13.56)$$

and the trial function, presented in Eq. (13.33)

$$G_0^e(y) = \exp[-\alpha_0^e(y + w/2)^2] + s \exp[-\alpha_0^e(y - w/2)^2] \qquad (13.57)$$

where $s = 1$ for the even mode e and -1 for the odd mode 0. Figure 13.33 shows some index profile and corresponding normal mode plots. The power transfer efficiency η is defined as

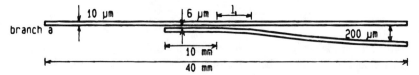

FIGURE 13.31 Design of a directional coupler. (*Source*: Albert, J. and Uip, G. L. 1988. Wide single-mode channels and directional coupler by two-step ion-exchange in glass. *IEEE J. Lightwave Technol.* 6:552–563. With permission.)

Design Methodology For Guided-Wave Photonic Devices

FABRICATION STEPS

- Cleaning
- A1 deposition
- Photoresist coating
- Exposition to UV light

- Development of photoresist
- Liquid etching of A1

- Ion exchange

- A1 removal
- Ion exchange

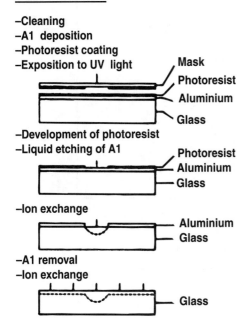

FIGURE 13.32 Diagram of fabrication procedure. (*Source*: Albert, J. and Uip, G. L. 1988. Wide single-mode channels and directional coupler by two-step ion-exchange in glass. *IEEE J. Lightwave Technol.* 6:552–563. With permission.)

$$\eta = P_b/(P_a + P_b) \tag{13.58}$$

where P_a and P_b are the respective powers at the outputs of channels a and b. P_a and P_b can be calculated theoretically and also measured. The design optimization in our context means calculating the performance parameters of this device, e.g., η, for a set of waveguide design parameters, in an iterative manner, until the best desired performance parameters are achieved. The device is, then, ready for fabrication.

The device shown in Fig. 13.31 was fabricated at 385°C by the two-step, ion-exchange process depicted in Fig. 13.32, keeping $t_1 + t_2 = 1$ hr. to maintain a single mode in the depth direction. The outputs of the two channel guides were measured either with a scanning optical power meter or a CCD camera at 0.63 μm. Some typical lateral output power profiles, as observed on a chart recorder, are presented in Fig. 13.34 for device samples fabricated with different t_1. The mode in each channel seems to be well confined. For (c) and (d), it is evident that the same sample could yield a different power transfer efficiency, depending upon the critical launching condition at the input. In Fig. 13.35, both the theoretically calculated and measured values of η for device samples fabricated with different t_1 are presented, and the overall agreement appears good.

A Widened X-Branch Wavelength Demultiplexer at 1.3 μm and 1.55 μm by K⁺-Ion Exchange in Glass

The device configuration, c shown in Fig. 13.36, consists of input and output waveguides in the form of two identical, tapered, directional couplers of length l_t connected adiabatically to a two-mode waveguide of length L. Each channel guide in the tapered region has a width w, whereas the central waveguide has a width of $2w$. This device works on the principle of interference between two normal modes (TMI) Papuchon et al., 1977; Neyer, 1983, supported by the structure, the symmetric mode Ψ_s and the antisymmetric mode Ψ_a, as sketched in Fig. 13.37. The total field in the structure is given by

$$\Psi(x, y, z) = \alpha_s \Psi_s(x, y, z) + \alpha_a \Psi_a(x, y, z) \tag{13.59}$$

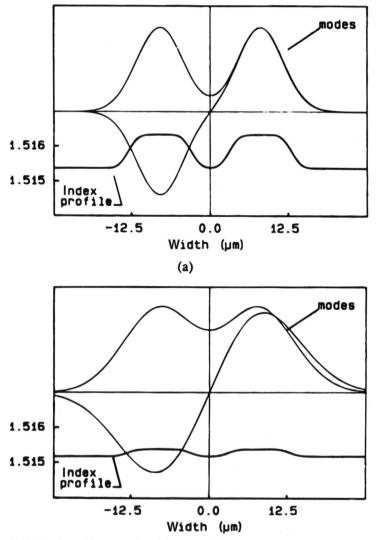

FIGURE 13.33 Two examples of the normal lateral. (*Source*: Albert, J. and Uip, G. L. 1988. Wide single-mode channels and directional coupler by two-step ion-exchange in glass. *IEEE J. Lightwave Technol.* 6:552–563. With permission.)

where α_s and α_a are the relative excitation amplitudes of the symmetric (even) mode and antisymmetric (odd) mode, respectively, so that $\alpha_s^2 + \alpha_a^2 = 1$.

In the tapered regions ($z < -L/2$ or $z > L/2$), the symmetric and anti-symmetric modes can be considered as two different combinations of the two fundamental modes of the two input channel guides taken separately (Marom and Ruschin, 1984; Forber and Marom, 1986). Hence,

$$\Psi_s(x, y, z) = [a_1(x, y) + a_2(x, y)]\, e^{-j\beta_s^z} \qquad (13.60\text{a})$$

and

$$\Psi_a(x, y, z) = [a_1(x, y) - a_2(x, y)]\, e^{-j\beta_a^z} \qquad (13.60\text{b})$$

where $a_1(x, y) = a(x, y - p/2)$ and $a_2(x, y) = a(x, y + p/2)$ and $a(x, y)$ is the normal ($\int\int |a(x,$

Design Methodology For Guided-Wave Photonic Devices

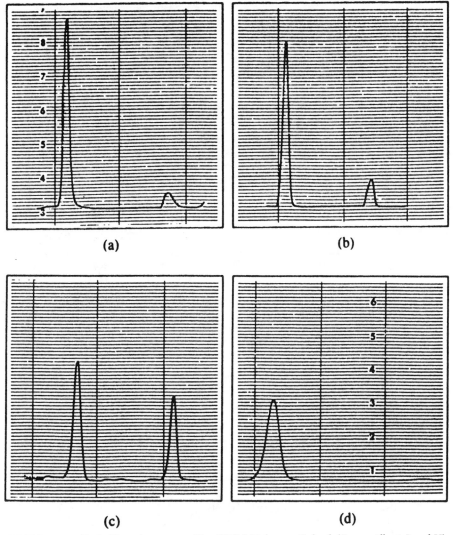

FIGURE 13.34 Typical lateral output profiles. *IEEE J. Lightwave Technol.* (*Source*: Albert, J. and Uip, G. L. 1988. Wide single-mode channels and directional coupler by two-step ion-exchange in glass. *IEEE J. Lightwave Technol.* 6:552–563. With permission.)

$y)|^2 \, dx \, dy = 1$) fundamental mode of a single waveguide of width w, centered $y = 0$. The two propagation constants β_s and β_a depend on the spacing between the input or output waveguides and, hence, on z.

In the central waveguide of width $2w$, β_s and β_a are the propagation constants of the fundamental mode Ψ_s (even) and the second-order mode Ψ_a (odd), respectively.

Due to mode orthogonality, Ψ_s and Ψ_a do not exchange energy along the entire device. However, because $\beta_s \neq \beta_a$, the two modes accumulate a relative phase difference $\phi = \int (\beta_s - \beta_a) \, dz$ along the device length. Combining Eqs. (13.59) and (13.60), the power output at ports 1 and 2 can be evaluated to yield (Forber and Marom, 1986)

$$\text{at output port 1; } (y < 0) : P = P_{in}\cos^2(\phi/2), \qquad (13.61a)$$

and

$$\text{at output port 2; } (y < 0) : P_x = P_{in}\sin^2(\phi/2), \qquad (13.61b)$$

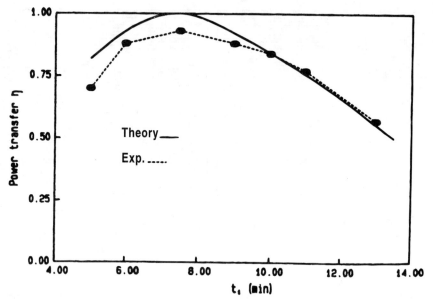

FIGURE 13.35 Power transfer efficiency η as a function. (*Source*: Albert, J. and Uip, G. L. 1988. Wide single-mode channels and directional coupler by two-step ion-exchange in glass. *IEEE J. Lightwave Technol.* 6:552–563. With permission.)

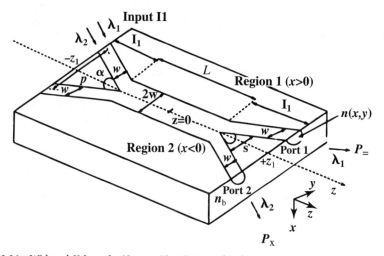

FIGURE 13.36 Widened X-branch. (*Source*: Yip, G. L. and Babin, L. 1995. Design optimization and fabrication of a widened x-branch demultiplexer by ion-exchanger in glass. In *Proc. Guided-Wave Optoelectronics: Device Characterization, Analysis, and Design*. Tamer, T., Griffel, G., and Bertoni, H. L., eds. pp. 221–229. Plenum Press, New York. With permission.)

where P_{in} is the input power at port 1 for either λ_1 (1.31 μm) or λ_2 (1.55 μm), as shown in Fig. 13.36, and ϕ is the accumulated phase difference over the two tapered and the central regions. P_{in} can be set to the normalized value of unity. The operating conditions, which must be satisfied, for demultiplexing are (Chung et al., 1989)

$$\phi(\lambda_1) = \Delta\beta_c(\lambda_1)L + 2\phi_t(\lambda_1) = m\pi \tag{13.62a}$$

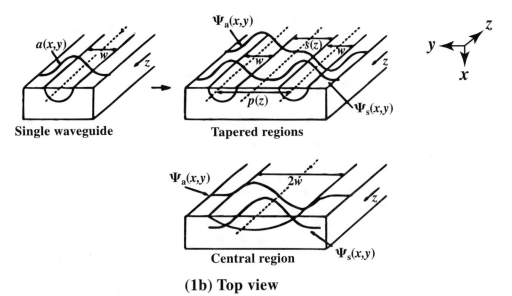

(1b) Top view

FIGURE 13.37 The symmetric and antisymmetric modes. (*Source*: Babin, L. 1993. Optimization and fabrication of a widened x-branch optical demultiplexer in glass. M. Eng. thesis. Dept. of Electrical Engineering, McGill University, Montreal, Quebec, Canada. With permission.)

and

$$\phi(\lambda_2) = \Delta\beta_c(\lambda_2)L + 2\phi_t(\lambda_2) = (m-1)\pi \tag{13.62b}$$

where $\Delta\beta_c = \beta_s - \beta_a$ over the central region and is constant, whereas $\phi_t = \int_{z1}^{z2} \Delta\beta_t(z)\,dz$ over either tapered region, z_1 and z_2 being either $-z_t$ and $-L/2$ or $L/2$ and z_t and $\Delta\beta_t(z) = \beta_s(z) - \beta_a(z)$.

If $m(=2n)$ is an even integer, using Eqs. (13.61a) and (13.61b), we can show that $P_=(\lambda_1) = 1$, $P_x(\lambda_1) = 0$, and $P_=(\lambda_2) = 0$, and $P_x(\lambda_2) = 1$. Hence, the light at λ_1 is received at the output port 1, and λ_2 is received at the output port 2. If $m(=2n-1)$ is an odd integer, the situation is reversed so that the light at λ_1 is received at the output port 2, and λ_2 at the output port 1. The extinction ratios (ER) at both wavelengths are defined and calculated as

$$\text{ER}(\lambda) = 10\log\left|\frac{P_x(\lambda)}{P_=(\lambda)}\right| = 10\log\left|\tan^2\left(\frac{\phi}{2}\right)\right| \tag{13.63}$$

Theoretically, $\text{ER}(\lambda_1) = \text{ER}(\lambda_2) = -\infty$ because $\tan[\phi(\lambda_{1,2})/2] = 0$ when $\phi = 2n\pi$ or $(2n+1)\pi$. Any deviation of ϕ from these values, denoted by $\Delta\phi$, due to waveguide parameter errors, caused by computational errors in design, for example, an error in L, or fabrication errors, for example, an error Δw in the guide width w in the tapered regions or $\Delta(2w)$ in the central guide width $2w$, could result in a seriously degraded extinction ratio.

To see this, we can get, from Eq. (13.63), near $\text{ER}(\lambda) = -\infty$ and assuming $\Delta\phi$ to be small,

$$\text{ER}(\lambda) = 10\log|\tan^2(\Delta\phi(\lambda)/2)|$$
$$\approx 20\log|\Delta\phi(\lambda)/2|. \tag{13.64}$$

For example, a small error of $\Delta\phi/2 \approx 0.1$ rad. (6°) would lead to a degradation of $\text{ER} \approx 20\log 0.1 \approx -20$ dB.

Using a reduced parameter R, similar to that defined by Cheng and Ramaswamy (1991), we can start our design problem by solving Eqs. (13.62a and b) for L.

$$R = \frac{m\pi - 2\phi_t(\lambda_1)}{(m-1)\pi - 2\phi_t(\lambda_2)} = \frac{\Delta\beta_c(\lambda_1)}{\Delta\beta_c(\lambda_2)}$$

$$= R_1^m \qquad\qquad\qquad = R_2. \qquad (13.65)$$

It is important to point out that $\Delta\beta_c$ depends on the wavelength λ, the waveguide width $2w$, hence, w, the waveguide diffusion depth d, hence, t_d (recalling $d = (D_e\, t_d)^{1/2}$ in Section 13.3). The phase ϕ_t also depends on these parameters, but, in addition, significantly, on the branching angle α. Hence, $\Delta\beta_c$ can be expressed as $\Delta\beta_c(\lambda, w, t_d)$ and ϕ_t as $\phi_t(\lambda, w, t_d, \alpha)$. If the parameters m, w and α are all fixed and λ_1 and λ_2 are set to 1.31 μm and 1.55 μm, respectively, t_d can be found from Eq. (13.65) by a root-search technique, namely, by seeking the solution of

$$R_1^m(t_d) - R_2(t_d) = 0 \qquad (13.66)$$

After t_d has been found, L can be determined from either Equation (13.62a) or (13.62b), namely,

$$L = \frac{m\pi - 2\phi_t(\lambda_1 = 1.31\mu m, \alpha, w, t_d)}{\Delta\beta_c(\lambda_1 = 1.31\mu m, w, t_d)}$$

$$= \frac{(m-1)\pi - 2\phi_t(\lambda_2 = 1.55\mu m, \alpha, w, t_d)}{\Delta\beta_c(\lambda_2 = 1.55\mu m, w, t_d)}. \qquad (13.67)$$

Most of the phase shift needed to effect demultiplexing is achieved over the central waveguide region. The contribution from ϕ_t over either of the tapered regions is very small. The design procedure can be represented by the flow chart in Fig. 13.38.

Using the procedure outlined in Fig. 13.38, one can calculate the lateral effective index profile of a single channel and a directional coupler as presented in Section 4.1. Hence, the lateral index profile of the X-branch could be similarly established as shown in Fig. 13.39a and b. Applying the 2-D FD-VBPM presented in Section 5 to the side-diffused model in Fig. 13.39b, the BPM simulations of the X-branch demultiplexer were performed. In Eq. (13.65), $\Delta\beta_c$ and $\Delta\beta_t$ can be computed, employing a procedure described in Feit and Fleck (1980). Using Eqs. (13.66) and (13.67), t_d and L are plotted versus w for different values for m in Fig. 13.40 as design curves for both TM and TE modes. The optical fields, propagating through the device at 1.31 μm and 1.55 μm, are depicted in Fig. 13.41 for the TM modes, where all the relevant parameters are also specified. From our numerous computational data, we have observed that, although the lengths L for the TM and TE modes could be about the same, the corresponding diffusion times t_d required are quite different. Hence, a device optimized for the TM mode propagation is *not* optimized for the TE mode operation and vice versa.

Some caution should be exercised in choosing the excitation field at Input I1. In general, there is no exact analytical expression for the fundamental mode of a diffused channel guide. In our BPM simulations, the fundamental mode of an input branch with a width w and a lateral step-index profile, taken as $N_{\text{eff}}(y = 0)$, was used as an approximation due to the availability of its analytical expression. After causing small radiation in the first few propagation steps, the approximate input mode will eventually excite the fundamental mode in our diffused input branch.

To calculate the extinction ratios ER and the radiation loss, it is necessary to evaluate the power in each output branch. The output field can be written in each branch as

Design Methodology For Guided-Wave Photonic Devices

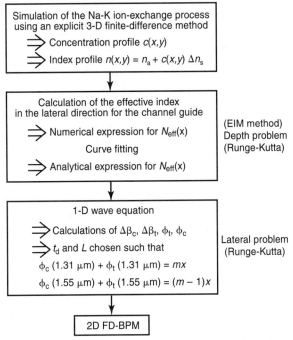

FIGURE 13.38 Design procedure flow chart. (*Source*: Babin, L. 1993. Optimization and fabrication of a widened x-branch optical demultiplexer in glass. M. Eng. thesis. Dept. of Electrical Engineering, McGill University, Montreal, Quebec, Canada. With permission.)

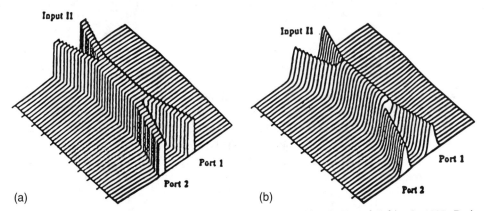

FIGURE 13.39 Lateral effective index. Plenum Press. (*Source*: Yip, G. L. and Babin, L. 1995. Design optimization and fabrication of a widened x-branch demultiplexer by ion-exchanger in glass. In *Proc. Guided-Wave Optoelectronics: Device Characterization, Analysis, and Design*. Tamer, T., Griffel, G., and Bertoni, H. L., eds. pp. 221–229. Plenum Press, New York. With permission.)

$$\psi_{out} = C_g \phi_g + \sum_i C_i \phi_i \qquad (13.68)$$

where ϕ_g is the normalized fundamental guided mode of either output branch, and ϕ_i's are the corresponding radiative modes. The output power normalized to the input power in the rth branch ($r = 1$ or 2) is given by $|C_{g,r}|^2$. Using the orthogonality between the guided and radiative modes, it is possible to derive, from Equation (13.68), that

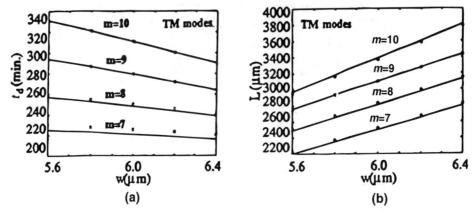

FIGURE 13.40 (a) t_d and (b) L vs. w w calculated by the Runge-Kutta method. Note c,d deleted by author. (*Source*: Babin, L. 1993. Optimization and fabrication of a widened x-branch optical demultiplexer in glass. M. Eng. thesis. Dept. of Electrical Engineering, McGill University, Montreal, Quebec, Canada. With permission.)

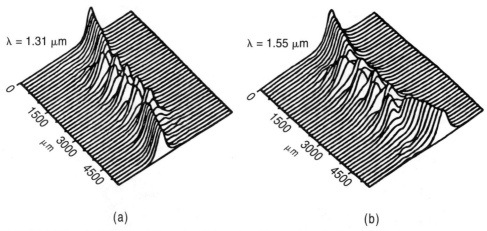

FIGURE 13.41 a,b BPM simulations for TM modes. (*Source*: Yip, G. L. and Babin, L. 1995. Design optimization and fabrication of a widened x-branch demultiplexer by ion-exchanger in glass. In *Proc. Guided-Wave Optoelectronics: Device Characterization, Analysis, and Design.* Tamer, T., Griffel, G., and Bertoni, H. L., eds. pp. 221–229. Plenum Press, New York. With permission.)

$$P_{g,r} = \left(\int_r \psi_{out} \cdot \phi_{g,r}^* \, dy \right)^2 \quad (13.69)$$

where $P_{g,1} = P_=$ and $P_{g,2} = P_x$. $\phi_{g,r}^*$ represents the BPM field calculated for each output branch, which is identical to that for the input branch. Because the BPM output field Ψ_{out} eventually approaches $\phi_{g,r}$, we can replace $\phi_{g,r}^*$ by Ψ_{out}^* in Equation (13.69). The radiation loss can, therefore, be expressed as

$$P_{rad} = 1 - (P_= + P_x). \quad (13.70)$$

The radiation loss is represented in dB as follows:

$$L_R = 10 \log(1 - P_{rad}), \quad (13.71)$$

so that $L_R = 0$ dB when $P_{rad} = 0$ and $L_R = -\infty$ dB when $P_{rad} = 1$

Design Methodology For Guided-Wave Photonic Devices

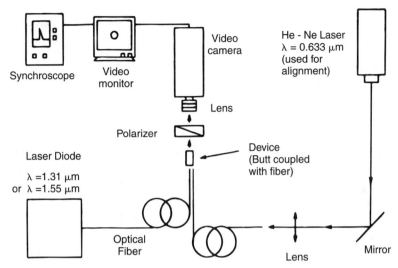

FIGURE 13.42 The measurement setup. (*Source*: Yip, G. L. and Babin, L. 1995. Design optimization and fabrication of a widened x-branch demultiplexer by ion-exchanger in glass. In *Proc. Guided-Wave Optoelectronics: Device Characterization, Analysis, and Design*. Tamer, T., Griffel, G., and Bertoni, H. L., eds. pp. 221–229. Plenum Press, New York. With permission.)

With the results of the BPM design simulation, a photomask containing several devices with different dimensions can be designed and used to fabricate several devices simultaneously on the same glass substrate. Several samples, each containing several devices, were prepared as depicted in Fig. 13.32, and the device performance, such as the extinction ratios (ER) were measured as indicated in the schematic setup in Fig. 13.42. Figures 13.43a,b show the near-field light spots at the two output arms for 1.31 μm and 1.55 μm, respectively, for the TM modes in one particular design. The corresponding measured extinction ratio (ER) is 20 dB at 1.31 μm and 15 dB at 1.523 μm, compared with the theoretical values of 35 dB at both wavelengths shown in Fig. 13.41.

Figures 13.40a, b are the results obtained assuming no errors in photolithography, namely, $w_t = w$, $2w_c = 2w$. Assuming an error of -0.4 μm in both $2w_c$ and w_t due to imperfect photolithography, namely, $2w_c = 11.6$ μm, $w_t = 5.6$ μm, and $2w_c \neq 2w_t$, a BPM simulation was carried out separately (because the results in Fig. 13.40 could not be used for this case) with $m = 9$, $\alpha = 0.5°$, $t_d = 270$ min., and $L = 2930$ μm for the TM modes. The corresponding theoretical ER's for this design are 35 dB at both 1.31 μm and 1.523 μm, whereas the measured ER's are 20 dB at both wavelengths.

To date, the findings reported represent the best theoretical and experimental results obtained by us. Although, there are still significant discrepancies between the design and measurement

(a) (b)

FIGURE 13.43 a,b Near-field output light spots. (*Source*: Yip, G. L. and Babin, L. 1995. Design optimization and fabrication of a widened x-branch demultiplexer by ion-exchanger in glass. In *Proc. Guided-Wave Optoelectronics: Device Characterization, Analysis, and Design*. Tamer, T., Griffel, G., and Bertoni, H. L., eds. pp. 221–229. Plenum Press, New York. With permission.)

ER's, the general agreement appears reasonable. Improvements are still possible, especially, in terms of fabrication and measurement.

An Electro-Optic Ti-diffused LiNbO$_3$, Ridge Waveguide, Mode-Confinement Modulator

The device configuration of the line modulator is shown in Fig. 13.44. The device consists of two channel ridge waveguides in regions 1 and 3 connected by a slab waveguide in region 2. The guided modal field in the input ridge guide, on entering the slab region, begins to diverge, so that, at the mouth of the output guide, the overlap between the transmitted field and the guided modal field of the output guide is reduced, resulting in a fraction of the input power being transmitted into the output guide. An increase in the lateral confinement of the modes in the slab waveguide can be achieved by electro-optically inducing a channel waveguide in region 2, producing an increased modal field overlap and, hence, increased transmission between the input and output channel guides. If the voltage polarity between the electrodes is reversed, the optical waves in region 2 will be scattered away from the output guide, yielding a transmission reduction. Hence, with the electrodes, the device performs as an intensity modulator for a voltage signal applied across the electrodes. The branch structure at the end of the slab region helps to guide away the scattered fields at the output junction.

Because the BPM can treat both the guided and radiation (scattered) modes, it is well suited for calculating the scattered modes in region 2 of the modulator. A theoretical analysis and design of this device requires the calculation of the transmitted power across the slab region into the output guide as a function of the applied electrode voltage. This can be accomplished with a combination of the EIM and FFT-BPM as discussed in Section 5 (Sekerka–Bajbus et al., 1990).

As discussed in Sections 13.2 and 13.3, the lateral effective-index distribution for the device can be found by replacing each lateral point, along the y direction of the crystal surface with a homogeneous slab guide of the same effective index $n_{\text{eff}}(y, z)$ as that of the diffused guide and solving the following WKB relationship:

$$2k_0 \int_{x_0}^{x_t} [n(x, y, z)^2 - n_{\text{eff}}^2(y, z)]^{1/2} \, dx = \left(2m + \frac{3}{2}\right)\pi, \quad m = 0, 1 \cdots \quad (13.72)$$

where $m = 0$ for a single mode waveguide and $n(x_t, y, z) = n_{\text{eff}}(y, z)$. The lower integration

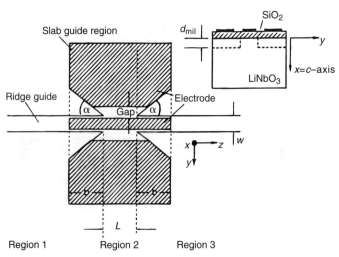

FIGURE 13.44 Device configuration. (*Source*: Sekerka-Bajbus, M. K., Yip, G. L., and Goto, N. 1990. BPM design-optimization and experimental improvements of a Ti:LiNbo$_3$ ridge waveguide linear-mode confinement modulator. *IEEE J. Lightwave Technol.* 8:1742–1749.)

limit x_0 is zero for the ridge channel guide but equal to the milling depth d_{mil} in the etched regions (by ion milling). The index distribution $n(x, y, z)$ is assumed to be gaussian along the x axis and, below the electrodes, is perturbed by the electro-optical effect. Hence,

$$n(x, y, z) = n_b + \Delta n_s \exp(-x^2/d^2) + \left[\frac{1}{2}\gamma_{33}n_b^3 E_x(x, y, z)\right] \quad (13.73)$$

where γ_{33} is the electro-optical coefficient along the x axis (C axis of $LiNbO_3$), E_x is the applied electric field in the x direction, and all other parameters have been defined previously. Given any voltage applied across the electrodes, E_x can be calculated by using a method as described in Section 4.5 (Thylén and Gravestrand, 1985), taking into consideration the attenuating influence of the SiO_2 buffer layer. The effective-index distribution in the slab region (region 2) for the on- and off-state voltage polarities are shown in Fig. 13.45. The optical-field distribution in the device can, then, be calculated by propagating an eigenmode along the two-dimensional effective-index models in regions 1, 2, and 3, using the FFT-BPM. The output power in a particular mode can be determined from the BPM data by overlapping the output optical field in region 3 with the specific normalized guide mode.

The effective index of the ridge structure can be controlled accurately through control of the milling depth d_{mil}, which helps to determine the number of lateral modes $(m + 1)$ a ridge guide of width w can support. Using the well known relationship [Equation 2.1.24 in Tamir (1988)],

$$m = INT\left[2\frac{W}{\lambda}\sqrt{(n_{eff}^{ridge})^2 - (n_{eff}^{milled})^2}\right], \quad (13.74)$$

$\Delta N = [(n_{eff}^{ridge})^2 - (n_{eff}^{milled})^2]^{1/2}$ is plotted versus λ/w in Fig. 13.46 with m as a parameter. The $m = 1$ can be used to estimate the maximum permissible value of ΔN for single-mode operation. N_{eff}^{ridge} is fixed by the diffusion conditions and independent of the subsequent milling. It can be determined by using Eqs. (13.72) and (13.73). The desired N_{eff}^{milled}, determined from Figure 13.46a, can be substituted in Equation (13.72) to obtain the required d_{mil}, iteratively. The results are plotted in Figure 13.46b for a gaussian diffusion profile with a surface-index increase of $\Delta n_s =$

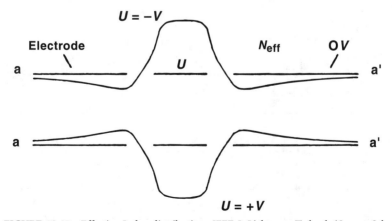

FIGURE 13.45 Effective-Index distribution. *IEEE J. Lightwave Technol.* (*Source*: Sekerka-Bajbus, M. K., Yip, G. L., and Goto, N. 1990. BPM design-optimization and experimental improvements of a Ti:LiNbo₃ ridge waveguide linear-mode confinement modulator. *IEEE J. Lightwave Technol.* 8:1742–1749.)

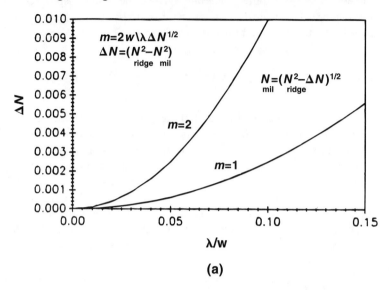

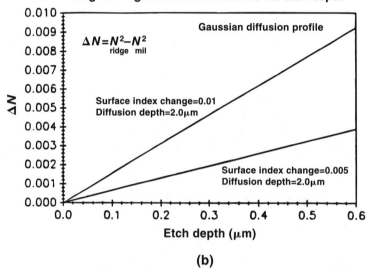

FIGURE 13.46 a,b Ridge waveguide design curves. (*Source*: Sekerka-Bajbus, M. K., Yip, G. L., and Goto, N. 1990. BPM design-optimization and experimental improvements of a Ti:LiNbo$_3$ ridge waveguide linear-mode confinement modulator. *IEEE J. Lightwave Technol.* 8:1742–1749.)

0.01 and 0.05, respectively, and a diffusion depth of $d = 2.0$ μm at $\lambda = 0.6328$ μm. The curves in Figs. 13.46a and b are very useful in designing ridge waveguides supporting $(m + 1)$ lateral modes.

The BPM simulations of the optical fields for a TM mode, propagating through a multimode modulator at an applied voltage of +5 and −5 V (side-electrodes grounded), respectively, are shown in Figure 13.47. The device's design parameters are given in Table 13.7 and are confirmed by the parameters of an actual fabricated device (sample 2 to be discussed later). The optical mode confinement in the induced channel guide in region 2 is clearly observed in Figure 13.47a. The scattered power along the branch structure, away from the mouth of the output ridge guide,

Design Methodology For Guided-Wave Photonic Devices

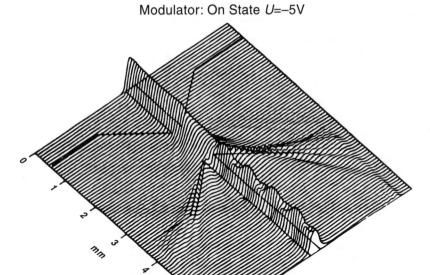

(a)

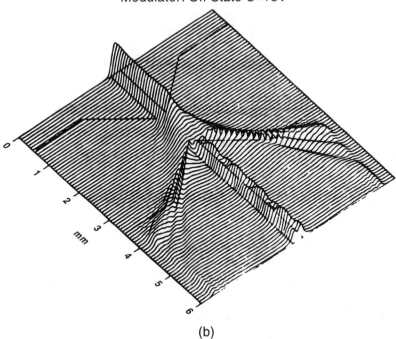

(b)

FIGURE 13.47 a,b Optical field distribution in the modulator. (*Source*: Sekerka-Bajbus, M. K., Yip, G. L., and Goto, N. 1990. BPM design-optimization and experimental improvements of a Ti:LiNbo$_3$ ridge waveguide linear-mode confinement modulator. *IEEE J. Lightwave Technol.* 8:1742–1749.)

TABLE 13.7 Device Parameters

Sample Number	1		2
Surface index change Δn_s	0.01		0.01
Diffusion depth D	2.0 μm		2.0 μm
Ridge height d^{mil}	0.3 μm		0.3 μm
Buffer layer thickness d	0.116 μm		0.08 μm
Modulation length L		1.0 mm	
Branch angle a		2.58°	
Branch length b		1.0 mm	
Ridge width w		10.0 μm	
Center electrode width		8.0 μm	
Inter electrode gap		4.0 μm	
Number of modes supported in-/output guide	3		3
NB: electrode axial misalignment	<3 μm		negligible

(*Source*: Yip, G. L. and Babin, L. 1995. Design optimization and fabrication of a widened x-branch demultiplexer by ion-exchanger in glass. In *Proc. Guided-Wave Optoelectronics: Device Characterization, Analysis, and Design.* Tamer, T., Griffel, G., and Bertoni, H. L., eds. pp. 221–229. Plenum Press, New York. With permission.)

is also clearly visible, especially for the off state. The presence of the second and third-order modes in the output ridge guides can also be seen for both voltage polarities. About 10–15% of the transmitted power propagates in these higher order modes. This amount of power can be calculated by overlapping the output modal field in region 3 with its normalized guided mode of the appropriate order m.

The theoretical model and analysis, just presented, makes it possible to study the effects of varying certain design parameters on the modulation characteristics. Such design calculations are very informative in obtaining a modulator design with some desired performance figures before proceeding on to the actual experimental work. Many BPM simulations have been performed to improve the modulator design for single-mode operations and to achieve important device characteristics such as (1) good modulation depth (>90%) and linearity of the output intensity versus the applied voltage, (2) a small drive voltage ($<\pm 10$ V, say), and (3) small device dimensions, etc. Fabrication tolerances, however, usually impose practical limitations on an achievable device design. In general, increasing the modulation length reduces the drive voltage required to switch off the power in the output guide in the off state. Reducing the branch angle α has the advantage of better separating the guided and scattered modes along the branch structure, so that large radiation intensity will not appear near the output field, especially in the device's off state. It is found that device performance is critically dependent on milling depth, but its proper value can be determined within the range permitted for single-mode operation for a particular device design.

Our design methodology is now used to improve the performance of a similar device fabricated previously with a modulation depth [$(I_{max} - I_{min})/I_{max}$] of 67% and a drive voltage of ± 20 V (Bélanger and Yip, 1986), by optimizing its photomask-independent parameters, while using the same photomask as before. With a buffer layer thickness of 0.1 μm, $\Delta n_s = 0.01$ and $d = 2$ μm, the influence of the milling depth d_{mil} is shown in Figure 13.48 for a device supporting up to three lateral modes. The device performance is, ideally, best under single-mode operation. However, the required milling depth of 0.05 μm is difficult to achieve experimentally, and coupling into such a shallow ridge guide using an input prism is also not easy. To avoid these difficulties, a milling depth of 0.3 μm is chosen. The resulting design supports three lateral modes and provides good modulation depth and linearity with a relatively small drive voltage of ± 10 V. These results are supported by experimental measurements to be presented.

Two line modulators with a ridge width of 10 μm were fabricated from the photomask designed previously without using the BPM (Sekerka–Bajbus et al., 1990). A 115-Å thick Ti film was

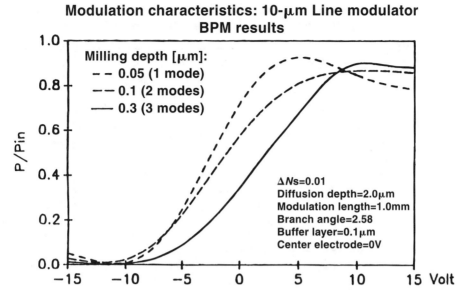

FIGURE 13.48 Modulation characteristics of a 10-μm modulator. (*Source*: Babin, L. 1993. Optimization and fabrication of a widened x-branch optical demultiplexer in glass. M. Eng. thesis. Dept. of Electrical Engineering, McGill University, Montreal, Quebec, Canada. With permission.)

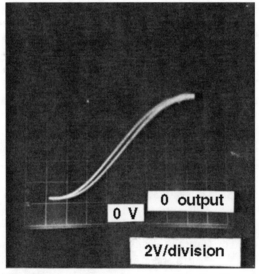

FIGURE 13.49 Measured output intensity. (*Source*: Babin, L. 1993. Optimization and fabrication of a widened x-branch optical demultiplexer in glass. M. Eng. thesis. Dept. of Electrical Engineering, McGill University, Montreal, Quebec, Canada. With permission.)

thermally evaporated onto a cleaned $LiNbO_3$ z-cut substrate. The titanium was diffused into the crystal at 975 °C for 4 hours in a flowing argon atmosphere and a further 2 hours in flowing oxygen to prevent lithium from diffusing out. This process generated a single-mode, slab waveguide with an estimated depth d and surface-index charge Δn_s of 2.0 μm and 0.01, respectively (using data as presented in section 13.3). By photolithography, the appropriate ridge structure of height d_{mil} was formed out of the slab waveguide by ion-beam milling. After milling, the sample was annealed at 500 °C for 2 hours to repair the surface damage caused by the milling process. The

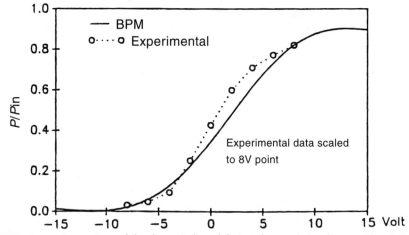

FIGURE 13.50 Comparison of the theoretical modulation characteristics. (*Source*: Yip, G. L. and Babin, L. 1995. Design optimization and fabrication of a widened x-branch demultiplexer by ion-exchanger in glass. In *Proc. Guided-Wave Optoelectronics: Device Characterization, Analysis, and Design.* Tamer, T., Griffel, G., and Bertoni, H. L., eds. pp. 221–229. Plenum Press, New York. With permission.)

TABLE 13.8 Comparison Between Experimental and Theoretical Results

	Drive Voltage Range [V]		Modulation Depth	
	theor.	exp.	theor.	exp.
Sample 1	−15 V, +20 V	−15 V, +15 V	95%	98%
Sample 2	−10 V, +10 V	−8 V, +8 V	99%	97%

(*Source*: Babin, L. 1993. Optimization and fabrication of a widened x-branch optical demultiplexer in glass. M. Eng. thesis. Dept. of Electrical Engineering, McGill University, Montreal, Quebec, Canada. With permission.)

output end was, then, polished to permit end fire coupling. Next, a thin SiO_2 buffer layer was RF sputtered onto the sample, followed by further annealing in oxygen for 2 hours. Finally, aluminum electrodes were deposited on top of the buffer layer by thermal evaporation, again through photolithography. The fabricated devices were, then, ready for measurements.

A He-Ne laser beam at 0.6328 μm was TM polarized and focused through a microscope objective lens onto a prism which coupled the light into the input ridge guide. At the output end, light leaving the ridge guide was focused through a second microscope lens onto a photodetector connected to an oscilloscope. The drive voltage applied to the electrodes was a slowly varying ramp from a function generator. Figure 13.49 shows the measured intensity versus the drive voltage for sample 2. Table 13.8 summarizes the measured along with the theoretically predicted performance. The higher drive voltage of ±15 V required for sample 1 is partly due to the slightly thicker buffer layer compared to sample 2 and an electrode misalignment. The much reduced drive voltage to ±8 V and the modulation depth of 97% for sample 2 constitute considerable improvements over previously published results (Bélanger and Yip, 1986). This is primarily due to a thinner buffer layer and a reduced ridge height, hence, reducing the number of lateral modes. Figure 13.50 compares the measured modulation characteristics with the theoretical design calculations, using the BPM. The good agreement confirms the usefulness of the BPM simulations in device design optimization.

Design and Fabrication of a Y-Branch TE-TM Mode Splitter in LiNbO$_3$ by Proton Exchange and Ti Diffusion

The geometry of this device on a z-cut LiNbO$_3$ substrate (Goto et al., 1989) is shown in Fig. 13.51. The input waveguide and arm 1 are fabricated by Ti diffusion (TI) whereas arm 2 is by proton exchange (PE). Because PE increases only the extraordinary index while having very little effect on the ordinary index, arm 2 is "seen" only by the E^x (TM) modes. If the effective index in arm 2 is higher than in arm 1, the fundamental E^x_{11} mode will propagate into arm 2. On the other hand, the existence of arm 2 has negligible effect on the E^y (TE) modes, and the fundamental E^y_{11} mode propagates along arm 1, hence, the TE-TM mode splitting. The taper angle θ_1 and branch angle θ_2 should be sufficiently small to avoid an appreciable mode conversion. They were chosen to be 0.0032 and 0.01 rad., respectively.

Again, the design must start with predetermined fabrication conditions, which are to be optimized with the design outcome in an iterative manner. Hence, the fabrication conditions are briefly described here. After Ti was diffused followed by one hour in flowing oxygen, a Ta mask pattern for PE was formed. The PE arm was formed by immersing the sample in phosphoric acid at 200°C for about 20 min., and its effective index adjusted by postannealing at 300°C for 20–60 min. After the PE process, the Ta film was removed, and the end facets of the device were polished for experimental characterization. The experimental results are presented in Table 13.9.

The effective index of diffused channel guides can be calculated by solving the dispersion equation for an equivalent homogeneous slab waveguide. The index distribution in the depth direction is assumed to be gaussian for both the TI and APE (PE followed by annealing) waveguides. In the lateral direction, the simple step-index model can be used. The optical field distribution in the device can, then, be calculated by propagating an eigenmode along this effective-index

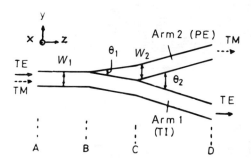

FIGURE 13.51 Comparison of the theoretical modulation characteristics. (*Source*: Yip, G. L. and Babin, L. 1995. Design optimization and fabrication of a widened x-branch demultiplexer by ion-exchanger in glass. In *Proc. Guided-Wave Optoelectronics: Device Characterization, Analysis, and Design.* Tamer, T., Griffel, G., and Bertoni, H. L., eds. pp. 221–229. Plenum Press, New York. With permission.)

TABLE 13.9 Experimental Results of TE-TM Mode Splitter

Sample No.	Fabrication Conditions	Effective Index	Extinction Ratio
1	Ti-diffusion: Ti: 107 Å Proton exchange: 20 min. at 200 °C Annealing: 15 min. at 300 °C	Arm 1 (TI): 2.2889 + 0.0005 (E^y_{11}) 2.2008 + 0.0005 (E^x_{11}) Arm 2 (PE): 2.2024 + 0.0005 (E^x_{11})	TE: >20 dB TM: = 20 dB
2	Ti-diffusion: Ti: 115 Å Proton exchange: 18 min. at 200 °C Annealing: 60 min. at 300 °C	Arm 1 (TI): 2.2950 + 0.0005 (E^y_{11}) 2.2015 + 0.0005 (E^x_{11}) Arm 2 (PE): 2.2015 + 0.0005 (E^x_{11})	TE: >20 dB TM: = −1.6 dB

(*Source*: Yip, G. L. and Babin, L. 1995. Design optimization and fabrication of a widened x-branch demultiplexer by ion-exchanger in glass. In *Proc. Guided-Wave Optoelectronics: Device Characterization, Analysis, and Design.* Tamer, T., Griffel, G., and Bertoni, H. L., eds. pp. 221–229. Plenum Press, New York. With permission.)

model, using the FFT-BPM. The power in each branch can be calculated by overlapping the optical field in the branch with a normalized eigenmode. The optical field distributions in the device are presented in Figures 13.52a and b for the TE and TM modes, respectively, using the data in Table 13.9. In Figure 13.52b the coupling of the TM modes into the PE waveguide along the taper is clearly visible. Most of the TM mode energy is transferred into the PE arm, whereas a small amount of energy is radiated into the substrate.

The amount of TM mode energy propagation into the PE arm is critically dependent on the difference in the effective indices between the two arms, $\Delta N (=N_{Pe} - N_{Ti})$. Fig. 13.53 is presented the extinction ratio for the E^z mode as a function of ΔN. Clearly, an extinction ratio of more than 20 dB for a single-mode device can be achieved only for a very narrow range of ΔN. This requires strict control of fabrication conditions and is consistent with experimental results obtained.

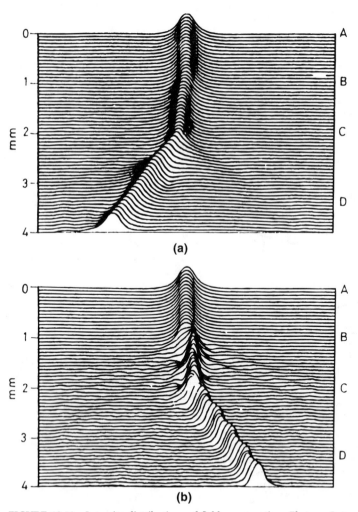

FIGURE 13.52 Intensity distributions of fields propagating. *Electron. Lett.* (*Source*: Yip, G. L. and Babin, L. 1995. Design optimization and fabrication of a widened x-branch demultiplexer by ion-exchanger in glass. In *Proc. Guided-Wave Optoelectronics: Device Characterization, Analysis, and Design.* Tamer, T., Griffel, G., and Bertoni, H. L., eds. pp. 221–229. Plenum Press, New York. With permission.)

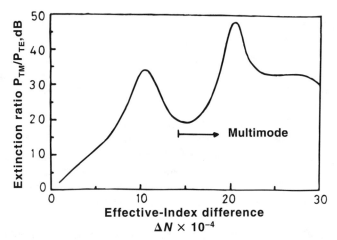

FIGURE 13.53 Extinction ratio of E^z wave. (*Source*: Yip, G. L. and Babin, L. 1995. Design optimization and fabrication of a widened x-branch demultiplexer by ion-exchanger in glass. In *Proc. Guided-Wave Optoelectronics: Device Characterization, Analysis, and Design*. Tamer, T., Griffel, G., and Bertoni, H. L., eds. pp. 221–229. Plenum Press, New York. With permission.)

The near-field patterns of the outputs in arm 1 and arm 2 and the intensity profiles are shown in Fig. 13.54. The extinction ratio of the outputs between the two arms in given in Table 6.5.1. Since the effective indices of the two arms of sample 2 are almost the same, the E^z (TM) wave is seen to split into both arms. However, sample 1 shows good TE-TM splitting because, in this case, ΔN at 16×10^{-4} has a proper value. These experimental results are in good agreement with theoretical predictions in Figure 13.53.

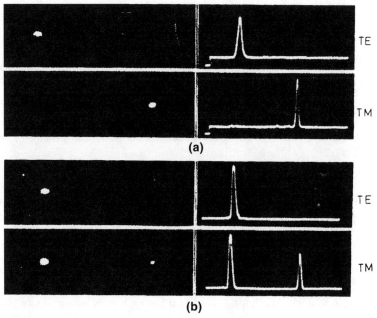

FIGURE 13.54 Near-Field patterns and their intensity profiles. (*Source*: Yip, G. L. and Babin, L. 1995. Design optimization and fabrication of a widened x-branch demultiplexer by ion-exchanger in glass. In *Proc. Guided-Wave Optoelectronics: Device Characterization, Analysis, and Design*. Tamer, T., Griffel, G., and Bertoni, H. L., eds. pp. 221–229. Plenum Press, New York. With permission.)

13.7 Conclusions

In this chapter, we have presented the theoretical and experimental characterizations of slab and channel waveguides by K^+-ion exchange in glass and Ti diffusion into $LiNbO_3$ and the modeling of their index profiles. These serve as illustrations for waveguides fabricated by diffusion-related technologies. The information obtained is important to the analysis, design, and implementation of planar waveguide devices, made by such technologies and employing slab and channel guides as basic structures. Once the index profiles are accurately modeled, the design procedure, involving a combination of the effective-index method and beam propagation method as presented in Sections 6.1, 6.3–6.5, or a combination of the effective-index method and variational method as presented in Section 6.2, then leads to design calculations of device performance parameters, which are well supported by measured performance parameters of the fabricated devices in the illustrative examples given.

We conclude that the key to accurate index modeling of a particular waveguide structure lies in a detailed and clear understanding of the physical processes going on in the fabrication stage with prescribed fabrication conditions, for example, diffusion and annealing temperature and time. Only after establishing an accurate index model, would clever mathematics play a significant role in the design stage, eventually leading to an optimized design. The methodology presented here should be generally applicable to all planar waveguide devices fabricated by the diffusion process, for example, ion exchange in glass involving other ion sources and substrates, Ti diffusion into $LiNbO_3$ and $LiTa_2O_3$ and proton exchange in the two latter substrates. In a wider sense, the design philosophy should be equally applicable to waveguides made from III–V compounds and polymers. Regrettably, we cannot elaborate here.

With the inclusion of sessions devoted to "Modeling, Numerical Simulation and Theory" since 1989, in such an important technology-related topical meeting (IEEE) as "Integrated Photonics Research" and the numerous workshops and symposia on "Design and Simulation of Guided-Wave Optoelectronic Devices", time seems to have arrived to achieve devices, which do not simply work on clever ideas but which have also been design optimized. Design predictions can be matched with the experimental performance of fabricated devices. In the process, much experimental effort can be saved.

Acknowledgements

The author would sincerely like to thank Dr. M. Gupta for his kind invitation to contribute this chapter. Former research associates and graduate students, J. Albert, M. Bélanger, L. Babin, J. Y. Chen, J. Finak, N. Goto, M. Haruna, K. Kishioka, P. C. Noutsios, M. K. Sekerka-Bajbus, and F. Xiang have contributed to the materials presented in this chapter. Our research was supported by NSERC (Natural Sciences and Engineering Research Council of Canada) operating and strategic grants. P. Jorgensen, L. Chen, and A. Yip typed the manuscript. Last but not least, the author would like to express his appreciation to his wife, Alice, and children, Gwendolyn, Alexandra, and Santa, for their understanding and support.

References

Abou-el-Liel, M., and Leonberger, F. 1988. Model for ion-exchanged waveguides in glass. *J. Amer. Cer. Soc.* 71, 497.

Albert, J., and Yip, G. L. 1988. Wide single-mode channels and directional coupler by two-step ion-exchange in glass, *IEEE J. of Lightwave Technol.* 6, 552.

Alferness, R. C. Titanium-Diffused Lithium Niobate Waveguide Devices. Chapter 4 In *Guided-Wave Optoelectronics*, Tamir, T. ed. Springer-Verlag, 1988, pp. 145–210.

Babin, L. Optimization and fabrication of a widened x-branch optical demultiplexer in glass. M. Eng. thesis, McGill University, 1993.

Babin, L. J. M., and Yip, G. L. Design and realization of a widened x-branch demultiplexer at 1.31 μm and 1.55 μm by K^+-Na^+ ion exchange in a glass substrate, OSA Topical Meeting on Integrated Photonics Research, February 1994, San Francisco, Calif., Tech. Dig., pp. 241–243.

Bélanger, M. and Yip, G. L. 1986. A novel Ti:LiNbO$_3$ ridge waveguide linear mode confinement modulator fabricated by reactive ion-beam etching. *IEEE J. Lightwave Technol.* 22, 252.

Burns, W. K., Klein, P. H., West, E. J., and Plew, L. E. 1979. Ti diffusion in Ti:LiNbO$_3$ planar and channel optical waveguides. *J. Appl. Phys.* 50, 6175.

Chartier, G. H., Jaussaud, P., de Oliveira, A. D., and Parriaux, O. 1978. Optical waveguides fabricated by electric-field controlled ion exchange in glass. *Electron. Lett.* 14, 132.

Chartier, G., Collier, P., Guez, A., Jaussaud, P., and Won, Y. 1980. Guided-Index surface or buried waveguides by ion-exchange in glass. *Appl. Opt.* 19, 1092.

Chen, T. R., and Yang, Z. L. 1985. Modes of a planar waveguide with Fermi index profile. *Appl. Opt.* 24, 2809.

Cheng, H. C., and Ramaswamy, R. V. 1991. Symmetrical directional coupler as a wavelength multiplexer-demultiplexer: Theory and experiment. *IEEE J. Quantum Electron.* QE-27, 567.

Crank, J. *The Mathematics of Diffusion*, 2nd Ed., Clarendon Press, London, 1975.

Čtyroký, J., Hofman, M., Janta, J., and Schröfel, J. 1984. 3-D analysis of Ti:LiNbO$_3$ channel waveguides and directional couplers. *IEEE J. of Quantum Electron.* QE-20, 400.

Chung, Y., 1989. Analysis of a tunable multi-channel two-mode-interference wavelength division multiplexer/demultiplexer. *IEEE J. Lightwave Technol.* 7, 766.

Chung, Y., and Dagli, N. 1990. Explicit finite difference beam propagation method: application to semiconductor rib waveguide Y-junction analysis. *Electron. Lett.* 26, 711.

Doremus, R. H. Ion exchange in glass, in *Ion-Exchange*, vol. 2, Marinsky, J. A., (ed.) Marcel Dekker, New York, 1969.

Feit, M. D., and Fleck, J. A. 1978. Light propagation in graded-index optical fiber. *Appl. Opt.* 17, 3990.

Feit, M. D., and Fleck, J. A., Jr. 1980. Computation of mode eigenfunctions in graded-index optical fibers by the propagation beam method. *Appl. Opt.* 19, 2240.

Feit, M. D., Fleck, J. A., and McCaughen, L. 1983. Comparison of calculated and measured performance of diffused channel-waveguide couplers. *J. Opt. Soc. Am.* 73, 1296.

Fleck, J. A., Morris J. R., and Feit, M. D. 1976. Time-dependent propagation of higher energy laser beams through the atmosphere. *Appl. Phys.* 10, 129.

Forber, R. A., and Marom, E. 1986. Symmetric directional coupler switches. *IEEE J. Quantum Electron.* 2E-22, 911.

Fukuma, M., Noda, J., and Iwasaki, H. 1978. Optical properties in titanium-diffused LiNbO$_3$ strip waveguides. *J. Appl. Phys.* 49, 3693.

Gerald, G. F. *Applied Numerical Analysis*, Addison-Wesley, 1970.

Giallorenzi, T. G., West, E. J., Kirk, R., Ginther, R., and Andrews, R. A. 1973. Optical waveguides formed by the thermal migration of ions in glass. *Appl. Opt.* 12, 1240.

Gill, S. 1951. A process for the step by step integration of differential equations in an automatic digital computing machine. *Proc. Camb. Philos. Soc.* 47, 96.

Goto, N., Sekerka–Bajbus, M. R., and Yip, G. L. 1989. BPM analysis of Y-branch TE-TM mode splitter in LiNbO$_3$ by proton exchange and Ti diffusion. *Electron. Lett.* 25, 1732.

Hadley, G. R, 1991. Transparent boundary conditions for beam propagation. *Opt. Lett.* 16, 624.

Haruna, M., Bélanger, M., and Yip, G. L. 1985. Passive 3-branch optical power dividers by K^+-ion exchange in glass. *Electron. Lett.* 21, 535.

Hocker, G. B., and Burns, W. K. 1977. Mode dispersion in diffused channel waveguide by the effective index method. 16, 113.

Huang, W. P., Xu. C. L., Chu, S. T., and Chaudhuri, S. K. 1992. The finite difference vector beam propagation method: Analysis and assessment. *IEEE J. Lightwave Technol.* 10, 295.

Izawa T., and Nakagome, H. 1972. Optical waveguide formed by electrically induced migration of ions in glass plates. *Appl. Phys. Lett.* 21, 584.

Jackel, J. L., Rice, C. E., and Veselka, J. J. 1982. Proton-exchange for high index waveguides in $LiNbO_3$. *Appl. Phys. Lett.* 41, 607.

Kaminow, I. P. *An Introduction to Electro-optic Devices*, Springer-Verlag, Berlin, 1974.

Kishioka, K., 1991. Characterization of ion exchange waveguide made with diluted KNO_3. *SPIE Proc.* 1583, 19.

Knox, P. M., and Toulios, P. P., Integrated circuits for the millimeter through optical frequency range, Proc. for Symp. on Submillimeter Waves, J. Fox, (ed.), Poly. Inst. of Brooklyn, March, 1990, pp. 497–56.

Liu, P. L., and Li, B. J. 1991. Study of form birefringence in the waveguide devices using the semi-vector beam propagation method. *IEEE Photon. Techno. Lett.* 3, 913.

Marom, E., and Ruschin, S. 1984. Relation between normal mode and coupled mode-analysis of parallel waveguides. *IEEE J. Quantum Electron.* QE-20, 1311.

Miliou, A., Zhenguang, H., Cheng, H. C., Srivastava, R., and Ramaswamy, R. 1989. Fiber-compatible K^+-Na^+ ion-exchanged channel waveguides: fabrication and characterization. *IEEE J. Quantum Electron.* 25, 1889.

Minakata, M., Saito, S., Shibata, M., and Miyazama, S. 1978. Precise determination of refractive-index changes in Ti-diffused $LiNbO_3$ optical waveguides. *J. Appl. Phys.* 49, 4677.

Naitoh, H., Nuroshita, M., and Nakayama, T. 1977. Mode control of Ti-diffused $LiNbO_3$ slab optical waveguides. *Appl. Opt.* 16, 2546.

Neyer, A. 1983. Electro optic x-switch using single mode $LiNbO_3$ channel waveguides. *Electron. Lett.* 19, 553.

Nishihara, H., Haruna, M., and Suhara, T. *Optical Integrated Circuits.* McGraw-Hill, 1989, Chapter 2.

Optical Materials, Optics Guide 4, Melles Griot Corporation, Irvine, Calif. 1988, pp. 3–10.

Popuchon, M., Roy, A., and Ostrowsky, D. B. 1977. Electrically active bifurcation: BOA. *Appl. Phys. Lett.* 31, 266.

Ralston, A., and Wilf, H. S. *Mathematical Methods for Digital Computers.* John Wiley & Sons, 1960, Vol. 1, pp. 111–120.

Ramaswamy, R. V., and Najafi, S. I. 1986. Planar, buried, ion-exchanged glass waveguides: Diffusion characteristics. *IEEE J. Quantum Electron.* 22, 883.

Ramer, O. G. 1982. Integrated optic electrooptic modulator electrode analysis. *IEEE J. Quantum Electron.* QE-18, 386.

Saijonmaa, J., and Yevick, D. 1983. Beam propagation analysis of loss in bent optical waveguides and fibers. *J. Opt. Soc. Am.* 73, 1785.

Scarmozzino, R., and Osgood, R. M., Jr. 1991. Comparison of finite difference and Fourier transform solutions of the parabolic wave equation with emphasis on integrated optics applications. *J. Opt. Soc. Am.* 8, 724.

Schmidt, R. V., and Kaminow, I. P. 1974. Metal diffused optical waveguides in $LiNbO_3$. *Appl. Phys. Lett.* 25, 458.

Sekerka–Bajbus, M. K., Yip, G. L., and Goto, N. 1990. BPM design-optimization and experimental improvements of a Ti:$LiNbO_3$ ridge waveguide linear-mode confinement modulator. *IEEE J. Lightwave Technol.* 8, 1742.

Sharma, A., Sharma, E., Goyal, I. C., and Ghatak, A. K. 1980. Variational analysis of directional couplers with graded index profile. *Opt. Commun.* 34, 39.

Stewart, G., Miller, C. A., Laybourn, P. J. R., Wilkinson, C. D. W., and De La Rue, R. M. Planar optical waveguides formed by silver ion migration in glass, *IEEE J. Quantum Electron.* QE-13, 192.

Tamir, T., ed., *Guided-Wave Optoelectronics*. Springer-Verlag, 1988, Chapter 2.

Tamir, T., ed., *Guided-Wave Optoelectronics*. Springer-Verlag, 1988, p. 16.

Taylor, H. F. 1976. Dispersion characteristics of diffused channel waveguides. *IEEE J. Quantum Electron.* QE-12, 748.

Thylén L., and Gravestrand, P. 1986. Integrated optic electro-optic device electrode analysis: The influence of buffer layers. *J. Opt. Commun.* 7, 11.

Tien, P. K., and Ulrich, R. 1970. Theory of prism-film coupler and thin-film light guide, *J. Opt. Soc. Am.* 60, 1325.

Vandenbulcke, P., and Lagasse, P. E. 1974. Static field analysis of thin film electro-optic modulators and switches. *Wave Electron.* 1, 295.

Xiang F., and Yip, G. L. 1992. Simple technique for determining substrate indices of isotropic materials by a multisheet Brewster angle measurement. *Appl. Opt.* 31, 7570.

Yip, G. L., and Finak, J. 1984. Directional-coupler power divider by two-step K^+-ion exchange. *Opt. Lett.* 9, 423.

Yip, G. L., and Albert, J., 1985. Characterization of planar optical waveguides by K^+-ion exchange in glass, *Opt. Lett.* 10, 151.

Yip, G. L., and Sekerka–Bajbus, M. A. 1988. Design of symmetric and asymmetric passive 3-branch power dividers by beam propagation method. *Electron. Lett.* 25, 1584.

Yip, G. L., Noutsios, P. C., and Kishioka, K. 1990. Characteristics of optical waveguides made by electric-field-assisted K^+-ion exchange. *Opt. Lett.* 15, 789.

Yip, G. L., Kishioka, K., Xiang, F., and Chen, J. Y. Characterization of planar optical waveguides by K^+-ion exchange in glass at 1.152 and 1.523 um. SPIE Proc. on Integrated Optical Circuits, Sept. 1991, Vol. 1583, pp. 14–18.

Yip, G. L., and Babin, L. J. M. BPM design of a widened x-branch demultiplexer in a glass substrate by K^+ ion exchange for $\lambda = 1.31$ μm and $\lambda = 1.55$ μm, OSA Topical Meet. on Integrated Photonics Research, March 1993, Palm Springs, Calif., Tech. Dig. pp. 72–754.

PART C
PHOTONIC SYSTEMS

14

Analog Coherent Optical Processing

David Casasent
Carnegie Mellon University

14.1 Introduction 591
14.2 Basic Operations Achievable 593
    Optical Fourier Transform (FT) • Optical Correlator • Extension of Operations Achievable
14.3 Components 596
    Spatial Light Modulators (SLMs) • Computer-Generated Holograms (CGHs)
14.4 Optical Signal Processors 599
14.5 Optical Image Processors 601
14.6 Summary 603

14.1 Introduction

This chapter discusses coherent analog optical processors for signal and image data. This involves a description of the basic architectures, the components from which they are fabricated, the operations performed, and why such operations are useful.

Section 2 describes the basic two-dimensional (2-D) operations considered: the Fourier transform and correlation. The salient design equations are provided for typical versions of such processors. Emphasis is given to systems level descriptions, and, thus, lens design is not addressed. Increased flexibility in the operations achievable is provided by computer-generated holograms, which can provide arbitrary complex-valued 2-D transfer functions. An enumeration of various optical feature spaces that can be produced with such elements is provided as examples of the many operations possible. The image processing use of the different feature spaces is noted as well as the general nature of the operations they provide.

Section 3 discusses the light modulator devices available. These are generally accepted as the limiting element in all optical processors. Because Chapter 9 details these devices, emphasis, here, is given to discussing the specifications available and required and how these affect the viable architectures and their performance. Section 3 also discusses computer-generated holograms and their vital role in fabricating cost-effective optical processors. Binary optics details these optical elements, and, thus, the focus of this chapter is on the functions to be recorded on such elements.

Signal and image processing requirements are quite different, and, thus, they are separately addressed in Sections 4 and 5. Many commercial optical signal processing systems have been fabricated and a set of optical hardware modules for signal and image processing are presently under development (Casasent 1993, 1994) and various optical component hardware are available commercially and from a new optoelectronics consortium (Consortium, 1993). Because it is premature to provide a table of system level specifications at this time, the optical processor architectures and components used are noted and references are provided to the growing list of operations achievable. Section 4 describes three of the major optical signal processing architectures and their properties. The hardware modules being developed are also noted. Section 5 provides similar data on optical image processors, notes the various operations that this author feels are needed for such systems to be of general use, and provides available detail on how to achieve such operations.

Because final hardware is not yet available, the structure of this chapter was chosen to allow the reader to determine the operations that can be achieved, the architecture to use, the available components, the design of such systems, and references for further details.

14.2 Basic Operations Achievable

Optical Fourier Transform (FT)

This operation is the hallmark of coherent optics. The basic 2-D system is shown in Fig. 14.1. The Fourier transform (FT) $F(u, v)$ of the input data $f(x, y)$ at P_1 is produced in plane P_2 with a spherical lens L_1. The input P_1 data must be available as a transparency. If it is an actual object, a diffraction pattern is produced that contains related information, but not in the easily analyzed FT form. The 2-D input $f(x, y)$ contains various spatial frequencies (u, v) oriented horizontally (u), vertically (v) or at any angle. Each of these produces a spot of light in two symmetric quadrants of P_2. The relationship between the spatial locations (x_2, y_2) in P_2 and input spatial frequencies is given by

$$(x_2, y_2) = (\lambda f_L u, \lambda f_L v), \tag{14.1}$$

where λ is the wavelength of the input coherent light and f_L is the focal length of the lens used. An input horizontal (vertical) frequency $u(v)$ produces two peaks in P_2 on the $\pm x_2(\pm y_2)$ axis. Input frequencies at different orientations produce peaks at angles equal to the orientations of the input spatial frequencies. Thus, only half of P_2 is needed to determine the frequencies present in the input. The amplitudes of these output P_2 peaks indicate the amount of each input spatial frequency present. Because all input images are real and positive, they are present on a bias and the value of this bias or dc value of the input is present at the central on-axis point in P_2.

Plane P_2 is viewed either by eye or, in general, by a 2-D detector array, and, thus, the intensity of the complex-valued FT is the sensed output. This is a great advantage, because the loss of phase information provides this system with one form of shift invariance, i.e., the $|FT|^2$ sensed

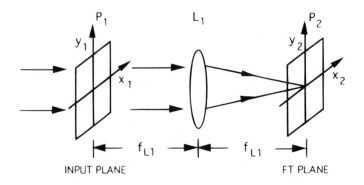

FIGURE 14.1 Basic 2-D optical Fourier transform system.

Analog Coherent Optical Processing

output is the same for an input object, regardless of its location in the input plane P_1. This is quite significant, because it allows one to determine the presence of an object (characterized by its $|FT|^2$) independent of its location. The relative locations of all spatial frequencies in the object are still present in the $|FT|^2$ output; only its position in P_1 is lost. This is advantageous, as it greatly simplifies the training required and the output analysis needed.

If the position of the input object f is desired, only the linear portion of the phase of the $|FT|^2$ output is needed to determine this information. To see this, we note that the FT of a 1-D shifted function $f(x - a)$ is related to the FT (described by the operator $\mathcal{F}$) $F(u)$ of the original unshifted function $f(x)$ by

$$\mathcal{F}[f(x - a)] = F(u) \exp(-j2\pi u a). \tag{14.2}$$

One can extract the linear phase component of P_2 by using a reference beam incident on P_2 and simply filtering the frequency of the recorded fringe pattern.

The following design equations are presented in 1-D for simplicity. For a given maximum input spatial frequency u_m, light leaves P_1 at an angle of diffraction ϕ, where

$$\text{Tan } \phi = \lambda u_m. \tag{14.3}$$

The size of lens aperture must be equal to

$$L + 2f_L \text{ Tan } \phi \tag{14.4}$$

where L is the 1-D size of the input P_1 data. The physical extent of P_1 is given by

$$2x_{2m} = 2\lambda f_L u_m. \tag{14.5}$$

The required size d of a P_2 detector is related to the input 1-D space bandwidth product (SBWP)

$$\text{SBWP} = L u_m \tag{14.6}$$

by

$$d = \lambda f_L u_m / \text{SBWP} = \lambda f_L / L. \tag{14.7}$$

This is half the width of the sinc function for an input P_1 aperture of width L. The number of P_2 detectors needed (in 1-D) is the 1-D SBWP of the input. These detectors are placed only in one half of the FT plane. This is also the number of spatial frequencies resolvable.

Equations (14.1) and (14.3)–(14.7) provide the major design equations for this basic 2-D FT system. Table 14.1 lists the major information available from the optical $|FT|^2$ output obtained. In practice, one, generally, places the input P_1 behind (to the right) the lens a distance d from P_2, with the L_1 to P_2 spacing still being f_L. This results in a scaling FT system with d replacing f_L in the prior equations. This is a practical expedient to reduce the size of the optical FT system

TABLE 14.1 Optical Fourier Transform Plane P_2 Detected Information

Central point gives the dc value of the input.
Peaks give the input spatial frequencies present.
Peak amplitudes provide the amount of each input spatial frequency present.
Location (x_2, y_2) of each peak provides orientation information on each input spatial frequency.
The detected intensity pattern is shift invariant.

and to match available detector sizes and λ values for input data with given maximum input spatial frequencies u_m and v_m.

Optical Correlator

The basic 2-D optical frequency plane correlator, shown in Fig. 14.2, is a cascade of two optical FT systems. The light amplitude distribution incident on P_1 is F (the FT of the input f). The transmittance of the filter at P_2 is $H^*(u, v)$ (the conjugate FT of $h(x, y)$. The P_3 amplitude output is the FT of FH^* which is the correlation (denoted by the symbol ⊗) of the two 2-D spatial functions f and h,

$$f \otimes h = \iint f(x, y)g(x - \xi, y - \eta) \, dx \, dy = c(\xi, \eta). \tag{14.8}$$

This is one of the most powerful operations in optical processing. Its use is well known in pattern recognition to locate multiple objects in an input scene with clutter present.

The fabrication of this architecture follows. The size of the P_2 filter plane in 1-D is given by Eq. (14.5) as $2x_{2m}$. If the lenses L_1 and L_2 have focal lengths f_{L1} and f_{L2} and are spaced from P_1, P_2, and P_3 by their f_L, as shown, then, the size of the correlation region of plane P_3 in 1-D is given by

$$(f_{L2}/f_{L1})(L_f + L_h) \tag{14.9}$$

where L_f and L_h are the physical extents of the f and h functions in 1-D.

Eqs. (14.4), (14.5), and (14.9) detail fabrication of this system. In practice, lens L_1 is placed in contact with P_1 or behind it, and a new second lens L is placed in contact with P_2. This results in a compact optical correlator (Davis et al. 1989).

The P_3 output contains peaks at the locations of all occurrences of the object h in the input image f. This is the powerful shift-invariance property of a correlator. It allows one to locate multiple versions of an object (or several objects) in the P_1 input plane.

Extension of Operations Achievable

Optical processors can produce many other operations beyond the basic linear system functions of the FT and correlation. These are generally achieved by using computer-generated hologram (CGH) elements (Cindrich and Lee 1991; Casasent 1985). These optical elements (Binary Optics, Chapter and Section 3), fabricated by integrated circuit techniques, allow production of an optical

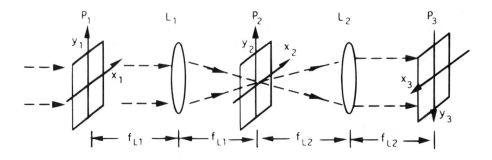

FIGURE 14.2 Basic frequency-plane optical correlator system.

Analog Coherent Optical Processing

TABLE 14.2 Selected List of Operations Achievable Using Computer-Generated Holograms

Transmittance Function	Output Produced	Use and Remarks	Reference
$\exp[jk(x^2 + y^2)/2f_1]$	Fourier transform (FT)	Signal and image processing	Cindrich and Lee 1991
Frequencies at different orientations in separate spatial regions	Wedge- and ring-sampled FT	Scale and rotation-invariant feature space	Casasent et al. 1986
Sum of orthogonal cylindrical lenses	Hough transform	Location, length, and parameters of input lines and curves	Casasent and Richards, 1988a
Polar-log spatially varying frequencies	Polar–Mellin transform	General coordinate transform Space-variant system Scale and rotation invariance	Casasent et al. 1987
Frequency-multiplexed monomials	Moments	Image processing	Casasent et al. 1982
		Multiple transfer functions Frequency-multiplexing	

element with any general 2-D amplitude and phase transmittance function. Table 14.2 lists a selected subset of the various optically generated feature spaces that can be produced using such elements. The optical system is generally that shown in Fig. 14.3, which is quite simple. Table 14.2 notes the transmittance function recorded on the CGH, the output produced, its use, and other remarks. References to detailed equations for the CGH functions are provided in Table 14.2. The operations noted are not a complete list. They are chosen to note the variety and flexibility of operations possible. Brief remarks follow on the different CGHs noted.

The first example in Table 14.2 notes that an FT lens can be fabricated as a CGH, with the advantages of size, weight, and cost. Because the number of FT features is large (the input SBWP), the FT has been sampled with wedge- and ring-shaped detector elements (one output value is provided for all $|FT|^2$ energy in a wedge- or ring-shaped region symmetric about the center of the FT plane). This reduces the dimensionality of the FT space and provides a set of in-plane, distortion-invariant features describing the input (the $|FT|^2$ output is shift-invariant, the wedge outputs are scale-invariant and the ring outputs are rotation-invariant). Numerous uses for this feature space have been detailed (Lendaris and Stanley 1970; Clark and Casasent 1988). The second example notes this CGH. The third CGH noted is a combination of multiple lenses. These integrate the P_1 input in different directions and produce a Hough transform (Duda and Hart 1972) output in which the locations and heights of peaks describe the locations and lengths of lines in the input P_1 image. This operation has many uses in product inspection (Casasent and Richards 1988b) and general image processing (Casasent and Richards 1988b; Gindi and Gmitro 1984).

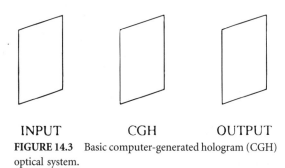

INPUT   CGH   OUTPUT

FIGURE 14.3 Basic computer-generated hologram (CGH) optical system.

In the fourth example chosen, the CGH produces a coordinate transform of the input P_1 data and an FT of the result. The specific example noted produces a polar and/or Mellin transform with distortion-invariant properties. This is an excellent example of the power of CGHs, in which each input point can be mapped to any desired output point (or to several output points). In the most general sense (Casasent and Psaltis 1979), this CGH produces an optical system with a different impulse response for each input point (and a reason for such a space-variant optical system). Examples 2 to 4 are excellent cases in which conventional lenses cannot provide the operations described.

The last example noted in Table 14.2 uses the optical system of Fig. 14.4 in which P_1 is imaged onto P_2 and the FT of the light leaving P_2 is produced at the P_3 output where it is sampled at one point. If the P_1 input is $f(x, y)$ and the transmittance of P_2 is $g(x, y)$, the P_3 output (on-axis) is given by

$$\iint f(x, y) g(x, y) \, dx \, dy. \tag{14.10}$$

This shows how an optical system can perform a point-by-point product of two functions (multiplication) and the parallel summation of all partial products. The P_2 function can contain the sum of N different functions g_n, each on a different spatial frequency carriers w_n (in x and y). This frequency-multiplexed property of an optical system allows parallel processing with multiple transfer functions. Specifically, the P_3 output is Eq. (14.10) with N different functions g_n with each scalar output located at a different P_3 position. To produce the moments,

$$m_{pq} = \iint f(x, y) x^p y^q \, dx \, dy, \tag{14.11}$$

of an input function $f(x, y)$, the different g_n functions are the monomials $x^p y^q$.

14.3 Components

The elements of an optical processing system generally consist of input, filter, and CGH planes; free space; lenses or CGH elements; a light source; and an output detector.

Spatial Light Modulators (SLMs)

The key element in any optical data processor that determines its practicality is, typically, the spatial light modulator (SLM) device used to enter data into the optical system. These elements

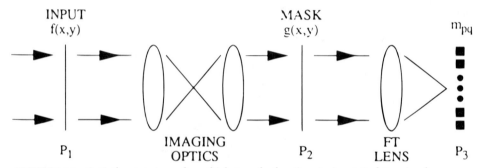

FIGURE 14.4 Optical system to apply multiple transfer functions to input P_1 data using frequency-multiplexed computer-generated holograms (CGHs) at P_2.

are essential for use at the input P_1 in Fig. 14.1–14.4. Table 14.3 lists the issues associated with the system level use of 2-D SLMs. Most parameters are self-explanatory. Specific values for various SLMs are given in the chapter on modulators.

The first and major parameter distinguishing different SLMs is whether the input data is optically or electronically written onto the device. Optically addressed SLMs are more restrictive (because the input scene must be imaged onto the device in the proper wavelength range) and require excellent sensitivity (because ambient light is required). Input data can be displayed on a CRT, and the CRT can be imaged onto the SLM, but this increases the cost and size of the system. Thus, electrically addressed SLMs are the most attractive ones, because they allow input from any sensor data source and, hence, result in much more general purpose systems.

For optically addressed SLMs, modulation and the available output light decreases as the resolution of the input data increases. Resolution at 50% of the maximum modulation is generally a useful measure, although operation at higher resolution with reduced modulation can be allowed in some applications if the output data can be measured at reduced light levels. Electrically addressed devices have more obvious resolution values. For optical processing, small-sized input pixels are necessary or long focal length lenses and associated large-sized processors result (because the size of the FT plane must be large enough to detect the highest input spatial frequency). The input SLM must have a gray scale, because images and signals do. One exception is a morphological processor (section 5). Sixteen to sixty-four gray levels are generally adequate for most applications. Linearity of the gray level (of the light leaving the device) with the input analog value is desirable but is not essential in image processing.

Optical efficiency is important because it determines the amount of optical power necessary and the sensitivity required by the output detector (i.e., the light budget of the processor). The uniformity of an SLM refers to how uniform is the output light across the device when the input data is constant. This value is often only 5–10%, but one can, generally, correct for this with fixed external masks. Similar remarks apply to optical flatness.

The speed of an SLM is generally of primary concern because high processing rates are necessary to compete with electronic processing. The data cannot, generally, be processed until the full frame of information has been written. Thus, the SLM must have some storage. It must also have an active erase mechanism or, equivalently, the device's output for the next frame of data must not be a function of the prior frame of data. Erasure must generally be very complete, i.e., erasure to the 10% point is rarely acceptable. These storage and erasure requirements are what

TABLE 14.3 Two-Dimensional Spatial Light Modulator (SLM) Parameters of Concern

Parameter	Remarks
Input data	Optically addressed SLMs required CRT or optical input and are generally less useful.
	Electrically addressed SLMs result in a more general-purpose system with input from any sensor.
Resolution	At 0.5 or 0.02 of maximum modulation (for optically addressed devices).
	The number of pixels in an electrically addressed device.
	Small pixel sizes are essential for compact processors.
Gray scale	Number of input gray levels resolvable.
	Linearity is desirable.
Contrast	Maximum to minimum output light level.
Efficiency	Useful amount of input light leaving the SLM.
Uniformity	Can generally be corrected for with fixed masks.
Dead spaces	Non useful region between input elements in an electrically addressed device.
Optical flatness	Generally $\lambda/4$–$\lambda/10$ is desired.
	This can, generally, be externally corrected.
Speed	Write, read, and erase cycle time.
	Cannot read and process until the full frame has been written.
	Is erasure active and complete?
Miscellaneous	Cost, size, weight, volume, and power dissipation.

distinguish SLMs from displays (optical flatness, the need for high spatial frequency or small pixels, and no general need for color are other differences). Display devices can, clearly, be modified for use as SLMs. Most presently available SLMs are only binary devices, but they can achieve frame rates of 1000–5000 2-D images per second. They can also operate as phase modulators (0 and π phase transmittance levels) and in some cases as ternary phase-amplitude devices (with transmittances of 0 and ± 1). As a result, a popular current approach is to apply many P_2 filters (in the system of Fig. 14.2) at very high rates (using a fast SLM at P_2) for a given P_1 input image. Simple filters can be designed that require only binary or ternary levels, however, this class of filters is very limited.

Acousto-Optic (AO) devices (see chapter on modulators) are a very different class of electrically addressed SLM. Input electrical signals are converted to sound waves that travel the length of an AO cell (they do not reflect from the far end of the device). Thus, they produce modulated traveling waves that vary in space and time. AO devices exist with multiple electrical inputs and, hence, with parallel channels of data. AO devices have excellent gray scales and operate at very high data rates. They are, generally, used in signal rather than in image processing. However, one very attractive AO architecture exists (Molley and Stalker 1990; Psaltis, 1984) that produces 2-D gray-scale correlations with electrically addressed and adaptive 2-D filter functions.

Computer-Generated Holograms (CGHs)

These optical elements are detailed elsewhere (Chapter 17), and some of their uses in optical processing were noted in section 2. In data processing, the different 2-D filter functions required in image processing can be fabricated as CGHs. In such cases, the CGHs have 2-D, complex-valued amplitude and phase transmittances. Amplitude modulation is generally achieved through halftoning and phase modulation by varying the optical thickness of the CGH; alternate encoding techniques exist (Cindrich and Lee 1991). The relevant CGH parameters are the number of amplitude and phase levels and the number of pixel elements (SBWP) possible in a given area. With e-beam and analog photoresist fabrication methods, 0.1-μm resolution and 64 phase levels are easily achieved. Thus, 2-D CGH spatial filters can easily be fabricated.

These CGH optical elements are fixed on film, rather than being adaptive. Thus, the architecture of Fig. 14.5 is a very attractive approach to utilizing a large bank of fixed optical filters in a 2-D correlator. In this architecture, a 2-D laser diode array is placed at P_0. The laser diode activated selects the spatial location at P_2 that is accessed (the FT of P_1 occurs at different P_2 locations, depending upon the P_0 laser diode activated). At each P_2 spatial location, several frequency-

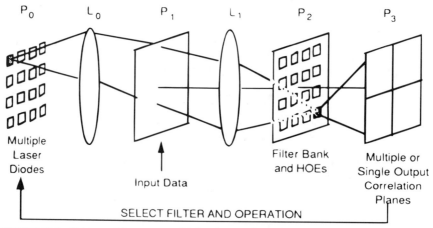

FIGURE 14.5 Space- and frequency-multiplexed, general-purpose, optical correlator using a fixed filter bank of CGHs.

Analog Coherent Optical Processing

multiplexed filters are placed, and the output P_3 correlation plane contains the correlation of the P_1 input data and the multiple filters at the selected P_2 spatial location (four such frequency-multiplexed filters is a reasonable number and four correlation planes are shown in Fig. 14.5). CGHs are necessary after each input P_0 laser diode. CGHs (lenses or holographic optical elements, HOEs) are also necessary on each P_2 spatially multiplexed filter to produce the output P_3 correlation plane in the same location (where the P_3 detector array is placed). This use of a fixed bank of space- and frequency-multiplexed CGH P_2 filters produces a very attractive general purpose optical processor, in which one optical architecture can achieve different operations depending upon the P_2 filter accessed. The major advantage of this Fig. 14.5 architecture is that it does not require a real-time P_2 SLM. Its speed is, thus, much greater and is limited (in practice) by the read-out rate of the P_3 detector. The read-out rate is generally limited to approximately 500 frames/second with present 2-D detector and electronic technology.

Note that a CGH can also produce multiple replicas of the FT of the input P_1 data at different locations in P_2. In such a case, different filter functions at different spatial P_2 locations can be accessed in parallel, and, again, multiple output correlation planes can be produced (producing four such parallel correlation planes is realistic with present detector technology).

14.4 Optical Signal Processors

Signal processing typically refers to the processing of 1-D waveforms (such as radar signals) at high frequencies (much above television rates). The basic operations required (Berg and Lee 1983) are FTs (to determine the presence of signals, their frequencies, and the velocity of objects) and correlations (to determine range and to demodulate data). When multiple antennae are used, the FT across the received signals determines the angle of a signal source (direction of arrival).

Optical systems with acousto-optic (AO) devices are generally used, because of the high bandwidth of these devices. Table 14.4 lists the AO cell and signal parameters used. Doppler (velocity) resolution is inversely proportional to the signal duration T_S, and range resolution is inversely proportional to the signal bandwidth BW_S. Thus, long duration signals with high bandwidths (a large $TBWP_S$) are preferable, and their processing requirements also become more demanding as AO cell bandwidth BW_A, etc., must increase. The signal delay T_D that can be handled in a processor is of concern, because T_D determines range. The T_D range required varies with the application. In communications, an infinite T_D search is necessary to acquire synchronization, whereas $T_D = 0$ during demodulation (when the receiver is in synchronization).

Table 14.5 notes the different optical processing modules presently being fabricated (Casasent 1993, 1994). For completeness, the 2-D optical image processing correlator module (a version of Figure 14.2) is included. The remaining modules are signal processing elements. Other commercial optical signal processing systems are noted elsewhere (Consortium, 1993). Brief descriptions of simplified versions of the major AO architectures are now provided. Details of single-sideband filtering and heterodyne detection are detailed elsewhere (Berg and Lee 1983).

TABLE 14.4 Acousto-Optic (AO) Cell and Signal Parameters

T_A	AO aperture time
	Time for a signal to travel the length of the AO cell
BW_A	AO cell bandwidth
$TBWP_A$	$T_A BW_A$, AO time-bandwidth product
T_S	Signal duration
BW_S	Signal bandwidth
$TBWP_S$	$T_S BW_S$, Signal TBWP
T_D	Time delay between transmitted and received signals

TABLE 14.5 Optical Processing Hardware Modules

Optical Architecture	Operation
2-D image correlator (Fig. 14.2)	2-D pattern recognition
AO time- and space-integrating processor	Produces SAR images
Multichannel AO 2-D FT (Fig. 14.1)	Determines signal direction of arrival
AO Fourier transform channelizer (Fig. 14.6)	Wideband spectrum analysis
AO time-integrating correlator (Figure 14.8)	Wideband signal correlator
AO triple-product processor	Range-Doppler radar processing
AO tapped-delay line	Adaptive phased-array nulling of jammers
AO space-integrating correlator (Figure 4.2)	Optical disk associative memory
Time-delay network	Optical phased-array radar input

An AO spectrum analyzer (Fig. 14.6) is similar to the 2-D FT system, except it is 1-D. Bandwidths in excess of several GHz are possible and 1000 frequencies ($TBWP_A$) can be handled with a dynamic range in excess of 60 dB. Phased-array radars have a number of separate receiving antennae. The received signals from N such antennae are fed to the N channels of a multichannel AO cell placed at P_1 of Figure 14.1. The 2-D FT of the contents of the cell provides an output peak whose location in 2-D denotes the frequency and direction of all input signals to the antennae.

The two basic types of AO correlators are the space-integrating (SI) and time-integrating (TI) architectures shown in simplified form in Figures 14.7 and 14.8 for the correlation of two signals s_1 and s_2. When $s_1(t)$ enters AO1 in Figure 14.7, its transmittance is described by $s_1(t - x/v)$, where v is the velocity of the sound wave in the cell. Thus, its transmittance varies with distance x along the cell and with time t (because the signal moves along the aperture of the cell). In Figure 14.7, AO1 is imaged onto AO2, with a transmittance $s_2(-t - x/v)$. The light leaving AO2 is, then, described by $s_1(t - x/v)s_2(-t - x/v)$. Lens L integrates this product in space (hence, the name space-integrating (SI) correlator) onto a single output detector. The time output from the detector (DET) at P_3 is thus of the form,

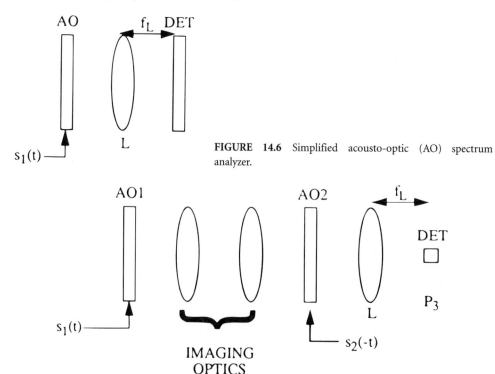

FIGURE 14.6 Simplified acousto-optic (AO) spectrum analyzer.

FIGURE 14.7 Simplified space-integrating acousto-optic (AO) correlator.

Analog Coherent Optical Processing

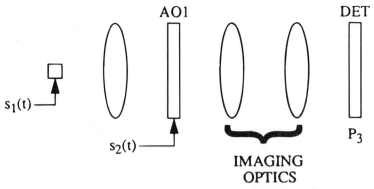

FIGURE 14.8 Simplified time-integrating acousto-optic (AO) correlator.

$$\int s_2(x')s_1(x'-t')\,dx', \qquad 14.12$$

which is the correlation of s_1 and s_2. The correlation is performed by actually sliding one signal by the other, by multiplication and spatial integration. The reference signal $s_2(t)$ being searched for is, generally, cyclically repeated and fed to AO2, and the received signal being processed is fed to AO1. The time at which a peak occurs on the output detector denotes when the reference signal is present and, hence, T_D for range, synchronization, etc., purposes. The parameters of the AO cells determine the range of signal parameters that can be handled. The signal duration T_S must be $<T_A$ (in practice $T_S + T_D < T_A$) and the signal bandwidth BW_S must be $<BW_A$. The T_D that one can search for and the length of signal T_S that one can process is, thus, limited in this system. The system does not require coherent light.

In the time-integrating architecture of Figure 14.8, the light from a point source (a laser diode, light emitting diode, or an AO cell) is modulated in time with $s_1(t)$ and collimated to uniformly illuminate AO1 which is fed with a signal $s_2(t)$. The light leaving AO1 is given by $s_1(t)s_2(t-x/v)$. AO1 is imaged onto a linear detector array (DET) at P_3 which time integrates (hence the name TI correlator) to give the correlation output in space x across the detector array

$$\int s_1(t)s_2(t-x')\,dt. \qquad 14.13$$

The TI correlator can process signals of nearly any length T_S (set by the integration time of the detectors and their dynamic range), with a bandwidth $<BW_A$ and can handle a $T_D < T_A$.

Many other AO systems exist. If chirp (linear frequency-modulated) signals are fed to a correlator, the system can perform the FT, and the bandwidth and resolution of the spectrum analyzer are controlled by the BW and the length of the chirp signals. A hybrid AO system employing both space and time integration is noted in Table 14.5. Several 2-D AO architectures exist. One of the most popular is the triple-product processor. This consists of a point modulator, one AO cell oriented vertically, and a second AO cell oriented horizontally with a 2-D output detector array. Three different signal inputs are possible. The two AO cells control the different axes in the 2-D output, which correspond to range and Doppler in the triple-product processor noted in Table 14.5.

14.5 Optical Image Processors

Optical feature extractor systems (Figure 14.3) are used in product inspection and for identifying one object in the field of view. When more than one object is present and when noise or clutter

TABLE 14.6 Hierarchical Image Processing Operations

Operation	Discussion
Clutter reduction	Reduce major noise in the scene
Detection	Locate positions of all candidate objects
False alarm reduction	Remove detection errors
Image enhancement	Improve contents of each detection region
Feature extraction	Describe the content of each detection region
Classification	Determine the object in each detection region

is severe, correlator architectures are preferable, because of their shift invariance and processing gain. However, classic, matched spatial filters are not sufficient for general image processing. Rather, a larger suite of image processing functions is necessary, and they should be realizable on the same basic architecture (for cost reasons). A discussion of such operations and their realization as different filter functions at P_2 of Figure 14.5 with references follow. This is also intended to describe how a number of different, fixed, P_2 filters would be used in Figure 14.5 for general image processing.

Different levels of computer vision have been defined. Table 14.6 provides one such list. Morphological processors are a general class of low-level vision processors (Maragos, 1987). They perform nonlinear operations, such as erosions and dilations. They are also achieved (Botha and Casasent 1989) on Fig. 14.5 with structuring element filters at P_2 (their size determines the size of image regions eroded away or filled in) and with high (erosion) or low (dilation) thresholds at P_3. They are performed on level-sliced versions of a gray-scale input image, and the binary output results are summed. Thus, they involve combinations of nonlinear operations. The filters required can be realized on ternary devices, and only binary P_1 and P_3 devices are required. Thus, they are easily achieved on simple SLMs. This class of operations performs clutter reduction and image enhancement. Another morphological operation, the hit-miss transform (Casasent et al. 1992), achieves detection. It involves correlating level-sliced input images with a hit filter and with a miss filter, forming the AND of the binarized hit-and-miss erosion outputs, and ORing the results for different level-sliced inputs. The hit-and-miss filters contain rectangles of different sizes, and the resultant output detects objects between the sizes of the small and large rectangles.

Two new and very attractive optical filters implement the wavelet transform and the Gabor transform (Szu and Caulfield 1992, 1994). These provide multiresolution analysis and the best, joint, space and frequency description of an image. These filters have been shown useful for detection. To reduce false alarms, different detection outputs are fused. Distortion-invariant filters are yet another class of filters for detection or recognition (Casasent and Chao 1994). Table 14.7 summarizes the different filter functions possible. When this suite of filters is placed at P_2 of Figure 14.5, this architecture is quite attractive for general purpose image processing.

TABLE 14.7 Suite of Optical Image Processing Filter Functions Achievable

Operation or Filter	Use
Morphology	Clutter reduction
	Image enhancement
Hit-miss transform	Detection
Wavelet transform	Detection
Gabor transform	Detection
Fusion	False alarm reduction
Distortion-invariant filters	Detection, recognition
Feature extraction	Describe image regions

14.6 Summary

Optical processors are a major application for various optoelectronic devices and components. Spatial light modulators are the key elements used to enter data into optical processors. Computer-generated holograms are the key elements used to fabricate optical processors. Their integrated circuit fabrication techniques allow different filter functions to be realized and different operations to be performed on optical processors. Laser diode arrays, an input spatial light modulator, and a set of computer-generated hologram filters allow fabrication of a general purpose, optical image processor. A wide variety of filtering algorithms exist covering the different levels of computer vision. Optical signal processing represents a rapidly maturing area with a number of hardware modules soon to be completed and tested. Many different architectures can be assembled, and many different operations can be achieved with the different optical components and algorithms available.

References

Berg, N., and J. Lee, eds. 1983. *Acousto-Optic Architectures and Applications*. New York: Marcel Dekker.

Botha, E., and Casasent, D. 1989. Applications of optical morphological transformations. *Optical Eng.* 28: 501–505.

Casasent, D. 1985. Computer generated holograms in pattern recognition: A review. *Optical Eng.* 24: 724.

Casasent, D., ed., 1993. *Transition of Optical Processors into Systems 1993, Proc. Soc. Photo Opt. Instrum. Eng.* 1958.

Casasent, D., ed., 1994. *Transition of Optical Processors into Systems 1994. Proc. Soc. Photo Opt. Instrum. Eng.* 2236.

Casasent, D., and Chao, T. H., eds. 1994. *Optical Pattern Recognition V, Proc. Soc. Photo Opt. Instrum. Engr.*, 2237; 1992 *Optical Eng.*, 31(5), Special issue on Optical Pattern Recognition.

Casasent, D., Cheatham, R. L., and Fetterly, D., 1982. Optical system to compute intensity moments: design, *Appl. Opt.* 21: 3292.

Casasent, D., Xia, S. F., Song, J. Z., and Lee, A. 1986. Diffraction pattern sampling using a computer-generated hologram. *Appl. Opt.* 25: 983.

Casasent, D., Xia, S. F., Lee, A., and Song, J. Z. 1987. Real-time deformation invariant optical pattern recognition using coordinate transformations. *Appl. Opt.* 26: 938.

Casasent, D., Schaefer, R., and Sturgill R. 1992. Optical hit-miss morphological transform, *Appl. Opt.* 31: 6255.

Casasent, D., and Psaltis, D. 1979. Deformation-invariant, space-variant optical pattern recognition". In *Progress in Optic*, Wolf, E., ed., XVI, New York: Holland, Chapter XVI.

Casasent, D., and Richards, J. 1988. Optical Hough and Fourier processors for product inspection. *Optical Eng.* 27(4): 258.

Casasent, D., and Richards, J. 1988. Industrial use of a real-time optical inspection system. *Appl. Opt.* 27: 4653.

Cindrich, I. and Lee, S. H., eds. 1991. *Computer and Optically Generated Holographic Optics (4th in a Series), Proc. Soc. Photo Opt. Instrum. Eng.* 1555.

Clark, D., and Casasent, D. 1988. Practical optical Fourier analysis for high-speed inspection, *Optical Eng.* 27(5): 365.

Consortium for Optical and Optoelectronic Technologies in Computing, George Mason University, 1993.

Davis, J. A., Waring, M. A., Bach, G. W., Lilly, R. A., and Cottrell, D. M. 1989. Compact optical correlator design. *Appl. Opt.* 28: 10.

Duda, R. O., and Hart, P. E. 1972. Use of the Hough transform to detect lines and curves in pictures, *Commun., Assoc. Computing Machinery* 15: 11.

Gindi, G. R., and Gmitro, A. F. 1984. Optical feature extraction via the Radon transform. *Optical Eng.* 23(5): 499.

Lendaris, S. and Stanley, G. 1970. Diffraction pattern sampling for automatic target recognition. *Proc. IEEE* 58: 198.

Maragos, P. 1987. Tutorial advances in morphological image processing and analysis. *Optical Eng.* 26: 623.

Molley, P. A., and Stalker, K. T. 1990. Acousto-Optic signal processing for real-time image recognition. *Optical Eng.* 29(9): 1073.

Psaltis, D. 1984. 2-D optical processing using 1-D input devices. *Proc. IEEE* 72: 962.

Szu, H. H., and Caulfield, H. J., eds. 1992, 1994. Special Issue on Wavelet Transforms and their Applications. *Optical Eng.* 31 and 33.

15
Optical Digital Computing

Jürgen Jahns
Fernuniversität-GH Hagen

15.1 Introduction 605
15.2 Communicating with Light Signals 607
15.3 Optoelectronic Devices for Processing and Interconnection. 608
15.4 Optical Interconnections 612
15.5 Architecture for an Optical Computer 617

> But let your communication be Yea, yea; Nea, nea:
> for whatsoever is more than these cometh of evil.[1]
> Matthew 5:37

15.1 Introduction

Today's computers have extraordinary processing powers and capabilities. For example, the Connection Machine which can consist of 64K processing elements is able to perform 10 GFLOPS (FLOP: floating point operation) (Maresca and Fountain, 1991). Further improvements can be expected in microelectronic fabrication technology, so that one can assume that the capabilities of individual processors will be further increased. Furthermore, there still exists a demand for even faster machines, for example, for simulations, real-time pattern recognition, and weather forecasting. However, the speed of today's high-performance electronic computers is increasingly limited by communication problems, such as the number and bandwidth of the interconnections, and by data storage and retrieval rates, rather than by processing power. Several limitations exist for the performance of all-electronic computers:

   a. architectural limitations, such as the often-cited "von Neumann bottleneck," (Backus, 1978; Huang, 1980), that limits the amount of data exchanged between processing unit and memory,
   b. topological limitations due to the two-dimensional layout of the wiring inherent in electronics (Goodman et al. 1984),
   c. physical limitations, such as a limited time-bandwidth product of electric wires that slows down the interconnection speed and requires more energy (Miller, 1989), and

[1] To my knowledge, the above verse has been used in the context of digital communications by the Bell Labs scientist, Claude Shannon.

d. technological limitations due to the lifetime of aluminum wires caused by effects such as electromigration (Keyes and Armstrong, 1987).

Optics, with its large bandwidth and interconnection capabilities, can offer interesting solutions to help alleviate these limitations. In particular, free-space optical propagation allows one to use the third spatial dimension to move signals to and from a chip. In addition, optical signal transmission can be more energy efficient because it is not necessary to charge a wire or a cable to transmit a signal. This is the basis of fiber optic communications for long and medium distances. However, even for very short distances, optical communications can be advantageous energy-wise compared to conventional electronic interconnections (Miller, 1989). Finally, optics offers various degrees of freedom, such as wavelength or polarization, that can be used for signal multiplexing to enhance the throughput of a communication channel.

For quite a while, there has been strong interest in exploring the capabilities of optics for computer technology. The field of optical computing, in general, started in the early sixties. At that time, optical computing was mainly synonymous with *analog* optical signal processing. Analog optical signal processing is based on the ability of a lens to perform a two-dimensional Fourier transformation (Goodman, 1968). This basic operation can be extended to implement optical convolutions and correlations useful for processing of radar signals or for pattern recognition (Van der Lugt, 1964). This field has been brought near perfection during the past three decades by improvement in the algorithms and hardware, such as temporal and spatial light modulators. An overview is given by Casasent (1994).

Digital computing also started in the early sixties when people used the newly invented laser to implement nonlinear logic operations. A collection of articles can be found in conference proceedings edited by J. T. Tippett et al. (1965) and in Basov et al. (1972). However, these early trials of optical digital computing were not too successful, and predictions of an everlasting inferiority of optics to electronics (Keyes and Armstrong, 1969) discouraged people from the subject.

A second effort started in the second half of the seventies. Two developments appear to be of importance: first, the work by Huang (1975, 1984) who pointed out that the limitations of electronic computers might be the strength of optics. Huang also introduced the residue number system to the optical computing community (Huang, 1975; Huang et al. 1979) which, for a while, was the subject of very active research for signal processing applications. The second development of importance was the discovery of optical bistability, first in gases (Gibbs, McCall, and Venkatesan, 1976). The development of nanoscale fabrication techniques, such as molecular beam epitaxy, later led to the investigation of multiple quantum-well structures in solid-state materials, in particular, in gallium arsenide (Miller et al. 1984).

A very important era in the still short life of optical digital computing was the time from 1985 to the early 1990s when activities in many research groups laid a groundwork in device research, interconnections, and systems and architectures for optical computers. Significant progress was made in understanding many details about using optics to perform digital logic and where the limitations still are. The fact that there are still limitations to optics has become evident during recent years when issues, such as speed limitations, complexity of algorithms, and difficulties with the packaging of free-space optics became apparent. Simultaneously, the advances of electronics are still impressive causing doubts whether optics will be able to improve what its big brother has already achieved.

The goal of this chapter is to review the work on optical digital computing in four areas: optical communications, optical switching optical processing, and computing architectures. The coverage here cannot be exhaustive; our main purpose is to explain, lexicographically some of the key words of optical digital computing. For more information, the reader is referred to the recent literature (Hinton and Midwinter, 1990; Midwinter, 1993; Jahns and Lee, 1993).

Optical Digital Computing

15.2 Communicating with Light Signals

The most stringent limits on current electronic computers exist in interconnection technology (Goodman et al. 1984). Although integration densities on chips and die sizes continue to increase, the number and capacity of connections cannot be raised at the same rate due to basic scaling rules (Reisman, 1983). This puts considerable stress on the communication channels at all levels, from intra chip to board-to-board. These become more and more evident as processing speeds increase becaue capacitances and inductivities increase and can cause problems such as cross talk, signal skew, and increasingly power-hungry line drivers. We have already mentioned material effects such as electromigration, which results from the formation of grain boundaries in Al wires at high current densities (Keyes, 1987).

Transmitting signals by photons and electrons is fundamentally different. Electrons are fermions, i.e., they carry charges and, therefore, interact strongly with each other. This is beneficial for implementing nonlinear devices to perform logic operations, for example, but not necessarily helpful for communications. Furthermore, electrons are confined to metallic wires which are one-dimensional (1-D). This is visualized in the left part of Fig. 15.1. Photons, on the other hand, are bosons or uncharged particles and, practically, do not interact with each other. This does not allow one to influence one light beam directly by another. Instead, it requires the use of some other medium to influence light beams by varying the refractive index or absorption. However, the lack of interaction of photons is beneficial for communications. Using fibers, waveguides, and free-space propagation, extremely large bandwidths are available. Whereas fibers and waveguides are limited to one or two dimensions, free-space optical communication offers the possibility of interconnecting chips through the third spatial dimension (Fig. 15.1).

Free-space optics is based on the imaging properties of a lens (Fig. 15.2). The theoretical limit of free-space, optically interconnecting densities is given as the ratio A/λ^2 (A - area of the cross-section, λ - wavelength). For example, for a wavelength of $1\mu m$ and an area of $1mm^2$ one would obtain a theoretical limit of 10^6 channels/mm^2. Practical values, however, are usually smaller due to the resolution of the imaging optics. The area of the point spread function or pixel formed by an aberration-free optical system with a numerical aperture NA = $n \sin \alpha$ (n - refractive index, α - half angle) $\propto (\lambda/NA)^2$. For a typical numerical aperture NA = 0.2 and the same values for A and λ as above, one would obtain 4×10^4 channels/mm^2, still an impressively large number. Another further limitation to interconnection density might be thermal dissipation on the chip which imposes a lower limit on the spatial separation of optical input/output devices.

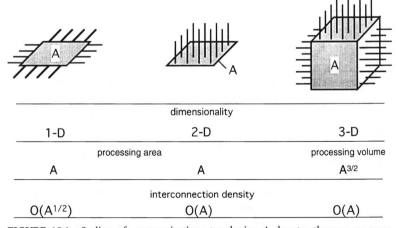

	dimensionality	
1-D	2-D	3-D
	processing area	processing volume
A	A	$A^{3/2}$
	interconnection density	
$O(A^{1/2})$	$O(A)$	$O(A)$

FIGURE 15.1 Scaling of communications topologies. A denotes the area or cross-section of a processor. The two-dimensional case provides the optimum situation for the interconnection density as compared to the chip area.

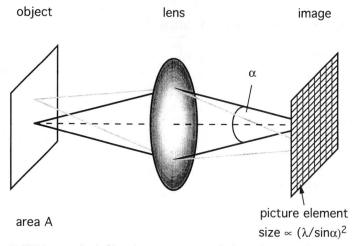

FIGURE 15.2 Optical imaging system. Due to the limited numerical aperture the resolution in the image plane is finite. In the absence of aberrations, the size of a focal spot is uniform across the image plane.

Because of the broadband nature of optical communication channels, the transmission capacity of optical systems is essentially limited by the performance of the transmission and detection components. Fiber optic communications has contributed significantly to the development of useful components and technologies. Laser diodes as optical transmitters can be modulated at extreme speeds because of their small size and are also highly efficient in converting electrical to optical energy. Depending on the semiconductor material system used (GaAs or InP), operating wavelengths range in the near infrared (above and below 850 nm) or in the mid-infrared around 1.3 μm and 1.55 μm. Modulation frequencies in the range of 10 GHz or more can be achieved. Optical output power is in the milliwatt range which is sufficient for communications.

Fast and sensitive photodiodes can be built as detectors. Silicon, an indirect semiconductor, can be used for near-infrared applications. Silicon is also the most important material for electronics, today, although more and more high-speed circuits are realized with GaAs technology despite technological difficulties with GaAs electronics. For example, incompatibilities of fabrication technologies for electronic and optoelectronic devices have still to be overcome, but approaches to fully integrated GaAs optoelectronic circuits (OEICs) are promising. For silicon based systems in the near-infrared, hybrid integration techniques are a way to combine detectors, drivers, and logic circuits with GaAs light sources and modulators.

15.3 Optoelectronic Devices for Processing and Interconnection

In the remainder of this chapter, we will concentrate on free-space optical interconnection and computing technology. For this purpose, devices are of interest that can be integrated on one chip as 1-D or 2-D arrays. Various modulator-based and emitter-based device types have been investigated during the past couple of years. All of them are GaAs devices operating in the near-infrared, mostly at a wavelength around 850 nm. As for modulators, self-electro-optic effect (SEED) devices (Miller et al. 1984; Lentine et al. 1988; Lentine and Miller, 1993) have been used in optical computing research projects at AT&T Bell Laboratories. Another device, promoted by NEC, is the VSTEP that was demonstrated to operate as a modulator and as a laser device (Kasahara, 1993). Vertical cavity, surface-emitting laser diodes (Iga et al. 1987) have become very important since they were demonstrated to be feasible at room temperature (Jewell et al. 1989; Morgan et al. 1991). For detailed information about the physics of the devices, the interested reader is referred to the literature and to those sections in this volume that deal with device

Optical Digital Computing

technology. Here, we want to consider the functional and system aspects of switching and interconnecting devices. As two typical examples, we will discuss the SEED device and the vertical cavity, surface-emitting laser diode (VCSEL).

SEEDs make use of the changes in optical absorption that can be caused by changes in an electric field normal to the quantum-well layers of a semiconductor material (Miller et al. 1984). If the quantum-well structure is placed in the intrinsic region of a reverse bias diode, one can change its absorption by varying the electric field. A light beam sent onto the device can, therefore, be modulated by applying an electrical signal to the device (Lentine, 1993).

The SEED can also be used as a detector, in which an incident light beam can generate a photocurrent. By placing two SEEDs in series, as shown in Fig. 15.3, one can build a structure with optical inputs that control optical outputs. The specific device shown is the symmetric SEED (S-SEED) that consists of two pin diodes with MQW structures in the intrinsic region. Each one serves as the load for the other. There are two optical inputs and two optical outputs. The switching of the device depends on the ratio of the two optical inputs, not on their absolute intensities (within practical limits). This feature is very desirable from a systems point of view, because the operation of the device is relatively insensitive to power variations. An S-SEED can be operated as a logic gate and as a latch, i.e., a memory device. The contrast ratio between the high and the low state for one of the output beams is, typically, 10:1.

Another type of device that is interesting for optical interconnections, in particular, is the vertical cavity, surface-emitting laser diode (VCSEL) (Jewell and Olbright, 1993). It consists, essentially, of a Fabry–Perot resonator between two integrated mirrors. The entire structure is fabricated by molecular beam epitaxy which allows one to achieve extremely high reflectivities for the mirrors and, therefore, very large gains. A typical "microlaser" structure is shown in Fig. 15.4. VCSELs have been demonstrated with current thresholds, typically, at voltages of 3–5 V. Modulation speeds are in the range of several gigahertz.

From a systems point of view, an important difference exists between passive (modulator-based) devices and active (emitter-based) devices. Because optical power has to be supplied to the modulator arrays whereas this is not necessary for active device arrays. This requires suitable schemes for splitting up the light from a single laser source and mechanisms for combining beams by polarization or other optical degrees of freedom (Fig. 15.5). The array illumination can be achieved with various techniques summarized by Streibl (1989, 1993). The beam combination can also be done by various techniques mostly discussed in McCormick (1993) and Cloonan (1993). One of the most common techniques is to use polarization beam splitters, as shown in Fig. 15.5b. We assume that a light beam is s-polarized as indicated in the figure. It is deflected

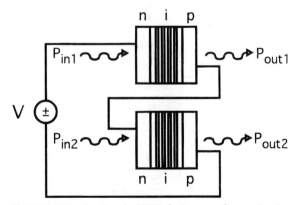

FIGURE 15.3 Symmetric SEED device according to Lentine (1993). Two individual, multiple, quantum-well modulators are connected in series to provide a switching characteristic independent of absolute light levels (within practical limits).

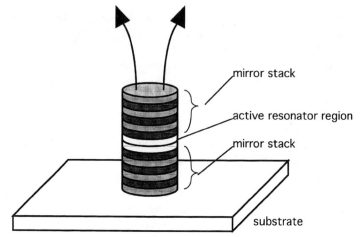

FIGURE 15.4 Vertical cavity, surface-emitting laser diode consisting of a quantum-well between two mirror stacks.

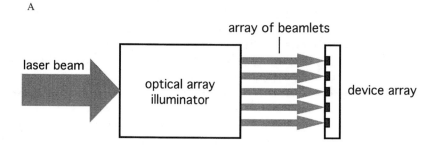

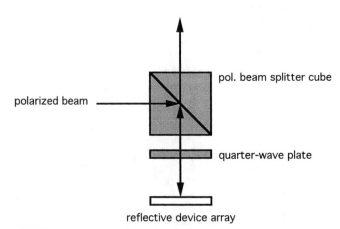

FIGURE 15.5 a) Principle of optical array illumination. A single laser beam is split up into a 1-D or 2-D array of equal intensity beams to illuminate an array of modulator devices. b) Beam combination using a beam splitter cube.

by a polarizing beam splitter and sent onto a device array that is assumed to be operating in reflection (as is usually the case for the SEED devices). The state of polarization of the reflected beam is now p-polarized by a quarter-wave plate and will be transmitted by the beam splitter cube.

A device that has gained significant importance is the "smart pixel" (Hinton, 1988; Cheng et al. 1993; D'Asaro et al. 1993; Kasahara, 1993b; Wilmsen et al. 1993). In the early work on digital optical computing systems in the mid-eighties, it was assumed that devices, such as the SEED, would operate as simple logic gates (AND, NOR, etc.) and that the optical interconnections would be used to directly connect gates in different device arrays. We will talk about this in a later section. However, this concept of low-level interconnections can have problems related to the gate count, for example, i.e., the number of logic gates required to perform a certain function may be high. For this reason and also for reasons related to the physics of the devices, a lot of interest is now focused on the smart pixel structure. Here, an optical switching device serves merely as an optical input and output device. It is surrounded by electronics which may be used to complement the function of the optical device (amplification, thresholding, . . .) or its function may be on a higher level (digital logic, signal routing, signal processing, . . .).

The use of the smart pixel has consequences for device technology and also for interconnection optics. In a conventional scheme, optical devices and electronic circuitry are separate (Fig. 15.6). For optical interconnections, this has the advantage that the optics is concentrated in a small area (typically 1 mm $\times$ 1 mm) and that individual devices are spaced very densely. This matches well with conventional imaging optics that provides high resolution over relatively small areas. The optical devices can be combined with electronic circuitry by hybrid integration mechanisms. The tight spacing of the devices, however, causes problems with the geometry of electric wires used to provide electric power and voltage to the individual devices. The density of the devices is ultimately limited by practical constraints related to the density of the electrical wiring.

In the smart pixel model, the optical device area is split up into many little pieces distributed over a much larger chip area (Fig. 15.7). The pitch of the devices may be 100 μm or more. The entire chip diameter may be several millimeters. For the optics this means that one has to use imaging systems that can handle a large optical field at high resolution. Promising approaches based on imaging with lenslet arrays will be discussed in a later section. The problem of the actual implementation of smart pixels might be more difficult to solve because they require integrating optoelectronic and electronic functions on the same chip. For an all-GaAs implementation, this brings with it the problems of process compatibility and yield. Despite promising

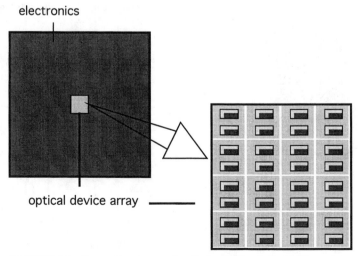

FIGURE 15.6 Conventional array of optical switching devices with supporting electronics. A close-up view is shown on the right of the figure.

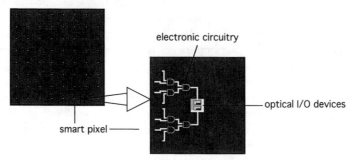

FIGURE 15.7 Array of smart pixels. An individual smart pixel is represented on the right.

approaches (Fang et al. 1990), integrating GaAs devices on Si is a difficult problem that has not yet been solved. An alternative might exist in the form of hybrid integration schemes such as flip-chip bonding which can provide high-density interconnections between two chips.

15.4 Optical Interconnections

The subject of free-space optical interconnects for digital optical computing systems is close to the field of switching networks, in particular, space-switching networks (Benes, 1965). A space-switching network consists of arrays of input and output devices and switching nodes with spatially separated inputs and outputs. They are interconnected through physical channels such as wires, waveguides, or free space. The specific function of the nodes is not specified at this point. It may vary with the specific application for which the network is used. For a sorting network, for example, the nodes may be compare-exchange units. Each node may, in fact, be a small network of its own.

An interconnect can be represented as a bipartite graph as shown in Fig. 15.8a. It consists of arrays of inputs and outputs, shown on the left and right, respectively, and lines indicating the interconnections. The specific network shown is a crossbar where every input is connected to

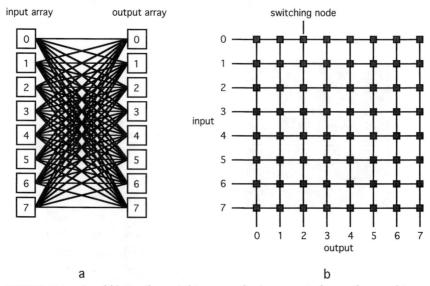

FIGURE 15.8 a) and b) Crossbar switching network, a) represents the crossbar as a bipartite graph.

every output. Another representation of the crossbar is shown in Fig. 15.8b where it becomes obvious that the number of switching nodes (the dark squares) is NM for N input devices and M output devices. The figure shows the special case where $N = M$. Two quantities to describe an interconnection network are the fan out of the switching nodes and the diameter of the network. The fan out F is the number of input positions and output positions of the nodes. The diameter D is the number of switching stages required to link any input to any output device. The crossbar represents an extreme because it consists of only one switching stage, where on the other hand, the devices have a fanout of $F = N$. The crossbar allows routing each of the N inputs to any of the M outputs or, in general, arbitrary combinations of outputs. The total number of interconnection patterns supported by a crossbar is N^M (or N^N in our special case). The crossbar is a nonblocking network, i.e., it allows connecting an input device to an output device—or to any combination of output devices—independent of already existing connections.

There are two disadvantages of the crossbar: one is the high cost in terms of the required number of switching nodes. In the area of telephone switching networks this has led to the invention of the Clos network (Clos, 1953). A second disadvantage is the large fan out that the switching nodes have to support. This feature is not desirable for optical devices as pointed out by Prise et al. (1988). Therefore, other types of interconnection networks have been considered for optical computing and switching systems. Of particular interest are multistage interconnection networks (MINs) which are usually constructed of 2×2 crossbar switches (i.e., $F = 2$). Permutation networks based on MINs, typically, consist of $\log_2 N$ stages (i.e., $D = \log_2 N$) where each stage has N switching nodes. MINs support all $N!$ permutations and some, but not all, broadcasting connections (Smith et al. 1993).

A well-known MIN is the Perfect Shuffle (Stone, 1971) (Fig. 15.9a). The name is derived from a technique for shuffling a deck of N cards by interleaving the upper and the lower halves with one another. After repeating the same operation $\log_2 N$ times, the initial order of the cards is reestablished. Other types of MINs are the Banyan (Goke and Lipovski, 1973) and the Crossover network (Wise, 1981; Jahns and Murdocca, 1989). These are shown in Figure 15.9b and c. The Perfect Shuffle, the Banyan, and the Crossover are topologically equivalent. This means that one can be transformed into the other by reordering and renaming the nodes. A formal condition for the equivalence of networks is the "buddy condition" (Agrawal, 1983) which was shown to apply for the Crossover and the Banyan by Jahns and Murdocca (1989).

For optically implementing the different types of interconnection networks, three fundamental approaches can be distinguished (Fig. 15.10):

   a. Matrix-vector multiplier setup (Goodman et al. 1978) for 1-D input and output arrays (Fig. 15.10a). Light from one input device is spread over a whole row of spatial light modulators which can be switched to be transmissive or absorptive. A specific output device receives all signals from a whole column in the spatial light modulator array. Thereby, arbitrary interconnections can be set up.
   b. For implementing MINs, such as the Perfect Shuffle, split-and-combine setups have been used most often. A specific implementation of the Perfect Shuffle proposed by Brenner and Huang (1988) is shown in Fig. 15.10b. A beam splitter cube is used to form two separate optical paths for the light beams emerging from the input array and to combine the arrays in the output plane. Mirrors with suitable tilts introduce a shift between the two output arrays. In the center of the output plane, one obtains the data in the right sequence, as indicated in the figure. Other optical implementations of the Perfect Shuffle network have been described, for example, by Lohmann et al. (1986), Eichmann and Li (1987), and Sawchuk and Glaser (1988). Optical setups for the Banyan network were described by Jahns (1990) and Cloonan and Herron (1989). Implementations of optical Crossover networks were demonstrated by Jahns and Murdocca (1988).

614   Handbook of Photonics

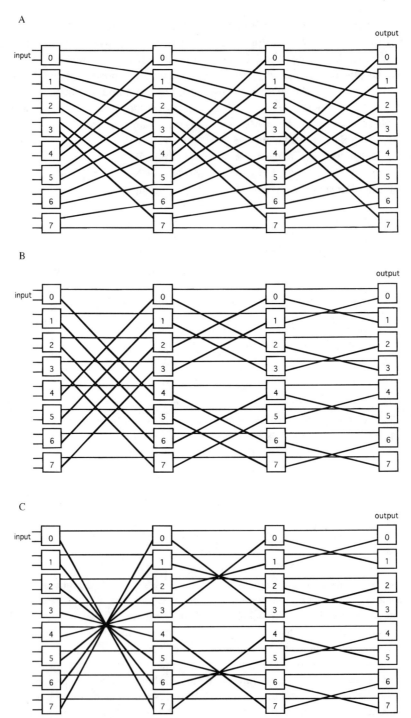

FIGURE 15.9 Three multistage interconnection networks: a) Perfect Shuffle; b) Banyan; c) Crossover network.

Optical Digital Computing

A

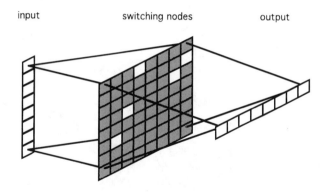

B

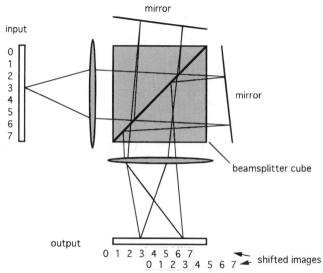

FIGURE 15.10 Optical implementation of switching networks: a) implementation of a crossbar switch using a matrix-vector multiplier arrangement; b) implementation of a Perfect Shuffle according to Brenner and Huang (1988); c) microchannel optics for implementing arbitrary interconnections. This setup can also be used to implement the crossbar or the stages of a MIN like the Perfect Shuffle. The channel density is determined by the aperture of the optical elements in the array.

c. Multifacet optics can be used to implement arbitrary interconnects. This can be achieved with microchannel optics as shown in Fig. 15.10c. Each channel consists of a separate, miniaturized, imaging system. An optical element is used to collimate and deflect the light emerging from the input node. A similar element is used to couple the light to the correspoding output node. A microoptical implementation of such an element using "binary optics"-type diffractive elements (Veldkamp, 1994) was demonstrated by Jahns and Däschner (1990) and with diffractive-refractive microoptics by Sauer et al. (1994). Volume holographic elements were used for this purpose by Robertson et al. (1991). It should be noted that in the case of an implementation with multifaceted optics, the interconnection density is limited by the size of the individual aperture. Typical values for δx might be on the order of 100 μm. Furthermore, for a given aperture size δx, the

C

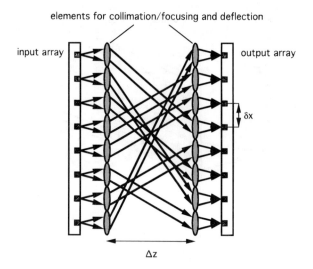

FIGURE 15.10 Continued.

interconnection length Δz, i.e., the separation between the input and output coupler is determined by diffraction effects (Leggatt, 1991; Sauer et al. 1994): $\Delta z \leq (\pi/\lambda)\delta x^2$ where λ denotes the wavelength of the light.

Irregular optical interconnections have been discussed in connection with optical interconnections for electronic computers. A scheme, shown in Fig. 15.11, was suggested by Goodman et al. (1984) and has been considered by various groups (Kostuk et al. 1985; Feldman et al. 1987). It is also the basis of the POEM (programmable optoelectronic multiprocessor) approach (Kiamilev et al. 1989). Several processors are interconnected on a wafer scale through free space by holograms. Intraprocessor communication is realized by electronics. Volume holographic elements can be used to multiplex many routing elements in a single component. Emitters and detectors are integrated with the processing elements on the same chip. A specific configuration of the POEM architecture based on photorefractive reconfigurable holograms is discussed by Ford et al. (1990).

A specific problem of optical interconnects is the aspect of system packaging. Conventional optical systems rely on the use of mechanical mounting which results in bulky and expensive

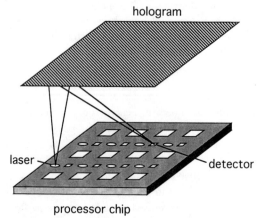

FIGURE 15.11 Optical interconnections for intrachip communications. Emitter and detector elements are distributed across the chip area.

Optical Digital Computing

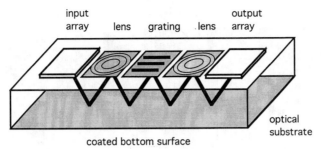

FIGURE 15.12 Priciple of planar optical packaging. Free-space interconnects are provided by substrate mode propagation of the light signals. The optics is fabricated by lithographic means.

setups. Microoptical integration schemes based on lithographic fabrication of the optics are an alternative. Two basic approaches were suggested in recent years: "stacked microoptics" (Iga et al. 1982; Brenner, 1991) and "planar optics" (Jahns and Huang, 1989). In the planar optics approach, microoptical elements are integrated on one or both sides of an optical flat (Fig. 15.12). Light signals propagate inside the substrate. Optoelectronic chips are mounted on the substrate by solder bump bonding (Jahns et al. 1992). Parallel interconnections with high interconnection densities (Jahns and Acklin, 1993), optical clock distribution (Walker and Jahns, 1992), and the implementation of space-variant interconnection networks (Jahns and Däschner, 1991; Song et al. 1992) have been demonstrated.

15.5 Architecture for an Optical Computer

Optical interconnections are being considered for various levels in computing and communications systems. It has been argued by Miller (1989) and Feldman et al. (1989) that optical communication should be used for all but the shortest distances. Table 15.1 shows which optical media are being used or investigated.

A model of an optical digital computer, developed at AT&T Bell Laboratories, was based on chip-to-chip interconnections (Smith et al. 1993). Optical connections were provided between individual logic gates (for example, OR gates) that were implemented as arrays of SEED devices (Fig. 15.13). Masks block light at selected positions to customize the interconnects so as to perform specific tasks, such as addition, subtraction, or sorting. Stages of $2 \log_2 N$ are used to process columns and rows, sequentially. Feedback may be provided to read the output data back into the system.

Programmable array logic is used for the design of the processor circuits (Mead and Conway, 1980; Murdocca et al. 1988). The positions of the logic gates in the array are fixed. All gates have a fan in and fan out of 2. All gates on the same chip perform the same logic operation such as

TABLE 15.1 Hierarchy of Optical Interconnections

Application	Distance (in m)	Medium
Satellite communication	10^6	Free space
Telecommunication	10^5–10^6	Fiber
Local area networks	10^2–10^3	Fiber
System-to-system	10^1–10^2	Fiber
Board-to-board	10^{-1}–10^0	Fiber
(intrasystem)		Waveguide
		Free space
Chip-to-chip	10^{-3}–10^{-2}	Waveguide
		Free space

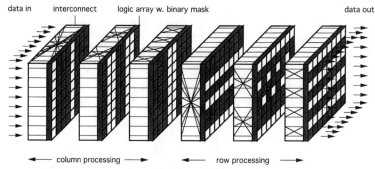

FIGURE 15.13 Architecture of a free-space optical computer according to Smith et al. (1993). The individual stages are separated in the figure for clarity. Optical device arrays are interconnected through free space. Masks in the device planes are used to customize the interconnections.

an OR or AND. The degrees of freedom available to the circuit designer are the positions of the inputs and the outputs and the selection of which mask positions should be open or closed. The general idea is first to generate all min terms of the input variables and then to select and combine the min terms required to implement specific functions. The generation of min terms of two variables x and y is shown in Fig. 15.14 (Murdocca et al. 1988). The squares in this figure represent AND gates. The black lines show connections used, whereas dashed lines indicate connections blocked by masks in the device planes. The design of complex circuits was described by Murdocca (1990).

One difficulty of the optical PLA approach is the large number of gates required to design a specific circuit. In the free-space optical approach, the expense for the interconnection optics between the stages is quite significant, at least, if conventional optomechanics is used. In general, the gate count for performing logic operations is relatively high, an issue which has led to doubts about the suitability of free-space interconnects at the gate level. Several improvements for reducing the gate count have been discussed (Smith et al. 1993).

A second approach to implementing optical logic is computational origami (Huang, 1992). This concept uses highly regular interconnections because they can be implemented with conventional imaging optics. Data processing is done by transforming the task into a sequence of five basic operations: NOR, Crossover, Bypass, Broadcast, and Don't Care. Figure 15.15a shows a simple

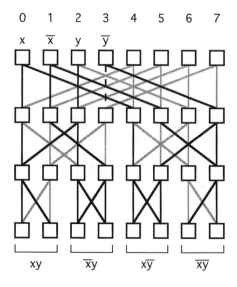

FIGURE 15.14 Interconnection network to generate min terms of two logic variables according to Murdocca et al. (1988).

Optical Digital Computing

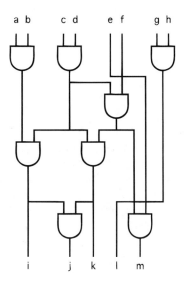

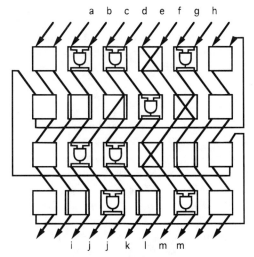

FIGURE 15.15 Computational origami: a) conventional electronic cicuit and b) regularized version in an origami processor according to Huang (1992).

conventional logic circuit consisting of eight AND gates. It is redrawn in a folded, regular way in Fig. 15.15b. A arrray of logic gates is used with a width of six and a depth of four.

Computational origami is a way of shaping a computational task into a form so that it becomes suitable to the hardware and communications requirements of a processor. The relevance to optics is that computational origami can be used to match the computational task with a highly regular interconnection scheme. It also illustrates how to perform computations with delay lines rather than with random-access memory. The function of memory in a computation is implemented by delaying the data stream until it is needed. Computational origami accomplishes using fixed amounts of time for the delays. This avoids random-access memory and, therefore, reduces the hardware complexity of a computer.

15.6 Conclusion

Today's electronic computers are becoming more and more limited by communications problems. The application of optics to computing may offer significant advantages over conventional computer technology and architecture. It is not clear, yet, how the merging of electronic and optical technology will be achieved, although the interconnection area is the most likely candidate where optics can be considered a serious alternative. Significant progress has been made in the past few years in all areas of optical computing. New technologies, such as molecular beam epitaxy, have permitted building better optical switching devices than considered possible ten years ago. Lithographic fabrication techniques have allowed optics to leave the conventional domain of optomechanics to build microoptic components and systems. Further progress can be expected in areas like smart pixel devices, which should also influence new work on architectures and algorithms, and system level packaging which appears to be of critical importance for making optics acceptable in the electronics world.

Acknowledgment

The author would like to thank his former colleagues in the Optical Computing Department at AT&T Bell Laboratories in Holmdel, NJ, for the interesting and stimulating atmosphere. In particular, I would like to express my gratitude to Alan Huang.

Also, I want to thank Stefan Sinzinger for his help in preparing this manuscript.

References

Agrawal, D. P. 1983. Graph theoretical analysis and design of multistage interconnection networks, *IEEE Trans. Comp.* C-32: 637.
Backus, J. 1982. Function-level computing. *Spectrum* 22.
Basov, N. G., Culver, W. H., and Shah, B. 1972. Applications of lasers to computers. In: *Laser Handbook*, F. T. Arecchi and E. O. Schulz–Dubois, eds. Amsterdam: North-Holland.
Benes, V. E. 1965. *Mathematical Theory of Connecting Networks and Telephone Traffic.* New York: Academic.
Brenner, K.-H. 3-D integration of digital optical systems. In Proc. Optical Computing Conf., Optical Soc. Am., Washington, DC. 1991, pp. 25–28.
Brenner, K.-H., and Huang, A. 1988. Optical implementation of the perfect shuffle interconnections. *Appl. Opt.* 27: 135–137.
Casasent, D. 1994. Analog optical data processing. In: *CRC Handbook on Photonics*, M. Gupta, ed., Boca Raton, Fla.: CRC.
Cheng, J., Zhou, P., Sun, S. Z., Hersee, S., Myers, D. R., Zolper, J., and Vawter, G. A. 1993. Surface-emitting laser-based smart pixels for two-dimensional optical logic and reconfigurable interconnections. *IEEE J. Quant. Electron.* 29: 741–756.
Cloonan, T. J. 1993. Architectural considerations for optical computing and photonic switching. In *Optical Computing Hardware*, Jahns, J. and Lee, S. H., Eds. Boston: Academic.
Cloonan, T. J., and Herron, M. J. 1989. Optical implementation and performance of one-dimensional trimmed inverse augmented data manipulator networks for multiprocessor computer systems. *Opt. Eng.* 28: 305–314.
Clos, C. 1953. A study of non-blocking switching networks, *Bell Syst. Tech. J.* 32: 406.

D'Asaro, L. A., Chirovski, L. M. F., Laskowski, E. J., Pei, S. S., Woodward, T. K., Lentine, A. L., Leibenguth, R. E., Focht, M. W., Freund, J. M., Guth, G. G., and Smith, L. E. 1993. Batch fabrication and operation of GaAs-Al$_x$Ga$_{1-x}$As field-effect transistor-self-electrooptic effect device (FET-SEED) smart pixel arrays. *IEEE J. Quant. Electron.* 29, 670–677.

Eichmann, G. and Li, Y. 1987. Compact optical generalized perfect shuffle. *Appl. Opt.* 26, 1167.

Fang, S. F., Adomi, K., Iyer, S., Morkoç, H., Zabel, H., Choi, C., and Otsuka, N. 1990. Gallium arsenide and other compound semiconductors on silicon. *J. Appl. Phys.* 68, R31–R58.

Feldman, M. R., Esener, S. C., Guest, C. C., and Lee, S. H. 1987. Comparison between optical and electrical interconnects based on power and speed considerations. *Appl. Opt.* 27: 1742–1751.

Feldman, M. R., Guest, C. C., Drabik, T. J., and Esener, S. C. 1989. Comparison between electrical and free space optical interconnects for fine grain processor arrays based on interconnect density capabilities. *Appl. Opt.* 28: 3820–3829.

Ford, J. E., Lee, S. H., and Fainman, Y. 1990. Application of photorefractive crystals to optical interconnection. In *Digital Optical Computing II*. Proc. SPIE 1215: 155–165.

Gibbs, H. M., McCall, S. L., and Venkatesan, T. N. C. 1976. Differential gain and bistability using a sodium-filled Fabry–Perot interferometer, *Phys. Rev. Lett.* 36: 1135.

Goke, L., and Lipovski, G. Banyan networks for partitioning multiprocessor systems. In 1st Annu. Symp. on Computer Architecture. New York: ACM/IEEE, 1973, p. 21.

Goodman, J. W. 1968. *Introduction to Fourier Optics*. San Francisco: McGraw-Hill.

Goodman, J. W., Dias, A. R., and Woody, L. M. 1978. Fully parallel, high speed incoherent optical method for performing discrete Fourier transforms. *Opt. Lett.* 2: 1.

Goodman, J. W., Leonberger, F. J., Kung, S., and Athale, R. A. 1984. Optical interconnections for VLSI systems. *Proc. IEEE* 72: 850–865.

Hinton, H. S. 1988. Architectural considerations for photonic switching networks. *IEEE J. Sel. Areas Commun.*, 6: 1209–1226.

Hinton, H. S., and Midwinter, J. E., eds. 1990. *Photonics in Switching*. New York: IEEE.

Huang, A. The implementation of a residue arithmetic unit via optical and other physical phenomena. Proc. Int. Optical Comp. Conf. Washington, DC, 1975, p. 14.

Huang, A. 1992. Computational origami: the folding of circuits and systems. *Appl. Opt.* 31: 5419–5422.

Huang, A. 1980. Design for an optical general purpose computer. In *Proc. 1980 Int. Optical Computing Conference (Book II)*, W. T. Rhodes, ed. SPIE, Washington, DC, Vol. 232, p. 119.

Huang, A., Tsunoda, Y., Goodman, J. W., and Ishihara, S. 1979. Optical computation using residue arithmetic. *Appl. Opt.* 18: 149.

Iga, K., Kinoshita, S., and Koyame, F. 1987. Microcavity GaAlAs/GaAs surface emitting lasers with I$_{th}$ = 6 mA. *Electron. Lett.* 23: 134–136.

Iga, K., Oikawa, M., Misawa, S., Banno, J., and Kokubun, Y. 1982. Stacked planar optics: An application of the planar microlens. *Appl. Opt.* 21: 3456–3460.

Jahns, J. 1990. Optical implementation of the Banyan network. *Opt. Commun.*, 76: 321–324.

Jahns, J., and Lee, S. H. 1993. *Optical Computing Hardware*. Boston: Academic.

Jahns, J., and Däschner, W. Integrated free-space optical permutation network. In Proc. Optical Computing Conf., Optical Society of America, Washington, DC, 1991, p. 29–32.

Jahns, J., and Acklin, B. 1993. Integrated planar optical imaging system with high interconnection density. *Opt. Lett.* 18: 1594–1596.

Jahns, J., and Däschner, W. 1990. Optical cyclic shifter using diffractive lenslet arrays. *Opt. Commun.*, 79: 407–410.

Jahns, J., and Huang, A. 1989. Planar integration of free-space optical components. *Appl. Opt.* 28: 1602–1605.

Jahns, J., and Murdocca, M. J. 1988. Crossover networks and their optical implementation. *Appl. Opt.* 27: 3155–3160.

Jahns, J., Morgan, R. A., Nguyen, H. N., Walker, J. A., Walker, S. J., and Wong, Y. M. 1992. Hybrid integration of surface-emitting microlaser chip and planar optics substrate for interconnection applications. *IEEE Phot. Tech. Lett.* 4: 1369–1372.

Jewell, J. L., and Olbright, G. R. 1993 Microlaser devices for optical computing. In *Optical Computing Hardware*, Jahns, J., and Lee, S. H. eds. Boston: Academic.

Jewell, J. L., Scherer, A., McCall, S. L., Lee, Y. H., Walker, S., Harbison, J. P., and Florez, L. T. 1989. Low-threshold electrically pumped vertical-cavity surface-emitting microlasers. *Electron. Lett.* 25: 1123–1124.

Kasahara, K. 1993. Vertical-to-surface transmission electrophotonic devices. In *Optical Computing Hardware*, Jahns, J., and Lee, S. H. eds. Boston: Academic.

Kasahara, K. 1993b. VSTEP-based smart pixels. *IEEE J. Quant. Electron.* 29: 757–768.

Keyes, R. W. 1987. *The Physics of VLSI Systems*. Wokingham, England: Addison-Wesley, p. 168.

Keyes, R. W., and Armstrong, J. A. 1969. Thermal limitations in optical logic. *Appl. Opt.* 8: 2549.

Kiamilev, F., Esener, S., Paturi, R., Fainman, Y., Mercier, P., Guest, C. C., and Lee, S. H. 1989. Programmable optoelectronic multiprocessors and their comparison with symbolic substitution for digital optical computing. *Optical Eng.* 28: 396–409.

Kostuk, R. K., Goodman, J. W., and Hesselink, L. 1985. Optical imaging applied to microelectronic chip-to-chip interconnections. *Appl. Opt.* 24: 2851–2858.

Leggatt, J. S. Silicon and photoresist microlenses for use with single mode fibres. In 3rd Microoptics Conference, Yokohama, Japan, 1991, p. 68.

Lentine, A. L. 1993. Self-electrooptic effect devices for optical information processing. In *Optical Computing Hardware*, Jahns, J., and Lee, S. H. eds. Boston: Academic.

Lentine, A. L., and Miller, D. A. B. 1993. Evolution of the SEED technology: Bistable logic gates to optoelectronic smart pixels. *IEEE J. Quant. Electron.* 29: 655–669.

Lentine, A. L., Hinton, H. S., Miller, D. A. B., Henry, J. E., Cunningham, J. E., and Chirovsky, L. M. F. 1988. Symmetric self-electrooptic effect device: Optical set-reset latch. *Appl. Phys. Lett.* 52: 1419.

Lohmann, A. W., Stork, W., and Stucke, G. 1986. Optical perfect shuffle. *Appl. Opt.* 25: 1530–1531.

Maresca, M., and Fountain, T. J. 1991. Massively parallel computers. *Proc. IEEE* 79: 395.

McCormick, F. B. 1993. Free-space interconnection techniques. In *Photonics in switching*, Midwinter, J. E., ed. Boston: Academic, Vol. II.

Mead, C., and Conway, L. 1980. *Introduction to VLSI Systems*. Reading, MA: Addison-Wesley.

Midwinter, J. E., ed. 1993. *Photonics in Switching*. Boston: Academic.

Miller, D. A. B. 1989. Optics for low-energy communication inside digital processors: Quantum detectors, sources, and modulators as efficient impedance converters. *Opt. Lett.* 14: 146–148.

Miller, D. A. B., Chemla, D. S., Damen, T. C., Gossard, A. C., Wiegmann, W., Wood, T. H., and Burrus, C. A. 1984. Novel hybrid optically bistable switch: The quantum well self-electrooptic effect device. *Appl. Phys. Lett.* 45: 13.

Morgan, R. A., Robinson, K. C., Chirovsky, L. M. F., Focht, M. F., Guth, G. D., Leibenguth, R. E., Glogovsky, K. G., Przybylek, G. J., and Smith, L. E. 1991. Uniform 64 × 1 arrays of individually addressed vertical cavity top surface emitting lasers. *Electron. Lett.* 16: 1400.

Murdocca, M. J. 1990. *A Digital Design Methodology for Optical Computing*. Cambridge, MA: MIT.

Murdocca, M. J., Huang, A., Jahns, J., and Streibl, N. 1988. Optical design of programmable logic arrays. *Appl. Opt.* 27: 1651–1660.

Prise, M. E., Streibl, N., and Downs, M. M. 1988. Optical considerations in the design of digital optical computers. *Optical Quantum Electron.* 20: 49–77.

Reisman, A. 1983. Device, circuit, and technology scaling to micron and submicron dimensions. *Proc. IEEE* 71: 550.

Robertson, B., Restall, E. J., Taghizadeh, M. R., and Walker, A. C. 1991. Space-variant holographic optical elements in dichromated gelatin. *Appl. Opt.* 30: 2368–2375.

Sauer, F., Jahns, J., Nijander, C. R., Feldblum, A. Y., and Townsend, W. P. 1994. Refractive-diffractive microoptics for permutation interconnects. *Optical Eng.* 33: 1550–1560.

Sawchuck, A. A. and Glaser, I. 1988. Geometries for optical implementations of the perfect shuffle. *Proc. SPIE* 88: 270.

Smith, D. E., Murdocca, M. J., and Stone, T. W. 1993. Parallel optical interconnections. In *Optical Computing Hardware*, Jahns, J., and Lee, S. H., eds. Boston: Academic.

Song, S. H., Lee, E. H., Carey, C. D., Selviah, D. R., and Midwinter, J. E. 1992. Planar optical implementation of crossover interconnects. *Opt. Lett.* 17: 1253–1255.

Stone, H. S. 1971. Parallel processing with the perfect shuffle. *IEEE Trans. Comp.* C-20: 153–161.

Streibl, N. 1989. Beam shaping with array generators. *J. Mod. Opt.* 12: 1559–1573.

Streibl, N. 1993. Multiple beamsplitters. In *Optical Computing Hardware*, Jahns, J., and Lee, S. H. eds. Boston: Academic.

Tippett, J. T. et al., eds. 1965. *Optical and Electro-Optical Information Processing*. Cambridge, MA: MIT.

Van der Lugt, A. B. 1964. Signal detection by complex spatial filtering. *IEEE Trans. Inform. Theory.* IT-10: 2.

Veldkamp, W. 1994. Binary optics. In *CRC Handbook on Photonics*, M. Gupta, ed. Boca Raton, Fla.: CRC.

Walker, S. J., and Jahns, J. 1992. Optical clock distribution using integrated free-space optics. *Opt. Commun.* 90: 359–371.

Wilmsen, C. W., Beyette, J., F. R., An, X., Feld, S. A., and Geib, K. M. 1993. Smart pixels using the light amplifying optical switch (LAOS). *IEEE J. Quantum Electron.* 29: 769–774.

Wise, D. S. 1981. Compact layout of Banyan/FFT networks. In *Proc. CU Conf. VLSI Systems and Computations*, Kung, Sproull, and Steele, eds. Rockville, MD: Computer Science, p. 186.

16
Optical Communications

16.1	Perspective	624
16.2	Review of Essential Components...............................	627
	Optical Fibers • Sources • Detectors • Filters • Various Components	
16.3	Basic Optical Systems..	646
	Modulation Formats • Detection Schemes • Signal and Noises • Performance Criteria • System Design • Point-to-Point and Multipoint Systems • Wireless Free-Space IR System	
16.4	Optical Amplifiers	667
	Fundamental Amplifier Characteristics • Semiconductor Amplifiers • Erbium-Doped Fiber Amplifiers	
16.5	Multichannel Systems	684
	Topologies and Architectures • Time-Division Multiplexing • Wavelength-Division Multiplexing • Space-Division Multiplexing • Code-Division Multiplexing	
16.6	Solitons...	701
16.7	Analog Communications	704
	Subcarrier Multiplexing • Multichannel Frequency-Division Multiplexing	
16.8	Future Trends...	709

Alan E. Willner
University of Southern California

16.1 Perspective

Optical communications is a very old form of data transfer. Line-of-sight primitive digital systems have included lighting bonfires on mountain tops to send a simple one-bit message, smoke signals to send a multiple-bit message, and ship-to-ship broad-incoherent-beam transmission of Morse-code messages. The inventions of the low-loss optical fiber (Marcuse, 1982) and the high-speed semiconductor laser (Kressel and Butler, 1978) have caused an explosion in the transmission capacity of optical systems (Li, 1983). In this chapter we will discuss the technologies behind sophisticated optical systems, some of which can transmit >10 billion digital bits of information per second across transpacific distances (Bergano et al. 1995). In addition to higher capacity, tethered systems have lower loss and higher security than untethered systems. We will, therefore, discuss mainly fiber-based systems (Kazovsky, Benedetto, and Willner, 1996) with some reference to other line-of-sight optical technologies.

Optical Communications

One of the most basic and undisputed applications of optical rather than electrical communication is data transmission on long-distance links. The reason optics is an obvious choice is because of the ultrawide bandwidth (>10 THz) of low-loss (<0.2 dB/km) transmission properties of an optical fiber (Marcuse, 1982). The lowest loss electrical wire is a coaxial cable which is nothing more than an electrical transmission line. To transmit information down a transmission line, the line must be powered up and, then, down. This must be performed extremely fast for a high-speed digital system. In an electrical system, powering of a transmission line is limited in speed by a time constant which is proportional to the frequency-dependent characteristic impedance. This impedance increases substantially at GHz speeds to >100 dB/mile (see Fig. 16.1), and therefore the electrical cable becomes quite lossy for high-speed transmission (Jordan and Penny, 1982). Because photons do not interact with each other, the optical fiber waveguide does not have any "impedance" and such time-constant-related losses do not exist in an optical fiber. The major figure-of-merit for a transmission system is the bit-rate/distance product, and we wish to maximize both of these quantities.

In addition to transmission of information from point-to-point, the rapid switching of information from a given input port to one of many output ports is extremely important for high-throughput networks which interconnect many high-speed users (Berztsekas and Gallager, 1987). Switching of an optical signal can be performed electronically, by detection and retransmission, or by purely optical means (Hinton, 1993; Midwinter, 1993). Detection and retransmission may have an inherent optoelectronic speed bottleneck and may consume much power. We will discuss some of the techniques for optical switching and multiplexing, but it is still not clear what the circumstances are for which optical switching will eventually be preferred.

A generic optical system can be described in a simple block diagram of Fig. 16.2. Because almost invariably, the signal origin and destination are in electronic form, the optical medium is simply used to connect the two electrical end points with low loss and high speed. An optoelectronic transmitter and receiver are required to interface the electronic and optical domains. Furthermore, an optical system can be used for transmission or for switching purposes. Unless

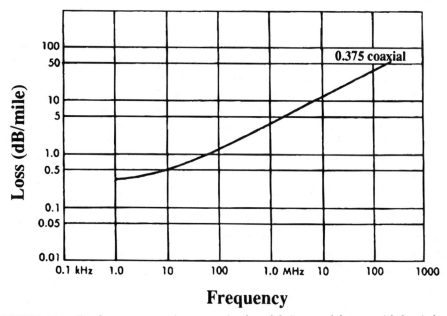

FIGURE 16.1 Signal power attenuation versus signal modulation speed for a coaxial electrical cable. (After Jordan and Penny, 1982.)

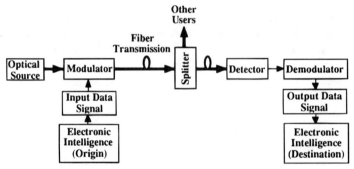

FIGURE 16.2 Block diagram of a generic optical communication system.

otherwise stated, fiber will be the medium for optical transmission, although a brief discussion of free-space systems will be included.

In any optical system, the ultimate measure of performance is the signal-to-noise ratio (SNR) of a recovered signal (Schwartz, 1980). The higher the SNR when recovering data, the lower will be the probability of error, or bit-error-rate (BER) in a digital system, or loss of fidelity in an analog system. Of course, various types of optical systems will differ in the SNR required for satisfactory system performance. We will devote much discussion throughout this chapter finding the signal and noise powers in many different optical communication systems to analyze the communications channel.

Analog and digital systems are two basic types communications. Analog can reproduce a given waveform exactly and is compatible with present cable-television systems, but the carrier-to-noise ratio required for near-error-free transmission must be extremely high (Haykin, 1983). On the other hand, digital systems are compatible with the way in which computers and the modern-telephone-network communicate, and one need only distinguish between a "0" and "1" bit (Roden, 1995). The signal-to-noise ratio required is much smaller than in analog, and therefore digital systems are easier to implement and maintain. We devote most of this chapter to digital systems.

The progress in optical communications over the past 15 years has been astounding. In 1980, AT&T could transmit 672 two-way conversations along a pair of optical fibers (MacMillan, 1994). In 1994, an AT&T network connecting Florida with the Virgin Islands could carry 320,000 two-way conversations along two pairs of optical fibers. We will be discussing the technology which has enabled such progress plus the technology which may increase the capacity once again by an order of magnitude within the next decade. Just as exciting as point-to-point transmission heroics is the technology which will allow future high-speed networking and switching of information. The so-called National Information Infrastructure (Kaminow et al. 1995) or "information super-highway," will most likely make use of many of the topics contained in this chapter.

The advances of fiber optical systems have progressed in three generations of technology (Henry et al. 1988, Chapter 21). The first generation systems used 0.8-μm GaAs semiconductor lasers and multimode fibers. These systems operated at 50–100 Mbit/s and the fiber links were only 10 km long (Agrawal, 1992). The second and third generation systems employed single-mode fibers and 1.3-μm InGaAs and 1.55-μm InGaAsP lasers. The next generation systems may include single-frequency, 1.55-μm lasers, Erbium-doped fiber amplifiers, and dispersion-shifted fiber. We will be discussing all of these technologies.

Optical communications is a fairly large and rapidly advancing scientific field. To treat many ideas in just one chapter, we will cover each topic fairly broadly without too much depth. Only the most basic material will be included, and many references are included to facilitate further study by the reader. Furthermore, because it is not clear which specific optical technology will be important in future systems, we will attempt to include a treatment of each possibly crucial technology. Finally, optical systems are different from electrical systems in several aspects, chief of which is the physical implementation. Consequently, the vast majority of this chapter deals with the physical optical technology and its systems implementations.

Read on, and find out more about the physical aspects which will enable future communications systems to connect a multitude of high-speed users throughout the globe. The "information superhighway" may be at hand, and allow your imagination to envision what will travel on this highway of the future!!

16.2 Review of Essential Components

As mentioned in the introduction, some photonic technologies which form the basis for optical communication systems are covered in other chapters in this handbook. They include fibers, sources, detectors, filters, couplers, and modulators. We will only briefly discuss the essence of these topics as they relate to optical communications. The reader should review these earlier chapters for more in-depth information about these areas.

Optical Fibers

As most readers already know, an optical beam can be thought of as an electromagnetic wave or as a large collection of photon particles (Born and Wolf, 1959). Either can be used for instructional purposes, with the choice depending on the specific topic. We will begin with the wave approach.

An optical wave propagates as a gaussian beam which has its peak intensity at the center of the beam and whose intensity decreases radially with Gaussian statistics (Kogelnik, 1965):

$$E(r, z) = A_0 \frac{w_0}{w} \exp\left(-\frac{r^2}{w^2}\right), \tag{16.1}$$

where $|E|$ is the absolute value of the electromagnetic field, A_0 is the peak amplitude at the beam center, w is the beam waist at the full width at half-maximum (FWHM), w_0 is the FWHM of the beam waist at the focal point and its minimum value, r is the radial distance from the beam center, and z is the direction of propagation. The wave also experiences divergence, i.e., radial spreading, as it propagates (Kogelnik, 1965):

$$w^2(z) = w_0^2\left[1 + \left(\frac{\lambda z}{\pi w_0^2 n}\right)^2\right], \tag{16.2}$$

where n is the index of refraction of the medium. Alternatively, we can employ a ray-optics approach to optical wave propagation. Such a treatment can be quite instructive and is a good approximation for some circumstances. Throughout this chapter, we will alternate between using any of the above three formalisms. For the sake of initially explaining optical fibers, we will use a ray-optics approach.

Reflection and refraction will occur when an optical ray, which is propagating in a dielectric medium of index of refraction n_1, impinges on a discontinuous boundary between this n_1 medium and another medium of index of refraction n_2. Snell's law describes this phenomenon of reflection and refraction, where the transmitted refracted beam follows the following relationship (Klein and Furtak, 1986):

$$n_1 \sin \theta_1 = n_2 \sin \theta_2, \tag{16.3}$$

where θ is the angle at which the optical ray hits the boundary with respect to the boundary

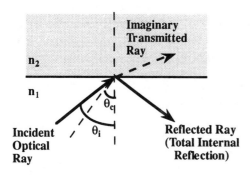

FIGURE 16.3 Total internal reflection at a dielectric boundary when the incident ray's angle exceeds the critical angle.

normal within each medium. The reflected beam "bounces" off the interface back into the original medium 1 at the same original angle with respect to the normal. However, the reflected beam now propagates above and away from the interface normal if the incident beam was below the normal (and vice-versa). The power in the transmitted and reflected beams must equal the total incident power if the process is lossless. Let us assume that the index of the medium in which the ray is propagating is larger than the second medium ($n_1 > n_2$). If the ray hits the boundary at a very grazing angle for which θ approaches 90°, then, there is an angle for which the inverse-sin of a number >1 is imaginary, thus, resulting in no real solution. If the angle of incidence is greater than this critical angle (θ_c), then: (i) the refracted beam itself is imaginary, (ii) no power penetrates the dielectric boundary, and (iii) total internal reflection occurs (see Fig. 16.3) (Gloge, 1971). No power has been lost upon reflection at a dielectric boundary.

If we surround the high-index medium of propagation with another medium of lower index, we can achieve lossless waveguiding of an optical beam as long as the beam strikes the boundary at an angle greater than the critical angle. However, there is a discrete set of angles greater than θ_c which will support waveguiding. These discrete angles are eigenvalue solutions to the wave propagation equations which define the waveguide. Figure 16.4 illustrates several possible propagating waves, each of which is called a spatial mode of the waveguide. The number of spatial modes M, which can be supported in a circular waveguide or optical fiber is given by (Gowar, 1984)

$$N = \frac{2\pi a}{\lambda} (n_1^2 - n_2^2), \quad (16.4)$$

where a is the core radius, n_1 is the core index, and n_2 is the cladding index; the key feature of Eq. (16.4) is that the number of modes increases when the index difference or the fiber radius is large. An optical fiber is thus "a high-index core surrounded by a lower-index cladding." Guided optical propagation provides a reliable, low loss, and secure medium of transmission.

Figure 16.4 illustrated the ray-model approach of many modes propagating through a multimode fiber whose core diameter is fairly large and usually ≥ 50 μm. The lowest order modes, i.e., the modes most easily generated, propagate nearly straight down the fiber at angles shallow

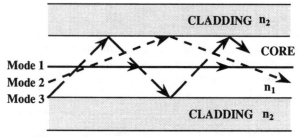

FIGURE 16.4 Ray model illustration of the propagation of many modes in a dielectric waveguide.

Optical Communications

to the core-cladding interface, and the higher order modes reflect off the core-cladding interface at steeper angles. The most important disadvantage of a multimode fiber is intermodal dispersion which arises for the following reason (see Fig. 16.5). A rectangular light pulse of temporal width T input to the multimode fiber will generate several modes, with the pulse's power distributed among the lower and higher order modes. However, because the different modes all propagate at the speed of light but at different angles down the fiber, they will also travel different absolute path-length distances even though they all propagate axially down the fiber for the same fixed fiber length. Therefore, each mode will arrive at the fiber end at different times, causing the initial light pulse to spread and disperse temporally, ΔT, by the time it reaches the fiber end. This type of dispersion limits the distance and bit rate which can be transmitted, because some finite time must exist until the subsequent light pulse can be transmitted without mutual interference. Intermodal dispersion can be minimized (but not eliminated) by using a graded-index (GRIN) multimode fiber (Keck, 1976), which was described in the chapter on optical fibers. In general, multimode fibers should be avoided for distances larger than several hundred meters and speeds >1 Gbit/s.

The only way to avoid intermodal dispersion is to reduce the core size and allow only one mode to propagate in a fiber. The typical single-mode fiber core diameter is ~8 μm, and the single mode propagates axially straight down the fiber. If we consider this fundamental mode as a real gaussian beam, the single mode will have most of its energy residing in the core with an evanescent field tail propagating alongside the core within the cladding region. In a single-mode fiber, another type of dispersion and is known as intramodal dispersion, will limit the optical transmission (Gallorenzi, 1978). This dispersion is a combination of material and waveguide dispersion in which the fiber material and fiber waveguide are slightly wavelength-dependent (Jones, 1988):

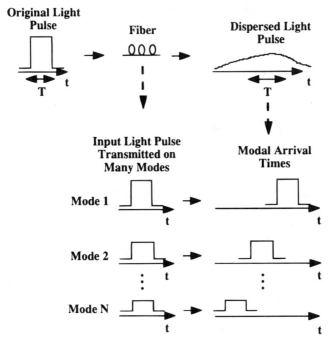

FIGURE 16.5 Illustration of intermodal dispersion in which an input optical pulse has its power transmitted on several different modes, each arriving at a different time.

$$\text{Intramodal Dispersion} = \text{Material Dispersion} + \text{Waveguide Dispersion}. \qquad (16.5)$$

The material dispersion is caused by the wavelength dependence of the fiber's index of refraction. The argument proceeds as follows. Light propagates down the fiber at the speed of light c divided by the wavelength-dependent index of refraction $n(\lambda)$. Because each light pulse contains some distribution of frequencies, these different frequencies (i.e., wavelengths) will travel at different speeds down the fiber and arrive at the fiber end at different times; the frequency distribution of a pulse of light is due to the laser linewidth and the information bandwidth and will be discussed later in section III.1. The waveguide dispersion arises because the fiber is, in reality, an optical transmission line for which the speed of the electromagnetic wave depends on the relationship between the geometric shape of the waveguide and the propagating wave. Although the fiber geometry itself is fixed, the light pulse will contain a spread of different wavelengths, each of which will travel at different speeds along the fiber. Figure 16.6 shows the dispersion coefficient D (ps/(nm · km)) of a conventional single-mode fiber with the material and waveguide contributions plotted separately (Agrawal, 1992, p. 44). For a given system, a pulse will disperse more in time for a wider frequency distribution of the light and for a longer length of fiber. Note that the fiber has an inherent material dispersion coefficient which changes sign and goes through a zero point, resulting in a total intramodal dispersion zero point near 1.3 μm for a conventional fiber.

Another key characteristic of a single-mode fiber is the power attenuation α per kilometer shown in Fig. 16.7 as a function of signal wavelength (Kaiser and Keck, 1988). The fundamental physical limits imposed on the fiber attenuation are due to scattering off the silica atoms at shorter wavelengths and the material absorption at longer wavelengths. There are two minima in the loss curve, one near 1.3 μm and an even lower one near 1.55 μm. The 1.55-μm loss minimum of ~0.2 dB/km (<1% light lost in 1 km!) is quite close to the theoretical limit with a bandwidth of ~25 THz. Low loss is extremely important because a light pulse must contain a certain minimum amount of power to be detected such that a "0" or "1" data bit can be unambiguously detected. If not for dispersion, we would clearly prefer to operate with 1.55-μm light due to its lower loss for long-distance systems.

We now have a trade-off for the widely installed conventional single-mode fiber in which the dispersion minimum is at 1.3 μm but the loss minimum is at 1.55 μm. The choice as to which wavelength to use depends on the practical limits of a particular system. However, there is another type of fiber, known as dispersion-shifted fiber (DSF), which has both zero dispersion and the loss minimum located at 1.55 μm. Because the waveguide dispersion depends on the waveguide's geometry, we can change the fiber doping and core (which effectively changes the waveguiding

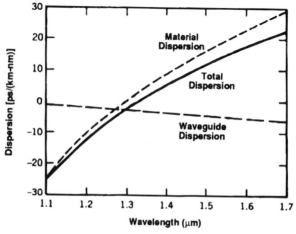

FIGURE 16.6 Dispersion coefficient D as a function of wavelength in a silica single-mode fiber. (After Agrawal, 1992.)

Optical Communications

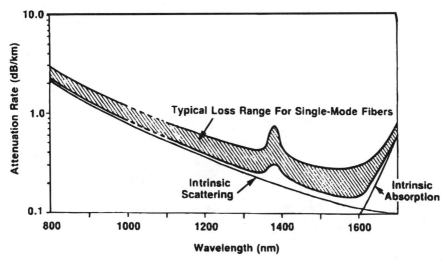

FIGURE 16.7 Attenuation versus wavelength in a single-mode silica fiber. (After Kaiser and Keck, 1988.)

geometry) to change the material dispersion and, thus, move the zero dispersion point to 1.55 μm. Figure 16.8 shows the dispersion curve (Kaiser and Keck, 1988, p. 35). One practical drawback of DSF is that the vast majority of already installed fiber is the conventional type, not DSF.

Coupling of light into and out of a small-core, single-mode fiber is much more difficult to achieve than coupling electrical signals in copper wires because (i) photons are weakly confined to the waveguide whereas electrons are tightly bound to the wire, and (ii) the core of a fiber is typically much smaller than the core of an electrical wire. Light must be coupled into the fiber from a diverging laser beam, and two fibers must be connected to each other. Let us first discuss coupling light into the fiber. Focusing the laser light into the fiber end must be performed with great care because the angle of acceptance cannot be greater than the critical angle for total internal reflection. We must excite only the lowest order fundamental mode in the 8-μm core, and this mode has a grazing angle with the core-cladding boundary. Therefore, the focusing must be gradual to ensure a small angle and must provide a small focal spot size similar in dimension to the single-mode core. The maximum limit of the acceptance angle for a single- or multimode fiber is known as the numerical aperture NA and is given by (Jones, 1987)

$$NA = \sin \theta_c = \sqrt{n_1^2 - n_2^2}, \qquad (16.6)$$

assuming that light is coupled into the fiber from air. The second issue of connecting two different

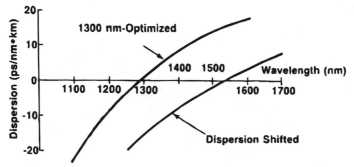

FIGURE 16.8 Dispersion coefficient versus wavelength in a dispersion-shifted fiber. (After Kaiser and Keck, 1988, p. 35)

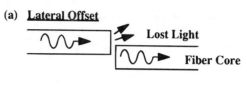
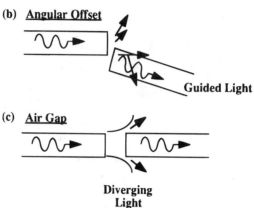

FIGURE 16.9 Sources of loss in a fiber-to-fiber connection.

fibers in a system must be performed with great care due to the small size of the cores. We wish to achieve connections exhibiting (i) low loss, (ii) low back reflection, (iii) repeatability, and (iv) reliability. As shown in Fig. 16.9, the following will cause losses: (i) an axial offset because the core areas do not fully overlap, causing light to couple into the second fiber's cladding, (ii) an angular offset because not all the light will be coupled at an angle above the critical angle, and (iii) an air gap in which both an index mismatch and beam divergence will limit coupling into the second fiber's core. Two popular methods for connections are the permanent splice and the mechanical connector. The permanent "fusion" splice can be accomplished by placing two fiber ends near each other, generating a high-voltage electric arc which melts the fiber ends, and "fusing" the fibers together. Losses and back reflection are extremely low being <0.1 dB and <-60 dB, respectively. Disadvantages are that the splice is delicate, must be protected, and is permanent. Alternatively, there are several types of mechanical connectors, such as Biconic, ST, and FC/PC (Young, 1989). For brevity, we will discuss only the general concept. Figure 16.10 shows that the polished fiber end is permanently placed at the center of a mechanical receptacle, two such receptacles are mechanically screwed together, and a (near) physical contact is made between the fiber cores (Carlisle, 1985). Losses and back reflection are still fairly good, typically, <0.3 dB and <-45 dB, respectively.

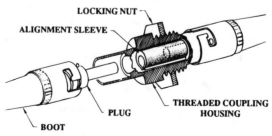

FIGURE 16.10 Diagram of a typical mechanical fiber splice. (After Carlisle, 1985.)

Optical Communications

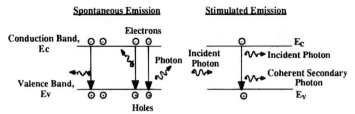

FIGURE 16.11 Spontaneous and stimulated emission in a semiconductor.

Sources

Photons have an energy which is dependent on the wavelength of the light λ (Siegman, 1986):

$$E = hc/\lambda = h\nu, \tag{16.7}$$

where c is the speed of light. Furthermore, semiconductors have an energy bandgap between the electron-rich valence band and the hole-rich conduction bands. If a photon is incident on a semiconductor, the photon can be absorbed if its energy is larger than the energy bandgap. In such a case, the photon's energy can be transferred to a valence electron pushing it up into the conduction band and freeing it to move through the semiconductor. This is known as stimulated absorption (Yariv, 1975).

Alternatively, photons can be emitted from a semiconductor if an electron in the conduction band drops down in energy, ΔE, into the valence band, thereby, combining with a hole in the valence band and emitting a photon of the same energy as ΔE. This process of photon emission can occur due to two different processes as illustrated in Fig. 16.11 (Yariv, 1975). The first process is called spontaneous emission in which a finite-lifetime electron in the conduction band randomly combines with a hole to emit a photon. These electrons exist in the conduction band due to prior pumping of them into the higher energy level, typically by electrical biasing (i.e., current injection). Because the electrons fill an energy well in the conduction band with a distribution in energy states (see Fig. 16.12), the energy drop in the electron, upon spontaneous recombination, will produce uncorrelated, incoherent photons at many different wavelengths (i.e., energies),

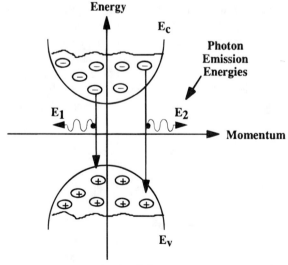

FIGURE 16.12 Energy-band diagram versus momentum in a direct-bandgap semiconductor. E_1 and E_2 represent possible energies of emitted photons.

producing a wide spectral bandwidth over which photon emission can occur. These random photons can be considered as noise in the optical system.

In the second process, called stimulated emission, a single photon of a given energy is incident on a semiconductor and causes electron-hole recombination. This stimulated recombination results in the emission of a photon of the same energy as the original incident photon, thus, producing two photons from an initial one. If the electron population in the conduction band is high enough to sustain continued stimulated emission, then, an incident photon at wavelength λ_i will produce two, four, etc., photons which are all coherent with each other and at the same wavelength as the original photon. This process produces gain and the medium is considered active. As the wave traverses through an active medium, it will be amplified as shown in Fig. 16.13. The gain G depends on the gain coefficient per unit length g and the length of the medium L (Yariv, 1976)

$$G(\lambda) = \frac{\text{Signal Output Power}}{\text{Signal Input Power}} = \exp[g(\lambda)L]. \qquad (16.8)$$

The relative rates for stimulated and spontaneous emission are determined by the electron populations in the various energy bands and the external pumping. (Optical amplification will be revisited in detail in Section 16.4.)

Before discussing lasers, it is useful, first, to introduce a less sophisticated optical source known as the light-emitting diode (LED). In general, light emission requires that carriers be easily pumped into the conduction band and, then, localized so that they will recombine to emit light. This can be accomplished by forming a p-n junction, hence, the word "diode" in the LED name. By selectively tailoring the layers surrounding the p-n junction and creating a potential well in the n region, the electrons and holes can be generated and, then, localized near the n-semiconductor's depletion region. Carriers are generated through an electrical contact and forward biasing of the p-n junction. These carriers recombine and spontaneously emit photons in the depletion region. The randomly generated photons will also experience some limited amplification (i.e., they will themselves cause stimulated emission). The generated light is broadband (~100 nm), incoherent, and not directionally well-confined. The light can be partially coupled into a fiber as shown in Fig. 16.14 (Burrus and Miller, 1971). Continued electrical pumping will sustain a light beam emanating from the p-n junction. Because the area of a typical LED is quite large (~250 μm), the capacitance time constant is also quite large, limiting the ON-OFF modulation speed to less than a few hundred mHz. In addition to this speed restriction, an LED has limited usefulness in high-speed long-distance communications due to: (i) the high material dispersion caused by the extremely broadband light, and (ii) the high intermodal dispersion because a large-core multi-mode fiber is typically used to maximize the capture of the ill-confined light.

A laser is a more complicated system because it involves coherent light generation and feedback over a small bandwidth. Let us first address the issue of feedback for laser oscillation. If we

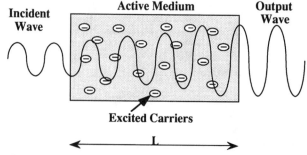

FIGURE 16.13 Amplification of a wave as it propagates through an active medium.

Optical Communications

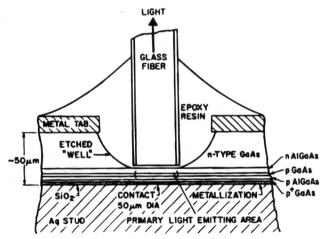

FIGURE 16.14 Light emanating from an LED coupled into an optical fiber. (After Burrus and Miller, 1971.)

surround a medium in which a light beam is propagating with two near-100% mirrors, then we have formed a frequency-selective resonant cavity which can be explained with the help of Fig. 16.15. An optical wave $L_1(t, z)$, having a given wavelength and phase, is propagating to the left inside this cavity. Upon reflection at the left mirror, a right-propagating wave will be generated, $R_1(t, z)$. Given a long coherence length for the optical wave (Born and Wolf, 1980, Chapter 10) the left and right propagating waves will be coherent with each other and have the same frequency ω, but they each have their own phase φ. This reflection process continues with a reflection of $R_1(t, z)$ at the right mirror which generates a second left-propagating wave of $L_2(t, z)$. If $L_1(t, z)$ and $L_2(t, z)$ are coherent, as would be the case in stimulated emission, then, they would interfere with each other at the left mirror. If they are in phase with each other (i.e., the phase difference equals $0, 2\pi, 4\pi, \ldots$), then, they will add constructively, and both waves will be mostly reflected and partially transmitted at the left boundary. If they are completely out of phase (i.e., the phase difference equals $\pi, 3\pi, 5\pi, \ldots$), then, they will add destructively, and no wave will be reflected.

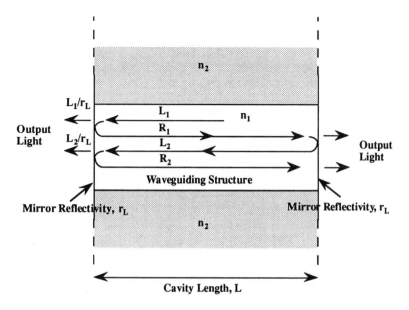

FIGURE 16.15 Multiple coherent reflections in a Fabry–Perot cavity.

In reality, there are an infinite number of reflections, and all of the reflections must add constructively in phase at the two mirror boundaries. Therefore, the round-trip distance of wave propagation between the two mirrors must be an integral number of 2π phase shifts or an integral number of wavelengths (Yariv, 1976, p. 64):

$$2L = m \frac{\lambda}{n_1}, \quad m = 1, 2, \ldots\ldots \quad (16.9)$$

This equation assumes that (i) the waves are confined to the plane of propagation, which can occur if we confine the light to a thin waveguide by using low-index surrounding cladding layers, and (ii) there is no angular dependence. There are an infinite number of integers which will satisfy the in-phase requirement. Therefore, the passband of this resonant cavity is periodic in wavelength, and the cavity is known as a Fabry–Perot cavity (Fabry and Perot, 1899). Note that the original wave can be continually generated by pumping-induced stimulated emission inside an active semiconductor. Additionally, the mirrors in a semiconductor wavelength-selective cavity are formed by the perfectly parallel, smooth, cleaved facets of the semiconductor ends. The light would, then, emanate from the cavity at these <100% mirrors and be useful as laser light.

The infinite number of reflections in the resonant cavity can be sustained by gain compensating for existing losses. There are two main sources of losses in a semiconductor cavity: (1) <100% reflectivity of the end mirrors (r_L and r_R being, typically, ~30% with semiconductor cleaved facets), and (2) nonradiative absorption of light within the semiconductor medium α_a. Both of these losses are, to first order, wavelength-independent. The semiconductor must provide gain within the active p-n junction to overcome these losses and sustain light oscillation in the resonant cavity (Casey and Panish, 1978).

Every laser must, therefore, fulfill three basic criteria:

1. gain to provide stimulated emission,
2. (gain > loss) to sustain reflections, and
3. wavelength-selective feedback

These three conditions are displayed graphically in Fig. 16.16 which shows that lasing will occur where are all the conditions are fulfilled, i.e., within the gain bandwidth, above the loss line, and at the wavelength resonances. In the Fabry–Perot diode laser, there are several frequency modes which satisfy all the lasing condition, and so, this laser is called a multimode laser with the all modes typically occupying several nanometers in bandwidth (Agrawal and Dutta, 1986). As we discussed in the previous section on fibers, we wish to have a very narrow frequency bandwidth to minimize intraspatially modal dispersion in the single-mode fiber. Therefore, multifrequency-mode lasers are useful only for shorter distance (<50 km) or lower speed (<1 GHz) links, but they are much more useful than the simple LED.

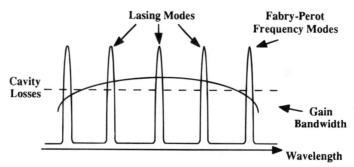

FIGURE 16.16 Illustration of the three lasing conditions of (i) a gain medium, (ii) a frequency-selective cavity filter, and (iii) gain > loss.

Because of intramodal dispersion, we are quite interested in single-frequency-mode lasers. This can be accomplished by referring back to Fig. 16.16 and providing a scenario in which only one frequency mode will satisfy all three lasing criteria. The gain bandwidth of the active medium is fixed, and we certainly do not wish to increase the losses in the cavity. The best method is to find a wavelength-selective cavity whose periodic resonance peaks are so far apart that only one passband resides within the active gain bandwidth and above the loss line. This requires a very efficient frequency-selective cavity with a high Q-factor, or high Finesse. One such cavity would integrate Bragg gratings to provide the feedback at the ends of the active region instead of using simple mirrors or cleaved facets (Saleh and Terch, 1991). Figure 16.17a illustrates the basic concept of the wavelength-selective reflectivity of these gratings. The gratings are formed by a modulation in either the refractive index or the gain; this modulation can exist outside of the waveguiding region and in the cladding region where it will affect the tail, or evanescent field, of the propagating wave. At each period in the grating, a small portion of the propagating wave is reflected. If all the small reflections are in phase, then, they will add constructively when they propagate in the reverse direction; of course, wavelengths which do not satisfy the in-phase requirement will not be reflected and will not, eventually, resonate and lase. This in-phase requirement for the gratings is called the Bragg condition (Saleh and Terch, 1991):

$$\Lambda = \frac{\lambda}{2n}, \qquad (16.10)$$

where Λ is the spacing of the Bragg corrugation; this equation assumes that there is no angular dependence if the wave is confined to the horizontal waveguide.

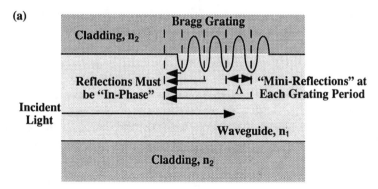

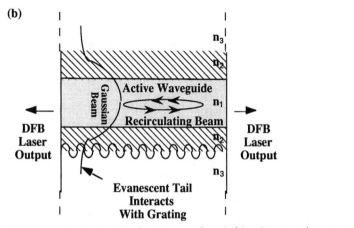

FIGURE 16.17 (a) Small reflections at each period in a Bragg grating can cause frequency-selective reflections. (b) Simple diagram of a distributed feedback laser.

Not only can the Bragg reflecting regions be placed at the end of the active region, but they can actually be placed within (i.e., above or below) the active region itself. Such intracavity, distributed, Bragg reflection satisfies the eigenvalues of coupled-mode theory to provide efficient frequency-selective feedback. This laser, simplistically shown in Fig. 16.17b, is called a distributed-feedback (DFB) laser (Kogelnik and Shank, 1972).

The light output L of a single- or multimode laser is a function of the input current I. As the current increases, the gain increases until it is equal to the losses. At this point, known as the laser threshold current (I_{th}), lasing will occur because all three lasing conditions will be satisfied (Siegman, 1986). A typical L-I curve is shown in Fig. 16.18. The laser light output can be modulated by directly modulating the bias current injected into the semiconductor active medium. This modulation can replicate the digital bits of some data stream giving rise to light "ON" and "OFF" corresponding to "1" and "0" data bits, respectively.

One final note is that the laser light is generated from stimulated emission in which all the photons are coherent and at the same wavelength. However, this is not strictly true because there are some quantum fluctuations in the energy of the photons emitted by stimulated emission compared to the original photons. These energy fluctuations correspond to fluctuations in the wavelength, giving rise to a laser linewidth, which, for a DFB laser, is typically 10 MHz of uncertainty. For high-speed Gbit/s, direct-detection systems, this linewidth is negligible, but this linewidth may cause difficulties for coherent systems (see Detection Schemes).

Detectors

Transmitted optical data must be unambiguously recovered, and, therefore, photodetection is an extremely important process. Photodetection is quite similar to photogeneration, except in reverse. We have already established that an incident photon will be absorbed by a semiconductor if this photon has enough energy for an electron to absorb the energy, overcome the energy bandgap, and become a free carrier in the conduction band. We must be able to measure the number of electrons generated by the incident photons, i.e., the photon-generated current. This requires separating and measuring the generated electron-hole pairs before they can recombine. A simple reverse-biased p-n junction can provide the necessary conditions, as shown in Fig. 16.19. The depletion region of a p-n junction has a built-in electric field due to bound, separated, ionized cores. If photons generate free electron-hole pairs in the depletion region, then, the carriers will be separated by the built-in electric field and can, then, be measured as a current before recombination occurs. The amount of generated photocurrent I_{ph}, is given by (Campbell, 1989)

$$I_{ph} = \frac{\eta q P}{h\nu}, \qquad (16.11)$$

where η is the quantum efficiency of a given photon producing one electron, P is the incident optical power, $h\nu$ is the photon energy, and q is the electronic charge.

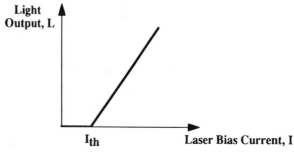

FIGURE 16.18 A typical curve of the laser light output versus input bias current.

Optical Communications

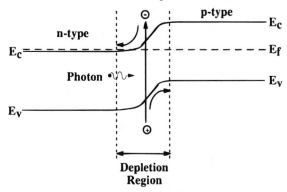

FIGURE 16.19 An energy-band diagram of a p-n junction in which photogenerated carriers are separated and measured as a photogenerated current.

Even if the incident optical power is constant, two main noise sources generated in the detector are thermal noise σ_{th}^2 and shot noise σ_{sh}^2 (Lee and Li, 1974). Three important characteristics about these noises are that

(i) they have a statistical variance,
(ii) they cover all possible frequencies (i.e., white noise) which are supported by the system's electrical detection bandwidth, and
(iii) they can be approximated as having a gaussian amplitude distribution centered around the intended photocurrent mean. The gaussian distribution is centered at a high current level for a "1" bit and a low current level for a "0" bit.

Thermal noise is caused by thermal energy in the detector. The thermal energy is randomly absorbed by electrons which are pushed up into the conduction band and will be mistakenly detected as photocurrent. This thermal noise power is independent of incident optical power and has a statistical variance (Lee and Li, 1979):

$$\sigma_{\text{th}}^2 = \frac{4kTB_e}{\Omega} \quad (16.12)$$

where B_e is the low-pass-filter electrical bandwidth of the detector, k is Boltzman's constant, T is the detector temperature, and Ω is the detector resistance. The shot noise is caused by the quantum randomness of generating carriers in a detector at random times. It is the most fundamental quantum limit of photodetection because one can never eliminate this noise term. The shot noise variance power is given by (Lee and Li, 1979)

$$\sigma_{\text{sh}}^2 = 2q\left(\frac{q}{h\nu}\right)PB_e \quad (16.13)$$

and is proportional to the absorbed optical power. In direct-detection systems, thermal noise usually dominates shot noise. Later, we will discuss the effects these noise sources have on the signal-to-noise ratio of the recovered data.

Although we have described the detector as a simple p-n junction, in reality, there are two different types of detectors which dominate commercial use: the p-i-n detector and the avalanche photodiode (APD). These are shown in Fig. 16.20a and b (Palais, 1988). As its name implies, the p-i-n detector is a p-n junction with a long, intrinsic, undoped region between the two ionized halves of the depletion region (Lee and Li, 1979). The electric field remains high throughout the long, intrinsic region giving it the ability to efficiently separate photogenerated carriers and produce photocurrent. The effect of the p-i-n detector is to have a much larger region in which photons can be absorbed; in a simple p-n junction the depletion region by itself is too thin to absorb much of the incoming light beam because absorption in a semiconductor has a characteristic and exponentially-decaying absorption depth. Similarly, the purpose of the APD is to increase the photocurrent-generating efficiency of the detector. This is accomplished by adding a region of very highly doped material within the p-i-n detector, thereby, creating a small region of extremely high electric field (Palais, 1988). Photogenerated electrons within this high-electric-field region will be accelerated in velocity and, consequently, gain much energy. The moving electron can collide with a valence-band, bound electron and transfer its new excess energy to it, thereby, ionizing and freeing it to move; note that this excess energy must be greater than the ionization energy of an electron. The newly freed electron resides within an intense electric field and will, itself, begin to move in the same direction as the original electron. These two electrons are now moving as photocurrent where only one existed before. This process can continue as long as the electrons are in the highly doped region, achieving a multiplication factor of as large as ~ 10.

Although the p-i-n and the APD detectors allow the measurement of a photocurrent, more functions must usually be performed to ensure adequate data recovery. These functions may include (Personick, 1977)

    a. electrical amplification by using a transimpedance amplifier,
    b. clock recovery to synchronize bit recovery of an arriving data packet,
    c. decision circuitry to judge if a "1" or "0" bit has arrived, and
    d. electrical filtering to limit the high frequency noise.

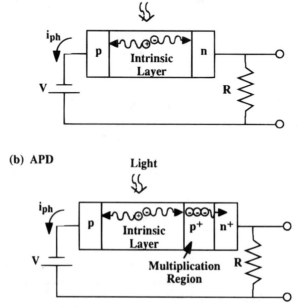

FIGURE 16.20 Diagram of a (a) p-i-n and (b) APD photodetectors.

Optical Communications

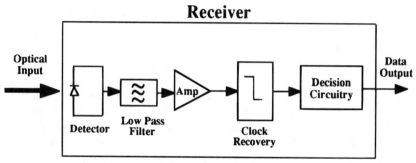

FIGURE 16.21 Block diagram of functions performed in advanced receiver packages.

All of these functions would be performed in an optical receiver package (see Fig. 16.21). At present, high-speed GHz receivers can be commercially purchased.

Filters

Optical filters, like electrical filters, can play a major role in an optical communication system by transmitting a desired wavelength but blocking all others. Two common filter applications are

(1) a demultiplexer when many different signals are transmitted along a single optical fiber with each signal being on a different wavelength (Willner, 1990) and

(2) to limit the transmitted bandwidth, as would be required at the output of an optical amplifier which generates broadband, amplified, spontaneous emission noise (see Section 16.4).

The Fabry–Perot optical filter operates in much the same manner as the Fabry–Perot resonant laser cavity. Figure 16.22a shows three media separated by two partially reflecting mirrors which form a wavelength-selective, Fabry–Perot optical filter cavity. An incident right-propagating optical wave at angle θ to the normal of the left mirror is partially transmitted. After this transmitted

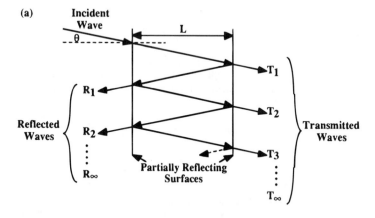

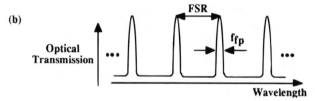

FIGURE 16.22 (a) Basic operation of a Fabry–Perot filter. (b) Wavelength-selective optical transmission of a Fabry–Perot filter.

right-propagating wave traverses the cavity interior, it experiences partial transmission T_1 and reflection when it encounters the right mirror. The new, left-propagating, reflected wave will experience reflection and transmission at the left mirror, and, subsequently, the new, right-propagating, reflected wave will encounter the right mirror and experience partial transmission T_2 and reflection. This process continues so that there are an infinite number of transmitted waves at the right mirror, with each higher number transmitted wave having a smaller intensity due to any losses. Assuming that the cavity length is much smaller than the coherence length of the laser light, all these transmitted waves will be coherent because they originate from the same coherent incident source and will interfere with each other as they are transmitted through the right mirror boundary. If these transmitted waves are in-phase with each other (i.e., the phase difference equals 0, 2π, 4π, . . .) when they encounter the right mirror boundary, then, they will all add constructively and the original incident wave will be transmitted through the FP optical filter cavity. If the transmitted waves are completely out of phase (i.e., the phase difference equals π, 3π, 5π, . . .), then, the waves will destroy each other, thereby blocking the original wave from passing through the cavity and reflecting it back away from the filter. If the transmitted waves are slightly out of phase, then, they will not add constructively, but mostly destructively, with little optical power of the original wave passing through the filter. Because all the waves must be in phase for filter transmission, the round-trip distance of wave propagation inside the cavity must be an integral number of wavelengths (Yariv, 1976):

$$L = \frac{m\lambda}{2n \cos \theta}, \qquad m = 1, 2, \ldots \ldots, \tag{16.14}$$

where n is the refractive index inside the cavity, L is the distance between the two mirrors, and λ is the wavelength of light to be transmitted through the filter. There are an infinite number of integers which will satisfy the in-phase requirement, and the passband of the filter is periodic with wavelength (Fig. 16.22b). The optical filter has a transmission resonance bandwidth f_{FP}, a free spectral range between any two resonances FSR, and a filter Finesse $\Im$, in which (Yariv, 1976)

$$\Im = \frac{\text{FSR}}{f_{FP}} = \frac{C}{2nLf_{FP}}. \tag{16.15}$$

The Finesse is a figure-of-merit for the optical filter. A high Finesse corresponds to a more efficient filter. The reflectivity of the mirrors affects the Finesse according to the following relationship (Yariv, 1976).

$$\Im = \frac{\pi\sqrt{r}}{1-r}, \tag{16.16}$$

where r is the power reflectivity at each of the two mirrors. As evident from Eq. (16.14), wavelength selectivity depends on the angle of incidence of the incoming wave because it affects the round-trip propagation distance inside the cavity required for phase matching of all the transmitted waves. Therefore, this simple filter can be wavelength-tuned by tilting the angle of the filter relative to the incoming wave. To first order, angle tuning effectively shifts all the periodic resonances, in unison, to different transmitting wavelengths.

Various Components

Directional Couplers

An optical signal must frequently be split from one input to two output ports for signal distribution or monitoring; note that the reverse function of combining is also desirable. This can be accom-

Optical Communications

plished passively by using a directional coupler (Tamir, 1979) as shown in Fig. 16.23. Two separate, single-mode waveguides are brought close together for some interaction length and are, then, separated. As mentioned previously, the gaussian-shaped, single-mode wave propagating in a waveguide will have most of its energy residing in the core accompanied by an evanescent field tail propagating alongside the core within the cladding region. A wave propagating in waveguide 1 will have an evanescent tail which partially falls within waveguide 2 while the wave is propagating along the interaction length. Due to coupled-mode theory (Tamir, 1979), this evanescent tail will excite an optical wave within waveguide 2. Power will gradually be transferred from waveguide 1 into waveguide 2. This coupling is resonant and periodic, so that light will oscillate back and forth between these two waveguides if the interaction length is sufficiently long. A 3-dB 50% power splitter can be fabricated by tailoring the interaction length to provide only coupling of half the power from waveguide 1 into waveguide 2. After the waveguides are separated so that no more coupling occurs, light can be independently transmitted from the output of two output ports.

Crossbar Switches

Directional couplers are passive devices in which the ratio of coupling is determined by the interaction length. However, by placing an electrode over the interaction region, we can induce a voltage-controlled electric field. This electric field changes the physical properties and refractive index of the waveguide material. Such a change in the index will alter the speed and phase retardation of the optical wave which, consequently, changes the effective interaction length and the coupling ratio (Alferness, 1982). By tuning the voltage, we can tune the optical output ratio between ports 1 and 2. Such a tunable device, which can switch an incoming signal between two different output ports, is known as a crossbar switch.

Modulators

We have mentioned one method of modulating an optical wave with digital data. That method is directly modulating the bias current of the laser which will modulate the carrier concentration within the active laser semiconductor medium. However, modulating the carrier concentration has a deleterious side effect because the refractive index of the active medium depends slightly on the carrier concentration. Changing the index will alter the effective optical length of the cavity between the feedback (or distributed feedback) mirrors. The wavelength selectivity of the cavity depends critically on the phase-matching conditions that the round-trip distance between the cavity mirrors is an integral number of wavelengths. Therefore, a change in the index will alter the spectral location of the cavity resonances. As the carrier concentration is changed upon direct bias-current modulation, the cavity will shift slightly and laser light will be produced over a spread of slightly different wavelengths. The wavelength spreading is known as modulation chirp, and is shown in Fig. 16.24 for amplitude-shift-keyed (ASK) (i.e., "ON"-"OFF") modulation of the light (Koch and Boroers, 1984). The chirp can have a 3-dB bandwidth as wide as 10 GHz and will limit ultralong-distance communications due to intramodal dispersion.

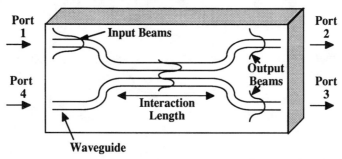

FIGURE 16.23 Optical waveguide interaction in a four-port directional coupler.

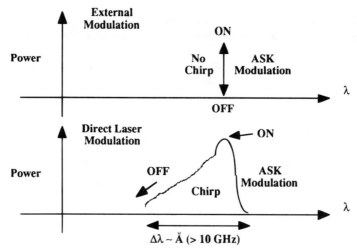

FIGURE 16.24 Optical spectrum showing the chirping of the output wavelength of a directly modulated, semiconductor laser.

One method of avoiding chirp is to have the laser emit a constant (continuous-wave) light beam and, then, externally modulate the light without producing any additive chirp. One type of external modulator is a lithium-niobate Mach–Zehnder interferometer (Korotky, 1984), shown in Fig. 16.25. An input single-mode waveguide is divided into two separate arms and, then, recombined into one waveguide after some distance. Assuming a short propagation distance, the light beams in the two separate arms are coherent because they originate from the same coherent source. These two beams will, then, interfere with each other when combining into the output waveguide. If the distance of propagation is the same in the two arms, then, the two waves are in phase and will add constructively resulting in light appearing at the output waveguide. Alternatively, if the distance in the two arms is different by one-half wavelength, then, the two waves will be π radians out of phase, the waves will add destructively, and no light will appear at the output waveguide. Let us begin with equal absolute distance in the two waveguide arms of the interferometer. By placing an electrode over only one of the arms, we can induce a voltage-controlled electric field in that arm which will change the refractive index, retard the wave, and change the phase delay of that arm relative to the other arm. For a sufficiently high applied voltage, we can induce a π phase shift in one arm, causing the two separate waves to be π out of phase when they recombine at the output waveguide. This will result in destructive interference and will cut off the optical transmission at a specific signal wavelength. The ON/OFF transmission characteristics, as a function of voltage, follow a raised, cosine-squared curve. These modulators can provide a contrast ratio >20 dB, are fairly expensive, and reduce the chirp to a negligible value.

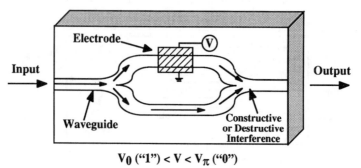

FIGURE 16.25 Mach–Zehnder interferometer acting as an optical modulator.

Isolators

Although an optical fiber itself can accommodate bidirectional optical propagation, many portions of an optical system will require that the light signal propagate only in one direction. This unidirectionality is usually governed by the functions which must be performed by various optical components, i.e., signal generation, detection, filtering, etc. In addition to these basic operations, unidirectional, signal propagation is also desired when we wish to severely limit reflections which can appear throughout the system and, typically, are due to slight discontinuities in refractive index at physical interfaces; note that nonlinear scattering will be briefly discussed later in Section 16.5. Most systems require some form of optical isolation which will allow light to pass in one direction but will not allow any reflected light to pass in the other direction. Reflections are detrimental to proper system operation for several reasons, including that

  a. reflections will slightly change the lasing frequency and threshold current of single-frequency lasers,
  b. reflections impair the ability of the local oscillator to track the incoming signal in coherent detection,
  c. if two reflections are close enough so that they are within the coherence length of the signal laser light, then, these two reflecting surfaces will establish a self-contained Fabry–Perot cavity with periodic resonances which will modulate the original signal, and
  d. reflections will increase the noise figure of an optical amplifier because the gain-producing carriers are now being used to amplify an undesired reflected wave. Moreover, two strong reflections surrounding the active region of a high-gain optical amplifier may create a cavity and initiate lasing inside the amplifier.

The operation of an optical isolator (Green, 1993) can be explained by following the light propagation shown in Fig. 16.26. Any light beam has a given polarization (i.e., direction of oscillation of the wave amplitude within the plane of propagation), whether it be linearly polarized (such as at the output to a semiconductor laser) or circularly polarized (such as within a fiber).

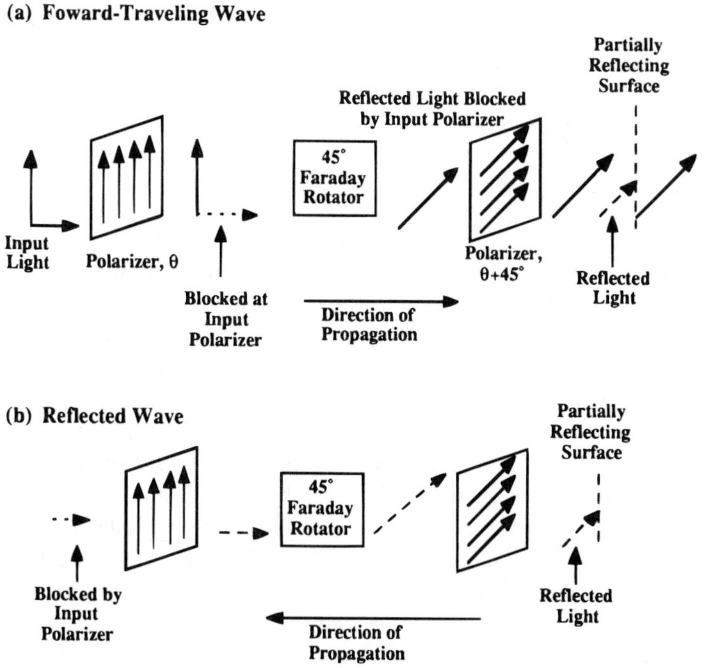

FIGURE 16.26 Illustration of the operation of an optical isolator. The isolator includes polarizers and a 45° Faraday rotator.

Even circularly polarized light can be decomposed into its two linearly polarized component parts. For this simple explanation, let us consider the case of linear polarization. The isolator consists of a 45° Faraday rotator surrounded on each side by a polarizer. Each polarizer passes only the light that matches its specific polarization direction. These polarizers are arranged so that the second (i.e., right) polarizer's transmission direction is shifted by +45° relative to the first (i.e., left) polarizer's transmission direction. The Faraday rotator takes any input optical beam and shifts its polarization by +45° independent of the direction of propagation through the rotator. We can explain the isolator operation by piecing together the above conditions. The left polarizer is aligned so that its transmission polarization direction matches the polarization of a right-propagating optical signal beam, for example, 0° to the horizontal plane. Subsequently, the transmitted optical beam passes through the Faraday rotator which shifts the beam's polarization by +45°. The right polarizer is aligned so that its transmission polarization direction matches the polarization of the light emanating from the rotator, i.e., +45°. This original incident light, thus, passes through the isolator without attenuation. However, if a reflection should occur after the isolator, then, this left-propagating reflected wave must pass the isolator in the reverse direction. This is extremely unlikely to happen. In the worst case, let us assume that the reflected wave has a polarization which matches the output right polarizer, thereby, passing the reflected wave in the reverse direction. When the reflected wave propagates through the rotator, its polarization is again rotated by +45°, now at +90° to the horizontal. Because the input left polarizer was set to pass only light at 0°, then, the reflected wave will be blocked by this polarizer, effectively optically isolating the input side from any reflected wave. In a single isolator, reflections can, typically, be <-40 dB below the original signal.

This simple isolator can also be achieved in a polarization-independent fiber-compatible in-line form even though we do not necessarily know the exact polarization of the light inside a fiber. Such isolation is accomplished by separating the incoming wave into its two component polarization parts with a polarization beam splitter. We, then, pass each polarization through separate linear-polarization isolators which function in parallel to isolate the two different polarizations. Then, the two light polarizations are combined and coupled into a single output (Green, 1993). Generically, separating polarizations and operating on each one independently is known as polarization diversity.

16.3 Basic Optical Systems

Modulation Formats

Information can be classified in two general forms, analog and digital (see Figure 16.27). Analog signals represent quantities which can take on any value within an infinite continuum of amplitude and time. This kind of signal is typical of many measured quantities, such as temperature, volume, frequency, dimensions, etc. In theory, Analog signals are ultimately accurate. However, to achieve a high degree of accuracy in signal recovery, the noise must be kept extremely low compared to the signal power (~ -50 dB). Alternatively, digital signals can represent many quantities which may require only measuring a limited number of discrete levels at specific bit-time intervals. It is only necessary to transmit and detect the difference between discrete levels (i.e., between "0" and "1" levels). Consequently, the noise relative to the signal that can be tolerated for error-free transmission is much higher than in analog transmission and, typically, is ~ -20 dB.

We would prefer to transmit digital signals due to the less stringent requirements of the signal-to-noise ratio (SNR). Even if a signal is originally in analog form, it is fairly straightforward to convert any analog signal into a digital signal and then transmit it. Inherently, we will lose some accuracy because we are taking a continuous-time signal of infinite amplitude levels and forcing it to be represented by a finite number of discrete levels measured at discrete time intervals. The analog signal is periodically sampled at a high enough sampling rate (i.e., above the Nyquist rate

Optical Communications

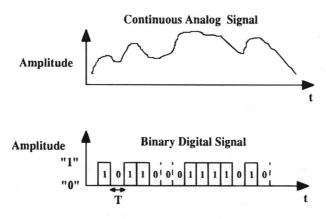

FIGURE 16.27 Analog and digital waveforms.

(Couch, 1983)) to adequately recover the highest frequency components in the signal; the highest frequency components in a signal are those portions that experience the fastest transitions in amplitude as a function of time. Each digital amplitude measurement of the analog signal is that value which corresponds most closely to one of the discrete digital levels. Each discrete level can be represented by a string of numbers, typically, a binary set of "0"s and "1"s; note that multilevel digital representation can also be used, but binary is the overwhelmingly popular choice. This is called pulse-code modulation (Couch, 1983) and is represented in Fig. 16.28. The more levels and the smaller the time interval in sampling, then, the more data bits which must be transmitted per unit time. This means that a higher bandwidth channel is required to transmit the data. In

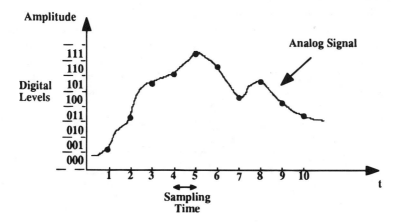

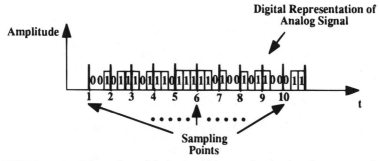

FIGURE 16.28 Pulse-code modulation in which an analog signal can be represented by a string of digital bits.

the limit, infinite bandwidth digital signals correspond to an analog signal. Of course, most quantities do not require such accuracy in representation. Because of the less stringent requirements of digital transmission and reception, we will almost exclusively discuss digital transmission except in section 16.7 which deals with subcarrier analog transmission.

Although there are several types of modulation coding formats for transmitting binary digital signals, we will discuss only the three most common. Fig. 16.29 shows non-return-to-zero (NRZ), return-to-zero (RZ), and biphase formats (Roden, 1995). NRZ is the simplest format in which the amplitude level is high during the bit-time, if a "1" is transmitted, and is low if a "0" is transmitted. If two or several "1"s are transmitted in sequence, then, the level remains high producing high-frequency components for isolated "1"s and low-frequency components for a string of "1"s. RZ format requires that a "1" always return to the low state during the bit-time even when two "1"s are transmitted in sequence, whereas the "0" bit remains at a low level. This eliminates the possibility that a long string of "1"s will produce a constant high level but does not prevent a long string of "0"s from producing a constant low level. Biphase transmission, in which a "1" or a "0" requires a transition in the middle of a bit-time even if several "1"s or "0"s are transmitted in sequence, eliminates the possibility of any long constant levels, i.e., no low-frequency components exist in the data bit stream. This is important only if the components in the system do not have a sufficiently wide bandwidth to accommodate all possible frequencies in the signal. Because NRZ is by far the simplest and most popular format, our discussion will be mainly limited to NRZ format requiring the system to accommodate the full digital signal bandwidth. However, note that RZ format is used for ultrahigh-speed pulse transmission and for soliton systems.

We have mentioned the notion of frequency components in a digital signal and the bandwidth which this signal occupies. Figures 16.30a, b, and c show graphically what we mean by the frequency components of a digital signal. As is well known, a square pulse in time can be represented in frequency space by taking the Fourier transform. The Fourier transform of the square pulse is a sinc function showing that the vast majority of the energy in the pulse resides in the center lobe (Roden, 1995). However, because an ideal square pulse has an infinitely fast rise and fall transition, some energy of the pulse exists at infinite frequencies; recall that there are infinite frequency components in a delta function. The center of the top flat portion of a pulse represents the lowest frequency components of the pulse because this portion has the

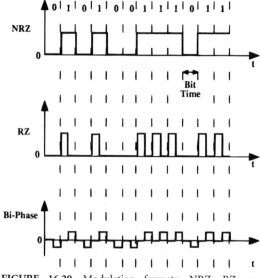

FIGURE 16.29 Modulation formats: NRZ, RZ, and biphase.

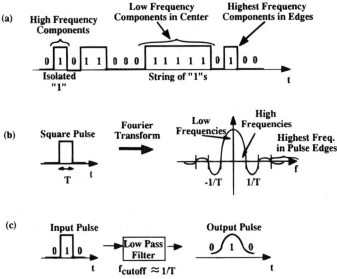

FIGURE 16.30 (a) An illustrative example of the frequency components in a time-based string of bits. (b) A square pulse and its Fourier transform. (c) A square pulse after passing through a low-pass-filter.

slowest transitions. We can retain most of the energy in a pulse and lose the energy in the highest frequencies if we pass a pulse through an electrical low-pass filter of a cutoff frequency which is half the bit rate. In such a case, we have retained only the first term(s) of the infinite Fourier series composed of cosines at the discrete harmonics of the signal. The shape of a pulse when passing through such a low-pass filter becomes rounded at the edges but the center of the pulse still reaches its maximum amplitude. Because the digital decision of a receiver circuit occurs at the center of the bit-time, there is little power penalty from decision ambiguity due to this low-pass filtering. Two reasons why we discuss the low-pass filtering of a digital signal are (i) the unintentional limiting of the system bandwidth by limited component bandwidth, and (ii) the intentional limiting of the bandwidth to restrict the excess noise at high frequencies (we will discuss noise later).

If we return to consider a random NRZ signal which can have isolated "1"s and strings of "1"s, we see that the highest indispensable frequency components of the data stream f_H are in the isolated "1"s and the first and last "1"s in a long sequence. The lowest indispensable frequency components f_L are in the center of the longest string of "1"s where the transition rate is the slowest:

$$f_H \approx \frac{1}{T},$$

and

$$f_L \approx \frac{1}{(\text{maximum \# of consecutive ``1''s})*T}. \tag{16.17}$$

(Later, we will see that the highest indispensable frequency is approximately $f_H/2$, which corresponds to the Nyquist sampling rate.) To have good transmission performance, the system must have a uniform response covering all the relevant frequency components in the digital bit stream. Note that the transmitted optical spectrum of an intensity-modulated random NRZ signal has a certain information, full-width, half-maximum, frequency bandwidth which is roughly equal to the bit-rate R_b (Roden, 1995).

The digital signal, which may be modulated at approximately Gbit/s rates, is being transmitted on an optical carrier wave whose frequency is in the multi-THz regime. The optical carrier is used because the optical wave has very low losses in the fiber whereas lower microwave frequencies have extremely high transmission losses in an optical fiber (or in a coaxial cable at Gbit/s speeds). This optical carrier wave $A(t)$ has an intensity amplitude A_0, an angular frequency ω_c, and a phase φ (Killen, 1991):

$$A(t) = A_0 \cos(\omega_c + \phi). \tag{16.18}$$

A binary digital signal implies transmitting two different quantities of anything which can subsequently be detected as representing a "1" and "0", i.e., we can transmit blue and red, and this can represent "1" and "0" in the receiver electronics if blue and red can be distinguished. We can, therefore, modulate either the amplitude, frequency, or phase of the optical carrier between two different values to represent either a "1" or "0", known, respectively, as amplitude-, frequency-, and phase-shift keying (ASK, FSK, and PSK), while the other two variables remain constant (Killen, 1991):

$$A_{\text{ASK}}(t) = [A_0 + m(\Delta A)]\cos(\omega_c t + \phi), \qquad m = \begin{cases} +1, & \text{"1"} \\ -1, & \text{"0"} \end{cases} \tag{16.19a}$$

$$A_{\text{FSK}}(t) = A_0 \cos\{[\omega_c + m(\Delta\omega)]t + \phi\}, \tag{16.19b}$$

and

$$A_{\text{PSK}}(t) = A_0 \cos[\omega_c t + (\phi + m\pi)], \tag{16.19c}$$

where ΔA is the amplitude modulation and is less than A_0, and $\Delta\omega$ is the FSK frequency deviation. ASK has two different light amplitude levels, FSK has two different optical carrier wavelengths, and PSK has two different phases which can be detected as an amplitude change in the center of the bit-time for which a "1" or "0" bit can be determined. Figure 16.31 shows the impression of a simple digital signal on the optical carrier. These three formats can be implemented by appropriately changing the optical source, by modulating the light amplitude, laser output wavelength, or using an external phase shifter. ASK is important because it is the simplest to implement, FSK is important because a smaller chirp is incurred when direct modulation of a laser is used, and PSK is important because, in theory, it requires the least amount of optical power to enjoy error-free data recovery.[14] It should be mentioned that ASK, by far the most common form, is called On-Off-Keying (OOK) if the "0" level is really at zero amplitude. Note that the frequency of the optical carrier is so high that the optical detector, whose electronic bandwidth is usually <<100 GHz, will not detect it and only the envelope of the Gbit/s data will be electrically recovered.

Detection Schemes

Once a signal has been modulated and transmitted, it must be accurately detected. Two detection schemes are direct detection and heterodyne (i.e., coherent) detection (Senior, 1985; Betti et al. 1995). Direct detection can be used with ASK and FSK modulation, and ASK, FSK, and PSK can be used with heterodyne detection. [Note that homodyne coherent detection is also possible, but, presently, is not considered the detection scheme of choice due to the relative difficulty of implementing it (Stephens and Nicholson, 1987; Kahn et al. 1990). Due to the limited space in this chapter, we will leave homodyne detection to the reader to examine the reference material.]

Optical Communications

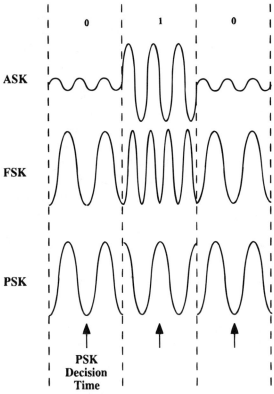

FIGURE 16.31 ASK, FSK, and PSK time modulation while employing an optical carrier wave.

Direct detection, by far the simpler of the two schemes, involves detecting the amount of optical power incident on the optical detector. Direct detection of an ASK (or OOK) signal, i.e., light "ON" or light "OFF", is extremely simple to accomplish using a detector and a high-bandwidth power meter (see Fig. 16.32a). The ASK signal is recovered by a detector of a certain electrical low-pass-filtering bandwidth. The electrical spectrum of the recovered ASK signal, which is sent in random NRZ format and can be measured on an RF spectrum analyzer, is shown in Fig. 16.32b. As mentioned in the previous section, only the first lobe is necessary for recovering

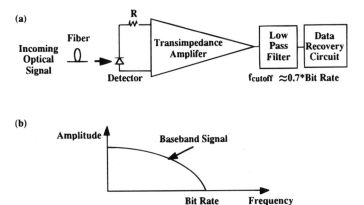

FIGURE 16.32 (a) Direct detection optical system. (b) Baseband signal of a directly detected, NRZ signal.

the data because only the transition edges would be affected by cutting off the higher lobes. The first lobe is considered the baseband signal representing the data stream (Roden, 1995).

Directly detecting an FSK signal is a little more complicated because direct detection can measure only optical power, but cannot resolve the incoming optical wavelengths which are at carrier frequencies much higher than the detector bandwidth. Direct detection of an FSK signal can be facilitated by using an optical filter to convert the FSK signal into ASK format before detection (Tamir, 1979). To see how this is achieved, Fig. 16.33 shows the optical spectrum of an FSK signal. An FSK spectrum is composed of two separate wavelengths, each being "ON" when the other is "OFF". Given FSK with a wide deviation between the two wavelengths, each spectrum appears like an independent ASK signal, i.e., instead of blue-red-blue it can be thought of as two separate signals of blue-"OFF"-blue and "OFF"-red-"OFF". If we tune an optical filter to pass only the wavelength representing the "1" bits and block the wavelength for the "0" bits, the output signal appears to be a single ASK signal. Therefore, a narrow optical filter placed before the detector effectively converts an FSK signal into ASK format which is, then, straightforward to detect. Note that the optical filter has a lorentzian-line-shaped pass-band, blocking most of the optical power from the rejected wavelength, but allowing some of its optical power to pass. This unwanted optical power will raise the "0" level from zero amplitude to some finite value, thereby, reducing the "1"-to-"0" contrast ratio and increasing the system power penalty.

Heterodyne detection is more difficult to accomplish, but is considered a more sensitive detection scheme (Stephens and Nicholson, 1987). Heterodyne detection relies on the ability of two waves to "mix" with each other when they both have the same polarization and are incident on a detector. Because an optoelectronic detector is a p-n junction which follows a "square law" between input power and generated current, the detector itself mixes two independent optical waves. The input optical field is squared by the detector, generating square terms and cross-term products. Figure 16.34 shows the combining of two optical waves into a single detector by using a partially reflecting beam splitter, this function can also be performed by a simple fiber directional coupler. The two optical waves are (i) the modulated transmitted signal $E_s(t)$ and (ii) a continuous-wave (CW) local oscillator E_{LO}. By invoking trigonometric relationships, the generated output photocurrent from the detector is given by (Linke et al. 1988)

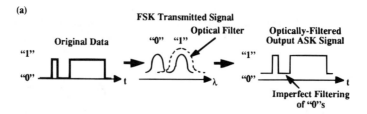

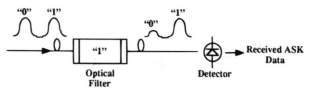

FIGURE 16.33 (a) FSK-to-ASK conversion. The illustration shows the optical spectrum of a transmitted FSK signal and the resultant ASK signal after passing through an optical filter. (b) The hardware implementation showing an optical filter and direct detection.

Optical Communications

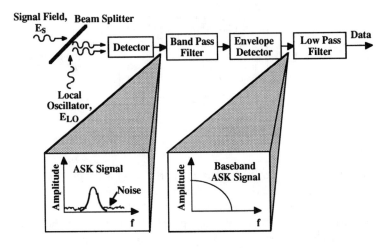

FIGURE 16.34 Heterodyne detection of an incoming signal.

$$I_{ph} = [E_s(t) + E_{LO}][E_s(t) + E_{LO}]^*$$
$$= E_s^2(t) + E_{LO}^2 + 2|E_s||E_{LO}|\cos(\omega_s t)\cos(\omega_{LO} t)$$
$$= P_s(t) + P_{LO} + \sqrt{P_s P_{LO}}\{\cos[(\omega_s + \omega_{LO})t] + \cos[(\omega_s - \omega_{LO})t]\}, \quad (16.20)$$

where the superscript * denotes the complex conjugate of the waves, E_s is the signal electric field and equals $[|E_s|\cos(\omega_s + \varphi)]$, E_{LO} is the signal electric field and equals $[|E_{LO}|\cos(\omega_{LO} + \varphi)]$, P represents optical power (or the square of the electromagnetic field), P_{LO} is the dc (or CW) local oscillator power, and P_s is the modulated signal power. (We have neglected to write the phases of the signal and local oscillator because we consider that they do not change significantly relative to each other over a short time; this requirement means that we must have a narrow linewidth source and narrow linewidth local oscillator to maintain coherence.) We can simplify this equation when considering that (i) the local oscillator power is dc and can be filtered out, (ii) the local oscillator power can and will be made much greater than the incoming weak signal which has been transmitted over some distance allowing us to eliminate the P_s^2 term, and (iii) the term which includes the sum of the two frequencies can be eliminated because the sum frequency is so high as to be unrecoverable by a detector of limited bandwidth or can be easily filtered out in relation to the difference-frequency term. The simplified equation is, then,

$$I_{ph} \approx \sqrt{P_s P_{LO}} \cos[(\omega_s - \omega_{LO})t]. \quad (16.21)$$

We can now see the advantages of this method. Instead of direct detection, which can achieve a recovered signal power of P_s, we can adjust the local oscillator power in heterodyne detection to be as high as possible, with the *effective* signal power recovered being increased to the square root of $P_s P_{LO}$. Our sensitivity has, therefore, been increased significantly. Although we can increase the local oscillator power to increase the effective signal power, there is a limit to its effectiveness because we will be limited at some point by the high-power-induced shot noise; as the signal power increases, the noise power will increase at the same rate, thereby, saturating the signal-to-noise ratio. Shot noise will be discussed further in the next section.

The center of the recovered "beat" signal's electrical spectrum will appear at the frequency difference between the signal and local oscillator. Therefore, the local oscillator must be wavelength tunable so that the recovered electrical signal can be placed at an intermediate frequency which is within the bandwidth of the detector. Note that even a 1-nm separation corresponds to a beat

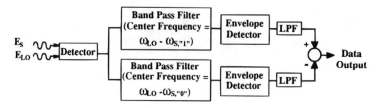

FIGURE 16.35 Heterodyne recovery of an FSK signal.

frequency >100 GHz and will not be recovered by the detector. Figure 16.34 shows the full heterodyne receiver which includes (i) an electrical band-pass filter following the photodetector to allow only the recovered electrical signal to pass and block other channels and other noise-containing frequencies, (ii) an electrical envelope detector to recover only the data modulation which will, then, appear as a baseband electrical spectrum, and (iii) a low-pass filter to limit the high frequency noise and recover only the data (Linke and Gnauck, 1988).

Heterodyne detection can also be used to detect FSK signals. This can be accomplished similarly to direct detection, except that the conversion from FSK-to-ASK using a filter is now accomplished in the electrical domain, not in the optical domain (Betti et al. 1995). Figure 16.35 shows the recovery of an FSK signal by heterodyning. The two wavelengths representing the "1" and "0" bits are detected by the photodetector and appear at two different locations in the electrical spectrum corresponding to the frequency difference between that bit's wavelength and the local oscillator wavelength. The electrical band-pass filter will, then, discriminate between the two beat frequency signals, passing the "1" bit difference frequency and rejecting the "0" bit difference frequency. Then, the envelope detector can easily recover the resultant ASK digital signal.

It is important to mention, briefly two important issues with heterodyne reception. One involves the fact that half the optical power is wasted when coupling the signal with the local oscillator. A balanced heterodyne receiver, as shown in Fig. 16.36a can be used, in which the two waves are combined in a 2 × 2 passive, fiber directional coupler (Abbas et al. 1985). Each output port contains the mixed waves. If each port is separately passed through a heterodyne detector, we can combine the two resultant digital signals to enhance the power received in the "1" bits, thereby, enhancing the receiver sensitivity. Another key issue is that the polarization of the signal and local oscillator must match. This is difficult because the polarization of light in a fiber is difficult to control and will wander and drift. Therefore, a polarization diversity receiver can be employed in which the two signal polarizations are separated at the receiver input and, then,

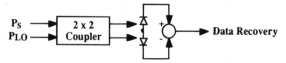

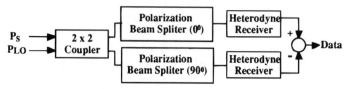

FIGURE 16.36 (a) A balanced heterodyne receiver. (b) A polarization diversity heterodyne receiver.

separately detected in two parallel heterodyne detectors whose local oscillator polarizations are 90° out of phase with respect to each other (see Fig. 36b) (Glance, 1987).

PSK signals can also be detected by heterodyne reception (Betti et al. 1995). What is necessary is the ability to detect the phase of the wave, 0 or π, by mixing the signal with a local oscillator. The difference in these two phases corresponds to a wave whose electric field is either positive or negative in the center of the bit where the digital decision occurs. We can repeat the formalism for PSK heterodyne detection as was done with ASK heterodyne detection. Because we have a generated photocurrent for the "1" and "0" bits, both of which can be used for data recovery, the "effective" detected photocurrent I_{PSK} is given by (Jones, 1988)

$$I_{PSK} = 4\sqrt{P_s P_{LO}} \cos(\phi_s - \phi_{LO}). \tag{16.22}$$

A detection circuit must be used to distinguish the two amplitude levels. Note that data recovery is extremely sensitive to the phase of the signal and local oscillator and that these two phases remain constant relative to each other. As mentioned in section II.2, the laser has a finite linewidth due to quantum fluctuations in the stimulated emission. This linewidth will cause the phase to wander and make it impossible to recover a signal. Therefore, PSK requires extremely narrow linewidth lasers, much narrower than needed even for FSK heterodyne detection.

Although different systems exist, the majority are NRZ ASK using direct detection. Unless otherwise stated, we will be discussing this generic type of system. Furthermore, due to the brevity of this chapter, we will not be able to discuss the other types of systems in much more detail except when particularly necessary.

Signal and Noises

Signal power is defined as the mean power in the modulated signal, and noise power is the statistical variance in the modulated signal (Schwartz, 1980). To detect a digital bit and accurately decide if it is a "1" or "0" bit, the signal power must be much larger than the noise power. If the noise is too large, then, a false "0" or false "1" may be detected. The signal-to-noise ratio (SNR) is considered an indispensable quantity when attempting to evaluate the performance of a system.

The electrically measured optical signal power mean is directly related to the square of the generated photocurrent in a photodetector (Lee and Li, 1979):

$$S = I_{ph}^2 = \left(\frac{\eta q P_s}{h\nu}\right)^2. \tag{16.23}$$

The signal power is proportional to the amount of power available for a decision circuit to decide if a "1" or "0" bit was transmitted. Therefore, the effective signal power is the high level in a "1" bit relative to the low level in a "0" bit; this relationship of ($P_{"1"}/P_{"0"}$) is known as the contrast ratio and is optimally equal to infinity (see Fig. 16.37a). If OOK modulation is used in which the "0" bit is transmitted with 0 power, then Eq. (16.1) describes the signal power level. In a case in which the light was not completely turned "OFF" during a "0" bit, then, there is a reduction in the contrast ratio between the "1" and "0" bits. Such a reduction affects the effective signal power as a subtraction:

$$S = (P_{"1"} - P_{"0"})^2. \tag{16.24}$$

A reduction in the contrast ratio from an increase in the "0" level can arise from several sources. Three of the most common are (Roden, 1995)

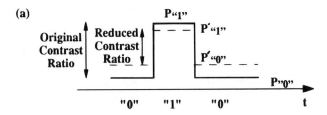

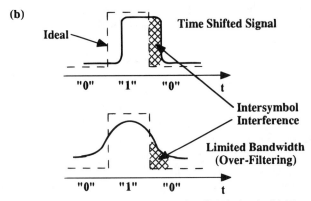

FIGURE 16.37 (a) Contrast ratio of a digital signal. (b) Two generic and simple examples of intersymbol interference in a digital stream of bits.

a. a nonzero power level when transmitting a "0" bit,
b. some optical power from another source incident on the detector, as is the case when another channel at another wavelength is not completely blocked by an optical filter, and
c. some of the rejected wavelength in an FSK signal leaks through the optical filter, when using direct detection to perform FSK-to-ASK conversion.

The effective signal power can also be reduced by simple attenuation of a "1" bit due to any number of types of losses (i.e. fiber attenuation, off-peak transmission through an optical filter, insufficient bandwidth in the signal source or detector, etc.).

Another fairly common source for reducing the contrast ratio is called intersymbol interference, shown graphically and in simple form in Fig. 16.37b. Intersymbol interference, essentially, means that a bit transmitted in one time slot will have some of its energy appear in an adjacent time slot, potentially creating a higher power level for a "0" bit and a lower power level for a "1" bit; this will reduce the signal contrast ratio. This interference can be caused by several mechanisms, but some common causes include (Henry et al. 1988) (i) insufficient system bandwidth which introduces phase shifts for the bits whose frequency components are outside of the systems 3-dB bandwidth, (ii) system dispersion (especially due to fiber) which spreads the pulse width and causes a pulse's power to appear in the two adjacent bit-times, and (iii) system nonlinearities which cause a power-dependent bit transmission velocity.

A reduction in the effective signal power, either by a lowering of the "1" bit level or a raising of the "0" bit level, causes a system power penalty, $PP_{CR}(dB)$, due to a reduced contrast ratio. Note that a system power penalty is the increase in the amount of signal power necessary to transmit a data bit unambiguously given an otherwise well-performing system. PP_{CR} is given by (Willner, 1990)

$$PP_{CR} = 10 \log \left[\frac{P_{"1"} - P_{"0"}}{P'_{"1"} - P'_{"0"}} \right]^2, \qquad (16.25)$$

where $P'_{"1"}$ represents the lowered value detected in the "1" bit due to losses and $P'_{"0"}$ is the raised value detected in the "0" bits due to incomplete suppression of the "0" bit or because of intersymbol interference. Unless otherwise stated, we will consider P_s as the effective signal power and that $P'_{"0"}$ is ~0.

The noises generated in a typical system are different for direct detection and for heterodyne detection. We will discuss each detection method separately. Furthermore, noise in a p-i-n detector differs from noise in an APD detector. For brevity, we will discuss only the p-i-n detector with the comment that, although current is multiplied in an APD, the noise is also increased.

We consider all the noise terms to be the power in the signal variance. Because these noise terms represent independent, incoherent, statistical variations in the detector current, we can decouple each term and simply add them together in a total noise term, σ^2_{tot}. The three noise terms in simple direct detection are the shot, thermal, and circuit noise. The circuit noise is simply variances of electron generation in the receiver circuit. However, due to advances in circuit design, we typically assume that the thermal noise dominates the circuit noise. Retrieving the equations for the thermal and shot noises from section II.3 on photodetection, σ^2_{tot} is expressed as (Campbell, 1989)

$$\sigma^2_{tot} = \left[\frac{4kT}{\Omega} + 2q\left(\frac{q}{h\nu}\right)P_s\right]B_e. \quad (16.26)$$

Typically, thermal noise dominates over shot noise in direct detection, giving rise to the term thermal-noise-limited system reception. These noise powers increase with bandwidth and cover all possible frequencies, and the noise powers are limited only by limiting the frequencies which are supported by the system. The receiver bandwidth must be at least 50–70% of the bit-rate to adequately recover the transmitted bits without incurring a power penalty; recall that a receiver bandwidth of approximately half the bit-rate will smooth the bit transitions but the center of each bit, where the digital decision is performed, is relatively unaffected. In a well-operated system, any higher bandwidth would not increase the recovered signal much but would allow more noise to be recovered. The receiver will almost always have a low-pass filter to limit the unwanted high-frequency noise. As an example, a 2 Gbit/s signal would require twice the electrical bandwidth of a 1 Gbit/s signal for detection. Therefore, twice the noise is produced, and twice the signal power is necessary to achieve the same SNR, thereby, incurring a 3-dB system decrease in sensitivity. When combining all the above, the SNR (in decibels (dB)) for direct detection is given by

$$\text{SNR} = 10 \log\left\{\frac{(P'_{"1"} - P'_{"0"})^2}{\left[\frac{4kT}{\Omega} + 2q\left(\frac{q}{h\nu}\right)P_s\right]B_e}\right\}. \quad (16.27)$$

The SNR for heterodyne coherent detection differs from the SNR for direct detection because, now, a local oscillator is impinging on the detector at the same time as the signal. As outlined in the previous section, the effective signal power has been enhanced by the local oscillator power (Linke and Gnauck, 1988):

$$S = P_s P_{LO}. \quad (16.28)$$

[Note that a factor of 2 would be inserted into the right side of Eq. (16.28) if a balanced heterodyne receiver was used.] The thermal noise is independent of optical power and, therefore, is not changed due to a local oscillator in heterodyne detection. However, the shot noise is proportional to the total absorbed optical power and is now much larger due to the local oscillator (Linke and Gnauck, 1988):

$$\sigma_{SL}^2 = 2q\left(\frac{q}{h\nu}\right)(\sqrt{P_s P_{LO}} + P_s + P_{LO})B_e. \quad (16.29)$$

Therefore, given that P_{LO} is much greater than P_s, the SNR for a coherent system can be approximated as (Agrawal, 1992; p. 230)

$$SNR = \frac{2P_s P_{LO}}{\left[2qP_{LO} + \frac{4kT}{\Omega}\right]B_e}. \quad (16.30)$$

We can make the effective signal power as large as we want by increasing the local-oscillator power. However, the shot-noise power also increases when the local-oscillator power is increased. When the total noise is dominated by the shot noise, the SNR does not continue increasing with increasing P_{LO}. The receiver is then shot-noise-limited, which is the optimal region for achieving the highest SNR in a heterodyne detection system.

Performance Criteria

The undisputed criterion for system performance is determining whether a data bit was correctly recovered as a "1" or "0". This is reflected in determining (and minimizing) the bit-error rate (BER) of a system. Fig. 16.38 shows the voltage or current levels generated from the recovered signal for a "1" and a "0" bit. The received noise has the effect of adding or subtracting electrons when either an optical pulse was transmitted or when no pulse was transmitted. The noise has an approximate Gaussian amplitude distribution around the expected signal value. This represents the variance in the expected number of generated electrons. A decision circuit must set a threshold current level I_{th}, above which is considered a "1" bit and below which is considered a "0" bit. If we assume, for simplicity, that the system is thermal-noise-limited and the noise is roughly equal on the "1" and "0" bits, then, the decision level will be roughly halfway between the two levels; note that the noises using an APD receiver are not equal for a "1" and "0" because the APD noises depend on the input signal power. Due to the Gaussian noise distribution around a signal mean, a recovered "0" bit will periodically appear above the threshold level with a finite probability, and a recovered "1" bit will also appear below the threshold level with a finite probability. These two scenarios represent a false "1" and false "0", respectively, and are considered errors in the system. The probability of error in the system, $P(e)$, is the sum of these two possible errors (Couch, 1993):

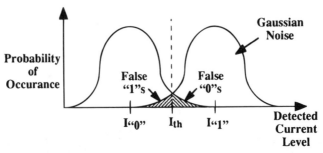

FIGURE 16.38 The probability of current detection when considering the recovered signal level and the Gaussian noise.

Optical Communications

$$P(e) = P(``1")P(e|``1") + P(``0")P(e|``0"), \qquad (16.31)$$

where $P(e|``1")$ and $P(e|``0")$ are the probabilities of error, given that a "1" and a "0" are transmitted, respectively, and $P(``1")$ and $P(``0")$ are respectively, the probabilities that a "1" and a "0", had originally been transmitted.

The key to deriving the BER, or probability of error, is to find the probability of a "0" being above I_{th} and a "1" being below I_{th}. All we need to do is to integrate the probability distribution of the noise from I_{th} to ∞ and from $-\infty$ to I_{th} for the "0" and "1" bits, respectively (Couch, 1993):

$$P(e) = \left\{ \frac{1}{\sqrt{2\pi\sigma^2_{``1"}}} \int_{-\infty}^{I_{th}} e^{-(I-I_{``1"})^2/2\sigma^2_{``1"}} \, dI \right\} + \left\{ \frac{1}{\sqrt{2\pi\sigma^2_{``0"}}} \int_{I_{th}}^{\infty} e^{-(I-I_{``0"})^2/2\sigma^2_{``0"}} \, dI \right\}. \qquad (16.32)$$

Such an integration involves integrating the gaussian noise distribution. By using some mathematical substitutions, integration of a gaussian function can be performed by computing the error function of the threshold level relative to the expected level (Spiegel, 1968). The error function $\text{erf}(x)$, where $x = (I_{``1"} - I_{th})$, finds the probability of error in a transmitted "1", and the complementary error function $\text{erfc}(x) = [1 - \text{erf}(x)]$, finds the probability of error in a transmitted "0". However, we can make appropriate substitutions and find the total probability of error in terms of $\text{erfc}(x)$ (Roden, 1995):

$$\text{BER} = \frac{1}{4} \left[\text{erfc}\left(\frac{I_{``1"} - I_{th}}{\sigma_{``1"}\sqrt{2}} \right) + \text{erfc}\left(\frac{I_{th} - I_{``0"}}{\sigma_{``0"}\sqrt{2}} \right) \right]. \qquad (16.33)$$

The BER is directly related to the SNR, with a higher signal or lower noise contributing to a lower probability of error. In a direct-detection system, the BER is a function of the SNR as follows (Roden, 1995):

$$\text{BER} = \frac{1}{2} \text{erfc}\left(\frac{Q}{\sqrt{2}} \right), \qquad (16.34)$$

where Q is related to the SNR and defined as

$$Q = \frac{I_{``1"} - I_{``0"}}{\sigma_{``1"} + \sigma_{``0"}}. \qquad (16.35)$$

The optimum threshold level i.e., decision level I_{th}^{opt}, is the average of the "0" and "1" currents weighted by the noises (Olsson, 1989):

$$I_{th}^{opt} = \frac{\sigma_{``0"} I_{``1"} + \sigma_{``1"} I_{``0"}}{\sigma_{``0"} + \sigma_{``1"}}. \qquad (16.36)$$

Figure 16.39 shows the BER as a function of the received optical power (Jones, 1988, p. 30). Note that it follows the form of an error function. This figure shows the sensitivity of a system, i.e., the amount of detected optical power necessary to achieve a certain bit-error rate. For this system, an SNR of ~ 16 dB is required for a BER $= 10^{-9}$. We wish to achieve an optical system exhibiting a very low BER, perhaps even 10^{-14}. The reasons for requiring such a low BER are that

   a. we wish to limit the retransmission of data upon error detection. Such retransmission incurs a significant time delay, which negates the benefits of a high-speed optical transmission system.

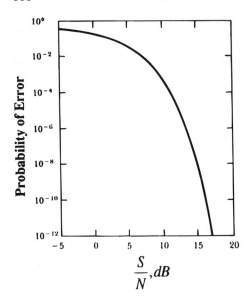

FIGURE 16.39 BER curve as a function of the signal power for a direct detection system. (After Jones, 1988.)

b. we wish to limit the amount of overhead in a transmitted stream of data which is reserved for error-correction coding. Such overhead bits are a waste of bandwidth, given an "error-free" system.

It may seem odd that the noise must be 16 dB below the signal power, roughly a factor of 50, to simply distinguish between a "1" and a "0" bit. However, the noise is a statistical quantity which may have a small variance but which will periodically achieve a very high value. In other words, although the noise may be small on average, sometimes a sharp noise spike, causing a severe increase or decrease in the electron count, will cause an error in a bit. This could be during the bit time of only one out of a billion transmitted bits, corresponding to the probability of error in the Gaussian noise distribution. Therefore the variance and noise powers must be kept extremely low.

The sensitivity and BER can also be found for asynchronous heterodyne receivers detecting ASK modulated signals (Betti et al. 1995):

$$\text{BER} = \frac{1}{2}\exp\left(\frac{\text{SNR}}{8}\right). \tag{16.37}$$

In practical systems, heterodyne detection is more sensitive than direct detection. Figure 16.40

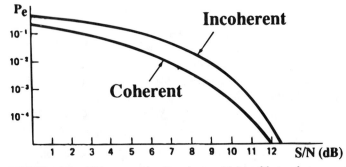

FIGURE 16.40 BER curves showing the sensitivity of heterodyne versus direct-detection schemes. (After Roden, 1995.)

Optical Communications

TABLE 16.1 Comparison of Theoretical and Practical Sensitivity Limits in the Number of Photons-Per-Bit Which Must be Recovered to Achieve a Bit-Error

Modulation/Receiver	Direct Detection	Heterodyne (Asynchronous)	Homodyne
OOK	10 (200)	40 (80)	18 (36)
FSK	–	40 (80)	–
PSK	–	20 (40)	9 (18)

shows two BER curves as examples of the sensitivity difference (Roden, 1995, p. 478). Results for direct-detection receivers hover near -30 dBm (i.e., -30 dB down from a milliwatt, or simply a microwatt in this case) to achieve a BER of 10^{-9} at 1 Gbit/s. Heterodyne receivers are typically 10 dB more sensitive, requiring only ~ -40 dBm to achieve a BER of 10^{-9} at 1 Gbit/s. We have discussed some of the difficulties in implementing heterodyne detection, and the key question is whether the increased difficulty is worth the added sensitivity. In section 16.4, we will discuss how direct detection using an optical amplifier is as sensitive as heterodyne detection using a local oscillator.

Table 16.1 gives a comparison of the theoretical limits of the various modulation and detection schemes (Henry et al. 1988, Chapter 21, p. 813). Actual systems rarely come close to these theoretical limits, and the challenge to reach the lower limits is considerable. Note that the limits are given in photons per bit, which can easily be calculated for any bit-rate by noting that the energy in a pulse is $h\nu$, the photon energy, multiplied by the number of photons.

Although the ultimate quantitative measure of performance is the BER, another key performance criterion which is easy to measure on an oscilloscope and gives a qualitative "feel" for system performance is the eye diagram (Couch, 1983). Figure 16.41 describes how an eye diagram is generated on an oscilloscope from a given pseudorandom NRZ data stream. To begin with, an oscilloscope trace of a series of distinct bits can be seen on an oscilloscope because the bit stream is periodic, repeating continuously so that a nanosecond bit (which cannot be seen as a single event by the human observer) is regenerated countless times and, thus, is visible on the oscilloscope

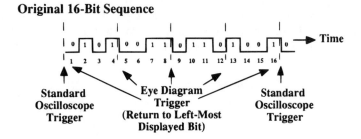

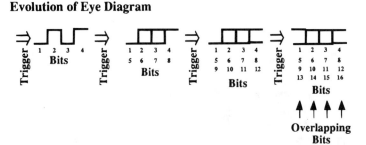

FIGURE 16.41 Generation of a sharp eye diagram.

screen. This continuous repetition of the bit stream occurs is because the entire bit sequence is repeated and each bit appears during each repetition in the same spatial (i.e., temporal) position on the oscilloscope screen; note that a truly random bit stream is not periodic, and so, only a predetermined pseudorandom bit sequence can be displayed. This repetition of the bit stream in which the same bit during each sequence appears in the same location on the screen is accomplished by the bit stream generator sending out a trigger pulse in addition to the clock pulses which synchronize the bits in a time slot. This trigger pulse forces the oscilloscope to begin its trace at the extreme left part of the screen for each repetition of the entire bit stream.

To generate an eye diagram, the bit-pattern generator triggers the oscilloscope to begin its trace much more often than at the end of the entire bit stream. As an illustrative example, Fig. 16.41 shows a 16-bit repetitive sequence which, normally, is triggered every 16 bits but is now triggered every 4 bits, with the trace beginning at the left part of the screen for bits 1, 5, 9, and 13. The oscilloscope traces and displays bits 1 through 4, returns to the left of the screen and, then, traces bits 5 through 9. There are now only four displayed bit time slots into which all the 16 bits must fit. Bits 1, 5, 9, and 13 are displayed in time slot #1, bits 2, 6, 10, and 14 are displayed in time slot #2, bits 3, 7, 11, and 15 are displayed in time slot #3, and bits 4, 8, 12, and 16 are displayed in time slot #4. In our example, each time slot will, sometimes, have a "1" bit and, sometimes, have a "0" bit displayed. The total display will eventually look like a series of boxes. As in all optical transmission systems, a low-pass filter is used to limit the high-frequency noise which exists above the necessary signal bandwidth. This filter will round off the sharp edges of transmitted rectangular bits. What remains is the "eye" diagram as shown in Fig. 16.42a, which literally resembles an "eye." The upper and lower rails have transitions when a "1" bit is followed by a "0" bit, and vice-versa. These rails also show no transitions between time slots if a string of "1"s or "0"s is transmitted. The rails also have some amplitude width corresponding to system noise in the vertical direction and timing jitter in the horizontal direction. Figure 42a shows an open eye diagram in which it is fairly easy to distinguish the difference between a "1" and "0" in any given time slot. The decision circuit will decide the identity of the bit at the center of the open eye, usually, in the middle of the time slot and at the halfway point between the upper and lower rails. Figure 42b shows how the eye closes if the contrast ratio is too low or if too much noise exists. Also shown in Fig. 16.42b is the timing jitter which is an uncertainty in the exact start of a bit-time (bit recovery time) and reduces the sensitivity of the receiver. The timing jitter can be caused by faulty circuitry or by various noise terms which result in phase variations. Although the eye diagram cannot represent a 10^{-9} BER because one bit out of a billion could

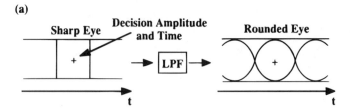

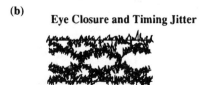

FIGURE 16.42 (a) A low-pass-filtered eye diagram. (b) A closed-eye diagram due to contrast-ratio degradation, amplitude noise, and timing jitter.

Optical Communications

never be seen on a screen, the eye diagram is still an easy and good qualitative indication of proper system performance.

System Design

A system must be designed to function properly and ensure near-error-free data recovery. This implies that enough signal power be detected (i.e., a sufficiently high SNR) to meet or exceed the sensitivity of the receiver. From a system design point of view, this is accomplished by adding or subtracting the various factors which would contribute to an increase or decrease in the signal power required for proper system performance.

We have discussed many quantities in an optical system in units of decibels (dB) for ease of system design. Decibels have the property that they compare two quantities, either an input divided by an output or one quantity relative to a standard value, in a logarithmic relationship (Spiegel, 1968):

$$dB = 10 \log\left[\frac{\text{output}}{\text{input}}\right] \text{ (power)}, \tag{16.38a}$$

$$dB = 20 \log\left[\frac{\text{output}}{\text{input}}\right] \text{ (voltage)}, \tag{16.38b}$$

and

$$dBm = 10 \log\left[\frac{x \text{ mW}}{1 \text{ mW}}\right] \text{ (power)}, \tag{16.38c}$$

where dBm compares the given value to a milliwatt of power. Although both power and voltage decibels were outlined in Eq. (16.38), we will restrict ourselves to discussing decibels in terms of power for the remainder of the chapter. An important point is that decibels are described in logarithms. Logarithms have the unique characteristic that the logarithm of one number minus the logarithm of another number is equal to the logarithm of the division of the two numbers (Spiegel, 1968):

$$10 \log\left[\frac{\text{output}}{\text{input}}\right] = 10 \log\left[\frac{\text{output}}{1\text{mW}}\right] - 10 \log\left[\frac{\text{input}}{1\text{mW}}\right], \tag{16.39}$$

Component Loss (dB) = Output (dBm) − Input (dBm).

Division and multiplication are transformed into subtraction and addition, respectively, which are much easier calculations to perform especially when concatenating many system elements for evaluation. If a device has either a loss or gain, this can be described by a subtraction or addition of the signal power in decibels. A system power penalty for the losses in a given component (i.e., output/input) would just be a subtraction of some number of decibels. Another elegant quality is that dB by itself is not an absolute number but describes a ratio which is simply a characteristic of a device. It is, therefore, independent of the absolute input or output power level of a device.

We will now design a simple optical system using the addition and subtraction of decibels (Senior, 1985). This simple ASK NRZ direct-detection system is shown in Fig. 16.43. The laser puts out a certain amount of optical power in dBm. Any aspect can provide a system addition or subtraction. The end receiver has a certain sensitivity in dBm which is the optical power necessary for near-error-free data recovery by the receiver. We must ensure that more power than this minimum sensitivity reaches the receiver. In fact, we usually speak of a system power margin, which represents an extra power buffer above the minimum power level required for unambiguous reception. This power margin, which is typically a few dB, allows for slight occasional degradations in system performance without affecting proper system performance. The designed system would be evaluated as follows:

Laser output	+3 dBm
Power penalty due to a 50% contrast ratio	−3 dB
Coupling loss into fiber	−3 dB
Fiber attenuation in 30-km link	−15 dB
Splitting loss in a 4 × 4 coupler	−6 dB
Insertion loss in an optical filter	−1.5 dB
Requested power margin	−4 dB
Direct-Detection receiver sensitivity (BER = 10^{-9} at 2.5 Gbit/s)	−29.5 dBm
Total balanced system	0 dBm

Note that any system with different components or characteristics would have different design requirements, i.e., designs would change if there were multimode fiber, wavelength dependencies, heterodyne reception, laser chirp, insertion losses, etc. If the losses were too large for the receiver sensitivity, then, optical amplification would be required somewhere in our system. In Section 16.4, we will discuss the particulars of incorporating optical amplifiers in an optical system because they not only provide gain but also introduce additional noise terms.

Point-to-Point and Multipoint Systems

We have been discussing the basic concepts of simple data transmission and detection. Figure 16.44 shows two generic types of optical systems, point-to-point and multipoint systems. Point-to-point links can be thought of as transmission of a data stream from point A to point B in which the design rules discussed in the previous section apply. No routing or switching occurs in the optical data path. The most obvious example of a point-to-point system is an undersea, long-distance, fiber-optic cable connecting two points which are very far apart. By the simplest of calculations, a 1 Gbit/s data stream, which is launched at 0 dBm into a fiber of attenuation 0.2 dB/km using a receiver of sensitivity −30 dBm, can propagate ~150 km if only attenuation losses and a thermal-noise-limited receiver are considered. In reality, the system would be limited to ~50 km by issues, such as power margin,

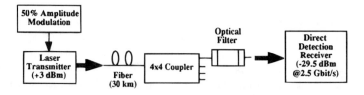

FIGURE 16.43 Example of designing a direct-detection optical system.

Optical Communications

(a) Point-to-Point System

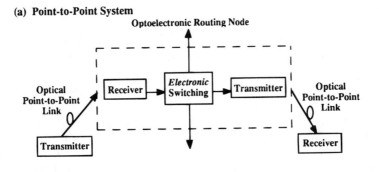

(b) Multipoint Systems

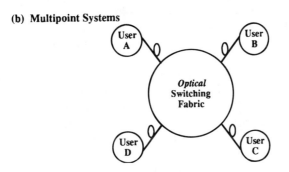

FIGURE 16.44 Point-to-point and multipoint optical systems.

dispersion, and insertion losses. Therefore, an ultralong-distance (>1000 km) point-to-point optical system would require that the signal be periodically detected and then retransmitted by an optical regenerator (Personick, 1985). This regenerator is composed of a receiver coupled with a transmitter. The regenerator must have electronics for clock recovery and electronic amplification, both of which are bit-rate specific to the incoming data stream and can accommodate only one signal wavelength at a given time. Regenerators are fairly expensive and not very flexible. To upgrade a system, these regenerators must be replaced, not very practical for an undersea cable system. In Section 16.4, we will discuss how an optical amplifier can replace regenerators without producing all their disadvantages.

Multipoint systems can be much more complex than point-to-point systems. A user is connected to many other users through some passive or active, optical switching medium. (Note that multipoint systems, in which all the switching is done electrically after an optical signal is detected, would be considered point-to-point *optical* systems.) The intercommunication can be unidirectional, as in a broadcast system, or bidirectional, as in a local-area network (LAN). It is the goal of future systems to have simultaneous transmission of many different signals through the system, thereby, increasing the aggregate system capacity. Signal recovery and design rules still apply, but may be much more complex. In addition, efficient routing, low interchannel cross talk, and output-port contention resolution are just some of the intricate issues involved. Multichannel systems are the subject of section 16.5, at which point we will have much more discussion on this topic.

Because of the high losses in electrical coaxial systems at speeds exceeding 100s of MHz, it is a foregone conclusion that low-loss, high-bandwidth, optical fiber transmission is the mode of choice for long-distance, point-to-point systems (Runge and Bergano, 1988). Although multipoint systems do hold the promise of extremely high-bandwidth communications among many users, however, the technology, network, and cost challenges have still not been definitively solved. The hope is that multipoint systems will become practical within the next few years.

Wireless Free-Space IR System

Wireless, free-space infrared (IR) transmission is a topic of optical communications which does not involve optical fiber but which is still an interesting and ever more important subject (Barry et al. 1991). These optical wireless systems are becoming common for relatively low-speed portable transmission within a confined area (Photonics Corp.). Wireless IR communication offers untethered and relatively secure communication between, say, computers and their peripherals or among computers. Therefore, we devote a small section to discuss it briefly.

We will treat this communication medium as a point-to-point system. Figure 16.45 shows a diagram of an enclosed room in which IR communications may occur. Note that IR light cannot penetrate walls and will not interfere with communications in another room; interroom interference is what makes wireless radio communications impractical for this application. An IR transmitter can communicate by line-of-sight without much difficulty. If the ceiling and/or walls of a room reflect some of the IR light, then, strict line-of-sight transmission would not be required. The light would, probably, be detected as a combination of line-of-sight and reflected beams. The possibility of "multiple paths" from transmitter to receiver causes spreading in the transmitted pulse because the different paths take different amounts of time to reach the receiver. This phenomenon, known as multipath dispersive noise, severely limits the bit-rate of transmission to ~10–100 Mbit/s, certainly well below Gbit/s speeds. The transmitter can be placed in any portable unit, or a single transmitter could be placed on the ceiling which would facilitate communications between rooms.

The transmitted signal is a baseband digital data stream. The proper technologies must be chosen to keep performance and speed high and cost low. The transmitters and receivers must transmit/accept light into/from a wide-angle aperture to allow communication from anywhere in a room. LEDs are probably adequate for diffuse-area transmitters, and wide-area hemispherical detectors would be adequate for receivers (Barry et al. 1991). Reflecting surfaces in the room must be used. Additionally, proper optical filtering in the form of a film covering the detector is necessary to filter away all the unwanted ambient background room light. These technologies must all be quite low in cost.

The near-term applications can already be found in many line-of-sight applications, i.e., portable digital assistants (Photonics Corp.). The more robust and portable systems are still in the research and development stage. However, conventional wisdom is that IR communications will have a major impact on connectivity in the future office.

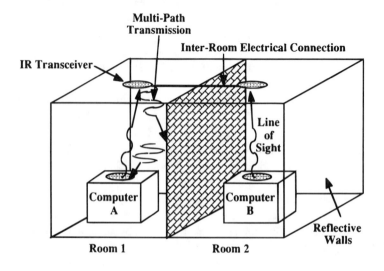

FIGURE 16.45 An enclosed room with IR optical communication links.

16.4 Optical Amplifiers

Fundamental Amplifier Characteristics

Introduction

Optical amplifiers have recently stolen center stage in the optical telecommunications world due to rapid device progress and revolutionary systems (Li, 1993). These devices enable new and exciting optical systems to be conceived and demonstrated. In fact, much of the most relevant recent advances in optical communications (i.e., long-distance NRZ and soliton systems, and wide-area and broadcast multichannel systems) can be traced to the incorporation of optical amplifiers. As a simple introduction, optical amplifiers can be thought of as lasers (gain media) with a low feedback mechanism and whose excited carriers amplify an incident signal but do not generate their own coherent signal (Yariv, 1991).

Similar to electronic amplifiers, optical amplifiers can be used to compensate for signal attenuation resulting from distribution, transmission, or component-insertion losses (Agrawal, 1992, Chapter 8; Green, 1993, Chapter 6). As shown in Fig. 16.46, both types of amplifiers provide signal gain G, but also introduce additive noise (variance = σ^2) into the system. Each amplifier requires some form of external power to provide the energy for amplification. A voltage source is required for the electrical amplifier, and a current or optical source is required for the optical amplifier. This current or optical source is used to pump carriers into a higher excited energy level. Given an incident signal photon, some of these carriers, then, experience stimulated emission and emit a photon at the input signal wavelength. We will discuss fundamental characteristics, systems issues, and potential applications associated with implementing optical amplifiers in optical communication systems.

The original motivation for the recent widespread research was to replace the regenerators in long-haul, transoceanic fiber-optic systems which were placed every ~50 km along the entire multimegameter span. Such regenerators correct for fiber attenuation and chromatic dispersion by detecting an optical signal and, then, retransmitting it as a new signal using its own internal laser. Regenerators (being a hybrid of optics and electronics) are expensive, electronically bit-rate and modulation-format specific, and waste much power and time in converting the signal from photons to electrons and back again to photons. In contrast, the optical amplifier is, ideally, a transparent box which provides gain and is also insensitive to the bit-rate, modulation-format, power, and wavelengths of the signal(s) passing through it. The signals remain in optical form during amplification, and optical amplifiers are potentially cheaper and more reliable than regenerators. However, the optical amplifier is not an ideal device because (a) it can provide only a limited amount of output power before the gain diminishes due to a finite number of excited carriers available to amplify an intense input signal, (b) the gain spectrum is not necessarily flat

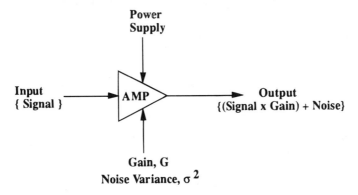

FIGURE 16.46 Basic amplifier characteristics.

over the entire region in which signals may be transmitted, (c) the additive noise causes a degradation in receiver sensitivity, and (d) fiber dispersion and nonlinear effects are allowed to accumulate unimpeded. We note here that, for ultralong-distance systems, (i) the signal wavelength must be near 1.55 μm for lowest attenuation, and (ii) dispersion-shifted fiber must be used so that the dispersion parameter will be close to zero for the ~1.55-μm signal wavelength.

The three basic system configurations envisioned for the incorporation of optical amplifiers are shown in Fig. 16.47 (Giles, 1992). The first configuration places the amplifier immediately following the laser transmitter to act as a "power", or "post-", amplifier. This boosts the signal power so that the detected signal is still above the thermal noise level of the receiver; note that any noise introduced by the power amplifier will be similarly attenuated together with the signal, as they are transmitted through the lossy system. The main figure-of-merit is for the amplifier to have a high-saturation output power. The second configuration places the amplifier "in-line" for incorporation at one or more places along the transmission path. The in-line amplifier(s) corrects for periodic signal attenuation due to fiber attenuation or network distribution-splitting losses (Hill et al. 1990). The third possibility places the amplifier directly before the receiver, thus, functioning as a "preamplifier." In this case, the signal has already been significantly attenuated along the transmission path. The main figures-of-merit are high gain and low additive noise because the entire amplifier output is immediately detected. As such, the receiver will be limited by the amplifier noise, not by the receiver thermal noise.

Before discussing generic details of amplified systems, we will briefly introduce the two most prominent (at present) optical amplifiers. Semiconductor optical amplifiers (SOA) (O'Mahony, 1988; Saitoh et al. 1986; Saitoh and Mukai, 1987; Simon, 1987; Fye, 1984) and erbium-doped, fiber-optic amplifiers (EDFA) (Mears et al. 1987; Desurvirre, 1987; Bjarklev, 1993) each consists of an active medium which has its "carriers" or "ions" inverted into an excited energy level. This population inversion enables an externally input optical field to initiate stimulated emission and experience coherent gain. The population inversion is achieved by the absorption of energy from a pump source. Furthermore, an external signal must be efficiently coupled into and out of the amplifier.

Figure 16.48 depicts the basic amplifier building blocks for the SOA and the EDFA. The traveling-wave (TW) SOA is nothing more than a semiconductor laser without facet reflections. An electrical current inverts the medium by transferring electrons from the valence band to the conduction band, producing spontaneous emission (fluorescence) and the potential for stimulated emission, if an external optical field is present. The stimulated emission is the signal gain. However,

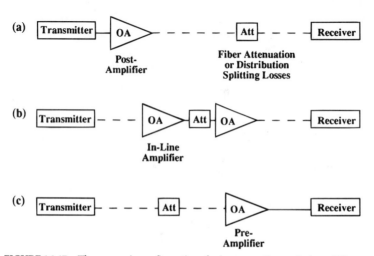

FIGURE 16.47 Three generic configurations for incorporating optical amplifiers into transmission or distribution systems.

Optical Communications

(a) Semiconductor Amplifier

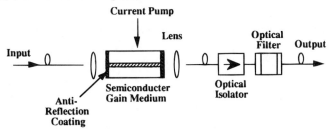

(b) Erbium-Doped Fiber Amplifier

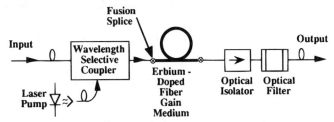

FIGURE 16.48 Block diagram of a semiconductor and a fiber amplifier. The optical isolator and optical filter are included although they may not be required under all circumstances.

the spontaneous emission is itself amplified (i.e., amplified spontaneous emission (ASE)) and is considered the randomly fluctuating noncoherent amplifier noise. If we are dealing with a circular-waveguide, fiber-based communication system (Stone and Burrus, 1973), an external signal must be coupled into and out of the amplifier's rectangular active region producing a mode field mismatch and, consequently, insertion losses. On the other hand, the fiber amplifier is a length of glass fiber which has been doped with a rare-earth metal, such as erbium ions. These ions are an active medium with the potential to experience a population inversion and emit spontaneous and stimulated emission light near a desirable signal wavelength. The pump is typically another light source whose wavelength is preferentially absorbed by the erbium ions. The pump and signal are combined and coupled into the erbium-doped fiber by a wavelength-selective coupler. The pump and signal may co- or counterpropagate with respect to each other inside the doped length of fiber. Therefore, light absorbed by the doped fiber at the pump wavelength will produce gain for a signal at a different wavelength. Note that the insertion losses are minimal because the transmission and the active medium are both fiber-based.

Both types of amplifiers are susceptible to external reflections that adversely affect the stimulated and spontaneous emission rates and the frequency-selectivity of the cavity. As a result, an optical isolator, which permits light to pass only in one direction and prevents reflections back into the amplifier, typically, is required for both types of amplifiers.

Amplifier Gain and Noise

In general, the total gain of the active medium G, is related to the gain coefficient per unit length g and the length of the active medium L:

$$G = \exp(gL). \qquad (16.40)$$

The gain coefficient represents the likelihood that a stimulated emission event occurs such that an excited carrier makes a transition from the upper energy level to the lower energy level, thus, causing a photon to be emitted at a wavelength corresponding to the energy level difference. The gain for an input signal occurs when an input photon produces a stimulated emission event, emitting a photon which is coherent and at the same wavelength as the original photon.

The gain of an amplifier will change depending on the input optical conditions and can become saturated. Essentially, a weak input optical signal can experience a certain amount of gain by initiating stimulated emission of the inverted carriers. However, if an incoming signal is so large that there simply are an insufficient number of inverted carriers to allow stimulated emission to occur for all the incoming photons, then, the total gain for this intense input signal will be less than for a weak input signal. For intense signals, the amplifier gain is diminished and the amplifier itself is considered saturated. Another way of thinking about saturation is that the absorbed pump power can provide only a maximum number of excited carriers, and, therefore, an incoming large signal cannot be amplified to the point where more power is cut out from the amplifier than was initially provided by the pump source. The saturation input (or output) power is usually considered to be that input (or output) power which will produce a reduction in the small-signal gain by 3 dB. The transcendental equation describing the gain is (Agrawal, 1992)

$$G = G_0 \exp\left[-\frac{(G-1)}{G}\frac{P_{out}}{P_{sat}}\right] = \exp(gz), \qquad (16.41)$$

where G_0 is the unsaturated total gain, Z is the gain-medium length, and P_{sat} represents the maximum (i.e., saturated) amount of optical power which can be output from the amplifier. P_{sat} is dependent on the active medium, the signal wavelength, the carrier lifetime, and the mode overlap between the optical field and the carriers. Figure 16.49 shows how the gain is reduced for an increase in the input optical signal power.

The noise in an amplifier is inherently due to the random, incoherent, spontaneous emission events of excited carriers. Each spontaneously decaying carrier can radiate a photon in any solid angle. The fraction of the spontaneous emission emitted within the critical angle of the waveguiding region and coupled into the optically guiding region will, itself, cause further stimulated emission producing amplified spontaneous emission (ASE). This ASE is quite broadband, occurring over the entire 50–100-nm material gain bandwidth. Additionally, because there is only a finite number of excited carriers, then, there is a trade-off between gain and noise, i.e., an increase in utilizing the carriers for ASE noise will result in fewer available carriers to provide signal gain. This is an

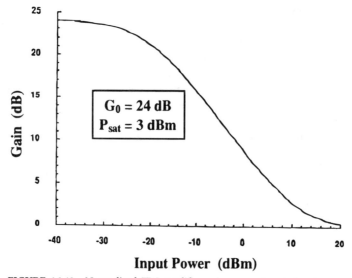

FIGURE 16.49 Normalized TW-amplifier gain versus input signal power demonstrating the effects of gain saturation.

additional reason why we wish to suppress forward or backward reflections into the active medium because any reflected wave will deplete the available gain and increase the noise component.

A fairly typical amplified-channel block diagram is shown in Fig. 16.50 (Kazovsky). The signal, P_{sig}, may initially pass through some lossy components, with a lumped insertion loss of L_{in}. The amplifier then provides gain G and adds noise of variance σ^2. Because the modulated input signal is typically narrowband whereas the ASE spectrum of the TW amplifier is quite broad (10s of nm), an optical filter of bandwidth B_0 will usually be present to pass the signal and block much of the ASE noise. We can also lump any insertion losses at the output of the amplifier in one term L_{out}. [We will not include L_{in} and L_{out} in the following analysis; the reader should note that the effect of L_{in} is to attenuate the optical signal power P_{sig} whereas L_{out} attenuates both P_{sig} and the spontaneous emission power P_{sp}.] The optical detector has a characteristic resistance Ω_r, and the entire receiver has an electrical bandwidth B_e to pass the modulated baseband signal and block all higher frequency noise terms. Note that the signal is in one polarization but the ASE occurs in both polarizations.

The equation describing the light power P along the length of an amplifier in one polarization is (Yariv, 1991)

$$P(z) = P_{sig}e^{gz} + n_{sp}h\nu(\Delta\nu)(e^{gz} - 1), \qquad (16.42)$$

where the first term describes the signal, the second term describes the noise in one polarization, z is the length along the amplifier, P_{sig} is the signal power input to the amplifier, N_2 is the carrier population in the higher energy level, N_1 is the population in the lower energy state, $h\nu$ is the photon energy, and $(\Delta\nu)$ is the optical bandwidth of the active medium. In Eq. (16.43), we define the spontaneous emission factor n_{sp} as a measure of the efficiency of the carrier population inversion in terms of (Olsson, 1989; Henry, 1986)

$$n_{sp} = \frac{N_2}{N_2 - N_1}. \qquad (16.43)$$

A higher value for n_{sp} implies that more ASE noise is generated in the amplifier in proportion the generated gain. Based on the definition, the quantum-limited minimum value for n_{sp} in any amplifier is equal to 1. The second term in Eq. (16.42) is more clearly understood when defining the total ASE noise power P_{sp} over the gain bandwidth $(\Delta\nu)$ in one polarization as (Olsson, 1989)

$$P_{sp} = n_{sp}(G - 1)\, h\nu(\Delta\nu). \qquad (16.44)$$

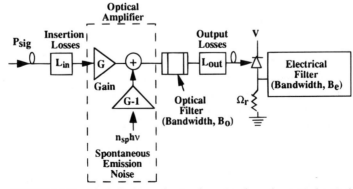

FIGURE 16.50 Block diagram of a signal passing through a typical optical amplifier and then being detected. (After Kazovsky.)

We have approximated that the spectral density of the ASE power is nearly uniform over the entire bandwidth Δv and equal to $[n_{sp}(G-1)hv]$. If we were to consider both polarizations, a simple factor of 2 multiplies the right side of Eq. (16.44);

$$P_{sp} = 2 n_{sp} (G-1) h v (\Delta v). \tag{16.45}$$

We are interested in the electrical noise which is ultimately generated in the optical detector and which governs the overall sensitivity of the system (Yamamoto, 1980). As discussed previously in section II.3, two noise terms which are common to all detectors are the shot noise σ_{sh}^2 and the thermal noise σ_{th}^2. The incoherent ASE noise terms generated in the amplifier are very broadband and are very dependent on the optical and electrical bandwidths of the corresponding filters.

Because the detector is inherently a square-law device which responds to the intensity (i.e., square) of the incoming optical field, a "beat" term is produced if two different optical waves are incident, $A(t)$ and $B(t)$. Squaring of $(A + B)$ produces A^2 plus B^2 plus the beat term of $2AB$; based on trigonometric identities, the beat term includes the cosine of the sum and difference frequencies between $A(t)$ and $B(t)$. Because the frequencies in the above equation are ultrahigh optical frequencies (THz) undetectable by the photodetector, the sum frequency term will not be detected and only the difference frequency will appear at the electrical output. The two waves in our system, impinging on the detector, are the signal and the ASE noise. Therefore, A^2 represents the signal power, B^2 represents the ASE noise power, and $2AB$ represents the signal-spontaneous electrical beat noise. However, the situation is more complicated because the ASE does not exist at a specific wavelength but is broadband consisting of an infinite number of incoherent waves each at a different frequency within the gain spectrum. Therefore, we must integrate over the entire ASE noise passing through the optical and electrical filters, and then beat (i.e., multiply) each thin bandwidth "slice" of ASE with the, approximately, single-frequency signal term. The resulting signal-spontaneous beat noise σ_{sig-sp}^2 that falls within the optical filter and electrical detector bandwidths is, then, given by (Olsson, 1989)

$$\sigma_{sig-sp}^2 = 4q\left(\frac{q}{hv}\right) P_{sig}G(G-1)n_{sp}B_e. \tag{16.46}$$

The B^2, or ASE noise power, term also must be evaluated because it is not at a single frequency or phase and therefore produces beat terms between one part of the ASE spectrum and another. After integration and convolution, the spontaneous-spontaneous electrical beat noise σ_{sp-sp}^2 is (Olsson, 1989) expressed by

$$\sigma_{sp-sp}^2 = 4q^2(G-1)^2 n_{sp}^2 B_e B_o. \tag{16.47}$$

Note that σ_{sig-sp}^2 and σ_{sp-sp}^2 can be reduced by small optical and electrical filter bandwidths but cannot be eliminated because some ASE must pass the through the filters within the same bandwidth as the signal. Additionally, the majority of the signal-spontaneous beat noise is generated from the low-frequency portion (within a few GHz) of the ASE immediately surrounding the signal and so cannot be reduced significantly.

The SNR for our optically amplified system is expressed as

$$\text{SNR} = \frac{(GP_{sig})^2}{\sigma_{sh}^2 + \sigma_{th}^2 + \sigma_{sig-sp}^2 + \sigma_{sp-sp}^2}. \tag{16.48}$$

We simplify this expression by making three generally valid assumptions: (1) the shot noise is small in comparison to all other terms, (2) the receiver noise will be dominated by the ASE noise, not by the thermal noise (typically true for systems employing a preamplifier before the detector), and (3) $G \gg 1$. After making the appropriate substitutions, the SNR can be approximated as

$$\text{SNR} = \frac{P_{\text{sig}}^2}{[2P_{\text{sig}}n_{\text{sp}}h\nu + 2n_{\text{sp}}^2(h\nu)^2 B_\text{o}]2B_\text{e}}. \tag{16.49}$$

Given all of our assumptions, the SNR does not change much with an increase in gain because the thermal noise has been overwhelmed by the ASE-generated terms and any higher gain will increase the signal and noise at nearly the same rate! Given the equation for the SNR, we can easily extend our analysis to include the calculation of the probability of error for a random digital bit stream (Marcuse, 1990) per the equations found in Section 16.3.

As with an electrical amplifier, an important parameter of an optical amplifier is the noise figure NF expressed by

$$\text{NF} = \frac{\text{SNR}_{\text{in}}}{\text{SNR}_{\text{out}}}, \tag{16.50}$$

where the SNR_{in} and SNR_{out} are the electrically equivalent SNRs of the optical wave going into and coming out of the amplifier. The absolute lowest (i.e., quantum-limited) NF is $2n_{\text{sp}}$. Because the minimum n_{sp} is 1 for complete inversion, the quantum-limited NF for an amplifier, given the approximations above, is 2 (i.e., 3 dB)! The typical noise figure for a semiconductor amplifier is 6–8 dB (Agrawal, 1992; Suzuki et al. 1993) and for an EDFA is ~4–5 dB.

Based on this analysis, we emphasize that an amplified system (excluding the "postamplifier" configuration) should be operated so that the receiver SNR is spontaneous-beat-noise limited and not thermal-beat-noise limited. We wish to increase the signal gain to the point where an increase in the signal gain will also proportionally increase the signal-spontaneous beat noise, which, then, provides us with the highest SNR possible for our system.

Semiconductor Amplifiers

The semiconductor active medium is a rectangular waveguide which provides gain to an optical signal that is propagating through it (Siegman, 1986, Chapter 30). Signal gain occurs when: (a) carriers are excited from the valence to the conduction band in this quasi-two-level energy system, and (b) an externally input signal initiates stimulated emission when propagating through the material. We can approximate that the bandwidth of the unsaturated spectral gain coefficient $g(\lambda)$ of the active medium is homogeneously broadened. An empirical formula describes the central portion of the curve as an inverse quadratic (Henning et al. 1985; Agrawal and Dutta, 1986, Chapter 2)

$$g = a(n - n_\text{o}) - \gamma(\lambda - \lambda_\text{m})^2, \tag{16.51}$$

in which n is the carrier population in the conduction band, n_o is the carrier density required to achieve transparency, a and γ are fitting constants, and λ_m is the gain peak wavelength and depends on carrier density. The center wavelength is typically designed to be near either 1.3 or 1.55 μm which correspond to the two fiber loss minima and are the most useful for optical systems. Because carriers have an energy distribution within each energy band, gain can occur over a wide range of wavelengths (~50–100 nm). The unsaturated semiconductor gain G_o depends on the level of pumping, the carrier lifetime, the saturation power, and the signal wavelength.

When considering an SOA, there may be reflections at the right and left boundaries, or facets, of the amplifier. These reflections have a profound impact on the gain achievable from the amplifier. A frequency-dependent gain expression in the presence of boundary reflections G_r is (O'Mahony, 1988)

$$G_r = \frac{(1 - R_1)(1 - R_2)G_o}{(1 - G_o\sqrt{R_1R_2})^2 + 4G_o\sqrt{R_1R_2}\sin^2\left[\dfrac{(\omega - \omega_o)L}{(c/n)}\right]}, \quad (16.52)$$

in which R_1 and R_2 are the power reflectivities at the two boundaries, and ω_o is the center frequency of the gain spectrum. The denominator in this equation is periodic, producing periodic Fabry–Perot resonances in the gain spectrum. If the reflections are suppressed and $R_1 = R_2 = 0$, then, the gain ripples disappear, and this device becomes a wideband, traveling-wave (TW) amplifier. Because FP amplifiers have fairly constant gain only over a narrowband, they are not considered as practical for optical systems as TW amplifiers because TW amplifiers can provide high gain for signals covering a wide range of wavelengths. As such, our discussion will deal exclusively with the characteristics of TW amplifiers. Three methods for suppressing facet reflections in an SOA are (i) using an antireflection coating which can substantially reduce boundary reflections (Vassallo, 1989) (ii) using a buried non guiding, passive, window region between the end of the central active layer and each of the facets in which the optical wave diverges before being reflected at the facet, thereby, coupling very little reflected energy back into the amplifying waveguide (Olsson et al. 1989) and (iii) using an angled waveguide relative to the facet normal in which any reflected waves will reflect at an angle larger than the critical angle; the reflected wave will not efficiently couple back into the amplifying waveguide (Zah et al. 1987).

Unlike fiber amplifiers, the semiconductor gain medium, being rectangular (not square) and having different crystal planes, has a polarization-dependent gain. This differential gain between the TE and TM waves can be as large as 8 dB. Recent work has attempted to minimize this differential by creating a nearly square, cross-sectional, active area thereby, making the TE and TM fill-factors similar. A difference of <1 dB has been produced by using this method and by using multiple-quantum-well and strained-layer material (Newkirk et al. 1993). If the semiconductor amplifier is integrated on the same chip following a fixed linear-polarization laser transmitter, then, only a single polarization passes through the amplifier and any polarization dependence is not a problem. Note that the fiber-based EDFA's are circular and the wave in a fiber is typically circularly polarized, producing a signal gain which is negligibly dependent on the polarization of the incoming signal.

In multichannel systems, the ideal amplifier will provide gain equally to all channels over a broad wavelength range without any other effects. However, a few factors inherent in the SOA make wavelength-division-multiplexed, multichannel systems more difficult to implement than single-channel systems. These include intermodulation distortion and saturation-induced cross talk. Intermodulation distortion and saturation-induced cross talk are both nonlinear effects.

Intermodulation distortion can be explained as follows. When two channels are incident into a closed amplifier system and their combined powers are near the amplifier saturation power, nonlinear effects occur which generate beat frequencies at the cross-product of the two optical carrier waves. The carrier density (i.e., gain) will be modulated by the interference between any two optical signals, and this modulation occurs at the sum and difference beat frequencies which are generated by all possible combinations of input channels. This carrier density modulation at the beat frequencies produces additional modulated signals which can interfere with the original desired signals. Figure 16.51 illustrates this scenario for two input signals which produce four output waves. Therefore, this nonlinear effect is called four-wave mixing, or, alternaly, intermodulation distortion (IMD) (Agrawal, 1987; Jopson and Darcie, 1988; Jopson et al. 1987). The amplitude

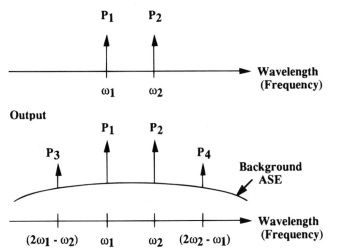

FIGURE 16.51 Example of four-wave mixing with two original signals and two newly produced signals.

of these products is proportional to the difference frequency and the carrier lifetime τ (Agrawal, 1992). In semiconductor amplifiers, the carrier lifetime is in the ns range. In general, when τ_s and $(\omega_1 - \omega_2)$ are small, significant power is transferred to the signal products. Conversely, if $(\omega_1 - \omega_2)$ or τ_s is large, then the beat modulation frequency is so great that the carrier population will be too slow to respond and transfer power to the products. Therefore, a spacing greater than several GHz is, typically, sufficient to quench any IMD production in SOAs.

Cross talk can also occur in a gain-saturated amplifier. If the intensity of the amplifier input signals increases beyond the saturation input power, then the gain will decrease. When the input signal intensity eventually drops, the gain will increase to its original unsaturated value. Therefore, the gain and input signal power are inverse functions of each other when the amplifier is saturated. This gain fluctuation occurs as rapidly as the carrier lifetime of the amplifier, again being ~1 ns in a typical SOA, and is comparable to the bit-time in a Gbit/s data stream. If we assume two input channels and a homogeneously broadened amplifier which becomes equally saturated across the entire gain bandwidth, then, an increase in the input intensity of one channel beyond the input saturation power will necessitate a decrease in the gain of both channels, thereby, causing cross talk in the second channel. If the gain can respond on the same time scale as a bit-time in a Gbit/s transmission system, then, as one channel is being ASK modulated, the second channel will also have its gain modulated within a bit-time, producing signal distortion and a system power penalty. This scenario is depicted in Fig. 16.52. [Note that the above two nonlinear effects are negligible for fiber amplifiers because their carrier lifetime is approximately 10 ms, far too long to produce any intermodulation distortion for any reasonably spaced channels and far too long to produce saturation-induced gain fluctuations on the time scale of an individual high-speed bit.]

There are several applications of SOAs in optical communication systems:

(i) Long-distance communications - Semiconductor amplifiers have been demonstrated successfully as power-, in-line, and preamplifiers (Jopson et al. 1988). However, erbium-doped, fiber-based amplifiers are more desirable for all three of these applications at 1.55 μm because of their superior characteristics (i.e., higher P_{sat}, lower noise figure, higher gain, and fiber compatibility). The real possible application of semiconductor amplifiers in long-distance communications is for 1.3-μm systems (Ryu et al. 1989).

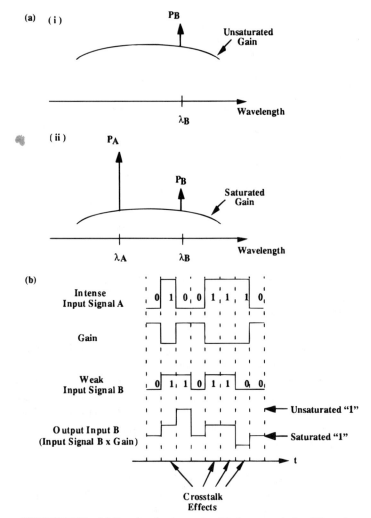

FIGURE 16.52 (a) Signal and gain spectra: (i) given a weak signal B producing no saturation, and (ii) given an intense signal A and a weak signal B producing gain saturation. (b) Bit-stream sequences for two signals propagating through a semiconductor amplifier. All pulse transitions are sharp because we assume that the response time of the amplifier gain is much greater than the bit-rate. If they are comparable, then, pulse-rounding effects will occur.

Because much of the fiber installed worldwide is conventional fiber with the dispersion zero at 1.3 μm, it is possible that some systems will operate at 1.3 μm and require amplification. Moreover, there is presently no practical fiber-based amplifier in the 1.3 μm range. Alternatively, recent research has demonstrated long-distance (>300 km) transmission at 1.55 μm with conventional fiber by using chromatic inversion and erbium-doped, fiber amplifiers (Gnauck et al. 1993).

(ii) Opto-Electronic Integrated Circuits (OEIC's) - The main advantages of using a semiconductor amplifier as opposed to a fiber amplifier include its small size, potential low cost, and integratability on a chip containing many other optoelectronic components (i.e., lasers and detectors). For instance, one can integrate a semiconductor amplifier in a photonic integrated circuit (PIC) where the polarization-dependent gain is of no consequence because the polarization is well defined on chip (Koch and Koven, 1991).

(iii) Photonic Switching Gates and Modulators - Beyond providing simple gain, a semiconductor amplifier can be used as a high-speed switching element in a photonic system because the semiconductor will (a) amplify if pumped, and (b) absorb if unpumped. The operation is simply to provide a current pump, when an optical data packet is to be passed, and discontinue the pump, when a data packet is to be blocked.

(iv) Broadband All-Optical Wavelength Shifter - In wavelength-division-multiplexed switching, network-routing nodes, the ability to shift the wavelength of an incoming packet from one wavelength to another may be critically important for network efficiency. Recent research has demonstrated the ability to shift the wavelength of a bit stream by several nm as quickly as several nanoseconds in an all-optical manner (Barnsley and Fiddyment, 1991). This method involves two beams which are copropagating through the amplifier, one, intense, ASK-modulated pump signal at wavelength #1 and one, weak, unmodulated probe beam at wavelength #2 (Glance et al. 1992; Wiesenfeld and Glance 1992; Durhuns et al. 1992; Joergensen et al. 1993). The operation of a wavelength shifter is based on the ability of an intense signal to reduce (i.e., saturate) the gain of an optical amplifier. If this intense signal at wavelength #1 is amplitude modulated and the gain can respond on fast time scales (as only SOAs can), then, the gain of the amplifier will also be modulated and will be an inverse function of the original modulated signal. The output on wavelength #2 will now have the modulation inverse that on wavelength #1.

Erbium-Doped Fiber Amplifiers

Recently, the erbium-doped, fiber amplifier (EDFA) has generated much excitement in the telecommunications world (Li, 1993; Bjarklev, 1993; Sunak, 1992). The main reasons are that (a) the erbium ions (Er^{3+}) emit light in the 1.55-μm loss-minimum band of optical fiber, (b) high gain and low noise is produced, and (c) a circular fiber-based amplifier is inherently compatible with a fiber-optic system. The EDFA has relatively few disadvantages making it an almost ideal critically important component for long-haul communications.

Fundamental differences between the SOA and the EDFA can be traced mainly to the following attributes:

(i) The semiconductor amplifier is essentially, a two-energy-level system whereas the erbium-doped fiber amplifier is a three-energy-level system (Armitage, 1988). In the EDFA, ions are excited from the ground state (population N_0) into an excited state (N_2). These ions quickly decay to the metastable level (N_1) from which both stimulated and spontaneous emission occur as they drop down to the ground state. Additionally, an obvious difference is that the population inversion in the semiconductor amplifier is achieved by a current source whereas the fiber amplifier is inverted by an optical source.

(ii) The length of the fiber amplifier is in meters whereas the length of the semiconductor amplifier is $\sim$1 mm. This dramatic difference in length assumes uniform inversion along the length of the amplifier valid only for the semiconductor amplifier and not for the fiber amplifier.

(iii) The fiber amplifier is circular, not rectangular, thus, eliminating (a) significant attenuation when coupling to a standard optical fiber and (b) any significant polarization dependence in the gain.

(iv) The carrier lifetime of erbium ions is in ms whereas the lifetime of semiconductor carriers is in ns. This difference significantly reduces the two nonlinear problems in multichannel systems of intermodulation distortion (four-wave mixing) and gain-saturation-induced cross talk. [Note that gain transients in highly-saturated EDFAs may be much faster than previously assumed, as fast at $n\mu$s. (Zyskind et al., 1996)]

To produce the amplifier gain medium, the silica fiber core of a standard single-mode fiber is doped with erbium ions. Because of the many different energy levels in erbium, several wavelengths will be absorbed by the ions. In general, absorption corresponds to a photon causing an ion to make a transition to a higher energy level of energy difference $\Delta E = h\nu$ matching the energy of the photon. Once a photon is absorbed and an ion is excited to a energy level higher than the first excited state, the carrier decays very rapidly to the first excited level. Once the carrier is in the first excited state, it has a very long lifetime of $\sim$10 ms (Mears et al. 1987), thereby, enabling us to consider the first excited level to be metastable. Depending on the external optical excitation signal, this ion will decay in a stimulated or spontaneous manner to the ground state and emit a photon. The absorption is not as strong for all possible wavelengths and is governed critically by the tendency of a pump photon to be absorbed as determined by the cross-section of the erbium ion with that photon. The two wavelengths having the strongest absorption coefficient are 0.98 μm and 1.48 μm. Fortunately, fabricating high-power, multimode, laser diodes for the 0.98 and 1.48-μm wavelengths can be accomplished by using strained-layer, quantum-well material, with output powers >100 mW achievable and commercially available (OKI Corp.). Laser diode pumps are attractive sources because they are compact, reliable, and potentially inexpensive.

Both the absorption and the emission spectra have an associated bandwidth. These bandwidths depend on the spread in wavelengths which can be absorbed or emitted from a given energy level. Such a spread in wavelengths is caused by Stark-splitting of the energy levels, allowing a deviation from an exact wavelength. This is highly desirable because (a) the exact wavelength of the pump laser may not be controllable and is impossible to fix for a multimode laser and (b) the input signal may be at one of several wavelengths, especially in a WDM system. Figure 16.53 shows the cross-sectional bandwidth of the 1.48-μm absorption and the 1.55-μm fluorescence (i.e., emission) spectrum of a typical erbium-doped fiber (Giles and Desurvire, 1991; Miniscalco, 1991). The fluorescence spectrum of Fig. 16.53 was taken from a fiber which not only contained erbium but was codoped with aluminum (Desurvire, 1991). Codoping the erbium fiber with another material allows for higher erbium doping concentrations and makes the gain bandwidth somewhat broader and more uniform. Note that the EDFA gain is, therefore, nonuniform because the gain corresponds very closely to the fluorescence spectra.

Several components may be required for proper operation of an EDFA. Because the erbium-doped gain medium must be pumped optically, the 0.98- or 1.48-μm pump light (usually in the form of a diode laser) and the 1.55-μm signal must be combined within the doped fiber to achieve

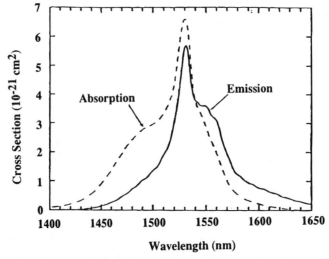

FIGURE 16.53 The absorption and fluorescence spectra for erbium near 1.5 μm. (© 1991 IEEE. After Miniscalco, 1991.)

pumping and signal gain. Pigtailed, grating-based, three-port, wavelength-division multiplexers (WDM) can perform this coupling function with <0.5 dB of insertion loss and >40 dB of return loss, even when combining wavelengths as close as 1.48 and 1.53 μm. Two other components which are not, strictly speaking, essential to an EDFA but which may be required to prevent system degradation are an optical filter and an optical isolator. An output optical filter of ~1–2 nm will limit the broadband spontaneous-spontaneous beat noise generated in the detector. Additionally, an optical isolator may be necessary to prevent reflections back into the amplifier which may cause the noise figure of the EDFA to increase and may even cause the EDFA to lase if the gain and reflections are sufficiently large.

In general, the wavelength-dependent gain $g(\lambda)$ and loss coefficients $\alpha(\lambda)$ of the EDFA can be described as follows (Desurvire, 1991):

$$g(\lambda) = \sigma_a(\lambda)\Gamma(\lambda)N_t \tag{16.53}$$

and

$$\alpha(\lambda) = \sigma_e(\lambda)\Gamma(\lambda)N_t \tag{16.54}$$

where σ_e and σ_a are the wavelength-dependent emission and absorption cross-sections of the erbium ions, Γ is the wavelength-dependent overlap integral between the optical mode and the erbium ions, and n_t is the density of erbium ions in the fiber. The total wavelength-dependent gain is the integral of the gain and loss over the entire length of the fiber:

$$G_k = \exp\left\{\int_0^L \left(g_k \frac{N_1}{N_t} - \alpha_k \frac{N_0}{N_t}\right) dz\right\}, \tag{16.55}$$

where k represents all the possible wavelengths to be considered, L is the erbium-doped fiber length, and N_1 and N_0 are the carrier densities in the first-metastable and ground states, respectively.

Three spatial-temporal equations which describe the evolution of the pump, signal, and ASE along the fiber are (Giles et al. 1989a)

$$dP_p^\pm(z, t) = \mp P_p^\pm \Gamma_p(\sigma_{pa}N_0 - \sigma_{pe}N_1) \mp \alpha_p P_p^\pm, \tag{16.56}$$

$$dP_s(z, t) = P_s\Gamma_s(\sigma_{se}N_1 - \sigma_{sa}N_0) - \alpha_p P_s, \tag{16.57}$$

and

$$\frac{dP_{ASE}^\pm(z, t)}{dz} = \pm P_{ASE}^\pm \Gamma_s(\sigma_{se}N_1 - \sigma_{sa}N_0) \pm 2\sigma_{se}N_1\Gamma_s h\nu_s(\Delta\nu) \mp \alpha_s P_{ASE}^\pm, \tag{16.58}$$

where the + and − represent propagation in the forward or backward direction and the subscripts p and s represent pump and signal, respectively, for the various power and absorption/emission cross-section terms; for simplicity, Eq. (16.56) neglects the possibility of pump absorption for carriers excited to another excited state. Note that the signal propagates only in the forward direction, the ASE propagates in both directions, and the pump can co- or counterpropagate.

An experimental curve of the gain dependence on pump power is shown in Fig. 16.54 (Desurvire, 1987, 1991) A minimum pump power is required for the gain to overcome the losses and achieve transparency. The gain, then, increases rapidly and is followed by a plateau region which represents the situation when nearly all the available carriers have already been inverted throughout the gain medium. The amplifier should be operated near this fully inverted (plateau) region to minimize n_{sp} and obtain a low noise figure.

The noise considerations of an EDFA (Li and Terch, 1991; Saleh et al. 1990) are quite similar to those for a semiconductor amplifier. One difference between the two types of amplifiers is the extreme wavelength dependence on the noise figure of the EDFA (Giles and DiGiovanni, 1990):

$$n_{sp} = \frac{1}{1 - \left[\dfrac{\sigma_{sa}(\lambda)\sigma_{pe}(\lambda)}{\sigma_{se}(\lambda)\sigma_{pa}(\lambda)}\right] - \left[\dfrac{\sigma_{sa}}{\sigma_{se}}\dfrac{P_p^{th}}{P_p}\right]}. \quad (16.59)$$

The second term in the denominator of Eq. (16.59) represents the relative absorption and emission cross-sections, and the third term in the denominator represents the contribution to noise due to inadequate pumping.

The individual spectra for the absorption and emission (fluorescence) of light near 1.55 μm in an EDFA was shown to be nonuniform and also to overlap with each other. This means that if a signal is incident on the amplifier at a wavelength containing significant absorption and emission cross-sections, then, the signal will experience both gain and absorption, thereby, contributing to some additional ASE. This overlap occurs on the short wavelength end of the gain spectra, and so, n_{sp} and the NF will be higher at shorter wavelengths than at longer wavelengths where the overlap is much less. Figure 16.55 shows the noise figure as a function of wavelength, and the noise figure is clearly seen to be higher at the shorter wavelengths (Giles et al. 1989b). Thus, it is advantageous to operate the signal wavelength, not at the gain peak of the amplifier, but at a higher wavelength which has a lower noise figure. Note that the noise figure is very close to 3 dB, which is the quantum-limited value, as determined earlier in this section.

As with semiconductor amplifiers, the EDFA gain can be compressed to a small value (Way et al. 1991) if the amplifier is saturated by an intense input signal (Willner and Desurvire, 1991). An amplifier figure-of-merit is to have a high output saturation power, which is especially desirable for power-amplifier applications in which we wish to boost the output of a laser diode to a value higher than a semiconductor medium can provide. The saturation output power is a function

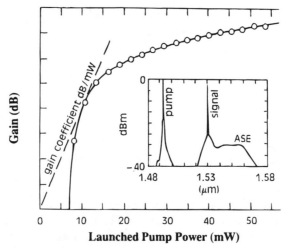

FIGURE 16.54 EDFA gain as a function of 1.48-μm pump power. (After Desurvire, 1991.)

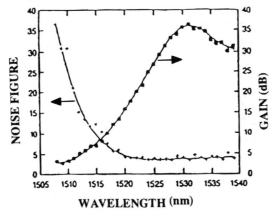

FIGURE 16.55 Noise figure as a function of wavelength in an EDFA pumped with 1.48-μm light. (© 1989 IEEE. After Giles et al. 1989b.)

of the erbium concentration, fiber length, and pump power because a higher output power will result from an increase in the inverted carriers. Of course, the limit is reached when all the erbium ions in the fiber length have been inverted. In Fig. 16.56 we plot the signal gain as a function of signal output power, showing the saturation output power of the amplifier (Zyskind et al. 1991a). Saturation output powers as high as ~20 dBm have been measured with extremely high pumping and an extremely large number of erbium ions (Laming et al. 1992).

Recall that two critical problems existed in multichannel operation in SOAs, namely, intermodulation distortion caused by four-wave mixing and gain-saturation induced cross talk. The relative effects of each of these can be traced to the carrier lifetime (i.e., gain response time) of an optical amplifier. Semiconductor amplifiers have lifetimes on the order of ns whereas EDFA lifetimes are in the ms range. Both of these two deleterious nonlinear effects are considered negligible in the EDFA due to its extremely long, gain response time. Gain-saturation-induced cross talk is considered negligible for frequencies faster than kHz speeds (see Fig. 16.57) (Giles et al. 1989a). [A recent paper has reported that the lifetime and gain transients in highly-saturated and cascaded EDFAs can be as fast as $n|\mu s$, much faster than previously assumed but still much slower than SOAs (Zyskind et al., 1996).]

EDFAs have highly desirable systems qualities which make it virtually certain that they will be implemented in actual systems in the very near future. The basic applications of EDFAs within

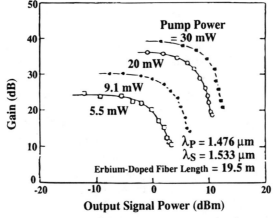

FIGURE 16.56 EDFA gain versus output signal power. (After Zyskind et al. 1991a.)

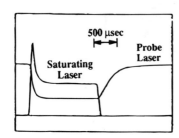

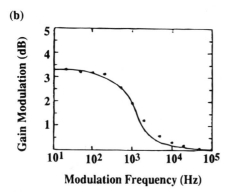

FIGURE 16.57 (a) Signal and gain transients in an EDFA. (b) Cross talk between two channels as a function of the rate of amplitude modulation on one of the channels. (After Giles et al. 1989a.)

optical communication systems (excluding WDM and soliton systems which will be addressed in later sections) include

(i) Amplifier Cascades and Long Distance Communications - Probably the one implementation of critical importance is long-distance terrestrial and transoceanic communications. The EDFA is an adequate substitute for expensive regenerators, even though EDFAs do not compensate for intramodal chromatic dispersion. A typical EDFA cascade has periodic introduction of gain, noise, and fiber attenuation losses. The operation of a cascade can be understood by considering the propagation of the signal and ASE as they traverse the system. The input of the amplifier is simply the signal, whereas the input to all subsequent amplifiers is the output of the previous amplifier which includes both signal and ASE power. The total output power $P_{out,i}$ of the ith amplifier in a cascade is given by (Giles and Desurvire, 1991a)

$$P_{out,i} = LG_iP_{out,i-1} + 2n_{sp}(G_i - 1)h\nu B_o, \qquad (16.60)$$

where L is the interamplifier fiber-attenuation and insertion losses, and the gain at each amplifier is not necessarily the small-signal gain but can be a saturated-gain value. Figure 16.58 shows the signal and ASE noise progression when traversing a cascade of many EDFAs (Giles and Desurvire, 1991a). As the signal propagates along the cascade, three phenomena occur: (1) the signal slowly decays, (2) the ASE noise slowly accumulates, (3) the SNR decreases slowly. A fundamental principle of cascaded amplifiers is that the signal and the ASE are both amplified from one amplifier to the next. In fact, because both the signal and the ASE from one amplifier are input to the following amplifier, amplifier saturation will quickly result, causing the total output power from each EDFA to equilibrate to the same output-saturation value.

Recent results for 5-Gb/s and 10-Gb/s NRZ transmission with 274 cascaded EDFAs have shown that nearly error-free transmission can be achieved for distances of 9,000 km, effectively conquering the barrier of repeaterless transmission for any possible distance around the globe (e.g., transpacific distances are ~9,000 km) (Bergano et al. 1992; Taga et al. 1993)!! Figure 16.59 shows the bit-

Optical Communications

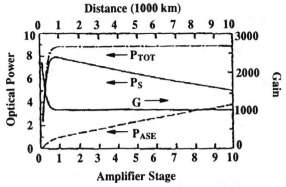

FIGURE 16.58 Signal and ASE noise powers along an EDFA cascade. (© 1991 IEEE. After Giles and Desurvire, 1991a.)

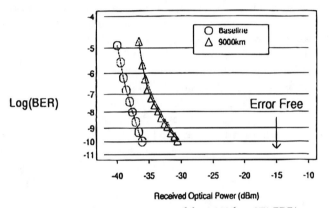

FIGURE 16.59 Bit-error-rate curve of the 9,000-km, 274-EDFA cascaded system. (After Bergano et al. 1992.)

error-rate curve for the results of this experiment. EDFAs will have a remarkable impact on future long-distance communications!!!

(ii) Preamplifiers and Power Amplifiers - EDFAs have been successfully used as preamplifiers in receivers (Gabla et al. 1991; Jacobs, 1990). The results are quite impressive, with a 10^{-9} bit error-rate sensitivity of -46 dB at 2 Gb/s for a single-stage experiment and -40 dB at 10 Gb/s for a double-stage EDFA experiment (Giles et al. 1989b; Laning et al. 1992). It now seems clear that direct detection with EDFAs can rival the high sensitivity of coherent detection with a high-power local oscillator. Additionally, EDFAs have shown their ability for impressive power boosting at the output of a laser transmitter. Results showing $+27$-dBm output power are unheard of with semiconductor amplifiers (Massicott et al. 1990).

(iii) Narrow Linewidth Sources - Certain communication systems require sources whose linewidths are quite narrow, perhaps, ~ 100 kHz. The linewidth of typical single-frequency lasers is >1 MHz, unless they are coupled to an external cavity. Because this arrangement is complex, a simpler solution may be an erbium-doped fiber laser, which lases near 1.55 μm. A circular length of erbium-doped fiber with a frequency-selective filter incorporated into the loop acts as a laser whose cavity is several meters long (Zyskind et al. 1991b). Because the cavity is circular, the feedback is provided, not by mirrors, but by circular return. Linewidths of 100 kHz have been achieved. This scheme is quite straightforward to implement, is fiber compatible, and can be tuned over the entire gain bandwidth of the EDFA.

In general, because most of the installed (i.e., existing) optical fiber is of the conventional type with a dispersion zero at 1.3 μm, it is highly desirable to have a fiber-based optical amplifier at 1.3 μm which has characteristics similar to the 1.55-μm EDFA. Doping a fiber with praseodymium will enable a fiber to experience fluorescence and, thus, amplify in the 1.3-μm range. However, because praseodymium has a four-level energy system, not a three-level system (as in erbium), the pump efficiency to achieve high gain is extremely low. In fact, hundreds of mWs of pump power are required to achieve a 20-dB gain, whereas <20 mW is required for EDFAs (Bhimuzu et al. 1993). The future of PDFAs hinges on the future of extremely high-power and inexpensive pump sources.

16.5 Multichannel Systems

Optics will unquestionably be used for high-speed, point-to-point transmission systems. It is hoped that we will also be able to use the high-speed, low-loss characteristics of optics to enable ultrahigh-capacity, multiuser systems in which the signals, usually, remain in optical form as they are switched through a network (Green, 1993 Chapter 1; Hinton, 1990; Kaminow, 1988). Much research has been dedicated to this problem and it is clear that such systems can be built. Photonic networks will most likely exist, but questions pertaining to performance and cost remain and will eventually determine if (and when) such systems will be implemented.

As we discuss multiuser systems, it should be noted that there is no "correct" network. Each option presents advantages and disadvantages, and each situation must be evaluated separately to find which method is the most practical and provides the highest performance. Moreover, because any given situation may have multiple solutions, it is likely that resulting networks will be hybrids of two or more methods (i.e., wavelength plus time multiplexing).

Topologies and Architectures

Network users can be interconnected in any number of ways. Fig. 16.60 shows the four most common topologies: bus, ring, star, tree (Acampora, 1994). The bus, the simplest configuration

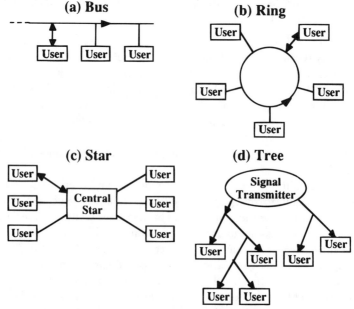

FIGURE 16.60 Four of the most common topologies for multichannel systems.

to implement, has all network nodes are connected to a common backbone. Optical tapping losses, typically, limit the number of users in a bus. Furthermore, the bus is not very reliable because a break in the fiber bus will isolate nodes on each side of the break. A ring can be thought of as a bus which closes on itself. Tapping losses are still an issue, but a break in the ring will not isolate any nodes because the direction of transmission can be bidirectional (i.e., two unidirectional rings which occupy the same path). The star configuration makes use of a central star in which each node is attached to a central point and can access all other nodes. A break in the fiber isolates only a single node, and the optical losses are more tolerable. By comparing the losses inherent in a bus and star, the optical power P_{bus}^M available at the Mth user node for a bus as compared to a star, P_{star}^M, for an M node network are given by (Henry et al. 1988).

$$P_{\text{Bus}}^M = P_T C[\beta(1 - C)]^{M-1} \qquad (16.61a)$$

and

$$P_{\text{star}}^M = \frac{P_T \gamma}{M}, \qquad (16.61b)$$

where C is the ratio of the amount of optical power tapped off at each node, β is the excess loss at each tapping node, and γ is the excess loss in an $M \times M$ star. When M is large, the difference in losses can be significant. Of course, optical amplifiers can be used to compensate for some of these losses. The tree configuration resembles the branches of a tree. Whereas the other topologies are intended to support bidirectional communication, this topology is useful for distributing information unidirectionally from a central point to a multitude of users, as in a cable television system.

Although networks connect many users, the users may exist near or far geographically. The architectural types of networks are the local-, metropolitan-, and wide-area networks (LAN, MAN, and WAN) for users that are, respectively, accessible locally (i.e., intra- and interbuilding), regionally (i.e., city-wide and between neighboring cities), and nationally (see Fig. 16.61) (Schwartz, 1987). Hybrid systems exist, and, typically, a wide-area network will consist of smaller local-area networks, with mixing and matching between the most practical topologies for a given system. For example, stars and rings may be desirable for LANs, whereas buses may be the only practical solution for WANs.

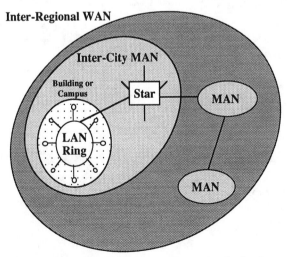

FIGURE 16.61 Wide-, metropolitan-, and local-area networks.

The two main methods of transmitting information in a network are circuit switching and packet switching, as described in Fig. 16.62a (Hui, 1990). In circuit switching, if node A wishes to communicate with node D, then, these two nodes must establish a link by appropriate "handshaking" and acknowledgment techniques. Once a link is confirmed, a dedicated path is established between the two nodes throughout the entire network. If this link requires routing along a high-bandwidth trunk line, then, these two nodes have reserved for themselves a small slice of this bandwidth for their own dedicated use; this slice may be some specific time slot in a time-multiplexed scheme, to be described in the next section. Even if no data is being transmitted, the line is still dedicated and, therefore, bandwidth is wasted.

Another major point is that the handshaking process between widely separated nodes may take a long time compared to the data transfer speed. If 100,000 bits are being transferred at 1 Gbit/s and the set-up time is 1 ms, then, the total time is 1.1 ms and is dominated by the long set-up time. Circuit switching emphasizes a transparent data pipe with a long set-up and tear-down time and is more appropriate for low-speed transmission over a long transmission time, such as a telephone call. Therefore, optical switching devices need only switch at ~ ms speeds. A circuit-switched network is quite easy to implement and control.

Alternatively, packet switching does not require a link with a dedicated and reserved bandwidth. Instead, data is bunched into packets. Each packet must have overhead bits in the header for providing information about its routing and destination for switching through a network. A flag (i.e., a unique set of bits) is, typically, at the beginning of the packet to synchronize data recovery and alert a receiver that a packet is arriving. The payload is the user-generated end-to-end data. Depending on the network and on the available bandwidth on different lines, any packet may take several possible routes to reach a destination. In such a case, a sequence number is included in the packet to ensure that the packets will not be misinterpreted if they arrive out of order. Packet switching efficiently uses the available bandwidth because bandwidth is used only for transmitting high-speed data and the line will not be forced idle if a given user is not transmitting data (as is the case for circuit switching). However, packet switching has much more stringent requirements for the switching technologies because switching must occur on the time scale of an individual packet, certainly not more than a μs time scale.

There are many network-level issues which must be addressed in a packet-switched network (Berthold, 1994). Some examples include

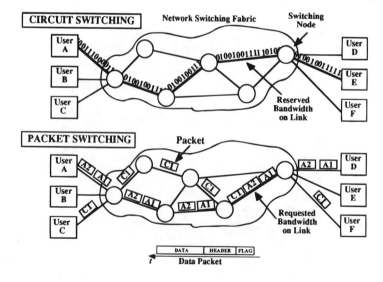

FIGURE 16.62 (a) Circuit and (b) packet switching.

i. Control - The packet routing and communication paths can be determined either from a central location or in a distributed fashion in which each node has some degree of autonomy.
ii. Routing - A packet can be routed actively or passively. Active routing implies that: (i) information in the packet headers helps determine the state of a given switch and (ii) the state must be changed as the dynamic network changes. Passive routing means that a routing path is established for a given set of conditions, but no active changing of a switch is required. An example of passive switching is a wavelength-selective switch in which all packets arriving on wavelength 1 are routed to output port A whereas all packets arriving on wavelength 2 are routed to the output port B (Brackett, 1994).
iii. Synchronization - High-Speed packets may be arriving from many different locations. Packets can arrive synchronously (i.e., at regular intervals in predetermined packet time slots) or asynchronously (i.e., at random time intervals, not in predetermined time slots).
iv. Clock Recovery - A clock must be recovered in order to (i) unambiguously detect digital data bits and (ii) anticipate packet detection if synchronous time slots are used. Global network clocks are difficult to achieve because the network may cover a very wide area and also because fiber dispersion will spread a narrow optical pulse.
v. Contention Resolution - As a packet traverses a network, it may converge on a 2×2 switch simultaneously with another packet from a different user. Both of these packets may request destination routing out the same output port, but only one can be transmitted to a single output port at a given time. Such packet output-port contention must be resolved rapidly to maintain high throughput and efficiency in our high-speed network (Cisneros et al. 1993).
vi. Buffering - If a packet cannot be sent immediately to its correct destination for any number of reasons, a decision must be made as to what to do with this packet. The choices for a packet which loses contention include (a) buffer the packet in a memory or delay line or (b) send the packet to another network node where it is detected and retransmitted, so that it will eventually route its way to the desired destination. This second scenario, which may be used to overcome technological shortcomings of the network, is known as a multihop network scenario, in which a packet may make several "hops" before arriving at its destination. One well-known example of a multihop system is called ShuffleNet (Acampora et al. 1987).

The following issues in packet switching do not exist in circuit switching: (i) contention resolution which is solved by the initial handshaking process, (ii) controlling the routing path which is predetermined and fixed, and (iii) packet switching delay because no switching decision must be made along the fixed path. Furthermore, depending on whether a LAN or WAN network architecture is used, key network issues such as delay, contention resolution, and synchronization have different challenges.

A standardized network protocol must be used to ensure that data packets are all formatted with recognizable routing information, so that the packets can be switched through the network with full global compatibility. The two standards (or their foreign equivalents) which show the most promise of full adoption for a global optical network is Synchronous Optical Network (SONET) (Ching, 1993) and Asynchronous Transfer Mode (ATM) (McEachern, 1992). Data and header information are bunched into small 53-byte ATM packets. These packets arrive at a switching node at random times and are grouped together into a large 125-μs SONET frame which makes its way in predetermined, synchronous, time slots through the network. The ATM packets are "unloaded" by the SONET frame when its direction is switched through the network and it can be placed into a different SONET frame. The analogy has been made that the ATM packets represent people randomly boarding a time-scheduled SONET train (Cisneros, private communication).

Time-Division Multiplexing

The basic multichannel network concept of optical time-division multiplexing (TDM) is the same as for traditional electrical networks and is shown in Fig. 16.63 (Oshima et al. 1985; Prucnal et al. 1986b). If there are N users who wish to share the same high-bandwidth optical medium, then we must divide a high-speed bit time T into N slots. Each user is assigned a specific time slot within each bit-time to transmit data. Photonic TDM is where the optical bits are multiplexed and demultiplexed in the optical domain. This implementation uses a fast photonic switch to time multiplex several lower speed channels into one higher speed channel. This signal is then transmitted on the fiber medium and, then, demultiplexed at a switching node or receiving end. The receiver would, then, recover a demultiplexed lower speed signal by receiving only a specified time slot within each bit-time.

Advantages of TDM include (i) no output-port contention problem because each data bit occupies its own time slot and there is only a single high-speed signal present at any given instant and (ii) the implementation for low-speed photonic networks is quite straightforward and similar to electronic networks. One major disadvantage of TDM is that this scheme requires ultrahigh-speed switching components if the individual signals are themselves high-speed and if there are many users. For example, if there are 10 users each transmitting at 2.5 Gbit/s, then, the photonic switches must have a 25-GHz bandwidth. It is obvious that this method will experience a capacity limitation because the bandwidth of photonic switches will not exceed 10s of GHz in the near future. Another difficulty of TDM is that network control, stability, and electronic processing become quite difficult (and expensive) to perform efficiently at very high speeds. Furthermore, the transmission of short pulses is extremely prone to fiber dispersion effects, unless soliton pulses are used for transmission (see section 16.6).

Some of the device technologies which may be critical to performing high-speed TDM include

   a. high-speed switches - these could be high-speed lithium niobate switches in which many 2×2 switches are cascaded to form any $N \times N$ combination (Tacker et al. 1987). Additionally, the nonlinear optical loop mirror may provide time-switching rates at speeds >100 GHz (Doran and Wood, 1988).
   b. generation of high-speed pulses - a mode-locked laser could produce picosecond pulses at GHz repetition rates. These pulses could be modulated to produce a high-speed TDM signal, and such a signal would be in RZ format.

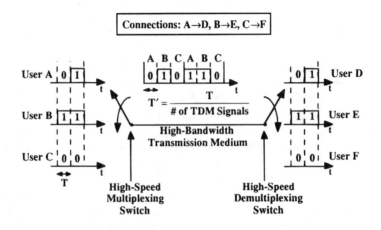

FIGURE 16.63 Concept of time-division multiplexing.

Wavelength-Division Multiplexing

One unique feature of an optical fiber is its extremely wide, low-loss bandwidth range (Midwinter, 1979). In the 1.55-μm, low-loss window, 25,000 GHz of bandwidth exist. It seems natural to dramatically increase the system capacity by transmitting several different wavelengths simultaneously down a fiber to more fully utilize this enormous bandwidth. Such wavelength-division multiplexing (WDM) has recently generated much excitement and research because it can have very high aggregate system capacity (Agrawal, 1992, Chapter 7; Brackett, 1990; Kaminow, 1990). At present, it is expected that WDM will be one of the methods of choice for future ultrahigh-bandwidth multichannel systems. Of course, this sentiment could change as technology evolves.

In WDM systems, many independent signals can be transmitted simultaneously on one fiber, with each signal located at a different wavelength. If wavelength-selective components are implemented, these signals can be routed and detected independently, with the wavelength determining the communication path. One example of the system capacity enhancement is the situation in which ten 2.5-Gbit/s signals can be transmitted on one fiber producing a system capacity of 25 Gbit/s. This wavelength parallelism circumvents the problem of typical optoelectronic devices which do not have bandwidths exceeding a few GHz unless they are exotic and expensive.

The wavelength of a signal is used for routing purposes and defines the data origin, destination, or path. A typical system is shown in Fig. 16.64. In this example, the wavelength of each transmitter is fixed and the wavelength of each receiver is tunable; the opposite arrangement could be implemented with nearly equivalent network functionality, although some differences do exist in specific cases. Figure 16.64 also shows the WDM system in three generic scenarios: (i) a quasi-point-to-point system useful for many channels transmitted over long distances, (ii) a star network topology, and (iii) a ring network topology.

Enabling Technologies

There has been much research to create efficient, high-performance, and low-cost WDM technologies. A diagram showing some of the devices which may be necessary in a system is shown in Fig. 16.65. In general, the following device parameters, where applicable, define the figures-of-merit for a WDM system: (i) wide wavelength tuning range, (ii) rapid wavelength tuning speed, (iii) rapid modulation bandwidth, (iv) narrow spectral characteristics, (v) high Finesse, (vi) multiuser capability, and (vii) potential low cost. Note that the wavelength tuning speed of the some of the WDM devices is critical for a packet-switched network requiring ~μs tuning speeds (or less) but is much less critical for a circuit-switched network which can tolerate ~ms switching speeds. Although the number of potential WDM devices being considered is quite large, we describe only a representative subset of these devices, each discussed in a short paragraph below:

Multiwavelength Transmitters. Multiwavelength transmitters are required if the wavelength of the transmitter must be tuned to different values corresponding to different destinations. Two leading contenders are (i) a multielectrode, wavelength-tunable, distributed Bragg-reflector (DBR) laser (Koch et al. 1988) (and the distributed-feedback laser) which can be tuned over some range of wavelengths and (ii) a multiwavelength laser array in which each laser emits light at a distinct, separate wavelength (see Fig. 16.66) (Zah et al. 1992). The wavelength tunability of a multielectrode laser is controlled by the current injected into the different electrodes. The tuning mechanism can be explained easily in the case of a two-electrode DBR laser in which one electrode is connected to the active medium and the second electrode is connected to the Bragg reflecting region. The cavity mirrors for this laser are provided on one side by a cleaved semiconductor facet and on the other side by the Bragg reflective grating. Current injection into the electrode attached to the active medium causes lasing. Current injection into the electrode attached to the Bragg region pumps this semiconductor region with carriers, thereby, slightly changing the index of refraction which changes the effective grating period for coherent reflection and, ultimately, the length of

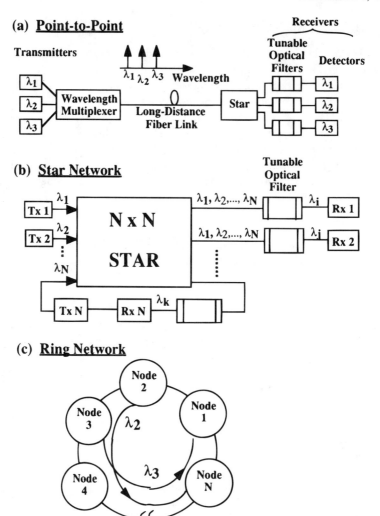

FIGURE 16.64 A generic WDM system for a (a) point-to-point link, (b) star network, and (c) ring network.

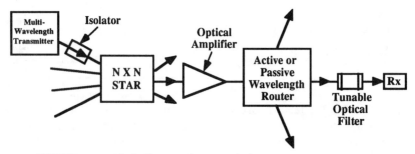

FIGURE 16.65 Block diagram of some typical components in a WDM system.

Optical Communications

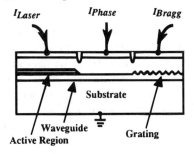

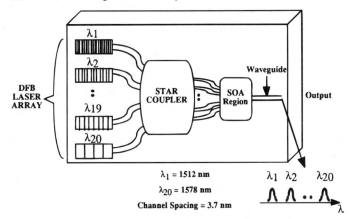

FIGURE 16.66 Schematic of (a) a multielectrode DBR laser [© IEE 1988. After Koch et al. 1988.] and (b) a multiwavelength laser array.

the optical lasing cavity defining the output wavelength. Many other multiwavelength transmitters have been proposed. However, each solution has its own advantages and disadvantages with no choice clearly superior presently.

Wavelength-Tunable Receivers. If the wavelength which the receiver detects a signal is tunable to pick up a signal from fixed-wavelength transmitters, then, we require a tunable filter. The filter is, typically, used to demultiplex many WDM channels, transmitting one and blocking all others (Kaminow et al. 1988; Willner et al. 1990b). Four possible tunable filters depicted in Fig. 16.67 and summarized in Table 16.2 are

(i) the semiconductor filter - a semiconductor with an intracavity Bragg grating which is extremely wavelength selective (Choa and Koch, 1992). Changing the current changes the index of refraction and the wavelength which will satisfy the Bragg condition. Tuning speed is extremely fast (ns), but the tuning range is limited to a few nm and a uniform passband is difficult to maintain over the entire wavelength range.

(ii) the fiber Fabry–Perot - highly reflective mirrors are deposited on the ends of a fiber. These mirrors define a wavelength-selective Fabry–Perot cavity which can be wavelength tuned by mechanically varying the distance between the two mirrors (Stone and Stulz, 1987). The tuning range is large (10s of nm) but switching speed is slow (ms).

(iii) the liquid-crystal Fabry–Perot - the Fabry–Perot cavity between two mirrors is filled with a liquid crystal (Patel, 1990). The orientation of the molecules in the liquid crystal can be realigned based on an externally applied voltage. This molecular alignment determines the index of refraction of the liquid crystal and, therefore, the effective length of the FP

(a) Semiconductor Filter

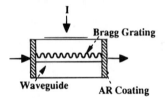

(b) Fiber Fabry-Perot

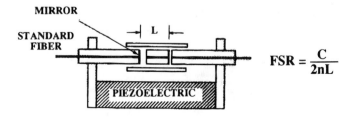

(c) Liquid-Crystal Fabry-Perot

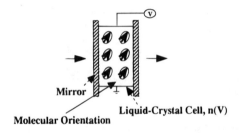

(d) Acousto-Optic Filter

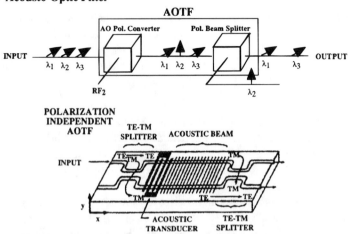

FIGURE 16.67 Four tunable optical filters: (a) the semiconductor filter, (b) the Fiber Fabry–Perot [© 1987 IEE. After Stone and Stulz, 1987.], (c) the liquid-crystal filter, and (d) the acousto-optic filter [© 1990 IEEE. After Smith et al. 1990.].

Optical Communications

TABLE 16.2 Comparison of Major Filter Characteristics

	Tuning Speed	Tuning Range (nm)	Advantage	Disadvantage
Fiber Fabry-Perot	ms	>10	Simple	Slow
Liquid-Crystal filter	ms	>10	Filter arrays	Nonlinear tuning
Acousto-Optic filter	μs	>10	Fast and multiple independent passbands	Potential high cost
Semiconductor filter	ns	>1	Fastest	Passband nonuniform with tuning

cavity. Tuning range is large but switching speed is slow (ms).

(iv) the acousto-optic filter (AOTF) - wavelength-selective Bragg gratings can be formed on a lithium niobate substrate by inducing an acoustic wave in the material. A shift in the polarization is induced in only that wavelength which matches the Bragg acoustic grating. An output polarizing beam splitter can be used to filter out only that wavelength which has had its polarization changed (Smith et al. 1990). The AOTF can accommodate multiple acoustic waves and, therefore, multiple independent passbands. Switching speed is moderate and is in the μs regime. Because the AOTF has two output ports, it can also be used as a wavelength-selective switching element.

Passive Stars. In star network topology, passive stars are necessary to allow all users to communicate with all other users. Each output port has access to all input ports, with the input power divided among all the outputs. Small $N \times N$ passive stars are typically formed by cascading 2×2 directional couplers. However, it would be impractical to have a 128×128 passive star composed of cascaded directional couplers. One clever planar solution is to have all input waveguides radiate their light into a free-space region which is then coupled into all the output waveguides (see Fig. 16.68) (Dragone, 1989). By tailoring the shape of the free-space region according to antenna theory and Fourier optics, the light from any input waveguide will radiate and couple equally into each of the output waveguides. This is known as the integrated star coupler.

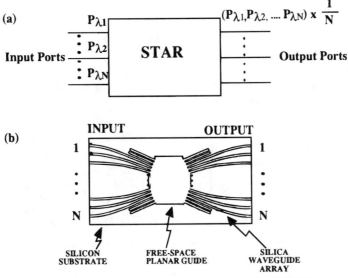

FIGURE 16.68 Integrated planar $N \times N$ star coupler (© 1989 IEEE. After Dragone, 1989.).

Wavelength-Division Multiplexers. If passive wavelength routing is required in the network, then $M \times N$ devices are needed facilitate the selective routing of a signal from a given input port to a given output port according to the input signal's wavelength. These devices typically contain a wavelength-selective grating. One straightforward example shown in Fig. 16.69 is a grating integrated with one input and several wavelength-specific outputs (or vice-versa) (Soole et al. 1991). Another highly acclaimed example is the frequency-selective integrated router (Dragone, 1991). This router is composed of two integrated star couplers which are interconnected by a series of waveguides of varying length. These staggered-length waveguides between the two stars form a phase grating which is the basis of the wavelength selectivity; it is beyond our scope to discuss the functionality of the frequency router, but the reader is encouraged to pursue this issue further. The passbands are <0.5 nm, and the number of input and output wavelengths can be >20. Many other examples exist, such as a narrowband channel-dropping filter (Haus and Lai, 1992).

Amplifiers. Erbium-Doped fiber amplifiers will most likely be used in fiber-based networks to compensate for fiber attenuation and optical splitting losses (Willner et al. 1990a). One major issue with using EDFA's in a WDM system is that the gain is wavelength-dependent whereas the losses are wavelength-independent, thereby, causing an SNR differential among many WDM channels. This problem can become quite severe when many amplifiers are cascaded in a system. The relatively flat shoulder region near 1.555 μm is the optimal spectral position for the channels. There are methods for equalizing the gain, but the system may perform adequately without equalization for at least several cascaded amplifiers and <20 channels.

System Guidelines

Figure 16.70 shows the spectrum of many WDM channels which must be demultiplexed, so that only one channel is received and the other channels are blocked. If we assume that the optical filter is a Fabry–Perot, the transmission function of the filter has a lorentzian-shaped, periodic passband (Born and Wolf, 1980, Chapter 1):

$$T(f) = \frac{1}{1 + \left[\dfrac{2(f - f_0)}{f_{FP}}\right]^2} \qquad (16.62)$$

where f_{FP} is the full-width, half-maximum of the filter, f_0 is the center frequency of a passband where the transmission is a maximum, and wavelength can be related to frequency by $\lambda = c/f$. Such a shape will allow some inter wavelength cross talk because not all of the optical power from the other channels will be blocked. This cross talk will have the effect of reducing the contrast ratio of the selected recovered signal. Therefore, a narrow filter is desirable because it

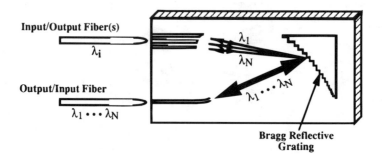

FIGURE 16.69 Integrated grating-based wavelength (de)multiplexer.

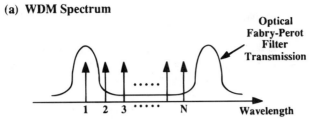

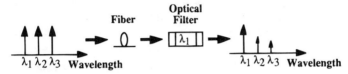

FIGURE 16.70. Spectrum of many WDM channels being demultiplexed by a Fabry–Perot optical filter.

will limit the amount of power leaking through the filter from neighboring channels. Furthermore, devices must also have a wide, wavelength-tuning range, so that the channels can be tuned and placed sufficiently far apart to avoid cross talk. Because the filter has a periodic passband, two different channels could be transmitted simultaneously through two different filter passbands causing ambiguous channel recovery. Therefore, all the channels must be located within one free-spectral range of the filter. The number of channels which can be accommodated in this WDM system is given by

$$N = \frac{\text{FSR}}{\Delta f}, \tag{16.63}$$

where N is the number of channels, FSR is the free-spectral range, and Δf is the channel spacing which will permit only negligible power leakage through the filter of the neighboring channels. In Eq. (16.63), we have neglected signal broadening due to the information bandwidth, laser chirp, and laser linewidth, all of which would necessitate a larger channel spacing and, consequently, fewer channels. The power penalty PP_r caused by cross-talk power from a neighboring channel leaking through the optical filter and reducing the contrast ratio of the selected signal is given by (Willner, 1990)

$$PP_r = 10 \log\left[1 + \left(\frac{f_{\text{Fp}} + f_{\text{ch}}}{2\Delta f}\right)^2\right], \tag{16.64}$$

where f_{ch} is the full-width half-maximum (FWHM) of the individual channels which takes into account channel bandwidth broadening effects; laser linewidth has been excluded because (i) it can be neglected for high-speed, direct-detection systems and (ii) the linewidth noise effects on the system must be handled differently, anyway.

It is important to mention a few key issues relating to system operation:

    i. Wavelength Chirp: Because the optical filter should reject all nonselected channels and all channels must fit within a finite bandwidth, the optical filter passband should not be too wide, so as not to make the channel separation large. Typically, the filter bandwidth should be ~3 times the high-speed information bandwidth in ASK direct-detection systems; note that the laser linewidth, typically <10 MHz for a DFB laser, would not be a factor for Gbit/s transmission. However, if the laser is directly modulated, so that the carrier density and, hence, the laser wavelength are chirped as the current is turned ON and OFF, then, the channel bandwidth may as wide as 10 GHz, necessitating a wider filter (Elrefaie et al. 1989). This chirp would also be deleterious for long-distance WDM transmission which would temporally spread the pulse due to fiber dispersion. If bandwidth is critical, then, external modulation, which produces minimal chirp, should be used.

ii. Frequency Stability: System operation requires that the signal wavelength and the wavelength-selectivity of devices be stable over time. If the signal bandwidth is small compared to the channel spacing, then, the system is fairly tolerant to small wavelength drifts. If tighter controls are needed, then, several techniques can be employed for stabilization: electronic feedback circuitry using passive optical tapping (Kaminow et al. 1989), rigid locking to an atomic laser transition frequency (Chung et al. 1988), and locking each channel to a different resonance of an FP periodic filter. Unless controls are used, it is unlikely that channel spacings less than 0.5 nm will be realized.

iii. Wavelength Reuse: In a simple WDM network, N users can be interconnected by N wavelengths, with each wavelength uniquely identifying each user. However, it is presently not practical to provide 1000 wavelengths for 1000 users in a large network. Therefore, due to technological limitations, it may be necessary to reuse the same wavelengths at different LANs within a large WAN. Wavelength reuse could exist with passive routing through the network, so that no two identical wavelengths appear at the same part of the network with the same destination. Wavelength-division multiplexers are key components, and N^2 users can be accommodated with N wavelengths (Brackett, 1994). If more flexibility is desired, all-optical wavelength shifting at key switching nodes can be employed. One method for active wavelength shifting was mentioned in Section 16.4 in which the signal from an intense pump wavelength is copied onto an unmodulated probe wavelength when both wavelengths are input to a saturated semiconductor optical amplifier.

iv. Fiber Nonlinearities: [The reader should be aware that fiber nonlinear effects in a WDM system is a fairly advanced subject and requires a substantial degree of explanation to adequately treat the subject. Due to the nature of this one brief chapter, we will only point out the most relevant features in an excruciatingly brief fashion.] The index of refraction of an optical fiber is nonlinear with signal power (Agrawal, 1990):

$$n = n_{\text{linear}} + \overline{n}_2 \left(\frac{P}{A_{\text{eff}}} \right), \tag{16.65}$$

where $\overline{n}_2$ is the nonlinear index coefficient and is (3.2×10^{-16} cm^2/W) for a silica fiber, n_{linear} is the linear (i.e., standard) index of refraction, P is the optical power, and A_{eff} is the mode cross-section. Although the nonlinear portion of the index is only 1 part in 10 billion, nonlinear effects can seriously affect high-speed, long-distance optical systems (Tkach, 1994). Four important nonlinear effects are

(1) four-wave mixing: Two channels (or three, etc.) will mix with each other producing optical power at the sum and difference beat frequencies. Four-wave mixing occurs when the channel is located near the dispersion zero wavelength in the fiber.

(2) cross-phase and self-phase modulation: The optical power will locally change the index of refraction, thus, creating a variation in the phase (i.e., wavelength) for different portions of a pulse. The pulse will temporally spread and deform.

(3) stimulated Raman scattering: Power from one channel will be transferred to another channel through a broadband (10s of nm), optical scattering process. Power is transferred from the shorter wavelength channels to the longer wavelength channels; shorter wavelengths experience loss and longer wavelengths experience gain.

(4) stimulated Brillouin scattering: Power from one channel will be transferred to another channel through narrowband acoustic phonon scattering. This process is so narrowband (~10 MHz) as to be fairly negligible for Gbit/s systems.

Power or modulation is transferred from one channel to another resulting in performance-degrading cross talk as the signals propagate down a length of fiber. These effects should be evaluated for higher power signals (>1 mW), longer fiber lengths (>10 km), and higher speeds (>5 Gbit/s). The theoretical limits are shown in Fig. 16.71, and the actual system limits may be much more constraining; this figure assumes a 10 GHz channel spacing and operation near 1.55 μm using dispersion-shifted fiber (Chraplyvy, 1990). The reader is encouraged to read Chraplyvy (1990) for more detail.

There are several examples of experimental WDM systems that have been built. Figure 16.72 shows two examples, a 100-channel experiment by Nippon Telephone and Telegraph in Japan (Inoue et al. 1991) and the LAMBDANET project at Bell Communications Research (Goodman et al. 1990). Additionally, the Rainbow project at IBM (Dono et al. 1990), an AT&T/Lincoln Laboratories system (Kaminow et al. 1994), and a WDM coherent heterodyne experiment (Glance et al. 1989) all represent advances in the state of the art of WDM system technology.

Space-Division Multiplexing

Space-division multiplexing (SDM) is a fairly mature technology in which each switching element uniquely defines an optical path and connection (Diaz et al. 1988). As an example, the two-dimensional optical crossbar switch of Fig. 16.73 is composed of N input fibers, N output fibers, and an $N \times N$ array of optical switching elements, such as spatial light modulators. Each input

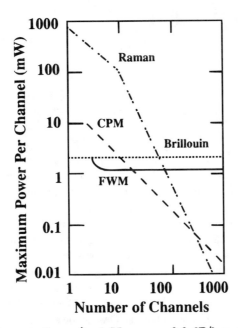

Assumptions: $\lambda = 1.55$ μm, $\alpha = 0.2$ dB/km, $\Delta f = 10$ GHz

FIGURE 16.71 Limitations on optical power per channel due to nonlinearities in a WDM system. (© 1990 IEEE. After Chraplyuy, 1990.)

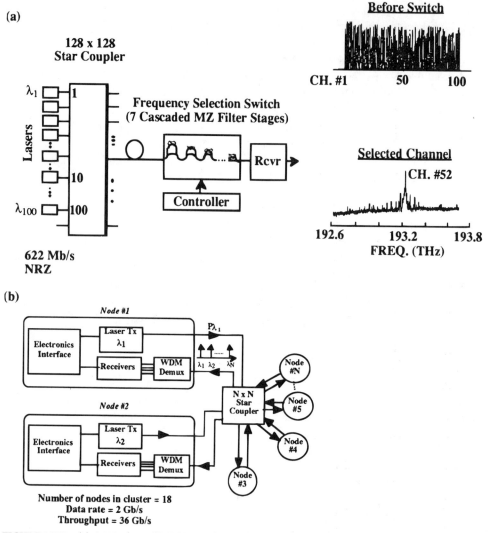

FIGURE 16.72 (a) A 100-channel WDM experiment. (© 1991 IEEE. After Inoue et al. 1991.) (b) LAMBDA NET experimental demonstration. (© 1990 IEEE. After Goodman et al. 1990.)

fiber is split into N branches, each branch connected to a different spatial element in a given row. Each output fiber is also split into N branches, each branch connected to a different spatial element in a given column. Because each spatial switching element can be independently chosen to transmit or block the light, then, a unique path connecting any input fiber to any output fiber can be established by allowing the appropriate spatial element to transmit. Although SDM is a simple technique, it suffers from some disadvantages: (i) the number of elements scales as the square of the number of users, making this technology limited to a small number of users due to fabrication limitations and (ii) optical splitting and recombining when using passive optical $N \times N$ star couplers will each incur a $1/N$ optical splitting loss which limits operation. We can compensate for these losses by using external EDFA's or by using an array of SOAs as the switching elements themselves (Fujiwara et al. 1991). Because many switching technologies (i.e., liquid crystal) offer slow ms switching speeds, fast switching SOAs can be employed to obviate the speed bottleneck. An alternate method for implementing SDM is to use an array of electro-optic, lithium niobate switches arranged in a multistage, planar orientation (Kondo et al. 1987). The switching speeds are high (GHz) but there is a practical fabrication-limited size associated with such switches.

Optical Communications

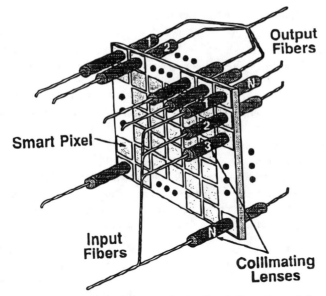

FIGURE 16.73 Space-division multiplexed optical crossbar switch.

Code-Division Multiplexing

A method of encoding, which predates optical communications, is code-division multiplexing (CDM) which establishes the unique communications link by encoding the destination inside of each bit-time (Prucnal et al. 1986a; Salehi et al. 1990). Fig. 16.74a shows how a single bit-time is divided into M chip slots. Each user is assigned a specific code of "1"s and "0"s within the chip times. N users would require N codes. Each transmitter must encode the correct sequence within each bit-time for a given destination. Each destination has a receiver which will decode the intended data with the inverse of the transmitter coding hardware but will not decode data intended for another destination. (Note that a fixed transmitter code and a tunable receiver code is another alternative.) If a decoder has the correct inverse hardware as the transmitter, then, the

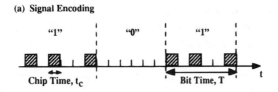

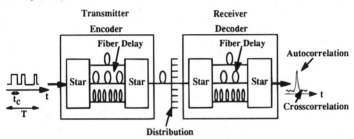

FIGURE 16.74 (a) An address-encoded bit composed of many chips. (b) Encoding (and decoding) of an optical bit with variable-length optical delay lines.

output will produce a peak in the optical signal called the autocorrelation. This autocorrelation signal will trigger a detected "1" bit in the receiver, if it is above a certain threshold level. If the receiver does not have the correct decoder, then, the cross-correlation signal, which appears as a noisy, temporally broad, background signal, will contain all the energy, and the autocorrelation will be small, producing only "0" bits. A straightforward method of encoding and decoding a chip sequence is to use a parallel array of different-length fiber delay lines (see Fig. 16.74b).

One advantage of CDM is increased security because only a receiver with the correct decoder can recover a data stream. The disadvantages include (i) a limitation in the number of users for which M chip slots will produce a high autocorrelation and a low cross-correlation signal when the decoder and encoder match and (ii) a limitation in the speed of the system because very short pulses are required within each bit-time, limiting the bit-rate for a finite-pulse-width transmitter.

16.6 Solitons

We have been discussing the basics of optical transmission, while always keeping in mind that there is some fundamental speed and distance limit when NRZ pulses are propagating. The limitations exist either because (i) fiber chromatic dispersion will temporally spread the pulse or (ii) in the case of zero dispersion, fiber nonlinearities will distort the pulse shape and cause pulses on different wavelengths to interact. However, a class of pulses exists which retains its shape indefinitely, allowing for the potential of ultrahigh-speed (>20 Gbit/s) optical transmission per channel over multimegameter distances. These pulses, known as solitons (Nasegawa and Tappert, 1973; Mollenauer et al. 1980), have few fundamental limits for a single channel and hold the exciting promise that the not so near future need for ever higher data rates anywhere in the world will be met by this technology. Presently there is no other technology which can make this claim. [More advanced limitation issues, such as the Gordon–Haus effect, will not be discussed in this chapter (Gordon and Haus, 1986)].

An optical pulse is nothing more than a wave which propagates along a fiber. This propagation in a single-mode fiber can be described by a nonlinear Schrödinger equation (Mollenauer et al. 1986):

$$-i\frac{\partial u}{\partial z} = \frac{\partial^2 u}{\partial^2 t} + |u|^2 u - i(\alpha/2)u, \tag{16.66}$$

where u represents the pulse envelope function, z the propagation distance along the fiber, t the retarded time, and α the energy-loss coefficient. The first term on the right side of the equation represents the fiber chromatic dispersion. The second, or nonlinear, term represents the dependence of the fiber index of refraction on the light intensity inside the fiber. The final term represents the fiber loss or gain. Dispersion broadens the optical pulse in time, whereas the nonlinear term broadens the pulse frequency spectrum. This nonlinear frequency broadening is known as self-phase modulation (SPM). If we employ optical amplifiers so that the fiber loss is compensated by the EDFA gain, then, we can initially ignore the third term in Eq. (16.66). One solution to the above equation is the soliton (Mollenauer et al. 1986):

$$u(z, t) = \sec h(t)e^{iz/2}. \tag{16.67}$$

The exponential term is simply a phase relationship which does not affect the shape of the pulse. Equation (16.67) shows that the fundamental (zeroth-order) soliton pulse has a hyperbolic secant shape which does not change with distance. This is due to a unique combination of factors for which the dispersion exactly cancels the SPM (i.e., these two effects have the same magnitude but are opposite in sign), making for a nondispersive pulse; note that dispersion can be positive

FIGURE 16.75 Cartoon illustration of a pulse conceptually composed of three runners, representing the pulse's leading edge, center, and trailing edge. (After Evangelides.)

or negative, and so, the soliton can exist only in the wavelength range for which the dispersion is opposite to the SPM. An intuitive way of understanding the soliton is that the pulse creates a index "well" which travels along with the pulse. The leading edge of the pulse, which would normally travel faster, is retarded by the intense pulse center, changing the index. The trailing edge, which would normally fall behind, speeds up due to the pulse center. This idea is shown in Fig. 16.75 in a cartoon of three joggers running on a mattress, a fast leading runner, a large center one, and a slow trailing one (Evangelides). Because the large middle jogger depresses the mattress, the speed of the fast jogger is reduced whereas the speed of the slow jogger is increased. The pulse and the relative positions of the joggers stay the same!

The soliton period has a characteristic length over which the various effects interact. The soliton length z_c is given by (Mollenauer et al. 1986)

$$z_c = K\left(\frac{2\pi c}{\lambda^2}\right)\left(\frac{\tau_{sol}^2}{D}\right), \qquad (16.68)$$

where τ_{sol} is the time interval between the two times associated with the soliton's FWHM intensity. This soliton length could be hundreds of kilometers for short pulses generated near the fiber dispersion minimum. The soliton also has a required peak power P_c which will allow exact cancellation between dispersion and SPM (Mollenauer et al. 1986):

$$P_c = \left(\frac{A_{eff}}{2\pi n_2}\right)\left(\frac{\lambda}{Z_c}\right), \qquad (16.69)$$

where A_{eff} is the effective area of the fiber core and n_2 is the fiber nonlinear coefficient. Figure 16.76 shows a typical soliton propagating along a fiber with no change in shape. Note that higher

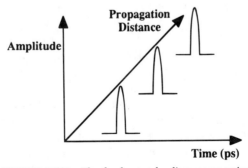

FIGURE 16.76 The fundamental soliton propagating along an optical fiber.

order solitons also exist in which the pulse shape oscillates over the soliton length, returning to the original, fundamental soliton shape after every period.

One of the truly amazing qualities of the soliton is that it is fairly robust to perturbations away from the ideal values. The soliton is the only stable solution to the nonlinear Schrödinger equation, and so, the pulse will evolve into a soliton as it propagates along the fiber, so long its shape and peak power are relatively close to the exact solution. It is, therefore, reasonable to generate a nonideal pulse from a laser which will evolve into a soliton over a short distance. This time evolution into a soliton is accompanied by the generation of low-intensity dispersive waves which are viewed as loss in our soliton system.

In our explanation of the basic soliton, we had assumed that the loss term was zero in the nonlinear Schrödinger equation. The robust soliton behavior is manifest in its ability to accommodate lumped amplification and in not requiring that the loss be absolutely zero at every point along the fiber. Lumped amplification can be employed as long as (i) the average path gain compensates for the average path loss and (ii) the pulse power remains within the range of the soliton solution. This is also true of the dispersion parameter of the fiber, which is allowed to have even, discontinuous changes over small lengths of fiber. Note that the changes in gain and dispersion must occur over distances much smaller than the soliton characteristic length z_0 (Mollenauer et al., 1986).

Transmission of solitons requires the generation of short pulses. This is typically accomplished with mode-locked lasers which can easily achieve <50-ps pulse generation at high repetition rates (Eisenstein et al. 1986). A train of pulses is produced, and a secondary data bit-stream controls an external modulator which either passes a pulse if a "1" bit is present or blocks the pulse if a "0" bit is present. Because solitons are inherently short pulses, the modulation format is return-to-zero, not nonreturn-to-zero (see Section 16.3).

Solitons could not exist if optical amplification were not used to compensate for the fiber loss. To this end, EDFAs have been the key enabling technology for long-distance soliton technology. EDFAs can be placed every 10–30 km along an ultralong-distance link and still provide the appropriate soliton power conditions mentioned above. As with a nonsoliton system, the amplified, spontaneous emission noise accumulates along the length of the link in a typical soliton system. However, an extremely clever technique has been demonstrated which reduces the accumulation of ASE. This method uses a cascade of sliding-frequency filters in which a transmission filter is placed after each EDFA, with each filter center frequency slightly offset from one filter to the next (Mecozzi et al. 1991; Kodama and Hasegawa, 1992). The nonlinear soliton "slides" in frequency with the changing filter centers, but the linear ASE does not translate frequency with the changing filter centers and, therefore, does not accumulate (i.e., experience as much gain) as rapidly as the soliton because part of the ASE under the soliton is always blocked by the next filter in the cascade.

Just as with standard nonsoliton NRZ transmission, we wish to increase the system capacity and take advantage of the broadband fiber to transmit many soliton signals simultaneously down the fiber on different wavelengths, i.e., WDM. Because soliton pulses on different wavelengths will propagate along the dispersive fiber at different speeds, two such WDM pulses will eventually collide with and pass through each other. This collision is characterized by an attraction between two such pulses. It is quite fortuitous that the solitons' attraction is symmetrical, causing one soliton to speed up and, then, slow down, with the opposite happening to the soliton on the other wavelength (Mollenauer et al. 1991). Almost "magically," the two solitons survive without much change after experiencing a collision, even though somewhat violent interactions take place during the collision. It should be noted that it is undesirable to have a large fraction of a collision occurring inside an amplifier because the two solitons' mutual interaction will be affected by the amplification and will no longer be symmetrical. Therefore, the length of the collision should be 2–3 times longer than the amplifier spacing to ensure that WDM soliton collisions are benign.

Figure 16.77 shows an example of an eight-channel WDM soliton experiment demonstrating the feasibility of WDM system enhancement while using soliton pulses (Nyman et al. 1995).

Given the above discussion concerning the wonderful attributes of soliton transmission, one may wonder why presently deployed systems are not using them. The main reason is that solitons are still more difficult to generate (i.e., the need for mode-locked lasers) and transmit (i.e., the requirements on soliton shape power, and collisions) than standard NRZ signals, and NRZ has achieved outstanding-eight-channel results of 5 Gbit/s per channel over multimegameter distances (Bergano et al. 1995). Although NRZ pulses, in theory, cannot achieve the outstanding performance of solitons, the performance of NRZ transmission is still more than adequate to fulfill data transfer needs for the next several years. It is, however, highly probable that the limit will eventually be reached for NRZ transmission, requiring the deployment of soliton systems for ultrahigh-capacity links.

16.7 Analog Communications

This chapter has dealt entirely with digital optical communications due to the potential for very high speed communications per channel. However, analog optical communications has some very appealing characteristics which make it a possible solution for some near-term (and, perhaps, long-term) general communications problems. These include cable television, video distribution, $\ll 1$ Gbit/s/channel local-area networks, and personal communication system interfaces. We will describe some of the basic elements of analog subcarrier multiplexing and its application to multichannel systems.

Subcarrier Multiplexing

The essence of subcarrier multiplexing (SCM) is taking all of the (de)modulating, (de)multiplexing, and routing functions, which could be performed optically, and, instead, performing them electrically (Kashima, 1993; Mestdagh, 1995; Darcie, 1987; Olshansky et al. 1989; Way, 1989). The only optical functions that remain are (i) optical generation using a laser, (ii) optical transmission over an optical fiber, (iii) optical detection using a photodetector, and (iv) some passive optical

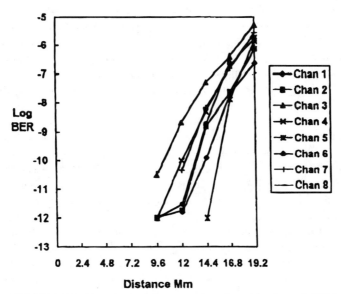

FIGURE 16.77 Experimental demonstration of an eight-channel WDM soliton transmission system. (After Nyman et al. 1995.)

coupling and splitting. The advantages of performing these functions electrically is that, under present circumstances (a) electrical components are cheaper and more reliable than optical components and (b) electrical filters can have an efficient and near-ideal multipole design, whereas optical filters are only single pole.

Figure 16.78 shows a single-channel SCM system. In digital transmission, the optical beam is a carrier wave which is directly modulated by the data. We now remove the data one step further from the carrier wave by using a ~GHz subcarrier electrical wave which is modulated by the data and which, in turn, modulates the ~THz optical carrier wave. The ~MHz data is electrically mixed with the electrical subcarrier, producing sum and difference frequencies as results in standard heterodyning (see Section 16.3); note that an electrical band-pass filter is used to allow only one product, typically, the sum frequency, to pass and be transmitted. The resultant modulation directly modulates either the current on a laser or the optical transmission function of an external modulator. The electrical subcarrier is a simple sinusoidal wave which has an amplitude, frequency (typically 100s of MHz or GHz), and phase, in which the optical wave P can be described by (Kashima, 1993)

$$P = P_c(1 + Am_{\text{data}}\cos(2\pi f_{\text{sc}}t + \varphi_{\text{sc}}), \tag{16.70}$$

where f_{sc} is the frequency of the subcarrier, A is the amplitude, m_{data} is the data modulation index, and P_c is the unmodulated optical output. The subcarrier will appear as a modulated tone in the electrical frequency spectrum before optical transmission. After transmission, the optical wave is detected at the receiving end. The optical carrier wave is far above the detector bandwidth, but the electrical subcarrier will fall within the detector bandwidth and be detected. We employ the same principle, as in the transmitter, to demultiplex the data from the subcarrier. At the receiver, the same subcarrier wave frequency of f_{sc} is mixed with the recovered transmitted signal, again producing sum and difference frequencies. The sum frequency is $(2f_{\text{sc}} + f_{\text{data}})$ and the difference frequency is simply f_{data}. A low-pass filter is, then, used to allow only the difference frequency to pass, allowing the data to be unambiguously recovered. This method allows all of the multiplexing and demultiplexing to occur in the electrical domain, and efficient electrical filters can be incorporated.

One of the key reasons for using analog optical transmission is that it is compatible with much of the analog transmission used today in video transmission of cable television (CATV) signals (Way, 1989). The modulation format in CATV is amplitude-modulation vestigial-sideband (AM-VSB), in which the subcarrier is amplitude-modulated and the AM sidebands created by the modulation are manipulated so as to occupy a minimum bandwidth without sacrificing transmission quality. Due to the compatibility issue, we will discuss only AM subcarrier multiplexing, although FM is also quite possible but requires a smaller signal power and larger bandwidth (Way et al. 1990).

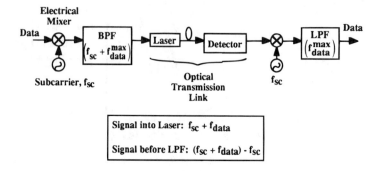

FIGURE 16.78 Basic diagram of subcarrier multiplexing.

Optical Communications

The key performance criterion is achieving a high carrier-to-noise ratio (CNR), compared to the SNR in digital communications. In a typical system, the receiver still contains random-shot noise and thermal noise which are generated in the photodetector. In addition, the laser itself produces what is called relative intensity noise (RIN) over a wide bandwidth (Sato, 1983). RIN is due to the random phase fluctuations in the laser light (i.e., stimulated emission) which are converted into amplitude fluctuations by reflections back into the laser or by multiple reflections along the transmission path; note that it is critically important to keep reflections low (<-65 dB) in an analog system by using high-quality connectors, splices, and isolators. Even fiber dispersion will have the affect of increasing the RIN. The RIN can be described by the statistical fluctuations in the photogenerated current (Sato, 1983):

$$\text{RIN} = \frac{\langle i_{\text{ph}}^2 \rangle}{\langle i_{\text{ph}} \rangle^2} \qquad (16.71)$$

and the RIN-generated noise power, σ_{RIN}^2 is given by

$$\sigma_{\text{RIN}}^2 = (\text{RIN})\bar{P}B_e, \qquad (16.72)$$

where $\bar{P}$ is the average received optical power, and B_e is the electrical bandwidth of the receiver. RIN is typically a very small value, but AM communications requires an extremely high CNR for good signal fidelity, so that the noise must be kept to a minimum and the optical power must be kept high. Because analog transmission requires a very high average optical power and because the RIN noise power is proportional to $\bar{P}^4$, RIN usually becomes the dominating noise factor. The entire equation for the CNR for AM optical transmission is (Agrawal, 1992)

$$\text{CNR} = \frac{(m\bar{P})^2/2}{\sigma_{\text{sh}}^2 + \sigma_{\text{th}}^2 + \sigma_{\text{RIN}}^2}. \qquad (16.73)$$

To achieve high fidelity and "error-free" signal recovery, the CNR typically must be >50 dB, much higher than the ~ 20 dB required by direct-detection ASK digital systems!

When an external modulator is not used, the laser acts to convert the electrical signal directly into an optical signal, with a modulation of the laser bias current producing a direct modulation of the laser light output. Any deviation in the linearity of the laser light as a function of the bias current will produce a decrease in the CNR because the electrical modulation is not exactly replicated by the optical signal output (see also intermodulation distortion mentioned in the next section) (Stubkjaer and Danielsen, 1980). Because the CNR must be extremely high in an analog system, we, therefore, require that the laser light output be extremely linear with the bias current.

There are several limitations in SCM systems, and we will mention just the issue of "clipping." In general, for a communications channel, it is desirable to produce as high a bias swing (i.e., light output swing) as possible in the laser to produce a high modulation index in the carrier power. Moreover, a large bias swing is needed to support many channels because each channel must individually contribute to a minimum modulation swing (see next section). However, there is a limit to the bias swing which can be supported by the laser because (i) the laser will produce no light when the bias current is below the laser threshold current and (ii) the laser linearity will degrade above a certain bias current (Saleh, 1989). If the electrical modulation falls below the threshold current, then, the SCM signal is "clipped", and the CNR will be in adequate for

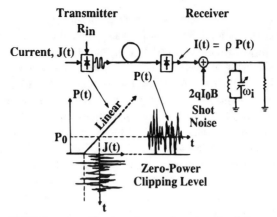

FIGURE 16.79 Clipping of an analog SCM signal due to the bias-current modulation appearing below the threshold current. (© 1989 IEE. After Saleh, 1989.)

signal recovery (see Fig. 16.79). If the modulation is above a certain current, then, nonlinearities will destroy the signal fidelity and produce intermodulation products and distortions in the presence of other channels.

Multichannel Frequency-Division Multiplexing

We have explained SCM in terms of single-channel transmission. However, the powerful advantage of SCM is to transmit many channels simultaneously on one laser, with each channel transmitted on its own subcarrier frequency (Darcie, 1987). This is known as frequency-division multiplexing (FDM) and, conceptually, is very analogous to WDM. Figure 16.80a shows a schematic of a multichannel SCM system, each channel mixed with a different subcarrier frequency and collec-

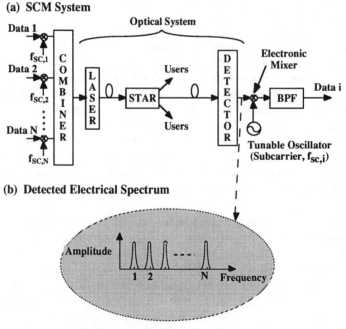

FIGURE 16.80 (a) Frequency-division multiplexed optical SCM system. (b) The electrical spectrum of the transmitted FDM system.

tively sharing the same laser transmitter. Each transmitted channel has its own electrical subcarrier generator, and the receiver can have a fixed or tunable oscillator which is used to recover one of the subcarrier frequencies, i.e., to selectively demultiplex the many channels. Shown in Fig. 16.80b is a rough representation of the electrical frequency spectrum of the multichannel transmission. One advantage of this SCM scheme is to have many channels share the cost of the same expensive optical hardware. One potential disadvantage is that the channels sharing the laser must all be in close proximity.

Another major disadvantage of the SCM scheme is that there is a significant limitation on the transmission speed of the individual channels and on the total number of channels due to the finite bandwidth of the laser transmitter and receiver. Reasonably priced optoelectronic devices do not, presently, exceed several GHz in bandwidth. If we assume that each channel occupies roughly its data rate and we separate the channels by twice the data rate to allow for filtering (and neglect the intricacies of nonlinear distortions), then, we have the following rough approximation for the total number of channels N:

$$N \approx \frac{B_e}{2B_{ch}}, \tag{16.74}$$

where B_e is the electrical bandwidth of the system, and B_{ch} is the information bandwidth of each channel. It is very difficult to have many 100 MHz channels sharing one laser, and GHz/channel transmission is not considered practical. However, transmitting many video channels (of a few MHz bandwidth each) is quite reasonable.

Because a real system is not perfectly linear in converting the electrical signal into an optical signal, another major limitation in the number of total channels is due to two and three channels mixing together nonlinearly, producing sum- and difference-frequency intermodulation distortion (IMD) products (Darcie et al. 1985). If the optical output as a function of bias current is not exactly linear, then, these IMD products will be produced. If we assume three channels, for simplicity, then, the second-order nonlinear products appear at $f_1 \pm f_2$, $f_1 \pm f_3$, and $f_2 \pm f_3$ and the third-order products appear at $f_1 \pm f_2 \pm f_3$, where the frequencies represents the subcarrier frequencies of the individual channels. Given several original channels, these products may appear at the same frequency as an original channel, decreasing the CNR as follows (Agrawal, 1992):

$$\text{CNR} = \frac{(m\overline{P})^2/2}{\sigma_{sh}^2 + \sigma_{th}^2 + \sigma_{RIN}^2 + \sigma_{IMD}^2}, \tag{16.75}$$

where σ_{IMD}^2 is the sum of all the IMD products contributing to system noise within the bandwidth of a single channel. The sum of all the second-order products and third-order products is called, respectively, the composite second order (CSO) and the composite triple beat (CTB). The effect of CSO and CTB is to reduce the number of channels which can be accommodated by an analog system.

It is quite likely that analog systems will incorporate EDFAs to compensate for losses. As was discussed in Section 16.4, the EDFA gain spectrum is not uniform with wavelength. If a directly modulated laser produces a chirp of several GHz of the output optical spectrum, then, the nonuniform EDFA gain will result in a nonlinear gain (i.e., transfer function) for the different frequency portions of the signal (Kuo, 1992). Higher CSO and CTB will result. There are several methods to reduce this problem, but the most straightforward solution is to use a low-chirp external modulator, using an external modulator also helps to reduce the adverse affects on CNR due to simple fiber dispersion.

Probably the most obvious application of analog SCM and FDM is for distribution of cable television (CATV) channels. Several channels can be transmitted on a single laser because the signals all are colocated at the CATV head-end. Furthermore, analog optical transmission of AM-VSB modulation format is completely compatible with the present-day CATV format. Figure 16.81 shows a CATV system in which fiber is used as a backbone, transmitting the channels to a user's local area by optical means and, then, transmitting the channels for the final short distance along a conventional coaxial-cable electrical system. The optical fiber has high bandwidth and low loss, significantly reducing the number of unreliable cascaded electrical amplifiers which must be traversed between head-end and user. Today's CATV systems use optical transmission. Future ultrahigh-capacity CATV systems may use more than one laser and EDFAs. One experiment demonstrated the potential distribution of 16 different wavelengths to 40 million users using EDFAs to compensate for the optical splitting losses (Hill et al. 1990). Another experiment demonstrated four wavelengths, each wavelength containing 50 AM-VSB channels (Uno et al. 1994).

Two final points should be mentioned. First, there is an anticipated explosion in wireless, personal communication systems (PCS). The PC system is expected to be analog, and it is possible that analog SCM optical transmission will be used to seamlessly connect the wireless system along a fiber-based system. Second, analog subcarriers can be used in conjunction with digital

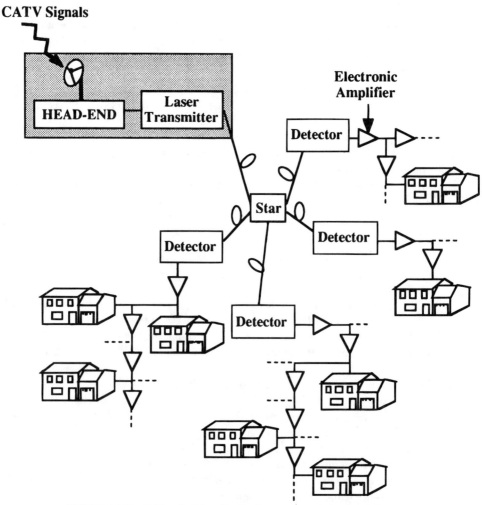

FIGURE 16.81 Cable television distribution system using optical fiber as a backbone.

baseband transmission. In fact, one optical network application is to use subcarrier signals to help control the network traffic and communication paths (Way et al. 1992).

16.8 Future Trends

In this chapter, we have covered many different topics involving high-speed optical communication systems. We have endeavored to treat the most important topics which will likely impact optical systems for years to come. These include WDM for multichannel communications, solitons for ultrahigh-speed communications, and analog transmission for CATV or wireless systems. The "glue" binding and impacting all of these areas is the recent and revolutionary introduction of the EDFA. Additionally, transmission limitations are being continually redefined. For example, by appropriately managing dispersion and nonlinearities, research results have demonstrated the ability to transmit high-speed signals over much longer distances of both conventional and dispersion-shifted fibers than was previously thought possible. (Tkach et al. 1995).

The progress in deployed optical systems has achieved an astounding rate of growth over the past 15 years with a significant impact on society. One may wish to speculate as to the bit-rate which will be available to each of us in the year 2000? 2010? 2020?

One fascinating point is that the overwhelming amount of scientific and technological progress has only occurred over the past 20 years. In fact, an interesting trend observed by Dr. Tingye Li of AT&T Bell Laboratories is that optical communications technology has enabled the doubling of the bit rate-distance transmission product every year (see Fig. 16.82)! (Li, 1988). This trend has been sustained by depending on different technologies, such as the inventions of the single-mode fiber, the single-frequency laser, and the EDFA. It is hard to imagine how optical communications will continue to make such astounding progress, but the striving continues due to the enormous inherent potential of the optical fiber. The limits are continually being pushed back. Although the sky isn't the limit, the fiber certainly is!

Acknowledgments

I wish to thank the following individuals for their kind and critical help in the preparation of this manuscript: Drs. Syang-Myau Hwang, James Leight, David Norte, Eugene Park, William Shieh, and Xing Yu Zou. Of course, Joshua Davis and Xiaoxin Qui were indispensable (at any hour!).

References

Abbas, G. L., V. W. S. Chan, and T. K. Yee. 1985. A dual detector optical heterodyne receiver for local oscillator noise suppression. *IEEE/OSA J. Lightwave Technol.* 3: 1110–1122.

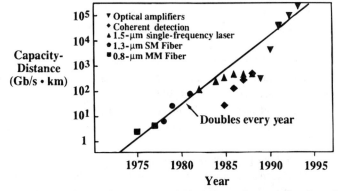

FIGURE 16.82 Bit rate-transmission distance product as a function of year for different optical technologies. (After Li, private communication.)

Acampora, A. S. 1994. *An Introduction to Broadband Networks*. New York: Plenum.

Acampora, A. S., M. J. Karol, M. G. Hluchyi. 1987. Terabit lightwave networks: The multihop approach. *AT&T Tech. J.* 66 (6): 21–34.

Agrawal, G. P., and N. K. Dutta. 1986. *Long-Wavelength Semiconductor Lasers*. New York: Van Nostrand Reinhold.

Agrawal, G. P. 1987. Four-wave mixing and phase conjunction in semiconductor laser media. *Opt. Lett.* 12: 260–262.

Agrawal G. P. 1990. *Nonlinear Fiber Optics*. New York: Academic.

Agrawal, G. P. 1992. *Fiber-Optic Communication Systems*. New York: John Wiley & Sons.

Alferness, R. C. 1982. Switches and Modulators, in *Integrated Optics: Theory and Technology*, G. Hunsperger, ed. New York: Springer-Verlag, Chapter 9.

Armitage, J. R. 1988. Three-Level fiber laser amplifier: A theoretical model. *Appl. Opt.* 27: 4831–4836.

Barnsley, P. E., and P. J. Fiddyment. 1991. Wavelength conversion from 1.3 to 1.55 μm using split contact optical amplifiers. *IEEE Photonics Technol. Lett.* 3: 256–258

Barry, J. R., J. M. Kahn, E. A. Lee, and D. G. Messerschmitt. High-Speed nondirective optical communication for wireless networks. *IEEE Network Mag.*, November, 1991, pp. 44–54.

Bergano, N. S., C. R. Davidson, G. M. Homsey, D. J. Kalmus, P. R. Trischitta, J. Aspell, D. A. Gray, R. L. Maybach, S. Yamamoto, H. Taga, N. Edagawa, Y. Yoshida, Y. Horiuchi, T. Kawazawa, Y. Namihira, and S. Akiba. 9000 km, 5 Gb/s NRZ transmission experiments using 274 erbium-doped fiber amplifiers. Topical Meeting on Optical Amplifiers and Their Applications '92, Tech. Dig., paper PD-11, Optical Soc. Am., Washington, D.C., 1992.

Bergano, N. S., C. R. Davidson, B. M. Nyman, S. G. Evangelides, J. M. Darcie, J. D. Evankow, P. C. Corbett, M. A. Mills, G. A. Ferguson, J. A. Nagel, J. L. Zyskind, J. W. Sulhoff, A. J. Lucero, and A. A. Klein. 40 Gb/s WDM transmission of eight 5 Gb/s data channels over transoceanic distances using the conventional NRZ modulation format. Conf. Optical Fiber Communications '95, paper PD19, San Diego, CA March, 1995, Optical Soc. Am. Wash., D.C., 1995.

Berthold, J. E. Networking fundamentals. Conf. on Optical Fiber Commun. San Jose, CA, Feb., 1994, Optical Soc. Am., Washington, D.C., 1994. Tutorial TuK. 3–20.

Berztsekas, D. P., and R. G. Gallager. 1987. *Data Networks*. Englewood Cliffs, NJ: Prentice-Hall.

Betti, S., G. de Marchis, and E. Iannone. 1995. *Coherent Optical Communication Systems*. New York: John Wiley & Sons, Chapter 5 and 6.

Bjarklev, A. 1993. *Optical Fiber Amplifiers*. Boston: Artech House, Chapter 10, p. 215.

Born, M., and E. Wolf. 1959. *Principles of Optics*, 6th Ed. New York: Pergamon.

Brackett, C. A. 1990. Dense wavelength division multiplexing: Principles and applications. *IEEE J. Selected Areas Commun.* 8: 948.

Brackett, C. A. Status and early results of the optical network technology consortium. Conf. on Optical Fiber Communications '94, Tech. Dig., paper WD3, San Jose, CA, Feb., 1994, Optical Soc. Am., Washington, D.C., 1994.

Burrus, C. A., and B. I. Miller. 1971. Small area double-heterostructure aluminum-gallium-arsenide electroluminescent sources for optical-fiber transmission lines. In *Optics Communications*, Amsterdam: Elvesier Science, Vol. 4, pp. 307–309.

Campbell, J. C. 1989. Photodetectors for long-wavelength lightwave systems. In *Optoelectronic Technology and Lightwave Communication Systems*. C. Lin, ed. Van Nostrand Reinhold, Chapter 14.

Carlisle, A. W. Small-Size high-performance lightguide connector for LANs. *Conf. Optical Fiber Commun.* '85, Proc. San Diego, CA, Optical Soc. Am., Wash., D.C., 1985.

Casey, H. C., and M. B. Panish. 1978. *Heterostructure Lasers*. New York: Academic.

Ching, Y.–C. SONET implementation. *IEEE Commun. Mag.*, Sept. 1993, pp. 34–40.

Choa, F. S., and T. L. Koch, 1991. Static and dynamical characteristics of narrow-band tunable resonant amplifiers as active filters and receivers. *IEEE/OSA J. Lightwave Technol.* 9: 73–83.

Chraplyvy, A. R. 1990. Limitations on lightwave communications imposed by optical fiber nonlinearities. *IEEE/OSA J. Lightwave Technol.* 8: 1548–1557.

Chung, Y. C., K. J. Pollock, P. J. Fitzgerald, B. Glance, R. W. Tkach, and A. R. Chraplyvy. 1988. *Electronics Lett.* 24: 1313.

Cisneros, A., S. F. Habiby, and A. E. Willner. Photonic contention resolution devices, a laboratory demonstration, Conf. Optical Fiber Communications '93 Technical Digest, San Jose, CA, Feb., 1993, Optical Soc. Am., Washington, D.C., 1993, 94–95.

Cisneros, A. Private communication.

Couch, L. W., II. 1983. *Digital and Analog Communication Systems.* New York: Macmillan.

Darcie, T. E. 1987. Subcarrier multiplexing for multiple-access lightwave networks. *IEEE J. Lightwave Technol.* 5: 1103–1110.

Darcie, T. E., R. S. Tucker, and G. J. Sullivan, 1985. *Electonics Lett.* 21: 665.

Desurvire, E., 1987. J. R. Simpson, and P. C. Becker. High-Gain erbium-doped traveling-wave fiber amplifier. *Opt. Lett.* 12: 888–890.

Desurvire, E. Erbium-Doped fiber amplifiers. Conf. Optical Fiber Communications '91, Tutorial Sessions, Optical Soc. Am., Washington, D.C., 1991.

Diaz, A. R., R. F. Kalman, J. W. Goodman, and A. A. Sawchuk. 1988. Fiber-Optic crossbar switch with broadcast capability. *Optical Eng.* 27: 1087–1095.

Dono, N. R., P. E. Green, K. Liu, R. Ramaswami, and F. F. Tong. 1990. A wavelength division multiple access network for computer communication. *IEEE J. Selected Areas Commun.* 8: 983–994.

Doran, N. J., and D. Wood. 1988. Nonlinear optical loop mirror. *Opt. Lett.* 13: 56–58.

Dragone, C. Efficient N X N couplers using Fourier optics. *IEEE/OSA J. Lightwave Technol.* 3: 467–471.

Dragone, C. 1991. An NxN optical multiplexer using a planar arrangement of two star couplers. *IEEE Photonics Technol. Lett.* 3: 812–815.

Durhuus, T., B. Mikkelsen, and K. E. Stubkjaer. 1992. Detailed dynamic model for semiconductor optical amplifiers and their cross talk and intermodulation distortion.*IEEE/OSA J. Lightwave Technol.* 10: 1056–1065.

Eisenstein, G., R. S. Tucker, U. Koren, and S. K. Korotky. 1986. Active mode-locking characteristics of InGaAsP single mode fiber composite cavity lasers. *IEEE J. Quantum Electron.* QE-22: 142–148.

Elrefaie, A., M. W. Maeda, and R. Guru. 1989. Impact of laser linewidth of optical channel spacing requirements for multichannel FSK and ASK systems. *IEEE Photon. Technol. Lett.* 1: 88–90.

Evangelides, S. G., private communication, In L. F. Mollenauer, Introduction to solitons. Conf. Optical Fiber Communications '94, Tech. Dig., Tutorial TuF, San Jose, CA, Feb. 1993, Optical Soc. Am. Washington, D.C., 1993.

Fabry, C., and A. Perot. 1899. Theorie et Applications d'Une Nouvelle de Spectroscopie Interferentielle. *Ann. Chim. Phys.* 16, 115.

Fujiwara, M., H. Nishimoto, T. Kajitani, M. Itoh, and S. Suzuki. 1991. Studies on semiconductor optical amplifiers for line capacity expansion in photonic space-division switching system. *IEEE/OSA J. Lightwave Technol.* 9: 155–160.

Fye, D. M. 1984. Practical limitations on optical amplifier performance. *IEEE/OSA J. Lightwave Tech.* 2: 403–406.

Gabla, P. M., E. Leclerc, and C. Coeurjolly. 1991. Practical implementation of a highly sensitive receiver using an erbium-doped fiber preamplifier. *IEEE Photonics Technol. Lett.* 3: 727–729.

Giallorenzi, T. G. 1978. Optical communications research and technology: Fiber optics, *Proc. IEEE* 66: 744–780.

Giles, C. R., E. Desurvire, and J. R. Simpson. 1989a. Transient gain and cross talk in erbium-doped fiber amplifiers. *Opt. Lett.* 14: 880–882.

Giles, C. R., E. Desurvire, J. L. Zyskind, and J. R. Simpson. 1989b. Noise performance of erbium-doped fiber amplifier pumped at 1.49 μm and application to signal preamplification at 1.8 Gbit/s. *IEEE Photonics Technol. Lett.* 1: 367–369.

Giles, C. R., and D. Di Giovanni. 1990. Spectral dependence of gain and noise in erbium-doped fiber amplifiers. *IEEE Photonics Technol. Lett.* 2: 797–800.

Giles, C. R., and E. Desurvire. 1991a. Propagation of signal and noise in concatenated erbium-doped fiber optical amplifiers. *IEEE/OSA J. Lightwave Technol.* 9, 147–154.

Giles, C. R., and E. Desurvire, 1991b. Modeling erbium-doped fiber amplifiers. *IEEE/OSA J. Lightwave Technol.* 9: 271–283.

Giles, C. R. System applications of optical amplifiers. Conf. Optical Fiber Communications '92, Tutorial Sessions, TuF, Optical Soc. Am. Washington., D.C., 1992.

Glance, B. 1987. Polarization independent coherent optical receiver. *IEEE/OSA J. Lightwave Technol.* 5: pp. 274–276.

Glance, B., T. L. Koch, O. Scaramucci, K. C. Reichmann, U. Koren, and C. A. Burrus. 1989. Densely spaced FDM coherent optical star network using monolithic widely frequency-tunable lasers. *Electron. Lett.* 25: 672–673.

Glance, B., J. M. Wiesenfeld, U. Koren, A. H. Gnauck, H. M. Presby, and A. Jourdon. 1992. High-Performance optical wavelength shifter. *Electron. Lett.* 28: 1714–1715.

Gloge, D. 1971. Weakly guiding fibers. *Appl. Opt.* 10: 2252–2258.

Gnauck, A. H., R. M. Jopson, and R. M. Derosier. 1993. 10-Gb/s 360 km transmission over dispersive fiber using midsystem spectral inversion. *IEEE Photonics Technol. Lett.* 5: 663–666.

Goodman, M. S., H. Kobrinski, M. P. Vecchi, R. M. Bulley, and J. L. Gimlett. 1990. The LAMBDANET multiwavelength network: Architecture, applications, and demonstrations. *IEEE J. Selected Areas Commun.* 8: 995–1004.

Gordon, J. P., and H. A. Haus. 1986. Random walk of coherently amplified solitons in optical fiber transmission. *Opt. Lett.* 11: 665–667.

Gowar, L. 1984. *Optical Communication Systems.* p. 135, Englewood Cliffs, NJ: Prentice-Hall.

Green, P. E., Jr. 1993. *Fiber Optic Networks,* Englewood Cliffs, NJ: Prentice Hall.

Haykin, S. 1983. *Communication Systems,* 2 Ed. John Wiley & Sons, New York.

Hasegawa, A., and F. Tappert. 1973. Transmission of stationary nonlinear optical pulses in dispersive dielectric fibers. *Appl. Phys. Lett.* 23: 142–144.

Haus, H. A., and Y. Lai, 1992. Narrow-Band optical channel-dropping filter. *IEEE/OSA J. Lightwave Technol.* 10: 57–62.

Henning, I. D., M. J. Adams, and J. V. Collins. 1985. Performance prediction from a new optical amplifier model, *IEEE J. Quantum Electron.* 2: 609–613.

Henry, C. H. 1986. Theory of spontaneous emission noise in open resonators and its application to lasers and amplifiers. *J. Lightwave Technol.* 4: 288–297.

Henry, P. S., R. A. Linke, and A. H. Gnauck. 1988. Introduction to lightwave systems. In *Optical Fiber Telecommunications II,* S. E. Miller and I. P. Kaminow, eds. New York: Academic.

Hill, A. M., R. Wyatt, J. F. Massicott, K. J. Blyth, D. S. Forrester, R. A. Lobbett, P. J. Smith, and D. B. Payne. 1990. Million-Way WDM broadcast network employing two stages of erbium-doped fiber amplifiers. *Electron. Lett.* 26: 1882–1884.

Hinton, H. S. 1990. Photonics in switching. *IEEE LTS* 26–35.

Hinton, H. S. 1993. *An Introduction to Photonic Switching Fabrics.* New York: Plenum.

Hui, J. Y. 1990. *Switching and Traffic Theory for Integrated Broadband Networks.* Boston: Kluwer Academic.

Inoue, K., H. Toba, and K. Nosu. 1991. Multichannel amplification utilizing an Er^{3+}-doped fiber amplifier. *IEEE/OSA J. Lightwave Technol.* 9: 368–374.

Jacobs, I. 1990. Effect of optical amplifier bandwidth on receiver sensitivity, *IEEE Trans. Commun.* 38: 1863–1864.

Joergensen, C., T. Durhuus, C. Braagaard, B. Mikkelsen, and K. E. Stubkjaer. 1993. Gb/s optical wavelength conversion using semiconductor optical amplifiers. *IEEE Photonics Technol. Lett.* 5: 657–660.

Jones, K. A. 1987. *Introduction to Optical Electronics*, Philadelphia, PA: Harper and Row, p. 68.

Jones, W. B., Jr., 1988. *Introduction to Optical Fiber Communication Systems*. New York: Holt, Rinehart, and Winston, p. 90.

Jopson, R. M. et al. 1987. Measurement of carrier density mediated intermodulation distortion in an optical amplifier. *Electron. Lett.* 23: 1394–1395.

Jopson, R. M., and T. E. Darcie. 1988. Calculation of multicarrier intermodulation distortion in semiconductor optical amplifier. *Electron. Lett.* 24: 1372–1374.

Jopson, R. M., A. H. Gnauck, B. L. Kasper, R. E. Tench, N. A. Olsson, C. A. Burrus, and A. R. Chraplyvy. 1988. 8 Gbit/s, 1.3 μm receiver using optical preamplifier. *Electron. Lett.* 25: 233–235.

Jordan, D. R., and P. L. Penney, eds. 1982. *Transmission Systems for Communications*, 5th. Ed. NJ: AT&T Bell Laboratories, p. 82.

Kahn, J. M., A. H. Gnauck, J. J. Veselka, S. K. Korotky, and B. L. Kasper. 1990. 4 Gbit/s PSK homodyne transmission system using phase-locked semiconductor lasers. *IEEE Photonics Technol. Lett.* 2: 285–287.

Kaiser, P., and D. B. Keck. 1988. Fiber types and their status. In *Optical Fiber Telecommunications II*, chapter 2, p. 40, S. E. Miller and I. P. Kaminow, eds. New York: Academic Chapter 2, p. 40.

Kaminow, I. P. 1988. Photonic local networks. In *Optical Fiber Telecommunications II*, S. E. Miller and I. P. Kaminow, eds. New York: Academic, Chapter 26, pp. 933–967.

Kaminow, I. P., P. P. Iannone, J. Stone, and L. W. Stulz. 1988. FDMA-FSK star network with a tunable optical fiber demultiplexer, *IEEE/OSA J. Lightwave Technol.* 6: 1406–1414.

Kaminow, I. P., P. P. Iannone, J. Stone, and L. W. Stulz. 1989. A tunable vernier fiber Fabry–Perot filter for FDM demultiplexing and detection. *IEEE Photonics Technol. Lett.* 1: 24–26.

Kaminow, I. P. 1990. FSK with direct detection in optical multiple-access FDM Networks. *ibid.*, 1005.

Kaminow, I. P., R. E. Thomas, and S. B. Alexander. Early results of the research consortium on wideband all-optical networks. Conf. Optical Fiber Communications '94. Tech. Dig., paper WD1, San Jose, CA, Feb. 1994, Optical Soc. Am. Washington, D.C., 1994.

Kaminow, I. P., R. E. Wagner, and A. E. Willner, eds. Technologies for a Global Information Infrastructure, Proc. IEEE LEOS Summer Topical, Keystone, CO, 1995. Piscataway, New Jersey: IEEE.

Kashima, N., 1993. *Optical Transmission for the Subscriber Loop*. Boston: Artech House, Chapters 3 and 9.

Kazovsky, L. G. Private communication.

Kazovsky, L. G., S. Benedetto, and A. E. Willner, 1996. Optical Fiber Communication Systems. Boston: Artech House.

Keck, D. B. 1976. Optical fiber waveguides. In *Fundamentals of Optical Fiber Communications*. M. K. Barnoski, ed. New York: Academic.

Killen, H. B. 1991. *Fiber Optic Communications*, Englewood Cliffs, NJ: Prentice-Hall, Chapter 3 and 4.

Klein, M. V., and T. E. Furtak. 1986. *Optics*, 2nd Ed. New York: John Wiley & Sons, p. 76.

Koch, T. L., and J. E. Bowers. 1984. Nature of wavelength chirping in directly modulated semiconductor lasers. *Electron. Lett.* 20: 1038–1039.

Koch, T. L., and U. Koren. 1991. Semiconductor photonic integrated circuits. *IEEE J. Quantum Electron.* 27: 641–653.

Koch, T. L., U. Koren, R. P. Gnall, C. A. Burrus, and B. I. Miller. 1988. Continuously tunable 1.5 μm multiple-quantum-well GaInAs/GaAsInP distributed-Bragg-reflector lasers. *Electron. Lett.* 21: 283–285.

Kodama, Y., and A. Hasegawa. 1992. Generation of asymptotically stable optical solitons and suppression of the Gordon–Haus effect. *Opt. Lett.* 17: 31.

Kogelnik, H. 1965. On the propagation of gaussian beams of light through lenslike media including those with a loss and gain variation. *Appl. Opt.* 4: 1562.

Kogelnik, H., and C. V. Shank. 1972. Coupled wave theory of distributed feedback lasers. *J. Appl. Phys.* 43: 2328.

Kondo, M., N. Takado, K. Komatsu, and Y. Ohta. 32 switch-elements integrated low cross talk Ti:LiNbO3 optical matrix switch. Conf. IOOC-ECOC '87, Tech. Dig., 1987, pp. 361–364.

Korotky, S. K. 1984. Three-Space representation of phase-mismatch switching in coupled two-state optical system. *IEEE J. Quantum Electron.* QE-22: 952–958.

Kressel, H., and J. K. Butler. 1978. *Semiconductor Lasers and Heterostructure LEDs.* New York, Academic.

Kuo, C. Y. 1992. *IEEE J. Lightwave Technol.* 10: 235.

Laming, R. I., A. H. Gnauck, C. R. Giles, M. N. Zervas, and D. N. Payne. High-Sensitivity optical preamplifier at 10 Gbit/s employing a low noise composite EDFA with 46 dB gain. Topical Meeting on Optical Amplifiers and Their Applications '92, Tech. Dig., paper PD-13, Optical Soc. Am., Washington, D.C., 1992.

Lee, T. P., and Tingye Li. 1979. Photodetectors. In *Optical Fiber Telecommunications I*, S. E. Miller and A. G. Chynoweth, eds. New York: Academic, Chapter 18.

Li, T. 1983. Advances in optical fiber communications: An historical perspective, IEEE J. Selected Areas Commun. 1: 356–372.

Li, T. private communication; modified from P. S. Henry, R. A. Linke, and A. H. Gnauck. 1988. Introduction to lightwave systems. In *Optical Fiber Telecommunications II*, S. E. Miller and I. P. Kaminow, eds. New York: Academic, Chapter 21, p. 782.

Li, T., and M. C. Teich. 1991. Bit-Error rate for a lightwave communication system incorporating an erbium-doped fibre amplifier. *Electron. Lett.* 27: 598–599.

Li, T. 1993. The impact of optical amplifiers on long-distance lightwave telecommunications. Proc. IEEE 81: 1568–1579.

Linke, R. A., and A. H. Gnauck. 1988. High capacity coherent lightwave systems. *IEEE/OSA J. Lightwave Technol.* 6: 1750–1769.

MacMillan, J. 1994. Advanced fiber optics, *U.S. News and World Report*, May 2, p. 58.

Marcuse, D. 1982. *Light Transmission Optics.* New York: Van Nostrand Reinhold.

Marcuse, D. 1990. Derivation of analytical expressions for the bit-error probability in lightwave systems with optical amplifiers. *IEEE/OSA J. Lightwave Technol.* 8: 1816–1823.

Massicott, J. F., R. Wyatt, B. J. Ainslie, and S. P. Craig-Ryan. 1990. Efficient, high power, high gain Er^{3+} doped silica fiber amplifier, *Electron. Lett.* 26: 1038–1039.

McEachern, J. A. Gigabit networking on the public transmission network *IEEE Commun. Mag.*, April 1992, pp. 70–78.

Mears, R. J., L. Reekie, I. M. Jauncy, and D. N. Payne. 1987. Low noise erbium-doped fiber amplifier operating at 1.54 μm. *Electron. Lett.* 23: 1026.

Mecozzi, A., J. D. Moores, H. A. Haus, and Y. Lai. 1991. Soliton transmission control. *Opt. Lett.* 16: 1841.

Mestdagh, D. J. G. 1995. *Fundamentals of Multiaccess Optical Fiber Networks.* Chapter 8, Boston: Artech House.

Midwinter, J. E. 1979. *Optical Fibers for Transmission.* New York: John Wiley & Sons.

Midwinter J. E. 1993. *Photonics in Switching*, New York: Academic, Vols. I and II.

Miniscalco, W. J. 1991. Erbium-Doped glasses for fiber amplifiers at 1500 nm. *IEEE J. Lightwave Technol.* 9: 234–250.

Mollenauer, L. F., R. H. Stolen, and J. P. Gordon. 1980. Experimental observation of picosecond pulse narrowing and solitons in optical fibers. *Phys. Rev. Lett.* 45: 1095–1097.

Mollenauer, L. F., J. P. Gordon, and M. N. Islam. 1986. Soliton propagation in long fibers with periodically compensated loss. *IEEE J. Quantum Electron.* QE-22: 157–173.

Mollenauer, L. F., S. G. Evangelides, and J. P. Gordon. 1991. Wavelength division multiplexing with solitons in ultra long distance transmission using lumped amplifiers. *IEEE J. Lightwave Technol.* 9: 362.

Newkirk, M. A., B. I. Miller, U. Koren, M. G. Young, M. Chien, R. M. Jopson, and C. A. Burrus. 1993. 1.5 μm multiquantum-well semiconductor optical amplifier with tensile and compressive strained wells for polarization-independent gain. *IEEE Photonics Technol. Lett.* 4: 406–408.

Nyman, B. M., S. G. Evangelides, G. T. Harvey, L. F. Mollenauer, P. V. Mamyshev, M. Saylors, S. K. Korotky, U. Koren, V. Mizrahi, T. A. Strasser, J. J. Veselka, J. D. Evankow, A. J. Lucero, J. A. Nagel, J. W. Sulhoff, J. L Zyskind, P. C. Corbett, M. A. Mills, and G. A. Ferguson. Soliton WDM transmission of 8 × 2.5 Gb/s, error free over 10 megameters. Conf. Optical Fiber Communications '94, Tech. Dig., paper PD21, San Diego, CA, March 1995, Optical Soc. Am., Washington, D.C., 1995.

OKI Corp. 1992. Product Brochure.

Olsson, N. A. 1989. Lightwave systems with optical amplifiers. *IEEE/OSA J. lightwave Technol.* 7: 1071–1082.

Olsson, N. A., R. F. Kazarinov, W. A. Nordland, C. H. Henry, M. G. Oberg, H. G. White, P. A. Garbinski, and A. Savage. 1989. Polarization independent optical amplifier with buried facets. *Electron. Lett.* 25: 1048.

O'Mahony, M. J. 1988. Semiconductor laser optical amplifiers for use in future fiber systems. *IEEE/OSA J. Lightwave Technol.* 6: 531–544.

Oshima, K., T. Kitayama, M. Yamaki, T. Matsui, and K. Ito. 1985. Fiber-Optic local area passive network using burst TDMA scheme. *IEEE/OSA J. Lightwave Technol.* 3: 502–510.

Olshansky, R., V. A. Lanzisera, and P. M. Hill. 1989. Subcarrier multiplexed lightwave systems for broad-band distribution. *IEEE J. Lightwave Technol.* 7: 1329–1342.

Palais, J. C. 1988. *Fiber Optic Communications*, 2nd ed. Englewood Cliffs, NJ: Prentice-Hall, pp. 152 and 159.

Patel, J. S., M. A. Saifi, D. W. Berreman, C. Lin, N. C. Andreadakis, and S. D. Lee. 1990. Electrically tunable optical filter for infrared wavelengths using liquid crystals in a Fabry–Perot etalon. *Appl. Phys. Lett.* 57: 1718–1720.

Personick, S. D. 1977. Receiver design for optical fiber systems. *Proc. IEEE*, 65: 1670–1678.

Personick, S. D. 1985. *Fiber Optics: Technology and Applications*. New York: Plenum.

Photonics Corp. 1994. *Product Catalog*, San Jose, CA.

Prucnal, P. R., M. A. Santoro, and T. R. Fran. 1986a. Spread spectrum fiber-optic local area network using optical processing, *IEEE/OSA J. Lightwave Technol.* 4: 547.

Prucnal, P. R., M. A. Santoro, and S. K. Sehgal. 1986b. Ultrafast all-optical synchronous multiple-access fiber networks. *IEEE J. Selected Areas Commun.* 4: 1484–1493.

Roden, M. S. 1995. *Analog and Digital Communication Systems*, 4th Ed., Upper Saddle River, NJ: Prentice-Hall.

Runge, P. K., and N. S. Bergano. 1988. Undersea cable transmission systems. In *Optical Fiber Telecommunications II*. S. E. Miller and I. P. Kaminow, eds. New York: Academic, Chapter 24.

Ryu, S., H. Taga, S. Yamamoto, K. Mochizuki, and H. Wakabayashi. 1989. *Electron. Lett.* 25: 1682.

Saitoh, T. et al. 1986. Recent progress in semiconductor laser amplifiers. *IEEE/OSA J. Lightwave Technol.* 6: 1656–1664.

Saitoh, T., and T. Mukai. 1987. 1.5 μm GaInAsP traveling-wave semiconductor laser amplifier. *IEEE J. Quantum Electron.* 23: 1010.

Saleh, A. A. M. 1989. Fundamental limit on number of channels in subcarrier-multiplexed lightwave CATV system. *Electron. Lett.* 25: 776–777.

Saleh, A. A. M., R. M. Jopson, J. D. Evankow, and J. Aspell. 1990. Modeling of gain in erbium-doped fiber amplifiers. *IEEE Photonics Technol. Lett.* 2: 714–717.

Saleh, B. E. A., and M. C. Teich. 1991. *Fundamentals of Photonics*. New York: John Wiley & Sons, p. 801.

Salehi, J. A., A. M. Weiner, and J. P. Heritage. 1990. *IEEE/OSA J. Lightwave Technol.* 8: 478.

Sato, K., 1983. Intensity noise of semiconductor laser diodes in fiber optic analog video transmission. *IEEE J. Quantum Electron.* QE-19: 1380.

Schwartz, M. 1980. *Information Transmission, Modulation, and Noise*, 3rd Ed. New York: McGraw-Hill.

Schwartz, M. 1987. *Telecommunication Networks, Protocols, Modeling, and Analysis*. New York: Addison-Wesley.

Senior, J. M. 1985. *Optical Fiber Communications*. New York: Prentice-Hall.

Shimizu, M. et al. 1993. 28.3 dB gain 1.3 μm-band, pr-doped fluoride fiber amplifier module pumped by 1.017 μm InGaAs-LDs, *IEEE Photonics Technol. Lett.* 5: 654–657.

Siegman, A. E. 1986. *Lasers*, Mill Valley, CA: University Science Books.

Simon, J. C. 1987. GaInAsP semiconductor laser amplifiers for single mode fiber communication. *IEEE/OSA J. Lightwave Technol.* 5: 1286–1295.

Smith, D. A., J. E. Baran, J. J. Johnson, and K.-W. Cheung. 1990. Integrated-Optic acoustically-tunable filters for WDM networks. *IEEE J. Selected Areas Commun.*, 8: 1151–1159.

Soole, J. B. D., A. Scherer, H. P. LeBlanc, N. C. Andreadakis, R. Bhat, and M. A. Koza. 1991. *Appl. Phys. Lett.* 58: 1949.

Spiegel, M. R. 1968. *Mathematical Handbook*. Shaum's Outline Series in Mathematics. New York: McGraw-Hill, p. 183.

Stephen, T. D., and G. Nicholson. 1987. Optical homodyne receiver with a six-port fiber coupler. *Electron. Lett.* 22: 1106–1107.

Stone, J., and C. A. Burrus. 1973. Neodymium doped silica lasers in end-pumped fiber geometry. *Appl. Phys. Lett.* 23: 388.

Stone, J., and L. W. Stulz. 1987. Pigtailed high-finesse tunable fiber Fabry–Perot interferometers with large, medium and small free spectral ranges. *Electron. Lett.* 23: 781–783.

Stubkjaer, K., and M. Danielsen, 1980. Nonlinearity of GaAlAs lasers-Harmonic distortion, *IEEE J. Quantum Electron.* QE-16: 531.

Sunak, H. R. D., Fundamentals of erbium-doped fiber amplifiers. Soc. Photo-Instrumentation Eng. OE/Fibers '92 Conf., Short Course, SC9, SPIE, Bellingham, WA, 1992.

Suzuki, Y., K. Magari, M. Ueki, T. Amano, O. Mikami, and M. Yamamoto. 1993 High-Gain, high-power 1.3 μm compressive strained MQW optical amplifier. *IEEE Photonics Tech. Lett.* 4: 404–406.

Taga, H., N. Edagawa, H. Tanaka, M. Suzuki, S. Yamamoto, H. Wakabayashi, N. S. Bergano, C. R. Davidson, G. M. Homsey, D. J. Kalmus, P. R. Trischitta, D. A. Gray, and R. L. Maybach. 10 Gb/s, 9000 km IM-DD transmission experiment using 274 erbium-doped fiber amplifier repeaters. Conf. Optical Fiber Commun. '93, paper PD-1, Optical Soc. Am., Washington, D.C., 1993.

Tamir, T. 1979. Beam and waveguide couplers. In *Integrated Optics, Topics in Applied Physics*, T. Tamir, ed. New York: Springer-Verlag, Vol. 7, chapter 3.

Tkach, R. W. Strategies for coping with fiber nonlinearities in lightwave systems. Conf. on Optical Fiber Communications '94, Technical Digest, Tutorial ThO1, San Jose, CA, Feb., 1994, Optical Soc. Am. Washington, D.C., 1994.

Tkach, R. W., R. M. Derosier, A. H. Gnauck, A. M. Vengsarkar, D. W. Peckham, J. L. Zyskind, J. W. Sulhoff, and A. R. Chraplyvy. 1995. Transmission of eight 20-Gb/s channels over 232 km of conventional single-mode fiber. *IEEE Photonics Technol. Lett.* 7: 1369–1371.

Tucker, R. S., G. Eisenstein, S. K., Korotky, G. Raybon, J. J. Veselka, L. L. Buhl, B. L. Kasper, R. C. Alferness. 1987. 16 Gbit/s fiber transmission experiment using optical time-division multiplexing. *Electron. Lett.* 23: (24):

Uno, T., M. Mitsuda. J. Ohya, Low distortion characteristics in amplified 4 × 50-channel WDM AM SCM transmission. Conf. on Optical Fiber Commun. '94, Tech. Dig., paper WM2, San Jose, CA, Feb. 1994, Optical Soc. Am. Washington, D.C., 1994.

Vassallo, C. 1989. Gain ripple minimization and higher-order modes in semiconductor optical amplifiers. *Electron. Lett.* 25: 789–790.

Way, W. I. 1989. Subcarrier multiplexed lightwave system design considerations for subscriber loop applications. *IEEE J. Lightwave Technol.* 7: 1806–1818.

Way, W. I., M. W. Maeda, A. Y. Yan, M. J. Andrejco, M. M. Choy, M. Saifi, and C. Lin, 1990. 160-Channel FM-Video transmission using FM/FDM and subcarrier multiplexing and an erbium doped optical fiber amplifier. *Electronics. Lett.* 26: 139.

Way, W. I., A. C. Von Lehman, M. J. Andrejco, M. A. Saifi, and C. Lin. Noise figure of a gain-saturated erbium-doped fiber amplifier pumped at 980 nm. Topical Meeting on Optical Amplifiers and Their Applications '91 Tech. Dig., paper TuB3, Monterey, CA, Optical Soc. Am., Washington, D.C., 1991.

Way, W. I., D. A. Smith, J. J. Johnson, and H. Izadpanah. 1992. A self-routing WDM high-capacity SONET ring network. *IEEE Phonics Technol. Lett.* 4: 402–405.

Wiesenfeld, J. M., and B. Glance. 1992. Cascadability and fanout of semiconductor optical amplifier wavelength shifter. *IEEE Photonics Technol. lett.* 4: 1168–1171.

Willner, A. E. 1990. Simplified model of a FSK-to-ASK direct-detection system using a Fabry-Perot demodulator. *IEEE Photonics Technol. Lett.* 2: 363–366.

Willner, A. E., E. Desurvire, H. M. Presby, C. A. Edwards, and J. Simpson, 1990a. LD-Pumped erbium-doped fiber preamplifiers with optimal noise filtering in a FDMA-FSK 1 Gb/s star network. *IEEE Photonics Technol. Lett.* 2: 669–672.

Willner, A. E., I. P. Kaminow, M. Kuznetsov, J. Stone, and L. W. Stulz. 1990. 1.2 Gb/s closely spaced FDMA-FSK direct-detection star network. *IEEE Photon. Technol. Lett.* 2: 223–226.

Willner, A. E., and E. Desurvire. 1991. Effect of gain saturated on receiver sensitivity in 1 Gb/s multichannel FSK direct-detection systems using erbium-doped fiber preamplifiers. *IEEE Photonics Technol. Lett.* 3: 259–261.

Yamamoto, Y. 1980. Noise and error rate performance of semiconductor laser amplifiers in PCM-IM optical transmission systems. *IEEE J. Quantum Electron.* QE-16: 1073–1081.

Yariv, A. 1975. *Quantum Electronics*, 2nd Ed. Wiley, New York: John Wiley & Sons, Chapter 8.

Yariv, A. 1976. *Introduction to Optical Electronics*, 2nd Ed. New York: Holt, Rinehart, and Winston, p. 98.

Yariv, A. 1991. Optical Electronics, 4th Ed. Philadelphia, Pa.: Holt, Rinehart, and Winston, Appendix D.

Young, W. C. 1989. Optical fiber connectors, splices, and joining technology. In *Optoelectronic Technology and Lightwave Communication Systems*, C. Lin, ed., Van Nostrand Reinhold, Chapter 6.

Zah, C. E., J. S. Osinski, C. Caneau, S. G. Menocal, L. A. Reith, J. Salzman, F. K. Shokoohi, and T. P. Lee. 1987. *Electron. Lett.* 23: 990.

Zyskind, J. L., D. J. DiGiovanni, J. W. Sulhoff, P. C. Becker, and C. H. Brito Cruz. High-Performance erbium-doped fiber amplifier pumped at 1.48 μm and 0.97 μm. Topical Meeting on Optical Amplifiers and Their Applications '91, Tech. Dig., paper PDP6, Monterey, CA, Optical Soc. Am. Washington, D.C., 1991.

Zyskind, J. L., J. W. Sulhoff, J. Stone, D. J. DiGiovanni, L. W. Stulz, H. M. Presby, A. Piccirilli, and P. E. Pramayon. 1991b. Diode-Pumped, electrically tunable erbium-doped fiber ring laser with fiber Fabry–Perot etalon. *Electron. Lett.* 27: 1950.

Zyskind, J. L. et al., Fast gain transients in EDFA cascades. Conf. Optical Fiber Communications, Post-Deadline Proceedings, San Jose, CA, Optical Soc. Am. Washington, D.C., 1996.

Zah, C. E., F. J. Favire, B. Pathak, R. Bhat, C. Caneau, P. S. D. Lin, A. S. Gozdz, N. C. Andreadakis, M. A. Koza, and T. P. Lee, 1992. Monolithic integration of multiwavelength compressive-strained multiquantum-well distributed-feedback laser array with star coupler and optical amplifiers, *Electron. Lett.* 28: 2361–2362.

The first of these disadvantages is being addressed by development work on integrated optics and micromechanical actuation devices, higher power solid-state lasers or semiconductor diode lasers at shorter wavelengths, and improved schemes for direct overwrite. Advances in these areas will benefit bit-by-bit serial recording approaches, if they arrive in time, or, in the longer term, they will be a boon for implementing compact parallel recording devices (for example, holographic systems). The cost issue is essentially a "chicken and egg" situation. Nothing drives cost down more effectively than volume production. However, to penetrate user applications on a massive scale requires clear performance advantages coupled with competitive costs, or else a clear cost advantage. To date, it has proven difficult to achieve costs/prices for read/write optical disk drives competitive enough to enable them to penetrate mass applications, such as personal computer drive slots. It may require either a production cost breakthrough or manufacturer risk-taking coupled with undeniably advantageous features or applications to push optical disk drives over the threshold to mass acceptance.

The noncontacting nature of optical recording is increasingly perceived as the key differentiating feature compared to magnetic storage systems, where the gradient of the dipole magnetic field requires that the recording transducer be in intimate proximity, or virtual contact, with the surface of the storage medium. The exceptional reliability of optical storage technology is inherent in the use of an optical stylus, which completely avoids any mechanical contact during the write/read process. The distance between the focusing lens and disk surface is on the order of a millimeter, which also enables removability and enhanced archivability. New multifunction optical disk drive systems, offering the choice of read-only (ROM), write-once-read-many (WORM), and rewritable disk memories are facilitating new applications not otherwise possible with strictly (rewritable) magnetic storage systems. The "borrowing" of components already designed, tested, and proven for magnetic hard drives is an effective strategy for closing the price gap and improving performance versus magnetic hard disk drives. Already, optical disk media are competitive with magnetic disk media on a cost per bit basis.

The mechanism of writing on an optical disk or tape involves focusing a semiconductor diode laser beam, whose light is absorbed and converted into thermal energy. The typical laser-beam power at the recording layer is about 10 mW, the focused spot size is about one micron, and the beam duration during writing is on the order of 50 nsec. This provides a power density of 1 MW/cm^2 or an energy density of 50 mJ/cm^2 at the recording layer. If all of the power is absorbed in a typical recording medium thickness of 100 nm, a temperature rise of several hundred °C can be generated in a few nsec. This rise in temperature induces a physical, chemical, or magnetic change in the film and produces an optical contrast between recorded and nonrecorded regions. This is the basic principle of optical information storage, whether in the form of disk, tape, card, etc. The physical change can be ablation of material, substrate deformation, bleaching, crystalline to amorphous phase change, or reversal of magnetic polarity, to name a few. Figure 17.1 schematically illustrates some important recording media configurations. In the case of ablation, if there is a metal reflector under the ablated film, the ablated region will have higher reflectivity (due to lower light absorption) than the adjacent area, and this difference in reflectivity indicates the location of marks. For ablative systems, organic dyes and some metals have been used for recording. In the case of phase change, if the coated film is crystalline, then the laser beam can form amorphous regions, which have lower reflectivity, again providing optical contrast for locating recorded marks. The phase change medium can be erasable if the reverse transition from amorphous to crystalline can be achieved in practical time scales of less than a microsecond. In magneto-optical (MO) recording, the writing process involves heating the irradiated region sufficiently close to a magnetic ordering temperature, e.g. T_{Curie} or T_{Neel}, so that the magnetic polarity can be switched by a small, reverse-applied magnetic field. This is the functional equivalent of thermally assisted magnetic recording and, like magnetic recording, has been shown to exhibit virtually unlimited rewrite cyclability. Readout contrast is achieved by exploiting the magneto-

17
Optical Data Storage

Charles F. Brucker
Eastman Kodak Company

Terry W. McDaniel
IBM Corporation

Mool C. Gupta
Eastman Kodak Company

17.1 Introduction 719
17.2 Design and Modeling 723
   Optical Thin-Film Design • Thermal Design and Modeling • Optical Design and Modeling
17.3 Optical Storage Media 733
   Compact Disk • Write-Once Ablative Media • Phase-Change Media • Magneto-Optical Media • Substrates • Optical Tape
17.4 Optical Storage Systems 748
   Optical Heads • Data Encoding and Decoding • ISO Standards
17.5 Future Optical Storage 755
   Heads and Systems • Media • Holographic Storage

17.1 Introduction

Optical disk storage technology for data storage typically finds application as a computer, video, or audio peripheral device that competes with other storage devices over a full spectrum of attributes including entry cost, media cost, media removability, form factor, power consumption, data rate, data accessing, reliability, ruggedness, interface support, and standardization. Optical tape drives and optical card readers would have a similar list of attributes. Different subsets of these attributes claim priority for different applications. By general consensus, the strongest attributes of optical disk storage technology are:

- Unlimited storage capacity because of media removability; cost per bit can become negligible as capacity per optical disk drive rises;
- Reliability—the media is very long-lived and the head-disk interface has extreme robustness because the separation is several orders of magnitude greater than that in magnetic recording;
- Random data access, providing near-line access to vast data capacity;
- Extremely robust servo systems and data error correction capability.

On the other hand, present-day optical disk drives suffer some disadvantages relative to other random-access data storage devices, specifically, hard disk drives that utilize magnetic recording:

- Optical disk drives are slower in data access because of larger head mass and lower disk rotation speed, and data throughput is less, due to unavailability or awkwardness of direct overwrite, limited writing laser power, and constraints on linear data density;
- Optical disk drive entry cost is somewhat higher because of drive cost, driven mainly by the cost of fabricating and precisely aligning bulky optical components, and also by the relatively low production volumes for optical disk drives.

Optical Data Storage

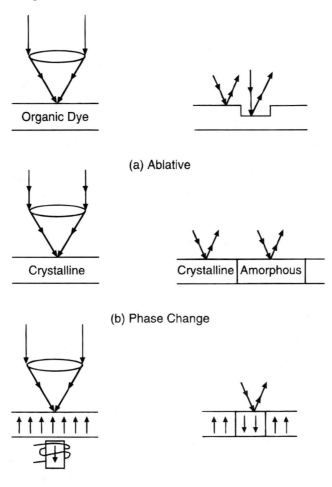

FIGURE 17.1 Schematic illustrations of (a) ablative write-once recording using an organic dye, (b) write-once recording using a crystalline-to-amorphous, phase-change medium, and (c) rewritable recording using magneto-optic media.

optical Kerr effect using polarization-sensitive detection optics. For all media types, the readout-beam power is chosen to be safely below the writing threshold, typically on the order of a milliwatt.

Organic dye-based and phase change WORM, magneto-optical rewritable, and other optical disk storage systems have been commercialized (Table 17.1) and are available in one or more of several formats designated by disk diameter: 64 mm (2-1/2″), 86 mm (3-1/2″), 120 mm (CD), 130 mm (5-1/4″), 304 mm (12″), and 356 mm (14″). A "low-end" storage system might consist of a CD-ROM drive and single CD-ROM disk. A typical CD has a track pitch of 1.6 μm, a recorded mark length range of about 1 to 3 μm, and a capacity (one side) of about 650 Mbytes. At the "high-end", complete library systems based on the 356-mm format are available with a disk capacity of about 15 GByte and a total system capacity exceeding 1 terabyte. Many choices are available to satisfy requirements for on-line capacity, total capacity, data rate, access performance, unalterability or rewritability, and cost. Among the smaller diameters (≤130 mm), it is well recognized that so many choices create confusion and raise costs due to lack of standardization. It will be interesting to see whether the need to be more interchangeable and manufacturing cost

TABLE 17.1 Mass Storage Matrix Comparing Basic System Attributes and Prices to the User for Several Optical and Magnetic Storage Formats (Prices as of Late 1995/Early 1996). Useful Sources for this Type of Information are the Freeman Reports (Santa Barbara, CA), the Disk/Trend Reports (Mountain View, CA), and the Santa Clara Consulting Group Reports (Santa Clara, CA)

Type of Storage	Form Factor	Capacity	Access Time (msec)	Data Transfer Rate* (MB/sec)	Drive Price ($US)	Recording Media Price ($US)	Media Price per MB ($US)
Rewritable optical disk (MO)	3.5"	128 MB	33–136	0.4–0.7	350	10	0.078
	5.25" (1×)	650 MB	25–77	0.6–2.1	1200	43	0.066
	5.25" (2×)	1.3 GB	36	1.6	1800	100	0.077
	5.25" (4×)	2.6 GB			2200	130	0.050
Write-Once optical disk	5.25"	650 MB	60–125	0.5–1.4	1,250	100	0.15
	12"	6.5 GB	100–750	0.3–1.6	19,000	800	0.12
	14"	14.8 GB	170	1	40,000	925	0.063
CD-ROM (4× drive)	120 mm (4.72")	640 MB	120–250	0.6	100	3	0.0047
CD-R (4× drive)	120 mm	640 MB	500	0.6	1400	10	0.016
Minidisc	2.5"	140 MB	300	0.15	400	20	0.14
PD	120 mm	650 MB	165	0.8	500	40	0.062
Optical tape reel	880 m (12" dia × 35 mm)	1 TB	65,000	3	250,000	10,000	0.010
Hard disk drive	3.5"	420 MB	15	1.5	100	NA	0.24
	5.25"	1–2 GB	12–16	2–4	600	NA	0.30
8-mm magnetic tape helical scan	5.25"	2.3–7 GB	30,000	0.25–0.5	2000	15	0.0021

*Sustained read

Optical Data Storage

effective will force convergence toward just one or two of these formats, e.g., the 120-mm form factor of the ubiquitous CD.

Historically, progress in the areal density achievable by optical recording has been closely related to laser development. A perspective on the technology migration path specific to 130-mm magneto-optic media in Figure 17.2 shows previous and anticipated improvement of areal density over time. In addition to decreasing laser wavelength, improvements in the substrate, format, recording media, and data channel can be expected to significantly impact user areal density. This chapter reviews both the current state of the art and newly proposed technologies for advancing optical storage solutions.

17.2 Design and Modeling

In this section, we give an introductory treatment of some areas of physical design in optical disk storage systems in which computer modeling has proven particularly useful. These topics can be associated with two of the primary components of the recording system-the storage medium and the recording/readout head. Because the currently predominant optical storage medium takes the form of a rotating disk, this section will deal exclusively with disk media formats. Nevertheless, many of the physical design issues related to fundamental optics, thin film optical behavior, and medium thermal performance can be easily translated to the case of nonrotating media, such as optical tape or cards. Examples of commercially successful optical

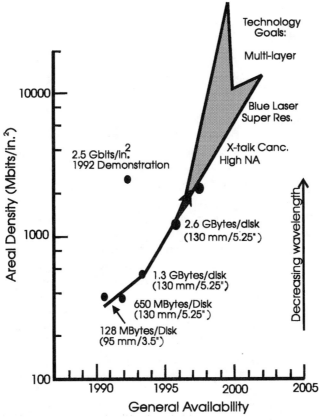

FIGURE 17.2 Technology migration path for magneto-optical recording, showing technology approaches for future density increases. After Lenth (1994).

storage media solutions made according to these design principles are discussed in section 17.3. Our discussion of optical heads is virtually independent of the physical configuration of the recording medium. Practical aspects of commercial head design and performance are discussed in Section 17.4.

The primary design issues associated with multilayer thin film stacks for optical data storage media relate to

(a) optical response, primarily for readout, but secondarily for writing and erasing;
(b) thermal response, primarily for writing and erasing, but secondarily for readout;
(c) material chemical and physical stability suitable for archival purposes; and
(d) cost and manufacturability.

Considerations (a) and (b) are particularly amenable to analysis and optimization with computer models, and we focus on them here, leaving discussion of items (c) and (d) to section 17.3 on media. Although we touch on most topics of importance, due to space constraints, we use references to more detailed treatments as an important part of our presentation for the reader who requires greater depth.

We would like to recommend a few general sources to supplement the abbreviated treatments of optical and thermal design. For optical response, the reader is referred to Chapters 5–8 in Mansuripur (1995); McDaniel and Bartholomeusz (1992); Marchant (1990), and Chapter 13 in Reference 4. For thermal design issues see Chapters 11 and 17 in Mansuripur (1995), McDaniel and Bartholomeusz (1996) and Marchant (1990), Chapters 5 and 13 in Hurst and McDaniel (1996), and Connell and Bloomberg (1988).

Optical Thin-Film Design

Beam-addressable, sequential bit-by-bit optical recording utilizes a focused laser directed onto a moving storage medium. The primary optical parameters of importance to the interaction of the beam with the medium for storage or readout are reflectance and absorption. A storage process encompasses writing and erasing, which, by definition, perform a physical alteration of the medium that may or may not be reversible. (Neither of these processes is performed by a user for read-only systems such as CD-ROM, a data format very similar to CD Audio. These media have information impressed on them at the time of manufacture.) A readout or reading process is intended to detect information nondestructively on the medium, and no irreversible physical change of the medium should occur. Thus, the processes of storage and readout are distinguished by different regimes of interaction strength between the laser and the medium.

Today, all optical disk writing and erasing processes are thermally driven, meaning that a temperature rise resulting from absorption of optical energy is responsible for the physical change induced in the medium. The only optical parameter adjusted by the user between storage and readout is the level of optical power directed onto the medium. It follows from this line of reasoning that the thermal effect induced in the medium during readout must be held below some threshold value. In terms of optical design, one must choose a medium absorption that allows the user to conveniently span the necessary thermal regimes with available laser power. Our discussion suggests that thermal effects can be ignored in analyzing readout (once we assure that reading power does not lift us over the threshold of medium thermal "damage"). Conversely, for storage processes, thermal analysis is paramount, and the only pertinent optical issue is whether enough of the available laser power is absorbed to induce the required heating for writing or erasure.

Turning to the principle optical issues in readout, we note that virtually all optical storage systems detect written or imprinted information as a modulation of reflected or recollected light power from the medium. Writing typically creates a "mark" (physical alteration) against a uniform background. The mark may have simple reflective contrast to the background, or it may be a

Optical Data Storage

topographical feature that is a diffractive element. In the case of MO media, the mark is a reversed magnetic domain which produces a modulated rotation of the direction of the incident linear polarization (MO Kerr effect). The modulated polarization state is converted to a detected power modulation by a suitable configuration of polarization-sensitive optics in the head (see Optical Storage Systems).

The analysis of readout optical response can be greatly simplified from the actual configuration used by considering only normally incident rays. Optical storage systems typically combine available low-cost laser diode radiation (λ = 650–850 nm) with intermediate numerical aperture optics (NA = 0.4–0.6) to achieve focused spot sizes in the submicron range (full width at half-maximum intensity FWHM = 0.65–1.0 μm). This implies that peripheral rays in a converging cone of a focused spot may approach 20° from the normal direction to the films deposited on the disk substrate (polycarbonate or glass, the incident medium adjacent to the films). However, this beam is formed from a Gaussian weighting of intensity across the focusing lens aperture, with the population of rays in the full beam centered about 0° (normal) incidence. Numerous studies (see Chapter 7 of Mansuripur, 1995 and Sarid, 1988) have established that, for this class of optical systems, the error in integrated beam effects associated with treating only normally incident rays and neglecting the effects of peripheral rays is very minor. Consequently, much of the detailed analysis of Chapters 5 and 7 of Mansuripur (1995) can be bypassed. The consideration of the interaction of a normally incident ray with a multilayer film stack is considerably simplified, and one can instead invest modeling effort in studying a wide range of film structures in search of optimally performing ones (McDaniel et al. 1991; Atkinson et al. 1992; McDaniel et al. 1994).

Reflectance R and absorption A are primary design parameters for an optical disk. They are not independent, however, because conservation of energy requires that $R + T + A = 1$, with T being transmittance. For single storage layer optical disks, T represents wasted light, so usually T will be made close to zero. T is practically zero for a quadrilayer MO disk, for example, and then $A = 1 - R$, leaving just one of the three parameters independent. Common constraints of the film stack design include the following:

- At the operating wavelength, R and A must be adjusted to allow writing/erasing with available write laser power and enable tracking with limited laser read power;
- The desired modulation of R between the background film and the written marks must be accommodated;
- One or more additional optical (or magneto-optical) parameters must also lie in a desired range for acceptable performance (e.g., Kerr rotation and ellipticity).

The problem of calculating the classical optical response of a thin film stack for normally incident rays is well-documented (Mansuripur, 1995, Chapter 5; Heavens, 1955). The physical issues are straightforward—given a set of film materials and their complex refractive indices, one needs to account for the cumulative effects of partial reflection and transmission at all film interfaces plus the effect of propagative phase delay through film thicknesses. Either matrix or iterative algebraic formulas are compatible with computer evaluation. Figure 17.3 illustrates a variety of typical responses from a four-layer (quadrilayer) film stack containing an MO storage layer. The MO signal from a differential detection head (see Optical Storage Systems) is expressed by $S_{MO} = cR \sin 2\theta_{K,max} \cos \Delta\varphi$, where c is a constant, $\theta_{K,max}$ is the maximum Kerr rotation (occurring at 0° phase shift), and $\Delta\varphi$ is the phase shift of the Kerr polarization component light relative to the unrotated, reflected Fresnel component. Alternatively, one can write $S_{MO} = cR \sin 2\theta_K \cos 2\epsilon_K$, where ϵ_K is the Kerr ellipticity. The MO Kerr effects and the phase shift $\Delta\varphi$ are thus interrelated by $\sin 2\theta_{K,max} \cos \Delta\varphi = \sin 2\theta_K \cos 2\epsilon_K$, an identity among the parameters of polarized light.

A quadrilayer film stack with an MO layer, as shown in Fig. 17.4, is a good example for discussion of optimization of an optical storage medium for readout. First, light is usually incident

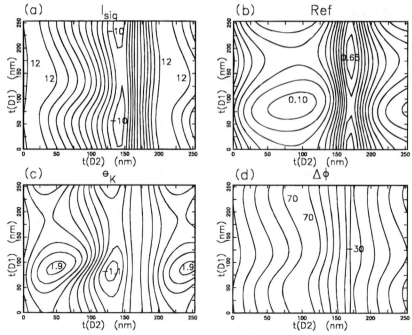

FIGURE 17.3 Optical and MO responses of a quadrilayer structure versus the variation of thicknesses of the encapsulating dielectrics, D1 and D2 (1.3X optical period); (a) MO signal current (arbitrary units); contours range from -10 to 12 in steps of 2; (b) reflectance in steps of 0.05 from 0.10 to 0.65; (c) Kerr rotation Θ_k in steps of $0.3°$ from $-1.1°$ to $1.9°$; (d) MO phase shift $\Delta\phi$ in steps of $20°$ from $-30°$ to $70°$.

on the storage films from an optically thick substrate; 0.6–1.2 mm is a common thickness range. Given that the depth of focus λ/NA^2 of intermediate NA optical heads is, at most, a few micrometers, the substrate serves as a protective cover keeping dust, debris, and handling damage, such as scratches away from the focal plane. The thickness of the first dielectric material deposited on the substrate is chosen to adjust the stack reflectance to an optimal value for good readout performance. The metallic storage film is chosen thick enough to produce adequate MO response (Kerr effect), but yet thin enough to have moderate transmittance (5–40%). This transmissive property is helpful in allowing some light to propagate onto the reflector film that caps the deposited stack, thereby creating a potentially resonant cavity in the stack. A useful resonance phenomenon in this application arises from multiple transmissions through the MO film to produce additional MO (Faraday) effects supplementing the reflective Kerr effect. (The reflector is thick enough–usually 50 nm or more–to insure that the stack transmittance $T \approx 0$.) Design analysis and experiment confirm that 1–5 dB signal enhancement relative to a single reflective MO layer can be realized by optimally tuning a "resonant" quadrilayer stack. A key part of the tuning is adjusting the dielectric thicknesses for optimal phasing for reflectance and the Kerr effect, which can result in an optimal value of signal or signal-to-noise ratio (SNR) (McDaniel and Bartholomeusz (1996); Atkinson et al. (1993)).

Finally, we note that the structure in Fig. 17.4 is well-suited for two additional purposes: (a) encapsulation of a highly chemically reactive MO material, often a rare earth-transition metal (RE-TM), to provide a stable archival storage medium, and (b) provision of a structure with considerable thermal design flexibility (next section). Most of the thin-film optical concepts introduced here, e.g., the use of antireflective overcoats and/or reflective and phase-adjusting backcoats, are applicable and useful for other optical storage thin-film designs (write once, phase change rewritable, etc.).

Optical Data Storage

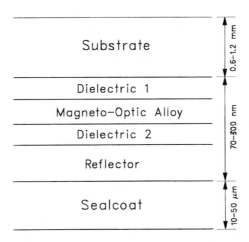

FIGURE 17.4 Generic quadrilayer structure with MO film.

Thermal Design and Modeling

Laser-beam heating of a moving film medium is well described by classical heat conduction and the Fourier law of heat transfer $Q = -k\nabla T$, where Q is heat flux, k is thermal conductivity, and T is temperature. The thermal performance of thin-film optical storage media must satisfy several requirements:

(a) adequate thermal efficiency to allow temperature elevation for writing and erasure at available laser powers;
(b) limited lateral heat spreading to minimize thermal crosstalk between adjacent marks or tracks; thermal crosstalk can contribute to unwanted intersymbol interference in written data;
(c) avoid excessive peak temperature rise to prevent medium damage or even excessive irreversible property changes;
(d) effectively dissipate heat imparted to the medium by the readout beam to prevent irreversible medium damage or partial erasure of written data.

Requirement (a) conflicts with (b) and (d), in general. Consequently, trade-offs and compromise are necessary. Item (b) is usually achieved by designing a low thermal resistance path from the MO film to the reflector, as in Fig. 17.4, and this tends to diminish the "efficiency" called for in (a).

The heating of thin-film structures with laser irradiation has been studied for applications ranging from laser annealing and machining to optical data storage (Madison and McDaniel (1989); Holtslag (1989)). Analytical methods are numerous (Carslaw and Jaeger (1988)), but their utility is usually limited to simple geometries and approximate estimates. Nevertheless, analytical results are very helpful for understanding the fundamental physics of heat flow and certain limiting conditions. One interesting benchmark is the adiabatic limit wherein one supposes that all of the energy absorbed from incident irradiation is converted into local thermal energy without diffusion loss. The assumption of no diffusion loss of heat from the incident radiation spot volume is very unrealistic, but this limit establishes an upper bound on estimated temperature rise for a given irradiation. The expression for temperature rise in this limit is $\Delta T = Q/C = P_{inc}(1 - R)\Delta t/cm = P_{inc}(1 - R)\Delta t/c\rho V$, where Q is the heat deposited, C is the heat capacity of the media layer(s), P_{inc} is the laser power incident on the disk, R is the reflectance, Δt is the laser pulse duration, c is the specific heat, m is the mass of the volume heated (without diffusion), ρ is the medium mass density, and V is the medium volume heated. Two other limits of interest in laser marking are discussed by Marchant (1990). The *static marking limit* applies when a laser pulse of finite duration is completed before the medium displacement under the beam is a significant fraction of the beam diameter. This is often expressed as $v\tau \ll s$, where v is the medium speed, τ is the pulse duration, and s is the measure of the beam diameter (often FWHM or $1/e^2$ width). The opposite limit is the *scanned marking limit* for which $v\tau \gg s$. Note that these limits can be approached by varying v or τ or both to achieve the desired product. Neither of these limits that Marchant reviews ignore diffusion effects. Simple analytical relations for the power requirement to achieve a marking condition are available for each limit. One particularly interesting relationship for a case of $v = 0$ and $\tau = \infty$, but including heat diffusion, is $\Delta T_{max} = P(1 - R)/2(2p)^{1/2}k\sigma_{eff}$, where ΔT_{max} is the peak temperature rise under the stationary beam, P is the beam power, R is reflectance, k is the thermal conductivity, and σ_{eff} is the (Gaussian) beam size parameter (McDaniel and Bartholomeusz, 1996).

It is interesting to compare the temperature rise ΔT predicted by the above formulas with that from a full computer simulation. We take common values $v = 0$ m/s, $P = P_{inc} = 20$ mW, $R = 0.20$, and $\Delta t = 10$ ns with material parameters k, c, and ρ as consistent average values for a quadrilayer MO film stack. In addition, we choose V and σ_{eff} consistent with a heated volume diameter, that is, beam diameter ≈ 1 μm and an assumed heating depth of 50 nm. The adiabatic limit ΔT_{ave} is 1330 K, the stationary CW (continuous wave) beam $\Delta T_{max} = 1140$ K, and the computer simulation predicts $\Delta T_{max} = 160$ K and $\Delta T_{ave} \approx 80$ K. The striking observation is how effective heat diffusion is in removing thermal energy. Notice that in the adiabatic case energy input is a finite 0.2 nJ, whereas CW heating has a continuous energy input of 20 mW = 20 mJ/s = 20 nJ/μs that results in a peak temperature achieved of nearly 200 K less. When the 10 ns pulse heats a realistically diffusive film structure, the temperature rise is nearly 20 times less than that for the adiabatic limit!

Because heat flow in a three-dimensional film structure, such as Fig. 17.4 in which the layer materials have a wide range of thermal properties (Table 17.2) is complicated, computer simulations are usually the only way to achieve accurate estimates of thermal profiles for arbitrary power pulsing of a moving laser beam (McDaniel and Mansuripur (1987); Shih, to be published). As an example of computer simulations that provide some insight into the behavior of temperature distribution in a moving optical disk during writing/erasing, consider the "thermal characteristic curves" shown in Fig. 17.5. Here, we use the simplifying approximation that medium marking is a threshold phenomenon that occurs instantaneously when a temperature threshold is passed. These plots show the laser-pulse power required to write a mark of a specified width (here equal to 0.666 μm) as a function of disk velocity and laser-pulse duration. An additional parameter that varies for the five panels is the baseline, continuous, laser power in which the writing pulse is embedded. These curves illustrate a wide, continuous range of behavior between the two limits discussed earlier ($v\tau$ ranges from 0 to ∞). An interesting variation of the characteristic curves is obtained if one employs "power cutting" before and after the pulsing event. Figure 17.6 gives an example of power cutting and shows that the effect raises and spreads the curves in power, because the medium experiences brief cooling intervals before and after the writing heat deposition. "Cut pulses" in writing serve as a convenient means of increasing the magnitude of temperature gradients that define the formation of mark boundaries, albeit at a cost of somewhat higher pulse power. High-temperature gradients are generally observed to improve the accuracy and quality (i.e., lower readout jitter) of written patterns in optical storage.

TABLE 17.2 Thermal Conductivities and Specific Heats of Common MO Disk Materials

Film Material	Thermal Conductivity (W/m-K)	Specific Heat (J/kg-K)
Polycarbonate	0.15–0.2	200–300
Glass	0.5–1.0	860
SiN_x	1.5–3.5	600–900
TbFeCo	6–9	320–400
Al or AlX	25–50	≈900

Optical Data Storage

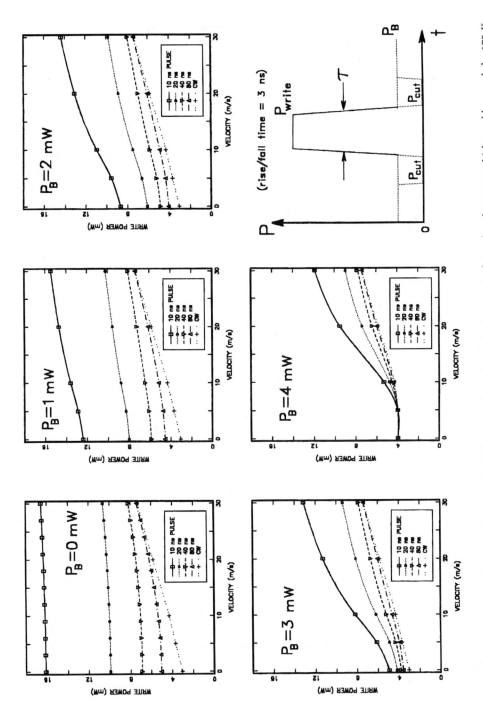

FIGURE 17.5 Thermal characteristic curves for MO media; the writing criterion was to determine the power which would expand the 175 K temperature rise contour to a width of 0.666 μm for constant mark width writing. Independent parameters are disk speed, write pulse duration, and baseline CW power. The 0 mW "cut powers" apply to the characteristic curves in the next figure.

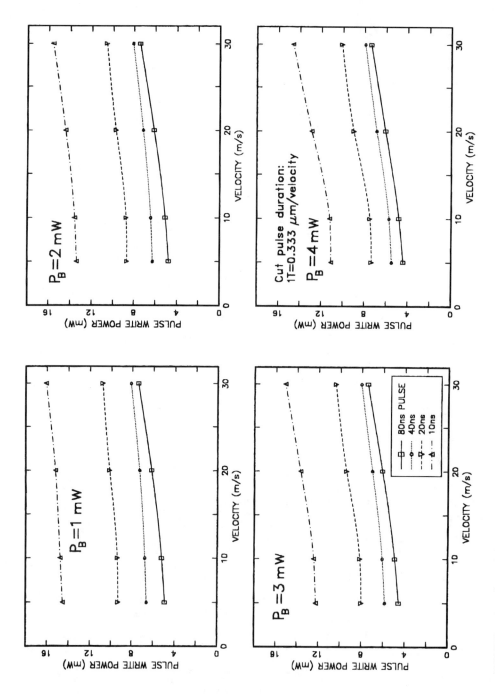

FIGURE 17.6 Thermal characteristic curves for MO media corresponding to the previous figure with the writing pulse modified to have a leading and trailing 0 mW cut power of one clock duration (see power-time profile in previous figure). When using cut powers, a zero baseline power is not allowed.

Thermal modeling is extremely valuable in designing the thermal response of the medium and in developing a writing or erasing strategy. Medium design for thermal performance clearly must be done in concert with optical design of the film stack and material selection (McDaniel and Sequeda, 1992). Fortunately, this design sequence can often occur serially. In a quadrilayer MO structure, for example, after the MO and top dielectric layer (D1) thicknesses are chosen primarily for optical and MO performance, the reflector and lower dielectric (D2) thicknesses, which play a secondary role optically, can be chosen to optimize the heat flow and overall power efficiency of the disk.

Hurst and McDaniel (1996) have an extensive discussion of the use of thermal models to assist in designing writing strategies for high-performance optical recording. The flexibility of a good computer simulation supplements experimental analysis by aiding in the understanding and optimization of pulsing strategies for write compensation. Without careful control of applied laser power, inaccurate or irregular mark shapes are written, which diminishes performance margins and, ultimately, limits achievable data density, particularly for pulsewidth modulation (PWM) recording.

Optical Design and Modeling

The literature on designing and modeling optical components and subsystems for optical data storage devices is vast. Our intent in this section is to raise the reader's awareness of this material and to provide a brief overview of how computer analytical tools have been instrumental in bringing insight into topics involving the propagation of electromagnetic radiation through the optical path of the recording head from the source to the detectors, including the interaction of the light with the storage medium.

A particularly versatile tool for optical path and component analysis is the program DIFFRACT developed at the University of Arizona by Professor Mansuripur. A recently published book (Mansuripur 1995, particularly, Chapter 8) by the program's developer contains a very complete review of these issues. Undoubtedly, other analytical programs exist, many not in the public domain, with similar capabilities. However, given the availability of the results from this program, and now the program itself (For information about DIFFRACT availability, contact MM Technology, Attention: Dr. Masud Mansuripur, Optical Sciences Center, University of Arizona, Tucson, AZ 85721), discussing DIFFRACT as a generic example is probably most useful.

DIFFRACT allows a user to specify the characteristics of a propagating beam and the optical components (lenses, apertures, mirrors, beam splitters, detectors, etc.) in the system path. Indeed, virtually any optical element of interest in an optical storage system can be defined and positioned in a beam for analysis. The program is based on scalar diffraction theory, treating each vector component of the electromagnetic field as a scalar field (Mansuripur, 1989). This analysis accounts for many vector field effects, even though it bypasses exact solutions of Maxwell's equations with boundary conditions. In spite of the absence of a rigorous vector diffraction treatment of light interaction with diffracting features, experience has shown that DIFFRACT accounts for all of the important phenomena observed experimentally in intermediate NA optical recording systems. The use of scalar theory provides great benefits in computational throughput because it is compatible with well-developed auxiliary analytical techniques, such as plane-wave decomposition and two-dimensional, fast Fourier transformation.

We next illustrate a few examples of the capability of DIFFRACT. A schematic of a generic optical head for an MO data storage device, shown in Fig. 17.7, will be helpful in representing the elements of a system which DIFFRACT can simulate. In the first example, Fig. 17.8 shows contour plots of the x, y, and z polarization components of the intensity of light in the focal plane of an objective lens. The light incident on the circularly symmetric objective lens was a plane wave with a Gaussian intensity profile over the x-y plane and linearly polarized along x. Notice that the dominant focused power is for x polarization as expected, but, interestingly,

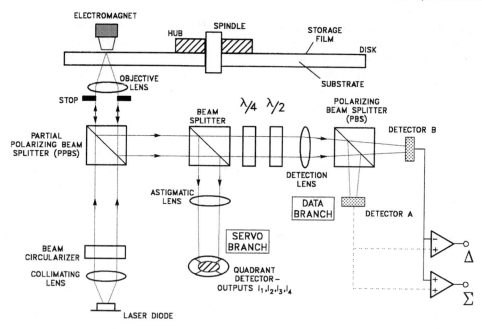

FIGURE 17.7 Generic schematic of an optical head for rewritable MO recording.

Fresnel refraction in passing through the objective lens results in noticeable y and z components with four- and twofold symmetry in the focal plane, respectively. This illustrates that vector component effects are available from this quasi-vector analysis. All parameters of the incident beam, the objective lens, and the detector are under the control of the modeler for design studies.

Figure 17.9 shows an example of predicted readout signals as a focused beam traverses both embossed pits on the surface of the substrate and written MO reverse domains placed sequentially down the centerline of an optical disk track. These signals are detected by the sum and difference channels of the data detection arm, respectively, illustrated in Fig. 17.7. Figure 17.10 shows track crossing signals seen in a servo tracking error signal (TES) channel utilizing a split detector difference signal which is sensitive to diffractive imbalance on the detector as the focused beam crosses embossed tracking pregrooves. Again, in both of these examples, the parameters of the incident beam, the diffractive features on the disk surface, the transmitting optics, and the signal-producing detectors are defined by the user. The modeling tool is a very powerful design aid and a device for understanding the physics governing the operation of the optical storage system.

17.3 Optical Storage Media

In this section, we present several examples of actual optical recording media products, which, for the most part, can be considered commercially successful. The theoretical underpinnings of optical media design have been reviewed in section 17.2; here we emphasize practical material, fabrication, and performance issues which have arisen during the development of real optical media solutions.

Compact Disk

Strictly speaking, read-only products, such as audio CD and CD-ROM, do not qualify as recording systems, but they are important to our discussion because they are forging the consumer optical disk marketplace and because they provide the technical base for more advanced optical recording

Optical Data Storage

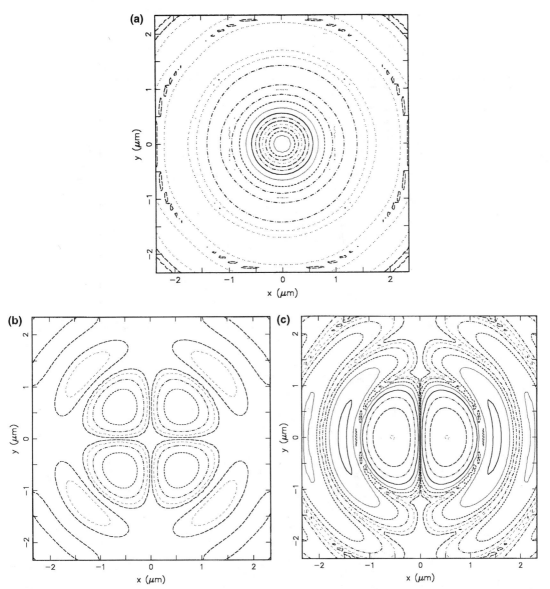

FIGURE 17.8 Contours of $I_i \propto |E_i|^2$, $i = x, y, z$ in (a),(b),(c) from DIFFRACT; in (a) the levels range from 0.9 (center) to 0.00003; in (b) and (c) the range is 10^{-2}, $10^{-2.5}$, ..., 10^{-6}; a Gaussian incident beam overfilled a circular lens aperture with a ratio of $d_{1/e2}/d_{\text{aper}} = 1.09$, with polarization along x; $\lambda = 780$ nm and NA = 0.55.

systems. CD systems also set challenging *de facto* standards for WORM and rewritable systems regarding (i) optical contrast, so that readout functions are as easy to deliver as in a read-only system, (ii) mark permanence, so that the medium can be used for archival storage and that repeated readout at reduced laser power will not alter the medium, and (iii) fidelity, so that mark sizes and shapes repeatably correspond to the areas of laser exposure. Information is "written" by replication of small physical pits on the polycarbonate substrate during the injection molding process. The depth of replicated pits is about 100 nm, carefully chosen to be about one-quarter of the wavelength of the laser light in polycarbonate, and the readout mechanism exploits the optical phase difference between light reflected from the mirror-like land surface and a molded pit.

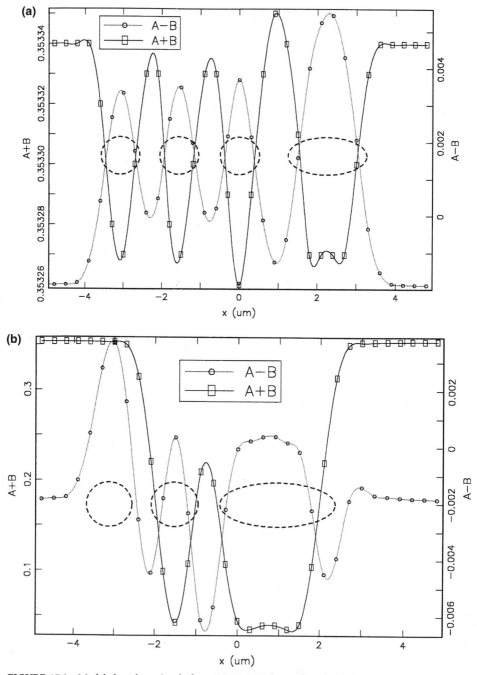

FIGURE 17.9 Modeled readout signals from DIFFRACT for reading down the centerline of embossed pits and/or reversed MO domains; Mark and spot sizes, data density, and track width are representative of second generation MO drive products. (a) four MO reversed domains; (b) one MO reversed domain, plus one small and one elongated embossed pit.

Optical Data Storage

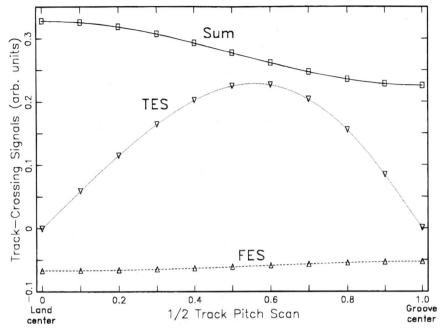

FIGURE 17.10 A tracking error (track crossing) signal, focus error signal, and quad detector sum signal for radial displacement of a readout beam over one-half period of modeled pregrooves in DIFFRACT. Note the diffraction loss in the sum signal over the tracking groove.

The original audio CD was one of the most successful new audio electronic products ever introduced. CD disk and drive technology advanced the state of audio art in every important respect—fidelity, convenience, and ruggedness. In 1990, eight years after its introduction, the CD had achieved a household penetration of 28 percent, with annual sales of 9.2 million players and 288 million disks in the United States (Pohlman, 1992). Annual worldwide demand for CDs passed the 1 billion mark in 1991. CD-ROM quickly started its own acceleration into the market of mass storage, and CD-I's future looks promising as an interactive audio-video medium. In 1988, the CD-R write-once standard was introduced, and, in 1990, the Photo CD became available. In 1994, prototype rewritable CDs began appearing at industry trade shows.

Here, we describe one of several fabrication processes for read-only CDs, such as audio CDs and CD-ROMs. Substrate manufacturing methods for read-only disks are used to make substrates for write-once and rewritable disks. Most CD pressing facilities utilize photoresist mastering and injection molding, as illustrated in Fig. 17.11. Disk mastering begins with a highly polished glass plate to which an adhesive chrome film or silane coupling agent is applied, followed by spin coating a layer of photoresist. The master glass plate coated with photoresist is placed on a "lathe" and exposed with a "cutting" laser to form a spiral track, with the laser intensity modulated according to a channel bit stream representing the original audio or data information. After exposure in the master cutter, the exposed areas are etched away using a developing fluid to create pits in the resist surface. Following development, a metal coating (usually silver or nickel) is evaporated onto the photoresist layer in preparation for electroforming. At this point, the disk master, which has a positive impression of the CD pit track, can be played on a master disk player to ascertain the accuracy of the disk information and pit geometry. After verification, the silvered master disk, which is now electrically conductive, is placed in a galvanic nickel electrolyte bath and a thick conformal Ni coating is gradually built up. The electroformed Ni plate is then separated from the glass master and rinsed. Because this usually damages the master's photoresist layer, masters may be used only once. The resulting nickel copy, called the "father", is a negative impression which, in cases of limited production, can be used to replicate CDs. More generally,

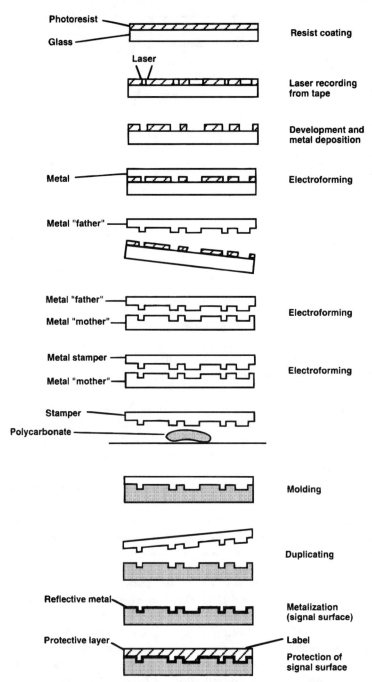

FIGURE 17.11 Schematic CD fabrication process utilizing photoresist mastering, injection molding, reflector metallization, and lacquer protection. After Pohlman (1992).

Optical Data Storage

the father can be used to galvanically generate four or five "mothers". Each mother, if acceptable, can, in turn, be used to generate four or five negative impression "sons" or "stampers" by the same process. When mounted to a die, stampers are used in replicating injection molding machines to produce CD disks.

The injection molding of CD disks presents considerable challenges. To achieve satisfactory mechanical flatness, optical homogeneity, and pit replication, consideration must be given to the quality and dryness of the polycarbonate resin, environmental cleanliness, nozzle and hopper temperature regulation, mold temperature regulation, adjustment and stability of injection volume and time, control of edge flashing due to seams between mold cavity moving parts, and stamper quality. Typically, the resin is heated to about 350°C for good fluidity and injected at high velocity and pressure into the mold, maintained at about 110°C. Injection molded polycarbonate disks tend to have large birefringence, or a polarization-dependent index of refraction, which can affect phase relationships during reading and lead to errors. Birefringence arises due to fluid shear during the injection process, which tends to align the polycarbonate chains in the plane of the disk. Because the polarizability, hence refractive index n, is higher when the E-field of the light is parallel to the polymer chains, n_ϕ and n_r tend to be much larger than n_z, where ϕ, r, and z correspond to E along the in-plane circumferential, in-plane radial, and normal-to-plane directions, respectively. Birefringence can be exacerbated by stress resulting from shrinkage, flow lines, and inclusions. Experience, based on a wide body of empirical knowledge, has shown that birefringence is usually worst near the outer disk radius and can be minimized, but not eliminated, with high melt temperature, rapid injection into the mold, low mold pressure, and short mold time. More dramatic improvements in this regard are being sought, primarily, through the development of new plastics.

The final steps in the CD fabrication process involve reflector metallization, lacquer protection, and label printing. The through-substrate reflectance coefficient in mirror areas is specified to be between 70 and 90%, which allows the use of Al, Cu, and Ag metals and Au as an expensive alternative. Typically, a 50–100-nm thick layer of Al or corrosion-stabilized Al alloy is magnetron sputter deposited. The reflector layer is, then, protected from physical abuse and oxidation by spin coating and curing a 6–7-μm thick plastic layer, e.g., UV curable acrylic with good adhesion to aluminum and high scratch resistance. Finally, a multicolor label is printed on the lacquer layer using sophisticated silk screening or pad printing machines. This label does not interfere with disk performance because light is incident from the polycarbonate side and what little light returns from the lacquer/label interface is thoroughly defocused. Basic CD physical parameters are summarized in Table 17.3.

Write-Once Ablative Media

As mentioned previously, the total power output of a diode laser in an optical recording system is not very great, but the power density of the optical stylus is impressively high. Light absorbed by a surface at such a high-power density can cause rapid alteration or damage by a variety of mechanisms. Actually, the fact that the energy arrives in the form of light is of secondary

TABLE 17.3 Basic Compact Disk Physical Parameters

Disk	Recording
120-mm diameter	Continuous spiral from center
1.2-mm thick	15,900 tracks per inch (1.6-μm track pitch)
4 layers:	
polycarbonate substrate	0.6-μm pit width
aluminum reflector	0.9 to 3.3-μm pit length
protective lacquer	6×10^8 bits/in.^{2} density
ink label	Constant linear velocity

importance, i.e., the important mechanisms are thermal. In ablative recording using thin metal films, surface energy differences between the film and substrate can lead to a thermodynamic tendency to bead up or form voids (Kivits et al. 1982). Because the continuous film is in a local energy minimum, a hole of some minimum size must be created before surface tension effects begin to dominate the fluid flow and create a stable void. For high-speed recording, localized melting may not be adequate to open a hole; usually, vaporization of the film surface dominates the initial hole-opening dynamics, hence the term *ablative* recording. Marks formed in light absorbing organic films also look like holes, but, in this case, the mark formation has little to do with surface tension. In fact, the surface energy may actually be increased by the recording process. Because the thermal conductivity of organic materials is much lower than that of metals, extremely high temperatures can be reached, and material ablation is the dominant process. Figure 17.12 shows a typical layer structure used for an organic recording medium (Gupta, 1988). The recording layer thickness is chosen for high absorption to provide high recording sensitivity and high read-out optical contrast. Numerical modeling of the reflectivity of the layer structure using optical constants of the various layers is extremely helpful in determining the optimum recording layer thickness. It may be desirable to contain the ablative debris using an encapsulating layer. Figure 17.13 shows such a structure and a typical plot of reflectivity versus layer thickness for an organic dye recording medium (Gupta, 1984).

Phase-Change Media

Some materials can exist stably in two or more structural phases which display different optical properties. Some phase-change materials of interest are group IV, V, and VI and low melting point elements, such as Se, Te, Ge, In, Sn, Sb, and their alloys. In *burn-bright* media, the heat of recording anneals the recording material from a metastable amorphous phase to a more favorable polycrystalline state, which is usually more reflective. Phase-change media can also exhibit the opposite transition, i.e., from crystalline to amorphous. The SbInSn phase-change, write-once medium developed by Eastman Kodak Company has demonstrated some of the highest performance in commercial optical disk and tape systems (Tyan et al. 1992). Figure 17.14a shows the reflectivity variation with film thickness for amorphous and crystalline SbInSn phases, and some actual recorded marks are shown in Fig. 17.14b. Because no net material motion is required for phase-change recording, i.e., the motion involves atomic reordering as opposed to macroscopic material displacement, in principle, it is the "cleanest" of the write-once mechanisms. It can also be difficult to control, because post recording phase change can occur in response to high storage temperature, contaminants, or autocatalytic reactions. Small volume changes may also occur upon phase reordering, leading to localized mechanical stresses in the recorded thin film.

Some phase-change materials can be reversibly converted between phases on time scales sufficiently short to permit practical rewritable media. Figure 17.15 shows a rewritable, phase-change, optical disk structure with a quadrilayer structure consisting of a bottom dielectric layer, active phase-change layer, upper dielectric layer, and reflection layer (Ohta et al. 1995). The bottom dielectric layer protects the polycarbonate substrate against thermal damage and is also used to control the recording sensitivity and cooling rate of the active layer. The use of amorphous-to-crystalline, phase-change materials for optical memory applications was first demonstrated by Ovshinsky et al. in 1968 using a TeGeSSb chalcogenide alloy system. This material undergoes a crystalline-to-amorphous transition on the order of 100 nsec, but the reverse (erase) recrystallization transition requires on the order of microseconds. Recent developments in materials have reduced the recrystallization time to the nanosecond time scale as shown in Figures 17.16. Tables 17.4 and 17.5 show refractive indices of materials typically used in rewritable, phase-change, disk structures and optical characteristics calculated from these indices. Using such materials and structures, single-pass overwriting has been demonstrated. The thicknesses of the layers are chosen to obtain the optimum combination of large reflectivity difference and large optical absorption.

Optical Data Storage

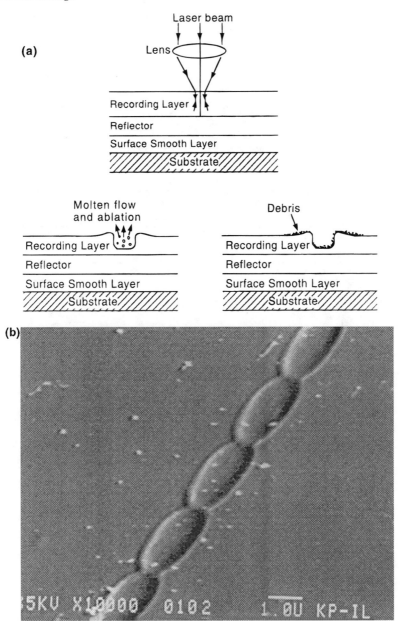

FIGURE 17.12 (a) Debris generation during laser recording process. (b) Debris particles on the organic dye surface due to a single recorded track. After Gupta (1984, 1988).

As shown in Table 17.5, the reflectivity difference between the amorphous and crystalline states is of the order of 20–30% at practical wavelengths (Ohta et al. 1995).

Rewritable, phase-change media in a sectored CD format are already commercially available, called phase-change dual (PD). Although rewritable magneto-optical media are perhaps a more mature media technology (see below), rewritable phase-change media offer easier compatibility with the large installed base of CD drives which detect data based on reflectance modulation (versus polarization modulation in the case of magneto-optic detection). PD drives can read current generation CD-ROM disks, but PD media cannot be read on current CD-ROM drives due to differing reflectance levels.

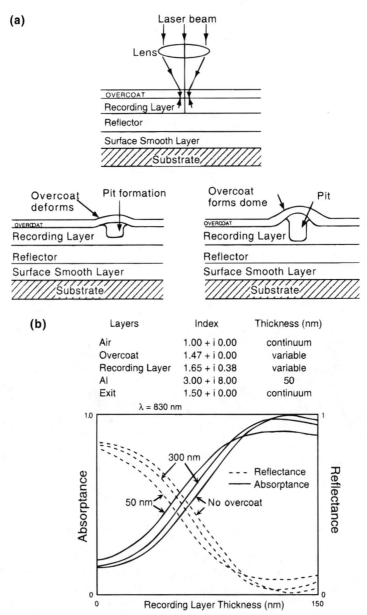

FIGURE 17.13 (a) Suppression of debris by a thin film coated over the laser recording layer. (b) Absorptance variation with organic dye thickness for different thicknesses of overcoat on the organic dye. Note that, in the table of indices, the sign of the imaginary component is a matter of convention. After Gupta (1984).

Magneto-Optical Media

Magneto-Optical (MO) media exploit the polar Kerr effect to achieve optical contrast between magnetic domains of "up" versus "down" polarity. Current generation media are based exclusively on rare earth-transition metal (RE-TM) amorphous alloys for the active MO layer. Generally, the RE-TM material is chemically unstable due to the high affinity of the rare-earth element for oxygen and is structurally metastable. As a consequence, the issue of long-term chemical and physical stability has been foremost in MO media design. Of necessity, the RE-TM layer is typically protected by sandwiching it between chemically stable dielectric and reflector layers which also

Optical Data Storage

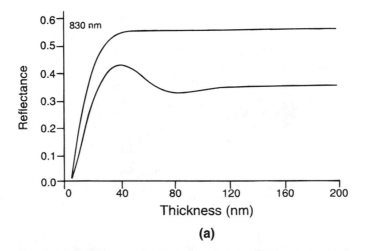

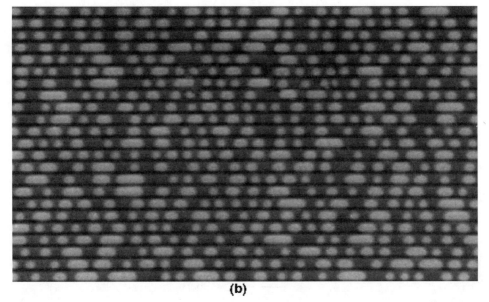

FIGURE 17.14 (a) Calculated reflectivity of an SbInSn thin film for amorphous and crystalline states as a function of film thickness. The refractive indices of the amorphous phase at 830 nm are about 5.07 + i2.62 and those of the crystalline phase 3.64 + i5.52. The reflectivity of the alloy, therefore, increases as a result of crystallization. After Tyan and co-workers (1992). (b) Photomicrograph or some laser-recorded marks in an SbInSn thin-film. The white features are the recorded marks, the track pitch is 1.6 μm, and the modulation code is EFM. After Tyan and coworkers (1992).

provide optical and thermal enhancement. Properly designed and fabricated, this optical layer stack provides virtually unlimited rewritability and an operational lifetime of at least a decade. On the other hand, the manufacturing complexity introduced by the need for a multilayer stack, in particular, thick dielectric layers, contributes significantly to cost due to high cycle time and fabricating yield loss. Rewritable phase-change media exhibit similar manufacturing complexity.

An example of an MO disk structure embodying the conventional "quadrilayer" structure is shown in Figure 17.17. The theoretical basis for the optical and thermal response of the basic quadrilayer structure has already been provided in section A. The quadrilayer forms an optical resonant cavity which provides near-optimum readback performance. The dielectric layers sandwiching the RE-TM layer are, typically, very stable, oxygen-free compounds, such as nitrides of

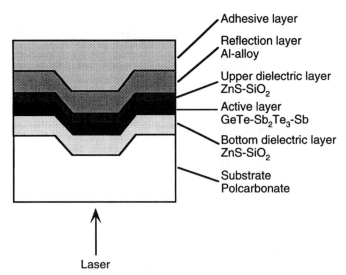

FIGURE 17.15 Cross-sectional view of a phase-change optical disk with rapid cooling structure. After Ohta and co-workers (1995).

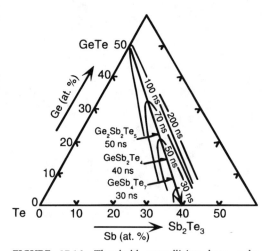

FIGURE 17.16 Threshold crystallizing laser pulse duration for Ge-Sb-Te ternary alloy films. The active layer films are sandwiched by protective layers. After Ohta and co-workers (1995).

TABLE 17.4 Refractive Indices of Materials

	Refractive index	
Material	830 nm	680 nm
TeGeSb amorphous	$4.61 - i\,1.05$	$4.39 - i\,1.53$
Crystal	$5.67 - i\,3.01$	$4.84 - i\,3.53$
ZnS-SiO$_2$	2.0	2.1
Al alloy	$2.2 - i\,7.5$	$1.8 - i\,6.1$
Polycarbonate	1.58	1.58

Optical Data Storage

TABLE 17.5 Calculated Optical Characteristics of the Disk

λ	ΔR	A_w	A_d
830 nm	27.7%	82.9%	67.4%
680 nm	19.8%	75.6%	62.0%

λ: wavelength of laser diode.

ΔR: reflectivity difference between amorphous state and crystalline state.

A_w: absorption of amorphous state.

A_d: absorption of crystalline state.

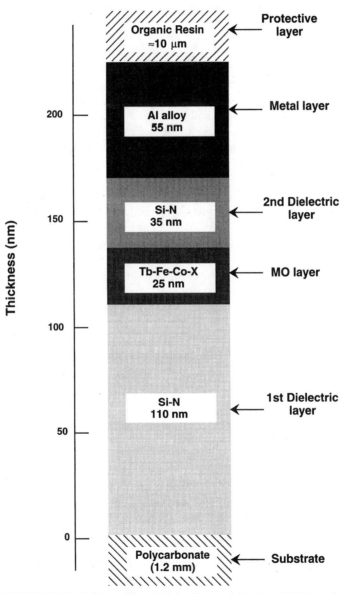

FIGURE 17.17 Commercial quadrilayer film stack based on RE-TM as the functional MO layer. Adapted from Ikeda and co-workers (1993).

Si or Al and are deposited with utmost care to eliminate permeation pathways for oxidation and corrosion agents. The metal reflector layer, typically, a corrosion stabilized alloy of aluminum, must also provide environmental protection in addition to a measure of optical and thermal tuning. Ideally, all light which is not otherwise absorbed in the RE-TM layer is returned to the detector. From an optical point of view, the practical design objective is to efficiently maximize θ_k while providing the level of reflectance required for adequate tracking and focusing of signals during readback. Thermally, the chief objectives are to provide adequate recording sensitivity, given the available power density in the writing beam, and to achieve appropriate temperature gradients at the leading and trailing edges of the written mark to freeze the domain wall position repeatably from mark to mark. A more in-depth discussion of practical design, fabrication, and performance issues for real MO recording media is provided in Brucker (1996).

It is rather remarkable that reliable MO disk systems with performance approaching that of magnetic hard disk drives have been achieved using alloys such as Tb-Fe-Co with an intrinsic Kerr rotation of only 0.2 to 0.3 degrees. Given the installed manufacturing capacity for RE-TM MO media, it is quite conceivable that RE-TM media will remain the mainstay in MO recording systems for some time to come, perhaps with additives to enhance performance at blue-green wavelengths. At the same time, it is clear that materials with superior magneto-optical properties and/or reduced cost could enable new applications in this media sector. A larger θ_k, for example, could increase the signal-to-noise-ratio, allowing greater information-carrying capacity in the data channel of which the storage medium is a part or could make practical the use of thin, semitransparent MO layers required for multilevel and near-field optical recording (see Section 17.5). Materials with larger θ_k are known, but few satisfy the practical prerequisites of appropriate magnetic ordering temperature, sufficient perpendicular magnetic anisotropy and coercivity, high reflectance, low medium noise, environmental stability, and process compatibility with other materials used in the disk structure. A listing of materials with θ_K comparable to or greater than TbFeCo is shown in Table 17.6. Those materials considered "leading candidates" for practical MO recording media are compared in Table 17.7. Garnets and Mn-based materials have strong Kerr effects, but have some fundamental shortcomings which must be overcome before they can be considered commercially viable. The Co/Pt multilayers possess all the requisite properties for high-density recording, are process compatible with temperature-sensitive polycarbonate substrate material, and have sufficient chemical stability to allow simpler stack structures and alternative dielectric materials, e.g., oxides. The measured carrier-to-noise ratio for Co/Pt disk media is comparable to that of RE-TM. (Zeper et al. 1992). Largely due to its excellent environmental stability, Co/Pt was the material of choice for the recent demonstration of 45 Gbits/in.[2] data density using near-field optical scanning microscopy (see Section 17.5).

Substrates

The optical disk substrate is the rigid mechanical structure which supports and protects the storage films. It serves several specific functions:

(a) It provides a rigid, stable platform to carry the delicate thin-film data storage structure, keeping it reasonably close to the laser focal point as the disk spins rapidly.

(b) It serves as a protective dust cover to keep dirt, dust, and debris far from the laser focal plane and to protect against handling damage; to some degree, it may also act as a barrier to external material contaminants.

(c) Considering function (b), the substrate is an important part of the system optical path, being the final element from the laser to the storage film; it follows the objective lens, whose function is to form a well-focused, unaberrated spot on the disk.

(d) It carries embossed servo and/or pre-format information to aid the servo and data channels in following, accessing, organizing, and interpreting data stored on the disk.

TABLE 17.6 Selected Materials with High Kerr Rotation

Material	θ_k(deg)	Corrosion Stability	Comments	References
Co (film, hexagonal)	0.35		For reference; $T_c = 1125\ °C$; $n + ik = 2.5 + i4.0$ at $\lambda = 633$ nm	Buschow et al. (1983); Hansen and Andenko Ward, (1991)
Fe	0.41		For reference; $T_c = 770\ °C$ (, p 669); $n + ik = 3.0 + i3.2$ at $\lambda = 633$ nm	Lynch and Hunter (1991)
$Tb_{22}Fe_{67}Co_{11}$	0.27	poor	$R = 0.6$, $n + ik = 3.2 + i3.5$ at $\lambda = 780$ nm; T_{prep} = room temp; $K = 7$ W/m-K	
Co-Pt alloy	0.40	good	$\lambda = 633$ nm; $T_{prep} = 300\ °C$	Weller et al. (1993)
Fe-Pt alloy	0.5–0.6	good	$\lambda = 633$ nm; $T_{prep} = 600\ °C$	Lairson et al. (1993)
Co-Pt multilayer	0.25	good	$R = 0.7$, $n + ik = 2.6 + i5.2$ at $\lambda = 780$ nm; T_{prep} = room temp; $K = 25$ W/m-K	
Fe-Ag bilayer	0.50		$\lambda = 633$ nm	Zhou et al. (1991) Buschow et al. (1983)
PtMnSb	0.93	good	$\lambda = 633$ nm	
MnBi	0.6–0.7	poor	$\lambda = 633$ nm; T_{prep} = room temp to 300 °C	
Garnet	high	excellent	$T_{prep} \geq 400\ °C$	
CeSb	14		$\lambda = 2480$ nm; $T_c(T_N) = -258\ °C\ (-270\ °C)$	Reim et al. (1986)
EuS	10		$\lambda = 520$ nm; $T_N = -268\ °C$	Gambino et al. (1993)
USbTe	9		$R = 0.5$ at $\lambda = 1770$ nm; $T_c = -69\ °C$	Reim et al. (1984)

θ_k is the intrinsic (bulk or thick film) Kerr rotation.
R is the reflectance.
$n + ik$ is the complex index of refraction.
λ is the measurement wavelength.
$T_c(T_N)$ is the Curie (Neel) magnetic ordering temperature.
T_{prep} is the highest temperature required to fabricate the material.
K is the coefficient of thermal conductivity.

TABLE 17.7 Comparison of "Leading Candidate" MO Thin Film Materials with Potential Application for Blue-Green Recording

Material	FOM $\times 10^{3*}$ (400–550 nm)	Comments
TbFeCoTa	9–11	"Industry standard," i.e., current generation MO media is based exclusively on RE-TM amorphous alloys for the active MO layer. Constrained by chemical instability. Sputtering targets difficult to fabricate.
Co/Pt	12–14	Recording performance comparable to RE-TM. Offers the distinct advantage of good chemical stability, permitting simpler optical stack structures and reduced media cost.
MnBi	13–15	High θ_K and strong perpendicular anisotropy. Problems include high grain boundary noise, structural phase instability upon write-erase cycling, and high film preparation temperature.
PtMnSb	16–25	Very high θ_K. Problems include narrow compositional tolerance and lack of perpendicular anisotropy.
BiFe Garnet	23–32	Outstanding structural and chemical stability, very high θ_K. Recording performance limited by media noise which appears to be process dependent. High film preparation temperature.

*The intrinsic magneto-optical figure-of-merit is defined by FOM $\equiv |\epsilon_{xy}|/(2Im\epsilon_{xx})$, where ϵ_{xx} and ϵ_{xy} are the diagonal and off-diagonal elements of the dielectric tensor of the MO material. The FOM is equal to the upper bound of the shot-noise-limited, signal-to-noise ratio in MO readout and can be approached in a properly designed quadrilayer structure. θ_K is related to the FOM through the (approximate) expression $\theta_K + i\epsilon_K = i\epsilon_{xy}/[\epsilon_{xx}^{1/2}(1 - \epsilon_{xx})]$, where ϵ_K is the Kerr ellipticity. After H. Fu and co-workers (1995).

Function (a) places stringent constraints on the shape and stability of substrate materials, which are, most commonly, injection molded polymers or tempered glass. Disk flatness and warp are controlled by specified limits on axial displacement, acceleration, and tilt angle (local slope in the radial and tangential directions) relative to a reference disk plane. Axial deflections under ±200–400 μm and tilt magnitudes under 5–10 milliradians are commonly required for data storage applications. Function (d) is often addressed through embossing concentric or spiral track pregrooves from a precision manufactured master disk, and the total radial runout of these tracks relative to the disk center is usually limited to 50–100 μm, peak-to-peak.

The quality of a disk substrate as a material diffusion barrier in function (b) depends on the material choice. In general, dense glasses are superior protective materials, whereas porous polymers tend to allow slow diffusion of some contaminant species, including water vapor. Water uptake in plastic substrates is known to contribute to warpage, and special steps to guard against this may be necessary. Fortunately, many optical disk designs consist of two laminated substrates with the deposited films positioned near the center of the sandwich. These balanced mechanical packages, are, to a first approximation, immune from curvature and warpage due to similar stressing on the two halves. However, inexpensive single-sided optical disks (e.g., CD formats) are ubiquitous, and the performance demands being placed on them are increasing rapidly, so that the material absorption issue is likely to become more important in the future. This is particularly so for the newly emerging, all-purpose DVD format (see Section 17.5) which will, some day, supplant CD.

Function (c) is extremely critical, as already alluded to in Section 17.2. Therefore, aberrations and distortions on the propagating optical wavefront, caused by the substrate, must be strictly limited to preserve the optical quality in the beam focused on the storage films. For this reason, the refractive index and thickness uniformity of the substrate must be carefully controlled to assure that a corrected objective lens can provide a low-aberration, focused beam. Indeed, the objective lens must compensate for the additional optical path represented by the substrate material, which, otherwise, would introduce excessive spherical aberration into the focused beam for an uncorrected objective. Even more troublesome is a variable and unpredictable aberration introduced by optical disk substrate birefringence. Birefringence is anisotropy in the refractive index of a transmitting optical material, and optical disk substrate materials, especially molded polymers, are prone to two anisotropies that manifest themselves in spot distortion and, thus, data signal and servo system degradation. Notice that this problem will be especially troubling for optical recording techniques utilizing polarized light, because different polarization states of propagating beams sense refractive indices along different directions. Lateral (in-plane) and vertical (axial) birefringence each cause signature degradations of signal or focused beam quality. These issues have been quite thoroughly analyzed (Mansuripur, 1995; McDaniel and Bartholomeusz, 1996; Marchant, 1990), and most of the burden for handling this problem falls on the manufacturer of disk substrates who must carefully control the material selection and injection molding processes to limit the two types of birefringence. Although the variation of refractive index with direction is still quite small in absolute terms (even in the most birefringent plastics, $\Delta n \approx 10^{-5}$ for the in-plane component, whereas $\Delta n \approx 10^{-4}$ to 10^{-3} for the vertical component), birefringence gives rise to an integrated effect on a propagating beam. Different polarization components experience relative phase delays (retardation), depending on the propagative direction of the light ray. The dephasing effects are equivalent to induced aberration and the wavefront quality of the focused beam is degraded.

Optical Tape

The need to archive the ever increasing amount of data being generated is challenging conventional methods of storage, such as paper, microfilm, magnetic tape, and magnetic and optical disks. Optical tape offers the potential of being a high-capacity, low-cost, and low-maintenance technology not only for storing but also for archiving data. Though magnetic tape is also economical, using it for long-term archiving is hardly ideal. Over time, the magnetic tape medium is degraded by creep, track deformation, and magnetic print-through in the tightly wound cartridge. Every few years, therefore, data should be transferred to fresh tape, an expensive inconvenience which does not arise with optical tape. The introduction of a commercial optical tape recorder (Gelbar, 1990) has demonstrated the technical feasibility of optical tape systems, but advances in hardware and media costs and performance are needed to make this technology widely accepted in the future.

Compared to disk data storage, tape storage enjoys exceptional volumetric density, because the support is so thin, at the price of greatly increased access time, because so much area must be linearly searched. Tape makes more effective use of the available surface area than disks, with the recorded area exceeding 80% of the total area available for half-inch tape versus about 65% for a 130-mm disk and 40% for a CD (single-sided). For the same areal density, optical tape has about 25 times the volumetric density of CD, and there is no fundamental reason why optical tape substrates cannot be reduced in thickness from the current 0.076 mm (0.003″) to 0.013 mm (0.0005″) or less. Just as optical storage brings higher areal density than magnetics to disks, so also with tape. Through the use of parallel read/write channels, optical tape can also support high data rates, on the order of 3 MB/s in current systems. Several companies worldwide are investigating optical tape not only for its storage and performance advantages, but also because the necessary recording materials, lasers, optics, channels, and tape handling methods are fairly straightforward extensions of existing optical disk and magnetic tape storage technologies.

The most serious drawbacks of optical tape have to do with reduced access time and engineering difficulties such as tape guiding, debris generation, contamination, and wear. For a cartridge holding several tens of meters of tape which can be searched at a rate of 10 m/s, the average access time is several seconds. For tape, access time is roughly proportional to cartridge capacity and inversely proportional to tape handling speed.

Just as with magnetic tape, optical tape in mass production would be an extremely low-cost medium because continuous web manufacture costs much less than the batch processing used to make disks (see Table 17.1). The cartridging, packaging, and testing would dominate the manufacturing cost, not the tape media itself. Optical tape is uniquely suited for archiving large quantities of data requiring relatively infrequent access, an application which takes good advantage of the economical nature of tape while not exacerbating its handling difficulties.

17.4 Optical Storage Systems

Optical Heads

In general, an optical head contains the following elements: (i) a semiconductor diode laser, (ii) focus and tracking actuators, (iii) a focusing/collector lens, (iv) optical components, and (v) detectors. Current diode lasers are made of GaAlAs/GaAs-based semiconductors emitting around 800-nm wavelength. The write power incident on the media is on the order of 10–50 mW and the read power is a few milliwatts. The laser has a single spatial mode and can be pulsed with a rise time of a few nanoseconds. Two actuators are used, one for maintaining focus on the rotating (and wobbling) disk surface and one for following the recorded track. While no part of an optical head has to "fly" next to the surface of the medium like a magnetic hard disk drive, the height of the objective lens must be maintained with just as much accuracy and precision given a typical depth of focus of ± 0.5 μm. The tracking tolerance is also very stringent, typically ± 0.1 μm, due to the extremely small track pitch of optical recording systems. The submicron focus and tracking tolerances are maintained by active servo systems composed of optical position sensors feedback coupled to high-bandwidth actuators. Maintaining focus and tracking on a rapidly spinning disk which invariably exhibits wobble and eccentricity is no mean task and represents challenges both to the design of servo systems capable of high radial and vertical acceleration and to the fabrication of disk media with high mechanical precision.

Practical focus sensors generally detect deviations from collimation of the reflected beam. Examples include half aperture and astigmatic focus sensors. Figure 17.18 shows the working principle of an astigmatic focus sensor. The astigmatic field lens has two different focal lengths along orthogonal axes and focuses the beam to two perpendicular line foci. A four-element detector positioned between these foci senses the beam profile as circular when in focus $A + D = B + C$ or elliptical when out of focus $A + D \neq B + C$. An important strength of this scheme is its relative insensitivity to lateral misalignment of the detector, because it reacts, primarily, to the shape of the spot, not its position. Tracking sensors rely on grooves, data, or preformatted servo fields to develop a corrective error signal. Examples include push-pull, outrigger, and sampled servo tracking sensors. Push-pull tracking requires continuous grooves or quasi-continuous tracks of pits to produce ± 1st order diffraction beams which are collected along with the 0th order reflected beam and sensed by a split detector, divided parallel to the groves. When centered exactly over a groove, the ± 1st orders interfere equally with the 0th order, but, when off-center, an error signal is generated by relative aperture illumination effects indicating which

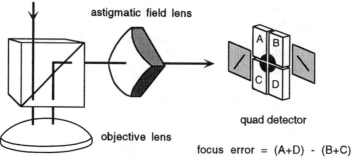

FIGURE 17.18 An astigmatic focus sensor showing conversion of defocus to a change in spot shape. After Marchant (1990).

side of the aperture is receiving more light. The optimum optical depth of pits on read-only disks for maximum push-pull tracking signal, in general, differs from the optimum phase depth of pits for maximum data contrast, requiring a design compromise. Also, whereas push-pull tracking is sensitive and convenient, false error signals can be created by out-of-focus defects or dirt and other effects which obscure the light in one area of the reflected beam.

Several alternative tracking techniques have been developed to avoid the difficulties of push-pull tracking, in which the tracking and focus systems are decoupled, by eliminating the continuous grooves. In the three-spot, or outrigger, tracking sensor, the laser beam is passed through a weak large-period phase grating before entering the objective lens. The $\pm$1st order diffraction beams form low-power secondary spots on either side of the main spot. The grating is aligned so that the outrigger spots are displaced by a fraction of the track pitch in the cross-track direction relative to the main beam. The difference between the two outrigger signals gives a direct indication of the main beam's tracking position. The outrigger tracking scheme is rugged, reliable, and relatively inexpensive, and has been implemented by Philips in a practical head for CD readout. Of course, the scheme requires the constant presence of a data signal and is, therefore, not useful in WORM or rewritable systems where blank, unrecorded tracks must be followed. This deficiency can be overcome by the sampled tracking method, analogous to the tracking scheme employed in high-end magnetic hard disk drives. Many small servo fields are molded into the substrate or servowritten after the disk is manufactured, guaranteeing that some "data" marks are always present. This scheme is simple to implement and, in effect, transfers the difficult job to the mastering station or servowriter where the servo fields are created. Sampled tracking has been implemented by Kodak in its 14-inch WORM media. The overhead requirement of many (several thousand) servo fields on each track, of course, impacts the storage capacity available to the user, especially for small disks.

Figure 17.7 shows a typical configuration for a differential detection MO head. The head serves these key functions:

- Prepare, propagate, and focus the laser beam in its path from the laser to the disk; this applies for write, erase, and read functions;
- Collect, propagate, and direct reflected (diffracted) light from the disk to the data detection branch;
- Collect, propagate, and direct reflected (diffracted) light from the disk to the servo data detection branch;
- Access any radial track position on the disk;
- Maintain servo control whereby the laser beam retains focus on the storage feature plane while accessing tracks or track following.

A magneto-optical head utilizes polarized light (usually linear; directed parallel or perpendicular to the track direction, depending on the design). Because the writing and erasing functions are required simply to deliver thermal energy to the medium, the polarization state is irrelevant for those functions. MO data readout is fundamentally dependent on the Kerr effect, an interaction between polarized light and material magnetization, so that the readout processing of polarized light reflected (diffracted) from the MO disk is critical. Servo information detection normally does not use polarization information, only the reflection (diffraction) of the unrotated Fresnel light component.

Because the "light-source arm" of the head is shared among the write, erase, and read functions, and the redirection of readout light into the data detection branch involves polarization, careful design consideration must be given to properties of the beam splitters, partially polarizing or otherwise. The issue is preserving efficiency in the power splitting for light passage in each direction for each function. Notice that the partially polarizing beam splitter (PPBS) in the light-source arm will waste a fraction of the light from the laser because the reflection of the PPBS for unrotated light must be nonzero on the return pass from the disk to send usable light on for detection. A typical unrotated transmission is 70% and its reflection is 30%. For rotated (Kerr) light returning from the disk, the reflectance is usually designed to be near 100%. It turns out that this intentional attenuation of the unrotated component effectively increases the Kerr rotation θ_K, although the overall effect on SNR must be considered in designing the splitting fractions of the PPBS (McDaniel et al. 1994).

Optical recording requires a very high-quality focused spot at the disk. Therefore, the beam aberration and wavefront distortion must be held to a very small fraction of a wavelength to have a small, symmetric focused beam at the disk. Excessive coma can degrade intersymbol interference, resolution, and adjacent track cross talk. Excessive astigmatism gives rise to focus offset errors and undesired trade-offs between data and servo signal quality (both focus and tracking) (Sugaya and Mansuripur, 1994).

Access (focus acquire, track seek) and servo (focusing, tracking) performance are critical head attributes. The mass of the moving components (objective lens and its mount and actuator only for "split optics" head designs) is a key determinant for the access speed and bandwidth of the servo actuation systems. Reduced mass is also important for better shock resistance and ruggedness. The capability of an optical disk servo system is seen to be very impressive when one considers that the reading or writing beam focal point (≈ 1 μm^3 volume) must be held in the storage layer while it translates radially and axially by as much as 50–100 μm in times of the order of milliseconds due to disk runout at several thousand rpm disk rotation rates.

The light-source arm has optics to condition a highly elliptical beam from the laser diode to a nearly circular gaussian beam. Adjustment of overfill at the objective lens aperture is an important factor in determining focused spot size and shape. In the servo branch, Fig. 17.7 shows an astigmatic lens coupled with a quadrant detector. As previously discussed, the astigmatic servo derives a focus error signal (FES) as FES $\propto (I_1 + I_3) - (I_2 + I_4)$. Diffraction from tracking grooves, as the beam moves radially on the disk, provides a tracking error signal (TES) as TES $\propto (I_1 + I_2) - (I_3 + I_4)$. Often, these signals are normalized by the sum signal from the four quadrants. In the data branch, a polarized beam splitter (PBS) is oriented with its fast axis at 45° to the unrotated polarization direction. The output beams to detectors A and B have amplitude $u + r$ and $u - r$, where u is the unrotated (Fresnel) and r is the rotated (Kerr) light amplitude at each detector. The difference signal is $\Delta \propto 4ur$ and the sum signal is $\Sigma \propto 2(u^2 + r^2)$. Remember that, due to the weakness of the MO Kerr effect, $r \ll u$. We see that the differential readout signal scales linearly with the unrotated light amplitude u, illustrating that the design of the PPBS polarization splitting is a factor in the data SNR.

Data Encoding and Decoding

The Institute of Electrical and Electronic Engineers defines a code as "a plan for representing each of a finite number of values or symbols as a particular arrangement or sequence of discrete conditions or events." Thus, coding is the process of transforming messages or signals in accordance with a definite set of rules. Binary *source* data is usually unconstrained, i.e., a randomly selected bit has an equal probability of being 1 or 0, and arbitrarily long sequences of all 1's or 0's may occur. Data emerging from the modulation encoder, on the other hand, is usually d, k constrained, where d and k specify the minimum and maximum number of 0's between any two 1's. Such d, k constrained *channel* data is said to be run-length-limited (RLL) and can be represented, strictly, by segments of binary data starting with a 1 and followed by at least d but not more than k contiguous 0's. The constraint d is used to control pulse crowding effects, whereas k is used to assure self-clocking ability. For a constant channel data rate to correspond to a constant source data rate, practical RLL codes are usually further constrained to map some m bits of source data onto n bits of channel data, where n > m. The ratio m/n (code rate) provides a first order measure of code efficiency. The need to be dc-free, essential for magnetic recording systems which

Optical Data Storage

sense domains inductively, is somewhat relaxed for optical recording systems. However, most codes still have problems maintaining an accurate decoding threshold in the presence of low-frequency content due to, e.g., dc instability. The code must also limit the maximum run length to maintain clock synchronization during readback and to facilitate decoding.

Foremost in the choice of modern modulation codes is efficiency, the ability to maximize the net linear density while maintaining satisfactory data reliability. Some simple examples illustrate the trade-offs the channel designer faces in achieving an efficient, yet, reliable system. The simplest way to represent binary information is to pulse the signal for each 1 and leave the signal low for each 0. This is referred to as return-to-zero (RZ) coding because the signal drops back to zero at the end of each bit interval. If the pulse length is equal to the bit cell, the signal remains high during consecutive 1's, known as non-return-to-zero (NRZ). In a related scheme, the signal level is switched whenever a 1 is encountered, but left in the prior high or low state for any 0 (NRZI). Although very efficient, these schemes are seldom used for recording. Their most serious deficiency is the lack of clock stability during readout. With arbitrarily long intervals between transitions permitted ($k = \infty$), a change in media velocity can desynchronize the clock. A simple solution to the timing problem is the biphase (or Manchester) code, for which the signal is high during one-half of each bit cell and low during the other half, with the phase determined by whether the data bit is 0 or 1. Biphase modulation is not only self-clocking but also dc-free, however, at the price of low data density. In essence, the source data stream is converted into a channel bit stream by defining two channel bits for each source data bit; a datum 1 becomes a 10, and 0 becomes a 01. A biphase encoded track ($m/n = 1/2$) can store only half as much information as NRZ or NRZI ($m/n = 1$). These and other selected codes are summarized in Table 17.8.

Simultaneously satisfying the objectives of efficiency, clock synchronization, and dc stability is a challenging undertaking, and a host of codes have been developed, each touted as superior to the others, at least, for its specific application. Two different modulation codes are specified for the first-generation, 325 MByte/side, 130-mm MO disk storage system (International Standard ISO/IEC 10089). In the first of these, the so-called "four-out-of-fifteen" (FOOF) 4/15 code (Steenbergen et al. 1985), each byte of data is mapped into 15 channel bits, exactly 4 of which are 1s, using a look-up table. In comparison with biphase, of which it is an elaborate variant, 4/15 has slightly improved data density and high, but uniform, dc content. FOOF code is designed with a spectral null at half the channel bit frequency to permit the use of a substrate with a buried clock in the form of tracking grooves with slight periodic width modulations. In this way, a separate clock signal, useful for determining the location of the optical stylus on the disk, can be filtered from the readout channel. Codes, such as FOOF 4/15, are known as block codes because they are based on fixed-length data blocks.

The second modulation code specified for 130-mm MO is the (2,7) RLL code developed by IBM to enhance the data rate and capacity of magnetic disk drives. Although the m/n efficiency is slightly less for (2,7) versus FOOF coding (Table 17.8), for various reasons it has become the format of choice for first-generation systems and, in fact, is the only code specified for second-generation, 654 MByte/side, 130-mm MO ($2\times$ MO ISO standard). The (2,7) RLL code is a variable length code in which source data segments, 2, 3, or 4 bits long, are mapped to channel

TABLE 17.8 Comparison of Selected Codes

Code	d	k	m	n	m/n	Max dc
NRZ	0	·	1	1	1	100%
bi-Δ	0	1	1	2	0.5	0
MFM	1	3	1	2	0.5	33
FOOF 4/15	0	16	8	15	0.533	47, uniform
RCA 3-Δ	1	7	2	3	0.667	56
IBM 2,7	2	7	1	2	0.5	40
EFM 8/14	2	10	8	17	0.471	0

data segments, 4, 6, or 8 bits long, such that the catenated channel segments obey (2,7) constraints. To satisfy further constraints on allowed source data patterns, padding bits may be added which result in variable length, channel data blocks.

The last code listed in Table 17.8 is eight-to-fourteen modulation (EFM), a block code designed for CD. In EFM, each 8-bit source byte is translated to a 14-bit channel word using a look-up table obeying (2,10) constraints. By design, pattern "uniqueness" is greater in the 14-bit words than in the original 8-bit symbols, providing a measure of error prevention. Blocks of 14 bits are subsequently linked by three merging bits, two of which (0s) are required to prevent the occurrence of successive 1s between serial words, a violation of the coding scheme. A major design objective for EFM was the elimination of problems with low-frequency content. The remaining merging bit is, thus, chosen to maintain the signal's average digital sum value at zero (at a cost of 6% efficiency). During demodulation, the three merging bits are discarded.

There are several useful references which the interested reader may consult to supplement this brief discussion of coding/decoding in the digital channel, as well as other key system attributes, such as track and sector format, measures of signal quality, error detection and correction strategies, and certification and verification to place guaranteed limits on the error distribution. Many of the cornerstones of the optical data storage channel were first developed for magnetic recording systems, as described by Patel (1988) and Jorgensen (1988). A comprehensive overview specific to optical recording systems is provided by Marchant (1990). In-depth and up-to-date discussions specific to CD and MO can be found in Pohlman (1992) and Howe (1996), respectively.

ISO Standards

A uniform industry standard is an important requirement for the success of many products. The floppy disk and the compact disk owe much of their acceptance to a single, industrywide standard. Historically, the optical drive industry has, on more than one occasion, simultaneously introduced incompatible disk formats and media types, which, undoubtedly, has confused customers, resulted in higher than necessary initial costs, and inhibited growth. On the one hand, this is hardly surprising given the great variety of materials systems that lend themselves to write-once or rewritable optical media (although the selection can be sharply narrowed when a constraint, such as backward compatibility with an existing format, is imposed). On the other hand, an awareness has developed that the entire industry can be stunted by zealous promotion of a company's unique technology without regard for interchangability and open systems.

The optical disk industry is learning its lesson in the area of standards and is making good progress in negotiating worldwide standards. Work on standards is organized primarily by disk diameter, including 2.5 inch, 3.5 inch, 120 mm, 130 mm, 12 inch, and 14 inch (Table 17.9). Development and formal approval of a standard is a lengthy process, usually taking at least a year including legal and procedural reviews and a period for public comment. The resultant standards document is a fairly exhaustive specification as to the dimensional and mechanical characteristics of the disk and case (if any), the interface between the disk and the drive, characteristics of the substrate and recording layer, common test procedures to ensure interchangability within a range of environmental conditions, and, more recently, the structure for organizing and labeling files on optical disk media. Normally, formal standards do not specify a necessary lifetime within certain operating or storage environments; rather, such numbers are treated as quality features specific to individual products. Life testing is often performed by the medium manufacturer, which can lead to difficulties if the results are not published in a scientific journal or if tests are performed on laboratory samples, rather than commercially available media. Nonvendor life testing by organizations such as the National Institute of Standards and Technology (NIST) and the Optical Storage Testing and Preservation within the Library of Congress is being encouraged to improve the standardization and objectivity of lifetime estimation.

TABLE 17.9 A. Survey of "basic" standards for CD and MD Media

Basic CD system	Color/ Designator	Year	Description	Standard "Owners"
CD-digital audio (CD-DA)	Red Book	1982	Establishes physical format standards for CD-DA digital audio disks so they can be physically readable on CD-DA drives made by different manufacturers.	Philips and Sony
CD-read only memory (CD-ROM)	Yellow Book	1985	Differs from Red Book in redefinition of the data into groups of 2352 bytes and an extra error detection and correction level.	Philips and Sony
CD-ROM volume and file structure	"High Sierra Group"	1988	Enables data placed on CD-ROM disks by different developers and manufacturers to be readable on different computers using different operating systems and configured with different CD-ROM drives. With enhancements, became "ECMA-119"/"ISO 9660" standard.	"High Sierra Group" ECMA-119 ISO 9660
CD rewritable (CD-MO)	Orange Book Part I	1990	CD-MO format gives the possibility for both audio and data recording (Obsolete).	Philips and Sony
CD write once (CD-WO or CD-R)	Orange Book Part II	1990	CD-DA backward compatible, used in Photo CD.	Philips and Sony
Mini disk (MD)	Rainbow Book	1991	Portable "CD" based on CD-DA with data compression, additional interleave on data, and use of semiconductor memory. If ATRAC Digital Audio Compression technology of the MD-format was applied to the standard CD disk, the playing time would be multiplied by a factor of 5 for a play time of 6 hours.	Sony
CD-R volume and file structure	"Frankfurt Group"	1992	Revision of CD-ROM High Sierra format for interchanging files for CD-ROM and CD-R media. Became "ECMA-168" standard.	"Frankfurt Group" ECMA-168

Especially encouraging is the gathering consensus on a single format for high density DVD. (The initials originally stood for Digital Video Disk, but that term has not been adopted because "video" is an incomplete description in view of audio and computer applications). The two contending formats, "MMCD" and "SD," are being successfully merged into a common format with the physical structure of the double-sided SD and the signal modulation technology originally developed for MMCD. An optical pickup with two switchable lenses can provide backward compatibility with existing CD-ROMs. Hopefully, a common, rewritable CD format can also be arrived at, although, so far, progress in negotiation between the rival CD-E and PD camps has not been very visible. Success (or lack thereof) in this effort could impact the eventual introduction of rewritability to DVD drives.

17.5 Future Optical Storage

Dramatic improvements in optical storage technology can be expected in storage capacity, access time, and cost. Much of the improvement will come through the use of shorter wavelength light

TABLE 17.9 B. Survey of "Application" Standards Based on CD Media Standards

Basic CD system	Color/Designator	Year	Description	Standard "Owners"
CD-interactive (CD-I)	Green Book	1989	Interactive real-time multimedia system (text, picture, sound, video) using dedicated real-time operating system (RTOS). Initially conceived and prototyped by Philips in 1983–1984. Final specification released to Philip's and Sony's 150 licensees in 1987.	Philips and Sony
CD-TV	Amiga CD	1990	Enhanced CD-ROM format for interactive multimedia or 60 minutes of full motion video.	Commodore
DVI		1990	60 minutes of full motion video.	Intel (RCA)
CD-ROM XA	Yellow Book Extension	1991	Extension of CD-ROM with ADPCM audio and specific computer video modes as specified in CD-I. Additional hardware needed to play XA disks.	Philips and Sony
Photo CD	CD-ROM XA Bridge Disk	1991	Blend of CD-DA, CD-ROM XA, CD-I, and CD-R. The CD bridge disk is a special type of CD-ROM XA disk with a CD-I application program.	Kodak and Philips
3DO multimedia format		1993	New CD-based concept for TV-oriented multimedia applications. Direct competition for CD-I. Based on RISC processor for optimum graphical performance.	3DO
Video CD, logical development of Karaoke CD launched in Japanese market	White Book	1993	CD bridge disk format for 74 minutes of full-motion video at VHS quality. The Video CD can be played on a CD-I player extended with a video decompression board ($\approx$$200 in 1994).	Philips, JVC, Sony and Matsushita
Kaleida's PC-based multimedia applications		1994	Blend of CD-DA, CD-ROM, CD-ROM XA, CD-I, and CD-R formats for interactive multimedia applications based on PC's. Under development.	Kaleida: Apple, IBM, Microsoft, . . .

TABLE 17.9 C. Developing Standards Based on CD

Basic CD system	Color/Designator	Year	Description	Standard "Owners"
PD phase-change disk			New CD-WORM standard using "Burn Bright" recording media and sectored data format. Not Red Book compatible.	
DVD-ROM			Areal density about 7× CD-ROM. Laminated construction, single-or double-sided, single or double layer. About 135 minutes of full-motion video per surface.	

Optical Data Storage

TABLE 17.9 D. Survey of Standards for 300 mm (12-inch) and 356 mm (14-inch) Optical Disks

International Standard No.	Description	Standard "Owner"
IS 13614	Nominal 300-mm diameter, double-sided, write-once, optical disk cartridge using sampled servo format (SSF) method of information interchange.	–
IS 13403	Nominal 300-mm diameter, double-sided, write-once optical disk cartridge using continuous composite servo (CCS) method of information interchange.	–
IS 10085	Nominal 356-mm diameter, double-sided, write-once, optical disk cartridge using sampled servo format (SSF) method of information interchange.	Eastman Kodak Co.

for increased storage capacity and enhanced data transfer rate and through the use of miniaturized optical heads. Innovations in substrate, format, recording media, and data channels will also impact future optical storage systems.

Heads and Systems

Integrated Heads

Current optical heads are bulky and heavy and require careful alignment and stabilization of the discrete optical components. As a consequence, optical recording, in general, has poorer access time and higher cost than competitive magnetic technology. The use of integrated optics is one approach to achieving the long sought goal of a compact and integrated optical head, but the technology is not yet practical. Ura and co-workers (1986) showed an integrated optical head whose schematic is shown in Figure 17.19. The laser beam is coupled to a waveguide device and outcoupled and focused on the disk surface using a focus grating coupler. The reflected light beam returning from the disk surface is coupled back into the waveguide. Focus, tracking, and RF signals are attained with an integrated optical head design. The difficulties in practical

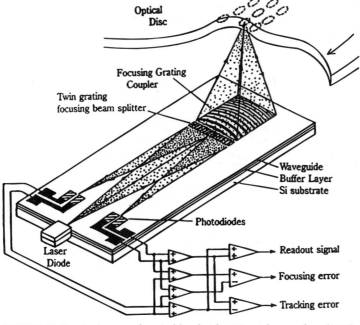

FIGURE 17.19 An integrated optical head. After Ura and co-workers (1986).

applications are loss of laser power in optical components, diode laser coupling, sensitivity to laser wavelength fluctuation, and beam quality. Improved integrated head designs have been discussed (Strasser and Gupta, 1993), but so far they have not reached a practical stage.

Short Wavelength Lasers

For optimum optical recording, the laser beam is focused to a diffraction-limited spot of size approximately 1.2 λ/NA (full beam width), where NA is the numerical aperture of the lens and λ is the wavelength of the light beam. For a given NA, the shorter the wavelength, the smaller the spot size. If the currently used 800 nm IR light is replaced by 400 nm blue light, then, the spot size will be halved, allowing an increase in the in-track storage density and data transfer rate ($\propto$NA/λ) by a factor of two. With smaller recorded marks, the tracks can also be closer by about a factor of two, giving a net increase in storage capacity by a factor of about four. The optimum recording power is proportional to $(\lambda/NA)^2$, so that recording at shorter wavelengths will require less power. The depth of focus, which varies as $\lambda/(NA)^2$, will decrease for shorter wavelengths, requiring tighter control on the focus actuator. Largely due to the inverse squared dependence on NA for depth of focus, it is far more desirable to reduce wavelength than increase NA to achieve a smaller spot.

Short wavelength sources will be very important in optical recording, and significant research and development has been directed toward their development (Risk, 1990). Short wavelength sources will also have many other applications, such as red, green and blue (RGB) sources for display and printing. Two basic approaches have been pursued for developing short wavelength laser sources, one based on direct conversion using II-VI or III-N materials and the other based on indirect conversion using existing GaAlAs/GaAs lasers. In indirect conversion, using second-harmonic generation in bulk $KNbO_3$ crystals, 54 mW of 428 nm output power has been demonstrated (Hurst and Kozlovsky, 1993) with an electrical to optical conversion efficiency of 12%. The frequency conversion approach using bulk crystals or waveguide approaches has been shown (Gupta et al. 1994; Yamada et al. 1993) to produce tens of mW of blue wavelength power, but these systems will be expensive and, hence, may find applications only in high cost products. Direct conversion lasers have not reached a commercial stage because of lifetime issues and will likely require several years of further development. Recently, direct conversion lasing has been demonstrated using II-VI materials (Haase et al. 1991) and III-N materials (Nakamura et al., to be published). The performance requirements for short wavelength laser sources are similar to the requirements for GaAs based IR lasers as shown in Table 17.10.

Optical Super Resolution

High-density magneto-optical disk recording has been demonstrated using superresolving optics. It is well known that, by introducing a shading band at the center of the collimated beam, the

TABLE 17.10 Laser Requirements for Optical Recording

Wavelength	400–800 nm
Wavelength variation	<0.5 nm/°C
	0.1 nm/mW
Pulsed power	20 mW
CW power	>2 mW
Spatial mode	Single
Div. ang. aspect ratio	<4:1
Wavefront quality	<0.02 RMS
Polarization ratio	>30:1
RIN (relative intensity noise)	<−120 dB/Hz
Modulation speed	>1 MHz
Rise/Fall time	<4 ns
Lifetime	10,000 h
Operating temperature	10–55 °C

focused main spot can be reduced to less than the diffraction-limited size, accompanied by the appearance of sidelobes. As the shading bandwidth increases, the main spot size decreases, but, at the same time, the main spot power decreases and the sidelobe intensity increases. A double rhomb prism in place of the shading band has been shown to provide a comparable reduction in focused spot size without appreciable light power loss. Using an optimized combination of a double rhomb prism and a readout slit to provide spatial rejection of undesirable sidelobe intensity, an improvement in linear recording density of about 1.2 times that possible with conventional optics has been demonstrated (Yamanaka et al. 1990).

Near-Field Methods

Another means of circumventing the classical diffraction limit to resolution in conventional optics is near-field optical techniques. Although the imaging process is not understood as well as in conventional far-field optical systems, near-field optical techniques have been applied to a wide range of disciplines including microscopy (Betzig and Trautman, 1992) and lithography (Froelich et al. 1992). More important for this discussion, the possibility of extremely high-density data storage has been demonstrated using near-field methods. One such approach utilizes the recently introduced solid immersion lens (SIL) (Mansfield and King, 1990), which exploits the reduced wavelength of light in a high index glass ($n = 1.83$) to achieve a spot size below the minimum achievable in air. By combining SIL optics with the air-bearing "slider" technology used in magnetic recording to fly magnetic heads above spinning disks, an effective spot size of about 360 nm has been demonstrated using 830-nm light based on read-back resolution from a conventional MO TbFeCo storage medium (Terris et al. 1995). For this demonstration, the beam was incident from the air side instead of the substrate side, and a Si wafer substrate was used to promote rapid cooling and avoid erasure during read. The SIL approach (and the NSOM approach described below) requires a small head-to-medium spacing to avoid evanescent decay of the light beam, which can be achieved by adapting the SIL optics to a flying slider. Based on time interval analysis of (2,7) coded pseudorandom data, a minimum practical mark length of 0.57 μm has been demonstrated. Data rate was limited only by the low linear velocity of 1.25 m/s needed to maintain a low flying height; a modified air bearing design would permit more competitive data rates.

An alternative near-field optical technique utilizes a small aperture between the sample and the light source. If the aperture-to-sample distance is kept smaller than a wavelength, resolution will be determined by the aperture size rather than by the diffraction limit. In particular, the use of a metallized tapered optical fiber, in the end of which is an aperture, has been adapted to a near-field scanning optical microscopy (NSOM) technique to image and record domains. In this technique, a subwavelength-sized source or detector of light is placed in close proximity to a sample and raster-scanned to generate an image. NSOM can provide imagery with polarization contrast. Using a thin, semitransparent MO film as a storage medium, resolution of 30–50 nm has been consistently obtained using NSOM in the imaging mode, whereas domains on the order of 60 nm have been reproducibly written in the recording mode (Betzig et al. 1992). Such small domain sizes correspond to data densities of ≈ 45 Gbits/in.^{2}, well in excess of current magneto-optic or magnetic technologies. These densities require a probe-disk separation of ≈ 10 nm, which, for a practical system, necessitates improvements in the surface cleanliness and flying height of the head. It has been shown that MO materials exist, e.g., Co/Pt multilayer films (see Section 17.3, Magneto-Optical Media) which can, in principle, support 100 Gbit/in.^{2} with a shot-noise-limited signal-to-noise-ratio adequate to achieve data rates in excess of 100 MHz (Kryder, 1995). It remains to be seen if a commercially viable combination of parameters can be found, but near-field optical disk storage is certainly well positioned to benefit from advances in both hard disk drive and optical media technologies. Some proposed recording methods using near-field optics mounted on flying air bearing sliders, similar to those used in magnetic hard disk drives are shown in Figure 17.20.

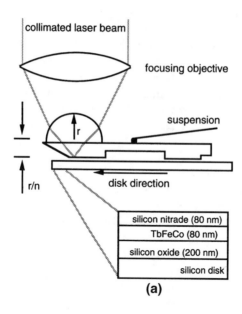

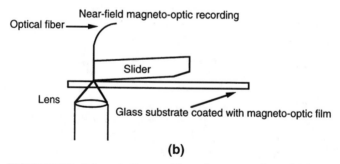

FIGURE 17.20 Schematic diagrams illustrating proposed near-field optical techniques and methods of coupling light to the disk surface using air bearing sliders. (a) Flying solid immersion lens and disk structure. After Terris and co-workers (1995). (b) Flying near-field aperture and collection lens on opposite sides of a semitransparent disk structure. After Kryder (1995).

Media

Double-Sided and Multiple Data Layer CDs

Existing CD products are restricted to two-dimensional storage on a single recording surface. Driven by the desire to store high-quality digital moving images on a small optical disk, double-sided and multiple data layer approaches, which extend optical storage into the third dimension and achieve dramatic increases in disk capacity, are being developed. The double-sided DVD format disk (Figure 17.21) is formed by face-to-face bonding of two half-thickness, 0.6-mm disks. The resulting DVD package is physically compatible with the current generation 1.2-mm CD for backward compatibility. The DVD concept in its single layer embodiment achieves up to 5 GB

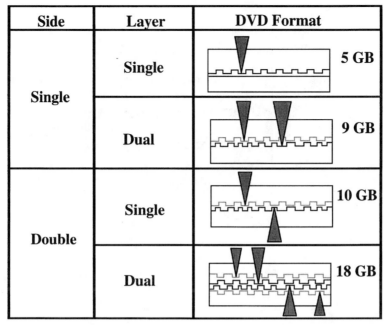

FIGURE 17.21 Schematic illustration of proposed DVD disk structure, with approximate read-only capacities indicated. Write-once and rewritable capacities will be somewhat less.

capacity per side, about 7.5 times the 650-MB capacity of a standard CD, or up to 10-GB total capacity. This capacity increase is anticipated using the combination of reduced wavelength to 650 nm, reduced track pitch to 0.73 μm, increased numerical aperture to 0.6, and more efficient signal modulation and error correction protocols. Use of digital image compression technology according to the MPEG-2 standard, and encoding at a high-speed variable transfer rate allows each side of a DVD disk to carry up to 142 minutes of high quality moving images and sound, enough for most full-length feature movies. The laminated DVD structure offers improved mechanical stability against environmental changes and improved margin against tilt-induced aberration due to the shorter optical path through polycarbonate. Disadvantages include slightly greater fabrication complexity, increased vulnerability to dirt and scratch-induced defocusing due to the thinner substrate, and the inability to access both data surfaces from one side, i.e., the disk must be flipped or two laser scanners must be used. Note that, in principle, single-side access could be achieved (see below), but at considerable increase in disk fabrication complexity. Of course the DVD format also lends itself to computer-related applications, as a very large capacity ROM, and to very long play, high-fidelity audio disks. Double-sided CD optical storage technology can expand bit capacity with minimal technical risk. Negotiation of industry wide interchange standards remains a key milestone.

The aluminized surface of an audio CD returns much of the incident light back to the read head, an important feature when laser power must be conserved. Stimulated by the realization that an unmetallized polycarbonate CD can provide adequate readback signal if enough laser power is available, an intriguing approach using multiple semitransparent data layers, all accessible from the same side, is also being actively investigated (Rubin et al. 1994) (Figure 17.22). Selective access to each of the data surfaces can be achieved simply by moving the laser focusing lens up and down. Continuous play between layers can be maintained by reversing the sense of the spiral groove on successive surfaces and by the use of a modest send-ahead data buffer during the laser refocus jump. Conversely, successive layers could be formatted differently, e.g., DVD on one layer and CD on the other. Given the short depth of focus of a 0.5 NA lens, a layer separation of less

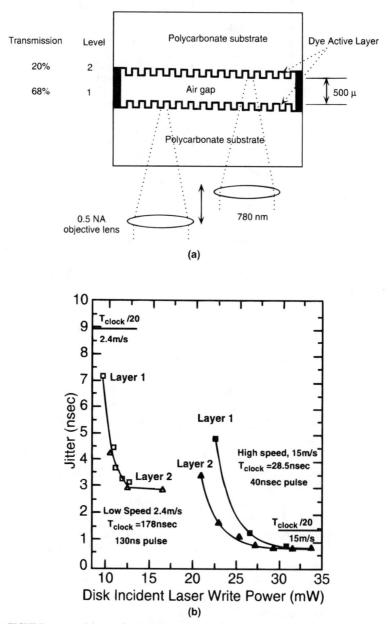

FIGURE 17.22 (a) Two-level multilevel optical storage disk structure built using an air gap between the levels. Level one was optimized to have high transmission of 68% and reasonable reflectivity of 18%. Level two was designed to have a high reflectivity of 30% to compensate for attenuation through level one during readback. (b) Read and write performance of two-level structure at low (2.4 m/s) and high (15 m/s) media speeds. The vertical axis is the jitter from the closely spaced 3T marks. The horizontal axis is the laser power incident on level one necessary to write on level one or two. After Rubin and co-workers (1994).

than 100 μm is sufficient for effective suppression of interlayer crosstalk during reading and writing. The number of usable recording surfaces is, ultimately, limited only by the available laser power. In the case of read-only applications, there appear to be no fundamental hurdles for disks with 10 or more surfaces. In the case of write-once applications, each recording surface must provide sufficient absorption for writing data. Four-level, write-once recording with product-level signal-to-noise-ratio has been demonstrated (Rubin et al. 1994). Challenges for multiple data layer media include the demonstration of an economical fabrication approach, particularly, regarding formatting (e.g., pregrooves) of each data surface and the commercial production of low reflectance coatings with adequate uniformity.

Magneto-Optic Direct Overwrite and Magnetically Induced Super-Resolution

Opportunities for future generation MO media arise not only from superior MO materials, as discussed previously, but also from the clever use of existing materials to enhance performance. One example of this is the functional separation of layers based on exchange-coupled multilayers (ECML), an approach for which conventional RE-TM alloys are particularly well-suited due to their ferrimagnetic ordering and subsequent compensation temperature behavior. The ability to utilize ECML sandwich structures of two or more layers with complementary properties has opened up new and exciting possibilities which would be difficult or impossible with a single MO layer. The most spectacular examples are the achievement of single-pass, direct overwrite (DOW) with laser modulation and magnetically induced superresolution (MSR). Figure 17.23(a) illustrates one such DOW scheme utilizing an ECML medium structure and an initialization magnet (Saito et al. 1987). Several ECML DOW structures have been described which meet or exceed the ISO CNR specification of ≥ 45 dB at 0.75-μm mark length. Some of these structures do not require the extra initialization magnet, which impedes downward compatibility and occupies precious drive space.

Figure 17.23(b) shows an ECML scheme for realizing MSR (Aratani et al. 1991). The principle of superresolution in microscopy indicates that resolving power exceeding the classical diffraction limit can be obtained by putting an aperture directly against the object, which intercepts all light outside a certain region. In MO MSR, a dynamic aperture is induced magnetically in an exchange-coupled readout layer. Dramatically improved mark length performance using 780-nm light has been demonstrated compared with conventional MO disk structures. Even though these ECML structures increase media complexity and cost, their great versatility will likely maintain their appeal for future product developments.

Land and Groove Recording

In most current generation optical disk media, a guardband is provided between successive data tracks to reduce crosstalk interference. This guardband represents space unused for data storage and limits the track density. Land and groove recording achieves virtually zero guardband width, i.e., roughly a doubling of the track density, using the optical phase difference between land and groove regions to cancel crosstalk from neighboring tracks. It has been demonstrated that a 0.6 μm track pitch, half that of current CDs, is feasible using this approach (Honma et al. 1994). The effectiveness of crosstalk cancellation is very sensitive, however, to small deviations in parameters, such as groove depth, domain length, and Kerr ellipticity.

Holographic Storage

Holographic storage provides a three-dimensional optical storage system where terabytes of information can be stored with high data rates in the Gb/s range (Psaltis and Mok, 1995; Hesselink and Bashaw, 1993). The other novel features of holographic storage are fast access time (microseconds), parallel processing of pages as one unit during writing and reading, and no

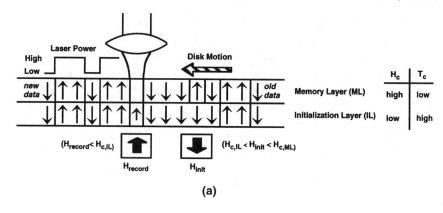

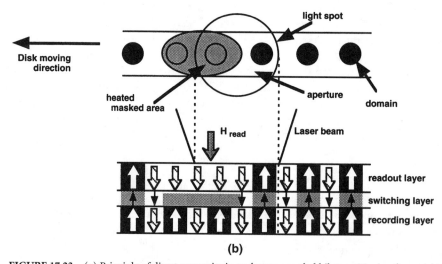

FIGURE 17.23 (a) Principle of direct overwrite in exchange-coupled bilayers. Moving from right to left, the initialization layer is dc-erased downward by the initialization magnet, with no effect on previously written domains in the memory layer. Under high laser power, both layers are heated near their T_c, and upward domains are first frozen into the initialization layer by the recording magnet and, subsequently, copied to the memory layer by exchange coupling upon further cooling. Under low laser power, only the memory layer is heated near its T_c, and downward domains are copied from the unchanged initialization layer to the memory layer upon cooling. After Saito and co-workers (1987).

(b) Principle of magnetically induced superresolution (MSR) in exchange-coupled multilayers using front aperture detection (FAD). In MSR-FAD, an intermediate switching layer with low Curie temperature $T_{c,switch}$ is used to control the effective exchange field $H_{exchange}$ between recording and readout layers. Domains are written in the recording layer by conventional light intensity modulation and transferred to the readout layer during cooling below $T_{c,switch}$ when the condition $H_{exchange} > H_{record} + H_{c,read}$ is reached. During readout, motion of the disk to the left results in a heated region downstream of the irradiated spot, and when $T_{c,switch}$ is exceeded, the readout layer magnetization is aligned with H_{read} (because $H_{exchange} = 0$ and $H_{read} > H_{c,read}$). This effectively dc-erases the heated region, which, thus, does not contribute to signal modulation and acts as a magnetic mask. The signal is modulated by unswitched domains in the remaining chevron-shaped aperture under the irradiated spot. After readout, recorded domains are restored in the readout layer by $H_{exchange}$, just as in the original recording process. After Aratani and co-workers (1991).

moving parts. In addition to data rate and capacity advantage, holographic storage can be readily adaptable to optical signal processing techniques, such as image recognition, etc.

Figure 17.24 compares recording with a volume holographic system and a current optical recording system (Asthana and Finkelstein, 1995). When two laser beams interfere with each other in a light-sensitive material, its optical properties are changed by the grating formed by the intersecting beams. A hologram of the page of data is created when the signal beam meets the reference beam in the photosensitive material. After data has been recorded, the page can be

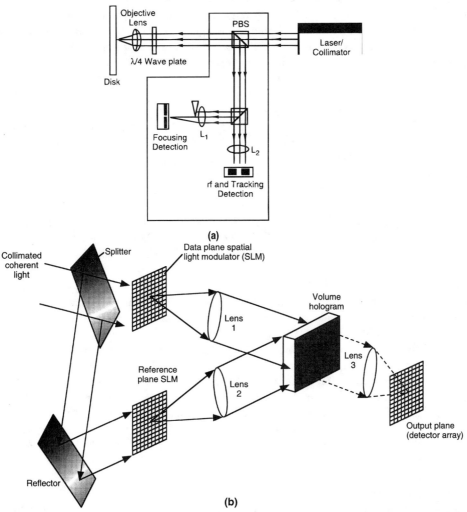

FIGURE 17.24 (a) Layout of a typical bulk optical head that is read-write capable. (b) A volume holographic system. During recording, a collimated laser beam is split into two parts, one for reference and the other for carrying data. The latter passes through a spatial light modulator, which reformats an input bit stream into a page form suitable for holographic storage and modulates it onto the beam. The modulated beam, then, penetrates the holographic material, interacting with the reference beam to form a fringe pattern which the material records. If a second spatial light modulator is used to change the reference beam's incident angle, more than one set of data can be recorded in the same spot. The recording medium can be a sandwich of thin holographic films and transparent buffer layers or a bulk single crystal. After Asthana and Finkelstein (1995).

holographically reconstructed by shining the reference beam into the crystal. The reference beam is diffracted, so that it creates the image of the original page. The diffracted beam is then projected onto a charge-coupled device (CCD) thereby reading the stored information. The major components in the holographic system are a spatial light modulator for data input, a laser diode source or two-dimensional laser array, a holographic recording medium, and a charge-coupled device (CCD) as a data readout device. Further improvements in components are required, such as (i) high-speed scanning devices, (ii) high bandwidth and large size spatial light modulators, (iii) high-speed parallel readout CCD devices, and (iv) high-power visible or near-infrared diode lasers. Currently, one of the major hindrances to commercialization of holographic storage has been appropriate recording media. Presently, there are two types of recording media used for holographic storage: photopolymers and iron-doped $LiNbO_3$ and $Sr_xBa_{1-x}Nb_2O_6$ (SBN). The issues with photopolymers are (i) data loss due to shrinkage of the photopolymer, (ii) limited thickness (≤ 40 mm) which restricts the number of angularly multiplexed holograms that can be stored, (iii) no erasability of the media, and (iv) low signal-to-noise-ratio and high bit-error rates due to low photorefractive response. The difficulties with iron-doped $LiNbO_3$ are slower recording response time and the need to "fix" the recorded data by a thermal process.

Intersecting signal and reference beams generate an interference pattern with regions of high intensity of the form $I = I_0[1 + m \cos(Kx)]$, in which I is the intensity, m is the modulation factor and K is the grating vector given by $K = 2\pi/\Lambda$, where Λ is grating period and x is transverse to the direction of propagation. The photons in the high intensity regions excite electrons into the conduction band, and electrons then diffuse into darker regions of the grating. The separation of electrons from positive ions establishes an electric field in the crystal. Through electro-optic effects, this electric field produces a change in refractive index. The refractive index variation grating, formed inside the crystal, causes the reference beam to diffract for image retrieval. In a holographic medium that does not display nonlinear beam coupling, the diffraction efficiency is given as $\eta = \sin^2(\pi m n^3 rE/2\lambda \cos \theta)$, where E is the space charge field, n is the index of refraction of the crystal, r is the electro-optic coefficient, λ is the wavelength of light, and θ is the angle of incidence.

The information capacity of 3-D data storage is approximately the holographic volume divided by the volume of the smallest feature. If we consider that a hologram is an information channel with W independent levels, and a resolving power of at least $1\backslash\lambda$ cycles per dimension, the storage capacity $\rho = \log_2 W/\lambda^3$ (bits/mm^3). For $W = 1024$ and $\lambda = 500$ nm, $\rho = 8 \times 10^{10}$ bits/mm^3. This calculation does not assume any particular multiplexing scheme. The theoretical limits are not realized because noise tends to reduce the usable capacity before geometric or wavelength limits are approached. The storage density can be also written as $\rho_1 = N_p^2 N_o/\omega^2$, where N_p is the number of pixels, N_o is the number of holograms, and ω is the beam size. The number of holograms by angular multiplexing is given by $h_1 = \Delta\theta_r/\Delta\theta$, where $\Delta\theta = \lambda \cos \theta_s/[L \sin (\theta_s + \theta_r)]$, where θ_r and θ_s are reference and signal beam angles and L is the thickness of the sample. For $\lambda = 500$ nm, $\theta_s = \theta_r = 20°$, $\Delta\theta_r = 4.2°$ and $L = 1$ cm, $h_1 = 1000$ holograms. In a demonstration, 10,000 pages of information have been stored at a single location in an iron-doped $LiNbO_3$ crystal (Psaltis et al.1995).

Holographic storage is most efficient when large data blocks or pages are being recorded. With data blocks as large as 1 Mbit, the recording of digital images is a natural application. If we couple this with very high-speed access to thousands of pages (using, e.g., acousto-optic deflectors with 10–50 μs access) and data rates in excess of 1 Gbit/s (e.g., a 1 Mbit page is parallel detected in 1 ms), this form of storage is particularly well suited for multiple user servicing. Because of this, a host of applications can be envisioned including video-on-demand, on-line catalogues and manuals, distribution of satellite images and maps, geophysical data, medical and radiological imagery, digital libraries, etc. Commercialization of holographic memories is still several years away. However, entry-level versions of this storage technology may be available by the turn of the century.

References

Aratani, K., A. Fukumoto, M. Ohta, M. Kaneko, and K. Watanabe. 1991. Magnetically induced super-resolution in novel magneto-optical Disk. *SPIE* 1499: 209.

Asthana, P., and B. Finkelstein. Superdense Optical Storage. *IEEE Spectrum* Aug. 1995, pp. 25–37.

Atkinson, R., I. W. Salter, and J. Xu. 1992. Design, fabrication, and performance of enhanced magneto-optic quadrilayers with controllable ellipticity. *Appl. Opt.* 31: 4847.

Atkinson, R., I. W. Salter, and J. Xu. 1993. Angular performance of phase-optimized magneto-optic quadrilayers. *Optical Eng.* 32: 3288.

Betzig, E., and J. K. Trautman. 1992. Near-Field optics: Microscopy, spectroscopy, and surface modification beyond the diffraction limit. *Science* 257: 189 and references cited therein.

Betzig, E., J. K. Trautman, R. Wolfe, E. M. Gyorgy, P. L. Finn, M. H. Kryder, and C.–H. Chang. 1992. Near-Field magneto-optics and high density data storage. *Appl. Phys. Lett.* 61: 142.

Brucker, C. F. 1996. Magneto-Optical Thin Film Recording Materials in Practice. In *Handbook of Magneto-Optical Data Recording - Materials, Subsystems, Techniques*, T. W. McDaniel and R. H. Victora, eds. New Jersey: Noyes.

Buschow, K. H. J., P. G. van Engen, and R. Jongebreur. 1983. *J. Magn. Magn. Mater.* 38: 1.

Carslaw, H. S., and J. C. Jaeger. 1988. *Conduction of Heat in Solids*, 2nd Ed. Oxford University.

Connell, G. A. N., and D. S. Bloomberg. 1988. Magnetooptical Recording. In *Magnetic Recording*, C. D. Mee and E. D. Daniel, eds. McGraw-Hill, Vol. II.

Deeter M. N., and D. Sarid. 1988. Effects of incident angle on readout in magnetooptic storage media. *Appl. Opt.* 27: 713.

Froelich, F., T. Milster, and R. Uber. 1992. High-Resolution optical lithography with a near-field scanning subwavelength aperture. *Proc. SPIE* 1751: 312.

Fu, H., Z. Yan, S. K. Lee, and M. Mansuripur. 1995. Dielectric tensor characterization and evaluation of several magneto-optical recording media. *J. Appl. Phys.* 78: 4076.

Gambino, R. J., P. Fumagali, and R. R. Ruf. 1993. Metastable films of giant magneto-optic rotators, *J. Magn. Soc. Jpn.* 17: 276.

Gelbar, D. 1990. Optical data storage. *SPIE* 1316: 65.

Gupta, M. C. 1984. Laser recording on an overcoated organic dye binder medium. *Appl. Opt.* 23: 3950.

Gupta, M. C. 1988. A study of laser marking of thin films. *J. Mater. Res.* 3: 1187.

Gupta, M. C., W. Kozlovsky, and A. C. G. Nutt, 1994. Second harmonic generation in bulk and waveguide $LiTaO_3$ with domain inversion induced by electron beam scanning, *Appl. Phys. Lett.* 64: 3210.

Haase, M. A., J. Qui, J. M. DePuydt, and H. Cheng. 1991. Blue green laser diodes. *Appl. Phys. Lett.* 59: 1272.

Hansen, M., and K. Anderko. 1958. *Constitution of Binary Alloys.* (New York: McGraw-Hill, 1958).

Heavens, O. 1995. *Optical Properties of Thin Solid Films.* Academic.

Hesselink, L., and M. C. Bashaw. 1993. Optical memories implemented with photorefractive media. *Optical and Quantum Electron.* 25: S611–S661.

Holtslag, A. H. M. 1989. Calculations on temperature profiles in optical recording. *J. Appl. Phys.* 66: 1530.

Honma, H., T. Iwanga, K. Kayamura, M. Nakada, R. Katayama, K. Kobayashi, S. Itoi, and H. Inada. High density land/groove recording using PRML technology. Optical Data Storage Topical Meeting 1994 WD1.

Howe, D. Data Reliability and Errors. 1996. In *Handbook of Magneto-Optical Data Recording - Materials, Subsystems, Techniques*, T. W. McDaniel and R. H. Victora, eds. New Jersey: Noyes.

Hurst, J. E., and W. Kozlovsky. 1993. Optical recording at 2.5 Gbit/in.^{2} using a frequency doubled diode laser. *Jpn. J. Appl. Phys.* 32: 5301.

Hurst, J. E. Jr., and T. W. McDaniel. 1996. Writing and Erasing in Magneto-Optical Recording. In *Handbook of Magneto-Optical Data Recording—Materials, Subsystems, Techniques*, T. W. McDaniel and R. H. Victora, eds. New Jersey: Noyes.

Ikeda, E., T. Tanaka, T. Chiba, and H. Yoshimura. 1993. The properties of Sony recordable mini disc. *J. Magn. Soc. Jpn.* 26: 335.

International Standard (ISO/IEC 10089: Information Technology—130 mm Rewritable Optical Disk Cartridge for Information Exchange).

Jorgensen, F. 1988. *The Complete Book of Magnetic Recording*, 3rd Ed. Blue Ridge Summit, PA; TAB Books.

Kivits, P., R. de Bont, B. Jacobs, and P. Zalm. 1982. The hole formation process in tellurium layers for optical data storage. *Thin Solid Films* 87: 215.

Kryder, M. H. 1995. Near-Field optical recording: An approach to 100 Gbit/in^2 recording. *Optoelectronics* 10: 297.

Lairson, B. M., M. R. Visokay, R. Sinclair, and B. M. Clemens. 1993. *J. Magn. Soc. Jpn.* 17: 40.

Lenth, W. Optical storage: A growing mass market for lasers. *Laser Focus World*, Dec. 1994, p. 87.

Lynch D. W., and W. R. Hunter. 1991. In *Handbook of Optical Constants of Solids II*, E. D. Palik, ed. Boston: Academic.

Madison M. R., and T. W. McDaniel. 1989. Temperature distributions produced in an N-Layer film structure by a static or scanning laser or electron beam with applications to magneto-optical media. *J. Appl. Phys.* 66: 5738.

Mansfield, S. M., and G. S. Kino. 1990. *Appl. Phys. Lett.* 57: 2615.

Mansuripur, M. 1986. Distribution of light at and near the focus of high numerical aperture objectives, *J. Optical Soc. Am.* A 3: 2086; 1989. Certain Computational Aspects of Vector Diffraction Problems. *J. Optical Soc. Am.* A 6: 786.

Mansuripur, M. 1995. *The Physical Principles of Magneto-Optical Recording*. United Kingdom; Cambridge University Press.

Marchant, A. B. 1990. *Optical Recording—A Technical Overview*. Addison-Wesley.

McDaniel, T. W., and B. Bartholomeusz. 1996. Modeling Magneto-Optical Recording Processes. In *Handbook of Magneto-Optical Data Recording—Materials, Subsystems, Techniques*, T. W. McDaniel and R. H. Victora, New Jersey: Noyes.

McDaniel T. W., and M. Mansuripur. 1987. Numerical simulation of thermomagnetic writing in RE-TM films. *IEEE Trans. Magn.* 23: 2943.

McDaniel, T. W., K. A. Rubin, and B. I. Finkelstein. 1994. Optimum design of optical storage media for drive compatibility, *IEEE Trans. Magn.* 30: 4413.

McDaniel T. W., and F. O. Sequeda. 1992. Design and material selection for a thin film magneto-optic disk. *Appl. Phys. Commun.* 11: 427.

McDaniel, T. W., F. O. Sequeda, W. McGahan, and J. A. Woollam. 1991. Optical and magneto-optical performance of optimized disk structures, Proc. MORIS '91, *J. Magn. Soc. Jpn.* 15: (S1) 361.

Nakamura, S., M. Senoh, S. Nagahama, N. Iwasa, T. Yamada, T. Matsushita, H. Kiyoku, and Y. Sugimoto. 1996. InGaN-based multi-quantum-well structure laser diodes. *Appl. Phys. Lett.* 68:3269.

Ohta, T., N. Akahira, and I. Satoh. 1995. High density phase change optical recording. *Optoelectronics Devices Technol.* 10: 361.

Ovshinsky, S. R. 1968. *Phys. Rev. Lett.* 21: 1450.

Patel, A. M. 1988. In *Magnetic Recording*, C. D. Mee and E. D. Daniel, eds. New York: McGraw-Hill, Vol. III.

Pohlman, K. C. 1992. *The Compact Disc Handbook*, 2nd ed. Madison, Wisc. A-R Editions.

Psaltis, D., and F. Mok. Holographic memories. *Sci. Am.*, Nov. 1995, pp. 70–76.

Reim, W., J. Schoenes, and P. Wachter. 1984. New high efficiency Kerr rotators. *IEEE Trans. Magn.* 20: 1045.

Reim, W., J. Schoenes, F. Hulliger, and O. Vogt. 1986. *J. Magn. Mater.* 54: 1401.

Risk, W. P. Compact Blue Laser Devices, *Optics and Photonics News*, May 1990, pp. 10–15.

Rubin, K., H. Rosen, T. Strand, W. Imaino, and W. Tang. 1994. Optical Data Storage Topical Meeting, *Tech. Dig. Series* 10: 104.

Saito, J., M. Sato, H. Matsumoto, and H. Akasawa. 1987. Direct overwrite by light power modulation on magneto-optical multilayered media. *Jpn. J. Appl. Phys.* 26: 155.

Shih, O. W. A steady-state heat conduction solution for optical disk recording using constant or periodically modulated laser radiation, to be published in *J. Appl. Phys.*

Steenbergen, C., D. Lou, and H. Verboom. 1985. Working document for development of a standard for a modulation code to be used with an optical media unit for digital information interchange using a 130 mm nominal diameter disk. ANSI submission X3B11/85-135.

Strasser, T. A., and M. C. Gupta. 1993. Integrated optic grating-coupler based optical head. *Appl. Optics* 32: 7454.

Sugaya, S., and M. Mansuripur. 1994. Effects of substrate birefringence on focusing and tracking servo signals in magneto-optical disk data storage. *Appl. Opt.* 33: 5073.

Terris, B. D., H. J. Mamin, and D. Rugar. 1995. Near-Field optical data storage, *Appl. Phys. Lett.* 68: 141.

Tyan, Y.-S., D. R. Preuss, G. R. Olin, F. Vazan, K.-C. Pan, and P. K. Raychaudhuri, Kodak phase-change media for optical tape application, Proc. Goddard Conf. Mass Storage Systems and Technologies II, B. Kobler and P. C. Hariharan, eds. NASA Goddard Space Flight Center, 1992, pp. 499–511.

Ura, S., T. Suhara, H. Nishihara, and J. Koyama. 1986. An integrated-optic disk pickup device. *IEEE J. Lightwave Technol.* LT-4: 913.

Ward, L. 1991. In *Handbook of Optical Constants of Solids II*, E. D. Palik, ed. Boston: Academic.

Weller, D., J. Hurst, H. Notarys, H. Brandle, R. F. C. Farrow, R. Marks, and G. Harp. 1993. *J. Magn. Soc. Jpn.* 17: 72.

Yamada, M., N. Nada, M. Saitoh, and K. Watanabe, *Appl. Phys. Lett.* 62: 435.

Yamanaka, Y., Y. Hirose, H. Fuji, and K. Kubota. 1990. High density recording by superresolution in an optical disk memory system. *Appl. Opt.* 29: 3046.

Zeper, W. B., A. P. J. Jongenelis, B. A. J. Jacobs, H. W. van Kesteren, and P. F. Carcia. *IEEE Trans. Magn.* 28: 2503.

Zhou, S. M., M. Lu, and H. R. Shai. 1991. *Phys. Status Solidi (b)* 168: 651.

18
Electronic Displays

Carlo Infante
CBI Technology Consultants

18.1 Introduction 768
   General Requirements of Electronic Displays • Elements of Photometry • System and Ergonomic Considerations • The Major Display Technologies
18.2 The Cathode-Ray Tube 773
   Overview • Monochrome CRTs • Color CRTs • Important Relationships • Other CRT Devices
18.3 Flat Panel Devices 782
   Liquid Crystal Devices (LCDs) • Electroluminescent Devices (EL) • Plasma Displays (PDPs) • Light Emitting Diodes (LEDs) • Other Flat Panel Technologies
18.4 Display Measurement and Performance Evaluation 794
18.5 Conclusions 795

18.1 Introduction

This chapter addresses the essential elements of display technology. Displays are the primary communication link between a variety of electronic devices and the ultimate consumer of information, the human being. An enormous variety of display devices have been developed over the years to fill many different needs. These range from the small, low-information content devices found in Personal Information Managers (PIMs), hand-held calculators, and games to the exotic and expensive mammoth displays used in command and control situations or in outdoor stadiums. Important devices are those optimized for television viewing in the home and those capable of reproducing one or more pages of text and graphics, such as used in personal computers, terminals and workstations. Specialized or "niche markets" include helmet-mounted, head-up displays (HUDs) and very high-resolution displays for viewing radiographic images.

It would take several books, not just a single chapter, to fully discuss the principal elements of display technology. The work presented here provides only a brief overview of the subject. Hopefully, the reader will be spurred to look deeper into the topic by consulting some of the excellent texts available.

General Requirements of Electronic Displays

Because displays can be used in a wide variety of applications, it is not surprising that no single technology is applicable to every situation. However, it is useful to pause for a minute and consider some general guidelines as to what a display should be able to do:

Electronic Displays

conform to the needs of the human visual system
meet application requirements for
    color
    gray scale
    size
    resolution
    electrical and physical interface
    size, weight, ambient
cost

Displays are often used for extended periods of time, thus, viewing should be comfortable and adapted to the needs of the user. This an obvious requirement but not always followed in practice.

Elements of Photometry

What follows is a short discussion on the fundamentals of photometry (Sherr, 1993; Infante, 1994). It is well known that the human eye does not respond equally to all wavelengths of luminous energy. Although there are differences between individuals, there is something to be said for adopting a standard or prototype eye. This was done some years ago and the result is shown in Fig. 18.1.

Given a source of electromagnetic radiation, the luminous flux is defined as

$$F = 683 \int_0^\infty V(\lambda)P(\lambda) \, d\lambda \text{ lumens,}$$

where $V(\lambda)$ is the photopic function above and $P(\lambda)$ is the power spectrum of the source (watts at the wavelength λ). *Illumination* or *illuminance* is the luminous flux per unit area incident on a surface. Illumination is measured in lux or lumens/m². *Luminance* is the luminous flux emitted from a surface per unit solid angle. Luminance correlates somewhat with the perceptual sensation of brightness and is measured in nits or candela/m² (cd/m²). Table 18.1 summarizes these terms.

An important fact, often neglected, is that, although we can measure time, distance, and frequency with enormous precision, we can generally measure luminance to no better than a few percent (ref). Thus, care is needed in writing specifications or drawing conclusions from measurements.

The luminance of an image *per se*, is less important than its contrast with respect to unwanted information. Terms in general use are

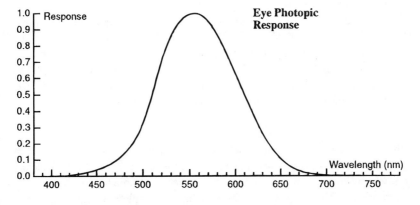

FIGURE 18.1 The response of the human visual system with wavelength.

TABLE 18.1 The Principal Photometric Units

Quantity	Defining Equation	SI Unit	US Unit	Conversion[1] Factor
Luminous flux	$F = K_m \int_0^\infty V(\lambda)P(\lambda)\,d\lambda$	lumens	lumens	1
Intensity	$I = dF/d\Omega$	candela (lumens/ steradian)	candela	1
Luminance	$L = dI/dA_e$	candela/m² (A_e is emitting area)	ft-lamberts (1 cd/ π ft²)	0.2919
Illuminance	$E = dF/dA_i$	lux (lum/m²) (A_i is illumin'd area)	Foot-candle (1 lum/ft²)	0.09294

[1] From metric into US units.

$$\text{Contrast} = \frac{L_S - L_B}{L_S + L_S}$$

and

$$\text{Contrast Ratio} = \frac{L_S}{L_B},$$

where L_S is the luminance of the "signal" and L_B is the luminance of the background.

System and Ergonomic Considerations

Overview

A description, however condensed, of the key aspects of display technology would not be complete without discussing those factors that make a display system useful. Generally, it is not obvious *a priori* what characteristics a display should have to provide useful information and comfortable viewing. This problem is the domain of the specialized science of ergonomics or human factors.

Naturally, the most important requirement is that the image be clearly visible: the luminance and contrast should fall within acceptable limits. This, in turn, depends in great part on the ambient illumination in which the display will operate, because of the limited accommodating power of the human eye and the finite reflectance of display materials. As an example, displays intended for use in the cockpit of an aircraft operate from unimpeded sunlight to near-total darkness, an enormous range, approximately 10^5 to 1 lux. Displays for the cockpit require special design techniques. Conversely, displays designed for the office operate in a much more restricted illumination range of the order of 500 to 2,000 Lux.

This has led to the promulgation and adoption of a number of standards issued by agencies, such as UL, VDE, FCC, ISO and ANSI. It is beyond the scope of this work to examine each in detail, but an overview and a discussion of some important issues may prove useful.

In the United States, the ANSI/HFS 100 Standard (ANSI 1988) published in 1988, has served as a foundation for a series of European requirements and contains a number of useful recommendations. The most important are summarized in Table 18.2.

Flicker

Flicker is the sensation we receive when looking at an image, such as that generated by a CRT, whose repetition frequency is insufficient. The critical fusion frequency (CFF) is defined as the frequency at which the flickering sensation just disappears. Although some of the fundamental principles governing CFF have been known for some time, a complete description of the phenomenon is still lacking. What is known is that CFF depends on a number of factors such as image

TABLE 18.2 Summary of the ANSI/HFS 100-1988 Recommendations[2]

CHARACTERISTIC	SPECIFICATION
Ambient illuminance	500 Lux Nominal
Display luminance	At least 35 cd/m^2
Luminance adjustment	Mandatory
Viewing distance	500 mm nominal
Character height	2.3 to 6.5 mm @ nom view distance
	3.1 mm preferred
Character width	92–93% of Height
Image polarity	Both acceptable
	(black-on-white & white-on-black)
Character format	7 × 9 minimum
Character modulation	(LM − Lm)/(LM + Lm) > 0.75
Luminance uniformity	Better than 50%
Jitter	Less than 0.1 mm @ nom view distance
Linearity (integral)	Better than 2%
Linearity (differential)	Better than 10%
Flicker	Flicker-Free for 90% of viewers

[2]The standard is currently undergoing revision.

luminance, ambient illumination, the angle the image forms with the eye, and the age and, possibly, even the gender of the observer. Confounding the problem is the fact that not all individuals are equally sensitive to flicker, so that any recommendation relating to flicker must be, of necessity, statistical. The most comprehensive work to date on the subject is due to Farrell et al. (1987) who published analytical equations that allow predicting flicker performance under a variety of conditions. These equations are at the basis of recommendations as to minimum refresh rates for CRT displays used in the office. Television displays are, generally, viewed at greater distances, thus, viewing angles are much smaller and flicker requirements are not nearly as severe. Thus, in television displays, the practice has been to choose the refresh rate equal to the power line frequency (60 Hz in the United States and 50 Hz in Europe) so as to minimize interference. To minimize the bandwidth over which the signal is transmitted, television systems generally employ interlace. This means that the picture is not scanned progressively from top to bottom. Rather, half the picture is scanned first (e.g., with 250 scanning lines) in 1/30 of a second. During the next "pass" or field, the remaining half of the picture is scanned, with the scanning lines halfway between the old ones.

Resolution, Image Quality

One of the most complex and controversial subjects in the ergonomics of displays in general and of CRT displays in particular, is that of image quality. Although this may be surprising, especially in view of the many excellent displays currently on the market, there are good reasons. In the first place, the shape of the scanning aperture is not square, as was shown. Thus, it is not at all obvious, *a priori*, just what is meant by "resolution". Furthermore, in an optimum display system, we expect the human eye to be the limiting factor, or, at least, a significant contributor to the resolution of the system. Thus even a first-order resolution theory must start with a description that encompasses both the behavior of gaussian spot profiles and that of the human eye.

This is exactly what has been done by a number of researchers. The description of the hardware aspects of the system are incorporated in the modulation transfer function (MTF) whereas the resolution properties of the human visual system are described by the contrast sensitivity function (CSF). Mathematical expressions have been developed by a number of researchers that allow prediction of image quality once the parameters of the system, such as luminance, viewing distance, and bandwidth are known. Currently, an algorithm known as SQRI (square root integral) has shown the best correlation between predicted values and subjective judgments. SQRI is being considered for inclusion in the revised version of the ANSI-HFS 100 Standard.

Radiation: X rays

A form of radiation that has been of some concern is X rays. X rays are emitted, of course, when energetic electrons strike a material, such as the phosphor screen. X rays penetrate quite deeply into matter including human tissue, especially at shorter wavelengths. In excessive doses, X rays pose a health hazard. The available energy of the electron beam limits the shortest wavelengths that can be produced. Reasonable design techniques, that limit the beam current together with the maximum beam energy even under failure conditions, insure that X-ray emissions produced by a CRT display will remain below limits imposed by appropriate regulatory agencies.

Radiation: Low Frequency Electric and Magnetic Fields

Recently, a number of articles in the popular media have expressed concerns over the potential health effects of the weak electric and magnetic fields produced by CRT monitors. At this time, no conclusive evidence exists that these weak fields do, in fact, cause adverse health effects. To the contrary, a major experiment conducted by the National Institute for Safety and Health (NIOSH) found no increase in miscarriages as a result of exposure to the weak electromagnetic fields emanating from CRT displays. Naturally, a variety of techniques exist that allow reducing of these fields. The techniques include coating the front glass surface of the CRT with a conductive layer, so as to eliminate the electrostatic charging of the front surface. Techniques to reduce dynamic electric and magnetic fields include canceling coils and electrodes, driven in synchronism with the horizontal sweep and arranged so as to cancel the stray flux. Irrespective of the physics of the situation, concerns have developed in the user community. As these market concerns coalesce, some form of regulation, similar in intent to that in effect for X-rays, will undoubtedly emerge.

The Major Display Technologies

A number of different display technologies have achieved a level of technological and market success over the years. These can be classified in a number of different ways. A traditional one has been that of separating the technologies according to whether they generate light (emissive) or simply reflect and modulate the light from separate sources (passive).

We prefer to classify the technologies, as shown in Table 18.3, according to the physical medium on which each depends.

Cathode-ray tubes (CRTs) and field emission devices (FEDs) use electron beams operating in a vacuum. Plasma display panels (PDPs) utilize the controlled electrical breakdown of gas molecules, whereas liquid crystal devices (LCDs) take advantage of the electro-optical properties of a class of materials known as liquid crystals that combine the properties of liquids together with the ordered structure inherent in crystals. Finally, electroluminescent devices (EL) generate light as a result of applying an electric field to a solid. Light emitting diodes (LEDs) are described fully in other parts of this volume. They are, of course, semiconductor devices which generate light by forward biasing a p-n junction with the appropriate material and doping it. Each of these technologies will now be discussed in some detail.

TABLE 18.3 The Major Display Technologies

Display Medium	Display Technology
Vacuum	CRT, FED
Gas	PDP
Liquid	LCD
Solid	EL, LED

18.2 The Cathode-Ray Tube

Overview

Although newer technologies have received a lot of attention, the cathode-ray tube (CRT) remains dominant both in terms of monetary value and the actual number of units shipped (Castellaxo, 1994). The reasons are many and include an established manufacturing base that routinely produces hundreds of millions of the devices at low cost and high yields and the superior ergonomics of CRTs. This situation has made CRTS pervasive in all sorts of applications including personal computers and television, to name just a few.

The block diagram of a CRT display is shown in Fig. 18.2. The word "monitor" is generally employed to describe this collection of waveform generators, power supplies together with the display device itself and a mechanical structure or housing. A television set consists of a monitor with the addition of a tuner, and audio and decoding functions.

The heart of the monitor is the CRT device itself shown in Fig. 18.3. It consists of an electron source (a thermionic cathode), beam forming and focusing electrodes, a deflection system, and a screen. Naturally, a glass envelope is required to sustain the required vacuum.

In short, electrons are emitted by the cathode, focused into a small bundle by some lenses and accelerated to several kilovolts before hitting the phosphor screen. There, the beam is converted into a luminous spot. The electron beam is modulated in *space* by a deflection field. In addition, the beam current can also be modulated in *intensity* by applying a modulating voltage between the CRT grid and cathode. As a result, a picture appears on the screen and is viewed by the observer. By far the most popular way of addressing a CRT is to repetitively cover the CRT screen with a "striped pattern" or raster as shown in Fig. 18.4. Linear deflection waveforms are applied along each axis at frequencies that are fixed and multiples of one another. The electron beam is turned on or off at the right time so that a picture is formed.

Monochrome CRTs

The simple monochrome CRT shown in Fig. 18.3 is capable of displaying only images of a single color. Although once dominant because of cost and simplicity, monochrome CRTs have been

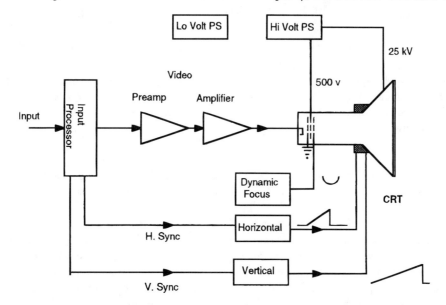

FIGURE 18.2 Monitor block diagram.

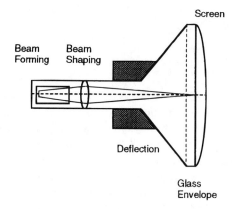

FIGURE 18.3 Monochrome CRT.

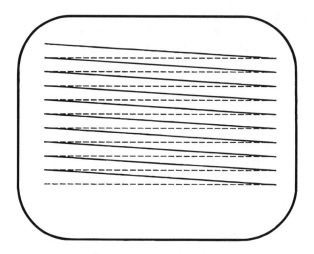

FIGURE 18.4 Raster scanning in a CRT.

largely supplanted in the marketplace by their more advanced offspring. Nonetheless, applications exist in which color is not a requirement. These include the display of complex mechanical drawings, simple alpha-numeric terminals, word processing, and X-ray imaging. For these, the superior resolution and ergonomics of the device, together with its inherently lower cost both at the device and at the system level, continue to be attractive. It is also a convenient way to begin the discussion of the technical aspects of CRTs.

Elements of Electron Optics

The region in which the beam is formed and focused in a CRT is called the electron gun. In the beam-forming region a filament and cathode, shown in Figs. 18.5a and b, generate an electron cloud.

These electrons are formed into a bundle called the crossover (ideally a point source) and one then accelerated and focused in the beam-shaping region. The electrons finally impinge on the phosphor screen where their energy is converted into light. This is shown diagrammatically in Fig. 18.6. A system of lenses is responsible for forming the beam into a focused spot. The details of the underlying physics are beyond the scope of this chapter. Suffice it to say that the electron paths are described by the same sort of equations that describe light rays. This means that concepts such as index of refraction, focal length, aberration and magnification all have an exact analogies

Electronic Displays

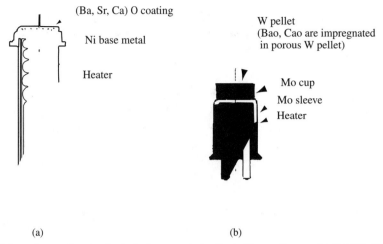

FIGURE 18.5 a. Conventional cathode. b. Impregnated cathode (*Source*: Yuamazaki, E. 1993. CRT Displays, *SID Seminar Notes*, II: F-5/1–47. With permission.)

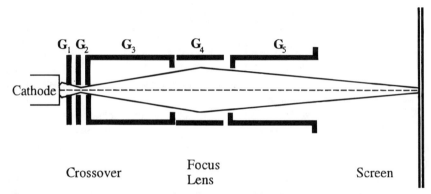

FIGURE 18.6 Operation of an electron gun: The electron cloud is formed into a crossover which is subsequently focused onto the screen.

in electron optics (El-Kareh and El-Kareh, 1970). More importantly, one can show that simple structures, such as a set of coaxial cylinders at different voltages, have properties very similar to those of an optical lens. Consider Fig. 18.7 in which we assume that the right-hand cylinder is at a higher potential.

The electron beam, entering from the left, is subject to the action of the electric field, proportional to the gradient of the potential. The gradient is, of course, perpendicular to the equipotentials at every point in space. Thus, on the *left* side, the beam sees a force deflecting it *toward* the axis. This has the effect of a converging lens. On the *right* hand side, the gradient is now directed *away* from the axis, thus, the effect is of a diverging lens. The net effect of the whole structure is that of a converging lens. Lenses with differing characteristics and focal lengths are obtained by changing the mechanical structure and the operating potentials. Real lenses, both optical and electron-optical, suffer from various imperfections commonly referred to as "aberrations". Spherical aberrations are particularly important in electron-optics. These increase rapidly as the lens diameter becomes smaller.

Deflection Systems

To generate a picture on the screen, the beam must be deflected in a controlled fashion. There are two known methods to deflect a beam: electrostatic and magnetic. In electrostatic deflection,

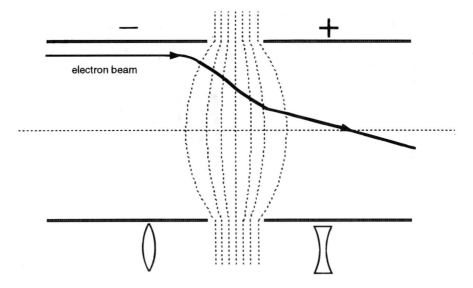

FIGURE 18.7 Operation of an electron-optical lens.

the beam is made to pass between two sets of conductive deflection plates, orthogonal to one another. The electric field resulting from voltages applied to the plates, allows positioning the beam anywhere on the screen. This technique is generally used in applications, such as oscilloscopes, in which deflection speed is a major requirement.

In applications in which a large deflection angle and high luminance are important requirements, magnetic deflection is preferred. Instead of deflection plates, one uses deflection coils, wound orthogonally to the CRT axis, as shown in Fig. 18.8. Current is made to flow in the coils and the resultant magnetic field deflects the beam.

It is important to note that, although the deflection voltage for electrostatic deflection increases *linearly* with the screen voltage E_b, in magnetic deflection, the required field B, hence, the deflection current, increases with the *square root* of E_b. Because display luminance is proportional to E_b, magnetically deflected systems are preferred in most applications.

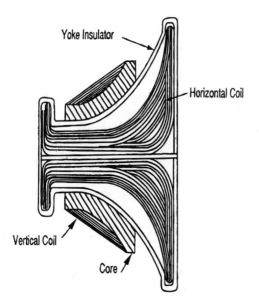

FIGURE 18.8 A CRT deflection yoke (Dasgupta, 1991).

One of the important characteristics of a deflection yoke is its effect on spot focus as the beam is deflected. If the magnetic field within the yoke is not uniform, different electrons within the beam will be deflected differently and an astigmatic spot results. The distribution of optimum turns required to generate a uniform field is a cosine function of the angle between the CRT axis and the coil loop. A good monochrome yoke is wound with the proper distribution so as to guarantee stability and consistency. This is achieved by using magnetic materials in which slots or teeth are provided and by carefully controlling manufacturing techniques. Any real deflection yoke has another effect on the spot focus. As the beam is deflected away from the axis, it encounters a fringing field and must travel a greater distance before reaching the screen. Both of these effects cause the beam to come to a focus before the screen. So-called dynamic focus techniques to compensate to this effect are often employed and involve changing the focus voltage as a function of the beam position.

Phosphors

After the deflection process, the electron beam reaches the screen. The screen consists of a glass faceplate, with the appropriate optical properties, on which a suitable phosphor material has been deposited. Phosphor materials are generally inorganic crystals, with particle sizes in the 5 to 15 μm range. Due to defects in the crystalline structure, the phosphor materials are able to absorb incident energy and convert it into light. This process as known as cathodoluminescence. Energetic electrons from the electron beam penetrate deep into the phosphor. As the beam is absorbed, it raises some of the phosphor electrons to an excited state. As sketched in Fig. 18.9, the excited electrons, then, return to the ground state, releasing the excess energy in the form of light with a wavelength and a persistence dependent on the particular phosphor and its preparation. Phosphors operated above a few kV are usually coated with a thin (150 to 300 nm) aluminum film that improves luminous efficiency by stabilizing the operating voltage and by reflecting more of the emitted light toward the observer. A wide variety of phosphors are available commercially with differing characteristics (Infante, 1986). For satisfactory operation, the right choice of phosphor material and deposition techniques must be made. The characteristics one usually looks for in a phosphor are efficiency, resolution, and good life, together with the appropriate color and persistence.

Color CRTs (Morrell et al. 1974)

Color is a major technological and marketing challenge for displays. Color is important both in computer displays, as it represents an additional information channel, and in the entertainment area, as color displays are much more natural and appealing than monochrome.

To display color, one takes advantage of Helmholtz's trichromaticity theory. Thus, each color can be represented by the addition of three independent primaries. As a result, much more information must be processed to display color. Practical devices achieve color by some form of sharing or separation mechanism in either space or in time. Picture elements in color are obtained

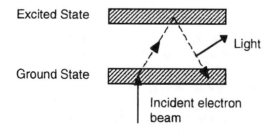

FIGURE 18.9 Phosphor operation.

by synthesizing either subpixels that are spatially separated or individual monochrome picture frames that are temporally distinct. Each subpixel or subframe is composed of a single primary color. The subpixel technique implies an inherent loss in spatial resolution, whereas the frame sequential technique requires a faster refresh rate to avoid flicker.

A number of technological solutions have been proposed for color CRTs including the CBS spinning disk, penetration CRTs, and the liquid crystal color shutter. None has achieved the market and technological success approaching that of the shadow mask CRT and its successors. The development of this device is linked to the name of David Sarnoff, the founder of RCA, and to the laboratory that bears his name.

Ergonomics of Color

Before exploring the hardware aspects of color, we pause for a moment to consider the ergonomic aspects of viewing color (Silverstein, 1994; Travis, 1991). While seeing color is a natural experience, the science of color is quite complex and as yet not totally developed. As an example of the complexity of the subject, the color of an object is dependent not only on the spectral content of the emitted light, but also on the characteristics of the surround and of the ambient illumination. Furthermore, a number of differing spectral representations exist for the same (perceived) color.

An essential problem is that, of a metric or space with which to represent color in a mathematical sense. There have been a number of proposals, none of them totally satisfactory. Currently, the CIE XYZ system is the most widespread. It is based on a set of color matching functions that define any color by three values.

Two classes of applications are often encountered. The first involves natural images, in which the fidelity of the color representation is essential. A second type of image has no correspondence in the real world, but uses color to distinguish or categorize information. Examples of the latter are color charts representing data or word processing documents in which color is used to highlight information.

Shadow-Mask CRTs

A number of alternative architectures for shadow-mask CRTs have evolved over the years. These are shown in Fig. 18.10. Although there are substantial variations among the devices, operating principles are similar. Three electron guns are employed and a thin mask or grille is interposed between the guns and the screen. The mask is made of stainless steel and "extrudes" the beams into beamlets insuring that each beamlet lands on the correct phosphor dot element. Naturally, a full frame of a single primary (e.g., red) should appear of uniform color to the eye (purity). For satisfactory performance, the registration between beamlet and phosphor element must be held to no more than a few microns. This is achieved by manufacturing each mask and screen as a matched pair (lighthouse exposure). A further requirement of a color display device is that small images (graphic lines or text characters) appear without fringing colors (convergence). This, in turn, requires that the centroid of each beam coincide with the other two beams over the entire screen to within 100 microns or so.

Delta Gun. The so-called delta gun is the oldest architecture and the simplest conceptually. Three individual guns are arranged in a triangle or delta inside the tube neck and angled (1° or so) with respect to the axis. The shadow mask consists of a thin (0.1 mm) sheet of steel with an aperture corresponding to each phosphor dot triad. The delta architecture makes good use of the available neck space, allowing relatively large electron-optical lenses. However, maintaining accurate positioning of the electron guns over time, temperature, and manufacturing process variables is very difficult. Delta guns require complex circuitry to compensate for these tolerances and have become practically obsolete.

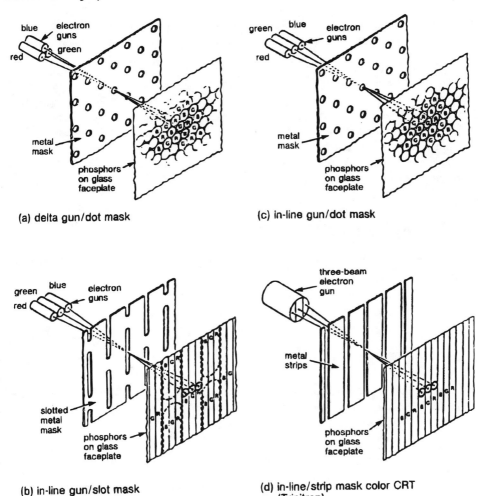

FIGURE 18.10 Shadow-Mask architectures (Silverstein and Merrifield, 1985)

In-Line Gun. In the in-line structure, the guns are arranged horizontally with the center (green) gun parallel to the axis. The outer guns are tilted inward by about 0.5°. In high resolution applications, a dot screen, similar to the delta dot screen is used, whereas, for lower resolution applications, a slotted mask is preferred for increased brightness. Eliminating the vertical tilt angle between guns and adopting a unitized mechanical structure greatly reduces purity and convergence difficulties to the point where these corrections are routinely built into the deflection yoke. A detailed description of self-converging yokes is beyond the scope of this work. The interested reader should consult the excellent literature on the subject (Dasgupta, 1991).

The Trinitron. A different approach to the problem was developed by Sony with the Trinitron. The guns are, again, arranged horizontally, but the beams go through a common focus lens. This achieves a space utilization factor even better than delta guns, by having a large common focus lens as shown in Fig. 18.10. For this to work, Sony had to use a more complex gun structure by adding a prefocus lens, so that most of the electron rays would go through the center of the lens as opposed to the periphery where aberrations are high. The faceplate is cylindrical and the mask is replaced by a grilled aperture. The phosphor is deposited on the screen in stripes. This eliminates the loss in resolution in the vertical direction. In moderate resolution applications, the grille allows higher transmission of the electron beam, thus, increasing luminance. This comes at the added expense and weight of a steel frame required to hold the grille under tension.

The Flat-Tension Mask. Yet another shadow-mask architecture is the flat-tension mask (FTM) that was developed by Zenith (Fig. 18.11). Instead of a domed shadow mask, the device uses a mask held under tension in both vertical and horizontal directions. This requires the use of a flat glass faceplate. Because the mask is under tension, as it absorbs power from the electron beam, the mask itself does not dilate, thereby, affecting purity, but remains flat. As a result, greater luminance levels can be achieved, together with other advantages, such as enabling interchangeable masks and screens.

Important Relationships

There are a number of mathematical relationships that apply to CRT systems. The most important are given below.

The luminance produced by a raster-scanned CRT system is given by

$$\text{Lum} = \frac{\eta T_G}{\pi} \frac{i_b E_b}{A} \delta,$$

where η is the phosphor efficiency in lumens/watt, T_G is the glass transmission, i_b and E_b are the beam current and voltage, and A is the raster area (m^2). δ is the raster duty cycle and Lum is the resulting luminance in nits (cd/m^2).

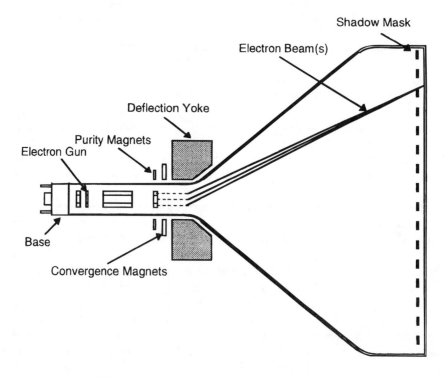

FIGURE 18.11 Structure of the flat-tension-mask CRT.

Electronic Displays

The deflection angle θ is given by

$$\sin \theta = C_F N_j \sqrt{\frac{q}{2mV_b}}$$

where C_F is a constant that depends on the geometry, N_i is the number of ampere-turns in the yoke, q and m are the charge and mass of the electron, respectively, and V_b is the accelerating voltage.

The size of the spot on the CRT screen is a complex function of the gun parameters and of the CRT geometry. In general, it decreases as V_b is made larger and the gun is made longer. For moderate values of beam current, the shape of the spot is gaussian.

Other CRT Devices

There have been a number of attempts at devising so-called flat CRTs, i.e., devices with all the advantages of the CRT but with greatly reduced depth. Currently, the so-called field emitting device (FED) (Spindt et al. 1976) is receiving considerable attention from a number of laboratories around the world. It is shown schematically in Fig. 18.12.

These devices do not employ thermionic cathodes like CRTs, but depend on cold cathode emission from small microtips. Because of the close proximity to the other electrodes and the

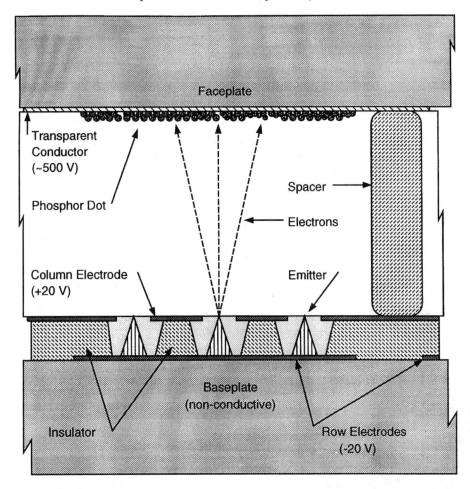

FIGURE 18.12 Structure of field Emitter display (Courtesy, Silicon Video).

shape of the tip, very high fields are present at the tip, which allows electron emission to take place. A large number of these tips can be built simultaneously, many per pixel. The tips are then addressed in a matrix form. Because of the small distance to the screen, there is no need for focusing structures. The electrons strike the screen, coated with phosphor, and light is emitted in much the same way as a CRT.

Clearly, there are many similarities between CRTs and FEDs. Both rely on vacuum, electron optics, and cathodoluminescence. The principal difference is that FEDs operate at much lower voltages, 300 to 1,000 V, as opposed to the 15–30 kV normally employed with CRTs. This not only makes very thin (a few millimeters) devices possible, but also enables low-voltage matrix addressing of the cathodes and eliminates X rays. It also means that many of the more common CRT phosphors are unusable and new phosphors such as those used with EL devices, must be chosen. There are number of technological and fabricating issues yet to be resolved. Chief among these are the viability of large area, low-cost devices and the demonstration of long-lived, low-voltage color phosphors with satisfactory chromaticity and luminous efficiency.

18.3 Flat Panel Devices

One of the principal limitations of the cathode-ray tube is its sheer bulk. This is due in large part to the difficulty of deflecting the electron beam by large angles. Thus, in applications in which size and weight are paramount, such as personal information managers, laptop computers, and pocket calculators, alternative display technologies are needed. Concurrently, it seems that something can be gained by abandoning the analog-addressing capability of the CRT, in favor of a digital approach.

To display just a limited set of alphanumeric characters, one can use economical displays featuring just seven segments, such as the one shown in Fig. 18.13

Complex images, however, require a more general arrangement such as the one shown in Fig 18.14.

To display an image, the row electrodes are addressed sequentially, beginning at the top. The appropriate voltages are applied to the column electrodes, thus, producing the image. Fig. 18.15 shows a simple matrix display consisting of 3 × 3 pixels, with the center pixel addressed (Scheffer, 1993).

It is apparent that, if simple addressing schemes are used, nonselected pixels are subjected to half the voltage of selected pixels. Unless the optical transfer characteristic of the display is very nonlinear, this can lead to poor contrast. Improved addressing schemes, such as the one shown in Fig. 18.15b, are possible. In the general case, one finds (Alt and Pleshko, 1974) that optimum performance is achieved when the voltages applied to the nonselect row electrodes and nonselect column electrodes are V_0/b and $2V_0/b$ respectively, where $b = \sqrt{N} + 1$ and N is the number of

FIGURE 18.13 Schematic diagram of a seven-segment display (Each segment has an independent connection).

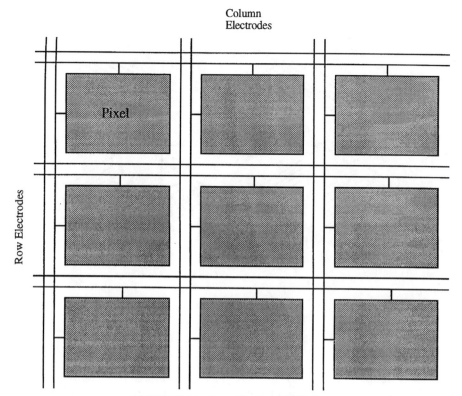

FIGURE 18.14 Generalized matrix display.

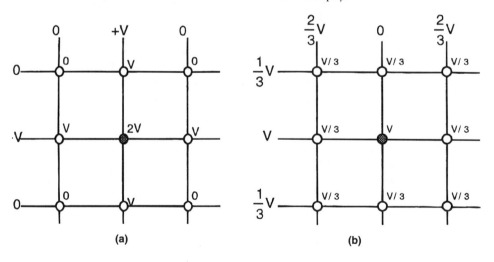

FIGURE 18.15 Matrix display addressing.

addressed lines. From this, it is clear that, as the resolution of the display increases, the ratio between select and nonselect voltage approaches unity. Thus, for high resolution images, display elements with sharp nonlinearities are required.

Liquid-Crystal Devices (LCDs)

Liquid-Crystal displays (LCDs) have been the subject of very intense development efforts in the last decades. Because of this and their inherent properties of low power requirement, low cost,

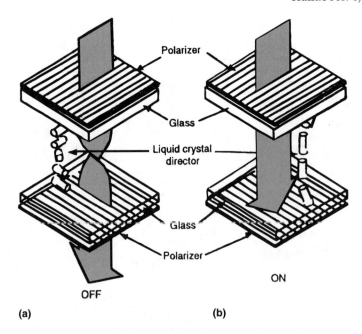

FIGURE 18.16 Operation of liquid-crystal displays (Scheffer 1993).

and the ability to display color, LCDs have become a major market presence. LCDs have been known for many years and were first applied to displays in the 1960s, at what was, then, called the RCA Laboratories. Liquid crystals are materials that combine the properties typical of ordered structures, such as crystals, while retaining the liquid form which makes them amenable to pouring, etc. There are a number of different material classes that have been developed over the years, such as nematic, cholesteric, and smectic (Chandrasekhar, 1992; Steemers, 1994). Currently the twisted nematic and its variants, such as supertwist nematic have become preeminent.

The fundamental property of LCs that make them useful as displays is that they are sensitive to an external electric field. This is shown in Fig. 18.16.

LC molecules rotate as a result of the electric field. The use of polarizers allows control of the optical transmission. As a result, an undriven cell will appear light, whereas the cell will appear dark when voltage is applied. The optical response is a function of the material and of the applied voltage, with a typical dependency shown in Fig. 18.17.

Because only a few volts are required for the cell to change states and the current drawn is negligible, the technology has found easy application in hand-held devices and laptop computers. The steepness of the response curve is finite, however. This is important in some applications, as will be seen.

Directly Addressed LCDs

Twisted Nematic Displays. To make a working display, the LC material is assembled in a cell, shown in Fig. 18.16. The cell consists of two glass plates, each coated with a transparent conductor. The cell is assembled so that the liquid-crystal molecules undergo a 90° twist from the top plate to the bottom plate. Polarizers are applied to the outside of each glass plate, with their polarization directions, usually, parallel to the LC direction of each plate.

The incident light on the cell is polarized linearly, as indicated. With an appropriate choice of LC material and cell gap, the light undergoes a 90° rotation in polarization as it passes through the liquid crystal, before exiting the bottom polarizer. In this mode of operation frequently used in LCDs, the cell is transmissive in its undriven state. When a voltage is applied to the liquid crystal, the molecules will align with the field, as shown in Fig. 18.16(b). The incident light does

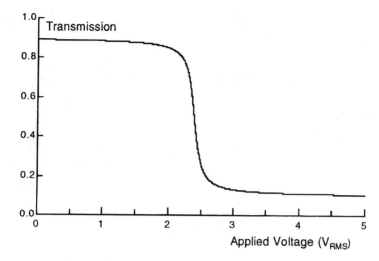

FIGURE 18.17 Transfer characteristic of a typical liquid-crystal cell [18].

not undergo rotation in polarization direction and is, therefore, absorbed in the exit polarizer. If the exit polarizer is rotated by 90°, the TN cell is opaque in the undriven state, and transmissive in the driven state. This is called the 'normally dark' mode, which is less frequently used. It is important to note that the cell responds to the RMS value of the driving voltage. Any dc component results in drastically reduced cell life.

Multiplexed Addressing. The straightforward addressing schemes of Fig. 18.15 are referred to as multiplexed addressing. With these schemes, the limited steepness of the electro-optic curve of available TN materials resulted in displays with satisfactory contrast and viewing angles only for very small number of elements.

Supertwisted Nematic Displays. Improving the electro-optic curve by brute force proved to be quite difficult. However, increasing the twist from the top plate to the bottom plate from 90° to 180° or even 270°, resulted in materials with much steeper curves. LC materials made in this way are called supertwist nematic (STN). Although the STN LCD made high-resolution displays possible, it has a number of limitations. Due to the very limited duty cycle of the addressing pulse for high-resolution displays, materials with response times of 150 to 300 ms are required. This, in turn, limits the display's ability to follow rapid changes in the image, such as mouse or cursor movements.

Another limitation is that the STN cell operates in a black and blue mode, as opposed to the black and white operation of a TN cell. This effect can be compensated for by adding second, undriven, STN cell with opposite liquid-crystal molecule rotation (called the double STN or DSTN cell) at some cost penalty.

Active Addressing. A new method, known as active addressing of STN displays, has been described in the literature (Scheffer, 1993). It involves optimizing the driving waveforms and appears to be a significant improvement in addressing technology. An STN display with a 60:1 contrast ratio and a 35-ms response time has been demonstrated.

Active-Matrix LCDs

Two and Three-Terminal Devices. For a liquid crystal to perform satisfactorily in directly addressed, high-resolution applications, it must demonstrate a steep electro-optic curve, as we have seen, in addition to offering wide viewing angles, good contrast, low-voltage operation, and

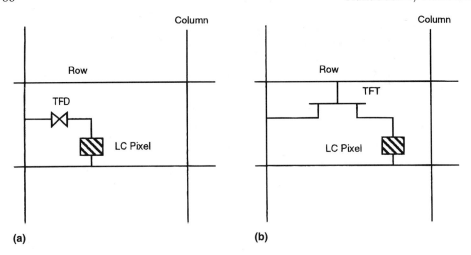

FIGURE 18.18 Active matrix liquid crystals.

a wide temperature operating range. A possible solution was proposed in the 1970s consisting of inserting a solid-state circuit device with strong nonlinearities (Morozumi, 1993). As Fig. 18.18 shows, the nonlinear device can be either a two-terminal device, such as a diode, or a three-terminal device, such as a transistor. The manufacturing processes to make thin-film devices using amorphous silicon have been proven compatible with large area, low temperature standard glass. Furthermore, the technology can provide satisfactory low-current performance. Thin-film *diodes* (TFDs) have the obvious advantage of being simpler devices that lead to higher yields in manufacturing. Thin-film *transistors* (TFTs), on the other hand, provide increased isolation between the control and signal lines and are less susceptible to manufacturing tolerances of the device. Currently, TFTs enjoy much greater popularity.

In summary, with AMLCDs, there is no inherent limit to the number of elements that can be addressed and, thus, to the resolution of the display. Furthermore, TN materials with good characteristics, such as contrast ratio and switching speed, can be used.

Manufacturing Issues (O'Mara, 1994). The validity of a technology is verified on the manufacturing floor. Although this is somewhat of a truism, the manufacture of LCD panels, both active and passive, poses a unique set of challenges and opportunities. The processes involved are quite complex, ranging from sputter deposition of transparent electrode material to amorphous silicon deposition by plasma-assisted chemical vapor deposition (PECVD) and including, of course, the most stringent clean room and process control techniques. Because a single short circuit or open conductor makes an entire row or column unusable, thus rejecting the entire panel, it is no surprise that manufacturing yields were very poor initially, of the order of 10% for a 10-in. color AMLCD. A series of technological and process advances, including adoption of class 10 clean rooms for lithography greater attention in combating electrostatic discharge, and the incorporation of second-generation manufacturing equipment resulted in the improvements summarized in Table 18.4.

TABLE 18.4 Historical Process Yield and Manufacturing Costs for 10-in. AMLCDs*

Year	Yield (approx)	Cost ($/module)
1991	10%	2,500
1992	32%	1,850
1993	53%	1,050

*O'Mara 1994.

Electronic Displays

Although the results are very impressive, yield and cost figures are still quite inferior to those achieved by older technologies, such as CRTs (approximately 90% and $60, respectively, for a 19-in. consumer-grade CRT). Only time will tell, of course, whether the continuing cost reductions demanded by the marketplace will be achieved. The continuing flow of massive investments bodes well for this aspect of the technology.

Color LCDs

An additive color approach is usually used in direct view displays, similar to the approach used in color CRTs. Pixels are grouped into triads and each aligned with a red, blue, or green color filter, processed on the cover sheet, as shown in Fig. 18.11. By controlling the transmission through each of these subpixels, a large number of colors can be rendered. A number of different color filter mosaics are used, with the 'stripe' mosaic commonly used for higher quality 'office automation' displays and the 'delta' mosaic often seen in 'audiovisual' displays.

To obtain good color performance, in addition to choosing filters with saturated primary colors, one also has to carefully select their transmissions to obtain a balanced white point. This requires careful optimization and good quality control practices in the filter materials.

Principles of LCD Lighting

As we seen previously, LDCs do not *generate* light but *modulate* it. Thus for satisfactory operation in a variety of ambients some form of supplementary lighting is necessary. Among the factors contributing to the great success of AMLCDs has been the development of backlighting technology (Lu et al. 1993; Lewin, 1993). Illuminating a liquid-crystal panel may seem simple, but the problem is far from trivial. The requirements of uniform illumination, low power consumption, long lamp life, compact physical dimensions, and, of course, low cost are difficult to achieve simultaneously. Initially, many LCDs were used in low-power applications and operated in the reflective mode, but the image quality was unsatisfactory. Currently, the vast majority of LCDs are operated in the transmissive mode and are backlit.

Early light sources for LCD backlighting were the powder EL lamps, which were very thin, but not very efficient, and short-lived. Currently, the most popular backlight technology uses cold cathode fluorescent lamps (CCFLs). These consist of a glass tube coated with phosphor and filled with low pressure gas. Applying a high voltage ionizes the gas and current flows, causing the emission of light, predominantly in the ultraviolet. The UV, in turn, excites the phosphor on the inside walls of the glass.

A typical backlight for a monochromatic STN LCD consists of a single CCFL illuminating a light pipe from the side as shown in Fig. 18.19.

The diffusing film at the top of the drawing insures, uniform resulting light, whereas the reflectors at the side and bottom increase the overall efficiency.

Because the optical transmission of black and white STN LCDs is about 20%, a backlight luminance of about 300 Cd/m^2 will result in a display luminance of 60 Cd/m^2, which is quite acceptable in an office.[3] This is routinely achieved with good uniformity and excellent life.

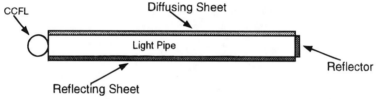

FIGURE 18.19 Schematic drawing of backlight for monochromatic LCDs (Lewin, 1993).

---

[3] See Table 18.2.

In the past, color LCDs had much lower optical transmission (about 3%). Thus, backlight designs for color LCDs used a number of CCFLs in a light box placed directly behind the display, with a consequent penalty in cost and power. Recent progress in color filter technology and improvements in TFT LCD design increased the optical transmission of color LCDs to about 4 to 5%. For a display luminance of 100 Cd/m^2, one needs a backlight luminance of 2,000 to 2,500 Cd/m^2. This is achieved with two modern, small diameter CCFLs (one on each end of the light pipe in Fig. 18.19) at a power consumption of about 6 watts.

Electroluminescent Devices (EL) (King, 1994)

Introduction

Electroluminescence is defined as nonthermal conversion of electrical energy into luminous energy. This would include the familiar light-emitting diode (LED) covered elsewhere. In *field electroluminescence devices*, commonly referred to as EL, light is generated by impact excitation of light-emitting centers in phosphor materials by high energy ($\approx$200 V) electrons. The electrons gain their high energy from an external electric field, thus, the name.

DC EL

Phosphorescent materials, such as ZnS, are known as efficient, low-voltage emitters, with a strong nonlinear relationship between the applied voltage and the resultant light. One can build simple flat panels in which the phosphor is directly addressed by conductors. The DCEL structure is shown in Fig. 18.20.

Although commercial devices have been made with this technology, it has gained a reputation for unreliability and limited life.

Thin Film EL (TFEL)

A more successful approach has been thin-film EL (TFEL) in which the phosphor is not directly in contact with the electrodes, but sandwiched between insulating layers as shown in Fig. 18.21.

Because of the capacitive nature of the coupling, excitation must be in AC form. Typical luminance-voltage characteristics are shown in Fig. 18.22. The highly nonlinear characteristic curves allow the device to be addressed electrically at a very high multiplexing ratio, while maintaining good contrast. This is what is required for the matrix addressing of flat panel displays with high information content, as was seen previously. Typical performance levels achieved by 640 × 480 TFEL units on the market are a luminance of 100 cd/m^2 and a contrast ratio of 15:1 in ambients typical of offices. These displays also have wide viewing angles ($\approx$160°).

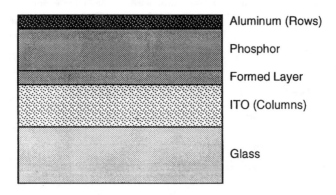

FIGURE 18.20 Structure of a dc flat panel.

Electronic Displays

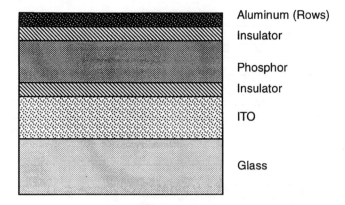

FIGURE 18.21 Structure of a TFEL flat panel.

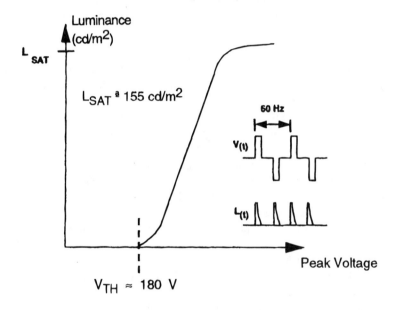

FIGURE 18.22 EO curve of EL.

EL Materials

Much of the performance of an EL display depends on the choice of phosphor material. In general, phosphors, whether used for EL or CRTs, consist of a host material doped with an activator which is the light emission center. The classical EL phosphor consists of a ZnS host lattice doped with Mn atoms for the light emission centers. To be a phosphor host lattice, a material must have a band gap large enough to emit visible light without absorption. This limits the class of possible materials to large-bandgap semiconductors (e.g., >2.5 eV) and insulators. The classical CRT phosphor host materials are the II–VI compounds and the rare-earth oxides and oxysulphides. To date, the oxide phosphors have been unsatisfactory, and only the II–VI materials, such as ZnS, SrS, and CaS, have demonstrated reasonable luminous efficiencies. This restricts the choice of materials for achieving color, as will be seen shortly.

Color EL

Any display technology must be capable of full color for long-term viability. This has posed a number of challenges to EL. One approach has been to develop a set of phosphors, similar to

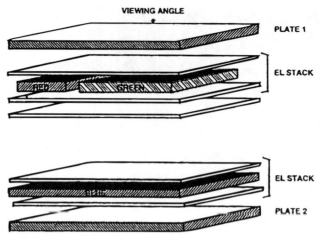

FIGURE 18.23 Construction of dual-substrate color EL (King, 1994).

what was done for CRTs, with emission spectra in the red, green, and blue, respectively. For satisfactory operation on a system level, this requires matching the *L-V* curves and achieving reasonable efficiencies so that the appropriate white is realized. This has caused considerable difficulty, particularly, with the blue phosphor.

An innovative solution has been reported recently and is shown in Fig. 18.23. The front (top) substrate consists of patterned red and green phosphors in an arrangement very similar to a conventional (monochromatic) EL layer structure. The rear substrate is a monochromatic blue-emitting EL panel with transparent top and bottom electrodes.

The dual panel structure also allows optimizing the drive electronics independently for each substrate. Furthermore, the threshold voltage of the blue phosphor does not need to match the threshold voltages for the red and green phosphors.

The performance reported for a complete panel based on the concept is summarized in Table 18.4.

A radically different approach is to use a single, broad-band phosphor that can then be filtered to produce an RGB display. This device structure is shown in Fig. 18.24. It has the advantage of maintaining the simple manufacturing process of a monochromatic TFEL display. The disadvantage is that it requires a very highly efficient broad-band phosphor.

Plasma Displays (PDPs)

Introduction

Plasma displays are one of the few display technologies that can readily be made in large sizes. Monochromatic versions, easily recognizable by their distinctive orange-reddish color, are often used in command and control applications in which their large size and shallow depth make them attractive alternatives to CRTs. PDPs can also achieve color as will be seen.

TABLE 18.5 Performance of Color TFEL Panel

Color Pixels		
		640×480
Displayed colors		16
Pixel pitch		0.31 mm
Subpixel size	Red	0.060×0.210 mm
	Green	0.170×0.210 mm
	Blue	0.210×0.210 mm

Electronic Displays

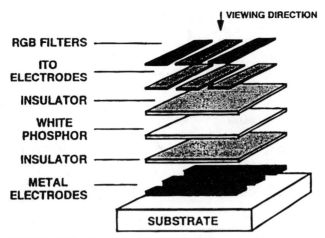

FIGURE 18.24 Construction of color EL employing fitters.

The operation of the device can best be understood by referring to Fig. 18.25 that shows the voltage–current relationship of a typical gas discharge found in plasma displays.

Note the large current range and the very strong nonlinearity at the firing voltage. This is a major attribute of gas discharges that allows matrix addressing. Plasma displays are generally operated at or near the junction of the normal and the abnormal glow regions.

PDP Addressing

Conceptually, there are a number of ways in which a plasma display can be addressed, DC, AC or a mixture of the two. A PDP can also be designed to support only segments or in a dot matrix form. Finally, PDPs may be designed to have inherent memory (i.e., the glow persists for some time after the excitation is removed). This has led to a variety of product and technological variations by a number of manufacturers who have produced PDPs with individual characteristics and enhancements. Because it is beyond the scope of this chapter to review each in detail, only the fundamental aspects will be covered in what follows. The interested reader is encouraged to pursue some of the excellent references on the subject (Weber, 1994).

DC PDPs

The simplest plasma product is a dc Plasma that is addressed along the lines of Fig. 18.4. A number of manufacturers make segmented displays that have achieved good reputations for cost and reliability under moderate ambient lighting conditions.

It is important to note that the current in a gas discharge must somehow be limited to a predictable value. DC plasma displays use an external resistor or a semiconductor current source. Furthermore, most dc plasma displays have a small amount of mercury added to the neon gas to extend the lifetime of the display by inhibiting the sputtering action of the gas discharge on the cathode.

AC PDPS

One undesirable characteristic of most dc displays is a background glow that reduces contrast. This has led to the development of ac plasma displays in which current limiting is achieved by a capacitor placed in series with each pixel. The driving waveforms are coupled capacitively to the gas discharge in much the same way as done in TFEL.

AC displays can achieve memory. When a voltage pulse is applied to an ac panel, the discharge deposits a charge on the wall that reduces the voltage across the gas. After a short time, the discharge will extinguish, and the light output will end until the applied voltage reverses polarity

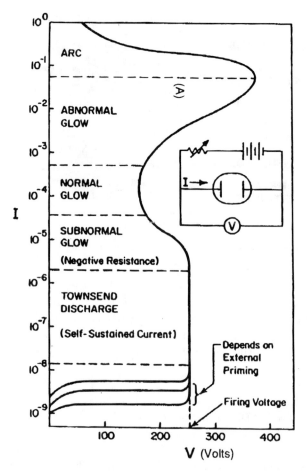

FIGURE 18.25 EO curve for plasma.

and a new discharge pulse occurs. The net effect of this memory mode, is a great increase in the effective duty cycle of the display, thus, an increased brightness, especially, in large displays.

Hybrid AC-DC

A very successful implementation of plasma displays is the hybrid ac-dc that was developed by SONY (Amano et al. 1982). This is, basically, a dc plasma panel with the addition of an ac trigger electrode buried under the cathodes. This trigger electrode is not used for display purposes but is used to create priming particles for the dc discharge.

The priming particles have two beneficial effects: they allow use of lower voltages to drive the normal dc address electrodes, which in turn require lower cost circuit drivers. Another advantage of this priming action is the shorter time delay between the start of the address pulse and the ignition of the discharge. Which, in turn, achieves a great increase in luminance uniformity.

Color Plasma Displays

After a long period of gestation, full color plasma displays have appeared on the market. These panels generate color by inserting RGB phosphors in the panel and exciting those phosphors with the ultraviolet light of the gas discharge. This creates a new set of problems that have been overcome in a variety ways, although intensive research is still occurring. One problem is, of course, that of luminous efficiency, due to the double energy conversion. A separate problem is the need for eliminating the cross talk between subpixels to insure purity.

Fig. 18.26 (Shinoda et al. 1993) shows a 21 in.-diagonal color PDP with 640 × 480 resolution. Note the barrier rib separators between each subpixel. The barrier ribs do not transmit the 147-nm radiation generated by the xenon gas used in most color plasma displays. A number of other manufacturers are working on similar approaches.

Conclusion

Some consider PDP a leading candidate for the large color displays required for the next generation of consumer television, in particular, for high-definition TV. There are a number of difficulties that must be resolved however. One is the high power requirements. These could be overcome by an increase in luminous efficiency. Another major obstacle is the high cost of the rather complex structures involved in PDPs.

Light Emitting Diodes (LEDs)

The physics of light-emitting diodes (LEDs) (Scherr, 1993) are addressed elsewhere in this volume. Some of the application considerations will be covered here. LEDs appear to be ideal candidates for matrix-driven displays. They are solid-state, low-voltage-driven, light-emitters with an intrinsic sharp, nonlinear threshold and a fast response. LEDs have been successfully fabricated to emit various colors including red, green, yellow, orange, amber, and blue. The luminous efficiencies that have been achieved depend on a number of factors, including the wavelength of the emitted light, and are of the order of a few lumens/watt or less.

Unfortunately, the fast response characteristic of LEDs has made them unsuitable for the high-resolution displays needed for personal computers and television applications. These displays, typically, require on the order of 0.1 to 1 million pixels. The resultant duty cycle, or on time, of the driving waveform for each pixel reduces the luminance and contrast to unacceptable levels.

Fortunately this does not apply to applications requiring 10 to 16 digits for which structures such as the seven-segment display are appropriate. This has made LEDs popular in applications, such as pocket calculators. Single-digit LED displays can also be made in moderately large sizes (2–6 in.) which can be assembled in message boards containing six lines or so of 80 characters.

Other Flat Panel Technologies

There are a number of flat panel technologies under development that deserve mention. *Electrophoretic display* is an intriguing concept that never quite fulfilled its promises. A cell, consisting of front and back glass plates, is filled with liquid containing electrophoretic particles in a suspension.

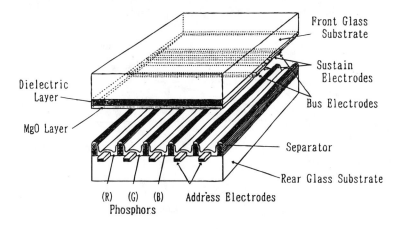

FIGURE 18.26 Example of color PDP.

The glass plates have conducting plates in contact with the liquid. The front plate is transparent and allows viewing of the image. The suspended particles, often made of TiO_2, become charged and are attracted to one of the electrodes by the action of an external electric field, thus, forming the image. Prototype displays with excellent contrast have been demonstrated, but keeping the particles from forming large aggregates or migrating to the bottom or the top of the cell over time due to imperfect gravity matching has proven quite difficult.

Another technique with promise is a derivative of liquid-crystal displays, the so-called *ferroelectric displays* (FLCD). Laboratory FLCDs have been made with high switching speeds, high contrast ratios and wide viewing angles. FLCDs displays are bistable: each pixel is driven into a white or black state. The pixel will remain in that state until it is switched again with the opposite polarity. This feature makes it feasible to address many lines at very low duty cycles and without loss of contrast. FLCDs require a very small cell gap (about 1.5 μm) for high performance, which makes mechanical ruggedness difficult. These displays have the potential of low cost because an active matrix is not needed. Large, high-resolution displays with FLCDs have been demonstrated, but commercial production is some years away.

18.4 Display Measurement and Performance Evaluation

The techniques that are the basis for setting, verifying, and monitoring specifications for displays form a subtechnology of their own. Some general principles and guidelines follow.

The most important characteristic of a display is its luminance, determined in great part, by the ambient conditions in which the product will be used. It is good to keep in mind that the human visual system responds logarithmically to luminance, so that an excessive precision in specifying or measuring luminance is inappropriate. Furthermore, the precision of most photometers is no better than a few percent (Infante, 1994).

Another characteristic of displays is the reflectance of the viewing surface that can play a large part in determining the usability of the display, particularly, in bright ambients. Reflectance can be specular or diffuse, and different measurement techniques are appropriate for each one. An example of a typical measurement process is shown in Fig. 18.27.

A very important characteristic of displays is the ability to present fine detail. This is often referred to as resolution, although the term is often misused and not distinguished from addressability, which is the characteristic of a display that measures how many pixels per unit length are presented. Resolution is a measure of the ability to resolve or distinguish fine detail. In many flat panels, the two are synonymous, whereas for CRTs, the two are quite distinct. A CRT display cannot resolve detail much smaller than the width of its own electron beam, although the beam itself might be stepped into much smaller steps. Measurement of resolution in a CRT

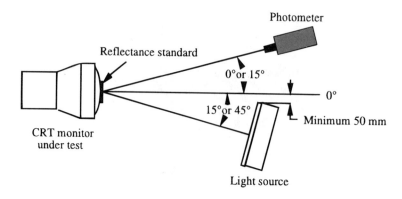

FIGURE 18.27 Reflectance measurement.

display is no simple matter because the profile of the beam itself is generally gaussian and its determination may be made complex by the shadow mask and screen structure in color CRTs. The effects of the latter will also contribute to the sharpness of the resulting image. Thus, it is not unusual for CRT displays to call out addressability, the size of the electron beam on the screen (at, say, 50% of the gaussian distribution), and the pitch of the shadow mask.

Another important parameter is a measure of the quality of the color presented. Although this may seem a trivial problem, measuring color is not quite as straightforward as it may seem, because of the influence of the ambient light, the surround, and the vision characteristics of the individual observer. Colorimeters with good accuracy and repeatability are available from a number of vendors.

Finally, the so-called geometric characteristics of a display can contribute a great deal to the perceived quality. These are factors, such as the apparent straightness of lines, the absence of distortions that transform squares into trapezoids, etc. These again are the province of special-purpose instrumentation.

18.5 Conclusions

The preceding discussion showed the variety of display technologies, all offering differing solutions to the common problem of transferring information to a human being. It should also be apparent that displays are supported by an enormous number of subtechnologies, ranging from detailed investigations of the properties of exotic materials to the very exact and demanding aspects of electron optics. The physical and psychophysical aspects of the human visual system are an underlying thread of all displays. What may not have been apparent is the scope of the continuing effort extant in the field. Older technologies, such as the CRT, continue to be perfected. Newer technologies, such as active matrix liquid crystals, achieve new performance levels on an almost daily basis. Finally brand-new technologies are being created. If an affordable solution to the age old problem of the "picture on the wall" still seems elusive, the continued intellectual and market growth of display technology seems assured for the foreseeable future.

References

Alt, P. M., and Pleshko, P. 1974. Scanning Limitations of Liquid Crystal Displays. *IEEE Trans. Electron Devices*, ED-21(7): 146–154.

Amano, Y., Yoshida, K., and Shionoya, T. 1982. High-Resolution DC plasma display panel. *SID Symp. Dig.* 160–161.

ANSI/HFS-100. 1988. American National Standard for Human Factors Engineering of Visual Display Terminal Workstation. Human Factors Soc.

Castellano, J. A. 1994. The CRT display market: Strategies and trends in the 1990s. *SID Symp. Dig.*, 25: 215–218.

Chandrasekhar, S. 1992. *Liquid Crystals*, 2nd ed. Cambridge, England: Cambridge University, p. 460.

Dasgupta, B. B. 1991. Deflection and convergence technology. *SID Seminar Lecture Notes* 1: 1–43.

El-Kareh, A. B., and El-Kareh, J. C. 1970. *Electron Beams, Lenses and Optics*. New York: Academic, Vols. I and II.

Farrell, J. E., Benson, B. L., and Haynie, C. R. 1987. Predicting Flicker Thresholds for Video Display Terminals. *Proceedings of the SID* 28(4): 449–453.

Infante, C. 1986. Introduction to CRT systems. *SID Seminar Lecture Notes* 1: 1–106.

Infante, C. 1994. CRT Display measurements and quality. *SID Seminar Lecture Notes* M-7: 1–54.

King, C. N. 1994. Electroluminescent Displays. *SID Seminar Lecture Notes* 1: M-9/1–38.

Lewin, I. 1993. Principles of LCD lighting. *SID Seminar Lecture Notes* 1: M-9/1–40.

Lu, S., Kuo, T., Huang, J., and Lin, F. 1993. Bright and Thin LCD Backlights for Monochrome and Color LCDs.

Morozumi, S. 1993. Active-Matrix LCDs. *SID Seminar Lecture Notes* 1: M-8/1–31.

Morrell, A. M., Law, H. B., Ramberg, E. G. and Herold, E. W. 1974. *Color Television Picture Tubes*. New York: Academic, p. 226.

O'Mara, W. C. 1994. AMLCD manufacturing. *SID Seminar Lecture Notes*: M-3/1–40.

Scheffer, T. 1993. Supertwisted-Nematic (STN) LCDs. *SID Seminar Lecture Notes* 1: M-7/1–63.

Sherr, S. 1993. *Electronic Displays*, 2nd ed. John Wiley & Sons, p. 624.

Shinoda, T., Wakitani, M., Nanto, T., Yoshikawa, K., et al. Development of technologies for large-area color AC plasma displays. *SID Symp. Dig.*, 161–164.

Silverstein, L. D., and Merrifield, R. M. The development and evaluation of color systems for airborne applications. DOT/FAA July 1985 PM-85-19.

Silverstein, L. D. 1994. Color in CRT and LC Displays. *SID Seminar Lecture Notes* F-3: 1–70.

Spindt, C. A., Brodie, I., Humphrey, L., and Westerbers, E. R. 1976. Physical properties of thin film field emission cathodes. *J. Appl. Phys.* 47: 5248.

Steemers, H. 1994. Fundamentals of liquid-crystal displays. *SIS Short Courses* S-3: 1–66.

Travis, D. 1991. *Effective Color Displays: Theory and Practice*. Academic.

Weber, L. F. 1994. Plasma Displays. *SID Seminar Lecture Notes* 1: M-8/1–35.

Yamazaki, E. 1993. CRT Displays. *SID Seminar Lecture Notes* II: F-5/1–47.

Index

Index

A

Abbe number, 470, 472
Above-bandgap absorption, 11
Absorbed light energy, 108
Absorber boundary condition, 561
Absorption, 5
Absorption coefficient, 5, 8–9
AC plasma displays, 791–792
Acoustic attenuation, 414
Acousto-optic devices, 411–418, 598
  differences from magnetostatic wave-optical devices, 420–421
Acousto-optic filters, 418, 693
Acousto-optic materials
  properties of, 414
Acousto-optic modulators, 415–417
Active Q switching, 378–381
Active-matrix liquid crystal displays, 785–787
Air quality monitoring, 57
AlGaAs material system
  absorption coefficient, 8–9
  bandgap, 8–9
  electronic applications, 7–8
  general characteristics, 7–8
  optoelectronic applications, 7–8
  refractive index, 8–9
AlGaInAs/InP material system
  absorption coefficient, 15
  direct bandgap, 15
  general characteristics, 14–15
  index, 15
  optoelectronic applications, 14–15
AlGaInP material system
  absorption coefficient, 16–17
  bandgaps, 16–17
  general characteristics, 15–17
  optoelectronic applications, 15–17
  refractive index, 16–17
AlGaP, 7
AlGaSb, 7
alpha SiC, 52

alpha-Alanine-doped TGS, 120
Amplified spontaneous emission, 669–670
Amplified-channel block diagram, 671
Amplifier cascades, 682
  and erbium-doped fiber amplifiers, 682
Amplifiers, in wavelength-division multiplexing, 694
Amplitude modulation, 300, 375
Amplitude-modulation vestigial-sideband, 704
Amplitude-shift keying, 650
Analog coherent optical processing, 591–602
  system components, 596–599
Analog optical communications, 703–709
Analog optical signal processing, 606
Analog signals, 646–648
Angle tuning, 136
Angular non-critical phase-matching, 235
Angular spectrum technique, 480, 497
ANSI 1988
  See also ANSI/HFS Standard, 770
ANSI/HFS 100 Standard, 770
  See also ANSI 1988, 770
Aperture guiding, 356
Area efficiency, 339
Aromaticity, 228–229
ASTEX CECR, 61
Asynchronous Transfer Mode
  See also ATM, 687
ATM
  See also Asynchronous Transfer Mode, 687
Attenuation in optical fibers, 443, 454
Avalanche multiplication, 311
Avalanche photodiodes, 302, 311–316, 639–640

B

Bacteriorhodopsin as optical modulator, 428
Band alignment, conversion of, 42
Bandgap energy, 8–9
  and room temperature, 9–10
Bandgap wavelength, 12–13
Bandwidth, 302

Banyan network, 613
Beam propagation method, 509, 560–561
Bessel functions, 440
Beta SiC
　See also 3C SiC, 52
Binary compounds, 5
Binary optic project cycle
　basic tasks, 461
　diagram, 462
Binary optics, 459–496
　advantages, 460–461
　applications, 461
　as microoptics, 471–473
　based on Maxwell's equations, 479–485
　based on scalar diffraction theory, 479–485
　fabrication, 463, 475
　　alternate methods, 495–496
　for monochromatic systems
　　design of, 468–470
　for thermal correction, 471–473
　for wideband systems
　　design of, 470–471
　geometrical optic designs
　　advantages, 464
　manufacturing tolerances, 478–479
　optical performance, 473–479
　ray tracing of, 463–479
　replication methods, 496
　using multiple masks, 490–495
　　processing steps, 492
Binary source data, 750
Biological materials as optical modulators, 428
Biphase code, 751
　See also Manchester code, 751
Biphase modulation coding format, 648
Birefringence, 91, 102, 111–113, 446–449, 454
Birefringence measurement, 510
Bismuth titanate crystals, 114–115, 122
Bit-error-rate, 625, 658–661
Blank fabrication
　processes, 446
Blue-green light emitting devices, 31–38
BNN crystals, 116–117, 119
Bohr radius, 27–28, 34, 36–37
Bond alternation, 227
Bound exciton recombination, 57
Bragg angle, 411
Bragg cells
　See also 1-D acousto-optic spatial light modulators, 415
Bragg condition, 637
Bragg grating, 516, 521, 689
Bragg regime, 411–412, 638
Brillouin scattering, 92, 454
Brillouin zone, 39
Broadband all-optical wavelength shifter
　semiconductor optical amplifiers in, 677
Bulk material, 7
Buried waveguide fabrication, 505
Burn-bright media, 738
Bus configuration in multichannel optical systems, 684–685

C

Cable television, 704
Cable television channel distribution, 707–708
Calcium titanate
　See Perovskites
Carrier-to-noise ratio, 705
Catastrophic optical damage, 301
Cathode ray tube systems
　mathematical relationships applying to, 780–781
Cathode ray tubes, 773–782
　block diagram of, 773
Cavity design, 350, 353–357
　Fabry-Perot cavities, 355–356
Cavity dumping, 383
CdSe/ZnTe heterostructures, 38
Charge-coupled device, 764
Chemical vapor deposition, 54, 99, 100
Cherenkov radiation, 137, 241
Chirp, 300
Christoffel matrix, 414
Chromatic correction, 470
　of purely diffractive systems, 471–473
　of wideband imaging systems, 470
Chromium, 346–348
CIE XYZ system, 778
Circuit switching, 686–687
Circular grating Bragg reflectors, 524
Cladding layer, 20
Cladding/electron confinement layers, 11, 16, 18
Clipping, 705
Cobalt, 349
Code, definition of, 750
Code-division multiplexing, 699–700
　advantage, 700
　disadvantages, 700
Coding, 750
Coercive field strength, 89
Coherent analog optical processors, 591
　operations, 592–596
Cold cathode fluorescent lamps, 787
Color
　ergonomic aspects, 778
　mathematical representation of, 778
Color cathode ray tubes, 777–781
Color plasma displays, 792–793
Column III elements
　compounds
　　aluminum, 5
　　gallium, 5
　　indium, 5
Column V elements
　compounds
　　antimony, 5
　　arsenic, 5
　　phosphorus, 5
Common anion, 27
Common anion rule, 27
Compact disks, 732–737
　fidelity in, 733
　injection molding of, 737

Index

mark permanence in, 733
optical contrast in, 733
Composite second order, 707
Composite triple beat, 707
Composites, 92
Computational origami, 618–620
Computer-generated holograms, 594–596, 598–599
Computer-generated waveguide hologram, 520
Conduction band, 18
Connection Machine, 605
Contact resistance, 34, 36
Continuous energy spectrum, 44
Contrast sensitivity function, 771
Corona poling, 165–166
Correction of monochromatic aberrations, 470–471
Coulomb interaction, 27–28
Coupled-cavity Q switches, 379–381
Coupled-wave equations, 130–131, 134
Cree Research SiC light emitting diode, 72
Critical angle, 436–437, 454
Critical fusion frequency, 770
Critical point energies, 6
Cross-correlation technique, 526
Cross-linked polymer networks, 269–270
Cross-phase modulation, 451–452, 696
Cross-relaxation, 362–363
Crossbar, 612–613
    disadvantages, 613
Crossbar switches, 643
Crossover, 774
Crossover network, 612
Crosstalk interference, 761
Crystallographic etching, 495
Curie point, 89–90
Curie-Weiss law, 90, 115
Current injection
    See also Electrical biasing, 633
Cutoff wavelength, 445

D

Dammann grating, 485, 497
Dark conductivity, 108
Data reduction at infinite dilution, 173
Data transmission, methods of, 686
    circuit switching, 686
    packet switching, 686
DC electric-field-induced, second harmonic generation, 167–171
DC plasma displays, 791
Deflection systems in cathode ray tubes, 775–777
Deflection yoke, 777
Degenerate four-wave mixing, 139
Delta gun, 778
Diamond turning, 495–496
Dielectric cladding, 553–555
Difference-frequency generation, 92, 118, 133–135
DIFFRACT, 731
Diffraction, 5
Diffraction efficiency, 473–478, 497

Digital signals, 646
Diluted magnetic semiconductor heterostructures, 42–45
Diode lasers, 7, 11
Direct bandgap energy, See also Emission energy, 5
Direct detection, 650–652
Direct gap materials, 6
Direct modulation of light sources, 425
Directional coupler power divider, 564
Directional couplers, 642–643
Dispersion-shifted fibers, 445, 454, 630–631
Display systems
    ergonomic considerations, 770–772
        flicker, 770–771
        image quality, 771
        low frequency electric and magnetic field radiation, 772
        resolution, 771
        X-ray radiation, 772
Display technologies, 772–794
Distributed Bragg reflector laser structure, 299
Distributed feedback laser structure, 299
Distributed-feedback laser, 638
Domain switching, 88
Domain wall, 88
Donor-to-acceptor pair recombination, 57
Dopants, 55–56
Doping, 6, 27, 33–34
Doppler-shift relationship, 413
Double ion exchange process, 506
Double positioning boundaries, 54
Double-sided compact disks, 758–761
Double-sided DVD format disk, 758
    disadvantages, 759
DVD format, 753

E

Edge-emitting lasers, 75
Effective-index method, 532–537
Eigenmodes, 90
    See also Linear polarized rays, 90
Eight-to-fourteen modulation code, 752
Elastooptic effect, 91
    See also Photoelasticity, Piezooptic effect, 91
Electric-field poling, 165–166
Electrical biasing
    See also Current injection, 633
Electrically addressed devices, 410
Electrically addressed modulators, 397
Electrically controlled birefringence, 111–113
Electrically controlled scattering, 113–114
Electrically controlled surface deformation, 114
Electro-optic devices
    analysis and design, 559–560
        numerical procedure in, 560
Electro-optic effect, 398–399
Electro-optic modulation, 166–167
Electro-optic modulators, 405–411
Electro-optic Ti-diffused LiNbO3 ridge waveguide, mode confinement modulator, 574–580

Electro-optic tuning, 368
Electro-optical index change, 555–559
Electrode matrix, 397
Electrode matrix devices, 410–411
Electroluminescence, 788
Electroluminescent devices, 788–790
Electromagnetic wave propagation, 556
Electromigration, 607
Electron acceptors, 174, 225
Electron beam, 775
Electron confinement, 34–35
Electron confinement layer, 20
Electron cyclotron resonance, 61
Electron donors, 174, 225
Electron gun, 775
Electron optics, 774–775
Electron-beam, 397
Electron-beam devices, 410
Electron-beam lithography, 517
Electron-hole plasma, 36–37
Electronic computers
   limitations, 605–606
Electronic displays, 768–795
   general requirements, 768–769
   performance evaluation, 794–795
Electrooptic devices, 398–411
Electrooptic effect, 101–104, 159
Electrooptic optical effects, 5
Electrooptic spatial light modulators
   performance of, 407–408
Electrophoretic display, 793
Electrostatic deflection, 775–776
Emission, 5
Emission energy, 6
   See also Direct bandgap energy, 5
Energy diffusion, 357–362
Erbium, 345–346
Erbium-doped fiber amplifiers
   and amplifier cascades, 682
   and narrow linewidth sources, 682
   and preamplifiers and power amplifiers, 683
Erbium-doped fiber-optic amplifiers, 668, 677–684
   differences from semiconductor optical amplifiers, 677
Erbium-doped YAG
   laser parameters of, 354
Etch pit densities, 54
Exchange-coupled multilayer direct overwrite structures, 761
Exchange-coupled multilayers, 761
Exciton absorption, 27–28, 32
Exciton recombination, 36
Excitons, 27
Extinction ratio, 573
Eye diagram, 661–663

F

Fabry-Perot cavities, 355–356, 409, 636, 641
Faraday rotation spatial light modulator devices, 419–420
Faraday rotator, 646

Fast Fourier transform, 481, 561
Fast Fourier transform-beam propagation method, 561
Femtosecond pulse propagation, 526
Fermi level, 33
Fermi-Dirac occupancy probability function, 299
Ferroelasticity, 98
Ferroelectric ceramics, 90
Ferroelectric crystals, 87
   material structure, 95–98
   photorefraction of, 105–109
   photorefractive applications of, 119–120
   preparation method, 95–98
Ferroelectric displays, 794
Ferroelectric domains, 88
Ferroelectric hysteresis loop, 89
Ferroelectric liquid crystals, 92
Ferroelectric materials, 87–123
   applications, 87
   basic concepts and characteristics, 87–92
   crystal growth methods, 93
   electrooptic applications, 90
   families of, 92
      composites, 92
      ferroelectric liquid crystals, 92
      lithium niobate, 92
      lithium tantalate, 92
      miscellaneous inorganic materials, 92
      perovskites, 92
      potassium titanyl phosphate, 92
      tungsten-bronze-type niobates, 92
      water-soluble crystals, 92
   general physical properties, 93
   general properties, 92–101
   growth techniques, 92–101
   nonlinear optic applications, 90
   optical properties, 101–115
      electrooptic effect, 101–104
      nonlinear, 104–105
      photorefractive properties, 105–109
      pyroelectric effect, 109
   photonic applications, 115–123
      electrooptic applications, 116–118
      ferroelectric structure-related applications, 121–123
      nonlinear optical applications, 118–119
      photorefractive applications, 119–120
      pyroelectric applications, 120–121
   pyroelectric properties of, 109
   structures, 92–101
Ferroelectric phase, 87, 89
Ferroelectric structure-related optical properties, 110–115
Ferroelectric thin films, 98–101
   applications, 100
   fabrication techniques, 99
   potential optical applications, 99
Ferroelectricity, 87, 89
Ferroelectricity, discovery of, 87
Fiber connections
   methods, 615
Fiber Fabry-Perot tunable filter, 691
Fiber optic communications, 607
Fiber optic systems, 435

Fiber-optic transmission, 435
Field emitting device, 781
Field-assisted ion exchange processes, 506, 539–542
Field-effect-transistor technology, 317
Filtering, 418
Final output beam, 396
Finite differences-beam propagation method, 561
Finite differences-vector beam propagation method, 561
Flat cathode ray tubes, 781–782
Flat panel devices, 782–783
Flat-tension mask, 780
Flicker, 770–771
Focus error signal, 750
Focused-ion-beam lithography, 517
Form birefringence, 488, 497
4H SiC, 52
4K data
  See also Low temperature data, 6
Four-level gain media, 335
Four-out-of-fifteen 4/15 code, 751
Four-wave mixing, 137, 451, 696
Fourier law of heat transfer, 727
Fourier transform technique, 559, 648
Franz-Keldysh effect, 405
Fraunhofer approximation, 484
Free exciton recombination, 57
Free-space optical interconnects, 612–617
Free-space optics, 606
Frequency doubling
  See also Second-harmonic generation, 92
Frequency modulation, 300
Frequency-division multiplexing, 706–709
Frequency-shift keying, 650
Fresnel approximation, 476, 480, 497, 561
Fresnel lenses, 502
Frohlich interaction, 28

G

GaAsP material system, 20–22
  absorption spectra, 22
  bandgaps, 22
  electroluminescence energies, 22
  general characteristics, 20–22
  optoelectronic applications, 20–22
Gabor transform, 602
Gadolinium molybdate crystals, 115, 122
Gain media
  laser parameters of, 351–354
  models, 333–334
Gain switching, 383
GaInAsP/InP material system
  absorption coefficient, 11–13
  direct bandgap, 11–13
  general characteristics, 10–14
  optoelectronic applications, 10–14
  refractive index, 11–13
Gallium nitride as optoelectronic material, 49–81
  doping, 66–68
  ohmic contacts to, 69

GaN lasers, 75, 79
GaN-based detectors, 80
  applications, 80
GaN-based light emitting diodes, 37, 72–74
Gas sensing, 57
Gas source molecular beam epitaxy, 54
GaSb material system
  absorption coefficients, 18–20
  bandgaps, 18–20
  general characteristics, 17–18
  optoelectronic applications, 17–18
  refractive indices, 18–20
Gaussian noise distribution, 658
Generalized grating model, 464–466, 497
Geometrical optics, 436
Geometrical optics theory, 461
Glass constant, 108
Glass integrated optic devices, 512–526
Glass waveguides with gratings, 515–519
Global optical networks, 687
Graded-index multimode fiber, 438, 629
Gradient index profile, 441, 443, 454
Gradient-index materials
  See also GRIN materials, 502
Grating beam splitters, 485–486, 497
Grating equation, 465, 480–483, 497
GRIN materials, 502
  applications, 502
GRIN-rod lenses, 502
Guest-host composites, 264–265
Guided-wave photonic devices
  design and fabrication, 561–583
  symbols, 530–532

H

Half-wave voltage, 102
Head-up displays, 768
Helmet-mounted displays, 768
Helmholtz equation, 560
Helmholtz's trichromaticity theory, 777
Heterodyne detection
  advantages of, 653
  See also Coherent detection, 650, 652–655, 657
Heterojunction bipolar transistors, 7
HgTe/CdTe type-III heterostructures, 39–42
High electronic mobility transistors, 7
High-speed photodetectors, 7
Hit-miss transform, 602
Holmium, 345
Holmium-doped YAG
  laser parameters of, 353
Holographic storage, 761–764
Homoepitaxy, 37
Host crystals
  properties of, 349–351
Hough transform, 595
Hybrid AC-DC plasma displays, 792
Hybridization, 67
Hyper-Rayleigh scattering, 172–173
Hyperpolarizability, 173–174, 225–229

I

IBM's Rainbow wavelength-division multiplexing project, 697
Idler wave, 135
Illuminance
  See also Illumination, 769–770
Illumination, 769–770
  See also Illuminance, 769–770
Image quality, 771
Impact ionization, 311–312
Impermeability tensor, 91
In situ combustion monitoring, 57
In-line gun, 779
Index ellipsoid, 90–91
  See also Indicatrix, 556
Indicatrix
  See also Index ellipsoid, 556
Indirect bandgap energy, 5
Information superhighway
  See also National Information Infrastructure, 626
Injection level, 79
Inorganic materials, 399–402
  linear electro-optic materials, 399–401
  optical modulation of, 399–401
  quadratic electro-optic materials, 401–402
Input optical beam, 396
Instantaneous polarization, 104
Integrated optical heads, 755–756
Integrated optical lenses, 502
Integrated star coupler, 693
Intensity-dependent refractive index, 139
Interconnection networks
  optical implementation of, 613–615
Intermodal dispersion, 629–630
Intermodulation distortion, 674–675, 681
Intersymbol interference, 656
Intracavity Q switches, 378–379
Intramodal dispersion, 630–631, 637
Intramode dispersion, 443
Inverse scattering theorem, 453
Inversion domain boundaries, 54
Inversion symmetry, 127
Ion exchange from molten salt, 503
Ion exchange in glass, 502–503
Ion exchange processes
  types of, 503–506
Ion implantation, 45, 55–56
Ion milling, 495–496
Ion-exchanged channel waveguides, 545–553
Ion-exchanged glass waveguides, 503–506
  basic fabrication processes, 503–506
  loss reduction in, 510–512
  losses in, 509–512
Ion-exchanged waveguides with ionic masking, 505
ISO Standards, 752–753

J

Jenkins model, 9
Johnson noise
  See also Thermal noise, Nyquist noise, 305, 307, 311

K

KDP water-soluble ferroelectric crystals, 96–97, 117
Kerr effect, 91, 749
Kerr electrooptic coefficient
  See also Quadratic electrooptic coefficient, 102
Kerr rotation, 750
Kleinman's symmetry, 127
KLN crystals, 117
KLTN crystals, 117
$KNbO_3$ crystals, 116, 118–119
Korteweg-de Vries equations, 453
KTP crystals, 117–119

L

Laguerre-Gauss functions, 441
LAMBDANET, 697–698
Lambrecht and Segall's local density function calculation, 77
Land and groove recording, 761
Laplace's equation, 559
Laser beams, 486–487
  reshaping and homogenization of, 486–487
Laser design
  factors, 350, 353–355
Laser efficiency, 338–339
Laser oscillation, 634–635
Laser parameters, 349–354
Lasers, multimode operation of, 357
LATGS crystals
  See also alpha-Alanine-doped TGS, 120
Lattice constants, 26
Lattice match, 7
Lead germanium oxide, 115
Lead lanthanum zirconate titanate, 95
Lead zirconate titanate, 95
Light confinement, 34–35
Light-emitting diodes, 20–22, 634, 793
  disadvantages, 634
LIGHTMOD/SIGHTMOD devices, 420
$LiNbO_3$ crystals, 116–120
$LiNbO_3$ waveguides, 542–545
  by proton exchange, 542–545
  by Ti diffusion, 542–545
    See also Ti:$LiNbO_3$ waveguides, 542–545, 555–560
Linear electro-optic materials, 399–401
Linear electro-optic effect, 91, 108, 399
  See Pockels effect, 91
Linear gate array, 122
Linear polarized rays
  See also Eigenmodes, 90
Linearly polarized mode approximation, 441
Liouville equation, 162
Liquid crystal color shutter, 778
Liquid crystal display lighting, 787–788
Liquid crystals, 402–403
Liquid phase epitaxy, 54, 99
Liquid-crystal displays, 783–784
Liquid-crystal Fabry-Perot tunable filter, 691–693

Index

Liquid-crystal televisions, 410
LISA device, 420
Lithium niobate, 92, 95
Lithium tantalate, 92, 95
Lithium-niobate Mach-Zehnder interferometer, 644
Local area networks, 685
Logic gates, 610
Long distance communications
   erbium-doped fiber amplifiers in, 682
Long-distance communications
   semiconductor optical amplifiers in, 675–676
Long-pulse operation
   See also Quasi-CW operation, 372–382
Lorenz-Lorentz model, 161
Low frequency electric field radiation, 772
   health aspects, 772
Low frequency magnetic field radiation, 772
   health aspects, 772
Low temperature data
   See also 4K data, 6
Low-energy electron beam irradiation, 66
Luminance, 769–770
Luminous flux, 769
Luneburg lenses, 502

M

Mach-Zehnder interferometers, 144, 166, 512–514
Macroscopic susceptibility, 156–160
Magnetic deflection, 775–776
Magnetically induced superresolution, 761
Magneto-optic direct overwrite, 761
Magneto-optic spatial light modulator devices, 418–421
   advantages, 420
   performance, 419
Magneto-optical heads, 749
Magneto-optical Kerr effect, 720–721, 725
Magneto-optical media, 740–744
Magneto-optical recording, 720
Magnetostatic wave interactions, 420–421
Magnetostatic wave-optical devices, 420–421
   differences from acousto-optic devices, 420–421
Magnetostatic waves, 419
Main-chain polymers, 267–268
   nonlinear optic applications, 268
Maker fringe method, 164
Maker fringes, 164
Manchester code
   See also Biphase code, 751
Manley-Rowe relations, 131
Material dispersion, 443, 629
Maxwell's equations, 454, 488–489
Maxwell's relaxation time, 108
Mean time between failure, 301
MEBES format, 494
Mechanical connector, 632
Mechanical modulation devices, 421–424
   advantages of, 421
Membrane modulation devices, 422–424
Metal-organic chemical vapor phase deposition, 25, 99

Metal-semiconductor-metal photodiode, 302, 316–322
Metalloorganic deposition, 99
Metalloorganic vapor phase epitaxy, 60
   disadvantages, 61
Metropolitan-area networks, 685
Microchannel plate spatial light modulators, 408
Microchip laser, 342–344
Micromachining, 495–497
   techniques, 495–496
Micromechanical spatial light modulators
   performance of, 422
Micropipes, 53
Microscopic susceptibility, 160–164
Microtwins, 54
Microwave plasma excitation, 61
Mid-infrared properties, 6
Miniature solid-state lasers
   components, 331–333
   frequency tuning, 366–372
      mode selection, 366–367
      mode shifting, 367–372
   fundamental-transverse-mode operation, 355–356
   linewidth, 365–366
   polarization control, 357
   symbols, 326–330
Miscellaneous inorganic ferroelectric crystals, 92, 97–98
   bismuth titanate, 97–98
   gadolinium molybdate, 97–98
   lead germanium oxide, 98
MMCD format, 753
Modal approach, 489
Mode index measurements, 545
Mode locking, 384–385
   AM mode locking, 384–385
   FM mode locking, 385
Modified chemical vapor deposition, 446, 454
Modulation chirp, 643–644
Modulation transfer function, 771
Modulators, 643–644
Molecular beam epitaxy, 14, 25, 42
Molecular vibrational resonance, 443
Monochrome CRTs, 773–777
   applications, 774
Monolithic Isolated Single-mode End-pumped Ring, 342
Morphological processors, 602
Multichannel optical networks
   architectural types of, 685
Multichannel optical systems, 684–700
   architectures, 684–687
   topologies, 684–687
Multifocal lenses, 486
Multimode optical fibers, 435, 441, 443, 628–629
Multipath dispersive noise, 666
Multiple data layer compact disks, 758–761
Multiple diffraction orders, 485–487
Multiple quantum well materials, 404–405
Multiple wavelength mixing, 451
Multiple wavelengths, 452
Multiplexed addressing, 785
Multipoint optical systems, 664–665
Multiquantum-well photodiodes, 316

Multistage interconnection networks, 613
Multiuser optical systems, 684
Multiwavelength transmitters, 689–691

N

National Information Infrastructure, 626
   See also Information superhighway, 626
National Institute for Safety and Health, 772
National Institute of Standards and Technology, 752
Near-field optical methods, 757
Near-field scanning optical microscopy, 757
Neodymium, 340–344
Neodymium-doped YAG
   laser parameters of, 351
Neodymium-doped YLF
   laser parameters of, 352
Neodymium-doped YVO4
   laser parameters of, 352
Network formers, 503
Network intermediates, 503
Network modifiers, 503
Nichia Chemical GaN light emitting diode, 72
Nichia Chemical InGaN light emitting diode, 72
Nitride alloys
   properties, 69–72
Nitride thin films, 64–65
Nitrogen doping, 21, 34
Nitrogen vacancies, 65
Noise equivalent power, 302, 304–305
Non-return-to-zero coding, 751
Non-return-to-zero modulation coding format, 648
Nonlinear dielectric susceptibility
   See also Nonlinear optical coefficient, 104
Nonlinear hydrodynamic wave, 453
Nonlinear optical coefficient
   See also Nonlinear dielectric susceptibility, 104
Nonlinear optical effects, 5
Nonlinear optical properties, 104–105
Nonlinear optical susceptibility, 126–129
   definition of, 126–127
   symmetry properties, 127–129
Nonlinear optical switches, 144–145
Nonlinear pulse propagation in fiber, 453
Nonlinear wave equation, 129–130
Nonlinearity
   and transparency, 226
Nonplanar ring, 342
Normal-mode operation
   See also Quasi-CW operation, Long-pulse operation, 372–382
NRZ pulses, 700–703
Nyquist noise
   See also Thermal noise, Johnson noise, 305, 307, 311

O

Ohmic contacts, 35–36
Oil film modulation devices, 424
On-off-keying, 650
1-D Acousto-optic spatial light modulators
   See also Bragg cells, 415
1550 NM window, 435
One-dimensional acousto-optic devices, 416–417
One-phonon scattering model, 28–29
Onsager model, 161
Optical addressing, 396
Optical amplifiers, 667–684
   basic system configurations, 668
   fundamental characteristics, 667–673
      gain, 669–673
      noise, 669–673
Optical beam, 627
Optical biaxial crystals, 90
Optical bistability, 606
Optical communications, 624–709
   generic system, diagram of, 625
   sources of, 633–638
   system design, 663–664
Optical computer
   architecture, 617–620
Optical confinement factor, 293
Optical constants, 6
Optical correlator, 594
Optical data storage, 719–764
   advantages, 719
   design and modeling, 723–732
   disadvantages, 719–720
   optical design and modeling, 731–732
Optical design, 461–463
   methods, 461–463
Optical detectors, 638–641
Optical digital computing, 605–620
   basic systems, 646–666
      modulation formats, 646–651
      performance criteria, 658–663
   history, 606
   system design, 663–664
Optical disk media
   thermal design and modeling, 727–731
   versus magnetic disk media, 720
Optical disk substrates, 744–747
   functions of, 744
Optical feature extractor systems
   See also Optical image processors, 601–602
Optical fiber amplification, 449–451
Optical fiber fabrication, 443
Optical fibers
   attenuation, 443
   multimode, 435, 441–443, 454
      in local area networks, 435
   See also Fiber optic systems, 435–454
   rare earth-doped, 435
   single-mode, 435, 443–445
Optical field distribution, 581–582
Optical filters, 641–642
   common applications of, 641
Optical Fourier transform, 592–594
Optical heads, 748–750, 755–757
Optical image processors, 600–602
   See also Optical feature extractor systems, 600–602

Optical intensity, 6
Optical interconnects, 608–617
  system packaging of, 616–617
Optical isolators, 645–646
Optical modulators, 7, 393–428
Optical noise power, 655
Optical nonlineriaties, 451–452
Optical parametric oscillator, 135, 236–237
Optical path, 438
Optical path and component analysis software, 732
Optical phase conjugation, 92
Optical processes
  absorption, 5
  diffraction, 5
  electrooptic optical effects, 5
  emission, 5
  nonlinear optical effects, 5
  reflection, 5
  waveguiding, 5
Optical processing
  hardware modules, 600
Optical propagation, 435
Optical properties, 6
Optical pulse, 700–703
  See Solitons, 700–703
Optical shutters, 121
Optical signal power, 655
Optical signal processors, 599–601
Optical signal transmission
  advantages, 606
Optical storage disks
  types of, 732–748
    compact disks, 732–737
    magneto-optical media, 740–744
    optical tape, 747–748
    phase-change media, 738–739
    substrates, 744–747
    write-once ablative media, 737–738
Optical storage systems, 748–753
Optical Storage Testing and Preservation, 752
Optical super resolution, 756–757
Optical tapes, 747–748
Optical thin films
  design, 724–726
Optical uniaxial crystals, 90
Optical wave propagation, 627
Optical wavelength control, 418
Optically addressed devices, 408–410
Optically anisotropic ferroelectrics, 90
Optically isotropic ferroelectrics, 90
Opto-electronic integrated circuits
  semiconductor optical amplifiers in, 676
Optoelectronic applications, 5–6
Optoelectronic devices, 291–322
Optoelectronic integrated circuits, 291
Optoelectronic interconnection devices, 608–612
Optoelectronic processing devices, 608–612
Organic materials, 155–271
  second-order nonlinear optical properties of, 156–164
Organic materials, characterization of
  experimental techniques, 164–173

data reduction, 173–174
DC electric-field-induced, 167–171
electric-field poling, 165–166
electro-optic modulation, 166–167
hyper-Rayleigh scattering, 172–173
second-harmonic generation, 164
solvatochromic method, 171–172
Organic materials, optical modulation of, 402
Organic nonlinear optical crystals
  phase-matching properties, 235–236
  physical properties, 236
  transparency constraint, 229, 235
Oscillation, 92
Output-coupling efficiency, 339
Outrigger tracking, 748–749
Outside vapor deposition, 446, 454
Oxygen octahedra ferroelectrics
  lithium niobate, 92
  lithium tantalate, 92
  perovskites, 92

P

p-type doping, 66
Packet switching, 686–687
  buffering in, 687
  clock recovery in, 687
  contention resolution in, 687
  control in, 687
  routing in, 687
  synchronization in, 687
Parabolic grading, 36
Parabolic index profile, 438
Paraelectric phase, 89
Parametric amplification, 92, 135
Partially polarizing beam splitter, 749–750
Passive mode locking, 385
Passive Q switching, 381–382
Passive stars, in wavelength-division multiplexing, 693
Passive three-branch power divider, 562–564
Penetration cathode ray tubes, 778
Perfect Shuffle network, 613
Permanent polarization, 122
Permanent splice, 632
Perovskites, 92, 95
Personal communication systems, 708
Personal Information Managers, 768
Personal ultraviolet exposure dosimetry, 57
Phase conjugation, 139
Phase function, 464–465
Phase-change dual optical storage media, 739
Phase-change media, 738–739
Phase-matching, 135–137
Phase-matching angle, 105
Phase-shift keying, 650
Phase-transition modulation devices, 425
Phonon coupling, 28
Phosphors, 777
Photo-cross-linking, 270
Photoassisted domain-switching effect, 114

Photoconductivity, 108
Photoconductors, 301
   properties of, 301
      bandwidth, 304
      noise equivalent power, 304–305
      quantum efficiency, 302–303
      responsivity, 303
Photodetection, 638–641
Photodetectors, 301–322
   principles, 301–302
Photodiodes, 301
Photoelasticity, 91
   See also Elastooptic effect, Piezooptic effect, 91
Photolithographic masks, 490–495
Photolithography, 494, 497
Photomasks, 490–495
Photometry
   elements of, 769–770
Photon emission, 633–634
Photon energy, 7
Photonic switching gates and modulators
   semiconductor optical amplifiers in, 676–677
Photorefractive effect, 92
   applications, 92
   mechanism in ferroelectric crystals, 105
   See also Self-pumped phase conjugation, 92
Photorefractive materials, 403–404
   defect chemistry of, 107
Photorefractive properties, 105–109
Photorefractivity, 92
   kinetics of, 108
Photoresist disk mastering, 735
Pi system, 227
Piezoelectric crystal, 87–88
Piezooptic effect, 91
   See also Photoelasticity, Elastooptic effect, 91
PIN photodiodes, 301–302, 307–311, 639–640
Planar microlenses, 502
Planar optics, 617
Planarity, 227
Plasma display addressing, 791
Plasma displays, 790–793
Plasma-assisted chemical vapor deposition, 786
PLZT materials, 110–114, 121
   electrooptic applications, 121–122
Pockels effect, 91, 108, 159, 399
Point-to-point optical systems, 664–665
Polar crystals, 87–88
Polarization, 357, 454
Polarization beam splitters, 609–610
Polarization diversity, 646
Polarization-dependent gain, 674
Polarization-mode dispersion, 446–449
Polarization-preserving fibers, 446–447
Poled polymers, 239, 261–271
   number density, 240
   optical loss, 262
   poling, 240–241
   potential nonlinearity, 241
   processibility, 261–262

   properties of, 241–261
   temporal stability of nonlinearity, 262–264
Poly(Bis(Trifluoroethoxy)Phosphazene), 424
Polymers, 92
Polytypism, 51
Population inversion, 331–332
Potassium dihydrogen phosphate
   See also KDP water-soluble ferroelectric crystals, 96–97
Potassium ion exchange, 506, 508–512, 514, 526
Potassium titanyl phosphate, 92, 97
Prism-coupler method, 537, 539
Profile dispersion, 443
Programmable array logic, 617–618
Programmable optoelectronic microprocessor approach, 616
Proton exchange, 581
Pulse breakup, 526
Pulse broadening, 443–444
Pulse build-up time, 378
Pulsed operation, 372–385
Pump efficiency, 339
Push-pull tracking, 748–749
Pyroelectric coefficient, 90
Pyroelectric effect, 90, 109

Q

Q-switched laser
   maximum peak power from, 375–376
   maximum pulse energy, 376
   minimum pulse width, 377
Q-switched rate equations, 374–375
Q-switching, 374
Quadratic effects, 91
   See Kerr effect, 91
Quadratic electro-optic materials, 401–402
Quadratic electrooptic coefficient
   See also Kerr electrooptic coefficient, 102
Quantum confined stark effect, 27
Quantum confinement, 27–31, 34–35
Quantum dots, 45
Quantum efficiency, 302–303, 339
Quantum size, 6–7
Quantum theory of solids, 91
Quantum wells, 27
   optical properties, 6–7
   type I heterostructures, 27
   type II heterostructures, 38–39
Quantum wires, 45
Quasi phase matching, 136–137
Quasi-CW operation
   See also Long-pulse operation, Normal-mode operation, 372–382
Quasi-fast Hankel transform, 484
Quasi-three-level gain media, 334
Quasilocalization, 44
Quaternary compounds, 5

Index

R

Raleigh-Ritz variational procedure, 537, 548
Raman media
  properties of, 148
Raman phonon energies, 6
Raman scattering, 92, 452
Random alloys, 16
Rare earth-doped fibers, 435
Rare earth-doped waveguides, 520–524
Rare-earth dopants, 339–346
Rare-earth transition-metal magneto-optical storage media, 740–744
Rate equations, 335–336
Rate-equation model, 335–339
Ray tracing, 463–479
Ray-optics approach, 627
Rayleigh scattering, 443, 454
Rayleigh-Sommerfeld model, 480
RCA Laboratories, 784
Reactive ion etching, 60, 495, 520
Read-only compact disks
  fabrication process for, 735–737
Read-only disk memory, 720
Readout optical beam, 396
Reflection, 5
Reflection mode Fourier transform infrared spectroscopy, 54
Refractive index, 5, 8–9
Regenerators, 667
Relative intensity noise, 705
Relaxation oscillations, 372–374
Remanent polarization, 88, 89
Residue number system, 606
Resolution, 771
Resonant tunneling diodes, 7
Responsivity, 303
Return-to-zero coding, 751
Return-to-zero modulation coding format, 648
Reverse current, 108
Rewritable disk memory, 720
Ridge waveguide laser, 297
Ring configuration in multichannel optical systems, 684–685
Ring resonators, 515, 521
Rochelle salt, 87
Room temperature bandgaps, 11, 16
Room temperature data, 6
Room temperature direct gap, 16
Room temperature indirect gap, 16
Ruby laser, 346
Runge-Kutta method, 537, 549
Russel, Scott, 452

S

Sampled servo tracking, 748–749
Sanyo SiC light emitting diode, 72
Sapphire
  properties of, 351
Sapphire substrates, 62
  buffer layer on, 63
Sarnoff, David, 778
Satellite-based missile plume detection, 57
Saturable absorbers, 7
Saturation-induced cross talk, 674–675, 681
SbInSn phase-change write-once optical storage, 738
SBN crystals, 117, 120–121
Scalar diffraction integrals, 483–485
Scalar diffraction theory, 461, 473–478, 479–485
Scalar wave equation, 438, 508, 561
Scanned marking limit, 728
Schottky barrier, 35, 56, 301
Schroedinger equation, 453
SD format, 753
Second-harmonic generation
  See also Frequency doubling, 92, 118, 133, 164
Second-order nonlinear optical materials
  dispersion parameters of, 140
  properties of, 138–139
Second-order nonlinearities, 129–137
Secondary-ion mass spectrometry, 545
SEED devices, 608–612
Self-electro-optic devices, 409
Self-electro-optic effect devices, 608–612
  See also SEED devices, 608–612
Self-focusing, 92, 141
Self-phase modulation, 452, 454, 696, 700
Self-pumped phase conjugation, 92
  See also Photorefractive effect, 92
Semiconducting, 40
Semiconductor cavities, 636
  sources of losses in, 636
Semiconductor diode lasers, 292–301
  i-V characteristic, 294–295
  lateral structure of, 296–297
  light-current characteristic, 295–296
  modulation of, 298–299
  optical spectrum, 297–298
  principles, 292
  reliability, 301
  spatial optical profiles, 297–298
  temperature effects, 300
  transverse structure of, 292–294
Semiconductor laser diodes
  failure mechanisms, 301
Semiconductor lasers, 524
Semiconductor optical amplifiers, 668, 673–677
  applications in optical communication systems, 675–677
  differences from erbium-doped fiber amplifiers, 677
Semiconductor tunable filter, 691
Semiconductor, quantum dot glass waveguides, 526
Semiconductors
  optical properties, 6
  II-IV semiconductors
    type II heterostructures, 38–39
  II-VI compounds, 24–46
  II-VI diluted magnetic semiconductors
    heterostructures, 42–45

II-VI semiconductors
  applications, 45–46
  heterostructures, 24–46
    design, 26
    unique features, 25
III-V compounds, 404–405
III-V nitride semiconductors, 59–68
  advantages for laser and photodetector applications, 59
  crystal growth, 60–62
  electrical properties of, 64–65
  epitaxy
    substrates for, 62
  physical properties of, 64
  polytypism in, 63–64
III-V semiconductors, 5, 20
Separate absorption-graded-multiplication avalanche photodiodes, 316
Shadow-mask cathode ray tubes, 778
Short wavelength lasers, 756
Shot noise, 639
  characteristics of, 639
SiC heterojunction bipolar transistor, 53
SiC light emitting diodes, 56–57
SiC photodiodes, 57–58
  ultraviolet optical applications, 57–58
SiC substrate crystal growth, 53
SiC/Si epilayers, 54
  defects, 54
Side-chain polymers, 265–267
Signal detection schemes, 650–655
Signal processing, 599
Signal wave, 135
Signal-to-noise ratio, 305, 625, 646, 655
Silicon carbide as optoelectronic material, 49–81
  advantages, 50
  ohmic contacts to, 56
Silicon mechanical devices, 424
Silver-film technique, 503
Single crystals, 229–239
  for electro-optics
    material requirements, 237–238, 237–239
    material status, 238
  for frequency conversion, 229, 235–237
  molecular strutures, 230–234
Single ion exchange, 506–508
Single-frequency lasers, 357, 363–364
Single-mode optical fibers, 435, 443–445, 454, 630
6H SiC, 52
Slab waveguides, 537–545
Slope efficiency, 338–339
  factors, 339
    area efficiency, 339
    output-coupling efficiency, 339
    pump efficiency, 339
    quantum efficiency, 339
Smart pixel, 610
Snell's law, 438, 627
SNLN crystals
  See also Strontium sodium lithium niobate, 121
Sol-gel processing, 99

advantages, 100
Solar cells, 7
Solid ferroelectrics, 90
  optically anisotropic, 90
  optically isotropic, 90
Solid immersion lens, 757
Solid state physics, 6
Solid-state gain media, 339–350
  rare-earth dopants, 339–340
    erbium, 345–346
    holmium, 345
    neodymium, 340–344
    thulium, 344
    ytterbium, 344
  transition-metal dopants, 346–349
    chromium, 346–348
    cobalt, 349
    titanium, 348
Soliton period, 453
Soliton propagation, 143–144
  in optical fibers, 451–452
Soliton pulses in optical fibers, 452–453
Soliton transmission in fiber, 435
Solitons, 700–703
Solvatochromic method, 171–172
SONET
  See also Synchronous Optical Network, 689
Sony Corp., 779–780
sp-d exchange interaction, 42
Space harmonic approach, 489
Space-division multiplexing, 697–698
Space-integrating architectures, 600
Space-switching networks, 612
Spatial hole burning, 358–362
Spatial light modulators, 394–395, 406–410, 596–598
  features of, 597
Spatial modulation, 427–428
  coherent-light approach, 427
  incoherent light approach, 427
Spatial modulators, 416–418
  one-dimensional devices, 416–417
  two-dimensional devices, 416–417
Spatial-temporal equations, 679
Spectral hole burning, 362–363
Spin superlattice, 45
Spontaneous emission, 332, 450
Spontaneous lifetime, 332
Spontaneous polarization, 87–88
Sputtering, 99
SQRI
  See also Square root integral, 771
Square root integral
  See also SQRI, 771
Stacked microoptics, 617
Stacking faults, 54
Stacking sequence, 52
Star configuration in multichannel optical systems, 684–685
Static marking limit, 727–728
Step-index model
  for channel waveguides, 545–553

Index

Step-index profile, 440, 444–445, 453
Stimulated absorption, 633
Stimulated Brillouin scattering, 149–150, 451–452, 696
   applications of, 150
Stimulated emission, 331–333, 634
Stimulated light scattering, 146–150
   general characteristics, 146–147
Stimulated Raman scattering, 147–149, 451–452, 696
Strain
   causes of, 7
Strained materials, 6–7
Strained quantum wells, 7, 9
Stress tuning, 367–368
Strontium sodium lithium niobate
   See also SNLN crystals, 121
Subcarrier multiplexing, 703–706
   advantage, 707
   disadvantages, 707
Sublimation, 53
Substitution pattern, 225–226
Sum-frequency generation, 92, 118, 130–133
Superlattice avalanche photodiodes, 316
Supertwist nematic liquid crystal displays, 785
   active addressing of, 785
Surface acoustic waves, 411
Surface emitting lasers, 75
Surface relief, 460
Sverdlov, Dr. Boris, 72
Sweatt model, 465–466, 496
Symmetric SEED
   See also S-SEED, 609–612
Synchronous Optical Network
   See also SONET, 687

T

Tap power divider, 517
TE-TM mode splitting, 581
Temperature tuning, 136
Temporal modulation, 405–406, 425–427
Temporal modulators, 415–416
   unique features, 416
Ternary compounds, 5
TGS water-soluble ferroelectric crystals, 97, 120
Thermal cross-linking, 269–270
Thermal decomposition, 53
Thermal guiding, 355–356
Thermal ion exchange, 506
Thermal modulation devices, 424–425
Thermal noise, 639
   characteristics of, 639
   See also Johnson noise, Nyquist noise, 305, 307, 311
Thermal spatial light modulators
   performance, 426
Thermal tuning, 367
Thermoplastics as optical modulators, 425
Thin-film diodes, 786
Thin-film electroluminescent devices, 788

Thin-film transistors, 786
Third-harmonic generation, 92, 137
Third-order nonlinear materials, 145–146
   properties of, 146
Third-order nonlinearities, 137–146
Third-order susceptibility, 142
3C SiC, 52
   See also beta SiC, 52
Three-level gain media, 333
Threshold inversion, 337
Thulim-holmium YLF
   laser parameters of, 354
Thulium, 344
Thulium-doped YAG
   laser parameters of, 353
Thulium-holmium YAG
   laser parameters of, 353
Ti diffusion, 581
Ti-diffused waveguides, 542–545
Ti:LiNbO3 waveguides, 542–545, 555–560
Time-division multiplexing, 688
   advantages of, 688
   and generation of high-speed pulses, 688
   and high-speed switches, 688
Time-integrating architectures, 600–601
Titanium, 348
Titanium-doped sapphire
   laser parameters of, 354
Total internal reflection, 436, 533
Total internal reflection modulator, 410
Tracking error signal, 732, 750
Transition-metal dopants, 346–349
Transition-metal impurities, 443
Transparency
   and nonlinearity, 226
Transparent boundary condition, 562
Transparent ferroelectric ceramics, 95
Transverse electrooptic effect, 102
Transverse waveguide, 293
Tranverse effective electrooptic coefficient, 102
Traveling wave addressing, 397, 411
Traveling-wave semiconductor optical amplifiers, 668, 674
Tree configuration in multichannel optical systems, 684–685
Trichromaticity theory, 777
Triglycine sulfate
   See also TGS water-soluble ferroelectric crystals, 97
Trinitron, 779–780
Tungsten-bronze-type niobates, 92, 96
   barium sodium niobate, 96
   potassium lithium niobate, 96
   potassium sodium strontium barium niobate, 96
   strontium barium niobate, 96
Twisted nematic liquid crystal displays, 784–785
Twisted-nematic, liquid crystal devices, 408
Two-beam coupling, 108
Two-dimensional acousto-optic devices, 417
Two-dimensional graded-index waveguides, 535–537
Two-dimensional step-index waveguides, 532–535
Two-mode interference cross-channel waveguides, 561
Two-photon absorption, 142

(2,7) RLL code, 751
Two-step ion-exchange process, 564–565

U

Undepleted pump regime, 131–132, 134
Upconversion, 132–133

V

Valasek, 87
van Cittert deconvolution algorithm, 495
Vapor-axial deposition, 446
Varshni expression, 9–10, 12
Vector diffraction theory, 461
Vector regime, 488
Vector wave equation, 439
Verdet constant, 418
Vertical cavity surface emitting lasers, 428, 608
Vibrational resonances of the OH ion, 443
VLSI-based fabrication, 460
VSTEP, 608–612

W

Water-soluble crystals, 92, 96–97
Wave function, 43–44
Wave function mapping, 43–44
Wave-front quality, 473
Waveguide dispersion, 629
Waveguides
    characterization of
        by field-assisted ion exchange in glass, 539–542
        by ion exchange in glass, 537–539
Waveguiding, 5
Wavelength chirp, 695–696
Wavelength filters, 512
Wavelength multi/demultiplexer, 516–517
Wavelength multiplexers, 512
Wavelength-division multiplexers, 694
Wavelength-division multiplexing, 689–697
    fiber nonlinearities in, 696
    frequency stability in, 696
    in a quasi-point-to-point system, 689–690
    in a ring network topology, 689–690
    in a star network topology, 689–690
    multiwavelength transmitters in, 689–691
    passive stars in, 693
    system guidelines, 694–697
    wavelength-tunable receivers in, 690–693
Wavelength-tunable receivers, 691–693
    and acousto-optic filter, 693
    and fiber Fabry-Perot, 691
    and liquid-crystal Fabry-Perot, 691–693
    and semiconductor filter, 691
Wavelet transform, 602
Wavemat MPDR 610, 61
Wedge method, 164
Wet chemical etching, 495
Wide area networks, 685
Widened X-branch wavelength demultiplexer, 565–574
Window/cap layer, 18
Wireless free-space infrared transmission system, 666
WKB approximation, 442–443, 508
WKB dispersion relationship, 537
Write-once ablative media, 737–738
Write-once-read-many disk memory, 720

X

X-ray microanalyzer method, 545
X-ray photoelectron spectroscopy, 510
X-ray radiation, 772
    health aspects, 772

Y

Y-branch TE-TM mode splitter in LiNbO3, 581–583
Ytterbium, 344
Ytterbium-doped YAG
    laser parameters of, 352
Yttrium aluminum garnet
    properties of, 349
Yttrium lithium fluoride
    properties of, 350
Yttrium orthovanadate
    properties of, 350

Z

Zeeman splitting, 42, 44
    as tool for wave function mapping, 44
ZnCdSe alloys, 33
ZnSe
    doping, 33–34
ZnSe quantum wells, 24–38
ZnSe-based laser diodes, 34–36
    defect expansion in, 37
    defect formation in, 37
    lasing mechanism in, 36–37
    operating lifetimes, 37
ZnSe/ZnTe short period superlattice, 38